To the Student

As you begin, you may feel anxious about the number of theorems, definitions, procedures, and equations. You may wonder if you can learn it all in time. Don't worry–your concerns are normal. This textbook was written with you in mind. If you attend class, work hard, and read and study this text, you will build the knowledge and skills you need to be successful. Here's how you can use the text to your benefit.

Read Carefully

When you get busy, it's easy to skip reading and go right to the problems. Don't … the text has a large number of examples and clear explanations to help you break down the mathematics into easy-to-understand steps. Reading will provide you with a clearer understanding, beyond simple memorization. Read before class (not after) so you can ask questions about anything you didn't understand. You'll be amazed at how much more you'll get out of class if you do this.

Use the Features

I use many different methods in the classroom to communicate. Those methods, when incorporated into the text, are called "features." The features serve many purposes, from providing timely review of material you learned before (just when you need it) to providing organized review sessions to help you prepare for quizzes and tests. Take advantage of the features and you will master the material.

To make this easier, we've provided a brief guide to getting the most from this text. Refer to "Prepare for Class," "Practice," and "Review" at the front of the text. Spend fifteen minutes reviewing the guide and familiarizing yourself with the features by flipping to the page numbers provided. Then, as you read, use them. This is the best way to make the most of your text.

Please do not hesitate to contact me through Pearson Education, with any questions, comments, or suggestions for improving this text. I look forward to hearing from you, and good luck with all of your studies.

Best Wishes!

Michael Sullivan

Prepare for Class "Read the Book"

Feature	Description	Benefit	Page
Every Chapter Opener begins with . . .			
Chapter-Opening Topic & Project	Each chapter begins with a discussion of a topic of current interest and ends with a related project.	The project lets you apply what you learned to solve a problem related to the topic.	101
Internet-Based Projects	The projects allow for the integration of spreadsheet technology that you will need to be a productive member of the workforce.	The projects give you an opportunity to collaborate and use mathematics to deal with issues of current interest.	188
Every Section begins with . . .			
LEARNING OBJECTIVES 2	Each section begins with a list of objectives. Objectives also appear in the text where the objective is covered.	These focus your study by emphasizing what's most important and where to find it.	102
Sections contain . . .			
PREPARING FOR THIS SECTION	Most sections begin with a list of key concepts to review with page numbers.	Ever forget what you've learned? This feature highlights previously learned material to be used in this section. Review it, and you'll always be prepared to move forward.	115
Now Work the 'Are You Prepared?' Problems	Problems that assess whether you have the prerequisite knowledge for the upcoming section.	Not sure you need the Preparing for This Section review? Work the 'Are You Prepared?' problems. If you get one wrong, you'll know exactly what you need to review and where to review it!	115, 127
Now Work PROBLEMS	These follow most examples and direct you to a related exercise.	We learn best by doing. You'll solidify your understanding of examples if you try a similar problem right away, to be sure you understand what you've just read.	117, 127
WARNING	Warnings are provided in the text.	These point out common mistakes and help you to avoid them.	117
Exploration and Seeing the Concept	These graphing utility activities foreshadow a concept or solidify a concept just presented.	You will obtain a deeper and more intuitive understanding of theorems and definitions.	68, 151
In Words	These provide alternative descriptions of select definitions and theorems.	Does math ever look foreign to you? This feature translates math into plain English.	116
Calculus	These appear next to information essential for the study of calculus.	Pay attention–if you spend extra time now, you'll do better later!	104, 129
SHOWCASE EXAMPLES	These examples provide "how-to" instruction by offering a guided, step-by-step approach to solving a problem.	With each step presented on the left and the mathematics displayed on the right, you can immediately see how each step is used.	153
Model It! Examples and Problems	These examples and problems require you to build a mathematical model from either a verbal description or data. The homework Model It! problems are marked by purple headings.	It is rare for a problem to come in the form "*Solve the following equation.*" Rather, the equation must be developed based on an explanation of the problem. These problems require you to develop models to find a solution to the problem.	108, 112
NEW! Need to Review?	These margin notes provide a just-in-time reminder of a concept needed now, but covered in an earlier section of the book. Each note is back-referenced to the chapter, section and page where the concept was originally discussed.	Sometimes as you read, you encounter a word or concept you know you've seen before, but don't remember exactly what it means. This feature will point you to where you first learned the word or concept. A quick review now will help you see the connection to what you are learning for the first time and make remembering easier the next time.	102

Practice "Work the Problems"

Feature	Description	Benefit	Page
'Are You Prepared?' Problems	These assess your retention of the prerequisite material you'll need. Answers are given at the end of the section exercises. This feature is related to the Preparing for This Section feature.	Do you always remember what you've learned? Working these problems is the best way to find out. If you get one wrong, you'll know exactly what you need to review and where to review it!	147, 156
Concepts and Vocabulary	These short-answer questions, mainly Fill-in-the-Blank, Multiple-Choice and True/False items, assess your understanding of key definitions and concepts in the current section.	It is difficult to learn math without knowing the language of mathematics. These problems test your understanding of the formulas and vocabulary.	157
Skill Building	Correlated with section examples, these problems provide straightforward practice.	It's important to dig in and develop your skills. These problems provide you with ample opportunity to do so.	157–159
Applications and Extensions	These problems allow you to apply your skills to real-world problems. They also allow you to extend concepts learned in the section.	You will see that the material learned within the section has many uses in everyday life.	159–161
NEW! Challenge Problems	These problems have been added in most sections and appear at the end of the Applications and Extensions exercises. They are intended to be thought-provoking, requiring some ingenuity to solve.	Are you a student who likes being challenged? Then the Challenge Problems are for you! Your professor might also choose to assign a challenge problem as a group project. The ability to work with a team is a highly regarded skill in the working world.	161
Explaining Concepts: Discussion and Writing	"Discussion and Writing" problems are colored red. They support class discussion, verbalization of mathematical ideas, and writing and research projects.	To verbalize an idea, or to describe it clearly in writing, shows real understanding. These problems nurture that understanding. Many are challenging, but you'll get out what you put in.	161
Retain Your Knowledge	These problems allow you to practice content learned earlier in the course.	Remembering how to solve all the different kinds of problems that you encounter throughout the course is difficult. This practice helps you remember.	161
Now Work PROBLEMS	Many examples refer you to a related homework problem. These related problems are marked by a pencil and orange numbers.	If you get stuck while working problems, look for the closest Now Work problem, and refer to the related example to see if it helps.	148, 149, 157, 158
Review Exercises	Every chapter concludes with a comprehensive list of exercises to practice. Use the list of objectives to determine the objective and examples that correspond to the problems.	Work these problems to ensure that you understand all the skills and concepts of the chapter. Think of it as a comprehensive review of the chapter.	183–186

Review "Study for Quizzes and Tests"

Feature	Description	Benefit	Page
The Chapter Review at the end of each chapter contains . . .			
Things to Know	A detailed list of important theorems, formulas, and definitions from the chapter.	Review these and you'll know the most important material in the chapter!	181–183
You Should Be Able to . . .	Contains a complete list of objectives by section, examples that illustrate the objective, and practice exercises that test your understanding of the objective.	Do the recommended exercises and you'll have mastered the key material. If you get something wrong, go back and work through the objective listed and try again.	183, 184
Review Exercises	These provide comprehensive review and practice of key skills, matched to the Learning Objectives for each section.	Practice makes perfect. These problems combine exercises from all sections, giving you a comprehensive review in one place.	184–186
Chapter Test	About 15–20 problems that can be taken as a Chapter Test. Be sure to take the Chapter Test under test conditions—no notes!	Be prepared. Take the sample practice test under test conditions. This will get you ready for your instructor's test. If you get a problem wrong, you can watch the Chapter Test Prep Video.	186, 187
Cumulative Review	These problem sets appear at the end of each chapter, beginning with Chapter 2. They combine problems from previous chapters, providing an ongoing cumulative review. When you use them in conjunction with the Retain Your Knowledge problems, you will be ready for the final exam.	These problem sets are really important. Completing them will ensure that you are not forgetting anything as you go. This will go a long way toward keeping you primed for the final exam.	187
Chapter Projects	The Chapter Projects apply to what you've learned in the chapter. Additional projects are available on the Instructor's Resource Center (IRC).	The Chapter Projects give you an opportunity to use what you've learned in the chapter to the opening topic. If your instructor allows, these make excellent opportunities to work in a group, which is often the best way to learn math.	188
Internet-Based Projects	In selected chapters, a Web-based project is given.	These projects give you an opportunity to collaborate and use mathematics to deal with issues of current interest by using the Internet to research and collect data.	188

To the Memory of Joe and Rita
and My Sister Maryrose

Trigonometry
A Unit Circle Approach

Eleventh Edition

Michael Sullivan

Chicago State University

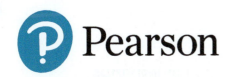 Pearson

Director, Portfolio Management: *Anne Kelly*
Senior Portfolio Management Analyst: *Dawn Murrin*
Portfolio Management Administrator: *Joseph Colella*
Manager, Courseware QA: *Mary Durnwald*
VP, Production & Digital Studio: *Ruth Berry*
Manager Producer: *Vicki Dreyfus*
Associate Producer: *Stacey Miller*
Manager, Content Development: *Kristina Evans*
Senior Content Developer: *Megan M. Burns*
Managing Producer: *Scott Disanno*
Content Producer: *Peggy McMahon*
Product Marketing Director: *Erin Kelly*

Product Marketer for Precalculus: *Stacey Sveum*
Product Marketing Assistant: *Shannon McCormack*
Field Marketing Manager: *Peggy Lucas*
Senior Publishing Services Analyst, Author Support: *Joe Vetere*
Manager, Rights/Permissions: *Gina Cheselka*
Manager Design: *Blair Brown*
Manufacturing Buyer: *Carol Melville, LSC Communications*
Cover Illustration: *Tamara Newnam*
Cover Design: *Jerilyn Bockorick, Pearson CSC*
Full Service Vendor: *Pearson, CSC*
Full Service Project Management: *Pearson CSC, Rose Kernan*
Composition: *Pearson CSC*

Library of Congress Cataloging-in-Publication Data

Names: Sullivan, Michael, 1942- author.
Title: Trigonometry : a unit circle approach / Michael Sullivan (Chicago State University).
Description: Eleventh edition. | Hoboken, NJ : Pearson, c2020. | Includes index.
Identifiers: LCCN 2019015166 | ISBN 9780135181133 (annotated instructor's ed.)
Subjects: LCSH: Trigonometry--Textbooks.
Classification: LCC QA531 .S87 2020 | DDC 516.24--dc23
LC record available at https://lccn.loc.gov/2019015166

About the Cover:
The image on this book's cover was inspired by a talk given by Michael Sullivan III: *Is Mathematical Talent Overrated?*

The answer is yes. In mathematics, innate talent plays a much smaller role than grit and motivation as you work toward your goal. If you put in the time and hard work, you can succeed in your math course—just as an athlete must work to medal in their sport.

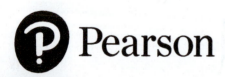 **Pearson**

10 2023

ISBN 10: **0135181135**
ISBN-13: **9780135181133**

Contents

1   Graphs and Functions   1

Three Distinct Series

Students have different goals, learning styles, and levels of preparation. Instructors have different teaching philosophies, styles, and techniques. Rather than write one series to fit all, the Sullivans have written three distinct series. All share the same goal—to develop a high level of mathematical understanding and an appreciation for the way mathematics can describe the world around us. The manner of reaching that goal, however, differs from series to series.

Flagship Series, Eleventh Edition

The Flagship Series is the most traditional in approach yet modern in its treatment of precalculus mathematics. In each text, needed review material is included, and is referenced when it is used. Graphing utility coverage is optional and can be included or excluded at the discretion of the instructor: *College Algebra, Algebra & Trigonometry, Trigonometry: A Unit Circle Approach, Precalculus.*

Enhanced with Graphing Utilities Series, Seventh Edition

This series provides a thorough integration of graphing utilities into topics, allowing students to explore mathematical concepts and encounter ideas usually studied in later courses. Many examples show solutions using algebra side-by-side with graphing techniques. Using technology, the approach to solving certain problems differs from the Flagship Series, while the emphasis on understanding concepts and building strong skills is maintained: *College Algebra, Algebra & Trigonometry, Precalculus.*

Concepts through Functions Series, Fourth Edition

This series differs from the others, utilizing a functions approach that serves as the organizing principle tying concepts together. Functions are introduced early in various formats. The approach supports the Rule of Four, which states that functions can be represented symbolically, numerically, graphically, and verbally. Each chapter introduces a new type of function and then develops all concepts pertaining to that particular function. The solutions of equations and inequalities, instead of being developed as stand-alone topics, are developed in the context of the underlying functions. Graphing utility coverage is optional and can be included or excluded at the discretion of the instructor: *College Algebra; Precalculus, with a Unit Circle Approach to Trigonometry; Precalculus, with a Right Triangle Approach to Trigonometry.*

The Flagship Series

College Algebra, Eleventh Edition

This text provides a contemporary approach to college algebra, with three chapters of review material preceding the chapters on functions. Graphing calculator usage is provided, but is optional. After completing this book, a student will be adequately prepared for trigonometry, finite mathematics, and business calculus.

Algebra & Trigonometry, Eleventh Edition

This text contains all the material in *College Algebra*, but also develops the trigonometric functions using a right triangle approach and shows how it relates to the unit circle approach. Graphing techniques are emphasized, including a thorough discussion of polar coordinates, parametric equations, and conics using polar coordinates. Vectors in the plane, sequences, induction, and the binomial theorem are also presented. Graphing calculator usage is provided, but is optional. After completing this book, a student will be adequately prepared for finite mathematics, business calculus, and engineering calculus.

Precalculus, Eleventh Edition

This text contains one review chapter before covering the traditional precalculus topics of polynomial, rational, exponential, and logarithmic functions and their graphs. The trigonometric functions are introduced using a unit circle approach and showing how it relates to the right triangle approach. Graphing techniques are emphasized, including a thorough discussion of polar coordinates, parametric equations, and conics using polar coordinates. Vectors in the plane and in space, including the dot and cross products, sequences, induction, and the binomial theorem are also presented. Graphing calculator usage is provided, but is optional. The final chapter provides an introduction to calculus, with a discussion of the limit, the derivative, and the integral of a function. After completing this book, a student will be adequately prepared for finite mathematics, business calculus, and engineering calculus.

Trigonometry: a Unit Circle Approach, Eleventh Edition

This text, designed for stand-alone courses in trigonometry, develops the trigonometric functions using a unit circle approach and shows how it relates to the right triangle approach. Vectors in the plane and in space, including the dot and cross products, are presented. Graphing techniques are emphasized, including a thorough discussion of polar coordinates, parametric equations, and conics using polar coordinates. Graphing calculator usage is provided, but is optional. After completing this book, a student will be adequately prepared for finite mathematics, business calculus, and engineering calculus.

Preface to the Instructor

As a professor of mathematics at an urban public university for 35 years, I understand the varied needs of students taking trigonometry. Students range from being underprepared with little mathematical background and a fear of mathematics, to being highly prepared and motivated. For some, this is their final course in mathematics. For others, it is preparation for future mathematics courses. I have written this text with both groups in mind.

A tremendous benefit of authoring a successful series is the broad-based feedback I receive from instructors and students who have used previous editions. I am sincerely grateful for their support. Virtually every change to this edition is the result of their thoughtful comments and suggestions. I hope that I have been able to take their ideas and, building upon a successful foundation of the tenth edition, make this series an even better learning and teaching tool for students and instructors.

Features in the Eleventh Edition

A descriptive list of the many special features of *Trigonometry* can be found on the endpapers in the front of this text. This list places the features in their proper context, as building blocks of an overall learning system that has been carefully crafted over the years to help students get the most out of the time they put into studying. Please take the time to review it and to discuss it with your students at the beginning of your course. My experience has been that when students use these features, they are more successful in the course.

- **Updated! Retain Your Knowledge Problems** These problems, which were new to the previous edition, are based on the article *"To Retain New Learning, Do the Math,"* published in the *Edurati Review*. In this article, Kevin Washburn suggests that "the more students are required to recall new content or skills, the better their memory will be." The Retain Your Knowledge problems were so well received that they have been expanded in this edition. Moreover, while the focus remains to help students maintain their skills, in most sections, problems were chosen that preview skills required to succeed in subsequent sections or in calculus. These are easily identified by the calculus icon ($\int$). All answers to Retain Your Knowledge problems are given in the back of the text and all are assignable in MyLab Math.

- **Guided Lecture Notes** Ideal for online, emporium/redesign courses, inverted classrooms, or traditional lecture classrooms. These lecture notes help students take thorough, organized, and understandable notes as they watch the Author in Action videos. They ask students to complete definitions, procedures, and examples based on the content of the videos and text. In addition, experience suggests that students learn by doing and understanding the why/how of the concept or property. Therefore, many

sections will have an exploration activity to motivate student learning. These explorations introduce the topic and/or connect it to either a real-world application or a previous section. For example, when the vertical-line test is discussed in Section 1.3, after the theorem statement, the notes ask the students to explain why the vertical-line test works by using the definition of a function. This challenge helps students process the information at a higher level of understanding.

- **Illustrations** Many of the figures have captions to help connect the illustrations to the explanations in the body of the text.

- **Graphing Utility Screen Captures** In several instances we have added Desmos screen captures along with the TI-84 Plus C screen captures. These updated screen captures provide alternate ways of visualizing concepts and making connections between equations, data and graphs in full color.

- **Chapter Projects**, which apply the concepts of each chapter to a real-world situation, have been enhanced to give students an up-to-the-minute experience. Many of these projects are new requiring the student to research information online in order to solve problems.

- **Exercise Sets** The exercises in the text have been reviewed and analyzed, some have been removed, and new ones have been added. All time-sensitive problems have been updated to the most recent information available. The problem sets remain classified according to purpose.

  The *'Are You Prepared?'* problems have been improved to better serve their purpose as a just-in-time review of concepts that the student will need to apply in the upcoming section.

  The *Concepts and Vocabulary* problems have been expanded to cover each objective of the section. These multiple-choice, fill-in-the-blank, and True/False exercises have been written to also serve as reading quizzes.

  Skill Building problems develop the student's computational skills with a large selection of exercises that are directly related to the objectives of the section. Within these exercises, *Mixed Practice* problems offer a comprehensive assessment of skills that relate to more than one objective. Often these require skills learned earlier in the course.

  Applications and Extensions problems have been updated. Further, many new application-type exercises have been added, especially ones involving information and data drawn from sources the student will recognize, to improve relevance and timeliness.

  At the end of Applications and Extensions, we have a collection of one or more *Challenge Problems*. These problems, as the title suggests, are intended to be thought-provoking, requiring some ingenuity to solve. They can be used for group work or to challenge students. At the end of the Annotated Instructor's

Edition and in the online Instructor's Solutions Manual, we have provided solutions to all these problems.

The ***Explaining Concepts: Discussion and Writing*** exercises provide opportunity for classroom discussion and group projects.

Updated! ***Retain Your Knowledge*** has been improved and expanded. The problems are based on material learned earlier in the course, especially calculus-related material. They serve to keep information that has already been learned "fresh" in the mind of the student. Answers to all these problems appear in the Student Edition.

NEW Need to Review? These margin notes provide a just-in-time reminder of a concept needed now, but covered in an earlier section of the book. Each note includes a reference to the chapter, section, and page where the concept was originally discussed.

Content Changes to the 11th edition

- ***Challenge Problems*** have been added in most sections at the end of the Applications and Extensions exercises. Challenge Problems are intended to be thought-provoking problems that require some ingenuity to solve. They can be used to challenge students or for group work.

- **Need to Review?** These margin notes provide a just-in-time review for a concept needed now, but covered earlier. Each note is back-referenced to the chapter, section, and page where the concept was originally discussed.

- Additional **Retain Your Knowledge** exercises, whose purpose is to keep learned material fresh in a student's mind, have been added to each section. Many of these new problems preview skills required for calculus or for concepts needed in subsequent sections.

- **Desmos** screen captures have been added throughout the text. This was done to recognize that graphing technology expands beyond graphing calculators.

- Examples and exercises throughout the text have been augmented to reflect a broader selection of STEM applications.

- Concepts and Vocabulary exercises have been expanded to cover each objective of a section.

- Skill building exercises have been expanded to assess a wider range of difficulty.

- Applied problems and those based on real data have been updated where appropriate.

Appendix A

- Section A.7 Objective 3 now includes rationalizing the numerator
  - NEW Example 6 Rationalizing Numerators
  - Problems 69–76 provide practice.
- Section A.7 Exercises now include more practice in simplifying radicals

- Section A.8 has been reorganized to treat the slope-intercept form of the equation of a line before finding an equation of a line using two points.

Chapter 1

- NEW Section 1.2 Example 9 Testing an Equation for Symmetry
- NEW Section 1.3 Objective 1 Describe a Relation
- NEW Section 1.3 Example 1 Describing a Relation demonstrates using the Rule of Four to express a relation numerically, as a mapping, and graphically given a verbal description.
- NEW Section 1.5 Example 4 Analyzing a Piecewise-defined Function
- Section 1.7 now finds and verifies inverse functions analytically and graphically.

Chapter 2

- NEW Section 2.1 Example 6 Field Width of a Digital Lens Reflex Camera Lens
- Sections 2.4 and 2.5 were reorganized for increased clarity.

Chapter 3

- Sections 3.1 and 3.2 were reorganized for increased clarity.

Chapter 5

- Section 5.3 The complex plane; DeMoivre's Theorem, was rewritten to support the exponential form of a complex number.
  - Euler's Formula is introduced to express a complex number in exponential form.
  - The exponential form of a complex number is used to compute products and quotients.
  - DeMoivre's Theorem is expressed using the exponential form of a complex number.
  - The exponential form is used to find complex roots.

Using the Eleventh Edition Effectively with Your Syllabus

To meet the varied needs of diverse syllabi, this text contains more content than is likely to be covered in a *Trigonometry* course. As the chart illustrates, this text has been organized with flexibility of use in mind. Within a given chapter, certain sections are optional (see the details that follow the figure below) and can be omitted without loss of continuity.

Appendix A Review
This chapter consists of review material. It may be used as the first part of the course or later as a just-in-time review when the content is required. Specific references to this chapter occur throughout the text to assist in the review process.

Chapter 1 Graphs and Functions
This chapter lays the foundation for Chapters 2, 6, and 7.

Chapter 2 Trigonometric Functions
Section 2.6 may be omitted in a brief course.

Chapter 3 Analytic Trigonometry
Section 3.7 may be omitted in a brief course.

Chapter 4 Applications of Trigonometric Functions
Sections 4.4 and 4.5 may be omitted in a brief course.

Chapter 5 Polar Coordinates; Vectors
Sections 5.1–5.3 and Sections 5.4–5.7 are independent and may be covered separately.

Chapter 6 Analytic Geometry
Sections 6.1–6.4 follow in sequence. Sections 6.5, 6.6, and 6.7 are independent of each other, but each requires Sections 6.1–6.4.

Chapter 7 Exponential and Logarithmic Functions
Sections 7.1–7.4 follow in sequence. Sections 7.5, 7.6, and 7.7 are optional.

Acknowledgments

Textbooks are written by authors, but evolve from an idea to final form through the efforts of many people. It was Don Dellen who first suggested this text and series to me. Don is remembered for his extensive contributions to publishing and mathematics.

Thanks are due to the following people for their assistance and encouragement to the preparation of this edition:

- From Pearson Education: Anne Kelly for her substantial contributions, ideas, and enthusiasm; Dawn Murrin, for her unmatched talent at getting the details right; Joseph Colella for always getting the reviews and pages to me on time; Peggy McMahon for directing the always difficult production process; Rose Kernan for handling liaison between the compositor and author; Peggy Lucas and Stacey Sveum for their genuine interest in marketing this text. Marcia Horton for her continued support and genuine interest; Paul Corey for his leadership and commitment to excellence; and the Pearson Sales team, for their continued confidence and personal support of Sullivan texts.

- Accuracy checkers: C. Brad Davis who read the entire manuscript and accuracy checked answers. His attention to detail is amazing; Timothy Britt, for creating the Solutions Manuals; and Kathleen Miranda and Pamela Trim for accuracy checking answers.

Finally, I offer my grateful thanks to the dedicated users and reviewers of my texts, whose collective insights form the backbone of each textbook revision.

James Africh, College of DuPage
Steve Agronsky, Cal Poly State University
Gererdo Aladro, Florida International University
Grant Alexander, Joliet Junior College
Dave Anderson, South Suburban College
Wes Anderson, Northwest Vista College
Richard Andrews, Florida A&M University
Joby Milo Anthony, University of Central Florida
James E. Arnold, University of Wisconsin-Milwaukee
Adel Arshaghi, Center for Educational Merit
Carolyn Autray, University of West Georgia
Agnes Azzolino, Middlesex County College
Wilson P. Banks, Illinois State University
Sudeshna Basu, Howard University
Timothy Bayer, Virginia Western CC
Dale R. Bedgood, East Texas State University
Beth Beno, South Suburban College
Carolyn Bernath, Tallahassee Community College
Rebecca Berthiaume, Edison State College
William H. Beyer, University of Akron
Annette Blackwelder, Florida State University
Richelle Blair, Lakeland Community College
Kevin Bodden, Lewis and Clark College
Jeffrey Boerner, University of Wisconsin-Stout
Connie Booker, Owensboro Community and Technical College
Barry Booten, Florida Atlantic University
Laurie Boudreaux, Nicholls State University
Larry Bouldin, Roane State Community College
Bob Bradshaw, Ohlone College

Trudy Bratten, Grossmont College
Tim Bremer, Broome Community College
Tim Britt, Jackson State Community College
Holly Broesamle, Oakland CC-Auburn Hills
Michael Brook, University of Delaware
Timothy Brown, Central Washington University
Joanne Brunner, Joliet Junior College
Warren Burch, Brevard Community College
Mary Butler, Lincoln Public Schools
Melanie Butler, West Virginia University
Jim Butterbach, Joliet Junior College
Roberto Cabezas, Miami Dade College
William J. Cable, University of Wisconsin-Stevens Point
Lois Calamia, Brookdale Community College
Jim Campbell, Lincoln Public Schools
Roger Carlsen, Moraine Valley Community College
Elena Catoiu, Joliet Junior College
Mathews Chakkanakuzhi, Palomar College
Tim Chappell, Penn Valley Community College
John Collado, South Suburban College
Amy Collins, Northwest Vista College
Alicia Collins, Mesa Community College
Nelson Collins, Joliet Junior College
Rebecca Connell, Troy University
Jim Cooper, Joliet Junior College
Denise Corbett, East Carolina University
Carlos C. Corona, San Antonio College
Theodore C. Coskey, South Seattle Community College
Rebecca Connell, Troy University
Donna Costello, Plano Senior High School

Rebecca Courter, Pasadena City College
Garrett Cox, The University of Texas at San Antonio
Paul Crittenden, University of Nebraska at Lincoln
John Davenport, East Texas State University
Faye Dang, Joliet Junior College
Antonio David, Del Mar College
Stephanie Deacon, Liberty University
Duane E. Deal, Ball State University
Jerry DeGroot, Purdue North Central
Timothy Deis, University of Wisconsin-Platteville
Joanna DelMonaco, Middlesex Community College
Vivian Dennis, Eastfield College
Deborah Dillon, R. L. Turner High School
Guesna Dohrman, Tallahassee Community College
Cheryl Doolittle, Iowa State University
Karen R. Dougan, University of Florida
Jerrett Dumouchel, Florida Community College at Jacksonville
Louise Dyson, Clark College
Paul D. East, Lexington Community College
Don Edmondson, University of Texas-Austin
Erica Egizio, Joliet Junior College
Jason Eltrevoog, Joliet Junior College
Christopher Ennis, University of Minnesota
Kathy Eppler, Salt Lake Community College
Ralph Esparza, Jr., Richland College
Garret J. Etgen, University of Houston
Scott Fallstrom, Shoreline Community College
Pete Falzone, Pensacola Junior College
Arash Farahmand, Skyline College

Said Fariabli, San Antonio College
W.A. Ferguson, University of Illinois-Urbana/Champaign
Iris B. Fetta, Clemson University
Mason Flake, student at Edison Community College
Timothy W. Flood, Pittsburg State University
Robert Frank, Westmoreland County Community College
Merle Friel, Humboldt State University
Richard A. Fritz, Moraine Valley Community College
Dewey Furness, Ricks College
Mary Jule Gabiou, North Idaho College
Randy Gallaher, Lewis and Clark College
Tina Garn, University of Arizona
Dawit Getachew, Chicago State University
Wayne Gibson, Rancho Santiago College
Loran W. Gierhart, University of Texas at San Antonio and Palo Alto College
Robert Gill, University of Minnesota Duluth
Nina Girard, University of Pittsburgh at Johnstown
Sudhir Kumar Goel, Valdosta State University
Adrienne Goldstein, Miami Dade College, Kendall Campus
Joan Goliday, Sante Fe Community College
Lourdes Gonzalez, Miami Dade College, Kendall Campus
Frederic Gooding, Goucher College
Donald Goral, Northern Virginia Community College
Sue Graupner, Lincoln Public Schools
Mary Beth Grayson, Liberty University
Jennifer L. Grimsley, University of Charleston
Ken Gurganus, University of North Carolina
Igor Halfin, University of Texas-San Antonio
James E. Hall, University of Wisconsin-Madison
Judy Hall, West Virginia University
Edward R. Hancock, DeVry Institute of Technology
Julia Hassett, DeVry Institute, Dupage
Christopher Hay-Jahans, University of South Dakota
Michah Heibel, Lincoln Public Schools
LaRae Helliwell, San Jose City College
Celeste Hernandez, Richland College
Gloria P. Hernandez, Louisiana State University at Eunice
Brother Herron, Brother Rice High School
Robert Hoburg, Western Connecticut State University
Lynda Hollingsworth, Northwest Missouri State University
Deltrye Holt, Augusta State University
Charla Holzbog, Denison High School
Lee Hruby, Naperville North High School
Miles Hubbard, St. Cloud State University
Kim Hughes, California State College-San Bernardino
Stanislav, Jabuka, University of Nevada, Reno
Ron Jamison, Brigham Young University
Richard A. Jensen, Manatee Community College
Glenn Johnson, Middlesex Community College
Sandra G. Johnson, St. Cloud State University
Tuesday Johnson, New Mexico State University
Susitha Karunaratne, Purdue University North Central
Moana H. Karsteter, Tallahassee Community College
Donna Katula, Joliet Junior College
Arthur Kaufman, College of Staten Island
Thomas Kearns, North Kentucky University
Jack Keating, Massasoit Community College
Shelia Kellenbarger, Lincoln Public Schools

Rachael Kenney, North Carolina State University
Penelope Kirby, Florida State University
John B. Klassen, North Idaho College
Debra Kopcso, Louisiana State University
Lynne Kowski, Raritan Valley Community College
Yelena Kravchuk, University of Alabama at Birmingham
Ray S. Kuan, Skyline College
Keith Kuchar, Manatee Community College
Tor Kwembe, Chicago State University
Linda J. Kyle, Tarrant Country Jr. College
H.E. Lacey, Texas A & M University
Darren Lacoste, Valencia College-West Campus
Harriet Lamm, Coastal Bend College
James Lapp, Fort Lewis College
Matt Larson, Lincoln Public Schools
Christopher Lattin, Oakton Community College
Julia Ledet, Lousiana State University
Wayne Lee, St. Phillips CC
Adele LeGere, Oakton Community College
Kevin Leith, University of Houston
JoAnn Lewin, Edison College
Jeff Lewis, Johnson County Community College
Janice C. Lyon, Tallahassee Community College
Jean McArthur, Joliet Junior College
Virginia McCarthy, Iowa State University
Karla McCavit, Albion College
Michael McClendon, University of Central Oklahoma
Tom McCollow, DeVry Institute of Technology
Marilyn McCollum, North Carolina State University
Jill McGowan, Howard University
Will McGowant, Howard University
Angela McNulty, Joliet Junior College
Lisa Meads, College of the Albemarle
Laurence Maher, North Texas State University
Jay A. Malmstrom, Oklahoma City Community College
Rebecca Mann, Apollo High School
Lynn Marecek, Santa Ana College
Sherry Martina, Naperville North High School
Ruby Martinez, San Antonio College
Alec Matheson, Lamar University
Nancy Matthews, University of Oklahoma
James Maxwell, Oklahoma State University-Stillwater
Marsha May, Midwestern State University
James McLaughlin, West Chester University
Judy Meckley, Joliet Junior College
David Meel, Bowling Green State University
Carolyn Meitler, Concordia University
Samia Metwali, Erie Community College
Rich Meyers, Joliet Junior College
Eldon Miller, University of Mississippi
James Miller, West Virginia University
Michael Miller, Iowa State University
Kathleen Miranda, SUNY at Old Westbury
Chris Mirbaha, The Community College of Baltimore County
Val Mohanakumar, Hillsborough Community College
Thomas Monaghan, Naperville North High School
Miguel Montanez, Miami Dade College, Wolfson Campus
Maria Montoya, Our Lady of the Lake University
Susan Moosai, Florida Atlantic University
Craig Morse, Naperville North High School
Samad Mortabit, Metropolitan State University
Pat Mower, Washburn University

Tammy Muhs, University of Central Florida
A. Muhundan, Manatee Community College
Jane Murphy, Middlesex Community College
Richard Nadel, Florida International University
Gabriel Nagy, Kansas State University
Bill Naegele, South Suburban College
Karla Neal, Lousiana State University
Lawrence E. Newman, Holyoke Community College
Dwight Newsome, Pasco-Hernando Community College
Denise Nunley, Maricopa Community Colleges
James Nymann, University of Texas-El Paso
Mark Omodt, Anoka-Ramsey Community College
Seth F. Oppenheimer, Mississippi State University
Leticia Oropesa, University of Miami
Linda Padilla, Joliet Junior College
Sanja Pantic, University of Illinois at Chicago
E. James Peake, Iowa State University
Kelly Pearson, Murray State University
Dashamir Petrela, Florida Atlantic University
Philip Pina, Florida Atlantic University
Charlotte Pisors, Baylor University
Michael Prophet, University of Northern Iowa
Laura Pyzdrowski, West Virginia University
Carrie Quesnell, Weber State University
Neal C. Raber, University of Akron
Thomas Radin, San Joaquin Delta College
Aibeng Serene Radulovic, Florida Atlantic University
Ken A. Rager, Metropolitan State College
Traci Reed, St. Johns River State College
Kenneth D. Reeves, San Antonio College
Elsi Reinhardt, Truckee Meadows Community College
Jose Remesar, Miami Dade College, Wolfson Campus
Jane Ringwald, Iowa State University
Douglas F. Robertson, University of Minnesota, MPLS
Stephen Rodi, Austin Community College
William Rogge, Lincoln Northeast High School
Howard L. Rolf, Baylor University
Mike Rosenthal, Florida International University
Phoebe Rouse, Lousiana State University
Edward Rozema, University of Tennessee at Chattanooga
Dennis C. Runde, Manatee Community College
Paul Runnion, Missouri University of Science and Technology
Amit Saini, University of Nevada-Reno
Laura Salazar, Northwest Vista College
Alan Saleski, Loyola University of Chicago
Susan Sandmeyer, Jamestown Community College
Brenda Santistevan, Salt Lake Community College
Linda Schmidt, Greenville Technical College
Ingrid Scott, Montgomery College
A.K. Shamma, University of West Florida
Zachery Sharon, University of Texas at San Antonio
Joshua Shelor, Virginia Western CC
Martin Sherry, Lower Columbia College
Carmen Shershin, Florida International University
Tatrana Shubin, San Jose State University
Anita Sikes, Delgado Community College
Timothy Sipka, Alma College
Charlotte Smedberg, University of Tampa
Lori Smellegar, Manatee Community College
Gayle Smith, Loyola Blakefield
Cindy Soderstrom, Salt Lake Community College

Leslie Soltis, Mercyhurst College
John Spellman, Southwest Texas State
   University
Karen Spike, University of North Carolina
Rajalakshmi Sriram, Okaloosa-Walton
   Community College
Katrina Staley, North Carolina Agricultural
   and Technical State University
Becky Stamper, Western Kentucky University
Judy Staver, Florida Community
   College-South
Robin Steinberg, Pima Community College
Neil Stephens, Hinsdale South High School
Sonya Stephens, Florida A&M Univeristy
Patrick Stevens, Joliet Junior College
John Sumner, University of Tampa
Matthew TenHuisen, University of North
   Carolina, Wilmington
Christopher Terry, Augusta State University
Diane Tesar, South Suburban College
Tommy Thompson, Brookhaven College
Martha K. Tietze, Shawnee Mission Northwest
   High School

Richard J. Tondra, Iowa State University
Florentina Tone, University of West Florida
Suzanne Topp, Salt Lake Community College
Marilyn Toscano, University of Wisconsin,
   Superior
Marvel Townsend, University of Florida
Jim Trudnowski, Carroll College
David Tseng, Miami Dade College, Kendall
   Campus
Robert Tuskey, Joliet Junior College
Mihaela Vajiac, Chapman University-Orange
Julia Varbalow, Thomas Nelson Community
   College-Leesville
Richard G. Vinson, University of South
   Alabama
Jorge Viola-Prioli, Florida Atlantic University
Mary Voxman, University of Idaho
Jennifer Walsh, Daytona Beach Community
   College
Donna Wandke, Naperville North High School
Timothy L. Warkentin, Cloud County
   Community College
Melissa J. Watts, Virginia State University

Hayat Weiss, Middlesex Community College
Kathryn Wetzel, Amarillo College
Darlene Whitkenack, Northern Illinois
   University
Suzanne Williams, Central Piedmont
   Community College
Larissa Williamson, University of Florida
Christine Wilson, West Virginia University
Brad Wind, Florida International University
Anna Wiodarczyk, Florida International
   University
Mary Wolyniak, Broome Community
   College
Canton Woods, Auburn University
Tamara S. Worner, Wayne State College
Terri Wright, New Hampshire Community
   Technical College, Manchester
Rob Wylie, Carl Albert State College
Aletheia Zambesi, University of West Florida
George Zazi, Chicago State University
Loris Zucca, Lone Star College-Kingwood
Steve Zuro, Joliet Junior College

Michael Sullivan

Get the Most Out of
MyLab Math

Math courses are continuously evolving to help today's students succeed. It's more challenging than ever to support students with a wide range of backgrounds, learner styles, and math anxieties. The flexibility to build a course that fits instructors' individual course formats—with a variety of content options and multimedia resources all in one place—has made MyLab Math the market-leading solution for teaching and learning mathematics since its inception.

Preparedness

One of the biggest challenges in College Algebra, Trigonometry, and Precalculus is making sure students are adequately prepared with prerequisite knowledge. For a student, having the essential algebra skills upfront in this course can dramatically increase success.

- **MyLab Math with Integrated Review** can be used in corequisite courses, or simply to help students who enter without a full understanding of prerequisite skills and concepts. **Integrated Review** provides videos on review topics with a corresponding worksheet, along with premade, assignable skills-check quizzes and personalized review homework assignments. **Integrated Review** is now available within all Sullivan 11th Edition MyLab Math courses.

Resources for Success

Pearson
MyLab

Instructor Resources

Online resources can be downloaded from www.pearson.com, or hardcopy resources can be ordered from your sales representative.

Annotated Instructor's Edition
Trigonometry, 11th Edition
ISBN – 0135229065 / 9780135229064
Shorter answers are on the page beside the exercises. Longer answers are in the back of the text.

Instructor's Solutions Manual
ISBN – 0135228999 / 9780135228999
Includes fully worked solutions to all exercises in the text.

Learning Catalytics Question Library
Questions written by Michael Sullivan III are available within MyLab Math to deliver through Learning Catalytics to engage students in your course.

Powerpoint® Lecture Slides
Fully editable slides correlate to the textbook.

Mini Lecture Notes
Includes additional examples and helpful teaching tips, by section.

Testgen®
TestGen (www.pearsoned.com/testgen) enables instructors to build, edit, print, and administer tests using a computerized bank of questions developed to cover all the objectives of the text.

Online Chapter Projects
Additional projects that give students an opportunity to apply what they learned in the chapter.

Student Resources

Additional resources to enhance student success.

Lecture Video
Author in Action videos are actual classroom lectures with fully worked out examples presented by Michael Sullivan, III. All video is assignable within MyLab Math.

Chapter Test Prep Videos
Students can watch instructors work through step-by-step solutions to all chapter test exercises from the text. These are available in MyLab Math and on YouTube.

Student's Solutions Manual
ISBN - 0134999312 / 9780134999319
Provides detailed worked-out solutions to odd-numbered exercises.

Guided Lecture Notes
ISBN – 0135244749 / 9780135244746
These lecture notes assist students in taking thorough, organized, and understandable notes while watching Author in Action videos. Students actively participate in learning the how/why of important concepts through explorations and activities. The Guided Lecture Notes are available as PDF's and customizable Word files in MyLab Math. They can also be packaged with the text and the MyLab Math access code.

Algebra Review
ISBN: 0131480065 / 9780131480063
Four printed chapters of Intermediate Algebra review available. Perfect for a corequisite course or for individual review.

Applications Index

Graphs and Functions

1

Choosing a Data Plan

When selecting a data plan for a device, most consumers choose a service provider first and then select an appropriate data plan from that provider. The choice as to the type of plan selected depends on your use of the device. For example, is online gaming important? Do you want to stream audio or video? The mathematics learned in this chapter can help you decide what plan is best suited to your particular needs.

 —See the Internet-based Chapter Project—

← A Look Back

Appendix A reviews skills from intermediate algebra.

A Look Ahead →

Here we connect algebra and geometry using the rectangular coordinate system to graph equations in two variables. Then we look at a special type of equation involving two variables called a *function*. This chapter deals with what a function is, how to graph functions, properties of functions, and how functions are used in applications. The word *function* apparently was introduced by René Descartes in 1637. For him, a function was simply any positive integral power of a variable *x*. Gottfried Wilhelm Leibniz (1646–1716), who always emphasized the geometric side of mathematics, used the word *function* to denote any quantity associated with a curve, such as the coordinates of a point on the curve. Leonhard Euler (1707–1783) used the word to mean any equation or formula involving variables and constants. His idea of a function is similar to the one most often seen in courses that precede calculus. Later, the use of functions in investigating heat flow equations led to a very broad definition that originated with Lejeune Dirichlet (1805–1859), which describes a function as a correspondence between two sets. That is the definition used in this text.

PREPARING FOR THIS SECTION *Before getting started, review the following:*

- Algebra Essentials (Section A.1, pp. A1–A10)
- Geometry Essentials (Section A.2, pp. A14–A19)

Now Work the 'Are You Prepared?' problems on page 5.

OBJECTIVES **1** Use the Distance Formula (p. 3)
            2 Use the Midpoint Formula (p. 5)

Figure 1 *xy*-Plane

Figure 2

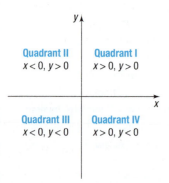

Figure 3

Rectangular Coordinates

We locate a point on the real number line by assigning it a single real number, called the *coordinate of the point*. For work in a two-dimensional plane, we locate points by using two numbers.

Begin with two real number lines located in the same plane: one horizontal and the other vertical. The horizontal line is called the **x-axis**, the vertical line the **y-axis**, and the point of intersection the **origin O**. See Figure 1. Assign coordinates to every point on these number lines using a convenient scale. In mathematics, we usually use the same scale on each axis, but in applications, different scales appropriate to the application may be used.

The origin *O* has a value of 0 on both the *x*-axis and the *y*-axis. Points on the *x*-axis to the right of *O* are associated with positive real numbers, and those to the left of *O* are associated with negative real numbers. Points on the *y*-axis above *O* are associated with positive real numbers, and those below *O* are associated with negative real numbers. In Figure 1, the *x*-axis and *y*-axis are labeled as *x* and *y*, respectively, and an arrow at the end of each axis is used to denote the positive direction.

The coordinate system described here is called a **rectangular** or **Cartesian*** **coordinate system**. The *x*-axis and *y*-axis lie in a *plane* called the **xy-plane**, and the *x*-axis and *y*-axis are referred to as the **coordinate axes**.

Any point *P* in the *xy*-plane can be located by using an **ordered pair** (x, y) of real numbers. Let *x* denote the signed distance of *P* from the *y*-axis (*signed* means that if *P* is to the right of the *y*-axis, then $x > 0$, and if *P* is to the left of the *y*-axis, then $x < 0$); and let *y* denote the signed distance of *P* from the *x*-axis. The ordered pair (x, y), also called the **coordinates** of *P*, gives us enough information to locate the point *P* in the plane.

For example, to locate the point whose coordinates are $(-3, 1)$, go 3 units along the *x*-axis to the left of *O* and then go straight up 1 unit. We **plot** this point by placing a dot at this location. See Figure 2, in which the points with coordinates $(-3, 1)$, $(-2, -3)$, $(3, -2)$, and $(3, 2)$ are plotted.

The origin has coordinates $(0, 0)$. Any point on the *x*-axis has coordinates of the form $(x, 0)$, and any point on the *y*-axis has coordinates of the form $(0, y)$.

If (x, y) are the coordinates of a point *P*, then *x* is called the **x-coordinate**, or **abscissa**, of *P*, and *y* is the **y-coordinate**, or **ordinate**, of *P*. We identify the point *P* by its coordinates (x, y) by writing $P = (x, y)$. Usually, we will simply say "the point (x, y)" rather than "the point whose coordinates are (x, y)."

The coordinate axes partition the *xy*-plane into four sections called **quadrants**, as shown in Figure 3. In quadrant I, both the *x*-coordinate and the *y*-coordinate of all points are positive; in quadrant II, *x* is negative and *y* is positive; in quadrant III, both *x* and *y* are negative; and in quadrant IV, *x* is positive and *y* is negative. Points on the coordinate axes belong to no quadrant.

Now Work PROBLEM 15

---

*Named after René Descartes (1596–1650), a French mathematician, philosopher, and theologian.

COMMENT On a graphing calculator, you can set the scale on each axis. Once this has been done, you obtain the **viewing rectangle**. See Figure 4 for a typical viewing rectangle. You should now read Section B.1, *The Viewing Rectangle*.

Figure 4 TI-84 Plus C Standard Viewing Rectangle

1 Use the Distance Formula

If the same units of measurement (such as inches, centimeters, and so on) are used for both the *x*-axis and *y*-axis, then all distances in the *xy*-plane can be measured using this unit of measurement.

EXAMPLE 1

Finding the Distance between Two Points

Find the distance *d* between the points $(1, 3)$ and $(5, 6)$.

Solution

First plot the points $(1, 3)$ and $(5, 6)$ and connect them with a line segment. See Figure 5(a). To find the length *d*, begin by drawing a horizontal line segment from $(1,3)$ to $(5,3)$ and a vertical line segment from $(5,3)$ to $(5,6)$, forming a right triangle, as shown in Figure 5(b). One leg of the triangle is of length 4 (since $|5 - 1| = 4$), and the other is of length 3 (since $|6 - 3| = 3$). By the Pythagorean Theorem, the square of the distance *d* that we seek is

$$d^2 = 4^2 + 3^2 = 16 + 9 = 25$$
$$d = \sqrt{25} = 5$$

Need to Review?

The Pythagorean Theorem and its converse are discussed in Section A.2, pp. A14–A15.

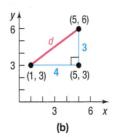

(a)   (b)

Figure 5

The **distance formula** provides a straightforward method for computing the distance between two points.

In Words

To compute the distance between two points, find the difference of the *x*-coordinates, square it, and add this to the square of the difference of the *y*-coordinates. The square root of this sum is the distance.

THEOREM Distance Formula

The distance between two points $P_1 = (x_1, y_1)$ and $P_2 = (x_2, y_2)$, denoted by $d(P_1, P_2)$, is

$$d(P_1, P_2) = \sqrt{(x_2 - x_1)^2 + (y_2 - y_1)^2} \tag{1}$$

EXAMPLE 2

Using the Distance Formula

Find the distance d between the points $(-4, 5)$ and $(3, 2)$.

Solution Using the distance formula, equation (1), reveals that the distance d is

$$d = \sqrt{[3 - (-4)]^2 + (2 - 5)^2} = \sqrt{7^2 + (-3)^2}$$
$$= \sqrt{49 + 9} = \sqrt{58} \approx 7.62$$

──── **Now Work** PROBLEMS **19** AND **23**

The distance between two points $P_1 = (x_1, y_1)$ and $P_2 = (x_2, y_2)$ is never a negative number. Also, the distance between two points is 0 only when the points are identical—that is, when $x_1 = x_2$ and $y_1 = y_2$. And, because $(x_2 - x_1)^2 = (x_1 - x_2)^2$ and $(y_2 - y_1)^2 = (y_1 - y_2)^2$, it makes no difference whether the distance is computed from P_1 to P_2 or from P_2 to P_1; that is, $d(P_1, P_2) = d(P_2, P_1)$.

The introduction to this chapter mentioned that algebra and geometry are connected by the rectangular coordinate system. The next example shows how algebra (the distance formula) can be used to solve geometry problems.

EXAMPLE 3

Using Algebra to Solve a Geometry Problem

Consider the three points $A = (-2, 1)$, $B = (2, 3)$, and $C = (3, 1)$.

(a) Plot each point and form the triangle ABC.
(b) Find the length of each side of the triangle.
(c) Show that the triangle is a right triangle.
(d) Find the area of the triangle.

Solution
(a) Figure 6 shows the points A, B, C and the triangle ABC.

(b) To find the length of each side of the triangle, use the distance formula, equation (1).

$$d(A, B) = \sqrt{[2 - (-2)]^2 + (3 - 1)^2} = \sqrt{16 + 4} = \sqrt{20} = 2\sqrt{5}$$
$$d(B, C) = \sqrt{(3 - 2)^2 + (1 - 3)^2} = \sqrt{1 + 4} = \sqrt{5}$$
$$d(A, C) = \sqrt{[3 - (-2)]^2 + (1 - 1)^2} = \sqrt{25 + 0} = 5$$

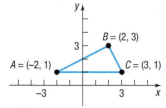

Figure 6

(c) If the sum of the squares of the lengths of two of the sides equals the square of the length of the third side, then the triangle is a right triangle. Looking at Figure 6, it seems reasonable to conjecture that the angle at vertex B might be a right angle. We shall check to see whether

$$[d(A, B)]^2 + [d(B, C)]^2 = [d(A, C)]^2$$

Using the results in part (b) yields

$$[d(A, B)]^2 + [d(B, C)]^2 = (2\sqrt{5})^2 + (\sqrt{5})^2$$
$$= 20 + 5 = 25 = [d(A, C)]^2$$

It follows from the converse of the Pythagorean Theorem that triangle ABC is a right triangle.

(d) Because the right angle is at vertex B, the sides AB and BC form the base and height of the triangle. Its area is

$$\text{Area} = \frac{1}{2} \cdot \text{Base} \cdot \text{Height} = \frac{1}{2} \cdot 2\sqrt{5} \cdot \sqrt{5} = 5 \text{ square units}$$

──── **Now Work** PROBLEM **33**

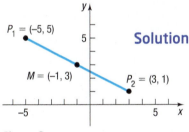

Figure 7

2 Use the Midpoint Formula

We now derive a formula for the coordinates of the **midpoint of a line segment**. Let $P_1 = (x_1, y_1)$ and $P_2 = (x_2, y_2)$ be the endpoints of a line segment, and let $M = (x, y)$ be the point on the line segment that is the same distance from P_1 as it is from P_2. See Figure 7. The triangles P_1AM and MBP_2 are congruent. [Do you see why? $d(P_1, M) = d(M, P_2)$ is given; also, $\angle AP_1M = \angle BMP_2$* and $\angle P_1MA = \angle MP_2B$. So, we have angle–side–angle.] Because triangles P_1AM and MBP_2 are congruent, corresponding sides are equal in length. That is,

$$x - x_1 = x_2 - x \quad \text{and} \quad y - y_1 = y_2 - y$$
$$2x = x_1 + x_2 \qquad\qquad 2y = y_1 + y_2$$
$$x = \frac{x_1 + x_2}{2} \qquad\qquad y = \frac{y_1 + y_2}{2}$$

> **In Words**
> To find the midpoint of a line segment, average the x-coordinates of the endpoints, and average the y-coordinates of the endpoints.

THEOREM Midpoint Formula

The midpoint $M = (x, y)$ of the line segment from $P_1 = (x_1, y_1)$ to $P_2 = (x_2, y_2)$ is

$$M = (x, y) = \left(\frac{x_1 + x_2}{2}, \frac{y_1 + y_2}{2} \right) \qquad (2)$$

EXAMPLE 4

Finding the Midpoint of a Line Segment

Find the midpoint of the line segment from $P_1 = (-5, 5)$ to $P_2 = (3, 1)$. Plot the points P_1 and P_2 and their midpoint.

Solution

Use the midpoint formula (2) with $x_1 = -5$, $y_1 = 5$, $x_2 = 3$, and $y_2 = 1$. The coordinates (x, y) of the midpoint M are

$$x = \frac{x_1 + x_2}{2} = \frac{-5 + 3}{2} = -1 \quad \text{and} \quad y = \frac{y_1 + y_2}{2} = \frac{5 + 1}{2} = 3$$

That is, $M = (-1, 3)$. See Figure 8.

Figure 8

➤ **Now Work** PROBLEM 39

*A postulate from geometry states that the transversal $\overline{P_1P_2}$ forms congruent corresponding angles with the parallel line segments $\overline{P_1A}$ and $\overline{MB}$.

1.1 Assess Your Understanding

'Are You Prepared?' *Answers are given at the end of these exercises. If you get a wrong answer, read the pages listed in* red.

1. On the real number line, the origin is assigned the number _____. (p. A4)

2. If -3 and 5 are the coordinates of two points on the real number line, the distance between these points is _____. (pp. A5–A6)

3. If 3 and 4 are the legs of a right triangle, the hypotenuse is _____. (p. A14)

4. Use the converse of the Pythagorean Theorem to show that a triangle whose sides are of lengths 11, 60, and 61 is a right triangle. (pp. A14–A15)

5. The area A of a triangle whose base is b and whose altitude is h is $A =$ _____. (p. A15)

6. **True or False** Two triangles are congruent if two angles and the included side of one equals two angles and the included side of the other. (pp. A16–A17)

Concepts and Vocabulary

7. If (x, y) are the coordinates of a point P in the xy-plane, then x is called the _____ of P, and y is the _____ of P.

8. The coordinate axes partition the xy-plane into four sections called _____.

9. If three distinct points P, Q, and R all lie on a line, and if $d(P, Q) = d(Q, R)$, then Q is called the _____ of the line segment from P to R.

10. *True or False* The distance between two points is sometimes a negative number.

11. *True or False* The point $(-1, 4)$ lies in quadrant IV of the Cartesian plane.

12. *True or False* The midpoint of a line segment is found by averaging the x-coordinates and averaging the y-coordinates of the endpoints.

13. *Multiple Choice* Which of the following statements is true for a point (x, y) that lies in quadrant III?
(a) Both x and y are positive.
(b) Both x and y are negative.
(c) x is positive, and y is negative.
(d) x is negative, and y is positive.

14. *Multiple Choice* Choose the expression that equals the distance between two points (x_1, y_1) and (x_2, y_2).
(a) $\sqrt{(x_2 - x_1)^2 + (y_2 - y_1)^2}$
(b) $\sqrt{(x_2 + x_1)^2 - (y_2 + y_1)^2}$
(c) $\sqrt{(x_2 - x_1)^2 - (y_2 - y_1)^2}$
(d) $\sqrt{(x_2 + x_1)^2 + (y_2 + y_1)^2}$

Skill Building

In Problems 15 and 16, plot each point in the xy-plane. State which quadrant or on what coordinate axis each point lies.

15. (a) $A = (-3, 2)$ **(d)** $D = (6, 5)$
(b) $B = (6, 0)$ **(e)** $E = (0, -3)$
(c) $C = (-2, -2)$ **(f)** $F = (6, -3)$

16. (a) $A = (1, 4)$ **(d)** $D = (4, 1)$
(b) $B = (-3, -4)$ **(e)** $E = (0, 1)$
(c) $C = (-3, 4)$ **(f)** $F = (-3, 0)$

17. Plot the points $(2, 0)$, $(2, -3)$, $(2, 4)$, $(2, 1)$, and $(2, -1)$. Describe the set of all points of the form $(2, y)$, where y is a real number.

18. Plot the points $(0, 3)$, $(1, 3)$, $(-2, 3)$, $(5, 3)$, and $(-4, 3)$. Describe the set of all points of the form $(x, 3)$, where x is a real number.

In Problems 19–32, find the distance d between the points P_1 and P_2.

19.   **20.**   **21.**   **22.**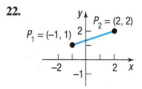

23. $P_1 = (3, -4)$; $P_2 = (5, 4)$

25. $P_1 = (-7, 3)$; $P_2 = (4, 0)$

27. $P_1 = (5, -2)$; $P_2 = (6, 1)$

29. $P_1 = (-0.2, 0.3)$; $P_2 = (2.3, 1.1)$

31. $P_1 = (a, b)$; $P_2 = (0, 0)$

24. $P_1 = (-1, 0)$; $P_2 = (2, 4)$

26. $P_1 = (2, -3)$; $P_2 = (4, 2)$

28. $P_1 = (-4, -3)$; $P_2 = (6, 2)$

30. $P_1 = (1.2, 2.3)$; $P_2 = (-0.3, 1.1)$

32. $P_1 = (a, a)$; $P_2 = (0, 0)$

In Problems 33–38, plot each point and form the triangle ABC. Show that the triangle is a right triangle. Find its area.

33. $A = (-2, 5)$; $B = (1, 3)$; $C = (-1, 0)$

35. $A = (-5, 3)$; $B = (6, 0)$; $C = (5, 5)$

37. $A = (4, -3)$; $B = (0, -3)$; $C = (4, 2)$

34. $A = (-2, 5)$; $B = (12, 3)$; $C = (10, -11)$

36. $A = (-6, 3)$; $B = (3, -5)$; $C = (-1, 5)$

38. $A = (4, -3)$; $B = (4, 1)$; $C = (2, 1)$

In Problems 39–46, find the midpoint of the line segment joining the points P_1 and P_2.

39. $P_1 = (3, -4)$; $P_2 = (5, 4)$

41. $P_1 = (-1, 4)$; $P_2 = (8, 0)$

43. $P_1 = (7, -5)$; $P_2 = (9, 1)$

45. $P_1 = (a, b)$; $P_2 = (0, 0)$

40. $P_1 = (-2, 0)$; $P_2 = (2, 4)$

42. $P_1 = (2, -3)$; $P_2 = (4, 2)$

44. $P_1 = (-4, -3)$; $P_2 = (2, 2)$

46. $P_1 = (a, a)$; $P_2 = (0, 0)$

Applications and Extensions

47. If the point $(2, 5)$ is shifted 3 units to the right and 2 units down, what are its new coordinates?

48. If the point $(-1, 6)$ is shifted 2 units to the left and 4 units up, what are its new coordinates?

49. Find all points having an x-coordinate of 3 whose distance from the point $(-2, -1)$ is 13.

   (a) By using the Pythagorean Theorem.

   (b) By using the distance formula.

50. Find all points having a y-coordinate of -6 whose distance from the point $(1, 2)$ is 17.

   (a) By using the Pythagorean Theorem.

   (b) By using the distance formula.

51. Find all points on the x-axis that are 6 units from the point $(4, -3)$.

52. Find all points on the y-axis that are 6 units from the point $(4, -3)$.

53. Suppose that $A = (2, 5)$ are the coordinates of a point in the xy-plane.

   (a) Find the coordinates of the point if A is shifted 3 units to the left and 4 units down.

   (b) Find the coordinates of the point if A is shifted 2 units to the left and 8 units up.

54. Plot the points $A = (-1, 8)$ and $M = (2, 3)$ in the xy-plane. If M is the midpoint of a line segment AB, find the coordinates of B.

55. The midpoint of the line segment from P_1 to P_2 is $(-1, 4)$. If $P_1 = (-3, 6)$, what is P_2?

56. The midpoint of the line segment from P_1 to P_2 is $(5, -4)$. If $P_2 = (7, -2)$, what is P_1?

57. Geometry The **medians** of a triangle are the line segments from each vertex to the midpoint of the opposite side (see the figure). Find the lengths of the medians of the triangle with vertices at $A = (0, 0)$, $B = (6, 0)$, and $C = (4, 4)$.

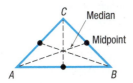

58. Geometry An **equilateral triangle** has three sides of equal length. If two vertices of an equilateral triangle are $(0, 4)$ and $(0, 0)$ find the third vertex. How many of these triangles are possible?

*In Problems 59–62, find the length of each side of the triangle determined by the three points P_1, P_2, and P_3. State whether the triangle is an isosceles triangle, a right triangle, neither of these, or both. (An **isosceles triangle** is one in which at least two of the sides are of equal length.)*

59. $P_1 = (2, 1)$; $P_2 = (-4, 1)$; $P_3 = (-4, -3)$

60. $P_1 = (-1, 4)$; $P_2 = (6, 2)$; $P_3 = (4, -5)$

61. $P_1 = (-2, -1)$; $P_2 = (0, 7)$; $P_3 = (3, 2)$

62. $P_1 = (7, 2)$; $P_2 = (-4, 0)$; $P_3 = (4, 6)$

63. Baseball A major league baseball "diamond" is actually a square 90 feet on a side (see the figure). What is the distance directly from home plate to second base (the diagonal of the square)?

64. Little League Baseball The layout of a Little League playing field is a square 60 feet on a side. How far is it directly from home plate to second base (the diagonal of the square)?

Source: 2018 Little League Baseball Official Regulations, Playing Rules, and Operating Policies

65. Baseball Refer to Problem 63. Overlay a rectangular coordinate system on a major league baseball diamond so that the origin is at home plate, the positive x-axis lies in the direction from home plate to first base, and the positive y-axis lies in the direction from home plate to third base.

   (a) What are the coordinates of first base, second base, and third base? Use feet as the unit of measurement.

   (b) If the right fielder is located at $(310, 15)$, how far is it from the right fielder to second base?

   (c) If the center fielder is located at $(300, 300)$, how far is it from the center fielder to third base?

66. Little League Baseball Refer to Problem 64. Overlay a rectangular coordinate system on a Little League baseball diamond so that the origin is at home plate, the positive x-axis lies in the direction from home plate to first base, and the positive y-axis lies in the direction from home plate to third base.

   (a) What are the coordinates of first base, second base, and third base? Use feet as the unit of measurement.

   (b) If the right fielder is located at $(180, 20)$, how far is it from the right fielder to second base?

   (c) If the center fielder is located at $(220, 220)$, how far is it from the center fielder to third base?

67. Distance between Moving Objects A Ford Focus and a Freightliner Cascadia truck leave an intersection at the same time. The Focus heads east at an average speed of 60 miles per hour, while the Cascadia heads south at an average speed of 45 miles per hour. Find an expression for their distance apart d (in miles) at the end of t hours.

68. Distance of a Moving Object from a Fixed Point A hot-air balloon, headed due east at an average speed of 15 miles per hour and at a constant altitude of 100 feet, passes over an intersection (see the figure). Find an expression for the distance d (measured in feet) from the balloon to the intersection t seconds later.

69. Drafting Error When a draftsman draws three lines that are to intersect at one point, the lines may not intersect as intended and subsequently will form an **error triangle**. If this error triangle is long and thin, one estimate for the location of the desired point is the midpoint of the shortest side. The figure shows one such error triangle.

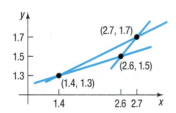

(a) Find an estimate for the desired intersection point.
(b) Find the distance from $(1.4, 1.3)$ to the midpoint found in part (a).

70. Net Sales The figure at the top of the next column illustrates the net sales growth of Costco Wholesale Corporation from 2013 through 2017. Use the midpoint formula to estimate the net sales of Costco Wholesale Corporation in 2015. How does your result compare to the reported value of $113.67 billion?
Source: Costco Wholesale Corporation 2017 Annual Report

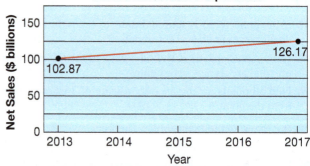

Costco Wholesale Corporation

71. Poverty Threshold Poverty thresholds are determined by the U.S. Census Bureau. A poverty threshold represents the minimum annual household income for a family not to be considered poor. In 2009, the poverty threshold for a family of four with two children under the age of 18 years was $21,756. In 2017, the poverty threshold for a family of four with two children under the age of 18 years was $24,858. Assuming that poverty thresholds increase in a straight-line fashion, use the midpoint formula to estimate the poverty threshold for a family of four with two children under the age of 18 in 2013. How does your result compare to the actual poverty threshold in 2013 of $23,624?
Source: U.S. Census Bureau

72. Challenge Problem Geometry Verify that the points $(0,0), (a,0)$, and $\left(\dfrac{a}{2}, \dfrac{\sqrt{3}a}{2}\right)$ are the vertices of an equilateral triangle. Then show that the midpoints of the three sides are the vertices of a second equilateral triangle.

73. Challenge Problem Geometry Find the midpoint of each diagonal of a square with side of length s. Draw the conclusion that the diagonals of a square intersect at their midpoints.
[**Hint:** Use $(0, 0), (0, s), (s, 0)$, and (s, s) as the vertices of the square.]

74. Challenge Problem Geometry A point P is equidistant from $(-5, 1)$ and $(4, -4)$. Find the coordinates of P if its y-coordinate is twice its x-coordinate.

75. Challenge Problem Geometry For any parallelogram, prove that the sum of the squares of the lengths of the sides equals the sum of the squares of the lengths of the diagonals.
[**Hint:** Use $(0, 0), (a, 0), (a + b, c)$, and (b, c) as the vertices of the parallelogram. Assume a, b, and c are positive.]

Explaining Concepts: Discussion and Writing

76. Write a paragraph that describes a Cartesian plane. Then write a second paragraph that describes how to plot points in the Cartesian plane. Your paragraphs should include the terms "coordinate axes," "ordered pair," "coordinates," "plot," "x-coordinate," and "y-coordinate."

'Are You Prepared?' Answers

1. 0   **2.** 8   **3.** 5   **4.** $11^2 + 60^2 = 121 + 3600 = 3721 = 61^2$ **5.** $\dfrac{1}{2}bh$ **6.** True

1.2 Graphs of Equations in Two Variables; Circles

PREPARING FOR THIS SECTION *Before getting started, review the following:*

- Solving Equations (Section A.4, pp. A27–A34)
- Complete the Square (Section A.3, pp. A24–A25)

✎ **Now Work** the 'Are You Prepared?' problems on page 19.

OBJECTIVES
1 Graph Equations by Plotting Points (p. 9)
2 Find Intercepts from a Graph (p. 11)
3 Find Intercepts from an Equation (p. 11)
4 Test an Equation for Symmetry with Respect to the *x*-Axis, the *y*-Axis, and the Origin (p. 12)
5 Know How to Graph Key Equations (p. 14)
6 Write the Standard Form of the Equation of a Circle (p. 16)
7 Graph a Circle (p. 17)
8 Work with the General Form of the Equation of a Circle (p. 18)

1 Graph Equations by Plotting Points

An **equation in two variables**, say *x* and *y*, is a statement in which two expressions involving *x* and *y* are equal. The expressions are called the **sides** of the equation. Since an equation is a statement, it may be true or false, depending on the value of the variables. Any pair of values for *x* and *y* that result in a true statement are said to **satisfy** the equation.

For example, the following are all equations in two variables *x* and *y*:

$$x^2 + y^2 = 5 \qquad 2x - y = 6 \qquad y = 2x + 5 \qquad x^2 = y$$

The first of these, $x^2 + y^2 = 5$, is satisfied for $x = 1, y = 2$, since $1^2 + 2^2 = 5$. Other choices of *x* and *y*, such as $x = -1, y = -2$, also satisfy this equation. It is not satisfied for $x = 2$ and $y = 3$, since $2^2 + 3^2 = 4 + 9 = 13 \neq 5$.

The **graph of an equation in two variables** *x* and *y* consists of the set of points in the *xy*-plane whose coordinates (x, y) satisfy the equation.

EXAMPLE 1 **Determining Whether a Point Is on the Graph of an Equation**

Determine if the following points are on the graph of the equation $2x - y = 6$.

(a) $(2, 3)$ (b) $(2, -2)$

Solution (a) For the point $(2, 3)$, check to see whether $x = 2, y = 3$ satisfies the equation $2x - y = 6$.

$$2x - y = 2 \cdot 2 - 3 = 4 - 3 = 1 \neq 6$$

The equation is not satisfied, so the point $(2, 3)$ is not on the graph of $2x - y = 6$.
(b) For the point $(2, -2)$,

$$2x - y = 2 \cdot 2 - (-2) = 4 + 2 = 6$$

The equation is satisfied, so the point $(2, -2)$ is on the graph of $2x - y = 6$.

Now Work PROBLEM 13

| EXAMPLE 2 | Graphing an Equation by Plotting Points |

Graph the equation: $y = 2x + 5$

Solution

The graph consists of all points (x, y) that satisfy the equation. To locate some of these points (and get an idea of the pattern of the graph), assign some numbers to x, and find corresponding values for y.

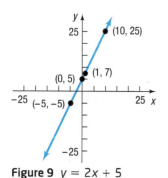

Figure 9 $y = 2x + 5$

If	Then	Point on Graph
$x = 0$	$y = 2 \cdot 0 + 5 = 5$	$(0, 5)$
$x = 1$	$y = 2 \cdot 1 + 5 = 7$	$(1, 7)$
$x = -5$	$y = 2 \cdot (-5) + 5 = -5$	$(-5, -5)$
$x = 10$	$y = 2 \cdot 10 + 5 = 25$	$(10, 25)$

By plotting these points and then connecting them, we obtain the graph (a *line*) of the equation $y = 2x + 5$, as shown in Figure 9.

| EXAMPLE 3 | Graphing an Equation by Plotting Points |

Graph the equation: $y = x^2$

Solution

Table 1 provides several points on the graph of $y = x^2$. Plotting these points and connecting them with a smooth curve gives the graph (a *parabola*) shown in Figure 10.

Table 1

x	$y = x^2$	(x, y)
-4	16	$(-4, 16)$
-3	9	$(-3, 9)$
-2	4	$(-2, 4)$
-1	1	$(-1, 1)$
0	0	$(0, 0)$
1	1	$(1, 1)$
2	4	$(2, 4)$
3	9	$(3, 9)$
4	16	$(4, 16)$

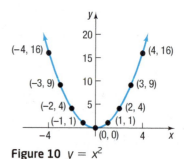

Figure 10 $y = x^2$

The graphs of the equations shown in Figures 9 and 10 do not show all points. For example, in Figure 9, the point $(20, 45)$ is a part of the graph of $y = 2x + 5$, but it is not shown. Since the graph of $y = 2x + 5$ can be extended out indefinitely, we use arrows to indicate that the pattern shown continues. It is important, when showing a graph, to present enough of the graph so that any viewer of the illustration will "see" the rest of it as an obvious continuation of what is actually there. This is referred to as a **complete graph**.

One way to obtain the complete graph of an equation is to plot enough points on the graph for a pattern to become evident. Then these points are connected with a smooth curve following the suggested pattern. But how many points are sufficient? Sometimes knowledge about the equation tells us. For example, we discuss in Section A.8 that if an equation is of the form $y = mx + b$, then its graph is a line. In this case, only two points are needed to obtain the complete graph.

One purpose of this text is to investigate the properties of equations in order to decide whether a graph is complete. Sometimes we shall graph equations by plotting points. Shortly, we shall investigate various techniques that will enable us to graph an equation without plotting so many points.

Two techniques that sometimes reduce the number of points required to graph an equation involve finding *intercepts* and checking for *symmetry*.

 COMMENT Another way to obtain the graph of an equation is to use a graphing utility. Read Section B.2, *Using a Graphing Utility to Graph Equations.* ∎

2 Find Intercepts from a Graph

The points, if any, at which a graph crosses or touches the coordinate axes are called the **intercepts** of the graph. See Figure 11. The *x*-coordinate of a point at which the graph crosses or touches the *x*-axis is an **x-intercept**, and the *y*-coordinate of a point at which the graph crosses or touches the *y*-axis is a **y-intercept**.

> **In Words**
> Intercepts are points (ordered pairs). An *x*-intercept or a *y*-intercept is a number. For example, the point (3, 0) is an intercept; the number 3 is an *x*-intercept.

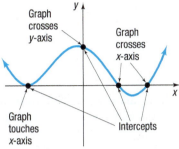

Figure 11

| EXAMPLE 4 | Finding Intercepts from a Graph |

Find the intercepts of the graph in Figure 12. What are its *x*-intercepts? What are its *y*-intercepts?

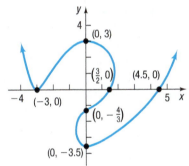

Figure 12

Solution The intercepts of the graph are the points

$$(-3, 0), \quad (0, 3), \quad \left(\frac{3}{2}, 0\right), \quad \left(0, -\frac{4}{3}\right), \quad (0, -3.5), \quad (4.5, 0)$$

The *x*-intercepts are -3, $\frac{3}{2}$, and 4.5; the *y*-intercepts are -3.5, $-\frac{4}{3}$, and 3.

In Example 4, notice that intercepts are listed as ordered pairs, and the *x*-intercepts and the *y*-intercepts are listed as numbers. We use this distinction throughout the text.

 Now Work PROBLEM **41(a)**

3 Find Intercepts from an Equation

The intercepts of a graph can be found from its equation by using the fact that points on the *x*-axis have *y*-coordinates equal to 0, and points on the *y*-axis have *x*-coordinates equal to 0.

COMMENT For many equations, finding intercepts may not be so easy. In such cases, a graphing utility can be used. Read the first part of Section B.3, *Using a Graphing Utility to Locate Intercepts and Check for Symmetry*, to find out how to locate intercepts using a graphing utility. ∎

> **Procedure for Finding Intercepts**
> - To find the *x*-intercept(s), if any, of the graph of an equation, let $y = 0$ in the equation and solve for *x*, where *x* is a real number.
> - To find the *y*-intercept(s), if any, of the graph of an equation, let $x = 0$ in the equation and solve for *y*, where *y* is a real number.

| EXAMPLE 5 | Finding Intercepts from an Equation |

Find the *x*-intercept(s) and the *y*-intercept(s) of the graph of $y = x^2 - 4$. Then graph $y = x^2 - 4$ by plotting points.

Solution To find the x-intercept(s), let $y = 0$ and obtain the equation

$$x^2 - 4 = 0 \quad \text{\textcolor{red}{$y = x^2 - 4$ with $y = 0$}}$$
$$(x + 2)(x - 2) = 0 \quad \text{\textcolor{red}{Factor.}}$$
$$x + 2 = 0 \quad \text{or} \quad x - 2 = 0 \quad \text{\textcolor{red}{Use the Zero-Product Property.}}$$
$$x = -2 \quad \text{or} \quad x = 2 \quad \text{\textcolor{red}{Solve.}}$$

The equation has two solutions, -2 and 2. The x-intercepts are -2 and 2.
To find the y-intercept(s), let $x = 0$ in the equation.

$$y = x^2 - 4$$
$$= 0^2 - 4 = -4$$

The y-intercept is -4.

Since $x^2 \geq 0$ for all x, we deduce from the equation $y = x^2 - 4$ that $y \geq -4$ for all x. This information, the intercepts, and the points from Table 2 enable us to graph $y = x^2 - 4$. See Figure 13.

Table 2

x	$y = x^2 - 4$	(x, y)
-3	5	$(-3, 5)$
-1	-3	$(-1, -3)$
1	-3	$(1, -3)$
3	5	$(3, 5)$

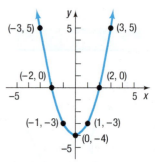

Figure 13 $y = x^2 - 4$

- **Now Work** PROBLEM 23

4 Test an Equation for Symmetry with Respect to the x-Axis, the y-Axis, and the Origin

Another helpful tool for graphing equations by hand involves *symmetry*, particularly symmetry with respect to the x-axis, the y-axis, and the origin.

Symmetry often occurs in nature. Consider the picture of the butterfly. Do you see the symmetry?

> **DEFINITION** Symmetry with Respect to the x-Axis
>
> A graph is **symmetric with respect to the x-axis** if, for every point (x, y) on the graph, the point $(x, -y)$ is also on the graph.

Figure 14 Symmetry with respect to the x-axis

Figure 14 illustrates the definition. Note that when a graph is symmetric with respect to the x-axis, the part of the graph above the x-axis is a reflection (or mirror image) of the part below it, and vice versa.

EXAMPLE 6 **Points Symmetric with Respect to the x-Axis**

If a graph is symmetric with respect to the x-axis, and the point $(3, 2)$ is on the graph, then the point $(3, -2)$ is also on the graph.

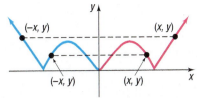

Figure 15 Symmetry with respect to the *y*-axis

> **DEFINITION** Symmetry with Respect to the *y*-Axis
>
> A graph is **symmetric with respect to the *y*-axis** if, for every point (x, y) on the graph, the point $(-x, y)$ is also on the graph.

Figure 15 illustrates the definition. When a graph is symmetric with respect to the *y*-axis, the part of the graph to the right of the *y*-axis is a reflection of the part to the left of it, and vice versa.

EXAMPLE 7   **Points Symmetric with Respect to the *y*-Axis**

If a graph is symmetric with respect to the *y*-axis and the point $(5, 8)$ is on the graph, then the point $(-5, 8)$ is also on the graph.

> **DEFINITION** Symmetry with Respect to the Origin
>
> A graph is **symmetric with respect to the origin** if, for every point (x, y) on the graph, the point $(-x, -y)$ is also on the graph.

Figure 16 Symmetry with respect to the origin

Figure 16 illustrates the definition. Symmetry with respect to the origin may be viewed in three ways:

- As a reflection about the *y*-axis, followed by a reflection about the *x*-axis
- As a projection along a line through the origin so that the distances from the origin are equal
- As half of a complete revolution about the origin

EXAMPLE 8   **Points Symmetric with Respect to the Origin**

If a graph is symmetric with respect to the origin, and the point $(4, -2)$ is on the graph, then the point $(-4, 2)$ is also on the graph.

 Now Work   PROBLEMS 31 AND 41(b)

When the graph of an equation is symmetric with respect to a coordinate axis or the origin, the number of points that you need to plot in order to see the pattern is reduced. For example, if the graph of an equation is symmetric with respect to the *y*-axis, then once points to the right of the *y*-axis are plotted, an equal number of points on the graph can be obtained by reflecting them about the *y*-axis. Because of this, before we graph an equation, we should first determine whether it has any symmetry. The following tests are used for this purpose.

> **Tests for Symmetry**
>
> To test the graph of an equation for symmetry with respect to the
>
> - *x*-Axis Replace *y* by $-y$ in the equation and simplify. If an equivalent equation results, the graph of the equation is symmetric with respect to the *x*-axis.
> - *y*-Axis Replace *x* by $-x$ in the equation and simplify. If an equivalent equation results, the graph of the equation is symmetric with respect to the *y*-axis.
> - *Origin* Replace *x* by $-x$ and *y* by $-y$ in the equation and simplify. If an equivalent equation results, the graph of the equation is symmetric with respect to the origin.

EXAMPLE 9   **Testing an Equation for Symmetry**

Test $4x^2 + 9y^2 = 36$ for symmetry.

Solution   *x-Axis:*   To test for symmetry with respect to the *x*-axis, replace *y* by $-y$. Since $4x^2 + 9(-y)^2 = 36$ is equivalent to $4x^2 + 9y^2 = 36$, the graph of the equation is symmetric with respect to the *x*-axis.

y-Axis: To test for symmetry with respect to the *y*-axis, replace *x* by $-x$. Since $4(-x)^2 + 9y^2 = 36$ is equivalent to $4x^2 + 9y^2 = 36$, the graph of the equation is symmetric with respect to the *y*-axis.

Origin: To test for symmetry with respect to the origin, replace *x* by $-x$ and *y* by $-y$. Since $4(-x)^2 + 9(-y)^2 = 36$ is equivalent to $4x^2 + 9y^2 = 36$, the graph of the equation is symmetric with respect to the origin.

EXAMPLE 10   **Testing an Equation for Symmetry**

Test $y = \dfrac{4x^2}{x^2 + 1}$ for symmetry.

Solution   *x-Axis:*   To test for symmetry with respect to the *x*-axis, replace *y* by $-y$. Since $-y = \dfrac{4x^2}{x^2 + 1}$ is not equivalent to $y = \dfrac{4x^2}{x^2 + 1}$, the graph of the equation is not symmetric with respect to the *x*-axis.

y-Axis: To test for symmetry with respect to the *y*-axis, replace *x* by $-x$. Since $y = \dfrac{4(-x)^2}{(-x)^2 + 1} = \dfrac{4x^2}{x^2 + 1}$ is equivalent to $y = \dfrac{4x^2}{x^2 + 1}$, the graph of the equation is symmetric with respect to the *y*-axis.

Origin: To test for symmetry with respect to the origin, replace *x* by $-x$ and *y* by $-y$.

$$-y = \frac{4(-x)^2}{(-x)^2 + 1} \qquad \textcolor{blue}{\textbf{Replace } x \textbf{ by } -x \textbf{ and } y \textbf{ by } -y.}$$

$$-y = \frac{4x^2}{x^2 + 1} \qquad \textcolor{blue}{\textbf{Simplify.}}$$

$$y = -\frac{4x^2}{x^2 + 1} \qquad \textcolor{blue}{\textbf{Multiply both sides by } -1.}$$

Since the result is not equivalent to the original equation, the graph of the equation $y = \dfrac{4x^2}{x^2 + 1}$ is not symmetric with respect to the origin.

Seeing the Concept

Figure 17 shows the graph of $4x^2 + 9y^2 = 36$ using Desmos. Do you see the symmetry with respect to the *x*-axis, the *y*-axis, and the origin?

Figure 17 $4x^2 + 9y^2 = 36$

Seeing the Concept

Figure 18 shows the graph of $y = \dfrac{4x^2}{x^2 + 1}$ using a TI-84 Plus C graphing calculator. Do you see the symmetry with respect to the *y*-axis?

Figure 18 $y = \dfrac{4x^2}{x^2 + 1}$

Now Work PROBLEM 61

5 Know How to Graph Key Equations

The next three examples use intercepts, symmetry, and point plotting to obtain the graphs of key equations. It is important to know the graphs of these key equations because we use them later. The first of these is $y = x^3$.

EXAMPLE 11

Graphing the Equation $y = x^3$ by Finding Intercepts, Checking for Symmetry, and Plotting Points

Graph the equation $y = x^3$ by plotting points. Find any intercepts and check for symmetry first.

Solution

First, find the intercepts. When $x = 0$, then $y = 0$; and when $y = 0$, then $x = 0$. The origin $(0, 0)$ is the only intercept. Now test for symmetry.

x-Axis: Replace y by $-y$. Since $-y = x^3$ is not equivalent to $y = x^3$, the graph is not symmetric with respect to the x-axis.

y-Axis: Replace x by $-x$. Since $y = (-x)^3 = -x^3$ is not equivalent to $y = x^3$, the graph is not symmetric with respect to the y-axis.

Origin: Replace x by $-x$ and y by $-y$. Since $-y = (-x)^3 = -x^3$ is equivalent to $y = x^3$ (multiply both sides by -1), the graph is symmetric with respect to the origin.

To graph $y = x^3$, use the equation to obtain several points on the graph. Because of the symmetry, we need to locate only points on the graph for which $x \geq 0$. See Table 3. Since $(1, 1)$ is on the graph, and the graph is symmetric with respect to the origin, the point $(-1, -1)$ is also on the graph. Plot the points from Table 3 and use the symmetry. Figure 19 shows the graph.

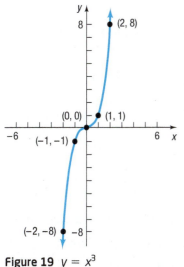

Figure 19 $y = x^3$

Table 3

x	y = x³	(x, y)
0	0	(0, 0)
1	1	(1, 1)
2	8	(2, 8)
3	27	(3, 27)

EXAMPLE 12

Graphing the Equation $x = y^2$

(a) Graph the equation $x = y^2$. Find any intercepts and check for symmetry first.
(b) Graph $x = y^2, y \geq 0$.

Solution

(a) The lone intercept is $(0, 0)$. The graph is symmetric with respect to the x-axis. (Do you see why? Replace y by $-y$.) Figure 20(a) shows the graph.

(b) If we restrict y so that $y \geq 0$, the equation $x = y^2, y \geq 0$, may be written equivalently as $y = \sqrt{x}$. The portion of the graph of $x = y^2$ in quadrant I is therefore the graph of $y = \sqrt{x}$. See Figure 20(b).

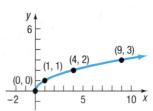

Figure 20 (a) $x = y^2$ **(b)** $y = \sqrt{x}$

Figure 21

COMMENT To see the graph of the equation $x = y^2$ on a graphing calculator, you will need to graph two equations: $Y_1 = \sqrt{x}$ and $Y_2 = -\sqrt{x}$. See Figure 21. ■

EXAMPLE 13

Graphing the Equation $y = \dfrac{1}{x}$

Graph the equation $y = \dfrac{1}{x}$. First, find any intercepts and check for symmetry.

Solution

Check for intercepts first. If we let $x = 0$, we obtain 0 in the denominator, which makes y undefined. We conclude that there is no y-intercept. If we let $y = 0$, we get the equation $\dfrac{1}{x} = 0$, which has no solution. We conclude that there is no x-intercept. The graph of $y = \dfrac{1}{x}$ does not cross or touch the coordinate axes.

Next check for symmetry:

x-Axis: Replacing y by $-y$ yields $-y = \dfrac{1}{x}$, which is not equivalent to $y = \dfrac{1}{x}$.

y-Axis: Replacing x by $-x$ yields $y = \dfrac{1}{-x} = -\dfrac{1}{x}$, which is not equivalent to $y = \dfrac{1}{x}$.

Origin: Replacing x by $-x$ and y by $-y$ yields $-y = -\dfrac{1}{x}$, which is equivalent to $y = \dfrac{1}{x}$. The graph is symmetric with respect to the origin.

Now set up Table 4, listing several points on the graph. Because of the symmetry with respect to the origin, we use only positive values of x. From Table 4 we infer that if x is a large and positive number, then $y = \dfrac{1}{x}$ is a positive number close to 0. We also infer that if x is a positive number close to 0, then $y = \dfrac{1}{x}$ is a large and positive number. Armed with this information, we can graph the equation.

Figure 22 illustrates some of these points and the graph of $y = \dfrac{1}{x}$. Observe how the absence of intercepts and the existence of symmetry with respect to the origin were utilized.

Table 4

x	$y = \dfrac{1}{x}$	(x, y)
$\dfrac{1}{10}$	10	$\left(\dfrac{1}{10}, 10\right)$
$\dfrac{1}{3}$	3	$\left(\dfrac{1}{3}, 3\right)$
$\dfrac{1}{2}$	2	$\left(\dfrac{1}{2}, 2\right)$
1	1	$(1, 1)$
2	$\dfrac{1}{2}$	$\left(2, \dfrac{1}{2}\right)$
3	$\dfrac{1}{3}$	$\left(3, \dfrac{1}{3}\right)$
10	$\dfrac{1}{10}$	$\left(10, \dfrac{1}{10}\right)$

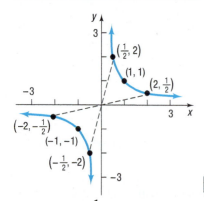

Figure 22 $y = \dfrac{1}{x}$

💾 **COMMENT** Refer to Example 2 in Section B.3, for the graph of $y = \dfrac{1}{x}$ using a graphing calculator. ∎

6 Write the Standard Form of the Equation of a Circle

One advantage of a coordinate system is that it enables us to translate a geometric statement into an algebraic statement, and vice versa. Consider, for example, the following geometric statement that defines a circle.

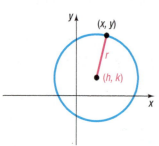

Figure 23
$(x - h)^2 + (y - k)^2 = r^2$

Need to Review?
The distance formula is discussed in Section 1.1, pp. 3–4.

> **DEFINITION   Circle**
>
> A **circle** is a set of points in the xy-plane that are a fixed distance r from a fixed point (h, k). The fixed distance r is called the **radius**, and the fixed point (h, k) is called the **center** of the circle.

Figure 23 shows the graph of a circle. To find the equation, let (x, y) represent the coordinates of any point on a circle with radius r and center (h, k). Then the distance between the points (x, y) and (h, k) must always equal r. That is, by the distance formula,

$$\sqrt{(x - h)^2 + (y - k)^2} = r$$

or, equivalently,

$$(x - h)^2 + (y - k)^2 = r^2$$

DEFINITION Standard Form of an Equation of a Circle

The **standard form of an equation of a circle** with radius r and center (h, k) is

$$(x - h)^2 + (y - k)^2 = r^2 \qquad\qquad \textbf{(1)}$$

THEOREM

The standard form of an equation of a circle of radius r with center at the origin $(0, 0)$ is

$$x^2 + y^2 = r^2$$

DEFINITION Unit Circle

If the radius $r = 1$, the circle whose center is at the origin is called the **unit circle** and has the equation

$$x^2 + y^2 = 1$$

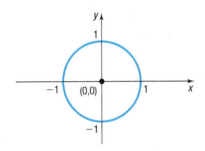

Figure 24
Unit circle $x^2 + y^2 = 1$

See Figure 24. Notice that the graph of the unit circle is symmetric with respect to the x-axis, the y-axis, and the origin.

EXAMPLE 14   **Writing the Standard Form of the Equation of a Circle**

Write the standard form of the equation of the circle with radius 5 and center $(-3, 6)$.

Solution   Substitute the values $r = 5$, $h = -3$, and $k = 6$ into equation (1).

$$(x - h)^2 + (y - k)^2 = r^2 \quad \textbf{Equation (1)}$$
$$[x - (-3)]^2 + (y - 6)^2 = 5^2$$
$$(x + 3)^2 + (y - 6)^2 = 25$$

━━━ **Now Work** PROBLEM **79**

---

7 Graph a Circle

EXAMPLE 15   **Graphing a Circle**

Graph the equation: $(x + 3)^2 + (y - 2)^2 = 16$

Solution   The given equation is in the standard form of an equation of a circle. To graph the circle, compare the equation to equation (1). The comparison gives information about the circle.

$$(x + 3)^2 + (y - 2)^2 = 16$$
$$(x - (-3))^2 + (y - 2)^2 = 4^2$$
$$(x - h)^2 \;+\; (y - k)^2 = r^2$$

We see that $h = -3$, $k = 2$, and $r = 4$. The circle has center $(-3, 2)$ and radius 4. To graph this circle, first plot the center $(-3, 2)$. Since the radius is 4, locate four points on the circle by plotting points 4 units to the left, to the right, up, and down from the center. These four points are then used as guides to obtain the graph. See Figure 25.

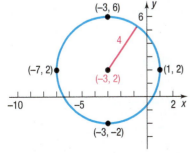

Figure 25 $(x + 3)^2 + (y - 2)^2 = 16$

━━━ **Now Work** PROBLEMS **95(a) AND (b)**

EXAMPLE 16

Finding the Intercepts of a Circle

Find the intercepts, if any, of the graph of the circle $(x + 3)^2 + (y - 2)^2 = 16$.

Solution

This is the equation graphed in Example 15. To find the x-intercepts, if any, let $y = 0$. Then

$$(x + 3)^2 + (y - 2)^2 = 16$$
$$(x + 3)^2 + (0 - 2)^2 = 16 \qquad \textcolor{blue}{y = 0}$$
$$(x + 3)^2 + 4 = 16 \qquad \textcolor{blue}{\text{Simplify.}}$$
$$(x + 3)^2 = 12 \qquad \textcolor{blue}{\text{Simplify.}}$$
$$x + 3 = \pm\sqrt{12} \qquad \textcolor{blue}{\text{Use the Square Root Method.}}$$
$$x = -3 \pm 2\sqrt{3} \quad \textcolor{blue}{\text{Solve for } x.}$$

> **In Words**
> The symbol $\pm$ is read "plus or minus." It means to add and subtract the quantity following the $\pm$ symbol. For example, 5 ± 2 means "$5 - 2 = 3$ or $5 + 2 = 7$."

The x-intercepts are $-3 - 2\sqrt{3} \approx -6.46$ and $-3 + 2\sqrt{3} \approx 0.46$.
To find the y-intercepts, if any, let $x = 0$. Then

$$(x + 3)^2 + (y - 2)^2 = 16$$
$$(0 + 3)^2 + (y - 2)^2 = 16$$
$$9 + (y - 2)^2 = 16$$
$$(y - 2)^2 = 7$$
$$y - 2 = \pm\sqrt{7}$$
$$y = 2 \pm \sqrt{7}$$

The y-intercepts are $2 - \sqrt{7} \approx -0.65$ and $2 + \sqrt{7} \approx 4.65$.

Look back at Figure 25 to verify the approximate locations of the intercepts.

—**Now Work** PROBLEM 95(C)

8 Work with the General Form of the Equation of a Circle

If we eliminate the parentheses from the standard form of the equation of the circle given in Example 15, we get

$$(x + 3)^2 + (y - 2)^2 = 16$$
$$x^2 + 6x + 9 + y^2 - 4y + 4 = 16$$

which simplifies to

$$x^2 + y^2 + 6x - 4y - 3 = 0$$

It can be shown that any equation of the form

$$x^2 + y^2 + ax + by + c = 0$$

has a graph that is a circle, is a point, or has no graph at all. For example, the graph of the equation $x^2 + y^2 = 0$ is the single point $(0, 0)$. The equation $x^2 + y^2 + 5 = 0$, or $x^2 + y^2 = -5$, has no graph, because sums of squares of real numbers are never negative.

> **DEFINITION** General Form of the Equation of a Circle
>
> When its graph is a circle, the equation
>
> $$x^2 + y^2 + ax + by + c = 0$$
>
> is the **general form of the equation of a circle.**

—**Now Work** PROBLEM 85

Need to Review?

Completing the square is discussed in Section A.3, pp. A24–A25.

If an equation of a circle is in general form, we use the method of completing the square to put the equation in standard form so that we can identify its center and radius.

EXAMPLE 17   **Graphing a Circle Whose Equation Is in General Form**

Graph the equation: $x^2 + y^2 + 4x - 6y + 12 = 0$

Solution   Group the terms involving x, group the terms involving y, and put the constant on the right side of the equation. The result is

$$(x^2 + 4x) + (y^2 - 6y) = -12$$

Next, complete the square of each expression in parentheses. Remember that any number added on the left side of the equation must also be added on the right.

$$(x^2 + 4x + 4) + (y^2 - 6y + 9) = -12 + 4 + 9$$

$$\left(\frac{4}{2}\right)^2 = 4 \qquad \left(\frac{-6}{2}\right)^2 = 9$$

$$(x + 2)^2 + (y - 3)^2 = 1 \quad \text{Factor.}$$

This equation is the standard form of the equation of a circle with radius 1 and center $(-2, 3)$. To graph the equation, use the center $(-2, 3)$ and the radius 1. See Figure 26.

Figure 26 $(x + 2)^2 + (y - 3)^2 = 1$

(Graph showing circle with center $(-2, 3)$, points labeled $(-2, 4)$, $(-1, 3)$, $(-3, 3)$, $(-2, 3)$, $(-2, 2)$, radius 1)

→ **Now Work** PROBLEM **99**

EXAMPLE 18   **Using a Graphing Utility to Graph a Circle**

Graph the equation: $x^2 + y^2 = 4$

Solution   This is the equation of a circle with center at the origin and radius 2. To graph this equation using a graphing utility, begin by solving for y.

$$x^2 + y^2 = 4$$
$$y^2 = 4 - x^2 \qquad \text{Subtract } x^2 \text{ from both sides.}$$
$$y = \pm\sqrt{4 - x^2} \qquad \text{Use the Square Root Method to solve for } y.$$

There are two equations to graph: first graph $Y_1 = \sqrt{4 - x^2}$ and then graph $Y_2 = -\sqrt{4 - x^2}$ on the same square screen. (Your circle will appear oval if you do not use a square screen.*) See Figure 27.

$Y_1 = \sqrt{4 - x^2}$

(Graphing calculator screen showing NORMAL FLOAT AUTO REAL RADIAN MP, with a circle graphed, axis markings 2.5, −4, 4, −2.5)

$Y_2 = -\sqrt{4 - x^2}$

Figure 27 $x^2 + y^2 = 4$

*The square screen ratio for the TI-84 Plus C graphing calculator is 8:5.

1.2 Assess Your Understanding

 'Are You Prepared?' *Answers are given at the end of these exercises. If you get a wrong answer, read the pages listed in* red.

1. To complete the square of $x^2 + 10x$, you would _____ (*add*/*subtract*) the number _____. (pp. A24–A25)

2. Use the Square Root Method to solve the equation $(x - 2)^2 = 9$. (p. A31)

Concepts and Vocabulary

3. The points, if any, at which a graph crosses or touches the coordinate axes are called _____.

4. If for every point (x, y) on the graph of an equation the point $(-x, y)$ is also on the graph, then the graph is symmetric with respect to the _____.

5. If the graph of an equation is symmetric with respect to the origin and $(3, -4)$ is a point on the graph, then _____ is also a point on the graph.

6. *True or False* To find the y-intercepts of the graph of an equation, let $x = 0$ and solve for y.

7. *True or False* If a graph is symmetric with respect to the x-axis, then it cannot be symmetric with respect to the y-axis.

8. For a circle, the _____ is the distance from the center to any point on the circle.

9. *True or False* The radius of the circle $x^2 + y^2 = 9$ is 3.

10. *Multiple Choice* Given that the intercepts of a graph are $(-4, 0)$ and $(0, 5)$, choose the statement that is true.
(a) The y-intercept is -4, and the x-intercept is 5.
(b) The y-intercepts are -4 and 5.
(c) The x-intercepts are -4 and 5.
(d) The x-intercept is -4, and the y-intercept is 5.

11. *True or False* The center of the circle
$$(x + 3)^2 + (y - 2)^2 = 13$$
is $(3, -2)$.

12. *Multiple Choice* The equation of a circle can be changed from general form to standard form by doing which of the following?
(a) completing the squares
(b) solving for x
(c) solving for y
(d) squaring both sides

Skill Building

In Problems 13–18, determine which of the given points are on the graph of the equation.

13. Equation: $y = x^4 - \sqrt{x}$
Points: $(0, 0)$; $(1, 1)$; $(2, 4)$

14. Equation: $y = x^3 - 2\sqrt{x}$
Points: $(0, 0)$; $(1, 1)$; $(1, -1)$

15. Equation: $y^2 = x^2 + 9$
Points: $(0, 3)$; $(3, 0)$; $(-3, 0)$

16. Equation: $y^3 = x + 1$
Points: $(1, 2)$; $(0, 1)$; $(-1, 0)$

17. Equation: $x^2 + y^2 = 4$
Points: $(0, 2)$; $(-2, 2)$; $(\sqrt{2}, \sqrt{2})$

18. Equation: $x^2 + 4y^2 = 4$
Points: $(0, 1)$; $(2, 0)$; $\left(2, \dfrac{1}{2}\right)$

In Problems 19–30, find the intercepts and graph each equation by plotting points. Be sure to label the intercepts.

19. $y = x + 2$

20. $y = x - 6$

21. $y = 2x + 8$

22. $y = 3x - 9$

23. $y = x^2 - 1$

24. $y = x^2 - 9$

25. $y = -x^2 + 4$

26. $y = -x^2 + 1$

27. $2x + 3y = 6$

28. $5x + 2y = 10$

29. $9x^2 + 4y = 36$

30. $4x^2 + y = 4$

In Problems 31–40, plot each point. Then plot the point that is symmetric to it with respect to (a) the x-axis; (b) the y-axis; (c) the origin.

31. $(3, 4)$

32. $(5, 3)$

33. $(-2, 1)$

34. $(4, -2)$

35. $(5, -2)$

36. $(-1, -1)$

37. $(-3, -4)$

38. $(4, 0)$

39. $(0, -3)$

40. $(-3, 0)$

In Problems 41–52, the graph of an equation is given. (a) Find the intercepts. (b) Indicate whether the graph is symmetric with respect to the x-axis, the y-axis, or the origin.

41.

42.

43.

44.

45.

46.

47.

48.

49. **50.** **51.** **52.**

In Problems 53–56, draw a complete graph so that it has the type of symmetry indicated.

53. *y*-axis **54.** *x*-axis **55.** Origin **56.** *y*-axis

In Problems 57–72, list the intercepts and test for symmetry.

57. $y^2 = x + 4$ **58.** $y^2 = x + 9$ **59.** $y = \sqrt[3]{x}$ **60.** $y = \sqrt[5]{x}$

61. $x^2 + y - 9 = 0$ **62.** $x^2 - y - 4 = 0$ **63.** $9x^2 + 4y^2 = 36$ **64.** $4x^2 + y^2 = 4$

65. $y = x^3 - 27$ **66.** $y = x^4 - 1$ **67.** $y = x^2 - 3x - 4$ **68.** $y = x^2 + 4$

69. $y = \dfrac{3x}{x^2 + 9}$ **70.** $y = \dfrac{x^2 - 4}{2x}$ **71.** $y = \dfrac{-x^3}{x^2 - 9}$ **72.** $y = \dfrac{x^4 + 1}{2x^5}$

In Problems 73–76, draw a quick sketch of the graph of each equation.

73. $y = x^3$ **74.** $x = y^2$ **75.** $y = \sqrt{x}$ **76.** $y = \dfrac{1}{x}$

77. If $(a, 4)$ is a point on the graph of $y = x^2 + 3x$, what is a?   **78.** If $(a, -5)$ is a point on the graph of $y = x^2 + 6x$, what is a?

In Problems 79–82, find the center and radius of each circle. Write the standard form of the equation.

79. **80.** **81.** **82.**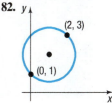

In Problems 83–92, write the standard form of the equation and the general form of the equation of each circle of radius r and center (h, k). *Graph each circle.*

83. $r = 2$; $(h, k) = (0, 0)$ **84.** $r = 3$; $(h, k) = (0, 0)$ **85.** $r = 2$; $(h, k) = (0, 2)$ **86.** $r = 3$; $(h, k) = (1, 0)$

87. $r = 5$; $(h, k) = (4, -3)$ **88.** $r = 4$; $(h, k) = (2, -3)$ **89.** $r = 4$; $(h, k) = (-2, 1)$ **90.** $r = 7$; $(h, k) = (-5, -2)$

91. $r = \dfrac{1}{2}$; $(h, k) = \left(\dfrac{1}{2}, 0\right)$ **92.** $r = \dfrac{1}{2}$; $(h, k) = \left(0, -\dfrac{1}{2}\right)$

In Problems 93–106, (a) find the center (h, k) *and radius r of each circle; (b) graph each circle; (c) find the intercepts, if any.*

93. $x^2 + y^2 = 4$ **94.** $x^2 + (y - 1)^2 = 1$ **95.** $2(x - 3)^2 + 2y^2 = 8$

96. $3(x + 1)^2 + 3(y - 1)^2 = 6$ **97.** $x^2 + y^2 - 2x - 4y - 4 = 0$ **98.** $x^2 + y^2 + 4x + 2y - 20 = 0$

🖊 **99.** $x^2 + y^2 + 4x - 4y - 1 = 0$ **100.** $x^2 + y^2 - 6x + 2y + 9 = 0$ **101.** $x^2 + y^2 - x + 2y + 1 = 0$

102. $x^2 + y^2 + x + y - \dfrac{1}{2} = 0$ **103.** $2x^2 + 2y^2 - 12x + 8y - 24 = 0$ **104.** $2x^2 + 2y^2 + 8x + 7 = 0$

105. $2x^2 + 8x + 2y^2 = 0$ **106.** $3x^2 + 3y^2 - 12y = 0$

In Problems 107–112, find the standard form of the equation of each circle.

107. Center at the origin and containing the point $(-2, 3)$ **108.** Center $(1, 0)$ and containing the point $(-3, 2)$

109. With endpoints of a diameter at $(1, 4)$ and $(-3, 2)$ **110.** With endpoints of a diameter at $(4, 3)$ and $(0, 1)$

111. Center $(2, -4)$ and circumference 16π **112.** Center $(-5, 6)$ and area 49π

Applications and Extensions

113. Given that the point $(1, 2)$ is on the graph of an equation that is symmetric with respect to the origin, what other point is on the graph?

114. If the graph of an equation is symmetric with respect to the y-axis and 6 is an x-intercept of this graph, name another x-intercept.

115. If the graph of an equation is symmetric with respect to the origin and -4 is an x-intercept of this graph, name another x-intercept.

116. If the graph of an equation is symmetric with respect to the x-axis and 2 is a y-intercept, name another y-intercept.

117. Microphones In studios and on stages, cardioid microphones are often preferred for the richness they add to voices and for their ability to reduce the level of sound from the sides and rear of the microphone. Suppose one such cardioid pattern is given by the equation $(x^2 + y^2 - x)^2 = x^2 + y^2$.

(a) Find the intercepts of the graph of the equation.
(b) Test for symmetry with respect to the x-axis, the y-axis, and the origin.

Source: www.notaviva.com

118. Solar Energy The solar electric generating systems at Kramer Junction, California, use parabolic troughs to heat a heat-transfer fluid to a high temperature. This fluid is used to generate steam that drives a power conversion system to produce electricity. For troughs 7.5 feet wide, an equation for the cross section is $16y^2 = 120x - 225$.

(a) Find the intercepts of the graph of the equation.

(b) Test for symmetry with respect to the x-axis, the y-axis, and the origin.

Source: U.S. Department of Energy

119. Find the area of the square in the figure.

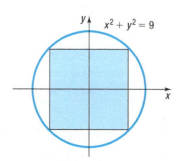

120. Find the area of the blue shaded region in the figure, assuming the quadrilateral inside the circle is a square.

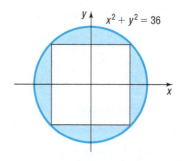

121. Ferris Wheel The original Ferris wheel was built in 1893 by 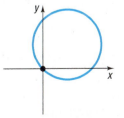 Pittsburgh, Pennsylvania bridge builder George W. Ferris. The Ferris wheel was originally built for the 1893 World's Fair in Chicago, but it was also later reconstructed for the 1904 World's Fair in St. Louis. It had a maximum height of 264 feet and a wheel diameter of 250 feet. Find an equation for the wheel if the center of the wheel is on the y-axis.
Source: guinnessworldrecords.com

122. Ferris Wheel The High Roller observation wheel in Las Vegas has a maximum height of 550 feet and a diameter of 520 feet, with one full rotation taking approximately 30 minutes. Find an equation for the wheel if the center of the wheel is on the y-axis.
Source: Las Vegas Review Journal

123. Vertically Circular Building Located in Al Raha, Abu Dhabi, the headquarters of property developing company Aldar is a vertically circular building with a diameter of 121 meters. The tip of the building is 110 meters aboveground. Find an equation for the building's outline if the center of the building is on the y-axis.

124. Vertically Circular Building The Sunrise Kempinski Hotel in Beijing, China, is a vertically circular building whose outline is described by the equation $x^2 + y^2 - 78y - 1843 = 0$ if the center of the building is on the y-axis. If x and y are in meters, what is the height of the building?

Problems 125–130 require the following discussion.

The **tangent line** to a circle may be defined as the line that intersects the circle in a single point, called the **point of tangency**. See the figure.

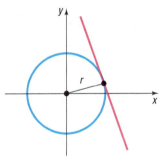

In Problems 125–128, find the standard form of the equation of each circle. (Refer to the preceding discussion).

125. Center $(2, 3)$ and tangent to the x-axis

126. Center $(-3, 1)$ and tangent to the y-axis

127. Center $(-1, 3)$ and tangent to the line $y = 2$

128. Center $(4, -2)$ and tangent to the line $x = 1$

129. *Challenge Problem* If the equation of a circle is $x^2 + y^2 = r^2$ and the equation of a tangent line is $y = mx + b$, show that:

(a) $r^2(1 + m^2) = b^2$

[**Hint:** The quadratic equation $x^2 + (mx + b)^2 = r^2$ has exactly one solution.]

(b) The point of tangency is $\left(\dfrac{-r^2 m}{b}, \dfrac{r^2}{b}\right)$.

(c) The tangent line is perpendicular to the line containing the center of the circle and the point of tangency.

130. *Challenge Problem* Refer to Problem 129.
The line $x - 2y + 4 = 0$ is tangent to a circle at $(0, 2)$. The line $y = 2x - 7$ is tangent to the same circle at $(3, -1)$. Find the center of the circle.

131. *Challenge Problem* If (x_1, y_1) and (x_2, y_2) are the endpoints of a diameter of a circle, show that an equation of the circle is

$$(x - x_1)(x - x_2) + (y - y_1)(y - y_2) = 0$$

132. *Challenge Problem* If $x^2 + y^2 + dx + ey + f = 0$ is the equation of a circle, show that

$$x_0 x + y_0 y + d\left(\frac{x + x_0}{2}\right) + e\left(\frac{y + y_0}{2}\right) + f = 0$$

is an equation of the tangent line to the circle at the point (x_0, y_0).

133. *Challenge Problem* Lemniscate For a nonzero constant a, find the intercepts of the graph of $(x^2 + y^2)^2 = a^2(x^2 - y^2)$. Then test for symmetry with respect to the x-axis, the y-axis, and the origin.

134. *Challenge Problem* Limaçon For nonzero constants a and b, find the intercepts of the graph of

$$(x^2 + y^2 - ax)^2 = b^2(x^2 + y^2)$$

Then test for symmetry with respect to the x-axis, the y-axis, and the origin.

Explaining Concepts: Discussion and Writing

135. Which of the following equations might have the graph shown? (More than one answer is possible.)

(a) $(x - 2)^2 + (y + 3)^2 = 13$
(b) $(x - 2)^2 + (y - 2)^2 = 8$
(c) $(x - 2)^2 + (y - 3)^2 = 13$
(d) $(x + 2)^2 + (y - 2)^2 = 8$
(e) $x^2 + y^2 - 4x - 9y = 0$
(f) $x^2 + y^2 + 4x - 2y = 0$
(g) $x^2 + y^2 - 9x - 4y = 0$
(h) $x^2 + y^2 - 4x - 4y = 4$

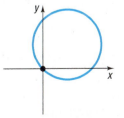

136. Which of the following equations might have the graph shown? (More than one answer is possible.)

(a) $(x - 2)^2 + y^2 = 3$
(b) $(x + 2)^2 + y^2 = 3$
(c) $x^2 + (y - 2)^2 = 3$
(d) $(x + 2)^2 + y^2 = 4$
(e) $x^2 + y^2 + 10x + 16 = 0$
(f) $x^2 + y^2 + 10x - 2y = 1$
(g) $x^2 + y^2 + 9x + 10 = 0$
(h) $x^2 + y^2 - 9x - 10 = 0$

137. (a) Graph $y = \sqrt{x^2}$, $y = x$, $y = |x|$, and $y = (\sqrt{x})^2$, noting which graphs are the same.
(b) Explain why the graphs of $y = \sqrt{x^2}$ and $y = |x|$ are the same.
(c) Explain why the graphs of $y = x$ and $y = (\sqrt{x})^2$ are not the same.
(d) Explain why the graphs of $y = \sqrt{x^2}$ and $y = x$ are not the same.

138. Draw a graph of an equation that contains two x-intercepts; at one the graph crosses the x-axis, and at the other the graph touches the x-axis.

139. Make up an equation with the intercepts $(2, 0)$, $(4, 0)$, and $(0, 1)$. Compare your equation with a friend's equation. Comment on any similarities.

140. Draw a graph that contains the points $(-2, -1)$, $(0, 1)$, $(1, 3)$, and $(3, 5)$. Compare your graph with those of other students. Are most of the graphs almost straight lines? How many are "curved"? Discuss the various ways in which these points might be connected.

141. An equation is being tested for symmetry with respect to the x-axis, the y-axis, and the origin. Explain why, if two of these symmetries are present, the remaining one must also be present.

142. Draw a graph that contains the points $(-2, 5)$, $(-1, 3)$, and $(0, 2)$ and is symmetric with respect to the y-axis. Compare your graph with those of other students; comment on any similarities. Can a graph contain these points and be symmetric with respect to the x-axis? the origin? Why or why not?

143. Explain how the center and radius of a circle can be used to graph the circle.

144. **What Went Wrong?** A student stated that the center and radius of the graph whose equation is $(x + 3)^2 + (y - 2)^2 = 16$ are $(3, -2)$ and 4, respectively. Why is this incorrect?

'Are You Prepared?' Answers

1. add; 25   **2.** $\{-1, 5\}$

1.3 Functions and Their Graphs

PREPARING FOR THIS SECTION *Before getting started, review the following:*

- Interval Notation (Section A.6, pp. A45–A47)
- Solving Inequalities (Section A.6, pp. A49–A51)
- Evaluating Algebraic Expressions, Domain of a Variable (Section A.1, pp. A6–A7)
- Rationalizing Denominators and Numerators (Section A.7, pp. A58–A59)

Now Work the 'Are You Prepared?' problems on page 37.

OBJECTIVES **1** Describe a Relation (p. 24)
2 Determine Whether a Relation Represents a Function (p. 26)
3 Use Function Notation; Find the Value of a Function (p. 29)
4 Find the Difference Quotient of a Function (p. 31)
5 Find the Domain of a Function Defined by an Equation (p. 32)
6 Identify the Graph of a Function (p. 34)
7 Obtain Information from or about the Graph of a Function (p. 35)

1 Describe a Relation

Often there are situations where one variable is somehow linked to another variable. For example, the price of a gallon of gas is linked to the price of a barrel of oil. A person can be associated to her telephone number(s). The volume V of a sphere depends on its radius R. The force F exerted by an object corresponds to its acceleration a. These are examples of a *relation*, a correspondence between two sets called the *domain* and the *range*.

DEFINITION **Relation**

A **relation** is a correspondence between two sets: a set X, called the **domain**, and a set Y, called the **range**. In a relation, each element from the domain corresponds to at least one element from the range.

Figure 28

If x is an element of the domain and y is an element of the range, and if a relation exists from x to y, then we say that y **corresponds** to x or that y **depends on** x, and we write $x \rightarrow y$. It is often helpful to think of x as the **input** and y as the **output** of the relation. See Figure 28.

Suppose an astronaut standing on the Moon throws a rock 20 meters up and starts a stopwatch as the rock begins to fall back down. The astronaut measures the height of the rock at 1, 2, 2.5, 3, 4, and 5 seconds and obtains heights of 19.2, 16.8, 15, 12.8, 7.2, and 0 meters, respectively. This is an example of a relation expressed **verbally**. The domain of the relation is the set $\{0, 1, 2, 2.5, 3, 4, 5\}$ and the range of the relation is the set $\{20, 19.2, 16.8, 15, 12.8, 7.2, 0\}$.

The astronaut could also express this relation *numerically*, *graphically*, or *algebraically*.

The relation can be expressed **numerically** using a table of numbers, as in Table 5, or by using a **set of ordered pairs**. Using ordered pairs, the relation is

$$\{(0, 20), (1, 19.2), (2, 16.8), (2.5, 15), (3, 12.8), (4, 7.2), (5, 0)\}$$

where the first element of each pair denotes the time and the second element denotes the height.

Suppose x represents the number of seconds on the stopwatch and y represents the height of the rock in meters. Then the relation can be expressed **graphically** by plotting the points (x, y). See Figure 29.

The relation can be represented as a **mapping** by drawing an arrow from an element in the domain to the corresponding element in the range. See Figure 30.

Table 5

Time (in seconds)	Height (in meters)
0	20
1	19.2
2	16.8
2.5	15
3	12.8
4	7.2
5	0

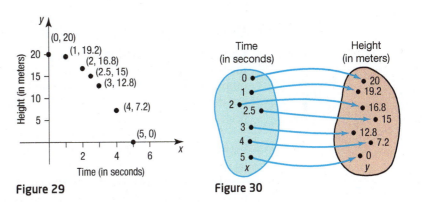

Figure 29   **Figure 30**

Finally, from physics, the relation can be expressed **algebraically** using the equation

$$y = 20 - 0.8x^2$$

EXAMPLE 1	**Describing a Relation**

A verbal description of a relation is given below.

The price of First Class U.S. postage stamps has changed over the years. To mail a letter in 2015 cost $0.49. In 2016 it cost $0.49 for part of the year and $0.47 for the rest of the year. In 2017 it cost $0.47 for part of the year and $0.49 for the rest of the year. In 2018 it cost $0.49 for part of the year and $0.50 for the rest of the year.

Using year as input and price as output,

(a) What is the domain and the range of the relation?

(b) Express the relation as a set of ordered pairs.

(c) Express the relation as a mapping.

(d) Express the relation as a graph.

Solution The relation establishes a correspondence between the input, year, and the output, price of a First Class U.S. postage stamp.

(a) The domain of the relation is { 2015, 2016, 2017, 2018 }. The range of the relation is { $0.47, $0.49, $0.50 }.

(b) The relation expressed as a set of ordered pairs is {(2015, $0.49), (2016, $0.47), (2016, $0.49), (2017, $0.47), (2017, $0.49), (2018, $0.49), (2018, $0.50)}

(c) See Figure 31 for the relation expressed as a mapping, using t for year and p for price.

(d) Figure 32 shows a graph of the relation.

Figure 31 First Class stamp price

Figure 32 First Class stamp price (2015–2018)

- **Now Work** PROBLEM 21

2 Determine Whether a Relation Represents a Function

Look back at the relation involving the height of a rock on the Moon described at the beginning of the section. Notice that each input, time, corresponds to exactly one output, height. Given a time, you could tell the exact height of the rock. But that is not the case with the price of stamps. Given the year 2018, you cannot determine the price of a stamp with certainty. It could be $0.49, or it could be $0.50.

Consider the mapping of the relation in Figure 33. It shows a correspondence between a substance and its specific heat. Notice that for each substance you can tell its specific heat with certainty.

The relation associating the time to the height of the rock is a *function*, and the relation associating a given substance to its specific heat is a function. But the relation associating the year to the price of a First Class postage stamp is *not* a function. To be a function, each input must correspond to *exactly one* output.

Figure 33 Specific heat of some common substances

> **DEFINITION Function**
>
> Let X and Y be two nonempty sets.* A **function** from X into Y is a relation that associates with each element of X exactly one element of Y.

The set X is called the **domain** of the function. For each element x in X, the corresponding element y in Y is called the **value** of the function at x, or the **image** of x. The set of all images of the elements in the domain is called the **range** of the function. See Figure 34.

Since there may be some elements in Y that are not the image of some x in X, it follows that the range of a function may be a proper subset of Y, as shown in Figure 34.

The idea behind a function is its certainty. If an input is given, we can use the function to determine the output. This is not possible if a relation is not a function. The requirement of "one output" provides a predictable behavior that is important when using mathematics to model or analyze the real world. It allows doctors to know

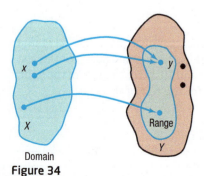

Figure 34

*The sets X and Y will usually be sets of real numbers, in which case a (real) function results. The two sets can also be sets of complex numbers, and then we have defined a complex function. In the broad definition (proposed by Lejeune Dirichlet), X and Y can be any two sets.

exactly how much medicine to give a patient, an engineer to determine the material to use in construction, a store manager to choose how many units to keep in stock, etc.

EXAMPLE 2

Determining Whether a Relation Given by a Mapping Is a Function

For each relation, state the domain and range. Then determine whether the relation is a function.

(a) See Figure 35. This relation shows a correspondence between an item on McDonald's $1-$2-$3 menu and its price.
Source: McDonald's Corporation, 2018

(b) See Figure 36. This relation shows a correspondence between activity level and daily calories needed for an individual with a basal metabolic rate (BMR) of 1975 kilocalories per day (kcal/day).

(c) See Figure 37. This relation shows a correspondence between gestation period (in days) and life expectancy (in years) for five animal species.

Figure 35

Figure 36

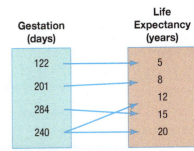

Figure 37

Solution

(a) The domain of the relation is {Cheeseburger, McChicken, Bacon McDouble, Triple Cheeseburger, Happy Meal}, and the range of the relation is { $1, $2, $3 }. The relation in Figure 35 is a function because each element in the domain corresponds to exactly one element in the range.

(b) The domain of the relation is { Sedentary, Light, Moderate, Intense }, and the range of the relation is { 2370, 2716, 3061, 3407 }. The relation in Figure 36 is a function because each element in the domain corresponds to exactly one element in the range.

(c) The domain of the relation is { 122, 201, 240, 284 }, and the range of the relation is { 5, 8, 12, 15, 20 }. The relation in Figure 37 is not a function because there is an element in the domain, 240, that corresponds to two elements in the range, 12 and 20. If a gestation period of 240 days is selected from the domain, a single life expectancy cannot be associated with it.

Now Work PROBLEM 23

We may also think of a function as a set of ordered pairs (x, y) in which no ordered pairs have the same first element and different second elements. The set of all first elements x is the domain of the function, and the set of all second elements y is its range. Each element x in the domain corresponds to exactly one element y in the range.

EXAMPLE 3

Determining Whether a Relation Given by a Set of Ordered Pairs Is a Function

For each relation, state the domain and range. Then determine whether the relation is a function.

(a) { (1, 4), (2, 5), (3, 6), (4, 7) }

(b) { (1, 4), (2, 4), (3, 5), (6, 10) }

(c) { (−3, 9), (−2, 4), (0, 0), (1, 1), (−3, 8) }

Solution

(a) The domain of this relation is $\{1, 2, 3, 4\}$, and its range is $\{4, 5, 6, 7\}$. This relation is a function because there are no ordered pairs with the same first element and different second elements.

(b) The domain of this relation is $\{1, 2, 3, 6\}$, and its range is $\{4, 5, 10\}$. This relation is a function because there are no ordered pairs with the same first element and different second elements.

(c) The domain of this relation is $\{-3, -2, 0, 1\}$, and its range is $\{0, 1, 4, 8, 9\}$. This relation is not a function because there are two ordered pairs, $(-3, 9)$ and $(-3, 8)$, that have the same first element and different second elements.

In Example 3(b), notice that 1 and 2 in the domain both have the same image in the range. This does not violate the definition of a function; two different first elements can have the same second element. The definition is violated when two ordered pairs have the same first element and different second elements, as in Example 3(c).

▬▬▬ **Now Work** PROBLEM **27**

Up to now we have shown how to identify when a relation is a function for relations defined by mappings (Example 2) and ordered pairs (Example 3). But relations can also be expressed as equations.

It is usually easiest to determine whether an equation, where y depends on x, is a function, when the equation is solved for y. If any value of x in the domain corresponds to more than one y, the equation does not define a function; otherwise, it does define a function.

EXAMPLE 4

Determining Whether an Equation Is a Function

Determine whether the equation $y = 2x - 5$ defines y as a function of x.

Solution

The equation tells us to take an input x, multiply it by 2, and then subtract 5. For any input x, these operations yield only one output y, so the equation is a function. For example, if $x = 1$, then $y = 2 \cdot 1 - 5 = -3$. If $x = 3$, then $y = 2 \cdot 3 - 5 = 1$.

The graph of the equation $y = 2x - 5$ is a line with slope 2 and y-intercept -5. The function is called a *linear function*. See Figure 38.

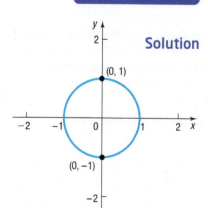

Figure 38 $y = 2x - 5$

EXAMPLE 5

Determining Whether an Equation Is a Function

Determine whether the equation $x^2 + y^2 = 1$ defines y as a function of x.

Solution

To determine whether the equation $x^2 + y^2 = 1$, which defines the unit circle, is a function, solve the equation for y.

$$x^2 + y^2 = 1$$
$$y^2 = 1 - x^2$$
$$y = \pm\sqrt{1 - x^2}$$

For values of x for which $-1 < x < 1$, two values of y result. For example, if $x = 0$, then $y = \pm 1$, so two different outputs result from the same input. This means that the equation $x^2 + y^2 = 1$ does not define a function. See Figure 39.

Figure 39 $x^2 + y^2 = 1$

▬▬▬ **Now Work** PROBLEM **41**

Input x
x
f
Output
y = f(x)

Figure 40 Input/output machine

In Words

If $y = f(x)$, then x is the input and y is the output corresponding to x.

WARNING The notation $y = f(x)$ denotes a function f. It does NOT mean "f times x." ∎

3 Use Function Notation; Find the Value of a Function

It is common practice to denote functions by letters such as f, g, F, G, and others. If f is a function, then for each number x in the domain, the corresponding number y in the range is designated by the symbol $f(x)$, read as "f of x," and we write $y = f(x)$. When a function is expressed in this way, we are using **function notation**.

We refer to $f(x)$ as the **value** of the function f at the number x. For example, the function in Example 4 may be written using function notation as $y = f(x) = 2x - 5$. Then $f(1) = -3$ and $f(3) = 1$.

Sometimes it is helpful to think of a function f as a machine that receives as input a number from the domain, manipulates it, and outputs a value in the range. See Figure 40.

The restrictions on this input/output machine are as follows:

* It accepts only numbers from the domain of the function.
* For each input, there is exactly one output (which may be repeated for different inputs).

Figure 41 illustrates some other functions. Notice that in every function, for each x in the domain, there is one corresponding value in the range.

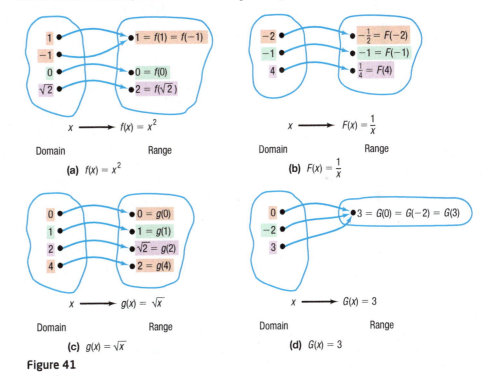

Figure 41

For a function $y = f(x)$, the variable x is called the **independent variable** because it can be assigned any number from the domain. The variable y is called the **dependent variable**, because its value depends on x.

Any symbols can be used to represent the independent and dependent variables. For example, if f is the *cube function*, then f can be given by $f(x) = x^3$ or $f(t) = t^3$ or $f(z) = z^3$. All three functions are the same. Each says to cube the independent variable to get the output. In practice, the symbols used for the independent and dependent variables are based on common usage, such as using t for time and a for acceleration.

The independent variable is also called the **argument** of the function. Thinking of the independent variable as an argument can sometimes make it easier to find the value of a function. For example, if f is the function defined by $f(x) = x^3$, then f tells us to cube the argument. Then $f(2)$ means to cube 2, $f(a)$ means to cube the number a, and $f(x + h)$ means to cube the quantity $x + h$.

EXAMPLE 6

Finding Values of a Function

For the function f defined by $f(x) = 2x^2 - 3x$, evaluate

(a) $f(3)$ (b) $f(x) + f(3)$ (c) $3f(x)$ (d) $f(-x)$

(e) $-f(x)$ (f) $f(3x)$ (g) $f(x + 3)$ (h) $f(x + h)$

Solution

(a) Substitute 3 for x in the equation for f, $f(x) = 2x^2 - 3x$, to get

$$f(3) = 2 \cdot 3^2 - 3 \cdot 3 = 18 - 9 = 9$$

The image of 3 is 9.

(b) $f(x) + f(3) = (2x^2 - 3x) + 9 = 2x^2 - 3x + 9$

(c) Multiply the equation for f by 3.

$$3f(x) = 3(2x^2 - 3x) = 6x^2 - 9x$$

(d) Substitute $-x$ for x in the equation for f and simplify.

$$f(-x) = 2(-x)^2 - 3(-x) = 2x^2 + 3x \quad \text{Notice the use of parentheses here.}$$

(e) $-f(x) = -(2x^2 - 3x) = -2x^2 + 3x$

(f) Substitute $3x$ for x in the equation for f and simplify.

$$f(3x) = 2(3x)^2 - 3 \cdot 3x = 2 \cdot 9x^2 - 9x = 18x^2 - 9x$$

(g) Substitute $x + 3$ for x in the equation for f and simplify.

$$f(x + 3) = 2(x + 3)^2 - 3(x + 3)$$
$$= 2(x^2 + 6x + 9) - 3x - 9$$
$$= 2x^2 + 12x + 18 - 3x - 9$$
$$= 2x^2 + 9x + 9$$

(h) Substitute $x + h$ for x in the equation for f and simplify.

$$f(x + h) = 2(x + h)^2 - 3(x + h)$$
$$= 2(x^2 + 2xh + h^2) - 3x - 3h$$
$$= 2x^2 + 4xh + 2h^2 - 3x - 3h$$

Notice in this example that $f(x + 3) \neq f(x) + f(3)$, $f(-x) \neq -f(x)$, and $3f(x) \neq f(3x)$.

Now Work PROBLEM 47

Most calculators have special keys that allow you to find the value of certain commonly used functions. Examples are keys for the square function $f(x) = x^2$, the square root function $f(x) = \sqrt{x}$, and the reciprocal function $f(x) = \dfrac{1}{x} = x^{-1}$.

EXAMPLE 7

Finding Values of a Function on a Calculator

Verify the following results on your calculator.

(a) $f(x) = x^2$ $f(1.234) = 1.234^2 = 1.522756$

(b) $F(x) = \dfrac{1}{x}$ $F(1.234) = \dfrac{1}{1.234} \approx 0.8103727715$

(c) $g(x) = \sqrt{x}$ $g(1.234) = \sqrt{1.234} \approx 1.110855526$

COMMENT Graphing calculators can be used to evaluate a function. Figure 42 shows the function $Y_1 = f(x) = 2x^2 - 3x$ evaluated at 3 on a TI-84 Plus C graphing calculator. Compare this result with the solution to Example 6(a).

```
NORMAL FLOAT AUTO REAL RADIAN MP

Plot1  Plot2  Plot3
■\Y₁∎2X²−3X

■\Y₂=
■\Y₃=
■\Y₄=
■\Y₅=
■\Y₆=
■\Y₇=
```

```
NORMAL FLOAT AUTO REAL RADIAN MP

Y₁(3)
                                    9
```

Figure 42 Evaluating $f(x) = 2x^2 - 3x$ for $x = 3$ ■

COMMENT A graphing calculator requires the explicit form of a function.   ■

Implicit Form of a Function

In general, when a function f is defined by an equation in x and y, we say that the function f is given **implicitly**. If it is possible to solve the equation for y in terms of x, then we write $y = f(x)$ and say that the function is given **explicitly**. For example,

Implicit Form	**Explicit Form**
$3x + y = 5$	$y = f(x) = -3x + 5$
$x^2 - y = 6$	$y = f(x) = x^2 - 6$
$xy = 4$	$y = f(x) = \dfrac{4}{x}$

SUMMARY

Important Facts about Functions

- For each x in the domain of a function f, there is exactly one image $f(x)$ in the range; however, more than one x in the domain can have the same image in the range.
- f is the symbol that we use to denote the function. It is symbolic of the equation (rule) that we use to get from x in the domain to $f(x)$ in the range.
- If $y = f(x)$, then x is the independent variable, or the argument of f, and y, or $f(x)$, is the dependent variable, or the value of f at x.

4 Find the Difference Quotient of a Function

An important concept in calculus involves using a certain quotient. For a function f, the inputs x and $x + h$, $h \neq 0$, result in the values $f(x)$ and $f(x + h)$. The quotient of their differences

$$\frac{f(x + h) - f(x)}{(x + h) - x} = \frac{f(x + h) - f(x)}{h}$$

with $h \neq 0$, is called the *difference quotient of f* at x.

> **DEFINITION** Difference Quotient
>
> The **difference quotient** of a function f at x is given by
>
> $$\frac{f(x + h) - f(x)}{h} \qquad h \neq 0 \tag{1}$$

The difference quotient is used in calculus to define the *derivative*, which leads to applications such as the velocity of an object and optimization of resources.

When finding a difference quotient, it is necessary to simplify expression (1) so that the h in the denominator can be cancelled.

| EXAMPLE 8 | **Finding the Difference Quotient of a Function** |

Find the difference quotient of each function.

(a) $f(x) = 2x^2 - 3x$ (b) $f(x) = \dfrac{4}{x}$ (c) $f(x) = \sqrt{x}$

Solution (a) $\dfrac{f(x + h) - f(x)}{h} = \dfrac{[2(x + h)^2 - 3(x + h)] - [2x^2 - 3x]}{h}$

$\uparrow$
$f(x + h) = 2(x + h)^2 - 3(x + h)$

$= \dfrac{2(x^2 + 2xh + h^2) - 3x - 3h - 2x^2 + 3x}{h}$ **Simplify.**

$= \dfrac{2x^2 + 4xh + 2h^2 - 3h - 2x^2}{h}$ **Distribute and combine like terms.**

$= \dfrac{4xh + 2h^2 - 3h}{h}$ **Combine like terms.**

$= \dfrac{h(4x + 2h - 3)}{h}$ **Factor out h.**

$= 4x + 2h - 3$ **Cancel the h's.**

(b) $\dfrac{f(x + h) - f(x)}{h} = \dfrac{\dfrac{4}{x + h} - \dfrac{4}{x}}{h}$ $f(x + h) = \dfrac{4}{x + h}$

$= \dfrac{\dfrac{4x - 4(x + h)}{x(x + h)}}{h}$ **Subtract.**

$= \dfrac{4x - 4x - 4h}{x(x + h)h}$ **Divide and distribute.**

$= \dfrac{-4h}{x(x + h)h}$ **Simplify.**

$= -\dfrac{4}{x(x + h)}$ **Cancel the h's.**

(c) When a function involves a square root, we rationalize the numerator of expression (1). After simplifying, the h's cancel.

$\dfrac{f(x + h) - f(x)}{h} = \dfrac{\sqrt{x + h} - \sqrt{x}}{h}$ $f(x + h) = \sqrt{x + h}$

$= \dfrac{\sqrt{x + h} - \sqrt{x}}{h} \cdot \dfrac{\sqrt{x + h} + \sqrt{x}}{\sqrt{x + h} + \sqrt{x}}$ **Rationalize the numerator.**

$= \dfrac{(\sqrt{x + h})^2 - (\sqrt{x})^2}{h(\sqrt{x + h} + \sqrt{x})}$ $(A - B)(A + B) = A^2 - B^2$

$= \dfrac{h}{h(\sqrt{x + h} + \sqrt{x})}$ $(\sqrt{x + h})^2 - (\sqrt{x})^2$
$= x + h - x = h$

$= \dfrac{1}{\sqrt{x + h} + \sqrt{x}}$ **Cancel the h's.**

Need to Review?
Rationalizing Numerators is discussed in Section A.7, pp. A58–A59.

━━▸ **Now Work** PROBLEM 71

5 Find the Domain of a Function Defined by an Equation

Often the domain of a function f is not specified; instead, only the equation defining the function is given. In such cases, the **domain of f** is the largest set of real numbers for which the value $f(x)$ is a real number. The domain of a function f is the same as the domain of the variable x in the expression $f(x)$.

EXAMPLE 9

Finding the Domain of a Function

Find the domain of each function.

(a) $f(x) = x^2 + 5x$

(b) $g(x) = \dfrac{3x}{x^2 - 4}$

(c) $h(t) = \sqrt{4 - 3t}$

(d) $F(x) = \dfrac{\sqrt{3x + 12}}{x - 5}$

Solution

(a) The function $f(x) = x^2 + 5x$ says to sum the square of a number and five times the number. Since these operations can be performed on any real number, the domain of f is the set of all real numbers.

(b) The function $g(x) = \dfrac{3x}{x^2 - 4}$ says to divide $3x$ by $x^2 - 4$. Since division by 0 is not defined, the denominator $x^2 - 4$ cannot be 0, so x cannot equal -2 or 2. The domain of the function g is $\{x \mid x \neq -2, x \neq 2\}$.

(c) The function $h(t) = \sqrt{4 - 3t}$ says to take the square root of $4 - 3t$. Since only nonnegative numbers have real square roots, the expression under the square root (the radicand) must be nonnegative (greater than or equal to zero). That is,

$$4 - 3t \geq 0$$
$$-3t \geq -4$$
$$t \leq \frac{4}{3}$$

The domain of h is $\left\{ t \,\middle|\, t \leq \dfrac{4}{3} \right\}$, or the interval $\left(-\infty, \dfrac{4}{3} \right]$.

(d) The function $F(x) = \dfrac{\sqrt{3x + 12}}{x - 5}$ says to take the square root of $3x + 12$ and divide the result by $x - 5$. This requires that $3x + 12 \geq 0$, so $x \geq -4$, and also that $x - 5 \neq 0$, so $x \neq 5$. Combining these two restrictions, the domain of F is

$$\{x \mid x \geq -4, x \neq 5\}$$

> **In Words**
> The domain of g found in Example 9(b) is
> $\{x \mid x \neq -2, x \neq 2\}$.
> This notation is read, "The domain of the function g is the set of all real numbers x such that x does not equal -2 and x does not equal 2."

The following steps may prove helpful for finding the domain of a function that is defined by an equation and whose domain is a subset of the real numbers.

Finding the Domain of a Function Defined by an Equation

- Start with the domain as the set of all real numbers.
- If the equation has a denominator, exclude any numbers for which the denominator is zero.
- If the equation has a radical with an even index, exclude any numbers for which the expression inside the radical (the radicand) is negative.

Now Work PROBLEM 59

We express the domain of a function using interval notation, set notation, or words, whichever is most convenient. If x is in the domain of a function f, we say that **f is defined at x**, or **$f(x)$ exists**. If x is not in the domain of f, we say that **f is not defined at x**, or **$f(x)$ does not exist**. For example, if $f(x) = \dfrac{x}{x^2 - 1}$, then $f(0)$ exists, but $f(1)$ and $f(-1)$ do not exist. (Do you see why?)

When a function is defined by an equation, it can be difficult to find its range unless we also have a graph of the function. So we are usually content to find only the domain.

When we use functions in applications, the domain may be restricted by physical or geometric considerations. For example, the domain of the function f defined by $f(x) = x^2$ is the set of all real numbers. However, if f represents the area of a square whose sides are of length x, the domain of f is restricted to the positive real numbers, since the length of a side can never be 0 or negative.

EXAMPLE 10

Finding the Domain of a Function Used in an Application

Express the area of a circle as a function of its radius. Find the domain.

Solution

Figure 43 Circle of radius r

See Figure 43. The formula for the area A of a circle of radius r is $A = \pi r^2$. Using r to represent the independent variable and A to represent the dependent variable, the function expressing this relationship is

$$A = A(r) = \pi r^2$$

In this application, the domain is $\{r \mid r > 0\}$. (Do you see why?)

Now Work PROBLEM 109

The Graph of a Function

In applications, a graph often demonstrates more clearly the relationship between two variables than, say, an equation or table. For example, Table 6 shows the average price of gasoline in the United States for the years 1991–2017 (adjusted for inflation, based on 2017 dollars). If we plot these data and then connect the points, we obtain Figure 44.

Table 6

Year	Price	Year	Price	Year	Price
1991	1.98	2000	2.11	2009	2.68
1992	1.90	2001	1.97	2010	3.12
1993	1.81	2002	1.83	2011	3.84
1994	1.78	2003	2.07	2012	3.87
1995	1.78	2004	2.40	2013	3.68
1996	1.87	2005	2.85	2014	3.48
1997	1.83	2006	3.13	2015	2.51
1998	1.55	2007	3.31	2016	2.19
1999	1.67	2008	3.70	2017	2.38

Source: U.S. Energy Information Administration

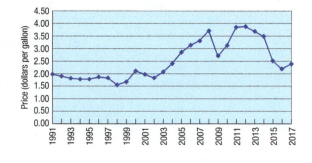

Figure 44 Average retail price of gasoline (2017 dollars)

We can see from the graph that the price of gasoline (adjusted for inflation) stayed roughly the same from 1993 to 1997 and was increasing from 2002 to 2008. The graph also shows that the lowest price occurred in 1998.

Look again at Figure 44. The graph shows that for each date on the horizontal axis, there is only one price on the vertical axis. The graph represents a function, although a rule for getting from date to price is not given.

When a function is defined by an equation in x and y, the **graph of the function** is the graph of the equation; that is, it is the set of points (x, y) in the xy-plane that satisfy the equation.

6 Identify the Graph of a Function

Not every collection of points in the xy-plane represents the graph of a function. Remember, for a function, each number x in the domain has exactly one image y in the range. This means that the graph of a function cannot contain two points with the same x-coordinate and different y-coordinates. Therefore, the graph of a function must satisfy the following *vertical-line test*.

In Words

If any vertical line intersects a graph at more than one point, the graph is not the graph of a function.

THEOREM Vertical-Line Test

A set of points in the xy-plane is the graph of a function if and only if every vertical line intersects the graph in at most one point.

EXAMPLE 11	Identifying the Graph of a Function

Which of the graphs in Figure 45 are graphs of functions?

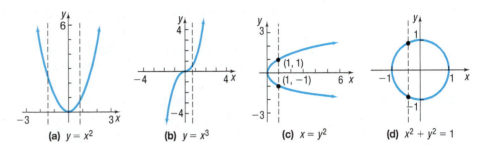

Figure 45

(a) $y = x^2$ (b) $y = x^3$ (c) $x = y^2$ (d) $x^2 + y^2 = 1$

Solution   The graphs in Figures 45(a) and 45(b) are graphs of functions, because every vertical line intersects each graph in at most one point. The graphs in Figures 45(c) and 45(d) are not graphs of functions, because there is a vertical line that intersects each graph in more than one point. Notice in Figure 45(c) that the input 1 corresponds to two outputs, -1 and 1. This is why the graph does not represent a function.

 Now Work PROBLEMS 87 AND 89

7 Obtain Information from or about the Graph of a Function

If (x, y) is a point on the graph of a function f, then y is the value of f at x; that is, $y = f(x)$. Also if $y = f(x)$, then (x, y) is a point on the graph of f. For example, if $(-2, 7)$ is on the graph of f, then $f(-2) = 7$, and if $f(5) = 8$, then the point $(5, 8)$ is on the graph of $y = f(x)$.

EXAMPLE 12	Obtaining Information from the Graph of a Function

Let f be the function whose graph is given in Figure 46. (The graph of f might represent the distance y that the bob of a pendulum is from its *at-rest* position at time x. Negative values of y mean that the pendulum is to the left of the at-rest position, and positive values of y mean that the pendulum is to the right of the at-rest position.)

(a) What are $f(0), f\left(\dfrac{3\pi}{2}\right)$, and $f(3\pi)$?

(b) What is the domain of f?

(c) What is the range of f?

(d) List the intercepts. (Recall that these are the points, if any, where the graph crosses or touches the coordinate axes.)

(e) How many times does the line $y = 2$ intersect the graph?

(f) For what values of x does $f(x) = -4$?

(g) For what values of x is $f(x) > 0$?

Figure 46

Solution   (a) Since $(0, 4)$ is on the graph of f, the y-coordinate 4 is the value of f at the x-coordinate 0; that is, $f(0) = 4$. In a similar way, when $x = \dfrac{3\pi}{2}$, then $y = 0$, so $f\left(\dfrac{3\pi}{2}\right) = 0$. When $x = 3\pi$, then $y = -4$, so $f(3\pi) = -4$.

(b) To determine the domain of f, notice that the points on the graph of f have x-coordinates between 0 and 4π, inclusive; and for each number x between 0 and 4π, there is a point $(x, f(x))$ on the graph. The domain of f is $\{x \mid 0 \le x \le 4\pi\}$ or the interval $[0, 4\pi]$.

(c) The points on the graph all have y-coordinates between -4 and 4, inclusive; and for each such number y, there is at least one number x in the domain. The range of f is $\{y \mid -4 \le y \le 4\}$ or the interval $[-4, 4]$.

(d) The intercepts are the points

$$(0, 4), \left(\frac{\pi}{2}, 0\right), \left(\frac{3\pi}{2}, 0\right), \left(\frac{5\pi}{2}, 0\right), \text{ and } \left(\frac{7\pi}{2}, 0\right)$$

(e) Draw the horizontal line $y = 2$ on the graph in Figure 46. Notice that the line intersects the graph four times.

(f) Since $(\pi, -4)$ and $(3\pi, -4)$ are the only points on the graph for which $y = f(x) = -4$, we have $f(x) = -4$ when $x = \pi$ and $x = 3\pi$.

(g) To determine where $f(x) > 0$, look at Figure 46 and determine the x-values from 0 to 4π for which the y-coordinate is positive. This occurs on $\left[0, \frac{\pi}{2}\right) \cup \left(\frac{3\pi}{2}, \frac{5\pi}{2}\right) \cup \left(\frac{7\pi}{2}, 4\pi\right]$. Using inequality notation, $f(x) > 0$ for $0 \le x < \frac{\pi}{2}$ or $\frac{3\pi}{2} < x < \frac{5\pi}{2}$ or $\frac{7\pi}{2} < x \le 4\pi$.

RECALL
The symbol $\cup$ stands for union. It means the set of elements that are in either of two sets.

When the graph of a function is given, its domain may be viewed as the shadow created by the graph on the x-axis by vertical beams of light. Its range can be viewed as the shadow created by the graph on the y-axis by horizontal beams of light. Try this technique with the graph given in Figure 46.

→ **Now Work** PROBLEM 83

EXAMPLE 13

Obtaining Information about the Graph of a Function

Consider the function: $f(x) = \dfrac{x + 1}{x + 2}$

(a) Find the domain of f.

(b) Is the point $\left(1, \dfrac{1}{2}\right)$ on the graph of f?

(c) If $x = 2$, what is $f(x)$? What point is on the graph of f?

(d) If $f(x) = 2$, what is x? What point is on the graph of f?

(e) What are the x-intercepts of the graph of f (if any)? What corresponding point(s) are on the graph of f?

Solution

(a) The domain of f is $\{x | x \ne -2\}$.

(b) When $x = 1$, then

$$f(1) = \frac{1 + 1}{1 + 2} = \frac{2}{3} \qquad f(x) = \frac{x + 1}{x + 2}$$

The point $\left(1, \dfrac{2}{3}\right)$ is on the graph of f; the point $\left(1, \dfrac{1}{2}\right)$ is not.

(c) If $x = 2$, then

$$f(2) = \frac{2 + 1}{2 + 2} = \frac{3}{4}$$

The point $\left(2, \dfrac{3}{4}\right)$ is on the graph of f.

(d) If $f(x) = 2$, then

$$\frac{x + 1}{x + 2} = 2 \qquad\qquad f(x) = 2$$
$$x + 1 = 2(x + 2) \qquad \text{Multiply both sides by } x + 2.$$
$$x + 1 = 2x + 4 \qquad \text{Distribute.}$$
$$x = -3 \qquad\qquad \text{Solve for } x.$$

NOTE In Example 13, -2 is not in the domain of f, so $x + 2$ is not zero and we can multiply both sides of the equation by $x + 2$. ∎

If $f(x) = 2$, then $x = -3$. The point $(-3, 2)$ is on the graph of f.

(e) The x-intercepts of the graph of f are the real solutions of the equation $f(x) = 0$ that are in the domain of f.

$$\frac{x+1}{x+2} = 0$$

$$x + 1 = 0 \qquad \text{Multiply both sides by } x + 2.$$

$$x = -1 \qquad \text{Subtract 1 from both sides.}$$

The only real solution of the equation $f(x) = \dfrac{x+1}{x+2} = 0$ is $x = -1$, so -1 is the only x-intercept. Since $f(-1) = 0$, the point $(-1, 0)$ is on the graph of f.

Now Work PROBLEM 99

SUMMARY

Function	• A relation between two sets of real numbers so that each number x in the first set, the domain, has corresponding to it exactly one number y in the second set, the range.
	• A set of ordered pairs (x, y) or $(x, f(x))$ in which no first element is paired with two different second elements.
	• The range is the set of y values of the function that are the images of the x values in the domain.
	• A function f may be defined implicitly by an equation involving x and y or explicitly by writing $y = f(x)$.
Unspecified Domain	If a function f is defined by an equation and no domain is specified, then the domain will be taken to be the largest set of real numbers for which the equation defines a real number.
Function Notation	• $y = f(x)$
	• f is a symbol for the function.
	• x is the independent variable, or argument.
	• y is the dependent variable.
	• $f(x)$ is the value of the function at x, or the image of x.
Graph of a Function	The collection of points (x, y) that satisfies the equation $y = f(x)$.
Vertical-Line Test	A collection of points is the graph of a function provided that every vertical line intersects the graph in at most one point.

1.3 Assess Your Understanding

'Are You Prepared?' *Answers are given at the end of these exercises. If you get a wrong answer, read the pages listed in red.*

1. The inequality $-1 < x < 3$ can be written in interval notation as _____. (pp. A45–A47)

2. If $x = -2$, the value of the expression $3x^2 - 5x + \dfrac{1}{x}$ is _____. (pp. A6–A7)

3. The domain of the variable in the expression $\dfrac{x-3}{x+4}$ is _____. (p. A7)

4. Solve the inequality: $3 - 2x > 5$. (pp. A49–A51)

5. To rationalize the denominator of $\dfrac{3}{\sqrt{5} - 2}$, multiply the numerator and denominator by _____. (pp. A58–A59)

6. A quotient is considered rationalized if its denominator has no _____. (pp. A58–A59)

Concepts and Vocabulary

7. If f is a function defined by the equation $y = f(x)$, then x is called the _____ variable, and y is the _____ variable.

8. *True or False* Every relation is a function.

9. *True or False* If no domain is specified for a function f, then the domain of f is taken to be the set of real numbers.

10. *True or False* The domain of the function $f(x) = \dfrac{x^2 - 4}{x}$ is $\{x | x \neq \pm 2\}$.

11. *Multiple Choice* The set of all images of the elements in the domain of a function is called the _____.
  (a) range (b) domain (c) solution set (d) function

12. *Multiple Choice* The independent variable is sometimes referred to as the _____ of the function.
  (a) range **(b)** value **(c)** argument **(d)** definition

13. *Multiple Choice* The expression $\dfrac{f(x + h) - f(x)}{h}$ is called the _____ of f.
  (a) radicand **(b)** image
  (c) correspondence **(d)** difference quotient

14. *Multiple Choice* When written as $y = f(x)$, a function is said to be defined _____.

  (a) explicitly **(b)** consistently
  (c) implicitly **(d)** rationally

15. A set of points in the xy-plane is the graph of a function if and only if every _____ line intersects the graph in at most one point.

16. If the point $(5, -3)$ is a point on the graph of f, then $f(\_\_) = \_\_\_$.

17. Find a so that the point $(-1, 2)$ is on the graph of $f(x) = ax^2 + 4$.

18. *True or False* The graph of a function $y = f(x)$ always crosses the y-axis.

19. *True or False* The y-intercept of the graph of the function $y = f(x)$, whose domain is all real numbers, is $f(0)$.

20. *Multiple Choice* The graph of a function $y = f(x)$ can have more than one of which type of intercept?

  (a) x-intercept **(b)** y-intercept
  (c) both **(d)** neither

Skill Building

In Problems 21 and 22, a relation expressed verbally is given.

(a) What is the domain and the range of the relation? (b) Express the relation using a mapping. (c) Express the relation as a set of ordered pairs.

21. The density of a gas under constant pressure depends on temperature. Holding pressure constant at 14.5 pounds per square inch, a chemist measures the density of an oxygen sample at temperatures of 0, 22, 40, 70, and 100°C and obtains densities of 1.411, 1.305, 1.229, 1.121, and 1.031 kg/m³, respectively.

22. A researcher wants to investigate how weight depends on height among adult males in Europe. She visits five regions in Europe and determines the average heights in those regions to be 1.80, 1.78, 1.77, 1.77, and 1.80 meters. The corresponding average weights are 87.1, 86.9, 83.0, 84.1, and 86.4 kg, respectively.

In Problems 23–34, state the domain and range for each relation. Then determine whether each relation represents a function.

23.

24.

25.

26.

27. $\{ (2, 6), (-3, 6), (4, 9), (2, 10) \}$ **28.** $\{ (-2, 5), (-1, 3), (3, 7), (4, 12) \}$ **29.** $\{ (1, 3), (2, 3), (3, 3), (4, 3) \}$

30. $\{ (0, -2), (1, 3), (2, 3), (3, 7) \}$ **31.** $\{ (-2, 4), (-2, 6), (0, 3), (3, 7) \}$ **32.** $\{ (-4, 4), (-3, 3), (-2, 2), (-1, 1), (-4, 0) \}$

33. $\{ (-2, 4), (-1, 1), (0, 0), (1, 1) \}$ **34.** $\{ (-2, 16), (-1, 4), (0, 3), (1, 4) \}$

In Problems 35–46, determine whether the equation defines y as a function of x.

35. $y = 2x^2 - 3x + 4$ **36.** $y = x^3$ **37.** $y = \dfrac{1}{x}$ **38.** $y = |x|$

39. $y^2 = 4 - x^2$ **40.** $y = \pm\sqrt{1 - 2x}$ **41.** $x = y^2$ **42.** $x + y^2 = 1$

43. $y = \sqrt[3]{x}$ **44.** $y = \dfrac{3x - 1}{x + 2}$ **45.** $2x^2 + 3y^2 = 1$ **46.** $x^2 - 4y^2 = 1$

In Problems 47–54, find the following for each function:

  (a) $f(0)$ *(b)* $f(1)$ *(c)* $f(-1)$ *(d)* $f(-x)$ *(e)* $-f(x)$ *(f)* $f(x + 1)$ *(g)* $f(2x)$ *(h)* $f(x + h)$

47. $f(x) = 3x^2 + 2x - 4$ **48.** $f(x) = -2x^2 + x - 1$ **49.** $f(x) = \dfrac{x}{x^2 + 1}$ **50.** $f(x) = \dfrac{x^2 - 1}{x + 4}$

51. $f(x) = |x| + 4$ **52.** $f(x) = \sqrt{x^2 + x}$ **53.** $f(x) = \dfrac{2x + 1}{3x - 5}$ **54.** $f(x) = 1 - \dfrac{1}{(x + 2)^2}$

In Problems 55–70, find the domain of each function.

55. $f(x) = -5x + 4$

56. $f(x) = x^2 + 2$

57. $f(x) = \dfrac{x}{x^2 + 1}$

58. $f(x) = \dfrac{x^2}{x^2 + 1}$

59. $g(x) = \dfrac{x}{x^2 - 16}$

60. $h(x) = \dfrac{2x}{x^2 - 4}$

61. $F(x) = \dfrac{x - 2}{x^3 + x}$

62. $G(x) = \dfrac{x + 4}{x^3 - 4x}$

63. $h(x) = \sqrt{3x - 12}$

64. $G(x) = \sqrt{1 - x}$

65. $p(x) = \sqrt{\dfrac{2}{x - 1}}$

66. $f(x) = \dfrac{4}{\sqrt{x - 9}}$

67. $f(x) = \dfrac{x}{\sqrt{x - 4}}$

68. $f(x) = \dfrac{-x}{\sqrt{-x - 2}}$

69. $P(t) = \dfrac{\sqrt{t - 4}}{3t - 21}$

70. $h(z) = \dfrac{\sqrt{z + 3}}{z - 2}$

 In Problems 71–82, find the difference quotient of f; that is, find $\dfrac{f(x + h) - f(x)}{h}, h \ne 0,$ *for each function. Be sure to simplify.*

71. $f(x) = 4x + 3$

72. $f(x) = -3x + 1$

73. $f(x) = x^2 - 4$

74. $f(x) = 3x^2 + 2$

75. $f(x) = x^2 - x + 4$

76. $f(x) = 3x^2 - 2x + 6$

77. $f(x) = \dfrac{1}{x^2}$

78. $f(x) = \dfrac{1}{x + 3}$

79. $f(x) = \dfrac{2x}{x + 3}$

80. $f(x) = \dfrac{5x}{x - 4}$

81. $f(x) = \sqrt{x - 2}$

82. $f(x) = \sqrt{x + 1}$

[**Hint:** Rationalize the numerator.]

83. Use the given graph of the function f to answer parts (a)–(n).

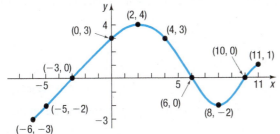

(a) Find $f(0)$ and $f(-6)$.
(b) Find $f(6)$ and $f(11)$.
(c) Is $f(3)$ positive or negative?
(d) Is $f(-4)$ positive or negative?
(e) For what values of x is $f(x) = 0$?
(f) For what values of x is $f(x) > 0$?
(g) What is the domain of f?
(h) What is the range of f?
(i) What are the x-intercepts?
(j) What is the y-intercept?
(k) How often does the line $y = \dfrac{1}{2}$ intersect the graph?
(l) How often does the line $x = 5$ intersect the graph?
(m) For what values of x does $f(x) = 3$?
(n) For what values of x does $f(x) = -2$?

84. Use the given graph of the function f to answer parts (a)–(n).

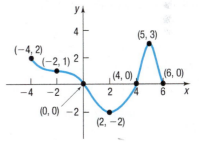

(a) Find $f(0)$ and $f(6)$.
(b) Find $f(2)$ and $f(-2)$.
(c) Is $f(3)$ positive or negative?
(d) Is $f(-1)$ positive or negative?
(e) For what values of x is $f(x) = 0$?
(f) For what values of x is $f(x) < 0$?
(g) What is the domain of f?
(h) What is the range of f?
(i) What are the x-intercepts?
(j) What is the y-intercept?
(k) How often does the line $y = -1$ intersect the graph?
(l) How often does the line $x = 1$ intersect the graph?
(m) For what value of x does $f(x) = 3$?
(n) For what value of x does $f(x) = -2$?

In Problems 85–96, determine whether the graph is that of a function by using the vertical-line test. If it is, use the graph to find:
(a) The domain and range   (b) The intercepts, if any   (c) Any symmetry with respect to the x-axis, the y-axis, or the origin

85.

86.

87.

88.

89.

90.

91.

92.

93.

94.

95.

96.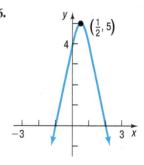

In Problems 97–102, answer the questions about the given function.

97. $f(x) = 2x^2 - x - 1$
  (a) Is the point $(-1, 2)$ on the graph of f?
  (b) If $x = -2$, what is $f(x)$? What point is on the graph of f?
  (c) If $f(x) = -1$, what is x? What point(s) are on the graph of f?
  (d) What is the domain of f?
  (e) List the x-intercepts, if any, of the graph of f.
  (f) List the y-intercept, if there is one, of the graph of f.

98. $f(x) = -3x^2 + 5x$
  (a) Is the point $(-1, 2)$ on the graph of f?
  (b) If $x = -2$, what is $f(x)$? What point is on the graph of f?
  (c) If $f(x) = -2$, what is x? What point(s) are on the graph of f?
  (d) What is the domain of f?
  (e) List the x-intercepts, if any, of the graph of f.
  (f) List the y-intercept, if there is one, of the graph of f.

99. $f(x) = \dfrac{x + 2}{x - 6}$
  (a) Is the point $(3, 14)$ on the graph of f?
  (b) If $x = 4$, what is $f(x)$? What point is on the graph of f?
  (c) If $f(x) = 2$, what is x? What point(s) are on the graph of f?
  (d) What is the domain of f?
  (e) List the x-intercepts, if any, of the graph of f.
  (f) List the y-intercept, if there is one, of the graph of f.

100. $f(x) = \dfrac{x^2 + 2}{x + 4}$
  (a) Is the point $\left(1, \dfrac{3}{5}\right)$ on the graph of f?
  (b) If $x = 0$, what is $f(x)$? What point is on the graph of f?
  (c) If $f(x) = \dfrac{1}{2}$, what is x? What point(s) are on the graph of f?
  (d) What is the domain of f?
  (e) List the x-intercepts, if any, of the graph of f.
  (f) List the y-intercept, if there is one, of the graph of f.

101. $f(x) = \dfrac{2x^2}{x^4 + 1}$
  (a) Is the point $(-1, 1)$ on the graph of f?
  (b) If $x = 2$, what is $f(x)$? What point is on the graph of f?
  (c) If $f(x) = 1$, what is x? What point(s) are on the graph of f?
  (d) What is the domain of f?
  (e) List the x-intercepts, if any, of the graph of f.
  (f) List the y-intercept, if there is one, of the graph of f.

102. $f(x) = \dfrac{2x}{x - 2}$
  (a) Is the point $\left(\dfrac{1}{2}, -\dfrac{2}{3}\right)$ on the graph of f?
  (b) If $x = 4$, what is $f(x)$? What point is on the graph of f?
  (c) If $f(x) = 1$, what is x? What point(s) are on the graph of f?
  (d) What is the domain of f?
  (e) List the x-intercepts, if any, of the graph of f.
  (f) List the y-intercept, if there is one, of the graph of f.

Applications and Extensions

103. If $f(x) = x^2 - 2x + 3$, find the value(s) of x so that $f(x) = 11$.

104. If $f(x) = \dfrac{5}{6}x - \dfrac{3}{4}$, find the value(s) of x so that $f(x) = -\dfrac{7}{16}$.

105. If $f(x) = 2x^3 + Ax^2 + 4x - 5$ and $f(2) = 5$, what is the value of A?

106. If $f(x) = 3x^2 - Bx + 4$ and $f(-1) = 12$, what is the value of B?

107. If $f(x) = \dfrac{3x + 8}{2x - A}$ and $f(0) = 2$, what is the value of A?

108. If $f(x) = \dfrac{2x - B}{3x + 4}$ and $f(2) = \dfrac{1}{2}$, what is the value of B?

109. Geometry Express the area A of a rectangle as a function of the length x if the length of the rectangle is twice its width.

110. Geometry Express the area A of an isosceles right triangle as a function of the length x of one of the two equal sides.

111. Constructing Functions Express the gross salary G of a person who earns \$16 per hour as a function of the number x of hours worked.

112. Constructing Functions Tiffany, a commissioned salesperson, earns \$100 base pay plus \$10 per item sold. Express her gross salary G as a function of the number x of items sold.

113. **Effect of Gravity on Earth** If a rock falls from a height of 20 meters on Earth, the height H (in meters) after x seconds is approximately

$$H(x) = 20 - 4.9x^2$$

   (a) What is the height of the rock when $x = 1$ second? $x = 1.1$ seconds? $x = 1.2$ seconds? $x = 1.3$ seconds?
   (b) When is the height of the rock 15 meters? When is it 10 meters? When is it 5 meters?
   (c) When does the rock strike the ground?

114. **Effect of Gravity on Jupiter** If a rock falls from a height of 20 meters on the planet Jupiter, its height H (in meters) after x seconds is approximately

$$H(x) = 20 - 13x^2$$

   (a) What is the height of the rock when $x = 1$ second? $x = 1.1$ seconds? $x = 1.2$ seconds?
   (b) When is the height of the rock 15 meters? When is it 10 meters? When is it 5 meters?
   (c) When does the rock strike the ground?

115. **Cost of Trans-Atlantic Travel** A Boeing 747 crosses the Atlantic Ocean (3000 miles) with an airspeed of 500 miles per hour. The cost C (in dollars) per passenger is given by

$$C(x) = 100 + \frac{x}{10} + \frac{36,000}{x}$$

   where x is the ground speed (airspeed $\pm$ wind).
   (a) What is the cost per passenger for quiescent (no wind) conditions?
   (b) What is the cost per passenger with a head wind of 50 miles per hour?
   (c) What is the cost per passenger with a tail wind of 100 miles per hour?
   (d) What is the cost per passenger with a head wind of 100 miles per hour?

116. **Cross-sectional Area** The cross-sectional area of a beam cut from a log with radius 1 foot is given by the function $A(x) = 4x\sqrt{1 - x^2}$, where x represents the length, in feet, of half the base of the beam. See the figure. Determine the cross-sectional area of the beam if the length of half the base of the beam is as follows:
   (a) One-third of a foot
   (b) One-half of a foot
   (c) Two-thirds of a foot

$A(x) = 4x\sqrt{1 - x^2}$

117. **Free-throw Shots** According to physicist Peter Brancazio, the key to a successful foul shot in basketball lies in the arc of the shot. Brancazio determined the optimal angle of the arc from the free-throw line to be 45 degrees. The arc also depends on the velocity with which the ball is shot. If a player shoots a foul shot, releasing the ball at a 45-degree angle from a position 6 feet above the floor, then the path of the ball can be modeled by the function

$$h(x) = -\frac{44x^2}{v^2} + x + 6$$

   where h is the height of the ball above the floor, x is the forward distance of the ball in front of the foul line, and v is the initial velocity with which the ball is shot in feet per second. Suppose a player shoots a ball with an initial velocity of 28 feet per second.

   (a) Determine the height of the ball after it has traveled 8 feet in front of the foul line.
   (b) Determine the height of the ball after it has traveled 12 feet in front of the foul line.
   (c) Find additional points and graph the path of the basketball.
   (d) The center of the hoop is 10 feet above the floor and 15 feet in front of the foul line. Will the ball go through the hoop? Why or why not? If not, with what initial velocity must the ball be shot in order for the ball to go through the hoop?

   Source: The Physics of Foul Shots, Discover, Vol. 21, No. 10, October 2000

118. **Granny Shots** The last player in the NBA to use an underhand foul shot (a "granny" shot) was Hall of Fame forward Rick Barry, who retired in 1980. Barry believes that current NBA players could increase their free-throw percentage if they were to use an underhand shot. Since underhand shots are released from a lower position, the angle of the shot must be increased. If a player shoots an underhand foul shot, releasing the ball at a 70-degree angle from a position 3.5 feet above the floor, then the path of the ball can be modeled by the function $h(x) = -\dfrac{136x^2}{v^2} + 2.7x + 3.5$, where h is the height of the ball above the floor, x is the forward distance of the ball in front of the foul line, and v is the initial velocity with which the ball is shot in feet per second.

   (a) The center of the hoop is 10 feet above the floor and 15 feet in front of the foul line. Determine the initial velocity with which the ball must be shot in order for the ball to go through the hoop.
   (b) Write the function for the path of the ball using the velocity found in part (a).
   (c) Determine the height of the ball after it has traveled 9 feet in front of the foul line.
   (d) Find additional points and graph the path of the basketball.

   Source: The Physics of Foul Shots, Discover, Vol. 21, No. 10, October 2000

119. **Motion of a Golf Ball** A golf ball is hit with an initial velocity of 130 feet per second at an inclination of 45° to the horizontal. In physics, it is established that the height h of the golf ball is given by the function

$$h(x) = \frac{-32x^2}{130^2} + x$$

(continued)

where x is the horizontal distance that the golf ball has traveled.

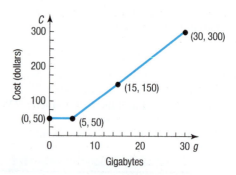

(a) Determine the height of the golf ball after it has traveled 100 feet.

(b) What is the height after it has traveled 300 feet?

(c) What is the height after it has traveled 500 feet?

(d) How far was the golf ball hit?

(e) Use a graphing utility to graph the function $h = h(x)$.

(f) Use a graphing utility to determine the distance that the ball has traveled when the height of the ball is 90 feet.

(g) Create a TABLE with TblStart = 0 and ΔTbl = 25. To the nearest 25 feet, how far does the ball travel before it reaches a maximum height? What is the maximum height?

(h) Adjust the value of ΔTbl until you determine the distance, to within 1 foot, that the ball travels before it reaches its maximum height.

120. **Reading and Interpreting Graphs** Let C be the function whose graph is given below. This graph represents the cost C of manufacturing q computers in a day.

(a) Determine $C(0)$. Interpret this value.

(b) Determine $C(10)$. Interpret this value.

(c) Determine $C(50)$. Interpret this value.

(d) What is the domain of C? What does this domain imply in terms of daily production?

(e) Describe the shape of the graph.

(f) The point (30, 32 000) is called an *inflection point*. Describe the behavior of the graph around the inflection point.

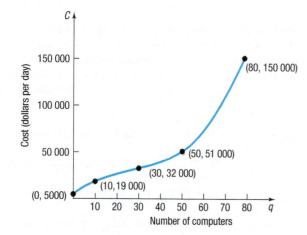

121. **Reading and Interpreting Graphs** Let C be the function whose graph is given below. This graph represents the cost C of using g gigabytes of data in a month for a data-only plan.

(a) Determine $C(0)$. Interpret this value.

(b) Determine $C(5)$. Interpret this value.

(c) Determine $C(15)$. Interpret this value.

(d) What is the domain of C? What does this domain imply in terms of the number of gigabytes?

(e) Describe the shape of the graph.

122. Some functions f have the property that

$$f(a + b) = f(a) + f(b)$$

for all real numbers a and b. Which of the following functions have this property?

(a) $h(x) = 2x$ (b) $g(x) = x^2$

(c) $F(x) = 5x - 2$ (d) $G(x) = \dfrac{1}{x}$

123. *Challenge Problem* Find the difference quotient of the function $f(x) = \sqrt[3]{x}$.

(**Hint**: Factor using $a^3 - b^3 = (a - b)(a^2 + ab + b^2)$ with $a = \sqrt[3]{x + h}$ and $b = \sqrt[3]{x}$.)

124. *Challenge Problem* If $f\left(\dfrac{x + 4}{5x - 4}\right) = 3x^2 - 2$, find $f(1)$.

125. *Challenge Problem* Suppose $f(x) = x^2 - 4x + c$ and
$g(x) = \dfrac{f(x)}{3} - 4$. Find $f(3)$ if $g(-2) = 5$.

126. *Challenge Problem* Suppose $f(x) = \sqrt{x} + 2$ and $g(x) = x^2 + n$. If $f(g(5)) = 4$, what is the value of $g(n)$?

Explaining Concepts: Discussion and Writing

127. Describe how you would find the domain and range of a function if you were given its graph. How would your strategy change if you were given the equation defining the function instead of its graph?

128. How many x-intercepts can the graph of a function have? How many y-intercepts can the graph of a function have?

129. Is a graph that consists of a single point the graph of a function? Can you write the equation of such a function?

130. Match each of the following functions with the graph that best describes the situation.
  (a) The cost of building a house as a function of its square footage
  (b) The height of an egg dropped from a 300-foot building as a function of time
  (c) The height of a human as a function of time
  (d) The demand for Big Macs as a function of price
  (e) The height of a child on a swing as a function of time

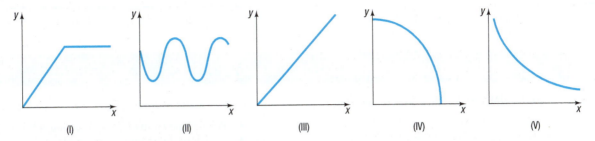

131. Match each of the following functions with the graph that best describes the situation.
  (a) The temperature of a bowl of soup as a function of time
  (b) The number of hours of daylight per day over a 2-year period
  (c) The population of Florida as a function of time
  (d) The distance traveled by a car going at a constant velocity as a function of time
  (e) The height of a golf ball hit with a 7-iron as a function of time

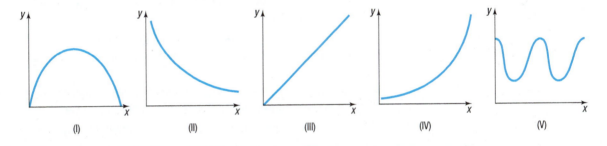

132. Consider the following scenario: Barbara decides to take a walk. She leaves home, walks 2 blocks in 5 minutes at a constant speed, and realizes that she forgot to lock the door. So Barbara runs home in 1 minute. While at her doorstep, it takes her 1 minute to find her keys and lock the door. Barbara walks 5 blocks in 15 minutes and then decides to jog home. It takes her 7 minutes to get home. Draw a graph of Barbara's distance from home (in blocks) as a function of time.

133. Consider the following scenario: Jayne enjoys riding her bicycle through the woods. At the forest preserve, she gets on her bicycle and rides up a 2000-foot incline in 10 minutes. She then travels down the incline in 3 minutes. The next 5000 feet is level terrain, and she covers the distance in 20 minutes. She rests for 15 minutes. Jayne then travels 10,000 feet in 30 minutes. Draw a graph of Jayne's distance traveled (in feet) as a function of time.

134. The following sketch represents the distance d (in miles) that Kevin was from home as a function of time t (in hours). Answer the questions by referring to the graph. In parts (a)–(g), how many hours elapsed and how far was Kevin from home during this time?

(a) From $t = 0$ to $t = 2$
(b) From $t = 2$ to $t = 2.5$
(c) From $t = 2.5$ to $t = 2.8$
(d) From $t = 2.8$ to $t = 3$
(e) From $t = 3$ to $t = 3.9$
(f) From $t = 3.9$ to $t = 4.2$
(g) From $t = 4.2$ to $t = 5.3$
(h) What is the farthest distance that Kevin was from home?
(i) How many times did Kevin return home?

135. The following sketch represents the speed v (in miles per hour) of Michael's car as a function of time t (in minutes).

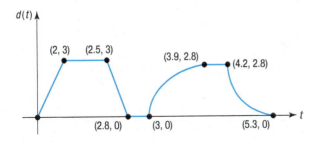

(a) Over what interval of time was Michael traveling fastest?
(b) Over what interval(s) of time was Michael's speed zero?
(c) What was Michael's speed between 0 and 2 minutes?
(d) What was Michael's speed between 4.2 and 6 minutes?
(e) What was Michael's speed between 7 and 7.4 minutes?
(f) When was Michael's speed constant?

136. Draw the graph of a function whose domain is $\{x | -3 \le x \le 8,\ x \ne 5\}$ and whose range is $\{y | -1 \le y \le 2,\ y \ne 0\}$. What point(s) in the rectangle $-3 \le x \le 8,$ $-1 \le y \le 2$ cannot be on the graph? Compare your graph with those of other students. What differences do you see?

137. Is there a function whose graph is symmetric with respect to the x-axis? Explain.

138. Explain why the vertical-line test works.

'Are You Prepared?' Answers

1. $(-1, 3)$ **2.** 21.5 **3.** $\{x | x \ne -4\}$ **4.** $\{x | x < -1\}$ **5.** $\sqrt{5} + 2$ **6.** radicals

---

1.4 Properties of Functions

PREPARING FOR THIS SECTION *Before getting started, review the following:*

- Intervals (Section A.6, pp. A45–A47)
- Intercepts (Section 1.2, pp. 11–12)
- Slope of a Line (Section A.8, pp. A64–A67)
- Point-Slope Form of a Line (Section A.8, pp. A68–A69)
- Symmetry (Section 1.2, pp. 12–14)

✎ **Now Work** the **'Are You Prepared?'** problems on page 52.

OBJECTIVES
1. Identify Even and Odd Functions from a Graph (p. 44)
2. Identify Even and Odd Functions from an Equation (p. 45)
3. Use a Graph to Determine Where a Function Is Increasing, Decreasing, or Constant (p. 46)
4. Use a Graph to Locate Local Maxima and Local Minima (p. 47)
5. Use a Graph to Locate the Absolute Maximum and the Absolute Minimum (p. 48)
6. Use a Graphing Utility to Approximate Local Maxima and Local Minima and to Determine Where a Function Is Increasing or Decreasing (p. 50)
7. Find the Average Rate of Change of a Function (p. 50)

To obtain the graph of a function $y = f(x)$, it is often helpful to know properties of the function and the impact of these properties on the graph of the function.

1 Identify Even and Odd Functions from a Graph

The words *even* and *odd*, when discussing a function f, describe the symmetry of the graph of the function.

A function f is even if and only if, whenever the point (x, y) is on the graph of f, the point $(-x, y)$ is also on the graph. Using function notation, we define an even function as follows:

DEFINITION **Even Function**

A function f is **even** if, for every number x in its domain, the number $-x$ is also in the domain and

$$f(-x) = f(x)$$

A function f is odd if and only if, whenever the point (x, y) is on the graph of f, the point $(-x, -y)$ is also on the graph. Using function notation, we define an odd function as follows:

DEFINITION **Odd Function**

A function f is **odd** if, for every number x in its domain, the number $-x$ is also in the domain and

$$f(-x) = -f(x)$$

Refer to page 13, where the tests for symmetry are listed. The results below follow.

> **THEOREM Graphs of Even and Odd Functions**
>
> - A function is even if and only if its graph is symmetric with respect to the y-axis.
> - A function is odd if and only if its graph is symmetric with respect to the origin.

EXAMPLE 1   Identifying Even and Odd Functions from a Graph

Determine whether each graph given in Figure 47 is the graph of an even function, an odd function, or a function that is neither even nor odd.

Figure 47   **(a)**   **(b)**   **(c)**

Solution   (a) The graph in Figure 47(a) is that of an even function, because the graph is symmetric with respect to the y-axis.

(b) The function whose graph is given in Figure 47(b) is neither even nor odd, because the graph is neither symmetric with respect to the y-axis nor symmetric with respect to the origin.

(c) The function whose graph is given in Figure 47(c) is odd, because its graph is symmetric with respect to the origin.

 Now Work PROBLEMS 25(a), (b), AND (d)

2 Identify Even and Odd Functions from an Equation

EXAMPLE 2   Identifying Even and Odd Functions Algebraically

Determine whether each of the following functions is even, odd, or neither. Then determine whether the graph is symmetric with respect to the y-axis, with respect to the origin, or neither.

(a) $f(x) = x^2 - 5$ (b) $g(x) = x^3 - 1$

(c) $h(x) = 5x^3 - x$ (d) $F(x) = |x|$

Solution   (a) To determine whether f is even, odd, or neither, replace x by $-x$ in $f(x) = x^2 - 5$.

$$f(-x) = (-x)^2 - 5 = x^2 - 5 = f(x)$$

Since $f(-x) = f(x)$, the function is even, and the graph of f is symmetric with respect to the y-axis.

(b) Replace x by $-x$ in $g(x) = x^3 - 1$.

$$g(-x) = (-x)^3 - 1 = -x^3 - 1$$

Since $g(-x) \neq g(x)$ and $g(-x) \neq -g(x) = -(x^3 - 1) = -x^3 + 1$, the function is neither even nor odd. The graph of g is not symmetric with respect to the y-axis, nor is it symmetric with respect to the origin.

(c) Replace x by $-x$ in $h(x) = 5x^3 - x$.

$$h(-x) = 5(-x)^3 - (-x) = -5x^3 + x = -(5x^3 - x) = -h(x)$$

Since $h(-x) = -h(x)$, h is an odd function, and the graph of h is symmetric with respect to the origin.

(d) Replace x by $-x$ in $F(x) = |x|$.

$$F(-x) = |-x| = |-1| \cdot |x| = |x| = F(x)$$

Since $F(-x) = F(x)$, F is an even function, and the graph of F is symmetric with respect to the y-axis.

━━━━▶ **Now Work** PROBLEM 37

3 Use a Graph to Determine Where a Function Is Increasing, Decreasing, or Constant

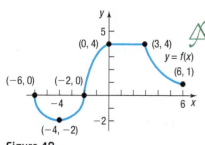

Figure 48

Consider the graph given in Figure 48. If you look from left to right along the graph of the function, you will notice that parts of the graph are going up, parts are going down, and parts are horizontal. In such cases, the function is described as *increasing*, *decreasing*, or *constant*, respectively.

DEFINITIONS Increasing Function, Decreasing Function, Constant Function

- A function f is **increasing** on an interval I if, for any choice of x_1 and x_2 in I, with $x_1 < x_2$, then $f(x_1) < f(x_2)$.

- A function f is **decreasing** on an interval I if, for any choice of x_1 and x_2 in I, with $x_1 < x_2$, then $f(x_1) > f(x_2)$.

- A function f is **constant** on an interval I if, for all choices of x in I, the values $f(x)$ are equal.

In Words

- If a function is increasing, then as the values of x get bigger, the values of the function also get bigger.
- If a function is decreasing, then as the values of x get bigger, the values of the function get smaller.
- If a function is constant, then as the values of x get bigger, the values of the function remain unchanged.

Need to Review?

Interval Notation is discussed in Section A.6, pp. A45–A47.

Figure 49 illustrates the definitions. The graph of an increasing function goes up from left to right, the graph of a decreasing function goes down from left to right, and the graph of a constant function remains at a fixed height. The interval I on which a function is increasing, decreasing, or constant may be open, closed, or half-open/half-closed depending on whether the endpoints of the interval satisfy the required inequality or not.

Figure 49

(a) For $x_1 < x_2$ in I, $f(x_1) < f(x_2)$; f is increasing on I.

(b) For $x_1 < x_2$ in I, $f(x_1) > f(x_2)$; f is decreasing on I.

(c) For all x in I, the values of f are equal; f is constant on I.

EXAMPLE 3

Determining Where a Function Is Increasing, Decreasing, or Constant from Its Graph

Determine the values of x for which the function in Figure 50 is increasing. Where is it decreasing? Where is it constant?

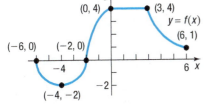

Figure 50

Solution When determining where a function is increasing, where it is decreasing, and where it is constant, we use intervals involving the independent variable x.

The function whose graph is given in Figure 50 is decreasing on the interval $[-6, -4]$ because for any choice of x_1 and x_2 in the interval for which $x_1 < x_2$, we have $f(x_1) > f(x_2)$. For example, -6 is included in the interval where f is decreasing because if x is any number for which $-6 < x \leq -4$, then $f(-6) > f(x)$.

Similarly, f is increasing on the interval $[-4, 0]$ because for any choice of x_1 and x_2 in the interval for which $x_1 < x_2$, we have $f(x_1) < f(x_2)$.

Finally, the function f is constant on the interval $[0, 3]$ and is decreasing on the interval $[3, 6]$.

Now Work PROBLEMS **13, 15, 17**, AND **25(c)**

4 Use a Graph to Locate Local Maxima and Local Minima

Suppose f is a function defined on an open interval I containing c. If the value of f at c is greater than or equal to the other values of f on I, then f has a *local maximum* at c.* See Figure 51(a).

If the value of f at c is less than or equal to the other values of f on I, then f has a *local minimum* at c. See Figure 51(b).

NOTE "Maxima" is the plural of "maximum"; "minima" is the plural of "minimum." ∎

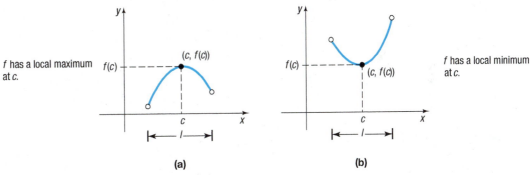

Figure 51 Local maximum and local minimum

DEFINITIONS Local Maximum/Minimum

Let f be a function defined on some interval I and let c be a number in I.

- A function f has a **local maximum** at c if there is an open interval in I containing c so that $f(c) \geq f(x)$ for all x in this open interval. The number $f(c)$ is called a **local maximum value of** f.
- A function f has a **local minimum** at c if there is an open interval in I containing c so that $f(c) \leq f(x)$ for all x in this open interval. The number $f(c)$ is called a **local minimum value of** f.

If f has a local maximum at c, then the value of f at c is greater than or equal to the values of f near c. If f has a local minimum at c, then the value of f at c is less than or equal to the values of f near c. The word *local* is used to suggest that it is only near c, not necessarily over the entire domain, that the value $f(c)$ has these properties.

*Some texts use the term *relative* instead of *local*.

EXAMPLE 4

Finding Local Maxima and Local Minima from the Graph of a Function and Determining Where the Function Is Increasing, Decreasing, or Constant

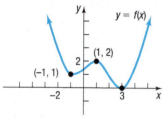

Figure 52

Figure 52 shows the graph of a function f.

(a) At what numbers x, if any, does f have a local maximum? List the local maximum value(s).

(b) At what numbers x, if any, does f have a local minimum? List the local minimum value(s).

(c) Find the intervals on which f is increasing. Find the intervals on which f is decreasing.

Solution

The domain of f is the set of real numbers.

(a) f has a local maximum at 1, since for all x close to 1, we have $f(x) \le f(1)$. The local maximum value is $f(1) = 2$.

(b) f has local minima at -1 and at 3. The local minimum values are $f(-1) = 1$ and $f(3) = 0$.

(c) The function whose graph is given in Figure 52 is increasing on the intervals $[-1, 1]$ and $[3, \infty)$, or for $-1 \le x \le 1$ and $x \ge 3$. The function is decreasing on the intervals $(-\infty, -1]$ and $[1, 3]$, or for $x \le -1$ and $1 \le x \le 3$. ∎

WARNING The y-value is the local maximum value or local minimum value, and it occurs at some number x. For example, in Figure 52, we say f has a local maximum at 1 and the local maximum value is 2. ∎

Now Work PROBLEMS **19** AND **21**

5 Use a Graph to Locate the Absolute Maximum and the Absolute Minimum

Look at the graph of the function f given in Figure 53. The domain of f is the closed interval $[a, b]$. Also, the largest value of f is $f(u)$ and the smallest value of f is $f(v)$. These are called, respectively, the *absolute maximum* and the *absolute minimum* of f on $[a, b]$.

domain: $[a, b]$
for all x in $[a, b]$, $f(x) \le f(u)$
for all x in $[a, b]$, $f(x) \ge f(v)$
absolute maximum: $f(u)$
absolute minimum: $f(v)$

Figure 53

> **DEFINITIONS** **Absolute Maximum and Absolute Minimum**
>
> Let f be a function defined on some interval I.
>
> • If there is a number u in I for which $f(u) \ge f(x)$ for all x in I, then f has an **absolute maximum at** u, and the number $f(u)$ is the **absolute maximum of f on I.**
>
> • If there is a number v in I for which $f(v) \le f(x)$ for all x in I, then f has an **absolute minimum at** v, and the number $f(v)$ is the **absolute minimum of f on I.**

The absolute maximum and absolute minimum of a function f are sometimes called the **absolute extrema** or the **extreme values** of f on I.

The absolute maximum or absolute minimum of a function f may not exist. Let's look at some examples.

EXAMPLE 5   **Finding the Absolute Maximum and the Absolute Minimum from the Graph of a Function**

For the graph of each function $y = f(x)$ in Figure 54, find the absolute maximum and the absolute minimum, if they exist. Also, find any local maxima or local minima.

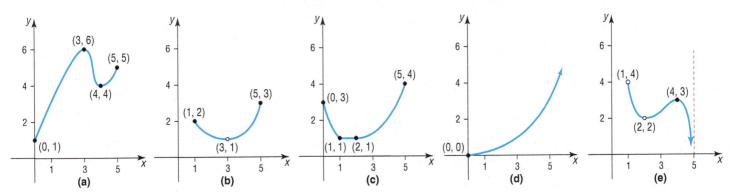

Figure 54

Solution   (a) The function f whose graph is given in Figure 54(a) has the closed interval $[0, 5]$ as its domain. The largest value of f is $f(3) = 6$, the absolute maximum. The smallest value of f is $f(0) = 1$, the absolute minimum. The function has a local maximum value of 6 at $x = 3$ and a local minimum value of 4 at $x = 4$.

(b) The function f whose graph is given in Figure 54(b) has domain $\{x \mid 1 \le x \le 5, x \neq 3\}$. Note that we exclude 3 from the domain because of the "hole" at $(3, 1)$. The largest value of f on its domain is $f(5) = 3$, the absolute maximum. There is no absolute minimum. Do you see why? As you trace the graph, getting closer to the point $(3, 1)$, there is no single smallest value. [As soon as you claim a smallest value, we can trace closer to $(3, 1)$ and get a smaller value!] The function has no local maxima or minima.

(c) The function f whose graph is given in Figure 54(c) has the interval $[0, 5]$ as its domain. The absolute maximum of f is $f(5) = 4$. The absolute minimum is 1. Notice that the absolute minimum occurs at any number in the interval $[1, 2]$. The function has a local minimum value of 1 at every x in the interval $[1, 2]$, but it has no local maximum.

(d) The function f given in Figure 54(d) has the interval $[0, \infty)$ as its domain. The function has no absolute maximum; the absolute minimum is $f(0) = 0$. The function has no local maxima or local minima.

(e) The function f in Figure 54(e) has domain $\{x \mid 1 < x < 5, x \neq 2\}$. The function has no absolute maximum and no absolute minimum. Do you see why? The function has a local maximum value of 3 at $x = 4$, but no local minimum value.

In calculus, there is a theorem with conditions that guarantee a function will have an absolute maximum and an absolute minimum.

THEOREM Extreme Value Theorem

If a function f is continuous* on a closed interval $[a, b]$, then f has an absolute maximum and an absolute minimum on $[a, b]$.

The absolute maximum (minimum) can be found by selecting the largest (smallest) value of f from the following list:

- The values of f at any local maxima or local minima of f in $[a, b]$.
- The value of f at each endpoint of $[a, b]$—that is, $f(a)$ and $f(b)$.

*Although a precise definition requires calculus, we'll agree for now that a function is continuous if its graph has no gaps or holes and can be traced without lifting the pencil from the paper.

For example, the graph of the function f given in Figure 54(a) is continuous on the closed interval $[0, 5]$. The Extreme Value Theorem guarantees that f has extreme values on $[0, 5]$. To find them, we list

- The value of f at the local extrema: $f(3) = 6, f(4) = 4$
- The value of f at the endpoints: $f(0) = 1, f(5) = 5$

The largest of these, 6, is the absolute maximum; the smallest of these, 1, is the absolute minimum.

 Now Work PROBLEM 49

6 Use a Graphing Utility to Approximate Local Maxima and Local Minima and to Determine Where a Function Is Increasing or Decreasing

 To locate the exact value at which a function f has a local maximum or a local minimum usually requires calculus. However, a graphing utility may be used to approximate these values using the MAXIMUM and MINIMUM features.

| EXAMPLE 6 | Using a Graphing Utility to Approximate Local Maxima and Minima and to Determine Where a Function Is Increasing or Decreasing |

(a) Use a graphing utility to graph $f(x) = 6x^3 - 12x + 5$ for $-2 \le x \le 2$. Approximate where f has a local maximum and where f has a local minimum.

(b) Determine where f is increasing and where it is decreasing.

Solution (a) Graphing utilities have a feature that finds the maximum or minimum point of a graph within a given interval. Graph the function f for $-2 \le x \le 2$. Using MAXIMUM on a TI-84 Plus C, we find that the local maximum value is 11.53 and that it occurs at $x = -0.82$, rounded to two decimal places. See Figure 55(a). Using MINIMUM, we find that the local minimum value is -1.53 and that it occurs at $x = 0.82$, rounded to two decimal places. See Figure 55(b). Figure 55(c) shows the two extrema in Desmos.

Figure 55 (a) Local maximum   (b) Local minimum   (c) Local extrema

(b) Looking at Figure 55, we see that f is increasing on the intervals $[-2, -0.82]$ and $[0.82, 2]$, or for $-2 \le x \le -0.82$ and $0.82 \le x \le 2$. And f is decreasing on the interval $[-0.82, 0.82]$, or for $-0.82 \le x \le 0.82$.

 Now Work PROBLEM 57

7 Find the Average Rate of Change of a Function

In Section A.8, we said that the slope of a line can be interpreted as the average rate of change. To find the average rate of change of a function between any two points on its graph, calculate the slope of the line containing the two points.

> **DEFINITION** Average Rate of Change
>
> If a and b, $a \neq b$, are in the domain of a function $y = f(x)$, the **average rate of change of f** from a to b is defined as
>
> $$\text{Average rate of change} = \frac{\Delta y}{\Delta x} = \frac{f(b) - f(a)}{b - a} \qquad a \neq b \qquad \textbf{(1)}$$

The symbol Δy in equation (1) is the "change in y," and Δx is the "change in x." The average rate of change of f is the change in y divided by the change in x.

EXAMPLE 7   **Finding the Average Rate of Change**

Find the average rate of change of $f(x) = 3x^2$:

(a) From 1 to 3   (b) From 1 to 5   (c) From 1 to 7

Solution (a) The average rate of change of $f(x) = 3x^2$ from 1 to 3 is

$$\frac{\Delta y}{\Delta x} = \frac{f(3) - f(1)}{3 - 1} = \frac{27 - 3}{3 - 1} = \frac{24}{2} = 12$$

(b) The average rate of change of $f(x) = 3x^2$ from 1 to 5 is

$$\frac{\Delta y}{\Delta x} = \frac{f(5) - f(1)}{5 - 1} = \frac{75 - 3}{5 - 1} = \frac{72}{4} = 18$$

(c) The average rate of change of $f(x) = 3x^2$ from 1 to 7 is

$$\frac{\Delta y}{\Delta x} = \frac{f(7) - f(1)}{7 - 1} = \frac{147 - 3}{7 - 1} = \frac{144}{6} = 24$$

See Figure 56 for a graph of $f(x) = 3x^2$. The function f is increasing on the interval $[0, \infty)$. The fact that the average rate of change is positive for any $x_1, x_2, x_1 \neq x_2$, in the interval $[1, 7]$ indicates that f is increasing on $1 \leq x \leq 7$.

Now Work PROBLEM **65**

Figure 56 $f(x) = 3x^2$

(Graph showing points (0,0), (1,3), (3,27), (5,75), (7,147) with Average rate of change = 24, Average rate of change = 18, Average rate of change = 12)

The Secant Line

The average rate of change of a function has an important geometric interpretation. Look at the graph of $y = f(x)$ in Figure 57. Two points are labeled on the graph: $(a, f(a))$ and $(b, f(b))$. The line containing these two points is called a **secant line**; its slope is

$$m_{\text{sec}} = \frac{f(b) - f(a)}{b - a}$$

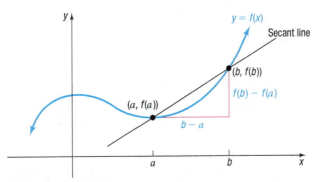

Figure 57 Secant line

> **THEOREM** **Slope of a Secant Line**
>
> The average rate of change of a function from a to b equals the slope of the secant line containing the two points $(a, f(a))$ and $(b, f(b))$ on its graph.

EXAMPLE 8

Finding an Equation of a Secant Line

Suppose that $g(x) = 3x^2 - 2x + 3$.

(a) Find the average rate of change of g from -2 to 1.

(b) Find an equation of the secant line containing $(-2, g(-2))$ and $(1, g(1))$.

 (c) Using a graphing utility, draw the graph of g and the secant line obtained in part (b) on the same screen.

Solution

(a) The average rate of change of $g(x) = 3x^2 - 2x + 3$ from -2 to 1 is

$$\text{Average rate of change} = \frac{g(1) - g(-2)}{1 - (-2)}$$

$$= \frac{4 - 19}{3} \qquad \begin{aligned} g(1) &= 3(1)^2 - 2 \cdot 1 + 3 = 4 \\ g(-2) &= 3(-2)^2 - 2(-2) + 3 = 19 \end{aligned}$$

$$= -\frac{15}{3} = -5$$

(b) The slope of the secant line containing $(-2, g(-2)) = (-2, 19)$ and $(1, g(1)) = (1, 4)$ is $m_{\text{sec}} = -5$. Use the point-slope form to find an equation of the secant line.

$$y - y_1 = m_{\text{sec}}(x - x_1) \qquad \textcolor{blue}{\text{Point-slope form of a secant line}}$$

$$y - 19 = -5(x - (-2)) \qquad \textcolor{blue}{x_1 = -2, y_1 = g(-2) = 19, m_{\text{sec}} = -5}$$

$$y - 19 = -5x - 10 \qquad \textcolor{blue}{\text{Distribute.}}$$

$$y = -5x + 9 \qquad \textcolor{blue}{\text{Slope-intercept form of the secant line}}$$

Figure 58 Graph of g and the secant line

 (c) Figure 58 shows the graph of g along with the secant line $y = -5x + 9$ on a TI-84 Plus C.

✏ **Now Work** PROBLEM **71**

1.4 Assess Your Understanding

'Are You Prepared?' *Answers are given at the end of these exercises. If you get a wrong answer, read the pages listed in* red.

1. The interval $(2, 5)$ can be written as the inequality _____. (pp. A45–A47)

2. The slope of the line containing the points $(-2, 3)$ and $(3, 8)$ is _____. (pp. A64–A67)

3. Test the equation $y = 5x^2 - 1$ for symmetry with respect to the x-axis, the y-axis, and the origin. (pp. 12–14)

4. Write the point-slope form of the line with slope 5 containing the point $(3, -2)$. (pp. A68–A69)

5. The intercepts of the graph of $y = x^2 - 9$ are _____. (pp. 11–12)

Concepts and Vocabulary

6. A function f is _____ on an interval I if, for any choice of x_1 and x_2 in I, with $x_1 < x_2$, then $f(x_1) < f(x_2)$.

7. A(n) _____ function f is one for which $f(-x) = f(x)$ for every x and $-x$ in the domain of f; a(n) _____ function f is one for which $f(-x) = -f(x)$ for every x and $-x$ in the domain of f.

8. *True or False* A function f is decreasing on an interval I if, for any choice of x_1 and x_2 in I, with $x_1 < x_2$, then $f(x_1) > f(x_2)$.

9. *True or False* A function f has a local minimum at c if there is an open interval I containing c so that $f(c) \leq f(x)$ for all x in this open interval.

10. *True or False* Even functions have graphs that are symmetric with respect to the origin.

11. *Multiple Choice* An odd function is symmetric with respect to _____.

(a) the *x*-axis   (b) the *y*-axis
(c) the origin   (d) the line *y* = *x*

12. *Multiple Choice* A function that is continuous on the interval _____ is guaranteed to have both an absolute maximum and an absolute minimum.

(a) (*a, b*)   (b) (*a, b*]   (c) [*a, b*)   (d) [*a, b*]

Skill Building

In Problems 13–24, use the graph on the right of the function f.

13. Is *f* increasing on the interval [−8, −2]?

14. Is *f* decreasing on the interval [−8, −4]?

15. Is *f* increasing on the interval [−2, 6]?

16. Is *f* decreasing on the interval [2, 5]?

17. List the interval(s) on which *f* is increasing.

18. List the interval(s) on which *f* is decreasing.

19. Is there a local maximum at 2? If yes, what is it?

20. Is there a local maximum at 5? If yes, what is it?

21. List the number(s) at which *f* has a local maximum. What are the local maximum values?

22. List the number(s) at which *f* has a local minimum. What are the local minimum values?

23. Find the absolute minimum of *f* on [−10, 7].

24. Find the absolute maximum of *f* on [−10, 7].

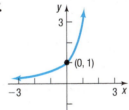

In Problems 25–32, the graph of a function is given. Use the graph to find:

(a) *The intercepts, if any*
(b) *The domain and range*
(c) *The intervals on which the function is increasing, decreasing, or constant*
(d) *Whether the function is even, odd, or neither*

25.

26.

27.

28.

29.

30.

31.

32.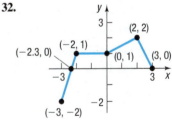

In Problems 33–36, the graph of a function f is given. Use the graph to find:

(a) *The numbers, if any, at which f has a local maximum. What are the local maximum values?*
(b) *The numbers, if any, at which f has a local minimum. What are the local minimum values?*

33.

34.

35.

36.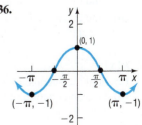

In Problems 37–48, determine algebraically whether each function is even, odd, or neither.

37. $f(x) = 4x^3$

38. $f(x) = 2x^4 - x^2$

39. $g(x) = 10 - x^2$

40. $h(x) = 3x^3 + 5$

41. $F(x) = \sqrt[3]{4x}$

42. $G(x) = \sqrt{x}$

43. $f(x) = x + |x|$

44. $f(x) = \sqrt[3]{2x^2 + 1}$

45. $g(x) = \dfrac{1}{x^2 + 8}$

46. $h(x) = \dfrac{x}{x^2 - 1}$

47. $h(x) = \dfrac{-x^3}{3x^2 - 9}$

48. $F(x) = \dfrac{2x}{|x|}$

In Problems 49–56, for each graph of a function $y = f(x)$, find the absolute maximum and the absolute minimum, if they exist. Identify any local maximum values or local minimum values.

49.

50.

51.

52.

53.

54.

55.

56.
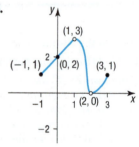

In Problems 57–64, use a graphing utility to graph each function over the indicated interval and approximate any local maximum values and local minimum values. Determine where the function is increasing and where it is decreasing. Round answers to two decimal places.

57. $f(x) = x^3 - 3x + 2$ $[-2, 2]$

58. $f(x) = x^3 - 3x^2 + 5$ $[-1, 3]$

59. $f(x) = x^5 - x^3$ $[-2, 2]$

60. $f(x) = x^4 - x^2$ $[-2, 2]$

61. $f(x) = -0.2x^3 - 0.6x^2 + 4x - 6$ $[-6, 4]$

62. $f(x) = -0.4x^3 + 0.6x^2 + 3x - 2$ $[-4, 5]$

63. $f(x) = 0.25x^4 + 0.3x^3 - 0.9x^2 + 3$ $[-3, 2]$

64. $f(x) = -0.4x^4 - 0.5x^3 + 0.8x^2 - 2$ $[-3, 2]$

65. Find the average rate of change of $f(x) = -2x^2 + 4$:
(a) From 0 to 2 (b) From 1 to 3 (c) From 1 to 4

66. Find the average rate of change of $f(x) = -x^3 + 1$:
(a) From 0 to 2 (b) From 1 to 3 (c) From −1 to 1

67. Find the average rate of change of $g(x) = x^3 - 4x + 7$:
(a) From −3 to −2 (b) From −1 to 1 (c) From 1 to 3

68. Find the average rate of change of $h(x) = x^2 - 2x + 3$:
(a) From −1 to 1 (b) From 0 to 2 (c) From 2 to 5

69. $f(x) = 5x - 2$
(a) Find the average rate of change from 1 to 3.
(b) Find an equation of the secant line containing $(1, f(1))$ and $(3, f(3))$.

70. $f(x) = -4x + 1$
(a) Find the average rate of change from 2 to 5.
(b) Find an equation of the secant line containing $(2, f(2))$ and $(5, f(5))$.

71. $g(x) = x^2 - 2$
(a) Find the average rate of change from −2 to 1.
(b) Find an equation of the secant line containing $(-2, g(-2))$ and $(1, g(1))$.

72. $g(x) = x^2 + 1$
(a) Find the average rate of change from −1 to 2.
(b) Find an equation of the secant line containing $(-1, g(-1))$ and $(2, g(2))$.

73. $h(x) = x^2 - 2x$
(a) Find the average rate of change from 2 to 4.
(b) Find an equation of the secant line containing $(2, h(2))$ and $(4, h(4))$.

74. $h(x) = -2x^2 + x$
(a) Find the average rate of change from 0 to 3.
(b) Find an equation of the secant line containing $(0, h(0))$ and $(3, h(3))$.

75. Mixed Practice $g(x) = x^3 - 27x$

   (a) Determine whether g is even, odd, or neither.

   (b) There is a local minimum value of -54 at 3. Determine the local maximum value.

76. Mixed Practice $f(x) = -x^3 + 12x$

   (a) Determine whether f is even, odd, or neither.

   (b) There is a local maximum value of 16 at 2. Determine the local minimum value.

Applications and Extensions

77. $F(x) = -x^4 + 8x^2 + 9$

   (a) Determine whether F is even, odd, or neither.

   (b) There is a local maximum value of 25 at $x = 2$. Find a second local maximum value.

   (c) Suppose the area of the region enclosed by the graph of F and the x-axis between $x = 0$ and $x = 3$ is 50.4 square units. Using the result from (a), determine the area of the region enclosed by the graph of F and the x-axis between $x = -3$ and $x = 0$.

78. $G(x) = -x^4 + 32x^2 + 144$

   (a) Determine whether G is even, odd, or neither.

   (b) There is a local maximum value of 400 at $x = 4$. Find a second local maximum value.

   (c) Suppose the area of the region enclosed by the graph of G and the x-axis between $x = 0$ and $x = 6$ is 1612.8 square units. Using the result from (a), determine the area of the region enclosed by the graph of G and the x-axis between $x = -6$ and $x = 0$.

79. Minimum Average Cost The average cost per hour in dollars, $\overline{C}$, of producing x riding lawn mowers can be modeled by the function

$$\overline{C}(x) = 0.3x^2 + 21x - 251 + \frac{2500}{x}$$

   (a) Use a graphing utility to graph $\overline{C} = \overline{C}(x)$.

   (b) Determine the number of riding lawn mowers to produce in order to minimize average cost.

   (c) What is the minimum average cost?

80. Medicine Concentration The concentration C of a medication in the bloodstream t hours after being administered is modeled by the function

$$C(t) = -0.002t^4 + 0.039t^3 - 0.285t^2 + 0.766t + 0.085$$

   (a) After how many hours will the concentration be highest?

   (b) A woman nursing a child must wait until the concentration is below 0.5 before she can feed him. After taking the medication, how long must she wait before feeding her child?

81. E. coli Growth A strain of E. coli Beu 397-recA441 is placed into a nutrient broth at 30° Celsius and allowed to grow. The data shown in the table are collected. The population is measured in grams and the time in hours. Since population P depends on time t, and each input corresponds to exactly one output, we can say that population is a function of time, so $P(t)$ represents the population at time t.

Time (hours), t	Population (grams), P
0	0.09
2.5	0.18
3.5	0.26
4.5	0.35
6	0.50

   (a) Find the average rate of change of the population from 0 to 2.5 hours.

   (b) Find the average rate of change of the population from 4.5 to 6 hours.

   (c) What is happening to the average rate of change as time passes?

82. National Debt The size of the total debt owed by the United States federal government continues to grow. In fact, according to the Department of the Treasury, the debt per person living in the United States is approximately \$63,720 (or over \$172,000 per U.S. household). The following data represent the U.S. debt for the years 2007–2017. Since the debt D depends on the year y, and each input corresponds to exactly one output, the debt is a function of the year. So $D(y)$ represents the debt for each year y.

Year	Debt (billions of dollars)
2007	9008
2008	10,025
2009	11,910
2010	13,562
2011	14,790
2012	16,066
2013	16,738
2014	17,824
2015	18,151
2016	19,573
2017	20,245

Source: www.treasurydirect.gov

   (a) Plot the points (2007, 9008), (2008, 10 025), and so on.

   (b) Draw a line segment from the point (2007, 9008) to (2012, 16 066). What does the slope of this line segment represent?

   (c) Find the average rate of change of the debt from 2008 to 2010.

   (d) Find the average rate of change of the debt from 2011 to 2013.

   (e) Find the average rate of change of the debt from 2014 to 2016.

   (f) What appears to be happening to the average rate of change as time passes?

83. For the function $f(x) = x^2$, compute the average rate of change:

   (a) From 0 to 1   **(b)** From 0 to 0.5   **(c)** From 0 to 0.1

   (d) From 0 to 0.01   **(e)** From 0 to 0.001

   (f) Use a graphing utility to graph each of the secant lines along with f.

   (g) What do you think is happening to the secant lines?

   (h) What is happening to the slopes of the secant lines? Is there some number that they are getting closer to? What is that number?

84. For the function $f(x) = x^2$, compute the average rate of change:
(a) From 1 to 2
(b) From 1 to 1.5
(c) From 1 to 1.1
(d) From 1 to 1.01
(e) From 1 to 1.001

 (f) Use a graphing utility to graph each of the secant lines along with f.
(g) What do you think is happening to the secant lines?
(h) What is happening to the slopes of the secant lines? Is there some number that they are getting closer to? What is that number?

Problems 85–92 require the following discussion of a secant line. The slope of the secant line containing the two points $(x, f(x))$ and $(x + h, f(x + h))$ on the graph of a function $y = f(x)$ may be given as

$$m_{\text{sec}} = \frac{f(x + h) - f(x)}{(x + h) - x} = \frac{f(x + h) - f(x)}{h} \qquad h \neq 0$$

(a) Express the slope of the secant line of each function in terms of x and h. Be sure to simplify your answer.
(b) Find m_{sec} for $h = 0.5, 0.1,$ and 0.01 at $x = 1$. What value does m_{sec} approach as h approaches 0?
(c) Find an equation for the secant line at $x = 1$ with $h = 0.01$.
(d) Use a graphing utility to graph f and the secant line found in part (c) in the same viewing window.

85. $f(x) = 2x + 5$ 　　　　**86.** $f(x) = -3x + 2$ 　　　　**87.** $f(x) = x^2 + 2x$ 　　　　**88.** $f(x) = 2x^2 + x$

89. $f(x) = 2x^2 - 3x + 1$ 　　**90.** $f(x) = -x^2 + 3x - 2$ 　　**91.** $f(x) = \dfrac{1}{x}$ 　　　　**92.** $f(x) = \dfrac{1}{x^2}$

93. *Challenge Problem* Mean Value Theorem Suppose $f(x) = x^3 + 2x^2 - x + 6$. From calculus, the Mean Value Theorem guarantees that there is at least one number in the open interval $(-1, 2)$ at which the value of the derivative of f, given by $f'(x) = 3x^2 + 4x - 1$, is equal to the average rate of change of f on the interval. Find all such numbers x in the interval.

94. *Challenge Problem* If f is an odd function, determine whether $g(x) = -2f\left(-\dfrac{x}{3}\right)$ is even, odd, or neither.

Explaining Concepts: Discussion and Writing

95. Draw the graph of a function that has the following properties: domain: all real numbers; range: all real numbers; intercepts: $(0, -3)$ and $(3, 0)$; a local maximum value of -2 at -1; a local minimum value of -6 at 2. Compare your graph with those of others. Comment on any differences.

96. Redo Problem 95 with the following additional information: increasing on $(-\infty, -1], [2, \infty)$; decreasing on $[-1, 2]$. Again compare your graph with others and comment on any differences.

97. How many x-intercepts can a function defined on an interval have if it is increasing on that interval? Explain.

98. Suppose that a friend of yours does not understand the idea of increasing and decreasing functions. Provide an explanation, complete with graphs, that clarifies the idea.

99. Can a function be both even and odd? Explain.

100. Using a graphing utility, graph $y = 5$ on the interval $[-3, 3]$. Use MAXIMUM to find the local maximum values on $[-3, 3]$. Comment on the result provided by the graphing utility.

101. A function f has a positive average rate of change on the interval $[2, 5]$. Is f increasing on $[2, 5]$? Explain.

102. Show that a constant function $f(x) = b$ has an average rate of change of 0. Compute the average rate of change of $y = \sqrt{4 - x^2}$ on the interval $[-2, 2]$. Explain how this can happen.

'Are You Prepared?' Answers

1. $2 < x < 5$ 　　　**2.** 1 　　　**3.** symmetric with respect to the y-axis 　　　**4.** $y + 2 = 5(x - 3)$ 　　　**5.** $(-3, 0), (3, 0), (0, -9)$

1.5 Library of Functions; Piecewise-defined Functions

PREPARING FOR THIS SECTION *Before getting started, review the following:*

- Intercepts (Section 1.2, pp. 11–12)
- Lines (Section A.8, pp. A64–A75)
- Graphs of Key Equations (Section 1.2: Example 3, p. 10; Examples 11–13, pp. 15–16)

✎ **Now Work** the 'Are You Prepared?' problems on page 63.

OBJECTIVES 1 Graph the Functions Listed in the Library of Functions (p. 57)
2 Analyze a Piecewise-defined Function (p. 61)

1 Graph the Functions Listed in the Library of Functions

Figure 59 shows the graph of $y = \sqrt{x}$. Based on the Vertical-Line Test, $y = f(x) = \sqrt{x}$ is a function, called the **square root function**. From the graph, we have the following properties:

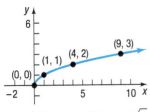

Figure 59 $y = f(x) = \sqrt{x}$

> **Properties of $f(x) = \sqrt{x}$**
>
> - The domain and the range of $f(x) = \sqrt{x}$ are the set of nonnegative real numbers.
> - The x-intercept of the graph of $f(x) = \sqrt{x}$ is 0. The y-intercept of the graph of $f(x) = \sqrt{x}$ is also 0.
> - $f(x) = \sqrt{x}$ is neither even nor odd.
> - $f(x) = \sqrt{x}$ is increasing on the interval $[0, \infty)$.
> - $f(x) = \sqrt{x}$ has an absolute minimum of 0 at $x = 0$.

EXAMPLE 1

Graphing the Cube Root Function

(a) Determine whether $f(x) = \sqrt[3]{x}$ is even, odd, or neither. State whether the graph of f is symmetric with respect to the y-axis, symmetric with respect to the origin, or neither.

(b) Find the intercepts, if any, of the graph of $f(x) = \sqrt[3]{x}$.

(c) Graph $f(x) = \sqrt[3]{x}$.

Solution

(a) Because

$$f(-x) = \sqrt[3]{-x} = -\sqrt[3]{x} = -f(x)$$

the cube root function is odd. The graph of f is symmetric with respect to the origin.

(b) The y-intercept is $f(0) = \sqrt[3]{0} = 0$. The x-intercept is found by solving the equation $f(x) = 0$.

$$f(x) = 0$$
$$\sqrt[3]{x} = 0 \qquad \textcolor{blue}{f(x) = \sqrt[3]{x}}$$
$$x = 0 \qquad \textcolor{blue}{\text{Cube both sides of the equation.}}$$

The x-intercept is also 0.

(c) Use the function to form Table 7 on the next page and obtain some points on the graph. Because of the symmetry with respect to the origin, we find only points (x, y) for which $x \geq 0$. Figure 60 on the next page shows the graph of $f(x) = \sqrt[3]{x}$.

Table 7

x	$y = f(x) = \sqrt[3]{x}$	(x, y)
0	0	(0, 0)
$\frac{1}{8}$	$\frac{1}{2}$	$\left(\frac{1}{8}, \frac{1}{2}\right)$
1	1	(1, 1)
8	2	(8, 2)

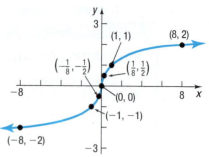

Figure 60 Cube Root Function

From the results of Example 1 and Figure 60, we have the following properties of the cube root function.

Properties of the Cube Root Function $f(x) = \sqrt[3]{x}$

- The domain and the range of $f(x) = \sqrt[3]{x}$ are the set of all real numbers.
- The x-intercept of the graph of $f(x) = \sqrt[3]{x}$ is 0. The y-intercept of the graph of $f(x) = \sqrt[3]{x}$ is also 0.
- $f(x) = \sqrt[3]{x}$ is odd. The graph is symmetric with respect to the origin.
- $f(x) = \sqrt[3]{x}$ is increasing on the interval $(-\infty, \infty)$.
- $f(x) = \sqrt[3]{x}$ does not have any local minima or any local maxima.

EXAMPLE 2 **Graphing the Absolute Value Function**

(a) Determine whether $f(x) = |x|$ is even, odd, or neither. State whether the graph of f is symmetric with respect to the y-axis, symmetric with respect to the origin, or neither.

(b) Determine the intercepts, if any, of the graph of $f(x) = |x|$.

(c) Graph $f(x) = |x|$.

Solution (a) Because

$$f(-x) = |-x|$$
$$= |x| = f(x)$$

the absolute value function is even. The graph of f is symmetric with respect to the y-axis.

(b) The y-intercept is $f(0) = |0| = 0$. The x-intercept is found by solving the equation $f(x) = |x| = 0$. The x-intercept is 0.

(c) Use the function to form Table 8 and obtain some points on the graph. Because of the symmetry with respect to the y-axis, we only need to find points (x, y) for which $x \geq 0$. Figure 61 shows the graph of $f(x) = |x|$.

Table 8

x	$y = f(x) = |x|$	(x, y)
0	0	(0, 0)
1	1	(1, 1)
2	2	(2, 2)
3	3	(3, 3)

Figure 61 Absolute Value Function

From Example 2 and Figure 61, we have the following properties of the absolute value function.

> **Properties of the Absolute Value Function $f(x) = |x|$**
>
> - The domain of $f(x) = |x|$ is the set of all real numbers. The range of f is $\{y | y \geq 0\}$.
> - The x-intercept of the graph of $f(x) = |x|$ is 0. The y-intercept of the graph of $f(x) = |x|$ is also 0.
> - $f(x) = |x|$ is even. The graph is symmetric with respect to the y-axis.
> - $f(x) = |x|$ is decreasing on the interval $(-\infty, 0]$. It is increasing on the interval $[0, \infty)$.
> - $f(x) = |x|$ has an absolute minimum of 0 at $x = 0$.

Below is a list of the key functions that we have discussed. In going through this list, pay special attention to the properties of each function, particularly to the domain of each function and to the shape of each graph. Knowing these graphs, along with key points on each graph, lays the foundation for further graphing techniques.

> **Constant Function**
>
> $$f(x) = b \qquad b \text{ is a real number}$$

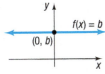

Figure 62 Constant Function

See Figure 62.

The domain of a **constant function** is the set of all real numbers; its range is the set consisting of a single number b. Its graph is a horizontal line whose y-intercept is b. The constant function is an even function.

> **Identity Function**
>
> $$f(x) = x$$

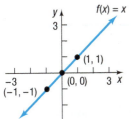

Figure 63 Identity Function

See Figure 63.

The domain and the range of the **identity function** are the set of all real numbers. Its graph is a line with slope 1 and y-intercept 0. The line consists of all points for which the x-coordinate equals the y-coordinate. The identity function is an odd function and is increasing over its domain. Note that the graph bisects quadrants I and III.

> **Square Function**
>
> $$f(x) = x^2$$

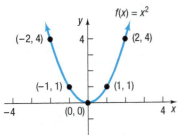

Figure 64 Square Function

See Figure 64.

The domain of the **square function** is the set of all real numbers; its range is the set of nonnegative real numbers. The graph of the square function is a parabola whose intercept is at $(0, 0)$. The square function is an even function that is decreasing on the interval $(-\infty, 0]$ and increasing on the interval $[0, \infty)$.

Figure 65 Cube Function

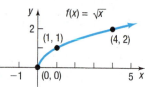

Figure 66 Square Root Function

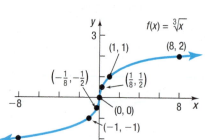

Figure 67 Cube Root Function

Figure 68 Reciprocal Function

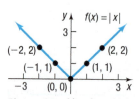

Figure 69 Absolute Value Function

Cube Function

$$f(x) = x^3$$

See Figure 65.

The domain and the range of the **cube function** are the set of all real numbers. The intercept of the graph is at $(0, 0)$. The cube function is odd and is increasing on the interval $(-\infty, \infty)$.

Square Root Function

$$f(x) = \sqrt{x}$$

See Figure 66.

The domain and the range of the **square root function** are the set of nonnegative real numbers. The intercept of the graph is at $(0,0)$. The square root function is neither even nor odd and is increasing on the interval $[0, \infty)$.

Cube Root Function

$$f(x) = \sqrt[3]{x}$$

See Figure 67.

The domain and the range of the **cube root function** are the set of all real numbers. The intercept of the graph is at $(0,0)$. The cube root function is an odd function that is increasing on the interval $(-\infty, \infty)$.

Reciprocal Function

$$f(x) = \frac{1}{x}$$

Refer to Example 13, page 16, for a discussion of the equation $y = \frac{1}{x}$. See Figure 68.

The domain and the range of the **reciprocal function** are the set of all nonzero real numbers. The graph has no intercepts. The reciprocal function is decreasing on the intervals $(-\infty, 0)$ and $(0, \infty)$ and is an odd function.

Absolute Value Function

$$f(x) = |x|$$

See Figure 69.

The domain of the **absolute value function** is the set of all real numbers; its range is the set of nonnegative real numbers. The intercept of the graph is at $(0, 0)$. If $x \geq 0$, then $f(x) = x$, and the graph of f is part of the line $y = x$; if $x < 0$, then $f(x) = -x$, and the graph of f is part of the line $y = -x$. The absolute value function is an even function; it is decreasing on the interval $(-\infty, 0]$ and increasing on the interval $[0, \infty)$.

The functions discussed so far are basic. Whenever you encounter one of them, you should see a mental picture of its graph. For example, if you encounter the function $f(x) = x^2$, you should see in your mind's eye a picture of a parabola.

Now Work PROBLEMS 11 THROUGH 18

2 Analyze a Piecewise-defined Function

Sometimes a function is defined using different equations on different parts of its domain. For example, the absolute value function $f(x) = |x|$ is actually defined by two equations: $f(x) = x$ if $x \geq 0$ and $f(x) = -x$ if $x < 0$. See Figure 69. For convenience, these equations are generally combined into one expression as

$$f(x) = |x| = \begin{cases} x & \text{if } x \geq 0 \\ -x & \text{if } x < 0 \end{cases}$$

When a function is defined by different equations on different parts of its domain, it is called a **piecewise-defined** function.

EXAMPLE 3

Analyzing a Piecewise-defined Function

A piecewise-defined function f is defined as

$$f(x) = \begin{cases} -2x & \text{if } x < 0 \\ \sqrt{x} & \text{if } x \geq 0 \end{cases}$$

(a) Find $f(-4)$, $f(0)$, and $f(4)$.

(b) Find the domain of f.

(c) Locate any intercepts.

(d) Graph f.

(e) Use the graph to find the range of f.

Solution

(a) When evaluating a piecewise-defined function, we first must identify which equation should be used in the evaluation. We do this by identifying in which part of the domain the value for the independent variable lies.

- To find $f(-4)$, observe that $-4 < 0$ so $x = -4$ lies in the domain of the first equation. This means that when $x = -4$, the equation for f is $f(x) = -2x$. Then

$$f(-4) = -2(-4) = 8$$

- To find $f(0)$, observe that when $x = 0$, the equation for f is $f(x) = \sqrt{x}$. Then

$$f(0) = \sqrt{0} = 0$$

- To find $f(4)$, observe that when $x = 4$, the equation for f is $f(x) = \sqrt{x}$. Then

$$f(4) = \sqrt{4} = 2$$

(b) The domain of the first equation of the function f is the set of all negative real numbers, $\{x \mid x < 0\}$, or $(-\infty, 0)$ in interval notation. The domain of the second equation of the function f is the set of all nonnegative real numbers, $\{x \mid x \geq 0\}$, or $[0, \infty)$ in interval notation. The domain of the function is the union of the domains of the individual equations. Since

$$(-\infty, 0) \cup [0, \infty) = (-\infty, \infty)$$

the domain of f is the set of all real numbers.

(c) The y-intercept of the graph of the function is $f(0)$. Because the equation for f when $x = 0$ is $f(x) = \sqrt{x}$, the y-intercept is $f(0) = \sqrt{0} = 0$.   (*continued*)

The x-intercepts of the graph of a function are the real solutions to the equation $f(x) = 0$. To find the x-intercepts of f, solve $f(x) = 0$ for each equation of the function, and then determine which values of x, if any, satisfy the domain of the equation.

$$f(x) = 0 \quad x < 0 \qquad\qquad f(x) = 0 \quad x \geq 0$$
$$-2x = 0 \qquad\qquad\qquad \sqrt{x} = 0$$
$$x = 0 \qquad\qquad\qquad\quad x = 0$$

The result on the left $(x = 0)$ is discarded because it does not satisfy the condition $x < 0$. The result on the right $(x = 0)$ is obtained under the condition $x \geq 0$, so 0 is an x-intercept. The only intercept is $(0, 0)$.

(d) To graph f, graph each equation. First graph the line $y = -2x$ and keep only the part for which $x < 0$. Then graph the square root function $y = \sqrt{x}$ for $x \geq 0$. See Figure 70.

(e) From the graph, we conclude that the range of f is the set of nonnegative real numbers, or the interval $[0, \infty)$.

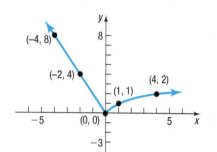

Figure 70

EXAMPLE 4

Analyzing a Piecewise-defined Function

A piecewise-defined function f is defined as

$$f(x) = \begin{cases} -2x + 1 & \text{if } -3 \leq x < 1 \\ 2 & \text{if } x = 1 \\ x^2 & \text{if } x > 1 \end{cases}$$

(a) Find $f(-2)$, $f(1)$, and $f(2)$.

(b) Find the domain of f.

(c) Locate any intercepts.

(d) Graph f.

(e) Use the graph to find the range of f.

Solution

(a) To find $f(-2)$, observe that $-3 \leq -2 < 1$, so when $x = -2$, the equation for f is given by $f(x) = -2x + 1$. Then

$$f(-2) = -2(-2) + 1 = 5 \quad f(x) = -2x + 1 \quad \text{if} \quad -3 \leq x < 1$$

To find $f(1)$, observe that when $x = 1$, the equation for f is $f(x) = 2$. So,

$$f(1) = 2 \quad f(x) = 2 \text{ if } x = 1$$

When $x = 2$, the equation for f is $f(x) = x^2$. Then

$$f(2) = 2^2 = 4 \quad f(x) = x^2 \quad \text{if} \quad x > 1$$

(b) The domain of f is the union of the domains of each equation in the piecewise-defined function. So the domain of f is $[-3, 1) \cup \{1\} \cup (1, \infty) = [-3, \infty)$. The domain of f is $[-3, \infty)$ in interval notation, or $\{x \mid x \geq -3\}$ in set notation.

(c) The y-intercept of the graph of the function is $f(0)$. Because the equation for f when $x = 0$ is $f(x) = -2x + 1$, the y-intercept is $f(0) = -2 \cdot 0 + 1 = 1$. The x-intercepts of the graph of a function f are the real solutions of the equation $f(x) = 0$. To find the x-intercepts of f, solve $f(x) = 0$ for each equation of the function, and then determine what values of x, if any, are in the domain of the equation.

$$f(x) = 0 \quad -3 \leq x < 1 \qquad f(x) = 0 \quad x = 1 \qquad f(x) = 0 \quad x > 1$$
$$-2x + 1 = 0 \qquad\qquad\qquad 2 = 0 \qquad\qquad\qquad x^2 = 0$$
$$-2x = -1 \qquad\qquad\qquad \text{No solution} \qquad\qquad x = 0$$
$$x = \frac{1}{2}$$

Figure 71

The first potential x-intercept, $x = \dfrac{1}{2}$, satisfies the condition $-3 \le x < 1$, so $x = \dfrac{1}{2}$ is an x-intercept. The second potential x-intercept, $x = 0$, does not satisfy the condition $x > 1$, so we discard it. The only x-intercept is $\dfrac{1}{2}$. The intercepts are $(0, 1)$ and $\left(\dfrac{1}{2}, 0\right)$.

(d) To graph f, first graph the line $y = -2x + 1$ and keep only the part for which $-3 \le x < 1$. When $x = 1, f(x) = 2$, so plot the point $(1, 2)$. Finally, graph the parabola $y = x^2$ and keep only the part for which $x > 1$. See Figure 71. Notice we use open circles at the points $(1, -1)$ and $(1, 1)$ to indicate these points are not part of the graph.

(e) From the graph, we conclude that the range of f is $\{y | y > -1\}$, or the interval $(-1, \infty)$.

Now Work PROBLEM 31

1.5 Assess Your Understanding

'Are You Prepared?' *Answers are given at the end of these exercises. If you get a wrong answer, read the pages listed in red.*

1. Graph $y = \sqrt{x}$. (p. 15)

2. Graph $y = \dfrac{1}{x}$. (p. 16)

3. Find the intercepts of the equation $y = x^3 - 8$. (pp. 11–12)

Concepts and Vocabulary

4. The function $f(x) = x^2$ is decreasing on the interval _____.

5. When functions are defined by more than one equation, they are called _____ functions.

6. *True or False* The cube function is odd and is increasing on the interval $(-\infty, \infty)$.

7. *True or False* The cube root function is odd and is decreasing on the interval $(-\infty, \infty)$.

8. *True or False* The domain and the range of the reciprocal function are the set of all real numbers.

9. *Multiple Choice* Which of the following functions has a graph that is symmetric about the y-axis?
   (a) $y = \sqrt{x}$ (b) $y = |x|$ (c) $y = x^3$ (d) $y = \dfrac{1}{x}$

10. *Multiple Choice* Consider the following function.
$$f(x) = \begin{cases} 3x - 2 & \text{if} & x < 2 \\ x^2 + 5 & \text{if} & 2 \le x < 10 \\ 3 & \text{if} & x \ge 10 \end{cases}$$
   Which expression(s) should be used to find the y-intercept?
   (a) $3x - 2$ (b) $x^2 + 5$ (c) 3 (d) all three

Skill Building

In Problems 11–18, match each graph to its function.

   A. *Constant function*
   E. *Square root function*

   B. *Identity function*
   F. *Reciprocal function*

   C. *Square function*
   G. *Absolute value function*

   D. *Cube function*
   H. *Cube root function*

11.

12.

13.

14.

15.

16.

17.

18.

In Problems 19–26, graph each function. Be sure to label three points on the graph.

19. $f(x) = x$ **20.** $f(x) = x^2$ **21.** $f(x) = x^3$ **22.** $f(x) = \sqrt{x}$

23. $f(x) = \dfrac{1}{x}$ **24.** $f(x) = |x|$ **25.** $f(x) = \sqrt[3]{x}$ **26.** $f(x) = 3$

27. If $f(x) = \begin{cases} -x^2 & \text{if } x < 0 \\ 4 & \text{if } x = 0 \\ 3x - 2 & \text{if } x > 0 \end{cases}$ find:

  (a) $f(-3)$ **(b)** $f(0)$ **(c)** $f(3)$

28. If $f(x) = \begin{cases} -3x & \text{if } x < -1 \\ 0 & \text{if } x = -1 \\ 2x^2 + 1 & \text{if } x > -1 \end{cases}$ find:

  (a) $f(-2)$ **(b)** $f(-1)$ **(c)** $f(0)$

29. If $f(x) = \begin{cases} 2x + 4 & \text{if } -3 \le x \le 1 \\ x^3 - 1 & \text{if } 1 < x \le 5 \end{cases}$ find:

  (a) $f(-2)$ **(b)** $f(0)$ **(c)** $f(1)$ **(d)** $f(3)$

30. If $f(x) = \begin{cases} x^3 & \text{if } -2 \le x < 1 \\ 3x + 2 & \text{if } 1 \le x \le 4 \end{cases}$ find:

  (a) $f(-1)$ **(b)** $f(0)$ **(c)** $f(1)$ **(d)** $f(3)$

In Problems 31–42:

  (a) Find the domain of each function.   *(b) Locate any intercepts.*   *(c) Graph each function.*
  (d) Based on the graph, find the range.

31. $f(x) = \begin{cases} 2x & \text{if } x \ne 0 \\ 1 & \text{if } x = 0 \end{cases}$

32. $f(x) = \begin{cases} 3x & \text{if } x \ne 0 \\ 4 & \text{if } x = 0 \end{cases}$

33. $f(x) = \begin{cases} -2x + 3 & \text{if } x < 1 \\ 3x - 2 & \text{if } x \ge 1 \end{cases}$

34. $f(x) = \begin{cases} x + 3 & \text{if } x < -2 \\ -2x - 3 & \text{if } x \ge -2 \end{cases}$

35. $f(x) = \begin{cases} x + 3 & \text{if } -2 \le x < 1 \\ 5 & \text{if } x = 1 \\ -x + 2 & \text{if } x > 1 \end{cases}$

36. $f(x) = \begin{cases} 2x + 5 & \text{if } -3 \le x < 0 \\ -3 & \text{if } x = 0 \\ -5x & \text{if } x > 0 \end{cases}$

37. $f(x) = \begin{cases} 1 + x & \text{if } x < 0 \\ x^2 & \text{if } x \ge 0 \end{cases}$

38. $f(x) = \begin{cases} \dfrac{1}{x} & \text{if } x < 0 \\ \sqrt[3]{x} & \text{if } x \ge 0 \end{cases}$

39. $f(x) = \begin{cases} |x| & \text{if } -2 \le x < 0 \\ x^3 & \text{if } x > 0 \end{cases}$

40. $f(x) = \begin{cases} 2 - x & \text{if } -3 \le x < 1 \\ \sqrt{x} & \text{if } x > 1 \end{cases}$

41. $f(x) = \begin{cases} x^2 & \text{if } 0 < x \le 2 \\ x + 2 & \text{if } 2 < x < 5 \\ 7 & \text{if } x \ge 5 \end{cases}$

42. $f(x) = \begin{cases} 3x + 5 & \text{if } -3 \le x < 0 \\ 5 & \text{if } 0 \le x \le 2 \\ x^2 + 1 & \text{if } x > 2 \end{cases}$

In Problems 43–46, the graph of a piecewise-defined function is given. Write a definition for each function.

43.

44.

45.

46.
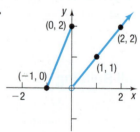

Applications and Extensions

47. (a) Graph $f(x) = \begin{cases} (x - 1)^2 & \text{if } 0 \le x < 2 \\ -2x + 10 & \text{if } 2 \le x \le 6 \end{cases}$

  (b) Find the domain of f.

  (c) Find the absolute maximum and the absolute minimum, if they exist.

48. (a) Graph $f(x) = \begin{cases} -x + 1 & \text{if } -2 \le x < 0 \\ 2 & \text{if } x = 0 \\ x + 1 & \text{if } 0 < x \le 2 \end{cases}$

  (b) Find the domain of f.

  (c) Find the absolute maximum and the absolute minimum, if they exist.

49. Tablet Service A monthly tablet plan costs $34.99. It includes 3 gigabytes of data and charges $15 per gigabyte for additional gigabytes. The following function is used to compute the monthly cost for a subscriber.

$$C(x) = \begin{cases} 34.99 & \text{if } 0 \le x \le 3 \\ 15x - 10.01 & \text{if } x > 3 \end{cases}$$

Compute the monthly cost for each of the following gigabytes of use.

(a) 2   **(b)** 5   **(c)** 13

50. Parking at O'Hare International Airport The short-term (no more than 24 hours) parking fee F (in dollars) for parking x hours on a weekday at O'Hare International Airport's main parking garage can be modeled by the function

$$F(x) = \begin{cases} 3 + x & \text{if } 0 < x \le 4 \\ 4x + 10 & \text{if } 4 < x \le 9 \\ 74 & \text{if } 9 < x \le 24 \end{cases}$$

where fractions of an hour are rounded up to the next hour. Determine the fee for parking in the short-term parking garage for

(a) 2 hours   **(b)** 7 hours

(c) 15 hours   **(d)** 8 hours and 24 minutes

Source: O'Hare International Airport

51. Cost of Natural Gas In March 2018, Spire, Inc. had the following rate schedule for natural gas usage in single-family residences.

Monthly service charge	$23.44
Delivery charge	
First 30 therms	$0.91686/therm
Over 30 therms	$0
Natural gas cost	
First 30 therms	$0.26486/therm
Over 30 therms	$0.50897/therm

(a) What is the charge for using 20 therms in a month?

(b) What is the charge for using 150 therms in a month?

(c) Develop a function that models the monthly charge C for x therms of gas.

(d) Graph the function found in part (c).

Source: Spire, Inc.

52. Cost of Natural Gas In April 2018, Nicor Gas had the following rate schedule for natural gas usage in small businesses.

Monthly customer charge	$90.00
Distribution charge	
1st 150 therms	$0.1201/therm
Next 4850 therms	$0.0549/therm
Over 5000 therms	$0.0482/therm
Gas supply charge	$0.35/therm

(a) What is the charge for using 1000 therms in a month?

(b) What is the charge for using 6000 therms in a month?

(c) Develop a function that models the monthly charge C for x therms of gas.

(d) Graph the function found in part (c).

Source: Nicor Gas

53. Federal Income Tax Two 2018 Tax Rate Schedules are given in the accompanying table. If x equals taxable income and y equals the tax due, construct a function $y = f(x)$ for Schedule X.

2018 Tax Rate Schedules										
Schedule X—Single					**Schedule Y-1—Married Filing Jointly or Qualified Widow(er)**					
If Taxable Income is Over	But Not Over	The Tax is This Amount	Plus This %	Of the Excess Over	If Taxable Income is Over	But Not Over	The Tax is This Amount		Plus This %	Of the Excess Over
$0	$9,525	$0 +	10%	$0	$0	$19,050	$0	+	10%	$0
9,525	38,700	952.50 +	12%	9,525	19,050	77,400	1,905	+	12%	19,050
38,700	82,500	4,453.50 +	22%	38,700	77,400	165,000	8,907.00	+	22%	77,400
82,500	157,500	14,089.50 +	24%	82,500	165,000	315,000	28,179.00	+	24%	165,000
157,500	200,000	32,089.50 +	32%	157,500	315,000	400,000	64,179.00	+	32%	315,000
200,000	500,000	45,689.50 +	35%	200,000	400,000	600,000	91,379.00	+	35%	400,000
500,000	–	150,689.50 +	37%	500,000	600,000	–	161,379.00	+	37%	600,000

54. Federal Income Tax Refer to the 2018 tax rate schedules. If x equals taxable income and y equals the tax due, construct a function $y = f(x)$ for Schedule Y-1.

55. Cost of Transporting Goods A trucking company transports goods between Chicago and New York, a distance of 960 miles. The company's policy is to charge, for each pound, $0.50 per mile for the first 100 miles, $0.40 per mile for the next 300 miles, $0.25 per mile for the next 400 miles, and no charge for the remaining 160 miles.

(a) Graph the relationship between the per-pound cost of transportation in dollars and mileage over the entire 960-mile route.

(b) Find the cost as a function of mileage for hauls between 100 and 400 miles from Chicago.

(c) Find the cost as a function of mileage for hauls between 400 and 800 miles from Chicago.

56. Car Rental Costs An economy car rented in Florida from Enterprise® on a weekly basis costs $185 per week. Extra days cost $37 per day until the day rate exceeds the weekly rate, in which case the weekly rate applies. Also, any part of a day used counts as a full day. Find the cost C of renting an economy car as a function of the number of days used x, where $7 \leq x \leq 14$. Graph this function.

57. Mortgage Fees Fannie Mae charges a loan-level price adjustment (LLPA) on all mortgages, which represents a fee homebuyers seeking a loan must pay. The rate paid depends on the credit score of the borrower, the amount borrowed, and the loan-to-value (LTV) ratio. The LTV ratio is the ratio of amount borrowed to appraised value of the home. For example, a homebuyer who wishes to borrow $250,000 with a credit score of 730 and an LTV ratio of 80% will pay 0.75% (0.0075) of $250,000, or $1875. The table shows the LLPA for various credit scores and an LTV ratio of 80%.

Credit Score	Loan-Level Price Adjustment Rate
≤ 659	3.00%
660–679	2.75%
680–699	1.75%
700–719	1.25%
720–739	0.75%
≥ 740	0.50%

Source: Fannie Mae

(a) Construct a function $C = C(s)$, where C is the loan-level price adjustment (LLPA) and s is the credit score of an individual who wishes to borrow $300,000 with an 80% LTV ratio.

(b) What is the LLPA on a $300,000 loan with an 80% LTV ratio for a borrower whose credit score is 725?

(c) What is the LLPA on a $300,000 loan with an 80% LTV ratio for a borrower whose credit score is 670?

58. Minimum Payments for Credit Cards Holders of credit cards issued by banks, department stores, oil companies, and so on, receive bills each month that state minimum amounts that must be paid by a certain due date. The minimum due depends on the total amount owed. One such credit card company uses the following rules: For a bill of less than $10, the entire amount is due. For a bill of at least $10 but less than $500, the minimum due is $10. A minimum of $30 is due on a bill of at least $500 but less than $1000, a minimum of $50 is due on a bill of at least $1000 but less than $1500, and a minimum of $70 is due on bills of $1500 or more. Find the function f that describes the minimum payment due on a bill of x dollars. Graph f.

59. Wind Chill The wind chill factor represents the air temperature at a standard wind speed that would produce the same heat loss as the given temperature and wind speed. One formula for computing the equivalent temperature is

$$W = \begin{cases} t & 0 \leq v < 1.79 \\ 33 - \dfrac{(10.45 + 10\sqrt{v} - v)(33 - t)}{22.04} & 1.79 \leq v \leq 20 \\ 33 - 1.5958(33 - t) & v > 20 \end{cases}$$

where v represents the wind speed (in meters per second) and t represents the air temperature (°C). Compute the wind chill for the following:

(a) An air temperature of 10°C and a wind speed of 1 meter per second (m/sec)

(b) An air temperature of 10°C and a wind speed of 5 m/sec

(c) An air temperature of 10°C and a wind speed of 15 m/sec

(d) An air temperature of 10°C and a wind speed of 25 m/sec

(e) Explain the physical meaning of the equation corresponding to $0 \leq v < 1.79$.

(f) Explain the physical meaning of the equation corresponding to $v > 20$.

60. Wind Chill Redo Problem 59(a)–(d) for an air temperature of -10°C.

61. First-class Mail In 2018 the U.S. Postal Service charged $1.00 postage for certain first-class mail retail flats (such as an 8.5″ by 11″ envelope) weighing up to 1 ounce, plus $0.21 for each additional ounce up to 13 ounces. First-class rates do not apply to flats weighing more than 13 ounces. Develop a model that relates C, the first-class postage charged, for a flat weighing x ounces. Graph the function.

Source: United States Postal Service

62. Challenge Problem Pool Depth Develop a model for the depth of the swimming pool shown below as a function of the distance from the wall on the left.

63. Challenge Problem Find the sum function $(f + g)(x)$ if

$$f(x) = \begin{cases} 2x + 3 & \text{if } x < 2 \\ x^2 + 5x & \text{if } x \geq 2 \end{cases}$$

and

$$g(x) = \begin{cases} -4x + 1 & \text{if } x \leq 0 \\ x - 7 & \text{if } x > 0 \end{cases}$$

Explaining Concepts: Discussion and Writing

In Problems 64–71, use a graphing utility.

64. Exploration Graph $y = x^2$. Then on the same screen graph $y = x^2 + 2$, followed by $y = x^2 + 4$, followed by $y = x^2 - 2$. What pattern do you observe? Can you predict the graph of $y = x^2 - 4$? Of $y = x^2 + 5$?

65. Exploration Graph $y = x^2$. Then on the same screen graph $y = (x - 2)^2$, followed by $y = (x - 4)^2$, followed by $y = (x + 2)^2$. What pattern do you observe? Can you predict the graph of $y = (x + 4)^2$? Of $y = (x - 5)^2$?

66. Exploration Graph $y = |x|$. Then on the same screen graph $y = 2|x|$, followed by $y = 4|x|$, followed by $y = \frac{1}{2}|x|$. What pattern do you observe? Can you predict the graph of $y = \frac{1}{4}|x|$? Of $y = 5|x|$?

67. Exploration Graph $y = x^2$. Then on the same screen graph $y = -x^2$. Now try $y = |x|$ and $y = -|x|$. What do you conclude?

68. Exploration Graph $y = \sqrt{x}$. Then on the same screen graph $y = \sqrt{-x}$. Now try $y = 2x + 1$ and $y = 2(-x) + 1$. What do you conclude?

69. Exploration Graph $y = x^3$. Then on the same screen graph $y = (x - 1)^3 + 2$. Could you have predicted the result?

70. Exploration Graph $y = x^2$, $y = x^4$, and $y = x^6$ on the same screen. What do you notice is the same about each graph? What do you notice is different?

71. Exploration Graph $y = x^3$, $y = x^5$, and $y = x^7$ on the same screen. What do you notice is the same about each graph? What do you notice is different?

72. Consider the equation

$$y = \begin{cases} 1 & \text{if } x \text{ is rational} \\ 0 & \text{if } x \text{ is irrational} \end{cases}$$

Is this a function? What is its domain? What is its range? What is its y-intercept, if any? What are its x-intercepts, if any? Is it even, odd, or neither? How would you describe its graph?

73. Define some functions that pass through $(0, 0)$ and $(1, 1)$ and are increasing for $x \geq 0$. Begin your list with $y = \sqrt{x}$, $y = x$, and $y = x^2$. Can you propose a general result about such functions?

'Are You Prepared?' Answers

1.

2.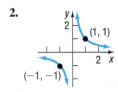

3. $(0, -8)$, $(2, 0)$

1.6 Graphing Techniques: Transformations

> **OBJECTIVES** **1** Graph Functions Using Vertical and Horizontal Shifts (p. 67)
> **2** Graph Functions Using Compressions and Stretches (p. 71)
> **3** Graph Functions Using Reflections about the *x*-Axis and the *y*-Axis (p. 73)

At this stage, you should be able to quickly graph any of the functions

$$y = x \qquad y = x^2 \qquad y = x^3 \qquad y = \sqrt{x} \qquad y = \sqrt[3]{x} \qquad y = \frac{1}{x} \qquad y = |x|$$

with key points plotted. If necessary, review the previous section, Figures 63 through 69.

Sometimes we are asked to graph a function that is "almost" like one that we already know how to graph. In this section, we develop techniques for graphing such functions. Collectively, these techniques are referred to as **transformations**.

1 Graph Functions Using Vertical and Horizontal Shifts

EXAMPLE 1 **Vertical Shift Up**

Use the graph of $f(x) = x^2$ to obtain the graph of $g(x) = x^2 + 3$. Find the domain and range of g.

Solution The function $g(x) = x^2 + 3$ is basically a square function. In fact, notice that $g(x) = f(x) + 3$. Create a table of values to obtain some points on the graphs of f and g. For example, when $x = 0$, then $y = f(0) = 0$ and $y = g(0) = 3$. When $x = 1$, then $y = f(1) = 1$ and $y = g(1) = 4$. Table 9 on the next page lists these and a few other points on each graph. Notice that each y-coordinate of a point on the graph of g is 3 units larger than the y-coordinate of the corresponding point on the graph of f. We conclude that the graph of g is identical to that of f, except that it is shifted vertically up 3 units. See Figure 72 on the next page.

Table 9

x	y = f(x) = x²	y = g(x) = x² + 3
-2	4	7
-1	1	4
0	0	3
1	1	4
2	4	7

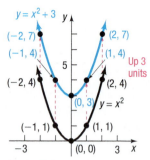

Figure 72

The domain of g is all real numbers, or $(-\infty, \infty)$. The range of g is $[3, \infty)$.

EXAMPLE 2   **Vertical Shift Down**

Use the graph of $f(x) = x^2$ to obtain the graph of $g(x) = x^2 - 4$. Find the domain and range of g.

Solution   The function $g(x) = x^2 - 4$ is basically a square function. Table 10 lists some points on the graphs of f and g. Notice that each y-coordinate of g is 4 units less than the corresponding y-coordinate of f. Also notice that $g(x) = f(x) - 4$.

To obtain the graph of g from the graph of f, subtract 4 from each y-coordinate on the graph of f. The graph of g is identical to that of f except that it is shifted down 4 units. See Figure 73.

Table 10

x	y = f(x) = x²	y = g(x) = x² - 4
-2	4	0
-1	1	-3
0	0	-4
1	1	-3
2	4	0

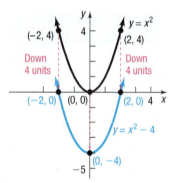

Figure 73

The domain of g is all real numbers, or $(-\infty, \infty)$. The range of g is $[-4, \infty)$.

A vertical shift affects only the range of a function, not the domain. For example, the range of $f(x) = x^2$ is $[0, \infty)$. In Example 1 the range of $g(x) = f(x) + 3$ is $[3, \infty)$, whereas in Example 2 the range of $g(x) = f(x) - 4$ is $[-4, \infty)$. The domain for all three functions is all real numbers.

Figure 74

Exploration On the same screen, graph each of the following functions:

$$Y_1 = x^2 \qquad Y_2 = x^2 + 2 \qquad Y_3 = x^2 - 2$$

Figure 74 illustrates the graphs using Desmos. You should have observed a general pattern. With $Y_1 = x^2$ on the screen, the graph of $Y_2 = x^2 + 2$ is identical to that of $Y_1 = x^2$, except that it is shifted vertically up 2 units. The graph of $Y_3 = x^2 - 2$ is identical to that of $Y_1 = x^2$, except that it is shifted vertically down 2 units.

We are led to the following conclusions:

> **In Words**
> For $y = f(x) + k$, $k > 0$, add k to each y-coordinate on the graph of $y = f(x)$ to shift the graph up k units.
> For $y = f(x) - k$, $k > 0$, subtract k from each y-coordinate to shift the graph down k units.

Vertical Shifts

- If a positive real number k is added to the output of a function $y = f(x)$, the graph of the new function $y = f(x) + k$ is the graph of f **shifted vertically up** k units.
- If a positive real number k is subtracted from the output of a function $y = f(x)$, the graph of the new function $y = f(x) - k$ is the graph of f **shifted vertically down** k units.

Now Work PROBLEM **39**

EXAMPLE 3 | **Horizontal Shift to the Right**

Use the graph of $f(x) = \sqrt{x}$ to obtain the graph of $g(x) = \sqrt{x - 2}$. Find the domain and range of g.

Solution | The function $g(x) = \sqrt{x - 2}$ is basically a square root function. Table 11 lists some points on the graphs of f and g. Note that when $f(x) = 0$, then $x = 0$, but when $g(x) = 0$, then $x = 2$. Also, when $f(x) = 2$, then $x = 4$, but when $g(x) = 2$, then $x = 6$. The x-coordinates on the graph of g are 2 units larger than the corresponding x-coordinates on the graph of f for any given y-coordinate. Also notice that $g(x) = f(x - 2)$. We conclude that the graph of g is identical to that of f, except that it is shifted horizontally 2 units to the right. See Figure 75.

Table 11

x	$y = f(x)$ $= \sqrt{x}$	x	$y = g(x)$ $= \sqrt{x - 2}$
0	0	2	0
1	1	3	1
4	2	6	2
9	3	11	3

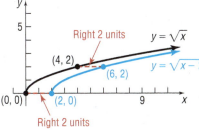

Figure 75

The domain of g is $[2, \infty)$ and the range of g is $[0, \infty)$.

EXAMPLE 4 | **Horizontal Shift to the Left**

Use the graph of $f(x) = \sqrt{x}$ to obtain the graph of $g(x) = \sqrt{x + 4}$. Find the domain and range of g.

Solution | The function $g(x) = \sqrt{x + 4}$ is basically a square root function. Table 12 lists some points on the graphs of f and g. Note that when $f(x) = 0$, then $x = 0$, but when $g(x) = 0$, then $x = -4$. Also, when $f(x) = 2$, then $x = 4$, but when $g(x) = 2$, then $x = 0$. The x-coordinates on the graph of g are 4 units smaller than the corresponding x-coordinates on the graph of f for any given y-coordinate. Also notice that $g(x) = f(x + 4)$. We conclude that the graph of g is identical to that of f, except that it is shifted horizontally 4 units to the left. See Figure 76.

Table 12

x	$y = f(x)$ $= \sqrt{x}$	x	$y = g(x)$ $= \sqrt{x + 4}$
0	0	-4	0
1	1	-3	1
4	2	0	2
9	3	5	3

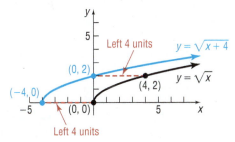

Figure 76

The domain of g is $[-4, \infty)$ and the range of g is $[0, \infty)$.

Now Work PROBLEM **43**

A horizontal shift affects only the domain of a function, not the range. For example, the domain of $f(x) = \sqrt{x}$ is $[0, \infty)$. In Example 3 the domain of $g(x) = f(x - 2)$ is $[2, \infty)$, whereas in Example 4 the domain of $g(x) = f(x + 4)$ is $[-4, \infty)$. The range for all three functions is $[0, \infty)$.

Figure 77

Exploration On the same screen, graph each of the following functions:

$$Y_1 = x^2 \qquad Y_2 = (x - 3)^2 \qquad Y_3 = (x + 2)^2$$

Figure 77 illustrates the graphs using Desmos.

You should have observed the following pattern. With the graph of $Y_1 = x^2$ on the screen, the graph of $Y_2 = (x - 3)^2$ is identical to that of $Y_1 = x^2$, except that it is shifted horizontally to the right 3 units. The graph of $Y_3 = (x + 2)^2$ is identical to that of $Y_1 = x^2$, except that it is shifted horizontally to the left 2 units.

We are led to the following conclusions:

In Words

For $y = f(x - h), h > 0$, add h to each x-coordinate on the graph of $y = f(x)$ to shift the graph right h units. For $y = f(x + h), h > 0$, subtract h from each x-coordinate on the graph of $y = f(x)$ to shift the graph left h units.

Horizontal Shifts

- If the argument x of a function f is replaced by $x - h, h > 0$, the graph of the new function $y = f(x - h)$ is the graph of f **shifted horizontally right** h units.

- If the argument x of a function f is replaced by $x + h, h > 0$, the graph of the new function $y = f(x + h)$ is the graph of f **shifted horizontally left** h units.

Observe the distinction between vertical and horizontal shifts. The graph of $f(x) = x^3 + 2$ is obtained by shifting the graph vertically because we evaluate the cube function first and then add 2. The graph of $g(x) = (x + 2)^3$ is obtained by shifting the graph of $y = x^3$ horizontally because we add 2 to x before we evaluate the cube function.

Another way to think of the distinction is to note where the shift occurs. If the shift occurs *outside* the basic function, as is the case with $f(x) = x^3 + 2$, then there is a vertical shift. If the shift occurs *inside* the basic function, as is the case with $f(x) = (x + 2)^3$, then there is a horizontal shift.

The graph of a function can be moved anywhere in the coordinate plane by combining vertical and horizontal shifts.

EXAMPLE 5   **Combining Vertical and Horizontal Shifts**

Graph the function $f(x) = |x + 3| - 5$. Find the domain and range of f.

Solution   We graph f in steps. First, note that f is basically an absolute value function, so begin with the graph of $y = |x|$ as shown in Figure 78(a). Next, to get the graph of $y = |x + 3|$, shift the graph of $y = |x|$ horizontally 3 units to the left. See Figure 78(b). Finally, to get the graph of $y = |x + 3| - 5$, shift the graph of $y = |x + 3|$ vertically down 5 units. See Figure 78(c).

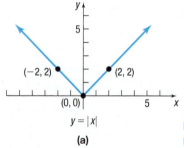

$y = |x|$

(a)

Replace x by $x + 3$;
Horizontal shift
left 3 units

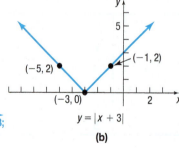

$y = |x + 3|$

(b)

Subtract 5;
Vertical shift
down 5 units

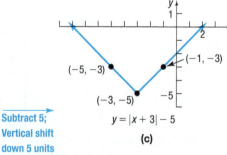

$y = |x + 3| - 5$

(c)

Figure 78

The domain of f is all real numbers, or $(-\infty, \infty)$. The range of f is $[-5, \infty)$.

Note the points plotted on each graph in Figure 78. Using key points can be helpful in keeping track of the transformations that have been made.

In Example 5, if the vertical shift had been done first, followed by the horizontal shift, the final graph would have been the same. Try it for yourself.

Now Work PROBLEMS **45** AND **69**

2 Graph Functions Using Compressions and Stretches

EXAMPLE 6	**Vertical Stretch**

Use the graph of $f(x) = \sqrt{x}$ to obtain the graph of $g(x) = 2\sqrt{x}$.

Solution To see the relationship between the graphs of f and g, we list points on each graph, as shown in Table 13. For each x, the y-coordinate of a point on the graph of g is 2 times as large as the corresponding y-coordinate on the graph of f. That is, $g(x) = 2f(x)$. The graph of $f(x) = \sqrt{x}$ is vertically stretched by a factor of 2 to obtain the graph of $g(x) = 2\sqrt{x}$. For example, $(1, 1)$ is on the graph of f, but $(1, 2)$ is on the graph of g. See Figure 79.

Table 13

x	$y = f(x) = \sqrt{x}$	$y = g(x) = 2\sqrt{x}$
0	0	0
1	1	2
4	2	4
9	3	6

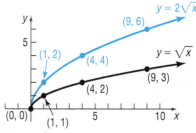

Figure 79

EXAMPLE 7	**Vertical Compression**

Use the graph of $f(x) = |x|$ to obtain the graph of $g(x) = \dfrac{1}{2}|x|$.

Solution For each x, the y-coordinate of a point on the graph of g is $\dfrac{1}{2}$ as large as the corresponding y-coordinate on the graph of f. That is, $g(x) = \dfrac{1}{2}f(x)$. The graph of $f(x) = |x|$ is vertically compressed by a factor of $\dfrac{1}{2}$ to obtain the graph of $g(x) = \dfrac{1}{2}|x|$. For example, $(2, 2)$ is on the graph of f, but $(2, 1)$ is on the graph of g. See Table 14 and Figure 80.

Table 14

x	$y = f(x) = |x|$	$y = g(x) = \dfrac{1}{2}|x|$
-2	2	1
-1	1	$\dfrac{1}{2}$
0	0	0
1	1	$\dfrac{1}{2}$
2	2	1

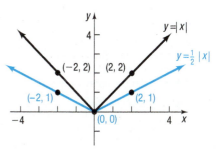

Figure 80

In Words

For $y = af(x), a > 0$, the factor a is "outside" the function, so it affects the y-coordinates. Multiply each y-coordinate on the graph of $y = f(x)$ by a.

Vertical Compression or Stretch

If a function $y = f(x)$ is multiplied by a positive number a, then the graph of the new function $y = af(x)$ is obtained by multiplying each y-coordinate of the graph of $y = f(x)$ by a.

- If $0 < a < 1$, a **vertical compression by a factor of a** results.
- If $a > 1$, a **vertical stretch by a factor of a** results.

Now Work PROBLEM 47

What happens if the argument x of a function $y = f(x)$ is multiplied by a positive number a, creating a new function $y = f(ax)$? To find the answer, look at the following Exploration.

Exploration On the same screen, graph each of the following functions:

$$Y_1 = f(x) = \sqrt{x} \qquad Y_2 = f(2x) = \sqrt{2x} \qquad Y_3 = f\left(\frac{1}{2}x\right) = \sqrt{\frac{1}{2}x} = \sqrt{\frac{x}{2}}$$

Create a table of values to explore the relation between the x- and y-coordinates of each function.

Result You should have obtained the graphs in Figure 81. Look at Table 15(a). Note that (1, 1), (4, 2), and (9, 3) are points on the graph of $Y_1 = \sqrt{x}$. Also, (0.5, 1), (2, 2), and (4.5, 3) are points on the graph of $Y_2 = \sqrt{2x}$. For a given y-coordinate, the x-coordinate on the graph of Y_2 is $\frac{1}{2}$ of the x-coordinate on Y_1.

Table 15

NORMAL FLOAT AUTO REAL RADIAN MP
PRESS ENTER TO EDIT

X	Y1	Y2
0	0	0
.5	.70711	1
1	1	1.4142
2	1.4142	2
4	2	2.8284
4.5	2.1213	3
8	2.8284	4
9	3	4.2426
16	4	5.6569
12.5	3.5355	5
25	5	7.0711

Y2=√2X

(a)

NORMAL FLOAT AUTO REAL RADIAN MP
PRESS ENTER TO EDIT

X	Y1	Y3
0	0	0
1	1	.70711
2	1.4142	1
4	2	1.4142
8	2.8284	2
9	3	2.1213
16	4	2.8284
18	4.2426	3
25	5	3.5355
32	5.6569	4
50	7.0711	5

Y3=√X/2

(b)

$y_1 = \sqrt{x}$

$y_2 = \sqrt{2x}$

$y_3 = \sqrt{\frac{x}{2}}$

desmos

Figure 81

We conclude that the graph of $Y_2 = \sqrt{2x}$ is obtained by multiplying the x-coordinate of each point on the graph of $Y_1 = \sqrt{x}$ by $\frac{1}{2}$. The graph of $Y_2 = \sqrt{2x}$ is the graph of $Y_1 = \sqrt{x}$ *compressed* horizontally.

Look at Table 15(b). Notice that (1, 1), (4, 2), and (9, 3) are points on the graph of $Y_1 = \sqrt{x}$. Also notice that (2, 1), (8, 2), and (18, 3) are points on the graph of $Y_3 = \sqrt{\frac{x}{2}}$. For a given y-coordinate, the x-coordinate on the graph of Y_3 is 2 times the x-coordinate on Y_1. We conclude that the graph of $Y_3 = \sqrt{\frac{x}{2}}$ is obtained by multiplying the x-coordinate of each point on the graph of $Y_1 = \sqrt{x}$ by 2. The graph of $Y_3 = \sqrt{\frac{x}{2}}$ is the graph of $Y_1 = \sqrt{x}$ *stretched* horizontally.

Based on the Exploration, we have the following result:

In Words

For $y = f(ax), a > 0$, the factor a is "inside" the function, so it affects the x-coordinates. Multiply each x-coordinate on the graph of $y = f(x)$ by $\frac{1}{a}$.

Horizontal Compression or Stretch

If the argument of a function $y = f(x)$ is multiplied by a positive number a, then the graph of the new function $y = f(ax)$ is obtained by multiplying each x-coordinate of the graph of $y = f(x)$ by $\frac{1}{a}$.

- If $a > 1$, a **horizontal compression by a factor of $\frac{1}{a}$** results.
- If $0 < a < 1$, a **horizontal stretch by a factor of $\frac{1}{a}$** results.

| EXAMPLE 8 | **Graphing Using Stretches and Compressions** |

The graph of $y = f(x)$ is given in Figure 82. Use this graph to find the graphs of

(a) $y = 2f(x)$ (b) $y = f(3x)$

Solution (a) Because the 2 is "outside" the function f, the graph of $y = 2f(x)$ is obtained by multiplying each y-coordinate of $y = f(x)$ by 2. See Figure 83.

(b) Because the 3 is "inside" the function f, the graph of $y = f(3x)$ is obtained from the graph of $y = f(x)$ by multiplying each x-coordinate of $y = f(x)$ by $\frac{1}{3}$. See Figure 84.

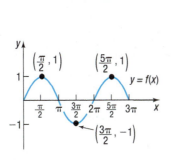

Figure 82 $y = f(x)$

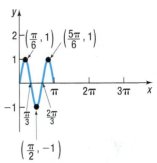

Figure 83 $y = 2f(x)$; Vertical stretch by a factor of 2

Figure 84 $y = f(3x)$; Horizontal compression by a factor of $\frac{1}{3}$

Now Work PROBLEMS 63(e) AND (g)

3 Graph Functions Using Reflections about the *x*-Axis and the *y*-Axis

| EXAMPLE 9 | **Reflection about the *x*-Axis** |

Graph the function $f(x) = -x^2$. Find the domain and range of f.

Solution Note that f is basically a square function, so begin with the graph of $y = x^2$, as shown in black in Figure 85. For each point (x, y) on the graph of $y = x^2$, the point $(x, -y)$ is on the graph of $y = -x^2$, as indicated in Table 16. Draw the graph of $y = -x^2$ by reflecting the graph of $y = x^2$ about the x-axis. See Figure 85.

Figure 85

Table 16

x	$y = x^2$	$y = -x^2$
-2	4	-4
-1	1	-1
0	0	0
1	1	-1
2	4	-4

The domain of f is all real numbers, or $(-\infty, \infty)$. The range of f is $(-\infty, 0]$.

> **Reflection about the *x*-Axis**
>
> When a function f is multiplied by -1, the graph of the new function $y = -f(x)$ is the **reflection about the *x*-axis** of the graph of the function f.

Now Work PROBLEM 49

EXAMPLE 10   **Reflection about the *y*-Axis**

Graph the function $f(x) = \sqrt{-x}$. Find the domain and range of f.

Solution   To graph $f(x) = \sqrt{-x}$, begin with the graph of $y = \sqrt{x}$, as shown in black in Figure 86. For each point (x, y) on the graph of $y = \sqrt{x}$, the point $(-x, y)$ is on the graph of $y = \sqrt{-x}$, as listed in Table 17. Obtain the graph of $y = \sqrt{-x}$ by reflecting the graph of $y = \sqrt{x}$ about the *y*-axis. See Figure 86.

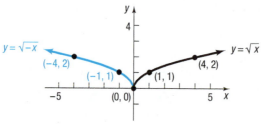

Figure 86

Table 17

x	$y = \sqrt{x}$	*x*	$y = \sqrt{-x}$
0	0	0	0
1	1	−1	1
4	2	−4	2
9	3	−9	3

The domain of f is $(-\infty, 0]$. The range of f is the set of all nonnegative real numbers, or $[0, \infty)$.

In Words
For $y = -f(x)$, multiply each *y*-coordinate on the graph of $y = f(x)$ by -1.
For $y = f(-x)$, multiply each *x*-coordinate by -1.

Reflection about the *y*-Axis

When the graph of the function f is known, the graph of the new function $y = f(-x)$ is the **reflection about the *y*-axis** of the graph of the function f.

SUMMARY OF GRAPHING TECHNIQUES

Each graphing technique has a different effect on the graph of a function. Compressions and stretches change the proportions of a graph, and reflections change the orientation, but not its proportions. Vertical and horizontal shifts change the location of the graph, without changing its proportions or orientation.

To Graph:	Draw the Graph of f and:	Functional Change to $f(x)$
Vertical shifts		
$y = f(x) + k, \quad k > 0$	Shift the graph of f up k units.	Add k to $f(x)$.
$y = f(x) - k, \quad k > 0$	Shift the graph of f down k units.	Subtract k from $f(x)$.
Horizontal shifts		
$y = f(x + h), \quad h > 0$	Shift the graph of f to the left h units.	Replace x by $x + h$.
$y = f(x - h), \quad h > 0$	Shift the graph of f to the right h units.	Replace x by $x - h$.
Compressing or stretching		
$y = af(x), \quad a > 0$	Multiply each *y*-coordinate of $y = f(x)$ by a. Stretch the graph of f vertically if $a > 1$. Compress the graph of f vertically if $0 < a < 1$.	Multiply $f(x)$ by a.
$y = f(ax), \quad a > 0$	Multiply each *x*-coordinate of $y = f(x)$ by $\dfrac{1}{a}$. Stretch the graph of f horizontally if $0 < a < 1$. Compress the graph of f horizontally if $a > 1$.	Replace x by ax.
Reflection about the *x*-axis		
$y = -f(x)$	Reflect the graph of f about the *x*-axis.	Multiply $f(x)$ by -1.
Reflection about the *y*-axis		
$y = f(-x)$	Reflect the graph of f about the *y*-axis.	Replace x by $-x$.

EXAMPLE 11

Determining the Function Obtained from a Series of Transformations

Find the function that is finally graphed after the following sequence of transformations are applied to the graph of $y = |x|$.

1. Shift left 2 units
2. Shift up 3 units
3. Reflect about the y-axis

Solution

1. Shift left 2 units: Replace x by $x + 2$. $y = |x + 2|$
2. Shift up 3 units: Add 3. $y = |x + 2| + 3$
3. Reflect about the y-axis: Replace x by $-x$. $y = |-x + 2| + 3$

Now Work PROBLEM 31

EXAMPLE 12

Using Graphing Techniques

Graph the function $R(x) = \dfrac{3}{x - 2} + 1$. Find the domain and range of R.

Solution

It is helpful to write R as $R(x) = 3 \cdot \dfrac{1}{x - 2} + 1$. Now use the following steps to obtain the graph of R.

HINT: Although the order in which transformations are performed can be altered, consider using the following order for consistency:
1. Horizontal shift
2. Reflections
3. Compressions and stretches
4. Vertical shift

STEP 1: $y = \dfrac{1}{x}$ **Reciprocal function**

STEP 2: $y = \dfrac{1}{x - 2}$ **Replace x by $x - 2$; horizontal shift to the right 2 units.**

STEP 3: $y = 3 \cdot \dfrac{1}{x - 2} = \dfrac{3}{x - 2}$ **Multiply by 3; vertical stretch by a factor of 3.**

STEP 4: $y = \dfrac{3}{x - 2} + 1$ **Add 1; vertical shift up 1 unit.**

See Figure 87.

Figure 87

(a) $y = \dfrac{1}{x}$ Replace x by $x - 2$; Horizontal shift right 2 units

(b) $y = \dfrac{1}{x-2}$ Multiply by 3; Vertical stretch

(c) $y = \dfrac{3}{x-2}$ Add 1; Vertical shift up 1 unit

(d) $y = \dfrac{3}{x-2} + 1$

The domain of $y = \dfrac{1}{x}$ is $\{x \mid x \neq 0\}$ and its range is $\{y \mid y \neq 0\}$. Because we shifted right 2 units and up 1 unit to obtain R, the domain of R is $\{x \mid x \neq 2\}$ and its range is $\{y \mid y \neq 1\}$.

Other orderings of the steps shown in Example 12 would also result in the graph of R. For example, try this one:

STEP 1: $y = \dfrac{1}{x}$ Reciprocal function

STEP 2: $y = 3 \cdot \dfrac{1}{x} = \dfrac{3}{x}$ Multiply by 3; vertical stretch by a factor of 3.

STEP 3: $y = \dfrac{3}{x - 2}$ Replace x by x − 2; horizontal shift to the right 2 units.

STEP 4: $y = \dfrac{3}{x - 2} + 1$ Add 1; vertical shift up 1 unit.

EXAMPLE 13   **Using Graphing Techniques**

Graph the function $f(x) = \sqrt{1 - x} + 2$. Find the domain and range of f.

Solution   Begin by rewriting $f(x)$ as $f(x) = \sqrt{1 - x} + 2 = \sqrt{-x + 1} + 2$. Now use the following steps.

STEP 1: $y = \sqrt{x}$ Square root function

STEP 2: $y = \sqrt{x + 1}$ Replace x by x + 1; horizontal shift to the left 1 unit.

STEP 3: $y = \sqrt{-x + 1} = \sqrt{1 - x}$ Replace x by − x; reflect about the y-axis.

STEP 4: $y = \sqrt{1 - x} + 2$ Add 2; vertical shift up 2 units.

See Figure 88.

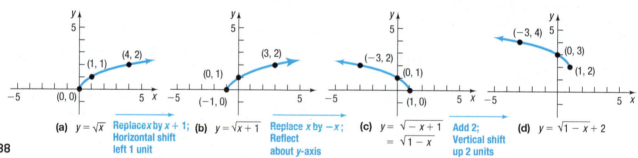

(a) $y = \sqrt{x}$ Replace x by x + 1; Horizontal shift left 1 unit   (b) $y = \sqrt{x + 1}$ Replace x by − x; Reflect about y-axis   (c) $y = \sqrt{-x + 1}$ $= \sqrt{1 - x}$ Add 2; Vertical shift up 2 units   (d) $y = \sqrt{1 - x} + 2$

Figure 88

The domain of f is $(-\infty, 1]$ and the range is $[2, \infty)$.

Now Work PROBLEM 55

Asymptotes

Look back at Figure 87(d). Notice that as the values of x become more negative, that is, as x becomes **unbounded in the negative direction** ($x \rightarrow -\infty$, read as "x **approaches negative infinity**"), the values $R(x)$ approach 1. Also, as the values of x become **unbounded in the positive direction** ($x \rightarrow \infty$, read as "x **approaches infinity**"), the values of $R(x)$ also approach 1. That is,

$$\text{As } x \rightarrow -\infty, R(x) \rightarrow 1.$$
$$\text{As } x \rightarrow \infty, R(x) \rightarrow 1.$$

This behavior of the graph is depicted by the horizontal line $y = 1$, which is called a *horizontal asymptote* of the graph of R.

DEFINITION Horizontal Asymptote

Let R denote a function. If, as $x \rightarrow -\infty$ or as $x \rightarrow \infty$, the values of $R(x)$ approach some fixed number L, then the line $y = L$ is a **horizontal asymptote** of the graph of R. [Refer to Figures 89(a) and (b).]

Now look again at Figure 87(d). Notice that as x gets closer to 2, but remains less than 2, denoted $x \rightarrow 2^-$, the values of R are becoming unbounded in the negative direction. That is, as $x \rightarrow 2^-$ (from the left), $R(x) \rightarrow -\infty$. And as x gets closer to 2, but remains larger than 2, denoted $x \rightarrow 2^+$, the values of R are becoming unbounded in the positive direction. That is, as $x \rightarrow 2^+$ (from the right), $R(x) \rightarrow \infty$. We conclude that

$$\text{As } x \rightarrow 2, \ |R(x)| \rightarrow \infty.$$

This behavior of the graph is depicted by the vertical line $x = 2$, which is called a *vertical asymptote* of the graph of R.

DEFINITION Vertical Asymptote

If, as x approaches some number c, the values $|R(x)| \rightarrow \infty$ [that is, $R(x) \rightarrow \infty$ or $R(x) \rightarrow -\infty$.], then the line $x = c$ is a **vertical asymptote** of the graph of R. [Refer to Figures 89(c) and (d).]

Even though the asymptotes of a function are not part of the graph of the function, they provide information about how the graph looks. Figure 89 illustrates some of the possibilities.

Figure 89

(a) End behavior:
As $x \rightarrow \infty$, $R(x) \rightarrow L$.
That is, the points on the graph of R are getting closer to the line $y = L$; $y = L$ is a horizontal asymptote.

(b) End behavior:
As $x \rightarrow -\infty$, $R(x) \rightarrow L$.
That is, the points on the graph of R are getting closer to the line $y = L$; $y = L$ is a horizontal asymptote.

(c) As $x \rightarrow c^-$, $R(x) \rightarrow \infty$;
As $x \rightarrow c^+$, $R(x) \rightarrow \infty$.
That is, the points on the graph of R are getting closer to the line $x = c$; $x = c$ is a vertical asymptote.

(d) As $x \rightarrow c^-$, $R(x) \rightarrow -\infty$;
As $x \rightarrow c^+$, $R(x) \rightarrow \infty$.
That is, the points on the graph of R are getting closer to the line $x = c$; $x = c$ is a vertical asymptote.

A horizontal asymptote, when it occurs, describes the end behavior of the graph as $x \rightarrow \infty$ or as $x \rightarrow -\infty$. **The graph of a function may intersect a horizontal asymptote.** A vertical asymptote, when it occurs, describes the behavior of the graph when x is close to some number c. **The graph of a function will never intersect a vertical asymptote.**

1.6 Assess Your Understanding

Concepts and Vocabulary

1. Suppose the graph of a function f is known. Then the graph of $y = f(x - 2)$ is obtained by a _____ shift of the graph of f to the _____ a distance of 2 units.

2. Suppose the graph of a function f is known. Then the graph of $y = f(-x)$ is a reflection about the _____-axis of the graph of the function $y = f(x)$.

3. *True or False* The graph of $y = \frac{1}{3}g(x)$ is the graph of $y = g(x)$ vertically stretched by a factor of 3.

4. *True or False* The graph of $y = -f(x)$ is the reflection about the x-axis of the graph of $y = f(x)$.

5. *True or False* The graph of a function may intersect a vertical asymptote.

6. *Multiple Choice* Which function has a graph that is the graph of $y = \sqrt{x}$ shifted down 3 units?
   (a) $y = \sqrt{x + 3}$ (b) $y = \sqrt{x - 3}$
   (c) $y = \sqrt{x} + 3$ (d) $y = \sqrt{x} - 3$

7. *Multiple Choice* Which function has a graph that is the graph of $y = f(x)$ horizontally stretched by a factor of 4?
   (a) $y = \frac{1}{4}f(x)$ (b) $y = f\left(\frac{1}{4}x\right)$
   (c) $y = 4f(x)$ (d) $y = f(4x)$

8. *True or False* The graph of a function may intersect a horizontal asymptote.

Skill Building

In Problems 9–20, match each graph to one of the following functions:

A. $y = x^2 + 2$
B. $y = -x^2 + 2$
C. $y = |x| + 2$

D. $y = -|x| + 2$
E. $y = (x - 2)^2$
F. $y = -(x + 2)^2$

G. $y = |x - 2|$
H. $y = -|x + 2|$
I. $y = 2x^2$

J. $y = -2x^2$
K. $y = 2|x|$
L. $y = -2|x|$

9.

10.

11.

12.

13.

14.

15.

16.

17.

18.

19.

20.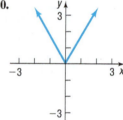

In Problems 21–30, write the function whose graph is the graph of $y = x^3$, but is:

21. Shifted to the right 4 units

22. Shifted to the left 4 units

23. Shifted up 4 units

24. Shifted down 4 units

25. Reflected about the y-axis

26. Reflected about the x-axis

27. Vertically stretched by a factor of 5

28. Horizontally stretched by a factor of 4

29. Horizontally compressed by a factor of $\frac{1}{2}$

30. Vertically compressed by a factor of $\frac{1}{4}$

In Problems 31–34, find the function that is finally graphed after each of the following transformations is applied to the graph of $y = \sqrt{x}$ in the order stated.

31. **(1)** Shift up 2 units
    (2) Reflect about the x-axis
    (3) Reflect about the y-axis

32. **(1)** Reflect about the x-axis
    (2) Shift right 3 units
    (3) Shift down 2 units

33. **(1)** Vertical stretch by a factor of 3
    (2) Shift up 4 units
    (3) Shift left 5 units

34. **(1)** Shift up 2 units
    (2) Reflect about the y-axis
    (3) Shift left 3 units

35. If $(3, 6)$ is a point on the graph of $y = f(x)$, which of the following points must be on the graph of $y = -f(x)$?
    (a) $(6, 3)$ **(b)** $(6, -3)$
    (c) $(3, -6)$ **(d)** $(-3, 6)$

36. If $(3, 6)$ is a point on the graph of $y = f(x)$, which of the following points must be on the graph of $y = f(-x)$?
    (a) $(6, 3)$ **(b)** $(6, -3)$
    (c) $(3, -6)$ **(d)** $(-3, 6)$

37. If $(1, 3)$ is a point on the graph of $y = f(x)$, which of the following points must be on the graph of $y = 2f(x)$?
    (a) $\left(1, \dfrac{3}{2}\right)$ **(b)** $(2, 3)$
    (c) $(1, 6)$ **(d)** $\left(\dfrac{1}{2}, 3\right)$

38. If $(4, 2)$ is a point on the graph of $y = f(x)$, which of the following points must be on the graph of $y = f(2x)$?
    (a) $(4, 1)$ **(b)** $(8, 2)$
    (c) $(2, 2)$ **(d)** $(4, 4)$

In Problems 39–62, graph each function using the techniques of shifting, compressing, stretching, and/or reflecting. Start with the graph of the basic function (for example, $y = x^2$) and show all the steps. Be sure to show at least three key points. Find the domain and the range of each function.

39. $f(x) = x^2 - 1$

40. $f(x) = x^2 + 4$

41. $g(x) = \sqrt{3x}$

42. $g(x) = \sqrt[3]{\frac{1}{2}x}$

43. $h(x) = \sqrt{x + 2}$

44. $h(x) = \sqrt{x + 1}$

45. $f(x) = (x - 1)^3 + 2$

46. $f(x) = (x + 2)^3 - 3$

47. $g(x) = 4\sqrt{x}$

48. $g(x) = \frac{1}{2}\sqrt{x}$

49. $f(x) = -\sqrt[3]{x}$

50. $f(x) = -\sqrt{x}$

51. $f(x) = 2(x + 1)^2 - 3$

52. $f(x) = 3(x - 2)^2 + 1$

53. $g(x) = 2\sqrt{x - 2} + 1$

54. $g(x) = 3|x + 1| - 3$

55. $h(x) = \sqrt{-x} - 2$

56. $h(x) = \frac{4}{x} + 2$

57. $f(x) = -(x + 1)^3 - 1$

58. $f(x) = -4\sqrt{x - 1}$

59. $g(x) = 2|1 - x|$

60. $g(x) = 4\sqrt{2 - x}$

61. $h(x) = \frac{1}{2x}$

62. $h(x) = \sqrt[3]{x - 1} + 3$

In Problems 63–66, the graph of a function f is illustrated. Use the graph of f as the first step toward graphing each of the following functions:

(a) $F(x) = f(x) + 3$

(b) $G(x) = f(x + 2)$

(c) $P(x) = -f(x)$

(d) $H(x) = f(x + 1) - 2$

(e) $Q(x) = \frac{1}{2}f(x)$

(f) $g(x) = f(-x)$

(g) $h(x) = f(2x)$

63.

64.

65.

66.
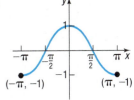

Mixed Practice *In Problems 67–74, complete the square of each quadratic expression. Then graph each function using graphing techniques. (If necessary, refer to Section A.3 to review completing the square.)*

67. $f(x) = x^2 + 2x$

68. $f(x) = x^2 - 6x$

69. $f(x) = x^2 - 8x + 1$

70. $f(x) = x^2 + 4x + 2$

71. $f(x) = 2x^2 - 12x + 19$

72. $f(x) = 3x^2 + 6x + 1$

73. $f(x) = -3x^2 - 12x - 17$

74. $f(x) = -2x^2 - 12x - 13$

Applications and Extensions

75. Suppose that the x-intercepts of the graph of $y = f(x)$ are -5 and 3.

   (a) What are the x-intercepts of the graph of $y = f(x + 2)$?

   (b) What are the x-intercepts of the graph of $y = f(x - 2)$?

   (c) What are the x-intercepts of the graph of $y = 4f(x)$?

   (d) What are the x-intercepts of the graph of $y = f(-x)$?

76. Suppose that the x-intercepts of the graph of $y = f(x)$ are -8 and 1.

   (a) What are the x-intercepts of the graph of $y = f(x + 4)$?

   (b) What are the x-intercepts of the graph of $y = f(x - 3)$?

   (c) What are the x-intercepts of the graph of $y = 2f(x)$?

   (d) What are the x-intercepts of the graph of $y = f(-x)$?

77. Suppose that the function $y = f(x)$ is increasing on the interval $[-1, 5]$.

   (a) Over what interval is the graph of $y = f(x + 2)$ increasing?

   (b) Over what interval is the graph of $y = f(x - 5)$ increasing?

   (c) Is the graph of $y = -f(x)$ increasing, decreasing, or neither on the interval $[-1, 5]$?

   (d) Is the graph of $y = f(-x)$ increasing, decreasing, or neither on the interval $[-5, 1]$?

78. Suppose that the function $y = f(x)$ is decreasing on the interval $[-2, 7]$.

   (a) Over what interval is the graph of $y = f(x + 2)$ decreasing?

   (b) Over what interval is the graph of $y = f(x - 5)$ decreasing?

   (c) Is the graph of $y = -f(x)$ increasing, decreasing, or neither on the interval $[-2, 7]$?

   (d) Is the graph of $y = f(-x)$ increasing, decreasing, or neither on the interval $[-7, 2]$?

79. The graph of a function f is illustrated in the figure.

   (a) Graph $y = |f(x)|$.

   (b) Graph $y = f(|x|)$.

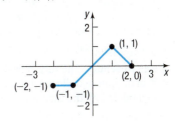

80. The graph of a function f is illustrated in the figure.

(a) Graph $y = |f(x)|$.

(b) Graph $y = f(|x|)$.

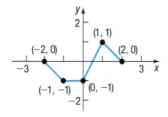

81. Suppose $(1, 3)$ is a point on the graph of $y = f(x)$.

(a) What point is on the graph of $y = f(x + 3) - 5$?

(b) What point is on the graph of $y = -2f(x - 2) + 1$?

(c) What point is on the graph of $y = f(2x + 3)$?

82. Suppose $(-3, 5)$ is a point on the graph of $y = g(x)$.

(a) What point is on the graph of $y = g(x + 1) - 3$?

(b) What point is on the graph of $y = -3g(x - 4) + 3$?

(c) What point is on the graph of $y = g(3x + 9)$?

83. (a) Graph $f(x) = |x - 3| - 3$ using transformations.

(b) Find the area of the region that is bounded by f and the x-axis and lies below the x-axis.

84. (a) Graph $f(x) = -2|x - 4| + 4$ using transformations.

(b) Find the area of the region that is bounded by f and the x-axis and lies above the x-axis.

85. Thermostat Control Energy conservation experts estimate that homeowners can save 5% to 10% on winter heating bills by programming their thermostats 5 to 10 degrees lower while sleeping. In the graph below, the temperature T (in degrees Fahrenheit) of a home is given as a function of time t (in hours after midnight) over a 24-hour period.

(a) At what temperature is the thermostat set during daytime hours? At what temperature is the thermostat set overnight?

(b) The homeowner reprograms the thermostat to $y = T(t) - 2$. Explain how this affects the temperature in the house. Graph this new function.

(c) The homeowner reprograms the thermostat to $y = T(t + 1)$. Explain how this affects the temperature in the house. Graph this new function.

Source: Roger Albright, 547 Ways to Be Fuel Smart, *2000*

86. Digital Music Revenues The total worldwide digital music revenues R, in billions of dollars, for the years 2012 through 2017 can be modeled by the function

$$R(x) = 0.15x^2 - 0.03x + 5.46$$

where x is the number of years after 2012.

(a) Find $R(0)$, $R(3)$, and $R(5)$ and explain what each value represents.

(b) Find $r(x) = R(x - 2)$.

(c) Find $r(2)$, $r(5)$ and $r(7)$ and explain what each value represents.

(d) In the model $r = r(x)$, what does x represent?

(e) Would there be an advantage in using the model r when estimating the projected revenues for a given year instead of the model R?

Source: IFPI Digital Music Report 2017

87. Temperature Measurements The relationship between the Celsius (°C) and Fahrenheit (°F) scales for measuring temperature is given by the equation

$$F = \frac{9}{5}C + 32$$

The relationship between the Celsius (°C) and Kelvin (K) scales is $K = C + 273$. Graph the equation $F = \frac{9}{5}C + 32$ using degrees Fahrenheit on the y-axis and degrees Celsius on the x-axis. Use the techniques introduced in this section to obtain the graph showing the relationship between Kelvin and Fahrenheit temperatures.

88. Period of a Pendulum The period T (in seconds) of a simple pendulum is a function of its length l (in feet) defined by the equation

$$T = 2\pi\sqrt{\frac{l}{g}}$$

where $g \approx 32.2$ feet per second per second is the acceleration due to gravity.

(a) Use a graphing utility to graph the function $T = T(l)$.

(b) Now graph the functions $T = T(l + 1)$, $T = T(l + 2)$, and $T = T(l + 3)$.

(c) Discuss how adding to the length l changes the period T.

(d) Now graph the functions $T = T(2l)$, $T = T(3l)$, and $T = T(4l)$.

(e) Discuss how multiplying the length l by factors of 2, 3, and 4 changes the period T.

89. The equation $y = (x - c)^2$ defines a *family of parabolas*, one parabola for each value of c. On one set of coordinate axes, graph the members of the family for $c = 0$, $c = 3$, and $c = -2$.

90. Repeat Problem 89 for the family of parabolas $y = x^2 + c$.

91. *Challenge Problem* If a function f is increasing on the intervals $[-3, 3]$ and $[11, 19]$ and decreasing on the interval $[3, 11]$, determine the interval(s) on which $g(x) = -3f(2x - 5)$ is increasing.

92. *Challenge Problem* In statistics, the standard normal density function is given by $f(x) = \frac{1}{\sqrt{2\pi}} \cdot \exp\left[-\frac{x^2}{2}\right]$. This function can be transformed to describe any general normal distribution with mean, μ, and standard deviation, σ. A general normal density function is given by $f(x) = \frac{1}{\sqrt{2\pi} \cdot \sigma} \cdot \exp\left[-\frac{(x - \mu)^2}{2\sigma^2}\right]$. Describe the transformations needed to get from the graph of the standard normal function to the graph of a general normal function.

Explaining Concepts: Discussion and Writing

93. Suppose that the graph of a function f is known. Explain how the graph of $y = 4f(x)$ differs from the graph of $y = f(4x)$.

94. Suppose that the graph of a function f is known. Explain how the graph of $y = f(x) - 2$ differs from the graph of $y = f(x - 2)$.

95. The area under the curve $y = \sqrt{x}$ bounded from below by the x-axis and on the right by $x = 4$ is $\dfrac{16}{3}$ square units. Using

the ideas presented in this section, what do you think is the area under the curve of $y = \sqrt{-x}$ bounded from below by the x-axis and on the left by $x = -4$? Justify your answer.

96. Explain how the range of the function $f(x) = x^2$ compares to the range of $g(x) = f(x) + k$.

97. Explain how the domain of $g(x) = \sqrt{x}$ compares to the domain of $g(x - k)$, where $k \geq 0$.

1.7 One-to-One Functions; Inverse Functions

PREPARING FOR THIS SECTION *Before getting started, review the following:*

- Functions and Their Graphs (Section 1.3, pp. 24–37)
- Increasing/Decreasing Functions (Section 1.4, pp. 46–47)

✏️ **Now Work** the 'Are You Prepared?' problems on page 89.

OBJECTIVES **1** Determine Whether a Function Is One-to-One (p. 81)
     2 Obtain the Graph of the Inverse Function from the Graph of a One-to-One Function (p. 84)
     3 Verify an Inverse Function (p. 85)
     4 Find the Inverse of a Function Defined by an Equation (p. 86)

1 Determine Whether a Function Is One-to-One

Section 1.3 presented five ways to represent a function: (1) verbally, (2) with a mapping, (3) as a set of ordered pairs, (4) with a graph, and (5) with an equation. For example, Figures 90 and 91 illustrate two different functions represented as mappings. The function in Figure 90 shows the correspondence between states and their populations (in millions). The function in Figure 91 shows a correspondence between animals and life expectancies (in years).

Figure 90

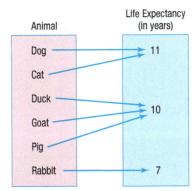

Figure 91

Suppose several people are asked to name a state that has a population of 0.9 million based on the function in Figure 90. Everyone will respond "South Dakota." Now, if the same people are asked to name an animal whose life expectancy is 11 years based on the function in Figure 91, some may respond "dog," while others may respond "cat." What is the difference between the functions in Figures 90 and 91? In Figure 90, no two elements in the domain correspond to the same element in the range. In Figure 91, this is not the case: Different elements in the domain correspond to the same element in the range. Functions such as the one in Figure 90 are given a special name.

> **DEFINITION** One-to-One
>
> A function is **one-to-one** if any two different inputs in the domain correspond to two different outputs in the range. That is, if x_1 and x_2 are two different inputs of a function f, then f is one-to-one if $f(x_1) \neq f(x_2)$.

In Words

A function is not one-to-one if two different inputs correspond to the same output.

Put another way, a function f is one-to-one if no y in the range is the image of more than one x in the domain. A function is not one-to-one if any two (or more) different elements in the domain correspond to the same element in the range. So the function in Figure 91 is not one-to-one because two different elements in the domain, *dog* and *cat*, both correspond to 11 (and also because three different elements in the domain correspond to 10).

Figure 92 illustrates the distinction among one-to-one functions, functions that are not one-to-one, and relations that are not functions.

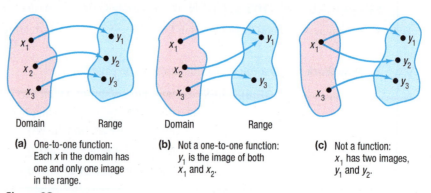

(a) One-to-one function: Each x in the domain has one and only one image in the range.

(b) Not a one-to-one function: y_1 is the image of both x_1 and x_2.

(c) Not a function: x_1 has two images, y_1 and y_2.

Figure 92

EXAMPLE 1

Determining Whether a Function Is One-to-One

Determine whether the following functions are one-to-one.

(a) For the following function, the domain represents the ages of five males, and the range represents their HDL (good) cholesterol scores (mg/dL).

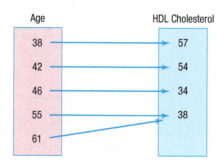

(b) $\{(-2, 6), (-1, 3), (0, 2), (1, 5), (2, 8)\}$

Solution

(a) The function is not one-to-one because there are two different inputs, 55 and 61, that correspond to the same output, 38.

(b) The function is one-to-one because every distinct input corresponds to a different output.

✏ **Now Work** PROBLEMS **13** AND **17**

For functions defined by an equation $y = f(x)$ and for which the graph of f is known, there is a simple test, called the **horizontal-line test**, to determine whether f is one-to-one.

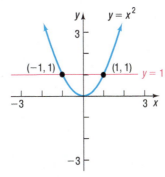

Figure 93
$f(x_1) = f(x_2) = h$ and $x_1 \neq x_2$;
f is not a one-to-one function.

> **THEOREM** Horizontal-line Test
>
> If every horizontal line intersects the graph of a function f in at most one point, then f is one-to-one.

The reason why this test works can be seen in Figure 93, where the horizontal line $y = h$ intersects the graph at two distinct points, (x_1, h) and (x_2, h). Since h is the image of both x_1 and x_2 and $x_1 \neq x_2$, f is not one-to-one. Based on Figure 93, we can state the horizontal-line test in another way: If the graph of any horizontal line intersects the graph of a function f at more than one point, then f is not one-to-one.

EXAMPLE 2   **Using the Horizontal-line Test**

For each function, use its graph to determine whether the function is one-to-one.

(a) $f(x) = x^2$ (b) $g(x) = x^3$

Solution   (a) Figure 94(a) illustrates the horizontal-line test for $f(x) = x^2$. The horizontal line $y = 1$ intersects the graph of f twice, at $(1, 1)$ and at $(-1, 1)$, so f is not one-to-one.

(b) Figure 94(b) illustrates the horizontal-line test for $g(x) = x^3$. Because every horizontal line intersects the graph of g exactly once, the function g is one-to-one.

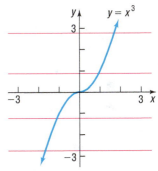

(a) A horizontal line intersects the graph twice; f is not one-to-one.

(b) Every horizontal line intersects the graph exactly once; g is one-to-one.

Figure 94

Now Work PROBLEM 21

Look more closely at the one-to-one function $g(x) = x^3$. This function is an increasing function. Because an increasing (or decreasing) function always has different y-values for unequal x-values, it follows that a function that is increasing (or decreasing) over its domain is also a one-to-one function.

> **THEOREM**
>
> • A function that is increasing on an interval I is a one-to-one function on I.
> • A function that is decreasing on an interval I is a one-to-one function on I.

One-to-one functions $y = f(x)$ have an important property. Corresponding to each x in the domain of f, there is exactly one y in the range because f is a function. And corresponding to each y in the range of f, there is exactly one x in the domain because f is one-to-one. The function represented by the correspondence from the range back to the domain is called the *inverse function of f*.

In Words

Suppose that f is a one-to-one function so that the input 5 corresponds to the output 10. For the inverse function f^{-1}, the input 10 will correspond to the output 5.

┌───

DEFINITION Inverse Function

Suppose $y = f(x)$ is a one-to-one function. The correspondence from the range of f to the domain of f is called the **inverse function of f.** The symbol f^{-1} is used to denote the inverse function of f. In other words, if $y = f(x)$ is a one-to-one function, then f has an inverse function f^{-1} and $x = f^{-1}(y)$.

───┘

2 Obtain the Graph of the Inverse Function from the Graph of a One-to-One Function

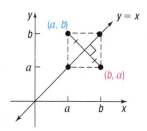

Figure 95

Suppose (a, b) is a point on the graph of a one-to-one function f defined by $y = f(x)$. Then $b = f(a)$. This means that $a = f^{-1}(b)$, so (b, a) is a point on the graph of the inverse function f^{-1}. The relationship between the point (a, b) on f and the point (b, a) on f^{-1} is shown in Figure 95. The line segment with endpoints (a, b) and (b, a) is perpendicular to the line $y = x$ and is bisected by the line $y = x$. (Do you see why?) It follows that the point (b, a) on f^{-1} is the reflection about the line $y = x$ of the point (a, b) on f.

┌───

THEOREM

The graph of a one-to-one function f and the graph of its inverse function f^{-1} are symmetric with respect to the line $y = x$.

───┘

Figure 96 illustrates the theorem. Once the graph of f is known, the graph of f^{-1} can be obtained by reflecting the graph of f about the line $y = x$.

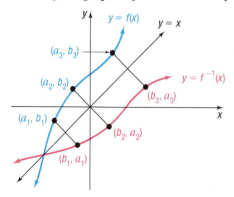

Figure 96

EXAMPLE 3

Graphing an Inverse Function

The graph in Figure 97(a) shows a one-to-one function $y = f(x)$. Draw the graph of its inverse.

Solution

First add the graph of $y = x$ to Figure 97(a). Since the points $(-2, -1)$, $(-1, 0)$, and $(2, 1)$ are on the graph of f, the points $(-1, -2)$, $(0, -1)$, and $(1, 2)$ must be on the graph of f^{-1}. Keeping in mind that the graph of f^{-1} is the reflection about the line $y = x$ of the graph of f, graph f^{-1}. See Figure 97(b).

Figure 97   (a)   (b)

Now Work PROBLEM 27

3 Verify an Inverse Function

Domain of f   **Range of f**

f

x

f⁻¹

Range of f⁻¹   **Domain of f⁻¹**

Figure 98

WARNING Be careful! f^{-1} is a symbol for the inverse function of f. The -1 used in f^{-1} is not an exponent. That is, f^{-1} does not mean the reciprocal of f; $f^{-1}(x)$ is not equal to $\dfrac{1}{f(x)}$. ∎

Suppose f is a one-to-one function. Then f has an inverse function f^{-1}. Figure 98 shows the relationship between the domain and range of f and the domain and range of f^{-1}. As Figure 98 illustrates,

> • Domain of f = Range of f^{-1} • Range of f = Domain of f^{-1}

Look again at Figure 98 to visualize the relationship. Starting with x, applying f, and then applying f^{-1} gets x back again. Starting with x, applying f^{-1}, and then applying f gets the number x back again. To put it simply, what f does, f^{-1} undoes, and vice versa. See the illustration that follows.

$$\boxed{\textbf{Input } x \textbf{ from domain of } f} \xrightarrow{\textbf{Apply } f} \boxed{f(x)} \xrightarrow{\textbf{Apply } f^{-1}} \boxed{f^{-1}(f(x)) = x}$$

$$\boxed{\text{Input } x \text{ from domain of } f^{-1}} \xrightarrow{\textbf{Apply } f^{-1}} \boxed{f^{-1}(x)} \xrightarrow{\textbf{Apply } f} \boxed{f(f^{-1}(x)) = x}$$

In other words,

> • $f^{-1}(f(x)) = x$ where x is in the domain of f
>
> • $f(f^{-1}(x)) = x$ where x is in the domain of f^{-1}

Consider the function $f(x) = 2x$, which multiplies the argument x by 2. The inverse function f^{-1} undoes whatever f does. So the inverse function of f is $f^{-1}(x) = \dfrac{1}{2}x$, which divides the argument by 2. For example, $f(3) = 2 \cdot 3 = 6$ and $f^{-1}(6) = \dfrac{1}{2} \cdot 6 = 3$, so f^{-1} undoes f. This is verified by showing that $f^{-1}(f(x)) = f^{-1}(2x) = \dfrac{1}{2} \cdot 2x = x$ and $f(f^{-1}(x)) = f\left(\dfrac{1}{2}x\right) = 2 \cdot \dfrac{1}{2}x = x$.

See Figure 99.

$f^{-1}(2x) = \frac{1}{2} \cdot 2x = x$

Figure 99

EXAMPLE 4	**Verifying Inverse Functions**

(a) Verify that the inverse of $g(x) = x^3$ is $g^{-1}(x) = \sqrt[3]{x}$.

(b) Verify that the inverse of $f(x) = 2x + 3$ is $f^{-1}(x) = \dfrac{1}{2}(x - 3)$.

Solution   (a) $g^{-1}(g(x)) = g^{-1}(x^3) = \sqrt[3]{x^3} = x$ for all x in the domain of g

$g(g^{-1}(x)) = g(\sqrt[3]{x}) = (\sqrt[3]{x})^3 = x$ for all x in the domain of g^{-1}

(b) $f^{-1}(f(x)) = f^{-1}(2x + 3) = \dfrac{1}{2}[(2x + 3) - 3] = \dfrac{1}{2} \cdot 2x = x$ for all x in the domain of f

$f(f^{-1}(x)) = f\left(\dfrac{1}{2}(x - 3)\right) = 2 \cdot \dfrac{1}{2}(x - 3) + 3 = x$ for all x in the domain of f^{-1}

EXAMPLE 5

Verifying Inverse Functions

Verify that the inverse of $f(x) = \dfrac{1}{x - 1}$ is $f^{-1}(x) = \dfrac{1}{x} + 1$. For what values of x is $f^{-1}(f(x)) = x$? For what values of x is $f(f^{-1}(x)) = x$?

Solution

The domain of f is $\{x | x \neq 1\}$ and the domain of f^{-1} is $\{x | x \neq 0\}$. Now

$$f^{-1}(f(x)) = f^{-1}\left(\frac{1}{x - 1}\right) = \frac{1}{\frac{1}{x - 1}} + 1 = x - 1 + 1 = x \quad \text{provided } x \neq 1$$

$$f(f^{-1}(x)) = f\left(\frac{1}{x} + 1\right) = \frac{1}{\left(\frac{1}{x} + 1\right) - 1} = \frac{1}{\frac{1}{x}} = x \quad \text{provided } x \neq 0$$

➤ **Now Work** PROBLEMS 33 AND 37

4 Find the Inverse of a Function Defined by an Equation

The fact that the graphs of a one-to-one function f and its inverse function f^{-1} are symmetric with respect to the line $y = x$ tells us more. It says that we can obtain f^{-1} by interchanging the roles of x and y in f. If f is defined by the equation

$$y = f(x)$$

then f^{-1} is defined by the equation

$$x = f(y)$$

Need to Review?
Implicit functions are
discussed in Section 1.3,
p. 31.

The equation $x = f(y)$ defines f^{-1} *implicitly*. If we can solve this equation for y, we will have the *explicit* form of f^{-1}, that is,

$$y = f^{-1}(x)$$

Let's use this procedure to find the inverse of $f(x) = 2x + 3$. Because f is a linear function and is increasing, f is one-to-one and so has an inverse function.

EXAMPLE 6

Finding the Inverse of a Function Defined by an Equation

Find the inverse of $f(x) = 2x + 3$. Graph f and f^{-1} on the same coordinate axes.

Step-by-Step Solution

Step 1 Replace $f(x)$ with y. Then interchange the variables x and y. This equation defines the inverse function f^{-1} implicitly.

Replace $f(x)$ with y in $f(x) = 2x + 3$ and obtain $y = 2x + 3$. Now interchange the variables x and y to obtain

$$x = 2y + 3$$

This equation defines the inverse function f^{-1} implicitly.

Step 2 If possible, solve the implicit equation for y in terms of x to obtain the explicit form of f^{-1}, $y = f^{-1}(x)$.

To find the explicit form of the inverse, solve $x = 2y + 3$ for y.

$$x = 2y + 3$$
$$2y + 3 = x$$
$$2y = x - 3 \qquad \text{\textbf{\textcolor{blue}{Subtract 3 from both sides.}}}$$
$$y = \frac{1}{2}(x - 3) \qquad \text{\textbf{\textcolor{blue}{Multiply both sides by $\frac{1}{2}$.}}}$$

The explicit form of the inverse function f^{-1} is

$$f^{-1}(x) = \frac{1}{2}(x - 3)$$

Step 3 Check the result by showing that $f^{-1}(f(x)) = x$ and $f(f^{-1}(x)) = x$.

We verified that f and f^{-1} are inverses in Example 4(b).

The graphs of $f(x) = 2x + 3$ and its inverse $f^{-1}(x) = \dfrac{1}{2}(x - 3)$ are shown in Figure 100. Note the symmetry of the graphs with respect to the line $y = x$.

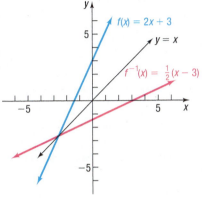

Figure 100

Steps for Finding the Inverse of a One-to-One Function

STEP 1: In $y = f(x)$, interchange the variables x and y to obtain

$$x = f(y)$$

This equation defines the inverse function f^{-1} implicitly.

STEP 2: If possible, solve the implicit equation for y in terms of x to obtain the explicit form of f^{-1}:

$$y = f^{-1}(x)$$

STEP 3: Check the result by showing that

$$f^{-1}(f(x)) = x \quad \text{and} \quad f(f^{-1}(x)) = x$$

EXAMPLE 7 **Finding the Inverse of a Function Defined by an Equation**

The function

$$f(x) = \frac{2x + 1}{x - 1} \qquad x \neq 1$$

is one-to-one. Find its inverse function and check the result.

Solution **STEP 1:** Replace $f(x)$ with y and interchange the variables x and y in

$$y = \frac{2x + 1}{x - 1}$$

to obtain

$$x = \frac{2y + 1}{y - 1}$$

STEP 2: Solve for y.

$$x = \frac{2y + 1}{y - 1}$$

$x(y - 1) = 2y + 1$ **Multiply both sides by y − 1.**

$xy - x = 2y + 1$ **Use the Distributive Property.**

$xy - 2y = x + 1$ **Subtract 2y from both sides; add x to both sides.**

$(x - 2)y = x + 1$ **Factor.**

$y = \dfrac{x + 1}{x - 2}$ **Divide by x − 2.**

The inverse function is

$$f^{-1}(x) = \frac{x + 1}{x - 2} \qquad x \neq 2 \quad \textbf{Replace y by } f^{-1}(x).$$

STEP 3: ✔ Check:

$$f^{-1}(f(x)) = f^{-1}\left(\frac{2x+1}{x-1}\right) = \frac{\dfrac{2x+1}{x-1}+1}{\dfrac{2x+1}{x-1}-2} = \frac{2x+1+x-1}{2x+1-2(x-1)} = \frac{3x}{3} = x, \; x \neq 1$$

$$f(f^{-1}(x)) = f\left(\frac{x+1}{x-2}\right) = \frac{2 \cdot \dfrac{x+1}{x-2}+1}{\dfrac{x+1}{x-2}-1} = \frac{2(x+1)+x-2}{x+1-(x-2)} = \frac{3x}{3} = x, \; x \neq 2$$

Exploration

In Example 7, we found that if $f(x) = \dfrac{2x+1}{x-1}$, then $f^{-1}(x) = \dfrac{x+1}{x-2}$. Compare the vertical and horizontal asymptotes of f and f^{-1}.

Result The vertical asymptote of f is $x = 1$, and the horizontal asymptote is $y = 2$. The vertical asymptote of f^{-1} is $x = 2$, and the horizontal asymptote is $y = 1$.

⟶ **Now Work** PROBLEMS **45** AND **59**

If a function is not one-to-one, it has no inverse function. Sometimes, though, an appropriate restriction on the domain of such a function yields a new function that is one-to-one. Then the function defined on the restricted domain has an inverse function.

EXAMPLE 8 ### Finding the Inverse of a Domain-restricted Function

Find the inverse of $y = f(x) = x^2$ if $x \geq 0$. Graph f and f^{-1}.

Solution The function $y = x^2$ is not one-to-one. [Refer to Example 2(a).] However, restricting the domain of this function to $x \geq 0$, as indicated, results in a new function that is increasing and therefore is one-to-one. Consequently, the function defined by $y = f(x) = x^2, x \geq 0$, has an inverse function, f^{-1}.
Follow the steps to find f^{-1}.

STEP 1: In the equation $y = x^2, x \geq 0$, interchange the variables x and y. The result is

$$x = y^2 \qquad y \geq 0$$

This equation defines the inverse function implicitly.

Figure 101

STEP 2: Solve for y to get the explicit form of the inverse. Because $y \geq 0$, only one solution for y is obtained: $y = \sqrt{x}$. So $f^{-1}(x) = \sqrt{x}$.

STEP 3: ✔ Check: $f^{-1}(f(x)) = f^{-1}(x^2) = \sqrt{x^2} = |x| = x$ because $x \geq 0$

$$f(f^{-1}(x)) = f(\sqrt{x}) = (\sqrt{x})^2 = x$$

Figure 101 illustrates the graphs of $f(x) = x^2, x \geq 0$, and $f^{-1}(x) = \sqrt{x}$.

SUMMARY

- If a function f is one-to-one, then it has an inverse function f^{-1}.
- Domain of f = Range of f^{-1}; Range of f = Domain of f^{-1}.
- To verify that f^{-1} is the inverse of f, show that $f^{-1}(f(x)) = x$ for every x in the domain of f and that $f(f^{-1}(x)) = x$ for every x in the domain of f^{-1}.
- The graphs of f and f^{-1} are symmetric with respect to the line $y = x$.

1.7 Assess Your Understanding

'Are You Prepared?' *Answers are given at the end of these exercises. If you get a wrong answer, read the pages listed in red.*

1. Suppose $f(x) = 4x - 2$. Find $f\left(\dfrac{x+2}{4}\right)$. (pp. 29–30)

2. Where is the function $f(x) = x^2$ increasing? Where is it decreasing? (pp. 46–47)

3. What is the domain of $f(x) = \dfrac{x+5}{x^2 + 3x - 18}$? (pp. 32–34)

4. Find the function $y = f(x)$ that is defined implicitly by the equation $x - 3y = -8$. (p. 31)

Concepts and Vocabulary

5. If x_1 and x_2 are any two different inputs of a function f, then f is one-to-one if _____.

6. If every horizontal line intersects the graph of a function f at no more than one point, then f is a(n) _____ function.

7. If f is a one-to-one function and $f(3) = 8$, then $f^{-1}(8) = $ ____.

8. If f^{-1} is the inverse of a function f, then the graphs of f and f^{-1} are symmetric with respect to the line _____.

9. If the domain of a one-to-one function f is $[4, \infty)$, then the range of its inverse function f^{-1} is _____.

10. *True or False* If f and g are inverse functions, then the domain of f is the same as the range of g.

11. *Multiple Choice* If $(-2, 3)$ is a point on the graph of a one-to-one function f, which of the following points is on the graph of f^{-1}?

 (a) $(3, -2)$ **(b)** $(2, -3)$ **(c)** $(-3, 2)$ **(d)** $(-2, -3)$

12. *Multiple Choice* Suppose f is a one-to-one function with a domain of $\{x \mid x \neq 3\}$ and a range of $\left\{y \,\middle|\, y \neq \dfrac{2}{3}\right\}$. Which of the following is the domain of f^{-1}?

 (a) $\{x \mid x \neq 3\}$ **(b)** All real numbers

 (c) $\left\{x \,\middle|\, x \neq \dfrac{2}{3}, x \neq 3\right\}$ **(d)** $\left\{x \,\middle|\, x \neq \dfrac{2}{3}\right\}$

Skill Building

In Problems 13–20, determine whether the function is one-to-one.

13.

14.

15.

16.

17. $\{(2, 6), (-3, 6), (4, 9), (1, 10)\}$

18. $\{(-2, 5), (-1, 3), (3, 7), (4, 12)\}$

19. $\{(0, 0), (1, 1), (2, 16), (3, 81)\}$

20. $\{(1, 2), (2, 8), (3, 18), (4, 32)\}$

In Problems 21–26, the graph of a function f is given. Use the horizontal-line test to determine whether f is one-to-one.

21.

22.

23.

24.

25.

26.

In Problems 27–32, the graph of a one-to-one function f is given. Draw the graph of the inverse function f^{-1}.

27.

28.

29.

30.

31.

32.

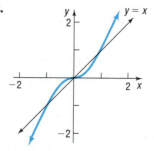

In Problems 33–42, verify that the functions f and g are inverses of each other by showing that $f(g(x)) = x$ and $g(f(x)) = x$. Give any values of x that need to be excluded from the domain of f and the domain of g.

33. $f(x) = 3x + 4$; $g(x) = \dfrac{1}{3}(x - 4)$

34. $f(x) = 3 - 2x$; $g(x) = -\dfrac{1}{2}(x - 3)$

35. $f(x) = 4x - 8$; $g(x) = \dfrac{x}{4} + 2$

36. $f(x) = 2x + 6$; $g(x) = \dfrac{1}{2}x - 3$

37. $f(x) = x^3 - 8$; $g(x) = \sqrt[3]{x + 8}$

38. $f(x) = (x - 2)^2, x \geq 2$; $g(x) = \sqrt{x} + 2$

39. $f(x) = \dfrac{1}{x}$; $g(x) = \dfrac{1}{x}$

40. $f(x) = x$; $g(x) = x$

41. $f(x) = \dfrac{2x + 3}{x + 4}$; $g(x) = \dfrac{4x - 3}{2 - x}$

42. $f(x) = \dfrac{x - 5}{2x + 3}$; $g(x) = \dfrac{3x + 5}{1 - 2x}$

In Problems 43–54, the function f is one-to-one. (a) Find its inverse function f^{-1} and check your answer. (b) Find the domain and the range of f and f^{-1}. (c) Graph f, f^{-1}, and $y = x$ on the same coordinate axes.

43. $f(x) = 3x$

44. $f(x) = -4x$

45. $f(x) = 4x + 2$

46. $f(x) = 1 - 3x$

47. $f(x) = x^3 - 1$

48. $f(x) = x^3 + 1$

49. $f(x) = x^2 + 4, \ x \geq 0$

50. $f(x) = x^2 + 9, \ x \geq 0$

51. $f(x) = \dfrac{4}{x}$

52. $f(x) = -\dfrac{3}{x}$

53. $f(x) = \dfrac{1}{x - 2}$

54. $f(x) = \dfrac{4}{x + 2}$

In Problems 55–72, the function f is one-to-one. (a) Find its inverse function f^{-1} and check your answer. (b) Find the domain and the range of f and f^{-1}.

55. $f(x) = \dfrac{2}{3+x}$

56. $f(x) = \dfrac{4}{2-x}$

57. $f(x) = \dfrac{3x}{x+2}$

58. $f(x) = -\dfrac{2x}{x-1}$

59. $f(x) = \dfrac{2x}{3x-1}$

60. $f(x) = -\dfrac{3x+1}{x}$

61. $f(x) = \dfrac{3x+4}{2x-3}$

62. $f(x) = \dfrac{2x-3}{x+4}$

63. $f(x) = \dfrac{2x+3}{x+2}$

64. $f(x) = \dfrac{-3x-4}{x-2}$

65. $f(x) = \dfrac{x^2-4}{2x^2}, \quad x > 0$

66. $f(x) = \dfrac{x^2+3}{3x^2}, \quad x > 0$

67. $f(x) = x^{\frac{2}{3}} - 4, \quad x \geq 0$

68. $f(x) = x^{\frac{3}{2}} + 5$

69. $f(x) = \sqrt[3]{x^5 - 2}$

70. $f(x) = \sqrt[5]{x^3 + 13}$

71. $f(x) = \dfrac{1}{9}(x-1)^2 + 2, \quad x \geq 1$

72. $f(x) = 2\sqrt{x+3} - 5$

Applications and Extensions

73. Use the graph of $y = f(x)$ given in Problem 27 to evaluate the following:

    (a) $f(-1)$ **(b)** $f(1)$ **(c)** $f^{-1}(1)$ **(d)** $f^{-1}(2)$

74. Use the graph of $y = f(x)$ given in Problem 28 to evaluate the following:

    (a) $f(2)$ **(b)** $f(1)$ **(c)** $f^{-1}(0)$ **(d)** $f^{-1}(-1)$

75. If $f(7) = 13$ and f is one-to-one, what is $f^{-1}(13)$?

76. If $g(-5) = 3$ and g is one-to-one, what is $g^{-1}(3)$?

77. The domain of a one-to-one function f is $[5, \infty)$, and its range is $[-2, \infty)$. State the domain and the range of f^{-1}.

78. The domain of a one-to-one function f is $[0, \infty)$, and its range is $[5, \infty)$. State the domain and the range of f^{-1}.

79. The domain of a one-to-one function g is $(-\infty, 0]$, and its range is $[0, \infty)$. State the domain and the range of g^{-1}.

80. The domain of a one-to-one function g is $[0, 15]$, and its range is $(0, 8)$. State the domain and the range of g^{-1}.

81. A function $y = f(x)$ is increasing on the interval $[0, 5]$. What conclusions can you draw about the graph of $y = f^{-1}(x)$?

82. A function $y = f(x)$ is decreasing on the interval $[0, 5]$. What conclusions can you draw about the graph of $y = f^{-1}(x)$?

83. Find the inverse of the linear function

$$f(x) = mx + b, \quad m \neq 0$$

84. Find the inverse of the function

$$f(x) = \sqrt{r^2 - x^2}, \quad 0 \leq x \leq r$$

85. A function f has an inverse function f^{-1}. If the graph of f lies in quadrant I, in which quadrant does the graph of f^{-1} lie?

86. A function f has an inverse function f^{-1}. If the graph of f lies in quadrant II, in which quadrant does the graph of f^{-1} lie?

87. The function $f(x) = |x|$ is not one-to-one. Find a suitable restriction on the domain of f so that the new function that results is one-to-one. Then find the inverse of the new function.

88. The function $f(x) = x^4$ is not one-to-one. Find a suitable restriction on the domain of f so that the new function that results is one-to-one. Then find the inverse of the new function.

In applications, the symbols used for the independent and dependent variables are often based on common usage. So, rather than using $y = f(x)$ to represent a function, an applied problem might use $C = C(q)$ to represent the cost C of manufacturing q units of a good. Because of this, the inverse notation f^{-1} used in a pure mathematics problem is not used when finding inverses of applied problems. Rather, the inverse of a function such as $C = C(q)$ will be $q = q(C)$. So $C = C(q)$ is a function that represents the cost C as a function of the number q of units manufactured, and $q = q(C)$ is a function that represents the number q as a function of the cost C. Problems 89–96 illustrate this idea.

89. Vehicle Stopping Distance Taking into account reaction time, the distance d (in feet) that a car requires to come to a complete stop while traveling r miles per hour is given by the function

$$d(r) = 6.97r - 90.39$$

    (a) Express the speed r at which the car is traveling as a function of the distance d required to come to a complete stop.

    (b) Verify that $r = r(d)$ is the inverse of $d = d(r)$ by showing that $r(d(r)) = r$ and $d(r(d)) = d$.

    (c) Predict the speed that a car was traveling if the distance required to stop was 300 feet.

90. Height and Head Circumference The head circumference C of a child is related to the height H of the child (both in inches) through the function

$$H(C) = 2.15C - 10.53$$

    (a) Express the head circumference C as a function of height H.

    (b) Verify that $C = C(H)$ is the inverse of $H = H(C)$ by showing that $H(C(H)) = H$ and $C(H(C)) = C$.

    (c) Predict the head circumference of a child who is 26 inches tall.

91. Ideal Body Weight One model for the ideal body weight W for men (in kilograms) as a function of height h (in inches) is given by the function

$$W(h) = 50 + 2.3(h - 60)$$

    (a) What is the ideal weight of a 6-foot male?

    (b) Express the height h as a function of weight W.

    (c) Verify that $h = h(W)$ is the inverse of $W = W(h)$ by showing that $h(W(h)) = h$ and $W(h(W)) = W$.

    (d) What is the height of a male who is at his ideal weight of 80 kilograms?

92. Temperature Conversion The function $F(C) = \dfrac{9}{5}C + 32$ converts a temperature from C degrees Celsius to F degrees Fahrenheit.
  (a) Express the temperature in degrees Celsius C as a function of the temperature in degrees Fahrenheit F.
  (b) Verify that $C = C(F)$ is the inverse of $F = F(C)$ by showing that $C(F(C)) = C$ and $F(C(F)) = F$.
  (c) What is the temperature in degrees Celsius if it is 70 degrees Fahrenheit?

93. Income Taxes The function
$$T(g) = 4453.05 + 0.22(g - 38,700)$$
represents the 2018 federal income tax T (in dollars) due for a "single" filer whose modified adjusted gross income is g dollars, where $38,700 < g \le 82,500$.
  (a) What is the domain of the function T?
  (b) Given that the tax due T is an increasing linear function of modified adjusted gross income g, find the range of the function T.
  (c) Find adjusted gross income g as a function of federal income tax T. What are the domain and the range of this function?

94. Income Taxes The function
$$T(g) = 1905 + 0.12(g - 19,050)$$
represents the 2018 federal income tax T (in dollars) due for a "married filing jointly" filer whose modified adjusted gross income is g dollars, where $19,050 < g \le 77,400$.
  (a) What is the domain of the function T?
  (b) Given that the tax due T is an increasing linear function of modified adjusted gross income g, find the range of the function T.
  (c) Find adjusted gross income g as a function of federal income tax T. What are the domain and the range of this function?

95. Gravity on Earth If a rock falls from a height of 100 meters above Earth, the height H (in meters) after t seconds is approximately
$$H(t) = 100 - 4.9t^2$$
  (a) In general, quadratic functions are not one-to-one. However, the function H is one-to-one. Why?
  (b) Find the inverse of H and verify your result.
  (c) How long will it take a rock to fall 80 meters?

96. Period of a Pendulum The period T (in seconds) of a simple pendulum as a function of its length l (in feet) is given by
$$T(l) = 2\pi\sqrt{\dfrac{l}{32.2}}$$
  (a) Express the length l as a function of the period T.
  (b) How long is a pendulum whose period is 3 seconds?

97. *Challenge Problem* Given
$$f(x) = \dfrac{ax + b}{cx + d}$$
find $f^{-1}(x)$. If $c \ne 0$, under what conditions on $a, b, c,$ and d is $f = f^{-1}$?

98. *Challenge Problem* If $h(x) = f(g(x))$, find h^{-1} in terms of f^{-1} and g^{-1}.

99. *Challenge Problem* For $f(x) = \begin{cases} 2x + 3, & x < 0 \\ 3x + 4, & x \ge 0 \end{cases}$
  (a) Find the domain and range of f.
  (b) Find f^{-1}.
  (c) Find the domain and range of f^{-1}.

Explaining Concepts: Discussion and Writing

100. Can a one-to-one function and its inverse be equal? What must be true about the graph of f for this to happen? Give some examples to support your conclusion.

101. Draw the graph of a one-to-one function that contains the points $(-2, -3)$, $(0, 0)$, and $(1, 5)$. Now draw the graph of its inverse. Compare your graph to those of other students. Discuss any similarities. What differences do you see?

102. Give an example of a function whose domain is the set of real numbers and that is neither increasing nor decreasing on its domain, but is one-to-one.
[**Hint:** Use a piecewise-defined function.]

103. Is every odd function one-to-one? Explain.

104. Suppose that $C(g)$ represents the cost C, in dollars, of manufacturing g cars. Explain what $C^{-1}(800,000)$ represents.

105. Explain why the horizontal-line test can be used to identify one-to-one functions from a graph.

106. Explain why a function must be one-to-one in order to have an inverse that is a function. Use the function $y = x^2$ to support your explanation.

'Are You Prepared?' Answers

1. x
2. Increasing on $[0, \infty)$; decreasing on $(-\infty, 0]$
3. $\{x \mid x \ne -6, x \ne 3\}$
4. $y = \dfrac{x + 8}{3}$ or $y = \dfrac{1}{3}x + \dfrac{8}{3}$

Chapter Review

Library of Functions

Constant function (p. 59)

$$f(x) = b$$

The graph is a horizontal line with y-intercept b.

Identity function (p. 59)

$$f(x) = x$$

The graph is a line with slope 1 and y-intercept 0.

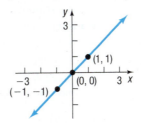

Linear function

$$f(x) = mx + b$$

The graph is a line with slope m and y-intercept b.

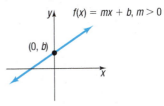

Square function (p. 59)

$$f(x) = x^2$$

The graph is a parabola with intercept at $(0, 0)$.

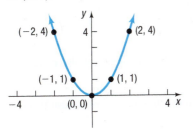

Cube function (p. 60)

$$f(x) = x^3$$

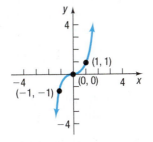

Square root function (p. 60)

$$f(x) = \sqrt{x}$$

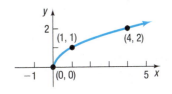

Cube root function (p. 60)

$$f(x) = \sqrt[3]{x}$$

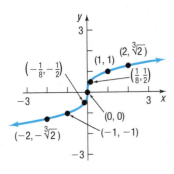

Reciprocal function (p. 60)

$$f(x) = \frac{1}{x}$$

Absolute value function (p. 60)

$$f(x) = |x|$$

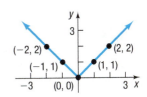

Things to Know

Formulas

Distance formula (p. 3)

$$d = \sqrt{(x_2 - x_1)^2 + (y_2 - y_1)^2}$$

Midpoint formula (p. 5)

$$(x, y) = \left(\frac{x_1 + x_2}{2}, \frac{y_1 + y_2}{2} \right)$$

Equations of Circles

Standard form of the equation of a circle (p. 17)

$(x - h)^2 + (y - k)^2 = r^2$; r is the radius of the circle, (h, k) is the center of the circle.

Equation of the unit circle (p. 17)

$$x^2 + y^2 = 1$$

General form of the equation of a circle (p. 18)

$x^2 + y^2 + ax + by + c = 0$ with restrictions on a, b, and c

Function (pp. 26–28)

- A relation between two sets so that each element x in the first set, the domain, has corresponding to it exactly one element y in the second set, the range. The range is the set of y values of the function for the x values in the domain.
- A function can also be characterized as a set of ordered pairs (x, y) in which no first element is paired with two different second elements.

Function notation (pp. 29–31)

- $y = f(x)$
- f is a symbol for the function.
- x is the argument, or independent variable.
- y is the dependent variable.
- $f(x)$ is the value of the function at x, or the image of x.
- A function f may be defined implicitly by an equation involving x and y or explicitly by writing $y = f(x)$.

Difference quotient of f (p. 31)

$$\frac{f(x + h) - f(x)}{h} \quad h \neq 0$$

Domain (pp. 32–33)

If unspecified, the domain of a function f defined by an equation is the largest set of real numbers for which $f(x)$ is a real number.

Vertical-line test (p. 34)

A set of points in the plane is the graph of a function if and only if every vertical line intersects the graph in at most one point.

Even function (p. 44)

$f(-x) = f(x)$ for every x in the domain ($-x$ must also be in the domain).

Odd function (p. 44)

$f(-x) = -f(x)$ for every x in the domain ($-x$ must also be in the domain).

Increasing function (p. 46)

A function f is increasing on an interval I if, for any choice of x_1 and x_2 in I, with $x_1 < x_2$, we have $f(x_1) < f(x_2)$.

Decreasing function (p. 46)

A function f is decreasing on an interval I if, for any choice of x_1 and x_2 in I, with $x_1 < x_2$, we have $f(x_1) > f(x_2)$.

Constant function (p. 46)

A function f is constant on an interval I if, for all choices of x in I, the values of $f(x)$ are equal.

Local maximum (p. 47)

A function f, defined on some interval I, has a local maximum at c if there is an open interval in I containing c such that, for all x in this open interval, $f(x) \leq f(c)$.

Local minimum (p. 47)

A function f, defined on some interval I, has a local minimum at c if there is an open interval in I containing c such that, for all x in this open interval, $f(x) \geq f(c)$.

Absolute maximum and

Let f denote a function defined on some interval I.

Absolute minimum (p. 48)

- If there is a number u in I for which $f(x) \leq f(u)$ for all x in I, then f has an absolute maximum at u, and the number $f(u)$ is the absolute maximum of f on I.
- If there is a number v in I for which $f(x) \geq f(v)$, for all x in I, then f has an absolute minimum at v and the number $f(v)$ is the absolute minimum of f on I.

Average rate of change of a function (p. 51)

The average rate of change of f from a to b is

$$\frac{\Delta y}{\Delta x} = \frac{f(b) - f(a)}{b - a} \quad a \neq b$$

One-to-one function f (p. 82)

- A function for which any two different inputs in the domain correspond to two different outputs in the range
- For any choice of elements x_1, x_2 in the domain of f, if $x_1 \neq x_2$, then $f(x_1) \neq f(x_2)$

Horizontal-line test (p. 83)

If every horizontal line intersects the graph of a function f in at most one point, f is one-to-one.

Inverse function f^{-1} of f (pp. 84–85)

- Domain of f = range of f^{-1}; range of f = domain of f^{-1}
- $f^{-1}(f(x)) = x$ for all x in the domain of f
- $f(f^{-1}(x)) = x$ for all x in the domain of f^{-1}
- The graphs of f and f^{-1} are symmetric with respect to the line $y = x$.

Objectives

Section	You should be able to ...	Examples	Review Exercises
1.1	1 Use the distance formula (p. 3)	1–3	1(a)–3(a), 16, 17
	2 Use the midpoint formula (p. 5)	4	1(b)–3(b), 17
1.2	1 Graph equations by plotting points (p. 9)	1–3	4
	2 Find intercepts from a graph (p. 11)	4	5
	3 Find intercepts from an equation (p. 11)	5	6–10
	4 Test an equation for symmetry with respect to the x-axis, the y-axis, and the origin (p. 12)	6–9	6–10
	5 Know how to graph key equations (p. 14)	10–12	44, 45
	6 Write the standard form of the equation of a circle (p. 16)	13	11, 12, 17
	7 Graph a circle (p. 17)	14, 15	13–15
	8 Work with the general form of the equation of a circle (p. 18)	16	14, 15
1.3	1 Describe a relation (p. 24)	1	18
	2 Determine whether a relation represents a function (p. 26)	2–5	19, 20
	3 Use function notation; find the value of a function (p. 29)	6, 7	21–23, 54
	4 Find the difference quotient of a function (p. 31)	8	30
	5 Find the domain of a function defined by an equation (p. 32)	9, 10	24–29
	6 Identify the graph of a function (p. 34)	11	42, 43
	7 Obtain information from or about the graph of a function (p. 35)	12, 13	31(a)–(e), 32(a), 32(e), 32(g)
1.4	1 Determine even and odd functions from a graph (p. 44)	1	32(f)
	2 Identify even and odd functions from the equation (p. 45)	2	33–36
	3 Use a graph to determine where a function is increasing, decreasing, or constant (p. 46)	3	32(b)
	4 Use a graph to locate local maxima and local minima (p. 47)	4	32(c)
	5 Use a graph to locate the absolute maximum and the absolute minimum (p. 48)	5	32(d)
	6 Use a graphing utility to approximate local maxima and local minima and to determine where a function is increasing or decreasing (p. 50)	6	37, 38
	7 Find the average rate of change of a function (p. 50)	7, 8	39–41

Section	You should be able to . . .	Examples	Review Exercises
1.5	**1** Graph the functions listed in the library of functions (p. 57)	1, 2	44, 45
	2 Graph piecewise-defined functions (p. 61)	3, 4	52, 53
1.6	**1** Graph functions using vertical and horizontal shifts (p. 67)	1–5	31(f), 48–51
	2 Graph functions using compressions and stretches (p. 71)	6–8, 12	31(g), 47, 51
	3 Graph functions using reflections about the x-axis and the y-axis (p. 73)	9–11, 13	31(h), 47, 49, 51
1.7	**1** Determine whether a function is one-to-one (p. 81)	1, 2	55(a), 56
	2 Obtain the graph of the inverse function from the graph of the function (p. 85)	3	56
	3 Verify an inverse function (p. 85)	4–5	57–58
	4 Find the inverse of a function defined by an equation (p. 86)	6–8	59–62

Review Exercises

In Problems 1–3, find the following for each pair of points:
  (a) The distance between the points
  (b) The midpoint of the line segment connecting the points

1. $(0, 0)$; $(4, 2)$

2. $(1, -1)$; $(-2, 3)$

3. $(4, -4)$; $(4, 8)$

4. Graph $y = x^2 + 4$ by plotting points.

5. List the intercepts of the graph below.

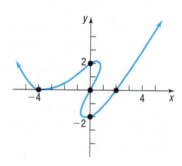

In Problems 6–10, list the intercepts and test for symmetry with respect to the x-axis, the y-axis, and the origin.

6. $2x = 3y^2$

7. $x^2 + 4y^2 = 16$

8. $y = x^4 + 2x^2 + 1$

9. $y = x^3 - x$

10. $x^2 + x + y^2 + 2y = 0$

In Problems 11 and 12, find the standard form of the equation of the circle whose center and radius are given.

11. $(h, k) = (-2, 3)$; $r = 4$

12. $(h, k) = (-1, -2)$; $r = 1$

In Problems 13–15, find the center and radius of each circle. Graph each circle. Find the intercepts, if any, of each circle.

13. $x^2 + (y - 1)^2 = 4$

14. $x^2 + y^2 - 2x + 4y - 4 = 0$

15. $3x^2 + 3y^2 - 6x + 12y = 0$

16. Show that the points $A = (3, 4)$, $B = (1, 1)$, and $C = (-2, 3)$ are the vertices of an isosceles triangle.

17. The endpoints of the diameter of a circle are $(-3, 2)$ and $(5, -6)$. Find the center and radius of the circle. Write the standard equation of this circle.

18. While shopping online for AA batteries, Masoud found that he could order a pack of 8 batteries for $6.30, a pack of 16 for $13.99, a pack of 20 for $12.32, or a pack of 24 for $13.99. Define a relation using number of batteries as input and price as output.
  (a) What are the domain and the range of the relation?
  (b) Express the relation as a set of ordered pairs.
  (c) Express the relation as a mapping.
  (d) Express the relation as a graph.

In Problems 19 and 20, determine whether each relation represents a function. For each function, state the domain and range.

19. $\{(-1, 0), (2, 3), (4, 0)\}$

20. $\{(4, -1), (2, 1), (4, 2)\}$

In Problems 21–23, find the following for each function:
   (a) $f(2)$ (b) $f(-2)$ (c) $f(-x)$ (d) $-f(x)$ (e) $f(x-2)$ (f) $f(2x)$

21. $f(x) = \dfrac{3x}{x^2 - 1}$
  22. $f(x) = \sqrt{x^2 - 4}$
  23. $f(x) = \dfrac{x^2 - 4}{x^2}$

In Problems 24–29, find the domain of each function.

24. $f(x) = \dfrac{x}{x^2 - 9}$
  25. $f(x) = \sqrt{2 - x}$
  26. $g(x) = \dfrac{|x|}{x}$

27. $f(x) = \dfrac{x}{x^2 + 2x - 3}$
  28. $f(x) = \dfrac{\sqrt{x+1}}{x^2 - 4}$
  29. $g(x) = \dfrac{x}{\sqrt{x+8}}$

30. Find the difference quotient of $f(x) = -2x^2 + x + 1$; that is, find $\dfrac{f(x+h) - f(x)}{h}, h \neq 0.$

31. Consider the graph of the function f on the right.
  (a) Find the domain and the range of f.
  (b) List the intercepts.
  (c) Find $f(-2)$.
  (d) For what value of x does $f(x) = -3$?
  (e) Solve $f(x) > 0$.
  (f) Graph $y = f(x - 3)$.
  (g) Graph $y = f\left(\dfrac{1}{2}x\right)$.
  (h) Graph $y = -f(x)$.

32. Use the graph of the function f shown to find:
  (a) The domain and the range of f.
  (b) The intervals on which f is increasing, decreasing, or constant.
  (c) The local minimum values and local maximum values.
  (d) The absolute maximum and absolute minimum.
  (e) Whether the graph is symmetric with respect to the x-axis, the y-axis, or the origin.
  (f) Whether the function is even, odd, or neither.
  (g) The intercepts, if any.

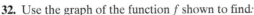

In Problems 33–36, determine (algebraically) whether the given function is even, odd, or neither.

33. $f(x) = x^3 - 4x$
  34. $g(x) = \dfrac{4 + x^2}{1 + x^4}$
  35. $G(x) = 1 - x + x^3$
  36. $f(x) = \dfrac{x}{1 + x^2}$

In Problems 37 and 38, use a graphing utility to graph each function over the indicated interval. Approximate any local maximum values and local minimum values. Determine where the function is increasing and where it is decreasing.

37. $f(x) = 2x^3 - 5x + 1 \quad [-3, 3]$
  38. $f(x) = 2x^4 - 5x^3 + 2x + 1 \quad [-2, 3]$

39. Find the average rate of change of $f(x) = 8x^2 - x$:
  (a) From 1 to 2   (b) From 0 to 1   (c) From 2 to 4

In Problems 40 and 41, find the average rate of change from 2 to 3 for each function f. Be sure to simplify.
40. $f(x) = 2 - 5x$
  41. $f(x) = 3x - 4x^2$

In Problems 42 and 43, is the graph shown the graph of a function?
42.

43.

In Problems 44 and 45, graph each function. Be sure to label at least three points.
  44. $f(x) = |x|$
  45. $f(x) = \sqrt{x}$

In Problems 46–51, graph each function using the techniques of shifting, compressing or stretching, and reflections. Identify any intercepts of the graph. State the domain and, based on the graph, find the range.

46. $F(x) = |x| - 4$

47. $g(x) = -2|x|$

48. $h(x) = \sqrt{x} - 1$

49. $f(x) = \sqrt{1 - x}$

50. $h(x) = (x - 1)^2 + 2$

51. $g(x) = -2(x + 2)^3 - 8$

In Problems 52 and 53:

   (a) Find the domain of each function.   *(b) Locate any intercepts.*   *(c) Graph each function.*

   (d) Based on the graph, find the range.

52. $f(x) = \begin{cases} 3x & \text{if } -2 < x \le 1 \\ x + 1 & \text{if } x > 1 \end{cases}$

53. $f(x) = \begin{cases} x & \text{if } -4 \le x < 0 \\ 1 & \text{if } x = 0 \\ 3x & \text{if } x > 0 \end{cases}$

54. A function f is defined by

$$f(x) = \frac{Ax + 5}{6x - 2}$$

   If $f(1) = 4$, find A.

55. (a) Verify that the function below is one-to-one; (b) find its inverse.

   $\{(1, 2), (3, 5), (5, 8), (6, 10)\}$

56. State why the graph of the function is one-to-one. Then draw the graph of the inverse function f^{-1}.

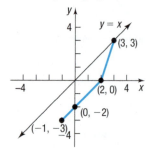

In Problems 57 and 58, verify that the functions f and g are inverses of each other by showing that $f(g(x)) = x$ and $g(f(x)) = x$. Give any values of x that need to be excluded from the domain of f and the domain of g

57. $f(x) = 5x - 10;\ g(x) = \dfrac{1}{5}x + 2$

58. $f(x) = \dfrac{x - 4}{x};\ g(x) = \dfrac{4}{1 - x}$

In Problems 59–62, each function is one-to-one. Find the inverse of each function and check your answer. State the domain and range of f and f^{-1}.

59. $f(x) = \dfrac{2x + 3}{5x - 2}$

60. $f(x) = \dfrac{1}{x - 1}$

61. $f(x) = \sqrt{x - 2}$

62. $f(x) = x^{1/3} + 1$

Chapter Test

The Chapter Test Prep Videos include step-by-step solutions to all chapter test exercises. These videos are available in MyLab™ Math, or on this text's YouTube Channel. Refer to the Preface for a link to the YouTube channel.

In Problems 1–2, use $P_1 = (-1, 3)$ and $P_2 = (5, -1)$.

1. Find the distance from P_1 to P_2.

2. Find the midpoint of the line segment joining P_1 and P_2.

3. Graph $y = x^2 - 9$ by plotting points.

4. Sketch the graph of $y^2 = x$.

5. List the intercepts and test for symmetry: $x^2 + y = 9$.

6. Write the general form of the circle with center $(4, -3)$ and radius 5.

7. Find the center and radius of the circle $x^2 + y^2 + 4x - 2y - 4 = 0$. Graph this circle.

8. Determine whether each relation represents a function. For each function, state the domain and the range.

   (a) $\{(2, 5), (4, 6), (6, 7), (8, 8)\}$

   (b) $\{(1, 3), (4, -2), (-3, 5), (1, 7)\}$

   (c)

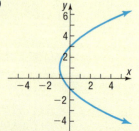

   (d)

In Problems 9–11, find the domain of each function and evaluate each function at x = −1.

9. $f(x) = \sqrt{4 - 5x}$

10. $g(x) = \dfrac{x + 2}{|x + 2|}$

11. $h(x) = \dfrac{x - 4}{x^2 + 5x - 36}$

12. Consider the graph of the function f:

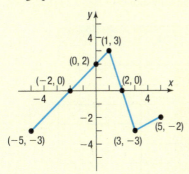

(a) Find the domain and the range of f.
(b) List the intercepts.
(c) Find $f(1)$.
(d) For what value(s) of x does $f(x) = -3$?
(e) Solve $f(x) < 0$.

 13. Use a graphing utility to graph the function $f(x) = -x^4 + 2x^3 + 4x^2 - 2$ on the interval $[-5, 5]$. Approximate any local maximum values and local minimum values rounded to two decimal places. Determine where the function is increasing and where it is decreasing.

14. Consider the function $g(x) = \begin{cases} 2x + 1 & \text{if } x < -1 \\ x - 4 & \text{if } x \geq -1 \end{cases}$

(a) Graph the function.
(b) List the intercepts.
(c) Find $g(-5)$.
(d) Find $g(2)$.

15. For the function $f(x) = 3x^2 - 2x + 4$, find the average rate of change of f from 3 to 4.

16. Graph each function using the techniques of shifting, compressing or stretching, and reflections. Start with the graph of the basic function and show all stages.

(a) $h(x) = -2(x + 1)^3 + 3$
(b) $g(x) = |x + 4| + 2$

17. Find the inverse of $f(x) = \dfrac{2}{3x - 5}$ and check your answer. State the domain and the range of f and f^{-1}.

18. If the point $(3, -5)$ is on the graph of a one-to-one function f, what point must be on the graph of f^{-1}?

Chapter Projects

 Internet-based Project

I. **Choosing a Data Plan** Collect information from your family, friends, or consumer agencies such as Consumer Reports. Then decide on a service provider, choosing the company that you feel offers the best service. Once you have selected a service provider, research the various types of individual plans offered by the company by visiting the provider's website. Many cellular providers offer family plans that include unlimited talk, text, and data. However, once a data cap has been reached, service may be slowed, which prevents media from being streamed. So, many customers still purchase data-only plans for devices such as tablets or laptops. The monthly cost is primarily determined by the amount of data used and the number of data-only devices.

1. Suppose you expect to use 10 gigabytes of data for a single tablet. What would be the monthly cost of each plan you are considering?

2. Suppose you expect to use 30 gigabytes of data and want a personal hotspot, but you still have only a single tablet. What would be the monthly cost of each plan you are considering?

3. Suppose you expect to use 20 gigabytes of data with three tablets sharing the data. What would be the monthly cost of each plan you are considering?

4. Suppose you expect to use 20 gigabytes of data with a single tablet and a personal hotspot. What would be the monthly cost of each plan you are considering?

5. Build a model that describes the monthly cost C, in dollars, as a function of the number g of data gigabytes used, assuming a single tablet and a personal hotspot for each plan you are considering.

6. Graph each function from Problem 5.

7. Based on your particular usage, which plan is best for you?

8. Now, develop an Excel spreadsheet to analyze the various plans you are considering. Suppose you want a plan that offers 50 gigabytes of shared data and costs $60 per month. Additional gigabytes of data cost $15 per gigabyte, extra tablets can be added to the plan for $10 each per month, and each hotspot or laptop costs $20 per month. Because these data plans have a cost structure based on piecewise-defined functions, we need an "if/then" statement within Excel to analyze the cost of the plan. Use the accompanying Excel spreadsheet as a guide in developing your spreadsheet. Enter into your spreadsheet a variety of possible amounts of data and various numbers of additional tablets, laptops, and hotspots.

	A	B	C	D
1				
2	Monthly fee	$60		
3	Allotted data per month (GB)	50		
4	Data used (GB)	12		
5	Cost per additional GB of data	$15		
6				
7	Monthly cost of hotspot or laptop	$20		
8	Number of hotspots or laptops	1		
9	Monthly cost of additional tablet	$10		
10	Number of additional tablets	2		
11				
12	Cost of data	=IF(B4<B3,B2,B2+B5*(B4-B3))		
13	Cost of additional devices/hotspots	=B8*B7+B10*B9		
14				
15	Total Cost	=B12+B13		
16				

9. Write a paragraph supporting the choice in plans that best meets your needs.

10. How are "if/then" loops similar to a piecewise-defined function?

Source: Excel © 2018 Microsoft Corporation. Used with Permission from Microsoft.

The following projects are available on the Instructor's Resource Center (IRC).

II. Project at Motorola: *Wireless Internet Service* Use functions and their graphs to analyze the total cost of various wireless Internet service plans.

III. Cost of Cable When government regulations and customer preference influence the path of a new cable line, the Pythagorean Theorem can be used to assess the cost of installation.

IV. Oil Spill Functions are used to analyze the size and spread of an oil spill from a leaking tanker.

Trigonometric Functions

Length of Day Revisited

The length of a day depends on the day of the year as well as on the latitude of the location. Latitude gives the location of a point on Earth north or south of the equator. In the Internet Project at the end of this chapter, we will find a model that describes the relation between the length of day and day of the year for a specific latitude.

 —*See the Internet-based Chapter Project I—*

← A Look Back

In Chapter 1, we began our discussion of functions. We defined domain and range and independent and dependent variables; we found the value of a function and graphed functions. We continued our study of functions by listing properties of functions, such as being even or odd, and we created a library of functions, naming key functions and listing their properties, including the graph.

A Look Ahead →

In this chapter, we define the trigonometric functions, six functions that have wide application. We find their domain and range, evaluate them, graph them, and develop a list of their properties.

There are two widely accepted approaches to the development of the trigonometric functions: one uses right triangles; the other uses circles, especially the unit circle. In this text, we develop the trigonometric functions using the unit circle. In Chapter 4, we present right triangle trigonometry.

Outline

2.1 Angles, Arc Length, and Circular Motion

PREPARING FOR THIS SECTION *Before getting started, review the following:*

• Circumference and Area of a Circle (Section A.2, p. A16)

Now Work the *'Are You Prepared?'* problems on page 110.

OBJECTIVES 1 Angles and Degree Measure (p. 102)
2 Convert between Decimal and Degree, Minute, Second Measures for Angles (p. 104)
3 Find the Length of an Arc of a Circle (p. 105)
4 Convert from Degrees to Radians and from Radians to Degrees (p. 106)
5 Find the Area of a Sector of a Circle (p. 108)
6 Find the Linear Speed of an Object Traveling in Circular Motion (p. 109)

1 Angles and Degree Measure

A **ray**, or **half-line**, is a portion of a line that starts at a point V on the line and extends indefinitely in one direction. The starting point V of a ray is called its **vertex**. See Figure 1.

When two rays are drawn with a common vertex, they form an **angle**. We call one ray of an angle the **initial side** and the other the **terminal side**. The angle formed is identified by showing the direction and amount of rotation from the initial side to the terminal side. If the rotation is in the counterclockwise direction, the angle is **positive**; if the rotation is clockwise, the angle is **negative**. See Figure 2.

Figure 1 A ray or half-line

  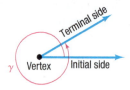

Figure 2
(a) Counterclockwise rotation; Positive angle
(b) Clockwise rotation; Negative angle
(c) Counterclockwise rotation; Positive angle

Lowercase Greek letters, such as α (alpha), β (beta), γ (gamma), and θ (theta), are often used to denote angles. Notice in Figure 2(a) that the angle α is positive because the direction of the rotation from the initial side to the terminal side is counterclockwise. The angle β in Figure 2(b) is negative because the rotation is clockwise. The angle γ in Figure 2(c) is positive. Notice that the angle α in Figure 2(a) and the angle γ in Figure 2(c) have the same initial side and the same terminal side. However, α and γ are unequal, because the amount of rotation required to go from the initial side to the terminal side is greater for angle γ than for angle α.

An angle θ is said to be in **standard position** if its vertex is at the origin of a rectangular coordinate system and its initial side coincides with the positive x-axis. See Figure 3.

Need to Review?
Rectangular Coordinates are discussed in Section 1.1, p. 2.

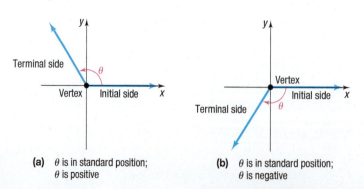

Figure 3 Standard position of an angle
(a) θ is in standard position; θ is positive
(b) θ is in standard position; θ is negative

When an angle θ is in standard position, either the terminal side will lie in a quadrant, in which case we say that θ **lies in that quadrant**, or the terminal side will lie on the x-axis or the y-axis, in which case we say that θ is a **quadrantal angle**. For example, the angle θ in Figure 4(a) lies in quadrant II, the angle θ in Figure 4(b) lies in quadrant IV, and the angle θ in Figure 4(c) is a quadrantal angle.

Figure 4   **(a)** θ lies in quadrant II   **(b)** θ lies in quadrant IV   **(c)** θ is a quadrantal angle

Angles are measured by determining the amount of rotation needed for the initial side to coincide with the terminal side. The two commonly used measures for angles are *degrees* and *radians*.

Degree Measure

HISTORICAL NOTE One counterclockwise rotation was said to measure 360° because the Babylonian year had 360 days. ■

The angle formed by rotating the initial side exactly once in the counterclockwise direction until it coincides with itself (1 revolution) is said to measure 360 degrees, abbreviated 360°. **One degree, 1°**, is $\frac{1}{360}$ revolution. A **right angle** is an angle that measures 90°, or $\frac{1}{4}$ revolution; a **straight angle** is an angle that measures 180°, or $\frac{1}{2}$ revolution. See Figure 5. As Figure 5(b) shows, it is customary to indicate a right angle by using the symbol ⌐.

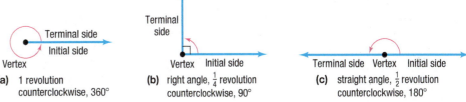

Figure 5   **(a)** 1 revolution counterclockwise, 360°   **(b)** right angle, $\frac{1}{4}$ revolution counterclockwise, 90°   **(c)** straight angle, $\frac{1}{2}$ revolution counterclockwise, 180°

It is also customary to refer to an angle that measures θ degrees as an angle *of θ degrees*.

EXAMPLE 1 **Drawing an Angle**

Draw each angle in standard position.

(a) 45°   (b) −90°   (c) 225°   (d) 405°

Solution   (a) An angle of 45° is $\frac{1}{2}$ of a right angle. See Figure 6.

(b) An angle of −90° is $\frac{1}{4}$ revolution in the clockwise direction. See Figure 7.

(c) An angle of 225° consists of a rotation through 180° followed by a rotation through 45°. See Figure 8.

(d) An angle of 405° consists of 1 revolution (360°) followed by a rotation through 45°. See Figure 9.

Figure 6 45° angle   **Figure 7** −90° angle   **Figure 8** 225° angle   **Figure 9** 405° angle

Now Work PROBLEM 11

2 Convert between Decimal and Degree, Minute, Second Measures for Angles

Although subdivisions of a degree may be obtained by using decimals, the notion of *minutes* and *seconds* may also be used. **One minute**, denoted by **1′**, is defined as $\dfrac{1}{60}$ degree. **One second**, denoted by **1″**, is defined as $\dfrac{1}{60}$ minute or, equivalently, $\dfrac{1}{3600}$ degree. An angle of, say, 30 degrees, 40 minutes, 10 seconds is written compactly as 30°40′10″. To summarize:

> 1 counterclockwise revolution = 360°
>
> 1° = 60′   1′ = 60″   **(1)**

It is sometimes necessary to convert from the degree, minute, second notation (D°M′S″) to a decimal form, and vice versa.

EXAMPLE 2

Converting between Decimal Form and the Degree, Minute, Second Form

(a) Convert 50°6′21″ to a decimal in degrees. Round the answer to four decimal places.

(b) Convert 21.256° to the degree, minute, second notation. Round the answer to the nearest second.

Solution

(a) Because $1' = \left(\dfrac{1}{60}\right)^{\circ}$ and $1'' = \left(\dfrac{1}{60}\right)' = \left(\dfrac{1}{60} \cdot \dfrac{1}{60}\right)^{\circ}$, convert as follows:

$$50°6'21'' = 50° + 6' + 21'' = 50° + 6 \cdot 1' + 21 \cdot 1''$$

$$= 50° + 6 \cdot \left(\dfrac{1}{60}\right)^{\circ} + 21 \cdot \left(\dfrac{1}{60} \cdot \dfrac{1}{60}\right)^{\circ} \quad \text{\color{blue}\textbf{Convert minutes and seconds to degrees.}}$$

$$\approx 50° + 0.1° + 0.0058°$$

$$= 50.1058°$$

(b) Because 1° = 60′ and 1′ = 60″, proceed as follows:

$$21.256° = 21° + 0.256°$$

$$= 21° + 0.256 \cdot 1°$$

$$= 21° + 0.256 \cdot 60' \quad \text{\color{blue}\textbf{Convert fraction of degree to minutes; 1° = 60′.}}$$

$$= 21° + 15.36'$$

$$= 21° + 15' + 0.36'$$

$$= 21° + 15' + 0.36 \cdot 1'$$

$$= 21° + 15' + 0.36 \cdot 60'' \quad \text{\color{blue}\textbf{Convert fraction of minute to seconds; 1′ = 60″.}}$$

$$= 21° + 15' + 21.6''$$

$$\approx 21°15'22'' \quad \text{\color{blue}\textbf{Round to the nearest second.}}$$

COMMENT Graphing calculators (and some scientific calculators) have the ability to convert from *degree, minute, second* to decimal form, and vice versa. Consult your owner's manual. ∎

 Now Work PROBLEMS **59** AND **65**

In many applications, such as describing the exact location of a star or the precise position of a ship at sea, angles measured in degrees, minutes, and even seconds are used. For calculation purposes, these are transformed to decimal form. In other applications, especially those in calculus, angles are measured using *radians*.

Radian Measure

A **central angle** is a positive angle whose vertex is at the center of a circle. The rays of a central angle subtend (intersect) an arc on the circle. If the radius of the circle is r

and the length of the arc subtended by the central angle is also r, then the measure of the angle is **1 radian**. See Figure 10(a).

For a circle of radius 1, the rays of a central angle with measure 1 radian subtend an arc of length 1. For a circle of radius 3, the rays of a central angle with measure 1 radian subtend an arc of length 3. See Figure 10(b).

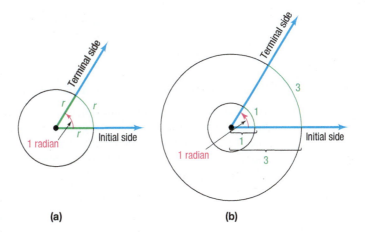

Figure 10　　　(a)　　　　　　　　　(b)

3 Find the Length of an Arc of a Circle

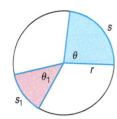

Figure 11 $\dfrac{\theta}{\theta_1} = \dfrac{s}{s_1}$

Now consider a circle of radius r and two central angles, θ and θ_1, measured in radians. Suppose that these central angles subtend arcs of lengths s and s_1, respectively, as shown in Figure 11. From geometry, the ratio of the measures of the angles equals the ratio of the corresponding lengths of the arcs subtended by these angles; that is,

$$\frac{\theta}{\theta_1} = \frac{s}{s_1} \tag{2}$$

Suppose that $\theta_1 = 1$ radian. Refer again to Figure 10(a). The length s_1 of the arc subtended by the central angle $\theta_1 = 1$ radian equals the radius r of the circle. Then $s_1 = r$, so equation (2) reduces to

$$\frac{\theta}{1} = \frac{s}{r} \quad \text{or} \quad s = r\theta \tag{3}$$

THEOREM Arc Length

For a circle of radius r, a central angle of θ radians subtends an arc whose length s is

$$s = r\theta \tag{4}$$

NOTE Formulas must be consistent with the units used. In formula (4), we write

$$s = r\theta$$

To see the units, use equation (3) and write

$$\frac{\theta \text{ radians}}{1 \text{ radian}} = \frac{s \text{ length units}}{r \text{ length units}}$$

$$s \text{ length units} = r \text{ length units} \frac{\theta \text{ radians}}{1 \text{ radian}}$$

The radians cancel, leaving

$$s \text{ length units} = (r \text{ length units})\theta \qquad s = r\theta$$

where θ appears to be "dimensionless" but, in fact, is measured in radians. So, in the formula $s = r\theta$, the dimension for θ is radians, and any convenient unit of length (such as inches or meters) can be used for s and r. ▪

EXAMPLE 3

Finding the Length of an Arc of a Circle

Find the length of the arc of a circle of radius 2 meters subtended by a central angle of 0.25 radian.

Solution   Use formula (4) with $r = 2$ meters and $\theta = 0.25$. The length s of the arc is

$$s = r\theta = 2 \cdot 0.25 = 0.5 \text{ meter}$$

 Now Work PROBLEM 71

4 Convert from Degrees to Radians and from Radians to Degrees

With two ways to measure angles, it is important to be able to convert from one measure to the other. Consider a circle of radius r. A central angle of 1 revolution subtends an arc equal to the circumference of the circle. See Figure 12. Because the circumference of a circle of radius r equals $2\pi r$, substitute $2\pi r$ for s in formula (4) to find that, for an angle θ of 1 revolution,

$$s = r\theta$$
$$2\pi r = r\theta \qquad \textbf{\textit{θ}} = \textbf{1 revolution; } s = \textbf{2πr}$$
$$\theta = 2\pi \text{ radians} \qquad \textbf{Solve for θ.}$$

Figure showing: $s = 2\pi r$, 1 revolution, r

Figure 12 1 revolution = 2π radians

From this, we have

$$\boxed{1 \text{ revolution} = 2\pi \text{ radians}} \qquad \textbf{(5)}$$

Since 1 revolution = 360°, we have

$$360° = 2\pi \text{ radians}$$

Dividing both sides by 2 yields

$$\boxed{180° = \pi \text{ radians}} \qquad \textbf{(6)}$$

Divide both sides of equation (6) by 180. Then

$$1 \text{ degree} = \frac{\pi}{180} \text{ radian}$$

Divide both sides of equation (6) by π. Then

$$\frac{180}{\pi} \text{ degrees} = 1 \text{ radian}$$

We have the following two conversion formulas:*

$$\boxed{1 \text{ degree} = \frac{\pi}{180} \text{ radian} \qquad 1 \text{ radian} = \frac{180}{\pi} \text{ degrees}} \qquad \textbf{(7)}$$

EXAMPLE 4

Converting from Degrees to Radians

Convert each angle in degrees to radians.

(a) 60° (b) 150° (c) −45° (d) 90° (e) 107°

---

*Some students prefer instead to use the proportion $\dfrac{\text{Degrees}}{180°} = \dfrac{\text{Radians}}{\pi}$. Then substitute for what is given and solve for the measurement sought.

Solution (a) $60° = 60 \cdot 1 \text{ degree} = 60 \cdot \dfrac{\pi}{180} \text{ radian} = \dfrac{\pi}{3} \text{ radians}$

(b) $150° = 150 \cdot 1° = 150 \cdot \dfrac{\pi}{180} \text{ radian} = \dfrac{5\pi}{6} \text{ radians}$

(c) $-45° = -45 \cdot \dfrac{\pi}{180} \text{ radian} = -\dfrac{\pi}{4} \text{ radian}$

(d) $90° = 90 \cdot \dfrac{\pi}{180} \text{ radian} = \dfrac{\pi}{2} \text{ radians}$

(e) $107° = 107 \cdot \dfrac{\pi}{180} \text{ radian} \approx 1.868 \text{ radians}$

Example 4, parts (a)–(d), illustrates that angles that are "nice" fractions of a revolution are expressed in radian measure as fractional multiples of π, rather than as decimals. For example, a right angle, as in Example 4(d), is left in the form $\dfrac{\pi}{2}$ radians, which is exact, rather than using the approximation $\dfrac{\pi}{2} \approx \dfrac{3.1416}{2} = 1.5708$ radians. When the fractions are not "nice," use the decimal approximation of the angle, as in Example 4(e).

✏️ **Now Work** PROBLEMS 23 AND 49

EXAMPLE 5 **Converting from Radians to Degrees**

Convert each angle in radians to degrees.

(a) $\dfrac{\pi}{6} \text{ radian}$ (b) $\dfrac{3\pi}{2} \text{ radians}$ (c) $-\dfrac{7\pi}{4} \text{ radians}$ (d) $\dfrac{7\pi}{3} \text{ radians}$ (e) 3 radians

Solution (a) $\dfrac{\pi}{6} \text{ radian} = \dfrac{\pi}{6} \cdot 1 \text{ radian} = \dfrac{\pi}{6} \cdot \dfrac{180}{\pi} \text{ degrees} = 30°$

(b) $\dfrac{3\pi}{2} \text{ radians} = \dfrac{3\pi}{2} \cdot \dfrac{180}{\pi} \text{ degrees} = 270°$

(c) $-\dfrac{7\pi}{4} \text{ radians} = -\dfrac{7\pi}{4} \cdot \dfrac{180}{\pi} \text{ degrees} = -315°$

(d) $\dfrac{7\pi}{3} \text{ radians} = \dfrac{7\pi}{3} \cdot \dfrac{180}{\pi} \text{ degrees} = 420°$

(e) $3 \text{ radians} = 3 \cdot \dfrac{180}{\pi} \text{ degrees} \approx 171.89°$

✏️ **Now Work** PROBLEM 35

Table 1 lists the degree and radian measures of some common angles. You should learn to feel equally comfortable using either measure.

Table 1

Degrees	0°	30°	45°	60°	90°	120°	135°	150°	180°
Radians	0	$\dfrac{\pi}{6}$	$\dfrac{\pi}{4}$	$\dfrac{\pi}{3}$	$\dfrac{\pi}{2}$	$\dfrac{2\pi}{3}$	$\dfrac{3\pi}{4}$	$\dfrac{5\pi}{6}$	π
Degrees		210°	225°	240°	270°	300°	315°	330°	360°
Radians		$\dfrac{7\pi}{6}$	$\dfrac{5\pi}{4}$	$\dfrac{4\pi}{3}$	$\dfrac{3\pi}{2}$	$\dfrac{5\pi}{3}$	$\dfrac{7\pi}{4}$	$\dfrac{11\pi}{6}$	2π

| EXAMPLE 6 | **Field Width of a DSLR Camera Lens** |

For small angles, the length of the arc subtended by a central angle is approximately equal to the length of the chord that is subtended. Use this fact to approximate the field width (the width of scenery the lens can image) of a 400mm camera lens at a distance of 750 feet if the viewing angle of the lens is 6°12′.

Solution

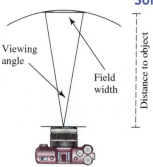

See Figure 13. The measure of the central angle is 6°12′, but remember that the angle in formula (4) must be in radians. So begin by converting the angle to radians.

$$\theta = 6°12' = 6.2° = 6.2 \cdot \frac{\pi}{180} \text{ radian} \approx 0.108 \text{ radian}$$

$$\underset{\textbf{12}' = \textbf{0.2}°}{\uparrow}$$

Use $\theta = 0.108$ radian and $r = 750$ feet in formula (4). The field width for the lens at a distance of 750 feet from the camera is approximately

$$s = r\theta = 750 \cdot 0.108 = 81 \text{ feet}$$

Figure 13 Camera viewing angle

NOTE If the measure of an angle is given as 5, it is understood to mean 5 radians; if the measure of an angle is given as 5°, it means 5 degrees. ∎

When an angle is measured in degrees, the degree symbol is always shown. However, when an angle is measured in radians, we usually omit the word *radians*. So if the measure of an angle is given as $\frac{\pi}{6}$, it is understood to mean $\frac{\pi}{6}$ radian.

Now Work PROBLEM 107

5 Find the Area of a Sector of a Circle

Consider a circle of radius r. Suppose that θ, measured in radians, is a central angle of this circle. See Figure 14. We seek a formula for the area A of the sector (shown in blue) formed by the angle θ.

Consider a circle of radius r and two central angles θ and θ_1, both measured in radians. See Figure 15. From geometry, the ratio of the measures of the angles equals the ratio of the corresponding areas of the sectors formed by these angles. That is,

$$\frac{\theta}{\theta_1} = \frac{A}{A_1}$$

Figure 14 Sector of a Circle

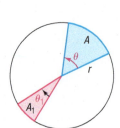

Now suppose that $\theta_1 = 2\pi$ radians. Then $A_1 =$ area of the circle $= \pi r^2$. Solving for A, we find

$$A = A_1 \frac{\theta}{\theta_1} = \pi r^2 \frac{\theta}{2\pi} = \frac{1}{2} r^2 \theta$$

$$\underset{\textbf{A}_1 = \pi \textbf{r}^2; \, \theta_1 = 2\pi}{\uparrow}$$

Figure 15 $\dfrac{\theta}{\theta_1} = \dfrac{A}{A_1}$

THEOREM **Area of a Sector**

The area A of the sector of a circle of radius r formed by a central angle of θ radians is

$$A = \frac{1}{2} r^2 \theta \qquad \qquad (8)$$

| EXAMPLE 7 | Finding the Area of a Sector of a Circle |

Find the area of the sector of a circle of radius 2 feet formed by an angle of 30°. Round the answer to two decimal places.

Solution   Use formula (8) with $r = 2$ feet and $\theta = 30° = \dfrac{\pi}{6}$ radian. [Remember, in formula (8), θ must be in radians.]

$$A = \frac{1}{2}r^2\theta = \frac{1}{2} \cdot 2^2 \cdot \frac{\pi}{6} = \frac{\pi}{3} \approx 1.05$$

The area A of the sector is 1.05 square feet, rounded to two decimal places.

 Now Work PROBLEM 79

6 Find the Linear Speed of an Object Traveling in Circular Motion

The average speed of an object is the distance traveled divided by the elapsed time. For motion along a circle, we distinguish between *linear speed* and *angular speed*.

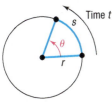

Figure 16 $v = \dfrac{s}{t}$

> **DEFINITION Linear Speed**
>
> Suppose that an object moves on a circle of radius r at a constant speed. If s is the distance traveled in time t on this circle, then the **linear speed** v of the object is defined as
>
> $$v = \frac{s}{t} \tag{9}$$

As this object travels on the circle, suppose that θ (measured in radians) is the central angle swept out in time t. See Figure 16.

> **DEFINITION Angular Speed**
>
> The **angular speed** ω (the Greek lowercase letter omega) of an object is the angle θ (measured in radians) swept out, divided by the elapsed time t; that is,
>
> $$\omega = \frac{\theta}{t} \tag{10}$$

Angular speed is the way the turning rate of an engine is described. For example, an engine idling at 900 rpm (revolutions per minute) is one that rotates at an angular speed of

$$900 \; \frac{\text{revolutions}}{\text{minute}} = 900 \; \frac{\text{revolutions}}{\text{minute}} \cdot 2\pi \; \frac{\text{radians}}{\text{revolution}} = 1800\pi \; \frac{\text{radians}}{\text{minute}}$$

There is an important relationship between linear speed and angular speed:

$$\text{linear speed} = v = \underset{\underset{(9)}{\uparrow}}{\frac{s}{t}} = \underset{\underset{s\,=\,r\theta}{\uparrow}}{\frac{r\theta}{t}} = r \cdot \underset{\underset{(10)}{\uparrow}}{\frac{\theta}{t}} = r \cdot \omega$$

> $$v = r\omega \tag{11}$$

where ω is measured in radians per unit time.

When using formula (11), $v = r\omega$, remember that $v = \dfrac{s}{t}$ (the linear speed) has the dimensions of length per unit of time (such as feet per second or miles per hour), r (the radius of the circular motion) has the same length dimension as s, and ω (the angular speed) has the dimension of radians per unit of time. If the angular speed is given in terms of *revolutions* per unit of time (as is often the case), be sure to convert it to *radians* per unit of time, using the fact that 1 revolution $= 2\pi$ radians, before using formula (11).

EXAMPLE 8

Finding Linear Speed

A child is spinning a rock at the end of a 2-foot rope at the rate of 180 revolutions per minute (rpm). Find the linear speed of the rock when it is released.

Solution

Look at Figure 17. The rock is moving around a circle of radius $r = 2$ feet. The angular speed ω of the rock is

$$\omega = 180 \; \frac{\text{revolutions}}{\text{minute}} = 180 \; \frac{\text{revolutions}}{\text{minute}} \cdot 2\pi \; \frac{\text{radians}}{\text{revolution}} = 360\pi \; \frac{\text{radians}}{\text{minute}}$$

From formula (11), $v = r\omega$, the linear speed v of the rock is

$$v = r\omega = 2 \text{ feet} \cdot 360\pi \; \frac{\text{radians}}{\text{minute}} = 720\pi \; \frac{\text{feet}}{\text{minute}} \approx 2262 \; \frac{\text{feet}}{\text{minute}}$$

Figure 17

The linear speed of the rock when it is released is 2262 ft/min $\approx$ 25.7 mi/h.

Now Work PROBLEM 99

2.1 Assess Your Understanding

'Are You Prepared?' *Answers are given at the end of these exercises. If you get a wrong answer, read the pages listed in* red.

1. What is the formula for the circumference C of a circle of radius r? (p. A16)

2. The area of a semicircle of radius 2 is _____. (p. A16)

Concepts and Vocabulary

3. An angle θ is in _____ _____ if its vertex is at the origin of a rectangular coordinate system and its initial side coincides with the positive x-axis.

4. A _____ _____ is a positive angle whose vertex is at the center of a circle.

5. *Multiple Choice* If the radius of a circle is r and the length of the arc subtended by a central angle is also r, then the measure of the angle is 1 _____.
   (a) degree
   (b) minute
   (c) second
   (d) radian

6. On a circle of radius r, a central angle of θ radians subtends an arc of length $s = $ _____; the area of the sector formed by this angle θ is $A = $ _____.

7. *Multiple Choice* $180° = $ _____ radians
   (a) $\dfrac{\pi}{2}$
   (b) π
   (c) $\dfrac{3\pi}{2}$
   (d) 2π

8. An object travels on a circle of radius r with constant speed. If s is the distance traveled in time t on the circle and θ is the central angle (in radians) swept out in time t, then the linear speed of the object is $v = $ _____ and the angular speed of the object is $\omega = $ _____.

9. *True or False* The angular speed ω of an object traveling on a circle of radius r is the angle θ (measured in radians) swept out, divided by the elapsed time t.

10. *True or False* For circular motion on a circle of radius r, linear speed equals angular speed divided by r.

Skill Building

In Problems 11–22, draw each angle in standard position.

11. 30°   **12.** 60°   **13.** 135°   **14.** −120°   **15.** 450°   **16.** 540°

17. $\dfrac{3\pi}{4}$ **18.** $\dfrac{4\pi}{3}$ **19.** $-\dfrac{\pi}{6}$ **20.** $-\dfrac{2\pi}{3}$ **21.** $\dfrac{16\pi}{3}$ **22.** $\dfrac{21\pi}{4}$

In Problems 23–34, convert each angle in degrees to radians. Express your answer as a multiple of π.

23. 30°   **24.** 120°   **25.** 495°   **26.** 330°   **27.** −60°   **28.** −30°

29. 540°   **30.** 270°   **31.** −240°   **32.** −225°   **33.** −90°   **34.** −180°

In Problems 35–46, convert each angle in radians to degrees.

35. $\dfrac{\pi}{3}$ **36.** $\dfrac{5\pi}{6}$ **37.** $-\dfrac{13\pi}{6}$ **38.** $-\dfrac{2\pi}{3}$ **39.** $\dfrac{9\pi}{2}$ **40.** 4π

41. $\dfrac{3\pi}{20}$ **42.** $\dfrac{5\pi}{12}$ **43.** $-\dfrac{\pi}{2}$ **44.** $-\pi$ **45.** $-\dfrac{17\pi}{15}$ **46.** $-\dfrac{3\pi}{4}$

In Problems 47–52, convert each angle in degrees to radians. Express your answer in decimal form, rounded to two decimal places.

47. 17°   **48.** 73°   **49.** −40°   **50.** −51°   **51.** 125°   **52.** 350°

In Problems 53–58, convert each angle in radians to degrees. Express your answer in decimal form, rounded to two decimal places.

53. 3.14   **54.** 0.75   **55.** 7

56. 3   **57.** 9.28   **58.** $\sqrt{2}$

In Problems 59–64, convert each angle to a decimal in degrees. Round your answer to two decimal places.

59. 40°10′25″   **60.** 61°42′21″   **61.** 50°14′20″

62. 73°40′40″   **63.** 9°9′9″   **64.** 98°22′45″

In Problems 65–70, convert each angle to D°M′S″ form. Round your answer to the nearest second.

65. 40.32°   **66.** 61.24°   **67.** 18.255°   **68.** 29.411°   **69.** 19.99°   **70.** 44.01°

In Problems 71–78, s denotes the length of the arc of a circle of radius r subtended by the central angle θ. Find the missing quantity. Round answers to three decimal places.

71. $r = 10$ meters, $\theta = \dfrac{1}{2}$ radian, $s = ?$ **72.** $r = 6$ feet, $\theta = 2$ radians, $s = ?$

73. $\theta = \dfrac{2}{3}$ radian, $s = 8$ feet, $r = ?$ **74.** $\theta = \dfrac{1}{4}$ radian, $s = 6$ centimeters, $r = ?$

75. $r = 10$ miles, $s = 9$ miles, $\theta = ?$ **76.** $r = 6$ meters, $s = 8$ meters, $\theta = ?$

77. $r = 2$ inches, $\theta = 30°$, $s = ?$ **78.** $r = 3$ meters, $\theta = 120°$, $s = ?$

In Problems 79–86, A denotes the area of the sector of a circle of radius r formed by the central angle θ. Find the missing quantity. Round answers to three decimal places.

79. $r = 10$ meters, $\theta = \dfrac{1}{2}$ radian, $A = ?$ **80.** $r = 6$ feet, $\theta = 2$ radians, $A = ?$

81. $\theta = \dfrac{1}{3}$ radian, $A = 2$ square feet, $r = ?$ **82.** $\theta = \dfrac{1}{4}$ radian, $A = 6$ square centimeters, $r = ?$

83. $r = 5$ miles, $A = 3$ square miles, $\theta = ?$ **84.** $r = 6$ meters, $A = 8$ square meters, $\theta = ?$

85. $r = 2$ inches, $\theta = 30°$, $A = ?$ **86.** $r = 3$ meters, $\theta = 120°$, $A = ?$

In Problems 87–90, find the length s and the area A. Round answers to three decimal places.

87.

88.

89.

90.

Applications and Extensions

91. Movement of a Minute Hand The minute hand of a clock is 6 inches long. How far does the tip of the minute hand move in 15 minutes? How far does it move in 25 minutes? Round answers to two decimal places.

92. Movement of a Pendulum A pendulum swings through an angle of 20° each second. If the pendulum is 40 inches long, how far does its tip move each second? Round answers to two decimal places.

93. Area of a Sector Find the area of the sector of a circle of radius 4 meters formed by an angle of 45°. Round the answer to two decimal places.

94. Area of a Sector Find the area of the sector of a circle of radius 3 centimeters formed by an angle of 60°. Round the answer to two decimal places.

95. Watering a Lawn A water sprinkler sprays water over a distance of 30 feet while rotating through an angle of 135°. What area of lawn receives water?

135°
30 ft

96. Designing a Water Sprinkler An engineer is asked to design a water sprinkler that will cover a field of 100 square yards that is in the shape of a sector of a circle of radius 15 yards. Through what angle should the sprinkler rotate?

97. Windshield Wiper The arm and blade of a windshield wiper have a total length of 34 inches. If the blade is 25 inches long and the wiper sweeps out an angle of 120°, how much window area can the blade clean?

98. Windshield Wiper The arm and blade of a windshield wiper have a total length of 30 inches. If the blade is 24 inches long and the wiper sweeps out an angle of 125°, how much window area can the blade clean?

99. Motion on a Circle An object is traveling on a circle with a radius of 5 centimeters. If in 20 seconds a central angle of $\frac{1}{3}$ radian is swept out, what is the angular speed of the object? What is its linear speed?

100. Ferris Wheels A neighborhood carnival has a Ferris wheel with a radius of 30 feet. You measure the time it takes for one revolution to be 70 seconds. What is the linear speed (in feet per second) of this Ferris wheel? What is the angular speed in radians per second?

101. Amusement Park Ride A gondola on an amusement park ride, similar to the Spin Cycle at Silverwood Theme Park, spins at a rate of 13 revolutions per minute. If the gondola is 25 feet from the ride's center, what is the linear speed of the gondola in miles per hour?

102. Amusement Park Ride A centrifugal force ride, similar to the Gravitron, spins at a rate of 22 revolutions per minute. If the diameter of the ride is 13 meters, what is the linear speed of the passengers in kilometers per hour?

103. Wind Turbine As of January 2018, the world's tallest wind turbine was located in Gaildorf, Germany, with a hub height of 178 meters and a rotor diameter of 137 meters. If the blades turn at a rate of 14 revolutions per minute, what is the linear speed of the blade tip, in km/h?

104. Blu-ray Drive A Blu-ray drive has a maximum speed of 10,000 revolutions per minute. If a Blu-ray disc has a diameter of 12 cm, what is the linear speed, in km/h, of a point 4 cm from the center if the disc is spinning at a rate of 8000 revolutions per minute?

105. Bicycle Wheels The diameter of each wheel of a bicycle is 26 inches. If you are traveling at a speed of 35 miles per hour on this bicycle, through how many revolutions per minute are the wheels turning?

106. Car Wheels The radius of each wheel of a car is 15 inches. If the wheels are turning at the rate of 3 revolutions per second, how fast is the car moving? Express your answer in inches per second and in miles per hour.

107. Photography If the viewing angle for a 600mm lens is 4°6′, use arc length to approximate the field width of the lens at a distance of 860 feet.

108. Photography If the viewing angle for an 800mm lens is 1°42′, use arc length to approximate the field width of the lens at a distance of 920 feet.

In Problems 109–110, the latitude of a location L is the angle formed by a ray drawn from the center of Earth to the equator and a ray drawn from the center of Earth to L. See the figure.

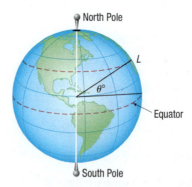

North Pole

L

$\theta°$

Equator

South Pole

109. Linear Speed on Earth Earth rotates on an axis through its poles. The distance from the axis to a location on Earth at 30° north latitude is about 3429.5 miles. Therefore, a location on Earth at 30° north latitude is spinning on a circle of radius 3429.5 miles. Compute the linear speed on the surface of Earth at 30° north latitude.

110. Linear Speed on Earth Earth rotates on an axis through its poles. The distance from the axis to a location on Earth at 40° north latitude is about 3033.5 miles. Therefore, a location on Earth at 40° north latitude is spinning on a circle of radius 3033.5 miles. Compute the linear speed on the surface of Earth at 40° north latitude.

111. **Speed of the Moon** The mean distance of the moon from Earth is 2.39×10^5 miles. Assuming that the orbit of the moon around Earth is circular and that 1 revolution takes 27.3 days, find the linear speed of the moon. Express your answer in miles per hour.

112. **Speed of Earth** The mean distance of Earth from the Sun is 9.29×10^7 miles. Assuming that the orbit of Earth around the Sun is circular and that 1 revolution takes 365 days, find the linear speed of Earth. Express your answer in miles per hour.

113. **Pulleys** Two pulleys, one with radius 2 inches and the other with radius 8 inches, are connected by a belt. (See the figure.) If the 2-inch pulley is caused to rotate at 3 revolutions per minute, determine the revolutions per minute of the 8-inch pulley.

[**Hint:** The linear speeds of the pulleys are the same; both equal the speed of the belt.]

114. **Pulleys** Two pulleys, one with radius r_1 and the other with radius r_2, are connected by a belt. The pulley with radius r_1 rotates at ω_1 revolutions per minute, whereas the pulley with radius r_2 rotates at ω_2 revolutions per minute. Show that

$$\frac{r_1}{r_2} = \frac{\omega_2}{\omega_1}$$

Use this result to rework Problem 113.

115. **Computing the Speed of a River Current** To approximate the speed of the current of a river, a circular paddle wheel with radius 4 feet is lowered into the water. If the current causes the wheel to rotate at a speed of 10 revolutions per minute, what is the speed of the current? Express your answer in miles per hour.

116. **Spin Balancing Tires** A spin balancer rotates the wheel of a car at 480 revolutions per minute. If the diameter of the wheel is 26 inches, what road speed is being tested? Express your answer in miles per hour. At how many revolutions per minute should the balancer be set to test a road speed of 80 miles per hour?

117. **The Cable Cars of San Francisco** At the Cable Car Museum you can see the four cable lines that are used to pull cable cars up and down the hills of San Francisco. Each cable travels at a speed of 9.55 miles per hour, driven by a rotating wheel whose diameter is 8.5 feet. How fast is the wheel rotating? Express your answer in revolutions per minute.

118. **Difference in Time of Sunrise** Naples, Florida, is about 90 miles due west of Ft. Lauderdale. How much sooner would a person in Ft. Lauderdale first see the rising Sun than a person in Naples? See Hint, top right.

[**Hint:** Consult the figure. When a person at Q sees the first rays of the Sun, a person at P is still in the dark. The person at P sees the first rays after Earth has rotated until P is at the location Q. Now use the fact that at the latitude of Ft. Lauderdale, in 24 hours an arc of length $2\pi(3559)$ miles is subtended.]

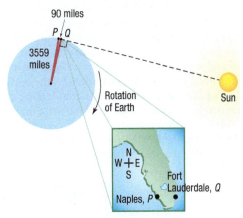

119. **Keeping Up with the Sun** How fast would you have to travel on the surface of Earth at the equator to keep up with the Sun (that is, so that the Sun would appear to remain in the same position in the sky)?

120. **Nautical Miles** A **nautical mile** equals the length of the arc subtended by a central angle of 1 minute on a great circle[†] on the surface of Earth. See the figure. If the radius of Earth is taken as 3960 miles, express 1 nautical mile in terms of ordinary, or **statute**, miles.

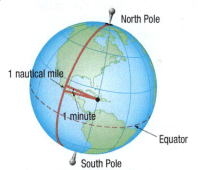

121. **Approximating the Circumference of Earth** Eratosthenes of Cyrene (276–195 BC) was a Greek scholar who lived and worked in Cyrene and Alexandria. One day while visiting in Syene, he noticed that the Sun's rays shone directly down a well. On this date 1 year later, in Alexandria, which is 500 miles due north of Syene, he measured the angle of the Sun to be about 7.2 degrees. See the figure. Use this information to approximate the radius and circumference of Earth.

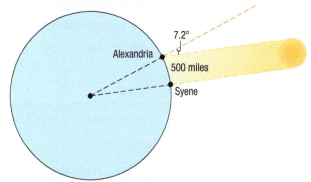

[†]Any circle drawn on the surface of Earth that divides Earth into two equal hemispheres.

122. Designing a Little League Field For a 60-foot Little League Baseball field, the distance from home base to the nearest fence (or other obstruction) in fair territory should be a minimum of 200 feet. The commissioner of parks and recreation is making plans for a new 60-foot field. Because of limited ground availability, he will use the minimum required distance to the outfield fence. To increase safety, however, he plans to include a 10-foot wide warning track on the inside of the fence. To further increase safety, the fence and warning track will extend both directions into foul territory. In total, the arc formed by the outfield fence (including the extensions into the foul territories) will be subtended by a central angle at home plate measuring 96°, as illustrated.

(a) Determine the length of the outfield fence.
(b) Determine the area of the warning track.

Source: www.littleleague.org

[**Note:** There is a 90° angle between the two foul lines. Then there are two 3° angles between the foul lines and the dashed lines shown. The angle between the two dashed lines outside the 200-foot foul lines is 96°.]

123. Challenge Problem Let the Dog Roam A dog is attached to a 9-foot rope fastened to the outside corner of a fenced-in garden that measures 6 feet by 10 feet. Assuming that the dog cannot enter the garden, compute the exact area that the dog can wander. Write the exact area in square feet.*

124. Challenge Problem Geometry See the figure. The measure of arc $\overset{\frown}{BE}$ is 2π. Find the exact area of the portion of the rectangle $ABCD$ that falls outside of the circle whose center is at A.*

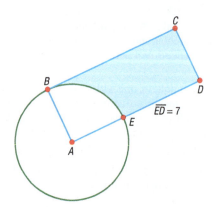

125. Challenge Problem Cycling A bicycle has a pedal drive wheel with radius 5.2 inches and a rear cog wheel with radius 1.8 inches. See the figure. How many revolutions will the pedals need to make to move the bicycle 50 feet if the wheels have a diameter of 30 inches? Round to the nearest tenth.

*Courtesy of the Joliet Junior College Mathematics Department

Explaining Concepts: Discussion and Writing

126. Do you prefer to measure angles using degrees or radians? Provide justification and a rationale for your choice.

127. What is 1 radian? What is 1 degree?

128. Which angle has the larger measure: 1 degree or 1 radian? Or are they equal?

129. Explain the difference between linear speed and angular speed.

130. For a circle of radius r, a central angle of θ degrees subtends an arc whose length s is $s = \dfrac{\pi}{180} r\theta$. Discuss whether this statement is true or false. Defend your position.

131. Discuss why ships and airplanes use nautical miles to measure distance. Explain the difference between a nautical mile and a statute mile.

132. Investigate the way that speed bicycles work. In particular, explain the differences and similarities between 5-speed and 9-speed derailleurs. Be sure to include a discussion of linear speed and angular speed.

Retain Your Knowledge

Problems 133–142 are based on material learned earlier in the course. The purpose of these problems is to keep the material fresh in your mind so that you are better prepared for the final exam.

133. Find the zero of $f(x) = 3x + 7$.

134. Solve: $5x^2 + 2 = 5 - 14x$

135. Write the function that is finally graphed if the following transformations are applied in order to the graph of $y = |x|$.
  1. Shift left 3 units.   2. Reflect about the x-axis.   3. Shift down 4 units.

136. Find the horizontal and vertical asymptotes of $R(x) = \dfrac{3x^2 - 12}{x^2 - 5x - 14}$.

137. Find c so the points $(2, c)$ and $(-1, 4)$ are on a line perpendicular to $2x - y = 5$.

138. Solve: $2\sqrt{x - 3} + 5 = 8$

139. Find the domain of $h(x) = \dfrac{3x}{x^2 - 9}$.

140. Find the difference quotient of $f(x) = 2x^3 - 5$.

141. Multiply: $(3x - 2)^3$

142. Use a graphing utility to determine the interval(s) where $g(x) = 3.2x^4 - 5.3x^2 + 2x - 1$ is decreasing.

'Are You Prepared?' Answers

1. $C = 2\pi r$ **2.** 2π

2.2 Trigonometric Functions: Unit Circle Approach

PREPARING FOR THIS SECTION *Before getting started, review the following:*

- Geometry Essentials (Section A.2, pp. A14–A19)
- Unit Circle (Section 1.2, p. 17)
- Symmetry (Section 1.2, pp. 12–14)
- Functions (Section 1.3, pp. 24–37)

Now Work the 'Are You Prepared?' problems on page 127.

OBJECTIVES
1. Find the Exact Values of the Trigonometric Functions Using a Point on the Unit Circle (p. 117)
2. Find the Exact Values of the Trigonometric Functions of Quadrantal Angles (p. 118)
3. Find the Exact Values of the Trigonometric Functions of $\dfrac{\pi}{4} = 45°$ (p. 120)
4. Find the Exact Values of the Trigonometric Functions of $\dfrac{\pi}{6} = 30°$ and $\dfrac{\pi}{3} = 60°$ (p. 121)
5. Find the Exact Values of the Trigonometric Functions for Integer Multiples of $\dfrac{\pi}{6} = 30°$, $\dfrac{\pi}{4} = 45°$, and $\dfrac{\pi}{3} = 60°$ (p. 123)
6. Use a Calculator to Approximate the Value of a Trigonometric Function (p. 125)
7. Use a Circle of Radius r to Evaluate the Trigonometric Functions (p. 126)

We now introduce the trigonometric functions using the unit circle.

The Unit Circle

Recall that the unit circle is a circle whose radius is 1 and whose center is at the origin of a rectangular coordinate system. Also recall that any circle of radius r has circumference of length $2\pi r$. Therefore, the unit circle (radius = 1) has a circumference of length 2π. In other words, for 1 revolution around the unit circle the length of the arc is 2π units.

The following discussion sets the stage for defining the trigonometric functions using the unit circle.

Let t be any real number. Position the t-axis so that it is vertical with the positive direction up. Place this t-axis in the xy-plane so that $t = 0$ is located at the point $(1, 0)$ in the xy-plane.

(a)

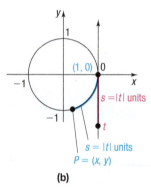

(b)

Figure 18

In Words

The point $P = (x, y)$ on the unit circle corresponding to a real number t is given by $(\cos t, \sin t)$.

If $t \geq 0$, let s be the distance from the origin to t on the t-axis. See the red portion of Figure 18(a). Beginning at the point $(1, 0)$ on the unit circle, travel $s = t$ units in the counterclockwise direction along the circle, to arrive at the point $P = (x, y)$. In this sense, the length $s = t$ units is being **wrapped** around the unit circle.

If $t < 0$, we begin at the point $(1, 0)$ on the unit circle and travel $s = |t|$ units in the clockwise direction to arrive at the point $P = (x, y)$. See Figure 18(b).

If $t > 2\pi$ or if $t < -2\pi$, it will be necessary to travel around the unit circle more than once before arriving at the point P. Do you see why?

Let's describe this process another way. Picture a string of length $s = |t|$ units being wrapped around a circle of radius 1 unit. Start wrapping the string around the circle at the point $(1, 0)$. If $t \geq 0$, wrap the string in the counterclockwise direction; if $t < 0$, wrap the string in the clockwise direction. The point $P = (x, y)$ is the point where the string ends.

This discussion tells us that, for any real number t, we can locate a unique point $P = (x, y)$ on the unit circle. We call P **the point on the unit circle that corresponds to t.** This is the important idea here. No matter what real number t is chosen, there is a unique point P on the unit circle corresponding to it. The coordinates of the point $P = (x, y)$ on the unit circle corresponding to the real number t are used to define the **six trigonometric functions of t.**

DEFINITION Trigonometric Functions of a Real Number

Let t be a real number and let $P = (x, y)$ be the point on the unit circle that corresponds to t.

The **sine function** associates with t the y-coordinate of P and is denoted by

$$\sin t = y$$

The **cosine function** associates with t the x-coordinate of P and is denoted by

$$\cos t = x$$

If $x \neq 0$, the **tangent function** associates with t the ratio of the y-coordinate to the x-coordinate of P and is denoted by

$$\tan t = \frac{y}{x}$$

If $y \neq 0$, the **cosecant function** is defined as

$$\csc t = \frac{1}{y}$$

If $x \neq 0$, the **secant function** is defined as

$$\sec t = \frac{1}{x}$$

If $y \neq 0$, the **cotangent function** is defined as

$$\cot t = \frac{x}{y}$$

Notice in these definitions that if $x = 0$, that is, if the point P is on the y-axis, then the tangent function and the secant function are undefined. Also, if $y = 0$, that is, if the point P is on the x-axis, then the cosecant function and the cotangent function are undefined.

Because we use the unit circle in these definitions of the trigonometric functions, they are sometimes referred to as **circular functions**.

1 Find the Exact Values of the Trigonometric Functions Using a Point on the Unit Circle

> EXAMPLE 1

Finding the Values of the Six Trigonometric Functions Using a Point on the Unit Circle

Let t be a real number and let $P = \left(-\frac{1}{2}, \frac{\sqrt{3}}{2}\right)$ be the point on the unit circle that corresponds to t. Find the values of $\sin t$, $\cos t$, $\tan t$, $\csc t$, $\sec t$, and $\cot t$.

Solution

See Figure 19. We follow the definition of the six trigonometric functions, using

$$P = \left(-\frac{1}{2}, \frac{\sqrt{3}}{2}\right) = (x, y). \text{ Then, with } x = -\frac{1}{2} \text{ and } y = \frac{\sqrt{3}}{2}, \text{ we have}$$

$$\sin t = y = \frac{\sqrt{3}}{2} \qquad \cos t = x = -\frac{1}{2} \qquad \tan t = \frac{y}{x} = \frac{\frac{\sqrt{3}}{2}}{-\frac{1}{2}} = -\sqrt{3}$$

$$\csc t = \frac{1}{y} = \frac{1}{\frac{\sqrt{3}}{2}} = \frac{2\sqrt{3}}{3} \qquad \sec t = \frac{1}{x} = \frac{1}{-\frac{1}{2}} = -2 \qquad \cot t = \frac{x}{y} = \frac{-\frac{1}{2}}{\frac{\sqrt{3}}{2}} = -\frac{\sqrt{3}}{3}$$

Figure 19

WARNING When writing the values of the trigonometric functions, do not forget the argument of the function.

$$\sin t = \frac{\sqrt{3}}{2} \quad \textbf{correct}$$

$$\sin = \frac{\sqrt{3}}{2} \quad \textbf{incorrect} \qquad \blacksquare$$

— **Now Work** PROBLEM **13**

Trigonometric Functions of Angles

Let $P = (x, y)$ be the point on the unit circle corresponding to the real number t. See Figure 20(a). Let θ be the angle in standard position, measured in radians, whose terminal side is the ray from the origin through P and whose arc length is $|t|$. See Figure 20(b). Since the unit circle has radius 1 unit, if $s = |t|$ units, then from the arc length formula $s = r|\theta|$, we have $\theta = t$ radians. See Figures 20(c) and (d).

Figure 20

(a)

(b)

(c)

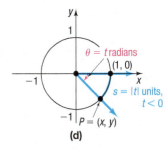

(d)

The point $P = (x, y)$ on the unit circle that corresponds to the real number t is also the point P on the terminal side of the angle $\theta = t$ radians. As a result, we can say that

$$\sin t = \sin \theta$$

Real number $\theta = t$ **radians**

and so on. We can now define the trigonometric functions of the angle θ.

DEFINITION Trigonometric Functions of an Angle θ

If $\theta = t$ radians, the **six trigonometric functions of the angle θ** are defined as

$$\sin\theta = \sin t \qquad \cos\theta = \cos t \qquad \tan\theta = \tan t$$
$$\csc\theta = \csc t \qquad \sec\theta = \sec t \qquad \cot\theta = \cot t$$

Even though the trigonometric functions can be viewed both as functions of real numbers and as functions of angles, it is customary to refer to trigonometric functions of real numbers and trigonometric functions of angles collectively as the *trigonometric functions*. We shall follow this practice from now on.

If an angle θ is measured in degrees, we shall use the degree symbol when writing a trigonometric function of θ, as, for example, in sin 30° and tan 45°. If an angle θ is measured in radians, then no symbol is used when writing a trigonometric function of θ, as, for example, in $\cos \pi$ and $\sec\dfrac{\pi}{3}$.

Finally, since the values of the trigonometric functions of an angle θ are determined by the coordinates of the point $P = (x, y)$ on the unit circle corresponding to θ, the units used to measure the angle θ are irrelevant. For example, it does not matter whether we write $\theta = \dfrac{\pi}{2}$ radians or $\theta = 90°$. The point on the unit circle corresponding to this angle is $P = (0, 1)$. As a result,

$$\sin\frac{\pi}{2} = \sin 90° = 1 \quad \text{and} \quad \cos\frac{\pi}{2} = \cos 90° = 0$$

2 Find the Exact Values of the Trigonometric Functions of Quadrantal Angles

To find the exact value of a trigonometric function of an angle θ or a real number t requires that we locate the point $P = (x, y)$ on the unit circle that corresponds to t. This is not always easy to do. In the examples that follow, we will evaluate the trigonometric functions of certain angles or real numbers for which this process is relatively easy. A calculator will be used to evaluate the trigonometric functions of most other angles.

EXAMPLE 2

Finding the Exact Values of the Six Trigonometric Functions of Quadrantal Angles

Find the exact values of the six trigonometric functions of each of the following angles:

(a) $\theta = 0 = 0°$

(b) $\theta = \dfrac{\pi}{2} = 90°$

(c) $\theta = \pi = 180°$

(d) $\theta = \dfrac{3\pi}{2} = 270°$

Solution

(a) The point on the unit circle that corresponds to $\theta = 0 = 0°$ is $P = (1, 0)$. See Figure 21(a). Then using $x = 1$ and $y = 0$

$$\sin 0 = \sin 0° = y = 0 \qquad \cos 0 = \cos 0° = x = 1$$

$$\tan 0 = \tan 0° = \frac{y}{x} = 0 \qquad \sec 0 = \sec 0° = \frac{1}{x} = 1$$

Since the y-coordinate of P is 0, $\csc 0$ and $\cot 0$ are not defined.

Figure 21 (a) $\theta = 0 = 0°$

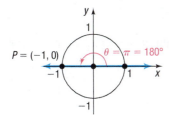

Figure 21 (b) $\theta = \dfrac{\pi}{2} = 90°$

Figure 21 (c) $\theta = \pi = 180°$

$\theta = \dfrac{3\pi}{2} = 270°$

$P = (0, -1)$

Figure 21 (d) $\theta = \dfrac{3\pi}{2} = 270°$

(b) The point on the unit circle that corresponds to $\theta = \dfrac{\pi}{2} = 90°$ is $P = (0, 1)$. See Figure 21(b). Then

$$\sin\frac{\pi}{2} = \sin 90° = y = 1 \qquad \cos\frac{\pi}{2} = \cos 90° = x = 0$$

$$\csc\frac{\pi}{2} = \csc 90° = \frac{1}{y} = 1 \qquad \cot\frac{\pi}{2} = \cot 90° = \frac{x}{y} = 0$$

Since the x-coordinate of P is 0, $\tan\dfrac{\pi}{2}$ and $\sec\dfrac{\pi}{2}$ are not defined.

(c) The point on the unit circle that corresponds to $\theta = \pi = 180°$ is $P = (-1, 0)$. See Figure 21(c). Then

$$\sin\pi = \sin 180° = y = 0 \qquad \cos\pi = \cos 180° = x = -1$$

$$\tan\pi = \tan 180° = \frac{y}{x} = 0 \qquad \sec\pi = \sec 180° = \frac{1}{x} = -1$$

Since the y-coordinate of P is 0, $\csc\pi$ and $\cot\pi$ are not defined.

(d) The point on the unit circle that corresponds to $\theta = \dfrac{3\pi}{2} = 270°$ is $P = (0, -1)$. See Figure 21(d). Then

$$\sin\frac{3\pi}{2} = \sin 270° = y = -1 \qquad \cos\frac{3\pi}{2} = \cos 270° = x = 0$$

$$\csc\frac{3\pi}{2} = \csc 270° = \frac{1}{y} = -1 \qquad \cot\frac{3\pi}{2} = \cot 270° = \frac{x}{y} = 0$$

Since the x-coordinate of P is 0, $\tan\dfrac{3\pi}{2}$ and $\sec\dfrac{3\pi}{2}$ are not defined.

Table 2 summarizes the values of the trigonometric functions found in Example 2.

Table 2

	Quadrantal Angles						
θ (Radians)	θ (Degrees)	$\sin\theta$	$\cos\theta$	$\tan\theta$	$\csc\theta$	$\sec\theta$	$\cot\theta$
0	0°	0	1	0	Not defined	1	Not defined
$\dfrac{\pi}{2}$	90°	1	0	Not defined	1	Not defined	0
π	180°	0	-1	0	Not defined	-1	Not defined
$\dfrac{3\pi}{2}$	270°	-1	0	Not defined	-1	Not defined	0

There is no need to memorize Table 2. To find the value of a trigonometric function of a quadrantal angle, draw the angle and apply the definition, as we did in Example 2.

EXAMPLE 3   **Finding Exact Values of the Trigonometric Functions of Quadrantal Angles**

Find the exact value of each of the following angles:

(a) $\sin(3\pi)$ (b) $\cos(-270°)$

Solution   (a) See Figure 22(a). The point P on the unit circle that corresponds to $\theta = 3\pi$ is $P = (-1, 0)$, so $\sin(3\pi) = y = 0$.   (b) See Figure 22(b). The point P on the unit circle that corresponds to $\theta = -270°$ is $P = (0, 1)$, so $\cos(-270°) = x = 0$.

Figure 22   **(a)** $\theta = 3\pi$

(b) $\theta = -270°$

 Now Work PROBLEMS **21** AND **61**

3 Find the Exact Values of the Trigonometric Functions of $\dfrac{\pi}{4} = 45°$

EXAMPLE 4   **Finding the Exact Values of the Trigonometric Functions of $\dfrac{\pi}{4} = 45°$**

Find the exact values of the six trigonometric functions of $\dfrac{\pi}{4} = 45°$.

Solution   We seek the coordinates of the point $P = (x, y)$ on the unit circle that corresponds to $\theta = \dfrac{\pi}{4} = 45°$. See Figure 23. First, observe that P lies on the line $y = x$. (Do you see why? Since $\theta = 45° = \dfrac{1}{2} \cdot 90°$, P must lie on the line that bisects quadrant I.) Since $P = (x, y)$ also lies on the unit circle, $x^2 + y^2 = 1$, it follows that

$$x^2 + y^2 = 1$$
$$x^2 + x^2 = 1 \quad \color{blue}{y = x, x > 0, y > 0}$$
$$2x^2 = 1$$
$$x = \frac{1}{\sqrt{2}} = \frac{\sqrt{2}}{2} \qquad y = \frac{\sqrt{2}}{2}$$

Figure 23 $\theta = \dfrac{\pi}{4} = 45°$

Then

$$\sin\frac{\pi}{4} = \sin 45° = \frac{\sqrt{2}}{2} \qquad \cos\frac{\pi}{4} = \cos 45° = \frac{\sqrt{2}}{2} \qquad \tan\frac{\pi}{4} = \tan 45° = \frac{\frac{\sqrt{2}}{2}}{\frac{\sqrt{2}}{2}} = 1$$

$$\csc\frac{\pi}{4} = \csc 45° = \frac{1}{\frac{\sqrt{2}}{2}} = \sqrt{2} \qquad \sec\frac{\pi}{4} = \sec 45° = \frac{1}{\frac{\sqrt{2}}{2}} = \sqrt{2} \qquad \cot\frac{\pi}{4} = \cot 45° = \frac{\frac{\sqrt{2}}{2}}{\frac{\sqrt{2}}{2}} = 1$$

EXAMPLE 5   **Finding the Exact Value of a Trigonometric Expression**

Find the exact value of each expression.

(a) $\sin 45° \cos 180°$ (b) $\tan \dfrac{\pi}{4} - \sin \dfrac{3\pi}{2}$ (c) $\left(\sec \dfrac{\pi}{4} \right)^2 + \csc \dfrac{\pi}{2}$

Solution   (a) $\sin 45° \cos 180° = \dfrac{\sqrt{2}}{2} \cdot (-1) = -\dfrac{\sqrt{2}}{2}$

        ↑   ↑

     From Example 4   **From Table 2**

(b) $\tan \dfrac{\pi}{4} - \sin \dfrac{3\pi}{2} = 1 - (-1) = 2$

       ↑   ↑

     From Example 4   **From Table 2**

(c) $\left(\sec \dfrac{\pi}{4} \right)^2 + \csc \dfrac{\pi}{2} = (\sqrt{2})^2 + 1 = 2 + 1 = 3$

━━━━━▶ **Now Work** PROBLEM 35

4 Find The Exact Values of the Trigonometric Functions of $\dfrac{\pi}{6} = 30°$ and $\dfrac{\pi}{3} = 60°$

Consider a right triangle in which one of the angles is $\dfrac{\pi}{6} = 30°$. It then follows that the third angle is $\dfrac{\pi}{3} = 60°$. Figure 24(a) illustrates such a triangle with hypotenuse of length 1. Our problem is to determine a and b.

Begin by placing next to the triangle in Figure 24(a) another triangle congruent to the first, as shown in Figure 24(b). Notice that we now have a triangle whose three angles each equal 60°. This triangle is therefore equilateral, so each side is of length 1.

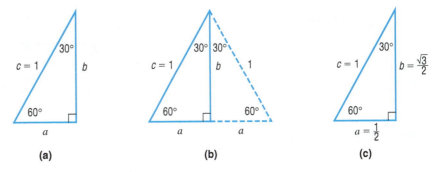

Figure 24 30°–60°–90° triangle   **(a)**   **(b)**   **(c)**

This means the base is $2a = 1$, and so $a = \dfrac{1}{2}$. By the Pythagorean Theorem, b satisfies the equation $a^2 + b^2 = c^2$, so we have

$$a^2 + b^2 = c^2$$

$$\dfrac{1}{4} + b^2 = 1 \qquad\qquad a = \dfrac{1}{2}, c = 1$$

$$b^2 = 1 - \dfrac{1}{4} = \dfrac{3}{4}$$

$$b = \dfrac{\sqrt{3}}{2} \qquad\qquad \text{\textit{b} > 0 because \textit{b} is the length of the side of a triangle.}$$

This results in Figure 24(c).

EXAMPLE 6   Finding the Exact Values of the Trigonometric Functions of $\frac{\pi}{3} = 60°$

Find the exact values of the six trigonometric functions of $\frac{\pi}{3} = 60°$.

Solution   Position the triangle in Figure 24(c) so that the 60° angle is in standard position. See Figure 25. The point on the unit circle that corresponds to $\theta = \frac{\pi}{3} = 60°$ is $P = \left(\frac{1}{2}, \frac{\sqrt{3}}{2}\right)$. Then

$$\sin \frac{\pi}{3} = \sin 60° = \frac{\sqrt{3}}{2} \qquad \cos \frac{\pi}{3} = \cos 60° = \frac{1}{2}$$

$$\csc \frac{\pi}{3} = \csc 60° = \frac{1}{\frac{\sqrt{3}}{2}} = \frac{2}{\sqrt{3}} = \frac{2\sqrt{3}}{3} \qquad \sec \frac{\pi}{3} = \sec 60° = \frac{1}{\frac{1}{2}} = 2$$

$$\tan \frac{\pi}{3} = \tan 60° = \frac{\frac{\sqrt{3}}{2}}{\frac{1}{2}} = \sqrt{3} \qquad \cot \frac{\pi}{3} = \cot 60° = \frac{\frac{1}{2}}{\frac{\sqrt{3}}{2}} = \frac{1}{\sqrt{3}} = \frac{\sqrt{3}}{3}$$

Figure 25 $\theta = \frac{\pi}{3} = 60°$

EXAMPLE 7   Finding the Exact Values of the Trigonometric Functions of $\frac{\pi}{6} = 30°$

Find the exact values of the trigonometric functions of $\frac{\pi}{6} = 30°$.

Solution   Position the triangle in Figure 24(c) so that the 30° angle is in standard position. See Figure 26. The point on the unit circle that corresponds to $\theta = \frac{\pi}{6} = 30°$ is $P = \left(\frac{\sqrt{3}}{2}, \frac{1}{2}\right)$. Then

$$\sin \frac{\pi}{6} = \sin 30° = \frac{1}{2} \qquad \cos \frac{\pi}{6} = \cos 30° = \frac{\sqrt{3}}{2}$$

$$\csc \frac{\pi}{6} = \csc 30° = \frac{1}{\frac{1}{2}} = 2 \qquad \sec \frac{\pi}{6} = \sec 30° = \frac{1}{\frac{\sqrt{3}}{2}} = \frac{2}{\sqrt{3}} = \frac{2\sqrt{3}}{3}$$

$$\tan \frac{\pi}{6} = \tan 30° = \frac{\frac{1}{2}}{\frac{\sqrt{3}}{2}} = \frac{1}{\sqrt{3}} = \frac{\sqrt{3}}{3} \qquad \cot \frac{\pi}{6} = \cot 30° = \frac{\frac{\sqrt{3}}{2}}{\frac{1}{2}} = \sqrt{3}$$

Figure 26 $\theta = \frac{\pi}{6} = 30°$

Table 3 summarizes the information just derived for $\frac{\pi}{6} = 30°, \frac{\pi}{4} = 45°$, and $\frac{\pi}{3} = 60°$. Until you memorize the entries in Table 3, you should draw an appropriate diagram to determine the values given in the table.

Table 3

θ (Radians)	θ (Degrees)	$\sin \theta$	$\cos \theta$	$\tan \theta$	$\csc \theta$	$\sec \theta$	$\cot \theta$
$\dfrac{\pi}{6}$	30°	$\dfrac{1}{2}$	$\dfrac{\sqrt{3}}{2}$	$\dfrac{\sqrt{3}}{3}$	2	$\dfrac{2\sqrt{3}}{3}$	$\sqrt{3}$
$\dfrac{\pi}{4}$	45°	$\dfrac{\sqrt{2}}{2}$	$\dfrac{\sqrt{2}}{2}$	1	$\sqrt{2}$	$\sqrt{2}$	1
$\dfrac{\pi}{3}$	60°	$\dfrac{\sqrt{3}}{2}$	$\dfrac{1}{2}$	$\sqrt{3}$	$\dfrac{2\sqrt{3}}{3}$	2	$\dfrac{\sqrt{3}}{3}$

 Now Work PROBLEM 41

EXAMPLE 8

Constructing a Rain Gutter

A rain gutter is to be constructed of aluminum sheets 12 inches wide. After marking off a length of 4 inches from each edge, this length is bent up at an angle θ. See Figure 27. The area A of the opening may be expressed as a function of θ as

$$A(\theta) = 16 \sin \theta (\cos \theta + 1)$$

Find the area A of the opening for $\theta = 30°$, $\theta = 45°$, and $\theta = 60°$.

Figure 27

Solution For $\theta = 30°$: $\quad A(30°) = 16 \sin 30° (\cos 30° + 1)$

$$= 16 \cdot \frac{1}{2} \cdot \left(\frac{\sqrt{3}}{2} + 1 \right) = 4\sqrt{3} + 8 \approx 14.93$$

The area of the opening for $\theta = 30°$ is about 14.93 square inches.

For $\theta = 45°$: $\quad A(45°) = 16 \sin 45° (\cos 45° + 1)$

$$= 16 \cdot \frac{\sqrt{2}}{2} \cdot \left(\frac{\sqrt{2}}{2} + 1 \right) = 8 + 8\sqrt{2} \approx 19.31$$

The area of the opening for $\theta = 45°$ is about 19.31 square inches.

For $\theta = 60°$: $\quad A(60°) = 16 \sin 60° (\cos 60° + 1)$

$$= 16 \cdot \frac{\sqrt{3}}{2} \cdot \left(\frac{1}{2} + 1 \right) = 12\sqrt{3} \approx 20.78$$

The area of the opening for $\theta = 60°$ is about 20.78 square inches.

5 Find the Exact Values of the Trigonometric Functions for Integer Multiples of $\dfrac{\pi}{6} = 30°$, $\dfrac{\pi}{4} = 45°$, and $\dfrac{\pi}{3} = 60°$

We know the exact values of the trigonometric functions of $\dfrac{\pi}{4} = 45°$. Using symmetry, we can find the exact values of the trigonometric functions of $\dfrac{3\pi}{4} = 135°$, $\dfrac{5\pi}{4} = 225°$, and $\dfrac{7\pi}{4} = 315°$. The point on the unit circle corresponding to $\dfrac{\pi}{4} = 45°$ is $\left(\dfrac{\sqrt{2}}{2}, \dfrac{\sqrt{2}}{2} \right)$.

See Figure 28. Using symmetry with respect to the y-axis, the point $\left(-\dfrac{\sqrt{2}}{2}, \dfrac{\sqrt{2}}{2} \right)$ is the point on the unit circle that corresponds to the angle $\dfrac{3\pi}{4} = 135°$. Similarly, using symmetry with respect to the origin, the point $\left(-\dfrac{\sqrt{2}}{2}, -\dfrac{\sqrt{2}}{2} \right)$ is the point on the unit

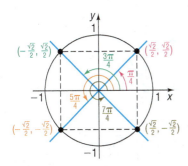

Figure 28

circle that corresponds to the angle $\frac{5\pi}{4} = 225°$. Finally, using symmetry with respect to the *x*-axis, the point $\left(\frac{\sqrt{2}}{2}, -\frac{\sqrt{2}}{2}\right)$ is the point on the unit circle that corresponds to the angle $\frac{7\pi}{4} = 315°$.

EXAMPLE 9

Finding Exact Values for Multiples of $\frac{\pi}{4} = 45°$

Find the exact value of each expression.

(a) $\cos\frac{5\pi}{4}$ (b) $\sin 135°$ (c) $\tan 315°$ (d) $\sin\left(-\frac{\pi}{4}\right)$ (e) $\cos\frac{11\pi}{4}$

Solution

(a) From Figure 28, we see the point $\left(-\frac{\sqrt{2}}{2}, -\frac{\sqrt{2}}{2}\right)$ corresponds to $\frac{5\pi}{4}$, so

$$\cos\frac{5\pi}{4} = x = -\frac{\sqrt{2}}{2}.$$

(b) Since $135° = \frac{3\pi}{4}$, the point $\left(-\frac{\sqrt{2}}{2}, \frac{\sqrt{2}}{2}\right)$ corresponds to $135°$, so $\sin 135° = \frac{\sqrt{2}}{2}$.

(c) Since $315° = \frac{7\pi}{4}$, the point $\left(\frac{\sqrt{2}}{2}, -\frac{\sqrt{2}}{2}\right)$ corresponds to $315°$, so

$$\tan 315° = \frac{-\dfrac{\sqrt{2}}{2}}{\dfrac{\sqrt{2}}{2}} = -1.$$

(d) The point $\left(\frac{\sqrt{2}}{2}, -\frac{\sqrt{2}}{2}\right)$ corresponds to $-\frac{\pi}{4}$, so $\sin\left(-\frac{\pi}{4}\right) = -\frac{\sqrt{2}}{2}$.

(e) The point $\left(-\frac{\sqrt{2}}{2}, \frac{\sqrt{2}}{2}\right)$ corresponds to $\frac{11\pi}{4}$, so $\cos\frac{11\pi}{4} = -\frac{\sqrt{2}}{2}$.

Now Work PROBLEMS **51** AND **55**

The use of symmetry also provides information about certain integer multiples of the angles $\frac{\pi}{6} = 30°$ and $\frac{\pi}{3} = 60°$. See Figures 29 and 30.

Figure 29

Figure 30

EXAMPLE 10

Finding Exact Values for Multiples of $\dfrac{\pi}{6} = 30°$ or $\dfrac{\pi}{3} = 60°$

Based on Figures 29 and 30, we see that

(a) $\cos 210° = \cos \dfrac{7\pi}{6} = -\dfrac{\sqrt{3}}{2}$

(b) $\sin(-60°) = \sin\left(-\dfrac{\pi}{3}\right) = -\dfrac{\sqrt{3}}{2}$

(c) $\tan \dfrac{5\pi}{3} = \dfrac{-\dfrac{\sqrt{3}}{2}}{\dfrac{1}{2}} = -\sqrt{3}$

(d) $\cos \dfrac{8\pi}{3} = \cos \dfrac{2\pi}{3} = -\dfrac{1}{2}$

 Now Work PROBLEM **47**

6 Use a Calculator to Approximate the Value of a Trigonometric Function

WARNING On your calculator the second functions $\sin^{-1}$, $\cos^{-1}$, and $\tan^{-1}$ do not represent the reciprocal of sin, cos, and tan. ∎

Before getting started, you must first decide whether to enter the angle in the calculator using radians or degrees and then set the calculator to the correct MODE. Check your instruction manual to find out how your calculator handles degrees and radians. Your calculator has keys marked $\boxed{\sin}$, $\boxed{\cos}$, and $\boxed{\tan}$. To find the values of the remaining three trigonometric functions, secant, cosecant, and cotangent, use the fact that, if $P = (x, y)$ is a point on the unit circle on the terminal side of θ, then

$$\sec \theta = \frac{1}{x} = \frac{1}{\cos \theta} \qquad \csc \theta = \frac{1}{y} = \frac{1}{\sin \theta} \qquad \cot \theta = \frac{x}{y} = \frac{1}{\dfrac{y}{x}} = \frac{1}{\tan \theta}$$

EXAMPLE 11

Using a Calculator to Approximate the Value of a Trigonometric Function

Use a calculator to find the approximate value of:

(a) $\cos 48°$ (b) $\csc 21°$ (c) $\tan \dfrac{\pi}{12}$

Express your answer rounded to two decimal places.

Solution

(a) First, set the MODE to receive degrees. Rounded to two decimal places,

$$\cos 48° = 0.6691306 \approx 0.67$$

(b) Most calculators do not have a csc key. The manufacturers assume that the user knows some trigonometry. To find the value of $\csc 21°$, use the fact that $\csc 21° = \dfrac{1}{\sin 21°}$. Rounded to two decimal places,

$$\csc 21° \approx 2.79$$

(c) Set the MODE to receive radians. Figure 31 shows the solution using a TI-84 Plus C graphing calculator. Rounded to two decimal places,

$$\tan \dfrac{\pi}{12} \approx 0.27$$

Figure 31

 Now Work PROBLEM **65**

7 Use a Circle of Radius r to Evaluate the Trigonometric Functions

Until now, finding the exact value of a trigonometric function of an angle θ required that we locate the corresponding point $P = (x, y)$ on the unit circle. In fact, though, any circle whose center is at the origin can be used.

Let θ be any nonquadrantal angle placed in standard position. Let $P = (x, y)$ be the point on the circle $x^2 + y^2 = r^2$ that corresponds to θ, and let $P^* = (x^*, y^*)$ be the point on the unit circle that corresponds to θ. See Figure 32, where θ is shown in quadrant II.

Notice that the triangles OA^*P^* and OAP are similar; as a result, the ratios of corresponding sides are equal.

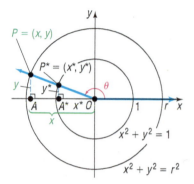

Figure 32

$$\frac{y^*}{1} = \frac{y}{r} \qquad \frac{x^*}{1} = \frac{x}{r} \qquad \frac{y^*}{x^*} = \frac{y}{x}$$

$$\frac{1}{y^*} = \frac{r}{y} \qquad \frac{1}{x^*} = \frac{r}{x} \qquad \frac{x^*}{y^*} = \frac{x}{y}$$

These results lead us to formulate the following theorem:

> **THEOREM**
>
> For an angle θ in standard position, let $P = (x, y)$ be the point on the terminal side of θ that is also on the circle $x^2 + y^2 = r^2$. Then
>
> $$\sin \theta = \frac{y}{r} \qquad\qquad \cos \theta = \frac{x}{r} \qquad\qquad \tan \theta = \frac{y}{x} \quad x \neq 0$$
>
> $$\csc \theta = \frac{r}{y} \quad y \neq 0 \qquad \sec \theta = \frac{r}{x} \quad x \neq 0 \qquad \cot \theta = \frac{x}{y} \quad y \neq 0$$

EXAMPLE 12 **Finding the Exact Values of the Six Trigonometric Functions**

Find the exact values of each of the six trigonometric functions of an angle θ if $(4, -3)$ is a point on its terminal side in standard position.

Solution See Figure 33. The point $(4, -3)$ is on a circle with center at the origin that has a radius of $r = \sqrt{4^2 + (-3)^2} = \sqrt{16 + 9} = \sqrt{25} = 5$.

For the point $(x, y) = (4, -3)$, we have $x = 4$ and $y = -3$. Since $r = 5$, we find

$$\sin \theta = \frac{y}{r} = -\frac{3}{5} \qquad \cos \theta = \frac{x}{r} = \frac{4}{5} \qquad \tan \theta = \frac{y}{x} = -\frac{3}{4}$$

$$\csc \theta = \frac{r}{y} = -\frac{5}{3} \qquad \sec \theta = \frac{r}{x} = \frac{5}{4} \qquad \cot \theta = \frac{x}{y} = -\frac{4}{3}$$

Figure 33

Now Work PROBLEM **77**

Historical Feature

The name *sine* for the sine function arose from a medieval confusion. Its origin is from the Sanskrit word *jīva* (meaning "chord"), first used in India by Araybhata the Elder (AD 510). He really meant half-chord, but abbreviated it. This was brought into Arabic as *jība*, which was meaningless. Because the proper Arabic word *jaib* would be written the same way (short vowels are not written out in Arabic), *jība* was pronounced as *jaib*, which meant "bosom" or "hollow," and *jaib* remains as the Arabic word for sine to this day. Scholars translating the Arabic works into Latin found that the word *sinus* also meant "bosom" or "hollow," and from *sinus* we get *sine*.

The name *tangent*, due to Thomas Finck (1583), can be understood by looking at Figure 34. The line segment $\overline{DC}$ is tangent to the circle at C. If $d(O, B) = d(O, C) = 1$, then the length of the line segment $\overline{DC}$ is

$$d(D, C) = \frac{d(D, C)}{1} = \frac{d(D, C)}{d(O, C)} = \tan \alpha$$

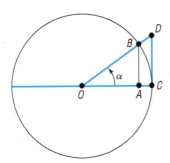

Figure 34

The old name for the tangent is *umbra versa* (meaning turned shadow), referring to the use of the tangent in solving height problems with shadows.

The names of the remaining functions came about as follows. If α and β are complementary angles, then $\cos \alpha = \sin \beta$. Because β is the complement of α, it was natural to write the cosine of α as sin co α. Probably for reasons involving ease of pronunciation, the *co* migrated to the front, and then cosine received a three-letter abbreviation to match sin, sec, and tan. The two other cofunctions were similarly treated, except that the long forms *cotan* and *cosec* survive to this day in some countries.

2.2 Assess Your Understanding

'Are You Prepared?' *Answers are given at the end of these exercises. If you get a wrong answer, read the pages listed in* red.

1. In a right triangle, with legs a and b and hypotenuse c, the Pythagorean Theorem states that _____ . (p. A14)

2. The value of the function $f(x) = 3x - 7$ at 5 is ___. (pp. 29–31)

3. **True or False** For a function $y = f(x)$, for each x in the domain, there is exactly one element y in the range. (pp. 26–28)

4. If two triangles are similar, then corresponding angles are _____ and the lengths of corresponding sides are _____ . (pp. A16–A19)

5. What point is symmetric with respect to the y-axis to the point $\left(\dfrac{1}{2}, \dfrac{\sqrt{3}}{2}\right)$? (pp. 12–14)

6. If (x, y) is a point on the unit circle in quadrant IV and if $x = \dfrac{\sqrt{3}}{2}$, what is y? (p. 17)

Concepts and Vocabulary

7. **Multiple Choice** Which function takes as input a real number t that corresponds to a point $P = (x, y)$ on the unit circle and outputs the x-coordinate?
   (a) sine **(b)** cosine **(c)** tangent **(d)** secant

8. The point on the unit circle that corresponds to $\theta = \dfrac{\pi}{2}$ is $P = $ _____ .

9. The point on the unit circle that corresponds to $\theta = \dfrac{\pi}{4}$ is

   $P = $ _____ .

10. **True or False** Exact values can be found for the sine of any angle.

11. For any angle θ in standard position, let $P = (x, y)$ be the point on the terminal side of θ that is also on the circle $x^2 + y^2 = r^2$. Then, $\sin \theta = $ __ and $\cos \theta = $ __.

12. **Multiple Choice** The point on the unit circle that corresponds to $\theta = \dfrac{\pi}{3}$ is

   (a) $\left(\dfrac{1}{2}, \dfrac{\sqrt{3}}{2}\right)$ **(b)** $\left(\dfrac{\sqrt{2}}{2}, \dfrac{\sqrt{2}}{2}\right)$

   (c) $\left(\dfrac{\sqrt{3}}{2}, \dfrac{1}{2}\right)$ **(d)** $\left(\sqrt{3}, \dfrac{2\sqrt{3}}{3}\right)$

Skill Building

In Problems 13–20, $P = (x, y)$ is the point on the unit circle that corresponds to a real number t. Find the exact values of the six trigonometric functions of t.

13. $\left(\dfrac{\sqrt{3}}{2}, \dfrac{1}{2}\right)$

14. $\left(\dfrac{1}{2}, -\dfrac{\sqrt{3}}{2}\right)$

15. $\left(-\dfrac{2}{5}, \dfrac{\sqrt{21}}{5}\right)$

16. $\left(-\dfrac{1}{5}, \dfrac{2\sqrt{6}}{5}\right)$

17. $\left(-\dfrac{\sqrt{2}}{2}, \dfrac{\sqrt{2}}{2}\right)$

18. $\left(\dfrac{\sqrt{2}}{2}, \dfrac{\sqrt{2}}{2}\right)$

19. $\left(\dfrac{2\sqrt{2}}{3}, -\dfrac{1}{3}\right)$

20. $\left(-\dfrac{\sqrt{5}}{3}, -\dfrac{2}{3}\right)$

In Problems 21–30, find the exact value. Do not use a calculator.

21. $\sin \dfrac{11\pi}{2}$ **22.** $\cos(7\pi)$ **23.** $\tan(6\pi)$ **24.** $\cot \dfrac{7\pi}{2}$ **25.** $\csc \dfrac{11\pi}{2}$

26. $\sec(8\pi)$ **27.** $\cos\left(-\dfrac{3\pi}{2}\right)$ **28.** $\sin(-3\pi)$ **29.** $\sec(-\pi)$ **30.** $\tan(-3\pi)$

In Problems 31–46, find the exact value of each expression. Do not use a calculator.

31. $\sin 45° + \cos 60°$ **32.** $\sin 30° - \cos 45°$ **33.** $\sin 90° + \tan 45°$ **34.** $\cos 180° - \sin 180°$

35. $\sin 45° \cos 45°$ **36.** $\tan 45° \cos 30°$ **37.** $\csc 45° \tan 60°$ **38.** $\sec 30° \cot 45°$

39. $4 \sin 90° - 3 \tan 180°$ **40.** $5 \cos 90° - 8 \sin 270°$ **41.** $2 \sin \dfrac{\pi}{3} - 3 \tan \dfrac{\pi}{6}$ **42.** $2 \sin \dfrac{\pi}{4} + 3 \tan \dfrac{\pi}{4}$

43. $2 \sec \dfrac{\pi}{4} + 4 \cot \dfrac{\pi}{3}$ **44.** $3 \csc \dfrac{\pi}{3} + \cot \dfrac{\pi}{4}$ **45.** $\csc \dfrac{\pi}{2} + \cot \dfrac{\pi}{2}$ **46.** $\sec \pi - \csc \dfrac{\pi}{2}$

In Problems 47–64, find the exact values of the six trigonometric functions of the given angle. If any are not defined, say "not defined."
Do not use a calculator.

47. $\dfrac{2\pi}{3}$ **48.** $\dfrac{5\pi}{6}$ **49.** $210°$ **50.** $240°$ **51.** $\dfrac{3\pi}{4}$ **52.** $\dfrac{11\pi}{4}$

53. $\dfrac{8\pi}{3}$ **54.** $\dfrac{13\pi}{6}$ **55.** $405°$ **56.** $390°$ **57.** $-\dfrac{\pi}{6}$ **58.** $-\dfrac{\pi}{3}$

59. $-135°$ **60.** $-240°$ **61.** $\dfrac{5\pi}{2}$ **62.** 5π **63.** $-\dfrac{14\pi}{3}$ **64.** $-\dfrac{13\pi}{6}$

In Problems 65–76, use a calculator to find the approximate value of each expression rounded to two decimal places.

65. $\sin 28°$ **66.** $\cos 14°$ **67.** $\sec 21°$ **68.** $\cot 70°$

69. $\tan \dfrac{\pi}{10}$ **70.** $\sin \dfrac{\pi}{8}$ **71.** $\cot \dfrac{\pi}{12}$ **72.** $\csc \dfrac{5\pi}{13}$

73. $\sin 1$ **74.** $\tan 1$ **75.** $\sin 1°$ **76.** $\tan 1°$

In Problems 77–84, a point on the terminal side of an angle θ in standard position is given. Find the exact value of each of the six
trigonometric functions of θ.

77. $(-3, 4)$ **78.** $(5, -12)$ **79.** $(2, -3)$ **80.** $(-1, -2)$

81. $(-2, -2)$ **82.** $(-1, 1)$ **83.** $\left(\dfrac{1}{3}, \dfrac{1}{4}\right)$ **84.** $(0.3, 0.4)$

---

85. Find the exact value of:

     $\sin 45° + \sin 135° + \sin 225° + \sin 315°$

86. Find the exact value of:

     $\tan 60° + \tan 150°$

87. Find the exact value of:

     $\sin 40° + \sin 130° + \sin 220° + \sin 310°$

88. Find the exact value of:

     $\tan 40° + \tan 140°$

89. If $f(\theta) = \sin \theta = 0.1$, find $f(\theta + \pi)$.

90. If $f(\theta) = \cos \theta = 0.3$, find $f(\theta + \pi)$.

91. If $f(\theta) = \tan \theta = 3$, find $f(\theta + \pi)$.

92. If $f(\theta) = \cot \theta = -2$, find $f(\theta + \pi)$.

93. If $\sin \theta = \dfrac{1}{5}$, find $\csc \theta$.

94. If $\cos \theta = \dfrac{2}{3}$, find $\sec \theta$.

In Problems 95–106, $f(\theta) = \sin \theta$ and $g(\theta) = \cos \theta$. Find the exact value of each function below if $\theta = 60°$. Do not use a calculator.

95. $f(\theta)$ **96.** $g(\theta)$ **97.** $f\left(\dfrac{\theta}{2}\right)$ **98.** $g\left(\dfrac{\theta}{2}\right)$

99. $[f(\theta)]^2$ **100.** $[g(\theta)]^2$ **101.** $f(2\theta)$ **102.** $g(2\theta)$

103. $2f(\theta)$ **104.** $2g(\theta)$ **105.** $f(-\theta)$ **106.** $g(-\theta)$

Mixed Practice *In Problems 107–116,* $f(x) = \sin x, g(x) = \cos x, h(x) = 2x,$ *and* $p(x) = \dfrac{x}{2}.$ *Find the value of each of the following:*

107. $(f + g)(30°)$

108. $(f - g)(60°)$

109. $(f \cdot g)\left(\dfrac{3\pi}{4}\right)$

110. $(f \cdot g)\left(\dfrac{4\pi}{3}\right)$

111. $f\left(h\left(\dfrac{\pi}{6}\right)\right)$

112. $g(p(60°))$

113. $p(g(315°))$

114. $h\left(f\left(\dfrac{5\pi}{6}\right)\right)$

115. (a) Find $f\left(\dfrac{\pi}{4}\right)$. What point is on the graph of f?

    (b) Assuming $0 \le x \le \dfrac{\pi}{2}$, f is one-to-one.* Use the result of part (a) to find a point on the graph of f^{-1}.

    (c) What point is on the graph of $y = f\left(x + \dfrac{\pi}{4}\right) - 3$ if $x = \dfrac{\pi}{4}$?

116. (a) Find $g\left(\dfrac{\pi}{6}\right)$. What point is on the graph of g?

    (b) Assuming $0 \le x \le \dfrac{\pi}{2}$, g is one-to-one.* Use the result of part (a) to find a point on the graph of g^{-1}.

    (c) What point is on the graph of $y = 2g\left(x - \dfrac{\pi}{6}\right)$ if $x = \dfrac{\pi}{6}$?

Applications and Extensions

117. Find two negative and three positive angles, expressed in radians, for which the point on the unit circle that corresponds to each angle is $\left(\dfrac{1}{2}, \dfrac{\sqrt{3}}{2}\right)$.

118. Find two negative and three positive angles, expressed in radians, for which the point on the unit circle that corresponds to each angle is $\left(-\dfrac{\sqrt{2}}{2}, \dfrac{\sqrt{2}}{2}\right)$.

119. Use a calculator in radian mode to complete the following table.

What do you conjecture about the value of $f(\theta) = \dfrac{\sin \theta}{\theta}$ as θ approaches 0?

θ	0.5	0.4	0.2	0.1	0.01	0.001	0.0001	0.00001
$\sin \theta$								
$f(\theta) = \dfrac{\sin \theta}{\theta}$								

120. Use a calculator in radian mode to complete the following table.

What do you conjecture about the value of $g(\theta) = \dfrac{\cos \theta - 1}{\theta}$ as θ approaches 0?

θ	0.5	0.4	0.2	0.1	0.01	0.001	0.0001	0.00001
$\cos \theta - 1$								
$g(\theta) = \dfrac{\cos \theta - 1}{\theta}$								

For Problems 121–124 on the next page, use the following discussion.

Projectile Motion *The path of a projectile fired at an inclination θ to the horizontal with initial speed v_0 is a parabola (see the figure).*

v_0 = Initial speed

Height, H

θ

Range, R

The range R of the projectile, that is, the horizontal distance that the projectile travels, is found by using the function

$$R(\theta) = \dfrac{v_0^2 \sin(2\theta)}{g}$$

where $g \approx 32.2$ feet per second per second ≈ 9.8 meters per second per second is the acceleration due to gravity. The maximum height H of the projectile is given by the function

$$H(\theta) = \dfrac{v_0^2 (\sin \theta)^2}{2g}$$

*In Section 3.1, we discuss the necessary domain restriction so that the function is one-to-one.

In Problems 121–124, find the range R and maximum height H. Round answers to two decimal places. See the discussion on the previous page.

121. The projectile is fired at an angle of 45° to the horizontal with an initial speed of 100 feet per second.

122. The projectile is fired at an angle of 30° to the horizontal with an initial speed of 150 meters per second.

123. The projectile is fired at an angle of 25° to the horizontal with an initial speed of 500 meters per second.

124. The projectile is fired at an angle of 50° to the horizontal with an initial speed of 200 feet per second.

125. Inclined Plane See the figure. If friction is ignored, the time t (in seconds) required for a block to slide down an inclined plane is given by the function

$$t(\theta) = \sqrt{\frac{2a}{g \sin \theta \cos \theta}}$$

where a is the length (in feet) of the base and $g \approx 32$ feet per second per second is the acceleration due to gravity. How long does it take a block to slide down an inclined plane with base $a = 10$ feet when:

(a) $\theta = 30°$?
(b) $\theta = 45°$?
(c) $\theta = 60°$?

126. Piston Engines In a certain piston engine, the distance x (in centimeters) from the center of the drive shaft to the head of the piston is given by the function

$$x(\theta) = \cos \theta + \sqrt{16 + 0.5 \cos (2\theta)}$$

where θ is the angle between the crank and the path of the piston head. See the figure. Find x when $\theta = 30°$ and when $\theta = 45°$.

127. Calculating the Time of a Trip Two oceanfront homes are located 8 miles apart on a straight stretch of beach, each a distance of 1 mile from a paved road that parallels the ocean. See the figure.

Sally can jog 8 miles per hour along the paved road, but only 3 miles per hour in the sand on the beach. Because of a river directly between the two houses, it is necessary to jog in the sand to the road, continue on the road, and then jog directly back in the sand to get from one house to the other. The time T to get from one house to the other as a function of the angle θ shown in the figure is

$$T(\theta) = 1 + \frac{2}{3 \sin \theta} - \frac{1}{4 \tan \theta}, \qquad 0° < \theta < 90°$$

(a) Calculate the time T for $\theta = 30°$. How long is Sally on the paved road?
(b) Calculate the time T for $\theta = 45°$. How long is Sally on the paved road?
(c) Calculate the time T for $\theta = 60°$. How long is Sally on the paved road?
(d) Calculate the time T for $\theta = 90°$. Describe the path taken. Why can't the formula for T be used?

128. Designing Fine Decorative Pieces A designer of decorative art plans to market solid gold spheres encased in clear crystal cones. Each sphere is of fixed radius R and will be enclosed in a cone of height h and radius r. See the figure. Many cones can be used to enclose the sphere, each having a different slant angle θ. The volume V of the cone can be expressed as a function of the slant angle θ of the cone as

$$V(\theta) = \frac{1}{3} \pi R^3 \frac{(1 + \sec \theta)^3}{(\tan \theta)^2} \qquad 0° < \theta < 90°$$

What volume V is required to enclose a sphere of radius 2 centimeters in a cone whose slant angle θ is 30°? 45°? 60°?

Use the following to answer Problems 129–132. The viewing angle, θ, of an object is the angle the object forms at the lens of the viewer's eye. This is also known as the perceived or angular size of the object. The viewing angle is related to the object's height, H, and distance from the viewer, D, through the formula $\tan \dfrac{\theta}{2} = \dfrac{H}{2D}.$

129. Tailgating While driving, Arletha observes the car in front of her with a viewing angle of 22°. If the car is 6 feet wide, how close is Arletha to the car in front of her? Round your answer to one decimal place.

130. Viewing Distance The Washington Monument in Washington, D.C. is 555 feet tall. If a tourist sees the monument with a viewing angle of 8°, how far away, to the nearest foot, is she from the monument?

131. Tree Height A forest ranger views a tree that is 200 feet away with a viewing angle of 20°. How tall is the tree to the nearest foot?

132. Radius of the Moon An astronomer observes the moon with a viewing angle of 0.52°. If the moon's average distance from Earth is 384,400 km, what is its radius to the nearest kilometer?

133. Projectile Distance An object is fired at an angle θ to the horizontal with an initial speed of v_0 feet per second. Ignoring air resistance, the length of the projectile's path is given by

$$L(\theta) = \frac{v_0^2}{32}\left[\sin\theta - (\cos\theta)^2 \cdot \left(\ln\left[\tan\left(\frac{\pi - 2\theta}{4}\right)\right]\right)\right]$$

where $0 < \theta < \frac{\pi}{2}$.*

(a) Find the length of the object's path for angles $\theta = \frac{\pi}{6}, \frac{\pi}{4}$, and $\frac{\pi}{3}$ if the initial velocity is 128 feet per second.

(b) Using a graphing utility, determine the angle required for the object to have a path length of 550 feet if the initial velocity is 128 feet per second.

(c) What angle will result in the longest path? How does this angle compare to the angle that results in the longest range? (See Problems 121–124.)

134. Photography The length L of the chord joining the endpoints of an arc on a circle of radius r subtended by a central angle θ, $0 < \theta \le \pi$, is given by $L = r\sqrt{2 - 2\cos\theta}$. Use this fact to approximate the field width (the width of scenery the lens can image) of a 450mm camera lens at a distance of 920 feet if the viewing angle of the lens is $\frac{\pi}{30}$.

135. Projectile Motion An object is propelled upward at an angle θ, $45° < \theta < 90°$, to the horizontal with an initial velocity of v_0 feet per second from the base of an inclined plane that makes an angle of 45° with the horizontal. See the figure.

If air resistance is ignored, the distance R that it travels up the inclined plane as a function of θ is given by

$$R(\theta) = \frac{v_0^2\sqrt{2}}{32}\left[\sin(2\theta) - \cos(2\theta) - 1\right]$$

(a) Find the distance R that the object travels along the inclined plane if the initial velocity is 32 feet per second and $\theta = 60°$.

(b) Graph $R = R(\theta)$ if the initial velocity is 32 feet per second.

(c) What value of θ makes R largest?

*Requires Section 7.2 Logarithmic Functions

136. If θ, $0 < \theta < \pi$, is the angle between the positive x-axis and a nonhorizontal, nonvertical line L, show that the slope m of L equals $\tan\theta$. The angle θ is called the **inclination** of L.

[**Hint:** See the figure, where we have drawn the line M parallel to L and passing through the origin. Use the fact that M intersects the unit circle at the point $(\cos\theta, \sin\theta)$.]

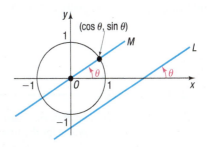

In Problems 137 and 138, use the figure to approximate the value of the six trigonometric functions at t to the nearest tenth. Then use a calculator to approximate each of the six trigonometric functions at t.

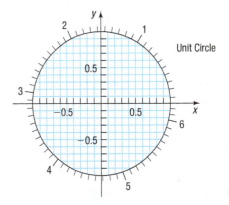

137. (a) $t = 1$ **(b)** $t = 5.1$

138. (a) $t = 2$ **(b)** $t = 4$

139. Challenge Problem Let θ be the measure of an angle, in radians, in standard position with $\pi < \theta < \frac{3\pi}{2}$. Find the exact y-coordinate of the intersection of the terminal side of θ with the unit circle, given $\cos\theta + \sin^2\theta = \frac{41}{49}$. State the answer as a single fraction, completely simplified, with rationalized denominator.

140. Challenge Problem Let θ be the measure of an angle, in radians, in standard position with $\frac{\pi}{2} < \theta < \pi$. Find the exact x-coordinate of the intersection of the terminal side of θ with the unit circle, given $\cos^2\theta - \sin\theta = -\frac{1}{9}$. State the answer as a single fraction, completely simplified, with rationalized denominator.

141. Challenge Problem If the terminal side of an angle contains the point $(5n, -12n)$ with $n > 0$, find $\sin\theta$.

Explaining Concepts: Discussion and Writing

142. Write a brief paragraph that explains how to quickly compute the trigonometric functions of 30°, 45°, and 60°.

143. Write a brief paragraph that explains how to quickly compute the trigonometric functions of 0°, 90°, 180°, and 270°.

144. How would you explain the meaning of the sine function to a fellow student who has just completed college algebra?

145. Draw a unit circle. Label the angles $0, \frac{\pi}{6}, \frac{\pi}{4}, \frac{\pi}{3}, \dots, \frac{7\pi}{4}, \frac{11\pi}{6}, 2\pi$ and the coordinates of the points on the unit circle that correspond to each of these angles. Explain how symmetry can be used to find the coordinates of points on the unit circle for angles whose terminal sides are in quadrants II, III, and IV.

Retain Your Knowledge

Problems 146–153 are based on material learned earlier in the course. The purpose of these problems is to keep the material fresh in your mind so that you are better prepared for the final exam.

146. If the polynomial function $P(x) = x^4 - 5x^3 - 9x^2 + 155x - 250$ has zeros of $4 + 3i$ and 2, find the remaining zeros of the function.

147. Find the remainder when $P(x) = 8x^4 - 2x^3 + x - 8$ is divided by $x + 2$.

148. Sidewalk Area A sidewalk with a uniform width of 3 feet is to be placed around a circular garden with a diameter of 24 feet. Find the exact area of the sidewalk.

149. Find the real zeros of $f(x) = 3x^2 - 7x - 9$.

150. If $f(x) = x^2 - 3$ and $g(x) = -x + 3$, determine where $g(x) \geq f(x)$.

151. Solve $\operatorname{int}(x + 3) = -2$.

152. If the point $(3, -4)$ is on the graph of $y = f(x)$, what corresponding point must be on the graph of $\frac{1}{2}f(x - 3)$?

153. If $g(x) = \frac{x^2}{4} - \frac{1}{x^2}$, simplify $\sqrt{1 + [g(x)]^2}$.

'Are You Prepared?' Answers

1. $c^2 = a^2 + b^2$ **2.** 8   **3.** True   **4.** equal; proportional   **5.** $\left(-\frac{1}{2}, \frac{\sqrt{3}}{2}\right)$ **6.** $-\frac{1}{2}$

2.3 Properties of the Trigonometric Functions

PREPARING FOR THIS SECTION *Before getting started, review the following:*

- Functions (Section 1.3, pp. 24–37)
- Identity (Section A.4, p. A27)
- Even and Odd Functions (Section 1.4, pp. 44–46)

Now Work the **'Are You Prepared?'** problems on page 143.

OBJECTIVES **1** Determine the Domain and the Range of the Trigonometric Functions (p. 133)

2 Determine the Period of the Trigonometric Functions (p. 135)

3 Determine the Signs of the Trigonometric Functions in a Given Quadrant (p. 136)

4 Find the Values of the Trigonometric Functions Using Fundamental Identities (p. 137)

5 Find the Exact Values of the Trigonometric Functions of an Angle Given One of the Functions and the Quadrant of the Angle (p. 139)

6 Use Even-Odd Properties to Find the Exact Values of the Trigonometric Functions (p. 142)

1 Determine the Domain and the Range of the Trigonometric Functions

Let θ be an angle in standard position, and let $P = (x, y)$ be the point on the unit circle that corresponds to θ. See Figure 35. Then, by the definition given earlier,

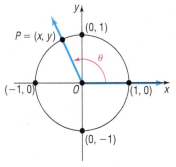

$$\sin \theta = y \qquad \cos \theta = x \qquad \tan \theta = \frac{y}{x} \quad x \neq 0$$

$$\csc \theta = \frac{1}{y} \quad y \neq 0 \qquad \sec \theta = \frac{1}{x} \quad x \neq 0 \qquad \cot \theta = \frac{x}{y} \quad y \neq 0$$

Figure 35

For $\sin \theta$ and $\cos \theta$, there is no concern about dividing by 0, so θ can be any angle. It follows that the domain of the sine function and cosine function is the set of all real numbers.

- The domain of the sine function is the set of all real numbers.
- The domain of the cosine function is the set of all real numbers.

For the tangent function and secant function, the x-coordinate of $P = (x, y)$ cannot be 0 since this results in division by 0. See Figure 35. On the unit circle, there are two such points, $(0, 1)$ and $(0, -1)$. These two points correspond to the angles $\frac{\pi}{2}$ $(90°)$ and $\frac{3\pi}{2}$ $(270°)$ or, more generally, to any angle that is an odd integer multiple of $\frac{\pi}{2}$ $(90°)$, such as $\pm \frac{\pi}{2}$ $(\pm 90°)$, $\pm \frac{3\pi}{2}$ $(\pm 270°)$, $\pm \frac{5\pi}{2}$ $(\pm 450°)$, and so on. Such angles must be excluded from the domain of the tangent function and secant function.

- The domain of the tangent function is the set of all real numbers, except odd integer multiples of $\frac{\pi}{2}$ $(90°)$.
- The domain of the secant function is the set of all real numbers, except odd integer multiples of $\frac{\pi}{2}$ $(90°)$.

For the cotangent function and cosecant function, the y-coordinate of $P = (x, y)$ cannot be 0 since this results in division by 0. See Figure 35. On the unit circle, there are two such points, $(1, 0)$ and $(-1, 0)$. These two points correspond to the angles $0(0°)$ and $\pi(180°)$ or, more generally, to any angle that is an integer multiple of $\pi(180°)$, such as $0(0°)$, $\pm \pi(\pm 180°)$, $\pm 2\pi(\pm 360°)$, $\pm 3\pi(\pm 540°)$, and so on. Such angles must therefore be excluded from the domain of the cotangent function and cosecant function.

- The domain of the cotangent function is the set of all real numbers, except integer multiples of $\pi(180°)$.
- The domain of the cosecant function is the set of all real numbers, except integer multiples of $\pi(180°)$.

Next we determine the range of each of the six trigonometric functions. Refer again to Figure 35. Let $P = (x, y)$ be the point on the unit circle that corresponds to the angle θ. It follows that $-1 \leq x \leq 1$ and $-1 \leq y \leq 1$. Since $\sin \theta = y$ and $\cos \theta = x$, we have

$$-1 \leq \sin \theta \leq 1 \qquad -1 \leq \cos \theta \leq 1$$

The range of both the sine function and the cosine function consists of all real numbers between -1 and 1, inclusive. Using absolute value notation, we have $|\sin \theta| \leq 1$ and $|\cos \theta| \leq 1$.

If θ is not an integer multiple of $\pi (180°)$, then $\csc \theta = \dfrac{1}{y}$. Since $y = \sin \theta$ and $|y| = |\sin \theta| \leq 1$, it follows that $|\csc \theta| = \dfrac{1}{|\sin \theta|} = \dfrac{1}{|y|} \geq 1 \left(\dfrac{1}{y} \leq -1 \text{ or } \dfrac{1}{y} \geq 1 \right).$ Since $\csc \theta = \dfrac{1}{y}$, the range of the cosecant function consists of all real numbers less than or equal to -1 or greater than or equal to 1. That is,

$$\csc \theta \leq -1 \quad \text{or} \quad \csc \theta \geq 1$$

If θ is not an odd integer multiple of $\dfrac{\pi}{2}(90°)$, then $\sec \theta = \dfrac{1}{x}$. Since $x = \cos \theta$ and $|x| = |\cos \theta| \leq 1$, it follows that $|\sec \theta| = \dfrac{1}{|\cos \theta|} = \dfrac{1}{|x|} \geq 1 \left(\dfrac{1}{x} \leq -1 \text{ or } \dfrac{1}{x} \geq 1 \right).$ Since $\sec \theta = \dfrac{1}{x}$, the range of the secant function consists of all real numbers less than or equal to -1 or greater than or equal to 1. That is,

$$\sec \theta \leq -1 \quad \text{or} \quad \sec \theta \geq 1$$

The range of both the tangent function and the cotangent function is the set of all real numbers.

$$-\infty < \tan \theta < \infty \qquad -\infty < \cot \theta < \infty$$

You are asked to prove this in Problems 123 and 124.
Table 4 summarizes these results.

Table 4

Function	Symbol	Domain	Range
sine	$f(\theta) = \sin \theta$	All real numbers	All real numbers from -1 to 1, inclusive
cosine	$f(\theta) = \cos \theta$	All real numbers	All real numbers from -1 to 1, inclusive
tangent	$f(\theta) = \tan \theta$	All real numbers, except odd integer multiples of $\dfrac{\pi}{2}$ ($90°$)	All real numbers
cosecant	$f(\theta) = \csc \theta$	All real numbers, except integer multiples of π ($180°$)	All real numbers less than or equal to -1 or greater than or equal to 1
secant	$f(\theta) = \sec \theta$	All real numbers, except odd integer multiples of $\dfrac{\pi}{2}$ ($90°$)	All real numbers less than or equal to -1 or greater than or equal to 1
cotangent	$f(\theta) = \cot \theta$	All real numbers, except integer multiples of π ($180°$)	All real numbers

Now Work PROBLEM **97**

2 Determine the Period of the Trigonometric Functions

Look at Figure 36. This figure shows that for an angle of $\dfrac{\pi}{3}$ radians the corresponding point P on the unit circle is $\left(\dfrac{1}{2}, \dfrac{\sqrt{3}}{2}\right)$. Notice that, for an angle of $\dfrac{\pi}{3} + 2\pi$ radians, the corresponding point P on the unit circle is also $\left(\dfrac{1}{2}, \dfrac{\sqrt{3}}{2}\right)$. Then

$$\sin\frac{\pi}{3} = \frac{\sqrt{3}}{2} \quad \text{and} \quad \sin\left(\frac{\pi}{3} + 2\pi\right) = \frac{\sqrt{3}}{2}$$

$$\cos\frac{\pi}{3} = \frac{1}{2} \quad \text{and} \quad \cos\left(\frac{\pi}{3} + 2\pi\right) = \frac{1}{2}$$

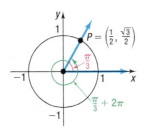

Figure 36 $\sin\left(\dfrac{\pi}{3} + 2\pi\right) = \sin\dfrac{\pi}{3}$;

$\cos\left(\dfrac{\pi}{3} + 2\pi\right) = \cos\dfrac{\pi}{3}$

This example illustrates a more general situation. For a given angle θ, measured in radians, suppose that we know the corresponding point $P = (x, y)$ on the unit circle. Now add 2π to θ. The point on the unit circle corresponding to $\theta + 2\pi$ is identical to the point P on the unit circle corresponding to θ. See Figure 37. The values of the trigonometric functions of $\theta + 2\pi$ are equal to the values of the corresponding trigonometric functions of θ.

If we add (or subtract) integer multiples of 2π to θ, the values of the sine and cosine function remain unchanged. That is, for all θ

$$\sin(\theta + 2\pi k) = \sin\theta \qquad \cos(\theta + 2\pi k) = \cos\theta \qquad \textbf{(1)}$$

where k is any integer.

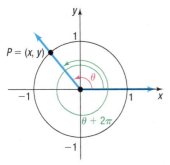

Figure 37 $\sin(\theta + 2\pi k) = \sin\theta$;

$\cos(\theta + 2\pi k) = \cos\theta$

Functions that exhibit this kind of behavior are called *periodic functions*.

> **DEFINITION Periodic Function and Fundamental Period**
>
> A function f is called **periodic** if there is a positive number p with the property that whenever θ is in the domain of f, so is $\theta + p$, and
>
> $$f(\theta + p) = f(\theta)$$
>
> If there is a smallest such number p, this smallest value is called the **(fundamental) period** of f.

Based on equation (1), the sine and cosine functions are periodic. In fact, the sine and cosine functions have period 2π. You are asked to prove this fact in Problems 125 and 126. The secant and cosecant functions are also periodic with period 2π, and the tangent and cotangent functions are periodic with period π. You are asked to prove these statements in Problems 127 through 130.

These facts are summarized as follows:

> **Periodic Properties**
>
> $$\sin(\theta + 2\pi) = \sin\theta \quad \cos(\theta + 2\pi) = \cos\theta \quad \tan(\theta + \pi) = \tan\theta$$
> $$\csc(\theta + 2\pi) = \csc\theta \quad \sec(\theta + 2\pi) = \sec\theta \quad \cot(\theta + \pi) = \cot\theta$$

In Words

Tangent and cotangent have period π; the others have period 2π.

Because the sine, cosine, secant, and cosecant functions have period 2π, once we know their values over an interval of length 2π, we know all their values; similarly, since the tangent and cotangent functions have period π, once we know their values over an interval of length π, we know all their values.

Finding Exact Values Using Periodic Properties

Find the exact value of each of the following angles:

(a) $\sin \dfrac{17\pi}{4}$ (b) $\cos(5\pi)$ (c) $\tan \dfrac{5\pi}{4}$

Solution

(a) It is best to sketch the angle first, as shown in Figure 38(a). Since the period of the sine function is 2π, each full revolution can be ignored leaving the angle $\dfrac{\pi}{4}$. Then

$$\sin \frac{17\pi}{4} = \sin\left(\frac{\pi}{4} + 4\pi\right) = \sin \frac{\pi}{4} = \frac{\sqrt{2}}{2}$$

(b) See Figure 38(b). Since the period of the cosine function is 2π, each full revolution can be ignored leaving the angle π. Then

$$\cos(5\pi) = \cos(\pi + 4\pi) = \cos \pi = -1$$

(c) See Figure 38(c). Since the period of the tangent function is π, each half-revolution can be ignored leaving the angle $\dfrac{\pi}{4}$. Then

$$\tan \frac{5\pi}{4} = \tan\left(\frac{\pi}{4} + \pi\right) = \tan \frac{\pi}{4} = 1$$

Figure 38   (a)   (b)   (c)

The periodic properties of the trigonometric functions will be very helpful to us when we study their graphs later in the chapter.

Now Work PROBLEM 11

3 Determine the Signs of the Trigonometric Functions in a Given Quadrant

Let $P = (x, y)$ be the point on the unit circle that corresponds to the angle θ. If we know in which quadrant the point P lies, then we can determine the signs of the trigonometric functions of θ. For example, if $P = (x, y)$ lies in quadrant IV, as shown in Figure 39, then we know that $x > 0$ and $y < 0$. Consequently,

$$\sin \theta = y < 0 \qquad \cos \theta = x > 0 \qquad \tan \theta = \frac{y}{x} < 0$$

$$\csc \theta = \frac{1}{y} < 0 \qquad \sec \theta = \frac{1}{x} > 0 \qquad \cot \theta = \frac{x}{y} < 0$$

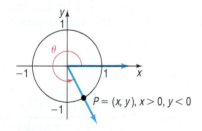

Figure 39
θ in quadrant IV, $x > 0$, $y < 0$

Table 5 lists the signs of the six trigonometric functions for each quadrant. Figure 40 provides two illustrations.

Table 5

Quadrant of P	sin θ, csc θ	cos θ, sec θ	tan θ, cot θ
I	Positive	Positive	Positive
II	Positive	Negative	Negative
III	Negative	Negative	Positive
IV	Negative	Positive	Negative

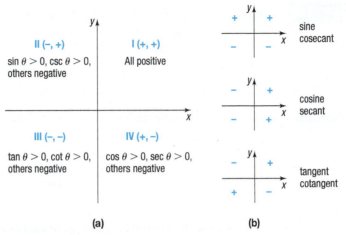

Figure 40 Signs of the trigonometric functions

| EXAMPLE 2 | **Finding the Quadrant in Which an Angle θ Lies** |

If $\sin \theta < 0$ and $\cos \theta < 0$, name the quadrant in which the angle θ lies.

Solution Let $P = (x, y)$ be the point on the unit circle corresponding to θ. Then $\sin \theta = y < 0$ and $\cos \theta = x < 0$. Because points in quadrant III have $x < 0$ and $y < 0$, θ lies in quadrant III.

Now Work PROBLEM 27

4 Find the Values of the Trigonometric Functions Using Fundamental Identities

If $P = (x, y)$ is the point on the unit circle corresponding to θ, then

$$\sin \theta = y \qquad \cos \theta = x \qquad \tan \theta = \frac{y}{x} \quad \text{if } x \neq 0$$

$$\csc \theta = \frac{1}{y} \quad \text{if } y \neq 0 \qquad \sec \theta = \frac{1}{x} \quad \text{if } x \neq 0 \qquad \cot \theta = \frac{x}{y} \quad \text{if } y \neq 0$$

Based on these definitions, we have the **reciprocal identities**:

> **Reciprocal Identities**
>
> $$\csc \theta = \frac{1}{\sin \theta} \qquad \sec \theta = \frac{1}{\cos \theta} \qquad \cot \theta = \frac{1}{\tan \theta} \qquad \text{(2)}$$

Two other fundamental identities are the **quotient identities**.

> **Quotient Identities**
>
> $$\tan \theta = \frac{\sin \theta}{\cos \theta} \qquad \cot \theta = \frac{\cos \theta}{\sin \theta} \qquad \text{(3)}$$

The proofs of identities (2) and (3) follow from the definitions of the trigonometric functions. (See Problems 131 and 132.)

If $\sin \theta$ and $\cos \theta$ are known, identities (2) and (3) make it easy to find the values of the remaining trigonometric functions.

EXAMPLE 3

Finding Exact Values Using Identities When Sine and Cosine Are Given

Given $\sin \theta = \dfrac{\sqrt{5}}{5}$ and $\cos \theta = \dfrac{2\sqrt{5}}{5}$, find the exact values of the four remaining trigonometric functions of θ using identities.

Solution Based on a quotient identity from (3), we have

$$\tan \theta = \frac{\sin \theta}{\cos \theta} = \frac{\dfrac{\sqrt{5}}{5}}{\dfrac{2\sqrt{5}}{5}} = \frac{1}{2}$$

Then we use the reciprocal identities from (2) to get

$$\csc \theta = \frac{1}{\sin \theta} = \frac{1}{\dfrac{\sqrt{5}}{5}} = \frac{5}{\sqrt{5}} = \sqrt{5} \qquad \sec \theta = \frac{1}{\cos \theta} = \frac{1}{\dfrac{2\sqrt{5}}{5}} = \frac{5}{2\sqrt{5}} = \frac{\sqrt{5}}{2} \qquad \cot \theta = \frac{1}{\tan \theta} = \frac{1}{\dfrac{1}{2}} = 2$$

━━━ **Now Work** PROBLEM 35

The equation of the unit circle is $x^2 + y^2 = 1$ or, equivalently,

$$y^2 + x^2 = 1$$

If $P = (x, y)$ is the point on the unit circle that corresponds to the angle θ, then $y = \sin \theta$ and $x = \cos \theta$, so we have

$$(\sin \theta)^2 + (\cos \theta)^2 = 1 \qquad (4)$$

It is customary to write $\sin^2 \theta$ instead of $(\sin \theta)^2$, $\cos^2 \theta$ instead of $(\cos \theta)^2$, and so on. With this notation, we can rewrite identity (4) as

$$\sin^2 \theta + \cos^2 \theta = 1 \qquad (5)$$

If $\cos \theta \neq 0$, we can divide each side of identity (5) by $\cos^2 \theta$.

$$\frac{\sin^2 \theta}{\cos^2 \theta} + \frac{\cos^2 \theta}{\cos^2 \theta} = \frac{1}{\cos^2 \theta}$$

$$\left(\frac{\sin \theta}{\cos \theta}\right)^2 + 1 = \left(\frac{1}{\cos \theta}\right)^2$$

Now use identities (2) and (3) to get

$$\tan^2 \theta + 1 = \sec^2 \theta \qquad (6)$$

Similarly, if $\sin \theta \neq 0$, we can divide both sides of identity (5) by $\sin^2 \theta$ and use identities (2) and (3) to get $1 + \cot^2 \theta = \csc^2 \theta$, which we write as

$$\cot^2 \theta + 1 = \csc^2 \theta \qquad (7)$$

The identities in (5), (6), and (7) are referred to as the **Pythagorean identities**. Collectively, the identities in (2), (3), and (5)−(7) are referred to as the **Fundamental Identities**.

Fundamental Identities

- $\tan \theta = \dfrac{\sin \theta}{\cos \theta}$
- $\cot \theta = \dfrac{\cos \theta}{\sin \theta}$

- $\csc \theta = \dfrac{1}{\sin \theta}$
- $\sec \theta = \dfrac{1}{\cos \theta}$
- $\cot \theta = \dfrac{1}{\tan \theta}$

- $\sin^2 \theta + \cos^2 \theta = 1$
- $\tan^2 \theta + 1 = \sec^2 \theta$
- $\cot^2 \theta + 1 = \csc^2 \theta$

EXAMPLE 4

Finding the Exact Value of a Trigonometric Expression Using Identities

Find the exact value of each expression. Do not use a calculator.

(a) $\tan 20° - \dfrac{\sin 20°}{\cos 20°}$

(b) $\sin^2 \dfrac{\pi}{12} + \dfrac{1}{\sec^2 \dfrac{\pi}{12}}$

Solution

(a) $\tan 20° - \underset{\underset{\dfrac{\sin \theta}{\cos \theta} = \tan \theta}{\uparrow}}{\dfrac{\sin 20°}{\cos 20°}} = \tan 20° - \tan 20° = 0$

(b) $\sin^2 \dfrac{\pi}{12} + \underset{\underset{\cos \theta = \dfrac{1}{\sec \theta}}{\uparrow}}{\dfrac{1}{\sec^2 \dfrac{\pi}{12}}} = \underset{\underset{\sin^2 \theta + \cos^2 \theta = 1}{\uparrow}}{\sin^2 \dfrac{\pi}{12} + \cos^2 \dfrac{\pi}{12}} = 1$

Now Work PROBLEM 79

---

5 Find the Exact Values of the Trigonometric Functions of an Angle Given One of the Functions and the Quadrant of the Angle

Many problems require finding the exact values of the remaining trigonometric functions when the value of one of them is known and the quadrant in which θ lies can be found. There are two approaches to solving such problems. One approach uses a circle of radius r; the other uses identities.

When using identities, sometimes a rearrangement is required. For example, the Pythagorean identity

$$\sin^2 \theta + \cos^2 \theta = 1$$

can be solved for $\sin \theta$ in terms of $\cos \theta$ (or vice versa) as follows:

$$\sin^2 \theta = 1 - \cos^2 \theta$$

$$\sin \theta = \pm \sqrt{1 - \cos^2 \theta}$$

where the $+$ sign is used if $\sin \theta > 0$ and the $-$ sign is used if $\sin \theta < 0$. Similarly, in $\tan^2 \theta + 1 = \sec^2 \theta$, we can solve for $\tan \theta$ (or $\sec \theta$), and in $\cot^2 \theta + 1 = \csc^2 \theta$, we can solve for $\cot \theta$ (or $\csc \theta$).

EXAMPLE 5   **Finding Exact Values Given One Value and the Sign of Another**

Given that $\sin\theta = \dfrac{1}{3}$ and $\cos\theta < 0$, find the exact value of each of the remaining five trigonometric functions.

Solution
Option 1
Using a Circle

Suppose that $P = (x, y)$ is the point on a circle that corresponds to θ. Since $\sin\theta = \dfrac{1}{3} > 0$ and $\cos\theta < 0$, the point $P = (x, y)$ is in quadrant II. Because $\sin\theta = \dfrac{1}{3} = \dfrac{y}{r}$, we let $y = 1$ and $r = 3$. The point $P = (x, y) = (x, 1)$ that corresponds to θ lies on the circle of radius 3 centered at the origin: $x^2 + y^2 = 9$. See Figure 41.

To find x, we use the fact that $x^2 + y^2 = 9$, $y = 1$, and P is in quadrant II (so $x < 0$).

Figure 41

$$x^2 + y^2 = 9$$
$$x^2 + 1^2 = 9 \qquad \textbf{y = 1}$$
$$x^2 = 8$$
$$x = -2\sqrt{2} \quad \textbf{x < 0}$$

Since $x = -2\sqrt{2}$, $y = 1$, and $r = 3$, we find that

$$\cos\theta = \frac{x}{r} = -\frac{2\sqrt{2}}{3} \qquad \tan\theta = \frac{y}{x} = \frac{1}{-2\sqrt{2}} = -\frac{\sqrt{2}}{4}$$

$$\csc\theta = \frac{r}{y} = \frac{3}{1} = 3 \qquad \sec\theta = \frac{r}{x} = \frac{3}{-2\sqrt{2}} = -\frac{3\sqrt{2}}{4} \qquad \cot\theta = \frac{x}{y} = \frac{-2\sqrt{2}}{1} = -2\sqrt{2}$$

Option 2
Using Identities

First, solve identity (5), $\sin^2\theta + \cos^2\theta = 1$, for $\cos\theta$.

$$\sin^2\theta + \cos^2\theta = 1$$
$$\cos^2\theta = 1 - \sin^2\theta$$
$$\cos\theta = \pm\sqrt{1 - \sin^2\theta}$$

Because $\cos\theta < 0$, choose the minus sign and use the fact that $\sin\theta = \dfrac{1}{3}$.

$$\cos\theta = -\sqrt{1 - \sin^2\theta} = -\sqrt{1 - \frac{1}{9}} = -\sqrt{\frac{8}{9}} = -\frac{2\sqrt{2}}{3}$$

$$\uparrow$$
$$\textbf{sin }\theta = \frac{1}{3}$$

Now, use $\sin\theta = \dfrac{1}{3}$ and $\cos\theta = -\dfrac{2\sqrt{2}}{3}$ and fundamental identities.

$$\tan\theta = \frac{\sin\theta}{\cos\theta} = \frac{\frac{1}{3}}{-\frac{2\sqrt{2}}{3}} = \frac{1}{-2\sqrt{2}} = -\frac{\sqrt{2}}{4} \qquad \cot\theta = \frac{1}{\tan\theta} = -2\sqrt{2}$$

$$\sec\theta = \frac{1}{\cos\theta} = \frac{1}{-\frac{2\sqrt{2}}{3}} = \frac{-3}{2\sqrt{2}} = -\frac{3\sqrt{2}}{4} \qquad \csc\theta = \frac{1}{\sin\theta} = \frac{1}{\frac{1}{3}} = 3$$

Finding the Values of the Trigonometric Functions of θ When the Value of One Function Is Known and the Quadrant of θ Is Known

Given the value of one trigonometric function and the quadrant in which θ lies, the exact value of each of the remaining five trigonometric functions can be found in either of two ways.

Option 1 Using a Circle of Radius r

STEP 1: Draw a circle centered at the origin showing the location of the angle θ and the point $P = (x, y)$ that corresponds to θ. The radius of the circle that contains $P = (x, y)$ is $r = \sqrt{x^2 + y^2}$.

STEP 2: Assign a value to two of the three variables x, y, r based on the value of the given trigonometric function and the location of P.

STEP 3: Use the fact that P lies on the circle $x^2 + y^2 = r^2$ to find the value of the missing variable.

STEP 4: Apply the theorem on page 126 to find the values of the remaining trigonometric functions.

Option 2 Using Identities

Use appropriate identities to find the value of each remaining trigonometric function.

EXAMPLE 6

Given the Value of One Trigonometric Function and the Sign of Another, Find the Values of the Remaining Ones

Given that $\tan \theta = \dfrac{1}{2}$ and $\sin \theta < 0$, find the exact value of each of the remaining five trigonometric functions of θ.

Solution
Option 1
Using a Circle

STEP 1: Since $\tan \theta = \dfrac{1}{2} > 0$ and $\sin \theta < 0$, the point $P = (x, y)$ that corresponds to θ lies in quadrant III. See Figure 42.

STEP 2: Since $\tan \theta = \dfrac{1}{2} = \dfrac{y}{x}$ and θ lies in quadrant III, let $x = -2$ and $y = -1$.

STEP 3: With $x = -2$ and $y = -1$, then $r = \sqrt{x^2 + y^2} = \sqrt{(-2)^2 + (-1)^2} = \sqrt{5}$, so P lies on the circle $x^2 + y^2 = 5$.

STEP 4: Apply the theorem on page 126 using $x = -2$, $y = -1$, and $r = \sqrt{5}$.

$$\sin \theta = \frac{y}{r} = \frac{-1}{\sqrt{5}} = -\frac{\sqrt{5}}{5} \qquad \cos \theta = \frac{x}{r} = \frac{-2}{\sqrt{5}} = -\frac{2\sqrt{5}}{5}$$

$$\csc \theta = \frac{r}{y} = \frac{\sqrt{5}}{-1} = -\sqrt{5} \qquad \sec \theta = \frac{r}{x} = \frac{\sqrt{5}}{-2} = -\frac{\sqrt{5}}{2} \qquad \cot \theta = \frac{x}{y} = \frac{-2}{-1} = 2$$

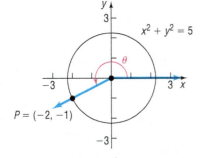

Figure 42 $\tan \theta = \dfrac{1}{2}$; θ in quadrant III

$x^2 + y^2 = 5$

$P = (-2, -1)$

Option 2
Using Identities

Because the value of $\tan \theta$ is known, use the Pythagorean identity that involves $\tan \theta$, that is, $\tan^2 \theta + 1 = \sec^2 \theta$. Since $\tan \theta = \dfrac{1}{2} > 0$ and $\sin \theta < 0$, then θ lies in quadrant III, where $\sec \theta < 0$.

$$\tan^2 \theta + 1 = \sec^2 \theta \qquad \textcolor{blue}{\textbf{Pythagorean identity}}$$

$$\left(\frac{1}{2}\right)^2 + 1 = \sec^2 \theta \qquad \textcolor{blue}{\tan \theta = \frac{1}{2}}$$

$$\sec^2 \theta = \frac{1}{4} + 1 = \frac{5}{4} \qquad \textcolor{blue}{\textbf{Proceed to solve for } \sec \theta.}$$

$$\sec \theta = -\frac{\sqrt{5}}{2} \qquad \textcolor{blue}{\sec \theta < 0}$$

Now we know $\tan \theta = \dfrac{1}{2}$ and $\sec \theta = -\dfrac{\sqrt{5}}{2}$. Using reciprocal identities, we find

$$\cos \theta = \frac{1}{\sec \theta} = \frac{1}{-\dfrac{\sqrt{5}}{2}} = -\frac{2}{\sqrt{5}} = -\frac{2\sqrt{5}}{5}$$

$$\cot \theta = \frac{1}{\tan \theta} = \frac{1}{\dfrac{1}{2}} = 2$$

To find $\sin \theta$, use the following reasoning:

$$\tan \theta = \frac{\sin \theta}{\cos \theta} \quad \text{so} \quad \sin \theta = (\tan \theta)(\cos \theta) = \left(\frac{1}{2}\right) \cdot \left(-\frac{2\sqrt{5}}{5}\right) = -\frac{\sqrt{5}}{5}$$

$$\csc \theta = \frac{1}{\sin \theta} = \frac{1}{-\dfrac{\sqrt{5}}{5}} = -\frac{5}{\sqrt{5}} = -\sqrt{5}$$

━━━━━ **Now Work** PROBLEM 43

6 Use Even-Odd Properties to Find the Exact Values of the Trigonometric Functions

Recall that a function f is even if $f(-\theta) = f(\theta)$ for all θ in the domain of f; a function f is odd if $f(-\theta) = -f(\theta)$ for all θ in the domain of f. We will now show that the trigonometric functions sine, tangent, cotangent, and cosecant are odd functions and the functions cosine and secant are even functions.

In Words
Cosine and secant are even functions; the others are odd functions.

THEOREM **Even-Odd Properties**

$$\sin(-\theta) = -\sin \theta \qquad \cos(-\theta) = \cos \theta \qquad \tan(-\theta) = -\tan \theta$$
$$\csc(-\theta) = -\csc \theta \qquad \sec(-\theta) = \sec \theta \qquad \cot(-\theta) = -\cot \theta$$

Proof Let $P = (x, y)$ be the point on the unit circle that corresponds to the angle θ. See Figure 43. Using symmetry, the point Q on the unit circle that corresponds to the angle $-\theta$ will have coordinates $(x, -y)$. Using the definition of the trigonometric functions, we have

$$\sin \theta = y \qquad \sin(-\theta) = -y \qquad \cos \theta = x \qquad \cos(-\theta) = x$$

so

$$\sin(-\theta) = -y = -\sin \theta \qquad \cos(-\theta) = x = \cos \theta$$

Now, using these results and some of the fundamental identities, we have

$$\tan(-\theta) = \frac{\sin(-\theta)}{\cos(-\theta)} = \frac{-\sin \theta}{\cos \theta} = -\tan \theta \qquad \cot(-\theta) = \frac{1}{\tan(-\theta)} = \frac{1}{-\tan \theta} = -\cot \theta$$

$$\sec(-\theta) = \frac{1}{\cos(-\theta)} = \frac{1}{\cos \theta} = \sec \theta \qquad \csc(-\theta) = \frac{1}{\sin(-\theta)} = \frac{1}{-\sin \theta} = -\csc \theta$$

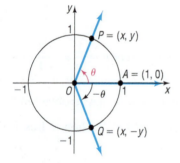

Figure 43

| EXAMPLE 7 | **Finding Exact Values Using Even-Odd Properties** |

Find the exact value of each of the following angles:

(a) $\sin(-45°)$ 　　(b) $\cos(-\pi)$ 　　(c) $\cot\left(-\dfrac{3\pi}{2}\right)$ 　　(d) $\tan\left(-\dfrac{37\pi}{4}\right)$

Solution 　(a) $\sin(-45°) = -\sin 45° = -\dfrac{\sqrt{2}}{2}$ 　　(b) $\cos(-\pi) = \cos\pi = -1$
　　　　　　　　　　$\uparrow$ 　　　　　　　　　　　　　　　　$\uparrow$
　　　　　　　Odd function 　　　　　　　　　　**Even function**

(c) $\cot\left(-\dfrac{3\pi}{2}\right) = -\cot\dfrac{3\pi}{2} = 0$
　　　　　$\uparrow$
　　Odd function

(d) $\tan\left(-\dfrac{37\pi}{4}\right) = -\tan\dfrac{37\pi}{4} = -\tan\left(\dfrac{\pi}{4} + 9\pi\right) = -\tan\dfrac{\pi}{4} = -1$
　　　　　　　　$\uparrow$ 　　　　　　　　　　　　　　　　　$\uparrow$
　　　Odd function 　　　　　　　　　　　**Period is π**

⟋ **Now Work** PROBLEM **59**

2.3 Assess Your Understanding

'Are You Prepared?' *Answers are given at the end of these exercises. If you get a wrong answer, read the pages listed in* red.

1. The domain of the function $f(x) = \dfrac{x+1}{2x+1}$ is _____.
   (pp. 32–34)

2. A function for which $f(x) = f(-x)$ for all x in the domain of f is called a(n) _____ function. (pp. 44–46)

3. **True or False** The function $f(x) = \sqrt{x}$ is even. (pp. 44–46)

4. **True or False** The equation $x^2 + 2x = (x+1)^2 - 1$ is an identity. (p. A27)

Concepts and Vocabulary

5. The sine, cosine, cosecant, and secant functions have period ____; the tangent and cotangent functions have period __.

6. The domain of the tangent function is _____.

7. **Multiple Choice** Which of the following is not in the range of the sine function?

   (a) $\dfrac{\pi}{4}$ 　(b) $\dfrac{3}{2}$ 　(c) -0.37 　(d) -1

8. **Multiple Choice** Which of the following functions is even?
   (a) cosine 　(b) sine 　(c) tangent 　(d) cosecant

9. $\sin^2\theta + \cos^2\theta = $ __

10. **True or False** $\sec\theta = \dfrac{1}{\sin\theta}$

Skill Building

In Problems 11–26, use the fact that the trigonometric functions are periodic to find the exact value of each expression. Do not use a calculator.

11. $\sin 405°$ 　　12. $\cos 420°$ 　　13. $\tan 405°$ 　　14. $\sin 390°$ 　　15. $\csc 450°$ 　　16. $\sec 540°$

17. $\cot 390°$ 　　18. $\sec 420°$ 　　19. $\cos\dfrac{33\pi}{4}$ 　　20. $\sin\dfrac{9\pi}{4}$ 　　21. $\tan(21\pi)$ 　　22. $\csc\dfrac{9\pi}{2}$

23. $\sec\dfrac{17\pi}{4}$ 　　24. $\cot\dfrac{17\pi}{4}$ 　　25. $\tan\dfrac{19\pi}{6}$ 　　26. $\sec\dfrac{25\pi}{6}$

In Problems 27–34, name the quadrant in which the angle θ lies.

27. $\sin\theta > 0$, 　$\cos\theta < 0$ 　　28. $\sin\theta < 0$, 　$\cos\theta > 0$ 　　29. $\sin\theta < 0$, 　$\tan\theta < 0$ 　　30. $\cos\theta > 0$, 　$\tan\theta > 0$

31. $\cos\theta > 0$, 　$\tan\theta < 0$ 　　32. $\cos\theta < 0$, 　$\tan\theta > 0$ 　　33. $\sec\theta < 0$, 　$\sin\theta > 0$ 　　34. $\csc\theta > 0$, 　$\cos\theta < 0$

In Problems 35–42, sin θ and cos θ are given. Find the exact value of each of the four remaining trigonometric functions.

35. $\sin\theta = -\dfrac{3}{5}, \quad \cos\theta = \dfrac{4}{5}$

36. $\sin\theta = \dfrac{4}{5}, \quad \cos\theta = -\dfrac{3}{5}$

37. $\sin\theta = \dfrac{2\sqrt{5}}{5}, \quad \cos\theta = \dfrac{\sqrt{5}}{5}$

38. $\sin\theta = -\dfrac{\sqrt{5}}{5}, \quad \cos\theta = -\dfrac{2\sqrt{5}}{5}$

39. $\sin\theta = \dfrac{1}{2}, \quad \cos\theta = \dfrac{\sqrt{3}}{2}$

40. $\sin\theta = \dfrac{\sqrt{3}}{2}, \quad \cos\theta = \dfrac{1}{2}$

41. $\sin\theta = -\dfrac{1}{3}, \quad \cos\theta = \dfrac{2\sqrt{2}}{3}$

42. $\sin\theta = \dfrac{2\sqrt{2}}{3}, \quad \cos\theta = -\dfrac{1}{3}$

In Problems 43–58, find the exact value of each of the remaining trigonometric functions of θ.

43. $\sin\theta = \dfrac{12}{13}, \quad \theta$ in quadrant II

44. $\cos\theta = \dfrac{3}{5}, \quad \theta$ in quadrant IV

45. $\cos\theta = -\dfrac{4}{5}, \quad \theta$ in quadrant III

46. $\sin\theta = -\dfrac{5}{13}, \quad \theta$ in quadrant III

47. $\sin\theta = \dfrac{5}{13}, \quad 90° < \theta < 180°$

48. $\cos\theta = \dfrac{4}{5}, \quad 270° < \theta < 360°$

49. $\cos\theta = -\dfrac{1}{3}, \quad \dfrac{\pi}{2} < \theta < \pi$

50. $\sin\theta = -\dfrac{2}{3}, \quad \pi < \theta < \dfrac{3\pi}{2}$

51. $\sin\theta = \dfrac{2}{3}, \quad \tan\theta < 0$

52. $\cos\theta = -\dfrac{1}{4}, \quad \tan\theta > 0$

53. $\sec\theta = 2, \quad \sin\theta < 0$

54. $\csc\theta = 3, \quad \cot\theta < 0$

55. $\tan\theta = \dfrac{3}{4}, \quad \sin\theta < 0$

56. $\cot\theta = \dfrac{4}{3}, \quad \cos\theta < 0$

57. $\tan\theta = -\dfrac{1}{3}, \quad \sin\theta > 0$

58. $\sec\theta = -2, \quad \tan\theta > 0$

In Problems 59–76, use the even-odd properties to find the exact value of each expression. Do not use a calculator.

59. $\sin(-60°)$ **60.** $\cos(-30°)$ **61.** $\tan(-30°)$ **62.** $\sin(-135°)$

63. $\sec(-60°)$ **64.** $\csc(-30°)$ **65.** $\sin(-90°)$ **66.** $\cos(-270°)$

67. $\tan\left(-\dfrac{\pi}{4}\right)$ **68.** $\sin(-\pi)$ **69.** $\cos\left(-\dfrac{\pi}{4}\right)$ **70.** $\sin\left(-\dfrac{\pi}{3}\right)$

71. $\tan(-\pi)$ **72.** $\sin\left(-\dfrac{3\pi}{2}\right)$ **73.** $\csc\left(-\dfrac{\pi}{4}\right)$ **74.** $\sec(-\pi)$

75. $\sec\left(-\dfrac{\pi}{6}\right)$ **76.** $\csc\left(-\dfrac{\pi}{3}\right)$

In Problems 77–88, use properties of the trigonometric functions to find the exact value of each expression. Do not use a calculator.

77. $\sin^2 40° + \cos^2 40°$ **78.** $\sec^2 18° - \tan^2 18°$ **79.** $\sin 80° \csc 80°$ **80.** $\tan 10° \cot 10°$

81. $\tan 40° - \dfrac{\sin 40°}{\cos 40°}$ **82.** $\cot 20° - \dfrac{\cos 20°}{\sin 20°}$ **83.** $\cos 400° \cdot \sec 40°$ **84.** $\tan 200° \cdot \cot 20°$

85. $\sin\left(-\dfrac{\pi}{12}\right)\csc\dfrac{25\pi}{12}$ **86.** $\sec\left(-\dfrac{\pi}{18}\right)\cdot\cos\dfrac{37\pi}{18}$ **87.** $\dfrac{\sin(-20°)}{\cos 380°} + \tan 200°$ **88.** $\dfrac{\sin 70°}{\cos(-430°)} + \tan(-70°)$

89. If $\sin\theta = 0.3$, find the value of:
$$\sin\theta + \sin(\theta + 2\pi) + \sin(\theta + 4\pi)$$

90. If $\cos\theta = 0.2$, find the value of:
$$\cos\theta + \cos(\theta + 2\pi) + \cos(\theta + 4\pi)$$

91. If $\tan\theta = 3$, find the value of:
$$\tan\theta + \tan(\theta + \pi) + \tan(\theta + 2\pi)$$

92. If $\cot\theta = -2$, find the value of:
$$\cot\theta + \cot(\theta - \pi) + \cot(\theta - 2\pi)$$

93. Find the exact value of:
$$\sin 1° + \sin 2° + \sin 3° + \cdots + \sin 358° + \sin 359°$$

94. Find the exact value of:
$$\cos 1° + \cos 2° + \cos 3° + \cdots + \cos 358° + \cos 359°$$

95. What is the domain of the sine function?

96. What is the domain of the cosine function?

97. For what numbers θ is $f(\theta) = \tan\theta$ not defined?

98. For what numbers θ is $f(\theta) = \cot\theta$ not defined?

99. For what numbers θ is $f(\theta) = \sec\theta$ not defined?

100. For what numbers θ is $f(\theta) = \csc\theta$ not defined?

101. What is the range of the sine function?

102. What is the range of the cosine function?

103. What is the range of the tangent function?

104. What is the range of the cotangent function?

105. What is the range of the secant function?

106. What is the range of the cosecant function?

107. Is the sine function even, odd, or neither? Is its graph symmetric? With respect to what?

108. Is the cosine function even, odd, or neither? Is its graph symmetric? With respect to what?

109. Is the tangent function even, odd, or neither? Is its graph symmetric? With respect to what?

110. Is the cotangent function even, odd, or neither? Is its graph symmetric? With respect to what?

111. Is the secant function even, odd, or neither? Is its graph symmetric? With respect to what?

112. Is the cosecant function even, odd, or neither? Is its graph symmetric? With respect to what?

Applications and Extensions

In Problems 113–118, use the periodic and even-odd properties.

113. If $f(\theta) = \sin \theta$ and $f(a) = \dfrac{1}{3}$, find the exact value of:

   (a) $f(-a)$ **(b)** $f(a) + f(a + 2\pi) + f(a + 4\pi)$

114. If $f(\theta) = \cos \theta$ and $f(a) = \dfrac{1}{4}$, find the exact value of:

   (a) $f(-a)$ **(b)** $f(a) + f(a + 2\pi) + f(a - 2\pi)$

115. If $f(\theta) = \tan \theta$ and $f(a) = 2$, find the exact value of:

   (a) $f(-a)$ **(b)** $f(a) + f(a + \pi) + f(a + 2\pi)$

116. If $f(\theta) = \cot \theta$ and $f(a) = -3$, find the exact value of:

   (a) $f(-a)$ **(b)** $f(a) + f(a + \pi) + f(a + 4\pi)$

117. If $f(\theta) = \sec \theta$ and $f(a) = -4$, find the exact value of:

   (a) $f(-a)$ **(b)** $f(a) + f(a + 2\pi) + f(a + 4\pi)$

118. If $f(\theta) = \csc \theta$ and $f(a) = 2$, find the exact value of:

   (a) $f(-a)$ **(b)** $f(a) + f(a + 2\pi) + f(a + 4\pi)$

119. Calculating the Time of a Trip From a parking lot, you want to walk to a house on the beach. The house is located 1500 feet down a paved path that parallels the ocean, which is 500 feet away. See the figure. Along the path you can walk 300 feet per minute, but in the sand on the beach you can only walk 100 feet per minute.

    The time T to get from the parking lot to the beach house expressed as a function of the angle θ shown in the figure is

$$T(\theta) = 5 - \frac{5}{3 \tan \theta} + \frac{5}{\sin \theta}, \qquad 0 < \theta < \frac{\pi}{2}$$

Calculate the time T if you walk directly from the parking lot to the house.

[**Hint:** $\tan \theta = \dfrac{500}{1500}$.]

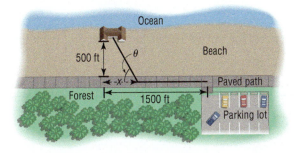

120. Calculating the Time of a Trip Two oceanfront homes are located 8 miles apart on a straight stretch of beach, each a distance of 1 mile from a paved path that parallels the ocean. Sally can jog 8 miles per hour on the paved path, but only 3 miles per hour in the sand on the beach. Because a river flows directly between the two houses, it is necessary to jog in the sand to the road, continue on the path, and then jog directly back in the sand to get from one house to the other. See the figure. The time T to get from one house to the other as a function of the angle θ shown in the figure is

$$T(\theta) = 1 + \frac{2}{3 \sin \theta} - \frac{1}{4 \tan \theta} \qquad 0 < \theta < \frac{\pi}{2}$$

   (a) Calculate the time T for $\tan \theta = \dfrac{1}{4}$.

   (b) Describe the path taken.

   (c) Explain why θ must be larger than 14°.

121. Predator Population In predator–prey relationships, the populations of the predator and prey are often cyclical. In a conservation area, rangers monitor the red fox population and have determined that the population can be modeled by the function

$$P(t) = 40 \cos\left(\frac{\pi t}{6}\right) + 110$$

where t is the number of months from the time monitoring began. Use the model to estimate the population of red foxes in the conservation area after 10 months, 20 months, and 30 months.

122. Lung Volume Normal resting lung volume V, in mL, for adult men varies over the breathing cycle and can be approximated by the model

$$V(t) = 250 \sin\left[\frac{2\pi(t - 1.25)}{5}\right] + 2650$$

where t is the number of seconds after breathing begins. Use the model to estimate the volume of air in a man's lungs after 2.5 seconds, 10 seconds, and 17 seconds.

123. Show that the range of the tangent function is the set of all real numbers.

124. Show that the range of the cotangent function is the set of all real numbers.

125. Show that the period of $f(\theta) = \sin\theta$ is 2π.

[**Hint:** Assume that $0 < p < 2\pi$ exists so that $\sin(\theta + p) = \sin\theta$ for all θ. Let $\theta = 0$ to find p. Then let $\theta = \dfrac{\pi}{2}$ to obtain a contradiction.]

126. Show that the period of $f(\theta) = \cos\theta$ is 2π.

127. Show that the period of $f(\theta) = \sec\theta$ is 2π.

128. Show that the period of $f(\theta) = \csc\theta$ is 2π.

129. Show that the period of $f(\theta) = \tan\theta$ is π.

130. Show that the period of $f(\theta) = \cot\theta$ is π.

131. Prove the reciprocal identities given in identity (2).

132. Prove the quotient identities given in identity (3).

133. Establish the identity:

$$(\sin\theta\cos\phi)^2 + (\sin\theta\sin\phi)^2 + \cos^2\theta = 1$$

134. ***Challenge Problem*** If $2\sin^2\theta + 3\cos^2\theta = 3\sin\theta\cos\theta + 1$ with θ in quadrant I, find the possible values for $\cot\theta$.

135. ***Challenge Problem*** If $\tan\theta = 3 - \sec\theta$ with θ in quadrant I, what is $\sin\theta + \cos\theta$?

136. ***Challenge Problem*** If $\sin(4\theta) = \cos(2\theta)$ and $0 < 4\theta < \dfrac{\pi}{2}$, find the exact value of $\sin(8\theta) + \cot(4\theta) - 2$.

137. ***Challenge Problem*** Find the exact value of $\sin\theta - \cos\theta$ if $\cos\theta - 8\sin\theta = 7$ and $180° < \theta < 270°$.

Explaining Concepts: Discussion and Writing

138. Write down five properties of the tangent function. Explain the meaning of each.

139. Describe your understanding of the meaning of a periodic function.

140. Explain how to find the value of $\sin 390°$ using periodic properties.

141. Explain how to find the value of $\cos(-45°)$ using even-odd properties.

142. Explain how to find the value of $\sin 390°$ and $\cos(-45°)$ using the unit circle.

Retain Your Knowledge

Problems 143–150 are based on material learned earlier in the course. The purpose of these problems is to keep the material fresh in your mind so that you are better prepared for the final exam.

143. Given $f(x) = x^2 - 3$ and $g(x) = x - 7$, find $f(g(x))$.

144. Graph $f(x) = -2x^2 + 12x - 13$ using transformations. Find the vertex and the axis of symmetry.

145. Find the real zeros of $f(x) = x^3 - 9x^2 + 3x - 27$.

146. Solve: $\sqrt{x + 2} - \sqrt{x - 5} = 2$

147. If the real zeros of $g(x)$ are -2 and 3, what are the real zeros of $g(x + 6)$?

148. Find c so that $f(x) = 6x^2 - 28x + c$ has a minimum value of $\dfrac{7}{3}$.

149. Find the intercepts of the graph of $-3x + 5y = 15$.

150. Find the difference quotient for $f(x) = \dfrac{3}{2}x^2 - 5x + 1$.

'Are You Prepared?' Answers

1. $\left\{x \middle| x \neq -\dfrac{1}{2}\right\}$ **2.** even   **3.** False   **4.** True

2.4 Graphs of the Sine and Cosine Functions

PREPARING FOR THIS SECTION *Before getting started, review the following:*

- Graphing Techniques: Transformations (Section 1.6, pp. 67–77)

Now Work the **'Are You Prepared?'** problems on page 156.

OBJECTIVES 1 Graph the Sine Function $y = \sin x$ and Functions of
the Form $y = A\sin(\omega x)$ (p. 147)
2 Graph the Cosine Function $y = \cos x$ and Functions of
the Form $y = A\cos(\omega x)$ (p. 149)
3 Determine the Amplitude and Period of Sinusoidal Functions (p. 150)
4 Graph Sinusoidal Functions Using Key Points (p. 152)
5 Find an Equation for a Sinusoidal Graph (p. 156)

We want to graph the trigonometric functions in the *xy*-plane. So we use the traditional symbols *x* for the independent variable (or argument) and *y* for the dependent variable for each function. Then the six trigonometric functions are written as

$$y = f(x) = \sin x \qquad y = f(x) = \cos x \qquad y = f(x) = \tan x$$
$$y = f(x) = \csc x \qquad y = f(x) = \sec x \qquad y = f(x) = \cot x$$

 Here the independent variable *x* represents an angle, measured in radians. However, in calculus, *x* will usually be treated as a real number. As noted earlier, these are equivalent ways of viewing *x*.

1 Graph the Sine Function $y = \sin x$ and Functions of the Form $y = A\sin(\omega x)$

Because the sine function has period 2π, it is only necessary to graph $y = \sin x$ on the interval $[0, 2\pi]$. The remainder of the graph will consist of repetitions of this portion of the graph.

To begin, consider Table 6, which lists some points on the graph of $y = \sin x$, $0 \le x \le 2\pi$. As the table shows, the graph of $y = \sin x$, $0 \le x \le 2\pi$, begins at the origin. As *x* increases from 0 to $\dfrac{\pi}{2}$, the value of $y = \sin x$ increases from 0 to 1; as *x* increases from $\dfrac{\pi}{2}$ to π to $\dfrac{3\pi}{2}$, the value of *y* decreases from 1 to 0 to -1; as *x* increases from $\dfrac{3\pi}{2}$ to 2π, the value of *y* increases from -1 to 0. Plotting the points listed in Table 6 and connecting them with a smooth curve yields the graph shown in Figure 44.

Table 6

x	$y = \sin x$	(x, y)
0	0	$(0, 0)$
$\dfrac{\pi}{6}$	$\dfrac{1}{2}$	$\left(\dfrac{\pi}{6}, \dfrac{1}{2}\right)$
$\dfrac{\pi}{2}$	1	$\left(\dfrac{\pi}{2}, 1\right)$
$\dfrac{5\pi}{6}$	$\dfrac{1}{2}$	$\left(\dfrac{5\pi}{6}, \dfrac{1}{2}\right)$
π	0	$(\pi, 0)$
$\dfrac{7\pi}{6}$	$-\dfrac{1}{2}$	$\left(\dfrac{7\pi}{6}, -\dfrac{1}{2}\right)$
$\dfrac{3\pi}{2}$	-1	$\left(\dfrac{3\pi}{2}, -1\right)$
$\dfrac{11\pi}{6}$	$-\dfrac{1}{2}$	$\left(\dfrac{11\pi}{6}, -\dfrac{1}{2}\right)$
2π	0	$(2\pi, 0)$

Figure 44 $y = \sin x,\ 0 \le x \le 2\pi$

The graph in Figure 44 is one period, or **cycle**, of the graph of $y = \sin x$. To obtain a more complete graph of $y = \sin x$, continue the graph in each direction, as shown in Figure 45.

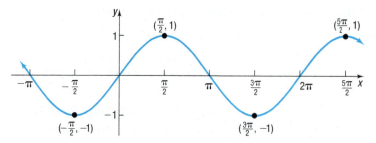

Figure 45 $y = \sin x$, $-\infty < x < \infty$

The graph of $y = \sin x$ illustrates some properties of the sine function.

Properties of the Sine Function $y = \sin x$

- The domain is the set of all real numbers.
- The range consists of all real numbers from -1 to 1, inclusive.
- The sine function is an odd function, as the symmetry of the graph with respect to the origin indicates.
- The sine function is periodic, with period 2π.
- The x-intercepts are $\ldots, -2\pi, -\pi, 0, \pi, 2\pi, 3\pi, \ldots$; the y-intercept is 0.
- The maximum value is 1 and occurs at $x = \ldots, -\dfrac{3\pi}{2}, \dfrac{\pi}{2}, \dfrac{5\pi}{2}, \dfrac{9\pi}{2}, \ldots$;

  the minimum value is -1 and occurs at $x = \ldots, -\dfrac{\pi}{2}, \dfrac{3\pi}{2}, \dfrac{7\pi}{2}, \dfrac{11\pi}{2}, \ldots$.

Now Work PROBLEM 11

EXAMPLE 1

Graphing Functions of the Form $y = A \sin x$ Using Transformations

Graph $y = 3 \sin x$ using transformations. Use the graph to determine the domain and the range of the function.

Solution Figure 46 illustrates the steps.

Figure 46

The domain of $y = 3 \sin x$ is the set of all real numbers, or $(-\infty, \infty)$. The range is $\{y \mid -3 \le y \le 3\}$, or $[-3, 3]$.

| EXAMPLE 2 | **Graphing Functions of the Form $y = A\sin(\omega x)$ Using Transformations** |

Graph $y = -\sin(2x)$ using transformations. Use the graph to determine the domain and the range of the function. Identify the period of the function $y = -\sin(2x)$.

Solution Figure 47 illustrates the steps.

(a) $y = \sin x$
(b) $y = -\sin x$
(c) $y = -\sin(2x)$

Figure 47

The domain of $y = -\sin(2x)$ is the set of all real numbers, or $(-\infty, \infty)$. The range is $\{y \mid -1 \le y \le 1\}$, or $[-1, 1]$.

The period of the function $y = -\sin(2x)$ is π because of the horizontal compression of the original period 2π by a factor of $\dfrac{1}{2}$. See Figure 47(c).

Now Work PROBLEM 39 USING TRANSFORMATIONS

2 Graph the Cosine Function $y = \cos x$ and Functions of the Form $y = A\cos(\omega x)$

The cosine function also has period 2π. To graph $y = \cos x$, begin by constructing Table 7, which lists some points on the graph of $y = \cos x, 0 \le x \le 2\pi$. As the table shows, the graph of $y = \cos x, 0 \le x \le 2\pi$, begins at the point $(0, 1)$.

As x increases from 0 to $\dfrac{\pi}{2}$ to π, the value of y decreases from 1 to 0 to -1; as x increases from π to $\dfrac{3\pi}{2}$ to 2π, the value of y increases from -1 to 0 to 1. As before, plot the points in Table 7 to get one period or cycle of the graph. See Figure 48.

Table 7

x	$y = \cos x$	(x, y)
0	1	$(0, 1)$
$\dfrac{\pi}{3}$	$\dfrac{1}{2}$	$\left(\dfrac{\pi}{3}, \dfrac{1}{2}\right)$
$\dfrac{\pi}{2}$	0	$\left(\dfrac{\pi}{2}, 0\right)$
$\dfrac{2\pi}{3}$	$-\dfrac{1}{2}$	$\left(\dfrac{2\pi}{3}, -\dfrac{1}{2}\right)$
π	-1	$(\pi, -1)$
$\dfrac{4\pi}{3}$	$-\dfrac{1}{2}$	$\left(\dfrac{4\pi}{3}, -\dfrac{1}{2}\right)$
$\dfrac{3\pi}{2}$	0	$\left(\dfrac{3\pi}{2}, 0\right)$
$\dfrac{5\pi}{3}$	$\dfrac{1}{2}$	$\left(\dfrac{5\pi}{3}, \dfrac{1}{2}\right)$
2π	1	$(2\pi, 1)$

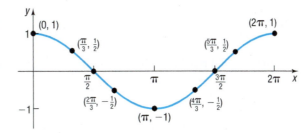

Figure 48 $y = \cos x, 0 \le x \le 2\pi$

A more complete graph of $y = \cos x$ is obtained by continuing the graph in each direction, as shown in Figure 49.

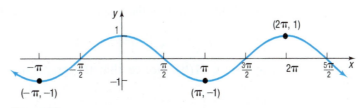

Figure 49 $y = \cos x, -\infty < x < \infty$

The graph of $y = \cos x$ illustrates some properties of the cosine function.

Properties of the Cosine Function

- The domain is the set of all real numbers.
- The range consists of all real numbers from -1 to 1, inclusive.
- The cosine function is an even function, as the symmetry of the graph with respect to the y-axis indicates.
- The cosine function is periodic, with period 2π.
- The x-intercepts are $\ldots, -\dfrac{3\pi}{2}, -\dfrac{\pi}{2}, \dfrac{\pi}{2}, \dfrac{3\pi}{2}, \dfrac{5\pi}{2}, \ldots$; the y-intercept is 1.
- The maximum value is 1 and occurs at $x = \ldots, -2\pi, 0, 2\pi, 4\pi, 6\pi, \ldots$; the minimum value is -1 and occurs at $x = \ldots, -\pi, \pi, 3\pi, 5\pi, \ldots$.

EXAMPLE 3

Graphing Functions of the Form $y = A\cos(\omega x)$ Using Transformations

Graph $y = 2\cos(3x)$ using transformations. Use the graph to determine the domain and the range of the function. Identify the period of the function $y = 2\cos(3x)$.

Solution Figure 50 shows the steps.

(a) $y = \cos x$ — Multiply by 2; Vertical stretch by a factor of 2 → (b) $y = 2\cos x$ — Replace x by $3x$; Horizontal compression by a factor of $\frac{1}{3}$ → (c) $y = 2\cos(3x)$

Figure 50

The domain of $y = 2\cos(3x)$ is the set of all real numbers, or $(-\infty, \infty)$. The range is $\{y \mid -2 \le y \le 2\}$, or $[-2, 2]$.

The period of the function $y = 2\cos(3x)$ is $\dfrac{2\pi}{3}$ because of the compression of the original period 2π by the factor of $\dfrac{1}{3}$. See Figure 50(c).

Now Work PROBLEM 43 USING TRANSFORMATIONS

3 Determine the Amplitude and Period of Sinusoidal Functions

The sine function and cosine function are referred to as **sinusoidal functions**. The discussion below provides the rationale for this definition.

Begin by shifting the graph of $y = \cos x$ to the right $\dfrac{\pi}{2}$ units to obtain the graph of $y = \cos\left(x - \dfrac{\pi}{2}\right)$. See Figure 51(a). Now look at the graph of $y = \sin x$ in Figure 51(b). Notice that the graph of $y = \sin x$ is the same as the graph of $y = \cos\left(x - \dfrac{\pi}{2}\right)$.

(a) $y = \cos x$ $y = \cos\left(x - \frac{\pi}{2}\right)$

(b) $y = \sin x$

Figure 51

Based on Figure 51, we conjecture that

<table>
<tr><td>

Seeing the Concept

Graph $Y_1 = \sin x$ and $Y_2 = \cos\left(x - \frac{\pi}{2}\right)$.

How many graphs do you see?

</td><td>

$$\sin x = \cos\left(x - \frac{\pi}{2}\right)$$

</td></tr>
</table>

We prove this in Chapter 3. Because of this relationship, the graphs of functions of the form $y = A\sin(\omega x)$ or $y = A\cos(\omega x)$ are referred to as **sinusoidal graphs**.

Figure 52 uses transformations to obtain the graph of $y = 2\cos x$. Note that the values of $y = 2\cos x$ lie between -2 and 2, inclusive.

(a) $y = \cos x$ Multiply by 2;
Vertical stretch
by a factor of 2   **(b)** $y = 2\cos x$

Figure 52

<table>
<tr><td>

In Words

The amplitude determines the vertical stretch or compression of the graph of $y = \sin x$ or $y = \cos x$.

</td><td>

The values of the functions $y = A\sin x$ and $y = A\cos x$, where $A \neq 0$, satisfy the inequalities

$$-|A| \le A\sin x \le |A| \quad \text{and} \quad -|A| \le A\cos x \le |A|$$

respectively. The number $|A|$ is called the **amplitude** of $y = A\sin x$ or $y = A\cos x$. See Figure 53.

</td></tr>
</table>

Figure 53 $y = A\sin x$, $A > 0$; period $= 2\pi$

If $\omega > 0$, the functions $y = \sin(\omega x)$ and $y = \cos(\omega x)$ have period $T = \dfrac{2\pi}{\omega}$. To see why, recall that the graph of $y = \sin(\omega x)$ is obtained from the graph of $y = \sin x$ by performing a horizontal compression or stretch by a factor $\dfrac{1}{\omega}$. This horizontal compression replaces the interval $[0, 2\pi]$, which contains one period of the graph of $y = \sin x$, by the interval $\left[0, \dfrac{2\pi}{\omega}\right]$, which contains one period of the graph of $y = \sin(\omega x)$. This is why the function $y = 2\cos(3x)$, graphed in Figure 50(c), with $\omega = 3$, has period $\dfrac{2\pi}{\omega} = \dfrac{2\pi}{3}$.

One period of the graph of $y = \sin(\omega x)$ or $y = \cos(\omega x)$ is called a **cycle**. Figure 54 illustrates the general situation. The blue portion of the graph is one cycle.

Figure 54 $y = A \sin(\omega x)$, $A > 0$, $\omega > 0$; period $= \dfrac{2\pi}{\omega}$

If $\omega > 0$ and $y = \sin(-\omega x)$ or $y = \cos(-\omega x)$, we use the even-odd properties of the sine and cosine functions, which are

$$\sin(-\omega x) = -\sin(\omega x) \quad \text{and} \quad \cos(-\omega x) = \cos(\omega x)$$

This gives us an equivalent form in which the coefficient of x is positive. For example,

$$\sin(-2x) = -\sin(2x) \quad \text{and} \quad \cos(-\pi x) = \cos(\pi x)$$

Because of this, we can assume that $\omega > 0$.

THEOREM

If $\omega > 0$, the amplitude and period of $y = A \sin(\omega x)$ and $y = A \cos(\omega x)$ are given by

$$\text{Amplitude} = |A| \qquad \text{Period} = T = \dfrac{2\pi}{\omega} \qquad \textbf{(1)}$$

EXAMPLE 4 **Finding the Amplitude and Period of a Sinusoidal Function**

Determine the amplitude and period of $y = 3 \sin(4x)$.

Solution Comparing $y = 3 \sin(4x)$ to $y = A \sin(\omega x)$, note that $A = 3$ and $\omega = 4$. From equation (1),

$$\text{Amplitude} = |A| = 3 \qquad \text{Period} = T = \dfrac{2\pi}{\omega} = \dfrac{2\pi}{4} = \dfrac{\pi}{2}$$

━━━━ **Now Work** PROBLEM **17**

4 Graph Sinusoidal Functions Using Key Points

So far, we have graphed functions of the form $y = A \sin(\omega x)$ or $y = A \cos(\omega x)$ using transformations. We now introduce another method that can be used to graph these functions.

Figure 55 shows one cycle of the graphs of $y = \sin x$ and $y = \cos x$ on the interval $[0, 2\pi]$. Each graph consists of four parts corresponding to the four subintervals:

$$\left[0, \frac{\pi}{2}\right], \quad \left[\frac{\pi}{2}, \pi\right], \quad \left[\pi, \frac{3\pi}{2}\right], \quad \left[\frac{3\pi}{2}, 2\pi\right]$$

Each subinterval is of length $\dfrac{\pi}{2}$ (the period 2π divided by 4), and the endpoints of these intervals $x = 0$, $x = \dfrac{\pi}{2}$, $x = \pi$, $x = \dfrac{3\pi}{2}$, $x = 2\pi$ give rise to five key points on each graph, as shown in Figure 55.

(a) $y = \sin x$

(b) $y = \cos x$

Figure 55

The next example illustrates how these five key points can be used to obtain the graph of a sinusoidal function.

EXAMPLE 5   **Graphing a Sinusoidal Function Using Key Points**

Graph $y = 3\sin(4x)$ using key points.

Step-by-Step Solution

Step 1 Determine the amplitude and period of the sinusoidal function.

Comparing $y = 3\sin(4x)$ to $y = A\sin(\omega x)$, note that $A = 3$ and $\omega = 4$, so the amplitude is $|A| = 3$ and the period is $\dfrac{2\pi}{\omega} = \dfrac{2\pi}{4} = \dfrac{\pi}{2}$. Because the amplitude is 3, the graph of $y = 3\sin(4x)$ lies between -3 and 3 on the y-axis. Because the period is $\dfrac{\pi}{2}$, one cycle begins at $x = 0$ and ends at $x = \dfrac{\pi}{2}$.

Step 2 Divide the interval $\left[0, \dfrac{2\pi}{\omega}\right]$ into four subintervals of the same length.

Divide the interval $\left[0, \dfrac{\pi}{2}\right]$ into four subintervals, each of length $\dfrac{\pi}{2} \div 4 = \dfrac{\pi}{8}$, as follows:

$$\left[0, \frac{\pi}{8}\right] \quad \left[\frac{\pi}{8}, \frac{\pi}{8} + \frac{\pi}{8}\right] = \left[\frac{\pi}{8}, \frac{\pi}{4}\right] \quad \left[\frac{\pi}{4}, \frac{\pi}{4} + \frac{\pi}{8}\right] = \left[\frac{\pi}{4}, \frac{3\pi}{8}\right] \quad \left[\frac{3\pi}{8}, \frac{3\pi}{8} + \frac{\pi}{8}\right] = \left[\frac{3\pi}{8}, \frac{\pi}{2}\right]$$

The endpoints of the subintervals are $0, \dfrac{\pi}{8}, \dfrac{\pi}{4}, \dfrac{3\pi}{8}, \dfrac{\pi}{2}$. These numbers represent the x-coordinates of the five key points on the graph.

Step 3 Use the endpoints of the subintervals from Step 2 to obtain five key points on the graph.

NOTE The five key points (x, y) also can be found using the five endpoints (Step 2) as the x-coordinates. Then the y-coordinates are the product of $A = 3$ and the y-coordinates of the five key points of $y = \sin x$. ∎

To obtain the y-coordinates of the five key points of $y = 3\sin(4x)$, evaluate $y = 3\sin(4x)$ at each endpoint found in Step 2. The five key points are then

$$(0, 0) \quad \left(\frac{\pi}{8}, 3\right) \quad \left(\frac{\pi}{4}, 0\right) \quad \left(\frac{3\pi}{8}, -3\right) \quad \left(\frac{\pi}{2}, 0\right)$$

Step 4 Plot the five key points and draw a sinusoidal graph to obtain the graph of one cycle. Extend the graph in each direction.

Plot the five key points obtained in Step 3, and fill in the graph of the sine curve as shown in Figure 56(a). Extend the graph in each direction to obtain the complete graph shown in Figure 56(b). Notice that additional key points appear every $\dfrac{\pi}{8}$ radian.

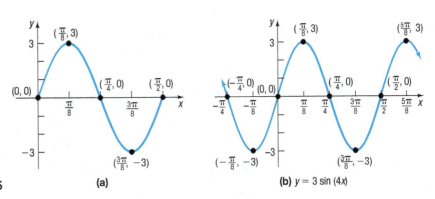

Figure 56   (a)   (b) $y = 3\sin(4x)$

Now Work PROBLEM **35** USING KEY POINTS

Graph $y = 3\sin(4x)$ using transformations. Which graphing method do you prefer?

SUMMARY

Steps for Graphing a Sinusoidal Function of the Form $y = A \sin(\omega x)$ or $y = A \cos(\omega x)$ Using Key Points

STEP 1: Determine the amplitude and period of the sinusoidal function.

STEP 2: Divide the interval $\left[0, \dfrac{2\pi}{\omega}\right]$ into four subintervals of the same length.

STEP 3: Use the endpoints of these subintervals to obtain five key points on the graph.

STEP 4: Plot the five key points, and draw a sinusoidal graph to obtain the graph of one cycle. Extend the graph in each direction to make it complete.

EXAMPLE 6

Graphing a Sinusoidal Function Using Key Points

Graph $y = 2 \sin\left(-\dfrac{\pi}{2}x\right)$ using key points.

Solution

Since the sine function is odd, use the equivalent form:

$$y = -2 \sin\left(\dfrac{\pi}{2}x\right)$$

STEP 1: Compare $y = -2 \sin\left(\dfrac{\pi}{2}x\right)$ to $y = A \sin(\omega x)$. Then $A = -2$ and $\omega = \dfrac{\pi}{2}$, so the amplitude is $|A| = |-2| = 2$ and the period is $T = \dfrac{2\pi}{\omega} = \dfrac{2\pi}{\dfrac{\pi}{2}} = 4$. The graph of $y = -2 \sin\left(\dfrac{\pi}{2}x\right)$ lies between -2 and 2 on the y-axis. One cycle begins at $x = 0$ and ends at $x = 4$.

STEP 2: Divide the interval $[0, 4]$ into four subintervals, each of length $4 \div 4 = 1$. The x-coordinates of the five key points are

0	$0 + 1 = 1$	$1 + 1 = 2$	$2 + 1 = 3$	$3 + 1 = 4$
1st **x-coordinate**	**2nd** **x-coordinate**	**3rd** **x-coordinate**	**4th** **x-coordinate**	**5th** **x-coordinate**

STEP 3: Evaluate $y = -2 \sin\left(\dfrac{\pi}{2}x\right)$ at each of the five x-coordinates.

- at $x = 0$, $y = -2 \sin 0 = 0$ • at $x = 1$, $y = -2 \sin \dfrac{\pi}{2} = -2$
- at $x = 2$, $y = 0$ • at $x = 3$, $y = 2$ • at $x = 4$, $y = 0$

The five key points on the graph are

$$(0,0) \quad (1, -2) \quad (2, 0) \quad (3, 2) \quad (4, 0)$$

STEP 4: Plot these five points, and fill in the graph of the sine function as shown in Figure 57(a). Extend the graph in each direction to obtain Figure 57(b).

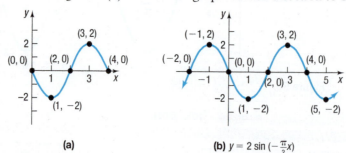

Figure 57

COMMENT To graph a sinusoidal function of the form $y = A \sin(\omega x)$ or $y = A \cos(\omega x)$ using a graphing utility, use the amplitude to set $Y_{\min}$ and $Y_{\max}$, and use the period to set $X_{\min}$ and $X_{\max}$. ∎

Now Work PROBLEM 39 USING KEY POINTS

If the function to be graphed is of the form $y = A \sin(\omega x) + B$ [or $y = A \cos(\omega x) + B$], first graph $y = A \sin(\omega x)$ [or $y = A \cos(\omega x)$] and then use a vertical shift.

EXAMPLE 7   **Graphing a Sinusoidal Function Using Key Points**

Graph $y = -4\cos(\pi x) - 2$ using key points. Use the graph to determine the domain and the range of $y = -4\cos(\pi x) - 2$.

Solution   Begin by graphing the function $y = -4\cos(\pi x)$.

Comparing $y = -4\cos(\pi x)$ to $y = A\cos(\omega x)$, note that $A = -4$ and $\omega = \pi$. The amplitude is $|A| = |-4| = 4$, and the period is $T = \dfrac{2\pi}{\omega} = \dfrac{2\pi}{\pi} = 2$.

The graph of $y = -4\cos(\pi x)$ lies between -4 and 4 on the y-axis. One cycle begins at $x = 0$ and ends at $x = 2$.

Divide the interval $[0, 2]$ into four subintervals, each of length $2 \div 4 = \dfrac{1}{2}$. The x-coordinates of the five key points are

$$0 \qquad 0 + \frac{1}{2} = \frac{1}{2} \qquad \frac{1}{2} + \frac{1}{2} = 1 \qquad 1 + \frac{1}{2} = \frac{3}{2} \qquad \frac{3}{2} + \frac{1}{2} = 2$$

| **1st** | **2nd** | **3rd** | **4th** | **5th** |
| **x-coordinate** | **x-coordinate** | **x-coordinate** | **x-coordinate** | **x-coordinate** |

Now evaluate $y = -4\cos(\pi x)$ at each of the five x-coordinates listed above.

$$(0, -4) \quad \left(\frac{1}{2}, 0\right) \quad (1, 4) \quad \left(\frac{3}{2}, 0\right) \quad (2, -4)$$

Plot these five points, and fill in the graph of the cosine function as shown in Figure 58(a). Extend the graph in each direction to obtain Figure 58(b), the graph of $y = -4\cos(\pi x)$.

A vertical shift down 2 units gives the graph of $y = -4\cos(\pi x) - 2$, as shown in Figure 58(c).

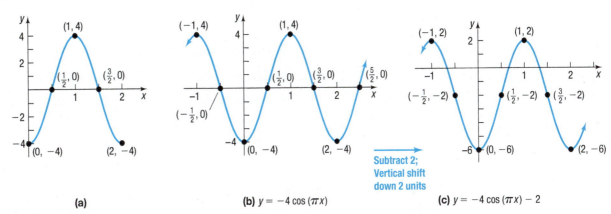

(a)

(b) $y = -4\cos(\pi x)$

(c) $y = -4\cos(\pi x) - 2$

Figure 58

The domain of $y = -4\cos(\pi x) - 2$ is the set of all real numbers, or $(-\infty, \infty)$. The range of $y = -4\cos(\pi x) - 2$ is $\{y \mid -6 \le y \le 2\}$, or $[-6, 2]$.

Now Work PROBLEM 49

5 Find an Equation for a Sinusoidal Graph

EXAMPLE 8	**Using a Graph to Find an Equation for a Sinusoidal Function**

Find an equation for the sinusoidal graph shown in Figure 59.

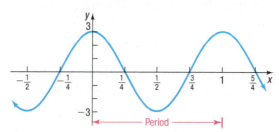

Figure 59

Solution

The graph has the characteristics of a cosine function. Do you see why? The maximum value, 3, occurs at $x = 0$. So the equation can be viewed as a cosine function $y = A \cos(\omega x)$ with $A = 3$ and period $T = 1$. Then $\dfrac{2\pi}{\omega} = 1$, so $\omega = 2\pi$. The cosine function whose graph is shown in Figure 59 is

$$y = A \cos(\omega x) = 3\cos(2\pi x)$$

EXAMPLE 9	**Using a Graph to Find an Equation for a Sinusoidal Function**

Find an equation for the sinusoidal graph shown in Figure 60.

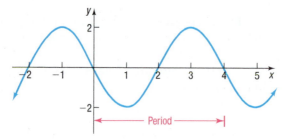

Figure 60

Solution

NOTE The equation could also be viewed as a cosine function with a horizontal shift, but viewing it as a sine function is easier. ∎

The graph is sinusoidal, with amplitude $|A| = 2$. The period is 4, so $\dfrac{2\pi}{\omega} = 4$, or $\omega = \dfrac{\pi}{2}$. Since the graph passes through the origin but is decreasing near the origin, the graph is that of a sine function reflected about the x-axis. This requires that $A = -2$. The sine function whose graph is given in Figure 60 is

$$y = A \sin(\omega x) = -2\sin\left(\frac{\pi}{2}x\right)$$

✏️–**Now Work** PROBLEMS **57** AND **61**

2.4 Assess Your Understanding

'Are You Prepared?' *Answers are given at the end of these exercises. If you get a wrong answer, read the pages listed in red.*

1. Use transformations to graph $y = 3x^2$. (pp. 67–77)

2. Use transformations to graph $y = \sqrt{2x}$. (pp. 67–77)

Concepts and Vocabulary

3. The maximum value of $y = \sin x$, $0 \le x \le 2\pi$, is _____
and occurs at $x =$ _____.

4. If the function $y = A \sin(\omega x)$, $A > 0$, has amplitude 3 and period 2, then $A =$ _____ and $\omega =$ _____.

5. The function $y = -3\cos(6x)$ has amplitude _____ and period _____.

6. *True or False* The graphs of $y = \sin x$ and $y = \cos x$ are identical except for a horizontal shift.

7. *True or False* For $y = 2 \sin(\pi x)$, the amplitude is 2 and the period is $\dfrac{\pi}{2}$.

8. *True or False* The graph of the sine function has infinitely many x-intercepts.

9. *Multiple Choice* One period of the graph of $y = \sin(\omega x)$ or $y = \cos(\omega x)$ is called a(n) ____.
(a) amplitude
(b) phase shift
(c) transformation
(d) cycle

10. *Multiple Choice* To graph $y = 3\sin(-2x)$ using key points, the equivalent form ____ could be graphed instead.
(a) $y = -3\sin(-2x)$
(b) $y = -2\sin(3x)$
(c) $y = 3\sin(2x)$
(d) $y = -3\sin(2x)$

Skill Building

11. $f(x) = \sin x$
  (a) What is the y-intercept of the graph of f?
  (b) For what numbers x, $-\pi \le x \le \pi$, is the graph of f increasing?
  (c) What is the absolute maximum of f?
  (d) For what numbers x, $0 \le x \le 2\pi$, does $f(x) = 0$?
  (e) For what numbers x, $-2\pi \le x \le 2\pi$, does $f(x) = 1$? Where does $f(x) = -1$?
  (f) For what numbers x, $-2\pi \le x \le 2\pi$, does $f(x) = -\dfrac{1}{2}$?
  (g) What are the x-intercepts of f?

12. $g(x) = \cos x$
  (a) What is the y-intercept of the graph of g?
  (b) For what numbers x, $-\pi \le x \le \pi$, is the graph of g decreasing?
  (c) What is the absolute minimum of g?
  (d) For what numbers x, $0 \le x \le 2\pi$, does $g(x) = 0$?
  (e) For what numbers x, $-2\pi \le x \le 2\pi$, does $g(x) = 1$? Where does $g(x) = -1$?
  (f) For what numbers x, $-2\pi \le x \le 2\pi$, does $g(x) = \dfrac{\sqrt{3}}{2}$?
  (g) What are the x-intercepts of g?

In Problems 13–22, determine the amplitude and period of each function without graphing.

13. $y = 5 \sin x$

14. $y = 3 \cos x$

15. $y = -3 \cos(4x)$

16. $y = -\sin\left(\dfrac{1}{2}x\right)$

17. $y = 6 \sin(\pi x)$

18. $y = -3 \cos(3x)$

19. $y = -\dfrac{1}{7}\cos\left(\dfrac{7}{2}x\right)$

20. $y = \dfrac{4}{3}\sin\left(\dfrac{2}{3}x\right)$

21. $y = \dfrac{10}{9}\sin\left(-\dfrac{2\pi}{5}x\right)$

22. $y = \dfrac{9}{5}\cos\left(-\dfrac{3\pi}{2}x\right)$

In Problems 23–32, match the given function to one of the graphs (A)–(J).

23. $y = 2 \sin\left(\dfrac{\pi}{2}x\right)$

24. $y = 2 \cos\left(\dfrac{\pi}{2}x\right)$

25. $y = 2 \cos\left(\dfrac{1}{2}x\right)$

26. $y = 3 \cos(2x)$

27. $y = -3 \sin(2x)$

28. $y = 2 \sin\left(\dfrac{1}{2}x\right)$

29. $y = -2 \cos\left(\dfrac{1}{2}x\right)$

30. $y = -2 \cos\left(\dfrac{\pi}{2}x\right)$

31. $y = 3 \sin(2x)$

32. $y = -2 \sin\left(\dfrac{1}{2}x\right)$

(A)

(B)

(C)

(D)

(E)

(F)

(G)

(H)

(I)

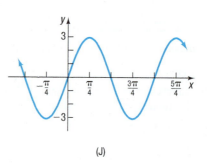

(J)

In Problems 33–56, graph each function using transformations or the method of key points. Be sure to label key points and show at least two cycles. Use the graph to determine the domain and the range of each function.

33. $y = 4\cos x$

34. $y = 3\sin x$

35. $y = -4\sin x$

36. $y = -3\cos x$

37. $y = \cos(4x)$

38. $y = \sin(3x)$

39. $y = \sin(-2x)$

40. $y = \cos(-2x)$

41. $y = 2\sin\left(\dfrac{1}{2}x\right)$

42. $y = 2\cos\left(\dfrac{1}{4}x\right)$

43. $y = -\dfrac{1}{2}\cos(2x)$

44. $y = -4\sin\left(\dfrac{1}{8}x\right)$

45. $y = 2\sin x + 3$

46. $y = 3\cos x + 2$

47. $y = 5\cos(\pi x) - 3$

48. $y = 4\sin\left(\dfrac{\pi}{2}x\right) - 2$

49. $y = -6\sin\left(\dfrac{\pi}{3}x\right) + 4$

50. $y = -3\cos\left(\dfrac{\pi}{4}x\right) + 2$

51. $y = 5 - 3\sin(2x)$

52. $y = 2 - 4\cos(3x)$

53. $y = \dfrac{5}{3}\sin\left(-\dfrac{2\pi}{3}x\right)$

54. $y = \dfrac{9}{5}\cos\left(-\dfrac{3\pi}{2}x\right)$

55. $y = -\dfrac{3}{2}\cos\left(\dfrac{\pi}{4}x\right) + \dfrac{1}{2}$

56. $y = -\dfrac{1}{2}\sin\left(\dfrac{\pi}{8}x\right) + \dfrac{3}{2}$

In Problems 57–60, write the equation of a sine function that has the given characteristics.

57. Amplitude: 3
Period: π

58. Amplitude: 2
Period: 4π

59. Amplitude: 3
Period: 2

60. Amplitude: 4
Period: 1

In Problems 61–74, find an equation for each graph.

61.

62.

63.

64.

65.

66.

67.

68.

69.

70.

71.

72.

73.

74.

Mixed Practice In Problems 75–78, find the average rate of change of f from 0 to $\dfrac{\pi}{2}$.

75. $f(x) = \sin x$ **76.** $f(x) = \cos x$ **77.** $f(x) = \sin \dfrac{x}{2}$ **78.** $f(x) = \cos(2x)$

Mixed Practice In Problems 79–82, find $f(g(x))$ and $g(f(x))$, and graph each of these functions.

79. $f(x) = \sin x$
$g(x) = 4x$

80. $f(x) = \cos x$
$g(x) = \dfrac{1}{2}x$

81. $f(x) = -2x$
$g(x) = \cos x$

82. $f(x) = -3x$
$g(x) = \sin x$

Applications and Extensions

83. Graph $f(x) = \begin{cases} \sin x & \text{if } 0 \le x < \dfrac{5\pi}{4} \\ \cos x & \text{if } \dfrac{5\pi}{4} \le x \le 2\pi \end{cases}$

84. Graph $g(x) = \begin{cases} 2\sin x & \text{if } 0 \le x \le \pi \\ \cos x + 1 & \text{if } \pi < x \le 2\pi \end{cases}$

85. Graph $y = |\sin x|, \ -2\pi \le x \le 2\pi$.

86. Graph $y = |\cos x|, \ -2\pi \le x \le 2\pi$.

87. Alternating Current (ac) Circuits The current I, in amperes, flowing through an ac (alternating current) circuit at time t, in seconds, is

$$I(t) = 220 \sin(60\pi t) \qquad t \ge 0$$

What is the period? What is the amplitude? Graph this function over two periods.

88. Alternating Current (ac) Circuits The current I, in amperes, flowing through an ac (alternating current) circuit at time t, in seconds, is

$$I(t) = 120 \sin(30\pi t) \qquad t \ge 0$$

What is the period? What is the amplitude? Graph this function over two periods.

89. Alternating Current (ac) Generators The voltage V produced by an ac generator is sinusoidal. As a function of time, the voltage V is

$$V(t) = V_0 \sin(2\pi ft)$$

where f is the **frequency**, the number of complete oscillations (cycles) per second. [In the United States and Canada, f is 60 hertz (Hz).] The **power** P delivered to a resistance R at any time t is defined as

$$P(t) = \frac{[V(t)]^2}{R}$$

(a) Show that $P(t) = \dfrac{V_0^2}{R}\sin^2(2\pi ft)$.

(b) The graph of P is shown in the figure. Express P as a sinusoidal function.

Power in an ac generator

(c) Deduce that

$$\sin^2(2\pi ft) = \frac{1}{2}[1 - \cos(4\pi ft)]$$

90. Tunnel Clearance A one-lane highway runs through a tunnel in the shape of one-half a sine curve cycle. The opening is 28 feet wide at road level and is 15 feet tall at its highest point.

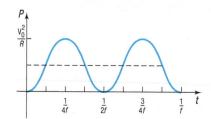

15 ft
14 ft
28 ft

(a) Find an equation for the sine curve that fits the opening. Place the origin at the left end of the opening.

(b) If the road is 14 feet wide with 7-foot shoulders on each side, what is the height of the tunnel at the edge of the road?

Source: en.wikipedia.org/wiki/Interstate_Highway_standards and *Ohio Revised Code*

91. Modeling Blood Pressure Several research papers use a sinusoidal graph to model blood pressure. Suppose an individual's blood pressure is modeled by the function

$$P(t) = 20 \sin\left(\frac{7\pi t}{3}\right) + 100$$

where the maximum value of P is the **systolic pressure**, which is the pressure when the heart contracts (beats), the minimum value is the **diastolic pressure**, and t is time, in seconds. The heart rate is the number of beats per minute.

(a) What is the individual's systolic pressure?

(b) What is the individual's diastolic pressure?

(c) What is the individual's heart rate?

92. Modeling Tides The function below models the water height H, in feet, at a monitoring station in Charleston, South Carolina.

$$H(t) = 2.91 \sin\left(\frac{24\pi}{149}t + 1.360\right) + 2.97$$

where t is the number of hours after midnight.
(a) What is the height of the water at high tide?
(b) What is the height of the water at low tide?
(c) What is the time between high and low tide?

93. Modeling Average Monthly Temperature The function below models the average monthly temperature T, in °F, for Indianapolis, Indiana.

$$T(x) = 23.65 \sin\left(\frac{\pi}{6}x - \frac{2\pi}{3}\right) + 51.75$$

where x is the month (January = 1, February = 2, etc.).
(a) What is the highest average monthly temperature?
(b) What is the lowest average monthly temperature?
(c) What is the time between the highest and lowest average temperatures?

94. Modeling Hours of Daylight The function below models the number of hours of daylight in Miami, Florida.

$$D(x) = 1.615 \sin\left(\frac{2\pi}{365}x - 1.39\right) + 12.135$$

where x is the day of the year.
(a) How many hours of daylight are there on the longest day?
(b) How many hours of daylight are there on the shortest day?
(c) What is the time between the longest and shortest days?

95. Ferris Wheel The function

$$h(t) = -100 \cos\left(\frac{\pi t}{50}\right) + 105$$

represents the height h, in feet, of a seat on a Ferris wheel as a function of time t, where t is measured in seconds.
(a) How high does a seat on the Ferris wheel go?
(b) How close to the ground does a seat get?
(c) If a ride lasts for 5 minutes, how many times will a passenger go around?
(d) What is the linear speed of the Ferris wheel in miles per hour? Round to one decimal place.

96. Holding Pattern The function

$$d(t) = 50 \cos\left(\frac{\pi t}{39}\right) + 60$$

represents the distance d, in miles, from the airport after t minutes of an airplane asked to fly in a circular holding pattern.
(a) What is the plane's average distance from the airport over one cycle?
(b) How long does it take the plane to complete one cycle in the holding pattern?
(c) What is the plane's speed, in miles per hour, while in the holding pattern?
(d) If the plane travels 1.8 miles per gallon of fuel, how much fuel is used in one cycle of the holding pattern?

97. Biorhythms In the theory of biorhythms, a sine function of the form

$$P(t) = 50 \sin(\omega t) + 50$$

is used to measure the percent P of a person's potential at time t, where t is measured in days and $t = 0$ is the day the person is born. Three characteristics are commonly measured:

 Physical potential: period of 23 days

 Emotional potential: period of 28 days

 Intellectual potential: period of 33 days

    (a) Find ω for each characteristic.

 (b) Using a graphing utility, graph all three functions on the same screen.

    (c) Is there a time t when all three characteristics have 100% potential? When is it?

 (d) Suppose that you are 20 years old today ($t = 7305$ days). Describe your physical, emotional, and intellectual potential for the next 30 days.

98. *Challenge Problem* If $y = A\sin(Bx - C) + D, A \neq 0$, for what values of D will the graph lie completely below the x-axis?

99. *Challenge Problem* If $A \neq 0$, find the intercepts of the graph of

$$y = A\cos[B(x - C)] + A$$

Explaining Concepts: Discussion and Writing

100. Explain how you would scale the x-axis and y-axis before graphing $y = 3\cos(\pi x)$.

101. Explain the term *amplitude* as it relates to the graph of a sinusoidal function.

102. Explain the term *period* as it relates to the graph of a sinusoidal function.

103. Explain how the amplitude and period of a sinusoidal graph are used to establish the scale on each coordinate axis.

104. Find an application in your major field that leads to a sinusoidal graph. Write a summary of your findings.

Retain Your Knowledge

Problems 105–111 are based on material learned earlier in the course. The purpose of these problems is to keep the material fresh in your mind so that you are better prepared for the final exam.

105. If $f(x) = x^2 - 5x + 1$, find $\dfrac{f(x + h) - f(x)}{h}$.

106. Find the vertex of the graph of $g(x) = -3x^2 + 12x - 7$.

107. Find the intercepts of the graph of $h(x) = 3|x + 2| - 1$.

108. Solve: $3x - 2(5x + 16) = -3x + 4(8 + x)$

109. Find the oblique asymptote of $g(x) = \dfrac{4x^3 + 6x^2 - 3x + 1}{2x^2 - 4x + 3}$.

110. Determine the interval on which $f(x) = -6x^2 - 19x + 7$ is decreasing.

111. Write the set $\left\{ x \,\middle|\, x \leq -2 \text{ or } x > \dfrac{4}{3} \right\}$ using interval notation.

'Are You Prepared?' Answers

1. Vertical stretch by a factor of 3

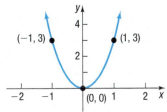

2. Horizontal compression by a factor of $\dfrac{1}{2}$

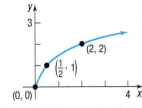

2.5 Graphs of the Tangent, Cotangent, Cosecant, and Secant Functions

PREPARING FOR THIS SECTION *Before getting started, review the following:*

• Vertical Asymptotes (Section 1.6, pp. 76–77)

Now Work the 'Are You Prepared?' problems on page 167.

OBJECTIVES 1 Graph the Tangent Function $y = \tan x$ and the Cotangent Function $y = \cot x$ (p. 162)

2 Graph Functions of the Form $y = A\tan(\omega x) + B$ and $y = A\cot(\omega x) + B$ (p. 164)

3 Graph the Cosecant Function $y = \csc x$ and the Secant Function $y = \sec x$ (p. 165)

4 Graph Functions of the Form $y = A\csc(\omega x) + B$ and $y = A\sec(\omega x) + B$ (p. 166)

1 Graph the Tangent Function $y = \tan x$ and the Cotangent Function $y = \cot x$

Because the tangent function has period π, we only need to determine the graph over some interval of length π. The rest of the graph consists of repetitions of that graph. Because the tangent function is not defined at $\ldots, -\dfrac{3\pi}{2}, -\dfrac{\pi}{2}, \dfrac{\pi}{2}, \dfrac{3\pi}{2}, \ldots,$ we concentrate on the interval $\left(-\dfrac{\pi}{2}, \dfrac{\pi}{2}\right)$, of length π, and construct Table 8, which lists some points on the graph of $y = \tan x$, $-\dfrac{\pi}{2} < x < \dfrac{\pi}{2}$. We plot the points in the table and connect them with a smooth curve. See Figure 61 for a partial graph of $y = \tan x$, where $-\dfrac{\pi}{3} \le x \le \dfrac{\pi}{3}$.

Table 8

x	$y = \tan x$	(x, y)
$-\dfrac{\pi}{3}$	$-\sqrt{3} \approx -1.73$	$\left(-\dfrac{\pi}{3}, -\sqrt{3}\right)$
$-\dfrac{\pi}{4}$	-1	$\left(-\dfrac{\pi}{4}, -1\right)$
$-\dfrac{\pi}{6}$	$-\dfrac{\sqrt{3}}{3} \approx -0.58$	$\left(-\dfrac{\pi}{6}, -\dfrac{\sqrt{3}}{3}\right)$
0	0	$(0, 0)$
$\dfrac{\pi}{6}$	$\dfrac{\sqrt{3}}{3} \approx 0.58$	$\left(\dfrac{\pi}{6}, \dfrac{\sqrt{3}}{3}\right)$
$\dfrac{\pi}{4}$	1	$\left(\dfrac{\pi}{4}, 1\right)$
$\dfrac{\pi}{3}$	$\sqrt{3} \approx 1.73$	$\left(\dfrac{\pi}{3}, \sqrt{3}\right)$

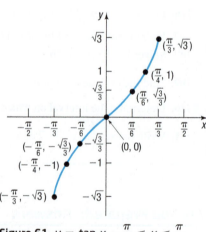

Figure 61 $y = \tan x$, $-\dfrac{\pi}{3} \le x \le \dfrac{\pi}{3}$

To complete one period of the graph of $y = \tan x$, investigate the behavior of the function as x approaches $-\dfrac{\pi}{2}$ and $\dfrac{\pi}{2}$. Be careful, though, because $y = \tan x$ is not defined at these numbers. To determine this behavior, use the identity

$$\tan x = \frac{\sin x}{\cos x}$$

See Table 9. If x is close to $\frac{\pi}{2} \approx 1.5708$ but remains less than $\frac{\pi}{2}$, then $\sin x$ is close to 1, and $\cos x$ is positive and close to 0. (To see this, refer to the graphs of the sine function and the cosine function.) The ratio $\frac{\sin x}{\cos x}$ is positive and large, so, as $x \to \frac{\pi}{2}$ from the left, then $\tan x \to \infty$. In other words, the vertical line $x = \frac{\pi}{2}$ is a vertical asymptote of the graph of $y = \tan x$.

Table 9

x	$\sin x$	$\cos x$	$y = \tan x$
$\frac{\pi}{3} \approx 1.05$	$\frac{\sqrt{3}}{2}$	$\frac{1}{2}$	$\sqrt{3} \approx 1.73$
1.5	0.9975	0.0707	14.1
1.57	0.9999	7.96×10^{-4}	1255.8
1.5707	0.9999	9.6×10^{-5}	10,381
$\frac{\pi}{2} \approx 1.5708$	1	0	Undefined

If x is close to $-\frac{\pi}{2}$ but remains greater than $-\frac{\pi}{2}$, then $\sin x$ is close to -1, $\cos x$ is positive and close to 0, and the ratio $\frac{\sin x}{\cos x}$ is negative and large. That is, as $x \to -\frac{\pi}{2}$ from the right, then $\tan x \to -\infty$. In other words, the vertical line $x = -\frac{\pi}{2}$ is also a vertical asymptote of the graph of $y = \tan x$.

With these observations, one period of the graph can be completed. Obtain the complete graph of $y = \tan x$ by repeating this period, as shown in Figure 62.

✓ **Check:** Graph $Y_1 = \tan x$ and compare the result with Figure 62. Use TRACE to see what happens as x gets close to $\frac{\pi}{2}$ but remains less than $\frac{\pi}{2}$.

Figure 62 $y = \tan x$, $-\infty < x < \infty$, x not equal to odd multiples of $\frac{\pi}{2}$, $-\infty < y < \infty$

The graph of $y = \tan x$ in Figure 62 illustrates the following properties.

Properties of the Tangent Function

- The domain is the set of all real numbers, except odd multiples of $\frac{\pi}{2}$.
- The range is the set of all real numbers.
- The tangent function is an odd function, as the symmetry of the graph with respect to the origin indicates.
- The tangent function is periodic, with period π.
- The x-intercepts are $\ldots$, $-2\pi, -\pi, 0, \pi, 2\pi, 3\pi, \ldots$; the y-intercept is 0.
- Vertical asymptotes occur at $x = \ldots , -\frac{3\pi}{2}, -\frac{\pi}{2}, \frac{\pi}{2}, \frac{3\pi}{2}, \ldots$.

━━━ **Now Work** PROBLEMS **7** AND **15**

The Graph of the Cotangent Function $y = \cot x$

The graph of $y = \cot x$ can be obtained in the same way as the graph of $y = \tan x$. The period of $y = \cot x$ is π. Because the cotangent function is not defined for integer multiples of π, concentrate on the interval $(0, \pi)$. Table 10 lists some points on the graph of $y = \cot x$, $0 < x < \pi$. To determine the behavior of $y = \cot x$ near 0 and π, use the identity $\cot x = \dfrac{\cos x}{\sin x}$. As x approaches 0 but remains greater than 0, the value of $\cos x$ is close to 1, and the value of $\sin x$ is positive and close to 0. The ratio $\dfrac{\cos x}{\sin x}$ is positive and large; so as $x \to 0$ from the right, then $\cot x \to \infty$. Similarly, as x approaches π but remains less than π, the value of $\cos x$ is close to -1, and the value of $\sin x$ is positive and close to 0. The ratio $\dfrac{\cos x}{\sin x} = \cot x$ is negative and large; so as $x \to \pi$ from the left, then $\cot x \to -\infty$. Figure 63 shows the graph.

Table 10

x	$y = \cot x$	(x, y)
$\dfrac{\pi}{6}$	$\sqrt{3}$	$\left(\dfrac{\pi}{6}, \sqrt{3}\right)$
$\dfrac{\pi}{4}$	1	$\left(\dfrac{\pi}{4}, 1\right)$
$\dfrac{\pi}{3}$	$\dfrac{\sqrt{3}}{3}$	$\left(\dfrac{\pi}{3}, \dfrac{\sqrt{3}}{3}\right)$
$\dfrac{\pi}{2}$	0	$\left(\dfrac{\pi}{2}, 0\right)$
$\dfrac{2\pi}{3}$	$-\dfrac{\sqrt{3}}{3}$	$\left(\dfrac{2\pi}{3}, -\dfrac{\sqrt{3}}{3}\right)$
$\dfrac{3\pi}{4}$	-1	$\left(\dfrac{3\pi}{4}, -1\right)$
$\dfrac{5\pi}{6}$	$-\sqrt{3}$	$\left(\dfrac{5\pi}{6}, -\sqrt{3}\right)$

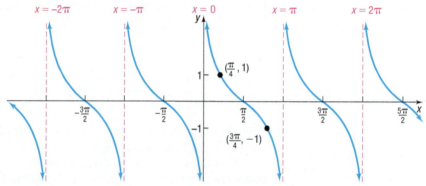

Figure 63 $y = \cot x$, $-\infty < x < \infty$, x not equal to integer multiples of π, $-\infty < y < \infty$

2 Graph Functions of the Form $y = A \tan(\omega x) + B$ and $y = A \cot(\omega x) + B$

For tangent functions, there is no concept of amplitude since the range of the tangent function is $(-\infty, \infty)$. The role of A in $y = A \tan(\omega x) + B$ is to provide the magnitude of the vertical stretch. The period of $y = \tan x$ is π, so the period of $y = A \tan(\omega x) + B$ is $\dfrac{\pi}{\omega}$, caused by the horizontal compression of the graph by a factor of $\dfrac{1}{\omega}$. Finally, the presence of B indicates a vertical shift.

EXAMPLE 1 **Graphing a Function of the Form $y = A \tan(\omega x) + B$**

Graph $y = 2 \tan x - 1$. Use the graph to determine the domain and the range of the function $y = 2 \tan x - 1$.

Solution Figure 64 shows the steps using transformations.

Figure 64

(a) $y = \tan x$ **Multiply by 2; Vertical stretch by a factor of 2**   (b) $y = 2 \tan x$ **Subtract 1; Vertically shift down 1 unit**   (c) $y = 2 \tan x - 1$

The domain of $y = 2 \tan x - 1$ is $\left\{ x \middle| x \neq \dfrac{k\pi}{2}, k \text{ is an odd integer} \right\}$, and the range is the set of all real numbers, or $(-\infty, \infty)$.

〰️ **Now Work** PROBLEM 21

The graph of $y = A \cot(\omega x) + B$ has characteristics similar to those of the tangent function. The cotangent function $y = A \cot(\omega x) + B$ has period $\dfrac{\pi}{\omega}$. The cotangent function has no amplitude. The role of A is to provide the magnitude of the vertical stretch; the presence of B indicates a vertical shift.

EXAMPLE 2 **Graphing a Function of the Form $y = A \cot(\omega x) + B$**

Graph $y = 3 \cot(2x)$. Use the graph to determine the domain and the range of $y = 3 \cot(2x)$.

Solution Figure 65 shows the steps using transformations.

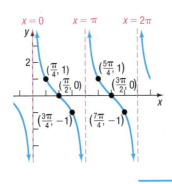
(a) $y = \cot x$
**Multiply by 3;
Vertical stretch by
a factor of 3**

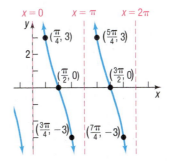
(b) $y = 3 \cot x$
**Replace x by $2x$;
Horizontal compression
by a factor of $\frac{1}{2}$**

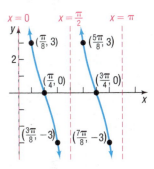
(c) $y = 3 \cot(2x)$

Figure 65

The domain of $y = 3 \cot(2x)$ is $\left\{ x \middle| x \neq \dfrac{k\pi}{2}, k \text{ is an integer} \right\}$, and the range is the set of all real numbers, or $(-\infty, \infty)$.

Note in Figure 65(c) that the period of $y = 3 \cot(2x)$ is $\dfrac{\pi}{2}$ because of the compression of the original period π by a factor of $\dfrac{1}{2}$. Notice that the asymptotes are $x = -\dfrac{\pi}{2}, x = 0, x = \dfrac{\pi}{2}, x = \pi, x = \dfrac{3\pi}{2}$, and so on, also because of the compression.

〰️ **Now Work** PROBLEM 23

3 Graph the Cosecant Function $y = \csc x$ and the Secant Function $y = \sec x$

The cosecant and secant functions, sometimes referred to as **reciprocal functions**, are graphed by making use of the reciprocal identities

$$ \csc x = \frac{1}{\sin x} \quad \text{and} \quad \sec x = \frac{1}{\cos x} $$

For example, the value of the cosecant function $y = \csc x$ at a given number x equals the reciprocal of the corresponding value of the sine function, provided that the value of the sine function is not 0. If the value of $\sin x$ is 0, then x is an integer multiple of π. At those numbers, the cosecant function is not defined. In fact, the graph of the cosecant function has a vertical asymptote at each integer multiple of π. Figure 66 shows the graph.

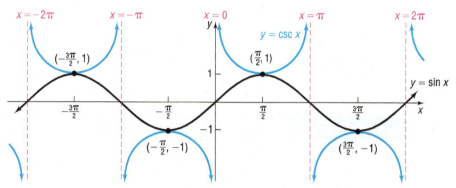

Figure 66 $y = \csc x$, $-\infty < x < \infty$, x not equal to integer multiples of π, $|y| \geq 1$

Using the idea of reciprocals, the graph of $y = \sec x$ is obtained in a similar manner. See Figure 67.

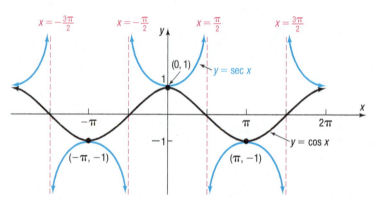

Figure 67 $y = \sec x$, $-\infty < x < \infty$, x not equal to odd multiples of $\frac{\pi}{2}$, $|y| \geq 1$

4 Graph Functions of the Form $y = A\csc(\omega x) + B$ and $y = A\sec(\omega x) + B$

The role of A in these functions is to set the range. The range of $y = \csc x$ is $\{y | y \leq -1 \text{ or } y \geq 1\}$; the range of $y = A\csc x$ is $\{y | y \leq -|A| \text{ or } y \geq |A|\}$ because of the vertical stretch of the graph by a factor of $|A|$. Just as with the sine and cosine functions, the period of $y = \csc(\omega x)$ and $y = \sec(\omega x)$ becomes $\dfrac{2\pi}{\omega}$ because of the horizontal compression of the graph by a factor of $\dfrac{1}{\omega}$. The presence of B indicates that a vertical shift is required.

EXAMPLE 3

Graphing a Function of the Form $y = A\csc(\omega x) + B$

Graph $y = 2\csc x - 1$. Use the graph to determine the domain and the range of $y = 2\csc x - 1$.

Solution   We use transformations. Figure 68 shows the required steps.

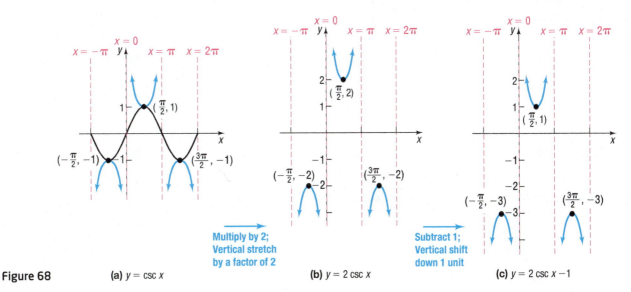

Figure 68   **(a)** $y = \csc x$ **(b)** $y = 2 \csc x$ **(c)** $y = 2 \csc x - 1$

The domain of $y = 2 \csc x - 1$ is $\{x \mid x \neq k\pi, k$ an integer$\}$. The range is $\{y \mid y \leq -3$ or $y \geq 1\}$, or using interval notation, $(-\infty, -3] \cup [1, \infty)$.

Now Work PROBLEM 29

2.5 Assess Your Understanding

'Are You Prepared?' *Answers are given at the end of these exercises. If you get a wrong answer, read the pages listed in* red.

1. The graph of $y = \dfrac{3x - 6}{x - 4}$ has a vertical asymptote. What is it? (pp. 76–77)

2. **True or False** If $x = 3$ is a vertical asymptote of the graph of a rational function R, then as $x \to 3, |R(x)| \to \infty$. (pp. 76–77)

Concepts and Vocabulary

3. The graph of $y = \tan x$ is symmetric with respect to the _____ and has vertical asymptotes at _____.

4. The graph of $y = \sec x$ is symmetric with respect to the _____ and has vertical asymptotes at _____.

5. **Multiple Choice** It is easiest to graph $y = \sec x$ by first sketching the graph of _____.
   (a) $y = \sin x$ **(b)** $y = \cos x$ **(c)** $y = \tan x$ **(d)** $y = \csc x$

6. **True or False** The graphs of $y = \tan x, y = \cot x, y = \sec x,$ and $y = \csc x$ each have infinitely many vertical asymptotes.

Skill Building

In Problems 7–16, if necessary, refer to the graphs of the functions to answer each question.

7. What is the y-intercept of $y = \tan x$?

8. What is the y-intercept of $y = \cot x$?

9. What is the y-intercept of $y = \sec x$?

10. What is the y-intercept of $y = \csc x$?

11. For what numbers $x, -2\pi \leq x \leq 2\pi$, does $\sec x = 1$? For what numbers x does $\sec x = -1$?

12. For what numbers $x, -2\pi \leq x \leq 2\pi$, does $\csc x = 1$? For what numbers x does $\csc x = -1$?

13. For what numbers $x, -2\pi \leq x \leq 2\pi$, does the graph of $y = \sec x$ have vertical asymptotes?

14. For what numbers $x, -2\pi \leq x \leq 2\pi$, does the graph of $y = \csc x$ have vertical asymptotes?

15. For what numbers $x, -2\pi \leq x \leq 2\pi$, does the graph of $y = \tan x$ have vertical asymptotes?

16. For what numbers $x, -2\pi \leq x \leq 2\pi$, does the graph of $y = \cot x$ have vertical asymptotes?

In Problems 17–40, graph each function. Be sure to label key points and show at least two cycles. Use the graph to determine the domain and the range of each function.

17. $y = 3 \tan x$

18. $y = -2 \tan x$

19. $y = 4 \cot x$

20. $y = -3 \cot x$

21. $y = \tan\left(\dfrac{\pi}{2}x\right)$

22. $y = \tan\left(\dfrac{1}{2}x\right)$

23. $y = \cot\left(\dfrac{1}{4}x\right)$

24. $y = \cot\left(\dfrac{\pi}{4}x\right)$

25. $y = 2 \sec x$

26. $y = \dfrac{1}{2}\csc x$

27. $y = -3 \csc x$

28. $y = -4 \sec x$

29. $y = 4 \sec\left(\dfrac{1}{2}x\right)$

30. $y = \dfrac{1}{2}\csc(2x)$

31. $y = -2 \csc(\pi x)$

32. $y = -3 \sec\left(\dfrac{\pi}{2}x\right)$

33. $y = \tan\left(\dfrac{1}{4}x\right) + 1$

34. $y = 2 \cot x - 1$

35. $y = \sec\left(\dfrac{2\pi}{3}x\right) + 2$

36. $y = \csc\left(\dfrac{3\pi}{2}x\right)$

37. $y = \dfrac{1}{2}\tan\left(\dfrac{1}{4}x\right) - 2$

38. $y = 3 \cot\left(\dfrac{1}{2}x\right) - 2$

39. $y = 2 \csc\left(\dfrac{1}{3}x\right) - 1$

40. $y = 3 \sec\left(\dfrac{1}{4}x\right) + 1$

Mixed Practice *In Problems 41–44, find the average rate of change of f from 0 to $\dfrac{\pi}{6}$.*

41. $f(x) = \tan x$

42. $f(x) = \sec x$

43. $f(x) = \tan(2x)$

44. $f(x) = \sec(2x)$

Mixed Practice *In Problems 45–48, find $f(g(x))$ and $g(f(x))$, and graph each of these functions.*

45. $f(x) = \tan x$
$g(x) = 4x$

46. $f(x) = 2 \sec x$
$g(x) = \dfrac{1}{2}x$

47. $f(x) = -2x$
$g(x) = \cot x$

48. $f(x) = \dfrac{1}{2}x$
$g(x) = 2 \csc x$

Mixed Practice *In Problems 49 and 50, graph each function.*

49. $f(x) = \begin{cases} \tan x & \text{if } 0 \le x < \dfrac{\pi}{2} \\ 0 & \text{if } x = \dfrac{\pi}{2} \\ \sec x & \text{if } \dfrac{\pi}{2} < x \le \pi \end{cases}$

50. $g(x) = \begin{cases} \csc x & \text{if } 0 < x < \pi \\ 0 & \text{if } x = \pi \\ \cot x & \text{if } \pi < x < 2\pi \end{cases}$

Applications and Extensions

51. Carrying a Ladder around a Corner Two hallways, one of width 3 feet, the other of width 4 feet, meet at a right angle. See the figure.

(a) Show that the length L of the ladder shown as a function of the angle θ is

$$L(\theta) = 3 \sec \theta + 4 \csc \theta$$

 (b) Graph $L = L(\theta), 0 < \theta < \dfrac{\pi}{2}$.

(c) For what value of θ is L the least?

(d) What is the length of the longest ladder that can be carried around the corner? Why is this also the least value of L?

52. A Rotating Beacon Suppose that a fire truck is parked in front of a building as shown in the figure.

The beacon light on top of the fire truck is located 10 feet from the wall and has a light on each side. If the beacon light rotates 1 revolution every 2 seconds, then a model for determining the distance d, in feet, that the beacon of light is from point A on the wall after t seconds is given by

$$d(t) = |10 \tan(\pi t)|$$

(a) Graph $d(t) = |10 \tan(\pi t)|$ for $0 \le t \le 2$.
(b) For what values of t is the function undefined? Explain what this means in terms of the beam of light on the wall.

(c) Fill in the following table.

t	0	0.1	0.2	0.3	0.4
$d(t) = 10 \tan(\pi t)$					

(d) Compute $\dfrac{d(0.1) - d(0)}{0.1 - 0}$, $\dfrac{d(0.2) - d(0.1)}{0.2 - 0.1}$, and so on, for each consecutive value of t. These are called **first differences**.

(e) Interpret the first differences found in part (d). What is happening to the speed of the beam of light as d increases?

53. Exploration Graph

$$y = \tan x \quad \text{and} \quad y = -\cot\left(x + \frac{\pi}{2}\right)$$

Do you think that $\tan x = -\cot\left(x + \dfrac{\pi}{2}\right)$?

54. Challenge Problem What are the domain and the range of $f(x) = \dfrac{1}{\sqrt{|\sin x|}}$? Find any vertical asymptotes.

Retain Your Knowledge

Problems 55–61 are based on material learned earlier in the course. The purpose of these problems is to keep the material fresh in your mind so that you are better prepared for the final exam.

55. Factor: $125p^3 - 8q^6$

56. Painting a Room Hazel can paint a room in 2 hours less time than her friend Gwyneth. Working together, they can paint the room in 2.4 hours. How long does it take each woman to paint the room by herself?

57. Use the slope and the y-intercept to graph the linear function

$$f(x) = \frac{1}{4}x - 3$$

58. If $f(x) = x^2 - 3x$, find $\dfrac{f(x) - f(c)}{x - c}$.

59. Find the intercepts of the graph of the function

$$f(x) = \frac{2x^2 + x - 6}{x + 3}$$

60. Complete the square in x to write $\sqrt{x^2 + 2x + 26}$ in the form $\sqrt{u^2 + a^2}$.

61. Find the domain of $f(x) = \sqrt[4]{5x - 2} - 3$.

'Are You Prepared?' Answers

1. $x = 4$ **2.** True

2.6 Phase Shift; Sinusoidal Curve Fitting

OBJECTIVES **1** Graph Sinusoidal Functions of the Form $y = A\sin(\omega x - \phi) + B$ (p. 169)
2 Build Sinusoidal Models from Data (p. 173)

1 Graph Sinusoidal Functions of the Form $y = A\sin(\omega x - \phi) + B$

Figure 69 One cycle of $y = A\sin(\omega x)$, $A > 0$, $\omega > 0$

We have seen that the graph of $y = A\sin(\omega x)$, $\omega > 0$, has amplitude $|A|$ and period $T = \dfrac{2\pi}{\omega}$. One cycle can be drawn as x varies from 0 to $\dfrac{2\pi}{\omega}$, or, equivalently, as ωx varies from 0 to 2π. See Figure 69.

Now consider the graph of

$$y = A \sin(\omega x - \phi)$$

which may also be written as

$$y = A \sin\left[\omega\left(x - \frac{\phi}{\omega}\right)\right]$$

NOTE The beginning and end of the period can also be found by solving the inequality:

$$0 \leq \omega x - \phi \leq 2\pi$$
$$\phi \leq \omega x \leq 2\pi + \phi$$
$$\frac{\phi}{\omega} \leq x \leq \frac{2\pi}{\omega} + \frac{\phi}{\omega} \qquad \blacksquare$$

where $\omega > 0$ and ϕ (the Greek letter phi) are real numbers. The graph is a sine curve with amplitude $|A|$. As $\omega x - \phi$ varies from 0 to 2π, one period is traced out. This period begins when

$$\omega x - \phi = 0 \quad \text{or} \quad x = \frac{\phi}{\omega}$$

and ends when

$$\omega x - \phi = 2\pi \quad \text{or} \quad x = \frac{\phi}{\omega} + \frac{2\pi}{\omega}$$

See Figure 70.

Notice that the graph of $y = A \sin(\omega x - \phi) = A \sin\left[\omega\left(x - \frac{\phi}{\omega}\right)\right]$ is the same

Figure 70 One cycle of
$y = A \sin(\omega x - \phi)$,
$A > 0, \omega > 0, \phi > 0$

as the graph of $y = A \sin(\omega x)$, except that it has been shifted $\left|\frac{\phi}{\omega}\right|$ units (to the right if $\phi > 0$ and to the left if $\phi < 0$). This number $\frac{\phi}{\omega}$ is called the **phase shift** of the graph of $y = A \sin(\omega x - \phi)$.

For the graphs of $y = A \sin(\omega x - \phi)$ or $y = A \cos(\omega x - \phi)$, $\omega > 0$,

$$\text{Amplitude} = |A| \qquad \text{Period} = T = \frac{2\pi}{\omega} \qquad \text{Phase shift} = \frac{\phi}{\omega}$$

The phase shift is to the left if $\phi < 0$ and to the right if $\phi > 0$.

EXAMPLE 1 | **Finding the Amplitude, Period, and Phase Shift of a Sinusoidal Function and Graphing It**

Find the amplitude, period, and phase shift of $y = 3 \sin(2x - \pi)$, and graph the function.

Solution   Use the same four steps used to graph sinusoidal functions of the form $y = A \sin(\omega x)$ or $y = A \cos(\omega x)$ given on page 154.

STEP 1: Comparing

$$y = 3 \sin(2x - \pi) = 3 \sin\left[2\left(x - \frac{\pi}{2}\right)\right]$$

to

$$y = A \sin(\omega x - \phi) = A \sin\left[\omega\left(x - \frac{\phi}{\omega}\right)\right]$$

note that $A = 3$, $\omega = 2$, and $\phi = \pi$. The graph is a sine curve with amplitude $|A| = 3$, period $T = \frac{2\pi}{\omega} = \frac{2\pi}{2} = \pi$, and phase shift $= \frac{\phi}{\omega} = \frac{\pi}{2}$.

STEP 2: The graph of $y = 3 \sin(2x - \pi)$ lies between -3 and 3 on the y-axis. One cycle begins at $x = \frac{\phi}{\omega} = \frac{\pi}{2}$ and ends at $x = \frac{\phi}{\omega} + \frac{2\pi}{\omega} = \frac{\pi}{2} + \pi = \frac{3\pi}{2}$.

NOTE The interval defining one cycle can also be found by solving the inequality

$$0 \le 2x - \pi \le 2\pi$$

Then

$$\pi \le 2x \le 3\pi$$
$$\frac{\pi}{2} \le x \le \frac{3\pi}{2}$$

∎

To find the five key points, divide the interval $\left[\dfrac{\pi}{2}, \dfrac{3\pi}{2}\right]$ into four subintervals, each of length $\pi \div 4 = \dfrac{\pi}{4}$, by finding the following values of x:

$\frac{\pi}{2}$	$\frac{\pi}{2} + \frac{\pi}{4} = \frac{3\pi}{4}$	$\frac{3\pi}{4} + \frac{\pi}{4} = \pi$	$\pi + \frac{\pi}{4} = \frac{5\pi}{4}$	$\frac{5\pi}{4} + \frac{\pi}{4} = \frac{3\pi}{2}$
1st x-coordinate	**2nd** x-coordinate	**3rd** x-coordinate	**4th** x-coordinate	**5th** x-coordinate

STEP 3: Use these values of x to determine the five key points on the graph:

$$\left(\frac{\pi}{2}, 0\right) \quad \left(\frac{3\pi}{4}, 3\right) \quad (\pi, 0) \quad \left(\frac{5\pi}{4}, -3\right) \quad \left(\frac{3\pi}{2}, 0\right)$$

STEP 4: Plot these five points and fill in the graph of the sine function as shown in Figure 71(a). Extend the graph in each direction to obtain Figure 71(b).

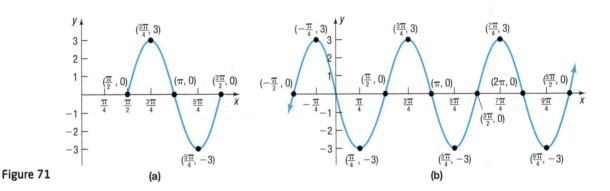

Figure 71

(a)

(b)

The graph of $y = 3\sin(2x - \pi) = 3\sin\left[2\left(x - \dfrac{\pi}{2}\right)\right]$ may also be obtained using transformations. See Figure 72.

Figure 72

(a) $y = \sin x$

Multiply by 3; Vertical stretch by a factor of 3

(b) $y = 3\sin x$

Replace x by $2x$; Horizontal compression by a factor of $\frac{1}{2}$

(c) $y = 3\sin(2x)$

Replace x by $x - \frac{\pi}{2}$; Shift right $\frac{\pi}{2}$ units

(d) $y = 3\sin\left[2\left(x - \frac{\pi}{2}\right)\right]$
$= 3\sin(2x - \pi)$

To graph a sinusoidal function of the form $y = A \sin(\omega x - \phi) + B$, first graph the function $y = A \sin(\omega x - \phi)$ and then apply a vertical shift.

EXAMPLE 2 **Finding the Amplitude, Period, and Phase Shift of a Sinusoidal Function and Graphing It**

Find the amplitude, period, and phase shift of $y = 2\cos(4x + 3\pi) + 1$, and graph the function.

Solution   **STEP 1:** Compare

$$y = 2\cos(4x + 3\pi) = 2\cos\left[4\left(x + \frac{3\pi}{4}\right)\right]$$

to

$$y = A\cos(\omega x - \phi) = A\cos\left[\omega\left(x - \frac{\phi}{\omega}\right)\right]$$

Note that $A = 2$, $\omega = 4$, and $\phi = -3\pi$. The graph is a cosine curve with amplitude $|A| = 2$, period $T = \dfrac{2\pi}{\omega} = \dfrac{2\pi}{4} = \dfrac{\pi}{2}$, and phase shift $= \dfrac{\phi}{\omega} = -\dfrac{3\pi}{4}$.

NOTE The interval defining one cycle can also be found by solving the inequality

$$0 \le 4x + 3\pi \le 2\pi$$

Then

$$-3\pi \le 4x \le -\pi$$

$$-\frac{3\pi}{4} \le x \le -\frac{\pi}{4} \qquad \blacksquare$$

STEP 2: The graph of $y = 2\cos(4x + 3\pi)$ lies between -2 and 2 on the y-axis. One cycle begins at $x = \dfrac{\phi}{\omega} = -\dfrac{3\pi}{4}$ and ends at $x = \dfrac{\phi}{\omega} + \dfrac{2\pi}{\omega} = -\dfrac{3\pi}{4} + \dfrac{\pi}{2} = -\dfrac{\pi}{4}$.

To find the five key points, divide the interval $\left[-\dfrac{3\pi}{4}, -\dfrac{\pi}{4}\right]$ into four subintervals, each of length $\dfrac{\pi}{2} \div 4 = \dfrac{\pi}{8}$, by finding the following values of x:

$-\dfrac{3\pi}{4}$	$-\dfrac{3\pi}{4} + \dfrac{\pi}{8} = -\dfrac{5\pi}{8}$	$-\dfrac{5\pi}{8} + \dfrac{\pi}{8} = -\dfrac{\pi}{2}$	$-\dfrac{\pi}{2} + \dfrac{\pi}{8} = -\dfrac{3\pi}{8}$	$-\dfrac{3\pi}{8} + \dfrac{\pi}{8} = -\dfrac{\pi}{4}$
1st x-coordinate	**2nd** x-coordinate	**3rd** x-coordinate	**4th** x-coordinate	**5th** x-coordinate

STEP 3: The five key points on the graph of $y = 2\cos(4x + 3\pi)$ are

$$\left(-\frac{3\pi}{4}, 2\right) \quad \left(-\frac{5\pi}{8}, 0\right) \quad \left(-\frac{\pi}{2}, -2\right) \quad \left(-\frac{3\pi}{8}, 0\right) \quad \left(-\frac{\pi}{4}, 2\right)$$

STEP 4: Plot these five points and fill in the graph of the cosine function as shown in Figure 73(a). Extend the graph in each direction to obtain Figure 73(b), the graph of $y = 2\cos(4x + 3\pi)$.

STEP 5: A vertical shift up 1 unit gives the final graph. See Figure 73(c).

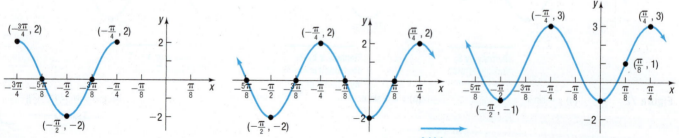

Figure 73   **(a)**   **(b)** $y = 2\cos(4x + 3\pi)$ Add 1; Vertical shift up 1 unit   **(c)** $y = 2\cos(4x + 3\pi) + 1$

The graph of $y = 2\cos{(4x + 3\pi)} + 1 = 2\cos\left[4\left(x + \dfrac{3\pi}{4}\right)\right] + 1$ may also be obtained using transformations. See Figure 74.

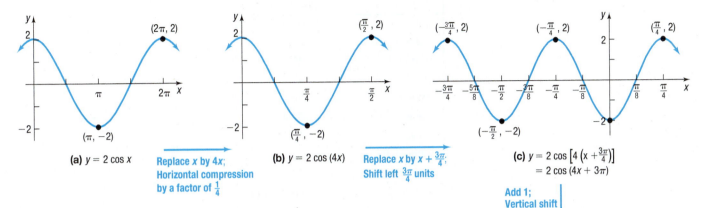

(a) $y = 2\cos x$

Replace x by $4x$;
Horizontal compression
by a factor of $\frac{1}{4}$

(b) $y = 2\cos{(4x)}$

Replace x by $x + \frac{3\pi}{4}$;
Shift left $\frac{3\pi}{4}$ units

(c) $y = 2\cos\left[4\left(x + \frac{3\pi}{4}\right)\right]$
$= 2\cos{(4x + 3\pi)}$

Add 1;
Vertical shift
up 1 unit

(d) $y = 2\cos{(4x + 3\pi)} + 1$

Figure 74

Now Work PROBLEM 3

SUMMARY

Steps for Graphing Sinusoidal Functions $y = A\sin{(\omega x - \phi)} + B$ or $y = A\cos{(\omega x - \phi)} + B$

STEP 1: Find the amplitude $|A|$, period $T = \dfrac{2\pi}{\omega}$, and phase shift $\dfrac{\phi}{\omega}$.

STEP 2: Determine the starting point of one cycle of the graph, $\dfrac{\phi}{\omega}$. Determine the ending point of one cycle of the graph, $\dfrac{\phi}{\omega} + \dfrac{2\pi}{\omega}$. Divide the interval $\left[\dfrac{\phi}{\omega}, \dfrac{\phi}{\omega} + \dfrac{2\pi}{\omega}\right]$ into four subintervals, each of length $\dfrac{2\pi}{\omega} \div 4$.

STEP 3: Use the endpoints of the subintervals to find the five key points on the graph.

STEP 4: Plot the five key points, and connect them with a sinusoidal graph to obtain one cycle of the graph. Extend the graph in each direction to make it complete.

STEP 5: If $B \neq 0$, apply a vertical shift.

2 Build Sinusoidal Models from Data

Scatter plots of data sometimes resemble the graph of a sinusoidal function. For example, the data given in Table 11 on the next page represent the average monthly temperatures in Denver, Colorado. Since the data represent *average* monthly temperatures collected over many years, the data will not vary much from year to year and so will essentially repeat each year. In other words, the data are periodic. Figure 75 shows a scatter plot of the data, where $x = 1$ represents January, $x = 2$ represents February, and so on.

Notice that the scatter plot looks like the graph of a sinusoidal function. We choose to fit the data to a sine function of the form

$$y = A\sin{(\omega x - \phi)} + B$$

where A, B, ω, and ϕ are constants.

Table 11

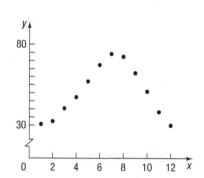

Month, x	Average Monthly Temperature, °F
January, 1	30.7
February, 2	32.5
March, 3	40.4
April, 4	47.4
May, 5	57.1
June, 6	67.4
July, 7	74.2
August, 8	72.5
September, 9	63.4
October, 10	50.9
November, 11	38.3
December, 12	30.0

Source: U.S. National Oceanic and Atmospheric Administration

Figure 75 Denver average monthly temperature

EXAMPLE 3 **Finding a Sinusoidal Function from Temperature Data**

Fit a sine function to the data in Table 11.

Solution Begin with a scatter plot of the data for one year. See Figure 76. The data will be fitted to a sine function of the form

$$y = A \sin(\omega x - \phi) + B$$

STEP 1: To find the amplitude A, compute

$$\text{Amplitude} = \frac{\text{largest data value} - \text{smallest data value}}{2}$$

$$= \frac{74.2 - 30.0}{2} = 22.1$$

To see the remaining steps in this process, superimpose the graph of the function $y = 22.1 \sin x$, where x represents months, on the scatter plot.

Figure 77 shows the two graphs. To fit the data, the graph needs to be shifted vertically, shifted horizontally, and stretched horizontally.

STEP 2: Determine the vertical shift by finding the average of the highest and lowest data values.

$$\text{Vertical shift} = \frac{74.2 + 30.0}{2} = 52.1$$

Now superimpose the graph of $y = 22.1 \sin x + 52.1$ on the scatter plot. See Figure 78.

We see that the graph needs to be shifted horizontally and stretched horizontally.

STEP 3: It is easier to find the horizontal stretch factor first. Since the temperatures repeat every 12 months, the period of the function is $T = 12$. Because $T = \dfrac{2\pi}{\omega} = 12$, then

$$\omega = \frac{2\pi}{12} = \frac{\pi}{6}$$

Figure 76

Figure 77

Figure 78

Figure 79

Figure 80

Now superimpose the graph of $y = 22.1 \sin\left(\dfrac{\pi}{6}x\right) + 52.1$ on the scatter plot. See Figure 79, where the graph still needs to be shifted horizontally.

STEP 4: To determine the horizontal shift, use the period $T = 12$ and divide the interval $[0, 12]$ into four subintervals of length $12 \div 4 = 3$:

$$[0, 3], \quad [3, 6], \quad [6, 9], \quad [9, 12]$$

The sine curve is increasing on the interval $[0, 3]$ and is decreasing on the interval $[3, 9]$, so a local maximum occurs at $x = 3$. The data indicate that a maximum occurs at $x = 7$ (corresponding to July's temperature), so the graph of the function must be shifted 4 units to the right by replacing x by $x - 4$. Doing this yields

$$y = 22.1 \sin\left(\dfrac{\pi}{6}(x - 4)\right) + 52.1$$

Distributing reveals that a sine function of the form $y = A \sin(\omega x - \phi) + B$ that fits the data is

$$y = 22.1 \sin\left(\dfrac{\pi}{6}x - \dfrac{2\pi}{3}\right) + 52.1$$

The graph of $y = 22.1 \sin\left(\dfrac{\pi}{6}x - \dfrac{2\pi}{3}\right) + 52.1$ and the scatter plot of the data are shown in Figure 80.

To summarize, these are the steps used to fit a sine function

$$y = A \sin(\omega x - \phi) + B$$

to sinusoidal data.

Steps for Fitting a Sine Function $y = A\sin(\omega x - \phi) + B$ to Data

STEP 1: Determine A, the amplitude of the function.

$$\text{Amplitude} = \frac{\text{largest data value} - \text{smallest data value}}{2}$$

STEP 2: Determine B, the vertical shift of the function.

$$\text{Vertical shift} = \frac{\text{largest data value} + \text{smallest data value}}{2}$$

STEP 3: Determine ω. Since the period T, the time it takes for the data to repeat, is $T = \dfrac{2\pi}{\omega}$, we have

$$\omega = \frac{2\pi}{T}$$

STEP 4: Determine the horizontal shift of the function by using the period of the data. Divide the period into four subintervals of equal length. Determine the x-coordinate for the maximum of the sine function and the x-coordinate for the maximum value of the data. Use this information to determine the value of the phase shift, $\dfrac{\phi}{\omega}$.

Now Work PROBLEM 25(a)–(c)

Let's look at another example. Since the number of hours of sunlight in a day cycles annually, the number of hours of sunlight in a day for a given location can be modeled by a sinusoidal function.

The longest day of the year (in terms of hours of sunlight) occurs on the day of the summer solstice. For locations in the Northern Hemisphere, the summer solstice is the time when the Sun is farthest north. In 2018, the summer solstice occurred on June 21 (the 172nd day of the year) at 6:07 AM EDT. The shortest day of the year occurs on the day of the winter solstice, the time when the Sun is farthest south (for locations in the Northern Hemisphere). In 2018, the winter solstice occurred on December 21 (the 355th day of the year) at 5:23 PM (EST).

EXAMPLE 4   **Finding a Sinusoidal Function for Hours of Daylight**

According to the *Old Farmer's Almanac*, the number of hours of sunlight in Boston on the day of the summer solstice is 15.28, and the number of hours of sunlight on the day of the winter solstice is 9.07.

(a) Find a sinusoidal function of the form $y = A \sin(\omega x - \phi) + B$ that fits the data.

(b) Use the function found in part (a) to predict the number of hours of sunlight in Boston on April 1, the 91st day of the year.

(c) Graph the function found in part (a).

(d) Look up the number of hours of sunlight for April 1 in the *Old Farmer's Almanac* and compare it to the results found in part (b).

Source: The Old Farmer's Almanac, www.almanac.com/rise

Solution (a) **STEP 1:** $\text{Amplitude} = \dfrac{\text{largest data value} - \text{smallest data value}}{2}$

$$= \dfrac{15.28 - 9.07}{2} = 3.105$$

STEP 2: $\text{Vertical shift} = \dfrac{\text{largest data value} + \text{smallest data value}}{2}$

$$= \dfrac{15.28 + 9.07}{2} = 12.175$$

STEP 3: The data repeat every 365 days. Since $T = \dfrac{2\pi}{\omega} = 365$, we find

$$\omega = \dfrac{2\pi}{365}$$

So far, we have $y = 3.105 \sin\left(\dfrac{2\pi}{365} x - \phi\right) + 12.175$.

STEP 4: To determine the horizontal shift, use the period $T = 365$ and divide the interval $[0, 365]$ into four subintervals of length $365 \div 4 = 91.25$:

$$[0, 91.25], \quad [91.25, 182.5], \quad [182.5, 273.75], \quad [273.75, 365]$$

The sine curve is increasing on the interval $[0, 91.25]$ and is decreasing on the interval $[91.25, 273.75]$, so a local maximum occurs at $x = 91.25$. Since the maximum occurs on the summer solstice at $x = 172$, we must shift the graph of the function $172 - 91.25 = 80.75$ units to the right by replacing x by $x - 80.75$. Doing this yields

$$y = 3.105 \sin\left(\dfrac{2\pi}{365}(x - 80.75)\right) + 12.175$$

Next, multiply out to obtain the form $y = A \sin(\omega x - \phi) + B$.

$$y = 3.105 \sin\left(\dfrac{2\pi}{365} x - \dfrac{323\pi}{730}\right) + 12.175$$

(b) To predict the number of hours of daylight on April 1, let $x = 91$ in the function found in part (a) and obtain

$$y = 3.105 \sin\left(\frac{2\pi}{365} \cdot 91 - \frac{323\pi}{730}\right) + 12.175$$

$$\approx 12.72$$

The prediction is that there will be about 12.72 hours = 12 hours 43 minutes of sunlight on April 1 in Boston.

(c) The graph of the function found in part (a) is given in Figure 81.

(d) According to the *Old Farmer's Almanac*, there will be 12 hours 45 minutes of sunlight on April 1 in Boston.

Figure 81

- **Now Work** PROBLEM **31**

Certain graphing utilities (such as the TI-83, TI-84 Plus C, and TI-89) have the capability of finding the sine function of best fit for sinusoidal data. At least four data points are required for this process.

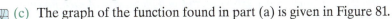

EXAMPLE 5

Finding the Sine Function of Best Fit

Use a graphing utility to find the sine function of best fit for the data in Table 11. Graph this function with the scatter plot of the data.

Solution

Enter the data from Table 11 and execute the SINe REGression program. The result is shown in Figure 82.

The output that the utility provides shows the equation

$$y = a \sin(bx + c) + d$$

The sinusoidal function of best fit is

$$y = 21.54 \sin(0.56x - 2.44) + 51.77$$

where x represents the month and y represents the average monthly temperature. Figure 83 shows the graph of the sinusoidal function of best fit on the scatter plot.

Figure 82

Figure 83

- **Now Work** PROBLEM **25(d)** AND **(e)**

2.6 Assess Your Understanding

Concepts and Vocabulary

1. For the graph of $y = A \sin(\omega x - \phi)$, the number $\frac{\phi}{\omega}$ is called the _____ _____.

2. **True or False** A graphing utility requires only two data points to find the sine function of best fit.

Skill Building

In Problems 3–18, find the amplitude (if one exists), period, and phase shift of each function. Graph each function. Be sure to label key points. Show at least two periods.

3. $y = 4 \sin(2x - \pi)$

4. $y = 3 \sin(3x - \pi)$

5. $y = 2 \cos\left(3x + \frac{\pi}{2}\right)$

6. $y = 3 \cos(2x + \pi)$

7. $y = -3 \sin\left(2x + \frac{\pi}{2}\right)$

8. $y = -2 \cos\left(2x - \frac{\pi}{2}\right)$

9. $y = 4 \sin(\pi x + 2) - 5$

10. $y = 2 \cos(2\pi x + 4) + 4$

11. $y = 3 \cos(\pi x - 2) + 5$

12. $y = 2 \cos(2\pi x - 4) - 1$

13. $y = -3 \sin\left(-2x + \frac{\pi}{2}\right)$

14. $y = -3 \cos\left(-2x + \frac{\pi}{2}\right)$

15. $y = 2 \tan(4x - \pi)$

16. $y = \frac{1}{2} \cot(2x - \pi)$

17. $y = 3 \csc\left(2x - \frac{\pi}{4}\right)$

18. $y = \frac{1}{2} \sec(3x - \pi)$

In Problems 19–22, write an equation of a sine function that has the given characteristics.

19. Amplitude: 2
    Period: π
    Phase shift: $\frac{1}{2}$

20. Amplitude: 3
    Period: $\frac{\pi}{2}$
    Phase shift: 2

21. Amplitude: 3
    Period: 3π
    Phase shift: $-\frac{1}{3}$

22. Amplitude: 2
    Period: π
    Phase shift: -2

Applications and Extensions

23. **Alternating Current (ac) Circuits** The current I, in amperes, flowing through an ac (alternating current) circuit at time t, in seconds, is

    $$I(t) = 120 \sin\left(30\pi t - \frac{\pi}{3}\right) \qquad t \geq 0$$

    What is the period? What is the amplitude? What is the phase shift? Graph this function over two periods.

24. **Alternating Current (ac) Circuits** The current I, in amperes, flowing through an ac (alternating current) circuit at time t, in seconds, is

    $$I(t) = 220 \sin\left(60\pi t - \frac{\pi}{6}\right) \qquad t \geq 0$$

    What is the period? What is the amplitude? What is the phase shift? Graph this function over two periods.

25. **Hurricanes** Hurricanes are categorized using the Saffir-Simpson Hurricane Scale, with winds 111–130 miles per hour (mph) corresponding to a category 3 hurricane, winds 131–155 mph corresponding to a category 4 hurricane, and winds in excess of 155 mph corresponding to a category 5 hurricane. The data on the right represent the number of major hurricanes in the Atlantic Basin (category 3, 4, or 5) each decade from 1921 to 2010.
    (a) Draw a scatter plot of the data.
    (b) Find a sinusoidal function of the form
        $y = A \sin(\omega x - \phi) + B$ that models the data.
    (c) Draw the sinusoidal function found in part (b) on the scatter plot.
    (d) Use a graphing utility to find the sinusoidal function of best fit.
    (e) Graph the sinusoidal function of best fit on a scatter plot of the data.

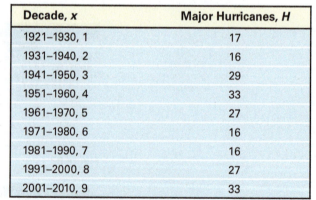

Decade, x	Major Hurricanes, H
1921–1930, 1	17
1931–1940, 2	16
1941–1950, 3	29
1951–1960, 4	33
1961–1970, 5	27
1971–1980, 6	16
1981–1990, 7	16
1991–2000, 8	27
2001–2010, 9	33

Source: U.S. National Oceanic and Atmospheric Administration

26. **Monthly Temperature** The data on the next page represent the average monthly temperatures for Washington, D.C.
    (a) Draw a scatter plot of the data for one period.
    (b) Find a sinusoidal function of the form
        $y = A \sin(\omega x - \phi) + B$ that models the data.
    (c) Draw the sinusoidal function found in part (b) on the scatter plot.
    (d) Use a graphing utility to find the sinusoidal function of best fit.
    (e) Graph the sinusoidal function of best fit on a scatter plot of the data.

Month, x	Average Monthly Temperature, °F
January, 1	36.0
February, 2	39.0
March, 3	46.8
April, 4	56.8
May, 5	66.0
June, 6	75.2
July, 7	79.8
August, 8	78.1
September, 9	71.0
October, 10	59.5
November, 11	49.6
December, 12	39.7

Source: U.S. National Oceanic and
Atmospheric Administration

27. Monthly Temperature The following data represent the average monthly temperatures for Indianapolis, Indiana.

Month, x	Average Monthly Temperature, °F
January, 1	28.1
February, 2	32.1
March, 3	42.2
April, 4	53.0
May, 5	62.7
June, 6	72.0
July, 7	75.4
August, 8	74.2
September, 9	66.9
October, 10	55.0
November, 11	43.6
December, 12	31.6

Source: U.S. National Oceanic and
Atmospheric Administration

(a) Draw a scatter plot of the data for one period.
(b) Find a sinusoidal function of the form

$$y = A \sin(\omega x - \phi) + B$$

that models the data.
(c) Draw the sinusoidal function found in part (b) on the scatter plot.
(d) Use a graphing utility to find the sinusoidal function of best fit.
(e) Graph the sinusoidal function of best fit on a scatter plot of the data.

28. Monthly Temperature The following data represent the average monthly temperatures for Baltimore, Maryland.
(a) Draw a scatter plot of the data for one period.
(b) Find a sinusoidal function of the form
$y = A \sin(\omega x - \phi) + B$ that models the data.

(c) Draw the sinusoidal function found in part (b) on the scatter plot.
(d) Use a graphing utility to find the sinusoidal function of best fit.
(e) Graph the sinusoidal function of best fit on a scatter plot of the data.

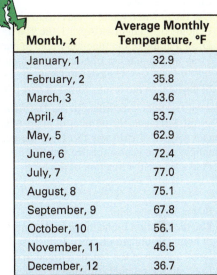

Month, x	Average Monthly Temperature, °F
January, 1	32.9
February, 2	35.8
March, 3	43.6
April, 4	53.7
May, 5	62.9
June, 6	72.4
July, 7	77.0
August, 8	75.1
September, 9	67.8
October, 10	56.1
November, 11	46.5
December, 12	36.7

Source: U.S. National Oceanic and
Atmospheric Administration

29. Tides The length of time between consecutive high tides is 12 hours and 25 minutes. According to the National Oceanic and Atmospheric Administration, on Saturday, April 21, 2018, in Charleston, South Carolina, high tide occurred at 12:25 AM (0.42 hours) and low tide occurred at 7:05 AM (7.08 hours). Water heights are measured as the amounts above or below the mean lower low water. The height of the water at high tide was 5.88 feet, and the height of the water at low tide was 0.06 foot.
(a) Approximately when did the next high tide occur?
(b) Find a sinusoidal function of the form

$$y = A \sin(\omega x - \phi) + B$$

that models the data.
(c) Use the function found in part (b) to predict the height of the water at 5 PM.

30. Tides The length of time between consecutive high tides is 12 hours and 25 minutes. According to the National Oceanic and Atmospheric Administration, on Saturday, April 21, 2018, in Sitka, Alaska, high tide occurred at 4:51 AM (4.85 hours) and low tide occurred at 11:50 AM (11.83 hours). Water heights are measured as the amounts above or below the mean lower low water. The height of the water at high tide was 10.03 feet, and the height of the water at low tide was −0.46 feet.
(a) Approximately when did the next high tide occur?
(b) Find a sinusoidal function of the form

$$y = A \sin(\omega x - \phi) + B$$

that models the data.
(c) Use the function found in part (b) to predict the height of the water at 3 PM.

31. Hours of Daylight According to the *Old Farmer's Almanac*, in Miami, Florida, the number of hours of sunlight on the summer solstice of 2018 was 13.75, and the number of hours of sunlight on the winter solstice was 10.52.

(a) Find a sinusoidal function of the form

$$y = A \sin(\omega x - \phi) + B$$

that models the data.

(b) Use the function found in part (a) to predict the number of hours of sunlight on April 1, the 91st day of the year.

(c) Draw a graph of the function found in part (a).

(d) Look up the number of hours of sunlight for April 1 in the *Old Farmer's Almanac*, and compare the actual hours of daylight to the results found in part (b).

32. Hours of Daylight According to the *Old Farmer's Almanac*, in Detroit, Michigan, the number of hours of sunlight on the summer solstice of 2018 was 15.27, and the number of hours of sunlight on the winter solstice was 9.07.

(a) Find a sinusoidal function of the form

$$y = A \sin(\omega x - \phi) + B$$

that models the data.

(b) Use the function found in part (a) to predict the number of hours of sunlight on April 1, the 91st day of the year.

(c) Draw a graph of the function found in part (a).

(d) Look up the number of hours of sunlight for April 1 in the *Old Farmer's Almanac*, and compare the actual hours of daylight to the results found in part (b).

33. Hours of Daylight According to the *Old Farmer's Almanac*, in Anchorage, Alaska, the number of hours of sunlight on the summer solstice of 2018 was 19.37, and the number of hours of sunlight on the winter solstice was 5.45.

(a) Find a sinusoidal function of the form

$$y = A \sin(\omega x - \phi) + B$$

that models the data.

(b) Use the function found in part (a) to predict the number of hours of sunlight on April 1, the 91st day of the year.

(c) Draw a graph of the function found in part (a).

(d) Look up the number of hours of sunlight for April 1 in the *Old Farmer's Almanac*, and compare the actual hours of daylight to the results found in part (b).

34. Hours of Daylight According to the *Old Farmer's Almanac*, in Honolulu, Hawaii, the number of hours of sunlight on the summer solstice of 2018 was 13.42, and the number of hours of sunlight on the winter solstice was 10.83.

(a) Find a sinusoidal function of the form

$$y = A \sin(\omega x - \phi) + B$$

that models the data.

(b) Use the function found in part (a) to predict the number of hours of sunlight on April 1, the 91st day of the year.

(c) Draw a graph of the function found in part (a).

(d) Look up the number of hours of sunlight for April 1 in the *Old Farmer's Almanac*, and compare the actual hours of daylight to the results found in part (b).

35. *Challenge Problem* Coaster Motion A wooden roller coaster at Six Flags contains a run in the shape of a sinusoidal curve, with a series of hills. The crest of each hill is 106 feet above the ground. If it takes a car 1.8 seconds to go from the top of a hill to the bottom (4 feet off the ground), find a sinusoidal function of the form

$$y = A \sin(\omega t - \phi) + B$$

that models the motion of the coaster train during this run starting at the top of a hill.

Discussion and Writing

36. Explain how the amplitude and period of a sinusoidal graph are used to establish the scale on each coordinate axis.

37. Find an application in your major field that leads to a sinusoidal graph. Write an account of your findings.

Retain Your Knowledge

Problems 38–46 are based on material learned earlier in the course. The purpose of these problems is to keep the material fresh in your mind so that you are better prepared for the final exam.

38. Given $f(x) = \dfrac{4x + 9}{2}$, find $f^{-1}(x)$.

39. Solve: $0.25(0.4x + 0.8) = 3.7 - 1.4x$

40. Multiply: $(8x + 15y)^2$

41. Find the exact distance between the points $(4, -1)$ and $(10, 3)$.

42. Solve: $|3x + 4| = |5x - 7|$

43. Given $y = x\sqrt{x + 4}$, let $u = x + 4$ and express y in terms of u.

44. If $x = a \sin t, a > 0$, and $-\dfrac{\pi}{2} \le t \le \dfrac{\pi}{2}$, find $\cos t$.

45. A rectangular garden is enclosed by 54 feet of fencing. If the length of the garden is 3 feet more than twice the width, what are the dimensions of the garden?

46. Find the vertical asymptotes, if any, of the graph of $R(x) = \dfrac{x^2 - 25}{x^2 - 2x - 15}$.

Chapter Review

Things to Know

Definitions

Angle in standard position (p. 102)	Vertex is at the origin; initial side is along the positive x-axis.
1 Degree (1°) (p. 103)	$1° = \dfrac{1}{360}$ revolution
1 Radian (pp. 104–105)	The measure of a central angle of a circle whose rays subtend an arc whose length is equal to the radius of the circle.
Trigonometric functions (p. 116)	$P = (x, y)$ is the point on the unit circle corresponding to $\theta = t$ radians.

$$\sin t = \sin \theta = y \qquad \cos t = \cos \theta = x \qquad \tan t = \tan \theta = \dfrac{y}{x}, x \neq 0$$

$$\csc t = \csc \theta = \dfrac{1}{y}, y \neq 0 \quad \sec t = \sec \theta = \dfrac{1}{x}, x \neq 0 \quad \cot t = \cot \theta = \dfrac{x}{y}, y \neq 0$$

Trigonometric functions using a circle of radius r (p. 126)	For an angle θ in standard position, $P = (x, y)$ is the point on the terminal side of θ that is also on the circle $x^2 + y^2 = r^2$.

$$\sin \theta = \dfrac{y}{r} \qquad \cos \theta = \dfrac{x}{r} \qquad \tan \theta = \dfrac{y}{x}, x \neq 0$$

$$\csc \theta = \dfrac{r}{y}, y \neq 0 \qquad \sec \theta = \dfrac{r}{x}, x \neq 0 \qquad \cot \theta = \dfrac{x}{y}, y \neq 0$$

Periodic function (p. 135)	A function f is periodic if for some number $p > 0$ for which $\theta + p$ is in the domain of f whenever θ is, then $f(\theta + p) = f(\theta)$. The smallest such p is the fundamental period.

Formulas

1 counterclockwise revolution = 360° (p. 104) = 2π radians (p. 106)

$1° = \dfrac{\pi}{180}$ radian (p. 106); 1 radian = $\dfrac{180}{\pi}$ degrees (p. 106)

Arc length: $s = r\theta$ (p. 105) — θ is measured in radians; s is the length of the arc subtended by the central angle θ of the circle of radius r.

Area of a sector: $A = \dfrac{1}{2}r^2\theta$ (p. 108) — A is the area of the sector of a circle of radius r formed by a central angle of θ radians.

Linear speed: $v = \dfrac{s}{t}$ (p. 109) — v is linear speed, distance per unit time.

Angular speed: $\omega = \dfrac{\theta}{t}$ (p. 109) — ω is angular speed measured in radians per unit time.
$v = r\omega$ (p. 109)

Table of Values (pp. 119 and 123)

θ (Radians)	θ (Degrees)	$\sin \theta$	$\cos \theta$	$\tan \theta$	$\csc \theta$	$\sec \theta$	$\cot \theta$
0	0°	0	1	0	Not defined	1	Not defined
$\dfrac{\pi}{6}$	30°	$\dfrac{1}{2}$	$\dfrac{\sqrt{3}}{2}$	$\dfrac{\sqrt{3}}{3}$	2	$\dfrac{2\sqrt{3}}{3}$	$\sqrt{3}$
$\dfrac{\pi}{4}$	45°	$\dfrac{\sqrt{2}}{2}$	$\dfrac{\sqrt{2}}{2}$	1	$\sqrt{2}$	$\sqrt{2}$	1
$\dfrac{\pi}{3}$	60°	$\dfrac{\sqrt{3}}{2}$	$\dfrac{1}{2}$	$\sqrt{3}$	$\dfrac{2\sqrt{3}}{3}$	2	$\dfrac{\sqrt{3}}{3}$
$\dfrac{\pi}{2}$	90°	1	0	Not defined	1	Not defined	0
π	180°	0	−1	0	Not defined	−1	Not defined
$\dfrac{3\pi}{2}$	270°	−1	0	Not defined	−1	Not defined	0

The Unit Circle (pp. 115–125)

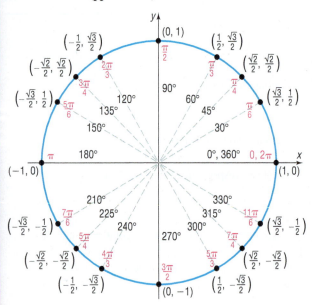

Fundamental identities (pp. 137–139)

Quotient identities: $\tan \theta = \dfrac{\sin \theta}{\cos \theta}$ $\cot \theta = \dfrac{\cos \theta}{\sin \theta}$

Reciprocal identities: $\csc \theta = \dfrac{1}{\sin \theta}$ $\sec \theta = \dfrac{1}{\cos \theta}$ $\cot \theta = \dfrac{1}{\tan \theta}$

Pythagorean identities: $\sin^2 \theta + \cos^2 \theta = 1$ $\tan^2 \theta + 1 = \sec^2 \theta$ $\cot^2 \theta + 1 = \csc^2 \theta$

Properties of the trigonometric functions (pp. 133–135, 142)

$y = \sin x$ (p. 148) Domain: $-\infty < x < \infty$

Range: $-1 \le y \le 1$

Periodic: period $= 2\pi \, (360°)$

Odd function

$y = \cos x$ (pp. 149–150) Domain: $-\infty < x < \infty$

Range: $-1 \le y \le 1$

Periodic: period $= 2\pi \, (360°)$

Even function

$y = \tan x$ (p. 163) Domain: $-\infty < x < \infty$, except odd integer

multiples of $\dfrac{\pi}{2} \, (90°)$

Range: $-\infty < y < \infty$

Periodic: period $= \pi \, (180°)$

Odd function

Vertical asymptotes at odd integer multiples of $\dfrac{\pi}{2}$

$y = \cot x$ (p. 164) Domain: $-\infty < x < \infty$, except integer multiples

of $\pi \, (180°)$

Range: $-\infty < y < \infty$

Periodic: period $= \pi \, (180°)$

Odd function

Vertical asymptotes at integer multiples of π

$y = \csc x$ (p. 166)	Domain: $-\infty < x < \infty$, except integer multiples of π (180°)	
	Range: $\lvert y \rvert \geq 1$ ($y \leq -1$ or $y \geq 1$)	
	Periodic: period $= 2\pi$ (360°)	
	Odd function	
	Vertical asymptotes at integer multiples of π	

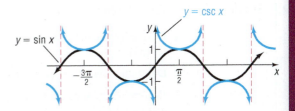

$y = \sec x$ (p. 166)	Domain: $-\infty < x < \infty$, except odd integer multiples of $\dfrac{\pi}{2}$ (90°)	
	Range: $\lvert y \rvert \geq 1$ ($y \leq -1$ or $y \geq 1$)	
	Periodic: period $= 2\pi$ (360°)	
	Even function	
	Vertical asymptotes at odd integer multiples of $\dfrac{\pi}{2}$	

Sinusoidal Graphs

$y = A \sin(\omega x) + B, \quad \omega > 0$

$y = A \cos(\omega x) + B, \quad \omega > 0$

$y = A \sin(\omega x - \phi) + B = A \sin\left[\omega\left(x - \dfrac{\phi}{\omega}\right)\right] + B$

$y = A \cos(\omega x - \phi) + B = A \cos\left[\omega\left(x - \dfrac{\phi}{\omega}\right)\right] + B$

Period $= \dfrac{2\pi}{\omega}$ (pp. 152, 170)

Amplitude $= \lvert A \rvert$ (pp. 152, 170)

Phase shift $= \dfrac{\phi}{\omega}$ (p. 170)

Objectives

Section	You should be able to...	Example(s)	Review Exercises
2.1	**1** Angles and degree measure (p. 102)	1	
	2 Convert between decimal and degree, minute, second measures for angles (p. 104)	2	48
	3 Find the length of an arc of a circle (p. 105)	3	49, 50
	4 Convert from degrees to radians and from radians to degrees (p. 106)	4–6	1–4
	5 Find the area of a sector of a circle (p. 108)	7	49
	6 Find the linear speed of an object traveling in circular motion (p. 109)	8	51, 52
2.2	**1** Find the exact values of the trigonometric functions using a point on the unit circle (p. 117)	1	45, 55
	2 Find the exact values of the trigonometric functions of quadrantal angles (p. 118)	2, 3	9, 55
	3 Find the exact values of the trigonometric functions of $\dfrac{\pi}{4} = 45°$ (p. 120)	4, 5	5, 6
	4 Find the exact values of the trigonometric functions of $\dfrac{\pi}{6} = 30°$ and $\dfrac{\pi}{3} = 60°$ (p. 121)	6–8	5, 6
	5 Find the exact values of the trigonometric functions for integer multiples of $\dfrac{\pi}{6} = 30°$, $\dfrac{\pi}{4} = 45°$, and $\dfrac{\pi}{3} = 60°$ (p. 123)	9, 10	7, 8, 10, 55
	6 Use a calculator to approximate the value of a trigonometric function (p. 125)	11	41, 42
	7 Use a circle of radius r to evaluate the trigonometric functions (p. 126)	12	46

(continued)

Section	You should be able to...	Example(s)	Review Exercises
2.3	**1** Determine the domain and the range of the trigonometric functions (p. 133)	pp. 133–134	47
	2 Determine the period of the trigonometric functions (p. 135)	1	47
	3 Determine the signs of the trigonometric functions in a given quadrant (p. 136)	2	43, 44
	4 Find the values of the trigonometric functions using fundamental identities (p. 137)	3, 4	11–15
	5 Find the exact values of the trigonometric functions of an angle given one of the functions and the quadrant of the angle (p. 139)	5, 6	16–23
	6 Use even-odd properties to find the exact values of the trigonometric functions (p. 142)	7	13–15
2.4	**1** Graph the sine function $y = \sin x$ and functions of the form $y = A \sin(\omega x)$ (p. 147)	1, 2	24
	2 Graph the cosine function $y = \cos x$ and functions of the form $y = A \cos(\omega x)$ (p. 149)	3	25
	3 Determine the amplitude and period of sinusoidal functions (p. 150)	4	33–38
	4 Graph sinusoidal functions using key points (p. 152)	5–7	24, 25, 35
	5 Find an equation for a sinusoidal graph (p. 156)	8, 9	39, 40
2.5	**1** Graph the tangent function $y = \tan x$ and the cotangent function $y = \cot x$ (p. 162)		26, 28
	2 Graph functions of the form $y = A \tan(\omega x) + B$ and $y = A \cot(\omega x) + B$ (p. 164)	1, 2	27, 32
	3 Graph the cosecant function $y = \csc x$ and the secant function $y = \sec x$ (p. 165)		30
	4 Graph functions of the form $y = A \csc(\omega x) + B$ and $y = A \sec(\omega x) + B$ (p. 166)	3	29
2.6	**1** Graph sinusoidal functions of the form $y = A \sin(\omega x - \phi) + B$ (p. 169)	1, 2	31, 35–38, 53(d)
	2 Build sinusoidal models from data (p. 173)	3–5	54

Review Exercises

In Problems 1 and 2, convert each angle in degrees to radians. Express your answer as a multiple of π.

1. $135°$

2. $18°$

In Problems 3 and 4, convert each angle in radians to degrees.

3. $\dfrac{3\pi}{4}$

4. $-\dfrac{5\pi}{2}$

In Problems 5–15, find the exact value of each expression. Do not use a calculator.

5. $\tan \dfrac{\pi}{4} - \sin \dfrac{\pi}{6}$

6. $3 \sin 45° - 4 \tan \dfrac{\pi}{6}$

7. $6 \cos \dfrac{3\pi}{4} + 2 \tan\left(-\dfrac{\pi}{3}\right)$

8. $\sec\left(-\dfrac{\pi}{3}\right) - \cot\left(-\dfrac{5\pi}{4}\right)$

9. $\tan \pi + \sin \pi$

10. $\cos 540° - \tan(-405°)$

11. $\sin^2 20° + \dfrac{1}{\sec^2 20°}$

12. $\sec 50° \cos 50°$

13. $\dfrac{\cos(-40°)}{\cos 40°}$

14. $\dfrac{\sin(-40°)}{\sin 40°}$

15. $\sin 310° \csc(-50°)$

In Problems 16–23, find the exact value of each of the remaining trigonometric functions.

16. $\sin \theta = \dfrac{4}{5}$, θ is acute

17. $\tan \theta = \dfrac{12}{5}$, $\sin \theta < 0$

18. $\sec \theta = -\dfrac{5}{4}$, $\tan \theta < 0$

19. $\sin \theta = \dfrac{12}{13}$, θ in quadrant II

20. $\sin \theta = -\dfrac{5}{13}$, $\dfrac{3\pi}{2} < \theta < 2\pi$

21. $\tan \theta = \dfrac{1}{3}$, $180° < \theta < 270°$

22. $\sec \theta = 3$, $\dfrac{3\pi}{2} < \theta < 2\pi$

23. $\cot \theta = -2$, $\dfrac{\pi}{2} < \theta < \pi$

In Problems 24–32, graph each function. Each graph should contain at least two periods. Use the graph to determine the domain and the range of each function.

24. $y = 2\sin(4x)$

25. $y = -3\cos(2x)$

26. $y = \tan(x + \pi)$

27. $y = -2\tan(3x)$

28. $y = \cot\left(x + \dfrac{\pi}{4}\right)$

29. $y = 4\sec(2x)$

30. $y = \csc\left(x + \dfrac{\pi}{4}\right)$

31. $y = 4\sin(2x + 4) - 2$

32. $y = 5\cot\left(\dfrac{x}{3} - \dfrac{\pi}{4}\right)$

In Problems 33 and 34, determine the amplitude and period of each function without graphing.

33. $y = \sin(2x)$

34. $y = -2\cos(3\pi x)$

In Problems 35–38, find the amplitude, period, and phase shift of each function. Graph each function. Show at least two periods.

35. $y = 4\sin(3x)$

36. $y = -\cos\left(\dfrac{1}{2}x + \dfrac{\pi}{2}\right)$

37. $y = \dfrac{1}{2}\sin\left(\dfrac{3}{2}x - \pi\right)$

38. $y = -\dfrac{2}{3}\cos(\pi x - 6)$

In Problems 39 and 40, find a function whose graph is given.

39.

40.

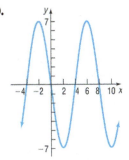

41. Use a calculator to approximate $\sin\dfrac{\pi}{8}$. Round the answer to two decimal places.

42. Use a calculator to approximate $\sec 10°$. Round the answer to two decimal places.

43. Determine the signs of the six trigonometric functions of an angle θ whose terminal side is in quadrant III.

44. Name the quadrant θ lies in if $\cos \theta > 0$ and $\tan \theta < 0$.

45. Find the exact values of the six trigonometric functions of t if $P = \left(-\dfrac{1}{3}, \dfrac{2\sqrt{2}}{3}\right)$ is the point on the unit circle that corresponds to t.

46. Find the exact value of $\sin t$, $\cos t$, and $\tan t$ if $P = (-2, 5)$ is the point on a circle that corresponds to t.

47. What are the domain and the range of the secant function? What is the period?

48. (a) Convert the angle $32°20'35''$ to a decimal in degrees. Round the answer to two decimal places.

(b) Convert the angle $63.18°$ to D°M′S″ form. Express the answer to the nearest second.

49. Find the length of the arc subtended by a central angle of $30°$ on a circle of radius 2 feet. What is the area of the sector?

50. The minute hand of a clock is 8 inches long. How far does the tip of the minute hand move in 30 minutes? How far does it move in 20 minutes?

51. Angular Speed of a Race Car A race car is driven around a circular track at a constant speed of 180 miles per hour. If the diameter of the track is $\dfrac{1}{2}$ mile, what is the angular speed of the car? Express your answer in revolutions per hour (which is equivalent to laps per hour).

52. Lighthouse Beacons The Montauk Point Lighthouse on Long Island has dual beams (two light sources opposite each other). Ships at sea observe a blinking light every 5 seconds. What rotation speed is required to achieve this?

53. Alternating Current The current I, in amperes, flowing through an ac (alternating current) circuit at time t is

$$I(t) = 220 \sin\left(30\pi t + \frac{\pi}{6}\right) \qquad t \geq 0$$

(a) What is the period?
(b) What is the amplitude?
(c) What is the phase shift?
(d) Graph this function over two periods.

54. Monthly Temperature The data below represent the average monthly temperatures for Phoenix, Arizona.

(a) Draw a scatter plot of the data for one period.
(b) Find a sinusoidal function of the form
$y = A \sin(\omega x - \phi) + B$ that models the data.
(c) Draw the sinusoidal function found in part (b) on the scatter plot.
(d) Use a graphing utility to find the sinusoidal function of best fit.
(e) Graph the sinusoidal function of best fit on the scatter plot.

Month, x	Average Monthly Temperature, °F
January, 1	56
February, 2	60
March, 3	65
April, 4	73
May, 5	82
June, 6	91
July, 7	95
August, 8	94
September, 9	88
October, 10	77
November, 11	64
December, 12	55

Source: U.S. National Oceanic and Atmospheric Administration

55. Unit Circle On the given unit circle, fill in the missing angles $(0 \leq \theta \leq 2\pi)$ and the corresponding points P.

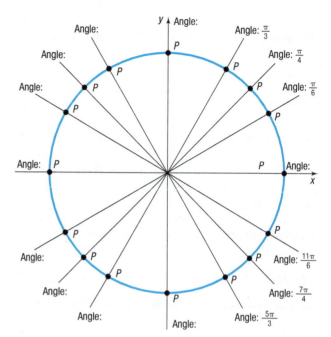

Chapter Test

CHAPTER
Test Prep
VIDEOS

The Chapter Test Prep Videos include step-by-step solutions to all chapter test exercises. These videos are available in MyLab™ Math, or on this text's YouTube Channel. Refer to the Preface for a link to the YouTube channel.

In Problems 1–3, convert each angle in degrees to radians. Express your answer as a multiple of π.

1. $260°$ **2.** $-400°$ **3.** $13°$

In Problems 4–6, convert each angle in radians to degrees.

4. $-\dfrac{\pi}{8}$ **5.** $\dfrac{9\pi}{2}$ **6.** $\dfrac{3\pi}{4}$

In Problems 7–12, find the exact value of each expression.

7. $\sin\dfrac{\pi}{6}$ **8.** $\cos\left(-\dfrac{5\pi}{4}\right) - \cos\dfrac{3\pi}{4}$

9. $\cos(-120°)$ **10.** $\tan 330°$

11. $\sin\dfrac{\pi}{2} - \tan\dfrac{19\pi}{4}$ **12.** $2\sin^2 60° - 3\cos 45°$

In Problems 13–16, use a calculator to evaluate each expression. Round your answer to three decimal places.

13. $\sin 17°$ **14.** $\cos\dfrac{2\pi}{5}$

15. $\sec 229°$ **16.** $\cot\dfrac{28\pi}{9}$

17. Fill in each table entry with the sign of each function.

	sin θ	cos θ	tan θ	sec θ	csc θ	cot θ
θ in QI						
θ in QII						
θ in QIII						
θ in QIV						

18. If $f(x) = \sin x$ and $f(a) = \dfrac{3}{5}$, find $f(-a)$.

In Problems 19–21, find the value of the remaining five trigonometric functions of θ.

19. $\sin \theta = \dfrac{5}{7}$, θ in quadrant II   **20.** $\cos \theta = \dfrac{2}{3}$, $\dfrac{3\pi}{2} < \theta < 2\pi$

21. $\tan \theta = -\dfrac{12}{5}$, $\dfrac{\pi}{2} < \theta < \pi$

In Problems 22–24, the point (x, y) is on the terminal side of angle θ in standard position. Find the exact value of the given trigonometric function.

22. $(2, 7)$, $\sin \theta$ **23.** $(-5, 11)$, $\cos \theta$

24. $(6, -3)$, $\tan \theta$

In Problems 25 and 26, graph the function.

25. $y = 2 \sin\left(\dfrac{x}{3} - \dfrac{\pi}{6}\right)$ **26.** $y = \tan\left(-x + \dfrac{\pi}{4}\right) + 2$

27. Write an equation for a sinusoidal graph with the following properties:

$$A = -3 \qquad \text{period} = \dfrac{2\pi}{3} \qquad \text{phase shift} = -\dfrac{\pi}{4}$$

28. Logan has a garden in the shape of a sector of a circle; the outer rim of the garden is 25 feet long and the central angle of the sector is 50°. She wants to add a 3-foot-wide walk to the outer rim. How many square feet of paving blocks will she need to build the walk?

29. Hungarian Adrian Annus won the gold medal for the hammer throw at the 2004 Olympics in Athens with a winning distance of 83.19 meters.* The event consists of swinging a 16-pound weight attached to a wire 190 centimeters long in a circle and then releasing it. Assuming his release is at a 45° angle to the ground, the hammer will travel a distance of $\dfrac{v_0^2}{g}$ meters, where $g = 9.8$ meters/second^2 and v_0 is the linear speed of the hammer when released. At what rate (rpm) was he swinging the hammer upon release?

*Annus was stripped of his medal after refusing to cooperate with postmedal drug testing.

Cumulative Review

1. Find the real solutions, if any, of the equation
$$2x^2 + x - 1 = 0$$

2. Find an equation for the line with slope -3 containing the point $(-2, 5)$.

3. Find an equation for a circle of radius 4 and center at the point $(0, -2)$.

4. Describe the equation $2x - 3y = 12$. Graph it.

5. Describe the equation $x^2 + y^2 - 2x + 4y - 4 = 0$. Graph it.

6. Use transformations to graph the function
$$y = (x - 3)^2 + 2$$

7. Sketch a graph of each of the following functions. Label at least three points on each graph.
   (a) $y = x^2$ **(b)** $y = x^3$
   (c) $y = \sin x$ **(d)** $y = \tan x$

8. Find the inverse function of $f(x) = 3x - 2$.

9. Find the exact value of $(\sin 14°)^2 + (\cos 14°)^2 - 3$.

10. Graph $y = 3 \sin(2x)$.

11. Find the exact value of $\tan \dfrac{\pi}{4} - 3 \cos \dfrac{\pi}{6} + \csc \dfrac{\pi}{6}$.

12. Find a sinusoidal function for the following graph.

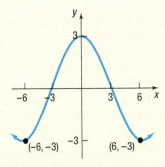

13. (a) Find a linear function that contains the points $(-2, 3)$ and $(1, -6)$. What is the slope? What are the intercepts of the function? Graph the function. Be sure to label the intercepts.
   (b) Find a quadratic function with vertex $(1, -6)$ that contains the point $(-2, 3)$. What are the intercepts of the function? Graph the function.

Chapter Projects

 Internet-based Project

I. **Length of Day Revisited** Go to *http://en.wikipedia.org/wiki/latitude* and read about latitude. Then go to *http://www.orchidculture.com/COD/daylength.html*.

1. For a particular latitude, record in a table the length of day for the various days of the year. For January 1, use 1 as the day, for January 16, use 16 as the day, for February 1, use 32 as the day, and so on. Enter the data into an Excel spreadsheet using column B for the day of the year and column C for the length of day.

2. Draw a scatter plot of the data with day of the year as the independent variable and length of day as the dependent variable using Excel.

3. Determine the sinusoidal function of best fit, $y = A \sin(Bx + C) + D$, as follows:

    (a) Enter initial guesses for the values of $A, B, C,$ and D into column A with the value of A in cell A1, B in cell A2, C in cell A3, and D in cell A4.

    (b) Enter "=A\$1*sin(A\$2*B1+A\$3)+A\$4" into cell D1. Copy this cell entry into the cells below D1 for as many rows as there are data. For example, if column C goes to row 23, then column D should also go to row 23.

    (c) Enter "=(D1−C1)^2" into cell E1. Copy this entry below cell E1 as described in part 3(b).

    (d) The idea behind curve fitting is to make the sum of the squared differences between what is predicted and actual observations as small as possible. Enter "=sum(E1..E#)" into cell A6, where # represents the row number of the last data point. For example, if you have 23 rows of data, enter "=sum(E1..E23)" in cell A6.

    (e) Now, install the Solver feature of Excel. To do this, click the File tab, and then select Options. Select Add-Ins. In the drop-down menu entitled "Manage," choose Excel Add-ins, and then click <u>G</u>o . . . . Check

the box entitled "Solver Add-in" and click OK. The Solver add-in is now available in the Data tab. Choose Solver. Fill in the screen as shown below.

Click Solve. The values for $A, B, C,$ and D are located in cells A1–A4. What is the sinusoidal function of best fit?

Citation: Excel © 2018 Microsoft Corporation. Used with permission from Microsoft.

4. Determine the longest day of the year according to your model. What is the day length on the longest day of the year? Determine the shortest day of the year according to your model. What is the day length on the shortest day of the year?

5. On which days is the day length exactly 12 hours according to your model?

6. Look up the day on which the vernal equinox and autumnal equinox occur. How do they match up with the results obtained in part 5?

7. Do you think your model accurately describes the relation between day of the year and length of day?

8. Use your model to predict the hours of daylight for the latitude you selected for various days of the year. Go to the *Old Farmer's Almanac* or other website (such as *http://astro.unl.edu/classaction/animations/coordsmotion/daylighthoursexplorer.html*) to determine the hours of daylight for the latitude you selected. How do the two compare?

The following projects are available on the Instructor's Resource Center (IRC):

II. **Tides** Data from a tide table are used to build a sine function that models tides.

III. **Project at Motorola** **_Digital Transmission over the Air_** Learn how Motorola Corporation transmits digital sequences by modulating the phase of the carrier waves.

IV. **Identifying Mountain Peaks in Hawaii** The visibility of a mountain is affected by its altitude, its distance from the viewer, and the curvature of Earth's surface. Trigonometry can be used to determine whether a distant object can be seen.

V. **CBL Experiment** Technology is used to model and study the effects of damping on sound waves.

Analytic Trigonometry

3

Mapping Your Mind

The ability to organize material in your mind is key to understanding. You have been exposed to a lot of concepts at this point in the course, and it is a worthwhile exercise to organize the material. In the past, we might organize material using index cards or an outline. But in today's digital world, we can use interesting software that enables us to digitally organize the material that is in our mind and share it with anyone on the Web.

 —See the Internet-based Chapter Project I—

← A Look Back

Chapter 1 introduced inverse functions and developed their properties, particularly the relationship between the domain and range of a function and those of its inverse. We learned that only functions that are one-to-one have an inverse function. But sometimes an appropriate restriction on the domain of a function that is not one-to-one yields a new function that is one-to-one. Then the function defined on the restricted domain has an inverse function. Also, we saw that the graphs of a function and its inverse are symmetric with respect to the line $y = x$.

Chapter 2 defined the six trigonometric functions and discussed their properties.

A Look Ahead →

The first two sections of this chapter define the six inverse trigonometric functions and investigate their properties. Section 3.3 discusses equations that contain trigonometric functions. Sections 3.4 through 3.7 continue the derivation of identities. These identities play an important role in calculus, the physical and life sciences, and economics, where they are used to simplify complicated expressions.

Outline

3.1 The Inverse Sine, Cosine, and Tangent Functions

PREPARING FOR THIS SECTION *Before getting started, review the following:*

- Inverse Functions (Section 1.7, pp. 81–88)
- Values of the Trigonometric Functions (Section 2.2, pp. 117–126)
- Properties of the Sine, Cosine, and Tangent Functions (Section 2.3, pp. 132–143)
- Graphs of the Sine, Cosine, and Tangent Functions (Sections 2.4, pp. 147–150 and 2.5, pp. 162–163)

Now Work the 'Are You Prepared?' problems on page 199.

OBJECTIVES 1 Define the Inverse Sine Function (p. 190)
2 Find the Value of an Inverse Sine Function (p. 191)
3 Define the Inverse Cosine Function (p. 193)
4 Find the Value of an Inverse Cosine Function (p. 194)
5 Define the Inverse Tangent Function (p. 194)
6 Find the Value of an Inverse Tangent Function (p. 196)
7 Use Properties of Inverse Functions to Find Exact Values of Certain Composite Functions (p. 196)
8 Find the Inverse Function of a Trigonometric Function (p. 198)
9 Solve Equations Involving Inverse Trigonometric Functions (p. 199)

In Section 1.7 we discussed inverse functions, and we concluded that if a function is one-to-one, it will have an inverse function. We also observed that if a function is not one-to-one, it may be possible to restrict its domain in some suitable manner so that the restricted function is one-to-one. For example, the function $y = x^2$ is not one-to-one; however, if the domain is restricted to $x \geq 0$, the new function is one-to-one.

Other properties of a one-to-one function f and its inverse function f^{-1} that were discussed in Section 1.7 are summarized next.

- $f^{-1}(f(x)) = x$ for every x in the domain of f.
- $f(f^{-1}(x)) = x$ for every x in the domain of f^{-1}.
- The domain of f = the range of f^{-1}, and the range of f = the domain of f^{-1}.
- The graph of a one-to-one function f and the graph of its inverse f^{-1} are symmetric with respect to the line $y = x$.
- If a function $y = f(x)$ has an inverse function, the implicit equation of the inverse function is $x = f(y)$. If we solve this equation for y, we obtain the explicit equation $y = f^{-1}(x)$.

1 Define the Inverse Sine Function

Figure 1 shows the graph of $y = \sin x$. Because every horizontal line $y = b$, where b is between -1 and 1, inclusive, intersects the graph of $y = \sin x$ infinitely many times, it follows from the horizontal-line test that the function $y = \sin x$ is not one-to-one.

Figure 1 $y = \sin x$, $-\infty < x < \infty$, $-1 \leq y \leq 1$

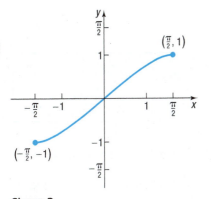

Figure 2
$y = \sin x, -\dfrac{\pi}{2} \le x \le \dfrac{\pi}{2}, -1 \le y \le 1$

NOTE Because the restricted domain of the sine function is $\left[-\dfrac{\pi}{2}, \dfrac{\pi}{2}\right]$, the range of the inverse sine function is $\left[-\dfrac{\pi}{2}, \dfrac{\pi}{2}\right]$, and because the range of the sine function is $[-1, 1]$, the domain of the inverse sine function is $[-1, 1]$.

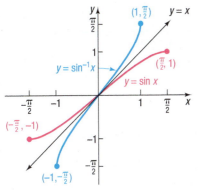

Figure 3
$y = \sin^{-1} x, -1 \le x \le 1,$
$-\dfrac{\pi}{2} \le y \le \dfrac{\pi}{2}$

However, if the domain of $y = \sin x$ is restricted to the interval $\left[-\dfrac{\pi}{2}, \dfrac{\pi}{2}\right]$, the restricted function

$$y = \sin x \qquad -\dfrac{\pi}{2} \le x \le \dfrac{\pi}{2}.$$

is one-to-one and has an inverse function.* See Figure 2.

An equation for the inverse of $y = f(x) = \sin x$ is obtained by interchanging x and y. The implicit form of the inverse function is $x = \sin y, -\dfrac{\pi}{2} \le y \le \dfrac{\pi}{2}$. The explicit form is called the **inverse sine** of x and is symbolized by $y = f^{-1}(x) = \sin^{-1} x$.

> **DEFINITION Inverse Sine Funtion**
>
> $$y = \sin^{-1} x \quad \text{if and only if} \quad x = \sin y$$
> $$\text{where} \quad -1 \le x \le 1 \quad \text{and} \quad -\dfrac{\pi}{2} \le y \le \dfrac{\pi}{2} \qquad \textbf{(1)}$$

The inverse sine of x, $y = \sin^{-1} x$, is read as "y is the angle or real number whose sine equals x," or, alternatively, is read "y is the inverse sine of x." Be careful about the notation used. The superscript -1 that appears in $y = \sin^{-1} x$ is not an exponent but the symbol used to denote the inverse function f^{-1} of f. (To avoid confusion, some texts use the notation $y = \arcsin x$ instead of $y = \sin^{-1} x$.)

The inverse of a function f receives as input an element from the range of f and returns as output an element in the domain of f. The restricted sine function, $y = f(x) = \sin x$, receives as input an angle or real number x in the interval $\left[-\dfrac{\pi}{2}, \dfrac{\pi}{2}\right]$ and outputs a real number in the interval $[-1, 1]$. Therefore, the inverse sine function, $y = \sin^{-1} x$, receives as input a real number in the interval $[-1, 1]$ or $-1 \le x \le 1$, its domain, and outputs an angle or real number in the interval $\left[-\dfrac{\pi}{2}, \dfrac{\pi}{2}\right]$ or $-\dfrac{\pi}{2} \le y \le \dfrac{\pi}{2}$, its range.

The graph of the inverse sine function can be obtained by reflecting the restricted portion of the graph of $y = f(x) = \sin x$ about the line $y = x$, as shown in Figure 3.

Now Work PROBLEM 5

2 Find the Value of an Inverse Sine Function

For some numbers x, it is possible to find the exact value of $y = \sin^{-1} x$.

EXAMPLE 1   **Finding the Exact Value of an Inverse Sine Function**

Find the exact value of: $\sin^{-1} 1$

Solution   Let $\theta = \sin^{-1} 1$. Then θ is the angle, $-\dfrac{\pi}{2} \le \theta \le \dfrac{\pi}{2}$, whose sine equals 1.

$$\theta = \sin^{-1} 1 \qquad -\dfrac{\pi}{2} \le \theta \le \dfrac{\pi}{2}$$

$$\sin \theta = 1 \qquad -\dfrac{\pi}{2} \le \theta \le \dfrac{\pi}{2} \quad \textbf{By definition of } y = \sin^{-1} x$$

(continued)

*Although there are many other ways to restrict the domain and obtain a one-to-one function, mathematicians have agreed to use the interval $\left[-\dfrac{\pi}{2}, \dfrac{\pi}{2}\right]$, to define the inverse of $y = \sin x$.

Now look at Table 1 and Figure 4.

Table 1

θ	$-\dfrac{\pi}{2}$	$-\dfrac{\pi}{3}$	$-\dfrac{\pi}{4}$	$-\dfrac{\pi}{6}$	0	$\dfrac{\pi}{6}$	$\dfrac{\pi}{4}$	$\dfrac{\pi}{3}$	$\dfrac{\pi}{2}$
$\sin\theta$	-1	$-\dfrac{\sqrt{3}}{2}$	$-\dfrac{\sqrt{2}}{2}$	$-\dfrac{1}{2}$	0	$\dfrac{1}{2}$	$\dfrac{\sqrt{2}}{2}$	$\dfrac{\sqrt{3}}{2}$	1

The only angle θ in the interval $\left[-\dfrac{\pi}{2}, \dfrac{\pi}{2}\right]$ whose sine is 1 is $\dfrac{\pi}{2}$. (Note that $\sin\dfrac{5\pi}{2}$ also equals 1, but $\dfrac{5\pi}{2}$ lies outside the interval $\left[-\dfrac{\pi}{2}, \dfrac{\pi}{2}\right]$, which is not allowed.) So

$$\sin^{-1} 1 = \frac{\pi}{2}$$

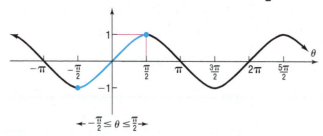

Figure 4

━━ **Now Work** PROBLEM 11

EXAMPLE 2   **Finding the Exact Value of an Inverse Sine Function**

Find the exact value of: $\sin^{-1}\left(-\dfrac{1}{2}\right)$

Solution   Let $\theta = \sin^{-1}\left(-\dfrac{1}{2}\right)$. Then θ is the angle, $-\dfrac{\pi}{2} \le \theta \le \dfrac{\pi}{2}$, whose sine equals $-\dfrac{1}{2}$.

$$\theta = \sin^{-1}\left(-\frac{1}{2}\right) \qquad -\frac{\pi}{2} \le \theta \le \frac{\pi}{2}$$

$$\sin\theta = -\frac{1}{2} \qquad -\frac{\pi}{2} \le \theta \le \frac{\pi}{2}$$

(Refer to Table 1 and Figure 4, if necessary.) The only angle in the interval $\left[-\dfrac{\pi}{2}, \dfrac{\pi}{2}\right]$ whose sine is $-\dfrac{1}{2}$ is $-\dfrac{\pi}{6}$, so

$$\sin^{-1}\left(-\frac{1}{2}\right) = -\frac{\pi}{6}$$

━━ **Now Work** PROBLEM 17

For most numbers x, the value $y = \sin^{-1} x$ must be approximated.

EXAMPLE 3   **Finding an Approximate Value of an Inverse Sine Function**

Find an approximate value of each of the following functions:

(a) $\sin^{-1}\dfrac{1}{3}$ (b) $\sin^{-1}\left(-\dfrac{1}{4}\right)$

Express the answer in radians rounded to two decimal places.

Solution

(a) Because the angle is to be measured in radians, first set the mode of the calculator to radians.* Rounded to two decimal places, $\sin^{-1}\frac{1}{3} = 0.34$.

(b) Figure 5(a) shows the solution using a TI-84 Plus C graphing calculator in radian mode. Figure 5(b) shows the solution using Desmos. Rounded to two decimal places, $\sin^{-1}\left(-\frac{1}{4}\right) = -0.25$.

Figure 5 (a) TI-84 Plus C   (b) Desmos

Now Work PROBLEM 27

3 Define the Inverse Cosine Function

Figure 6 shows the graph of $y = \cos x$. Because every horizontal line $y = b$, where b is between -1 and 1, inclusive, intersects the graph of $y = \cos x$ infinitely many times, it follows that the cosine function is not one-to-one.

Figure 6 $y = \cos x, -\infty < x < \infty, -1 \le y \le 1$

However, if the domain of $y = \cos x$ is restricted to the interval $[0, \pi]$, the restricted function

$$y = \cos x \qquad 0 \le x \le \pi$$

is one-to-one and has an inverse function.† See Figure 7.

An equation for the inverse of $y = f(x) = \cos x$ is obtained by interchanging x and y. The implicit form of the inverse function is $x = \cos y, 0 \le y \le \pi$. The explicit form is called the **inverse cosine** of x and is symbolized by $y = f^{-1}(x) = \cos^{-1} x$ (or by $y = \arccos x$).

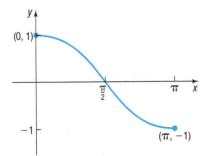

Figure 7
$y = \cos x, 0 \le x \le \pi, -1 \le y \le 1$

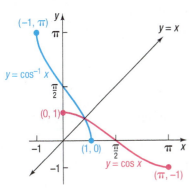

Figure 8
$y = \cos^{-1} x, -1 \le x \le 1, 0 \le y \le \pi$

> **DEFINITION** Inverse Cosine Function
>
> $$y = \cos^{-1}x \quad \text{if and only if} \quad x = \cos y$$
> $$\text{where} \quad -1 \le x \le 1 \quad \text{and} \quad 0 \le y \le \pi \tag{2}$$

Here y is the angle whose cosine is x. Because the range of the cosine function, $y = \cos x$, is $-1 \le y \le 1$, the domain of the inverse function $y = \cos^{-1}x$ is $-1 \le x \le 1$. Because the restricted domain of the cosine function, $y = \cos x$, is $0 \le x \le \pi$, the range of the inverse function $y = \cos^{-1}x$ is $0 \le y \le \pi$.

The graph of $y = \cos^{-1}x$ can be obtained by reflecting the restricted portion of the graph of $y = \cos x$ about the line $y = x$, as shown in Figure 8.

Now Work PROBLEM 7

*On most calculators, the inverse sine is obtained by pressing $\boxed{\text{SHIFT}}$ or $\boxed{\text{2}^\text{nd}}$, followed by $\boxed{\sin}$. On some calculators, $\boxed{\sin^{-1}}$ is pressed first, then 1/3 is entered; on others, this sequence is reversed. Consult your owner's manual for the correct sequence.
†This is the generally accepted restriction to define the inverse cosine function.

4 Find the Value of an Inverse Cosine Function

| EXAMPLE 4 | **Find the Exact Value of an Inverse Cosine Function** |

Find the exact value of: $\cos^{-1} 0$

Solution Let $\theta = \cos^{-1} 0$. Then θ is the angle, $0 \leq \theta \leq \pi$, whose cosine equals 0.

$$\theta = \cos^{-1} 0 \qquad 0 \leq \theta \leq \pi$$
$$\cos \theta = 0 \qquad 0 \leq \theta \leq \pi$$

Look at Table 2 and Figure 9.

Table 2

θ	$\cos \theta$
0	1
$\dfrac{\pi}{6}$	$\dfrac{\sqrt{3}}{2}$
$\dfrac{\pi}{4}$	$\dfrac{\sqrt{2}}{2}$
$\dfrac{\pi}{3}$	$\dfrac{1}{2}$
$\dfrac{\pi}{2}$	0
$\dfrac{2\pi}{3}$	$-\dfrac{1}{2}$
$\dfrac{3\pi}{4}$	$-\dfrac{\sqrt{2}}{2}$
$\dfrac{5\pi}{6}$	$-\dfrac{\sqrt{3}}{2}$
π	-1

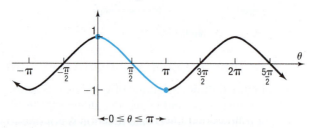

Figure 9

The only angle θ within the interval $[0, \pi]$ whose cosine is 0 is $\dfrac{\pi}{2}$. [Note that $\cos \dfrac{3\pi}{2}$ and $\cos\left(-\dfrac{\pi}{2}\right)$ also equal 0, but they lie outside the interval $[0, \pi]$, so these values are not allowed.] Therefore,

$$\cos^{-1} 0 = \frac{\pi}{2}$$

| EXAMPLE 5 | **Finding the Exact Value of an Inverse Cosine Function** |

Find the exact value of: $\cos^{-1}\left(-\dfrac{\sqrt{2}}{2}\right)$

Solution Let $\theta = \cos^{-1}\left(-\dfrac{\sqrt{2}}{2}\right)$. Then θ is the angle, $0 \leq \theta \leq \pi$, whose cosine equals $-\dfrac{\sqrt{2}}{2}$.

$$\theta = \cos^{-1}\left(-\frac{\sqrt{2}}{2}\right) \qquad 0 \leq \theta \leq \pi$$

$$\cos \theta = -\frac{\sqrt{2}}{2} \qquad 0 \leq \theta \leq \pi$$

Figure 10

Look at Table 2 and Figure 10.

The only angle θ within the interval $[0, \pi]$ whose cosine is $-\dfrac{\sqrt{2}}{2}$ is $\dfrac{3\pi}{4}$, so

$$\cos^{-1}\left(-\frac{\sqrt{2}}{2}\right) = \frac{3\pi}{4}$$

Now Work PROBLEM 21

5 Define the Inverse Tangent Function

Figure 11 on the next page shows the graph of $y = \tan x$. Because every horizontal line intersects the graph infinitely many times, it follows that the tangent function is not one-to-one.

However, if the domain of $y = \tan x$ is restricted to the interval $\left(-\dfrac{\pi}{2}, \dfrac{\pi}{2}\right)$, the restricted function

$$y = \tan x \qquad -\frac{\pi}{2} < x < \frac{\pi}{2}$$

is one-to-one and so has an inverse function.* See Figure 12.

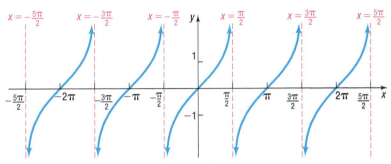

Figure 11
$y = \tan x$, $-\infty < x < \infty$, x not equal to odd multiples of $\dfrac{\pi}{2}$, $-\infty < y < \infty$

Figure 12
$y = \tan x$, $-\dfrac{\pi}{2} < x < \dfrac{\pi}{2}$, $-\infty < y < \infty$

An equation for the inverse of $y = f(x) = \tan x$ is obtained by interchanging x and y. The implicit form of the inverse function is $x = \tan y$, $-\dfrac{\pi}{2} < y < \dfrac{\pi}{2}$. The explicit form is called the **inverse tangent** of x and is symbolized by $y = f^{-1}(x) = \tan^{-1} x$ (or by $y = \arctan x$).

DEFINITION Inverse Tangent Function

$$y = \tan^{-1} x \quad \text{if and only if} \quad x = \tan y$$

$$\text{where} \quad -\infty < x < \infty \quad \text{and} \quad -\frac{\pi}{2} < y < \frac{\pi}{2}$$

(3)

Here y is the angle whose tangent is x. The domain of the function $y = \tan^{-1} x$ is $-\infty < x < \infty$, and its range is $-\dfrac{\pi}{2} < y < \dfrac{\pi}{2}$. The graph of $y = \tan^{-1} x$ can be obtained by reflecting the restricted portion of the graph of $y = \tan x$ about the line $y = x$, as shown in Figure 13.

Now Work PROBLEM **9**

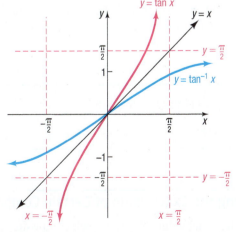

Figure 13 $y = \tan^{-1} x$, $-\infty < x < \infty$, $-\dfrac{\pi}{2} < y < \dfrac{\pi}{2}$

*This is the generally accepted restriction.

6 Find the Value of an Inverse Tangent Function

EXAMPLE 6

Finding the Exact Value of an Inverse Tangent Function

Find the exact value of:

(a) $\tan^{-1} 1$ (b) $\tan^{-1}\left(-\sqrt{3}\right)$

Solution

(a) Let $\theta = \tan^{-1} 1$. Then θ is the angle, $-\dfrac{\pi}{2} < \theta < \dfrac{\pi}{2}$, whose tangent equals 1.

$$\theta = \tan^{-1} 1 \qquad -\dfrac{\pi}{2} < \theta < \dfrac{\pi}{2}$$
$$\tan \theta = 1 \qquad -\dfrac{\pi}{2} < \theta < \dfrac{\pi}{2}$$

Table 3

θ	$\tan \theta$
$-\dfrac{\pi}{2}$	Undefined
$-\dfrac{\pi}{3}$	$-\sqrt{3}$
$-\dfrac{\pi}{4}$	-1
$-\dfrac{\pi}{6}$	$-\dfrac{\sqrt{3}}{3}$
0	0
$\dfrac{\pi}{6}$	$\dfrac{\sqrt{3}}{3}$
$\dfrac{\pi}{4}$	1
$\dfrac{\pi}{3}$	$\sqrt{3}$
$\dfrac{\pi}{2}$	Undefined

Look at Table 3. The only angle θ within the interval $\left(-\dfrac{\pi}{2}, \dfrac{\pi}{2}\right)$ whose tangent is 1 is $\dfrac{\pi}{4}$, so

$$\tan^{-1} 1 = \dfrac{\pi}{4}$$

(b) Let $\theta = \tan^{-1}\left(-\sqrt{3}\right)$. Then θ is the angle, $-\dfrac{\pi}{2} < \theta < \dfrac{\pi}{2}$, whose tangent equals $-\sqrt{3}$.

$$\theta = \tan^{-1}\left(-\sqrt{3}\right) \qquad -\dfrac{\pi}{2} < \theta < \dfrac{\pi}{2}$$
$$\tan \theta = -\sqrt{3} \qquad -\dfrac{\pi}{2} < \theta < \dfrac{\pi}{2}$$

Look at Table 3. The only angle θ within the interval $\left(-\dfrac{\pi}{2}, \dfrac{\pi}{2}\right)$ whose tangent is $-\sqrt{3}$ is $-\dfrac{\pi}{3}$, so

$$\tan^{-1}\left(-\sqrt{3}\right) = -\dfrac{\pi}{3}$$

 Now Work PROBLEM 15

7 Use Properties of Inverse Functions to Find Exact Values of Certain Composite Functions

Recall from the discussion of functions and their inverses in Section 1.7 that $f^{-1}(f(x)) = x$ for all x in the domain of f and that $f(f^{-1}(x)) = x$ for all x in the domain of f^{-1}. In terms of the sine function and its inverse, these properties are of the form

$$f^{-1}(f(x)) = \sin^{-1}(\sin x) = x \qquad \text{where } -\dfrac{\pi}{2} \le x \le \dfrac{\pi}{2} \qquad \textbf{(4a)}$$

$$f(f^{-1}(x)) = \sin(\sin^{-1} x) = x \qquad \text{where } -1 \le x \le 1 \qquad \textbf{(4b)}$$

EXAMPLE 7

Finding the Exact Value of Certain Composite Functions

Find the exact value, if any, of each composite function.

(a) $\sin^{-1}\left(\sin \dfrac{\pi}{8}\right)$ (b) $\sin^{-1}\left(\sin \dfrac{5\pi}{8}\right)$

Solution (a) The composite function $\sin^{-1}\left(\sin \dfrac{\pi}{8}\right)$ follows the form of property (4a).

Because $\dfrac{\pi}{8}$ is in the interval $\left[-\dfrac{\pi}{2}, \dfrac{\pi}{2}\right]$, use property (4a) to conclude

$$\sin^{-1}\left(\sin \dfrac{\pi}{8}\right) = \dfrac{\pi}{8}$$

(b) The composite function $\sin^{-1}\left(\sin \dfrac{5\pi}{8}\right)$ follows the form of property (4a). But

because $\dfrac{5\pi}{8}$ is not in the interval $\left[-\dfrac{\pi}{2}, \dfrac{\pi}{2}\right]$, $\sin^{-1}\left(\sin \dfrac{5\pi}{8}\right) \neq \dfrac{5\pi}{8}$.

To find $\sin^{-1}\left(\sin \dfrac{5\pi}{8}\right)$, first find an angle θ in the interval $\left[-\dfrac{\pi}{2}, \dfrac{\pi}{2}\right]$ for

which $\sin \theta = \sin \dfrac{5\pi}{8}$.

Figure 14 illustrates that $\sin \dfrac{5\pi}{8} = y = \sin \dfrac{3\pi}{8}$. Since $\dfrac{3\pi}{8}$ is in the

interval $\left[-\dfrac{\pi}{2}, \dfrac{\pi}{2}\right]$, we can use 4(a).

$$\sin^{-1}\left(\sin \dfrac{5\pi}{8}\right) = \sin^{-1}\left(\sin \dfrac{3\pi}{8}\right) = \dfrac{3\pi}{8}$$

↑
Use property (4a).

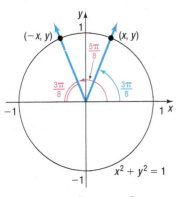

Figure 14 $\sin \dfrac{5\pi}{8} = \sin \dfrac{3\pi}{8}$

 Now Work PROBLEM 43

| EXAMPLE 8 | **Finding the Exact Value of Certain Composite Functions** |

Find the exact value, if any, of each composite function.

(a) $\sin\left(\sin^{-1} 0.5\right)$ (b) $\sin\left(\sin^{-1} 1.8\right)$

Solution (a) The composite function $\sin\left(\sin^{-1} 0.5\right)$ follows the form of property (4b). Because 0.5 is in the interval $[-1, 1]$, we can use 4(b).

$$\sin\left(\sin^{-1} 0.5\right) = 0.5$$

(b) The composite function $\sin\left(\sin^{-1} 1.8\right)$ follows the form of property (4b). But since 1.8 is not in the domain of the inverse sine function, $[-1, 1]$, $\sin^{-1} 1.8$ is not defined. Therefore, $\sin\left(\sin^{-1} 1.8\right)$ is also not defined.

 Now Work PROBLEM 51

For the cosine function and its inverse, the following properties hold.

$$f^{-1}(f(x)) = \cos^{-1}(\cos x) = x \qquad \text{where } 0 \leq x \leq \pi \qquad \textbf{(5a)}$$
$$f(f^{-1}(x)) = \cos(\cos^{-1} x) = x \qquad \text{where } -1 \leq x \leq 1 \qquad \textbf{(5b)}$$

| EXAMPLE 9 | **Using Properties of Inverse Functions to Find the Exact Value of Certain Composite Functions** |

Find the exact value, if any, of each composite function.

(a) $\cos^{-1}\left(\cos \dfrac{\pi}{12}\right)$ (b) $\cos\left[\cos^{-1}(-0.4)\right]$

(c) $\cos^{-1}\left[\cos\left(-\dfrac{2\pi}{3}\right)\right]$ (d) $\cos\left(\cos^{-1} \pi\right)$

Solution

(a) $\cos^{-1}\left(\cos\dfrac{\pi}{12}\right) = \dfrac{\pi}{12}$ $\dfrac{\pi}{12}$ is in the interval $[0, \pi]$; use property (5a).

(b) $\cos[\cos^{-1}(-0.4)] = -0.4$ -0.4 is in the interval $[-1, 1]$; use property (5b).

(c) The angle $-\dfrac{2\pi}{3}$ is not in the interval $[0, \pi]$, so property (5a) cannot be used. However, because the cosine function is even, $\cos\left(-\dfrac{2\pi}{3}\right) = \cos\dfrac{2\pi}{3}$. Because $\dfrac{2\pi}{3}$ is in the interval $[0, \pi]$, property (5a) can be used, and

$$\cos^{-1}\left[\cos\left(-\dfrac{2\pi}{3}\right)\right] = \cos^{-1}\left(\cos\dfrac{2\pi}{3}\right) = \dfrac{2\pi}{3}$$

(d) Because π is not in the interval $[-1, 1]$, the domain of the inverse cosine function, $\cos^{-1}\pi$ is not defined. This means the composite function $\cos(\cos^{-1}\pi)$ is also not defined.

Now Work PROBLEMS 39 AND 55

For the tangent function and its inverse, the following properties hold.

$$f^{-1}(f(x)) = \tan^{-1}(\tan x) = x \quad \text{where } -\dfrac{\pi}{2} < x < \dfrac{\pi}{2}$$

$$f(f^{-1}(x)) = \tan(\tan^{-1}x) = x \quad \text{where } -\infty < x < \infty$$

Now Work PROBLEM 45

8 Find the Inverse Function of a Trigonometric Function

EXAMPLE 10   **Finding the Inverse Function of a Trigonometric Function**

(a) Find the inverse function f^{-1} of $f(x) = 2\sin x - 1$, $-\dfrac{\pi}{2} \le x \le \dfrac{\pi}{2}$.

(b) Find the range of f and the domain and range of f^{-1}.

Solution

(a) The function f is one-to-one and so has an inverse function. Follow the steps on page 87 for finding the inverse function.

$$y = 2\sin x - 1$$
$$x = 2\sin y - 1 \quad \text{Interchange } x \text{ and } y.$$
$$x + 1 = 2\sin y \quad \text{Solve for } y.$$
$$\sin y = \dfrac{x+1}{2}$$
$$y = \sin^{-1}\dfrac{x+1}{2} \quad \text{Definition of inverse sine function}$$

The inverse function is $f^{-1}(x) = \sin^{-1}\dfrac{x+1}{2}$.

(b) To find the range of f, use the fact that the domain of f^{-1} equals the range of f. Since the domain of the inverse sine function is the interval $[-1, 1]$, the argument $\dfrac{x+1}{2}$ must be in the interval $[-1, 1]$.

$$-1 \le \dfrac{x+1}{2} \le 1$$
$$-2 \le x + 1 \le 2 \quad \text{Multiply by 2.}$$
$$-3 \le x \le 1 \quad \text{Subtract 1 from each part.}$$

NOTE The range of f also can be found using transformations. The range of $y = \sin x$ is $[-1, 1]$. the range of $y = 2 \sin x$ is $[-2, 2]$ due to the vertical stretch by a factor of 2. The range of $f(x) = 2 \sin x - 1$ is $[-3, 1]$ due to the shift down of 1 unit. ■

The domain of f^{-1} is $\{x \mid -3 \le x \le 1\}$, or the interval $[-3, 1]$. So the range of f is the interval $[-3, 1]$. The range of f^{-1} equals the domain of f. So the range of f^{-1} is $\left[-\dfrac{\pi}{2}, \dfrac{\pi}{2}\right]$.

━━━ **Now Work** PROBLEM **61**

9 Solve Equations Involving Inverse Trigonometric Functions

Equations that contain inverse trigonometric functions are called **inverse trigonometric equations**.

EXAMPLE 11

Solving an Inverse Trigonometric Equation

Solve the equation: $3 \sin^{-1} x = \pi$

Solution

To solve an equation involving a single inverse trigonometric function, first isolate the inverse trigonometric function.

$$3 \sin^{-1} x = \pi$$

$$\sin^{-1} x = \frac{\pi}{3} \qquad \textcolor{blue}{\textbf{Divide both sides by 3.}}$$

$$x = \sin \frac{\pi}{3} \qquad \textcolor{blue}{\textbf{y} = \sin^{-1} x \textbf{ if and only if } x = \sin y.}$$

$$x = \frac{\sqrt{3}}{2}$$

The solution set is $\left\{\dfrac{\sqrt{3}}{2}\right\}$.

━━━ **Now Work** PROBLEM **67**

3.1 Assess Your Understanding

'Are You Prepared?' *Answers are given at the end of these exercises. If you get a wrong answer, read the pages listed in* red.

1. What are the domain and the range of $y = \sin x$? (pp. 133–134)

2. If the domain of a one-to-one function is $[3, \infty)$, the range of its inverse is _____. (pp. 81–88)

3. *True or False* The graph of $y = \cos x$ is decreasing on the interval $[0, \pi]$. (pp. 149–150)

4. $\tan \dfrac{\pi}{4} =$ _____; $\sin \dfrac{\pi}{3} =$ _____; $\sin\left(-\dfrac{\pi}{6}\right) =$ _____;
   $\cos \pi =$ _____. (pp. 117–126)

Concepts and Vocabulary

5. $y = \sin^{-1} x$ if and only if _____, where $-1 \le x \le 1$ and $-\dfrac{\pi}{2} \le y \le \dfrac{\pi}{2}$.

6. $\cos^{-1}(\cos x) = x$ for all numbers x for which _____.

7. *True or False* The domain of $y = \cos^{-1} x$ is $-1 \le x \le 1$.

8. *True or False* $\sin(\sin^{-1} 0) = 0$ and $\cos(\cos^{-1} 0) = 0$.

9. *True or False* $y = \tan^{-1} x$ if and only if $x = \tan y$, where $-\infty < x < \infty$ and $-\dfrac{\pi}{2} < y < \dfrac{\pi}{2}$.

10. *Multiple Choice* $\sin^{-1}(\sin x) = x$ for all numbers x for which
   (a) $-\infty < x < \infty$
   (b) $0 \le x \le \pi$
   (c) $-1 \le x \le 1$
   (d) $-\dfrac{\pi}{2} \le x \le \dfrac{\pi}{2}$

Skill Building

In Problems 11–26, find the exact value of each expression.

11. $\sin^{-1} 0$

12. $\cos^{-1} 1$

13. $\sin^{-1}(-1)$

14. $\cos^{-1}(-1)$

15. $\tan^{-1} 0$

16. $\tan^{-1}(-1)$

17. $\sin^{-1} \dfrac{\sqrt{2}}{2}$

18. $\tan^{-1} \dfrac{\sqrt{3}}{3}$

19. $\tan^{-1} \sqrt{3}$

20. $\sin^{-1}\left(-\dfrac{\sqrt{3}}{2}\right)$

21. $\cos^{-1}\left(-\dfrac{\sqrt{3}}{2}\right)$

22. $\sin^{-1}\left(-\dfrac{\sqrt{2}}{2}\right)$

23. $\cos^{-1} \dfrac{\sqrt{2}}{2}$

24. $\cos^{-1}\left(-\dfrac{1}{2}\right)$

25. $\tan^{-1}\left(-\dfrac{\sqrt{3}}{3}\right)$

26. $\sin^{-1} \dfrac{1}{2}$

In Problems 27–38, use a calculator to find the approximate value of each expression rounded to two decimal places.

27. $\sin^{-1} 0.1$

28. $\cos^{-1} 0.6$

29. $\tan^{-1} 5$

30. $\tan^{-1} 0.2$

31. $\cos^{-1} \dfrac{7}{8}$

32. $\sin^{-1} \dfrac{1}{8}$

33. $\tan^{-1}(-0.4)$

34. $\tan^{-1}(-3)$

35. $\sin^{-1}(-0.12)$

36. $\cos^{-1}(-0.44)$

37. $\cos^{-1} \dfrac{\sqrt{2}}{3}$

38. $\sin^{-1} \dfrac{\sqrt{3}}{5}$

In Problems 39–58, find the exact value, if any, of each composite function. If there is no value, state it is "not defined." Do not use a calculator.

39. $\cos^{-1}\left(\cos \dfrac{4\pi}{5}\right)$

40. $\sin^{-1}\left[\sin\left(-\dfrac{\pi}{10}\right)\right]$

41. $\tan^{-1}\left[\tan\left(-\dfrac{3\pi}{8}\right)\right]$

42. $\sin^{-1}\left[\sin\left(-\dfrac{3\pi}{7}\right)\right]$

43. $\sin^{-1}\left(\sin \dfrac{9\pi}{8}\right)$

44. $\cos^{-1}\left[\cos\left(-\dfrac{5\pi}{3}\right)\right]$

45. $\tan^{-1}\left(\tan \dfrac{4\pi}{5}\right)$

46. $\tan^{-1}\left[\tan\left(-\dfrac{2\pi}{3}\right)\right]$

47. $\cos^{-1}\left[\cos\left(-\dfrac{\pi}{4}\right)\right]$

48. $\sin^{-1}\left[\sin\left(-\dfrac{3\pi}{4}\right)\right]$

49. $\tan^{-1}\left[\tan\left(\dfrac{\pi}{2}\right)\right]$

50. $\tan^{-1}\left[\tan\left(-\dfrac{3\pi}{2}\right)\right]$

51. $\sin\left(\sin^{-1} \dfrac{1}{4}\right)$

52. $\cos\left[\cos^{-1}\left(-\dfrac{2}{3}\right)\right]$

53. $\tan\left(\tan^{-1} 4\right)$

54. $\tan\left[\tan^{-1}(-2)\right]$

55. $\cos\left(\cos^{-1} 1.2\right)$

56. $\sin\left[\sin^{-1}(-2)\right]$

57. $\tan\left(\tan^{-1}\pi\right)$

58. $\sin\left[\sin^{-1}(-1.5)\right]$

In Problems 59–66, find the inverse function f^{-1} of each function f. Find the range of f and the domain and range of f^{-1}.

59. $f(x) = 5\sin x + 2; \ -\dfrac{\pi}{2} \le x \le \dfrac{\pi}{2}$

60. $f(x) = 2\tan x - 3; \ -\dfrac{\pi}{2} < x < \dfrac{\pi}{2}$

61. $f(x) = -2\cos(3x); 0 \le x \le \dfrac{\pi}{3}$

62. $f(x) = 3\sin(2x); -\dfrac{\pi}{4} \le x \le \dfrac{\pi}{4}$

63. $f(x) = -\tan(x+1) - 3; -1 - \dfrac{\pi}{2} < x < \dfrac{\pi}{2} - 1$

64. $f(x) = \cos(x+2) + 1; -2 \le x \le \pi - 2$

65. $f(x) = 3\sin(2x+1); -\dfrac{1}{2} - \dfrac{\pi}{4} \le x \le -\dfrac{1}{2} + \dfrac{\pi}{4}$

66. $f(x) = 2\cos(3x+2); -\dfrac{2}{3} \le x \le -\dfrac{2}{3} + \dfrac{\pi}{3}$

In Problems 67–74, find the exact solution of each equation.

67. $4\sin^{-1} x = \pi$

68. $2\cos^{-1} x = \pi$

69. $3\cos^{-1}(2x) = 2\pi$

70. $-6\sin^{-1}(3x) = \pi$

71. $3\tan^{-1} x = \pi$

72. $-4\tan^{-1} x = \pi$

73. $4\cos^{-1} x - 2\pi = 2\cos^{-1} x$

74. $5\sin^{-1} x - 2\pi = 2\sin^{-1} x - 3\pi$

Applications and Extensions

In Problems 75–80, use the following discussion. The formula

$$D = 24\left[1 - \frac{\cos^{-1}(\tan i \tan \theta)}{\pi}\right]$$

can be used to approximate the number of hours of daylight D when the declination of the Sun is i° at a location θ° north latitude for any date between the vernal equinox and autumnal equinox. The declination of the Sun is defined as the angle i between the equatorial plane and any ray of light from the Sun. The latitude of a location is the angle θ between the Equator and the location on the surface of Earth, with the vertex of the angle located at the center of Earth. See the figure. To use the formula, $\cos^{-1}(\tan i \tan \theta)$ must be expressed in radians.

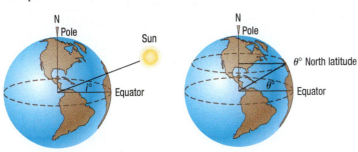

75. Approximate the number of hours of daylight in Houston, Texas (29°45′ north latitude), for the following dates:
    (a) Summer solstice ($i = 23.5°$)
    (b) Vernal equinox ($i = 0°$)
    (c) July 4 ($i = 22°48′$)

76. Approximate the number of hours of daylight in New York, New York (40°45′ north latitude), for the following dates:
    (a) Summer solstice ($i = 23.5°$)
    (b) Vernal equinox ($i = 0°$)
    (c) July 4 ($i = 22°48′$)

77. Approximate the number of hours of daylight in Honolulu, Hawaii (21°18′ north latitude), for the following dates:
    (a) Summer solstice ($i = 23.5°$)
    (b) Vernal equinox ($i = 0°$)
    (c) July 4 ($i = 22°48′$)

78. Approximate the number of hours of daylight in Anchorage, Alaska (61°10′ north latitude), for the following dates:
    (a) Summer solstice ($i = 23.5°$)
    (b) Vernal equinox ($i = 0°$)
    (c) July 4 ($i = 22°48′$)

79. Approximate the number of hours of daylight at the Equator (0° north latitude) for the following dates:
    (a) Summer solstice ($i = 23.5°$)
    (b) Vernal equinox ($i = 0°$)
    (c) July 4 ($i = 22°48′$)
    (d) What do you conclude about the number of hours of daylight throughout the year for a location at the Equator?

80. Approximate the number of hours of daylight for any location that is 66°30′ north latitude for the following dates:
    (a) Summer solstice ($i = 23.5°$)
    (b) Vernal equinox ($i = 0°$)
    (c) July 4 ($i = 22°48′$)
    (d) Thanks to the symmetry of the orbital path of Earth around the Sun, the number of hours of daylight on the winter solstice may be found by computing the number of

hours of daylight on the summer solstice and subtracting this result from 24 hours. Compute the number of hours of daylight for this location on the winter solstice. What do you conclude about daylight for a location at 66°30′ north latitude?

81. **Being the First to See the Rising Sun** Cadillac Mountain, elevation 1530 feet, is located in Acadia National Park, Maine, and is the highest peak on the east coast of the United States. It is said that a person standing on the summit will be the first person in the United States to see the rays of the rising Sun. How much sooner would a person atop Cadillac Mountain see the first rays than a person standing below, at sea level?

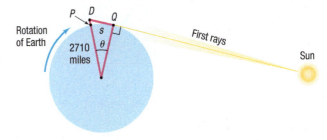

[**Hint:** Consult the figure. When the person at *D* sees the first rays of the Sun, the person at *P* does not. The person at *P* sees the first rays of the Sun only after Earth has rotated so that *P* is at location *Q*. Compute the length of the arc subtended by the central angle θ. Then use the fact that at the latitude of Cadillac Mountain, in 24 hours a length of $2\pi \cdot 2710 \approx 17027.4$ miles is subtended.]

82. **Movie Theater Screens** Suppose that a movie theater has a screen that is 28 feet tall. When you sit down, the bottom of the screen is 6 feet above your eye level. The angle formed by drawing a line from your eye to the bottom of the screen and another line from your eye to the top of the screen is called the **viewing angle**. In the figure, θ is the viewing angle. Suppose that you sit *x* feet from the screen. The viewing angle θ is given by the function

$$\theta(x) = \tan^{-1}\left(\frac{34}{x}\right) - \tan^{-1}\left(\frac{6}{x}\right)$$

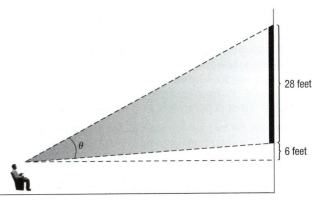

(a) What is your viewing angle if you sit 10 feet from the screen? 15 feet? 20 feet?
(b) If there are 5 feet between the screen and the first row of seats and there are 3 feet between each row and the row behind it, which row results in the largest viewing angle?

(continued)

(c) Using a graphing utility, graph

$$\theta(x) = \tan^{-1}\left(\frac{34}{x}\right) - \tan^{-1}\left(\frac{6}{x}\right)$$

What value of x results in the largest viewing angle?

83. Area under a Curve The area under the graph of $y = \dfrac{1}{1 + x^2}$ and above the x-axis between $x = a$ and $x = b$ is given by

$$\tan^{-1} b - \tan^{-1} a$$

See the figure.

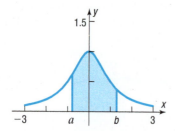

(a) Find the exact area under the graph of $y = \dfrac{1}{1 + x^2}$ and above the x-axis between $x = 0$ and $x = \sqrt{3}$.

(b) Find the exact area under the graph of $y = \dfrac{1}{1 + x^2}$ and above the x-axis between $x = -\dfrac{\sqrt{3}}{3}$ and $x = 1$.

84. Area under a Curve The area under the graph of $y = \dfrac{1}{\sqrt{1 - x^2}}$ and above the x-axis between $x = a$ and $x = b$ is given by

$$\sin^{-1} b - \sin^{-1} a$$

See the figure.

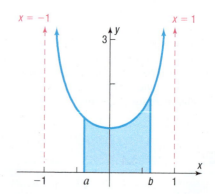

(a) Find the exact area under the graph of $y = \dfrac{1}{\sqrt{1 - x^2}}$ and above the x-axis between $x = 0$ and $x = \dfrac{\sqrt{3}}{2}$.

(b) Find the exact area under the graph of $y = \dfrac{1}{\sqrt{1 - x^2}}$ and above the x-axis between $x = -\dfrac{1}{2}$ and $x = \dfrac{1}{2}$.

Problems 85 and 86 require the following discussion:
The shortest distance between two points on Earth's surface can be determined from the latitude and longitude of the two locations. For example, if location 1 has $(\text{lat, lon}) = (\alpha_1, \beta_1)$ *and location 2 has* $(\text{lat, lon}) = (\alpha_2, \beta_2)$, *the shortest distance between the two locations is approximately*
$$d = r \cos^{-1}[\,(\cos\alpha_1 \cos\beta_1 \cos\alpha_2 \cos\beta_2) + (\cos\alpha_1 \sin\beta_1 \cos\alpha_2 \sin\beta_2) + (\sin\alpha_1 \sin\alpha_2)\,],$$
where $r = \text{radius of Earth} \approx 3960$ *miles and the inverse cosine function is expressed in radians.*
Also, N *latitude and* E *longitude are positive angles, and* S *latitude and* W *longitude are negative angles.*

City	Latitude	Longitude
Chicago, IL	41°50′N	87°37′W
Honolulu, HI	21°18′N	157°50′W
Melbourne, Australia	37°47′S	144°58′E

Source: www.infoplease.com

85. Shortest Distance from Chicago to Honolulu Find the shortest distance from Chicago, latitude 41°50′N, longitude 87°37′W, to Honolulu, latitude 21°18′N, longitude 157°50′W. Round your answer to the nearest mile.

86. Shortest Distance from Honolulu to Melbourne, Australia Find the shortest distance from Honolulu to Melbourne, Australia, latitude 37°47′S, longitude 144°58′E. Round your answer to the nearest mile.

87. Challenge Problem Solve:

$$\cos\left(\sin^{-1} x\right) = \tan\left(\cos^{-1}\frac{4}{5}\right)$$

88. Challenge Problem Find u in terms of x and r:

$$\tan\left(\cos^{-1}\frac{x}{r}\right) = \sin\left(\tan^{-1} u\right)$$

Retain Your Knowledge

Problems 89–96 are based on material learned earlier in the course. The purpose of these problems is to keep the material fresh in your mind so that you are better prepared for the final exam.

89. Solve: $|3x - 2| + 5 \le 9$

90. State why the graph of the function f shown to the right is one-to-one.
Then draw the graph of the inverse function f^{-1}.
Hint: The graph of $y = x$ is given.

91. Factor: $(2x + 1)^{-\frac{1}{2}}(x^2 + 3)^{-\frac{1}{2}} - (x^2 + 3)^{-\frac{3}{2}}x(2x + 1)^{\frac{1}{2}}$

92. The diameter of each wheel of a bicycle is 20 inches. If the wheels are turning at 336 revolutions per minute, how fast is the bicycle moving? Express the answer in miles per hour, rounded to the nearest integer.

93. Find the exact value of $\sin \dfrac{\pi}{3} \cos \dfrac{\pi}{3}$.

94. If $\cos \theta = \dfrac{24}{25}$, find the exact value of each of the remaining five trigonometric functions of acute angle θ.

95. If $\sin \theta > 0$ and $\cot \theta < 0$, name the quadrant in which the angle θ lies.

96. Find the average rate of change of $f(x) = \tan x$ from $\dfrac{\pi}{6}$ to $\dfrac{\pi}{4}$.

'Are You Prepared?' Answers

1. Domain: the set of all real numbers; Range: $-1 \le y \le 1$ **2.** $[3, \infty)$ **3.** True   **4.** $1; \dfrac{\sqrt{3}}{2}; -\dfrac{1}{2}; -1$

3.2 The Inverse Trigonometric Functions (Continued)

PREPARING FOR THIS SECTION *Before getting started, review the following:*

- Finding Exact Values of the Trigonometric Functions, Given the Value of a Trigonometric Function and the Quadrant of the Angle (Section 2.3, pp. 139–142)
- Graphs of the Secant, Cosecant, and Cotangent Functions (Section 2.5, pp. 164–167)

- Domain and Range of the Secant, Cosecant, and Cotangent Functions (Section 2.3, pp. 133–134)

Now Work the 'Are You Prepared?' problems on page 206.

OBJECTIVES   **1** Define the Inverse Secant, Cosecant, and Cotangent Functions (p. 203)
               2 Find the Value of Inverse Secant, Cosecant, and Cotangent Functions (p. 204)
               3 Find the Exact Value of Composite Functions Involving the Inverse Trigonometric Functions (p. 205)
               4 Write a Trigonometric Expression as an Algebraic Expression (p. 206)

1 Define the Inverse Secant, Cosecant, and Cotangent Functions

The inverse secant, inverse cosecant, and inverse cotangent functions are defined as follows:

> **DEFINITION Inverse Secant, Cosecant, and Cotangent Functions**
>
> - $y = \sec^{-1} x$ if and only if $x = \sec y$
>
>   where $|x| \ge 1$ and $0 \le y \le \pi$, $y \ne \dfrac{\pi}{2}$ *   **(1)**
>
> - $y = \csc^{-1} x$ if and only if $x = \csc y$
>
>   where $|x| \ge 1$ and $-\dfrac{\pi}{2} \le y \le \dfrac{\pi}{2}$, $y \ne 0$†   **(2)**
>
> - $y = \cot^{-1} x$ if and only if $x = \cot y$
>
>   where $-\infty < x < \infty$ and $0 < y < \pi$ **(3)**

Take time to review the graphs of the cotangent, cosecant, and secant functions in Figures 63, 66, and 67 in Section 2.5 to see the basis for these definitions.

Now Work PROBLEM 4

*Most texts use this definition. A few use the restriction $0 \le y < \dfrac{\pi}{2}, \pi \le y < \dfrac{3\pi}{2}$.

† Most texts use this definition. A few use the restriction $-\pi < y \le -\dfrac{\pi}{2}, 0 < y \le \dfrac{\pi}{2}$.

2 Find the Value of Inverse Secant, Cosecant, and Cotangent Functions

EXAMPLE 1

Finding the Exact Value of Inverse Secant, Cosecant, and Cotangent Functions

Find the exact value of each expression.

(a) $\sec^{-1} 1$ (b) $\csc^{-1} 2$ (c) $\cot^{-1}(-\sqrt{3})$

Solution

(a) Let $\theta = \sec^{-1} 1$. Then θ is the angle, $0 \le \theta \le \pi, \theta \ne \dfrac{\pi}{2}$, whose secant equals 1 (or, equivalently, whose cosine equals 1).

$$\theta = \sec^{-1} 1 \qquad 0 \le \theta \le \pi, \quad \theta \ne \frac{\pi}{2}$$

$$\sec \theta = 1 \qquad 0 \le \theta \le \pi, \quad \theta \ne \frac{\pi}{2}$$

The only angle θ for which $0 \le \theta \le \pi, \theta \ne \dfrac{\pi}{2}$, whose secant is 1 is 0, so $\sec^{-1} 1 = 0$.

(b) Let $\theta = \csc^{-1} 2$. Then θ is the angle, $-\dfrac{\pi}{2} \le \theta \le \dfrac{\pi}{2}, \theta \ne 0$, whose cosecant equals 2 $\left(\text{or, equivalently, whose sine equals } \dfrac{1}{2}\right)$.

$$\theta = \csc^{-1} 2 \qquad -\frac{\pi}{2} \le \theta \le \frac{\pi}{2}, \quad \theta \ne 0$$

$$\csc \theta = 2 \qquad -\frac{\pi}{2} \le \theta \le \frac{\pi}{2}, \quad \theta \ne 0$$

The only angle θ for which $-\dfrac{\pi}{2} \le \theta \le \dfrac{\pi}{2}, \theta \ne 0$, whose cosecant is 2 is $\dfrac{\pi}{6}$, so $\csc^{-1} 2 = \dfrac{\pi}{6}$.

(c) Let $\theta = \cot^{-1}(-\sqrt{3})$. Then θ is the angle, $0 < \theta < \pi$, whose cotangent equals $-\sqrt{3}$.

$$\theta = \cot^{-1}(-\sqrt{3}) \qquad 0 < \theta < \pi$$

$$\cot \theta = -\sqrt{3} \qquad 0 < \theta < \pi$$

The only angle θ for which $0 < \theta < \pi$ whose cotangent is $-\sqrt{3}$ is $\dfrac{5\pi}{6}$, so $\cot^{-1}(-\sqrt{3}) = \dfrac{5\pi}{6}$.

Now Work PROBLEM 11

Most calculators do not have keys that evaluate the inverse cotangent, cosecant, or secant functions. The easiest way to evaluate them is to convert each to an inverse trigonometric function whose range is the same as the one to be evaluated. In this regard, notice that $y = \cot^{-1} x$ and $y = \sec^{-1} x$, except where undefined, have the same range as $y = \cos^{-1} x$ and that $y = \csc^{-1} x$, except where undefined, has the same range as $y = \sin^{-1} x$.

NOTE Remember that the range of $y = \sin^{-1} x$ is $\left[-\dfrac{\pi}{2}, \dfrac{\pi}{2}\right]$ and that the range of $y = \cos^{-1} x$ is $[0, \pi]$. ∎

EXAMPLE 2

Approximating the Value of Inverse Trigonometric Functions

Use a calculator to approximate each expression in radians rounded to two decimal places.

(a) $\sec^{-1} 3$ (b) $\csc^{-1}(-4)$ (c) $\cot^{-1}\dfrac{1}{2}$ (d) $\cot^{-1}(-2)$

Solution   First, set the calculator to radian mode.

(a) Let $\theta = \sec^{-1} 3$. Then $\sec \theta = 3$ and $0 \leq \theta \leq \pi, \theta \neq \dfrac{\pi}{2}$. Now find $\cos \theta$ because $y = \cos^{-1} x$ has the same range as $y = \sec^{-1} x$, except where undefined. Because $\sec \theta = \dfrac{1}{\cos \theta} = 3$, this means $\cos \theta = \dfrac{1}{3}$. Then $\theta = \cos^{-1}\dfrac{1}{3}$, and

$$\sec^{-1} 3 = \theta = \cos^{-1}\frac{1}{3} \approx 1.23$$

↑
Use a calculator.

(b) Let $\theta = \csc^{-1}(-4)$. Then $\csc \theta = -4, -\dfrac{\pi}{2} \leq \theta \leq \dfrac{\pi}{2}, \theta \neq 0$. Now find $\sin \theta$ because $y = \sin^{-1} x$ has the same range as $y = \csc^{-1} x$, except where undefined. Because $\csc \theta = \dfrac{1}{\sin \theta} = -4$, this means $\sin \theta = -\dfrac{1}{4}$. Then $\theta = \sin^{-1}\left(-\dfrac{1}{4}\right)$, and

$$\csc^{-1}(-4) = \theta = \sin^{-1}\left(-\frac{1}{4}\right) \approx -0.25$$

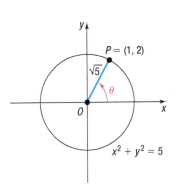

Figure 15 $\cot \theta = \dfrac{1}{2}, 0 < \theta < \pi$

(c) Let $\theta = \cot^{-1}\dfrac{1}{2}$. Then $\cot \theta = \dfrac{1}{2}, 0 < \theta < \pi$. Since $\cot \theta > 0$, it follows that θ lies in quadrant I. Now find $\cos \theta$ because $y = \cos^{-1} x$ has the same range as $y = \cot^{-1} x$, except where undefined. Use Figure 15 to find that $\cos \theta = \dfrac{1}{\sqrt{5}}, 0 < \theta < \dfrac{\pi}{2}$. So, $\theta = \cos^{-1}\left(\dfrac{1}{\sqrt{5}}\right)$, and

$$\cot^{-1}\frac{1}{2} = \theta = \cos^{-1}\left(\frac{1}{\sqrt{5}}\right) \approx 1.11$$

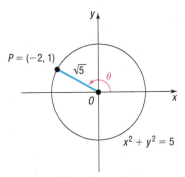

Figure 16 $\cot \theta = -2, 0 < \theta < \pi$

(d) Let $\theta = \cot^{-1}(-2)$. Then $\cot \theta = -2, 0 < \theta < \pi$. Since $\cot \theta < 0$, θ lies in quadrant II. Use Figure 16 to find that $\cos \theta = -\dfrac{2}{\sqrt{5}}, \dfrac{\pi}{2} < \theta < \pi$. This means $\theta = \cos^{-1}\left(-\dfrac{2}{\sqrt{5}}\right)$, and

$$\cot^{-1}(-2) = \theta = \cos^{-1}\left(-\frac{2}{\sqrt{5}}\right) \approx 2.68$$

Now Work PROBLEM 21

3 Find the Exact Value of Composite Functions Involving the Inverse Trigonometric Functions

EXAMPLE 3   **Finding the Exact Value of an Expression Involving Inverse Trigonometric Functions**

Find the exact value of: $\sin\left(\tan^{-1}\dfrac{1}{2}\right)$

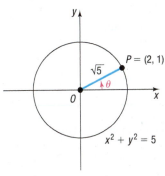

Figure 17 $\tan \theta = \dfrac{1}{2}$

Solution Let $\theta = \tan^{-1}\dfrac{1}{2}$. Then $\tan \theta = \dfrac{1}{2}$, where $-\dfrac{\pi}{2} < \theta < \dfrac{\pi}{2}$. Because $\tan \theta > 0$, it follows that $0 < \theta < \dfrac{\pi}{2}$, so θ lies in quadrant I. Because $\tan \theta = \dfrac{1}{2} = \dfrac{y}{x}$, let $x = 2$ and $y = 1$. Since $r = d(O, P) = \sqrt{2^2 + 1^2} = \sqrt{5}$, the point $P = (x, y) = (2, 1)$ is on the circle $x^2 + y^2 = 5$. See Figure 17. Then

$$\sin\left(\tan^{-1}\frac{1}{2}\right) = \sin \theta = \frac{1}{\sqrt{5}} = \frac{\sqrt{5}}{5}$$

↑
$\sin \theta = \dfrac{y}{r}$

EXAMPLE 4 **Finding the Exact Value of an Expression Involving Inverse Trigonometric Functions**

Find the exact value of: $\cos\left[\sin^{-1}\left(-\dfrac{1}{3}\right)\right]$

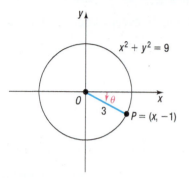

Figure 18 $\sin\theta = -\dfrac{1}{3}$

Solution Let $\theta = \sin^{-1}\left(-\dfrac{1}{3}\right)$. Then $\sin\theta = -\dfrac{1}{3}$ and $-\dfrac{\pi}{2} \le \theta \le \dfrac{\pi}{2}$. Because $\sin\theta < 0$, it follows that $-\dfrac{\pi}{2} \le \theta < 0$, so θ lies in quadrant IV. Since $\sin\theta = \dfrac{-1}{3} = \dfrac{y}{r}$, let $y = -1$ and $r = 3$. The point $P = (x, y) = (x, -1), x > 0$, is on a circle of radius 3, $x^2 + y^2 = 9$. See Figure 18. Then $x^2 = 8$ and $x = 2\sqrt{2}$ so

$$\cos\left[\sin^{-1}\left(-\dfrac{1}{3}\right)\right] = \cos\theta = \dfrac{2\sqrt{2}}{3}$$

$\uparrow$
$\cos\theta = \dfrac{x}{r}$

EXAMPLE 5 **Finding the Exact Value of an Expression Involving Inverse Trigonometric Functions**

Find the exact value of: $\tan\left[\cos^{-1}\left(-\dfrac{1}{3}\right)\right]$

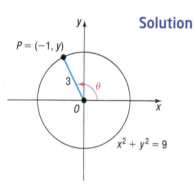

Figure 19 $\cos\theta = -\dfrac{1}{3}$

Solution Let $\theta = \cos^{-1}\left(-\dfrac{1}{3}\right)$. Then $\cos\theta = -\dfrac{1}{3}$ and $0 \le \theta \le \pi$. Because $\cos\theta < 0$, it follows that $\dfrac{\pi}{2} < \theta \le \pi$, so θ lies in quadrant II. Since $\cos\theta = \dfrac{-1}{3} = \dfrac{x}{r}$, let $x = -1$ and $r = 3$. The point $P = (x, y) = (-1, y), y > 0$, is on a circle of radius $r = 3$, $x^2 + y^2 = 9$. See Figure 19. Then $y^2 = 8$ and $y = 2\sqrt{2}$. Since $x = -1$, we have

$$\tan\left[\cos^{-1}\left(-\dfrac{1}{3}\right)\right] = \tan\theta = \dfrac{2\sqrt{2}}{-1} = -2\sqrt{2}$$

$\uparrow$
$\tan\theta = \dfrac{y}{x}$

— Now Work PROBLEMS **33** AND **51**

4 Write a Trigonometric Expression as an Algebraic Expression

EXAMPLE 6 **Writing a Trigonometric Expression as an Algebraic Expression**

Write $\sin(\tan^{-1} u)$ as an algebraic expression containing u.

Solution Let $\theta = \tan^{-1} u$ so that $\tan\theta = u$, $-\dfrac{\pi}{2} < \theta < \dfrac{\pi}{2}$, $-\infty < u < \infty$. This means $\sec\theta > 0$. Then

$$\sin(\tan^{-1} u) = \sin\theta = \sin\theta \cdot \dfrac{\cos\theta}{\cos\theta} = \tan\theta\cos\theta = \dfrac{\tan\theta}{\sec\theta} = \dfrac{\tan\theta}{\sqrt{1 + \tan^2\theta}} = \dfrac{u}{\sqrt{1 + u^2}}$$

$\uparrow$ $\uparrow$ $\uparrow$

Multiply by $1 = \dfrac{\cos\theta}{\cos\theta}$ $\dfrac{\sin\theta}{\cos\theta} = \tan\theta$ $\sec^2\theta = 1 + \tan^2\theta$
$\sec\theta > 0$

— Now Work PROBLEM **61**

3.2 Assess Your Understanding

'Are You Prepared?' *Answers are given at the end of these exercises. If you get a wrong answer, read the pages listed in* red.

1. What are the domain and the range of $y = \sec x$? (pp. 133–134)

2. *True or False* The graph of $y = \sec x$ is one-to-one on the set $\left[0, \dfrac{\pi}{2}\right) \cup \left(\dfrac{\pi}{2}, \pi\right]$. (pp. 164–167)

3. If $\tan\theta = \dfrac{1}{2}$, $-\dfrac{\pi}{2} < \theta < \dfrac{\pi}{2}$, then $\sin\theta =$ _____.
(pp. 139–142)

Concepts and Vocabulary

4. $y = \sec^{-1} x$ if and only if _____ , where $|x|$ _____ and ____ $\leq y \leq$ ____ , $y \neq \dfrac{\pi}{2}$.

5. To find the inverse secant of a real number x, $|x| \geq 1$, convert the inverse secant to an inverse _____ .

6. *True or False* It is impossible to obtain exact values for the inverse secant function.

7. *True or False* $\csc^{-1} 0.5$ is not defined.

8. *True or False* The domain of the inverse cotangent function is the set of real numbers.

Skill Building

In Problems 9–20, find the exact value of each expression.

9. $\cot^{-1} \sqrt{3}$

10. $\cot^{-1} 1$

11. $\csc^{-1}(-1)$

12. $\csc^{-1} \sqrt{2}$

13. $\sec^{-1} \dfrac{2\sqrt{3}}{3}$

14. $\sec^{-1}(-2)$

15. $\cot^{-1}\left(-\dfrac{\sqrt{3}}{3}\right)$

16. $\csc^{-1}\left(-\dfrac{2\sqrt{3}}{3}\right)$

17. $\sec^{-1}\left(-\sqrt{2}\right)$

18. $\cot^{-1}(-1)$

19. $\csc^{-1}\left(-\sqrt{2}\right)$

20. $\sec^{-1} 1$

In Problems 21–32, use a calculator to find the approximate value of each expression rounded to two decimal places.

21. $\sec^{-1} 4$

22. $\csc^{-1} 5$

23. $\cot^{-1} 2$

24. $\sec^{-1}(-3)$

25. $\csc^{-1}(-3)$

26. $\cot^{-1}\left(-\dfrac{1}{2}\right)$

27. $\cot^{-1}\left(-\sqrt{5}\right)$

28. $\cot^{-1}(-8.1)$

29. $\csc^{-1}\left(-\dfrac{3}{2}\right)$

30. $\sec^{-1}\left(-\dfrac{4}{3}\right)$

31. $\cot^{-1}\left(-\dfrac{3}{2}\right)$

32. $\cot^{-1}\left(-\sqrt{10}\right)$

In Problems 33–60, find the exact value of each expression.

33. $\cos\left(\sin^{-1} \dfrac{\sqrt{2}}{2}\right)$

34. $\sin\left(\cos^{-1} \dfrac{1}{2}\right)$

35. $\tan\left[\cos^{-1}\left(-\dfrac{\sqrt{3}}{2}\right)\right]$

36. $\tan\left[\sin^{-1}\left(-\dfrac{1}{2}\right)\right]$

37. $\sec\left(\cos^{-1} \dfrac{1}{2}\right)$

38. $\cot\left[\sin^{-1}\left(-\dfrac{1}{2}\right)\right]$

39. $\csc\left(\tan^{-1} 1\right)$

40. $\sec\left(\tan^{-1} \sqrt{3}\right)$

41. $\sin\left[\tan^{-1}(-1)\right]$

42. $\cos\left[\sin^{-1}\left(-\dfrac{\sqrt{3}}{2}\right)\right]$

43. $\sec\left[\sin^{-1}\left(-\dfrac{1}{2}\right)\right]$

44. $\csc\left[\cos^{-1}\left(-\dfrac{\sqrt{3}}{2}\right)\right]$

45. $\cos^{-1}\left(\sin \dfrac{5\pi}{4}\right)$

46. $\tan^{-1}\left(\cot \dfrac{2\pi}{3}\right)$

47. $\sin^{-1}\left[\cos\left(-\dfrac{7\pi}{6}\right)\right]$

48. $\cos^{-1}\left[\tan\left(-\dfrac{\pi}{4}\right)\right]$

49. $\tan\left(\sin^{-1} \dfrac{1}{3}\right)$

50. $\tan\left(\cos^{-1} \dfrac{1}{3}\right)$

51. $\sec\left(\tan^{-1} \dfrac{1}{2}\right)$

52. $\cos\left(\sin^{-1} \dfrac{\sqrt{2}}{3}\right)$

53. $\cot\left[\sin^{-1}\left(-\dfrac{\sqrt{2}}{3}\right)\right]$

54. $\csc\left[\tan^{-1}(-2)\right]$

55. $\sin\left[\tan^{-1}(-3)\right]$

56. $\cot\left[\cos^{-1}\left(-\dfrac{\sqrt{3}}{3}\right)\right]$

57. $\sec\left(\sin^{-1} \dfrac{2\sqrt{5}}{5}\right)$

58. $\csc\left(\tan^{-1} \dfrac{1}{2}\right)$

59. $\sin^{-1}\left(\cos \dfrac{3\pi}{4}\right)$

60. $\cos^{-1}\left(\sin \dfrac{7\pi}{6}\right)$

In Problems 61–70, write each trigonometric expression as an algebraic expression in u.

61. $\cos\left(\tan^{-1} u\right)$

62. $\sin\left(\cos^{-1} u\right)$

63. $\tan\left(\sin^{-1} u\right)$

64. $\tan\left(\cos^{-1} u\right)$

65. $\sin\left(\sec^{-1} u\right)$

66. $\sin\left(\cot^{-1} u\right)$

67. $\cos\left(\csc^{-1} u\right)$

68. $\cos\left(\sec^{-1} u\right)$

69. $\tan\left(\cot^{-1} u\right)$

70. $\tan\left(\sec^{-1} u\right)$

Mixed Practice In Problems 71–82, $f(x) = \sin x$, $-\dfrac{\pi}{2} \leq x \leq \dfrac{\pi}{2}$; $g(x) = \cos x$, $0 \leq x \leq \pi$; and $h(x) = \tan x$, $-\dfrac{\pi}{2} < x < \dfrac{\pi}{2}$. Find the exact value of each composite function.

71. $g\left(f^{-1}\left(\dfrac{12}{13}\right)\right)$

72. $f\left(g^{-1}\left(\dfrac{5}{13}\right)\right)$

73. $g^{-1}\left(f\left(-\dfrac{\pi}{4}\right)\right)$

74. $f^{-1}\left(g\left(\dfrac{5\pi}{6}\right)\right)$

75. $h\left(f^{-1}\left(-\dfrac{3}{5}\right)\right)$

76. $h\left(g^{-1}\left(-\dfrac{4}{5}\right)\right)$

77. $g\left(h^{-1}\left(\dfrac{12}{5}\right)\right)$

78. $f\left(h^{-1}\left(\dfrac{5}{12}\right)\right)$

79. $g^{-1}\left(f\left(-\dfrac{\pi}{3}\right)\right)$

80. $g^{-1}\left(f\left(-\dfrac{\pi}{6}\right)\right)$

81. $h\left(g^{-1}\left(-\dfrac{1}{4}\right)\right)$

82. $h\left(f^{-1}\left(-\dfrac{2}{5}\right)\right)$

Applications and Extensions

*Problems 83 and 84 require the following discussion: When granular materials are allowed to fall freely, they form conical (cone-shaped) piles. The naturally occurring angle, measured from the horizontal, at which the loose material comes to rest is called the **angle of repose** and varies for different materials. The angle of repose θ is related to the height h and the base radius r of the conical pile by the equation $\theta = \cot^{-1}\dfrac{r}{h}$. See the figure.*

83. Angle of Repose: De-icing Salt Due to potential transportation issues (for example, frozen waterways), de-icing salt used by highway departments in the Midwest must be ordered early and stored for future use. When de-icing salt is stored in a pile 14 feet high, the diameter of the base of the pile is 45 feet.

   (a) Find the angle of repose for de-icing salt.
   (b) What is the base diameter of a pile that is 17 feet high?
   (c) What is the height of a pile that has a base diameter of approximately 122 feet?

   Source: Salt Institute, *The Salt Storage Handbook*, 2015

84. Angle of Repose: Bunker Sand The steepness of sand bunkers on a golf course is affected by the angle of repose of the sand (a larger angle of repose allows for steeper bunkers). A freestanding pile of loose sand from a United States Golf Association (USGA) bunker had a height of 4 feet and a base diameter of approximately 6.68 feet.

   (a) Find the angle of repose for USGA bunker sand.

   (b) What is the height of such a pile if the diameter of the base is 8 feet?
   (c) A 6-foot-high pile of loose Tour Grade 50/50 sand has a base diameter of approximately 8.44 feet. Which type of sand (USGA or Tour Grade 50/50) would be better suited for steep bunkers?

   Source: Purdue University Turfgrass Science Program

85. Artillery A projectile fired into the first quadrant from the origin of a rectangular coordinate system will pass through the point (x, y) at time t according to the relationship $\cot \theta = \dfrac{2x}{2y + gt^2}$, where $\theta =$ the angle of elevation of the launcher and $g =$ the acceleration due to gravity = 32.2 feet/second^2. An artilleryman is firing at an enemy bunker located 2450 feet up the side of a hill that is 6175 feet away. He fires a round, and exactly 2.27 seconds later he scores a direct hit.

   (a) What angle of elevation did he use?
   (b) If the angle of elevation is also given by $\sec \theta = \dfrac{v_0 t}{x}$, where v_0 is the muzzle velocity of the weapon, find the muzzle velocity of the artillery piece he used.

   Source: www.egwald.com/geometry/projectile3d.php

86. Challenge Problem Find the exact value:

$$\cot\left[\sec^{-1}\left(\sin\frac{\pi}{3} + \tan\frac{\pi}{6}\right)\right]$$

87. Challenge Problem Write as an algebraic expression in x:

$$\sec\{\tan^{-1}[\sin(\cos^{-1}|x|)]\}$$

Explaining Concepts: Discussion and Writing

88. Explain in your own words how you would use your calculator to find the value of $\cot^{-1} 10$.

89. Consult three books on calculus, and then write down the definition in each of $y = \sec^{-1} x$ and $y = \csc^{-1} x$. Compare these with the definitions given in this text.

Retain Your Knowledge

Problems 90–98 are based on material learned earlier in the course. The purpose of these problems is to keep the material fresh in your mind so that you are better prepared for the final exam.

90. Find the complex zeros of $f(x) = x^4 + 21x^2 - 100$.

91. Determine algebraically whether $f(x) = x^3 + x^2 - x$ is even, odd, or neither.

92. Convert 315° to radians.

93. Find the length of the arc subtended by a central angle of 75° on a circle of radius 6 inches. Give both the exact length and an approximation rounded to two decimal places.

94. Consider $f(x) = -2x^2 - 10x + 3$.
   (a) Find the vertex.
   (b) Is the parabola concave up or concave down?
   (c) Find where f is increasing and where f is decreasing.

95. If $f(x) = \sqrt{x - 3}$ and $g(x) = \dfrac{x - 7}{x - 4}$, find the domain of $\left(\dfrac{f}{g}\right)(x)$.

96. Find the equation of a sine function with amplitude 4, period $\dfrac{\pi}{3}$, and phase shift 1.

97. Rationalize the numerator: $\dfrac{\sqrt{1 - x^2} - \sqrt{1 - c^2}}{x - c}$

98. Find the average rate of change of $f(x) = \sin x$ from $\dfrac{\pi}{2}$ to $\dfrac{4\pi}{3}$.

'Are You Prepared?' Answers

1. Domain: $\left\{x \mid x \neq \text{odd integer multiples of } \dfrac{\pi}{2}\right\}$; Range: $\{y \mid y \le -1 \text{ or } y \ge 1\}$ **2.** True   **3.** $\dfrac{\sqrt{5}}{5}$

3.3 Trigonometric Equations

PREPARING FOR THIS SECTION *Before getting started, review the following:*

- Solving Equations (Section A.4, pp. A27–A34)
- Values of the Trigonometric Functions (Section 2.2, pp. 117–126)

- Using a Graphing Utility to Solve Equations (Section B.4, pp. B6–B8)
- Fundamental Identities (Section 2.3, pp. 137–139)

Now Work the 'Are You Prepared?' problems on page 214.

OBJECTIVES 1 Solve Equations Involving a Single Trigonometric Function (p. 209)
2 Solve Trigonometric Equations Using a Calculator (p. 212)
3 Solve Trigonometric Equations Quadratic in Form (p. 213)
4 Solve Trigonometric Equations Using Fundamental Identities (p. 213)
5 Solve Trigonometric Equations Using a Graphing Utility (p. 214)

In this section, we discuss **trigonometric equations**—that is, equations involving trigonometric functions that are satisfied only by some values of the variable (or, possibly, are not satisfied by any values of the variable). The values that satisfy the equation are called **solutions** of the equation.

1 Solve Equations Involving a Single Trigonometric Function

EXAMPLE 1

Checking Whether a Given Number Is a Solution of a Trigonometric Equation

Determine whether $\theta = \dfrac{\pi}{4}$ is a solution of the equation $2 \sin \theta - 1 = 0$. Is $\theta = \dfrac{\pi}{6}$ a solution?

Solution

Replace θ by $\dfrac{\pi}{4}$ in the equation $2 \sin \theta - 1 = 0$. The result is

$$2 \sin \frac{\pi}{4} - 1 = 2 \cdot \frac{\sqrt{2}}{2} - 1 = \sqrt{2} - 1 \neq 0$$

Therefore, $\dfrac{\pi}{4}$ is not a solution.

Next replace θ by $\dfrac{\pi}{6}$ in the equation. The result is

$$2 \sin \frac{\pi}{6} - 1 = 2 \cdot \frac{1}{2} - 1 = 0$$

Therefore $\dfrac{\pi}{6}$ is a solution of the equation, $2 \sin \theta - 1 = 0$.

The equation $2 \sin \theta - 1 = 0$ in Example 1 has other solutions besides $\theta = \dfrac{\pi}{6}$. For example, $\theta = \dfrac{5\pi}{6}$ is also a solution, as is $\theta = \dfrac{13\pi}{6}$. (Check this for yourself.) In fact, the equation has an infinite number of solutions because of the periodicity of the sine function, as can be seen in Figure 20 on the next page, which shows the graph of $y = 2 \sin x - 1$. Each x-intercept of the graph represents a solution to the equation $2 \sin x - 1 = 0$.

Figure 20 $y = 2\sin x - 1$

Unless the domain of the variable is restricted, we need to find *all* the solutions of a trigonometric equation. As the next example illustrates, finding all the solutions can be accomplished by first finding solutions over an interval whose length equals the period of the function and then adding multiples of that period to the solutions found.

EXAMPLE 2 | **Finding All the Solutions of a Trigonometric Equation**

Solve the equation: $\cos\theta = \dfrac{1}{2}$

Give a general formula for all the solutions. List eight of the solutions.

Solution The period of the cosine function is 2π. In the interval $[0, 2\pi)$, there are two angles θ for which $\cos\theta = \dfrac{1}{2}$: $\theta = \dfrac{\pi}{3}$ and $\theta = \dfrac{5\pi}{3}$. See Figure 21. Because the cosine function has period 2π, all the solutions of $\cos\theta = \dfrac{1}{2}$ may be given by the general formula

$$\theta = \frac{\pi}{3} + 2k\pi \quad \text{or} \quad \theta = \frac{5\pi}{3} + 2k\pi \quad k \text{ any integer}$$

Eight of the solutions are

$$\underbrace{-\frac{5\pi}{3}, \quad -\frac{\pi}{3},}_{k = -1} \quad \underbrace{\frac{\pi}{3}, \quad \frac{5\pi}{3},}_{k = 0} \quad \underbrace{\frac{7\pi}{3}, \quad \frac{11\pi}{3},}_{k = 1} \quad \underbrace{\frac{13\pi}{3}, \quad \frac{17\pi}{3}}_{k = 2}$$

Figure 21

Figure 22

✔**Check:** To verify the solutions, graph $Y_1 = \cos x$ and $Y_2 = \dfrac{1}{2}$ and determine where the graphs intersect. (Be sure to graph in radian mode.) See Figure 22. The graph of Y_1 intersects the graph of Y_2 at $x = 1.05 \left(\approx \dfrac{\pi}{3} \right)$, $5.24 \left(\approx \dfrac{5\pi}{3} \right)$, $7.33 \left(\approx \dfrac{7\pi}{3} \right)$, and $11.52 \left(\approx \dfrac{11\pi}{3} \right)$, rounded to two decimal places.

Now Work PROBLEM 37

In most of the work we do, we shall be interested only in finding solutions of trigonometric equations for $0 \le \theta < 2\pi$.

EXAMPLE 3 | **Solving a Linear Trigonometric Equation**

Solve the equation: $2 \sin \theta + \sqrt{3} = 0, \quad 0 \le \theta < 2\pi$

Solution   First solve the equation for $\sin \theta$.

$$2 \sin \theta + \sqrt{3} = 0$$

$$2 \sin \theta = -\sqrt{3} \quad \text{Subtract } \sqrt{3} \text{ from both sides.}$$

$$\sin \theta = -\frac{\sqrt{3}}{2} \quad \text{Divide both sides by 2.}$$

In the interval $[0, 2\pi)$, there are two angles θ for which $\sin \theta = -\frac{\sqrt{3}}{2}$: $\theta = \frac{4\pi}{3}$ and $\theta = \frac{5\pi}{3}$. The solution set is $\left\{ \frac{4\pi}{3}, \frac{5\pi}{3} \right\}$.

> **Now Work** PROBLEM 13

When the argument of the trigonometric function in an equation is a multiple of θ, the general formula is required to solve the equation.

EXAMPLE 4 | **Solving a Trigonometric Equation Involving a Double Angle**

Solve the equation: $\sin(2\theta) = \frac{1}{2}, \quad 0 \le \theta < 2\pi$

Solution   In the interval $[0, 2\pi)$, the sine function equals $\frac{1}{2}$ at $\frac{\pi}{6}$ and $\frac{5\pi}{6}$. See Figure 23(a). In the interval $[0, 2\pi)$, the graph of $y = \sin(2\theta)$ completes two cycles, and the graph of $y = \sin(2\theta)$ intersects the graph of $y = \frac{1}{2}$ four times. See Figure 23(b). So, there are four solutions of the equation $\sin(2\theta) = \frac{1}{2}$ in the interval $[0, 2\pi)$. To find the solutions, use 2θ in the general formula that gives all the solutions.

(a)

(b)

Figure 23

$$2\theta = \frac{\pi}{6} + 2k\pi \quad \text{or} \quad 2\theta = \frac{5\pi}{6} + 2k\pi \quad k \text{ an integer} \quad \textbf{The argument is } 2\theta.$$

$$\theta = \frac{\pi}{12} + k\pi \quad \text{or} \quad \theta = \frac{5\pi}{12} + k\pi \qquad \textbf{Divide by 2.}$$

Then

$$\theta = \frac{\pi}{12} + (-1)\pi = \frac{-11\pi}{12} \quad \color{blue}{k = -1} \quad \theta = \frac{5\pi}{12} + (-1)\pi = \frac{-7\pi}{12}$$

$$\theta = \frac{\pi}{12} + 0 \cdot \pi = \frac{\pi}{12} \quad \color{blue}{k = 0} \quad \theta = \frac{5\pi}{12} + 0 \cdot \pi = \frac{5\pi}{12}$$

$$\theta = \frac{\pi}{12} + 1 \cdot \pi = \frac{13\pi}{12} \quad \color{blue}{k = 1} \quad \theta = \frac{5\pi}{12} + 1 \cdot \pi = \frac{17\pi}{12}$$

$$\theta = \frac{\pi}{12} + 2 \cdot \pi = \frac{25\pi}{12} \quad \color{blue}{k = 2} \quad \theta = \frac{5\pi}{12} + 2 \cdot \pi = \frac{29\pi}{12}$$

WARNING In solving a trigonometric equation for $\theta, 0 \le \theta < 2\pi$, in which the argument is not θ (as in Example 4), you must write down all the solutions first and then list those that are in the interval $[0, 2\pi)$. Otherwise, solutions may be lost. ∎

In the interval $[0, 2\pi)$, the solutions of $\sin(2\theta) = \frac{1}{2}$ are $\theta = \frac{\pi}{12}, \theta = \frac{5\pi}{12}, \theta = \frac{13\pi}{12}$, and $\theta = \frac{17\pi}{12}$. The solution set is $\left\{ \frac{\pi}{12}, \frac{5\pi}{12}, \frac{13\pi}{12}, \frac{17\pi}{12} \right\}$. The graph of $y = \sin(2\theta)$ intersects the graph of $y = \frac{1}{2}$ at the points $\left(\frac{\pi}{12}, \frac{1}{2} \right), \left(\frac{5\pi}{12}, \frac{1}{2} \right), \left(\frac{13\pi}{12}, \frac{1}{2} \right)$, and $\left(\frac{17\pi}{12}, \frac{1}{2} \right)$ in the interval $[0, 2\pi)$.

> **Now Work** PROBLEM 29

EXAMPLE 5

Solving a Trigonometric Equation

Solve the equation: $\tan\left(\theta - \dfrac{\pi}{2}\right) = 1, \quad 0 \le \theta < 2\pi$

Solution

The period of the tangent function is π. In the interval $[0, \pi)$, the tangent function has the value 1 when the argument is $\dfrac{\pi}{4}$. Because the argument is $\theta - \dfrac{\pi}{2}$ in the given equation, use it in the general formula that gives all the solutions.

$$\theta - \frac{\pi}{2} = \frac{\pi}{4} + k\pi \quad k \text{ any integer}$$

$$\theta = \frac{3\pi}{4} + k\pi$$

In the interval $[0, 2\pi)$, $\theta = \dfrac{3\pi}{4}$ and $\theta = \dfrac{3\pi}{4} + \pi = \dfrac{7\pi}{4}$ are the only solutions.

The solution set is $\left\{\dfrac{3\pi}{4}, \dfrac{7\pi}{4}\right\}$.

Now Work PROBLEM 27

2 Solve Trigonometric Equations Using a Calculator

The next example illustrates how to solve trigonometric equations using a calculator. Remember that the function keys on a calculator give only values consistent with the definition of the function.

EXAMPLE 6

Solving a Trigonometric Equation Using a Calculator

Use a calculator to solve the equation $\tan \theta = -2, 0 \le \theta < 2\pi$. Express any solutions in radians, rounded to two decimal places.

Solution

To solve $\tan \theta = -2$ on a calculator, first set the mode to radians. Then use the $\boxed{\tan^{-1}}$ key to obtain

$$\theta = \tan^{-1}(-2) \approx -1.1071487$$

Rounded to two decimal places, $\theta = \tan^{-1}(-2) = -1.11$ radian. Because of the definition of $y = \tan^{-1} x$, the angle θ that is obtained is the angle $-\dfrac{\pi}{2} < \theta < \dfrac{\pi}{2}$ for which $\tan \theta = -2$. But we want solutions for which $0 \le \theta < 2\pi$. Since the period of the tangent function is π, the angles $\pi - 1.11$ and $2\pi - 1.11$ are solutions that lie in the interval $[0, 2\pi)$. Note that the angle $3\pi - 1.11$ lies outside the interval and so is not a solution.

The solutions of the equation $\tan \theta = -2, 0 \le \theta < 2\pi$, are

$$\theta = \pi - 1.11 \approx 2.03 \text{ radians and } \theta = 2\pi - 1.11 \approx 5.17 \text{ radians}$$

The solution set is $\{2.03, 5.17\}$.

Figure 24 illustrates another way to obtain the solutions. Start with the angle $\theta = -1.11$. Then $\pi - 1.11$ is the angle in quadrant II, where $\tan \theta = -2$, and $2\pi - 1.11$ is the angle in quadrant IV where $\tan \theta = -2$.

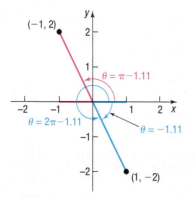

Figure 24 $\tan \theta = -2$

WARNING Example 6 illustrates that caution must be exercised when solving trigonometric equations on a calculator. Remember that the calculator supplies an angle only within the restrictions of the definition of the inverse trigonometric function. To find the remaining solutions, you must identify other quadrants, if any, in which a solution may be located. ∎

Now Work PROBLEM 49

3 Solve Trigonometric Equations Quadratic in Form

Many trigonometric equations can be solved using techniques that we already know, such as using the quadratic formula (if the equation is a second-degree polynomial) or factoring.

EXAMPLE 7

Solving a Trigonometric Equation Quadratic in Form

Solve the equation: $2 \sin^2 \theta - 3 \sin \theta + 1 = 0, \quad 0 \le \theta < 2\pi$

Solution

This equation is a quadratic equation in $\sin \theta$ that can be factored.

$$2 \sin^2 \theta - 3 \sin \theta + 1 = 0 \quad \text{\color{blue}} 2x^2 - 3x + 1 = 0, \quad x = \sin \theta$$
$$(2 \sin \theta - 1)(\sin \theta - 1) = 0 \quad \text{\color{blue}} (2x - 1)(x - 1) = 0$$
$$2 \sin \theta - 1 = 0 \quad \text{or} \quad \sin \theta - 1 = 0 \quad \text{\color{blue}Use the Zero-Product Property.}$$
$$\sin \theta = \frac{1}{2} \quad \text{or} \quad \sin \theta = 1$$

Solving each equation in the interval $[0, 2\pi)$ yields

$$\theta = \frac{\pi}{6} \qquad \theta = \frac{5\pi}{6} \qquad \theta = \frac{\pi}{2}$$

The solution set is $\left\{ \dfrac{\pi}{6}, \dfrac{\pi}{2}, \dfrac{5\pi}{6} \right\}$.

Now Work PROBLEM 63

4 Solve Trigonometric Equations Using Fundamental Identities

Often when a trigonometric equation contains more than one trigonometric function, identities can be used to obtain an equivalent equation that contains only one trigonometric function.

EXAMPLE 8

Solving a Trigonometric Equation Using Identities

Solve the equation: $3 \cos \theta + 3 = 2 \sin^2 \theta, \quad 0 \le \theta < 2\pi$

Solution

The equation contains a sine and a cosine. However, using the Pythagorean Identity, $\sin^2 \theta + \cos^2 \theta = 1$, the equation is transformed into an equivalent one containing only cosines.

$$3 \cos \theta + 3 = 2 \sin^2 \theta$$
$$3 \cos \theta + 3 = 2(1 - \cos^2 \theta) \quad \text{\color{blue}} \sin^2 \theta = 1 - \cos^2 \theta$$
$$3 \cos \theta + 3 = 2 - 2 \cos^2 \theta \quad \text{\color{blue}Quadratic in } \cos \theta$$
$$2 \cos^2 \theta + 3 \cos \theta + 1 = 0$$
$$(2 \cos \theta + 1)(\cos \theta + 1) = 0 \quad \text{\color{blue}Factor}$$
$$2 \cos \theta + 1 = 0 \quad \text{or} \quad \cos \theta + 1 = 0 \quad \text{\color{blue}Use the Zero-Product Property.}$$
$$\cos \theta = -\frac{1}{2} \quad \text{or} \quad \cos \theta = -1$$

Solving each equation in the interval $[0, 2\pi)$ yields

$$\theta = \frac{2\pi}{3} \qquad \theta = \frac{4\pi}{3} \qquad \theta = \pi$$

The solution set is $\left\{ \dfrac{2\pi}{3}, \pi, \dfrac{4\pi}{3} \right\}$.

Now Work PROBLEM 79

EXAMPLE 9	Solving a Trigonometric Equation Using Identities

Solve the equation: $\cos^2 \theta + \sin \theta = 2, \quad 0 \le \theta < 2\pi$

Solution This equation involves two trigonometric functions: sine and cosine. By using a Pythagorean Identity, we can express the equation in terms of just sine functions.

$$\cos^2 \theta + \sin \theta = 2$$

$$(1 - \sin^2 \theta) + \sin \theta = 2 \qquad \cos^2 \theta = 1 - \sin^2 \theta$$

$$\sin^2 \theta - \sin \theta + 1 = 0$$

This is a quadratic equation in $\sin \theta$. The discriminant is $b^2 - 4ac = 1 - 4 = -3 < 0$. Therefore, the equation has no real solution. The solution set is the empty set, $\varnothing$.

5 Solve Trigonometric Equations Using a Graphing Utility

The techniques introduced in this section apply only to certain types of trigonometric equations. Solutions for other types are usually studied in calculus, using numerical methods.

EXAMPLE 10	Solving a Trigonometric Equation Using a Graphing Utility

Solve: $5 \sin x + x = 3$

Express the solution(s) rounded to two decimal places.

Solution This trigonometric equation cannot be solved by previous methods. A graphing utility, though, can be used. Each solution of the equation is the x-coordinate of a point of intersection of the graphs of $Y_1 = 5 \sin x + x$ and $Y_2 = 3$. See Figure 25.

There are three points of intersection; the x-coordinates are the solutions of the equation. Use INTERSECT to find

$$x = 0.52 \qquad x = 3.18 \qquad x = 5.71$$

The solution set is $\{0.52, 3.18, 5.71\}$.

Figure 25

Now Work PROBLEM 85

3.3 Assess Your Understanding

'Are You Prepared?' *Answers are given at the end of these exercises. If you get a wrong answer, read the pages listed in* red.

1. Find the exact value of $\sec^2 \dfrac{\pi}{15} - \tan^2 \dfrac{\pi}{15}$. (pp. 137–139)

2. $\sin\left(\dfrac{\pi}{4}\right) = $ _____; $\cos\left(\dfrac{8\pi}{3}\right) = $ _____ (pp. 117–126)

3. Find the real solutions of $4x^2 - x - 5 = 0$. (pp. A27–A34)

4. Find the real solutions of $x^2 - x - 1 = 0$. (pp. A27–A34)

5. Find the real solutions of $(2x - 1)^2 - 3(2x - 1) - 4 = 0$. (pp. A27–A34)

6. Use a graphing utility to solve $5x^3 - 2 = x - x^2$. Round answers to two decimal places. (pp. B6–B8)

Concepts and Vocabulary

7. *True or False* Most trigonometric equations have unique solutions.

8. *True or False* Two solutions of the equation $\sin \theta = \dfrac{1}{2}$ are $\dfrac{\pi}{6}$ and $\dfrac{5\pi}{6}$.

9. *True or False* The solution set of the equation $\tan \theta = 1$ is given by $\left\{ \theta \middle| \theta = \dfrac{\pi}{4} + k\pi, k \text{ an integer} \right\}$.

10. *True or False* The equation $\sin \theta = 2$ has a real solution that can be found using a calculator.

11. *Multiple Choice* If all solutions of a trigonometric equation are given by the general formula $\theta = \dfrac{\pi}{6} + 2k\pi$ or $\theta = \dfrac{11\pi}{6} + 2k\pi$, where k is an integer, then which of the following is *not* a solution of the equation?

(a) $\dfrac{35\pi}{6}$ (b) $\dfrac{23\pi}{6}$ (c) $\dfrac{13\pi}{6}$ (d) $\dfrac{7\pi}{6}$

12. *Multiple Choice* Suppose $\theta = \dfrac{\pi}{2}$ is the only solution of a trigonometric equation in the interval $0 \le \theta < 2\pi$. Assuming a period of 2π, which of the following formulas gives all solutions of the equation, where k is an integer?

(a) $\theta = \dfrac{\pi}{2} + 2k\pi$ (b) $\theta = \dfrac{\pi}{2} + k\pi$

(c) $\theta = \dfrac{k\pi}{2}$ (d) $\theta = \dfrac{\pi + k\pi}{2}$

Skill Building

In Problems 13–36, solve each equation on the interval $0 \le \theta < 2\pi$.

13. $2 \sin \theta + 3 = 2$

14. $1 - \cos \theta = \dfrac{1}{2}$

15. $2 \sin \theta + 1 = 0$

16. $\cos \theta + 1 = 0$

17. $\tan \theta + 1 = 0$

18. $\sqrt{3} \cot \theta + 1 = 0$

19. $4 \sec \theta + 6 = -2$

20. $5 \csc \theta - 3 = 2$

21. $3\sqrt{2} \cos \theta + 2 = -1$

22. $4 \sin \theta + 3\sqrt{3} = \sqrt{3}$

23. $4 \cos^2 \theta = 1$

24. $\tan^2 \theta = \dfrac{1}{3}$

25. $2 \sin^2 \theta - 1 = 0$

26. $4 \cos^2 \theta - 3 = 0$

27. $\sin(3\theta) = -1$

28. $\tan \dfrac{\theta}{2} = \sqrt{3}$

29. $\cos(2\theta) = -\dfrac{1}{2}$

30. $\tan(2\theta) = -1$

31. $\sec \dfrac{3\theta}{2} = -2$

32. $\cot \dfrac{2\theta}{3} = -\sqrt{3}$

33. $\cos\left(2\theta - \dfrac{\pi}{2}\right) = -1$

34. $\sin\left(3\theta + \dfrac{\pi}{18}\right) = 1$

35. $\tan\left(\dfrac{\theta}{2} + \dfrac{\pi}{3}\right) = 1$

36. $\cos\left(\dfrac{\theta}{3} - \dfrac{\pi}{4}\right) = \dfrac{1}{2}$

In Problems 37–48, solve each equation. Give a general formula for all the solutions. List six solutions.

37. $\sin \theta = \dfrac{1}{2}$

38. $\tan \theta = 1$

39. $\tan \theta = -\dfrac{\sqrt{3}}{3}$

40. $\cos \theta = -\dfrac{\sqrt{3}}{2}$

41. $\cos \theta = 0$

42. $\sin \theta = \dfrac{\sqrt{2}}{2}$

43. $\sqrt{3} - \cot \theta = 0$

44. $2 - \sqrt{3} \csc \theta = 0$

45. $\cos(2\theta) = -\dfrac{1}{2}$

46. $\sin(2\theta) = -1$

47. $\sin \dfrac{\theta}{2} = -\dfrac{\sqrt{3}}{2}$

48. $\tan \dfrac{\theta}{2} = -1$

In Problems 49–60, use a calculator to solve each equation on the interval $0 \le \theta < 2\pi$. Round answers to two decimal places.

49. $\sin \theta = 0.4$

50. $\cos \theta = 0.6$

51. $\tan \theta = 5$

52. $\cot \theta = 2$

53. $\cos \theta = -0.9$

54. $\sin \theta = -0.2$

55. $\sec \theta = -4$

56. $\csc \theta = -3$

57. $5 \tan \theta + 9 = 0$

58. $4 \cot \theta = -5$

59. $3 \sin \theta - 2 = 0$

60. $4 \cos \theta + 3 = 0$

In Problems 61–84, solve each equation on the interval $0 \le \theta < 2\pi$.

61. $2 \cos^2 \theta + \cos \theta = 0$

62. $\sin^2 \theta - 1 = 0$

63. $2 \sin^2 \theta - \sin \theta - 1 = 0$

64. $2 \cos^2 \theta + \cos \theta - 1 = 0$

65. $(\tan \theta - 1)(\sec \theta - 1) = 0$

66. $(\cot \theta + 1)\left(\csc \theta - \dfrac{1}{2}\right) = 0$

67. $\sin^2 \theta - \cos^2 \theta = 1 + \cos \theta$

68. $\cos^2 \theta - \sin^2 \theta + \sin \theta = 0$

69. $\sin^2 \theta = 6(\cos(-\theta) + 1)$

70. $2 \sin^2 \theta = 3(1 - \cos(-\theta))$

71. $\cos \theta = -\sin(-\theta)$

72. $\cos \theta - \sin(-\theta) = 0$

73. $\tan \theta = 2 \sin \theta$

74. $\tan \theta = \cot \theta$

75. $1 + \sin \theta = 2 \cos^2 \theta$

76. $\sin^2 \theta = 2 \cos \theta + 2$

77. $2 \sin^2 \theta - 5 \sin \theta + 3 = 0$

78. $2 \cos^2 \theta - 7 \cos \theta - 4 = 0$

79. $3(1 - \cos \theta) = \sin^2 \theta$

80. $4(1 + \sin \theta) = \cos^2 \theta$

81. $\tan^2 \theta = \dfrac{3}{2} \sec \theta$

82. $\csc^2 \theta = \cot \theta + 1$

83. $\sec^2 \theta + \tan \theta = 0$

84. $\sec \theta = \tan \theta + \cot \theta$

In Problems 85–96, use a graphing utility to solve each equation. Express the solution(s) rounded to two decimal places.

85. $x + 5 \cos x = 0$

86. $x - 4 \sin x = 0$

87. $22x - 17 \sin x = 3$

88. $19x + 8 \cos x = 2$

89. $\sin x + \cos x = x$

90. $\sin x - \cos x = x$

91. $x^2 - 2 \cos x = 0$

92. $x^2 + 3 \sin x = 0$

93. $x^2 - 2 \sin(2x) = 3x$

94. $x^2 = x + 3 \cos(2x)$

95. $6 \sin x - e^x = 2, \quad x > 0$

96. $4 \cos(3x) - e^x = 1, \quad x > 0$

97. *Mixed Practice* What are the zeros of $f(x) = 4 \sin^2 x - 3$ on the interval $[0, 2\pi]$?

98. *Mixed Practice* What are the zeros of $f(x) = 2 \cos(3x) + 1$ on the interval $[0, \pi]$?

99. *Mixed Practice* $f(x) = 3 \sin x$
(a) Find the zeros of f on the interval $[-2\pi, 4\pi]$.
(b) Graph $f(x) = 3 \sin x$ on the interval $[-2\pi, 4\pi]$.
(c) Solve $f(x) = \dfrac{3}{2}$ on the interval $[-2\pi, 4\pi]$. What points are on the graph of f? Label these points on the graph drawn in part (b).
(d) Use the graph drawn in part (b) along with the results of part (c) to determine the values of x such that $f(x) > \dfrac{3}{2}$ on the interval $[-2\pi, 4\pi]$.

100. *Mixed Practice* $f(x) = 2 \cos x$
(a) Find the zeros of f on the interval $[-2\pi, 4\pi]$.
(b) Graph $f(x) = 2 \cos x$ on the interval $[-2\pi, 4\pi]$.
(c) Solve $f(x) = -\sqrt{3}$ on the interval $[-2\pi, 4\pi]$. What points are on the graph of f? Label these points on the graph drawn in part (b).
(d) Use the graph drawn in part (b) along with the results of part (c) to determine the values of x such that $f(x) < -\sqrt{3}$ on the interval $[-2\pi, 4\pi]$.

101. *Mixed Practice* $f(x) = 4 \tan x$
(a) Solve $f(x) = -4$.
(b) For what values of x is $f(x) < -4$ on the interval $\left(-\dfrac{\pi}{2}, \dfrac{\pi}{2}\right)$?

102. *Mixed Practice* $f(x) = \cot x$
(a) Solve $f(x) = -\sqrt{3}$.
(b) For what values of x is $f(x) > -\sqrt{3}$ on the interval $(0, \pi)$?

103. *Mixed Practice*
(a) Graph $f(x) = 3 \sin(2x) + 2$ and $g(x) = \dfrac{7}{2}$ on the same Cartesian plane for the interval $[0, \pi]$.
(b) Solve $f(x) = g(x)$ on the interval $[0, \pi]$, and label the points of intersection on the graph drawn in part (a).
(c) Solve $f(x) > g(x)$ on the interval $[0, \pi]$.
(d) Shade the region bounded by $f(x) = 3 \sin(2x) + 2$ and $g(x) = \dfrac{7}{2}$ between the two points found in part (b) on the graph drawn in part (a).

104. *Mixed Practice*
(a) Graph $f(x) = 2 \cos \dfrac{x}{2} + 3$ and $g(x) = 4$ on the same Cartesian plane for the interval $[0, 4\pi]$.
(b) Solve $f(x) = g(x)$ on the interval $[0, 4\pi]$, and label the points of intersection on the graph drawn in part (a).
(c) Solve $f(x) < g(x)$ on the interval $[0, 4\pi]$.
(d) Shade the region bounded by $f(x) = 2 \cos \dfrac{x}{2} + 3$ and $g(x) = 4$ between the two points found in part (b) on the graph drawn in part (a).

105. *Mixed Practice*
(a) Graph $f(x) = -4 \cos x$ and $g(x) = 2 \cos x + 3$ on the same Cartesian plane for the interval $[0, 2\pi]$.
(b) Solve $f(x) = g(x)$ on the interval $[0, 2\pi]$, and label the points of intersection on the graph drawn in part (a).
(c) Solve $f(x) > g(x)$ on the interval $[0, 2\pi]$.
(d) Shade the region bounded by $f(x) = -4 \cos x$ and $g(x) = 2 \cos x + 3$ between the two points found in part (b) on the graph drawn in part (a).

106. *Mixed Practice*
(a) Graph $f(x) = 2 \sin x$ and $g(x) = -2 \sin x + 2$ on the same Cartesian plane for the interval $[0, 2\pi]$.
(b) Solve $f(x) = g(x)$ on the interval $[0, 2\pi]$, and label the points of intersection on the graph drawn in part (a).
(c) Solve $f(x) > g(x)$ on the interval $[0, 2\pi]$.
(d) Shade the region bounded by $f(x) = 2 \sin x$ and $g(x) = -2 \sin x + 2$ between the two points found in part (b) on the graph drawn in part (a).

Applications and Extensions

107. Blood Pressure Several research papers use a sinusoidal graph to model blood pressure. Assuming that a person's heart beats 70 times per minute, the blood pressure P of an individual after t seconds can be modeled by the function

$$P(t) = 20 \sin\left(\frac{7\pi}{3}t\right) + 100$$

(a) In the interval $[0, 1]$, determine the times at which the blood pressure is 100 mmHg.
(b) In the interval $[0, 1]$, determine the times at which the blood pressure is 120 mmHg.
(c) In the interval $[0, 1]$, determine the times at which the blood pressure is between 100 and 105 mmHg.

108. The Ferris Wheel In 1893, George Ferris engineered the Ferris wheel. It was 250 feet in diameter. If a Ferris wheel makes 1 revolution every 40 seconds, then the function

$$h(t) = 125 \sin\left(0.157t - \frac{\pi}{2}\right) + 125$$

represents the height h, in feet, of a seat on the wheel as a function of time t, where t is measured in seconds. The ride begins when $t = 0$.
(a) During the first 40 seconds of the ride, at what time t is an individual on the Ferris wheel exactly 125 feet above the ground?
(b) During the first 80 seconds of the ride, at what time t is an individual on the Ferris wheel exactly 250 feet above the ground?
(c) During the first 40 seconds of the ride, over what interval of time t is an individual on the Ferris wheel more than 125 feet above the ground?

109. Holding Pattern An airplane is asked to stay within a holding pattern near Chicago's O'Hare International Airport. The function $d(x) = 70 \sin(0.65x) + 150$ represents the distance d, in miles, of the airplane from the airport at time x, in minutes.
(a) When the plane enters the holding pattern, $x = 0$, how far is it from O'Hare?
(b) During the first 20 minutes after the plane enters the holding pattern, at what time x is the plane exactly 100 miles from the airport?
(c) During the first 20 minutes after the plane enters the holding pattern, at what time x is the plane more than 100 miles from the airport?
(d) While the plane is in the holding pattern, will it ever be within 70 miles of the airport? Why?

110. Projectile Motion A golfer hits a golf ball with an initial velocity of 100 miles per hour. The range R of the ball as a function of the angle θ to the horizontal is given by $R(\theta) = 672 \sin(2\theta)$, where R is measured in feet.
(a) At what angle θ should the ball be hit if the golfer wants the ball to travel 450 feet (150 yards)?
(b) At what angle θ should the ball be hit if the golfer wants the ball to travel 540 feet (180 yards)?
(c) At what angle θ should the ball be hit if the golfer wants the ball to travel at least 480 feet (160 yards)?
(d) Can the golfer hit the ball 720 feet (240 yards)?

111. Heat Transfer In the study of heat transfer, the equation $x + \tan x = 0$ occurs. Graph $Y_1 = -x$ and $Y_2 = \tan x$ for $x \geq 0$. Conclude that there are an infinite number of

points of intersection of these two graphs. Now find the first two positive solutions of $x + \tan x = 0$ rounded to two decimal places.

112. Carrying a Ladder around a Corner Two hallways, one of width 3 feet, the other of width 4 feet, meet at a right angle. See the illustration. It can be shown that the length L of the ladder as a function of θ is $L(\theta) = 4 \csc \theta + 3 \sec \theta$.
(a) In calculus, you will be asked to find the length of the longest ladder that can turn the corner by solving the equation

$$3 \sec \theta \tan \theta - 4 \csc \theta \cot \theta = 0 \quad 0° < \theta < 90°$$

Solve this equation for θ.

(b) What is the length of the longest ladder that can be carried around the corner?
(c) Graph $L = L(\theta), 0° \leq \theta \leq 90°$, and find the angle θ that minimizes the length L.
(d) Compare the result with the one found in part (a). Explain why the two answers are the same.

113. Projectile Motion The horizontal distance that a projectile will travel in the air (ignoring air resistance) is given by the equation

$$R(\theta) = \frac{v_0^2 \sin(2\theta)}{g}$$

where v_0 is the initial velocity of the projectile, θ is the angle of elevation, and g is acceleration due to gravity (9.8 meters per second squared).
(a) If you can throw a baseball with an initial speed of 34.8 meters per second, at what angle of elevation θ should you direct the throw so that the ball travels a distance of 107 meters before striking the ground?
(b) Determine the maximum distance that you can throw the ball.
(c) Graph $R = R(\theta)$, with $v_0 = 34.8$ meters per second.
(d) Verify the results obtained in parts (a) and (b) using a graphing utility.

114. Projectile Motion Refer to Problem 113.
(a) If you can throw a baseball with an initial speed of 40 meters per second, at what angle of elevation θ should you direct the throw so that the ball travels a distance of 110 meters before striking the ground?
(b) Determine the maximum distance that you can throw the ball.
(c) Graph $R = R(\theta)$, with $v_0 = 40$ meters per second.
(d) Verify the results obtained in parts (a) and (b) using a graphing utility.

⚠ *The following discussion of Snell's Law of Refraction* (named after Willebrord Snell, 1580–1626) is needed for Problems 115–122. Light, sound, and other waves travel at different speeds, depending on the medium (air, water, wood, and so on) through which they pass. Suppose that light travels from a point A in one medium, where its speed is v_1, to a point B in another medium, where its speed is v_2. Refer to the figure, where the angle θ_1 is called the* angle of incidence *and the angle θ_2 is the* angle of refraction. *Snell's Law, which can be proved using calculus, states that*

$$\frac{\sin \theta_1}{\sin \theta_2} = \frac{v_1}{v_2}$$

The ratio $\dfrac{v_1}{v_2}$ is called the index of refraction. *Some values are given in the table shown below.*

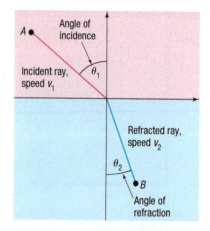

Some Indexes of Refraction	
Medium	**Index of Refraction†**
Water	1.33
Ethyl alcohol (20°C)	1.36
Carbon disulfide	1.63
Air (1 atm and 0°C)	1.00029
Diamond	2.42
Fused quartz	1.46
Glass, crown	1.52
Glass, dense flint	1.66
Sodium chloride	1.54

115. The index of refraction of light in passing from a vacuum into water is 1.33. If the angle of incidence is 40°, determine the angle of refraction.

116. The index of refraction of light in passing from a vacuum into dense flint glass is 1.66. If the angle of incidence is 50°, determine the angle of refraction.

117. Ptolemy, who lived in the city of Alexandria in Egypt during the second century AD, gave the measured values in the following table for the angle of incidence θ_1 and the angle of refraction θ_2 for a light beam passing from air into water. Do these values agree with Snell's Law? If so, what index of refraction results? (These data are of interest as the oldest recorded physical measurements.)

θ_1	θ_2	θ_1	θ_2
10°	8°	50°	35°0′
20°	15°30′	60°	40°30′
30°	22°30′	70°	45°30′
40°	29°0′	80°	50°0′

118. Bending Light The speed of yellow sodium light (wavelength, 589 nanometers) in a certain liquid is measured to be 1.92×10^8 meters per second. What is the index of refraction of this liquid, with respect to air, for sodium light? ‡

[**Hint:** The speed of light in air is approximately 2.998×10^8 meters per second.]

119. Bending Light A beam of light with a wavelength of 589 nanometers traveling in air makes an angle of incidence of 40° on a slab of transparent material, and the refracted beam makes an angle of refraction of 26°. Find the index of refraction of the material. ‡

120. Bending Light A light ray with a wavelength of 589 nanometers (produced by a sodium lamp) traveling through air makes an angle of incidence of 30° on a smooth, flat slab of crown glass. Find the angle of refraction. ‡

121. Brewster's Law If the angle of incidence and the angle of refraction are complementary angles, the angle of incidence is referred to as the Brewster angle θ_B. The Brewster angle is related to the indices of refraction of the two media, n_1 and n_2, by the equation $n_1 \sin \theta_B = n_2 \cos \theta_B$, where n_1 is the index of refraction of the incident medium and n_2 is the index of refraction of the refractive medium. Determine the Brewster angle for a light beam traveling through water (at 20°C) that makes an angle of incidence with a smooth, flat slab of crown glass.

122. Challenge Problem A light beam passes through a thick slab of material whose index of refraction is n_2. Show that the emerging beam is parallel to the incident beam. ‡

123. Challenge Problem Give the general formula for the solutions of the equation.

$$3 \sin \theta + \sqrt{3} \cos \theta = 0$$

124. Challenge Problem If $x^2 + (\tan \theta + \cot \theta)x + 1 = 0$ has two real solutions, $\{2 - \sqrt{3}, 2 + \sqrt{3}\}$, find $\sin \theta \cos \theta$.

---

*Because this law was also deduced by René Descartes in France, it is also known as Descartes' Law.

†For light of wavelength 589 nanometers, measured with respect to a vacuum. The index with respect to air is negligibly different in most cases.

‡Adapted from Halliday, Resnick, and Walker, *Fundamentals of Physics*, 10th ed., 2014, John Wiley & Sons.

Explaining Concepts: Discussion and Writing

125. Explain in your own words how you would use your calculator to solve the equation $\cos x = -0.6, 0 \le x < 2\pi$. How would you modify your approach to solve the equation $\cot x = 5, 0 < x < 2\pi$?

126. Explain why no further points of intersection (and therefore no further solutions) exist in Figure 25 for $x < -\pi$ or $x > 4\pi$.

Retain Your Knowledge

Problems 127–134 are based on material learned earlier in the course. The purpose of these problems is to keep the material fresh in your mind so that you are better prepared for the final exam.

127. Find the real zeros of $f(x) = 2x^2 - 9x + 8$.

128. If $\sin \theta = -\dfrac{\sqrt{10}}{10}$ and $\cos \theta = \dfrac{3\sqrt{10}}{10}$, find the exact value of each of the four remaining trigonometric functions.

129. Find the amplitude, period, and phase shift of the function $y = 2 \sin (2x - \pi)$. Graph the function. Show at least two periods.

130. Find the length of the arc of a circle of radius 15 centimeters subtended by a central angle of $36°$.

131. Find the value of a so that the line $ax - 3y = 10$ has slope 2.

132. Is the function $f(x) = \dfrac{3x}{5 - x^2}$ even, odd, or neither?

133. If $f(x) = \dfrac{8}{x^2}$, find an equation of the secant line containing the points $(1, f(1))$ and $(4, f(4))$.

134. Find the average rate of change of $f(x) = \cos^{-1}x$ from $\dfrac{1}{2}$ to 1.

'Are You Prepared?' Answers

1. 1   **2.** $\dfrac{\sqrt{2}}{2}; -\dfrac{1}{2}$ **3.** $\left\{-1, \dfrac{5}{4}\right\}$ **4.** $\left\{\dfrac{1 - \sqrt{5}}{2}, \dfrac{1 + \sqrt{5}}{2}\right\}$ **5.** $\left\{0, \dfrac{5}{2}\right\}$ **6.** {0.76}

3.4 Trigonometric Identities

PREPARING FOR THIS SECTION *Before getting started, review the following:*

- Fundamental Identities (Section 2.3, pp. 137–139)
- Even–Odd Properties (Section 2.3, pp. 142–143)

Now Work the 'Are You Prepared?' problems on page 224.

OBJECTIVES **1** Use Algebra to Simplify Trigonometric Expressions (p. 220)
                    2 Establish Identities (p. 221)

This section establishes additional identities involving trigonometric functions. First, let's define an *identity*.

DEFINITION **Identically Equal, Identity, and Conditional Equation**

Two functions f and g are **identically equal** if

$$f(x) = g(x)$$

for every value of x for which both functions are defined. Such an equation is referred to as an **identity**. An equation that is not an identity is called a **conditional equation**.

For example, the following are identities:

$$(x + 1)^2 = x^2 + 2x + 1 \qquad \sin^2 x + \cos^2 x = 1 \qquad \csc x = \frac{1}{\sin x}$$

The following are conditional equations:

$$2x + 5 = 0 \qquad \textbf{True only if } x = -\frac{5}{2}$$

$$\sin x = 0 \qquad \textbf{True only if } x = k\pi, \textbf{ } k \textbf{ an integer}$$

$$\sin x = \cos x \qquad \textbf{True only if } x = \frac{\pi}{4} + 2k\pi \textbf{ or } x = \frac{5\pi}{4} + 2k\pi, \textbf{ } k \textbf{ an integer}$$

Below are the trigonometric identities that have been established so far.

Quotient Identities

$$\tan\theta = \frac{\sin\theta}{\cos\theta} \qquad \cot\theta = \frac{\cos\theta}{\sin\theta}$$

Reciprocal Identities

$$\csc\theta = \frac{1}{\sin\theta} \qquad \sec\theta = \frac{1}{\cos\theta} \qquad \cot\theta = \frac{1}{\tan\theta}$$

Pythagorean Identities

$$\sin^2\theta + \cos^2\theta = 1 \qquad \tan^2\theta + 1 = \sec^2\theta$$

$$\cot^2\theta + 1 = \csc^2\theta$$

Even-Odd Identities

$$\sin(-\theta) = -\sin\theta \qquad \cos(-\theta) = \cos\theta \qquad \tan(-\theta) = -\tan\theta$$

$$\csc(-\theta) = -\csc\theta \qquad \sec(-\theta) = \sec\theta \qquad \cot(-\theta) = -\cot\theta$$

This list comprises what shall be referred to as the **basic trigonometric identities**. These identities should not merely be memorized, but should be *known* (just as you know your name rather than have it memorized). In fact, minor variations of a basic identity are often used. For example,

$$\sin^2\theta = 1 - \cos^2\theta \quad \text{or} \quad \cos^2\theta = 1 - \sin^2\theta$$

might be used instead of $\sin^2\theta + \cos^2\theta = 1$. For this reason, among others, it is very important to know these relationships and be comfortable with variations of them.

1 Use Algebra to Simplify Trigonometric Expressions

The ability to use algebra to manipulate trigonometric expressions is a key skill that one must have to establish identities. Four basic algebraic techniques are used to establish identities:

- Rewriting a trigonometric expression in terms of sine and cosine only
- Multiplying the numerator and denominator of a ratio by a "well-chosen 1"
- Writing sums of trigonometric ratios as a single ratio
- Factoring

EXAMPLE 1 **Using Algebraic Techniques to Simplify Trigonometric Expressions**

(a) Simplify $\dfrac{\cot \theta}{\csc \theta}$ by rewriting each trigonometric function in terms of sine and cosine functions.

(b) Show that $\dfrac{\cos \theta}{1 + \sin \theta} = \dfrac{1 - \sin \theta}{\cos \theta}$ by multiplying the numerator and denominator by $1 - \sin \theta$.

(c) Simplify $\dfrac{1 + \sin u}{\sin u} + \dfrac{\cot u - \cos u}{\cos u}$ by rewriting the expression as a single ratio.

(d) Simplify $\dfrac{\sin^2 v - 1}{\tan v \sin v - \tan v}$ by factoring.

Solution

(a) $\dfrac{\cot \theta}{\csc \theta} = \dfrac{\dfrac{\cos \theta}{\sin \theta}}{\dfrac{1}{\sin \theta}} = \dfrac{\cos \theta}{\sin \theta} \cdot \dfrac{\sin \theta}{1} = \cos \theta$

(b) $\dfrac{\cos \theta}{1 + \sin \theta} = \dfrac{\cos \theta}{1 + \sin \theta} \cdot \dfrac{1 - \sin \theta}{1 - \sin \theta} = \dfrac{\cos \theta (1 - \sin \theta)}{1 - \sin^2 \theta}$

$\uparrow$
Multiply by a well-chosen 1: $\dfrac{1 - \sin \theta}{1 - \sin \theta}$.

$= \dfrac{\cos \theta (1 - \sin \theta)}{\cos^2 \theta} = \dfrac{1 - \sin \theta}{\cos \theta}$

(c) $\dfrac{1 + \sin u}{\sin u} + \dfrac{\cot u - \cos u}{\cos u} = \dfrac{1 + \sin u}{\sin u} \cdot \dfrac{\cos u}{\cos u} + \dfrac{\cot u - \cos u}{\cos u} \cdot \dfrac{\sin u}{\sin u}$

$= \dfrac{\cos u + \sin u \cos u + \cot u \sin u - \cos u \sin u}{\sin u \cos u} = \dfrac{\cos u + \cot u \sin u}{\sin u \cos u}$

$= \dfrac{\cos u + \dfrac{\cos u}{\sin u} \cdot \sin u}{\sin u \cos u} = \dfrac{\cos u + \cos u}{\sin u \cos u} = \dfrac{2 \cos u}{\sin u \cos u} = \dfrac{2}{\sin u}$

$\uparrow$
$\cot u = \dfrac{\cos u}{\sin u}$

(d) $\dfrac{\sin^2 v - 1}{\tan v \sin v - \tan v} = \dfrac{(\sin v + 1)(\sin v - 1)}{\tan v (\sin v - 1)} = \dfrac{\sin v + 1}{\tan v}$

◢

Now Work PROBLEMS **11, 13,** AND **15**

2 Establish Identities

In the examples that follow, the directions read "Establish the identity . . . ." This is accomplished by starting with one side of the given equation (usually the side containing the more complicated expression) and, using appropriate basic identities and algebraic manipulations, arriving at the other side. The selection of appropriate basic identities to obtain the desired result is learned only through experience and lots of practice.

EXAMPLE 2 **Establishing an Identity**

Establish the identity: $\csc \theta \cdot \tan \theta = \sec \theta$

Solution Start with the left side, because it contains the more complicated expression. Then use a reciprocal identity and a quotient identity.

$$\csc \theta \cdot \tan \theta = \frac{1}{\sin \theta} \cdot \frac{\sin \theta}{\cos \theta} = \frac{1}{\cos \theta} = \sec \theta$$

The right side has been reached, so the identity is established.

➝ **Now Work** PROBLEM 21

EXAMPLE 3 **Establishing an Identity**

Establish the identity: $\sin^2(-\theta) + \cos^2(-\theta) = 1$

Solution Begin with the left side and, because the arguments are $-\theta$, use Even-Odd Identities.

$$\begin{aligned}
\sin^2(-\theta) + \cos^2(-\theta) &= [\sin(-\theta)]^2 + [\cos(-\theta)]^2 \\
&= (-\sin \theta)^2 + (\cos \theta)^2 &&\text{Even-Odd Identities} \\
&= (\sin \theta)^2 + (\cos \theta)^2 \\
&= 1 &&\text{Pythagorean Identity}
\end{aligned}$$

EXAMPLE 4 **Establishing an Identity**

Establish the identity: $\dfrac{\sin^2(-\theta) - \cos^2(-\theta)}{\sin(-\theta) - \cos(-\theta)} = \cos \theta - \sin \theta$

Solution The left side contains the more complicated expression. Also, the left side contains expressions with the argument $-\theta$, whereas the right side contains expressions with the argument θ. So start with the left side and use Even-Odd Identities.

$$\begin{aligned}
\frac{\sin^2(-\theta) - \cos^2(-\theta)}{\sin(-\theta) - \cos(-\theta)} &= \frac{[\sin(-\theta)]^2 - [\cos(-\theta)]^2}{\sin(-\theta) - \cos(-\theta)} \\
&= \frac{(-\sin \theta)^2 - (\cos \theta)^2}{-\sin \theta - \cos \theta} &&\text{Even-Odd Identities} \\
&= \frac{(\sin \theta)^2 - (\cos \theta)^2}{-\sin \theta - \cos \theta} &&\text{Simplify.} \\
&= \frac{(\sin \theta - \cos \theta)(\sin \theta + \cos \theta)}{-(\sin \theta + \cos \theta)} &&\text{Factor.} \\
&= \cos \theta - \sin \theta &&\text{Cancel and simplify.}
\end{aligned}$$

EXAMPLE 5 **Establishing an Identity**

Establish the identity: $\dfrac{1 + \tan u}{1 + \cot u} = \tan u$

Solution $\dfrac{1 + \tan u}{1 + \cot u} = \dfrac{1 + \tan u}{1 + \dfrac{1}{\tan u}} = \dfrac{1 + \tan u}{\dfrac{\tan u + 1}{\tan u}} = \dfrac{\tan u(1 + \tan u)}{\tan u + 1} = \tan u$

➝ **Now Work** PROBLEMS 25 AND 29

When sums or differences of quotients appear, it is usually best to rewrite them as a single quotient, especially if the other side of the identity consists of only one term.

EXAMPLE 6   **Establishing an Identity**

Establish the identity: $\dfrac{\sin\theta}{1+\cos\theta} + \dfrac{1+\cos\theta}{\sin\theta} = 2\csc\theta$

Solution   The left side is more complicated. Start with it and add.

$$\frac{\sin\theta}{1+\cos\theta} + \frac{1+\cos\theta}{\sin\theta} = \frac{\sin^2\theta + (1+\cos\theta)^2}{(1+\cos\theta)\cdot\sin\theta}$$ Add the quotients.

$$= \frac{\sin^2\theta + 1 + 2\cos\theta + \cos^2\theta}{(1+\cos\theta)\cdot\sin\theta}$$ Multiply out in the numerator.

$$= \frac{(\sin^2\theta + \cos^2\theta) + 1 + 2\cos\theta}{(1+\cos\theta)\cdot\sin\theta}$$ Regroup.

$$= \frac{2 + 2\cos\theta}{(1+\cos\theta)\cdot\sin\theta}$$ $\sin^2\theta + \cos^2\theta = 1$

$$= \frac{2(1+\cos\theta)}{(1+\cos\theta)\cdot\sin\theta}$$ Factor and cancel.

$$= \frac{2}{\sin\theta}$$

$$= 2\csc\theta$$ Reciprocal Identity

Now Work PROBLEM **51**

Sometimes it helps to write one side in terms of sine and cosine functions only.

EXAMPLE 7   **Establishing an Identity**

Establish the identity: $\dfrac{\tan v + \cot v}{\sec v \csc v} = 1$

Solution $\dfrac{\tan v + \cot v}{\sec v \csc v} = \dfrac{\dfrac{\sin v}{\cos v} + \dfrac{\cos v}{\sin v}}{\dfrac{1}{\cos v}\cdot\dfrac{1}{\sin v}} = \dfrac{\dfrac{\sin^2 v + \cos^2 v}{\cos v \sin v}}{\dfrac{1}{\cos v \sin v}}$

Change to sines and cosines.   Add the quotients in the numerator.

$$= \frac{1}{\cos v \sin v}\cdot\frac{\cos v \sin v}{1} = 1$$

Divide the quotients; $\sin^2 v + \cos^2 v = 1$.

Now Work PROBLEM **71**

Sometimes, multiplying the numerator and the denominator by an appropriate factor simplifies an expression.

| EXAMPLE 8 | Establishing an Identity |

Establish the identity: $\dfrac{1 - \sin \theta}{\cos \theta} = \dfrac{\cos \theta}{1 + \sin \theta}$

Solution Start with the left side and multiply the numerator and the denominator by $1 + \sin \theta$. (Alternatively, we could multiply the numerator and the denominator of the right side by $1 - \sin \theta$.)

$$\frac{1 - \sin \theta}{\cos \theta} = \frac{1 - \sin \theta}{\cos \theta} \cdot \frac{1 + \sin \theta}{1 + \sin \theta} \qquad \text{\textbf{Multiply the numerator and the}}$$
$$\text{\textbf{denominator by } } 1 + \sin \theta.$$

$$= \frac{1 - \sin^2 \theta}{\cos \theta (1 + \sin \theta)}$$

$$= \frac{\cos^2 \theta}{\cos \theta (1 + \sin \theta)} \qquad \mathbf{1 - \sin^2 \theta = \cos^2 \theta}$$

$$= \frac{\cos \theta}{1 + \sin \theta} \qquad \text{\textbf{Cancel.}}$$

Now Work PROBLEM 55

Although practice is the only real way to learn how to establish identities, the following guidelines should prove helpful.

WARNING Do not try to establish an identity by treating it as an equation. We cannot add or multiply both sides by the same expression because we do not know if the sides are equal. That is what we are trying to prove. ∎

> **Guidelines for Establishing Identities**
>
> - It is almost always preferable to start with the side containing the more complicated expression.
> - Rewrite sums or differences of quotients as a single quotient.
> - Sometimes it helps to rewrite one side in terms of sine and cosine functions only.
> - Always keep the goal in mind. As you manipulate one side of the expression, keep in mind the form of the expression on the other side.

3.4 Assess Your Understanding

'Are You Prepared?' *Answers are given at the end of these exercises. If you get a wrong answer, read the pages listed in* red.

1. *True or False* $\sin^2 \theta = 1 - \cos^2 \theta$. (p. 139)

2. *True or False* $\sin(-\theta) + \cos(-\theta) = \cos \theta - \sin \theta$. (p. 142)

Concepts and Vocabulary

3. Suppose that f and g are two functions with the same domain. If $f(x) = g(x)$ for every x in the domain, the equation is called a(n) _____. Otherwise, it is called a(n) _____ equation.

4. $\tan^2 \theta - \sec^2 \theta = $ _____.

5. $\cos(-\theta) - \cos \theta = $ ____.

6. *True or False* $\sin(-\theta) + \sin \theta = 0$ for any value of θ.

7. *True or False* In establishing an identity, it is often easiest to just multiply both sides by a well-chosen nonzero expression involving the variable.

8. *True or False* $\tan \theta \cdot \cos \theta = \sin \theta$ for any $\theta \ne (2k + 1) \dfrac{\pi}{2}$.

9. *Multiple Choice* Which of the following equations is *not* an identity?
   (a) $\cot^2 \theta + 1 = \csc^2 \theta$
   (b) $\tan(-\theta) = -\tan \theta$
   (c) $\tan \theta = \dfrac{\cos \theta}{\sin \theta}$
   (d) $\csc \theta = \dfrac{1}{\sin \theta}$

10. *Multiple Choice* The expression $\dfrac{1}{1 - \sin \theta} + \dfrac{1}{1 + \sin \theta}$ simplifies to which of the following?
    (a) $2 \cos^2 \theta$ (b) $2 \sec^2 \theta$ (c) $2 \sin^2 \theta$ (d) $2 \csc^2 \theta$

Skill Building

In Problems 11–20, simplify each trigonometric expression by following the indicated direction.

11. Rewrite in terms of sine and cosine functions:
$$\tan\theta \cdot \csc\theta.$$

12. Rewrite in terms of sine and cosine functions:
$$\cot\theta \cdot \sec\theta.$$

13. Multiply $\dfrac{\cos\theta}{1-\sin\theta}$ by $\dfrac{1+\sin\theta}{1+\sin\theta}$.

14. Multiply $\dfrac{\sin\theta}{1+\cos\theta}$ by $\dfrac{1-\cos\theta}{1-\cos\theta}$.

15. Rewrite as a single quotient:
$$\frac{\sin\theta+\cos\theta}{\cos\theta}+\frac{\cos\theta-\sin\theta}{\sin\theta}$$

16. Rewrite as a single quotient:
$$\frac{1}{1-\cos v}+\frac{1}{1+\cos v}$$

17. Multiply and simplify: $\dfrac{(\sin\theta+\cos\theta)(\sin\theta+\cos\theta)-1}{\sin\theta\cos\theta}$

18. Multiply and simplify: $\dfrac{(\tan\theta+1)(\tan\theta+1)-\sec^2\theta}{\tan\theta}$

19. Factor and simplify: $\dfrac{3\sin^2\theta+4\sin\theta+1}{\sin^2\theta+2\sin\theta+1}$

20. Factor and simplify: $\dfrac{\cos^2\theta-1}{\cos^2\theta-\cos\theta}$

In Problems 21–100, establish each identity.

21. $\csc\theta\cdot\cos\theta=\cot\theta$

22. $\sec\theta\cdot\sin\theta=\tan\theta$

23. $1+\tan^2(-\theta)=\sec^2\theta$

24. $1+\cot^2(-\theta)=\csc^2\theta$

25. $\cos\theta(\tan\theta+\cot\theta)=\csc\theta$

26. $\sin\theta(\cot\theta+\tan\theta)=\sec\theta$

27. $\tan u\cot u-\cos^2 u=\sin^2 u$

28. $\sin u\csc u-\cos^2 u=\sin^2 u$

29. $(\sec\theta-1)(\sec\theta+1)=\tan^2\theta$

30. $(\csc\theta-1)(\csc\theta+1)=\cot^2\theta$

31. $(\sec\theta+\tan\theta)(\sec\theta-\tan\theta)=1$

32. $(\csc\theta+\cot\theta)(\csc\theta-\cot\theta)=1$

33. $\cos^2\theta(1+\tan^2\theta)=1$

34. $(1-\cos^2\theta)(1+\cot^2\theta)=1$

35. $(\sin\theta+\cos\theta)^2+(\sin\theta-\cos\theta)^2=2$

36. $\tan^2\theta\cos^2\theta+\cot^2\theta\sin^2\theta=1$

37. $\sec^4\theta-\sec^2\theta=\tan^4\theta+\tan^2\theta$

38. $\csc^4\theta-\csc^2\theta=\cot^4\theta+\cot^2\theta$

39. $\sec u-\tan u=\dfrac{\cos u}{1+\sin u}$

40. $\csc u-\cot u=\dfrac{\sin u}{1+\cos u}$

41. $3\sin^2\theta+4\cos^2\theta=3+\cos^2\theta$

42. $9\sec^2\theta-5\tan^2\theta=5+4\sec^2\theta$

43. $1-\dfrac{\cos^2\theta}{1+\sin\theta}=\sin\theta$

44. $1-\dfrac{\sin^2\theta}{1-\cos\theta}=-\cos\theta$

45. $\dfrac{1+\tan v}{1-\tan v}=\dfrac{\cot v+1}{\cot v-1}$

46. $\dfrac{\csc v-1}{\csc v+1}=\dfrac{1-\sin v}{1+\sin v}$

47. $\dfrac{\sec\theta}{\csc\theta}+\dfrac{\sin\theta}{\cos\theta}=2\tan\theta$

48. $\dfrac{\csc\theta-1}{\cot\theta}=\dfrac{\cot\theta}{\csc\theta+1}$

49. $\dfrac{1+\sin\theta}{1-\sin\theta}=\dfrac{\csc\theta+1}{\csc\theta-1}$

50. $\dfrac{\cos\theta+1}{\cos\theta-1}=\dfrac{1+\sec\theta}{1-\sec\theta}$

51. $\dfrac{1-\sin v}{\cos v}+\dfrac{\cos v}{1-\sin v}=2\sec v$

52. $\dfrac{\cos v}{1+\sin v}+\dfrac{1+\sin v}{\cos v}=2\sec v$

53. $\dfrac{\sin\theta}{\sin\theta-\cos\theta}=\dfrac{1}{1-\cot\theta}$

54. $1-\dfrac{\sin^2\theta}{1+\cos\theta}=\cos\theta$

55. $\dfrac{1-\sin\theta}{1+\sin\theta}=(\sec\theta-\tan\theta)^2$

56. $\dfrac{1-\cos\theta}{1+\cos\theta}=(\csc\theta-\cot\theta)^2$

57. $\dfrac{\cos\theta}{1-\tan\theta}+\dfrac{\sin\theta}{1-\cot\theta}=\sin\theta+\cos\theta$

58. $\dfrac{\cot\theta}{1-\tan\theta}+\dfrac{\tan\theta}{1-\cot\theta}=1+\tan\theta+\cot\theta$

59. $\tan\theta+\dfrac{\cos\theta}{1+\sin\theta}=\sec\theta$

60. $\dfrac{\sin\theta\cos\theta}{\cos^2\theta-\sin^2\theta}=\dfrac{\tan\theta}{1-\tan^2\theta}$

61. $\dfrac{\tan\theta+\sec\theta-1}{\tan\theta-\sec\theta+1}=\tan\theta+\sec\theta$

62. $\dfrac{\sin\theta-\cos\theta+1}{\sin\theta+\cos\theta-1}=\dfrac{\sin\theta+1}{\cos\theta}$

63. $\dfrac{\tan\theta-\cot\theta}{\tan\theta+\cot\theta}=\sin^2\theta-\cos^2\theta$

64. $\dfrac{\sec\theta-\cos\theta}{\sec\theta+\cos\theta}=\dfrac{\sin^2\theta}{1+\cos^2\theta}$

65. $\dfrac{\tan u-\cot u}{\tan u+\cot u}+1=2\sin^2 u$

66. $\dfrac{\tan u-\cot u}{\tan u+\cot u}+2\cos^2 u=1$

67. $\dfrac{\sec\theta+\tan\theta}{\cot\theta+\cos\theta}=\tan\theta\sec\theta$

68. $\dfrac{\sec\theta}{1+\sec\theta}=\dfrac{1-\cos\theta}{\sin^2\theta}$

69. $\dfrac{1-\tan^2\theta}{1+\tan^2\theta}+1=2\cos^2\theta$

70. $\dfrac{1-\cot^2\theta}{1+\cot^2\theta}+2\cos^2\theta=1$

71. $\dfrac{\sec\theta-\csc\theta}{\sec\theta\csc\theta}=\sin\theta-\cos\theta$

72. $\dfrac{\sin^2\theta-\tan\theta}{\cos^2\theta-\cot\theta}=\tan^2\theta$

73. $\sec\theta-\cos\theta=\sin\theta\tan\theta$

74. $\tan\theta+\cot\theta=\sec\theta\csc\theta$

75. $\dfrac{1}{1-\sin\theta}+\dfrac{1}{1+\sin\theta}=2\sec^2\theta$

76. $\dfrac{1+\sin\theta}{1-\sin\theta}-\dfrac{1-\sin\theta}{1+\sin\theta}=4\tan\theta\sec\theta$

77. $\dfrac{\sec \theta}{1 - \sin \theta} = \dfrac{1 + \sin \theta}{\cos^3 \theta}$

78. $\dfrac{1 + \sin \theta}{1 - \sin \theta} = (\sec \theta + \tan \theta)^2$

79. $\dfrac{(\sec v - \tan v)^2 + 1}{\csc v (\sec v - \tan v)} = 2 \tan v$

80. $\dfrac{\sec^2 v - \tan^2 v + \tan v}{\sec v} = \sin v + \cos v$

81. $\dfrac{\sin \theta + \cos \theta}{\cos \theta} - \dfrac{\sin \theta - \cos \theta}{\sin \theta} = \sec \theta \csc \theta$

82. $\dfrac{\sin \theta + \cos \theta}{\sin \theta} - \dfrac{\cos \theta - \sin \theta}{\cos \theta} = \sec \theta \csc \theta$

83. $\dfrac{\sin^3 \theta + \cos^3 \theta}{\sin \theta + \cos \theta} = 1 - \sin \theta \cos \theta$

84. $\dfrac{\sin^3 \theta + \cos^3 \theta}{1 - 2 \cos^2 \theta} = \dfrac{\sec \theta - \sin \theta}{\tan \theta - 1}$

85. $\dfrac{\cos^2 \theta - \sin^2 \theta}{1 - \tan^2 \theta} = \cos^2 \theta$

86. $\dfrac{\cos \theta + \sin \theta - \sin^3 \theta}{\sin \theta} = \cot \theta + \cos^2 \theta$

87. $\dfrac{(2 \cos^2 \theta - 1)^2}{\cos^4 \theta - \sin^4 \theta} = 1 - 2 \sin^2 \theta$

88. $\dfrac{1 - 2 \cos^2 \theta}{\sin \theta \cos \theta} = \tan \theta - \cot \theta$

89. $\dfrac{1 + \sin \theta + \cos \theta}{1 + \sin \theta - \cos \theta} = \dfrac{1 + \cos \theta}{\sin \theta}$

90. $\dfrac{1 + \cos \theta + \sin \theta}{1 + \cos \theta - \sin \theta} = \sec \theta + \tan \theta$

91. $(a \sin \theta + b \cos \theta)^2 + (a \cos \theta - b \sin \theta)^2 = a^2 + b^2$

92. $(2a \sin \theta \cos \theta)^2 + a^2 (\cos^2 \theta - \sin^2 \theta)^2 = a^2$

93. $\dfrac{\tan \alpha + \tan \beta}{\cot \alpha + \cot \beta} = \tan \alpha \tan \beta$

94. $(\tan \alpha + \tan \beta)(1 - \cot \alpha \cot \beta) + (\cot \alpha + \cot \beta)(1 - \tan \alpha \tan \beta) = 0$

95. $(\sin \alpha + \cos \beta)^2 + (\cos \beta + \sin \alpha)(\cos \beta - \sin \alpha) = 2 \cos \beta (\sin \alpha + \cos \beta)$

96. $(\sin \alpha - \cos \beta)^2 + (\cos \beta + \sin \alpha)(\cos \beta - \sin \alpha) = -2 \cos \beta (\sin \alpha - \cos \beta)$

97. $\ln |\sec \theta| = -\ln |\cos \theta|$

98. $\ln |\tan \theta| = \ln |\sin \theta| - \ln |\cos \theta|$

99. $\ln |1 + \cos \theta| + \ln |1 - \cos \theta| = 2 \ln |\sin \theta|$

100. $\ln |\sec \theta + \tan \theta| + \ln |\sec \theta - \tan \theta| = 0$

In Problems 101–104, show that the functions f and g are identically equal.

101. $f(x) = \sin x \cdot \tan x \qquad g(x) = \sec x - \cos x$

102. $f(x) = \cos x \cdot \cot x \qquad g(x) = \csc x - \sin x$

103. $f(\theta) = \dfrac{1 - \sin \theta}{\cos \theta} - \dfrac{\cos \theta}{1 + \sin \theta} \qquad g(\theta) = 0$

104. $f(\theta) = \tan \theta + \sec \theta \qquad g(\theta) = \dfrac{\cos \theta}{1 - \sin \theta}$

105. Show that $\sqrt{16 + 6 \tan^2 \theta} = 4 \sec \theta$ if $-\dfrac{\pi}{2} < \theta < \dfrac{\pi}{2}$.

106. Show that $\sqrt{9 \sec^2 \theta - 9} = 3 \tan \theta$ if $\pi \le \theta < \dfrac{3\pi}{2}$.

Applications and Extensions

107. Searchlights A searchlight at the grand opening of a new car dealership casts a spot of light on a wall located 75 meters from the searchlight. The acceleration $\ddot{r}$ of the spot of light is found to be $\ddot{r} = 1200 \sec \theta (2 \sec^2 \theta - 1)$. Show that this is equivalent to $\ddot{r} = 1200 \left(\dfrac{1 + \sin^2 \theta}{\cos^3 \theta} \right)$.

Source: Adapted from Hibbeler, *Engineering Mechanics: Dynamics,* 13th ed., Pearson © 2013.

108. Optical Measurement Optical methods of measurement often rely on the interference of two light waves. If two light waves, identical except for a phase lag, are mixed together, the resulting intensity, or irradiance, is given by
$$I_t = 4A^2 \dfrac{(\csc \theta - 1)(\sec \theta + \tan \theta)}{\csc \theta \sec \theta}.$$
Show that this is equivalent to $I_t = (2A \cos \theta)^2$.

Source: Experimental Techniques, July/August 2002

109. Challenge Problem Prove: $\sin^{-1}(-x) = -\sin^{-1}x$

110. Challenge Problem Prove: $\cot^{-1}x = \tan^{-1}\left(\dfrac{1}{x}\right)$

Explaining Concepts: Discussion and Writing

111. Write a few paragraphs outlining your strategy for establishing identities.

112. Write down the three Pythagorean Identities.

113. Why do you think it is usually preferable to start with the side containing the more complicated expression when establishing an identity?

114. Make up an identity that is not a basic identity.

Retain Your Knowledge

Problems 115–123 are based on material learned earlier in the course. The purpose of these problems is to keep the material fresh in your mind so that you are better prepared for the final exam.

115. Determine whether $f(x) = -3x^2 + 120x + 50$ has a maximum or a minimum value, and then find the value.

116. If $f(x) = \dfrac{x + 1}{x - 2}$, find $f(3x - 4)$.

117. Find the exact values of the six trigonometric functions of an angle θ in standard position if $(-12, 5)$ is a point on its terminal side.

118. Find the average rate of change of $f(x) = \cos x$ from 0 to $\dfrac{\pi}{2}$.

119. Find the length of a line segment with endpoints $(-3, -4)$ and $(5, 8)$.

120. Find the area of the sector of a circle of radius 8 meters formed by an angle of $54°$.

121. **Kayaking** Ben paddled his kayak 8 miles upstream against a 1 mile per hour current and back again in 6 hours. How far could Ben have paddled in that time if there had been no current?

122. If an angle θ lies in quadrant III and $\cot \theta = \dfrac{8}{5}$, find $\sec \theta$.

123. Write the equation of the circle in standard form: $x^2 + y^2 - 12x + 4y + 31 = 0$

'Are You Prepared?' Answers

1. True   2. True

3.5 Sum and Difference Formulas

PREPARING FOR THIS SECTION *Before getting started, review the following:*

- Distance Formula (Section 1.1, pp. 3–4)
- Values of the Trigonometric Functions (Section 2.2, pp. 117–126)
- Congruent Triangles (Section A.2, pp. A16–A17)

- Finding Exact Values Given the Value of a Trigonometric Function and the Quadrant of the Angle (Section 2.3, pp. 139–142)

Now Work the 'Are You Prepared?' problems on page 236.

OBJECTIVES 1 Use Sum and Difference Formulas to Find Exact Values (p. 228)
2 Use Sum and Difference Formulas to Establish Identities (p. 232)
3 Use Sum and Difference Formulas Involving Inverse Trigonometric Functions (p. 233)
4 Solve Trigonometric Equations Linear in Sine and Cosine (p. 234)

This section continues the derivation of trigonometric identities by obtaining formulas that involve the sum or the difference of two angles, such as $\cos(\alpha + \beta)$, $\cos(\alpha - \beta)$, and $\sin(\alpha + \beta)$. These formulas are referred to as the *Sum and Difference Formulas*. We begin with the formulas for $\cos(\alpha + \beta)$ and $\cos(\alpha - \beta)$.

THEOREM Sum and Difference Formulas for the Cosine Function

$$\cos(\alpha + \beta) = \cos \alpha \cos \beta - \sin \alpha \sin \beta \qquad (1)$$

$$\cos(\alpha - \beta) = \cos \alpha \cos \beta + \sin \alpha \sin \beta \qquad (2)$$

In Words

Formula (1) states that the cosine of the sum of two angles equals the cosine of the first angle times the cosine of the second angle minus the sine of the first angle times the sine of the second angle.

Proof We prove formula (2) first. Although the formula is true for all numbers α and β, we assume in our proof that $0 < \beta < \alpha < 2\pi$. Begin with the unit circle and place the angles α and β in standard position, as shown in Figure 26(a) on the next page. The point P_1 lies on the terminal side of β, so its coordinates are $(\cos \beta, \sin \beta)$; and the point P_2 lies on the terminal side of α, so its coordinates are $(\cos \alpha, \sin \alpha)$.

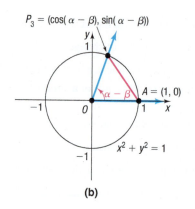

Figure 26   (a)   (b)

Now place the angle $\alpha - \beta$ in standard position, as shown in Figure 26(b). The point A has coordinates $(1, 0)$, and the point P_3 is on the terminal side of the angle $\alpha - \beta$, so its coordinates are $(\cos(\alpha - \beta), \sin(\alpha - \beta))$.

Looking at triangle OP_1P_2 in Figure 26(a) and triangle OAP_3 in Figure 26(b), note that these triangles are congruent. (Do you see why? SAS: two sides and the included angle, $\alpha - \beta$, are equal.) As a result, the unknown side of triangle OP_1P_2 and the unknown side of triangle OAP_3 must be equal; that is,

$$d(A, P_3) = d(P_1, P_2)$$

Now use the distance formula to obtain

$$\sqrt{[\cos(\alpha - \beta) - 1]^2 + [\sin(\alpha - \beta) - 0]^2} = \sqrt{(\cos\alpha - \cos\beta)^2 + (\sin\alpha - \sin\beta)^2} \quad \textbf{\textit{d}(A, P_3) = d(P_1, P_2)}$$

$$[\cos(\alpha - \beta) - 1]^2 + \sin^2(\alpha - \beta) = (\cos\alpha - \cos\beta)^2 + (\sin\alpha - \sin\beta)^2 \quad \textbf{Square both sides.}$$

$$\cos^2(\alpha - \beta) - 2\cos(\alpha - \beta) + 1 + \sin^2(\alpha - \beta) = \cos^2\alpha - 2\cos\alpha\cos\beta + \cos^2\beta \quad \textbf{Multiply out the squared}$$
$$+ \sin^2\alpha - 2\sin\alpha\sin\beta + \sin^2\beta \qquad \textbf{terms.}$$

Use a Pythagorean Identity (3 times).

$$2 - 2\cos(\alpha - \beta) = 2 - 2\cos\alpha\cos\beta - 2\sin\alpha\sin\beta$$

$$-2\cos(\alpha - \beta) = -2\cos\alpha\cos\beta - 2\sin\alpha\sin\beta \quad \textbf{Subtract 2 from both sides.}$$

$$\cos(\alpha - \beta) = \cos\alpha\cos\beta + \sin\alpha\sin\beta \quad \textbf{Divide both sides by } -2.$$

This is formula (2).

The proof of the Sum Formula for cosine follows from the Difference Formula for cosine and the Even-Odd Identities. Because $\alpha + \beta = \alpha - (-\beta)$, it follows that

$$\cos(\alpha + \beta) = \cos[\alpha - (-\beta)]$$
$$= \cos\alpha\cos(-\beta) + \sin\alpha\sin(-\beta) \quad \textbf{Use the Difference Formula for cosine.}$$
$$= \cos\alpha\cos\beta - \sin\alpha\sin\beta \quad \textbf{Use the Even-Odd Identities.}$$

■

1 Use Sum and Difference Formulas to Find Exact Values

One use of the Sum and Difference Formulas is to obtain the exact value of the cosine of an angle that can be expressed as the sum or difference of angles whose sine and cosine are known exactly.

EXAMPLE 1   **Using the Sum Formula to Find an Exact Value**

Find the exact value of $\cos 75°$.

Solution   Because $75° = 45° + 30°$, use formula (1) to obtain

$$\cos 75° = \cos(45° + 30°) = \cos 45° \cos 30° - \sin 45° \sin 30°$$

↑ **Sum Formula for cosine**

$$= \frac{\sqrt{2}}{2} \cdot \frac{\sqrt{3}}{2} - \frac{\sqrt{2}}{2} \cdot \frac{1}{2} = \frac{1}{4}(\sqrt{6} - \sqrt{2})$$

EXAMPLE 2   **Using the Difference Formula to Find an Exact Value**

Find the exact value of $\cos\dfrac{\pi}{12}$.

Solution
$$\cos\frac{\pi}{12} = \cos\left(\frac{3\pi}{12} - \frac{2\pi}{12}\right) = \cos\left(\frac{\pi}{4} - \frac{\pi}{6}\right)$$

$$= \cos\frac{\pi}{4}\cos\frac{\pi}{6} + \sin\frac{\pi}{4}\sin\frac{\pi}{6} \qquad \text{Use the Difference Formula for cosine.}$$

$$= \frac{\sqrt{2}}{2}\cdot\frac{\sqrt{3}}{2} + \frac{\sqrt{2}}{2}\cdot\frac{1}{2} = \frac{1}{4}\left(\sqrt{6} + \sqrt{2}\right)$$

— **Now Work** PROBLEMS **13** AND **19**

Another use of the Sum and Difference Formulas is to establish other identities. Two important identities, conjectured in Section 2.4, are given next.

$$\cos\left(\frac{\pi}{2} - \theta\right) = \sin\theta \qquad\qquad \textbf{(3a)}$$

$$\sin\left(\frac{\pi}{2} - \theta\right) = \cos\theta \qquad\qquad \textbf{(3b)}$$

Seeing the Concept

Graph $Y_1 = \cos\left(\dfrac{\pi}{2} - x\right)$ and $Y_2 = \sin x$ on the same screen. Does doing this provide evidence of result (3a)? How would you provide evidence of the result (3b)?

Proof To establish identity (3a), use the Difference Formula for $\cos(\alpha - \beta)$ with $\alpha = \dfrac{\pi}{2}$ and $\beta = \theta$.

$$\cos\left(\frac{\pi}{2} - \theta\right) = \cos\frac{\pi}{2}\cos\theta + \sin\frac{\pi}{2}\sin\theta$$

$$= 0\cdot\cos\theta + 1\cdot\sin\theta$$

$$= \sin\theta$$

To establish identity (3b), use the identity (3a) just established.

$$\sin\left(\frac{\pi}{2} - \theta\right) = \cos\left[\frac{\pi}{2} - \left(\frac{\pi}{2} - \theta\right)\right] = \cos\theta$$
$$\uparrow$$
$$\text{Use Identity (3a)}$$

Also, because the cosine function is even

$$\cos\left(\frac{\pi}{2} - \theta\right) = \cos\left[-\left(\theta - \frac{\pi}{2}\right)\right] = \cos\left(\theta - \frac{\pi}{2}\right)$$
$$\uparrow$$
$$\text{Even Property of Cosine}$$

and because

$$\cos\left(\frac{\pi}{2} - \theta\right) = \sin\theta$$
$$\uparrow$$
$$\text{Identity (3a)}$$

it follows that $\cos\left(\theta - \dfrac{\pi}{2}\right) = \sin\theta$. This means the graphs of $y = \cos\left(\theta - \dfrac{\pi}{2}\right)$ and $y = \sin\theta$ are identical.

Having established the identities (3a) and (3b), we now can derive the sum and difference formulas for $\sin(\alpha + \beta)$ and $\sin(\alpha - \beta)$.

Proof $\sin(\alpha + \beta) = \cos\left[\dfrac{\pi}{2} - (\alpha + \beta)\right]$ **Identity (3a)**

$$= \cos\left[\left(\dfrac{\pi}{2} - \alpha\right) - \beta\right]$$

$$= \cos\left(\dfrac{\pi}{2} - \alpha\right)\cos\beta + \sin\left(\dfrac{\pi}{2} - \alpha\right)\sin\beta \quad \textbf{Difference Formula for cosine}$$

$$= \sin\alpha\cos\beta + \cos\alpha\sin\beta \quad \textbf{Identities (3a) and (3b)}$$

$$\sin(\alpha - \beta) = \sin[\alpha + (-\beta)]$$

$$= \sin\alpha\cos(-\beta) + \cos\alpha\sin(-\beta) \quad \textbf{Use the Sum Formula for sine just obtained.}$$

$$= \sin\alpha\cos\beta + \cos\alpha(-\sin\beta) \quad \textbf{Even-Odd Identities}$$

$$= \sin\alpha\cos\beta - \cos\alpha\sin\beta \quad \blacksquare$$

In Words
Formula (4) states that the sine of the sum of two angles equals the sine of the first angle times the cosine of the second angle plus the cosine of the first angle times the sine of the second angle.

THEOREM **Sum and Difference Formulas for the Sine Function**

$$\sin(\alpha + \beta) = \sin\alpha\cos\beta + \cos\alpha\sin\beta \qquad (4)$$

$$\sin(\alpha - \beta) = \sin\alpha\cos\beta - \cos\alpha\sin\beta \qquad (5)$$

EXAMPLE 3 **Using the Sum Formula to Find an Exact Value**

Find the exact value of $\sin\dfrac{7\pi}{12}$.

Solution

$$\sin\dfrac{7\pi}{12} = \sin\left(\dfrac{3\pi}{12} + \dfrac{4\pi}{12}\right) = \sin\left(\dfrac{\pi}{4} + \dfrac{\pi}{3}\right)$$

$$= \sin\dfrac{\pi}{4}\cos\dfrac{\pi}{3} + \cos\dfrac{\pi}{4}\sin\dfrac{\pi}{3} \quad \textbf{Sum Formula for sine}$$

$$= \dfrac{\sqrt{2}}{2}\cdot\dfrac{1}{2} + \dfrac{\sqrt{2}}{2}\cdot\dfrac{\sqrt{3}}{2} = \dfrac{1}{4}(\sqrt{2} + \sqrt{6})$$

Now Work PROBLEM 21

EXAMPLE 4 **Using the Difference Formula to Find an Exact Value**

Find the exact value of $\sin 80°\cos 20° - \cos 80°\sin 20°$.

Solution The form of the expression $\sin 80°\cos 20° - \cos 80°\sin 20°$ is that of the right side of formula (5) for $\sin(\alpha - \beta)$ with $\alpha = 80°$ and $\beta = 20°$. That is,

$$\sin 80°\cos 20° - \cos 80°\sin 20° = \sin(80° - 20°) = \sin 60° = \dfrac{\sqrt{3}}{2}$$

Now Work PROBLEMS 27 AND 31

EXAMPLE 5 **Finding Exact Values**

If $\sin\alpha = \dfrac{4}{5}$, $\dfrac{\pi}{2} < \alpha < \pi$, and $\sin\beta = -\dfrac{2}{\sqrt{5}} = -\dfrac{2\sqrt{5}}{5}$, $\pi < \beta < \dfrac{3\pi}{2}$, find the exact value of each of the following.

(a) $\cos\alpha$ (b) $\cos\beta$ (c) $\cos(\alpha + \beta)$ (d) $\sin(\alpha + \beta)$

Solution

(a) Because $\sin \alpha = \dfrac{4}{5} = \dfrac{y}{r}$ and $\dfrac{\pi}{2} < \alpha < \pi$, let $y = 4$ and $r = 5$, and place α in quadrant II. The point $P = (x, y) = (x, 4), x < 0$, is on a circle of radius 5, $x^2 + y^2 = 25$. See Figure 27. Then

$$x^2 + y^2 = 25$$
$$x^2 + 16 = 25 \qquad \color{blue}{y = 4}$$
$$x^2 = 25 - 16 = 9$$
$$x = -3 \qquad \color{blue}{x < 0}$$

Then

$$\cos \alpha = \frac{x}{r} = -\frac{3}{5}$$

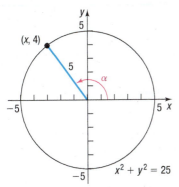

(*x*, 4)

$x^2 + y^2 = 25$

Figure 27 $\sin \alpha = \dfrac{4}{5}, \dfrac{\pi}{2} < \alpha < \pi$

Alternatively, we can use $\sin \alpha = \dfrac{4}{5}$ and the Pythagorean identity $\sin^2 \alpha + \cos^2 \alpha = 1$ to find $\cos \alpha$.

$$\cos \alpha = -\sqrt{1 - \sin^2 \alpha} = -\sqrt{1 - \frac{16}{25}} = -\sqrt{\frac{9}{25}} = -\frac{3}{5}$$

$\uparrow$
$\color{blue}{\alpha \text{ in quadrant II,}}$
$\color{blue}{\cos \alpha < 0}$

(b) Because $\sin \beta = \dfrac{-2}{\sqrt{5}} = \dfrac{y}{r}$ and $\pi < \beta < \dfrac{3\pi}{2}$, let $y = -2$ and $r = \sqrt{5}$, and place β in quadrant III. The point $P = (x, y) = (x, -2), x < 0$, is on a circle of radius $\sqrt{5}, x^2 + y^2 = 5$. See Figure 28. Then

$$x^2 + y^2 = 5$$
$$x^2 + 4 = 5 \qquad \color{blue}{y = -2}$$
$$x^2 = 1$$
$$x = -1 \qquad \color{blue}{x < 0}$$

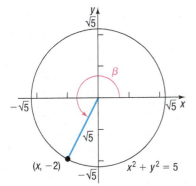

(*x*, −2)

$x^2 + y^2 = 5$

Figure 28 $\sin \beta = \dfrac{-2}{\sqrt{5}}, \pi < \beta < \dfrac{3\pi}{2}$

Then

$$\cos \beta = \frac{x}{r} = \frac{-1}{\sqrt{5}} = -\frac{\sqrt{5}}{5}$$

Alternatively, use $\sin \beta = -\dfrac{2\sqrt{5}}{5}$ and the Pythagorean identity $\sin^2 \beta + \cos^2 \beta = 1$ to find $\cos \beta$.

$$\cos \beta = -\sqrt{1 - \sin^2 \beta} = -\sqrt{1 - \frac{4}{5}} = -\sqrt{\frac{1}{5}} = -\frac{\sqrt{5}}{5}$$

(c) Use the results found in parts (a) and (b) and the Sum Formula for cosine.

$$\cos (\alpha + \beta) = \cos \alpha \cos \beta - \sin \alpha \sin \beta$$

$$= -\frac{3}{5}\left(-\frac{\sqrt{5}}{5}\right) - \frac{4}{5}\left(-\frac{2\sqrt{5}}{5}\right) = \frac{11\sqrt{5}}{25}$$

(d) Use the Sum Formula for sine.

$$\sin (\alpha + \beta) = \sin \alpha \cos \beta + \cos \alpha \sin \beta$$

$$= \frac{4}{5}\left(-\frac{\sqrt{5}}{5}\right) + \left(-\frac{3}{5}\right)\left(-\frac{2\sqrt{5}}{5}\right) = \frac{2\sqrt{5}}{25}$$

Now Work PROBLEM 35(a), (b), AND (c)

2 Use Sum and Difference Formulas to Establish Identities

EXAMPLE 6 **Establishing an Identity**

Establish the identity: $\dfrac{\cos(\alpha - \beta)}{\sin \alpha \sin \beta} = \cot \alpha \cot \beta + 1$

Solution

$$\frac{\cos(\alpha - \beta)}{\sin \alpha \sin \beta} = \frac{\cos \alpha \cos \beta + \sin \alpha \sin \beta}{\sin \alpha \sin \beta} \quad \text{Difference Formula for cosine}$$

$$= \frac{\cos \alpha \cos \beta}{\sin \alpha \sin \beta} + \frac{\sin \alpha \sin \beta}{\sin \alpha \sin \beta}$$

$$= \frac{\cos \alpha}{\sin \alpha} \cdot \frac{\cos \beta}{\sin \beta} + 1$$

$$= \cot \alpha \cot \beta + 1$$

Now Work PROBLEMS 49 AND 61

Use the identity $\tan \theta = \dfrac{\sin \theta}{\cos \theta}$ and the Sum Formulas for $\sin(\alpha + \beta)$ and $\cos(\alpha + \beta)$ to derive a formula for $\tan(\alpha + \beta)$.

Proof $\tan(\alpha + \beta) = \dfrac{\sin(\alpha + \beta)}{\cos(\alpha + \beta)} = \dfrac{\sin \alpha \cos \beta + \cos \alpha \sin \beta}{\cos \alpha \cos \beta - \sin \alpha \sin \beta}$

Now divide the numerator and the denominator by $\cos \alpha \cos \beta$.

$$\tan(\alpha + \beta) = \frac{\dfrac{\sin \alpha \cos \beta + \cos \alpha \sin \beta}{\cos \alpha \cos \beta}}{\dfrac{\cos \alpha \cos \beta - \sin \alpha \sin \beta}{\cos \alpha \cos \beta}} = \frac{\dfrac{\sin \alpha \cos \beta}{\cos \alpha \cos \beta} + \dfrac{\cos \alpha \sin \beta}{\cos \alpha \cos \beta}}{\dfrac{\cos \alpha \cos \beta}{\cos \alpha \cos \beta} - \dfrac{\sin \alpha \sin \beta}{\cos \alpha \cos \beta}}$$

$$= \frac{\dfrac{\sin \alpha}{\cos \alpha} + \dfrac{\sin \beta}{\cos \beta}}{1 - \dfrac{\sin \alpha}{\cos \alpha} \cdot \dfrac{\sin \beta}{\cos \beta}} = \frac{\tan \alpha + \tan \beta}{1 - \tan \alpha \tan \beta}$$

Proof Use the Sum Formula for $\tan(\alpha + \beta)$ and Even-Odd Properties to get the Difference Formula for the tangent function.

$$\tan(\alpha - \beta) = \tan[\alpha + (-\beta)] = \frac{\tan \alpha + \tan(-\beta)}{1 - \tan \alpha \tan(-\beta)} = \frac{\tan \alpha - \tan \beta}{1 + \tan \alpha \tan \beta}$$

$\tan(-\theta) = -\tan \theta$

We have proved the following results:

THEOREM Sum and Difference Formulas for the Tangent Function

$$\tan(\alpha + \beta) = \frac{\tan \alpha + \tan \beta}{1 - \tan \alpha \tan \beta} \tag{6}$$

$$\tan(\alpha - \beta) = \frac{\tan \alpha - \tan \beta}{1 + \tan \alpha \tan \beta} \tag{7}$$

Now Work PROBLEM 35(d)

EXAMPLE 7	**Establishing an Identity**

Establish the identity: $\tan(\theta + \pi) = \tan\theta$

Solution $\tan(\theta + \pi) = \dfrac{\tan\theta + \tan\pi}{1 - \tan\theta\tan\pi} = \dfrac{\tan\theta + 0}{1 - \tan\theta\cdot 0} = \tan\theta$

Example 7 verifies that the tangent function is periodic with period π.

EXAMPLE 8	**Establishing an Identity**

Establish the identity: $\tan\left(\theta + \dfrac{\pi}{2}\right) = -\cot\theta$

Solution   Formula (6) cannot be used because $\tan\dfrac{\pi}{2}$ is not defined. Instead, proceed as follows:

WARNING Be careful when using formulas (6) and (7). These formulas can be used only for angles α and β for which $\tan\alpha$ and $\tan\beta$ are defined. That is, they can be used for all angles except odd integer multiples of $\dfrac{\pi}{2}$. ■

$$\tan\left(\theta + \dfrac{\pi}{2}\right) = \dfrac{\sin\left(\theta + \dfrac{\pi}{2}\right)}{\cos\left(\theta + \dfrac{\pi}{2}\right)} = \dfrac{\sin\theta\cos\dfrac{\pi}{2} + \cos\theta\sin\dfrac{\pi}{2}}{\cos\theta\cos\dfrac{\pi}{2} - \sin\theta\sin\dfrac{\pi}{2}}$$

$$= \dfrac{(\sin\theta)\cdot 0 + (\cos\theta)\cdot 1}{(\cos\theta)\cdot 0 - (\sin\theta)\cdot 1} = \dfrac{\cos\theta}{-\sin\theta} = -\cot\theta$$

3 Use Sum and Difference Formulas Involving Inverse Trigonometric Functions

EXAMPLE 9	**Finding the Exact Value of an Expression Involving Inverse Trigonometric Functions**

Find the exact value of: $\sin\left(\cos^{-1}\dfrac{1}{2} + \sin^{-1}\dfrac{3}{5}\right)$

Solution   We want the sine of the sum of two angles, $\alpha = \cos^{-1}\dfrac{1}{2}$ and $\beta = \sin^{-1}\dfrac{3}{5}$. Then

$$\cos\alpha = \dfrac{1}{2} \quad 0 \le \alpha \le \pi \quad \text{and} \quad \sin\beta = \dfrac{3}{5} \quad -\dfrac{\pi}{2} \le \beta \le \dfrac{\pi}{2}$$

Use Pythagorean Identities to obtain $\sin\alpha$ and $\cos\beta$. Because $\sin\alpha \ge 0$ and $\cos\beta \ge 0$ (do you know why?), this means that

NOTE In Example 9 $\sin\alpha$ also can be found by using $\cos\alpha = \dfrac{1}{2} = \dfrac{x}{r}$, so $x = 1$ and $r = 2$. Then $y = \sqrt{3}$ and $\sin\alpha = \dfrac{y}{r} = \dfrac{\sqrt{3}}{2}$. Also, $\cos\beta$ can be found in a similar fashion. ■

$$\sin\alpha = \sqrt{1 - \cos^2\alpha} = \sqrt{1 - \dfrac{1}{4}} = \sqrt{\dfrac{3}{4}} = \dfrac{\sqrt{3}}{2}$$

$$\cos\beta = \sqrt{1 - \sin^2\beta} = \sqrt{1 - \dfrac{9}{25}} = \sqrt{\dfrac{16}{25}} = \dfrac{4}{5}$$

As a result,

$$\sin\left(\cos^{-1}\dfrac{1}{2} + \sin^{-1}\dfrac{3}{5}\right) = \sin(\alpha + \beta) = \sin\alpha\cos\beta + \cos\alpha\sin\beta$$

$$= \dfrac{\sqrt{3}}{2}\cdot\dfrac{4}{5} + \dfrac{1}{2}\cdot\dfrac{3}{5} = \dfrac{4\sqrt{3} + 3}{10}$$

✏️━━ **Now Work** PROBLEM **77**

EXAMPLE 10	**Writing a Trigonometric Expression as an Algebraic Expression**

Write $\sin(\sin^{-1}u + \cos^{-1}v)$ as an algebraic expression containing u and v (that is, without trigonometric functions). State the restrictions on u and v.

Solution

First, for $\sin^{-1} u$, the restriction on u is $-1 \le u \le 1$, and for $\cos^{-1} v$, the restriction on v is $-1 \le v \le 1$. Now let $\alpha = \sin^{-1} u$ and $\beta = \cos^{-1} v$. Then

$$\sin \alpha = u \qquad -\frac{\pi}{2} \le \alpha \le \frac{\pi}{2} \quad -1 \le u \le 1$$

$$\cos \beta = v \qquad 0 \le \beta \le \pi \quad -1 \le v \le 1$$

Because $-\dfrac{\pi}{2} \le \alpha \le \dfrac{\pi}{2}$, $\cos \alpha \ge 0$. So,

$$\cos \alpha = \sqrt{1 - \sin^2 \alpha} = \sqrt{1 - u^2}$$

Also, because $0 \le \beta \le \pi$, $\sin \beta \ge 0$. Then

$$\sin \beta = \sqrt{1 - \cos^2 \beta} = \sqrt{1 - v^2}$$

As a result,

$$\sin(\sin^{-1} u + \cos^{-1} v) = \sin(\alpha + \beta) = \sin \alpha \cos \beta + \cos \alpha \sin \beta$$

$$= uv + \sqrt{1 - u^2} \cdot \sqrt{1 - v^2}$$

✏️ **Now Work** PROBLEM 87

4 Solve Trigonometric Equations Linear in Sine and Cosine

Sometimes it is necessary to square both sides of an equation to obtain expressions that allow the use of identities. Remember, squaring both sides of an equation may introduce extraneous solutions. As a result, apparent solutions must be checked.

EXAMPLE 11

Solving a Trigonometric Equation Linear in Sine and Cosine

Solve the equation: $\sin \theta + \cos \theta = 1$, $0 \le \theta < 2\pi$

Option 1

Attempts to use available identities do not lead to equations that are easy to solve. (Try it yourself.) So, given the form of this equation, square both sides.

$$\sin \theta + \cos \theta = 1$$

$$(\sin \theta + \cos \theta)^2 = 1 \qquad \textbf{\textcolor{blue}{Square both sides.}}$$

$$\sin^2 \theta + 2 \sin \theta \cos \theta + \cos^2 \theta = 1 \qquad \textbf{\textcolor{blue}{Remove parentheses.}}$$

$$2 \sin \theta \cos \theta = 0 \qquad \textbf{\textcolor{blue}{$\sin^2 \theta + \cos^2 \theta = 1$}}$$

$$\sin \theta \cos \theta = 0$$

Setting each factor equal to zero leads to

$$\sin \theta = 0 \quad \text{or} \quad \cos \theta = 0$$

The apparent solutions are

$$\theta = 0 \qquad \theta = \pi \qquad \theta = \frac{\pi}{2} \qquad \theta = \frac{3\pi}{2}$$

Because both sides of the original equation were squared, these apparent solutions must be checked to see whether any are extraneous.

$$\theta = 0: \qquad \sin 0 + \cos 0 = 0 + 1 = 1 \qquad \textbf{\textcolor{blue}{A solution}}$$

$$\theta = \pi: \qquad \sin \pi + \cos \pi = 0 + (-1) = -1 \qquad \textbf{\textcolor{blue}{Not a solution}}$$

$$\theta = \frac{\pi}{2}: \qquad \sin \frac{\pi}{2} + \cos \frac{\pi}{2} = 1 + 0 = 1 \qquad \textbf{\textcolor{blue}{A solution}}$$

$$\theta = \frac{3\pi}{2}: \qquad \sin \frac{3\pi}{2} + \cos \frac{3\pi}{2} = -1 + 0 = -1 \qquad \textbf{\textcolor{blue}{Not a solution}}$$

The values $\theta = \pi$ and $\theta = \dfrac{3\pi}{2}$ are extraneous. The solution set is $\left\{0, \dfrac{\pi}{2}\right\}$.

Option 2 Start with the equation

$$\sin \theta + \cos \theta = 1$$

and divide both sides by $\sqrt{2}$. Then

$$\frac{1}{\sqrt{2}} \sin \theta + \frac{1}{\sqrt{2}} \cos \theta = \frac{1}{\sqrt{2}}$$

The left side now resembles the formula for the sine of the sum of two angles, one of which is θ. The other angle is unknown (call it ϕ.) Then

$$\sin (\theta + \phi) = \sin \theta \cos \phi + \cos \theta \sin \phi = \frac{1}{\sqrt{2}} \qquad \text{(8)}$$

Comparing (8) to $\dfrac{1}{\sqrt{2}} \sin \theta + \dfrac{1}{\sqrt{2}} \cos \theta = \dfrac{1}{\sqrt{2}}$, we see that

$$\cos \phi = \frac{1}{\sqrt{2}} = \frac{\sqrt{2}}{2} \qquad \sin \phi = \frac{1}{\sqrt{2}} = \frac{\sqrt{2}}{2} \qquad 0 \le \phi < 2\pi$$

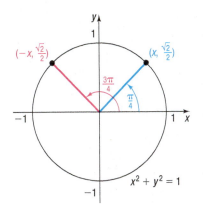

Figure 29

The angle ϕ is therefore $\dfrac{\pi}{4}$. As a result, equation (8) becomes

$$\sin\left(\theta + \frac{\pi}{4}\right) = \frac{\sqrt{2}}{2}$$

In the interval $[0, 2\pi)$, there are two angles whose sine is $\dfrac{\sqrt{2}}{2}$: $\dfrac{\pi}{4}$ and $\dfrac{3\pi}{4}$. See Figure 29. As a result,

$$\theta + \frac{\pi}{4} = \frac{\pi}{4} \quad \text{or} \quad \theta + \frac{\pi}{4} = \frac{3\pi}{4}$$

$$\theta = 0 \quad \text{or} \quad \theta = \frac{\pi}{2}$$

The solution set is $\left\{0, \dfrac{\pi}{2}\right\}$.

The second option can be used to solve any linear equation of the form $a \sin \theta + b \cos \theta = c$, where $a, b,$ and c are nonzero constants, by dividing both sides of the equation by $\sqrt{a^2 + b^2}$.

EXAMPLE 12 **Solving a Trigonometric Equation Linear in $\sin \theta$ and $\cos \theta$**

Solve:

$$a \sin \theta + b \cos \theta = c \qquad \text{(9)}$$

where $a, b,$ and c are constants and either $a \ne 0$ or $b \ne 0$.

Solution Divide both sides of equation (9) by $\sqrt{a^2 + b^2}$. Then

$$\frac{a}{\sqrt{a^2 + b^2}} \sin \theta + \frac{b}{\sqrt{a^2 + b^2}} \cos \theta = \frac{c}{\sqrt{a^2 + b^2}} \qquad \text{(10)}$$

There is a unique angle $\phi, 0 \le \phi < 2\pi$, for which

$$\cos \phi = \frac{a}{\sqrt{a^2 + b^2}} \quad \text{and} \quad \sin \phi = \frac{b}{\sqrt{a^2 + b^2}} \qquad \text{(11)}$$

Figure 30

Figure 30 shows the angle ϕ for $a > 0$ and $b > 0$.

(continued)

Equation (10) can be written as

$$\sin \theta \cos \phi + \cos \theta \sin \phi = \frac{c}{\sqrt{a^2 + b^2}}$$

or, equivalently,

$$\sin (\theta + \phi) = \frac{c}{\sqrt{a^2 + b^2}} \qquad \textbf{(12)}$$

where ϕ satisfies equation (11).

- If $|c| > \sqrt{a^2 + b^2}$, then $\sin (\theta + \phi) > 1$ or $\sin (\theta + \phi) < -1$, and equation (12) has no solution.
- If $|c| \leq \sqrt{a^2 + b^2}$, then the solutions of equation (12) are

$$\theta + \phi = \sin^{-1} \frac{c}{\sqrt{a^2 + b^2}} \quad \text{or} \quad \theta + \phi = \pi - \sin^{-1} \frac{c}{\sqrt{a^2 + b^2}}$$

Once the angle ϕ is determined by equations (11), the above are the solutions to equation (9).

Now Work PROBLEM 95

SUMMARY

Sum and Difference Formulas

$$\cos (\alpha + \beta) = \cos \alpha \cos \beta - \sin \alpha \sin \beta \qquad \cos (\alpha - \beta) = \cos \alpha \cos \beta + \sin \alpha \sin \beta$$

$$\sin (\alpha + \beta) = \sin \alpha \cos \beta + \cos \alpha \sin \beta \qquad \sin (\alpha - \beta) = \sin \alpha \cos \beta - \cos \alpha \sin \beta$$

$$\tan (\alpha + \beta) = \frac{\tan \alpha + \tan \beta}{1 - \tan \alpha \tan \beta} \qquad \tan (\alpha - \beta) = \frac{\tan \alpha - \tan \beta}{1 + \tan \alpha \tan \beta}$$

3.5 Assess Your Understanding

'Are You Prepared?' *Answers are given at the end of these exercises. If you get a wrong answer, read the pages listed in* red.

1. The distance d from the point $(2, -3)$ to the point $(5, 1)$ is _____. (pp. 3–4)

2. If $\sin \theta = \frac{4}{5}$ and θ is in quadrant II, then $\cos \theta =$ _____.
(pp. 139–142)

3. (a) $\sin \dfrac{\pi}{4} \cdot \cos \dfrac{\pi}{3} =$ _____ (pp. 120–123)

   (b) $\tan \dfrac{\pi}{4} - \sin \dfrac{\pi}{6} =$ _____ (pp. 120–123)

4. If $\sin \alpha = -\dfrac{4}{5}, \pi < \alpha < \dfrac{3\pi}{2}$, then $\cos \alpha =$ _____. (pp. 139–142)

5. Two triangles are _____ if the lengths of two corresponding sides are equal and the angles between the two sides have the same measure. (pp. A16–A17)

6. If $P = \left(-\dfrac{1}{3}, \dfrac{2\sqrt{2}}{3} \right)$ is a point on the unit circle that corresponds to a real number t, then $\sin t =$ _____,

   $\cos t =$ _____, and $\tan t =$ _____. (p. 117)

Concepts and Vocabulary

7. (a) $\cos (\alpha + \beta) = \cos \alpha \cos \beta$ _____ $\sin \alpha \sin \beta$
   (b) $\sin (\alpha - \beta) = \sin \alpha \cos \beta$ _____ $\cos \alpha \sin \beta$

8. *True or False* $\sin (\alpha + \beta) = \sin \alpha + \sin \beta + 2 \sin \alpha \sin \beta$

9. *True or False* $\cos \left(\dfrac{\pi}{2} - \theta \right) = \cos \theta$

10. *True or False* If $f(x) = \sin x$ and $g(x) = \cos x$, then
    $g(\alpha + \beta) = g(\alpha)g(\beta) - f(\alpha)f(\beta)$

11. *Multiple Choice* Choose the expression that completes the Sum Formula for tangent functions: $\tan(\alpha + \beta) =$ _____.

    (a) $\dfrac{\tan \alpha + \tan \beta}{1 - \tan \alpha \tan \beta}$ (b) $\dfrac{\tan \alpha - \tan \beta}{1 + \tan \alpha \tan \beta}$

    (c) $\dfrac{\tan \alpha + \tan \beta}{1 + \tan \alpha \tan \beta}$ (d) $\dfrac{\tan \alpha - \tan \beta}{1 - \tan \alpha \tan \beta}$

12. *Multiple Choice* Choose the expression that is equivalent to
    $$\sin 60° \cos 20° + \cos 60° \sin 20°$$

    (a) $\cos 40°$ (b) $\sin 40°$ (c) $\cos 80°$ (d) $\sin 80°$

Skill Building

In Problems 13–24, find the exact value of each expression.

13. $\cos 165°$ **14.** $\sin 105°$ **15.** $\tan 15°$ **16.** $\tan 195°$ **17.** $\sin \dfrac{5\pi}{12}$ **18.** $\sin \dfrac{\pi}{12}$

19. $\cos \dfrac{7\pi}{12}$ **20.** $\tan \dfrac{7\pi}{12}$ **21.** $\sin \dfrac{17\pi}{12}$ **22.** $\tan \dfrac{19\pi}{12}$ **23.** $\sec\left(-\dfrac{\pi}{12}\right)$ **24.** $\cot\left(-\dfrac{5\pi}{12}\right)$

In Problems 25–34, find the exact value of each expression.

25. $\sin 20° \cos 10° + \cos 20° \sin 10°$

26. $\sin 20° \cos 80° - \cos 20° \sin 80°$

27. $\cos 70° \cos 20° - \sin 70° \sin 20°$

28. $\cos 40° \cos 10° + \sin 40° \sin 10°$

29. $\dfrac{\tan 20° + \tan 25°}{1 - \tan 20° \tan 25°}$

30. $\dfrac{\tan 40° - \tan 10°}{1 + \tan 40° \tan 10°}$

31. $\sin \dfrac{\pi}{12} \cos \dfrac{7\pi}{12} - \cos \dfrac{\pi}{12} \sin \dfrac{7\pi}{12}$

32. $\cos \dfrac{5\pi}{12} \cos \dfrac{7\pi}{12} - \sin \dfrac{5\pi}{12} \sin \dfrac{7\pi}{12}$

33. $\cos \dfrac{\pi}{12} \cos \dfrac{5\pi}{12} + \sin \dfrac{5\pi}{12} \sin \dfrac{\pi}{12}$

34. $\sin \dfrac{\pi}{18} \cos \dfrac{5\pi}{18} + \cos \dfrac{\pi}{18} \sin \dfrac{5\pi}{18}$

In Problems 35–40, find the exact value of each of the following under the given conditions:

    (a) $\sin(\alpha + \beta)$ (b) $\cos(\alpha + \beta)$ (c) $\sin(\alpha - \beta)$ (d) $\tan(\alpha - \beta)$

35. $\sin \alpha = \dfrac{3}{5}, 0 < \alpha < \dfrac{\pi}{2};$ $\cos \beta = \dfrac{2\sqrt{5}}{5}, -\dfrac{\pi}{2} < \beta < 0$

36. $\cos \alpha = \dfrac{\sqrt{5}}{5}, 0 < \alpha < \dfrac{\pi}{2};$ $\sin \beta = -\dfrac{4}{5}, -\dfrac{\pi}{2} < \beta < 0$

37. $\tan \alpha = -\dfrac{4}{3}, \dfrac{\pi}{2} < \alpha < \pi;$ $\cos \beta = \dfrac{1}{2}, 0 < \beta < \dfrac{\pi}{2}$

38. $\tan \alpha = \dfrac{5}{12}, \pi < \alpha < \dfrac{3\pi}{2};$ $\sin \beta = -\dfrac{1}{2}, \pi < \beta < \dfrac{3\pi}{2}$

39. $\sin \alpha = \dfrac{5}{13}, -\dfrac{3\pi}{2} < \alpha < -\pi;$ $\tan \beta = -\sqrt{3}, \dfrac{\pi}{2} < \beta < \pi$

40. $\cos \alpha = \dfrac{1}{2}, -\dfrac{\pi}{2} < \alpha < 0;$ $\sin \beta = \dfrac{1}{3}, 0 < \beta < \dfrac{\pi}{2}$

41. If $\sin \theta = \dfrac{1}{3}, \theta$ in quadrant II, find the exact value of:

    (a) $\cos \theta$

    (b) $\sin\left(\theta + \dfrac{\pi}{6}\right)$

    (c) $\cos\left(\theta - \dfrac{\pi}{3}\right)$

    (d) $\tan\left(\theta + \dfrac{\pi}{4}\right)$

42. If $\cos \theta = \dfrac{1}{4}, \theta$ in quadrant IV, find the exact value of:

    (a) $\sin \theta$

    (b) $\sin\left(\theta - \dfrac{\pi}{6}\right)$

    (c) $\cos\left(\theta + \dfrac{\pi}{3}\right)$

    (d) $\tan\left(\theta - \dfrac{\pi}{4}\right)$

In Problems 43–48, use the figures to evaluate each function if $f(x) = \sin x$, $g(x) = \cos x$, and $h(x) = \tan x$.

43. $f(\alpha + \beta)$ **44.** $g(\alpha + \beta)$

45. $g(\alpha - \beta)$ **46.** $f(\alpha - \beta)$

47. $h(\alpha + \beta)$ **48.** $h(\alpha - \beta)$

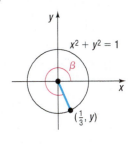

In Problems 49–74, establish each identity.

49. $\sin\left(\dfrac{\pi}{2} + \theta\right) = \cos \theta$

50. $\cos\left(\dfrac{\pi}{2} + \theta\right) = -\sin \theta$

51. $\sin(\pi - \theta) = \sin \theta$

52. $\cos(\pi - \theta) = -\cos \theta$

53. $\sin(\pi + \theta) = -\sin \theta$

54. $\cos(\pi + \theta) = -\cos \theta$

55. $\tan(\pi - \theta) = -\tan \theta$

56. $\tan(2\pi - \theta) = -\tan \theta$

57. $\sin\left(\dfrac{3\pi}{2} + \theta\right) = -\cos \theta$

58. $\cos\left(\dfrac{3\pi}{2} + \theta\right) = \sin \theta$

59. $\sin(\alpha + \beta) + \sin(\alpha - \beta) = 2 \sin \alpha \cos \beta$

60. $\cos(\alpha + \beta) + \cos(\alpha - \beta) = 2\cos\alpha\cos\beta$

61. $\dfrac{\sin(\alpha + \beta)}{\sin\alpha\cos\beta} = 1 + \cot\alpha\tan\beta$

62. $\dfrac{\sin(\alpha + \beta)}{\cos\alpha\cos\beta} = \tan\alpha + \tan\beta$

63. $\dfrac{\cos(\alpha + \beta)}{\cos\alpha\cos\beta} = 1 - \tan\alpha\tan\beta$

64. $\dfrac{\cos(\alpha - \beta)}{\sin\alpha\cos\beta} = \cot\alpha + \tan\beta$

65. $\dfrac{\sin(\alpha + \beta)}{\sin(\alpha - \beta)} = \dfrac{\tan\alpha + \tan\beta}{\tan\alpha - \tan\beta}$

66. $\dfrac{\cos(\alpha + \beta)}{\cos(\alpha - \beta)} = \dfrac{1 - \tan\alpha\tan\beta}{1 + \tan\alpha\tan\beta}$

67. $\cot(\alpha + \beta) = \dfrac{\cot\alpha\cot\beta - 1}{\cot\beta + \cot\alpha}$

68. $\cot(\alpha - \beta) = \dfrac{\cot\alpha\cot\beta + 1}{\cot\beta - \cot\alpha}$

69. $\sec(\alpha + \beta) = \dfrac{\csc\alpha\csc\beta}{\cot\alpha\cot\beta - 1}$

70. $\sec(\alpha - \beta) = \dfrac{\sec\alpha\sec\beta}{1 + \tan\alpha\tan\beta}$

71. $\sin(\alpha - \beta)\sin(\alpha + \beta) = \sin^2\alpha - \sin^2\beta$

72. $\cos(\alpha - \beta)\cos(\alpha + \beta) = \cos^2\alpha - \sin^2\beta$

73. $\sin(\theta + k\pi) = (-1)^k\sin\theta,\ k$ any integer

74. $\cos(\theta + k\pi) = (-1)^k\cos\theta,\ k$ any integer

In Problems 75–86, find the exact value of each expression.

75. $\sin\left(\sin^{-1}\dfrac{1}{2} + \cos^{-1}0\right)$

76. $\sin\left(\sin^{-1}\dfrac{\sqrt{3}}{2} + \cos^{-1}1\right)$

77. $\sin\left[\sin^{-1}\dfrac{3}{5} - \cos^{-1}\left(-\dfrac{4}{5}\right)\right]$

78. $\sin\left[\sin^{-1}\left(-\dfrac{4}{5}\right) - \tan^{-1}\dfrac{3}{4}\right]$

79. $\cos\left(\tan^{-1}\dfrac{4}{3} + \cos^{-1}\dfrac{5}{13}\right)$

80. $\cos\left[\tan^{-1}\dfrac{5}{12} - \sin^{-1}\left(-\dfrac{3}{5}\right)\right]$

81. $\cos\left(\sin^{-1}\dfrac{5}{13} - \tan^{-1}\dfrac{3}{4}\right)$

82. $\cos\left(\tan^{-1}\dfrac{4}{3} + \cos^{-1}\dfrac{12}{13}\right)$

83. $\tan\left(\sin^{-1}\dfrac{3}{5} + \dfrac{\pi}{6}\right)$

84. $\tan\left(\dfrac{\pi}{4} - \cos^{-1}\dfrac{3}{5}\right)$

85. $\tan\left(\sin^{-1}\dfrac{4}{5} + \cos^{-1}1\right)$

86. $\tan\left(\cos^{-1}\dfrac{4}{5} + \sin^{-1}1\right)$

In Problems 87–92, write each trigonometric expression as an algebraic expression containing u and v. Give the restrictions required on u and v.

87. $\cos(\cos^{-1}u + \sin^{-1}v)$

88. $\sin(\sin^{-1}u - \cos^{-1}v)$

89. $\sin(\tan^{-1}u - \sin^{-1}v)$

90. $\cos(\tan^{-1}u + \tan^{-1}v)$

91. $\tan(\sin^{-1}u - \cos^{-1}v)$

92. $\sec(\tan^{-1}u + \cos^{-1}v)$

In Problems 93–98, solve each equation on the interval $0 \le \theta < 2\pi$.

93. $\sin\theta - \sqrt{3}\cos\theta = 1$

94. $\sqrt{3}\sin\theta + \cos\theta = 1$

95. $\sin\theta + \cos\theta = \sqrt{2}$

96. $\sin\theta - \cos\theta = -\sqrt{2}$

97. $\tan\theta + \sqrt{3} = \sec\theta$

98. $\cot\theta + \csc\theta = -\sqrt{3}$

Applications and Extensions

99. Show that $\sin(\sin^{-1}v + \cos^{-1}v) = 1$.

100. Show that $\cos(\sin^{-1}v + \cos^{-1}v) = 0$.

101. Calculus Show that the difference quotient for $f(x) = \sin x$ is given by

$$\dfrac{f(x + h) - f(x)}{h} = \dfrac{\sin(x + h) - \sin x}{h}$$

$$= \cos x \cdot \dfrac{\sin h}{h} - \sin x \cdot \dfrac{1 - \cos h}{h}$$

102. Calculus Show that the difference quotient for $f(x) = \cos x$ is given by

$$\dfrac{f(x + h) - f(x)}{h} = \dfrac{\cos(x + h) - \cos x}{h}$$

$$= -\sin x \cdot \dfrac{\sin h}{h} - \cos x \cdot \dfrac{1 - \cos h}{h}$$

103. One, Two, Three
 (a) Show that $\tan(\tan^{-1}1 + \tan^{-1}2 + \tan^{-1}3) = 0$.
 (b) Conclude from part (a) that

$$\tan^{-1}1 + \tan^{-1}2 + \tan^{-1}3 = \pi$$

Source: College Mathematics Journal, Vol. 37, No. 3, May 2006

104. Electric Power In an alternating current (ac) circuit, the instantaneous power p at time t is given by

$$p(t) = V_mI_m\cos\phi\sin^2(\omega t) - V_mI_m\sin\phi\sin(\omega t)\cos(\omega t)$$

Show that this is equivalent to

$$p(t) = V_mI_m\sin(\omega t)\sin(\omega t - \phi)$$

Source: HyperPhysics, hosted by Georgia State University

105. Area of a Dodecagon Part I A regular dodecagon is a polygon with 12 sides of equal length. See the figure.

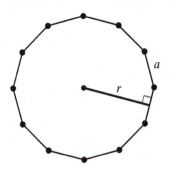

(a) The area A of a regular dodecagon is given by the formula $A = 12r^2 \tan \frac{\pi}{12}$, where r is the **apothem**, which is a line segment from the center of the polygon that is perpendicular to a side. Find the exact area of a regular dodecagon whose apothem is 10 inches.

(b) The area A of a regular dodecagon is also given by the formula $A = 3a^2 \cot \frac{\pi}{12}$, where a is the length of a side of the polygon. Find the exact area of a regular dodecagon if the length of a side is 15 centimeters.

106. Area of a Dodecagon Part II Refer to Problem 105. The figure shows that the interior angle of a regular dodecagon has measure 150°, and the *apothem* equals the radius of the inscribed circle.

(a) Find the exact area of a regular dodecagon with side $a = 5$ cm.

(b) Find the radius of the inscribed circle for the regular dodecagon from part (a).

(c) Find the exact area of the circle inscribed in a regular dodecagon with side $a = 5$ cm.

(d) Find the exact area of the region between the circle and the regular dodecagon.

107. Geometry: Angle between Two Lines Let L_1 and L_2 denote two nonvertical intersecting lines, and let θ denote the acute angle between L_1 and L_2 (see the figure). Show that

$$\tan \theta = \frac{m_2 - m_1}{1 + m_1 m_2}$$

where m_1 and m_2 are the slopes of L_1 and L_2, respectively. [**Hint**: Use the facts that $\tan \theta_1 = m_1$ and $\tan \theta_2 = m_2$.]

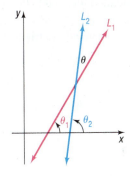

108. Challenge Problem Show that $\tan^{-1} v + \cot^{-1} v = \frac{\pi}{2}$.

109. Challenge Problem Show that $\tan^{-1}\left(\frac{1}{v}\right) = \frac{\pi}{2} - \tan^{-1} v$, if $v > 0$.

110. Challenge Problem If $\alpha + \beta + \gamma = 180°$ and

$$\cot \theta = \cot \alpha + \cot \beta + \cot \gamma \quad 0 < \theta < 90°$$

show that

$$\sin^3 \theta = \sin(\alpha - \theta) \sin(\beta - \theta) \sin(\gamma - \theta)$$

111. Challenge Problem If $\tan \alpha = x + 1$ and $\tan \beta = x - 1$, show that

$$2 \cot(\alpha - \beta) = x^2$$

Explaining Concepts: Discussion and Writing

112. Discuss the following derivation:

$$\tan\left(\theta + \frac{\pi}{2}\right) = \frac{\tan \theta + \tan \frac{\pi}{2}}{1 - \tan \theta \tan \frac{\pi}{2}} = \frac{\frac{\tan \theta}{\tan \frac{\pi}{2}} + 1}{\frac{1}{\tan \frac{\pi}{2}} - \tan \theta} = \frac{0 + 1}{0 - \tan \theta} = \frac{1}{-\tan \theta} = -\cot \theta$$

Can you justify each step?

113. Explain why formula (7) cannot be used to show that

$$\tan\left(\frac{\pi}{2} - \theta\right) = \cot \theta$$

Establish this identity by using formulas (3a) and (3b).

Retain Your Knowledge

Problems 114–121 are based on material learned earlier in the course. The purpose of these problems is to keep the material fresh in your mind so that you are better prepared for the final exam.

114. Determine the points of intersection of the graphs of
$f(x) = x^2 + 5x + 1$ and $g(x) = -2x^2 - 11x - 4$
by solving $f(x) = g(x)$.

115. Convert $\dfrac{17\pi}{6}$ to degrees.

116. Find the area of the sector of a circle of radius 6 meters formed by an angle of 45°. Give both the exact area and an approximation rounded to two decimal places.

117. Given $\tan\theta = -2$, $270° < \theta < 360°$, find the exact value of the remaining five trigonometric functions.

118. Write $f(x) = \dfrac{1}{4}x^2 + x - 2$ in vertex form.

119. Simplify: $(2x^2 y^3)^4 (3x^5 y)^2$

120. Solve: $\sqrt{3x - 2} - \sqrt{2x - 3} = 1$

121. Write $\dfrac{6x}{(x + 3)^{1/4}} + 8(x + 3)^{3/4}, x > -3$, as a single quotient with only positive exponents.

'Are You Prepared?' Answers

1. 5   **2.** $-\dfrac{3}{5}$ **3.** (a) $\dfrac{\sqrt{2}}{4}$ (b) $\dfrac{1}{2}$ **4.** $-\dfrac{3}{5}$ **5.** congruent   **6.** $\dfrac{2\sqrt{2}}{3}; -\dfrac{1}{3}; -2\sqrt{2}$

3.6 Double-angle and Half-angle Formulas

OBJECTIVES
1 Use Double-angle Formulas to Find Exact Values (p. 241)
2 Use Double-angle Formulas to Establish Identities (p. 241)
3 Use Half-angle Formulas to Find Exact Values (p. 244)

In this section, formulas for $\sin(2\theta)$, $\cos(2\theta)$, $\sin\left(\dfrac{1}{2}\theta\right)$, and $\cos\left(\dfrac{1}{2}\theta\right)$ are established in terms of $\sin\theta$ and $\cos\theta$. They are derived using the Sum Formulas.

In the Sum Formulas for $\sin(\alpha + \beta)$ and $\cos(\alpha + \beta)$, let $\alpha = \beta = \theta$. Then

$$\sin(\alpha + \beta) = \sin\alpha\cos\beta + \cos\alpha\sin\beta$$
$$\sin(\theta + \theta) = \sin\theta\cos\theta + \cos\theta\sin\theta$$
$$\sin(2\theta) = 2\sin\theta\cos\theta$$

and

$$\cos(\alpha + \beta) = \cos\alpha\cos\beta - \sin\alpha\sin\beta$$
$$\cos(\theta + \theta) = \cos\theta\cos\theta - \sin\theta\sin\theta$$
$$\cos(2\theta) = \cos^2\theta - \sin^2\theta$$

An application of the Pythagorean identity $\sin^2\theta + \cos^2\theta = 1$ results in two other ways to express $\cos(2\theta)$.

$$\cos(2\theta) = \cos^2\theta - \sin^2\theta = (1 - \sin^2\theta) - \sin^2\theta = 1 - 2\sin^2\theta$$

and

$$\cos(2\theta) = \cos^2\theta - \sin^2\theta = \cos^2\theta - (1 - \cos^2\theta) = 2\cos^2\theta - 1$$

The following theorem summarizes the *Double-angle Formulas.*

THEOREM Double-angle Formulas

$$\sin(2\theta) = 2\sin\theta\cos\theta \tag{1}$$

$$\cos(2\theta) = \cos^2\theta - \sin^2\theta \tag{2}$$

$$\cos(2\theta) = 1 - 2\sin^2\theta \tag{3}$$

$$\cos(2\theta) = 2\cos^2\theta - 1 \tag{4}$$

---

1 Use Double-angle Formulas to Find Exact Values

EXAMPLE 1 **Finding Exact Values Using Double-angle Formulas**

If $\sin\theta = \dfrac{3}{5}$, $\dfrac{\pi}{2} < \theta < \pi$, find the exact value of:

(a) $\sin(2\theta)$ (b) $\cos(2\theta)$

Solution

(a) Because $\sin(2\theta) = 2\sin\theta\cos\theta$ and because $\sin\theta = \dfrac{3}{5}$ is known, begin by finding $\cos\theta$. Since $\sin\theta = \dfrac{3}{5} = \dfrac{y}{r}$, $\dfrac{\pi}{2} < \theta < \pi$, let $y = 3$ and $r = 5$, and place θ in quadrant II. The point $P = (x, y) = (x, 3)$ is on a circle of radius 5, $x^2 + y^2 = 25$. See Figure 31. Then

$$x^2 + y^2 = 25$$
$$x^2 = 25 - 9 = 16 \quad \text{y = 3}$$
$$x = -4 \quad\quad\quad \text{x < 0}$$

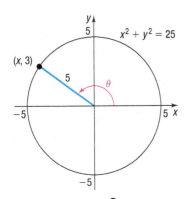

Figure 31 $\sin\theta = \dfrac{3}{5}$, $\dfrac{\pi}{2} < \theta < \pi$

This means that $\cos\theta = \dfrac{x}{r} = \dfrac{-4}{5}$. Now use Double-angle Formula (1) to obtain

$$\sin(2\theta) = 2\sin\theta\cos\theta = 2\cdot\frac{3}{5}\cdot\left(-\frac{4}{5}\right) = -\frac{24}{25}$$

(b) Because $\sin\theta = \dfrac{3}{5}$ is given, it is easiest to use Double-angle Formula (3) to find $\cos(2\theta)$.

$$\cos(2\theta) = 1 - 2\sin^2\theta = 1 - 2\cdot\frac{9}{25} = 1 - \frac{18}{25} = \frac{7}{25}$$

WARNING In finding $\cos(2\theta)$ in Example 1(b), a version of the Double-angle Formula, formula (3), was used. Note that it is not possible to use the Pythagorean identity $\cos(2\theta) = \pm\sqrt{1 - \sin^2(2\theta)}$, with $\sin(2\theta) = -\dfrac{24}{25}$, because there is no way of knowing which sign to choose. ∎

── **Now Work** PROBLEM 9(a) AND (b)

2 Use Double-angle Formulas to Establish Identities

EXAMPLE 2 **Establishing Identities**

(a) Develop a formula for $\tan(2\theta)$ in terms of $\tan\theta$.

(b) Develop a formula for $\sin(3\theta)$ in terms of $\sin\theta$ and $\cos\theta$.

Solution (a) In the Sum Formula for $\tan(\alpha + \beta)$, let $\alpha = \beta = \theta$. Then

$$\tan(\alpha + \beta) = \frac{\tan\alpha + \tan\beta}{1 - \tan\alpha\tan\beta}$$

$$\tan(\theta + \theta) = \frac{\tan\theta + \tan\theta}{1 - \tan\theta\tan\theta}$$

$$\tan(2\theta) = \frac{2\tan\theta}{1 - \tan^2\theta} \qquad (5)$$

(b) To find a formula for $\sin(3\theta)$, write 3θ as $2\theta + \theta$, and use the Sum Formula.

$$\sin(3\theta) = \sin(2\theta + \theta) = \sin(2\theta)\cos\theta + \cos(2\theta)\sin\theta$$

Now use the Double-angle Formulas to get

$$\sin(3\theta) = (2\sin\theta\cos\theta)(\cos\theta) + (\cos^2\theta - \sin^2\theta)(\sin\theta)$$
$$= 2\sin\theta\cos^2\theta + \sin\theta\cos^2\theta - \sin^3\theta$$
$$= 3\sin\theta\cos^2\theta - \sin^3\theta$$

The formula obtained in Example 2(b) also can be written as

$$\sin(3\theta) = 3\sin\theta\cos^2\theta - \sin^3\theta = 3\sin\theta(1 - \sin^2\theta) - \sin^3\theta$$
$$= 3\sin\theta - 4\sin^3\theta$$

That is, $\sin(3\theta)$ is a third-degree polynomial in the variable $\sin\theta$. In fact, $\sin(n\theta)$, n a positive odd integer, can always be written as a polynomial of degree n in the variable $\sin\theta$.*

Now Work PROBLEM 69

Rearranging the Double-angle Formulas (3) and (4) leads to other formulas that are used later and are important in calculus.

Begin with Double-angle Formula (3) and solve for $\sin^2\theta$.

$$\cos(2\theta) = 1 - 2\sin^2\theta$$

$$2\sin^2\theta = 1 - \cos(2\theta)$$

$$\sin^2\theta = \frac{1 - \cos(2\theta)}{2} \qquad (6)$$

Similarly, using Double-angle Formula (4), solve for $\cos^2\theta$.

$$\cos(2\theta) = 2\cos^2\theta - 1$$

$$2\cos^2\theta = 1 + \cos(2\theta)$$

$$\cos^2\theta = \frac{1 + \cos(2\theta)}{2} \qquad (7)$$

Formulas (6) and (7) can be used to develop a formula for $\tan^2\theta$.

$$\tan^2\theta = \frac{\sin^2\theta}{\cos^2\theta} = \frac{\dfrac{1 - \cos(2\theta)}{2}}{\dfrac{1 + \cos(2\theta)}{2}}$$

*Because of the work done by P. L. Chebyshëv, these polynomials are sometimes called *Chebyshëv polynomials.*

$$\tan^2 \theta = \frac{1 - \cos(2\theta)}{1 + \cos(2\theta)} \qquad \textbf{(8)}$$

Formulas (6) through (8) do not have to be memorized since their derivations are straightforward.

Formulas (6) and (7) are important in calculus. The next example illustrates a problem that arises in calculus requiring the use of formula (7).

EXAMPLE 3 Establishing an Identity

Write an equivalent expression for $\cos^4 \theta$ that does not involve any powers of sine or cosine greater than 1.

Solution The idea here is to use formula (7) twice.

$$\cos^4 \theta = (\cos^2 \theta)^2 = \left(\frac{1 + \cos(2\theta)}{2}\right)^2 \qquad \text{Formula (7)}$$

$$= \frac{1}{4}[1 + 2\cos(2\theta) + \cos^2(2\theta)]$$

$$= \frac{1}{4} + \frac{1}{2}\cos(2\theta) + \frac{1}{4}\cos^2(2\theta)$$

$$= \frac{1}{4} + \frac{1}{2}\cos(2\theta) + \frac{1}{4} \cdot \frac{1 + \cos(2 \cdot 2\theta)}{2} \qquad \text{Formula (7)}$$

$$= \frac{1}{4} + \frac{1}{2}\cos(2\theta) + \frac{1}{8}[1 + \cos(4\theta)]$$

$$= \frac{3}{8} + \frac{1}{2}\cos(2\theta) + \frac{1}{8}\cos(4\theta)$$

Now Work PROBLEM **43**

EXAMPLE 4 Solving a Trigonometric Equation Using Identities

Solve the equation: $\sin\theta \cos\theta = -\dfrac{1}{2}, \quad 0 \le \theta < 2\pi$

Solution The left side of the equation, except for a factor of 2, is in the form of the Double-angle Formula, $2\sin\theta\cos\theta = \sin(2\theta)$. Multiply both sides by 2.

$$\sin\theta\cos\theta = -\frac{1}{2}$$

$$2\sin\theta\cos\theta = -1 \qquad \textbf{Multiply both sides by 2.}$$

$$\sin(2\theta) = -1 \qquad \textbf{Double-angle Formula}$$

The argument is 2θ. Write the general formula that gives all the solutions of this equation, and then list those that are in the interval $[0, 2\pi)$. Because $\sin\left(\dfrac{3\pi}{2} + 2\pi k\right) = -1$, for any integer k, this means that

$$2\theta = \frac{3\pi}{2} + 2k\pi \qquad \textbf{\textit{k} an integer}$$

$$\theta = \frac{3\pi}{4} + k\pi$$

$$\theta = \frac{3\pi}{4} + (-1)\pi = -\frac{\pi}{4}, \quad \theta = \frac{3\pi}{4} + 0 \cdot \pi = \frac{3\pi}{4}, \quad \theta = \frac{3\pi}{4} + 1 \cdot \pi = \frac{7\pi}{4}, \quad \theta = \frac{3\pi}{4} + 2 \cdot \pi = \frac{11\pi}{4}$$

$$k = -1 \qquad\qquad k = 0 \qquad\qquad k = 1 \qquad\qquad k = 2$$

(continued)

The solutions in the interval $[0, 2\pi)$ are

$$\theta = \frac{3\pi}{4} \qquad \theta = \frac{7\pi}{4}$$

The solution set is $\left\{ \dfrac{3\pi}{4}, \dfrac{7\pi}{4} \right\}$.

Now Work PROBLEM 73

Figure 32

EXAMPLE 5   **Projectile Motion**

An object is propelled upward at an angle θ to the horizontal with an initial velocity of v_0 feet per second. See Figure 32. If air resistance is ignored, the **range** R—the horizontal distance that the object travels—is given by the function

$$R(\theta) = \frac{1}{16} v_0^2 \sin\theta \cos\theta$$

(a) Show that $R(\theta) = \dfrac{1}{32} v_0^2 \sin(2\theta)$.

(b) Find the angle θ for which R is a maximum.

Solution   (a) Rewrite the expression for the range using the Double-angle Formula $\sin(2\theta) = 2\sin\theta \cos\theta$. Then

$$R(\theta) = \frac{1}{16} v_0^2 \sin\theta \cos\theta = \frac{1}{16} v_0^2 \frac{2\sin\theta \cos\theta}{2} = \frac{1}{32} v_0^2 \sin(2\theta)$$

(b) In this form, the largest value for the range R can be found. For a fixed initial speed v_0, the angle θ of inclination to the horizontal determines the value of R. The largest value of a sine function is 1, which occurs when the argument 2θ is 90°. For maximum R, it follows that

$$2\theta = 90°$$
$$\theta = 45°$$

An inclination to the horizontal of 45° results in the maximum range.

3 Use Half-angle Formulas to Find Exact Values

Another important use of formulas (6) through (8) is to prove the *Half-angle Formulas*. In formulas (6) through (8), let $\theta = \dfrac{\alpha}{2}$. Then

$$\sin^2 \frac{\alpha}{2} = \frac{1 - \cos\alpha}{2} \qquad \cos^2 \frac{\alpha}{2} = \frac{1 + \cos\alpha}{2} \qquad \tan^2 \frac{\alpha}{2} = \frac{1 - \cos\alpha}{1 + \cos\alpha} \qquad \text{(9)}$$

Solving for the trigonometric functions on the left sides of equations (9) gives the Half-angle Formulas.

THEOREM   **Half-angle Formulas**

$$\sin \frac{\alpha}{2} = \pm \sqrt{\frac{1 - \cos\alpha}{2}} \qquad \text{(10)}$$

$$\cos \frac{\alpha}{2} = \pm \sqrt{\frac{1 + \cos\alpha}{2}} \qquad \text{(11)}$$

$$\tan \frac{\alpha}{2} = \pm \sqrt{\frac{1 - \cos\alpha}{1 + \cos\alpha}} \qquad \text{(12)}$$

where the $+$ or $-$ sign is determined by the quadrant of the angle $\dfrac{\alpha}{2}$.

EXAMPLE 6

Finding Exact Values Using Half-angle Formulas

Use a Half-angle Formula to find the exact value of:

(a) $\cos 15°$ (b) $\sin(-15°)$

Solution

(a) Because $15° = \dfrac{30°}{2}$, use the Half-angle Formula for $\cos \dfrac{\alpha}{2}$ with $\alpha = 30°$. Also, because $15°$ is in quadrant I, $\cos 15° > 0$, so choose the $+$ sign in using formula (11).

$$\cos 15° = \cos \frac{30°}{2} = \sqrt{\frac{1 + \cos 30°}{2}}$$

$$= \sqrt{\frac{1 + \sqrt{3}/2}{2}} = \sqrt{\frac{2 + \sqrt{3}}{4}} = \frac{\sqrt{2 + \sqrt{3}}}{2}$$

(b) Use the fact that $\sin(-15°) = -\sin 15°$, and then use formula (10).

$$\sin(-15°) = -\sin\frac{30°}{2} = -\sqrt{\frac{1 - \cos 30°}{2}}$$

$$= -\sqrt{\frac{1 - \sqrt{3}/2}{2}} = -\sqrt{\frac{2 - \sqrt{3}}{4}} = -\frac{\sqrt{2 - \sqrt{3}}}{2}$$

It is interesting to compare the answer found in Example 6(a) with the answer to Example 2 of Section 3.5. There it was calculated that

$$\cos\frac{\pi}{12} = \cos 15° = \frac{1}{4}\left(\sqrt{6} + \sqrt{2}\right)$$

Based on this and the result of Example 6(a),

$$\frac{1}{4}\left(\sqrt{6} + \sqrt{2}\right) \quad \text{and} \quad \frac{\sqrt{2 + \sqrt{3}}}{2}$$

are equal. (Since each expression is positive, you can verify this equality by squaring each expression.) Two very different-looking, yet correct, answers can be obtained, depending on the approach taken to solve a problem.

➤ **Now Work** PROBLEM 21

EXAMPLE 7

Finding Exact Values Using Half-angle Formulas

If $\cos\alpha = -\dfrac{3}{5}$, $\pi < \alpha < \dfrac{3\pi}{2}$, find the exact value of:

(a) $\sin\dfrac{\alpha}{2}$ (b) $\cos\dfrac{\alpha}{2}$ (c) $\tan\dfrac{\alpha}{2}$

Solution

First, observe that if $\pi < \alpha < \dfrac{3\pi}{2}$, then $\dfrac{\pi}{2} < \dfrac{\alpha}{2} < \dfrac{3\pi}{4}$. As a result, $\dfrac{\alpha}{2}$ lies in quadrant II.

(a) Because $\dfrac{\alpha}{2}$ lies in quadrant II, $\sin\dfrac{\alpha}{2} > 0$, so use the $+$ sign in formula (10) to get

$$\sin\frac{\alpha}{2} = \sqrt{\frac{1 - \cos\alpha}{2}} = \sqrt{\frac{1 - \left(-\frac{3}{5}\right)}{2}} = \sqrt{\frac{\frac{8}{5}}{2}} = \sqrt{\frac{4}{5}} = \frac{2}{\sqrt{5}} = \frac{2\sqrt{5}}{5}$$

(b) Because $\dfrac{\alpha}{2}$ lies in quadrant II, $\cos\dfrac{\alpha}{2} < 0$, so use the $-$ sign in formula (11) to get

$$\cos\frac{\alpha}{2} = -\sqrt{\frac{1 + \cos\alpha}{2}} = -\sqrt{\frac{1 + \left(-\frac{3}{5}\right)}{2}} = -\sqrt{\frac{\frac{2}{5}}{2}} = -\frac{1}{\sqrt{5}} = -\frac{\sqrt{5}}{5}$$

(continued)

(c) Because $\dfrac{\alpha}{2}$ lies in quadrant II, $\tan\dfrac{\alpha}{2} < 0$, so use the $-$ sign in formula (12) to get

$$\tan\frac{\alpha}{2} = -\sqrt{\frac{1-\cos\alpha}{1+\cos\alpha}} = -\sqrt{\frac{1-\left(-\dfrac{3}{5}\right)}{1+\left(-\dfrac{3}{5}\right)}} = -\sqrt{\frac{\dfrac{8}{5}}{\dfrac{2}{5}}} = -2$$

Another way to solve Example 7(c) is to use the results of parts (a) and (b).

$$\tan\frac{\alpha}{2} = \frac{\sin\dfrac{\alpha}{2}}{\cos\dfrac{\alpha}{2}} = \frac{\dfrac{2\sqrt{5}}{5}}{-\dfrac{\sqrt{5}}{5}} = -2$$

Now Work PROBLEMS **9(c)** AND **(d)**

There is a formula for $\tan\dfrac{\alpha}{2}$ that does not contain $+$ and $-$ signs, making it more useful than formula (12). To derive it, use the formulas

$$1-\cos\alpha = 2\sin^2\frac{\alpha}{2} \quad \textbf{Formula (9)}$$

and

$$\sin\alpha = \sin\left(2\cdot\frac{\alpha}{2}\right) = 2\sin\frac{\alpha}{2}\cos\frac{\alpha}{2} \quad \textbf{Double-angle Formula}$$

Then

$$\frac{1-\cos\alpha}{\sin\alpha} = \frac{2\sin^2\dfrac{\alpha}{2}}{2\sin\dfrac{\alpha}{2}\cos\dfrac{\alpha}{2}} = \frac{\sin\dfrac{\alpha}{2}}{\cos\dfrac{\alpha}{2}} = \tan\frac{\alpha}{2}$$

Because it also can be shown that

$$\frac{1-\cos\alpha}{\sin\alpha} = \frac{\sin\alpha}{1+\cos\alpha}$$

this results in the following two Half-angle Formulas:

> **Half-angle Formulas for $\tan\dfrac{\alpha}{2}$**
>
> $$\tan\frac{\alpha}{2} = \frac{1-\cos\alpha}{\sin\alpha} = \frac{\sin\alpha}{1+\cos\alpha} \qquad (13)$$

With this formula, the solution to Example 7(c) can be obtained as follows:

$$\cos\alpha = -\frac{3}{5} \quad \pi < \alpha < \frac{3\pi}{2}$$

$$\sin\alpha = -\sqrt{1-\cos^2\alpha} = -\sqrt{1-\frac{9}{25}} = -\sqrt{\frac{16}{25}} = -\frac{4}{5}$$

Then, by equation (13),

$$\tan\frac{\alpha}{2} = \frac{1-\cos\alpha}{\sin\alpha} = \frac{1-\left(-\dfrac{3}{5}\right)}{-\dfrac{4}{5}} = \frac{\dfrac{8}{5}}{-\dfrac{4}{5}} = -2$$

3.6 Assess Your Understanding

Concepts and Vocabulary

1. $\cos(2\theta) = \cos^2\theta - \underline{\hspace{1cm}} = \underline{\hspace{1cm}} - 1$
$= 1 - \underline{\hspace{1cm}}$

2. $\sin^2\underline{\hspace{0.5cm}} = \dfrac{1 - \cos\theta}{2}$

3. $\tan\dfrac{\theta}{2} = \dfrac{1 - \cos\theta}{\underline{\hspace{1cm}}}$

4. *True or False* $\tan(2\theta) = \dfrac{2\tan\theta}{1 - \tan^2\theta}$

5. *True or False* $\sin(2\theta)$ has two equivalent forms:
$2\sin\theta\cos\theta$ and $\sin^2\theta - \cos^2\theta$

6. *True or False* $\tan(2\theta) + \tan(2\theta) = \tan(4\theta)$

7. *Multiple Choice* Choose the expression that completes the Half-angle Formula for cosine functions: $\cos\dfrac{\alpha}{2} = \underline{\hspace{1cm}}$.

(a) $\pm\sqrt{\dfrac{1 - \cos\alpha}{2}}$ (b) $\pm\sqrt{\dfrac{1 + \cos\alpha}{2}}$

(c) $\pm\sqrt{\dfrac{\cos\alpha - \sin\alpha}{2}}$ (d) $\pm\sqrt{\dfrac{1 - \cos\alpha}{1 + \cos\alpha}}$

8. *Multiple Choice* If $\sin\alpha = \pm\sqrt{\dfrac{1 - \cos\theta}{2}}$, then which statement describes how θ is related to α?

(a) $\theta = \alpha$ (b) $\theta = \dfrac{\alpha}{2}$ (c) $\theta = 2\alpha$ (d) $\theta = \alpha^2$

Skill Building

In Problems 9–20, use the information given about the angle θ, $0 \le \theta < 2\pi$, to find the exact value of:

     (a) $\sin(2\theta)$ (b) $\cos(2\theta)$ (c) $\sin\dfrac{\theta}{2}$ (d) $\cos\dfrac{\theta}{2}$

9. $\sin\theta = \dfrac{3}{5}$, $0 < \theta < \dfrac{\pi}{2}$

10. $\cos\theta = \dfrac{3}{5}$, $0 < \theta < \dfrac{\pi}{2}$

11. $\tan\theta = \dfrac{4}{3}$, $\pi < \theta < \dfrac{3\pi}{2}$

12. $\tan\theta = \dfrac{1}{2}$, $\pi < \theta < \dfrac{3\pi}{2}$

13. $\cos\theta = -\dfrac{\sqrt{6}}{3}$, $\dfrac{\pi}{2} < \theta < \pi$

14. $\sin\theta = -\dfrac{\sqrt{3}}{3}$, $\dfrac{3\pi}{2} < \theta < 2\pi$

15. $\sec\theta = 3$, $\sin\theta > 0$

16. $\csc\theta = -\sqrt{5}$, $\cos\theta < 0$

17. $\cot\theta = -2$, $\sec\theta < 0$

18. $\sec\theta = 2$, $\csc\theta < 0$

19. $\tan\theta = -3$, $\sin\theta < 0$

20. $\cot\theta = 3$, $\cos\theta < 0$

In Problems 21–30, use Half-angle Formulas to find the exact value of each expression.

21. $\sin 22.5°$

22. $\cos 22.5°$

23. $\tan\dfrac{7\pi}{8}$

24. $\tan\dfrac{9\pi}{8}$

25. $\cos 165°$

26. $\sin 195°$

27. $\sec\dfrac{15\pi}{8}$

28. $\csc\dfrac{7\pi}{8}$

29. $\sin\left(-\dfrac{\pi}{8}\right)$

30. $\cos\left(-\dfrac{3\pi}{8}\right)$

In Problems 31–42, $f(x) = \sin x$, $g(x) = \cos x$, and $h(x) = \tan x$. Use the figures below to evaluate each function.

31. $f(2\theta)$

32. $g(2\theta)$

33. $g\left(\dfrac{\theta}{2}\right)$

34. $f\left(\dfrac{\theta}{2}\right)$

35. $h(2\theta)$

36. $h\left(\dfrac{\theta}{2}\right)$

37. $g(2\alpha)$

38. $f(2\alpha)$

39. $f\left(\dfrac{\alpha}{2}\right)$

40. $g\left(\dfrac{\alpha}{2}\right)$

41. $h\left(\dfrac{\alpha}{2}\right)$

42. $h(2\alpha)$

43. Show that $\sin^4\theta = \dfrac{3}{8} - \dfrac{1}{2}\cos(2\theta) + \dfrac{1}{8}\cos(4\theta)$.

44. Show that $\sin(4\theta) = (\cos\theta)(4\sin\theta - 8\sin^3\theta)$.

45. Show that $\sin^2\theta\cos^2\theta = \dfrac{1}{8} - \dfrac{1}{8}\cos(4\theta)$.

46. Show that $\sin^4\theta\cos^4\theta = \dfrac{3}{128} - \dfrac{1}{32}\cos(4\theta) + \dfrac{1}{128}\cos(8\theta)$.

47. Find an expression for $\cos(3\theta)$ as a third-degree polynomial in the variable $\cos\theta$.

48. Find an expression for $\cos(4\theta)$ as a fourth-degree polynomial in the variable $\cos\theta$.

49. Find an expression for $\sin(5\theta)$ as a fifth-degree polynomial in the variable $\sin\theta$.

50. Find an expression for $\cos(5\theta)$ as a fifth-degree polynomial in the variable $\cos\theta$.

In Problems 51–72, establish each identity.

51. $\cos^4\theta - \sin^4\theta = \cos(2\theta)$

52. $\dfrac{\cot\theta - \tan\theta}{\cot\theta + \tan\theta} = \cos(2\theta)$

53. $\cot(2\theta) = \dfrac{\cot^2\theta - 1}{2\cot\theta}$

54. $\cot(2\theta) = \dfrac{1}{2}(\cot\theta - \tan\theta)$

55. $\sec(2\theta) = \dfrac{\sec^2\theta}{2 - \sec^2\theta}$

56. $\csc(2\theta) = \dfrac{1}{2}\sec\theta\csc\theta$

57. $\cos^2(2u) - \sin^2(2u) = \cos(4u)$

58. $(4\sin u\cos u)(1 - 2\sin^2 u) = \sin(4u)$

59. $\dfrac{\cos(2\theta)}{1 + \sin(2\theta)} = \dfrac{\cot\theta - 1}{\cot\theta + 1}$

60. $\sin^2\theta\cos^2\theta = \dfrac{1}{4}\sin^2(2\theta)$

61. $\sec^2\dfrac{\theta}{2} = \dfrac{2}{1 + \cos\theta}$

62. $\csc^2\dfrac{\theta}{2} = \dfrac{2}{1 - \cos\theta}$

63. $\cot^2\dfrac{v}{2} = \dfrac{\sec v + 1}{\sec v - 1}$

64. $\tan\dfrac{v}{2} = \csc v - \cot v$

65. $\cos\theta = \dfrac{1 - \tan^2\dfrac{\theta}{2}}{1 + \tan^2\dfrac{\theta}{2}}$

66. $1 - \dfrac{1}{2}\sin(2\theta) = \dfrac{\sin^3\theta + \cos^3\theta}{\sin\theta + \cos\theta}$

67. $\dfrac{\sin(3\theta)}{\sin\theta} - \dfrac{\cos(3\theta)}{\cos\theta} = 2$

68. $\dfrac{\cos\theta + \sin\theta}{\cos\theta - \sin\theta} - \dfrac{\cos\theta - \sin\theta}{\cos\theta + \sin\theta} = 2\tan(2\theta)$

69. $\tan(3\theta) = \dfrac{3\tan\theta - \tan^3\theta}{1 - 3\tan^2\theta}$

70. $\tan\theta + \tan(\theta + 120°) + \tan(\theta + 240°) = 3\tan(3\theta)$

71. $\ln|\sin\theta| = \dfrac{1}{2}(\ln|1 - \cos(2\theta)| - \ln 2)$

72. $\ln|\cos\theta| = \dfrac{1}{2}(\ln|1 + \cos(2\theta)| - \ln 2)$

In Problems 73–82, solve each equation on the interval $0 \le \theta < 2\pi$.

73. $\cos(2\theta) + 6\sin^2\theta = 4$

74. $\cos(2\theta) = 2 - 2\sin^2\theta$

75. $\cos(2\theta) = \cos\theta$

76. $\sin(2\theta) = \cos\theta$

77. $\sin(2\theta) + \sin(4\theta) = 0$

78. $\cos(2\theta) + \cos(4\theta) = 0$

79. $3 - \sin\theta = \cos(2\theta)$

80. $\cos(2\theta) + 5\cos\theta + 3 = 0$

81. $\tan(2\theta) + 2\sin\theta = 0$

82. $\tan(2\theta) + 2\cos\theta = 0$

In Problems 83–94, find the exact value of each expression.

83. $\sin\left(2\sin^{-1}\dfrac{1}{2}\right)$

84. $\sin\left[2\sin^{-1}\dfrac{\sqrt{3}}{2}\right]$

85. $\cos\left(2\sin^{-1}\dfrac{3}{5}\right)$

86. $\cos\left(2\cos^{-1}\dfrac{4}{5}\right)$

87. $\tan\left[2\cos^{-1}\left(-\dfrac{3}{5}\right)\right]$

88. $\tan\left(2\tan^{-1}\dfrac{3}{4}\right)$

89. $\sin\left(2\cos^{-1}\dfrac{4}{5}\right)$

90. $\cos\left[2\tan^{-1}\left(-\dfrac{4}{3}\right)\right]$

91. $\sin^2\left(\dfrac{1}{2}\cos^{-1}\dfrac{3}{5}\right)$

92. $\cos^2\left(\dfrac{1}{2}\sin^{-1}\dfrac{3}{5}\right)$

93. $\sec\left(2\tan^{-1}\dfrac{3}{4}\right)$

94. $\csc\left[2\sin^{-1}\left(-\dfrac{3}{5}\right)\right]$

Applications and Extensions

In Problems 95–100, find the real zeros of each trigonometric function on the interval $0 \le \theta < 2\pi$.

95. $f(x) = \sin(2x) - \sin x$

96. $f(x) = \cos(2x) + \cos x$

97. $f(x) = \cos(2x) + \sin^2 x$

98. $f(x) = 2\sin^2 x - \sin(2x)$

99. $f(x) = \sin(2x) + \cos x$

100. $f(x) = \cos(2x) - 5\cos x - 2$

101. Area of an Octagon

(a) The area A of a regular octagon is given by the formula $A = 8r^2\tan\dfrac{\pi}{8}$, where r is the apothem, which is a line segment from the center of the octagon perpendicular to a side. See the figure. Find the exact area of a regular octagon whose apothem is 12 inches.

(b) The area A of a regular octagon is also given by the formula $A = 2a^2\cot\dfrac{\pi}{8}$, where a is the length of a side. Find the exact area of a regular octagon whose side is 9 centimeters.

102. **Constructing a Rain Gutter** A rain gutter is to be constructed of aluminum sheets 12 inches wide. After marking off a length of 4 inches from each edge, the builder bends this length up at an angle θ. See the figure. The area A of the opening as a function of θ is given by

$$A(\theta) = 16 \sin\theta\,(\cos\theta + 1) \quad 0° < \theta < 90°$$

 (a) In calculus, you will be asked to find the angle θ that maximizes A by solving the equation

$$\cos(2\theta) + \cos\theta = 0 \quad 0° < \theta < 90°$$

   Solve this equation for θ.
   (b) What is the maximum area A of the opening?
   (c) Graph $A = A(\theta), 0° \le \theta \le 90°$, and find the angle θ that maximizes the area A. Also find the maximum area.

103. **Laser Projection** In a laser projection system, the **optical angle** or **scanning angle** θ is related to the throw distance D from the scanner to the screen and the projected image width W by the equation

$$D = \frac{\frac{1}{2}W}{\csc\theta - \cot\theta}$$

   (a) Show that the projected image width is given by

$$W = 2D \tan\frac{\theta}{2}$$

   (b) Find the optical angle if the throw distance is 15 feet and the projected image width is 6.5 feet.

   Source: Pangolin Laser Systems, Inc.

104. **Product of Inertia** The **product of inertia** for an area about inclined axes is given by the formula

$$I_{uv} = I_x \sin\theta\cos\theta - I_y \sin\theta\cos\theta + I_{xy}(\cos^2\theta - \sin^2\theta)$$

   Show that this is equivalent to

$$I_{uv} = \frac{I_x - I_y}{2}\sin(2\theta) + I_{xy}\cos(2\theta)$$

   Source: Adapted from Hibbeler, *Engineering Mechanics: Statics*, 13th ed., Pearson © 2013.

105. **Projectile Motion** An object is propelled upward at an angle θ, $45° < \theta < 90°$, to the horizontal with an initial velocity of v_0 feet per second from the base of a plane that makes an angle of $45°$ with the horizontal. See the figure atop the right column. If air resistance is ignored, the distance R that it travels up the inclined plane is given by the function

$$R(\theta) = \frac{v_0^2\sqrt{2}}{16}\cos\theta\,(\sin\theta - \cos\theta)$$

   (a) Show that

$$R(\theta) = \frac{v_0^2\sqrt{2}}{32}\left[\sin(2\theta) - \cos(2\theta) - 1\right]$$

 (b) In calculus, you will be asked to find the angle θ that maximizes R by solving the equation

$$\sin(2\theta) + \cos(2\theta) = 0$$

   Solve this equation for θ.

   (c) What is the maximum distance R if $v_0 = 32$ feet per second?

   (d) Graph $R = R(\theta), 45° \le \theta \le 90°$, and find the angle θ that maximizes the distance R. Also find the maximum distance. Use $v_0 = 32$ feet per second. Compare the results with the answers found in parts (b) and (c).

106. **Sawtooth Curve** An oscilloscope often displays a sawtooth curve. This curve can be approximated by sinusoidal curves of varying periods and amplitudes. A first approximation to the sawtooth curve is given by

$$y = \frac{1}{2}\sin(2\pi x) + \frac{1}{4}\sin(4\pi x)$$

   Show that $y = \sin(2\pi x)\cos^2(\pi x)$.

107. **Area of an Isosceles Triangle** Show that the area A of an isosceles triangle whose equal sides are of length s, and where θ is the angle between them, is

$$A = \frac{1}{2}s^2\sin\theta$$

   [**Hint:** See the figure. The height h bisects the angle θ and is the perpendicular bisector of the base.]

108. Geometry A rectangle is inscribed in a semicircle of radius 1. See the figure.

(a) Express the area A of the rectangle as a function of the angle θ shown in the figure.
(b) Show that $A(\theta) = \sin(2\theta)$.
(c) Find the angle θ that results in the largest area A.
(d) Find the dimensions of this largest rectangle.

109. If $x = 2\tan\theta$, express $\sin(2\theta)$ as a function of x.

110. If $x = 2\tan\theta$, express $\cos(2\theta)$ as a function of x.

111. Find the value of the number C:

$$\frac{1}{2}\sin^2 x + C = -\frac{1}{4}\cos(2x)$$

112. Find the value of the number C:

$$\frac{1}{2}\cos^2 x + C = \frac{1}{4}\cos(2x)$$

113. If $z = \tan\dfrac{\alpha}{2}$, show that $\sin\alpha = \dfrac{2z}{1+z^2}$.

114. If $z = \tan\dfrac{\alpha}{2}$, show that $\cos\alpha = \dfrac{1-z^2}{1+z^2}$.

115. Graph $f(x) = \sin^2 x = \dfrac{1-\cos(2x)}{2}$ for $0 \le x \le 2\pi$ by using transformations.

116. Repeat Problem 115 for $g(x) = \cos^2 x$.

117. Use the fact that

$$\cos\frac{\pi}{12} = \frac{1}{4}\left(\sqrt{6} + \sqrt{2}\right)$$

to find $\sin\dfrac{\pi}{24}$ and $\cos\dfrac{\pi}{24}$.

118. Show that

$$\cos\frac{\pi}{8} = \frac{\sqrt{2+\sqrt{2}}}{2}$$

and use it to find $\sin\dfrac{\pi}{16}$ and $\cos\dfrac{\pi}{16}$.

119. *Challenge Problem* Show that

$$\sin^3\theta + \sin^3(\theta + 120°) + \sin^3(\theta + 240°) = -\frac{3}{4}\sin(3\theta)$$

120. *Challenge Problem* If $\tan\theta = a\tan\dfrac{\theta}{3}$, express $\tan\dfrac{\theta}{3}$ in terms of a.

121. *Challenge Problem*

If $\cos(2x) + (2m-1)\sin x + m - 1 = 0$, find m so that there is exactly one real solution for x, $-\dfrac{\pi}{2} \le x \le \dfrac{\pi}{2}$.†

†Courtesy of Joliet Junior College Mathematics Department

Explaining Concepts: Discussion and Writing

122. Research Chebyshëv polynomials. Write a report on your findings.

Retain Your Knowledge

Problems 123–130 are based on material learned earlier in the course. The purpose of these problems is to keep the material fresh in your mind so that you are better prepared for the final exam.

123. Find an equation of the line that contains the point $(2, -3)$ and is perpendicular to the line $y = -2x + 9$.

124. Graph $f(x) = -x^2 + 6x + 7$. Label the vertex and any intercepts.

125. Find the exact value of $\sin\dfrac{2\pi}{3} - \cos\dfrac{4\pi}{3}$.

126. Graph $y = -2\cos\left(\dfrac{\pi}{2}x\right)$. Show at least two periods.

127. Find a polynomial function of degree 3 whose real zeros are -5, -2, and 2. Use 1 for the leading coefficient.

128. The function $f(x) = \dfrac{3-x}{2x-5}$ is one-to-one. Find f^{-1}.

129. Find the distance between the vertices of the parabolas $f(x) = x^2 - 4x - 1$ and $g(x) = -x^2 - 6x - 2$.

130. Solve for D: $6x - 5xD - 5y + 4yD + 3 - 4D = 0$

3.7 Product-to-Sum and Sum-to-Product Formulas

OBJECTIVES 1 Express Products as Sums (p. 251)
2 Express Sums as Products (p. 252)

1 Express Products as Sums

Sum and Difference Formulas can be used to derive formulas for writing the products of sines and/or cosines as sums or differences. These identities are usually called the *Product-to-Sum Formulas*.

> **THEOREM Product-to-Sum Formulas**
>
> $$\sin \alpha \sin \beta = \frac{1}{2}[\cos(\alpha - \beta) - \cos(\alpha + \beta)] \tag{1}$$
>
> $$\cos \alpha \cos \beta = \frac{1}{2}[\cos(\alpha - \beta) + \cos(\alpha + \beta)] \tag{2}$$
>
> $$\sin \alpha \cos \beta = \frac{1}{2}[\sin(\alpha + \beta) + \sin(\alpha - \beta)] \tag{3}$$

These formulas do not have to be memorized. Instead, remember how they are derived. Then, when you want to use them, either look them up or derive them, as needed.

To derive Product-to-Sum Formulas (1) and (2), write down the Sum and Difference Formulas for cosine:

$$\cos(\alpha - \beta) = \cos \alpha \cos \beta + \sin \alpha \sin \beta \tag{4}$$
$$\cos(\alpha + \beta) = \cos \alpha \cos \beta - \sin \alpha \sin \beta \tag{5}$$

To derive formula (1), subtract equation (5) from equation (4)

$$\cos(\alpha - \beta) - \cos(\alpha + \beta) = 2 \sin \alpha \sin \beta$$

from which

$$\sin \alpha \sin \beta = \frac{1}{2}[\cos(\alpha - \beta) - \cos(\alpha + \beta)]$$

To derive formula (2), add equations (4) and (5)

$$\cos(\alpha - \beta) + \cos(\alpha + \beta) = 2 \cos \alpha \cos \beta$$

from which

$$\cos \alpha \cos \beta = \frac{1}{2}[\cos(\alpha - \beta) + \cos(\alpha + \beta)]$$

To derive Product-to-Sum Formula (3), use the Sum and Difference Formulas for sine in a similar way. (You are asked to do this in Problem 55.)

EXAMPLE 1 **Expressing Products as Sums**

Express each of the following products as a sum containing only sines or only cosines.

(a) $\sin(6\theta)\sin(4\theta)$ (b) $\cos(3\theta)\cos\theta$ (c) $\sin(3\theta)\cos(5\theta)$

Solution   (a) Use formula (1) to get

$$\sin(6\theta)\sin(4\theta) = \frac{1}{2}[\cos(6\theta - 4\theta) - \cos(6\theta + 4\theta)]$$

$$= \frac{1}{2}[\cos(2\theta) - \cos(10\theta)]$$

(b) Use formula (2) to get

$$\cos(3\theta)\cos\theta = \frac{1}{2}[\cos(3\theta - \theta) + \cos(3\theta + \theta)]$$

$$= \frac{1}{2}[\cos(2\theta) + \cos(4\theta)]$$

(c) Use formula (3) to get

$$\sin(3\theta)\cos(5\theta) = \frac{1}{2}[\sin(3\theta + 5\theta) + \sin(3\theta - 5\theta)]$$

$$= \frac{1}{2}[\sin(8\theta) + \sin(-2\theta)] = \frac{1}{2}[\sin(8\theta) - \sin(2\theta)]$$

━━━━ **Now Work** PROBLEM 7

2 Express Sums as Products

> **THEOREM** **Sum-to-Product Formulas**
>
> $$\sin\alpha + \sin\beta = 2\sin\frac{\alpha + \beta}{2}\cos\frac{\alpha - \beta}{2} \tag{6}$$
>
> $$\sin\alpha - \sin\beta = 2\sin\frac{\alpha - \beta}{2}\cos\frac{\alpha + \beta}{2} \tag{7}$$
>
> $$\cos\alpha + \cos\beta = 2\cos\frac{\alpha + \beta}{2}\cos\frac{\alpha - \beta}{2} \tag{8}$$
>
> $$\cos\alpha - \cos\beta = -2\sin\frac{\alpha + \beta}{2}\sin\frac{\alpha - \beta}{2} \tag{9}$$

Formula (6) is derived here. The derivations of formulas (7) through (9) are left as exercises (see Problems 56 through 58).

Proof

$$2\sin\frac{\alpha + \beta}{2}\cos\frac{\alpha - \beta}{2} = 2\cdot\frac{1}{2}\left[\sin\left(\frac{\alpha + \beta}{2} + \frac{\alpha - \beta}{2}\right) + \sin\left(\frac{\alpha + \beta}{2} - \frac{\alpha - \beta}{2}\right)\right]$$

↑
Product-to-Sum Formula (3)

$$= \sin\frac{2\alpha}{2} + \sin\frac{2\beta}{2} = \sin\alpha + \sin\beta$$

■

EXAMPLE 2   **Expressing Sums (or Differences) as Products**

Express each sum or difference as a product of sines and/or cosines.

(a) $\sin(5\theta) - \sin(3\theta)$ (b) $\cos(3\theta) + \cos(2\theta)$

Solution (a) Use formula (7) to get

$$\sin(5\theta) - \sin(3\theta) = 2 \sin \frac{5\theta - 3\theta}{2} \cos \frac{5\theta + 3\theta}{2}$$

$$= 2 \sin \theta \cos(4\theta)$$

(b) $\cos(3\theta) + \cos(2\theta) = 2 \cos \dfrac{3\theta + 2\theta}{2} \cos \dfrac{3\theta - 2\theta}{2}$ **Formula (8)**

$$= 2 \cos \frac{5\theta}{2} \cos \frac{\theta}{2}$$

-**Now Work** PROBLEM 17

3.7 Assess Your Understanding

Skill Building

In Problems 1–6, find the exact value of each expression.

1. $\sin 195° \cdot \cos 75°$

2. $\cos 285° \cdot \cos 195°$

3. $\sin 285° \cdot \sin 75°$

4. $\sin 75° + \sin 15°$

5. $\cos 255° - \cos 195°$

6. $\sin 255° - \sin 15°$

In Problems 7–16, express each product as a sum containing only sines or only cosines.

7. $\sin(4\theta) \sin(2\theta)$

8. $\cos(4\theta) \cos(2\theta)$

9. $\sin(4\theta) \cos(2\theta)$

10. $\sin(3\theta) \sin(5\theta)$

11. $\cos(3\theta) \cos(5\theta)$

12. $\sin(4\theta) \cos(6\theta)$

13. $\sin \theta \sin(2\theta)$

14. $\cos(3\theta) \cos(4\theta)$

15. $\sin \dfrac{3\theta}{2} \cos \dfrac{\theta}{2}$

16. $\sin \dfrac{\theta}{2} \cos \dfrac{5\theta}{2}$

In Problems 17–24, express each sum or difference as a product of sines and/or cosines.

17. $\sin(4\theta) - \sin(2\theta)$

18. $\sin(4\theta) + \sin(2\theta)$

19. $\cos(2\theta) + \cos(4\theta)$

20. $\cos(5\theta) - \cos(3\theta)$

21. $\sin \theta + \sin(3\theta)$

22. $\cos \theta + \cos(3\theta)$

23. $\cos \dfrac{\theta}{2} - \cos \dfrac{3\theta}{2}$

24. $\sin \dfrac{\theta}{2} - \sin \dfrac{3\theta}{2}$

In Problems 25–42, establish each identity.

25. $\dfrac{\sin \theta + \sin(3\theta)}{2 \sin(2\theta)} = \cos \theta$

26. $\dfrac{\cos \theta + \cos(3\theta)}{2 \cos(2\theta)} = \cos \theta$

27. $\dfrac{\sin(4\theta) + \sin(2\theta)}{\cos(4\theta) + \cos(2\theta)} = \tan(3\theta)$

28. $\dfrac{\cos \theta - \cos(3\theta)}{\sin(3\theta) - \sin \theta} = \tan(2\theta)$

29. $\dfrac{\cos \theta - \cos(3\theta)}{\sin \theta + \sin(3\theta)} = \tan \theta$

30. $\dfrac{\cos \theta - \cos(5\theta)}{\sin \theta + \sin(5\theta)} = \tan(2\theta)$

31. $\sin \theta [\sin \theta + \sin(3\theta)] = \cos \theta [\cos \theta - \cos(3\theta)]$

32. $\sin \theta [\sin(3\theta) + \sin(5\theta)] = \cos \theta [\cos(3\theta) - \cos(5\theta)]$

33. $\dfrac{\sin(4\theta) + \sin(8\theta)}{\cos(4\theta) + \cos(8\theta)} = \tan(6\theta)$

34. $\dfrac{\sin(4\theta) - \sin(8\theta)}{\cos(4\theta) - \cos(8\theta)} = -\cot(6\theta)$

35. $\dfrac{\sin(4\theta) + \sin(8\theta)}{\sin(4\theta) - \sin(8\theta)} = -\dfrac{\tan(6\theta)}{\tan(2\theta)}$

36. $\dfrac{\cos(4\theta) - \cos(8\theta)}{\cos(4\theta) + \cos(8\theta)} = \tan(2\theta) \tan(6\theta)$

37. $\dfrac{\sin \alpha + \sin \beta}{\sin \alpha - \sin \beta} = \tan \dfrac{\alpha + \beta}{2} \cot \dfrac{\alpha - \beta}{2}$

38. $\dfrac{\cos \alpha + \cos \beta}{\cos \alpha - \cos \beta} = -\cot \dfrac{\alpha + \beta}{2} \cot \dfrac{\alpha - \beta}{2}$

39. $\dfrac{\sin \alpha + \sin \beta}{\cos \alpha + \cos \beta} = \tan \dfrac{\alpha + \beta}{2}$

40. $\dfrac{\sin \alpha - \sin \beta}{\cos \alpha - \cos \beta} = -\cot \dfrac{\alpha + \beta}{2}$

41. $1 + \cos(2\theta) + \cos(4\theta) + \cos(6\theta) = 4 \cos \theta \cos(2\theta) \cos(3\theta)$

42. $1 - \cos(2\theta) + \cos(4\theta) - \cos(6\theta) = 4 \sin \theta \cos(2\theta) \sin(3\theta)$

43. Show that $\sin^4 \theta \cos^2 \theta = \dfrac{1}{16} - \dfrac{1}{32} \cos(2\theta) - \dfrac{1}{16} \cos(4\theta) + \dfrac{1}{32} \cos(6\theta)$.

44. Show that $\sin^2 \theta \cos^4 \theta = \dfrac{1}{16} + \dfrac{1}{32} \cos(2\theta) - \dfrac{1}{16} \cos(4\theta) - \dfrac{1}{32} \cos(6\theta)$.

45. Show that $\sin^6 \theta = \dfrac{5}{16} - \dfrac{15}{32} \cos(2\theta) + \dfrac{3}{16} \cos(4\theta) - \dfrac{1}{32} \cos(6\theta)$.

46. Show that $\cos^6 \theta = \dfrac{5}{16} + \dfrac{15}{32} \cos(2\theta) + \dfrac{3}{16} \cos(4\theta) + \dfrac{1}{32} \cos(6\theta)$.

In Problems 47–50, solve each equation on the interval $0 \leq \theta < 2\pi$.

47. $\sin(2\theta) + \sin(4\theta) = 0$

48. $\cos(2\theta) + \cos(4\theta) = 0$

49. $\cos(4\theta) - \cos(6\theta) = 0$

50. $\sin(4\theta) - \sin(6\theta) = 0$

Applications and Extensions

51. Touch-Tone Phones On a Touch-Tone phone, each button produces a unique sound. The sound produced is the sum of two tones, given by

$$y = \sin(2\pi l t) \quad \text{and} \quad y = \sin(2\pi h t)$$

where l and h are the low and high frequencies (cycles per second) shown on the illustration. For example, if you touch 7, the low frequency is $l = 852$ cycles per second and the high frequency is $h = 1209$ cycles per second. The sound emitted when you touch 7 is

$$y = \sin[2\pi(852)t] + \sin[2\pi(1209)t]$$

Touch-Tone phone

1209 cycles/sec   1477 cycles/sec

1336 cycles/sec

(a) Write this sound as a product of sines and/or cosines.
(b) Determine the maximum value of y.
(c) Graph the sound emitted when 7 is touched.

52. Touch-Tone Phones
(a) Write, as a product of sines and/or cosines, the sound emitted when the # key is touched.
(b) Determine the maximum value of y.
(c) Graph the sound emitted when the # key is touched.

53. Moment of Inertia The moment of inertia I of an object is a measure of how easy it is to rotate the object about some fixed point. In engineering mechanics, it is sometimes necessary to compute moments of inertia with respect to a set of rotated axes. These moments are given by the equations

$$I_u = I_x \cos^2\theta + I_y \sin^2\theta - 2I_{xy}\sin\theta\cos\theta$$
$$I_v = I_x \sin^2\theta + I_y \cos^2\theta + 2I_{xy}\sin\theta\cos\theta$$

Use Product-to-Sum Formulas to show that

$$I_u = \frac{I_x + I_y}{2} + \frac{I_x - I_y}{2}\cos(2\theta) - I_{xy}\sin(2\theta)$$

and

$$I_v = \frac{I_x + I_y}{2} - \frac{I_x - I_y}{2}\cos(2\theta) + I_{xy}\sin(2\theta)$$

Source: Adapted from Hibbeler, *Engineering Mechanics: Statics*, 13th ed., Pearson © 2013.

54. Projectile Motion The range R of a projectile propelled downward from the top of an inclined plane at an angle θ to the inclined plane is given by

$$R(\theta) = \frac{2v_0^2 \sin\theta \cos(\theta - \phi)}{g \cos^2\phi}$$

where v_0 is the initial velocity of the projectile, ϕ is the angle the plane makes with respect to the horizontal, and g is acceleration due to gravity.
(a) Show that for fixed v_0 and ϕ, the maximum range down the incline is given by $R_{\max} = \dfrac{v_0^2}{g(1 - \sin\phi)}$.
(b) Determine the maximum range if the projectile has an initial velocity of 50 meters/second, the angle of the plane is $\phi = 35°$, and $g = 9.8$ meters/second^2.

55. Derive formula (3).   **56.** Derive formula (7).

57. Derive formula (8).   **58.** Derive formula (9).

59. *Challenge Problem* If $\alpha + \beta + \gamma = \pi$, show that
$$\sin(2\alpha) + \sin(2\beta) + \sin(2\gamma) = 4\sin\alpha\sin\beta\sin\gamma$$

60. *Challenge Problem* If $\alpha + \beta + \gamma = \pi$, show that
$$\tan\alpha + \tan\beta + \tan\gamma = \tan\alpha\tan\beta\tan\gamma$$

Retain Your Knowledge

Problems 61–69 are based on material learned earlier in the course. The purpose of these problems is to keep the material fresh in your mind so that you are better prepared for the final exam.

61. For $y = 5\cos(4x - \pi)$, find the amplitude, the period, and the phase shift.

62. Find the exact value of $\cos\left(\csc^{-1}\dfrac{7}{5}\right)$.

63. Find the inverse function f^{-1} of $f(x) = 3\sin x - 5$, $-\dfrac{\pi}{2} \leq x \leq \dfrac{\pi}{2}$. Find the range of f and the domain and range of f^{-1}.

64. Find the exact value of $\tan\left(-\dfrac{\pi}{6}\right)$.

65. Complete the square to write the quadratic function $f(x) = \frac{1}{3}x^2 - 2x - 2$ in vertex form.

66. The figure shows two flywheels connected by a belt. If the 6-inch diameter flywheel spins at 2000 revolutions per minute, how fast does the 2.5-inch diameter flywheel spin?

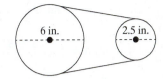
6 in. 2.5 in.

67. Solve the formula $A = \frac{1}{2}bh$ for h.

68. Given $f(x) = 2x^2 - 3x$ and $g(x) = 4x - 3$, determine where $f(x) \le g(x)$.

69. Find the difference quotient of $f(x) = \frac{2}{3}x + 9$.

Chapter Review

Things to Know

Definitions of the six inverse trigonometric functions

$y = \sin^{-1} x$ if and only if $x = \sin y$ where $-1 \le x \le 1$, $-\dfrac{\pi}{2} \le y \le \dfrac{\pi}{2}$ (p. 191)

$y = \cos^{-1} x$ if and only if $x = \cos y$ where $-1 \le x \le 1$, $0 \le y \le \pi$ (p. 193)

$y = \tan^{-1} x$ if and only if $x = \tan y$ where $-\infty < x < \infty$, $-\dfrac{\pi}{2} < y < \dfrac{\pi}{2}$ (p. 195)

$y = \sec^{-1} x$ if and only if $x = \sec y$ where $|x| \ge 1$, $0 \le y \le \pi$, $y \ne \dfrac{\pi}{2}$ (p. 203)

$y = \csc^{-1} x$ if and only if $x = \csc y$ where $|x| \ge 1$, $-\dfrac{\pi}{2} \le y \le \dfrac{\pi}{2}$, $y \ne 0$ (p. 203)

$y = \cot^{-1} x$ if and only if $x = \cot y$ where $-\infty < x < \infty$, $0 < y < \pi$ (p. 203)

Sum and Difference Formulas (pp. 227, 230, and 232)

$\cos(\alpha + \beta) = \cos\alpha \cos\beta - \sin\alpha \sin\beta$ $\cos(\alpha - \beta) = \cos\alpha \cos\beta + \sin\alpha \sin\beta$

$\sin(\alpha + \beta) = \sin\alpha \cos\beta + \cos\alpha \sin\beta$ $\sin(\alpha - \beta) = \sin\alpha \cos\beta - \cos\alpha \sin\beta$

$\tan(\alpha + \beta) = \dfrac{\tan\alpha + \tan\beta}{1 - \tan\alpha \tan\beta}$ $\tan(\alpha - \beta) = \dfrac{\tan\alpha - \tan\beta}{1 + \tan\alpha \tan\beta}$

Double-angle Formulas (pp. 241 and 242)

$\sin(2\theta) = 2\sin\theta \cos\theta$ $\cos(2\theta) = \cos^2\theta - \sin^2\theta$ $\tan(2\theta) = \dfrac{2\tan\theta}{1 - \tan^2\theta}$

$\cos(2\theta) = 2\cos^2\theta - 1$ $\cos(2\theta) = 1 - 2\sin^2\theta$

Half-angle Formulas (pp. 244 and 246)

$\sin^2\dfrac{\alpha}{2} = \dfrac{1 - \cos\alpha}{2}$ $\cos^2\dfrac{\alpha}{2} = \dfrac{1 + \cos\alpha}{2}$ $\tan^2\dfrac{\alpha}{2} = \dfrac{1 - \cos\alpha}{1 + \cos\alpha}$

$\sin\dfrac{\alpha}{2} = \pm\sqrt{\dfrac{1 - \cos\alpha}{2}}$ $\cos\dfrac{\alpha}{2} = \pm\sqrt{\dfrac{1 + \cos\alpha}{2}}$ $\tan\dfrac{\alpha}{2} = \pm\sqrt{\dfrac{1 - \cos\alpha}{1 + \cos\alpha}} = \dfrac{1 - \cos\alpha}{\sin\alpha} = \dfrac{\sin\alpha}{1 + \cos\alpha}$

where the $+$ or $-$ sign is determined by the quadrant of $\dfrac{\alpha}{2}$.

Product-to-Sum Formulas (p. 251)

$\sin\alpha \sin\beta = \dfrac{1}{2}[\cos(\alpha - \beta) - \cos(\alpha + \beta)]$

$\cos\alpha \cos\beta = \dfrac{1}{2}[\cos(\alpha - \beta) + \cos(\alpha + \beta)]$

$\sin\alpha \cos\beta = \dfrac{1}{2}[\sin(\alpha + \beta) + \sin(\alpha - \beta)]$

Sum-to-Product Formulas (p. 252)

$$\sin \alpha + \sin \beta = 2 \sin \frac{\alpha + \beta}{2} \cos \frac{\alpha - \beta}{2} \qquad \sin \alpha - \sin \beta = 2 \sin \frac{\alpha - \beta}{2} \cos \frac{\alpha + \beta}{2}$$

$$\cos \alpha + \cos \beta = 2 \cos \frac{\alpha + \beta}{2} \cos \frac{\alpha - \beta}{2} \qquad \cos \alpha - \cos \beta = -2 \sin \frac{\alpha + \beta}{2} \sin \frac{\alpha - \beta}{2}$$

Objectives

Section	You should be able to . . .	Example(s)	Review Exercises
3.1	**1** Define the inverse sine function (p. 190)	p. 191	1
	2 Find the value of an inverse sine function (p. 191)	1–3	7, 10
	3 Define the inverse cosine function (p. 193)	p. 193	2
	4 Find the value of an inverse cosine function (p. 194)	4, 5	8, 11
	5 Define the inverse tangent function (p. 194)	p. 195	3
	6 Find the value of an inverse tangent function (p. 196)	6	9, 12
	7 Use properties of inverse functions to find exact values of certain composite functions (p. 196)	7–9	15–23
	8 Find the inverse function of a trigonometric function (p. 198)	10	30, 31
	9 Solve equations involving inverse trigonometric functions (p. 199)	11	90, 91
3.2	**1** Define the inverse secant, cosecant, and cotangent functions (p. 203)	p. 203	4–6
	2 Find the value of inverse secant, cosecant, and cotangent functions (p. 204)	1, 2	13, 14
	3 Find the exact value of composite functions involving the inverse trigonometric functions (p. 205)	3–5	24–29
	4 Write a trigonometric expression as an algebraic expression (p. 206)	6	32, 33
3.3	**1** Solve equations involving a single trigonometric function (p. 209)	1–5	70–74
	2 Solve trigonometric equations using a calculator (p. 212)	6	75
	3 Solve trigonometric equations quadratic in form (p. 213)	7	78
	4 Solve trigonometric equations using fundamental identities (p. 213)	8, 9	76, 77, 79
	5 Solve trigonometric equations using a graphing utility (p. 214)	10	87–89
3.4	**1** Use algebra to simplify trigonometric expressions (p. 220)	1	34–50
	2 Establish identities (p. 221)	2–8	34–42
3.5	**1** Use Sum and Difference Formulas to find exact values (p. 228)	1–5	51–56, 59–63(a)–(d), 92
	2 Use Sum and Difference Formulas to establish identities (p. 232)	6–8	43, 44
	3 Use Sum and Difference Formulas involving inverse trigonometric functions (p. 233)	9, 10	64–67
	4 Solve trigonometric equations linear in sine and cosine (p. 234)	11, 12	81
3.6	**1** Use Double-angle Formulas to find exact values (p. 241)	1	59–63(e), (f), 68, 69, 93
	2 Use Double-angle Formulas to establish identities (p. 241)	2–5	46, 47, 80
	3 Use Half-angle Formulas to find exact values (p. 244)	6, 7	57, 58, 59–63(g), (h), 92
3.7	**1** Express products as sums (p. 251)	1	48
	2 Express sums as products (p. 252)	2	49, 50

Review Exercises

In Problems 1–6, state the domain and range of each function.

1. $y = \sin^{-1} x$

2. $y = \cos^{-1} x$

3. $y = \tan^{-1} x$

4. $y = \sec^{-1} x$

5. $y = \csc^{-1} x$

6. $y = \cot^{-1} x$

In Problems 7–14, find the exact value of each expression. Do not use a calculator.

7. $\sin^{-1} 1$

8. $\cos^{-1} 0$

9. $\tan^{-1} 1$

10. $\sin^{-1}\left(-\frac{1}{2}\right)$

11. $\cos^{-1}\left(-\frac{\sqrt{3}}{2}\right)$

12. $\tan^{-1}(-\sqrt{3})$

13. $\sec^{-1} \sqrt{2}$

14. $\cot^{-1}(-1)$

In Problems 15–29, find the exact value, if any, of each composite function. If there is no value, say it is "not defined." Do not use a calculator.

15. $\sin^{-1}\left(\sin\dfrac{3\pi}{8}\right)$

16. $\cos^{-1}\left(\cos\dfrac{3\pi}{4}\right)$

17. $\tan^{-1}\left(\tan\dfrac{2\pi}{3}\right)$

18. $\cos^{-1}\left(\cos\dfrac{15\pi}{7}\right)$

19. $\sin^{-1}\left[\sin\left(-\dfrac{8\pi}{9}\right)\right]$

20. $\sin\left(\sin^{-1}0.9\right)$

21. $\cos(\cos^{-1}0.6)$

22. $\tan\left[\tan^{-1}5\right]$

23. $\cos\left[\cos^{-1}(-1.6)\right]$

24. $\sin^{-1}\left(\cos\dfrac{2\pi}{3}\right)$

25. $\cos^{-1}\left(\tan\dfrac{3\pi}{4}\right)$

26. $\tan\left[\sin^{-1}\left(-\dfrac{\sqrt{3}}{2}\right)\right]$

27. $\sec\left(\tan^{-1}\dfrac{\sqrt{3}}{3}\right)$

28. $\sin\left(\cot^{-1}\dfrac{3}{4}\right)$

29. $\tan\left[\sin^{-1}\left(-\dfrac{4}{5}\right)\right]$

In Problems 30 and 31, find the inverse function f^{-1} of each function f. Find the range of f and the domain and range of f^{-1}.

30. $f(x)=2\sin(3x)\quad -\dfrac{\pi}{6}\le x\le\dfrac{\pi}{6}$

31. $f(x)=-\cos x+3\quad 0\le x\le\pi$

In Problems 32 and 33, write each trigonometric expression as an algebraic expression in u.

32. $\cos\left(\sin^{-1}u\right)$

33. $\tan\left(\csc^{-1}u\right)$

In Problems 34–50, establish each identity.

34. $\tan\theta\cot\theta-\sin^2\theta=\cos^2\theta$

35. $\sin^2\theta(1+\cot^2\theta)=1$

36. $5\cos^2\theta+3\sin^2\theta=3+2\cos^2\theta$

37. $\dfrac{1-\cos\theta}{\sin\theta}+\dfrac{\sin\theta}{1-\cos\theta}=2\csc\theta$

38. $\dfrac{\cos\theta}{\cos\theta-\sin\theta}=\dfrac{1}{1-\tan\theta}$

39. $\dfrac{\csc\theta}{1+\csc\theta}=\dfrac{1-\sin\theta}{\cos^2\theta}$

40. $\csc\theta-\sin\theta=\cos\theta\cot\theta$

41. $\dfrac{1-\sin\theta}{\sec\theta}=\dfrac{\cos^3\theta}{1+\sin\theta}$

42. $\dfrac{1-2\sin^2\theta}{\sin\theta\cos\theta}=\cot\theta-\tan\theta$

43. $\dfrac{\cos(\alpha+\beta)}{\cos\alpha\sin\beta}=\cot\beta-\tan\alpha$

44. $\dfrac{\cos(\alpha-\beta)}{\cos\alpha\cos\beta}=1+\tan\alpha\tan\beta$

45. $(1+\cos\theta)\tan\dfrac{\theta}{2}=\sin\theta$

46. $2\cot\theta\cot(2\theta)=\cot^2\theta-1$

47. $1-8\sin^2\theta\cos^2\theta=\cos(4\theta)$

48. $\dfrac{\sin(3\theta)\cos\theta-\sin\theta\cos(3\theta)}{\sin(2\theta)}=1$

49. $\dfrac{\sin(2\theta)+\sin(4\theta)}{\cos(2\theta)+\cos(4\theta)}=\tan(3\theta)$

50. $\dfrac{\cos(2\theta)-\cos(4\theta)}{\cos(2\theta)+\cos(4\theta)}-\tan\theta\tan(3\theta)=0$

In Problems 51–58, find the exact value of each expression.

51. $\sin 165°$

52. $\tan 105°$

53. $\cos\dfrac{5\pi}{12}$

54. $\sin\left(-\dfrac{\pi}{12}\right)$

55. $\cos 80°\cos 20°+\sin 80°\sin 20°$

56. $\sin 70°\cos 40°-\cos 70°\sin 40°$

57. $\tan\dfrac{\pi}{8}$

58. $\sin\dfrac{5\pi}{8}$

In Problems 59–63, use the information given about the angles α and β to find the exact value of:

(a) $\sin(\alpha+\beta)$

(b) $\cos(\alpha+\beta)$

(c) $\sin(\alpha-\beta)$

(d) $\tan(\alpha+\beta)$

(e) $\sin(2\alpha)$

(f) $\cos(2\beta)$

(g) $\sin\dfrac{\beta}{2}$

(h) $\cos\dfrac{\alpha}{2}$

59. $\sin\alpha=\dfrac{4}{5},0<\alpha<\dfrac{\pi}{2};\sin\beta=\dfrac{5}{13},\dfrac{\pi}{2}<\beta<\pi$

60. $\sin\alpha=-\dfrac{3}{5},\pi<\alpha<\dfrac{3\pi}{2};\cos\beta=\dfrac{12}{13},\dfrac{3\pi}{2}<\beta<2\pi$

61. $\tan\alpha=\dfrac{3}{4},\pi<\alpha<\dfrac{3\pi}{2};\tan\beta=\dfrac{12}{5},0<\beta<\dfrac{\pi}{2}$

62. $\sec\alpha=2,-\dfrac{\pi}{2}<\alpha<0;\sec\beta=3,\dfrac{3\pi}{2}<\beta<2\pi$

63. $\sin\alpha=-\dfrac{2}{3},\pi<\alpha<\dfrac{3\pi}{2};\cos\beta=-\dfrac{2}{3},\pi<\beta<\dfrac{3\pi}{2}$

In Problems 64–69, find the exact value of each expression.

64. $\cos\left(\sin^{-1}\dfrac{3}{5}-\cos^{-1}\dfrac{1}{2}\right)$

65. $\sin\left(\cos^{-1}\dfrac{5}{13}-\cos^{-1}\dfrac{4}{5}\right)$

66. $\tan\left[\sin^{-1}\left(-\dfrac{1}{2}\right) - \tan^{-1}\dfrac{3}{4}\right]$

67. $\cos\left[\tan^{-1}(-1) + \cos^{-1}\left(-\dfrac{4}{5}\right)\right]$

68. $\sin\left[2\cos^{-1}\left(-\dfrac{3}{5}\right)\right]$

69. $\cos\left(2\tan^{-1}\dfrac{4}{3}\right)$

In Problems 70–81, solve each equation on the interval $0 \le \theta < 2\pi$.

70. $\cos\theta = \dfrac{1}{2}$

71. $\tan\theta + \sqrt{3} = 0$

72. $\sin(2\theta) + 1 = 0$

73. $\tan(2\theta) = 0$

74. $\sec^2\theta = 4$

75. $0.2\sin\theta = 0.05$

76. $\sin\theta + \sin(2\theta) = 0$

77. $\sin(2\theta) - \cos\theta - 2\sin\theta + 1 = 0$

78. $2\sin^2\theta - 3\sin\theta + 1 = 0$

79. $4\sin^2\theta = 1 + 4\cos\theta$

80. $\sin(2\theta) = \sqrt{2}\cos\theta$

81. $\sin\theta - \cos\theta = 1$

In Problems 82–86, use a calculator to find an approximate value for each expression, rounded to two decimal places.

82. $\sin^{-1}0.7$

83. $\tan^{-1}(-2)$

84. $\cos^{-1}(-0.2)$

85. $\sec^{-1}3$

86. $\cot^{-1}(-4)$

In Problems 87–89, use a graphing utility to solve each equation on the interval $0 \le x \le 2\pi$. Approximate any solutions rounded to two decimal places.

87. $2x = 5\cos x$

88. $2\sin x + 3\cos x = 4x$

89. $\sin x = \ln x$

In Problems 90 and 91, find the exact solution of each equation.

90. $-3\sin^{-1}x = \pi$

91. $2\cos^{-1}x + \pi = 4\cos^{-1}x$

92. Use a Half-angle Formula to find the exact value of $\sin 15°$. Then use a Difference Formula to find the exact value of $\sin 15°$. Show that the answers you found are the same.

93. If you are given the value of $\cos\theta$ and want the exact value of $\cos(2\theta)$, what form of the Double-angle Formula for $\cos(2\theta)$ is most efficient to use?

Chapter Test

CHAPTER
Test Prep
VIDEOS

The Chapter Test Prep Videos include step-by-step solutions to all chapter test exercises. These videos are available in MyLab™ Math, or on this text's YouTube Channel. Refer to the Preface for a link to the YouTube channel.

In Problems 1–10, find the exact value of each expression. Express angles in radians.

1. $\sec^{-1}\left(\dfrac{2}{\sqrt{3}}\right)$

2. $\sin^{-1}\left(-\dfrac{\sqrt{2}}{2}\right)$

3. $\tan^{-1}(-\sqrt{3})$

4. $\cos^{-1}0$

5. $\cot^{-1}1$

6. $\csc^{-1}(-2)$

7. $\sin^{-1}\left(\sin\dfrac{11\pi}{5}\right)$

8. $\tan\left(\tan^{-1}\dfrac{7}{3}\right)$

9. $\cot\left(\csc^{-1}\sqrt{10}\right)$

10. $\sec\left(\cos^{-1}\left(-\dfrac{3}{4}\right)\right)$

In Problems 11–14, use a calculator to evaluate each expression. Express angles in radians rounded to two decimal places.

11. $\sin^{-1}0.382$

12. $\sec^{-1}1.4$

13. $\tan^{-1}3$

14. $\cot^{-1}5$

In Problems 15–20 establish each identity.

15. $\dfrac{\csc\theta + \cot\theta}{\sec\theta + \tan\theta} = \dfrac{\sec\theta - \tan\theta}{\csc\theta - \cot\theta}$

16. $\sin\theta\tan\theta + \cos\theta = \sec\theta$

17. $\tan\theta + \cot\theta = 2\csc(2\theta)$

18. $\dfrac{\sin(\alpha + \beta)}{\tan\alpha + \tan\beta} = \cos\alpha\cos\beta$

19. $\sin(3\theta) = 3\sin\theta - 4\sin^3\theta$

20. $\dfrac{\tan\theta - \cot\theta}{\tan\theta + \cot\theta} = 1 - 2\cos^2\theta$

In Problems 21–28 use sum, difference, product, or half-angle formulas to find the exact value of each expression.

21. $\cos 15°$

22. $\tan 75°$

23. $\sin\left(\dfrac{1}{2}\cos^{-1}\dfrac{3}{5}\right)$

24. $\tan\left(2\sin^{-1}\dfrac{6}{11}\right)$

25. $\cos\left(\sin^{-1}\dfrac{2}{3}+\tan^{-1}\dfrac{3}{2}\right)$

26. $\sin 75° \cos 15°$

27. $\sin 75° + \sin 15°$

28. $\cos 65° \cos 20° + \sin 65° \sin 20°$

In Problems 29–33, solve each equation on $0 \le \theta < 2\pi$.

29. $4\sin^2\theta - 3 = 0$

30. $-3\cos\left(\dfrac{\pi}{2} - \theta\right) = \tan\theta$

31. $\cos^2\theta + 2\sin\theta\cos\theta - \sin^2\theta = 0$

32. $\sin(\theta + 1) = \cos\theta$

33. $4\sin^2\theta + 7\sin\theta = 2$

Cumulative Review

1. Find the real solutions, if any, of the equation $3x^2 + x - 1 = 0$.

2. Find an equation for the line containing the points $(-2, 5)$ and $(4, -1)$. What is the distance between these points? What is their midpoint?

3. Test the equation $3x + y^2 = 9$ for symmetry with respect to the x-axis, y-axis, and origin. List the intercepts.

4. Use transformations to graph the equation $y = |x - 3| + 2$.

5. Use transformations to graph the equation

$$y = \cos\left(x - \dfrac{\pi}{2}\right) - 1$$

6. Graph each of the following functions. Label at least three points on each graph. Name the inverse function of each and show its graph.

   (a) $y = x^3$

   (b) $y = \sin x, \quad -\dfrac{\pi}{2} \le x \le \dfrac{\pi}{2}$

   (c) $y = \cos x, \quad 0 \le x \le \pi$

7. If $\sin\theta = -\dfrac{1}{3}$ and $\pi < \theta < \dfrac{3\pi}{2}$, find the exact value of:

   (a) $\cos\theta$ (b) $\tan\theta$ (c) $\sin(2\theta)$

   (d) $\cos(2\theta)$ (e) $\sin\left(\dfrac{1}{2}\theta\right)$ (f) $\cos\left(\dfrac{1}{2}\theta\right)$

8. Find the exact value of $\cos(\tan^{-1} 2)$.

9. If $\sin\alpha = \dfrac{1}{3}, \dfrac{\pi}{2} < \alpha < \pi$, and $\cos\beta = -\dfrac{1}{3}, \pi < \beta < \dfrac{3\pi}{2}$, find the exact value of:

   (a) $\cos\alpha$ (b) $\sin\beta$ (c) $\cos(2\alpha)$

   (d) $\cos(\alpha + \beta)$ (e) $\sin\dfrac{\beta}{2}$

Chapter Projects

 Internet-based Project

I. **Mapping Your Mind** The goal of this project is to organize the material learned in Chapters 2 and 3 in our minds. To do this, we will use mind-mapping software called Mindomo. Mindomo is free software that enables you to organize your thoughts digitally and share these thoughts with anyone on the Web. By organizing your thoughts, you can see the big picture and then communicate this big picture to others. You are also able to see how various concepts are related to each other.

1. Go to *http://www.mindomo.com* and register. Learn how to use Mindomo. A video on using Mindomo can be found at *http://www.screencast.com/t/ZPwJQDs4*
2. Use an Internet search engine to research Mind Mapping. Write a few paragraphs that explain the history and benefit of mind mapping.
3. Create a MindMap that explains the following:
   (a) The six trigonometric functions and their properties (including the inverses of these functions)
   (b) The fundamental trigonometric identities
      When creating your map, be creative! Perhaps you can share ideas about when a particular identity might be used, or when a particular identity cannot be used.
4. Share the MindMap so that students in your class can view it.

The following projects are available on the Instructor's Resource Center (IRC):

II. **Waves** Wave motion is described by a sinusoidal equation. The Principle of Superposition of two waves is discussed.

III. **Project at Motorola** *Sending Pictures Wirelessly* The electronic transmission of pictures is made practical by image compression, mathematical methods that greatly reduce the number of bits of data used to compose the picture.

IV. **Calculus of Differences** Finding consecutive difference quotients is called finding finite differences and is used to analyze the graph of an unknown function.

Applications of Trigonometric Functions

The Lewis and Clark Expedition

In today's world of GPS and smart phone apps that can precisely track one's location, it is difficult to fathom the magnitude of the challenge that confronted Meriwether Lewis and William Clark in 1804.

But Lewis and Clark managed. Commissioned by President Thomas Jefferson to explore the newly purchased Louisiana Territory, the co-captains led their expedition—the Corps of Discovery—on a journey that took nearly two and a half years and carried them more than 7000 miles. Starting at St. Louis, Missouri, they traveled up the Missouri River, across the Great Plains, over the Rocky Mountains, down the Columbia River to the Pacific Ocean, and then back. Along the way, using limited tools such as a compass and octant, they created more than 130 maps of the area with remarkable detail and accuracy.

 —See Chapter Project II—

← A Look Back

In Chapter 2, we defined the six trigonometric functions using the unit circle. In particular, we learned to evaluate the trigonometric functions. We also learned how to graph sinusoidal functions. In Chapter 3, we defined the inverse trigonometric functions and solved equations involving the trigonometric functions.

A Look Ahead →

In this chapter, we define the trigonometric functions using right triangles and then use the trigonometric functions to solve applied problems. The first four sections deal with applications involving right triangles and *oblique triangles*, triangles that do not have a right angle. To solve problems involving oblique triangles, we will develop the Law of Sines and the Law of Cosines. We will also develop formulas for finding the area of a triangle.

The final section deals with applications of sinusoidal functions involving simple harmonic motion and damped motion.

Outline

4.1 Right Triangle Trigonometry; Applications

PREPARING FOR THIS SECTION *Before getting started, review the following:*

- Pythagorean Theorem (Section A.2, pp. A14–A15)
- Trigonometric Equations (Section 3.3, pp. 209–214)

Now Work the '**Are You Prepared?**' problems on page 269.

OBJECTIVES 1 Find the Value of Trigonometric Functions of Acute Angles Using Right Triangles (p. 262)
2 Use the Complementary Angle Theorem (p. 264)
3 Solve Right Triangles (p. 264)
4 Solve Applied Problems (p. 265)

1 Find the Value of Trigonometric Functions of Acute Angles Using Right Triangles

A triangle in which one angle is a right angle (90°) is called a **right triangle**. Recall that the side opposite the right angle is called the **hypotenuse**, and the remaining two sides are called the **legs** of the triangle. In Figure 1(a), the hypotenuse is labeled as c to indicate that its length is c units, and, in a like manner, the legs are labeled as a and b. Because the triangle is a right triangle, the Pythagorean Theorem tells us that

$$a^2 + b^2 = c^2$$

Figure 1(a) also shows the angle θ. The angle θ is an **acute angle**: that is, $0° < \theta < 90°$ for θ measured in degrees and $0 < \theta < \dfrac{\pi}{2}$ for θ measured in radians. Place θ in standard position, as shown in Figure 1(b). Then the coordinates of the point P are (a, b). Also, P is a point on the terminal side of θ that is on the circle $x^2 + y^2 = c^2$. (Do you see why?)

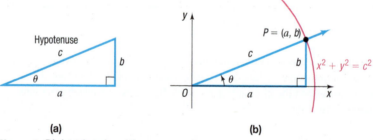

(a)   **(b)**

Figure 1 Right triangle with acute angle θ

Now apply the theorem on page 126 for evaluating trigonometric functions using a circle of radius c, $x^2 + y^2 = c^2$. By referring to the lengths of the sides of the triangle by the names *hypotenuse* (c), *opposite* (b), and *adjacent* (a), as indicated in Figure 2, the trigonometric functions of θ can be expressed as ratios of the sides of a right triangle.

Figure 2 Right triangle

$$\sin \theta = \frac{\text{Opposite}}{\text{Hypotenuse}} = \frac{b}{c} \qquad \csc \theta = \frac{\text{Hypotenuse}}{\text{Opposite}} = \frac{c}{b}$$

$$\cos \theta = \frac{\text{Adjacent}}{\text{Hypotenuse}} = \frac{a}{c} \qquad \sec \theta = \frac{\text{Hypotenuse}}{\text{Adjacent}} = \frac{c}{a} \qquad \textbf{(1)}$$

$$\tan \theta = \frac{\text{Opposite}}{\text{Adjacent}} = \frac{b}{a} \qquad \cot \theta = \frac{\text{Adjacent}}{\text{Opposite}} = \frac{a}{b}$$

Notice that each trigonometric function of the acute angle θ is positive.

EXAMPLE 1

Finding the Value of Trigonometric Functions from a Right Triangle

Find the exact value of the six trigonometric functions of the angle θ in Figure 3.

Solution

In Figure 3 the two given sides of the triangle are

$$c = \text{Hypotenuse} = 5 \qquad a = \text{Adjacent} = 3$$

To find the length of the opposite side, use the Pythagorean Theorem.

$$(\text{Adjacent})^2 + (\text{Opposite})^2 = (\text{Hypotenuse})^2$$
$$3^2 + (\text{Opposite})^2 = 5^2$$
$$(\text{Opposite})^2 = 25 - 9 = 16$$
$$\text{Opposite} = 4$$

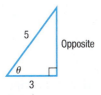

Figure 3

Now that the lengths of the three sides are known, use the ratios in (1) to find the value of each of the six trigonometric functions.

$$\sin \theta = \frac{\text{Opposite}}{\text{Hypotenuse}} = \frac{4}{5} \qquad \cos \theta = \frac{\text{Adjacent}}{\text{Hypotenuse}} = \frac{3}{5} \qquad \tan \theta = \frac{\text{Opposite}}{\text{Adjacent}} = \frac{4}{3}$$

$$\csc \theta = \frac{\text{Hypotenuse}}{\text{Opposite}} = \frac{5}{4} \qquad \sec \theta = \frac{\text{Hypotenuse}}{\text{Adjacent}} = \frac{5}{3} \qquad \cot \theta = \frac{\text{Adjacent}}{\text{Opposite}} = \frac{3}{4}$$

➤ **Now Work** PROBLEM 9

The values of the trigonometric functions of an acute angle are ratios of the lengths of the sides of a right triangle. This way of viewing the trigonometric functions leads to many applications and, in fact, was the point of view used by early mathematicians (before calculus) in studying the subject of trigonometry.

EXAMPLE 2

Constructing a Rain Gutter

A rain gutter is to be constructed of aluminum sheets 12 inches wide. See Figure 4(a). After marking off a length of 4 inches from each edge, the sides are bent up at an angle θ. See Figure 4(b).

(a) Express the area A of the opening as a function of θ.

(b) Graph $A = A(\theta)$. Find the angle θ that makes A largest. (This bend will allow the most water to flow through the gutter.)

Solution

(a) Look again at Figure 4(b). The area A of the opening is the sum of the areas of two congruent right triangles and one rectangle. Look at Figure 4(c), which shows the triangle on the right in Figure 4(b) redrawn. Note that

$$\cos \theta = \frac{a}{4}, \quad \text{so} \quad a = 4 \cos \theta \qquad \sin \theta = \frac{b}{4}, \quad \text{so} \quad b = 4 \sin \theta$$

The area of the triangle is

$$\text{area} = \frac{1}{2} \cdot \text{base} \cdot \text{height} = \frac{1}{2} ab = \frac{1}{2} \cdot 4 \cos \theta \cdot 4 \sin \theta = 8 \sin \theta \cos \theta$$

So the area of the two congruent triangles together is $2 \cdot 8 \sin \theta \cos \theta = 16 \sin \theta \cos \theta$. The rectangle has length 4 and height b, so its area is

$$\text{area of rectangle} = 4b = 4 \cdot 4 \sin \theta = 16 \sin \theta$$

Figure 4

(continued)

Figure 5

The area A of the opening is

$$A = \text{area of the two congruent triangles} + \text{area of the rectangle}$$
$$A(\theta) = 16 \sin \theta \cos \theta + 16 \sin \theta = 16 \sin \theta (\cos \theta + 1)$$

(b) Figure 5 shows the graph of $A = A(\theta)$ on a TI-84 Plus C. Using MAXIMUM, the angle θ that makes A largest is 60°.

2 Use the Complementary Angle Theorem

Two acute angles are called **complementary** if their sum is a right angle, or 90°. Because the sum of the angles of any triangle is 180°, it follows that, for a right triangle, the sum of the acute angles is 90°, so the two acute angles in every right triangle are complementary.

Refer now to Figure 6, which labels the angle opposite side b as B and the angle opposite side a as A. Notice that side b is adjacent to angle A and side a is adjacent to angle B. As a result,

Figure 6

$$\sin B = \frac{b}{c} = \cos A \qquad \cos B = \frac{a}{c} = \sin A \qquad \tan B = \frac{b}{a} = \cot A$$

$$\csc B = \frac{c}{b} = \sec A \qquad \sec B = \frac{c}{a} = \csc A \qquad \cot B = \frac{a}{b} = \tan A \tag{2}$$

Because of these relationships, the functions sine and cosine, tangent and cotangent, and secant and cosecant are called **cofunctions** of each other.

The identities (2) may be expressed in words as follows:

> **THEOREM Complementary Angle Theorem**
>
> Cofunctions of complementary angles are equal.

Examples of this theorem are given next:

Complementary angles   **Complementary angles**   **Complementary angles**

$$\sin 30° = \cos 60° \qquad \tan 40° = \cot 50° \qquad \sec 80° = \csc 10°$$

Cofunctions   **Cofunctions**   **Cofunctions**

EXAMPLE 3 **Using the Complementary Angle Theorem**

(a) $\sin 62° = \cos(90° - 62°) = \cos 28°$

(b) $\tan \dfrac{\pi}{12} = \cot\left(\dfrac{\pi}{2} - \dfrac{\pi}{12}\right) = \cot \dfrac{5\pi}{12}$

(c) $\sin^2 40° + \sin^2 50° = \sin^2 40° + \cos^2 40° = 1$

$$\sin 50° = \cos 40°$$

 Now Work PROBLEM 19

3 Solve Right Triangles

In the discussion that follows, a right triangle is always labeled so that side a is opposite angle A, side b is opposite angle B, and side c is the hypotenuse, as shown in Figure 7. **To solve a right triangle** means to find the lengths of its sides and the measurements of its angles. We express the lengths of the sides rounded to two decimal places and angle measures in degrees rounded to one decimal place. (Be sure that your calculator is in degree mode.)

Figure 7 Right triangle

To solve a right triangle, we need to know one of the acute angles A or B and a side, or else two sides (in which case the Pythagorean Theorem can be used). Also, because the sum of the measures of the angles of a triangle is 180°, the sum of the measures of angles A and B in a right triangle must be 90°.

THEOREM Properties of a Right Triangle

For the right triangle shown in Figure 7, we have

$$c^2 = a^2 + b^2 \qquad A + B = 90°$$

EXAMPLE 4 **Solving a Right Triangle**

Use Figure 8. If $b = 2$ and $A = 40°$, find $a, c,$ and B.

Solution Because $A = 40°$ and $A + B = 90°$, it follows that $B = 50°$. To find the sides a and c, use the facts that

$$\tan 40° = \frac{a}{2} \quad \text{and} \quad \cos 40° = \frac{2}{c}$$

Now solve for a and c.

$$a = 2\tan 40° \approx 1.68 \quad \text{and} \quad c = \frac{2}{\cos 40°} \approx 2.61$$

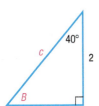

Figure 8

Now Work PROBLEM 29

EXAMPLE 5 **Solving a Right Triangle**

Use Figure 9. If $a = 3$ and $b = 2$, find $c, A,$ and B.

Solution Because $a = 3$ and $b = 2$, then, by the Pythagorean Theorem,

$$c^2 = a^2 + b^2 = 3^2 + 2^2 = 9 + 4 = 13$$
$$c = \sqrt{13} \approx 3.61$$

To find angle A, use the fact that

$$\tan A = \frac{3}{2} \quad \text{so} \quad A = \tan^{-1}\frac{3}{2}$$

Figure 9

NOTE To avoid round-off errors when using a calculator, we will store unrounded values in memory for use in subsequent calculations. ∎

Use a calculator with the mode set to degrees to find that $A = 56.3°$ rounded to one decimal place. Since $A + B = 90°$, this means that $B = 33.7°$.

Now Work PROBLEM 39

---

4 Solve Applied Problems*

In addition to developing models using right triangles, we can use right triangle trigonometry to measure heights and distances that are either awkward or impossible to measure by ordinary means. When using right triangles to solve these problems, pay attention to the known measures. This will indicate what trigonometric function to use. For example, if you know the measure of an angle and the length of the side adjacent to the angle, and wish to find the length of the opposite side, you would use the tangent function. Do you know why?

---

*In applied problems, it is important that answers be reported with both justifiable accuracy and appropriate significant figures. In this chapter we shall assume that the problem data are accurate to the number of significant digits resulting in sides being rounded to two decimal places and angles being rounded to one decimal place.

EXAMPLE 6

Finding the Width of a River

A surveyor can measure the width of a river by setting up a transit* at a point C on one side of the river and taking a sighting of a point A on the other side. Refer to Figure 10. After turning through an angle of 90° at C, the surveyor walks a distance of 200 meters to point B. Using the transit at B, the angle θ is measured and found to be 20°. What is the width of the river rounded to the nearest meter?

Figure 10

Solution

As seen in Figure 10, the width of the river is the length of side b, and a and θ are known. Use the facts that b is opposite θ and a is adjacent to θ and write

$$\tan \theta = \frac{b}{a}$$

which leads to

$$\tan 20° = \frac{b}{200}$$

$$b = 200 \tan 20° \approx 72.79 \text{ meters}$$

The width of the river is 73 meters, rounded to the nearest meter.

➤ **Now Work** PROBLEM 49

EXAMPLE 7

Finding the Inclination of a Mountain Trail

A straight trail leads from the Alpine Hotel, elevation 8000 feet, to a scenic overlook, elevation 11,100 feet. The length of the trail is 14,100 feet. What is the inclination (grade) of the trail? That is, what is the measure of angle B in Figure 11?

Solution

From Figure 11, the length of the side opposite angle B is $11,100 - 8000 = 3100$ feet, and the length of the hypotenuse is 14,100 feet. Angle B satisfies the equation

$$\sin B = \frac{3100}{14,100}$$

Using a calculator,

$$B = \sin^{-1} \frac{3100}{14,100} \approx 12.7°$$

The inclination (grade) of the trail is approximately 12.7°.

Figure 11

➤ **Now Work** PROBLEM 55

Vertical heights can sometimes be measured using either the *angle of elevation* or the *angle of depression*. If a person is looking up at an object, the acute angle measured from the horizontal to a line of sight to the object is called the **angle of elevation**. See Figure 12(a).

*An instrument used in surveying to measure angles.

(a) Angle of elevation

(b) Angle of depression

Figure 12

If a person is looking down at an object, the acute angle made by the line of sight to the object and the horizontal is called the **angle of depression**. See Figure 12(b).

EXAMPLE 8

Finding the Height of a Cloud

Meteorologists find the height of a cloud using an instrument called a **ceilometer**. A ceilometer consists of a **light projector** that directs a vertical light beam up to the cloud base and a **light detector** that scans the cloud to detect the light beam. See Figure 13(a). At Midway Airport in Chicago, a ceilometer was employed to find the height of the cloud cover. It was set up with its light detector 300 feet from its light projector. If the angle of elevation from the light detector to the base of the cloud was 75°, what was the height of the cloud cover?

Illuminated spot
on base of clouds

Vertical light beam

Cloud height h

h

Base b

θ

Light detector

Light projector

75°

300 ft

(a)

(b)

Figure 13

Solution

Figure 13(b) illustrates the situation. To find the height h, use the fact that $\tan 75° = \dfrac{h}{300}$, so

$$h = 300 \tan 75° \approx 1120 \text{ feet}$$

The ceiling (height to the base of the cloud cover) was approximately 1120 feet. ⌐

Now Work PROBLEM 51

The idea behind Example 8 can also be used to find the height of an object that is positioned above ground level.

EXAMPLE 9

Finding the Height of a Statue on a Building

Adorning the top of the Board of Trade building in Chicago is a statue of Ceres, the Roman goddess of wheat. From street level, two observations are taken 400 feet from the center of the building. The angle of elevation to the base of the statue is found to be 55.1°, and the angle of elevation to the top of the statue is 56.5°. See Figure 14(a) on the next page. What is the height of the statue?

Solution Figure 14(b) shows two triangles that replicate Figure 14(a). The height of the statue of Ceres will be $b' - b$. To find b and b', refer to Figure 14(b).

(a)

(b)

Figure 14

$$\tan 55.1° = \frac{b}{400} \qquad\qquad \tan 56.5° = \frac{b'}{400}$$

$$b = 400 \tan 55.1° \approx 573.39 \qquad b' = 400 \tan 56.5° \approx 604.33$$

The height of the statue is approximately $604.33 - 573.39 = 30.94$ feet ≈ 31 feet. ⌐

━━━ **Now Work** PROBLEM 71

EXAMPLE 10 **The Gibb's Hill Lighthouse, Southampton, Bermuda**

In operation since 1846, the Gibb's Hill Lighthouse stands 117 feet high on a hill 245 feet high, so its beam of light is 362 feet above sea level. A brochure states that the light can be seen on the horizon about 26 miles distant. Verify the accuracy of this statement.

Solution Figure 15 illustrates the situation. The central angle θ, positioned at the center of Earth, radius 3960 miles, obeys the equation

Figure 15

$$\cos \theta = \frac{3960}{3960 + \dfrac{362}{5280}} \approx 0.999982687 \qquad \text{1 mile = 5280 feet}$$

Solving for θ yields

$$\theta \approx \cos^{-1}(0.999982687) \approx 0.33715° \approx 20.23'$$

The brochure does not indicate whether the distance is measured in nautical miles or statute miles. Let's calculate both distances.

The distance s in nautical miles (refer to Problem 120, p. 113) is the measure of the angle θ in minutes, so $s \approx 20.23$ nautical miles.

The distance s in statute miles is given by the formula $s = r\theta$, where θ is measured in radians. Then, since

$$\theta \approx 0.33715° \approx 0.00588 \text{ radian}$$
$$\uparrow$$
$$1° = \frac{\pi}{180} \text{ radian}$$

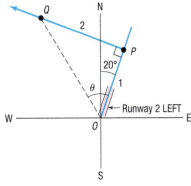

Figure 16

this means that

$$s = r\theta \approx 3960 \cdot 0.00588 \approx 23.3 \text{ miles}$$

In either case, it would seem that the brochure overstated the distance somewhat.

In navigation and surveying, the **direction** or **bearing** from a point O to a point P equals the acute angle θ between the ray OP and the vertical line through O, the north–south line.

Figure 16 illustrates some bearings. Notice that the bearing from O to P_1 is denoted by the symbolism N30°E, indicating that the bearing is 30° east of north. In writing the bearing from O to P, the direction north or south always appears first, followed by an acute angle, followed by east or west. In Figure 16, the bearing from O to P_2 is S50°W, and from O to P_3 it is N70°W.

EXAMPLE 11 **Finding the Bearing of an Object**

In Figure 16, what is the bearing from O to an object at P_4?

Solution The acute angle between the ray OP_4 and the north–south line through O is 20°. The bearing from O to P_4 is S20°E.

EXAMPLE 12 **Finding the Bearing of an Airplane**

A Boeing 777 aircraft takes off from O'Hare Airport on runway 2 LEFT, which has a bearing of N20°E.* After flying for 1 mile, the pilot of the aircraft requests permission to turn 90° and head toward the northwest. The request is granted. After the plane goes 2 miles in this direction, what bearing should the control tower use to locate the aircraft?

Solution Figure 17 illustrates the situation. After flying 1 mile from the airport O (the control tower), the aircraft is at P. After turning 90° toward the northwest and flying 2 miles, the aircraft is at the point Q. In triangle OPQ, the angle θ obeys the equation

$$\tan\theta = \frac{2}{1} = 2 \quad \text{so} \quad \theta = \tan^{-1}2 \approx 63.4°$$

The acute angle between north and the ray OQ is 63.4° − 20° = 43.4°. The bearing of the aircraft from O to Q is N43.4°W.

Now Work PROBLEM 63

Figure 17

4.1 Assess Your Understanding

'Are You Prepared?' *Answers are given at the end of these exercises. If you get a wrong answer, read the pages listed in red.*

1. In a right triangle, if the length of the hypotenuse is 5 and the length of one of the other sides is 3, what is the length of the third side? (pp. A14–A15)

2. If θ is an acute angle, solve the equation $\tan\theta = \frac{1}{2}$. Express your answer in degrees, rounded to one decimal place. (pp. 209–214)

3. If θ is an acute angle, solve the equation $\sin\theta = \frac{1}{2}$. (pp. 209–214)

*In air navigation, the term **azimuth** denotes the positive angle measured clockwise from the north (N) to a ray OP. In Figure 16, the azimuth from O to P_1 is 30°; the azimuth from O to P_2 is 230°; the azimuth from O to P_3 is 290°. In naming runways, the units digit is left off the azimuth. Runway 2 LEFT means the left runway with a direction of azimuth 20° (bearing N20°E). Runway 23 is the runway with azimuth 230° and bearing S50°W.

Concepts and Vocabulary

4. *True or False* sin 52° = cos 48°

5. *Multiple Choice* The sum of the measures of the two acute angles in a right triangle is _____.
(a) 45° **(b)** 90° **(c)** 180° **(d)** 360°

6. When you look up at an object, the acute angle measured from the horizontal to a line-of-sight observation of the object is called the _____ ___ _____.

7. *True or False* In a right triangle, if two sides are known, we can solve the triangle.

8. *True or False* In a right triangle, if we know the two acute angles, we can solve the triangle.

Skill Building

In Problems 9–18, find the exact value of the six trigonometric functions of the angle θ in each figure.

9.
10.
11.
12.
13.

14.
15.
16.
17.
18.

In Problems 19–28, find the exact value of each expression. Do not use a calculator.

19. sin 38° − cos 52°

20. tan 12° − cot 78°

21. $\dfrac{\cos 10°}{\sin 80°}$

22. $\dfrac{\cos 40°}{\sin 50°}$

23. $1 - \cos^2 20° - \cos^2 70°$

24. $1 + \tan^2 5° - \csc^2 85°$

25. $\tan 20° - \dfrac{\cos 70°}{\cos 20°}$

26. $\cot 40° - \dfrac{\sin 50°}{\sin 40°}$

27. cos 35° sin 55° + sin 35° cos 55°

28. sec 35° csc 55° − tan 35° cot 55°

In Problems 29–42, use the right triangle shown below. Then, using the given information, solve the triangle.

29. $b = 5$, $B = 20°$; find $a, c,$ and A

31. $a = 6$, $B = 40°$; find $b, c,$ and A

33. $b = 4$, $A = 10°$; find $a, c,$ and B

30. $b = 4$, $B = 10°$; find $a, c,$ and A

32. $a = 7$, $B = 50°$; find $b, c,$ and A

34. $b = 6$, $A = 20°$; find $a, c,$ and B

35. $a = 5$, $A = 25°$; find $b, c,$ and B

37. $c = 9$, $B = 20°$; find $b, a,$ and A

39. $a = 5$, $b = 3$; find $c, A,$ and B

41. $a = 2$, $c = 5$; find $b, A,$ and B

36. $a = 6$, $A = 40°$; find $b, c,$ and B

38. $c = 10$, $A = 40°$; find $b, a,$ and B

40. $a = 2$, $b = 8$; find $c, A,$ and B

42. $b = 4$, $c = 6$; find $a, A,$ and B

Applications and Extensions

43. Geometry The hypotenuse of a right triangle is 5 inches. If one leg is 2 inches, find the degree measure of each angle.

44. Geometry The hypotenuse of a right triangle is 3 feet. If one leg is 1 foot, find the degree measure of each angle.

45. Geometry A right triangle has a hypotenuse of length 8 inches. If one angle is 35°, find the length of each leg.

46. Geometry A right triangle has a hypotenuse of length 10 centimeters. If one angle is 40°, find the length of each leg.

47. Geometry A right triangle contains a 25° angle.
(a) If one leg is of length 5 inches, what is the length of the hypotenuse?
(b) There are two answers. How is this possible?

48. Geometry A right triangle contains an angle of $\dfrac{\pi}{8}$ radian.
(a) If one leg is of length 3 meters, what is the length of the hypotenuse?
(b) There are two answers. How is this possible?

49. Finding the Width of a Gorge Find the distance from A to C across the gorge illustrated in the figure.

50. Finding the Distance across a Pond Find the distance from A to C across the pond illustrated in the figure.

51. The Eiffel Tower The tallest tower built before the era of television masts, the Eiffel Tower was completed on March 31, 1889. Find the height of the Eiffel Tower (before a television mast was added to the top) using the information given in the figure.

52. Finding the Distance of a Ship from Shore A person in a small boat, offshore from a vertical cliff known to be 100 feet in height, takes a sighting of the top of the cliff. If the angle of elevation is found to be 25°, how far offshore is the boat?

53. Finding the Distance to a Plateau Suppose that you are headed toward a plateau 50 meters high. If the angle of elevation to the top of the plateau is 20°, how far are you from the base of the plateau?

54. Finding the Reach of a Ladder A 22-foot extension ladder leaning against a building makes a 70° angle with the ground. How far up the building does the ladder touch?

55. Finding the Angle of Elevation of the Sun At 10 AM on April 26, 2019, a building 300 feet high cast a shadow 50 feet long. What was the angle of elevation of the Sun?

56. Directing a Laser Beam A laser beam is to be directed through a small hole in the center of a circle of radius 10 feet. The origin of the beam is 35 feet from the circle (see the figure). At what angle of elevation should the beam be aimed to ensure that it goes through the hole?

57. Finding the Speed of a Truck A state trooper is hidden 30 feet from a highway. One second after a truck passes, the angle θ between the highway and the line of observation from the patrol car to the truck is measured. See the figure.

(a) If the angle measures 15°, how fast is the truck traveling? Express the answer in feet per second and in miles per hour.

(b) If the angle measures 20°, how fast is the truck traveling? Express the answer in feet per second and in miles per hour.

(c) If the speed limit is 55 miles per hour and a speeding ticket is issued for speeds of 5 miles per hour or more over the limit, for what angles should the trooper issue a ticket?

58. Security A security camera in a neighborhood bank is mounted on a wall 9 feet above the floor. What angle of depression should be used if the camera is to be directed to a spot 6 feet above the floor and 12 feet from the wall?

59. Parallax One method of measuring the distance from Earth to a star is the parallax method. The idea behind computing this distance is to measure the angle formed between the Earth and the star at two different points in time. Typically, the measurements are taken so that the side opposite the angle is as large as possible. Therefore, the optimal approach is to measure the angle when Earth is on opposite sides of the Sun, as shown in the figure.

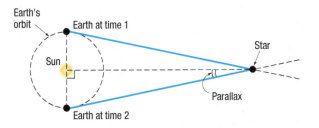

(a) Proxima Centauri is 4.22 light-years from Earth. If 1 light-year is about 5.9 trillion miles, how many miles is Proxima Centauri from Earth?

(b) The mean distance from Earth to the Sun is 93,000,000 miles. What is the parallax of Proxima Centauri?

60. Parallax See Problem 59. The star 61 Cygni, sometimes called Bessel's Star (after Friedrich Bessel, who measured the distance from Earth to the star in 1838), is a star in the constellation Cygnus.

(a) 61 Cygni is 11.14 light-years from Earth. If 1 light-year is about 5.9 trillion miles, how many miles is 61 Cygni from Earth?

(b) The mean distance from Earth to the Sun is 93,000,000 miles. What is the parallax of 61 Cygni?

61. Washington Monument The angle of elevation of the Sun is 35.1° at the instant the shadow cast by the Washington Monument is 789 feet long. Use this information to calculate the height of the monument.

62. Finding the Length of a Mountain Trail A straight trail with an inclination of 17° leads from a hotel at an elevation of 9000 feet to a mountain lake at an elevation of 11,200 feet. What is the length of the trail?

63. Finding the Bearing of an Aircraft A DC-9 aircraft leaves Midway Airport from runway 4 RIGHT, whose bearing is N40°E. After flying for $\frac{1}{2}$ mile, the pilot requests permission to turn 90° and head toward the southeast. The permission is granted. After the airplane goes 1 mile in this direction, what bearing should the control tower use to locate the aircraft?

64. Finding the Bearing of a Ship A ship leaves the port of Miami with a bearing of S80°E and a speed of 15 knots. After 1 hour, the ship turns 90° toward the south. After 2 hours, maintaining the same speed, what is the bearing to the ship from port?

65. Niagara Falls Incline Railway Situated between Portage Road and the Niagara Parkway directly across from the Canadian Horseshoe Falls, the Falls Incline Railway is a funicular that carries passengers up an embankment to Table Rock Observation Point. If the length of the track is 51.8 meters and the angle of inclination is 36°2′, determine the height of the embankment.

Source: www.niagaraparks.com

66. Willis Tower Willis Tower in Chicago is the second tallest building in the United States and is topped by a high antenna. A surveyor on the ground makes the following measurements:

1. The angle of elevation from his position to the top of the building is 34°.

2. The distance from his position to the top of the building is 2595 feet.

3. The distance from his position to the top of the antenna is 2760 feet.

(a) How far away from the (base of the) building is the surveyor located?

(b) How tall is the building?

(c) What is the angle of elevation from the surveyor to the top of the antenna?

(d) How tall is the antenna?

Source: Council on Tall Buildings and Urban Habitat

67. Constructing a Highway A highway whose primary directions are north–south is being constructed along the west coast of Florida. Near Naples, a bay obstructs the straight path of the road. Since the cost of a bridge is prohibitive, engineers decide to go around the bay. The figure shows the path that they decide on and the measurements taken. What is the length of highway needed to go around the bay?

68. Photography A camera is mounted on a tripod 4 feet high at a distance of 10 feet from George, who is 6 feet tall. See the figure. If the camera lens has angles of depression and elevation of 20°, will George's feet and head be seen by the lens? If not, how far back will the camera need to be moved to include George's feet and head?

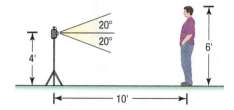

69. Finding the Distance between Two Objects A blimp, suspended in the air at a height of 500 feet, lies directly over a line from Soldier Field to the Adler Planetarium on Lake Michigan (see the figure). If the angle of depression from the blimp to the stadium is 32° and from the blimp to the planetarium is 23°, find the distance between Soldier Field and the Adler Planetarium.

70. Hot-Air Balloon While taking a ride in a hot-air balloon in Napa Valley, Francisco wonders how high he is. To find out, he chooses a landmark that is to the east of the balloon and measures the angle of depression to be 54°. A few minutes later, after traveling 100 feet east, the angle of depression to the same landmark is determined to be 61°. Use this information to determine the height of the balloon.

71. Mt. Rushmore To measure the height of Lincoln's caricature on Mt. Rushmore, two sightings 800 feet from the base of the mountain are taken. If the angle of elevation to the bottom of Lincoln's face is 32° and the angle of elevation to the top is 35°, what is the height of Lincoln's face?

72. The CN Tower The CN Tower, located in Toronto, Canada, is the tallest structure in the Americas. While visiting Toronto, a tourist wondered what the height of the tower above the top of the Sky Pod is. While standing 4000 feet from the tower, she measured the angle to the top of the Sky Pod to be 20.1°. At this same distance, the angle of elevation to the top of the tower was found to be 24.4°. Use this information to determine the height of the tower above the Sky Pod.

73. Chicago Skyscrapers The angle of inclination from the base of the John Hancock Center to the top of the main structure of the Willis Tower is approximately 10.3°. If the main structure of the Willis Tower is 1451 feet tall, how far apart are the two skyscrapers? Assume the bases of the two buildings are at the same elevation.

Source: www.emporis.com

74. Estimating the Width of the Mississippi River A tourist at the top of the Gateway Arch (height, 630 feet) in St. Louis, Missouri, observes a boat moored on the Illinois side of the Mississippi River 2070 feet directly across from the Arch. She also observes a boat moored on the Missouri side directly across from the first boat (see figure). Given that $B = \cot^{-1}\dfrac{67}{55}$, estimate the width of the Mississippi River at the St. Louis riverfront.

Source: U.S. Army Corps of Engineers

75. Finding the Pitch of a Roof A carpenter is preparing to put a roof on a garage that is 20 feet by 40 feet by 20 feet. A steel support beam 46 feet in length is positioned in the center of the garage. To support the roof, another beam will be attached to the top of the center beam (see the figure). At what angle of elevation is the new beam? In other words, what is the pitch of the roof?

76. Shooting Free Throws in Basketball The eyes of a basketball player are 6 feet above the floor. The player is at the free-throw line, which is 15 feet from the center of the basket rim (see the figure). What is the angle of elevation from the player's eyes to the center of the rim?

[**Hint:** The rim is 10 feet above the floor.]

77. Geometry Find the value of the angle θ in degrees rounded to the nearest tenth of a degree.

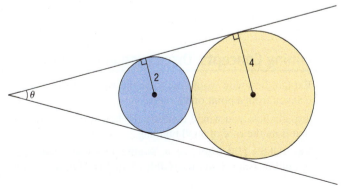

78. Surveillance Satellites A surveillance satellite circles Earth at a height of h miles above the surface. Suppose that d is the distance, in miles, on the surface of Earth that can be observed from the satellite. See the figure.

(a) Find an equation that relates the central angle θ to the height h.

(b) Find an equation that relates the observable distance d and θ.

(c) Find an equation that relates d and h.

(d) If d is to be 2500 miles, how high must the satellite orbit above Earth?

(e) If the satellite orbits at a height of 300 miles, what distance d on the surface can be observed?

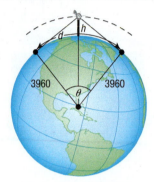

79. Calculating Pool Shots A pool player located at **X** wants to shoot the white ball off the top cushion and hit the red ball dead center. He knows from physics that the white ball will come off a cushion at the same angle as that at which it hit the cushion. If the deflection angle, θ, is 52°, where on the top cushion should he hit the white ball?

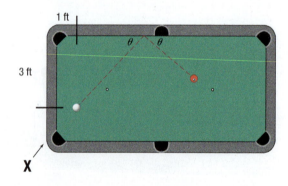

80. One World Trade Center One World Trade Center (1WTC) is the centerpiece of the rebuilding of the World Trade Center in New York City. The tower is 1776 feet tall (including its spire). The angle of elevation from the base of an office

building to the tip of the spire is 34°. The angle of elevation from the helipad on the roof of the office building to the tip of the spire is 20°.

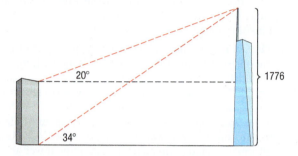

(a) How far away is the office building from One World Trade Center? Assume the side of the tower is vertical. Round to the nearest foot.

(b) How tall is the office building? Round to the nearest foot.

81. Challenge Problem Drive Wheel of an Engine The drive wheel of an engine is 13 inches in diameter, and the pulley on the rotary pump is 5 inches in diameter. If the shafts of the drive wheel and the pulley are 2 feet apart, what length of belt is required to join them as shown in the figure?

82. Challenge Problem Rework Problem 81 if the belt is crossed, as shown in the figure.

Explaining Concepts: Discussion and Writing

83. Explain how you would measure the width of the Grand Canyon from a point on its ridge.

84. Explain how you would measure the height of a TV tower that is on the roof of a tall building.

85. The Gibb's Hill Lighthouse, Southampton, Bermuda In operation since 1846, the Gibb's Hill Lighthouse stands

117 feet high on a hill 245 feet high, so its beam of light is 362 feet above sea level. A brochure states that ships 40 miles away can see the light and planes flying at 10,000 feet can see it 120 miles away. Verify the accuracy of these statements. What assumption did the brochure make about the height of the ship?

Retain Your Knowledge

Problems 86–94 are based on material learned earlier in the course. The purpose of these problems is to keep the material fresh in your mind so that you are better prepared for the final exam.

86. Determine whether $x - 3$ is a factor of $x^4 + 2x^3 - 21x^2 + 19x - 3$.

87. Find the exact value of $\sin \dfrac{\pi}{12}$.

88. If $f(x) = \sqrt{x}$, find $\dfrac{f(x) - f(4)}{x - 4}$, for $x = 5, 4.5,$ and 4.1. Round results to three decimal places.

89. Solve $2\sin^2\theta - \sin\theta + 5 = 6$ for $0 \le \theta < 2\pi$.

90. If the two triangles shown are similar, find x.

91. If a 4th degree polynomial function with real coefficients has zeros of 2, 7, and $3 - \sqrt{5}$, what is the remaining zero?

92. What is the remainder when $P(x) = 2x^4 - 3x^3 - x + 7$ is divided by $x + 2$?

93. Write the equation of a circle with radius $r = \sqrt{5}$ and center $(-4, 0)$ in standard from.

94. Find the domain of $g(x) = 3|x^2 - 1| - 5$.

'Are You Prepared?' Answers

1. 4   **2.** 26.6°   **3.** 30°

4.2 The Law of Sines

PREPARING FOR THIS SECTION *Before getting started, review the following:*

- Trigonometric Equations (Section 3.3, pp. 209–214)
- Difference Formula for the Sine Function (Section 3.5, p. 230)
- Geometry Essentials (Section A.2, pp. A14–A19)
- Approximating the Value of a Trigonometric Function (Section 2.2, p. 125)
- Approximating the Value of an Inverse Trigonometric Function (Section 3.2, pp. 204–205)

Now Work the 'Are You Prepared?' problems on page 282.

OBJECTIVES 1 Solve SAA or ASA Triangles (p. 276)
2 Solve SSA Triangles (p. 277)
3 Solve Applied Problems (p. 279)

If none of the angles of a triangle is a right angle, the triangle is called **oblique**. An oblique triangle will have either three acute angles or two acute angles and one obtuse angle (an angle measuring between 90° and 180°). See Figure 18.

Figure 18   **(a)** All angles are acute   **(b)** Two acute angles and one obtuse angle

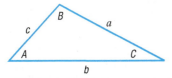

Figure 19 Oblique triangle

In the discussion that follows, an oblique triangle is always labeled so that side a is opposite angle A, side b is opposite angle B, and side c is opposite angle C, as shown in Figure 19.

WARNING Oblique triangles cannot be solved using the methods of Section 4.1. Do you know why? ∎

To **solve an oblique triangle** means to find the lengths of its sides and the measures of its angles. To do this, we need to know the length of one side,* along with (i) two angles, (ii) one angle and one other side, or (iii) the other two sides. There are four possibilities to consider.

> **CASE 1:** One side and two angles are known (ASA or SAA).
> **CASE 2:** Two sides and the angle opposite one of them are known (SSA).
> **CASE 3:** Two sides and the included angle are known (SAS).
> **CASE 4:** Three sides are known (SSS).

Figure 20 illustrates the four cases, where the known measurements are shown in blue.

Case 1: ASA

Case 1: SAA

Case 2: SSA

Case 3: SAS

Case 4: SSS

Figure 20

The **Law of Sines** is used to solve triangles for which Case 1 or 2 holds. Cases 3 and 4 are considered when we study the Law of Cosines in the next section.

> **THEOREM Law of Sines**
>
> For a triangle with sides a, b, c and opposite angles A, B, C, respectively,
>
> $$\frac{\sin A}{a} = \frac{\sin B}{b} = \frac{\sin C}{c} \qquad (1)$$

A proof of the Law of Sines is given at the end of this section.
The Law of Sines actually consists of three equalities:

$$\frac{\sin A}{a} = \frac{\sin B}{b} \qquad \frac{\sin A}{a} = \frac{\sin C}{c} \qquad \frac{\sin B}{b} = \frac{\sin C}{c}$$

Formula (1) is a compact way to write these three equations.
Typically, to solve triangles with the Law of Sines, we use the fact that the sum of the angles of any triangle equals 180°; that is,

$$A + B + C = 180° \qquad (2)$$

1 Solve SAA or ASA Triangles

The first two examples show how to solve a triangle when one side and two angles are known (Case 1: SAA or ASA).

EXAMPLE 1

Using the Law of Sines to Solve an SAA Triangle

Solve the triangle: $A = 40°, B = 60°, a = 4$

Solution Figure 21 shows the triangle to be solved. The third angle C is found using formula (2).

$$A + B + C = 180°$$
$$40° + 60° + C = 180°$$
$$C = 80°$$

* The length of one side must be known because knowing only the angles results in a family of *similar triangles*.

Figure 21

NOTE Although it is not a complete check, the reasonableness of answers can be verified by determining whether the longest side is opposite the largest angle and the shortest side is opposite the smallest angle. ■

Now use the Law of Sines (twice) to find the unknown sides b and c.

$$\frac{\sin A}{a} = \frac{\sin B}{b} \qquad \frac{\sin A}{a} = \frac{\sin C}{c}$$

Because $a = 4$, $A = 40°$, $B = 60°$, and $C = 80°$, we have

$$\frac{\sin 40°}{4} = \frac{\sin 60°}{b} \qquad \frac{\sin 40°}{4} = \frac{\sin 80°}{c}$$

Solving for b and c yields

$$b = \frac{4 \sin 60°}{\sin 40°} \approx 5.39 \qquad c = \frac{4 \sin 80°}{\sin 40°} \approx 6.13$$

Notice in Example 1 that b and c are found by working with the given side a. This is better than finding b first and working with a rounded value of b to find c.

- **Now Work** PROBLEM 11

EXAMPLE 2

Using the Law of Sines to Solve an ASA Triangle

Solve the triangle: $A = 35°$, $B = 15°$, $c = 5$

Solution

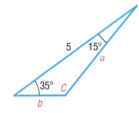

Figure 22

Figure 22 illustrates the triangle to be solved. Two angles are known ($A = 35°$ and $B = 15°$). Find the third angle using formula (2):

$$A + B + C = 180°$$
$$35° + 15° + C = 180°$$
$$C = 130°$$

Now the three angles and one side ($c = 5$) of the triangle are known. To find the remaining two sides a and b, use the Law of Sines (twice).

$$\frac{\sin A}{a} = \frac{\sin C}{c} \qquad\qquad \frac{\sin B}{b} = \frac{\sin C}{c}$$

$$\frac{\sin 35°}{a} = \frac{\sin 130°}{5} \qquad\qquad \frac{\sin 15°}{b} = \frac{\sin 130°}{5}$$

$$a = \frac{5 \sin 35°}{\sin 130°} \approx 3.74 \qquad b = \frac{5 \sin 15°}{\sin 130°} \approx 1.69$$

- **Now Work** PROBLEM 25

2 Solve SSA Triangles

Case 2 (SSA), which describes triangles for which two sides and the angle opposite one of them are known, is referred to as the **ambiguous case**, because the known information may result in one triangle, two triangles, or no triangle at all. Suppose that sides a and b and angle A are given, as illustrated in Figure 23. The key to determining how many triangles, if any, can be formed from the given information lies primarily with the relative size of side a, the height h, and the fact that $h = b \sin A$.

Figure 23 $h = b \sin A$

No Triangle If $a < h = b \sin A$, then side a is not long enough to form a triangle. See Figure 24.

One Right Triangle If $a = h = b \sin A$, then side a is just long enough to form one right triangle. See Figure 25.

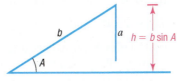

Figure 24 $a < h = b \sin A$

Figure 25 $a = h = b \sin A$

Two Triangles If $h = b \sin A < a$ and $a < b$, then two distinct triangles can be formed from the given information. See Figure 26.

One Triangle If $a \geq b$, only one triangle can be formed. See Figure 27.

Figure 26 $b \sin A < a$
and $a < b$

Figure 27 $a \geq b$

Fortunately, it is not necessary to rely on a figure or on complicated relationships to draw the correct conclusion in the ambiguous case. The Law of Sines leads us to the correct determination. Let's see how.

EXAMPLE 3

Using the Law of Sines to Solve an SSA Triangle (No Solution)

Solve the triangle: $a = 2, c = 1, C = 50°$

Solution

Because $a = 2, c = 1$, and $C = 50°$ are known, use the Law of Sines to find the angle A.

$$\frac{\sin A}{a} = \frac{\sin C}{c}$$

$$\frac{\sin A}{2} = \frac{\sin 50°}{1}$$

$$\sin A = 2 \sin 50° \approx 1.53$$

Since there is no angle A for which $\sin A > 1$, there is no triangle with the given measurements. Figure 28 illustrates the measurements given. Note that no matter how side c is positioned, it will never touch side b to form a triangle.

Figure 28

EXAMPLE 4

Using the Law of Sines to Solve an SSA Triangle (One Solution)

Solve the triangle: $a = 3, b = 2, A = 40°$

Solution

See Figure 29(a). Because $a = 3, b = 2$, and $A = 40°$ are known, use the Law of Sines to find the angle B.

$$\frac{\sin A}{a} = \frac{\sin B}{b}$$

Then

$$\frac{\sin 40°}{3} = \frac{\sin B}{2}$$

$$\sin B = \frac{2 \sin 40°}{3} \approx 0.43$$

Figure 29(a)

There are two angles $B, 0° < B < 180°$, for which $\sin B \approx 0.43$.

$$B_1 \approx 25.4° \quad \text{and} \quad B_2 \approx 180° - 25.4° = 154.6°$$

The angle $B_2 \approx 154.6°$ is discarded because the sum of the angles in a triangle must equal 180°, but $A + B_2 \approx 40° + 154.6° = 194.6° > 180°$. Using $B_1 \approx 25.4°$ gives

$$C = 180° - A - B_1 \approx 180° - 40° - 25.4° = 114.6°$$

NOTE The angle B_1 was determined by finding the value of $\sin^{-1}\left(\frac{2 \sin 40°}{3}\right)$. Using the rounded value and evaluating $\sin^{-1}(0.43)$ will yield a slightly different result. ∎

The third side c can now be determined using the Law of Sines.

$$\frac{\sin A}{a} = \frac{\sin C}{c}$$

$$\frac{\sin 40°}{3} = \frac{\sin 114.6°}{c}$$

$$c = \frac{3 \sin 114.6°}{\sin 40°} \approx 4.24$$

Figure 29(b)

Figure 29(b) illustrates the solved triangle.

EXAMPLE 5

Using the Law of Sines to Solve an SSA Triangle (Two Solutions)

Solve the triangle: $a = 6, b = 8, A = 35°$

Solution

Because $a = 6, b = 8$, and $A = 35°$ are known, use the Law of Sines to find the angle B.

$$\frac{\sin A}{a} = \frac{\sin B}{b}$$

Then

$$\frac{\sin 35°}{6} = \frac{\sin B}{8}$$

$$\sin B = \frac{8 \sin 35°}{6} \approx 0.76$$

$$B_1 \approx 49.9° \quad \text{or} \quad B_2 \approx 180° - 49.9° = 130.1°$$

Figure 30(a)

Both choices of B result in $A + B < 180°$. There are two triangles, one containing the angle $B_1 \approx 49.9°$ and the other containing the angle $B_2 \approx 130.1°$. See Figure 30(a). The third angle C is either

$$C_1 = 180° - A - B_1 \approx 95.1° \quad \text{or} \quad C_2 = 180° - A - B_2 \approx 14.9°$$

$$\uparrow \qquad\qquad\qquad\qquad\qquad \uparrow$$

$$A = 35°; B_1 = 49.9° \qquad\qquad\qquad A = 35°; B_2 = 130.1°$$

Now, use the Law of Sines to find the third side c.

Figure 30(b)

$$\frac{\sin A}{a} = \frac{\sin C_1}{c_1} \qquad\qquad \frac{\sin A}{a} = \frac{\sin C_2}{c_2}$$

$$\frac{\sin 35°}{6} = \frac{\sin 95.1°}{c_1} \qquad\qquad \frac{\sin 35°}{6} = \frac{\sin 14.9°}{c_2}$$

$$c_1 = \frac{6 \sin 95.1°}{\sin 35°} \approx 10.42 \qquad c_2 = \frac{6 \sin 14.9°}{\sin 35°} \approx 2.69$$

The two solved triangles are illustrated in Figure 30(b).

 Now Work PROBLEMS 27 AND 33

---

 3 Solve Applied Problems

EXAMPLE 6

Finding the Height of a Mountain

To measure the height of a mountain, a surveyor takes two sightings of the peak at a distance 900 meters apart on a direct line to the mountain.* See Figure 31(a) on the next page. The first observation results in an angle of elevation of 47°, and the second results in an angle of elevation of 35°. If the transit is 2 meters high, what is the height h of the mountain?

*For simplicity, assume that these sightings are at the same level.

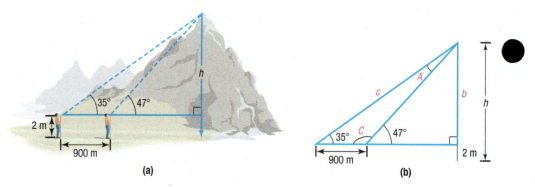

Figure 31

(a)

(b)

Solution Figure 31(b) shows the triangles that model the situation in Figure 31(a). Since $C + 47° = 180°$, this means that $C = 133°$. Also, since $A + C + 35° = 180°$, this means that $A = 180° - 35° - C = 145° - 133° = 12°$. Use the Law of Sines to find c.

$$\frac{\sin A}{a} = \frac{\sin C}{c} \qquad \textcolor{blue}{A = 12°, C = 133°, a = 900}$$

$$c = \frac{900 \sin 133°}{\sin 12°} \approx 3165.86$$

Using the larger right triangle gives

$$\sin 35° = \frac{b}{c}$$

$$b = 3165.86 \sin 35° \approx 1815.86 \approx 1816 \text{ meters}$$

The height of the peak from ground level is approximately $1816 + 2 = 1818$ meters.

 Now Work PROBLEM **39**

EXAMPLE 7 **Rescue at Sea**

Coast Guard Station Zulu is located 120 miles due west of Station X-ray. A ship at sea sends an SOS call that is received by each station. The call to Station Zulu indicates that the bearing of the ship from Zulu is N40°E (40° east of north). The call to Station X-ray indicates that the bearing of the ship from X-ray is N30°W (30° west of north).

(a) How far is each station from the ship?
(b) If a helicopter capable of flying 200 miles per hour is dispatched from the nearest station to the ship, how long will it take to reach the ship?

Solution (a) Figure 32 illustrates the situation. The angle C is found to be

$$C = 180° - 50° - 60° = 70°$$

The Law of Sines can now be used to find the two distances a and b that are needed.

$$\frac{\sin 50°}{a} = \frac{\sin 70°}{120}$$

$$a = \frac{120 \sin 50°}{\sin 70°} \approx 97.82 \text{ miles}$$

$$\frac{\sin 60°}{b} = \frac{\sin 70°}{120}$$

$$b = \frac{120 \sin 60°}{\sin 70°} \approx 110.59 \text{ miles}$$

Figure 32

Station Zulu is about 111 miles from the ship, and Station X-ray is about 98 miles from the ship.

(b) The time t needed for the helicopter to reach the ship from Station X-ray is found by using the formula

$$(\text{Rate}, r)\,(\text{Time}, t) = \text{Distance}, a$$

Then

$$t = \frac{a}{r} = \frac{97.82}{200} \approx 0.49 \text{ hour} \approx 29 \text{ minutes}$$

It will take about 29 minutes for the helicopter to reach the ship.

 Now Work PROBLEM 49

Proof of the Law of Sines To prove the Law of Sines, construct an altitude of length h from one of the vertices of a triangle. Figure 33(a) shows h for a triangle with three acute angles, and Figure 33(b) shows h for a triangle with an obtuse angle. In each case, the altitude is drawn from the vertex at B. Using either figure

$$\sin C = \frac{h}{a}$$

from which

$$h = a \sin C \qquad \text{(3)}$$

From Figure 33(a), it also follows that

$$\sin A = \frac{h}{c}$$

from which

$$h = c \sin A \qquad \text{(4)}$$

From Figure 33(b), it follows that

$$\sin (180° - A) = \sin A = \frac{h}{c}$$

$$\sin (180° - A) = \sin 180° \cos A - \cos 180° \sin A = \sin A$$

which again gives

$$h = c \sin A$$

So, whether the triangle has three acute angles or has two acute angles and one obtuse angle, equations (3) and (4) hold. As a result, the expressions for h in equations (3) and (4) are equal. That is,

$$a \sin C = c \sin A$$

from which

$$\frac{\sin A}{a} = \frac{\sin C}{c} \qquad \text{(5)}$$

In a similar manner, constructing the altitude h' from the vertex of angle A, as shown in Figure 34, reveals that

$$\sin B = \frac{h'}{c} \quad \text{and} \quad \sin C = \frac{h'}{b}$$

Equating the expressions for h' gives

$$h' = c \sin B = b \sin C$$

from which

$$\frac{\sin B}{b} = \frac{\sin C}{c} \qquad \text{(6)}$$

When equations (5) and (6) are combined, the result is the Law of Sines.

(a)

(b)

Figure 33

(a)

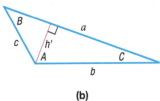

(b)

Figure 34

4.2 Assess Your Understanding

1. The difference formula for the sine function is
$\sin(A - B) =$ _____. (p. 230)

2. Solve $\sin A = \frac{1}{2}$ if $0 \le A \le \pi$. (pp. 209–212)

3. Approximate $\sin 40°$ and $\sin 80°$. (p. 125)

4. Approximate $\sin^{-1} 0.76$. Express the answer in degrees. (pp. 204–205)

Concepts and Vocabulary

5. *Multiple Choice* If none of the angles of a triangle is a right angle, the triangle is called _____.
(a) oblique **(b)** obtuse **(c)** acute **(d)** scalene

6. For a triangle with sides a, b, c and opposite angles A, B, C, the Law of Sines states that _____.

7. *Multiple Choice* If two angles of a triangle measure 48° and 93°, what is the measure of the third angle?
(a) 132° **(b)** 77° **(c)** 42° **(d)** 39°

8. *True or False* When two sides and an angle are given at least one triangle can be formed.

9. *True or False* The Law of Sines can be used to solve triangles where three sides are known.

10. Triangles for which two sides and the angle opposite one of them are known (SSA) are referred to as the _____ _____.

Skill Building

In Problems 11–18, solve each triangle.

11.

12.

13.

14.

15.

16.

17.

18.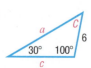

In Problems 19–26, solve each triangle.

19. $A = 55°$, $B = 25°$, $a = 4$

20. $A = 50°$, $C = 20°$, $a = 3$

21. $B = 64°$, $C = 47°$, $b = 6$

22. $A = 70°$, $B = 60°$, $c = 4$

23. $A = 110°$, $C = 30°$, $c = 3$

24. $B = 10°$, $C = 100°$, $b = 2$

25. $A = 40°$, $B = 40°$, $c = 2$

26. $B = 20°$, $C = 70°$, $a = 1$

In Problems 27–38, two sides and an angle are given. Determine whether the given information results in one triangle, two triangles, or no triangle at all. Solve any resulting triangle(s).

27. $a = 3$, $b = 2$, $A = 50°$

28. $b = 4$, $c = 3$, $B = 40°$

29. $b = 9$, $c = 4$, $B = 115°$

30. $a = 2$, $c = 1$, $A = 120°$

31. $a = 7$, $b = 14$, $A = 30°$

32. $b = 2$, $c = 3$, $B = 40°$

33. $b = 4$, $c = 6$, $B = 20°$

34. $a = 3$, $b = 7$, $A = 70°$

35. $a = 8$, $c = 3$, $C = 125°$

36. $b = 4$, $c = 5$, $B = 95°$

37. $a = 7$, $c = 3$, $C = 12°$

38. $b = 4$, $c = 5$, $B = 40°$

Applications and Extensions

39. Finding the Length of a Ski Lift Consult the figure. To find the length of the span of a proposed ski lift from P to Q, a surveyor measures $\angle DPQ$ to be 25° and then walks back a distance of 1000 feet to R and measures $\angle PRQ$ to be 15°. What is the distance from P to Q?

40. Finding the Height of a Mountain Use the figure in Problem 39 to find the height QD of the mountain.

41. Finding the Height of an Airplane An aircraft is spotted by two observers who are 1000 feet apart. As the airplane passes over the line joining them, each observer takes a sighting of the angle of elevation to the plane, as indicated in the figure. How high is the airplane?

42. Finding the Height of the Bridge over the Royal Gorge The highest bridge in the world is the bridge over the Royal Gorge of the Arkansas River in Colorado. Sightings to the same point at water level directly under the bridge are taken from each side of the 880-foot-long bridge, as indicated in the figure. How high is the bridge?

Source: Guinness Book of World Records

43. Land Dimensions A triangular plot of land has one side along a straight road measuring 200 feet. A second side makes a 50° angle with the road, and the third side makes a 43° angle with the road. How long are the other two sides?

44. Distance between Runners Two runners in a marathon determine that the angles of elevation of a news helicopter covering the race are 38° and 45°. If the helicopter is 1700 feet directly above the finish line, how far apart are the runners?

45. Landscaping Pat needs to determine the height of a tree before cutting it down to be sure that it will not fall on a nearby fence. The angle of elevation of the tree from one position on a flat path from the tree is 30°, and from a second position 40 feet farther along this path it is 20°. What is the height of the tree?

46. Construction A loading ramp 10 feet long that makes an angle of 18° with the horizontal is to be replaced by one that makes an angle of 12° with the horizontal. How long is the new ramp?

47. Commercial Navigation Adam must fly home to St. Louis from a business meeting in Oklahoma City. One flight option flies directly to St. Louis, a distance of about 461.1 miles. A second flight option flies first to Kansas City and then connects to St. Louis. The bearing from Oklahoma City to Kansas City is N29.6°E, and the bearing from Oklahoma City to St. Louis is N57.7°E. The bearing from St. Louis to Oklahoma City is S57.7°W, and the bearing from St. Louis to Kansas City is N79.4°W. How many more frequent flyer miles will Adam receive if he takes the connecting flight rather than the direct flight?

Source: www.landings.com

48. Time Lost to a Navigation Error In attempting to fly from city P to city Q, an aircraft followed a course that was 10° in error, as indicated in the figure. After flying a distance of 50 miles, the pilot corrected the course by turning at point R and flying 300 miles farther. If the constant speed of the aircraft was 250 miles per hour, how much time was lost due to the error?

49. Rescue at Sea Coast Guard Station Able is located 150 miles due south of Station Baker. A ship at sea sends an SOS call that is received by each station. The call to Station Able indicates the bearing of the ship is N55°E; the call to Station Baker indicates the bearing of the ship is S60°E.
(a) How far is each station from the ship?
(b) If a helicopter capable of flying 200 miles per hour is dispatched from the station nearest the ship, how long will it take to reach the ship?

50. Distance to the Moon
At exactly the same time, Tom and Alice measured the angle of elevation to the moon while standing exactly 300 km apart. The angle of elevation to the moon for Tom was 49.8974°, and the angle of elevation to the moon for Alice was 49.9312°. See the figure. To the nearest 1000 km, how far was the moon from Earth when the measurements were obtained?

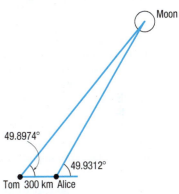

51. Finding the Lean of the Leaning Tower of Pisa The famous Leaning Tower of Pisa was originally 184.5 feet high.* At a distance of 123 feet from the base of the tower, the angle of elevation to the top of the tower is found to be 60°. Find $\angle RPQ$ indicated in the figure. Also, find the perpendicular distance from R to PQ.

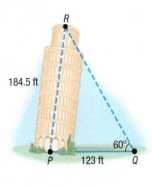

*On February 27, 1964, the government of Italy requested aid in preventing the tower from toppling. A multinational task force of engineers, mathematicians, and historians was assigned and met on the Azores islands to discuss stabilization methods. After over two decades of work on the subject, the tower was closed to the public in January 1990. During the time that the tower was closed, the bells were removed to relieve it of some weight, and cables were cinched around the third level and anchored several hundred meters away. Apartments and houses in the path of the tower were vacated for safety concerns. After a decade of corrective reconstruction and stabilization efforts, the tower was reopened to the public on December 15, 2001. Many methods were proposed to stabilize the tower, including the addition of 800 metric tons of lead counterweights to the raised end of the base. The final solution was to remove 38 cubic meters of soil from underneath the raised end. The tower has been declared stable for at least another 300 years.
Source: https://www.history.com/this-day-in-history/leaning-tower-needs-help

52. Crankshafts on Cars On a certain automobile, the crankshaft is 3 inches long and the connecting rod is 9 inches long (see the figure). At the time when ∠OPQ is 15°, how far is the piston (P) from the center (O) of the crankshaft?

53. Constructing a Highway U.S. 41, a highway whose primary direction is north–south, is being constructed along the west coast of Florida. Near Naples, a bay obstructs the straight path of the road. Since the cost of a bridge is prohibitive, engineers decide to go around the bay. The figure shows the path that they decide on and the measurements taken. What is the length of highway needed to go around the bay?

54. Calculating Distances at Sea The navigator of a ship at sea spots two lighthouses that she knows to be 3 miles apart along a straight seashore. She determines that the angles formed between two line-of-sight observations of the lighthouses and the line from the ship directly to shore are 15° and 35°. See the figure.
  (a) How far is the ship from lighthouse *P*?
  (b) How far is the ship from lighthouse *Q*?
  (c) How far is the ship from shore?

55. Designing an Awning An awning that covers a sliding glass door that is 88 inches tall forms an angle of 50° with the wall. The purpose of the awning is to prevent sunlight from entering the house when the angle of elevation of the Sun is more than 65°. See the figure. Find the length *L* of the awning.

56. Finding Distances A forest ranger is walking on a path inclined at 5° to the horizontal directly toward a 100-foot-tall fire observation tower. The angle of elevation from the path to the top of the tower is 40°. How far is the ranger from the tower at this time?

57. Great Pyramid of Cheops One of the original Seven Wonders of the World, the Great Pyramid of Cheops was built about 2580 BC. Its original height was 480 feet 11 inches, but owing to the loss of its topmost stones, it is now shorter. Find the current height of the Great Pyramid using the information given in the figure.

Source: *Guinness Book of World Records*

58. Determining the Height of an Aircraft Two sensors are spaced 700 feet apart along the approach to a small airport. When an aircraft is nearing the airport, the angle of elevation from the first sensor to the aircraft is 20°, and from the second sensor to the aircraft it is 15°. Determine how high the aircraft is at this time.

59. Mercury The distance from the Sun to Earth is approximately 149,600,000 kilometers (km). The distance from the Sun to Mercury is approximately 57,910,000 km. The **elongation angle** α is the angle formed between the line of sight from Earth to the Sun and the line of sight from Earth to Mercury. See the figure. Suppose that the elongation angle for Mercury is 15°. Use this information to find the possible distances between Earth and Mercury.

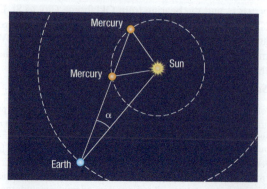

60. Venus The distance from the Sun to Earth is approximately149,600,000 km. The distance from the Sun to Venus is approximately 108,200,000 km. The elongation angle α is the angle formed between the line of sight from Earth to the Sun and the line of sight from Earth to Venus. Suppose that the elongation angle for Venus is 10°. Use this information to find the possible distances between Earth and Venus.

61. The Original Ferris Wheel George Washington Gale Ferris, Jr., designed the original Ferris wheel for the 1893 World's Columbian Exposition in Chicago, Illinois. The wheel had 36 equally spaced cars each the size of a school bus. The distance between adjacent cars was approximately 22 feet. Determine the diameter of the wheel to the nearest foot.

Source: Carnegie Library of Pittsburgh, www.clpgh.org

62. *Challenge Problem* Mollweide's Formula For any triangle, Mollweide's Formula (named after Karl Mollweide, 1774–1825) states that

$$\frac{a + b}{c} = \frac{\cos\left[\frac{1}{2}(A - B)\right]}{\sin\left(\frac{1}{2}C\right)}$$

Derive it.

[**Hint**: Use the Law of Sines and then a Sum-to-Product Formula.] Notice that this formula involves all six parts of a triangle. As a result, it is sometimes used to check the solution of a triangle.

63. *Challenge Problem* Mollweide's Formula Another form of Mollweide's Formula is

$$\frac{a - b}{c} = \frac{\sin\left[\frac{1}{2}(A - B)\right]}{\cos\left(\frac{1}{2}C\right)}$$

Derive it.

64. *Challenge Problem* For any triangle, derive the formula

$$a = b\cos C + c\cos B$$

65. *Challenge Problem* Law of Tangents For any triangle, derive the Law of Tangents:

$$\frac{a - b}{a + b} = \frac{\tan\left[\frac{1}{2}(A - B)\right]}{\tan\left[\frac{1}{2}(A + B)\right]}$$

66. *Challenge Problem* Circumscribing a Triangle Show that

$$\frac{\sin A}{a} = \frac{\sin B}{b} = \frac{\sin C}{c} = \frac{1}{2r}$$

See the figure where r is the radius of the circle circumscribing the triangle PQR whose sides are a, b, and c, and $PP' = 2r$ is a diameter of the circle.

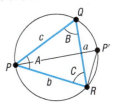

Explaining Concepts: Discussion and Writing

67. Make up three problems involving oblique triangles. One should result in one triangle, the second in two triangles, and the third in no triangle.

68. What do you do first if you are asked to solve a triangle and are given one side and two angles?

69. What do you do first if you are asked to solve a triangle and are given two sides and the angle opposite one of them?

70. Solve Example 6 using right-triangle geometry. Comment on which solution, using the Law of Sines or using right triangles, you prefer. Give reasons.

Retain Your Knowledge

Problems 71–78 are based on material learned earlier in the course. The purpose of these problems is to keep the material fresh in your mind so that you are better prepared for the final exam.

71. Solve: $3x^3 + 4x^2 - 27x - 36 = 0$

72. Find the exact distance between $P_1 = (-1, -7)$ and $P_2 = (2, -1)$. Then approximate the distance to two decimal places.

73. Find the exact value of $\tan\left[\cos^{-1}\left(-\frac{7}{8}\right)\right]$.

74. Graph $y = 4\sin\left(\frac{1}{2}x\right)$. Show at least two periods.

75. Find the horizontal asymptote of

$$h(x) = \frac{-3x^2 - 7x + 1}{4 - 9x^2}$$

76. Find an equation of the line perpendicular to $y = -\frac{2}{3}x + 7$ that contains the point $(-2, -5)$.

77. Determine whether $h(x) = 5x^3 - 4x + 1$ is even, odd, or neither.

78. Solve: $\frac{1}{3}(x - 6) + 4x > 0$

'Are You Prepared?' Answers

1. $\sin A \cos B - \cos A \sin B$ **2.** $\left\{\frac{\pi}{6}, \frac{5\pi}{6}\right\}$ **3.** $\sin 40° = 0.64; \sin 80° = 0.98$ **4.** 49.5°

4.3 The Law of Cosines

PREPARING FOR THIS SECTION *Before getting started, review the following:*

- Trigonometric Equations (Section 3.3, pp. 209–214)
- Distance Formula (Section 1.1, pp. 3–4)

Now Work the 'Are You Prepared?' problems on page 289.

OBJECTIVES 1 Solve SAS Triangles (p. 287)
2 Solve SSS Triangles (p. 287)
3 Solve Applied Problems (p. 288)

In the previous section, the Law of Sines was used to solve Case 1 (SAA or ASA) and Case 2 (SSA) of an oblique triangle. In this section, the Law of Cosines is derived and used to solve Cases 3 and 4.

> **CASE 3:** Two sides and the included angle are known (SAS).
>
> **CASE 4:** Three sides are known (SSS).

THEOREM **Law of Cosines**

For a triangle with sides a, b, c and opposite angles A, B, C, respectively,

$$c^2 = a^2 + b^2 - 2ab \cos C \tag{1}$$
$$b^2 = a^2 + c^2 - 2ac \cos B \tag{2}$$
$$a^2 = b^2 + c^2 - 2bc \cos A \tag{3}$$

Proof Only formula (1) is proved here. Formulas (2) and (3) can be proved using the same argument.

Begin by strategically placing a triangle on a rectangular coordinate system so that the vertex of angle C is at the origin and side b lies along the positive x-axis. Regardless of whether C is acute, as in Figure 35(a), or obtuse, as in Figure 35(b), the vertex of angle B has coordinates $(a \cos C, a \sin C)$. The vertex of angle A has coordinates $(b, 0)$.

Use the distance formula to compute c^2.

$$
\begin{aligned}
c^2 &= (b - a \cos C)^2 + (0 - a \sin C)^2 \\
&= b^2 - 2ab \cos C + a^2 \cos^2 C + a^2 \sin^2 C \\
&= b^2 - 2ab \cos C + a^2(\cos^2 C + \sin^2 C) \\
&= a^2 + b^2 - 2ab \cos C
\end{aligned}
$$

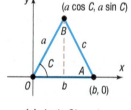

(a) Angle *C* is acute

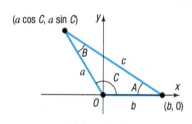

(b) Angle *C* is obtuse

Figure 35

Each of formulas (1), (2), and (3) may be stated in words as follows:

THEOREM **Law of Cosines**

The square of one side of a triangle equals the sum of the squares of the other two sides, minus twice their product times the cosine of their included angle.

Observe that if the triangle is a right triangle (so that, say, $C = 90°$), formula (1) becomes the familiar Pythagorean Theorem: $c^2 = a^2 + b^2$. That is, the Pythagorean Theorem is a special case of the Law of Cosines.

1 Solve SAS Triangles

The Law of Cosines is used to solve Case 3 (SAS), which applies to triangles for which two sides and the included angle are known.

EXAMPLE 1

Using the Law of Cosines to Solve an SAS Triangle

Solve the triangle: $a = 2$, $b = 3$, $C = 60°$

Solution

See Figure 36. Because two sides, a and b, and the included angle, $C = 60°$, are known, the Law of Cosines makes it easy to find the third side, c.

$$c^2 = a^2 + b^2 - 2ab \cos C$$
$$= 2^2 + 3^2 - 2 \cdot 2 \cdot 3 \cdot \cos 60° \quad \textcolor{blue}{a = 2, b = 3, C = 60°}$$
$$= 13 - 12 \cdot \frac{1}{2} = 7$$
$$c = \sqrt{7}$$

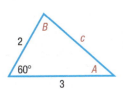

Figure 36

Side c is of length $\sqrt{7}$. To find the angles A and B, either the Law of Sines or the Law of Cosines may be used.

It is preferable to use the Law of Cosines because it will lead to an equation with *one* solution whether solving for A or B. Using the Law of Sines would lead to an equation with *two* solutions that would need to be checked to determine which solution fits the given data.*

We choose to use formulas (2) and (3) of the Law of Cosines to find A and B.

For A:

$$a^2 = b^2 + c^2 - 2bc \cos A$$
$$2bc \cos A = b^2 + c^2 - a^2$$
$$\cos A = \frac{b^2 + c^2 - a^2}{2bc} = \frac{9 + 7 - 4}{2 \cdot 3\sqrt{7}} = \frac{12}{6\sqrt{7}} = \frac{2\sqrt{7}}{7}$$
$$A = \cos^{-1} \frac{2\sqrt{7}}{7} \approx 40.9°$$

For B:

NOTE The angle B can also be found using $A + B + C = 180°$, so $B = 180° - 40.9° - 60° = 79.1°$. ∎

$$b^2 = a^2 + c^2 - 2ac \cos B$$
$$\cos B = \frac{a^2 + c^2 - b^2}{2ac} = \frac{4 + 7 - 9}{4\sqrt{7}} = \frac{2}{4\sqrt{7}} = \frac{\sqrt{7}}{14}$$
$$B = \cos^{-1} \frac{\sqrt{7}}{14} \approx 79.1°$$

Notice that $A + B + C = 40.9° + 79.1° + 60° = 180°$, as required.

━━ **Now Work** PROBLEMS **9** AND **17**

2 Solve SSS Triangles

The next example uses the Law of Cosines to solve a triangle when three sides are known, Case 4 (SSS).

---

*The Law of Sines can be used when seeking the angle opposite the smaller side, since it is acute. (In Figure 36, use the Law of Sines to find A, the angle opposite the smaller side.)

| EXAMPLE 2 | Using the Law of Cosines to Solve an SSS Triangle |

Solve the triangle: $a = 4, b = 3, c = 6$

Solution   See Figure 37. To find the angles $A, B,$ and C, use the Law of Cosines.

For A:

$$\cos A = \frac{b^2 + c^2 - a^2}{2bc} = \frac{9 + 36 - 16}{2 \cdot 3 \cdot 6} = \frac{29}{36}$$

$$A = \cos^{-1} \frac{29}{36} \approx 36.3°$$

For B:

$$\cos B = \frac{a^2 + c^2 - b^2}{2ac} = \frac{16 + 36 - 9}{2 \cdot 4 \cdot 6} = \frac{43}{48}$$

$$B = \cos^{-1} \frac{43}{48} \approx 26.4°$$

Now use A and B to find C:

$$C = 180° - A - B \approx 180° - 36.3° - 26.4° = 117.3°$$

Figure 37

⬤ **Now Work** PROBLEMS 15 AND 25

🌐 3 Solve Applied Problems

| EXAMPLE 3 | Correcting a Navigational Error |

A motorized sailboat leaves Naples, Florida, bound for Key West, 150 miles away. Maintaining a constant speed of 15 miles per hour, but encountering heavy crosswinds and strong currents, the crew finds, after 4 hours, that the sailboat is off course by 20°.

(a) How far is the sailboat from Key West at this time?

(b) Through what angle should the sailboat turn to correct its course?

(c) How much time has been added to the trip because of this? (Assume that the speed remains at 15 miles per hour.)

Solution   See Figure 38. With a speed of 15 miles per hour, the sailboat has gone 60 miles after 4 hours. The distance x of the sailboat from Key West is to be found, along with the angle θ that the sailboat should turn through to correct its course.

(a) To find x, use the Law of Cosines, because two sides and the included angle are known.

$$x^2 = 150^2 + 60^2 - 2 \cdot 150 \cdot 60 \cdot \cos 20° \approx 9185.53$$

$$x \approx 95.8$$

The sailboat is about 96 miles from Key West.

(b) With all three sides of the triangle now known, use the Law of Cosines again to find the angle A opposite the side of length 150 miles.

$$150^2 = 96^2 + 60^2 - 2 \cdot 96 \cdot 60 \cdot \cos A$$

$$9684 = -11{,}520 \cos A$$

$$\cos A \approx -0.8406$$

$$A \approx 147.2°$$

So,

$$\theta = 180° - A \approx 180° - 147.2° = 32.8°$$

The sailboat should turn through an angle of about 33° to correct its course.

Figure 38

(c) The total length of the trip is now 60 + 96 = 156 miles. The extra 6 miles will only require about 0.4 hour, or 24 minutes, more if the speed of 15 miles per hour is maintained.

> **Now Work** PROBLEM **47**

Historical Feature

The Law of Sines was known vaguely long before it was explicitly stated by Nasir Eddin (about AD 1250). Ptolemy (about AD 150) was aware of it in a form using a chord function instead of the sine function. But it was first clearly stated in Europe by Regiomontanus, writing in 1464.

The Law of Cosines appears first in Euclid's *Elements* (Book II), but in a well-disguised form in which squares built on the sides of triangles are added and a rectangle representing the cosine term is subtracted. It was thus known to all mathematicians because of their familiarity with Euclid's work. An early modern form of the Law of Cosines, that for finding the angle when the sides are known, was stated by François Viète (in 1593).

The Law of Tangents (see Problem 65 in Section 4.2) has become obsolete. In the past it was used in place of the Law of Cosines, because the Law of Cosines was very inconvenient for calculation with logarithms or slide rules. Mixing of addition and multiplication is now very easy on a calculator, however, and the Law of Tangents has been shelved along with the slide rule.

4.3 Assess Your Understanding

'Are You Prepared?' *Answers are given at the end of these exercises. If you get a wrong answer, read the pages listed in* red.

1. Write the formula for the distance d from $P_1 = (x_1, y_1)$ to $P_2 = (x_2, y_2)$. (p. 3)

2. If θ is an acute angle, solve the equation $\cos \theta = \dfrac{\sqrt{2}}{2}$. (pp. 209–214)

Concepts and Vocabulary

3. If three sides of a triangle are known, the Law of _____ is used to solve the triangle.

4. *Multiple Choice* If one side and two angles of a triangle are known, which law can be used to solve the triangle?
   (a) Law of Sines   (b) Law of Cosines
   (c) Either a or b   (d) The triangle cannot be solved.

5. *Multiple Choice* If two sides and the included angle of a triangle are known, which law can be used to solve the triangle?
   (a) Law of Sines   (b) Law of Cosines
   (c) Either a or b   (d) The triangle cannot be solved.

6. *True or False* Given only the three sides of a triangle, there is insufficient information to solve the triangle.

7. *True or False* The Law of Cosines states that the square of one side of a triangle equals the sum of the squares of the other two sides, minus twice their product.

8. *True or False* A special case of the Law of Cosines is the Pythagorean Theorem.

Skill Building

In Problems 9–16, solve each triangle.

9.

10.

11.

12.

13.

14.

15.

16.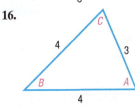

In Problems 17–32, solve each triangle.

17. $a = 3$, $b = 4$, $C = 40°$

18. $a = 2$, $c = 1$, $B = 10°$

19. $b = 2$, $c = 4$, $A = 75°$

20. $a = 6$, $b = 4$, $C = 60°$

21. $a = 5$, $c = 3$, $B = 105°$

22. $b = 4$, $c = 1$, $A = 120°$

23. $a = 2$, $b = 3$, $C = 70°$

24. $a = 3$, $c = 2$, $B = 90°$

25. $a = 20$, $b = 29$, $c = 21$

26. $a = 4$, $b = 5$, $c = 3$

27. $a = 2$, $b = 2$, $c = 2$

28. $a = 3$, $b = 3$, $c = 2$

29. $a = 6$, $b = 11$, $c = 12$

30. $a = 4$, $b = 3$, $c = 6$

31. $a = 15$, $b = 13$, $c = 3$

32. $a = 9$, $b = 7$, $c = 10$

Mixed Practice *In Problems 33–44, solve each triangle.*

33. $B = 20°$, $C = 75°$, $b = 5$

34. $A = 50°$, $B = 55°$, $c = 9$

35. $a = 6$, $b = 8$, $c = 9$

36. $a = 14$, $b = 7$, $A = 85°$

37. $B = 35°$, $C = 65°$, $a = 15$

38. $a = 4$, $c = 5$, $B = 55°$

39. $c = 8$, $A = 38°$, $B = 52°$

40. $a = 20$, $A = 73°$, $C = 17°$

41. $A = 10°$, $a = 3$, $b = 10$

42. $A = 65°$, $B = 72°$, $b = 7$

43. $b = 5$, $c = 12$, $A = 60°$

44. $a = 10$, $b = 10$, $c = 15$

Applications and Extensions

45. Distance to the Green A golfer hits an errant tee shot that lands in the rough. A marker in the center of the fairway is 150 yards from the center of the green. While standing on the marker and facing the green, the golfer turns 110° toward his ball. He then paces off 35 yards to his ball. See the figure. How far is the ball from the center of the green?

46. Navigation An airplane flies due north from Ft. Myers to Sarasota, a distance of 150 miles, and then turns through an angle of 50° and flies to Orlando, a distance of 100 miles. See the figure.

(a) How far is it directly from Ft. Myers to Orlando?

(b) What bearing should the pilot use to fly directly from Ft. Myers to Orlando?

47. Avoiding a Tropical Storm A cruise ship maintains an average speed of 15 knots in going from San Juan, Puerto Rico, to Barbados, West Indies, a distance of 600 nautical miles. To avoid a tropical storm, the captain heads out of San Juan in a direction 20° off a direct heading to

Barbados. The captain maintains the 15-knot speed for 10 hours, after which time the path to Barbados becomes clear of storms.

(a) Through what angle should the captain turn to head directly to Barbados?

(b) Once the turn is made, how long will it be before the ship reaches Barbados if the same 15-knot speed is maintained?

48. Revising a Flight Plan In attempting to fly from Chicago to Louisville, a distance of 330 miles, a pilot inadvertently took a course that was 10° in error, as indicated in the figure.

(a) If the aircraft maintains an average speed of 220 miles per hour, and if the error in direction is discovered after 15 minutes, through what angle should the pilot turn to head toward Louisville?

(b) What new average speed should the pilot maintain so that the total time of the trip is 90 minutes?

49. Major League Baseball Field A major league baseball diamond is actually a square 90 feet on a side. The pitching rubber is located 60.5 feet from home plate on a line joining home plate and second base.
(a) How far is it from the pitching rubber to first base?
(b) How far is it from the pitching rubber to second base?
(c) If a pitcher faces home plate, through what angle does he need to turn to face first base?

50. Little League Baseball Field According to Little League baseball official regulations, the diamond is a square 60 feet on a side. The pitching rubber is located 46 feet from home plate on a line joining home plate and second base.
(a) How far is it from the pitching rubber to first base?
(b) How far is it from the pitching rubber to second base?
(c) If a pitcher faces home plate, through what angle does he need to turn to face first base?

51. Finding the Length of a Guy Wire The height of a radio tower is 500 feet, and the ground on one side of the tower slopes upward at an angle of 10° (see the figure).
(a) How long should a guy wire be if it is to connect to the top of the tower and be secured at a point on the sloped side 100 feet from the base of the tower?
(b) How long should a second guy wire be if it is to connect to the middle of the tower and be secured at a position 100 feet from the base on the flat side?

52. Finding the Length of a Guy Wire A radio tower 500 feet high is located on the side of a hill with an inclination to the horizontal of 5°. See the figure. How long should two guy wires be if they are to connect to the top of the tower and be secured at two points 100 feet directly above and directly below the base of the tower?

53. Identifying Remains The Purkait triangle, located at the proximal end of the femur, has been used to identify the gender of fragmented skeletal remains. See the figure.
(a) Given $\overline{AB} = 30.1$ mm, $\overline{AC} = 51.4$ mm, and $A = 89.2°$, find the length of $\overline{BC}$.
(b) If the average length of $\overline{BC}$ is 59.4 mm for males and 53.3 mm for females, which gender would be identified for the measurements in part (a)?

54. Identifying Remains Like the Purkait triangle in Problem 53, the metric triangle is located at the proximal end of the femur and has been used to identify the gender of fragmented skeletal remains. See the figure.
(a) If $\overline{AC} = 48.8$ mm, $\overline{BC} = 62.2$ mm, and $C = 89°$, find the length of $\overline{AB}$.
(b) If $\overline{AB} < 80$ mm typically indicates a female and $\overline{AB} > 80$ mm typically indicates a male, which gender, if any, would be identified from the measurements in part (a)?

55. Soccer Angles A soccer goal is 8 yards wide. Suppose a goalie is standing on her line in the center of her goal as a striker from the opposing team moves the ball towards her. The near post angle, α, is formed by rays extending from the ball to the near post and the goalie. Similarly, the far post angle, β, is formed by rays extending from the ball to the far post and the goalie. See the figure.

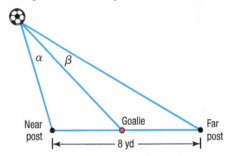

(a) Determine the near post angle and the far post angle when the striker is 20 yards from the near post and 24 yards from the far post.
(b) How far is the goalie from the ball?
(c) To cover the near post, the goalie moves toward the near post to make the near post angle and the far post angle equal. How far toward her near post does the goalie need to move?

56. Covering the Angles In soccer, a defending goalkeeper wants to take up a position which bisects the angle that needs to be covered. See the figure. The keeper stands square to the ball—that is, perpendicular to the line of bisection—at a point where the area covered (shaded) lies completely outside the goal. How far is the goalkeeper from the center of the goal line if an attacking striker is 24 yards from the near post and 30 yards from the far post?

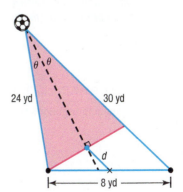

57. Wrigley Field, Home of the Chicago Cubs The distance from home plate to the fence in dead center in Wrigley Field is 400 feet (see the figure). How far is it from the fence in dead center to third base?

400 ft

90 ft   90 ft

58. Little League Baseball The distance from home plate to the fence in dead center at the Oak Lawn Little League field is 280 feet. How far is it from the fence in dead center to third base?

[**Hint**: The distance between the bases in Little League is 60 feet.]

59. Building a Swing Set Clint is building a wooden swing set for his children. Each supporting end of the swing set is to be an A-frame constructed with two 10-foot-long 4 by 4's joined at a 45° angle. To prevent the swing set from tipping over, Clint wants to secure the base of each A-frame to concrete footings. How far apart should the footings for each A-frame be?

60. Rods and Pistons See the figure (top, right). Rod *OA* rotates about the fixed point *O* so that point *A* travels on a circle of radius *r*. Connected to point *A* is another rod *AB* of length *L* > 2*r*, and point *B* is connected to a piston. Show that the distance *x* between point *O* and point *B* is given by

$$x = r\cos\theta + \sqrt{r^2\cos^2\theta + L^2 - r^2}$$

where θ is the angle of rotation of rod *OA*.

61. Challenge Problem Geometry Show that the length *d* of a chord of a circle of radius *r* is given by the formula

$$d = 2r\sin\frac{\theta}{2}$$

where θ, $0 < \theta < \pi$, is the central angle formed by the radii to the ends of the chord. See the figure. Use this result to derive the fact that $\sin\theta < \theta$, where θ is measured in radians.

62. Challenge Problem For any triangle, show that

$$\cos\frac{C}{2} = \sqrt{\frac{s(s-c)}{ab}}$$

where $s = \dfrac{1}{2}(a+b+c)$.

63. Challenge Problem For any triangle, show that

$$\sin\frac{C}{2} = \sqrt{\frac{(s-a)(s-b)}{ab}}$$

where $s = \dfrac{1}{2}(a+b+c)$.

64. Challenge Problem Use the Law of Cosines to prove the identity

$$\frac{\cos A}{a} + \frac{\cos B}{b} + \frac{\cos C}{c} = \frac{a^2 + b^2 + c^2}{2abc}$$

Explaining Concepts: Discussion and Writing

65. What do you do first if you are asked to solve a triangle and are given two sides and the included angle?

66. What do you do first if you are asked to solve a triangle and are given three sides?

67. Make up an applied problem that requires using the Law of Cosines.

68. Write down your strategy for solving an oblique triangle.

69. State the Law of Cosines in words.

Retain Your Knowledge

Problems 70–77 are based on material learned earlier in the course. The purpose of these problems is to keep the material fresh in your mind so that you are better prepared for the final exam.

70. Graph: $R(x) = \dfrac{2x+1}{x-3}$

71. If $\tan\theta = -\dfrac{2\sqrt{6}}{5}$ and $\cos\theta = -\dfrac{5}{7}$, find the exact value of each of the four remaining trigonometric functions.

72. Find an equation for the graph.

73. Find $f^{-1}(x)$ if $f(x) = \dfrac{A}{5x + 2}, A \neq 0$.

74. If $F(x) = -\dfrac{x^3}{3} + 3x + C$ and $[a, b] = [1, 2]$, find $F(b) - F(a)$.

75. Solve: $|4x - 3| \geq x + 1$

76. Convert $96°$ to radians.

77. If $f(x) = ax^2 - 2x + 5$ and $a < 0$, in which quadrant is the vertex located? How many x-intercepts does the graph of f have?

'Are You Prepared?' Answers

1. $d = \sqrt{(x_2 - x_1)^2 + (y_2 - y_1)^2}$ **2.** $\theta = 45°$ or $\dfrac{\pi}{4}$

4.4 Area of a Triangle

PREPARING FOR THIS SECTION *Before getting started, review the following:*

- Know Geometry Formulas (Section A.2, pp. A15–A16)
- Use Half-angle Formulas to Find Exact Values (Section 3.6, pp. 244–246)

Now Work the **'Are You Prepared?'** problems on page 296.

OBJECTIVES **1** Find the Area of SAS Triangles (p. 293)
                 2 Find the Area of SSS Triangles (p. 294)

In this section, several formulas for calculating the area of a triangle are derived.

NOTE Typically, *A* is used for area. However, because *A* is also used as the measure of an angle, *K* is used here for area to avoid confusion. ∎

THEOREM **Area of a Triangle**

The area K of a triangle is

$$K = \frac{1}{2}bh \tag{1}$$

where b is the base and h is the altitude drawn to that base.

Proof Look at the triangle in Figure 39. Around the triangle construct a rectangle of altitude h and base b, as shown in Figure 40.

    Triangles 1 and 2 in Figure 40 are equal in area, as are triangles 3 and 4. Consequently, the area of the triangle with base b and altitude h is exactly half the area of the rectangle, which is bh.

Figure 39

Figure 40

∎

1 Find the Area of SAS Triangles

If the base b and the altitude h to that base are known, then the area of the triangle can be found using formula (1). Usually, though, the information required to use formula (1) is not given. Suppose, for example, that two sides a and b and

Figure 41

the included angle C are known. See Figure 41. Then the altitude h can be found by noting that

$$\frac{h}{a} = \sin C$$

so

$$h = a \sin C$$

Using this fact in formula (1) produces

$$K = \frac{1}{2}bh = \frac{1}{2}b(a \sin C) = \frac{1}{2}ab \sin C$$

The area K of the triangle is given by the formula

> **THEOREM Area of an SAS Triangle**
>
> $$K = \frac{1}{2}ab \sin C \qquad (2)$$

Dropping altitudes from the other two vertices of the triangle leads to the following corresponding formulas:

> $$K = \frac{1}{2}bc \sin A \qquad (3)$$
>
> $$K = \frac{1}{2}ac \sin B \qquad (4)$$

It is easiest to remember these formulas by using the following wording:

> **THEOREM Area of an SAS Triangle**
>
> The area K of a triangle equals one-half the product of two of its sides times the sine of their included angle.

EXAMPLE 1   **Finding the Area of an SAS Triangle**

Find the area K of the triangle for which $a = 8$, $b = 6$, and $C = 30°$.

Solution   See Figure 42. Use formula (2) to get

$$K = \frac{1}{2}\, ab \sin C = \frac{1}{2} \cdot 8 \cdot 6 \cdot \sin 30° = 12 \text{ square units}$$

Figure 42

 Now Work PROBLEMS **9** AND **17**

2 Find the Area of SSS Triangles

If the three sides of a triangle are known, another formula, called *Heron's Formula* (named after Heron of Alexandria), can be used to find the area of a triangle.

> **THEOREM Heron's Formula**
>
> The area K of a triangle with sides $a, b,$ and c is
>
> $$K = \sqrt{s(s-a)(s-b)(s-c)} \qquad (5)$$
>
> where $s = \frac{1}{2}(a + b + c)$.

| EXAMPLE 2 | Finding the Area of an SSS Triangle |

Find the area of a triangle whose sides are 4, 5, and 7.

Solution Let $a = 4$, $b = 5$, and $c = 7$. Then

$$s = \frac{1}{2}(a + b + c) = \frac{1}{2}(4 + 5 + 7) = 8$$

Heron's Formula gives the area K as

$$K = \sqrt{s(s-a)(s-b)(s-c)} = \sqrt{8(8-4)(8-5)(8-7)}$$

$$= \sqrt{8 \cdot 4 \cdot 3 \cdot 1} = \sqrt{96} = 4\sqrt{6} \text{ square units}$$

Now Work PROBLEMS 15 AND 23

Proof of Heron's Formula The proof given here uses the Law of Cosines.
From the Law of Cosines,

$$c^2 = a^2 + b^2 - 2ab \cos C$$

and the Half-angle Formula,

$$\cos^2 \frac{C}{2} = \frac{1 + \cos C}{2}$$

it follows that

$$\cos^2 \frac{C}{2} = \frac{1 + \cos C}{2} = \frac{1 + \dfrac{a^2 + b^2 - c^2}{2ab}}{2}$$

$$= \frac{a^2 + 2ab + b^2 - c^2}{4ab} = \frac{(a+b)^2 - c^2}{4ab}$$

$$= \underset{\underset{\text{Difference of two squares}}{\uparrow}}{\frac{(a+b-c)(a+b+c)}{4ab}} = \underset{\underset{\substack{a+b-c = a+b+c-2c \\ = 2s - 2c = 2(s-c)}}{\uparrow}}{\frac{2(s-c) \cdot 2s}{4ab}} = \frac{s(s-c)}{ab} \quad \text{(6)}$$

Similarly, using $\sin^2 \dfrac{C}{2} = \dfrac{1 - \cos C}{2}$, it follows that

$$\sin^2 \frac{C}{2} = \frac{(s-a)(s-b)}{ab} \quad \text{(7)}$$

Now use formula (2) for the area.

$$K = \frac{1}{2}ab \sin C$$

$$= \frac{1}{2}ab \cdot 2 \sin \frac{C}{2} \cos \frac{C}{2} \qquad \sin C = \sin\left[2\left(\frac{C}{2}\right)\right] = 2 \sin \frac{C}{2} \cos \frac{C}{2}$$

$$= ab \sqrt{\frac{(s-a)(s-b)}{ab}} \sqrt{\frac{s(s-c)}{ab}} \qquad \text{Use equations (6) and (7).}$$

$$= \sqrt{s(s-a)(s-b)(s-c)}$$

Historical Feature

Heron's Formula (also known as *Hero's Formula*) was first expressed by Heron of Alexandria (first century AD), who had, besides his mathematical talents, engineering skills. In various temples, his mechanical devices produced effects that seemed supernatural and supposedly moved visitors to generosity. Heron's book *Metrica*, on making such devices, has survived and was discovered in 1896 in the city of Constantinople.

Heron's Formulas for the area of a triangle caused some mild discomfort in Greek mathematics, because a product with two factors was an area and one with three factors was a volume, but four factors seemed contradictory in Heron's time.

4.4 Assess Your Understanding

'Are You Prepared?' *The answers are given at the end of these exercises. If you get a wrong answer, read the pages listed in* red.

1. The area K of a triangle whose base is b and whose altitude is h is _____ . (p. A15)

2. **True or False** $\cos^2 \dfrac{\theta}{2} = \dfrac{1 + \sin \theta}{2}$ (pp. 244–246)

Concepts and Vocabulary

3. If two sides a and b and the included angle C are known in a triangle, then the area K is found using the formula $K =$ _____ .

4. The area K of a triangle with sides a, b, and c is

   $K =$ _____ ,

   where $s =$ _____ .

5. Find the area of the right triangle whose legs are of length 3 and 4.

6. **True or False** The area of a triangle equals one-half the product of the lengths of two of its sides times the sine of their included angle.

7. **Multiple Choice** Given two sides of a triangle, b and c, and the included angle A, the altitude h from angle B to side b is given by _____ .

   (a) $\dfrac{1}{2}ab \sin A$ (b) $b \sin A$ (c) $c \sin A$ (d) $\dfrac{1}{2}bc \sin A$

8. **Multiple Choice** Heron's Formula is used to find the area of _____ triangles.

   (a) ASA   (b) SAS   (c) SSS   (d) AAS

Skill Building

In Problems 9–16, find the area of each triangle. Round answers to two decimal places.

9.

10.

11.

12.

13.

14.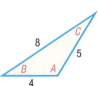

15.

16.

In Problems 17–28, find the area of each triangle. Round answers to two decimal places.

17. $a = 3$, $b = 4$, $C = 50°$
18. $a = 2$, $c = 1$, $B = 10°$
19. $b = 1$, $c = 8$, $A = 75°$
20. $a = 6$, $b = 4$, $C = 60°$
21. $a = 3$, $c = 2$, $B = 115°$
22. $b = 4$, $c = 1$, $A = 120°$
23. $a = 12$, $b = 35$, $c = 37$
24. $a = 4$, $b = 5$, $c = 3$
25. $a = 4$, $b = 4$, $c = 4$
26. $a = 3$, $b = 3$, $c = 2$
27. $a = 11$, $b = 14$, $c = 20$
28. $a = 4$, $b = 3$, $c = 6$

Applications and Extensions

29. **Area of an ASA Triangle** If two angles and the included side are given, the third angle is easy to find. Use the Law of Sines to show that the area K of a triangle with side a and angles A, B, and C is

$$K = \frac{a^2 \sin B \sin C}{2 \sin A}$$

30. **Area of a Triangle** Prove the two other forms of the formula for the area K of a triangle given in Problem 29.

$$K = \frac{b^2 \sin A \sin C}{2 \sin B} \quad \text{and} \quad K = \frac{c^2 \sin A \sin B}{2 \sin C}$$

In Problems 31–36, use the results of Problem 29 or 30 to find the area of each triangle. Round answers to two decimal places.

31. $A = 40°$, $B = 20°$, $a = 2$
32. $A = 50°$, $C = 20°$, $a = 3$
33. $B = 70°$, $C = 10°$, $b = 5$
34. $A = 70°$, $B = 60°$, $c = 4$
35. $A = 110°$, $C = 30°$, $c = 3$
36. $B = 10°$, $C = 100°$, $b = 2$

37. Area of a Segment Find the area of the segment (shaded in blue in the figure) of a circle whose radius is 8 feet, formed by a central angle of 70°.

[**Hint:** Subtract the area of the triangle from the area of the sector to obtain the area of the segment.]

38. Area of a Segment Find the area of the segment of a circle whose radius is 5 inches, formed by a central angle of 40°.

39. Cost of a Triangular Lot The dimensions of a triangular lot are 100 feet by 50 feet by 75 feet. If the price of the land is $3 per square foot, how much does the lot cost?

40. Amount of Material to Make a Tent A cone-shaped tent is made from a circular piece of canvas 24 feet in diameter by removing a sector with central angle 100° and connecting the ends. What is the surface area of the tent?

41. Fighter Jet Design The Eurofighter Typhoon has a canard-delta wing design that contains a large triangular main wing. Use the dimensions shown to approximate the area of one of the main wings.

42. Property Area A lot for sale in a subdivision has the shape of the quadrilateral shown in the figure. Find the area of the lot to the nearest square foot.

43. Dimensions of Home Plate The dimensions of home plate at any major league baseball stadium are shown. Find the area of home plate.

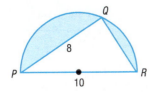

44. Computing Areas See the figure. Find the area of the shaded region enclosed in a semicircle of diameter 10 inches. The length of the chord PQ is 8 inches.

[**Hint:** Triangle PQR is a right triangle.]

45. Geometry See the figure, which shows a circle of radius r with center at O. Find the area K of the shaded region as a function of the central angle θ.

46. Approximating the Area of a Lake To approximate the area of a lake, a surveyor walks around the perimeter of the lake, taking the measurements shown in the figure. Using this technique, what is the approximate area of the lake?

[**Hint:** Use the Law of Cosines on the three triangles shown, and then find the sum of their areas.]

47. The Flatiron Building Completed in 1902 in New York City, the Flatiron Building is triangular shaped and bounded by 22nd Street, Broadway, and 5th Avenue. The building measures approximately 87 feet on the 22nd Street side, 190 feet on the Broadway side, and 173 feet on the 5th Avenue side. Approximate the ground area covered by the building.

Source: Sarah Bradford Landau and Carl W. Condit, *Rise of the New York Skyscraper: 1865–1913.* New Haven, CT: Yale University Press, 1996

48. Bermuda Triangle The Bermuda Triangle is roughly defined by Hamilton, Bermuda; San Juan, Puerto Rico; and Fort Lauderdale, Florida. The distances from Hamilton to Fort Lauderdale, Fort Lauderdale to San Juan, and San Juan to Hamilton are approximately 1028, 1046, and 965 miles, respectively. Ignoring the curvature of Earth, approximate the area of the Bermuda Triangle.

Source: www.worldatlas.com

49. Bretschneider's Formula There is a Heron-type formula that can be used to find the area of a general quadrilateral.

$$K = \sqrt{(s-a)(s-b)(s-c)(s-d) - abcd\cos^2\theta}$$

where a, b, c, and d are the side lengths, θ is half the sum of two opposite angles, and s is half the perimeter.

Show that if a triangle is considered a quadrilateral with one side equal to 0, Bretschneider's Formula reduces to Heron's Formula.

50. (a) Show that the area of a regular dodecagon (12-sided polygon) is given by $K = 3a^2\cot\frac{\pi}{12}$ or $K = 12r^2\tan\frac{\pi}{12}$, where a is the length of one of the sides and r is the radius of the inscribed circle.

(b) Given that each interior angle of a regular n-sided polygon $(n \geq 3)$ measures $\frac{(n-2)\cdot\pi}{n}$, generalize these formulas for any regular polygon.

51. Geometry Refer to the figure. If $|OA| = 1$, show that:

(a) Area $\triangle OAC = \dfrac{1}{2}\sin\alpha\cos\alpha$

(b) Area $\triangle OCB = \dfrac{1}{2}|OB|^2\sin\beta\cos\beta$

(c) Area $\triangle OAB = \dfrac{1}{2}|OB|\sin(\alpha+\beta)$

(d) $|OB| = \dfrac{\cos\alpha}{\cos\beta}$

(e) $\sin(\alpha+\beta)$
$= \sin\alpha\cos\beta + \cos\alpha\sin\beta$

[*Hint*: Area $\triangle OAB$ = Area $\triangle OAC$ + Area $\triangle OCB$]

52. Geometry Refer to the figure, in which a unit circle is drawn. The line segment DB is tangent to the circle and θ is acute.
(a) Express the area of $\triangle OBC$ in terms of $\sin\theta$ and $\cos\theta$.
(b) Express the area of $\triangle OBD$ in terms of $\sin\theta$ and $\cos\theta$.
(c) The area of the sector $\overset{\frown}{OBC}$ of the circle is $\dfrac{1}{2}\theta$, where θ is measured in radians. Use the results of parts (a) and (b) and the fact that

$$\text{Area }\triangle OBC < \text{Area }\overset{\frown}{OBC} < \text{Area }\triangle OBD$$

to show that

$$1 < \frac{\theta}{\sin\theta} < \frac{1}{\cos\theta}$$

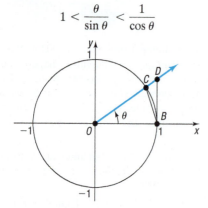

53. The Cow Problem* A cow is tethered to one corner of a square barn, 10 feet by 10 feet, with a rope 100 feet long. What is the maximum grazing area for the cow? See the figure.

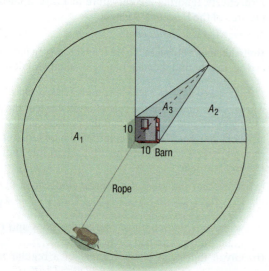

54. Another Cow Problem If the barn in Problem 53 is rectangular, 10 feet by 20 feet, what is the maximum grazing area for the cow?

55. Perfect Triangles A *perfect triangle* is one having integers for sides for which the area is numerically equal to the perimeter. Show that the triangles with the given side lengths are perfect.
(a) 9, 10, 17   **(b)** 6, 25, 29

Source: M.V. Bonsangue, G. E. Gannon, E. Buchman, and N. Gross, "In Search of Perfect Triangles," *Mathematics Teacher*, Vol. 92, No. 1, 1999: 56–61

56. If h_1, h_2, and h_3 are the altitudes dropped from P, Q, and R, respectively, in a triangle (see the figure), show that

$$\frac{1}{h_1} + \frac{1}{h_2} + \frac{1}{h_3} = \frac{s}{K}$$

where K is the area of the triangle and $s = \dfrac{1}{2}(a+b+c)$.

[**Hint:** $h_1 = \dfrac{2K}{a}$.]

57. Show that a formula for the altitude h from a vertex to the opposite side a of a triangle is

$$h = \frac{a\sin B\sin C}{\sin A}$$

58. Challenge Problem A triangle has vertices $A(0,0)$, $B(1,0)$, and C, where C is the point on the unit circle corresponding to an angle of $105°$ when it is drawn in standard position. Find the area of the triangle. State the answer in complete simplified form with a rationalized denominator.

Challenge Problems **Inscribed Circle** *For Problems 59–62, the lines that bisect each angle of a triangle meet in a single point O, and the perpendicular distance r from O to each side of the triangle is the same. The circle with center at O and radius r is called the inscribed circle of the triangle (see the figure).*

59. Use the formula from Problem 57 with triangle OPQ to show that

$$r = \frac{c\sin\dfrac{A}{2}\sin\dfrac{B}{2}}{\cos\dfrac{C}{2}}$$

60. Use the result of Problem 59 and the results of Problems 62 and 63 in Section 4.3 to show that

$$\cot \frac{C}{2} = \frac{s - c}{r}$$

where $s = \frac{1}{2}(a + b + c)$.

61. Show that

$$\cot \frac{A}{2} + \cot \frac{B}{2} + \cot \frac{C}{2} = \frac{s}{r}$$

62. Show that the area K of triangle PQR is $K = rs$, where $s = \frac{1}{2}(a + b + c)$. Then show that

$$r = \sqrt{\frac{(s - a)(s - b)(s - c)}{s}}$$

Explaining Concepts: Discussion and Writing

63. What do you do first if you are asked to find the area of a triangle and are given two sides and the included angle?

64. What do you do first if you are asked to find the area of a triangle and are given three sides?

65. State the formula for finding the area of an SAS triangle in words.

Retain Your Knowledge

Problems 66–74 are based on material learned earlier in the course. The purpose of these problems is to keep the material fresh in your mind so that you are better prepared for the final exam.

66. Without graphing, determine whether the quadratic function $f(x) = -3x^2 + 12x + 5$ has a maximum value or a minimum value, and then find the value.

67. Solve the inequality: $\dfrac{x + 1}{x^2 - 9} \le 0$

68. $P = \left(-\dfrac{\sqrt{7}}{3}, \dfrac{\sqrt{2}}{3}\right)$ is the point on the unit circle that corresponds to a real number t. Find the exact values of the six trigonometric functions of t.

69. Establish the identity: $\csc\theta - \sin\theta = \cos\theta\cot\theta$

70. A rectangle has a diagonal of length 12. Express the perimeter P as a function of its width, w.

71. List all potential rational zeros of $P(x) = 2x^3 - 5x^2 + 13x + 6$.

72. Solve: $|(5x - 7) - 5| \le 0.05$

73. Solve: $x(x - 7) = 18$

74. The slope m of the tangent line to the graph of $f(x) = 3x^4 - 7x^2 + 2$ at any number x is given by $m = f'(x) = 12x^3 - 14x$. Find an equation of the tangent line at $x = 1$.

'Are You Prepared?' Answers

1. $K = \dfrac{1}{2}bh$ **2.** False

4.5 Simple Harmonic Motion; Damped Motion; Combining Waves

PREPARING FOR THIS SECTION *Before getting started, review the following:*

- Sinusoidal Graphs (Section 2.4, pp. 150–156)
- Angular Speed (Section 2.1, pp. 109–110)

Now Work the 'Are You Prepared?' problems on page 305.

OBJECTIVES **1** Build a Model for an Object in Simple Harmonic Motion (p. 299)
 2 Analyze Simple Harmonic Motion (p. 301)
 3 Analyze an Object in Damped Motion (p. 302)
 4 Graph the Sum of Two Functions (p. 304)

1 Build a Model for an Object in Simple Harmonic Motion

Many physical phenomena can be described as simple harmonic motion. Radio and television waves, light waves, sound waves, and water waves exhibit motion that is simple harmonic.

Tuning fork

Figure 43 Coiled spring

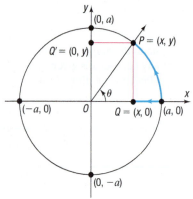

Figure 44

The swinging of a pendulum, the vibrations of a tuning fork, and the bobbing of a weight attached to a coiled spring are examples of vibrational motion. In this type of motion, an object swings back and forth over the same path. In Figure 43, the point B is the **equilibrium (rest) position** of the vibrating object. The **amplitude** is the distance from the object's rest position to its point of greatest displacement (either point A or point C in Figure 43). The **period** is the time required to complete one vibration—that is, the time it takes to go from, say, point A through B to C and back to A.

> **Simple harmonic motion** is a special kind of vibrational motion in which the acceleration a of the object is directly proportional to the negative of its displacement d from its rest position. That is, $a = -kd, k > 0$.

For example, when the mass hanging from the spring in Figure 43 is pulled down from its rest position B to the point C, the force of the spring tries to restore the mass to its rest position. Assuming that there is no frictional force to retard the motion, the amplitude will remain constant. The force increases in direct proportion to the distance that the mass is pulled from its rest position. Since the force increases directly, the acceleration of the mass of the object must do likewise, because (by Newton's Second Law of Motion) force is directly proportional to acceleration. As a result, the acceleration of the object varies directly with its displacement, and the motion is an example of simple harmonic motion.

Simple harmonic motion is related to circular motion. To see this relationship, consider a circle of radius a, with center at $(0, 0)$. See Figure 44. Suppose that an object initially placed at $(a, 0)$ moves counterclockwise around the circle at a constant angular speed ω. Suppose further that after time t has elapsed the object is at the point $P = (x, y)$ on the circle. The angle θ, in radians, swept out by the ray $\overrightarrow{OP}$ in this time t is

$$\theta = \omega t \quad \omega = \frac{\theta}{t}$$

The coordinates of the point P at time t are

$$x = a \cos \theta = a \cos(\omega t)$$
$$y = a \sin \theta = a \sin(\omega t)$$

Corresponding to each position $P = (x, y)$ of the object moving about the circle, there is the point $Q = (x, 0)$, called the **projection of P on the x-axis**. As P moves around the circle at a constant rate, the point Q moves back and forth between the points $(a, 0)$ and $(-a, 0)$ along the x-axis with a motion that is simple harmonic. Similarly, for each point P there is a point $Q' = (0, y)$, called the **projection of P on the y-axis**. As P moves around the circle, the point Q' moves back and forth between the points $(0, a)$ and $(0, -a)$ on the y-axis with a motion that is simple harmonic. Therefore, simple harmonic motion can be described as the projection of constant circular motion on a coordinate axis.

To illustrate, again consider a mass hanging from a spring where the mass is pulled down from its rest position to the point C and then released. See Figure 45(a). The graph shown in Figure 45(b) describes the displacement d of the object from its rest position as a function of time t, assuming that no frictional force is present.

Figure 45   (a)   (b)

> **THEOREM** **Simple Harmonic Motion**
>
> An object that moves on a coordinate axis so that the displacement d from its rest position at time t is given by either
>
> $$d(t) = a\cos(\omega t) \quad \text{or} \quad d(t) = a\sin(\omega t)$$
>
> where a and $\omega > 0$ are constants, moves with simple harmonic motion. The motion has amplitude $|a|$ and period $T = \dfrac{2\pi}{\omega}$.

The **frequency** f of an object in simple harmonic motion is the number of oscillations per unit time. Since the period is the time required for one oscillation, it follows that the frequency is the reciprocal of the period; that is,

$$f = \frac{1}{T} = \frac{\omega}{2\pi} \qquad \omega > 0$$

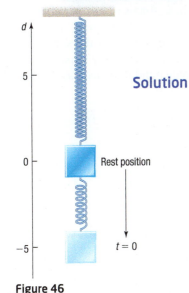

Figure 46

NOTE In the solution to Example 1, $a = -5$ because the object is initially pulled down. (If the initial direction is up, then use $a = 5$.) ∎

| EXAMPLE 1 | **Build a Model for an Object in Simple Harmonic Motion** |

Suppose that an object attached to a coiled spring is pulled down a distance of 5 inches from its rest position and then released. If the time for one oscillation is 3 seconds, develop a model that relates the displacement d of the object from its rest position after time t (in seconds). Assume no friction.

Solution The motion of the object is simple harmonic. See Figure 46. When the object is released $(t = 0)$, the displacement of the object from the rest position is -5 units (since the object was pulled down). Because $d = -5$ when $t = 0$, it is easier to use the cosine function

$$d(t) = a\cos(\omega t)$$

to describe the motion.* The amplitude is $|-5| = 5$ and the period is 3, so

$$a = -5 \quad \text{and} \quad \frac{2\pi}{\omega} = \text{period} = 3, \quad \text{so } \omega = \frac{2\pi}{3}$$

A function that models the motion of the object is

$$d(t) = -5\cos\left(\frac{2\pi}{3}t\right)$$

╺ **Now Work** PROBLEM 7

2 Analyze Simple Harmonic Motion

| EXAMPLE 2 | **Analyzing the Motion of an Object** |

Suppose that the displacement d (in meters) of an object at time t (in seconds) is given by the function

$$d(t) = 10\sin(5t)$$

(a) Describe the motion of the object.
(b) What is the maximum displacement from its rest position?
(c) What is the time required for one oscillation?
(d) What is the frequency?

---

*No phase shift is required if a cosine function is used.

Solution The function $d(t) = 10 \sin(5t)$ is of the form

$$d(t) = a \sin(\omega t)$$

where $a = 10$ and $\omega = 5$.

(a) The motion is simple harmonic.

(b) The maximum displacement of the object from its rest position is the amplitude: $|a| = 10$ meters.

(c) The time required for one oscillation is the period:

$$\text{Period} = T = \frac{2\pi}{\omega} = \frac{2\pi}{5} \text{ seconds}$$

(d) The frequency is the reciprocal of the period.

$$\text{Frequency} = f = \frac{1}{T} = \frac{5}{2\pi} \text{ oscillation per second}$$

Now Work PROBLEM 15

3 Analyze an Object in Damped Motion*

In the models discussed up to now, the motion was simple harmonic. That is, they assumed no force was retarding the motion. However, most physical phenomena are affected by friction or other resistive forces. These forces remove energy from a moving system and thereby damp its motion. For example, when a mass hanging from a spring is pulled down a distance a and released, the friction in the spring causes the distance the mass moves from its rest position to decrease over time. As a result, the amplitude of any real oscillating spring or swinging pendulum decreases with time due to air resistance, friction, or other forces. See Figure 47.

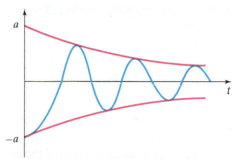

Figure 47 Damped motion

A model that describes this phenomenon maintains a sinusoidal component, but the amplitude of this component decreases with time to account for the damping effect. Moreover, the period of the oscillating component is affected by the damping. The next theorem, from physics, describes damped motion.

THEOREM Damped Motion

The displacement d of an oscillating object from its rest position at time t is given by

$$d(t) = ae^{-bt/(2m)} \cos\left(\sqrt{\omega^2 - \frac{b^2}{4m^2}}\, t\right)$$

where b is the **damping factor** or **damping coefficient** and m is the mass of the oscillating object. Here $|a|$ is the displacement at $t = 0$, and $\dfrac{2\pi}{\omega}$ is the period under simple harmonic motion (no damping).

*Requires Section 7.1, Exponential Functions

Notice that for $b = 0$ (zero damping), we have the formula for simple harmonic motion with amplitude $|a|$ and period $\dfrac{2\pi}{\omega}$.

EXAMPLE 3 | **Analyzing the Graph of an Object in Damped Motion**

Analyze the graph of an object in damped motion modeled by

$$d(t) = e^{-t/\pi} \cos t \quad t \geq 0$$

Solution

The displacement d is the product of $y = e^{-t/\pi}$ and $y = \cos t$. Using properties of absolute value and the fact that $|\cos t| \leq 1$, it follows that

$$|d(t)| = |e^{-t/\pi} \cos t| = |e^{-t/\pi}||\cos t| \leq |e^{-t/\pi}| = e^{-t/\pi}$$

$$\uparrow$$
$$\mathbf{e^{-t/\pi} > 0}$$

As a result,

$$-e^{-t/\pi} \leq d(t) \leq e^{-t/\pi}$$

This means that the graph of d lies between the graphs of $y = e^{-t/\pi}$ and $y = -e^{-t/\pi}$, called the **bounding curves** of d.

Also, the graph of d touches the graphs of the bounding curves when $|\cos t| = 1$; that is, when $t = 0, \pi, 2\pi$, and so on. The x-intercepts of the graph of d occur when $\cos t = 0$; that is, at $\dfrac{\pi}{2}, \dfrac{3\pi}{2}, \dfrac{5\pi}{2}$, and so on. See Table 1.

Table 1

t	0	$\dfrac{\pi}{2}$	π	$\dfrac{3\pi}{2}$	2π
$e^{-t/\pi}$	1	$e^{-1/2}$	e^{-1}	$e^{-3/2}$	e^{-2}
$\cos t$	1	0	-1	0	1
$d(t) = e^{-t/\pi} \cos t$	1	0	$-e^{-1}$	0	e^{-2}
Point on graph of d	$(0, 1)$	$\left(\dfrac{\pi}{2}, 0\right)$	$(\pi, -e^{-1})$	$\left(\dfrac{3\pi}{2}, 0\right)$	$(2\pi, e^{-2})$

Figure 48 shows the graph of $y = \cos t$, in black, the graphs of $y = e^{-t/\pi}$ and $y = -e^{-t/\pi}$, the bounding curves, in red, and the graph of $d(t) = e^{-t/\pi} \cos t$ in blue.

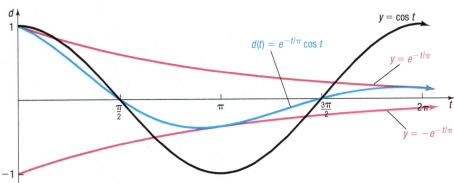

Figure 48 Damped vibration graph with bounding curves

Figure 49

Exploration

Graph $Y_1 = e^{-x/\pi} \cos x$, along with $Y_2 = e^{-x/\pi}$ and $Y_3 = -e^{-x/\pi}$, for $0 \leq x \leq 2\pi$. Determine where Y_1 has its first turning point (local minimum). Compare this to where Y_1 intersects Y_3.

Result Figure 49 shows the graphs of $Y_1 = e^{-x/\pi} \cos x$, $Y_2 = e^{-x/\pi}$, and $Y_3 = -e^{-x/\pi}$ on a TI-84 Plus C. Using MINIMUM, the first turning point occurs at $x \approx 2.83$; Y_1 INTERSECTS Y_3 at $x = \pi \approx 3.14$.

✏️ **Now Work** PROBLEM **23**

Situations also exist where external forces cause a vibrating system to oscillate at larger and larger amplitudes. This phenomenon, known as **resonance** from the Latin *resonare* (meaning "resound") or *resonantia* (meaning "echo"), occurs when external vibrations match the natural frequency of the vibrating system. Resonance can be destructive to bridges, buildings, or even mechanical devices. For example, bridges can be affected by soldiers marching in step, buildings can be affected by blowing winds, and automobiles can be affected by the vibrations of its tires. Engineers account for expected external vibrations in their designs and incorporate shock absorbers or dampers to counter the effect of resonance.

4 Graph the Sum of Two Functions

Many physical and biological applications require the graph of the sum of two functions, such as

$$f(x) = x + \sin x \quad \text{or} \quad g(x) = \sin x + \cos(2x)$$

For example, if two tones are emitted, the sound produced is the sum of the waves produced by the two tones. See Problem 51 in Section 3.7 for an explanation of Touch-Tone phones.

To graph the sum of two (or more) functions, add the *y*-coordinates that correspond to equal values of *x*.

EXAMPLE 4

Graphing the Sum of Two Functions

Graph $f(x) = x + \sin x$.

Solution

First, graph the component functions,

$$y = f_1(x) = x \qquad y = f_2(x) = \sin x$$

on the same coordinate axes. See Figure 50(a). Now, select several values of *x* say

$$x = 0 \qquad x = \frac{\pi}{2} \qquad x = \pi \qquad x = \frac{3\pi}{2} \qquad \text{and} \qquad x = 2\pi$$

and use them to compute $f(x) = f_1(x) + f_2(x)$. Table 2 shows the computations. Plot these points and connect them to get the graph, as shown in Figure 50(b).

Table 2

x	0	$\dfrac{\pi}{2}$	π	$\dfrac{3\pi}{2}$	2π
$y = f_1(x) = x$	0	$\dfrac{\pi}{2}$	π	$\dfrac{3\pi}{2}$	2π
$y = f_2(x) = \sin x$	0	1	0	-1	0
$f(x) = x + \sin x$	0	$\dfrac{\pi}{2} + 1 \approx 2.57$	π	$\dfrac{3\pi}{2} - 1 \approx 3.71$	2π
Point on graph of f	$(0, 0)$	$\left(\dfrac{\pi}{2}, 2.57\right)$	(π, π)	$\left(\dfrac{3\pi}{2}, 3.71\right)$	$(2\pi, 2\pi)$

✔ **Check:** Graph $Y_1 = x$, $Y_2 = \sin x$, and $Y_3 = x + \sin x$ and compare the result with Figure 50(b). Use INTERSECT to verify that the graphs of Y_1 and Y_3 intersect when $\sin x = 0$.

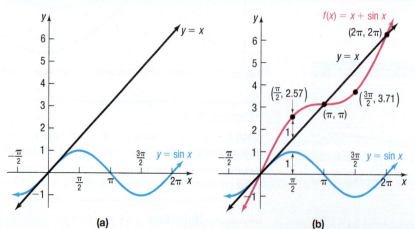

Figure 50

(a)   (b)

In Figure 50(b), notice that the graph of $f(x) = x + \sin x$ intersects the line $y = x$ whenever $\sin x = 0$. Also, notice that the graph of f is not periodic. The next example shows a periodic graph.

EXAMPLE 5 — **Graphing the Sum of Two Sinusoidal Functions**

Graph $f(x) = \sin x + \cos(2x)$.

Solution Graph f by adding the y-coordinates of $y = \sin x$ and $y = \cos(2x)$. Table 3 shows the steps for computing several points on the graph of f. Figure 51 illustrates the graphs of the component functions, $y = f_1(x) = \sin x$ (in blue), and $y = f_2(x) = \cos(2x)$ (in black), and the graph of $f(x) = \sin x + \cos(2x)$, which is shown in red.

Table 3

x	$-\dfrac{\pi}{2}$	0	$\dfrac{\pi}{2}$	π	$\dfrac{3\pi}{2}$	2π
$y = f_1(x) = \sin x$	-1	0	1	0	-1	0
$y = f_2(x) = \cos(2x)$	-1	1	-1	1	-1	1
$f(x) = \sin x + \cos(2x)$	-2	1	0	1	-2	1
Point on graph of f	$\left(-\dfrac{\pi}{2}, -2\right)$	$(0, 1)$	$\left(\dfrac{\pi}{2}, 0\right)$	$(\pi, 1)$	$\left(\dfrac{3\pi}{2}, -2\right)$	$(2\pi, 1)$

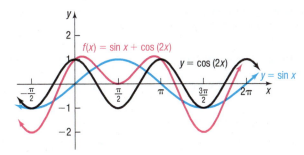

Figure 51

Notice that the function $f(x) = \sin x + \cos(2x)$ is periodic, with period 2π.

✓**Check:** Graph $Y_1 = \sin x$, $Y_2 = \cos(2x)$, and $Y_3 = \sin x + \cos(2x)$ and compare the result with Figure 51.

✏ **Now Work** PROBLEM 27

4.5 Assess Your Understanding

'Are You Prepared?' *The answers are given at the end of these exercises. If you get a wrong answer, read the pages listed in red.*

1. The amplitude A and period T of $f(x) = 5\sin(4x)$ are _____ and _____. (pp. 150–152)

2. Approximate the angular speed of the second hand on a clock in rad/sec. (Round to three decimal places.) (pp. 109–110)

3. Write an equation for a sine function with period 12 and amplitude 7. (p. 156)

Concepts and Vocabulary

4. The motion of an object is given by $d(t) = 4\cos(6t)$. Such motion is described as _____ _____. The number 4 is called the _____.

5. When a mass hanging from a spring is pulled down and then released, the motion is called _____ _____ if there is

no frictional force to retard the motion, and the motion is called _____ if there is friction.

6. **True or False** If the distance d of an object from its rest position at time t is given by a sinusoidal graph, the motion of the object is simple harmonic motion.

Skill Building

In Problems 7–10, an object attached to a coiled spring is pulled down a distance a from its rest position and then released. Assuming that the motion is simple harmonic with period T, find a function that relates the displacement d of the object from its rest position after t seconds. Assume that the positive direction of the motion is up.

7. $a = 5$; $T = 2$ seconds

8. $a = 10$; $T = 3$ seconds

9. $a = 7$; $T = 5\pi$ seconds

10. $a = 4$; $T = \dfrac{\pi}{2}$ seconds

11. Rework Problem 7 under the same conditions, except that at time $t = 0$, the object is at its rest position and moving down.

12. Rework Problem 8 under the same conditions, except that at time $t = 0$, the object is at its rest position and moving down.

13. Rework Problem 9 under the same conditions, except that at time $t = 0$, the object is at its rest position and moving down.

14. Rework Problem 10 under the same conditions, except that at time $t = 0$, the object is at its rest position and moving down.

In Problems 15–22, the displacement d (in meters) of an object at time t (in seconds) is given.
   (a) *Describe the motion of the object.*
   (b) *What is the maximum displacement from its rest position?*
   (c) *What is the time required for one oscillation?*
   (d) *What is the frequency?*

15. $d(t) = 5\sin(3t)$

16. $d(t) = 4\sin(2t)$

17. $d(t) = 8\cos(2\pi t)$

18. $d(t) = 5\cos\left(\dfrac{\pi}{2}t\right)$

19. $d(t) = -9\sin\left(\dfrac{1}{4}t\right)$

20. $d(t) = -2\cos(2t)$

21. $d(t) = 3 + 7\cos(3\pi t)$

22. $d(t) = 4 + 3\sin(\pi t)$

In Problems 23–26, graph each damped vibration curve for $0 \le t \le 2\pi$.

23. $d(t) = e^{-t/\pi}\cos(2t)$

24. $d(t) = e^{-t/2\pi}\cos(2t)$

25. $d(t) = e^{-t/2\pi}\cos t$

26. $d(t) = e^{-t/4\pi}\cos t$

In Problems 27–34, graph each function by adding y-coordinates.

27. $f(x) = x + \cos x$

28. $f(x) = x + \cos(2x)$

29. $f(x) = x - \sin x$

30. $f(x) = x - \cos x$

31. $f(x) = \sin x + \cos x$

32. $f(x) = \sin(2x) + \cos x$

33. $g(x) = \sin x + \sin(2x)$

34. $g(x) = \cos(2x) + \cos x$

Mixed Practice *In Problems 35–40, (a) use the Product-to-Sum Formulas to express each product as a sum, and (b) use the method of adding y-coordinates to graph each function on the interval $[0, 2\pi]$.*

35. $f(x) = \sin(2x)\sin x$

36. $F(x) = \sin(3x)\sin x$

37. $G(x) = \cos(4x)\cos(2x)$

38. $h(x) = \cos(2x)\cos(x)$

39. $H(x) = 2\sin(3x)\cos(x)$

40. $g(x) = 2\sin x\cos(3x)$

Applications and Extensions

In Problems 41–46, an object of mass m (in grams) attached to a coiled spring with damping factor b (in grams per second) is pulled down a distance a (in centimeters) from its rest position and then released. Assume that the positive direction of the motion is up and the period is T (in seconds) under simple harmonic motion.
   (a) *Find a function that relates the displacement d of the object from its rest position after t seconds.*
   (b) *Graph the function found in part (a) for 5 oscillations using a graphing utility.*

41. $m = 25$, $a = 10$, $b = 0.7$, $T = 5$

42. $m = 20$, $a = 15$, $b = 0.75$, $T = 6$

43. $m = 30$, $a = 18$, $b = 0.6$, $T = 4$

44. $m = 15$, $a = 16$, $b = 0.65$, $T = 5$

45. $m = 10$, $a = 5$, $b = 0.8$, $T = 3$

46. $m = 10$, $a = 5$, $b = 0.7$, $T = 3$

In Problems 47–52, the function d models the distance (in meters) of the bob of a pendulum of mass m (in kilograms) from its rest position at time t (in seconds) is given. The bob is released from the left of its rest position and represents a negative direction.

(a) *Describe the motion of the object. Be sure to give the mass and damping factor.*
(b) *What is the initial displacement of the bob? That is, what is the displacement at t = 0?*
(c) *Graph the motion using a graphing utility.*
(d) *What is the displacement of the bob at the start of the second oscillation?*
(e) *What happens to the displacement of the bob as time increases without bound?*

47. $d(t) = -20e^{-0.7t/40} \cos\left(\sqrt{\left(\frac{2\pi}{5}\right)^2 - \frac{0.49}{1600}}\, t\right)$

48. $d(t) = -20e^{-0.8t/40} \cos\left(\sqrt{\left(\frac{2\pi}{5}\right)^2 - \frac{0.64}{1600}}\, t\right)$

49. $d(t) = -30e^{-0.6t/80} \cos\left(\sqrt{\left(\frac{2\pi}{7}\right)^2 - \frac{0.36}{6400}}\, t\right)$

50. $d(t) = -30e^{-0.5t/70} \cos\left(\sqrt{\left(\frac{\pi}{2}\right)^2 - \frac{0.25}{4900}}\, t\right)$

51. $d(t) = -15e^{-0.9t/30} \cos\left(\sqrt{\left(\frac{\pi}{3}\right)^2 - \frac{0.81}{900}}\, t\right)$

52. $d(t) = -10e^{-0.8t/50} \cos\left(\sqrt{\left(\frac{2\pi}{3}\right)^2 - \frac{0.64}{2500}}\, t\right)$

---

53. Loudspeaker A loudspeaker diaphragm is oscillating in simple harmonic motion described by the function $d(t) = a \cos(\omega t)$ with a frequency of 520 hertz (cycles per second) and a maximum displacement of 0.80 millimeter. Find ω and then find a function that describes the movement of the diaphragm.

54. Colossus Added to Six Flags St. Louis in 1986, the Colossus is a giant Ferris wheel. Its diameter is 165 feet; it rotates at a rate of about 1.6 revolutions per minute; and the bottom of the wheel is 15 feet above the ground. Find a function that relates a rider's height h above the ground at time t. Assume the passenger begins the ride at the bottom of the wheel.

Source: Six Flags Theme Parks, Inc.

55. Tuning Fork The end of a tuning fork moves in simple harmonic motion described by the function $d(t) = a \sin(\omega t)$. If a tuning fork for the note A above middle C on an even-tempered scale (A_4, the tone by which an orchestra tunes itself) has a frequency of 440 hertz (cycles per second), find ω. If the maximum displacement of the end of the tuning fork is 0.01 millimeter, find a function that describes the movement of the tuning fork.

Source: David Lapp. Physics of Music and Musical Instruments. Medford, MA: Tufts University, 2003

56. Tuning Fork The end of a tuning fork moves in simple harmonic motion described by the function $d(t) = a \sin(\omega t)$. If a tuning fork for the note E above middle C on an even-tempered scale (E_4) has a frequency of approximately 329.63 hertz (cycles per second), find ω. If the maximum displacement of the end of the tuning fork is 0.025 millimeter, find a function that describes the movement of the tuning fork.

Source: David Lapp. Physics of Music and Musical Instruments. Medford, MA: Tufts University, 2003

57. Charging a Capacitor See the figure (top, right). If a charged capacitor is connected to a coil by closing a switch, energy is transferred to the coil and then back to the capacitor in an oscillatory motion. The voltage V (in volts) across the capacitor will gradually diminish to 0 with time t (in seconds).
(a) Graph the function relating V and t:

$$V(t) = e^{-t/3} \cos(\pi t) \qquad 0 \le t \le 3$$

(b) At what times t does the graph of V touch the graph of $y = e^{-t/3}$? When does the graph of V touch the graph of $y = -e^{-t/3}$?

(c) When is the voltage V between -0.4 and 0.4 volt?

58. The Sawtooth Curve An oscilloscope often displays a *sawtooth curve.* This curve can be approximated by sinusoidal curves of varying periods and amplitudes.
(a) Use a graphing utility to graph the following function, which can be used to approximate the sawtooth curve.

$$f(x) = \frac{1}{2}\sin(2\pi x) + \frac{1}{4}\sin(4\pi x) \qquad 0 \le x \le 4$$

(b) A better approximation to the sawtooth curve is given by

$$f(x) = \frac{1}{2}\sin(2\pi x) + \frac{1}{4}\sin(4\pi x) + \frac{1}{8}\sin(8\pi x)$$

Use a graphing utility to graph this function for $0 \le x \le 4$ and compare the result to the graph obtained in part (a).

(c) A third and even better approximation to the sawtooth curve is given by

$$f(x) = \frac{1}{2}\sin(2\pi x) + \frac{1}{4}\sin(4\pi x) + \frac{1}{8}\sin(8\pi x) + \frac{1}{16}\sin(16\pi x)$$

Use a graphing utility to graph this function for $0 \le x \le 4$ and compare the result to the graphs obtained in parts (a) and (b).

(d) What do you think the next approximation to the sawtooth curve is?

59. **A Clock Signal** A *clock signal* is a non-sinusoidal signal used to coordinate actions of a digital circuit. Such signals oscillate between two levels, high and low, "instantaneously" at regular intervals. The most common clock signal has the form of a *square wave* and can be approximated by the sum of simple harmonic sinusoidal waves, such as

$$f(x) = 2.35 + \sin x + \frac{\sin(3x)}{3} + \frac{\sin(5x)}{5} + \frac{\sin(7x)}{7} + \frac{\sin(9x)}{9}$$

Graph this function for $-4\pi \le x \le 4\pi$.

 60. **Non-Sinusoidal Waves** Both the sawtooth and square waves (see Problems 58 and 59) are examples of non-sinusoidal waves. Another type of non-sinusoidal wave is illustrated by the function

$$f(x) = 1.6 + \cos x + \frac{1}{9}\cos(3x) + \frac{1}{25}\cos(5x) + \frac{1}{49}\cos(7x)$$

Graph the function for $-5\pi \le x \le 5\pi$.

61. Graph the sound emitted by the * key on a Touch-Tone phone. See Problem 51 in Section 3.7.

 62. **CBL Experiment** The sound from a tuning fork is collected over time. A model of the form $y = A\cos[B(x - C)]$ is fitted to the data. Find the amplitude, frequency, and period of the graph.
(Activity 23, Real-World Math with the CBL System.)

 63. **CBL Experiment** Pendulum motion is analyzed to estimate simple harmonic motion. A plot is generated with the position of the pendulum over time. The graph is used to find a sinusoidal curve of the form $y = A\cos[B(x - C)] + D$. Find the amplitude, period, and frequency.
(Activity 16, Real-World Math with the CBL System.)

64. *Challenge Problem* **Beats** When two sinusoidal waves travel through the same medium, a third wave is formed that is the sum of the two original waves. If the two waves have slightly different frequencies, the sum of the waves results in an interference pattern known as a *beat*. Musicians use this idea when tuning an instrument with the aid of a tuning fork. If the instrument and the tuning fork play the same frequency, no beat is heard. Suppose two waves given by the functions, $y_1 = 3\cos(\omega_1 t)$ and $y_2 = 3\cos(\omega_2 t)$ where $\omega_1 > \omega_2$ pass through the same medium, and each has a maximum at $t = 0$ sec.
(a) How long does it take the sum function $y_3 = y_1 + y_2$ to equal 0 for the first time?
(b) If the periods of the two functions y_1 and y_2 are $T_1 = 19$ sec and $T_2 = 20$ sec, respectively, find the first time the sum $y_3 = y_1 + y_2 = 0$.
(c) Use the values from part (b) to graph y_3 over the interval $0 \le x \le 600$. Do the waves appear to be in tune?

Explaining Concepts: Discussion and Writing

 65. Graph the function $f(x) = \dfrac{\sin x}{x}, x > 0$. Based on the graph, what do you conjecture about the value of $\dfrac{\sin x}{x}$ for x close to 0?

66. Graph $y = x\sin x$, $y = x^2\sin x$, and $y = x^3\sin x$ for $x > 0$. What patterns do you observe?

 67. Graph $y = \dfrac{1}{x}\sin x$, $y = \dfrac{1}{x^2}\sin x$, and $y = \dfrac{1}{x^3}\sin x$ for $x > 0$. What patterns do you observe?

68. How would you explain simple harmonic motion to a friend? How would you explain damped motion?

Retain Your Knowledge

Problems 69–75 are based on material learned earlier in the course. The purpose of these problems is to keep the material fresh in your mind so that you are better prepared for the final exam.

69. The function $f(x) = \dfrac{x - 3}{x - 4}, x \ne 4$, is one-to-one. Find its inverse function.

70. If $\cos\alpha = \dfrac{4}{5}, 0 < \alpha < \dfrac{\pi}{2}$, find the exact value of:
(a) $\cos\dfrac{\alpha}{2}$ **(b)** $\sin\dfrac{\alpha}{2}$ **(c)** $\tan\dfrac{\alpha}{2}$

71. If $f(x) = \sqrt{3 - 5x}$ and $g(x) = x^2 + 7$, find $g(f(x))$ and its domain.

72. If $\cos\theta = \dfrac{5}{7}$ and $\tan\theta < 0$, what is the value of $\csc\theta$?

73. The normal line to a graph at a point is the line perpendicular to the tangent line of the graph at the point. If the tangent line is $y = \dfrac{2}{3}x - 1$ when $f(3) = 1$, find an equation of the normal line.

74. If $h(x)$ is a function with range $[-5, 8]$, what is the range of $h(2x + 3)$?

75. Solve: $x^2(5x - 3)(x + 2) \le 0$

'Are You Prepared?' Answers

1. $A = 5; T = \dfrac{\pi}{2}$ 2. 0.105 rad/sec   3. $y = 7\sin\left(\dfrac{\pi x}{6}\right)$

Chapter Review

Things to Know

Formulas

Law of Sines (p. 276)

$$\frac{\sin A}{a} = \frac{\sin B}{b} = \frac{\sin C}{c}$$

Law of Cosines (p. 286)

$$c^2 = a^2 + b^2 - 2ab \cos C$$
$$b^2 = a^2 + c^2 - 2ac \cos B$$
$$a^2 = b^2 + c^2 - 2bc \cos A$$

Area of a triangle (pp. 293–295)

$$K = \frac{1}{2}bh \quad K = \frac{1}{2}ab \sin C \quad K = \frac{1}{2}bc \sin A \quad K = \frac{1}{2}ac \sin B$$

$$K = \sqrt{s(s-a)(s-b)(s-c)} \quad \text{where} \quad s = \frac{1}{2}(a+b+c)$$

Objectives

Section		You should be able to . . .	Example(s)	Review Exercises
4.1	1	Find the value of trigonometric functions of acute angles using right triangles (p. 262)	1, 2	1, 2, 27
	2	Use the complementary angle theorem (p. 264)	3	3–5
	3	Solve right triangles (p. 264)	4, 5	6, 7, 27
	4	Solve applied problems (p. 265)	6–12	28–31, 36–38
4.2	1	Solve SAA or ASA triangles (p. 276)	1, 2	8, 19
	2	Solve SSA triangles (p. 277)	3–5	9, 10, 12, 16, 18
	3	Solve applied problems (p. 279)	6, 7	32, 33
4.3	1	Solve SAS triangles (p. 287)	1	11, 15, 20
	2	Solve SSS triangles (p. 287)	2	13, 14, 17
	3	Solve applied problems (p. 288)	3	34
4.4	1	Find the area of SAS triangles (p. 293)	1	21, 22, 26, 35
	2	Find the area of SSS triangles (p. 294)	2	23, 24
4.5	1	Build a model for an object in simple harmonic motion (p. 299)	1	39
	2	Analyze simple harmonic motion (p. 301)	2	40, 41
	3	Analyze an object in damped motion (p. 302)	3	42, 43
	4	Graph the sum of two functions (p. 304)	4, 5	44

Review Exercises

In Problems 1 and 2, find the exact value of the six trigonometric functions of the angle θ in each figure.

1.

2.

In Problems 3–5, find the exact value of each expression. Do not use a calculator.

3. $\cos 62° - \sin 28°$

4. $\dfrac{\sec 55°}{\csc 35°}$

5. $\cos^2 40° + \cos^2 50°$

In Problems 6 and 7, solve each triangle.

6.

7.

In Problems 8–20, find the remaining angle(s) and side(s) of each triangle, if it (they) exists. If no triangle exists, say "No triangle."

8. $A = 50°$, $B = 30°$, $a = 1$ **9.** $A = 100°$, $a = 5$, $c = 2$ **10.** $a = 3$, $c = 1$, $C = 110°$

11. $a = 3$, $c = 1$, $B = 100°$ **12.** $a = 3$, $b = 5$, $B = 80°$ **13.** $a = 2$, $b = 3$, $c = 1$

14. $a = 10$, $b = 7$, $c = 8$ **15.** $a = 1$, $b = 3$, $C = 40°$ **16.** $a = 5$, $b = 3$, $A = 80°$

17. $a = 1$, $b = \dfrac{1}{2}$, $c = \dfrac{4}{3}$ **18.** $a = 3$, $A = 10°$, $b = 4$ **19.** $a = 4$, $A = 20°$, $B = 100°$

20. $c = 5$, $b = 4$, $A = 70°$

In Problems 21–25, find the area of each triangle.

21. $a = 2$, $b = 3$, $C = 40°$ **22.** $b = 4$, $c = 10$, $A = 70°$ **23.** $a = 4$, $b = 3$, $c = 5$

24. $a = 4$, $b = 2$, $c = 5$ **25.** $A = 50°$, $B = 30°$, $a = 1$

---

26. Area of a Segment Find the area of the segment of a circle whose radius is 6 inches formed by a central angle of 50°.

27. Geometry The hypotenuse of a right triangle is 12 feet. If one leg is 8 feet, find the degree measure of each angle.

28. Finding the Width of a River Find the distance from *A* to *C* across the river illustrated in the figure.

29. Finding the Distance to Shore The Willis Tower in Chicago is 1454 feet tall and is situated about 1 mile inland from the shore of Lake Michigan, as indicated in the figure. An observer in a pleasure boat on the lake directly in front of the Willis Tower looks at the top of the tower and measures the angle of elevation as 5°. How far offshore is the boat?

30. Finding the Speed of a Glider From a glider 200 feet above the ground, two sightings of a stationary object directly in front are taken 1 minute apart (see the figure, top right). What is the speed of the glider?

31. Finding the Grade of a Mountain Trail A straight trail with a uniform inclination leads from a hotel, elevation 5000 feet, to a lake in a valley, elevation 4100 feet. The length of the trail is 4100 feet. What is the inclination (grade) of the trail?

32. Finding the Height of a Helicopter Two observers simultaneously measure the angle of elevation of a helicopter. One angle is measured as 25°, the other as 40° (see the figure). If the observers are 100 feet apart and the helicopter lies over the line joining them, how high is the helicopter?

33. Constructing a Highway A highway whose primary directions are north–south is being constructed along the west coast of Florida. Near Naples, a bay obstructs the straight path of the road. Since the cost of a bridge is prohibitive, engineers decide to go around the bay. The figure shows the path that they decide on and the measurements taken. What is the length of highway needed to go around the bay?

34. Correcting a Navigational Error A sailboat leaves St. Thomas bound for an island in the British West Indies, 200 miles away. Maintaining a constant speed of 18 miles per hour, but encountering heavy crosswinds and strong currents, the crew finds, after 4 hours, that the sailboat is off course by 15°.
(a) How far is the sailboat from the island at this time?
(b) Through what angle should the sailboat turn to correct its course?
(c) How much time has been added to the trip because of this? (Assume that the speed remains at 18 miles per hour.)

35. Approximating the Area of a Lake To approximate the area of a lake, Cindy walks around the perimeter of the lake, taking the measurements shown in the figure. Using this technique, what is the approximate area of the lake?

[**Hint:** Use the Law of Cosines on the three triangles shown and then find the sum of their areas.]

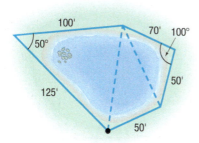

36. Finding the Bearing of a Ship The *Majesty* leaves the Port at Boston for Bermuda with a bearing of S80°E at an average speed of 10 knots. After 1 hour, the ship turns 90° toward the southwest. After 2 hours at an average speed of 20 knots, what is the bearing of the ship from Boston?

37. Frictional Force A box sitting on a flat surface has a coefficient of static friction of $\mu_s = 0.3$. If one end of the surface is raised, static friction prevents the box from sliding until the force of static friction is overcome. The critical angle at which the box begins to slide, θ_c, can be found from the equation $\tan \theta_c = \mu_s$.
(a) What is the critical angle for the box?
(b) If the box is 5 ft from the pivot point, at what height will the box begin to slide? See the figure.

38. Frictional Force (See Problem 37.) Once the box begins to slide and accelerate, kinetic friction acts to slow the box with a coefficient of kinetic friction $\mu_k = 0.1$. The raised end of the surface can be lowered to a point where the box continues sliding but does not accelerate. The critical angle at which this happens, $\theta_c{}'$, can be found from the equation $\tan \theta_c{}' = \mu_k$.
(a) What is this critical angle for the box?
(b) If the box is 5 ft from the pivot point, at what height will the box stop accelerating?

39. Simple Harmonic Motion An object attached to a coiled spring is pulled down a distance $a = 3$ units from its rest position and then released. Assuming that the motion is simple harmonic with period $T = 4$ seconds, find a model that relates the displacement d of the object from its rest position after t seconds. Also assume that the positive direction of the motion is up.

In Problems 40 and 41, the displacement d (in feet) of an object from its rest position at time t (in seconds) is given.
(a) Describe the motion of the object.
(b) What is the maximum displacement from its rest position?
(c) What is the time required for one oscillation?
(d) What is the frequency?

40. $d(t) = 6 \sin(2t)$ **41.** $d(t) = -2 \cos(\pi t)$

42. Damped Harmonic Motion An object of mass $m = 40$ grams attached to a coiled spring with damping factor $b = 0.75$ gram/second is pulled down a distance $a = 15$ centimeters from its rest position and then released. Assume that the positive direction of the motion is up and the period is $T = 5$ seconds under simple harmonic motion.
(a) Find a function that models the displacement d of the object from its rest position after t seconds.
(b) Graph the equation found in part (a) for 5 oscillations.

43. Damped Motion The displacement d (in meters) of the bob of a pendulum of mass 20 kilograms from its rest position at time t (in seconds) is given as

$$d(t) = -15e^{-0.6t/40} \cos\left(\sqrt{\left(\frac{2\pi}{5}\right)^2 - \frac{0.36}{1600}}\, t\right)$$

(a) Describe the motion of the object.
(b) What is the initial displacement of the bob? That is, what is the displacement at $t = 0$?
(c) Graph the function d using a graphing utility.
(d) What is the displacement of the bob at the start of the second oscillation?
(e) What happens to the displacement of the bob as time increases without bound?

44. Graph $y = 2 \sin x + \cos(2x)$ by adding y-coordinates.

Chapter Test

1. Find the exact value of the six trigonometric functions of the angle θ in the figure.

2. Find the exact value of sin 40° − cos 50°.

In Problems 3–5, use the given information to determine the three remaining parts of each triangle.

3.

4.

5.

In Problems 6–8, solve each triangle.

6. $A = 55°$, $C = 20°$, $a = 4$

7. $a = 3$, $b = 7$, $A = 40°$

8. $a = 8$, $b = 4$, $C = 70°$

9. Find the area of the triangle described in Problem 8.

10. Find the area of the triangle described in Problem 5.

11. A 12-foot ladder leans against a building. The top of the ladder leans against the wall 10.5 feet from the ground. What is the angle formed by the ground and the ladder?

12. A hot-air balloon is flying at a height of 600 feet and is directly above the Marshall Space Flight Center in Huntsville, Alabama. The pilot of the balloon looks down at the airport that is known to be 5 miles from the Marshall Space Flight Center. What is the angle of depression from the balloon to the airport?

13. Find the area of the shaded region enclosed in a semicircle of diameter 8 centimeters. The length of the chord AB is 6 centimeters.

 [**Hint:** Triangle ABC is a right triangle.]

14. Find the area of the quadrilateral shown.

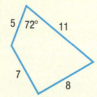

15. Madison wants to swim across Lake William from the fishing lodge (point A) to the boat ramp (point B), but she wants to know the distance first. Highway 20 goes right past the boat ramp and County Road 3 goes to the lodge. The two roads intersect at point C, 4.2 miles from the ramp and 3.5 miles from the lodge. Madison uses a transit to measure the angle of intersection of the two roads to be 32°. How far will she need to swim?

16. Given that $\triangle OAB$ is an isosceles triangle and the shaded sector is a semicircle, find the area of the entire region. Express your answer as a decimal rounded to two places.

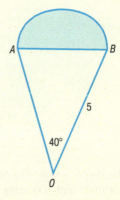

17. The area of the triangle shown below is $54\sqrt{6}$ square units. Find the lengths of the sides.

18. Logan is playing on her swing. One full swing (front to back to front) takes 6 seconds, and at the peak of her swing she is at an angle of 42° with the vertical. If her swing is 5 feet long, and we ignore all resistive forces, find a function that models her horizontal displacement (from the rest position) after time t.

Cumulative Review

1. Find the real solutions, if any, of the equation $3x^2 + 1 = 4x$.

2. Find an equation for the circle with center at the point $(-5, 1)$ and radius 3. Graph this circle.

3. Find the domain of the function

$$f(x) = \sqrt{x^2 - 3x - 4}$$

4. Graph the function: $y = 3\sin(\pi x)$

5. Graph the function: $y = -2\cos(2x - \pi)$

6. If $\tan\theta = -2$ and $\dfrac{3\pi}{2} < \theta < 2\pi$, find the exact value of:

   (a) $\sin\theta$ (b) $\cos\theta$ (c) $\sin(2\theta)$

   (d) $\cos(2\theta)$ (e) $\sin\left(\dfrac{1}{2}\theta\right)$ (f) $\cos\left(\dfrac{1}{2}\theta\right)$

7. Graph each of the following functions on the interval $[0, 4]$:

   (a) $y = \sin x$

   (b) $y = 2x + \sin x$

8. Graph each the following functions:

   (a) $y = x$ (b) $y = x^2$ (c) $y = \sqrt{x}$

   (d) $y = x^3$ (e) $y = \sin x$ (f) $y = \cos x$

   (g) $y = \tan x$

9. Solve the triangle for which side a is 20, side c is 15, and angle C is 40°.

10. On the interval $0 \le \theta < 2\pi$, solve the equation $\cos(3\theta) = 1$.

11. What is the length of the arc subtended by a central angle of 60° on a circle of radius 3 feet?

Chapter Projects

I. **Spherical Trigonometry** When the distance between two locations on the surface of Earth is small, we can treat Earth as a plane and compute the distance in statutory miles. Using this assumption, we can use the Law of Sines and the Law of Cosines to approximate distances and angles. However, the Earth is a sphere, so as the distance between two points on its surface increases, the linear distance is less accurate. Under this circumstance, we need to take into account the curvature of Earth when using the Law of Sines and the Law of Cosines.

1. See the figure. The points A, B, and C are the vertices of a spherical triangle with sides a, b, and c, a three-sided figure drawn on the surface of a sphere with center at the point O. Connect each vertex by a radius to the center O of the sphere. Now draw tangent lines to the sides a and b of the triangle that go through C. Extend the lines OA and OB to intersect the tangent lines at P and Q, respectively. List the plane right triangles. Find the measures of the central angles.

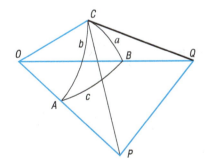

2. Use the Law of Cosines with triangles OPQ and CPQ to find two expressions for the length of PQ.

3. Subtract the expressions in part (2) from each other. Solve for the term containing $\cos c$.

4. Use the Pythagorean Theorem to find another value for $OQ^2 - CQ^2$ and $OP^2 - CP^2$. Now solve for $\cos c$.

5. Replacing the ratios in part (4) by the cosines of the sides of the spherical triangle, you should now have the Law of Cosines for spherical triangles:

$$\cos c = \cos a \cos b + \sin a \sin b \cos C$$

Source: For the spherical Law of Cosines, see *Mathematics from the Birth of Numbers* by Jan Gullberg. W. W. Norton & Co., Publishers, 1996, pp. 491–494.

II. **The Lewis and Clark Expedition** Lewis and Clark followed several rivers in their trek from what is now Great Falls, Montana, to the Pacific coast. First, they went down the Missouri and Jefferson rivers from Great Falls to Lemhi, Idaho. Because the two cities are at different longitudes and different latitudes, we must account for the curvature of Earth when computing the distance that they traveled. Assume that the radius of Earth is 3960 miles.

1. Great Falls is at approximately 47.5°N and 111.3°W. Lemhi is at approximately 45.5°N and 113.5°W. (We will assume that the rivers flow straight from Great Falls to Lemhi on the surface of Earth.) This line is
(continued)

called a geodesic line. Use the Law of Cosines for a spherical triangle [see Project I, part (5),] to find the angle between Great Falls and Lemhi. (The central angles are found by using the differences in the latitudes and longitudes of the towns. See the figure.) Then find the length of the arc joining the two towns. (Recall $s = r\theta$.)

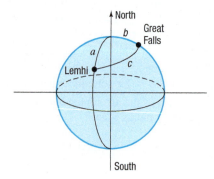

2. From Lemhi, they went up the Bitteroot River and the Snake River to what is now Lewiston and Clarkston on the border of Idaho and Washington. Although this is not really a side to a triangle, we will make a side that goes from Lemhi to Lewiston and Clarkston. If Lewiston and Clarkston are at about 46.5°N 117.0°W, find the distance from Lemhi using the Law of Cosines for a spherical triangle and the arc length.

3. How far did the explorers travel just to get that far?

4. Draw a plane triangle connecting the three towns. If the distance from Lewiston to Great Falls is 282 miles and the angle at Great Falls is 42° and the angle at Lewiston is 48.5°, find the distance from Great Falls to Lemhi and from Lemhi to Lewiston. How do these distances compare with the ones computed in parts (1) and (2)?

Source: For Lewis and Clark Expedition: *American Journey: The Quest for Liberty to 1877, Texas Edition.* Prentice Hall, 1992, p. 345.

Source: For map coordinates: *National Geographic Atlas of the World,* published by National Geographic Society, 1981, pp. 74–75.

Citation: Used with permission of *Technology Review,* from W. Roush, "From Lewis and Clark to Landsat: David Rumsey's Digital Maps Marry Past and Present," 108, no. 7, © 2005; permission conveyed through Copyright Clearance Center, Inc.

The following projects are available at the Instructor's Resource Center (IRC):

III. Project at Motorola: ***How Can You Build or Analyze a Vibration Profile?*** Fourier functions not only are important to analyze vibrations but also are what a mathematician would call interesting. Complete the project to see why.

IV. Leaning Tower of Pisa   Trigonometry is used to analyze the apparent height and tilt of the Leaning Tower of Pisa.

V. Locating Lost Treasure   Clever treasure seekers who know the Law of Sines are able to find a buried treasure efficiently.

VI. Jacob's Field   Angles of elevation and the Law of Sines are used to determine the height of the stadium wall and the distance from home plate to the top of the wall.

Polar Coordinates; Vectors

How Airplanes Fly

Four aerodynamic forces act on an airplane in flight: **lift**, **drag**, **thrust**, and **weight** (gravity).

Drag is the resistance of air molecules hitting the airplane (the *backward* force), thrust is the power of the airplane's engine (the *forward* force), lift is the *upward* force, and weight is the *downward* force. So for airplanes to fly and stay airborne, the thrust must be greater than the drag, and the lift must be greater than the weight.

This is certainly the case when an airplane takes off or climbs. However, when it is in straight and level flight, the opposing forces of lift, and weight are balanced. During a descent, weight exceeds lift, and to slow the airplane, drag has to overcome thrust.

Thrust is generated by the airplane's engine (propeller or jet), weight is created by the natural force of gravity acting on the airplane, and drag comes from friction as the plane moves through air molecules. Drag is also a *reaction* to lift, and this lift must be generated by the airplane in flight. This is done by the **wings** of the airplane.

A cross section of a typical airplane wing shows the top surface to be more curved than the bottom surface. This shaped profile is called an **airfoil** (or aerofoil), and the shape is used because an airfoil generates significantly more lift than opposing drag. In other words, it is very **efficient** at generating lift.

During flight, air naturally flows over and beneath the wing and is deflected upward over the top surface and downward beneath the lower surface. Any difference in deflection causes a difference in air pressure (pressure gradient), and because of the airfoil shape, the pressure of the deflected air is lower above the airfoil than below it. As a result the wing is "pushed" upward by the higher pressure beneath, or, you can argue, it is "sucked" upward by the lower pressure above.

> **Source**: Adapted from Pete Carpenter. How Airplanes Fly—The Basic Principles of Flight
> http://www.rc-airplane-world.com/how-airplanes-fly.html,
> accessed May 2018. © rc-airplane-world.com

—See Chapter Project I—

⬅ A Look Back, A Look Ahead ➡

This chapter is in two parts: Polar Coordinates (Sections 5.1–5.3) and Vectors (Sections 5.4–5.7). They are independent of each other and may be covered in either order.

Sections 5.1–5.3: In Chapter 1, we introduced rectangular coordinates (the *xy*-plane) and discussed the graph of an equation in two variables involving *x* and *y*. In Sections 5.1 and 5.2, we introduce polar coordinates, an alternative to rectangular coordinates, and discuss graphing equations that involve polar coordinates. In Section 5.3, we discuss raising a complex number to a real power. As it turns out, polar coordinates are useful for the discussion.

Sections 5.4–5.7: We have seen in many chapters that we are often required to solve an equation to obtain a solution to applied problems. In the last four sections of this chapter, we develop the notion of a vector and show how it can be used to model applied problems in physics and engineering.

Outline

5.1 Polar Coordinates

PREPARING FOR THIS SECTION *Before getting started, review the following:*

- Rectangular Coordinates (Section 1.1, pp. 2–3)
- Definition of the Trigonometric Functions (Section 2.2, pp. 116 and 126)
- The Distance Formula (Section 1.1, pp. 3–4)

- Inverse Tangent Function (Section 3.1, pp. 194–196)
- Completing the Square (Section A.3, pp. A24–A25)
- Angles; Degree Measure; Radian Measure (Section 2.1, pp. 102–107)

 Now Work the '**Are You Prepared?**' problems on page 323.

OBJECTIVES 1 Plot Points Using Polar Coordinates (p. 316)
2 Convert from Polar Coordinates to Rectangular Coordinates (p. 318)
3 Convert from Rectangular Coordinates to Polar Coordinates (p. 320)
4 Transform Equations between Polar and Rectangular Forms (p. 322)

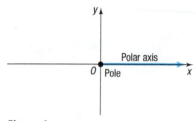

Figure 1

So far, we have always used a system of rectangular coordinates to plot points in the plane. Now we are ready to describe another system, called *polar coordinates*. In many instances, polar coordinates offer certain advantages over rectangular coordinates.

In a rectangular coordinate system, you will recall, a point in the plane is represented by an ordered pair of numbers (x, y), where x and y equal the signed distances of the point from the y-axis and the x-axis, respectively. In a polar coordinate system, we select a point, called the **pole**, and then a ray with vertex at the pole, called the **polar axis**. See Figure 1. Comparing the rectangular and polar coordinate systems, note that the origin in rectangular coordinates coincides with the pole in polar coordinates, and the positive x-axis in rectangular coordinates coincides with the polar axis in polar coordinates.

1 Plot Points Using Polar Coordinates

A point P in a polar coordinate system is represented by an ordered pair (r, θ) of numbers. If $r > 0$, then r is the distance of the point from the pole; θ is an angle (in degrees or radians) formed by the polar axis and a ray from the pole through the point. We call the ordered pair (r, θ) the **polar coordinates** of the point. See Figure 2.

As an example, suppose that a point P has polar coordinates $\left(2, \dfrac{\pi}{4}\right)$. Locate P by first drawing an angle of $\dfrac{\pi}{4}$ radian, placing its vertex at the pole and its initial side along the polar axis. Then go out a distance of 2 units along the terminal side of the angle to reach the point P. See Figure 3.

Figure 2

Figure 3

In using polar coordinates (r, θ), it is possible for r to be negative. When this happens, instead of the point being on the terminal side of θ, it is on the ray from the pole extending in the direction *opposite* the terminal side of θ at a distance $|r|$ units from the pole. See Figure 4 for an illustration.

For example, to plot the point $\left(-3, \dfrac{2\pi}{3}\right)$, use the ray in the opposite direction of $\dfrac{2\pi}{3}$ and go out $|-3| = 3$ units along that ray. See Figure 5.

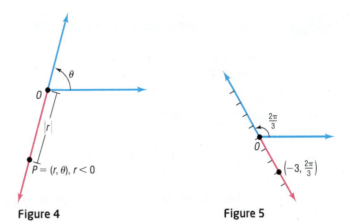

Figure 4   Figure 5

EXAMPLE 1   **Plotting Points Using Polar Coordinates**

Plot the points with the following polar coordinates:

(a) $\left(3, \dfrac{5\pi}{3}\right)$ (b) $\left(2, -\dfrac{\pi}{4}\right)$ (c) $(3, 0)$ (d) $\left(-2, \dfrac{\pi}{4}\right)$

Solution   Figure 6 shows the points.

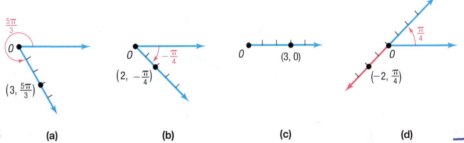

Figure 6   (a)   (b)   (c)   (d)

Now Work PROBLEMS 13, 25, AND 31

Recall that an angle measured counterclockwise is positive and an angle measured clockwise is negative. This convention has some interesting consequences related to polar coordinates.

EXAMPLE 2   **Finding Several Polar Coordinates of a Single Point**

Consider again the point P with polar coordinates $\left(2, \dfrac{\pi}{4}\right)$, as shown in Figure 7(a). Because $\dfrac{\pi}{4}, \dfrac{9\pi}{4}$, and $-\dfrac{7\pi}{4}$ all have the same terminal side, this point P also can be located by using the polar coordinates $\left(2, \dfrac{9\pi}{4}\right)$ or the polar coordinates $\left(2, -\dfrac{7\pi}{4}\right)$, as shown in Figures 7(b) and (c). The point $\left(2, \dfrac{\pi}{4}\right)$ can also be represented by the polar coordinates $\left(-2, \dfrac{5\pi}{4}\right)$. See Figure 7(d).

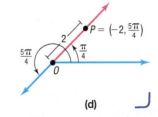

Figure 7   (a)   (b)   (c)   (d)

| EXAMPLE 3 | **Finding Other Polar Coordinates of a Given Point** |

Plot the point P with polar coordinates $\left(3, \dfrac{\pi}{6}\right)$, and find other polar coordinates (r, θ) of this same point for which:

(a) $r > 0$, $2\pi \le \theta < 4\pi$ (b) $r < 0$, $0 \le \theta < 2\pi$
(c) $r > 0$, $-2\pi \le \theta < 0$

Solution   The point $\left(3, \dfrac{\pi}{6}\right)$ is plotted in Figure 8.

$P = (3, \frac{\pi}{6})$

$\frac{\pi}{6}$

0

Figure 8

(a) Add 1 revolution (2π radians) to the angle $\dfrac{\pi}{6}$ to get

$$P = \left(3, \frac{\pi}{6} + 2\pi\right) = \left(3, \frac{13\pi}{6}\right).$$

See Figure 9.

(b) Add $\dfrac{1}{2}$ revolution (π radians) to the angle $\dfrac{\pi}{6}$, and replace 3 by -3 to

get $P = \left(-3, \dfrac{\pi}{6} + \pi\right) = \left(-3, \dfrac{7\pi}{6}\right)$. See Figure 10.

(c) Subtract 2π from the angle $\dfrac{\pi}{6}$ to get $P = \left(3, \dfrac{\pi}{6} - 2\pi\right) = \left(3, -\dfrac{11\pi}{6}\right)$. See
Figure 11.

$P = (3, \frac{13\pi}{6})$

$\frac{13\pi}{6}$

0

Figure 9

$\frac{7\pi}{6}$

$P = (-3, \frac{7\pi}{6})$

0

Figure 10

$-\frac{11\pi}{6}$

$P = (3, -\frac{11\pi}{6})$

0

Figure 11

These examples show a major difference between rectangular coordinates and polar coordinates. A point has exactly one pair of rectangular coordinates; however, a point has infinitely many pairs of polar coordinates.

SUMMARY

A point with polar coordinates (r, θ), θ in radians, can also be represented by either of the following:

$$(r, \theta + 2\pi k) \quad \text{or} \quad (-r, \theta + \pi + 2\pi k) \qquad k \text{ an integer}$$

The polar coordinates of the pole are $(0, \theta)$, where θ can be any angle.

— **Now Work** PROBLEM **35**

2 Convert from Polar Coordinates to Rectangular Coordinates

Sometimes it is necessary to convert coordinates or equations in rectangular form to polar form, and vice versa. To do this, recall that the origin in rectangular coordinates is the pole in polar coordinates and that the positive x-axis in rectangular coordinates is the polar axis in polar coordinates.

THEOREM **Conversion from Polar Coordinates to Rectangular Coordinates**

If P is a point with polar coordinates (r, θ), the rectangular coordinates (x, y) of P are given by

$$x = r \cos \theta \qquad y = r \sin \theta \tag{1}$$

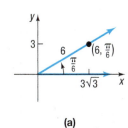

Figure 12

Proof Suppose that P has the polar coordinates (r, θ). We seek the rectangular coordinates (x, y) of P. Refer to Figure 12.

- If $r = 0$, then, regardless of θ, the point P is the pole, for which the rectangular coordinates are $(0, 0)$. Formula (1) is valid for $r = 0$.
- If $r > 0$, the point P is on the terminal side of θ, and $r = d(O, P) = \sqrt{x^2 + y^2}$. Because

$$\cos \theta = \frac{x}{r} \qquad \sin \theta = \frac{y}{r}$$

this means

$$x = r \cos \theta \qquad y = r \sin \theta$$

- If $r < 0$ and θ is in radians, then the point $P = (r, \theta)$ can be represented as $(-r, \pi + \theta)$, where $-r > 0$. Because

$$\cos(\pi + \theta) = -\cos \theta = \frac{x}{-r} \qquad \sin(\pi + \theta) = -\sin \theta = \frac{y}{-r}$$

this means

$$x = r \cos \theta \qquad y = r \sin \theta \qquad\blacksquare$$

EXAMPLE 4

Converting from Polar Coordinates to Rectangular Coordinates

Find the rectangular coordinates of the points with polar coordinates:

(a) $\left(6, \dfrac{\pi}{6}\right)$ (b) $\left(-4, -\dfrac{\pi}{4}\right)$

Solution

Use equations (1): $x = r \cos \theta$ and $y = r \sin \theta$.

(a) Figure 13(a) shows $\left(6, \dfrac{\pi}{6}\right)$ plotted. Notice that $\left(6, \dfrac{\pi}{6}\right)$ lies in quadrant I of the rectangular coordinate system. So both the x-coordinate and the y-coordinate will be positive. Substituting $r = 6$ and $\theta = \dfrac{\pi}{6}$ gives

$$x = r \cos \theta = 6 \cos \frac{\pi}{6} = 6 \cdot \frac{\sqrt{3}}{2} = 3\sqrt{3}$$

$$y = r \sin \theta = 6 \sin \frac{\pi}{6} = 6 \cdot \frac{1}{2} = 3$$

The rectangular coordinates of the point $\left(6, \dfrac{\pi}{6}\right)$ are $\left(3\sqrt{3}, 3\right)$, which lies in quadrant I, as expected.

(b) Figure 13(b) shows $\left(-4, -\dfrac{\pi}{4}\right)$ plotted. Notice that $\left(-4, -\dfrac{\pi}{4}\right)$ lies in quadrant II of the rectangular coordinate system. Substituting $r = -4$ and $\theta = -\dfrac{\pi}{4}$ gives

$$x = r \cos \theta = -4 \cos\left(-\frac{\pi}{4}\right) = -4 \cdot \frac{\sqrt{2}}{2} = -2\sqrt{2}$$

$$y = r \sin \theta = -4 \sin\left(-\frac{\pi}{4}\right) = -4\left(-\frac{\sqrt{2}}{2}\right) = 2\sqrt{2}$$

The rectangular coordinates of the point $\left(-4, -\dfrac{\pi}{4}\right)$ are $\left(-2\sqrt{2}, 2\sqrt{2}\right)$, which lies in quadrant II, as expected.

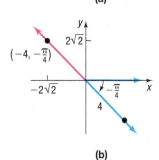

(a)

(b)

Figure 13

COMMENT Many calculators have the capability of converting from polar coordinates to rectangular coordinates. Consult your user's manual for the proper keystrokes. In most cases this procedure is tedious, so you will probably find using equations (1) is faster. ∎

Now Work PROBLEMS **43** AND **55**

3 Convert from Rectangular Coordinates to Polar Coordinates

Converting from rectangular coordinates (x, y) to polar coordinates (r, θ) is a little more complicated. Notice that each solution begins by plotting the given rectangular coordinates.

EXAMPLE 5

Converting from Rectangular Coordinates to Polar Coordinates where the Point Lies on a Coordinate Axis

Find polar coordinates of a point whose rectangular coordinates are $(0, 3)$.

Step-by-Step Solution

Step 1: Plot the point (x, y) and note the quadrant the point lies in or the coordinate axis the point lies on.

Plot the point $(0, 3)$ in a rectangular coordinate system. See Figure 14. The point lies on the positive y-axis.

Figure 14

Step 2: Find the distance r from the origin to the point.

The point $(0, 3)$ lies on the y-axis a distance of 3 units from the origin (pole), so $r = 3$.

Step 3: Determine θ.

A ray with vertex at the pole through $(0, 3)$ forms an angle $\theta = \dfrac{\pi}{2}$ with the polar axis.

Polar coordinates for this point can be given by $\left(3, \dfrac{\pi}{2}\right)$. Other possible representations include $\left(-3, -\dfrac{\pi}{2}\right)$ and $\left(3, \dfrac{5\pi}{2}\right)$.

Figure 15 shows polar coordinates of points that lie on either the x-axis or the y-axis. In each illustration, $a > 0$.

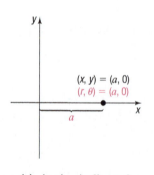

(a) $(x, y) = (a, 0), a > 0$

Figure 15

(b) $(x, y) = (0, a), a > 0$

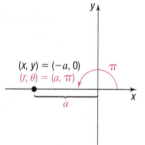

(c) $(x, y) = (-a, 0), a > 0$

(d) $(x, y) = (0, -a), a > 0$

──➤ **Now Work** PROBLEM **59**

EXAMPLE 6

Converting from Rectangular Coordinates to Polar Coordinates where the Point Lies in a Quadrant

Find the polar coordinates of a point whose rectangular coordinates are $(2, -2)$.

Step-by-Step Solution

Step 1: Plot the point (x, y) and note the quadrant the point lies in or the coordinate axis the point lies on.

Plot the point $(2, -2)$ in a rectangular coordinate system. See Figure 16. The point lies in quadrant IV.

Figure 16

Step 2: Find the distance r from the origin to the point.

$$r = \sqrt{x^2 + y^2} = \sqrt{2^2 + (-2)^2} = \sqrt{8} = 2\sqrt{2}$$

Step 3: Determine θ.

Find θ by recalling that $\tan\theta = \dfrac{y}{x}$, so $\theta = \tan^{-1}\dfrac{y}{x}, -\dfrac{\pi}{2} < \theta < \dfrac{\pi}{2}$. Because $(2, -2)$ lies in quadrant IV, this means that $-\dfrac{\pi}{2} < \theta < 0$. As a result,

$$\theta = \tan^{-1}\frac{y}{x} = \tan^{-1}\left(\frac{-2}{2}\right) = \tan^{-1}(-1) = -\frac{\pi}{4}$$

📟 **COMMENT** Many calculators have the capability of converting from rectangular coordinates to polar coordinates. Consult your user's manual for the proper keystrokes. ∎

Polar coordinates for the point $(2, -2)$ are $\left(2\sqrt{2}, -\dfrac{\pi}{4}\right)$. Other possible representations include $\left(2\sqrt{2}, \dfrac{7\pi}{4}\right)$ and $\left(-2\sqrt{2}, \dfrac{3\pi}{4}\right)$. ⌐

EXAMPLE 7

Converting from Rectangular Coordinates to Polar Coordinates

Find polar coordinates of a point whose rectangular coordinates are $\left(-1, -\sqrt{3}\right)$.

Solution

STEP 1: See Figure 17. The point lies in quadrant III.
STEP 2: The distance r from the origin to the point $\left(-1, -\sqrt{3}\right)$ is

$$r = \sqrt{(-1)^2 + \left(-\sqrt{3}\right)^2} = \sqrt{4} = 2$$

STEP 3: To find θ, use

$$\alpha = \tan^{-1}\frac{y}{x} = \tan^{-1}\frac{-\sqrt{3}}{-1} = \tan^{-1}\sqrt{3} = \frac{\pi}{3}, \quad -\frac{\pi}{2} < \alpha < \frac{\pi}{2}$$

$(x, y) = (-1, -\sqrt{3})$
Figure 17

Since the point $\left(-1, -\sqrt{3}\right)$ lies in quadrant III and the inverse tangent function gives an angle in quadrant I, add π to the result to obtain an angle in quadrant III. Then

$$\theta = \pi + \alpha = \pi + \tan^{-1}\sqrt{3} = \pi + \frac{\pi}{3} = \frac{4\pi}{3}$$

Polar coordinates for this point are $\left(2, \dfrac{4\pi}{3}\right)$. Other possible representations include $\left(-2, \dfrac{\pi}{3}\right)$ and $\left(2, -\dfrac{2\pi}{3}\right)$. ⌐

Figure 18 shows how to find polar coordinates of a point that lies in a quadrant when its rectangular coordinates (x, y) are given.

Figure 18 **(a)** $r = \sqrt{x^2 + y^2}$
$\theta = \tan^{-1}\dfrac{y}{x}$

(b) $r = \sqrt{x^2 + y^2}$
$\theta = \pi + \tan^{-1}\dfrac{y}{x}$

(c) $r = \sqrt{x^2 + y^2}$
$\theta = \pi + \tan^{-1}\dfrac{y}{x}$

(d) $r = \sqrt{x^2 + y^2}$
$\theta = \tan^{-1}\dfrac{y}{x}$

THEOREM Converting from Rectangular Coordinates to Polar Coordinates

If P is a point with rectangular coordinates (x, y), the polar coordinates (r, θ) of P are given by

$$r^2 = x^2 + y^2 \qquad \tan \theta = \frac{y}{x} \qquad \text{if } x \neq 0$$

$$r = y \qquad\qquad\quad \theta = \frac{\pi}{2} \qquad \text{if } x = 0$$

(2)

To use equations (2) effectively, follow these steps:

Steps for Converting from Rectangular to Polar Coordinates

STEP 1: Plot the point (x, y), as shown in Examples 5, 6, and 7.

STEP 2: If $x = 0$ or $y = 0$, use the graph to find r.
If $x \neq 0$ and $y \neq 0$, then $r = \sqrt{x^2 + y^2}$.

STEP 3: Find θ. If $x = 0$ or $y = 0$, use the graph to find θ.
If $x \neq 0$ and $y \neq 0$, note the quadrant in which the point lies.

$$\text{Quadrant I or IV:} \quad \theta = \tan^{-1}\frac{y}{x}$$

$$\text{Quadrant II or III:} \quad \theta = \pi + \tan^{-1}\frac{y}{x}$$

Now Work PROBLEM 63

4 Transform Equations between Polar and Rectangular Forms

Equations (1) and (2) can be used to transform equations from polar form to rectangular form, and vice versa. Two common techniques for transforming an equation from polar form to rectangular form are to

- Multiply both sides of the equation by r
- Square both sides of the equation

EXAMPLE 8

Transforming an Equation from Polar to Rectangular Form

Transform the equation $r = 6 \cos \theta$ from polar coordinates to rectangular coordinates, and identify the graph.

Solution

Multiplying both sides by r makes it easier to use equations (1) and (2).

$$r = 6 \cos \theta$$

$$r^2 = 6r \cos \theta \qquad \text{Multiply both sides by } r.$$

$$x^2 + y^2 = 6x \qquad r^2 = x^2 + y^2; x = r \cos \theta$$

This is the equation of a circle. Complete the square to obtain the standard form.

$$x^2 + y^2 = 6x$$

$$(x^2 - 6x) + y^2 = 0 \qquad \text{General form}$$

$$(x^2 - 6x + 9) + y^2 = 9 \qquad \text{Complete the square in } x.$$

$$(x - 3)^2 + y^2 = 9 \qquad \text{Factor.}$$

This is the standard form of the equation of a circle with center $(3, 0)$ and radius 3.

Now Work PROBLEM 79

| EXAMPLE 9 | **Transforming an Equation from Rectangular to Polar Form** |

Transform the equation $4xy = 9$ from rectangular coordinates to polar coordinates.

Solution   Use $x = r \cos \theta$ and $y = r \sin \theta$.

$$4xy = 9$$
$$4 (r \cos \theta) (r \sin \theta) = 9 \quad x = r \cos \theta, y = r \sin \theta$$
$$4r^2 \cos \theta \sin \theta = 9$$

This is the polar form of the equation. It can be simplified as follows:

$$2r^2 (2 \sin \theta \cos \theta) = 9 \quad \text{Factor out } 2r^2.$$
$$2r^2 \sin (2\theta) = 9 \quad \text{Double-angle Formula}$$

 Now Work PROBLEM 73

5.1 Assess Your Understanding

'Are You Prepared?' *Answers are given at the end of these exercises. If you get a wrong answer, read the pages listed in* red.

1. Plot the point whose rectangular coordinates are $(3, -1)$. What quadrant does the point lie in? (pp. 2–3)

2. The distance between two points $P_1 = (x_1, y_1)$ and $P_2 = (x_2, y_2)$ is $d(P_1, P_2) =$ _____. (pp. 3–4)

3. To complete the square of $x^2 + 6x$, add ____. (pp. A24–A25)

4. Draw the angle $\dfrac{5\pi}{6}$ in standard position. (pp. 102–107)

5. If $P = (x, y)$ is a point on the terminal side of the angle θ at a distance r from the origin, then $\tan \theta =$ _____. (p. 126)

6. $\tan^{-1} (-1) =$ _____. (pp. 194–196)

Concepts and Vocabulary

7. The origin in rectangular coordinates coincides with the _____ in polar coordinates; the positive x-axis in rectangular coordinates coincides with the _____ _____ in polar coordinates.

8. If P is a point with polar coordinates (r, θ), the rectangular coordinates (x, y) of P are given by $x =$ ____ and $y =$ ____.

9. *Multiple Choice* In a rectangular coordinate system, where does the point with polar coordinates $\left(1, -\dfrac{\pi}{2}\right)$ lie?

   (a) in quadrant IV
   (b) on the y-axis
   (c) in quadrant II
   (d) on the x-axis

10. *Multiple Choice* The point $\left(5, \dfrac{\pi}{6}\right)$ can also be represented by which polar coordinates?

   (a) $\left(5, -\dfrac{\pi}{6}\right)$
   (b) $\left(-5, \dfrac{13\pi}{6}\right)$
   (c) $\left(5, -\dfrac{5\pi}{6}\right)$
   (d) $\left(-5, \dfrac{7\pi}{6}\right)$

11. *True or False* In the polar coordinates (r, θ), r can be negative.

12. *True or False* The polar coordinates of a point are unique.

Skill Building

In Problems 13–20, match each point in polar coordinates with either A, B, C, or D on the graph.

13. $\left(2, -\dfrac{11\pi}{6}\right)$

14. $\left(-2, -\dfrac{\pi}{6}\right)$

15. $\left(-2, \dfrac{\pi}{6}\right)$

16. $\left(2, \dfrac{7\pi}{6}\right)$

17. $\left(2, \dfrac{5\pi}{6}\right)$

18. $\left(-2, \dfrac{5\pi}{6}\right)$

19. $\left(-2, \dfrac{7\pi}{6}\right)$

20. $\left(2, \dfrac{11\pi}{6}\right)$

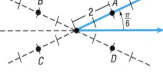

In Problems 21–34, plot each point given in polar coordinates.

21. $\left(3, \dfrac{\pi}{2}\right)$

22. $\left(4, \dfrac{3\pi}{2}\right)$

23. $(-2, 0)$

24. $(-3, \pi)$

25. $\left(6, \dfrac{\pi}{6}\right)$

26. $\left(5, \dfrac{5\pi}{3}\right)$

27. $\left(-2, \dfrac{3\pi}{4}\right)$

28. $\left(-3, \dfrac{2\pi}{3}\right)$

29. $\left(4, -\dfrac{2\pi}{3}\right)$

30. $\left(2, -\dfrac{5\pi}{4}\right)$

31. $\left(-1, -\dfrac{\pi}{3}\right)$

32. $\left(-3, -\dfrac{3\pi}{4}\right)$

33. $(-2, -\pi)$

34. $\left(-3, -\dfrac{\pi}{2}\right)$

In Problems 35–42, plot each point given in polar coordinates, and find other polar coordinates (r, θ) *of the point for which:*

(a) $r > 0$, $-2\pi \le \theta < 0$ (b) $r < 0$, $0 \le \theta < 2\pi$ (c) $r > 0$, $2\pi \le \theta < 4\pi$

35. $\left(5, \dfrac{2\pi}{3}\right)$ **36.** $\left(4, \dfrac{3\pi}{4}\right)$ **37.** $(-2, 3\pi)$ **38.** $(-3, 4\pi)$

39. $\left(1, \dfrac{\pi}{2}\right)$ **40.** $(2, \pi)$ **41.** $\left(-3, -\dfrac{\pi}{4}\right)$ **42.** $\left(-2, -\dfrac{2\pi}{3}\right)$

In Problems 43–58, polar coordinates of a point are given. Find the rectangular coordinates of each point.

43. $\left(3, \dfrac{\pi}{2}\right)$ **44.** $\left(4, \dfrac{3\pi}{2}\right)$ **45.** $(-2, 0)$ **46.** $(-3, \pi)$

47. $\left(6, \dfrac{5\pi}{6}\right)$ **48.** $\left(5, \dfrac{5\pi}{3}\right)$ **49.** $\left(-2, \dfrac{3\pi}{4}\right)$ **50.** $\left(-2, \dfrac{2\pi}{3}\right)$

51. $\left(-5, -\dfrac{\pi}{6}\right)$ **52.** $\left(-6, -\dfrac{\pi}{4}\right)$ **53.** $(-2, -\pi)$ **54.** $\left(-3, -\dfrac{\pi}{2}\right)$

55. $\left(7.5, \dfrac{11\pi}{18}\right)$ **56.** $\left(-3.1, \dfrac{91\pi}{90}\right)$ **57.** $(6.3, 3.8)$ **58.** $(8.1, 5.2)$

In Problems 59–70, the rectangular coordinates of a point are given. Find polar coordinates for each point.

59. $(3, 0)$ **60.** $(0, 2)$ **61.** $(-1, 0)$ **62.** $(0, -2)$

63. $(1, -1)$ **64.** $(-3, 3)$ **65.** $(5, 5\sqrt{3})$ **66.** $\left(-\dfrac{\sqrt{3}}{2}, -\dfrac{1}{2}\right)$

67. $(1.3, -2.1)$ **68.** $(-0.8, -2.1)$ **69.** $(8.3, 4.2)$ **70.** $(-2.3, 0.2)$

In Problems 71–78, the letters x and y represent rectangular coordinates. Write each equation using polar coordinates (r, θ).

71. $2x^2 + 2y^2 = 3$ **72.** $x^2 + y^2 = x$ **73.** $x^2 = 4y$ **74.** $y^2 = 2x$

75. $2xy = 1$ **76.** $4x^2 y = 1$ **77.** $x = 4$ **78.** $y = -3$

In Problems 79–86, the letters r and θ represent polar coordinates. Write each equation using rectangular coordinates (x, y).

79. $r = \cos \theta$ **80.** $r = \sin \theta + 1$ **81.** $r^2 = \cos \theta$ **82.** $r = \sin \theta - \cos \theta$

83. $r = 2$ **84.** $r = 4$ **85.** $r = \dfrac{4}{1 - \cos \theta}$ **86.** $r = \dfrac{3}{3 - \cos \theta}$

Applications and Extensions

87. Chicago In Chicago, the road system is set up like a Cartesian plane, where streets are indicated by the number of blocks they are from Madison Street and State Street. For example, Wrigley Field in Chicago is located at 1060 West Addison, which is 10 blocks west of State Street and 36 blocks north of Madison Street. Treat the intersection of Madison Street and State Street as the origin of a coordinate system, with east being the positive *x*-axis.
  (a) Write the location of Wrigley Field using rectangular coordinates.
  (b) Write the location of Wrigley Field using polar coordinates. Use the east direction for the polar axis. Express θ in degrees.
  (c) Guaranteed Rate Field, home of the White Sox, is located at 35th and Princeton, which is 3 blocks west of State Street and 35 blocks south of Madison. Write the location of Guaranteed Rate Field using rectangular coordinates.
  (d) Write the location of Guaranteed Rate Field using polar coordinates. Use the east direction for the polar axis. Express θ in degrees.

88. Show that the formula for the distance d between two points $P_1 = (r_1, \theta_1)$ and $P_2 = (r_2, \theta_2)$ is

$$d = \sqrt{r_1^2 + r_2^2 - 2r_1 r_2 \cos(\theta_2 - \theta_1)}$$

City of Chicago, Illinois

Addison Street

Addison Street

1 mile
1 km

↑N

Wrigley Field
1060 West Addison

Madison Street

Guaranteed Rate Field
35th and Princeton

35th Street

State Street

35th Street

89. *Challenge Problem* **Radar Detection** At 10:15 A.M., a radar station detects an aircraft at a point 80 miles away and 25 degrees north of due east. At 10:25 A.M., the aircraft is 110 miles away and 5 degrees south of due east.
  (a) Using the radar station as the pole and due east as the polar axis, write the two locations of the aircraft in polar coordinates.
  (b) Write the two locations of the aircraft in rectangular coordinates. Round answers to two decimal places.
  (c) What is the speed of the aircraft in miles per hour? Round the answer to one decimal place.

90. *Challenge Problem* **Radar station** A uses a coordinate system where A is located at the pole and due east is the polar axis. On this system, two other radar stations, B and C, are located at coordinates $(150, -24°)$ and $(100, 32°)$, respectively. If radar station B uses a coordinate system where B is located at the pole and due east is the polar axis, then what are the coordinates of radar stations A and C on this second system? Round answers to one decimal place.

Explaining Concepts: Discussion and Writing

91. In converting from polar coordinates to rectangular coordinates, what equations will you use?

92. Explain how to convert from rectangular coordinates to polar coordinates.

93. Is the street system in your town based on a rectangular coordinate system, a polar coordinate system, or some other system? Explain.

Retain Your Knowledge

Problems 94–101 are based on material learned earlier in the course. The purpose of these problems is to keep the material fresh in your mind so that you are better prepared for the final exam.

94. Use Descartes' Rule of Signs to determine the possible number of positive or negative real zeros for the function
$$f(x) = -2x^3 + 6x^2 - 7x - 8$$

95. Find the midpoint of the line segment connecting the points $(-3, 7)$ and $\left(\dfrac{1}{2}, 2\right)$.

96. Given that the point $(3, 8)$ is on the graph of $y = f(x)$, what is the corresponding point on the graph of $y = -2f(x + 3) + 5$?

97. If $z = 2 - 5i$ and $w = 4 + i$, find $z \cdot w$.

98. Solve the equation: $\quad 4 \sin \theta \cos \theta = 1, 0 \le \theta < 2\pi$

99. Solve the triangle: $\quad A = 65°, B = 37°, c = 10$

100. Find the exact value of $\sin \dfrac{7\pi}{12}$.

101. Show that $\sin^5 x = \sin x - 2 \cos^2 x \sin x + \cos^4 x \sin x$.

'Are You Prepared?' Answers

1. ; quadrant IV   **2.** $\sqrt{(x_2 - x_1)^2 + (y_2 - y_1)^2}$ **3.** 9   **4.**   **5.** $\dfrac{y}{x}$ **6.** $-\dfrac{\pi}{4}$

5.2 Polar Equations and Graphs

PREPARING FOR THIS SECTION *Before getting started, review the following:*

- Symmetry (Section 1.2, pp. 12–14)
- Circles (Section 1.2, pp. 16–19)
- Even-Odd Properties of Trigonometric Functions (Section 2.3, pp. 142–143)
- Difference Formulas for Sine and Cosine (Section 3.5, pp. 227 and 230)
- Values of the Sine and Cosine Functions at Certain Angles (Section 2.2, pp. 118–125)

Now Work the 'Are You Prepared?' problems on page 337.

OBJECTIVES **1** Identify and Graph Polar Equations by Converting to Rectangular Equations (p. 326)

2 Test Polar Equations for Symmetry (p. 329)

3 Graph Polar Equations by Plotting Points (p. 330)

Just as a rectangular grid may be used to plot points given by rectangular coordinates, such as the points $(-3, 1)$ and $(1, 2)$ shown in Figure 19(a), a grid consisting of concentric circles (with centers at the pole) and rays (with vertices at the pole) can be used to plot points given by polar coordinates, such as the points $\left(4, \dfrac{5\pi}{4}\right)$ and $\left(2, \dfrac{\pi}{4}\right)$ shown in Figure 19(b). Such **polar grids** are used to graph *polar equations*.

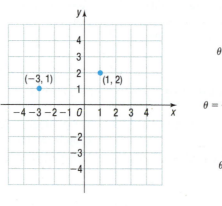

Figure 19   (a) Rectangular grid   (b) Polar grid

> ### DEFINITION Polar Equation
>
> An equation whose variables are polar coordinates is called a **polar equation**. The **graph of a polar equation** consists of all points whose polar coordinates satisfy the equation.

1 Identify and Graph Polar Equations by Converting to Rectangular Equations

One method that can be used to graph a polar equation is to convert the equation to rectangular coordinates. In the following discussion, (x, y) represents the rectangular coordinates of a point P, and (r, θ) represents polar coordinates of the point P.

EXAMPLE 1   **Identifying and Graphing a Polar Equation (Circle)**

Identify and graph the equation: $r = 3$

Solution   Convert the polar equation to a rectangular equation.

$$r = 3$$
$$r^2 = 9 \quad \textbf{Square both sides.}$$
$$x^2 + y^2 = 9 \quad \textbf{\textit{r}}^2 = \textbf{\textit{x}}^2 + \textbf{\textit{y}}^2$$

The graph of $r = 3$ is a circle, with center at the pole and radius 3. See Figure 20.

Figure 20 $r = 3$ or $x^2 + y^2 = 9$

Now Work   PROBLEM 15

EXAMPLE 2   **Identifying and Graphing a Polar Equation (Line)**

Identify and graph the equation: $\theta = \dfrac{\pi}{3}$

Solution   Convert the polar equation to a rectangular equation.

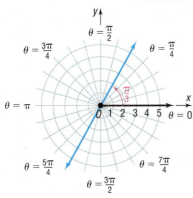

$$\theta = \frac{\pi}{3}$$

$$\tan \theta = \tan \frac{\pi}{3} \quad \textsf{Find the tangent of both sides.}$$

$$\frac{y}{x} = \sqrt{3} \quad \tan \theta = \frac{y}{x}; \tan \frac{\pi}{3} = \sqrt{3}$$

$$y = \sqrt{3}x$$

The graph of $\theta = \dfrac{\pi}{3}$ is a line passing through the pole making an angle of $\dfrac{\pi}{3}$ with the polar axis. See Figure 21.

Figure 21 $\theta = \dfrac{\pi}{3}$ or $y = \sqrt{3}x$

Now Work PROBLEM 17

EXAMPLE 3   **Identifying and Graphing a Polar Equation (Horizontal Line)**

Identify and graph the equation: $r \sin \theta = 2$

Solution   Because $y = r \sin \theta$, we can write the equation as

$$y = 2$$

Therefore, the graph of $r \sin \theta = 2$ is a horizontal line 2 units above the pole. See Figure 22.

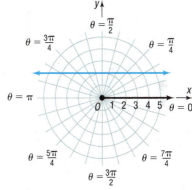

🖳 **COMMENT** A graphing utility can be used to graph polar equations. Read *Using a Graphing Utility to Graph a Polar Equation*, Section B.8. ■

Figure 22 $r \sin \theta = 2$ or $y = 2$

EXAMPLE 4   **Identifying and Graphing a Polar Equation (Vertical Line)**

Identify and graph the equation: $r \cos \theta = -3$

Solution   Since $x = r \cos \theta$, we can write the equation as

$$x = -3$$

Therefore, the graph of $r \cos \theta = -3$ is a vertical line 3 units to the left of the pole. See Figure 23.

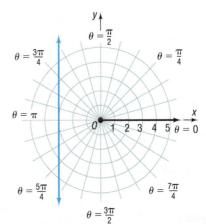

Figure 23 $r \cos \theta = -3$ or $x = -3$

Examples 3 and 4 on the previous page lead to the following results. (The proofs are left as exercises. See Problems 83 and 84.)

> **THEOREM**
>
> Let a be a real number. Then the graph of the equation
>
> $$r \sin \theta = a$$
>
> is a horizontal line. It lies a units above the pole if $a \geq 0$ and lies $|a|$ units below the pole if $a < 0$.
>
> The graph of the equation
>
> $$r \cos \theta = a$$
>
> is a vertical line. It lies a units to the right of the pole if $a \geq 0$ and lies $|a|$ units to the left of the pole if $a < 0$.

 Now Work PROBLEM 21

EXAMPLE 5 **Identifying and Graphing a Polar Equation (Circle)**

Identify and graph the equation: $r = 4 \sin \theta$

Solution To transform the polar equation to rectangular coordinates, multiply both sides by r.

$$r^2 = 4r \sin \theta$$

Now use the facts that $r^2 = x^2 + y^2$ and $y = r \sin \theta$. Then

$$x^2 + y^2 = 4y$$

$$x^2 + (y^2 - 4y) = 0$$

$$x^2 + (y^2 - 4y + 4) = 4 \qquad \text{Complete the square in } y.$$

$$x^2 + (y - 2)^2 = 4 \qquad \text{Factor.}$$

This is the standard equation of a circle with center $(0, 2)$ in rectangular coordinates and radius 2. See Figure 24.

Figure 24
$r = 4 \sin \theta$ or $x^2 + (y - 2)^2 = 4$

EXAMPLE 6 **Identifying and Graphing a Polar Equation (Circle)**

Identify and graph the equation: $r = -2 \cos \theta$

Solution To transform the polar equation to rectangular coordinates, multiply both sides by r.

$$r^2 = -2r \cos \theta$$

$$x^2 + y^2 = -2x \qquad r^2 = x^2 + y^2; \quad x = r \cos \theta$$

$$x^2 + 2x + y^2 = 0$$

$$(x^2 + 2x + 1) + y^2 = 1 \qquad \text{Complete the square in } x.$$

$$(x + 1)^2 + y^2 = 1 \qquad \text{Factor.}$$

This is the standard equation of a circle with center $(-1, 0)$ in rectangular coordinates and radius 1. See Figure 25.

Figure 25
$r = -2 \cos \theta$ or $(x + 1)^2 + y^2 = 1$

Based on Examples 5 and 6 and the Exploration to the left, we are led to the following results. (The proofs are left as exercises. See Problems 85–88.)

THEOREM

Suppose a is a positive real number. Then

Equation	Description
• $r = 2a\sin\theta$	Circle: radius a; center at $(0, a)$ in rectangular coordinates
• $r = -2a\sin\theta$	Circle: radius a; center at $(0, -a)$ in rectangular coordinates
• $r = 2a\cos\theta$	Circle: radius a; center at $(a, 0)$ in rectangular coordinates
• $r = -2a\cos\theta$	Circle: radius a; center at $(-a, 0)$ in rectangular coordinates

Each circle passes through the pole.

Now Work PROBLEM 23

The method of converting a polar equation to an identifiable rectangular equation to obtain the graph is not always helpful, nor is it always necessary. Usually, a table is created that lists several points on the graph. By checking for symmetry, it may be possible to reduce the number of points needed to draw the graph.

2 Test Polar Equations for Symmetry

In polar coordinates, the points (r, θ) and $(r, -\theta)$ are symmetric with respect to the polar axis (and to the x-axis). See Figure 26(a). The points (r, θ) and $(r, \pi - \theta)$ are symmetric with respect to the line $\theta = \dfrac{\pi}{2}$ (the y-axis). See Figure 26(b). The points (r, θ) and $(-r, \theta)$ are symmetric with respect to the pole (the origin). See Figure 26(c).

Figure 26

(a) Points symmetric with respect to the polar axis

(b) Points symmetric with respect to the line $\theta = \dfrac{\pi}{2}$

(c) Points symmetric with respect to the pole

The following tests are a consequence of these observations.

THEOREM Tests for Symmetry

- **Symmetry with Respect to the Polar Axis (x-Axis)**

  In a polar equation, replace θ by $-\theta$. If an equivalent equation results, the graph is symmetric with respect to the polar axis.

- **Symmetry with Respect to the Line $\theta = \dfrac{\pi}{2}$ (y-Axis)**

  In a polar equation, replace θ by $\pi - \theta$. If an equivalent equation results, the graph is symmetric with respect to the line $\theta = \dfrac{\pi}{2}$.

- **Symmetry with Respect to the Pole (Origin)**

  In a polar equation, replace r by $-r$ or θ by $\theta + \pi$. If an equivalent equation results, the graph is symmetric with respect to the pole.

The three tests for symmetry given on the previous page are *sufficient* conditions for symmetry, but they are not *necessary* conditions. That is, an equation may fail these tests and still have a graph that is symmetric with respect to the polar axis, the line $\theta = \dfrac{\pi}{2}$, or the pole. For example, the graph of $r = \sin(2\theta)$ turns out to be symmetric with respect to the polar axis, the line $\theta = \dfrac{\pi}{2}$, and the pole, but only the test for symmetry with respect to the pole (replace θ by $\theta + \pi$) works. See also Problems 93–95.

3 Graph Polar Equations by Plotting Points

EXAMPLE 7

Graphing a Polar Equation (Cardioid)

Graph the equation: $r = 1 - \sin\theta$

Solution

Check for symmetry first.

Polar Axis: Replace θ by $-\theta$. The result is

$$r = 1 - \sin(-\theta) = 1 + \sin\theta \qquad \textcolor{blue}{\sin(-\theta) = -\sin\theta}$$

The test fails, so the graph may or may not be symmetric with respect to the polar axis.

The Line $\theta = \dfrac{\pi}{2}$: Replace θ by $\pi - \theta$. The result is

$$r = 1 - \sin(\pi - \theta) = 1 - (\sin\pi\cos\theta - \cos\pi\sin\theta)$$
$$= 1 - [0 \cdot \cos\theta - (-1)\sin\theta] = 1 - \sin\theta$$

The test is satisfied, so the graph is symmetric with respect to the line $\theta = \dfrac{\pi}{2}$.

The Pole: Replace r by $-r$. Then the result is $-r = 1 - \sin\theta$, so $r = -1 + \sin\theta$. The test fails. Replace θ by $\theta + \pi$. The result is

$$r = 1 - \sin(\theta + \pi)$$
$$= 1 - [\sin\theta\cos\pi + \cos\theta\sin\pi]$$
$$= 1 - [\sin\theta \cdot (-1) + \cos\theta \cdot 0]$$
$$= 1 + \sin\theta$$

This test also fails, so the graph may or may not be symmetric with respect to the pole.

Next, identify points on the graph by assigning values to the angle θ and calculating the corresponding values of r. Due to the periodicity of the sine function and the symmetry with respect to the line $\theta = \dfrac{\pi}{2}$, just assign values to θ from $-\dfrac{\pi}{2}$ to $\dfrac{\pi}{2}$, as given in Table 1.

Now plot the points (r, θ) from Table 1 and trace out the graph, beginning at the point $\left(2, -\dfrac{\pi}{2}\right)$ and ending at the point $\left(0, \dfrac{\pi}{2}\right)$. Then reflect this portion of the graph about the line $\theta = \dfrac{\pi}{2}$ (the y-axis) to obtain the complete graph. See Figure 27.

Table 1

θ	$r = 1 - \sin\theta$
$-\dfrac{\pi}{2}$	$1 - (-1) = 2$
$-\dfrac{\pi}{3}$	$1 - \left(-\dfrac{\sqrt{3}}{2}\right) \approx 1.87$
$-\dfrac{\pi}{6}$	$1 - \left(-\dfrac{1}{2}\right) = \dfrac{3}{2}$
0	$1 - 0 = 1$
$\dfrac{\pi}{6}$	$1 - \dfrac{1}{2} = \dfrac{1}{2}$
$\dfrac{\pi}{3}$	$1 - \dfrac{\sqrt{3}}{2} \approx 0.13$
$\dfrac{\pi}{2}$	$1 - 1 = 0$

Figure 27 $r = 1 - \sin\theta$

The curve in Figure 27 is an example of a *cardioid* (a heart-shaped curve).

Exploration

Graph $r_1 = 1 + \sin\theta$. Clear the screen and graph $r_1 = 1 - \cos\theta$. Clear the screen and graph $r_1 = 1 + \cos\theta$. Do you see a pattern?

DEFINITION Cardiods

Cardioids are characterized by equations of the form

- $r = a(1 + \cos \theta)$ • $r = a(1 + \sin \theta)$
- $r = a(1 - \cos \theta)$ • $r = a(1 - \sin \theta)$

where $a > 0$. The graph of a cardioid passes through the pole.

Now Work PROBLEM 39

| EXAMPLE 8 | **Graphing a Polar Equation (Limaçon without an Inner Loop)** |

Graph the equation: $r = 3 + 2 \cos \theta$

Solution   Check for symmetry first.

Polar Axis: Replace θ by $-\theta$. The result is

$$r = 3 + 2 \cos(-\theta) = 3 + 2 \cos \theta \quad \cos(-\theta) = \cos \theta$$

The test is satisfied, so the graph is symmetric with respect to the polar axis.

The Line $\theta = \dfrac{\pi}{2}$: Replace θ by $\pi - \theta$. The result is

$$r = 3 + 2 \cos(\pi - \theta) = 3 + 2(\cos \pi \cos \theta + \sin \pi \sin \theta)$$
$$= 3 - 2 \cos \theta$$

The test fails, so the graph may or may not be symmetric with respect to the line $\theta = \dfrac{\pi}{2}$.

The Pole: Replace r by $-r$. The test fails, so the graph may or may not be symmetric with respect to the pole. Replace θ by $\theta + \pi$. The test fails, so the graph may or may not be symmetric with respect to the pole.

Next, identify points on the graph by assigning values to the angle θ and calculating the corresponding values of r. Due to the periodicity of the cosine function and the symmetry with respect to the polar axis, just assign values to θ from 0 to π, as given in Table 2.

Now plot the points (r, θ) from Table 2 and trace out the graph, beginning at the point $(5, 0)$ and ending at the point $(1, \pi)$. Then reflect this portion of the graph about the polar axis (the x-axis) to obtain the complete graph. See Figure 28.

Table 2

θ	$r = 3 + 2 \cos \theta$
0	$3 + 2 \cdot 1 = 5$
$\dfrac{\pi}{6}$	$3 + 2 \cdot \dfrac{\sqrt{3}}{2} \approx 4.73$
$\dfrac{\pi}{3}$	$3 + 2 \cdot \dfrac{1}{2} = 4$
$\dfrac{\pi}{2}$	$3 + 2 \cdot 0 = 3$
$\dfrac{2\pi}{3}$	$3 + 2\left(-\dfrac{1}{2}\right) = 2$
$\dfrac{5\pi}{6}$	$3 + 2\left(-\dfrac{\sqrt{3}}{2}\right) \approx 1.27$
π	$3 + 2(-1) = 1$

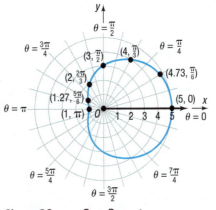

Figure 28 $r = 3 + 2 \cos \theta$

The curve in Figure 28 is an example of a *limaçon* (a French word for *snail*) *without an inner loop.*

Exploration

Graph $r_1 = 3 - 2 \cos \theta$. Clear the screen and graph $r_1 = 3 + 2 \sin \theta$. Clear the screen and graph $r_1 = 3 - 2 \sin \theta$. Do you see a pattern?

DEFINITION Limaçons without an Inner Loop

Limaçons without an inner loop are characterized by equations of the form

- $r = a + b \cos \theta$
- $r = a + b \sin \theta$
- $r = a - b \cos \theta$
- $r = a - b \sin \theta$

where $a > b > 0$. The graph of a limaçon without an inner loop does not pass through the pole.

▬▬▬▬ **Now Work** PROBLEM 45

EXAMPLE 9

Graphing a Polar Equation (Limaçon with an Inner Loop)

Graph the equation: $r = 1 + 2 \cos \theta$

Solution

First, check for symmetry.

Polar Axis: Replace θ by $-\theta$. The result is

$$r = 1 + 2 \cos(-\theta) = 1 + 2 \cos \theta$$

The test is satisfied, so the graph is symmetric with respect to the polar axis.

The Line $\theta = \dfrac{\pi}{2}$: Replace θ by $\pi - \theta$. The result is

$$r = 1 + 2 \cos(\pi - \theta) = 1 + 2(\cos \pi \cos \theta + \sin \pi \sin \theta)$$
$$= 1 - 2 \cos \theta$$

The test fails, so the graph may or may not be symmetric with respect to the line $\theta = \dfrac{\pi}{2}$.

The Pole: Replace r by $-r$. The test fails, so the graph may or may not be symmetric with respect to the pole. Replace θ by $\theta + \pi$. The test fails, so the graph may or may not be symmetric with respect to the pole.

Next, identify points on the graph of $r = 1 + 2 \cos \theta$ by assigning values to the angle θ and calculating the corresponding values of r. Due to the periodicity of the cosine function and the symmetry with respect to the polar axis, just assign values to θ from 0 to π, as given in Table 3.

Now plot the points (r, θ) from Table 3, beginning at $(3, 0)$ and ending at $(-1, \pi)$. See Figure 29(a). Finally, reflect this portion of the graph about the polar axis (the x-axis) to obtain the complete graph. See Figure 29(b).

Table 3

θ	$r = 1 + 2 \cos \theta$
0	$1 + 2 \cdot 1 = 3$
$\dfrac{\pi}{6}$	$1 + 2 \cdot \dfrac{\sqrt{3}}{2} \approx 2.73$
$\dfrac{\pi}{3}$	$1 + 2 \cdot \dfrac{1}{2} = 2$
$\dfrac{\pi}{2}$	$1 + 2 \cdot 0 = 1$
$\dfrac{2\pi}{3}$	$1 + 2\left(-\dfrac{1}{2}\right) = 0$
$\dfrac{5\pi}{6}$	$1 + 2\left(-\dfrac{\sqrt{3}}{2}\right) \approx -0.73$
π	$1 + 2(-1) = -1$

Figure 29 (a)   (b) $r = 1 + 2 \cos \theta$

Exploration

Graph $r_1 = 1 - 2 \cos \theta$. Clear the screen and graph $r_1 = 1 + 2 \sin \theta$. Clear the screen and graph $r_1 = 1 - 2 \sin \theta$. Do you see a pattern?

The curve in Figure 29(b) is an example of a *limaçon with an inner loop*.

> **DEFINITION** Limaçons with an Inner Loop
>
> **Limaçons with an inner loop** are characterized by equations of the form
>
> - $r = a + b \cos \theta$ • $r = a + b \sin \theta$
> - $r = a - b \cos \theta$ • $r = a - b \sin \theta$
>
> where $b > a > 0$. The graph of a limaçon with an inner loop passes through the pole twice.

──── Now Work PROBLEM 47

EXAMPLE 10 **Graphing a Polar Equation (Rose)**

Graph the equation: $r = 2 \cos(2\theta)$

Solution Check for symmetry.

Polar Axis: Replace θ by $-\theta$. The result is

$$r = 2 \cos[2(-\theta)] = 2 \cos(2\theta)$$

The test is satisfied, so the graph is symmetric with respect to the polar axis.

The Line $\theta = \dfrac{\pi}{2}$: Replace θ by $\pi - \theta$. The result is

$$r = 2 \cos[2(\pi - \theta)] = 2 \cos(2\pi - 2\theta) = 2 \cos(2\theta)$$

The test is satisfied, so the graph is symmetric with respect to the line $\theta = \dfrac{\pi}{2}$.

The Pole: Since the graph is symmetric with respect to both the polar axis and the line $\theta = \dfrac{\pi}{2}$, it must be symmetric with respect to the pole.

Next, construct Table 4. Because of the periodicity of the cosine function and the symmetry with respect to the polar axis, the line $\theta = \dfrac{\pi}{2}$, and the pole, only list values of θ from 0 to $\dfrac{\pi}{2}$.

Plot and connect these points as shown in Figure 30(a). Finally, because of symmetry, reflect this portion of the graph first about the polar axis (the x-axis) and then about the line $\theta = \dfrac{\pi}{2}$ (the y-axis) to obtain the complete graph. See Figure 30(b).

Table 4

θ	$r = 2 \cos(2\theta)$
0	$2 \cdot 1 = 2$
$\dfrac{\pi}{6}$	$2 \cdot \dfrac{1}{2} = 1$
$\dfrac{\pi}{4}$	$2 \cdot 0 = 0$
$\dfrac{\pi}{3}$	$2\left(-\dfrac{1}{2}\right) = -1$
$\dfrac{\pi}{2}$	$2(-1) = -2$

Exploration

Graph $r_1 = 2 \cos(4\theta)$; clear the screen and graph $r_1 = 2 \cos(6\theta)$. How many petals did each of these graphs have?

Clear the screen and graph, in order, each on a clear screen, $r_1 = 2 \cos(3\theta)$, $r_1 = 2 \cos(5\theta)$, and $r_1 = 2 \cos(7\theta)$. What do you notice about the number of petals?

Figure 30   (a)   (b) $r = 2 \cos(2\theta)$

The curve in Figure 30(b) is called a *rose* with four petals.

DEFINITION Rose

Rose curves are characterized by equations of the form

$$\bullet \ r = a\cos(n\theta) \qquad \bullet \ r = a\sin(n\theta) \qquad a \neq 0$$

and have graphs that are rose shaped.

If $n \neq 0$ is even, the rose has $2n$ petals; if $n \neq \pm 1$ is odd, the rose has n petals.

Now Work PROBLEM 51

EXAMPLE 11 Graphing a Polar Equation (Lemniscate)

Graph the equation: $r^2 = 4\sin(2\theta)$

Solution We leave it to you to verify that the graph is symmetric with respect to the pole. Because of the symmetry with respect to the pole, only list those values of θ between $\theta = 0$ and $\theta = \pi$. Note that for $\frac{\pi}{2} < \theta < \pi$ (quadrant II) there are no points on the graph since $r^2 < 0$ for such values. Table 5 lists points on the graph for values of $\theta = 0$ through $\theta = \frac{\pi}{2}$. The points from Table 5 where $r \geq 0$ are plotted in Figure 31(a). The remaining points on the graph may be obtained by using symmetry. Figure 31(b) shows the final graph drawn.

Table 5

θ	$r^2 = 4\sin(2\theta)$	r
0	$4 \cdot 0 = 0$	0
$\dfrac{\pi}{6}$	$4 \cdot \dfrac{\sqrt{3}}{2} = 2\sqrt{3}$	± 1.9
$\dfrac{\pi}{4}$	$4 \cdot 1 = 4$	± 2
$\dfrac{\pi}{3}$	$4 \cdot \dfrac{\sqrt{3}}{2} = 2\sqrt{3}$	± 1.9
$\dfrac{\pi}{2}$	$4 \cdot 0 = 0$	0

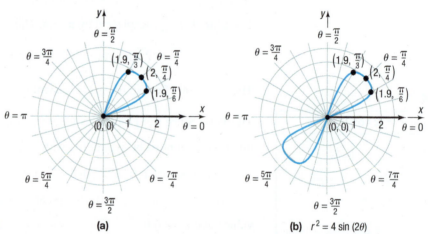

Figure 31 (a)   (b) $r^2 = 4\sin(2\theta)$

The curve in Figure 31(b) is an example of a *lemniscate* (from the Greek word for *ribbon*).

DEFINITION Lemniscates

Lemniscates are characterized by equations of the form

$$\bullet \ r^2 = a^2\sin(2\theta) \qquad \bullet \ r^2 = a^2\cos(2\theta)$$

where $a \neq 0$, and have graphs that are propeller shaped.

Now Work PROBLEM 55

EXAMPLE 12 Graphing a Polar Equation (Spiral)

Graph the equation: $r = e^{\theta/5}$

Solution The tests for symmetry with respect to the pole, the polar axis, and the line $\theta = \frac{\pi}{2}$ fail. Furthermore, there is no number θ for which $r = 0$, so the graph does not pass through the pole. From the equation, note that r is positive for all θ, r increases as θ increases, $r \to 0$ as $\theta \to -\infty$, and $r \to \infty$ as $\theta \to \infty$.

Table 6

θ	$r = e^{\theta/5}$
$-\dfrac{3\pi}{2}$	0.39
$-\pi$	0.53
$-\dfrac{\pi}{2}$	0.73
$-\dfrac{\pi}{4}$	0.85
0	1
$\dfrac{\pi}{4}$	1.17
$\dfrac{\pi}{2}$	1.37
π	1.87
$\dfrac{3\pi}{2}$	2.57
2π	3.51

With the help of a calculator, the values in Table 6 can be obtained. See Figure 32.

Figure 32 $r = e^{\theta/5}$

The curve in Figure 32 is called a **logarithmic spiral**, since its equation may be written as $\theta = 5 \ln r$ and it spirals infinitely both toward the pole and away from it.

Classification of Polar Equations

The equations of some lines and circles in polar coordinates and their corresponding equations in rectangular coordinates are given in Table 7. Also included are the names and graphs of a few of the more frequently encountered polar equations.

Table 7

Lines			
Description	Line passing through the pole making an angle α with the polar axis	Vertical line	Horizontal line
Rectangular equation	$y = (\tan \alpha)x$	$x = a$	$y = b$
Polar equation	$\theta = \alpha$	$r \cos \theta = a$	$r \sin \theta = b$
Typical graph			

Circles			
Description	Center at the pole, radius a	Passing through the pole, tangent to the line $\theta = \dfrac{\pi}{2}$, center on the polar axis, radius a	Passing through the pole, tangent to the polar axis, center on the line $\theta = \dfrac{\pi}{2}$, radius a
Rectangular equation	$x^2 + y^2 = a^2, \quad a > 0$	$x^2 + y^2 = \pm 2ax, \quad a > 0$	$x^2 + y^2 = \pm 2ay, \quad a > 0$
Polar equation	$r = a, \quad a > 0$	$r = \pm 2a \cos \theta, \quad a > 0$	$r = \pm 2a \sin \theta, \quad a > 0$
Typical graph			

(continued)

Table 7 (Continued)

Other Equations		
Name	Cardioid	Limaçon without inner loop	Limaçon with inner loop
Polar equations	$r = a \pm a \cos \theta, \quad a > 0$	$r = a \pm b \cos \theta, \quad a > b > 0$	$r = a \pm b \cos \theta, \quad b > a > 0$
	$r = a \pm a \sin \theta, \quad a > 0$	$r = a \pm b \sin \theta, \quad a > b > 0$	$r = a \pm b \sin \theta, \quad b > a > 0$
Typical graph			

Name	Lemniscate	Rose with three petals	Rose with four petals
Polar equations	$r^2 = a^2 \cos(2\theta), \quad a \neq 0$	$r = a \sin(3\theta), \quad a > 0$	$r = a \sin(2\theta), \quad a > 0$
	$r^2 = a^2 \sin(2\theta), \quad a \neq 0$	$r = a \cos(3\theta), \quad a > 0$	$r = a \cos(2\theta), \quad a > 0$
Typical graph			

Sketching Quickly

If a polar equation involves only a sine (or cosine) function, you can quickly obtain its graph by making use of Table 7, periodicity, and a short table.

EXAMPLE 13 **Sketching the Graph of a Polar Equation Quickly**

Graph the equation: $r = 2 + 2 \sin \theta$

Solution Because $a = b = 2$, the graph of this polar equation is a cardioid. The period of $\sin \theta$ is 2π, so form a table using $0 \leq \theta \leq 2\pi$, compute r, plot the points (r, θ), and sketch the graph of a cardioid as θ varies from 0 to 2π. See Table 8 and Figure 33.

Table 8

θ	$r = 2 + 2 \sin \theta$
0	$2 + 2 \cdot 0 = 2$
$\frac{\pi}{2}$	$2 + 2 \cdot 1 = 4$
π	$2 + 2 \cdot 0 = 2$
$\frac{3\pi}{2}$	$2 + 2(-1) = 0$
2π	$2 + 2 \cdot 0 = 2$

Figure 33 $r = 2 + 2 \sin \theta$

⚠️ **Calculus Comment** For those of you planning to study calculus, a comment about one important role of polar equations is in order.

In rectangular coordinates, the equation $x^2 + y^2 = 1$, whose graph is the unit circle, is not the graph of a function. In fact, it requires two functions to obtain the graph of the unit circle:

$$y_1 = \sqrt{1 - x^2} \quad \textcolor{blue}{\textbf{Upper semicircle}} \qquad y_2 = -\sqrt{1 - x^2} \quad \textcolor{blue}{\textbf{Lower semicircle}}$$

In polar coordinates, the equation $r = 1$, whose graph is also the unit circle, does define a function. For each choice of θ, there is only one corresponding value of r, that is, $r = 1$. Since many problems in calculus require the use of functions, the opportunity to express nonfunctions in rectangular coordinates as functions in polar coordinates becomes extremely useful.

Note also that the vertical-line test for functions is valid only for equations in rectangular coordinates.

Historical Feature

Jakob Bernoulli (1654–1705)

Polar coordinates seem to have been invented by Jakob Bernoulli (1654–1705) in about 1691, although, as with most such ideas, earlier traces of the notion exist. Early users of calculus remained committed to rectangular coordinates, and polar coordinates did not become widely used until the early 1800s. Even then, it was mostly geometers who used them for describing odd curves. Finally, about the mid-1800s, applied mathematicians realized the tremendous simplification that polar coordinates make possible in the description of objects with circular or cylindrical symmetry. From then on, their use became widespread.

5.2 Assess Your Understanding

'Are You Prepared?' *Answers are given at the end of these exercises. If you get a wrong answer, read the pages listed in* *red*.

1. If the rectangular coordinates of a point are $(4, -6)$, the point symmetric to it with respect to the origin is _____. (pp. 12–14)

2. The difference formula for cosine is $\cos(A - B) =$ _____. (p. 227)

3. The standard equation of a circle with center at $(-2, 5)$ and radius 3 is _____. (pp. 16–19)

4. Is the sine function even, odd, or neither? (pp. 142–143)

5. $\sin \dfrac{5\pi}{4} =$ _____ (pp. 118–125)

6. $\cos \dfrac{2\pi}{3} =$ _____ (pp. 118–125)

Concepts and Vocabulary

7. An equation whose variables are polar coordinates is called a(n) _____ _____.

8. *True or False* The tests for symmetry in polar coordinates are always conclusive.

9. To test whether the graph of a polar equation may be symmetric with respect to the polar axis, replace θ by _____.

10. To test whether the graph of a polar equation may be symmetric with respect to the line $\theta = \dfrac{\pi}{2}$, replace θ by _____.

11. Rose curves are characterized by equations of the form $r = a \cos(n\theta)$ or $r = a \sin(n\theta), a \neq 0$. If $n \neq 0$ is even, the rose has _____ petals; if $n \neq \pm 1$ is odd, the rose has _____ petals.

12. *True or False* A cardioid passes through the pole.

13. *Multiple Choice* For a positive real number a, which polar equation is a circle with radius a and center $(a, 0)$ in rectangular coordinates?
    (a) $r = 2a \sin \theta$
    (b) $r = -2a \sin \theta$
    (c) $r = 2a \cos \theta$
    (d) $r = -2a \cos \theta$

14. *Multiple Choice* In polar coordinates, the points (r, θ) and $(-r, \theta)$ are symmetric with respect to which of the following?
    (a) the polar axis (or x-axis)
    (b) the pole (or origin)
    (c) the line $\theta = \dfrac{\pi}{2}$ (or y-axis)
    (d) the line $\theta = \dfrac{\pi}{4}$ (or $y = x$)

Skill Building

In Problems 15–30, transform each polar equation to an equation in rectangular coordinates. Then identify and graph the equation.

15. $r = 4$

16. $r = 2$

17. $\theta = \dfrac{\pi}{3}$

18. $\theta = -\dfrac{\pi}{4}$

19. $r \sin \theta = 4$

20. $r \cos \theta = 4$

21. $r \cos \theta = -2$

22. $r \sin \theta = -2$

23. $r = 2\cos\theta$ **24.** $r = 2\sin\theta$ **25.** $r = -4\sin\theta$ **26.** $r = -4\cos\theta$

27. $r\sec\theta = 4$ **28.** $r\csc\theta = 8$ **29.** $r\csc\theta = -2$ **30.** $r\sec\theta = -4$

In Problems 31–38, match each of the graphs (A) through (H) to one of the following polar equations.

31. $r = 2$ **32.** $\theta = \dfrac{\pi}{4}$ **33.** $r = 2\cos\theta$ **34.** $r\cos\theta = 2$

35. $r = 1 + \cos\theta$ **36.** $r = 2\sin\theta$ **37.** $\theta = \dfrac{3\pi}{4}$ **38.** $r\sin\theta = 2$

(A)

(B)

(C)

(D)

(E)

(F)

(G)

(H)

In Problems 39–62, identify and graph each polar equation.

39. $r = 2 + 2\cos\theta$ **40.** $r = 1 + \sin\theta$ **41.** $r = 3 - 3\sin\theta$ **42.** $r = 2 - 2\cos\theta$

43. $r = 2 + \sin\theta$ **44.** $r = 2 - \cos\theta$ **45.** $r = 4 - 2\cos\theta$ **46.** $r = 4 + 2\sin\theta$

47. $r = 1 + 2\sin\theta$ **48.** $r = 1 - 2\sin\theta$ **49.** $r = 2 - 3\cos\theta$ **50.** $r = 2 + 4\cos\theta$

51. $r = 3\cos(2\theta)$ **52.** $r = 2\sin(3\theta)$ **53.** $r = 4\sin(5\theta)$ **54.** $r = 3\cos(4\theta)$

55. $r^2 = 9\cos(2\theta)$ **56.** $r^2 = \sin(2\theta)$ **57.** $r = 2^\theta$ **58.** $r = 3^\theta$

59. $r = 1 - \cos\theta$ **60.** $r = 3 + \cos\theta$ **61.** $r = 1 - 3\cos\theta$ **62.** $r = 4\cos(3\theta)$

Applications and Extensions

In Problems 63–66, the polar equation for each graph is either $r = a + b\cos\theta$ or $r = a + b\sin\theta$, $a > 0$. Select the correct equation and find the values of a and b.

63.

64.

65.

66.

In Problems 67–72, graph each pair of polar equations on the same polar grid. Find the polar coordinates of the point(s) of intersection and label the point(s) on the graph.

67. $r = 8 \cos \theta; \ r = 2 \sec \theta$

68. $r = 8 \sin \theta; \ r = 4 \csc \theta$

69. $r = \sin \theta; \ r = 1 + \cos \theta$

70. $r = 3; \ r = 2 + 2 \cos \theta$

71. $r = 1 + \sin \theta; \ r = 1 + \cos \theta$

72. $r = 1 + \cos \theta; \ r = 3 \cos \theta$

In Problems 73–82, graph each polar equation.

73. $r = \dfrac{2}{1 - \cos \theta}$ *(parabola)*

74. $r = \dfrac{2}{1 - 2 \cos \theta}$ *(hyperbola)*

75. $r = \dfrac{1}{3 - 2 \cos \theta}$ *(ellipse)*

76. $r = \dfrac{1}{1 - \cos \theta}$ *(parabola)*

77. $r = \theta, \ \ \theta \geq 0$ *(spiral of Archimedes)*

78. $r = \dfrac{3}{\theta}$ *(reciprocal spiral)*

79. $r = \csc \theta - 2, \ \ 0 < \theta < \pi$ *(conchoid)*

80. $r = \sin \theta \tan \theta$ *(cissoid)*

81. $r = \tan \theta, \ \ -\dfrac{\pi}{2} < \theta < \dfrac{\pi}{2}$ *(kappa curve)*

82. $r = \cos \dfrac{\theta}{2}$

83. Show that the graph of the equation $r \sin \theta = a$ is a horizontal line a units above the pole if $a \geq 0$ and $|a|$ units below the pole if $a < 0$.

84. Show that the graph of the equation $r \cos \theta = a$ is a vertical line a units to the right of the pole if $a \geq 0$ and $|a|$ units to the left of the pole if $a < 0$.

85. Show that the graph of the equation $r = 2a \sin \theta, a > 0$, is a circle of radius a with center $(0, a)$ in rectangular coordinates.

86. Show that the graph of the equation $r = -2a \sin \theta, a > 0$, is a circle of radius a with center $(0, -a)$ in rectangular coordinates.

87. Show that the graph of the equation $r = 2a \cos \theta, a > 0$, is a circle of radius a with center $(a, 0)$ in rectangular coordinates.

88. Show that the graph of the equation $r = -2a \cos \theta, a > 0$, is a circle of radius a with center $(-a, 0)$ in rectangular coordinates.

89. Sailing Polar plots provide attainable speeds of a specific sailboat sailing at different angles to a wind of given speed. See the figure. Use the plot to approximate the attainable speed of the sailboat for the given conditions.

(a) Sailing at a 140° angle to a 6-knot wind.

(b) Sailing at a 160° angle to a 10-knot wind.

(c) Sailing at a 80° angle to a 20-knot wind.

(d) If the wind blows at 20 knots, for what angles will the sailboat attain a speed of 10 knots or faster?

(e) If the wind blows at 10 knots, what is the maximum attainable speed of the sailboat? For what angle(s) does this speed occur?

Source: myhanse.com

90. *Challenge Problem* Show that $r = a \cos \theta + b \sin \theta$, with a, b not both zero, is the equation of a circle. Find the center and radius of the circle.

91. *Challenge Problem* Express $r^2 = \cos(2\theta)$ in rectangular coordinates free of radicals.

92. *Challenge Problem* Prove that the area of the triangle with vertices $(0, 0), (r_1, \theta_1)$, and $(r_2, \theta_2), 0 \leq \theta_1 < \theta_2 \leq \pi$, is

$$K = \frac{1}{2} r_1 r_2 \sin(\theta_2 - \theta_1)$$

Explaining Concepts: Discussion and Writing

93. Explain why the following test for symmetry is valid: Replace r by $-r$ and θ by $-\theta$ in a polar equation. If an equivalent equation results, the graph is symmetric with respect to the line $\theta = \dfrac{\pi}{2}$ (*y*-axis).

(a) Show that the test on page 329 fails for $r^2 = \cos \theta$, yet this new test works.

(b) Show that the test on page 329 works for $r^2 = \sin \theta$, yet this new test fails.

94. Write down two different tests for symmetry with respect to the polar axis. Find examples in which one test works and the other fails. Which test do you prefer to use? Justify your answer.

95. The tests for symmetry given on page 329 are sufficient, but not necessary. Explain what this means.

96. Explain why the vertical-line test used to identify functions in rectangular coordinates does not work for equations expressed in polar coordinates.

Retain Your Knowledge

Problems 97–105 are based on material learned earlier in the course. The purpose of these problems is to keep the material fresh in your mind so that you are better prepared for the final exam.

97. Solve: $\dfrac{5}{x-3} \geq 1$

98. Convert $\dfrac{7\pi}{3}$ radians to degrees.

99. Determine the amplitude and period of $y = -2\sin(5x)$ without graphing.

100. Find any asymptotes for the graph of

$$R(x) = \frac{x+3}{x^2 - x - 12}$$

101. Find the remainder when $3x^5 - 2x^3 + 7x - 5$ is divided by $x - 1$.

102. Find the area of a triangle with sides 6, 11, and 13.

103. Solve: $6x^2 + 7x = 20$

104. $m = f'(x) = 3x^2 + 8x$ gives the slope of the tangent line to the graph of $f(x) = x^3 + 4x^2 - 5$ at any number x. Find an equation of the tangent line to f at $x = -2$.

105. Show that $\cos^3 x = \cos x - \sin^2 x \cos x$.

'Are You Prepared?' Answers

1. $(-4, 6)$ **2.** $\cos A \cos B + \sin A \sin B$ **3.** $(x+2)^2 + (y-5)^2 = 9$ **4.** Odd **5.** $-\dfrac{\sqrt{2}}{2}$ **6.** $-\dfrac{1}{2}$

5.3 The Complex Plane; De Moivre's Theorem

PREPARING FOR THIS SECTION *Before getting started, review the following:*

- Complex Numbers (Section A.5, pp. A37–A42)
- Values of the Sine and Cosine Functions at Certain Angles (Section 2.2, pp. 118–125)
- Sum and Difference Formulas for Sine and Cosine (Section 3.5, pp. 227 and 230)
- Laws of Exponents (Section A.1, pp. A7–A9)

Now Work the **'Are You Prepared?'** problems on page 307.

OBJECTIVES 1 Plot Points in the Complex Plane (p. 340)
2 Convert a Complex Number between Rectangular Form and Polar Form or Exponential Form (p. 341)
3 Find Products and Quotients of Complex Numbers (p. 343)
4 Use De Moivre's Theorem (p. 344)
5 Find Complex Roots (p. 345)

1 Plot Points in the Complex Plane

Figure 34 Complex plane

Complex numbers are discussed in Section A.5. In that discussion, we were not prepared to give a geometric interpretation of a complex number. Now we are ready.

A complex number $z = x + yi$ can be interpreted geometrically as the point (x, y) in the xy-plane. Each point in the plane corresponds to a complex number, and conversely, each complex number corresponds to a point in the plane. The collection of such points is referred to as the **complex plane**. The x-axis is referred to as the **real axis**, because any point that lies on the real axis is of the form $z = x + 0i = x$, a real number. The y-axis is called the **imaginary axis**, because any point that lies on the imaginary axis is of the form $z = 0 + yi = yi$, a pure imaginary number. See Figure 34.

| EXAMPLE 1 | **Plotting a Point in the Complex Plane** |

Plot the point corresponding to $z = \sqrt{3} - i$ in the complex plane.

Solution The point corresponding to $z = \sqrt{3} - i$ has the rectangular coordinates $(\sqrt{3}, -1)$. This point, located in quadrant IV, is plotted in Figure 35.

Figure 35

DEFINITION Magnitude or Modulus

Suppose $z = x + yi$ is a complex number. The **magnitude** or **modulus** of z, denoted $|z|$, is the distance from the origin to the point (x, y). That is,

$$|z| = \sqrt{x^2 + y^2} \qquad \textbf{(1)}$$

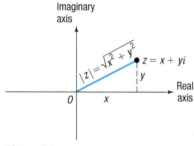

Figure 36

See Figure 36 for an illustration.

This definition for $|z|$ is consistent with the definition for the absolute value of a real number: If $z = x + yi$ is a real number then $z = x + 0i$ and

$$|z| = \sqrt{x^2 + 0^2} = \sqrt{x^2} = |x|$$

For this reason, the magnitude of z is sometimes called the **absolute value of z**.

Recall that if $z = x + yi$, then its **conjugate**, denoted $\bar{z}$, is $\bar{z} = x - yi$. Because $z\bar{z} = x^2 + y^2$ is a nonnegative real number, it follows from equation (1) that the magnitude of z can be written as

$$|z| = \sqrt{z\bar{z}} \qquad \textbf{(2)}$$

---

2 Convert a Complex Number between Rectangular Form and Polar Form or Exponential Form

When a complex number is written in the standard form $z = x + yi$, it is in **rectangular**, or **Cartesian**, **form**, because (x, y) are the rectangular coordinates of the corresponding point in the complex plane. Suppose that (r, θ) are polar coordinates of this point. Then

$$x = r \cos \theta \qquad y = r \sin \theta$$

DEFINITION Polar Form of a Complex Number

If $r \geq 0$ and $0 \leq \theta < 2\pi$, the complex number $z = x + yi$ can be written in **polar form** as

$$z = x + yi = r \cos \theta + (r \sin \theta)i = r(\cos \theta + i \sin \theta) \qquad \textbf{(3)}$$

$z = x + yi = r(\cos \theta + i \sin \theta)$,
$r \geq 0, 0 \leq \theta < 2\pi$

Figure 37

See Figure 37.

If $z = r(\cos \theta + i \sin \theta)$ is the polar form of a complex number,* the angle $\theta, 0 \leq \theta < 2\pi$, is called the **argument of z**.

*Some texts abbreviate the polar form using $z = r(\cos \theta + i \sin \theta) = r \operatorname{cis} \theta$.

Also, because $r \geq 0$, we have $r = \sqrt{x^2 + y^2}$. From equation (1), it follows that the magnitude of $z = r(\cos \theta + i \sin \theta)$ is

$$|z| = r$$

NOTE The number $e = 2.718281828\ldots$ is discussed in more detail in Section 7.1. ∎

Leonhard Euler (1707–1783) established a relationship, known as *Euler's Formula*, between complex numbers and the number e.

THEOREM Euler's Formula

For any real number θ,

$$e^{i\theta} = \cos \theta + i \sin \theta$$

The proof of Euler's Formula requires mathematics beyond the level of this text, so it is not included here.

Euler's Formula allows us to write the polar form of a complex number using exponential notation.

$$r(\cos \theta + i \sin \theta) = re^{i\theta} \qquad (4)$$

When a complex number is written in the form $z = re^{i\theta}$, it is said to be written in **exponential form**. Note in Euler's Formula that θ is a real number. That is, θ is in radians.

EXAMPLE 2 **Writing a Complex Number in Polar Form and in Exponential Form**

Write $z = \sqrt{3} - i$ in polar form and in exponential form.

Solution Because $x = \sqrt{3}$ and $y = -1$, it follows that

$$r = \sqrt{x^2 + y^2} = \sqrt{(\sqrt{3})^2 + (-1)^2} = \sqrt{4} = 2$$

so

$$\cos \theta = \frac{x}{r} = \frac{\sqrt{3}}{2} \qquad \sin \theta = \frac{y}{r} = \frac{-1}{2} \qquad 0 \leq \theta < 2\pi$$

The angle θ, $0 \leq \theta < 2\pi$, that satisfies both equations is $\theta = \dfrac{11\pi}{6}$. With $\theta = \dfrac{11\pi}{6}$ and $r = 2$, the polar form of $z = \sqrt{3} - i$ is

$$z = r(\cos \theta + i \sin \theta) = 2\left(\cos \frac{11\pi}{6} + i \sin \frac{11\pi}{6} \right)$$

The exponential form of $z = \sqrt{3} - i$ is

$$z = re^{i\theta} = 2e^{i \cdot 11\pi/6} \qquad \theta = \frac{11\pi}{6}, r = 2$$

✏ **Now Work PROBLEM 13**

EXAMPLE 3 **Plotting a Point in the Complex Plane and Converting It to Rectangular Form**

Plot the point corresponding to $z = 2\left(\cos \dfrac{\pi}{6} + i \sin \dfrac{\pi}{6} \right) = 2e^{i \cdot \pi/6}$ in the complex plane, and convert z to rectangular form.

Figure 38 $z = 2e^{i \cdot \pi/6}$

Solution To plot the complex number $z = 2\left(\cos\dfrac{\pi}{6} + i \sin\dfrac{\pi}{6}\right) = 2e^{i \cdot \pi/6}$, plot the point whose polar coordinates are $(r, \theta) = \left(2, \dfrac{\pi}{6}\right)$ as shown in Figure 38. To express z in rectangular form, expand $z = 2\left(\cos\dfrac{\pi}{6} + i \sin\dfrac{\pi}{6}\right)$.

$$z = 2\left(\cos\frac{\pi}{6} + i \sin\frac{\pi}{6}\right) = 2\left(\frac{\sqrt{3}}{2} + \frac{1}{2}i\right) = \sqrt{3} + i$$

Now Work PROBLEM 25

3 Find Products and Quotients of Complex Numbers

The exponential form of a complex number is particularly useful for finding products and quotients of complex numbers. The following theorem states that the laws of exponents can be used.

Need to Review?

The Laws of Exponents are discussed in Section A.1, pp. A7–A9.

THEOREM

Suppose $z_1 = r_1 e^{i\theta_1}$ and $z_2 = r_2 e^{i\theta_2}$ are two complex numbers. Then

$$z_1 z_2 = r_1 e^{i\theta_1} \cdot r_2 e^{i\theta_2} = r_1 r_2 e^{i(\theta_1 + \theta_2)} \qquad (5)$$

If $z_2 \neq 0$, then

$$\frac{z_1}{z_2} = \frac{r_1 e^{i\theta_1}}{r_2 e^{i\theta_2}} = \frac{r_1}{r_2} e^{i(\theta_1 - \theta_2)} \qquad (6)$$

Proof We prove formula (5). The proof of formula (6) is left as an exercise (see Problem 70).

$$
\begin{aligned}
z_1 z_2 &= r_1 e^{i\theta_1} \cdot r_2 e^{i\theta_2} \\
&= r_1(\cos\theta_1 + i \sin\theta_1) \cdot r_2(\cos\theta_2 + i \sin\theta_2) \\
&= r_1 r_2 (\cos\theta_1 + i \sin\theta_1)(\cos\theta_2 + i \sin\theta_2) \\
&= r_1 r_2 [\, (\cos\theta_1 \cos\theta_2 - \sin\theta_1 \cos\theta_2) + i(\sin\theta_1 \cos\theta_2 + \cos\theta_1 \sin\theta_2) \,] \\
&= r_1 r_2 [\cos(\theta_1 + \theta_2) + i \sin(\theta_1 + \theta_2)\,] \\
&= r_1 r_2 e^{i(\theta_1 + \theta_2)}
\end{aligned}
$$
∎

EXAMPLE 4 | **Finding Products and Quotients of Complex Numbers**

If $z = 3\left(\cos\dfrac{\pi}{9} + i \sin\dfrac{\pi}{9}\right)$ and $w = 5\left(\cos\dfrac{5\pi}{9} + i \sin\dfrac{5\pi}{9}\right)$, find

(a) zw (b) $\dfrac{z}{w}$

Express the answers in polar form and in exponential form.

Solution (a) $zw = 3\left(\cos\dfrac{\pi}{9} + i \sin\dfrac{\pi}{9}\right) \cdot 5\left(\cos\dfrac{5\pi}{9} + i \sin\dfrac{5\pi}{9}\right)$

$= 3e^{i \cdot \pi/9} \cdot 5e^{i \cdot 5\pi/9}$ **Write the complex numbers in exponential form.**

$= 3 \cdot 5 \cdot e^{i \cdot (\pi/9 + 5\pi/9)}$ **Use formula (5).**

$= 15e^{i \cdot 2\pi/3}$ **Exponential form of the product zw**

$= 15\left(\cos\dfrac{2\pi}{3} + i \sin\dfrac{2\pi}{3}\right)$ **Polar form of the product zw**

(continued)

(b) $\dfrac{z}{w} = \dfrac{3\left(\cos\dfrac{\pi}{9} + i\sin\dfrac{\pi}{9}\right)}{5\left(\cos\dfrac{5\pi}{9} + i\sin\dfrac{5\pi}{9}\right)} = \dfrac{3e^{i\cdot\pi/9}}{5e^{i\cdot5\pi/9}}$ **Write the complex numbers in exponential form.**

$= \dfrac{3}{5}e^{i\cdot(\pi/9 - 5\pi/9)} = \dfrac{3}{5}e^{i\cdot(-4\pi/9)}$ **Use formula (6).**

$= \dfrac{3}{5}\left[\cos\left(-\dfrac{4\pi}{9}\right) + i\sin\left(-\dfrac{4\pi}{9}\right)\right]$

$= \dfrac{3}{5}\left[\cos\dfrac{14\pi}{9} + i\sin\dfrac{14\pi}{9}\right]$ **The argument must be between 0 and 2π.**
Polar form of the quotient $\dfrac{z}{w}$

$= \dfrac{3}{5}e^{i\cdot14\pi/9}$ **Exponential form of the quotient $\dfrac{z}{w}$**

- **Now Work** PROBLEM **37**

Notice that the solution to Example 4(b) demonstrates that the argument of a complex number is periodic.

THEOREM

The argument of a complex number is periodic.

$$re^{i\theta} = re^{i(\theta + 2k\pi)} \quad k \text{ an integer}$$

You are asked to prove this theorem in Problem 71.

4 Use De Moivre's Theorem

We have seen that the four fundamental operations of addition, subtraction, multiplication, and division can be performed with complex numbers.

De Moivre's Theorem, stated by Abraham De Moivre (1667–1754) in 1730, but already known to many people by 1710, is important because it allows the last two fundamental algebraic operations, raising to a power and extracting roots, to be used with complex numbers.

De Moivre's Theorem, in its most basic form, is a formula for raising a complex number z to the power n, where $n \geq 1$ is an integer.

Suppose $z = re^{i\theta}$ is a complex number. Then

$$n = 2: \quad z^2 = (re^{i\theta})^2 = re^{i\theta} \cdot re^{i\theta} = r^2 e^{i(2\theta)} \qquad \text{Formula (5)}$$

$$n = 3: \quad z^3 = (re^{i\theta})^3 = r^2 e^{i(2\theta)} \cdot re^{i\theta} = r^3 e^{i(3\theta)} \qquad \text{Formula (5)}$$

$$n = 4: \quad z^4 = (re^{i\theta})^4 = r^3 e^{i(3\theta)} \cdot re^{i\theta} = r^4 e^{i(4\theta)} \qquad \text{Formula (5)}$$

Do you see the pattern? When written in exponential form, rules for exponents can be used to raise a complex number to a positive integer power.

THEOREM **De Moivre's Theorem**

If $z = re^{i\theta}$ is a complex number, then

$$z^n = r^n e^{i(n\theta)}$$

where $n \geq 1$ is an integer.

The proof of De Moivre's Theorem requires mathematical induction (which is not discussed in this text), so it is omitted here. The theorem is actually true for all integers n. You are asked to prove this in Problem 73.

EXAMPLE 5	**Using De Moivre's Theorem**

Express $\left[2\left(\cos\dfrac{\pi}{9} + i\sin\dfrac{\pi}{9}\right)\right]^3$ in exponential form $re^{i\theta}$ and in rectangular form $x + yi$.

Solution

$\left[2\left(\cos\dfrac{\pi}{9} + i\sin\dfrac{\pi}{9}\right)\right]^3 = \left(2e^{i\cdot\pi/9}\right)^3$ **Convert the complex number to exponential form.**

$\qquad\qquad = 2^3 e^{i\cdot(3\cdot\pi/9)} = 8e^{i\cdot\pi/3}$ **Use De Moivre's Theorem.**

$\qquad\qquad = 8\left(\cos\dfrac{\pi}{3} + i\sin\dfrac{\pi}{3}\right)$ **Convert the complex number to polar form.**

$\qquad\qquad = 8\left(\dfrac{1}{2} + \dfrac{\sqrt{3}}{2}i\right) = 4 + 4\sqrt{3}i$

So, $\left[2\left(\cos\dfrac{\pi}{9} + i\sin\dfrac{\pi}{9}\right)\right]^3 = 8e^{i\cdot\pi/3} = 4 + 4\sqrt{3}i$.

Now Work PROBLEM 45

EXAMPLE 6	**Using De Moivre's Theorem**

Express $(1 + i)^5$ in exponential form $re^{i\theta}$ and in rectangular form $x + yi$.

Solution

To use De Moivre's Theorem, first convert the complex number to exponential form. Since the magnitude of $1 + i$ is $\sqrt{1^2 + 1^2} = \sqrt{2}$, begin by writing

$$1 + i = \sqrt{2}\left(\dfrac{1}{\sqrt{2}} + \dfrac{1}{\sqrt{2}}i\right) = \sqrt{2}\left(\cos\dfrac{\pi}{4} + i\sin\dfrac{\pi}{4}\right) = \sqrt{2}e^{i\cdot\pi/4}$$

NOTE In the solution of Example 6, the approach used in Example 2 could also be used to write $1 + i$ in polar form. ∎

Then

$$(1 + i)^5 = \left(\sqrt{2}e^{i\cdot\pi/4}\right)^5 = \left(\sqrt{2}\right)^5 e^{i(5\cdot\pi/4)} = 4\sqrt{2}e^{i\cdot5\pi/4}$$

$$= 4\sqrt{2}\left(\cos\dfrac{5\pi}{4} + i\sin\dfrac{5\pi}{4}\right) = 4\sqrt{2}\left[-\dfrac{1}{\sqrt{2}} + \left(-\dfrac{1}{\sqrt{2}}\right)i\right] = -4 - 4i$$

So, $(1 + i)^5 = 4\sqrt{2}e^{i\cdot5\pi/4} = -4 - 4i$.

5 Find Complex Roots

Suppose w is a complex number, and $n \geq 2$ is a positive integer. Any complex number z that satisfies the equation

$$z^n = w$$

is a **complex nth root** of w. In keeping with previous usage, if $n = 2$, the solutions of the equation $z^2 = w$ are called **complex square roots** of w, and if $n = 3$, the solutions of the equation $z^3 = w$ are called **complex cube roots** of w.

> **THEOREM** **Finding Complex Roots**
>
> Suppose $w = re^{i\theta}$ is a complex number and $n \geq 2$ is an integer. If $w \neq 0$, there are n distinct complex roots of w, given by the formula
>
> $$z_k = \sqrt[n]{r}\, e^{i(1/n)(\theta + 2k\pi)} \qquad\qquad (7)$$
>
> where $k = 0, 1, 2, \ldots, n - 1$.

Proof (Outline) We do not prove this result in its entirety. Instead, we show only that each z_k in formula (7) satisfies the equation $z_k^n = w$, proving that each z_k is a complex nth root of w.

$$z_k^n = \left[\sqrt[n]{r}\, e^{i(1/n)(\theta + 2k\pi)} \right]^n = \left(\sqrt[n]{r} \right)^n e^{i\cdot [n\cdot(1/n)(\theta + 2k\pi)]} = r e^{i(\theta + 2k\pi)} = r e^{i\theta} = w$$

De Moivre's Theorem   The argument of a complex number is periodic.

So each z_k, $k = 0, 1, \ldots, n - 1$, is a complex nth root of w. To complete the proof, we need to show that each z_k, $k = 0, 1, \ldots, n - 1$, is, in fact, distinct and that there are no complex nth roots of w other than those given by formula (7). ∎

EXAMPLE 7

Finding Complex Cube Roots

Find the complex cube roots of $-1 + \sqrt{3}i$. Express the answers in exponential form.

Solution

First, express $-1 + \sqrt{3}i$ in exponential form.

$$-1 + \sqrt{3}i = 2\left(-\frac{1}{2} + \frac{\sqrt{3}}{2}i \right) = 2\left(\cos\frac{2\pi}{3} + i\sin\frac{2\pi}{3} \right) = 2\,e^{i\cdot 2\pi/3}$$

$\cos\theta = -\dfrac{1}{2}; \sin\theta = \dfrac{\sqrt{3}}{2} \Rightarrow \theta = \dfrac{2\pi}{3}$

Then using formula (7), the three complex cube roots of $-1 + \sqrt{3}i = 2\,e^{i\cdot 2\pi/3}$ are

$$z_k = \sqrt[3]{2}\, e^{i(1/3)(2\pi/3 + 6k\pi/3)} = \sqrt[3]{2}\, e^{i(2\pi + 6k\pi)/9} \quad k = 0, 1, 2$$

That is,

$$z_0 = \sqrt[3]{2}\, e^{i(2\pi + 6\cdot 0\cdot \pi)/9} = \sqrt[3]{2}\, e^{i\cdot 2\pi/9}$$

$$z_1 = \sqrt[3]{2}\, e^{i(2\pi + 6\cdot 1\cdot \pi)/9} = \sqrt[3]{2}\, e^{i\cdot 8\pi/9}$$

$$z_2 = \sqrt[3]{2}\, e^{i(2\pi + 6\cdot 2\cdot \pi)/9} = \sqrt[3]{2}\, e^{i\cdot 14\pi/9}$$

Notice that each complex root of $-1 + \sqrt{3}i$ has the same magnitude, $\sqrt[3]{2}$. This means that the point corresponding to each cube root is the same distance from the origin and lies on a circle with center at the origin and radius $\sqrt[3]{2}$. Furthermore, the arguments of these cube roots are $\dfrac{2\pi}{9}, \dfrac{8\pi}{9}$, and $\dfrac{14\pi}{9}$, and the difference of consecutive pairs is $\dfrac{1}{3} \cdot 2\pi = \dfrac{2\pi}{3}$. This means that the three points are equally spaced on the circle, as shown in Figure 39. These results are not coincidental. In fact, you are asked to show that these results hold for complex nth roots in Problems 67 through 69.

Figure 39

✏️ ─── **Now Work** PROBLEM **57**

Historical Feature

John Wallis

The Babylonians, Greeks, and Arabs considered square roots of negative quantities to be impossible and equations with complex solutions to be unsolvable. The first hint that there was some connection between real solutions of equations and complex numbers came when Girolamo Cardano (1501–1576) and Tartaglia (1499–1557) found *real* roots of cubic equations by taking cube roots of *complex* quantities. For centuries thereafter, mathematicians worked with complex numbers without much belief in their actual existence. In 1673, John Wallis appears to have been the first to suggest the graphical representation of complex numbers, a truly significant idea that was not pursued further until about 1800. Several people, including Karl Friedrich Gauss (1777–1855), then rediscovered the idea, and graphical representation helped to establish complex numbers as equal members of the number family. In practical applications, complex numbers have found their greatest uses in the study of alternating current, where they are a commonplace tool, and in the field of subatomic physics.

Historical Problems

1. The quadratic formula works perfectly well if the coefficients are complex numbers. Solve the following.
   (a) $z^2 - (2 + 5i)z - 3 + 5i = 0$
   (b) $z^2 - (1 + i)z - 2 - i = 0$

5.3 Assess Your Understanding

'Are You Prepared?' *Answers are given at the end of these exercises. If you get a wrong answer, read the pages listed in red.*

1. The conjugate of $-4 - 3i$ is _____ . (pp. A37–A42)

2. The Sum Formulas for the sine and cosine functions are:
   (a) $\sin(A + B) = $ _____ . (p. 230)
   (b) $\cos(A + B) = $ _____ . (p. 227)

3. $\sin\dfrac{2\pi}{3} = $ _____; $\cos\dfrac{4\pi}{3} = $ _____ (pp. 118–125)

4. Simplify: $e^2 \cdot e^5 = $ _____; $(e^4)^3 = $ _____ (pp. A7–A9)

Concepts and Vocabulary

5. In the complex plane, the x-axis is referred to as the _____ axis, and the y-axis is called the _____ axis.

6. When a complex number z is written in the polar form $z = r(\cos\theta + i\sin\theta)$, the nonnegative number r is the _____ or _____ of z, and the angle θ, $0 \le \theta < 2\pi$, is the _____ of z.

7. Suppose $z_1 = r_1 e^{i\theta_1}$ and $z_2 = r_2 e^{i\theta_2}$ are two complex numbers. Then $z_1 z_2 = $ _____ .

8. *True or False* If $z = re^{i\theta}$ is a complex number and n is an integer, then $z^n = r^n e^{i\theta}$.

9. Every nonzero complex number has exactly _____ distinct complex cube roots.

10. *True or False* The polar form of a nonzero complex number is unique.

11. *Multiple Choice* If $z = x + yi$ is a complex number, then the magnitude of z is:
    (a) $x^2 + y^2$ (b) $|x| + |y|$
    (c) $\sqrt{x^2 + y^2}$ (d) $\sqrt{|x| + |y|}$

12. *Multiple Choice* If $z_1 = r_1 e^{i\theta_1}$ and $z_2 = r_2 e^{i\theta_2}$ are complex numbers, then $\dfrac{z_1}{z_2}$, $z_2 \ne 0$, equals:
    (a) $\dfrac{r_1}{r_2} e^{i(\theta_1 - \theta_2)}$ (b) $\dfrac{r_1}{r_2} e^{i(\theta_1 \cdot \theta_2)}$
    (c) $\dfrac{r_1}{r_2} e^{i(\theta_1 + \theta_2)}$ (d) $\dfrac{r_1}{r_2} e^{i(\theta_1/\theta_2)}$

Skill Building

In Problems 13–24, plot each complex number in the complex plane and write it in polar form and in exponential form.

13. $1 + i$
14. $-1 + i$
15. $\sqrt{3} - i$
16. $1 - \sqrt{3}i$
17. $-3i$
18. -2
19. $4 - 4i$
20. $9\sqrt{3} + 9i$
21. $3 - 4i$
22. $2 + \sqrt{3}i$
23. $-2 + 3i$
24. $\sqrt{5} - i$

In Problems 25–36, write each complex number in rectangular form.

25. $2\left(\cos\dfrac{2\pi}{3} + i\sin\dfrac{2\pi}{3}\right)$
26. $3\left(\cos\dfrac{7\pi}{6} + i\sin\dfrac{7\pi}{6}\right)$
27. $4e^{i \cdot 7\pi/4}$

28. $2e^{i \cdot 5\pi/6}$
29. $3\left(\cos\dfrac{3\pi}{2} + i\sin\dfrac{3\pi}{2}\right)$
30. $4\left(\cos\dfrac{\pi}{2} + i\sin\dfrac{\pi}{2}\right)$

31. $7e^{i\pi}$
32. $3e^{i \cdot \pi/2}$
33. $0.2\left(\cos\dfrac{5\pi}{9} + i\sin\dfrac{5\pi}{9}\right)$

34. $0.4\left(\cos\dfrac{10\pi}{9} + i\sin\dfrac{10\pi}{9}\right)$
35. $2e^{i \cdot \pi/18}$
36. $3e^{i \cdot \pi/10}$

In Problems 37–44, find zw and $\dfrac{z}{w}$. Write each answer in polar form and in exponential form.

37. $z = 2\left(\cos\dfrac{2\pi}{9} + i\sin\dfrac{2\pi}{9}\right)$

$w = 4\left(\cos\dfrac{\pi}{9} + i\sin\dfrac{\pi}{9}\right)$

38. $z = \cos\dfrac{2\pi}{3} + i\sin\dfrac{2\pi}{3}$

$w = \cos\dfrac{5\pi}{9} + i\sin\dfrac{5\pi}{9}$

39. $z = 3e^{i\cdot 13\pi/18}$

$w = 4e^{i\cdot 3\pi/2}$

40. $z = 2e^{i\cdot 4\pi/9}$

$w = 6e^{i\cdot 10\pi/9}$

41. $z = 2\left(\cos\dfrac{\pi}{8} + i\sin\dfrac{\pi}{8}\right)$

$w = 2\left(\cos\dfrac{\pi}{10} + i\sin\dfrac{\pi}{10}\right)$

42. $z = 4\left(\cos\dfrac{3\pi}{8} + i\sin\dfrac{3\pi}{8}\right)$

$w = 2\left(\cos\dfrac{9\pi}{16} + i\sin\dfrac{9\pi}{16}\right)$

43. $z = 2 + 2i$

$w = \sqrt{3} - i$

44. $z = 1 - i$

$w = 1 - \sqrt{3}i$

In Problems 45–56, write each expression in rectangular form $x + yi$ and in exponential form $re^{i\theta}$.

45. $\left[4\left(\cos\dfrac{2\pi}{9} + i\sin\dfrac{2\pi}{9}\right)\right]^3$

46. $\left[3\left(\cos\dfrac{4\pi}{9} + i\sin\dfrac{4\pi}{9}\right)\right]^3$

47. $\left[2\left(\cos\dfrac{\pi}{10} + i\sin\dfrac{\pi}{10}\right)\right]^5$

48. $\left[\sqrt{2}\left(\cos\dfrac{5\pi}{16} + i\sin\dfrac{5\pi}{16}\right)\right]^4$

49. $\left[\sqrt{3}\left(\cos\dfrac{\pi}{18} + i\sin\dfrac{\pi}{18}\right)\right]^6$

50. $\left[\dfrac{1}{2}\left(\cos\dfrac{2\pi}{5} + i\sin\dfrac{2\pi}{5}\right)\right]^5$

51. $\left[\sqrt{5}e^{i\cdot 3\pi/16}\right]^4$

52. $\left[\sqrt{3}e^{i\cdot 5\pi/18}\right]^6$

53. $(1 - i)^5$

54. $(\sqrt{3} - i)^6$

55. $(\sqrt{2} - i)^6$

56. $(1 - \sqrt{5}i)^8$

In Problems 57–64, find all the complex roots. Write your answers in exponential form.

57. The complex cube roots of $1 + i$

58. The complex fourth roots of $\sqrt{3} - i$

59. The complex fourth roots of $4 - 4\sqrt{3}i$

60. The complex cube roots of $-8 - 8i$

61. The complex fourth roots of $-16i$

62. The complex cube roots of -8

63. The complex fifth roots of i

64. The complex fifth roots of $-i$

Applications and Extensions

65. Find the four complex fourth roots of unity, 1, and plot them.

66. Find the six complex sixth roots of unity, 1, and plot them.

67. Show that each complex nth root of a nonzero complex number w has the same magnitude.

68. Use the result of Problem 67 to draw the conclusion that each complex nth root lies on a circle with center at the origin. What is the radius of this circle?

69. Refer to Problem 68. Show that the complex nth roots of a nonzero complex number w are equally spaced on the circle.

70. Prove formula (6).

71. Prove $re^{i\theta} = re^{i(\theta + 2k\pi)}$, k an integer.

72. Euler's Identity Show that $e^{i\pi} + 1 = 0$.

73. Prove that De Moivre's Theorem is true for *all* integers n by assuming it is true for integers $n \geq 1$ and then showing it is true for 0 and for negative integers.

Hint: Multiply the numerator and the denominator by the conjugate of the denominator, and use even-odd properties.

74. Mandelbrot Sets

(a) Consider the expression $a_n = (a_{n-1})^2 + z$, where z is some complex number (called the **seed**) and $a_0 = z$. Compute $a_1 (=a_0^2 + z)$, $a_2 (=a_1^2 + z)$, $a_3 (=a_2^2 + z)$, a_4, a_5, and a_6 for the following seeds: $z_1 = 0.1 - 0.4i$, $z_2 = 0.5 + 0.8i$, $z_3 = -0.9 + 0.7i$, $z_4 = -1.1 + 0.1i$, $z_5 = 0 - 1.3i$, and $z_6 = 1 + 1i$.

(b) The dark portion of the graph represents the set of all values $z = x + yi$ that are in the Mandelbrot set. Determine which complex numbers in part (a) are in this set by plotting them on the graph. Do the complex numbers that are not in the Mandelbrot set have any common characteristics regarding the values of a_6 found in part (a)?

(c) Compute $|z| = \sqrt{x^2 + y^2}$ for each of the complex numbers in part (a). Now compute $|a_6|$ for each of the complex numbers in part (a). For which complex numbers is $|a_6| \leq |z|$ and $|z| \leq 2$? Conclude that the criterion for a complex number to be in the Mandelbrot set is that $|a_n| \leq |z|$ and $|z| \leq 2$.

Retain Your Knowledge

Problems 75–82 are based on material learned earlier in the course. The purpose of these problems is to keep the material fresh in your mind so that you are better prepared for the final exam.

75. Find the area of the triangle with $a = 8$, $b = 11$, and $C = 113°$.

76. Convert $240°$ to radians. Express your answer as a multiple of π.

77. Simplify: $\sqrt[3]{24x^2y^5}$

78. Determine whether $f(x) = 5x^2 - 12x + 4$ has a maximum value or a minimum value, and then find the value.

79. Solve the triangle: $a = 6$, $b = 8$, $c = 12$

80. Given $f(x) = 3x^2 - 4x$, find $f(5x^3)$.

81. Find an equation of the line perpendicular to the graph of $f(x) = \dfrac{2}{3}x - 5$ at $x = 6$.

82. Show that $\sqrt{16\sec^2 x - 16} = 4\tan x$.

'Are You Prepared?' Answers

1. $-4 + 3i$ **2. (a)** $\sin A \cos B + \cos A \sin B$ **(b)** $\cos A \cos B - \sin A \sin B$ **3.** $\dfrac{\sqrt{3}}{2}; -\dfrac{1}{2}$ **4.** $e^7; e^{12}$

5.4 Vectors

OBJECTIVES 1 Graph Vectors (p. 351)
2 Find a Position Vector (p. 352)
3 Add and Subtract Vectors Algebraically (p. 354)
4 Find a Scalar Multiple and the Magnitude of a Vector (p. 354)
5 Find a Unit Vector (p. 355)
6 Find a Vector from Its Direction and Magnitude (p. 356)
7 Model with Vectors (p. 357)

In simple terms, a **vector** (derived from the Latin *vehere*, meaning "to carry") is a quantity that has both magnitude and direction. It is customary to represent a vector by using an arrow. The length of the arrow represents the **magnitude** of the vector, and the arrowhead indicates the **direction** of the vector.

Many quantities in physics can be represented by vectors. For example, the velocity of an aircraft can be represented by an arrow that points in the direction of movement; the length of the arrow represents the speed. If the aircraft speeds up, we lengthen the arrow; if the aircraft changes direction, we introduce an arrow in the new direction. See Figure 40. Based on this representation, it is not surprising that vectors and *directed line segments* are somehow related.

Figure 40

Geometric Vectors

If P and Q are two distinct points in the xy-plane, there is exactly one line containing both P and Q [Figure 41(a)]. The points on that part of the line that joins P to Q, including P and Q, form what is called the **line segment** $\overline{PQ}$ [Figure 41(b)]. Ordering the points so that they proceed from P to Q results in a **directed line segment** from P to Q, or a **geometric vector**, denoted by $\overrightarrow{PQ}$. In a directed line segment $\overrightarrow{PQ}$, P is called the **initial point** and Q the **terminal point**, as indicated in Figure 41(c).

Figure 41

(a) Line containing P and Q (b) Line segment $\overline{PQ}$ (c) Directed line segment $\overrightarrow{PQ}$

NOTE In print, boldface letters are used to denote vectors, to distinguish them from numbers. For handwritten work, an arrow is placed over a letter to denote a vector. ∎

The magnitude of the directed line segment $\overrightarrow{PQ}$ is the distance from the point P to the point Q; that is, it is the length of the line segment. The direction of $\overrightarrow{PQ}$ is from P to Q. If a vector $\mathbf{v}$ has the same magnitude and the same direction as the directed line segment $\overrightarrow{PQ}$, then

$$\mathbf{v} = \overrightarrow{PQ}$$

The vector $\mathbf{v}$ whose magnitude is 0 is called the **zero vector**, **0**. The zero vector is assigned no direction.

Two vectors $\mathbf{v}$ and $\mathbf{w}$ are **equal**, written

$$\mathbf{v} = \mathbf{w}$$

if they have the same magnitude and the same direction.

For example, the three vectors shown in Figure 42 have the same magnitude and the same direction, so they are equal, even though they have different initial points and different terminal points. As a result, it is useful to think of a vector simply as an arrow, keeping in mind that two arrows (vectors) are equal if they have the same direction and the same magnitude (length).

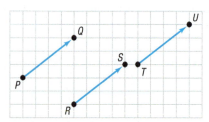

Figure 42 Equal vectors

Adding Vectors Geometrically

The **sum $\mathbf{v} + \mathbf{w}$** of two vectors is defined as follows: Position the vectors $\mathbf{v}$ and $\mathbf{w}$ so that the terminal point of $\mathbf{v}$ coincides with the initial point of $\mathbf{w}$, as shown in Figure 43. The vector $\mathbf{v} + \mathbf{w}$ is then the unique vector whose initial point coincides with the initial point of $\mathbf{v}$ and whose terminal point coincides with the terminal point of $\mathbf{w}$.

Figure 43 Adding vectors

Vector addition is **commutative**. That is, if $\mathbf{v}$ and $\mathbf{w}$ are any two vectors, then

$$\mathbf{v} + \mathbf{w} = \mathbf{w} + \mathbf{v}$$

Figure 44 illustrates this fact. (Observe that the commutative property is another way of saying that opposite sides of a parallelogram are equal and parallel.)

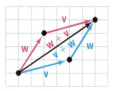

Figure 44 $\mathbf{v} + \mathbf{w} = \mathbf{w} + \mathbf{v}$

Vector addition is also **associative**. That is, if $\mathbf{u}$, $\mathbf{v}$, and $\mathbf{w}$ are vectors, then

$$\mathbf{u} + (\mathbf{v} + \mathbf{w}) = (\mathbf{u} + \mathbf{v}) + \mathbf{w}$$

Figure 45 illustrates the associative property for vectors.

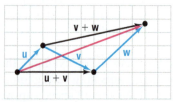

Figure 45
$(\mathbf{u} + \mathbf{v}) + \mathbf{w} = \mathbf{u} + (\mathbf{v} + \mathbf{w})$

The zero vector **0** has the property that

$$\mathbf{v} + \mathbf{0} = \mathbf{0} + \mathbf{v} = \mathbf{v}$$

for any vector $\mathbf{v}$.

If $\mathbf{v}$ is a vector, then $-\mathbf{v}$ is the vector that has the same magnitude as $\mathbf{v}$, but whose direction is opposite to $\mathbf{v}$, as shown in Figure 46.

Furthermore,

Figure 46 Opposite vectors

$$\mathbf{v} + (-\mathbf{v}) = \mathbf{0}$$

If $\mathbf{v}$ and $\mathbf{w}$ are two vectors, then the **difference $\mathbf{v} - \mathbf{w}$** is defined as

$$\mathbf{v} - \mathbf{w} = \mathbf{v} + (-\mathbf{w})$$

Figure 47 $\mathbf{v} - \mathbf{w} = \mathbf{v} + (-\mathbf{w})$

Figure 47 illustrates the relationships among $\mathbf{v}$, $\mathbf{w}$, and $\mathbf{v} - \mathbf{w}$.

Multiplying Vectors by Numbers Geometrically

When using vectors, real numbers are referred to as **scalars**. Scalars are quantities that have only magnitude. Examples of scalar quantities from physics are temperature, speed, and time. We now define how to multiply a vector by a scalar.

> **DEFINITION** Scalar Multiple
>
> If α is a scalar and $\mathbf{v}$ is a vector, the **scalar multiple** $\alpha\mathbf{v}$ is defined as follows:
>
> - If $\alpha > 0$, $\alpha\mathbf{v}$ is the vector whose magnitude is α times the magnitude of $\mathbf{v}$ and whose direction is the same as that of $\mathbf{v}$.
> - If $\alpha < 0$, $\alpha\mathbf{v}$ is the vector whose magnitude is $|\alpha|$ times the magnitude of $\mathbf{v}$ and whose direction is opposite that of $\mathbf{v}$.
> - If $\alpha = 0$ or if $\mathbf{v} = \mathbf{0}$, then $\alpha\mathbf{v} = \mathbf{0}$.

Figure 48 Scalar multiples

See Figure 48 for some illustrations.

For example, if $\mathbf{a}$ is the acceleration of an object of mass m due to a force $\mathbf{F}$ being exerted on it, then, by Newton's second law of motion, $\mathbf{F} = m\mathbf{a}$. Here, $m\mathbf{a}$ is the product of the scalar m and the vector $\mathbf{a}$.

Scalar multiples have the following properties:

> - $0\mathbf{v} = \mathbf{0}$ • $1\mathbf{v} = \mathbf{v}$ • $-1\mathbf{v} = -\mathbf{v}$
> - $(\alpha + \beta)\mathbf{v} = \alpha\mathbf{v} + \beta\mathbf{v}$ • $\alpha(\mathbf{v} + \mathbf{w}) = \alpha\mathbf{v} + \alpha\mathbf{w}$
> - $\alpha(\beta\mathbf{v}) = (\alpha\beta)\mathbf{v}$

1 Graph Vectors

EXAMPLE 1 **Graphing Vectors**

Use the vectors illustrated in Figure 49 to graph each of the following vectors:

(a) $\mathbf{v} - \mathbf{w}$ (b) $2\mathbf{v} + 3\mathbf{w}$ (c) $2\mathbf{v} - \mathbf{w} + \mathbf{u}$

Solution Figure 50 shows each graph.

Figure 49

(a) $\mathbf{v} - \mathbf{w}$

(b) $2\mathbf{v} + 3\mathbf{w}$

(c) $2\mathbf{v} - \mathbf{w} + \mathbf{u}$

Figure 50

Now Work PROBLEMS **11** AND **13**

Magnitude of Vectors

The symbol $\|\mathbf{v}\|$ represents the **magnitude** of a vector $\mathbf{v}$. Since $\|\mathbf{v}\|$ equals the length of a directed line segment, it follows that $\|\mathbf{v}\|$ has the following properties:

> **THEOREM Properties of the Magnitude $\|\mathbf{v}\|$ of a Vector v**
>
> If $\mathbf{v}$ is a vector and if α is a scalar, then
> (a) $\|\mathbf{v}\| \geq 0$ (b) $\|\mathbf{v}\| = 0$ if and only if $\mathbf{v} = \mathbf{0}$
> (c) $\|-\mathbf{v}\| = \|\mathbf{v}\|$ (d) $\|\alpha\mathbf{v}\| = |\alpha|\|\mathbf{v}\|$

Property (a) is a consequence of the fact that distance is a nonnegative number. Property (b) follows because the length of the directed line segment $\overrightarrow{PQ}$ is positive unless P and Q are the same point, in which case the length is 0. Property (c) follows because the length of the line segment $\overline{PQ}$ equals the length of the line segment $\overline{QP}$. Property (d) is a direct consequence of the definition of a scalar multiple.

> **DEFINITION Unit Vector**
>
> A vector $\mathbf{u}$ for which $\|\mathbf{u}\| = 1$ is called a **unit vector**.

2 Find a Position Vector

To find the magnitude and direction of a vector, an algebraic way of representing vectors is needed.

> **DEFINITION Algebraic Vector**
>
> An **algebraic vector** $\mathbf{v}$ is represented as
>
> $$\mathbf{v} = \langle a, b \rangle$$
>
> where a and b are real numbers (scalars) called the **components** of the vector $\mathbf{v}$.

Figure 51 Position vector v

A rectangular coordinate system is used to represent algebraic vectors in the plane. If $\mathbf{v} = \langle a, b \rangle$ is an algebraic vector whose initial point is at the origin, then $\mathbf{v}$ is called a **position vector**. See Figure 51. Notice that the terminal point of the position vector $\mathbf{v} = \langle a, b \rangle$ is the point $P = (a, b)$.

The next theorem states that any vector whose initial point is not at the origin is equal to a unique position vector.

> **THEOREM**
>
> Suppose that $\mathbf{v}$ is a vector with initial point $P_1 = (x_1, y_1)$, not necessarily the origin, and terminal point $P_2 = (x_2, y_2)$. If $\mathbf{v} = \overrightarrow{P_1 P_2}$, then $\mathbf{v}$ is equal to the position vector
>
> $$\mathbf{v} = \langle x_2 - x_1, y_2 - y_1 \rangle \qquad (1)$$

To see why this is true, look at Figure 52.

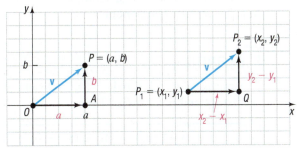

Figure 52 $\mathbf{v} = \langle a, b \rangle = \langle x_2 - x_1, y_2 - y_1 \rangle$

Triangle OPA and triangle $P_1 P_2 Q$ are congruent. [Do you see why? The line segments have the same magnitude, so $d(O, P) = d(P_1, P_2)$; and they have the same direction, so $\angle POA = \angle P_2 P_1 Q$. Since the triangles are right triangles, we have angle–side–angle.] It follows that corresponding sides are equal. As a result, $x_2 - x_1 = a$ and $y_2 - y_1 = b$, so $\mathbf{v}$ may be written as

$$\mathbf{v} = \langle a, b \rangle = \langle x_2 - x_1, y_2 - y_1 \rangle$$

Because of this result, any algebraic vector can be replaced by a unique position vector, and vice versa. This flexibility is one of the main reasons for the wide use of vectors.

| EXAMPLE 2 | **Finding a Position Vector** |

Find the position vector of the vector $\mathbf{v} = \overrightarrow{P_1 P_2}$ if $P_1 = (-1, 2)$ and $P_2 = (4, 6)$.

Solution By equation (1), the position vector equal to $\mathbf{v}$ is

$$\mathbf{v} = \langle 4 - (-1), 6 - 2 \rangle = \langle 5, 4 \rangle$$

See Figure 53.

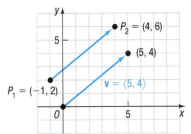

Figure 53

Two position vectors $\mathbf{v}$ and $\mathbf{w}$ are equal if and only if the terminal point of $\mathbf{v}$ is the same as the terminal point of $\mathbf{w}$. This leads to the following theorem:

THEOREM Equality of Vectors

Two vectors $\mathbf{v}$ and $\mathbf{w}$ are equal if and only if their corresponding components are equal. That is,

$$\text{If } \mathbf{v} = \langle a_1, b_1 \rangle \quad \text{and} \quad \mathbf{w} = \langle a_2, b_2 \rangle$$
$$\text{then} \quad \mathbf{v} = \mathbf{w} \quad \text{if and only if} \quad a_1 = a_2 \quad \text{and} \quad b_1 = b_2.$$

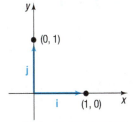

Figure 54 Unit vectors **i** and **j**

We now present an alternative representation of a vector in the plane that is common in the physical sciences. Let $\mathbf{i}$ denote the unit vector whose direction is along the positive x-axis; let $\mathbf{j}$ denote the unit vector whose direction is along the positive y-axis. Then $\mathbf{i} = \langle 1, 0 \rangle$ and $\mathbf{j} = \langle 0, 1 \rangle$, as shown in Figure 54. Any vector $\mathbf{v} = \langle a, b \rangle$ can be written using the unit vectors $\mathbf{i}$ and $\mathbf{j}$ as follows:

$$\mathbf{v} = \langle a, b \rangle = a \langle 1, 0 \rangle + b \langle 0, 1 \rangle = a\mathbf{i} + b\mathbf{j}$$

The quantities a and b are called the **horizontal** and **vertical components** of $\mathbf{v}$, respectively. For example, if $\mathbf{v} = \langle 5, 4 \rangle = 5\mathbf{i} + 4\mathbf{j}$, then 5 is the horizontal component and 4 is the vertical component.

─── **Now Work** PROBLEM 29

3 Add and Subtract Vectors Algebraically

The sum, difference, scalar multiple, and magnitude of algebraic vectors are defined in terms of their components.

> **DEFINITION**
>
> Suppose $\mathbf{v} = a_1\mathbf{i} + b_1\mathbf{j} = \langle a_1, b_1 \rangle$ and $\mathbf{w} = a_2\mathbf{i} + b_2\mathbf{j} = \langle a_2, b_2 \rangle$ are two vectors, and α is a scalar. Then
>
> $$\mathbf{v} + \mathbf{w} = (a_1 + a_2)\mathbf{i} + (b_1 + b_2)\mathbf{j} = \langle a_1 + a_2, b_1 + b_2 \rangle \tag{2}$$
> $$\mathbf{v} - \mathbf{w} = (a_1 - a_2)\mathbf{i} + (b_1 - b_2)\mathbf{j} = \langle a_1 - a_2, b_1 - b_2 \rangle \tag{3}$$
> $$\alpha\mathbf{v} = (\alpha a_1)\mathbf{i} + (\alpha b_1)\mathbf{j} = \langle \alpha a_1, \alpha b_1 \rangle \tag{4}$$
> $$\|\mathbf{v}\| = \sqrt{a_1^2 + b_1^2} \tag{5}$$

In Words

To add two vectors, add corresponding components. To subtract two vectors, subtract corresponding components.

These definitions are compatible with the geometric definitions given earlier in this section. See Figure 55.

Figure 55   **(a)** Illustration of property (2)   **(b)** Illustration of property (4), $\alpha > 0$ **(c)** Illustration of property (5):
$\|\mathbf{v}\| =$ Distance from O to P_1
$\|\mathbf{v}\| = \sqrt{a_1^2 + b_1^2}$

EXAMPLE 3   **Adding and Subtracting Vectors**

If $\mathbf{v} = 2\mathbf{i} + 3\mathbf{j} = \langle 2, 3 \rangle$ and $\mathbf{w} = 3\mathbf{i} - 4\mathbf{j} = \langle 3, -4 \rangle$, find:

(a) $\mathbf{v} + \mathbf{w}$ (b) $\mathbf{v} - \mathbf{w}$

Solution   (a) $\mathbf{v} + \mathbf{w} = (2\mathbf{i} + 3\mathbf{j}) + (3\mathbf{i} - 4\mathbf{j}) = (2 + 3)\mathbf{i} + (3 - 4)\mathbf{j} = 5\mathbf{i} - \mathbf{j}$

or

$\mathbf{v} + \mathbf{w} = \langle 2, 3 \rangle + \langle 3, -4 \rangle = \langle 2 + 3, 3 + (-4) \rangle = \langle 5, -1 \rangle$

(b) $\mathbf{v} - \mathbf{w} = (2\mathbf{i} + 3\mathbf{j}) - (3\mathbf{i} - 4\mathbf{j}) = (2 - 3)\mathbf{i} + [3 - (-4)]\mathbf{j} = -\mathbf{i} + 7\mathbf{j}$

or

$\mathbf{v} - \mathbf{w} = \langle 2, 3 \rangle - \langle 3, -4 \rangle = \langle 2 - 3, 3 - (-4) \rangle = \langle -1, 7 \rangle$

4 Find a Scalar Multiple and the Magnitude of a Vector

EXAMPLE 4   **Finding Scalar Multiples and Magnitudes of Vectors**

If $\mathbf{v} = 2\mathbf{i} + 3\mathbf{j} = \langle 2, 3 \rangle$ and $\mathbf{w} = 3\mathbf{i} - 4\mathbf{j} = \langle 3, -4 \rangle$, find:

(a) $3\mathbf{v}$ (b) $2\mathbf{v} - 3\mathbf{w}$ (c) $\|\mathbf{v}\|$

Solution (a) $3\mathbf{v} = 3(2\mathbf{i} + 3\mathbf{j}) = 6\mathbf{i} + 9\mathbf{j}$

or

$3\mathbf{v} = 3\langle 2, 3 \rangle = \langle 6, 9 \rangle$

(b) $2\mathbf{v} - 3\mathbf{w} = 2(2\mathbf{i} + 3\mathbf{j}) - 3(3\mathbf{i} - 4\mathbf{j}) = 4\mathbf{i} + 6\mathbf{j} - 9\mathbf{i} + 12\mathbf{j} = -5\mathbf{i} + 18\mathbf{j}$

or

$2\mathbf{v} - 3\mathbf{w} = 2\langle 2, 3 \rangle - 3\langle 3, -4 \rangle = \langle 4, 6 \rangle - \langle 9, -12 \rangle$

$\qquad\qquad = \langle 4 - 9, 6 - (-12) \rangle = \langle -5, 18 \rangle$

(c) $\|\mathbf{v}\| = \|2\mathbf{i} + 3\mathbf{j}\| = \sqrt{2^2 + 3^2} = \sqrt{13}$

 Now Work PROBLEMS 35 AND 43

5 Find a Unit Vector

Recall that a unit vector $\mathbf{u}$ is a vector for which $\|\mathbf{u}\| = 1$. In many applications, it is useful to be able to find a unit vector $\mathbf{u}$ that has the same direction as a given vector $\mathbf{v}$.

> **THEOREM** **Unit Vector in the Direction of v**
>
> For any nonzero vector $\mathbf{v}$, the vector
>
> $$\mathbf{u} = \frac{\mathbf{v}}{\|\mathbf{v}\|} \qquad\qquad (6)$$
>
> is a unit vector that has the same direction as $\mathbf{v}$.

Proof Let $\mathbf{v} = a\mathbf{i} + b\mathbf{j}$. Then $\|\mathbf{v}\| = \sqrt{a^2 + b^2}$ and

$$\mathbf{u} = \frac{\mathbf{v}}{\|\mathbf{v}\|} = \frac{a\mathbf{i} + b\mathbf{j}}{\sqrt{a^2 + b^2}} = \frac{a}{\sqrt{a^2 + b^2}}\mathbf{i} + \frac{b}{\sqrt{a^2 + b^2}}\mathbf{j}$$

The vector $\mathbf{u}$ has the same direction as $\mathbf{v}$, since $\|\mathbf{v}\| > 0$. Also,

$$\|\mathbf{u}\| = \sqrt{\frac{a^2}{a^2 + b^2} + \frac{b^2}{a^2 + b^2}} = \sqrt{\frac{a^2 + b^2}{a^2 + b^2}} = 1$$

That is, $\mathbf{u}$ is a unit vector that has the same direction as $\mathbf{v}$. ∎

The following is a consequence of this theorem.

> If $\mathbf{u}$ is a unit vector that has the same direction as a vector $\mathbf{v}$, then $\mathbf{v}$ can be expressed as
>
> $$\mathbf{v} = \|\mathbf{v}\|\mathbf{u} \qquad\qquad (7)$$

EXAMPLE 5 **Finding a Unit Vector**

Find a unit vector that has the same direction as $\mathbf{v} = 4\mathbf{i} - 3\mathbf{j}$.

Solution Find $\|\mathbf{v}\|$ first.

$$\|\mathbf{v}\| = \|4\mathbf{i} - 3\mathbf{j}\| = \sqrt{16 + 9} = 5$$

(continued)

Now multiply $\mathbf{v}$ by the scalar $\dfrac{1}{\|\mathbf{v}\|} = \dfrac{1}{5}$. A unit vector that has the same direction as $\mathbf{v}$ is

$$\frac{\mathbf{v}}{\|\mathbf{v}\|} = \frac{4\mathbf{i} - 3\mathbf{j}}{5} = \frac{4}{5}\mathbf{i} - \frac{3}{5}\mathbf{j}$$

✔ **Check:** This vector is a unit vector because

$$\left\|\frac{\mathbf{v}}{\|\mathbf{v}\|}\right\| = \sqrt{\left(\frac{4}{5}\right)^2 + \left(-\frac{3}{5}\right)^2} = \sqrt{\frac{16}{25} + \frac{9}{25}} = \sqrt{\frac{25}{25}} = 1$$

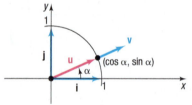

━━━ **Now Work** PROBLEM 53

6 Find a Vector from Its Direction and Magnitude

If a vector represents the speed and direction of an object, it is called a **velocity vector**. If a vector represents the direction and amount of a force acting on an object, it is called a **force vector**. In many applications, a vector is described in terms of its magnitude and direction, rather than in terms of its components. For example, a ball thrown with an initial speed of 25 miles per hour at an angle of 30° to the horizontal is a velocity vector.

Suppose that we are given the magnitude $\|\mathbf{v}\|$ of a nonzero vector $\mathbf{v}$ and the **direction angle** α, $0° \le \alpha < 360°$, between $\mathbf{v}$ and $\mathbf{i}$. To express $\mathbf{v}$ in terms of $\|\mathbf{v}\|$ and α, first find the unit vector $\mathbf{u}$ having the same direction as $\mathbf{v}$.

Look at Figure 56. The coordinates of the terminal point of $\mathbf{u}$ are $(\cos\alpha, \sin\alpha)$. Then $\mathbf{u} = \cos\alpha\,\mathbf{i} + \sin\alpha\,\mathbf{j}$ and, from equation (7),

$$\mathbf{v} = \|\mathbf{v}\|(\cos\alpha\,\mathbf{i} + \sin\alpha\,\mathbf{j}) \qquad (8)$$

Figure 56 $\mathbf{v} = \|\mathbf{v}\|(\cos\alpha\,\mathbf{i} + \sin\alpha\,\mathbf{j})$

where α is the direction angle between $\mathbf{v}$ and $\mathbf{i}$.

EXAMPLE 6   Finding a Vector When Its Magnitude and Direction Are Given

A ball is thrown with an initial speed of 25 miles per hour in a direction that makes an angle of 30° with the positive x-axis. Express the velocity vector $\mathbf{v}$ in terms of $\mathbf{i}$ and $\mathbf{j}$. What is the initial speed in the horizontal direction? What is the initial speed in the vertical direction?

Solution   The magnitude of $\mathbf{v}$ is $\|\mathbf{v}\| = 25$ miles per hour, and the angle between the direction of $\mathbf{v}$ and $\mathbf{i}$, the positive x-axis, is $\alpha = 30°$. By equation (8),

$$\mathbf{v} = \|\mathbf{v}\|(\cos\alpha\,\mathbf{i} + \sin\alpha\,\mathbf{j}) = 25(\cos 30°\mathbf{i} + \sin 30°\mathbf{j})$$

$$= 25\left(\frac{\sqrt{3}}{2}\mathbf{i} + \frac{1}{2}\mathbf{j}\right) = \frac{25\sqrt{3}}{2}\mathbf{i} + \frac{25}{2}\mathbf{j}$$

The initial speed of the ball in the horizontal direction is the horizontal component of $\mathbf{v}$, $\dfrac{25\sqrt{3}}{2} \approx 21.65$ miles per hour. The initial speed in the vertical direction is the vertical component of $\mathbf{v}$, $\dfrac{25}{2} = 12.5$ miles per hour. See Figure 57.

Figure 57

━━━ **Now Work** PROBLEM 61

EXAMPLE 7

Finding the Direction Angle of a Vector

Find the direction angle α of $\mathbf{v} = 4\mathbf{i} - 4\mathbf{j}$.

Solution

Figure 58 shows the vector $\mathbf{v}$ and its direction angle α. To find α, use the terminal point $(4, -4)$ and the fact that

$$\tan \alpha = \frac{-4}{4} = -1$$

Because $0° \leq \alpha < 360°$, the direction angle is $\alpha = 315°$.

Figure 58

🖉 **Now Work** PROBLEM 67

7 Model with Vectors

Because forces can be represented by vectors, two forces "combine" the way that vectors "add." If $\mathbf{F}_1$ and $\mathbf{F}_2$ are two forces simultaneously acting on an object, the vector sum $\mathbf{F}_1 + \mathbf{F}_2$ is the **resultant force**. The resultant force produces the same effect on the object as that obtained when the two forces $\mathbf{F}_1$ and $\mathbf{F}_2$ act on the object. See Figure 59.

Figure 59 Resultant force

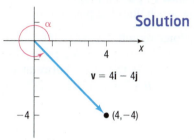

EXAMPLE 8

Finding the Actual Speed and Direction of an Aircraft

A Boeing 767 aircraft maintains a constant airspeed of 500 miles per hour headed due south. The jet stream is 80 miles per hour in the northeasterly direction.

(a) Express the velocity $\mathbf{v}_a$ of the 767 relative to the air and the velocity $\mathbf{v}_w$ of the jet stream in terms of $\mathbf{i}$ and $\mathbf{j}$.

(b) Find the velocity of the 767 relative to the ground.

(c) Find the actual speed and direction of the 767 relative to the ground.

Solution

(a) Set up a coordinate system in which north (N) is along the positive y-axis. See Figure 60. The velocity of the 767 relative to the air is $\mathbf{v}_a = -500\mathbf{j}$. The velocity of the jet stream $\mathbf{v}_w$ has magnitude 80 and direction NE (northeast), so the angle between $\mathbf{v}_w$ and $\mathbf{i}$ is 45°. Express $\mathbf{v}_w$ in terms of $\mathbf{i}$ and $\mathbf{j}$ as

$$\mathbf{v}_w = 80(\cos 45° \mathbf{i} + \sin 45° \mathbf{j}) = 80\left(\frac{\sqrt{2}}{2}\mathbf{i} + \frac{\sqrt{2}}{2}\mathbf{j}\right) = 40\sqrt{2}\,(\mathbf{i} + \mathbf{j})$$

(b) The velocity of the 767 relative to the ground $\mathbf{v}_g$ is

$$\mathbf{v}_g = \mathbf{v}_a + \mathbf{v}_w = -500\mathbf{j} + 40\sqrt{2}(\mathbf{i} + \mathbf{j}) = 40\sqrt{2}\mathbf{i} + \left(40\sqrt{2} - 500\right)\mathbf{j}$$

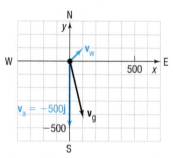

Figure 60

(c) The actual speed of the 767 is

$$\|\mathbf{v}_g\| = \sqrt{\left(40\sqrt{2}\right)^2 + \left(40\sqrt{2} - 500\right)^2} \approx 447 \text{ miles per hour}$$

To find the actual direction of the 767 relative to the ground, determine the direction angle of $\mathbf{v}_g$. The direction angle is found by solving

$$\tan \alpha = \frac{40\sqrt{2} - 500}{40\sqrt{2}}$$

Then $\alpha \approx -82.7°$. The 767 is traveling S7.3°E.

🖉 **Now Work** PROBLEM 79

| EXAMPLE 9 | **Finding the Weight of a Piano** |

Two movers require a force of magnitude 300 pounds to push a piano up a ramp inclined at an angle of 20° from the horizontal. How much does the piano weigh?

Solution Let $\mathbf{F}_1$ represent the force of gravity, $\mathbf{F}_2$ represent the force required to move the piano up the ramp, and $\mathbf{F}_3$ represent the force of the piano against the ramp. See Figure 61. The angle between the ground and the ramp is the same as the angle between $\mathbf{F}_1$ and $\mathbf{F}_3$ because triangles ABC and BDE are similar, so $\angle BAC = \angle DBE = 20°$. To find the magnitude of $\mathbf{F}_1$ (the weight of the piano), calculate

$$\sin 20° = \frac{\|\mathbf{F}_2\|}{\|\mathbf{F}_1\|} = \frac{300}{\|\mathbf{F}_1\|}$$

$$\|\mathbf{F}_1\| = \frac{300 \text{ lb}}{\sin 20°} \approx 877 \text{ lb}$$

The piano weighs approximately 877 pounds.

Figure 61

In Figure 61, the triangle formed by the force vectors (in blue) is called a **force diagram**.

An object is said to be in **static equilibrium** if the object is at rest and the sum of all forces acting on the object is zero—that is, if the resultant force is **0**.

| EXAMPLE 10 | **Analyzing an Object in Static Equilibrium** |

A box of supplies that weighs 1200 pounds is suspended by two cables attached to the ceiling, as shown in Figure 62. What are the tensions in the two cables?

Solution Draw a force diagram using the vectors as shown in Figure 63. The tensions in the cables are the magnitudes $\|\mathbf{F}_1\|$ and $\|\mathbf{F}_2\|$ of the force vectors $\mathbf{F}_1$ and $\mathbf{F}_2$. The magnitude of the force vector $\mathbf{F}_3$ equals 1200 pounds, the weight of the box. Now write each force vector in terms of the unit vectors $\mathbf{i}$ and $\mathbf{j}$. For $\mathbf{F}_1$ and $\mathbf{F}_2$, use $\mathbf{v} = \|\mathbf{v}\|(\cos\alpha\,\mathbf{i} + \sin\alpha\,\mathbf{j})$, where α is the angle between the vector and the positive x-axis.

$$\mathbf{F}_1 = \|\mathbf{F}_1\|(\cos 150°\mathbf{i} + \sin 150°\mathbf{j}) = \|\mathbf{F}_1\|\left(-\frac{\sqrt{3}}{2}\mathbf{i} + \frac{1}{2}\mathbf{j}\right) = -\frac{\sqrt{3}}{2}\|\mathbf{F}_1\|\mathbf{i} + \frac{1}{2}\|\mathbf{F}_1\|\mathbf{j}$$

$$\mathbf{v} = \|\mathbf{v}\|(\cos\alpha\,\mathbf{i} + \sin\alpha\,\mathbf{j})$$

Figure 62

$$\mathbf{F}_2 = \|\mathbf{F}_2\|(\cos 45°\mathbf{i} + \sin 45°\mathbf{j}) = \|\mathbf{F}_2\|\left(\frac{\sqrt{2}}{2}\mathbf{i} + \frac{\sqrt{2}}{2}\mathbf{j}\right) = \frac{\sqrt{2}}{2}\|\mathbf{F}_2\|\mathbf{i} + \frac{\sqrt{2}}{2}\|\mathbf{F}_2\|\mathbf{j}$$

$$\mathbf{v} = \|\mathbf{v}\|(\cos\alpha\,\mathbf{i} + \sin\alpha\,\mathbf{j})$$

$$\mathbf{F}_3 = -1200\mathbf{j}$$

For static equilibrium, the sum of the force vectors must equal the zero vector.

$$\mathbf{F}_1 + \mathbf{F}_2 + \mathbf{F}_3 = -\frac{\sqrt{3}}{2}\|\mathbf{F}_1\|\mathbf{i} + \frac{1}{2}\|\mathbf{F}_1\|\mathbf{j} + \frac{\sqrt{2}}{2}\|\mathbf{F}_2\|\mathbf{i} + \frac{\sqrt{2}}{2}\|\mathbf{F}_2\|\mathbf{j} - 1200\mathbf{j} = \mathbf{0}$$

Figure 63 Force diagram

The $\mathbf{i}$ component and $\mathbf{j}$ component must both equal zero. This results in the two equations

$$-\frac{\sqrt{3}}{2}\|\mathbf{F}_1\| + \frac{\sqrt{2}}{2}\|\mathbf{F}_2\| = 0 \tag{9}$$

$$\frac{1}{2}\|\mathbf{F}_1\| + \frac{\sqrt{2}}{2}\|\mathbf{F}_2\| - 1200 = 0 \tag{10}$$

Solve equation (9) for $\|\mathbf{F}_2\|$ to obtain

$$\|\mathbf{F}_2\| = \frac{\sqrt{3}}{\sqrt{2}}\|\mathbf{F}_1\| \tag{11}$$

Substituting into equation (10) and solving for $\|\mathbf{F}_1\|$ yields

$$\frac{1}{2}\|\mathbf{F}_1\| + \frac{\sqrt{2}}{2}\left(\frac{\sqrt{3}}{\sqrt{2}}\|\mathbf{F}_1\|\right) - 1200 = 0$$

$$\frac{1}{2}\|\mathbf{F}_1\| + \frac{\sqrt{3}}{2}\|\mathbf{F}_1\| - 1200 = 0$$

$$\frac{1 + \sqrt{3}}{2}\|\mathbf{F}_1\| = 1200$$

$$\|\mathbf{F}_1\| = \frac{2400}{1 + \sqrt{3}} \approx 878.5 \text{ pounds}$$

Substituting this value into equation (11) gives $\|\mathbf{F}_2\|$.

$$\|\mathbf{F}_2\| = \frac{\sqrt{3}}{\sqrt{2}}\|\mathbf{F}_1\| = \frac{\sqrt{3}}{\sqrt{2}} \cdot \frac{2400}{1 + \sqrt{3}} \approx 1075.9 \text{ pounds}$$

The left cable has tension of approximately 878.5 pounds, and the right cable has tension of approximately 1075.9 pounds.

Now Work PROBLEM 89

Historical Feature

Josiah Gibbs
(1839–1903)

The history of vectors is surprisingly complicated for such a natural concept. In the *xy*-plane, complex numbers do a good job of imitating vectors. About 1840, mathematicians became interested in finding a system that would do for three dimensions what the complex numbers do for two dimensions. Hermann Grassmann (1809–1877), in Germany, and William Rowan Hamilton (1805–1865), in Ireland, both attempted to find solutions.

Hamilton's system was the *quaternions*, which are best thought of as a real number plus a vector; they do for four dimensions what complex numbers do for two dimensions. In this system the order of multiplication matters; that is, $ab \neq ba$. Also, two products of vectors

emerged, the scalar product (or dot product) and the vector product (or cross product).

Grassmann's abstract style, although easily read today, was almost impenetrable during the nineteenth century, and only a few of his ideas were appreciated. Among those few were the same scalar and vector products that Hamilton had found.

About 1880, the American physicist Josiah Willard Gibbs (1839–1903) worked out an algebra involving only the simplest concepts: the vectors and the two products. He then added some calculus, and the resulting system was simple, flexible, and well adapted to expressing a large number of physical laws. This system remains in use essentially unchanged. Hamilton's and Grassmann's more extensive systems each gave birth to much interesting mathematics, but little of it is seen at elementary levels.

5.4 Assess Your Understanding

Concepts and Vocabulary

1. A _____ is a quantity that has both magnitude and direction.

2. If $\mathbf{v}$ is a vector, then $\mathbf{v} + (-\mathbf{v}) = $ ____.

3. A vector $\mathbf{u}$ for which $\|\mathbf{u}\| = 1$ is called a(n) _____ vector.

4. If $\mathbf{v} = \langle a, b \rangle$ is an algebraic vector whose initial point is the origin, then $\mathbf{v}$ is called a(n) _____ vector.

5. If $\mathbf{v} = a\mathbf{i} + b\mathbf{j}$, then a is called the _____ component of $\mathbf{v}$ and b is called the _____ component of $\mathbf{v}$.

6. If $\mathbf{F}_1$ and $\mathbf{F}_2$ are two forces acting on an object, the vector sum $\mathbf{F}_1 + \mathbf{F}_2$ is called the _____ force.

7. *True or False* Force is an example of a vector.

8. *True or False* Mass is an example of a vector.

9. *Multiple Choice* If $\mathbf{v}$ is a vector with initial point (x_1, y_1) and terminal point (x_2, y_2), then which of the following is the position vector that equals $\mathbf{v}$?

(a) $\langle x_2 - x_1, y_2 - y_1 \rangle$ **(b)** $\langle x_1 - x_2, y_1 - y_2 \rangle$

(c) $\left\langle \dfrac{x_2 - x_1}{2}, \dfrac{y_2 - y_1}{2} \right\rangle$ **(d)** $\left\langle \dfrac{x_1 + x_2}{2}, \dfrac{y_1 + y_2}{2} \right\rangle$

10. *Multiple Choice* If $\mathbf{v}$ is a nonzero vector with direction angle α, $0° \leq \alpha < 360°$, between $\mathbf{v}$ and $\mathbf{i}$, then $\mathbf{v}$ equals which of the following?

(a) $\|\mathbf{v}\|(\cos\alpha\mathbf{i} - \sin\alpha\mathbf{j})$ **(b)** $\|\mathbf{v}\|(\cos\alpha\mathbf{i} + \sin\alpha\mathbf{j})$

(c) $\|\mathbf{v}\|(\sin\alpha\mathbf{i} - \cos\alpha\mathbf{j})$ **(d)** $\|\mathbf{v}\|(\sin\alpha\mathbf{i} + \cos\alpha\mathbf{j})$

Skill Building

In Problems 11–18, use the vectors in the figure at the right to graph each of the following vectors.

11. v + w   **12. u + v**

13. 3v   **14. 2w**

15. v − w   **16. u − v**

17. 3v + u − 2w   **18. 2u − 3v + w**

In Problems 19–26, use the figure at the right. Determine whether each statement given is true or false.

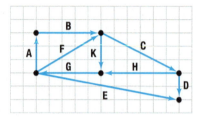

19. A + B = F   **20. K + G = F**

21. C = D − E + F   **22. G + H + E = D**

23. E + D = G + H   **24. H − C = G − F**

25. A + B + K + G = 0   **26. A + B + C + H + G = 0**

*In Problems 27–34, the vector **v** has initial point P and terminal point Q. Find its position vector. That is, express **v** in the form a**i** + b**j**.*

27. $P = (0,0);\ Q = (3,4)$ **28.** $P = (0,0);\ Q = (-3,-5)$

29. $P = (3,2);\ Q = (5,6)$ **30.** $P = (-3,2);\ Q = (6,5)$

31. $P(-2,-1);\ Q = (6,-2)$ **32.** $P = (-1,4);\ Q = (6,2)$

33. $P = (1,0);\ Q = (0,1)$ **34.** $P = (1,1);\ Q = (2,2)$

In Problems 35–42, find $\|\mathbf{v}\|$.

35. v = 3i − 4j   **36. v = −5i + 12j**   **37. v = i − j**

38. v = −i − j   **39. v = −2i + 3j**   **40. v = 6i + 2j**

41. v = cos θi + sin θj   **42. v = i + cot θj**

*In Problems 43–48, find each quantity if **v** = 3**i** − 5**j** and **w** = −2**i** + 3**j**.*

43. 2v + 3w   **44. 3v − 2w**   **45. $\|v − w\|$**

46. $\|v + w\|$ **47. $\|v\| − \|w\|$**   **48. $\|v\| + \|w\|$**

*In Problems 49–54, find the unit vector in the same direction as **v**.*

49. v = 5i   **50. v = −3j**   **51. v = 3i − 4j**

52. v = −5i + 12j   **53. v = i − j**   **54. v = 2i − j**

55. If $\|\mathbf{v}\| = 4$, what is the magnitude of $\frac{1}{2}\mathbf{v} + 3\mathbf{v}$?   **56.** If $\|\mathbf{v}\| = 2$, what is the magnitude of $-\frac{3}{4}\mathbf{v}$?

57. Find a vector **v** whose magnitude is 4 and whose component in the **i** direction is twice the component in the **j** direction.   **58.** Find a vector **v** whose magnitude is 3 and whose component in the **i** direction is equal to the component in the **j** direction.

59. If **v** = 2**i** − **j** and **w** = x**i** + 3**j**, find all numbers x for which $\|\mathbf{v} + \mathbf{w}\| = 5$.   **60.** If $P = (-3,1)$ and $Q = (x,4)$, find all numbers x so that the vector represented by $\overrightarrow{PQ}$ has length 5.

*In Problems 61–66, write the vector **v** in the form a**i** + b**j**, given its magnitude $\|\mathbf{v}\|$ and the angle α it makes with the positive x-axis.*

61. $\|\mathbf{v}\| = 5,\ \alpha = 60°$ **62.** $\|\mathbf{v}\| = 8,\ \alpha = 45°$ **63.** $\|\mathbf{v}\| = 14,\ \alpha = 120°$

64. $\|\mathbf{v}\| = 3,\ \alpha = 240°$ **65.** $\|\mathbf{v}\| = 25,\ \alpha = 330°$ **66.** $\|\mathbf{v}\| = 15,\ \alpha = 315°$

*In Problems 67–74, find the direction angle of **v**.*

67. v = 3i + 3j   **68. v = i + $\sqrt{3}$j**   **69. v = −3$\sqrt{3}$i + 3j**   **70. v = −5i − 5j**

71. v = 4i − 2j   **72. v = 6i − 4j**   **73. v = −i − 5j**   **74. v = −i + 3j**

Applications and Extensions

75. Force Vectors A child pulls a wagon with a force of 40 pounds. The handle of the wagon makes an angle of 30° with the ground. Express the force vector **F** in terms of **i** and **j**.

76. Force Vectors A man pushes a wheelbarrow up an incline of 20° with a force of 100 pounds. Express the force vector **F** in terms of **i** and **j**.

77. Resultant Force Two forces of magnitude 40 newtons (N) and 60 N act on an object at angles of 30° and −45° with the positive *x*-axis, as shown in the figure. Find the direction and magnitude of the resultant force; that is, find $\mathbf{F}_1 + \mathbf{F}_2$.

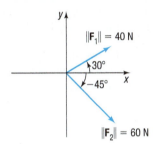

78. Resultant Force Two forces of magnitude 30 newtons (N) and 70 N act on an object at angles of 45° and 120° with the positive *x*-axis, as shown in the figure. Find the direction and magnitude of the resultant force; that is, find $\mathbf{F}_1 + \mathbf{F}_2$.

79. Finding the Actual Speed and Direction of an Aircraft A Boeing 787 Dreamliner maintains a constant airspeed of 550 miles per hour (mph) headed due north. The jet stream is 100 mph in the northeasterly direction.
  (a) Express the velocity $\mathbf{v}_a$ of the 787 relative to the air and the velocity $\mathbf{v}_w$ of the jet stream in terms of **i** and **j**.
  (b) Find the velocity of the 787 relative to the ground.
  (c) Find the actual speed and direction of the 787 relative to the ground.

80. Finding the Actual Speed and Direction of an Aircraft An Airbus A320 jet maintains a constant airspeed of 500 mph headed due west. The jet stream is 100 mph in the southeasterly direction.
  (a) Express the velocity $\mathbf{v}_a$ of the A320 relative to the air and the velocity $\mathbf{v}_w$ of the jet stream in terms of **i** and **j**.
  (b) Find the velocity of the A320 relative to the ground.
  (c) Find the actual speed and direction of the A320 relative to the ground.

81. Ground Speed and Direction of an Airplane An airplane has an airspeed of 500 kilometers per hour (km/h) bearing N45°E. The wind velocity is 60 km/h in the direction N30°W. Find the resultant vector representing the path of the plane relative to the ground. What is the groundspeed of the plane? What is its direction?

82. Ground Speed and Direction of an Airplane An airplane has an airspeed of 600 km/h bearing S30°E. The wind velocity is 40 km/h in the direction S45°E. Find the resultant vector representing the path of the plane relative to the ground. What is the groundspeed of the plane? What is its direction?

83. Weight of a Boat A force of magnitude 700 pounds is required to hold a boat and its trailer in place on a ramp whose incline is 10° to the horizontal. What is the combined weight of the boat and its trailer?

84. Weight of a Car A force of magnitude 1200 pounds is required to prevent a car from rolling down a hill whose incline is 15° to the horizontal. What is the weight of the car?

85. Correct Direction for Crossing a River A river has a constant current of 3 km/h. At what angle to a boat dock should a motorboat capable of maintaining a constant speed of 20 km/h be headed in order to reach a point directly opposite the dock? If the river is $\frac{1}{2}$ kilometer wide, how long will it take to cross?

86. Finding the Correct Compass Heading The pilot of an aircraft wishes to head directly east but is faced with a wind speed of 40 mph from the northwest. If the pilot maintains an airspeed of 250 mph, what compass heading should be maintained to head directly east? What is the actual speed of the aircraft?

87. Charting a Course A helicopter pilot needs to travel to a regional airport 25 miles away. She flies at an actual heading of N16.26°E with an airspeed of 120 mph, and there is a wind blowing directly east at 20 mph.
  (a) Determine the compass heading that the pilot needs to reach her destination.
  (b) How long will it take her to reach her destination? Round to the nearest minute.

88. Crossing a River A captain needs to pilot a boat across a river that is 2 km wide. The current in the river is 2 km/h and the speed of the boat in still water is 10 km/h. The desired landing point on the other side is 1 km upstream.
  (a) Find the direction that the captain should take.
  (b) How long will the trip take?

89. Static Equilibrium A weight of 1000 pounds is suspended from two cables, as shown in the figure. What are the tensions in the two cables?

90. Static Equilibrium A weight of 800 pounds is suspended from two cables, as shown in the figure. What are the tensions in the two cables?

91. Static Equilibrium A tightrope walker located at a certain point deflects the rope as indicated in the figure. If the weight of the tightrope walker is 150 pounds, how much tension is in each part of the rope?

92. Static Equilibrium Repeat Problem 91 if the angle on the left is 3.8°, the angle on the right is 2.6°, and the weight of the tightrope walker is 135 pounds.

93. Static Friction A 20-pound box sits at rest on a horizontal surface, and there is friction between the box and the surface. One side of the surface is raised slowly to create a ramp. The friction force **f** opposes the direction of motion and is proportional to the normal force $\mathbf{F_N}$ exerted by the surface on the box. The proportionality constant is called the **coefficient of friction**, μ. When the angle of the ramp, θ, reaches 20°, the box begins to slide. Find the value of μ to two decimal places.

94. Inclined Ramp A 2-pound weight is attached to a 3-pound weight by a rope that passes over an ideal pulley. The smaller weight hangs vertically, while the larger weight sits on a frictionless inclined ramp with angle θ. The rope exerts

a tension force **T** on both weights along the direction of the rope. Find the angle measure for θ that is needed to keep the larger weight from sliding down the ramp. Round your answer to the nearest tenth of a degree.

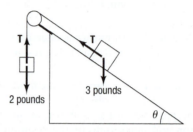

95. Inclined Ramp A box sitting on a horizontal surface is attached to a second box sitting on an inclined ramp by a rope that passes over an ideal pulley. The rope exerts a tension force **T** on both weights along the direction of the rope, and the coefficient of friction between the surface and boxes is 0.6 (see Problem 93). If the box on the right weighs 100 pounds and the angle of the ramp is 35°, how much must the box on the left weigh for the system to be in static equilibrium? Round your answer to two decimal places.

96. Muscle Force Two muscles exert force on a bone at the same point. The first muscle exerts a force of 800 N at a 10° angle with the bone. The second muscle exerts a force of 710 N at a 35° angle with the bone. What are the direction and magnitude of the resulting force on the bone?

97. Truck Pull At a county fair truck pull, two pickup trucks are attached to the back end of a monster truck as illustrated in the figure. One of the pickups pulls with a force of 2000 pounds, and the other pulls with a force of 3000 pounds. There is an angle of 45° between them. With how much force must the monster truck pull in order to remain unmoved?

[**Hint:** Find the resultant force of the two trucks.]

98. Removing a Stump A farmer wishes to remove a stump from a field by pulling it out with his tractor. Having removed many stumps before, he estimates that he will need 6 tons (12,000 pounds) of force to remove the stump. However, his tractor is only capable of pulling with a force of 7000 pounds, so he asks his neighbor to help. His neighbor's tractor can pull with a force of 5500 pounds. They attach the two tractors to the stump with a 40° angle between the forces, as shown in the figure.

(a) Assuming the farmer's estimate of a needed 6-ton force is correct, will the farmer be successful in removing the stump?

(b) Had the farmer arranged the tractors with a 25° angle between the forces, would he have been successful in removing the stump?

99. Computer Graphics The field of computer graphics utilizes vectors to compute translations of points. For example, if the point $(-3, 2)$, represented by vector $\mathbf{u} = \langle -3, 2 \rangle$, is to be translated by $\mathbf{v} = \langle 5, 2 \rangle$, then the new location will be $\mathbf{u}' = \mathbf{u} + \mathbf{v} = \langle -3, 2 \rangle + \langle 5, 2 \rangle = \langle 2, 4 \rangle$. So, the point $(-3, 2)$ is translated by $\mathbf{v}$ to $(2, 4)$ as shown in the figure.

(a) Determine the new coordinates of $(3, -1)$ if it is translated by $\mathbf{v} = \langle -4, 5 \rangle$.

(b) Illustrate this translation graphically.

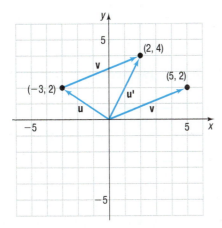

Source: Phil Dadd. Vectors and Matrices: A Primer. www .gamedev.net/articles/programming/math-and-physics/ vectors-and-matrices-a-primer-r1832/

100. Computer Graphics Refer to Problem 99. The points $(-3, 0)$, $(-1, -2)$, $(3, 1)$, and $(1, 3)$ are the vertices of a parallelogram $ABCD$.

(a) Find the vertices of a new parallelogram $A'B'C'D'$ if $ABCD$ is translated by $\mathbf{v} = \langle 3, -2 \rangle$.

(b) Find the vertices of a new parallelogram $A'B'C'D'$ if $ABCD$ is translated by $-\dfrac{1}{2}\mathbf{v}$.

101. Static Equilibrium Show on the following graph the force needed for the object at P to be in static equilibrium.

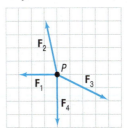

102. *Challenge Problem* Landscaping To drag a 500-pound boulder into place, Adam, Bill, and Chuck attach three ropes to the boulder as shown in the diagram. If Adam pulls with 240 pounds of force and Chuck pulls with 110 pounds of force, then Bill must pull with how much force in order for the boulder to move?

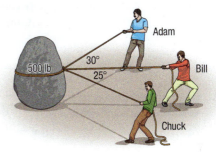

103. *Challenge Problem* See Problem 102. If Bill pulls due east with 200 pounds of force, then what direction does the boulder move?

Explaining Concepts: Discussion and Writing

104. Explain in your own words what a vector is. Give an example of a vector.

105. Write a brief paragraph comparing the algebra of complex numbers and the algebra of vectors.

106. Explain the difference between an algebraic vector and a position vector.

Retain Your Knowledge

Problems 107–116 are based on material learned earlier in the course. The purpose of these problems is to keep the material fresh in your mind so that you are better prepared for the final exam.

107. Solve: $\sqrt[3]{x - 2} = 3$

108. Factor $-3x^3 + 12x^2 + 36x$ completely.

109. Find the exact value of $\tan\left[\cos^{-1}\left(\dfrac{1}{2}\right)\right]$.

110. Find the amplitude, period, and phase shift of
$$y = \frac{3}{2}\cos(6x + 3\pi)$$
Graph the function, showing at least two periods.

111. Find the distance between the points $(-5, -8)$ and $(7, 1)$.

112. Write the equation of the circle in standard form:
$$x^2 + y^2 - 20x + 4y + 55 = 0$$

113. Find all the intercepts of the graph of
$$f(x) = x^3 + 2x^2 - 9x - 18$$

114. Solve: $4(x - 5)^2 + 9 = 53$

115. If $f(x) = x^4$, find $\dfrac{f(x) - f(3)}{x - 3}$.

116. If $f(\theta) = \sqrt{25 - \theta^2}$, show that $f(5\sin\theta) = 5\cos\theta$, $-\dfrac{\pi}{2} \le \theta \le \dfrac{\pi}{2}$.

5.5 The Dot Product

PREPARING FOR THIS SECTION *Before getting started, review the following:*

• Law of Cosines (Section 4.3, pp. 286–288)

Now Work the 'Are You Prepared?' problem on page 369.

OBJECTIVES   **1** Find the Dot Product of Two Vectors (p. 364)
                2 Find the Angle between Two Vectors (p. 365)
                3 Determine Whether Two Vectors Are Parallel (p. 366)
                4 Determine Whether Two Vectors Are Orthogonal (p. 366)
                5 Decompose a Vector into Two Orthogonal Vectors (p. 366)
                6 Compute Work (p. 368)

1 Find the Dot Product of Two Vectors

In Section 5.4, we defined the product of a scalar and a vector. Here we define the product of two vectors, called a *dot product*.

> **DEFINITION** **Dot Product**
>
> If $\mathbf{v} = a_1\mathbf{i} + b_1\mathbf{j}$ and $\mathbf{w} = a_2\mathbf{i} + b_2\mathbf{j}$ are two vectors, the **dot product v · w** is defined as
>
> $$\mathbf{v} \cdot \mathbf{w} = a_1 a_2 + b_1 b_2 \qquad (1)$$

EXAMPLE 1   **Finding Dot Products**

If $\mathbf{v} = 2\mathbf{i} - 3\mathbf{j}$ and $\mathbf{w} = 5\mathbf{i} + 3\mathbf{j}$, find:

(a) $\mathbf{v} \cdot \mathbf{w}$ (b) $\mathbf{w} \cdot \mathbf{v}$ (c) $\mathbf{v} \cdot \mathbf{v}$ (d) $\mathbf{w} \cdot \mathbf{w}$ (e) $\|\mathbf{v}\|$ (f) $\|\mathbf{w}\|$

Solution
(a) $\mathbf{v} \cdot \mathbf{w} = 2 \cdot 5 + (-3)3 = 1$ (b) $\mathbf{w} \cdot \mathbf{v} = 5 \cdot 2 + 3(-3) = 1$
(c) $\mathbf{v} \cdot \mathbf{v} = 2 \cdot 2 + (-3)(-3) = 13$ (d) $\mathbf{w} \cdot \mathbf{w} = 5 \cdot 5 + 3 \cdot 3 = 34$
(e) $\|\mathbf{v}\| = \sqrt{2^2 + (-3)^2} = \sqrt{13}$ (f) $\|\mathbf{w}\| = \sqrt{5^2 + 3^2} = \sqrt{34}$

COMMENT A scalar multiple $\alpha\mathbf{v}$ is a vector. A dot product $\mathbf{u} \cdot \mathbf{v}$ is a scalar (real number). ∎

Since the dot product $\mathbf{v} \cdot \mathbf{w}$ of two vectors $\mathbf{v}$ and $\mathbf{w}$ is a real number (a scalar), the dot product is sometimes referred to as the **scalar product**.

The results obtained in Example 1 suggest some general properties of the dot product.

> **THEOREM** **Properties of the Dot Product**
>
> If $\mathbf{u}$, $\mathbf{v}$, and $\mathbf{w}$ are vectors, then
>
> **Commutative Property**
>
> $$\mathbf{u} \cdot \mathbf{v} = \mathbf{v} \cdot \mathbf{u} \qquad (2)$$
>
> **Distributive Property**
>
> $$\mathbf{u} \cdot (\mathbf{v} + \mathbf{w}) = \mathbf{u} \cdot \mathbf{v} + \mathbf{u} \cdot \mathbf{w} \qquad (3)$$
>
> $$\mathbf{v} \cdot \mathbf{v} = \|\mathbf{v}\|^2 \qquad (4)$$
> $$\mathbf{0} \cdot \mathbf{v} = 0 \qquad (5)$$

Proof We prove properties (2) and (4) here and leave properties (3) and (5) as exercises (see Problems 38 and 39).

To prove property (2), let $\mathbf{u} = a_1\mathbf{i} + b_1\mathbf{j}$ and $\mathbf{v} = a_2\mathbf{i} + b_2\mathbf{j}$. Then

$$\mathbf{u} \cdot \mathbf{v} = a_1 a_2 + b_1 b_2 = a_2 a_1 + b_2 b_1 = \mathbf{v} \cdot \mathbf{u}$$

To prove property (4), let $\mathbf{v} = a\mathbf{i} + b\mathbf{j}$. Then

$$\mathbf{v} \cdot \mathbf{v} = a^2 + b^2 = \|\mathbf{v}\|^2$$

∎

2 Find the Angle between Two Vectors

One use of the dot product is to find the angle between two vectors.

Let $\mathbf{u}$ and $\mathbf{v}$ be two vectors with the same initial point A. Then the vectors $\mathbf{u}$, $\mathbf{v}$, and $\mathbf{u} - \mathbf{v}$ form a triangle. See Figure 64. The angle θ at vertex A of the triangle is the angle between the vectors $\mathbf{u}$ and $\mathbf{v}$. We wish to find a formula for calculating the angle θ.

The sides of the triangle have lengths $\|\mathbf{v}\|$, $\|\mathbf{u}\|$, and $\|\mathbf{u} - \mathbf{v}\|$, and θ is the included angle between the sides of length $\|\mathbf{v}\|$ and $\|\mathbf{u}\|$. The Law of Cosines (Section 4.3) can be used to find the cosine of the included angle.

$$\|\mathbf{u} - \mathbf{v}\|^2 = \|\mathbf{u}\|^2 + \|\mathbf{v}\|^2 - 2\|\mathbf{u}\|\|\mathbf{v}\| \cos\theta$$

Now use property (4) to rewrite this equation in terms of dot products.

$$(\mathbf{u} - \mathbf{v}) \cdot (\mathbf{u} - \mathbf{v}) = \mathbf{u} \cdot \mathbf{u} + \mathbf{v} \cdot \mathbf{v} - 2\|\mathbf{u}\|\|\mathbf{v}\| \cos\theta \qquad \textbf{(6)}$$

Then use the distributive property (3) twice on the left side of (6) to obtain

$$
\begin{aligned}
(\mathbf{u} - \mathbf{v}) \cdot (\mathbf{u} - \mathbf{v}) &= \mathbf{u} \cdot (\mathbf{u} - \mathbf{v}) - \mathbf{v} \cdot (\mathbf{u} - \mathbf{v}) \\
&= \mathbf{u} \cdot \mathbf{u} - \mathbf{u} \cdot \mathbf{v} - \mathbf{v} \cdot \mathbf{u} + \mathbf{v} \cdot \mathbf{v} \\
&= \mathbf{u} \cdot \mathbf{u} + \mathbf{v} \cdot \mathbf{v} - 2\,\mathbf{u} \cdot \mathbf{v} \qquad \textbf{(7)}
\end{aligned}
$$

↑
Commutative Property (2)

Combining equations (6) and (7) gives

$$\mathbf{u} \cdot \mathbf{u} + \mathbf{v} \cdot \mathbf{v} - 2\,\mathbf{u} \cdot \mathbf{v} = \mathbf{u} \cdot \mathbf{u} + \mathbf{v} \cdot \mathbf{v} - 2\|\mathbf{u}\|\|\mathbf{v}\| \cos\theta$$

$$\mathbf{u} \cdot \mathbf{v} = \|\mathbf{u}\|\|\mathbf{v}\| \cos\theta$$

> **THEOREM Angle between Vectors**
>
> If $\mathbf{u}$ and $\mathbf{v}$ are two nonzero vectors, the angle θ, $0 \le \theta \le \pi$, between $\mathbf{u}$ and $\mathbf{v}$ is determined by the formula
>
> $$\cos\theta = \frac{\mathbf{u} \cdot \mathbf{v}}{\|\mathbf{u}\|\,\|\mathbf{v}\|} \qquad \textbf{(8)}$$

u
u − v
θ
v
A

Figure 64

EXAMPLE 2

Finding the Angle θ between Two Vectors

Find the angle θ between $\mathbf{u} = 4\mathbf{i} - 3\mathbf{j}$ and $\mathbf{v} = 2\mathbf{i} + 5\mathbf{j}$.

Solution Find $\mathbf{u} \cdot \mathbf{v}$, $\|\mathbf{u}\|$, and $\|\mathbf{v}\|$.

$$\mathbf{u} \cdot \mathbf{v} = 4 \cdot 2 + (-3) \cdot 5 = -7$$
$$\|\mathbf{u}\| = \sqrt{4^2 + (-3)^2} = 5$$
$$\|\mathbf{v}\| = \sqrt{2^2 + 5^2} = \sqrt{29}$$

By formula (8), if θ is the angle between $\mathbf{u}$ and $\mathbf{v}$, then

$$\cos\theta = \frac{\mathbf{u} \cdot \mathbf{v}}{\|\mathbf{u}\|\,\|\mathbf{v}\|} = \frac{-7}{5\sqrt{29}} \approx -0.26$$

Therefore, $\theta \approx \cos^{-1}(-0.26) \approx 105°$. See Figure 65.

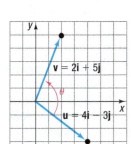

Figure 65 $\theta \approx 105°$

Now Work PROBLEM 9(a) AND (b)

3 Determine Whether Two Vectors Are Parallel

Two vectors **v** and **w** are said to be **parallel** if there is a nonzero scalar α so that $\mathbf{v} = \alpha\mathbf{w}$. In this case, the angle θ between **v** and **w** is 0 or π.

> **EXAMPLE 3**

Determining Whether Two Vectors Are Parallel

The vectors $\mathbf{v} = 3\mathbf{i} - \mathbf{j}$ and $\mathbf{w} = 6\mathbf{i} - 2\mathbf{j}$ are parallel, since $\mathbf{v} = \dfrac{1}{2}\mathbf{w}$. Furthermore, since

$$\cos\theta = \frac{\mathbf{v}\cdot\mathbf{w}}{\|\mathbf{v}\|\,\|\mathbf{w}\|} = \frac{18 + 2}{\sqrt{10}\,\sqrt{40}} = \frac{20}{\sqrt{400}} = 1$$

the angle θ between **v** and **w** is 0.

4 Determine Whether Two Vectors Are Orthogonal

If the angle θ between two nonzero vectors **v** and **w** is $\dfrac{\pi}{2}$, the vectors **v** and **w** are called **orthogonal**. See Figure 66.

Figure 66 **v** is orthogonal to **w**.

Since $\cos\dfrac{\pi}{2} = 0$, it follows from formula (8) that if the vectors **v** and **w** are orthogonal, then $\mathbf{v}\cdot\mathbf{w} = 0$.

On the other hand, if $\mathbf{v}\cdot\mathbf{w} = 0$, then $\mathbf{v} = \mathbf{0}$ or $\mathbf{w} = \mathbf{0}$ or $\cos\theta = 0$. If $\cos\theta = 0$, then $\theta = \dfrac{\pi}{2}$, and **v** and **w** are orthogonal. If **v** or **w** is the zero vector, then, since the zero vector has no specific direction, we adopt the convention that the zero vector is orthogonal to every vector.

NOTE *Orthogonal, perpendicular, and normal are all terms that mean "meet at a right angle." It is customary to refer to two vectors as being* orthogonal, *to two lines as being* perpendicular, *and to a line and a plane or a vector and a plane as being* normal. ∎

> **THEOREM**
>
> Two vectors **v** and **w** are orthogonal if and only if
>
> $$\mathbf{v}\cdot\mathbf{w} = 0$$

> **EXAMPLE 4**

Determining Whether Two Vectors Are Orthogonal

The vectors

$$\mathbf{v} = 2\mathbf{i} - \mathbf{j} \quad \text{and} \quad \mathbf{w} = 3\mathbf{i} + 6\mathbf{j}$$

are orthogonal, since

$$\mathbf{v}\cdot\mathbf{w} = 6 - 6 = 0$$

See Figure 67.

Figure 67

Now Work PROBLEM 9(c)

5 Decompose a Vector into Two Orthogonal Vectors

Figure 68

In many physical applications, it is necessary to find "how much" of a vector is applied in a given direction. Look at Figure 68. The force **F** due to gravity is pulling straight down (toward the center of Earth) on the block. To study the effect of gravity on the block, it is necessary to determine how much of **F** is actually pushing the block down the incline ($\mathbf{F}_1$) and how much is pressing the block against the incline ($\mathbf{F}_2$), at a right angle to the incline. Knowing the **decomposition** of **F** often enables us to determine when friction (the force holding the block in place on the incline) is overcome and the block will slide down the incline.

Figure 69

Suppose that **v** and **w** are two nonzero, nonorthogonal vectors with the same initial point P. We seek to decompose **v** into two vectors: $\mathbf{v}_1$, which is parallel to **w**, and $\mathbf{v}_2$, which is orthogonal to **w**. See Figure 69(a) and (b). The vector $\mathbf{v}_1$ is called the **vector projection of v onto w**.

The vector $\mathbf{v}_1$ is obtained by dropping a perpendicular from the terminal point of **v** to the line containing **w**. The vector $\mathbf{v}_1$ is the vector from P to the intersection of the line containing **w** and the perpendicular. The vector $\mathbf{v}_2$ is given by $\mathbf{v}_2 = \mathbf{v} - \mathbf{v}_1$. Note that $\mathbf{v} = \mathbf{v}_1 + \mathbf{v}_2$, the vector $\mathbf{v}_1$ is parallel to **w**, and the vector $\mathbf{v}_2$ is orthogonal to **w**. This is the decomposition of **v** that was sought.

Now we seek a formula for $\mathbf{v}_1$ that is based on a knowledge of the vectors **v** and **w**. Since $\mathbf{v} = \mathbf{v}_1 + \mathbf{v}_2$, we have

$$\mathbf{v} \cdot \mathbf{w} = (\mathbf{v}_1 + \mathbf{v}_2) \cdot \mathbf{w} = \mathbf{v}_1 \cdot \mathbf{w} + \mathbf{v}_2 \cdot \mathbf{w} \qquad (9)$$

Since $\mathbf{v}_2$ is orthogonal to **w**, we have $\mathbf{v}_2 \cdot \mathbf{w} = 0$. Since $\mathbf{v}_1$ is parallel to **w**, we have $\mathbf{v}_1 = \alpha \mathbf{w}$ for some scalar α. Equation (9) can be written as

$$\mathbf{v} \cdot \mathbf{w} = \alpha \mathbf{w} \cdot \mathbf{w} = \alpha \|\mathbf{w}\|^2 \qquad \mathbf{v}_1 = \alpha \mathbf{w}; \mathbf{v}_2 \cdot \mathbf{w} = 0$$

$$\alpha = \frac{\mathbf{v} \cdot \mathbf{w}}{\|\mathbf{w}\|^2}$$

Then

$$\mathbf{v}_1 = \alpha \mathbf{w} = \frac{\mathbf{v} \cdot \mathbf{w}}{\|\mathbf{w}\|^2} \mathbf{w}$$

> **THEOREM**
>
> If **v** and **w** are two nonzero, nonorthogonal vectors, the vector projection of **v** onto **w** is
>
> $$\mathbf{v}_1 = \frac{\mathbf{v} \cdot \mathbf{w}}{\|\mathbf{w}\|^2} \mathbf{w} \qquad (10)$$
>
> The decomposition of **v** into $\mathbf{v}_1$ and $\mathbf{v}_2$, where $\mathbf{v}_1$ is parallel to **w**, and $\mathbf{v}_2$ is orthogonal to **w**, is
>
> $$\mathbf{v}_1 = \frac{\mathbf{v} \cdot \mathbf{w}}{\|\mathbf{w}\|^2} \mathbf{w} \qquad \mathbf{v}_2 = \mathbf{v} - \mathbf{v}_1 \qquad (11)$$

EXAMPLE 5 **Decomposing a Vector into Two Orthogonal Vectors**

Find the vector projection of $\mathbf{v} = \mathbf{i} + 3\mathbf{j}$ onto $\mathbf{w} = \mathbf{i} + \mathbf{j}$. Decompose **v** into two vectors, $\mathbf{v}_1$ and $\mathbf{v}_2$, where $\mathbf{v}_1$ is parallel to **w**, and $\mathbf{v}_2$ is orthogonal to **w**.

Solution Use formulas (10) and (11).

$$\mathbf{v}_1 = \frac{\mathbf{v} \cdot \mathbf{w}}{\|\mathbf{w}\|^2} \mathbf{w} = \frac{1 + 3}{(\sqrt{2})^2} \mathbf{w} = 2\mathbf{w} = 2(\mathbf{i} + \mathbf{j})$$

$$\mathbf{v}_2 = \mathbf{v} - \mathbf{v}_1 = (\mathbf{i} + 3\mathbf{j}) - 2(\mathbf{i} + \mathbf{j}) = -\mathbf{i} + \mathbf{j}$$

See Figure 70.

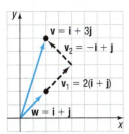

Figure 70

Now Work PROBLEM 21

EXAMPLE 6 **Finding the Force Required to Hold a Wagon on a Hill**

A wagon with two small children as occupants weighs 100 pounds and is on a hill with a grade of $20°$. What is the magnitude of the force that is required to keep the wagon from rolling down the hill?

Solution

Figure 71

See Figure 71. We need to find the magnitude of the force **v** that would cause the wagon to roll down the hill. A force with the same magnitude in the opposite direction of **v** will keep the wagon from rolling down the hill. The force of gravity is orthogonal to the level ground, so the force of the wagon due to gravity can be represented by the vector

$$\mathbf{F_g} = -100\mathbf{j}$$

Determine the vector projection of $\mathbf{F_g}$ onto **w**, which is the force parallel to the hill. The vector **w** is given by

$$\mathbf{w} = \cos 20°\mathbf{i} + \sin 20°\mathbf{j}$$

The vector projection of $\mathbf{F_g}$ onto **w** is

$$\mathbf{v} = \frac{\mathbf{F_g} \cdot \mathbf{w}}{\|\mathbf{w}\|^2}\,\mathbf{w}$$

$$= \frac{-100(\sin 20°)}{\left(\sqrt{\cos^2 20° + \sin^2 20°}\right)^2}\,(\cos 20°\mathbf{i} + \sin 20°\mathbf{j})$$

$$= -34.2\,(\cos 20°\mathbf{i} + \sin 20°\mathbf{j})$$

The magnitude of **v** is 34.2 pounds, so the magnitude of the force required to keep the wagon from rolling down the hill is 34.2 pounds.

6 Compute Work

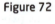

In elementary physics, the **work** W done by a constant force **F** in moving an object from a point A to a point B is defined as

$$W = (\text{magnitude of force})\,(\text{distance}) = \|\mathbf{F}\|\|\overrightarrow{AB}\|$$

Figure 72

Work is commonly measured in foot-pounds or in newton-meters (joules).

In this definition, it is assumed that the force **F** is applied along the line of motion. If the constant force **F** is not along the line of motion, but instead is at an angle θ to the direction of the motion, as illustrated in Figure 72, then the **work** W **done by F** in moving an object from A to B is defined as

$$W = \mathbf{F} \cdot \overrightarrow{AB} \qquad (12)$$

This definition is compatible with the force-times-distance definition, since

$$W = (\text{amount of force in the direction of } \overrightarrow{AB})\,(\text{distance})$$

$$= \|\text{projection of } \mathbf{F} \text{ on } AB\|\|\overrightarrow{AB}\| = \underset{\uparrow}{\frac{\mathbf{F} \cdot \overrightarrow{AB}}{\|\overrightarrow{AB}\|^2}}\|\overrightarrow{AB}\|\|\overrightarrow{AB}\| = \mathbf{F} \cdot \overrightarrow{AB}$$

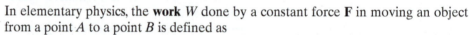

Use formula (10)

EXAMPLE 7   **Computing Work**

A girl is pulling a wagon with a force of 50 pounds. How much work is done in moving the wagon 100 feet if the handle makes an angle of 30° with the ground? See Figure 73(a).

Figure 73   **(a)**   **(b)**

Solution　Position the vectors in a coordinate system in such a way that the wagon is moved from $(0, 0)$ to $(100, 0)$. The motion is from $A = (0, 0)$ to $B = (100, 0)$, so $\overrightarrow{AB} = 100\mathbf{i}$. The force vector $\mathbf{F}$, as shown in Figure 73(b), is

$$\mathbf{F} = 50(\cos 30°\mathbf{i} + \sin 30°\mathbf{j}) = 50\left(\frac{\sqrt{3}}{2}\mathbf{i} + \frac{1}{2}\mathbf{j}\right) = 25(\sqrt{3}\,\mathbf{i} + \mathbf{j})$$

By formula (12), the work done is

$$W = \mathbf{F} \cdot \overrightarrow{AB} = 25(\sqrt{3}\,\mathbf{i} + \mathbf{j}) \cdot 100\mathbf{i} = 2500\sqrt{3} \text{ foot-pounds}$$

Now Work PROBLEM 29

Historical Feature

We stated in the Historical Feature in Section 5.4 that complex numbers were used as vectors in the plane before the general notion of a vector was clarified. Suppose that we make the correspondence

Vector $\longleftrightarrow$ Complex number

$a\mathbf{i} + b\mathbf{j} \longleftrightarrow a + bi$

$c\mathbf{i} + d\mathbf{j} \longleftrightarrow c + di$

Show that

$$(a\mathbf{i} + b\mathbf{j}) \cdot (c\mathbf{i} + d\mathbf{j}) = \text{real part } [(\overline{a + bi})(c + di)]$$

This is how the dot product was found originally. The imaginary part is also interesting. It is a determinant and represents the area of the parallelogram whose edges are the vectors. See Section 5.7. This is close to some of Hermann Grassmann's ideas and is also connected with the scalar triple product of three-dimensional vectors.

5.5 Assess Your Understanding

'Are You Prepared?' *The answer is given at the end of these exercises. If you get a wrong answer, read the pages listed in* red.

1. In a triangle with sides a, b, c and angles A, B, C, the Law of Cosines states that _____ . (p. 286)

Concepts and Vocabulary

2. If $\mathbf{v} = a_1\mathbf{i} + b_1\mathbf{j}$ and $\mathbf{w} = a_2\mathbf{i} + b_2\mathbf{j}$ are two vectors, then the _____ _____ is defined as $\mathbf{v} \cdot \mathbf{w} = a_1a_2 + b_1b_2$.

3. If $\mathbf{v} \cdot \mathbf{w} = 0$, then the two vectors $\mathbf{v}$ and $\mathbf{w}$ are _____.

4. If $\mathbf{v} = 3\mathbf{w}$, then the two vectors $\mathbf{v}$ and $\mathbf{w}$ are _____ .

5. *True or False* Given two nonzero, nonorthogonal vectors $\mathbf{v}$ and $\mathbf{w}$, it is always possible to decompose $\mathbf{v}$ into two vectors, one parallel to $\mathbf{w}$ and the other orthogonal to $\mathbf{w}$.

6. *True or False* Work is a physical example of a vector.

7. *Multiple Choice* The angle $\theta, 0 \leq \theta \leq \pi$, between two nonzero vectors $\mathbf{u}$ and $\mathbf{v}$ can be found using what formula?

(a) $\sin\theta = \dfrac{\|\mathbf{u}\|}{\|\mathbf{v}\|}$　　(b) $\cos\theta = \dfrac{\|\mathbf{u}\|}{\|\mathbf{v}\|}$

(c) $\sin\theta = \dfrac{\mathbf{u} \cdot \mathbf{v}}{\|\mathbf{u}\| \|\mathbf{v}\|}$　　(d) $\cos\theta = \dfrac{\mathbf{u} \cdot \mathbf{v}}{\|\mathbf{u}\| \|\mathbf{v}\|}$

8. *Multiple Choice* If two nonzero vectors $\mathbf{v}$ and $\mathbf{w}$ are orthogonal, then the angle between them has what measure?

(a) π　　(b) $\dfrac{\pi}{2}$　　(c) $\dfrac{3\pi}{2}$　　(d) 2π

Skill Building

In Problems 9–18, (a) find the dot product $\mathbf{v} \cdot \mathbf{w}$; *(b) find the angle between* $\mathbf{v}$ *and* $\mathbf{w}$; *(c) state whether the vectors are parallel, orthogonal, or neither.*

9. $\mathbf{v} = \mathbf{i} - \mathbf{j}$, $\mathbf{w} = \mathbf{i} + \mathbf{j}$

10. $\mathbf{v} = \mathbf{i} + \mathbf{j}$, $\mathbf{w} = -\mathbf{i} + \mathbf{j}$

11. $\mathbf{v} = 2\mathbf{i} + \mathbf{j}$, $\mathbf{w} = \mathbf{i} - 2\mathbf{j}$

12. $\mathbf{v} = 2\mathbf{i} + 2\mathbf{j}$, $\mathbf{w} = \mathbf{i} + 2\mathbf{j}$

13. $\mathbf{v} = \sqrt{3}\mathbf{i} - \mathbf{j}$, $\mathbf{w} = \mathbf{i} + \mathbf{j}$

14. $\mathbf{v} = \mathbf{i} + \sqrt{3}\mathbf{j}$, $\mathbf{w} = \mathbf{i} - \mathbf{j}$

15. $\mathbf{v} = 3\mathbf{i} + 4\mathbf{j}$, $\mathbf{w} = -6\mathbf{i} - 8\mathbf{j}$

16. $\mathbf{v} = 3\mathbf{i} - 4\mathbf{j}$, $\mathbf{w} = 9\mathbf{i} - 12\mathbf{j}$

17. $\mathbf{v} = 4\mathbf{i}$, $\mathbf{w} = \mathbf{j}$

18. $\mathbf{v} = \mathbf{i}$, $\mathbf{w} = -3\mathbf{j}$

19. Find a so that the vectors $\mathbf{v} = \mathbf{i} - a\mathbf{j}$ and $\mathbf{w} = 2\mathbf{i} + 3\mathbf{j}$ are orthogonal.

20. Find b so that the vectors $\mathbf{v} = \mathbf{i} + \mathbf{j}$ and $\mathbf{w} = \mathbf{i} + b\mathbf{j}$ are orthogonal.

In Problems 21–26, decompose $\mathbf{v}$ *into two vectors* $\mathbf{v}_1$ *and* $\mathbf{v}_2$, *where* $\mathbf{v}_1$ *is parallel to* $\mathbf{w}$, *and* $\mathbf{v}_2$ *is orthogonal to* $\mathbf{w}$.

21. $\mathbf{v} = 2\mathbf{i} - 3\mathbf{j}$, $\mathbf{w} = \mathbf{i} - \mathbf{j}$

22. $\mathbf{v} = -3\mathbf{i} + 2\mathbf{j}$, $\mathbf{w} = 2\mathbf{i} + \mathbf{j}$

23. $\mathbf{v} = \mathbf{i} - \mathbf{j}$, $\mathbf{w} = -\mathbf{i} - 2\mathbf{j}$

24. $\mathbf{v} = 2\mathbf{i} - \mathbf{j}$, $\mathbf{w} = \mathbf{i} - 2\mathbf{j}$

25. $\mathbf{v} = 3\mathbf{i} + \mathbf{j}$, $\mathbf{w} = -2\mathbf{i} - \mathbf{j}$

26. $\mathbf{v} = \mathbf{i} - 3\mathbf{j}$, $\mathbf{w} = 4\mathbf{i} - \mathbf{j}$

Applications and Extensions

27. Find a vector of magnitude 15 that is parallel to $4\mathbf{i} - 3\mathbf{j}$.

28. Find a vector of magnitude 5 that is parallel to $-12\mathbf{i} + 9\mathbf{j}$.

29. Computing Work Find the work done by a force of 3 pounds acting in a direction of 60° to the horizontal in moving an object 6 feet from $(0, 0)$ to $(6, 0)$.

30. Computing Work A wagon is pulled horizontally by exerting a force of 20 pounds on the handle at an angle of 30° with the horizontal. How much work is done in moving the wagon 100 feet?

31. Solar Energy The amount of energy collected by a solar panel depends on the intensity of the sun's rays and the area of the panel. Let the vector $\mathbf{I}$ represent the intensity, in watts per square centimeter, having the direction of the sun's rays. Let the vector $\mathbf{A}$ represent the area, in square centimeters, whose direction is the orientation of a solar panel. See the figure. The total number of watts collected by the panel is given by $W = |\mathbf{I} \cdot \mathbf{A}|$.

Suppose that $\mathbf{I} = \langle -0.02, -0.01 \rangle$ and $\mathbf{A} = \langle 300, 400 \rangle$.
(a) Find $\|\mathbf{I}\|$ and $\|\mathbf{A}\|$, and interpret the meaning of each.
(b) Compute W and interpret its meaning.
(c) If the solar panel is to collect the maximum number of watts, what must be true about $\mathbf{I}$ and $\mathbf{A}$?

32. Rainfall Measurement Let the vector $\mathbf{R}$ represent the amount of rainfall, in inches, whose direction is the inclination of the rain to a rain gauge. Let the vector $\mathbf{A}$ represent the area, in square inches, whose direction is the orientation of the opening of the rain gauge. See the figure. The volume of rain collected in the gauge, in cubic inches, is given by $V = |\mathbf{R} \cdot \mathbf{A}|$, even when the rain falls in a slanted direction or the gauge is not perfectly vertical.

Suppose that $\mathbf{R} = \langle 0.75, -1.75 \rangle$ and $\mathbf{A} = \langle 0.3, 1 \rangle$.
(a) Find $\|\mathbf{R}\|$ and $\|\mathbf{A}\|$, and interpret the meaning of each.
(b) Compute V and interpret its meaning.
(c) If the gauge is to collect the maximum volume of rain, what must be true about $\mathbf{R}$ and $\mathbf{A}$?

33. Braking Load A Toyota Sienna with a gross weight of 5300 pounds is parked on a street with an 8° grade. See the figure. Find the magnitude of the force required to keep the Sienna from rolling down the hill. What is the magnitude of the force perpendicular to the hill?

Weight = 5300 pounds

34. Braking Load A Chevrolet Silverado with a gross weight of 4500 pounds is parked on a street with a 10° grade. Find the magnitude of the force required to keep the Silverado from rolling down the hill. What is the magnitude of the force perpendicular to the hill?

35. Ramp Angle Billy and Timmy are using a ramp to load furniture into a truck. While rolling a 250-pound piano up the ramp, they discover that the truck is too full of other furniture for the piano to fit. Timmy holds the piano in place on the ramp while Billy repositions other items to make room for it in the truck. If the angle of inclination of the ramp is 20°, how many pounds of force must Timmy exert to hold the piano in position?

250 lb

36. Incline Angle A bulldozer exerts 1000 pounds of force to prevent a 5000-pound boulder from rolling down a hill. Determine the angle of inclination of the hill.

37. Find the acute angle that a constant unit force vector makes with the positive x-axis if the work done by the force in moving a particle from $(0, 0)$ to $(4, 0)$ equals 2.

38. Prove the distributive property:

$$\mathbf{u} \cdot (\mathbf{v} + \mathbf{w}) = \mathbf{u} \cdot \mathbf{v} + \mathbf{u} \cdot \mathbf{w}$$

39. Prove property (5): $\mathbf{0} \cdot \mathbf{v} = 0$.

40. If $\mathbf{v}$ is a unit vector and the angle between $\mathbf{v}$ and $\mathbf{i}$ is α, show that $\mathbf{v} = \cos \alpha \mathbf{i} + \sin \alpha \mathbf{j}$.

41. Suppose that $\mathbf{v}$ and $\mathbf{w}$ are unit vectors. If the angle between $\mathbf{v}$ and $\mathbf{i}$ is α and the angle between $\mathbf{w}$ and $\mathbf{i}$ is β, use the idea of the dot product $\mathbf{v} \cdot \mathbf{w}$ to prove that

$$\cos(\alpha - \beta) = \cos \alpha \cos \beta + \sin \alpha \sin \beta$$

42. Show that the projection of $\mathbf{v}$ onto $\mathbf{i}$ is $(\mathbf{v} \cdot \mathbf{i})\mathbf{i}$. Then show that we can always write a vector $\mathbf{v}$ as

$$\mathbf{v} = (\mathbf{v} \cdot \mathbf{i})\mathbf{i} + (\mathbf{v} \cdot \mathbf{j})\mathbf{j}$$

43. Let **v** and **w** denote two nonzero vectors. Show that the vectors $\|\mathbf{w}\|\mathbf{v} + \|\mathbf{v}\|\mathbf{w}$ and $\|\mathbf{w}\|\mathbf{v} - \|\mathbf{v}\|\mathbf{w}$ are orthogonal.

44. Let **v** and **w** denote two nonzero vectors. Show that the vector $\mathbf{v} - \alpha\mathbf{w}$ is orthogonal to **w** if $\alpha = \dfrac{\mathbf{v} \cdot \mathbf{w}}{\|\mathbf{w}\|^2}$.

45. Given vectors $\mathbf{u} = \mathbf{i} + 5\mathbf{j}$ and $\mathbf{v} = 4\mathbf{i} + y\mathbf{j}$, find y so that the angle between the vectors is $60°$.[†]

46. Given vectors $\mathbf{u} = x\mathbf{i} + 2\mathbf{j}$ and $\mathbf{v} = 7\mathbf{i} - 3\mathbf{j}$, find x so that the angle between the vectors is $30°$.

47. Given vectors $\mathbf{u} = 2x\mathbf{i} + 3\mathbf{j}$ and $\mathbf{v} = x\mathbf{i} - 8\mathbf{j}$, find x so that **u** and **v** are orthogonal.

48. In the definition of work given in this section, what is the work done if **F** is orthogonal to $\overrightarrow{AB}$?

49. *Challenge Problem*
(a) If **u** and **v** have the same magnitude, show that $\mathbf{u} + \mathbf{v}$ and $\mathbf{u} - \mathbf{v}$ are orthogonal.
(b) Use this to prove that an angle inscribed in a semicircle is a right angle (see the figure).

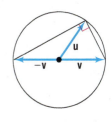

50. *Challenge Problem* Prove the **polarization identity**,

$$\|\mathbf{u} + \mathbf{v}\|^2 - \|\mathbf{u} - \mathbf{v}\|^2 = 4(\mathbf{u} \cdot \mathbf{v})$$

[†]Courtesy of the Joliet Junior College Mathematics Department

Explaining Concepts: Discussion and Writing

51. Create an application (different from any found in the text) that requires a dot product.

Retain Your Knowledge

Problems 52–60 are based on material learned earlier in the course. The purpose of these problems is to keep the material fresh in your mind so that you are better prepared for the final exam.

52. Find the average rate of change of $f(x) = x^3 - 5x^2 + 27$ from -3 to 2.

53. Find the exact value of $5\cos 60° + 2\tan\dfrac{\pi}{4}$. Do not use a calculator.

54. Establish the identity: $(1 - \sin^2\theta)(1 + \tan^2\theta) = 1$

55. Volume of a Box An open-top box is made from a sheet of metal by cutting squares from each corner and folding up the sides. The sheet has a length of 19 inches and a width of 13 inches. If x is the length of one side of each square to be cut out, write a function, $V(x)$, for the volume of the box in terms of x.

56. What is the function that is graphed after the graph of $y = \sqrt[3]{x}$ is shifted left 4 units and up 9 units?

57. Find all asymptotes of the graph of $f(x) = \dfrac{2x^2 - 5}{x^2 - 2x - 15}$.

58. Find the exact value of $\cos 80° \cos 70° - \sin 80° \sin 70°$.

59. Find the vertex and determine if the graph of $f(x) = \dfrac{2}{3}x^2 - 12x + 10$ is concave up or concave down.

60. If $f(x) = \dfrac{1}{(x^2 + 9)^{3/2}}$, show that
$$f(3\tan x) = \dfrac{1}{27\,|\sec^3 x|}$$

'Are You Prepared?' Answers

1. $c^2 = a^2 + b^2 - 2ab\cos C$

5.6 Vectors in Space

PREPARING FOR THIS SECTION *Before getting started, review the following:*

• Distance Formula (Section 1.1, pp. 3–4)

Now Work the 'Are You Prepared?' problem on page 379.

OBJECTIVES **1** Find the Distance between Two Points in Space (p. 372)
2 Find Position Vectors in Space (p. 373)
3 Perform Operations on Vectors (p. 374)
4 Find the Dot Product (p. 375)
5 Find the Angle between Two Vectors (p. 376)
6 Find the Direction Angles of a Vector (p. 377)

Rectangular Coordinates in Space

In the plane, each point is associated with an ordered pair of real numbers. In space, each point is associated with an ordered triple of real numbers. Through a fixed point, called the **origin** O, draw three mutually perpendicular lines: the x-axis, the y-axis, and the z-axis. On each of these axes, select an appropriate scale and the positive direction. See Figure 74.

The direction chosen for the positive z-axis in Figure 74 makes the system *right-handed*. This conforms to the *right-hand rule*, which states that if the index finger of the right hand points in the direction of the positive x-axis and the middle finger points in the direction of the positive y-axis, then the thumb will point in the direction of the positive z-axis. See Figure 75.

Figure 74

Figure 75

Associate with each point P an ordered triple (x, y, z) of real numbers, the **coordinates of P**. For example, the point $(2, 3, 4)$ is located by starting at the origin and moving 2 units along the positive x-axis, 3 units in the direction of the positive y-axis, and 4 units in the direction of the positive z-axis. See Figure 76.

Figure 76 also shows the location of the points $(2, 0, 0)$, $(0, 3, 0)$, $(0, 0, 4)$, and $(2, 3, 0)$. Points of the form $(x, 0, 0)$ lie on the x-axis, and points of the forms $(0, y, 0)$ and $(0, 0, z)$ lie on the y-axis and z-axis, respectively. Points of the form $(x, y, 0)$ lie in a plane called the **xy-plane**. Its equation is $z = 0$. Similarly, points of the form $(x, 0, z)$ lie in the **xz-plane** (equation $y = 0$), and points of the form $(0, y, z)$ lie in the **yz-plane** (equation $x = 0$). See Figure 77(a). By extension of these ideas, all points obeying the equation $z = 3$ will lie in a plane parallel to and 3 units above the xy-plane. The equation $y = 4$ represents a plane parallel to the xz-plane and 4 units to the right of the plane $y = 0$. See Figure 77(b).

Figure 76

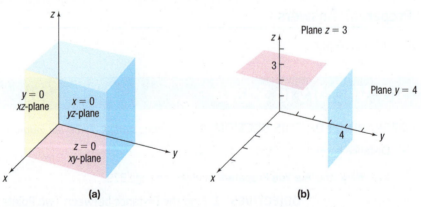

(a)   (b)

Figure 77

─── **Now Work** PROBLEM 9

1 Find the Distance between Two Points in Space

The formula for the distance between two points in space is an extension of the Distance Formula for points in the plane given in Section 1.1.

> **THEOREM Distance Formula in Space**
>
> If $P_1 = (x_1, y_1, z_1)$ and $P_2 = (x_2, y_2, z_2)$ are two points in space, the distance d from P_1 to P_2 is
>
> $$d = \sqrt{(x_2 - x_1)^2 + (y_2 - y_1)^2 + (z_2 - z_1)^2} \qquad \textbf{(1)}$$

The proof, which we omit, utilizes a double application of the Pythagorean Theorem.

EXAMPLE 1 **Using the Distance Formula**

Find the distance from $P_1 = (-1, 3, 2)$ to $P_2 = (4, -2, 5)$.

Solution $d = \sqrt{[4 - (-1)]^2 + [-2 - 3]^2 + [5 - 2]^2} = \sqrt{25 + 25 + 9} = \sqrt{59}$

　　　　　　 Now Work PROBLEM 15

2 Find Position Vectors in Space

To represent vectors in space, we introduce the unit vectors $\mathbf{i}, \mathbf{j}$, and $\mathbf{k}$ whose directions are along the positive x-axis, the positive y-axis, and the positive z-axis, respectively. If $\mathbf{v}$ is a vector with initial point at the origin O and terminal point at $P = (a, b, c)$, then we can represent $\mathbf{v}$ in terms of the vectors $\mathbf{i}, \mathbf{j}$, and $\mathbf{k}$ as

$$\mathbf{v} = a\mathbf{i} + b\mathbf{j} + c\mathbf{k}$$

Figure 78

See Figure 78.

The scalars a, b, and c are called the **components** of the vector $\mathbf{v} = a\mathbf{i} + b\mathbf{j} + c\mathbf{k}$, with a being the component in the direction $\mathbf{i}$, b the component in the direction $\mathbf{j}$, and c the component in the direction $\mathbf{k}$.

A vector whose initial point is at the origin is called a **position vector**. The next result states that any vector whose initial point is not at the origin is equal to a unique position vector.

> **THEOREM**
>
> Suppose that $\mathbf{v}$ is a vector with initial point $P_1 = (x_1, y_1, z_1)$, not necessarily the origin, and terminal point $P_2 = (x_2, y_2, z_2)$. If $\mathbf{v} = \overrightarrow{P_1 P_2}$, then $\mathbf{v}$ is equal to the position vector
>
> $$\mathbf{v} = (x_2 - x_1)\mathbf{i} + (y_2 - y_1)\mathbf{j} + (z_2 - z_1)\mathbf{k} \qquad \textbf{(2)}$$

Figure 79 illustrates this result.

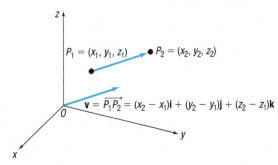

Figure 79

EXAMPLE 2

Finding a Position Vector

Find the position vector of the vector $\mathbf{v} = \overrightarrow{P_1P_2}$ if $P_1 = (-1, 2, 3)$ and $P_2 = (4, 6, 2)$.

Solution

By equation (2), the position vector equal to $\mathbf{v}$ is

$$\mathbf{v} = [4 - (-1)]\mathbf{i} + (6 - 2)\mathbf{j} + (2 - 3)\mathbf{k} = 5\mathbf{i} + 4\mathbf{j} - \mathbf{k}$$

Now Work PROBLEM 29

3 Perform Operations on Vectors

Equality, addition, subtraction, scalar product, and magnitude can be defined in terms of the components of a vector.

> **DEFINITION**
>
> Let $\mathbf{v} = a_1\mathbf{i} + b_1\mathbf{j} + c_1\mathbf{k}$ and $\mathbf{w} = a_2\mathbf{i} + b_2\mathbf{j} + c_2\mathbf{k}$ be two vectors, and let α be a scalar. Then
>
> - $\mathbf{v} = \mathbf{w}$ if and only if $a_1 = a_2, b_1 = b_2$, and $c_1 = c_2$
> - $\mathbf{v} + \mathbf{w} = (a_1 + a_2)\mathbf{i} + (b_1 + b_2)\mathbf{j} + (c_1 + c_2)\mathbf{k}$
> - $\mathbf{v} - \mathbf{w} = (a_1 - a_2)\mathbf{i} + (b_1 - b_2)\mathbf{j} + (c_1 - c_2)\mathbf{k}$
> - $\alpha\mathbf{v} = (\alpha a_1)\mathbf{i} + (\alpha b_1)\mathbf{j} + (\alpha c_1)\mathbf{k}$
> - $\|\mathbf{v}\| = \sqrt{a_1^2 + b_1^2 + c_1^2}$

These definitions are compatible with the geometric definitions given in Section 5.4 for vectors in a plane.

EXAMPLE 3

Adding and Subtracting Vectors

If $\mathbf{v} = 2\mathbf{i} + 3\mathbf{j} - 2\mathbf{k}$ and $\mathbf{w} = 3\mathbf{i} - 4\mathbf{j} + 5\mathbf{k}$, find:

(a) $\mathbf{v} + \mathbf{w}$ (b) $\mathbf{v} - \mathbf{w}$

Solution

(a) $\mathbf{v} + \mathbf{w} = (2\mathbf{i} + 3\mathbf{j} - 2\mathbf{k}) + (3\mathbf{i} - 4\mathbf{j} + 5\mathbf{k})$
$\qquad\quad = (2 + 3)\mathbf{i} + (3 - 4)\mathbf{j} + (-2 + 5)\mathbf{k}$
$\qquad\quad = 5\mathbf{i} - \mathbf{j} + 3\mathbf{k}$

(b) $\mathbf{v} - \mathbf{w} = (2\mathbf{i} + 3\mathbf{j} - 2\mathbf{k}) - (3\mathbf{i} - 4\mathbf{j} + 5\mathbf{k})$
$\qquad\quad = (2 - 3)\mathbf{i} + [3 - (-4)]\mathbf{j} + [-2 - 5]\mathbf{k}$
$\qquad\quad = -\mathbf{i} + 7\mathbf{j} - 7\mathbf{k}$

EXAMPLE 4

Finding Scalar Products and Magnitudes

If $\mathbf{v} = 2\mathbf{i} + 3\mathbf{j} - 2\mathbf{k}$ and $\mathbf{w} = 3\mathbf{i} - 4\mathbf{j} + 5\mathbf{k}$, find:

(a) $3\mathbf{v}$ (b) $2\mathbf{v} - 3\mathbf{w}$ (c) $\|\mathbf{v}\|$

Solution

(a) $3\mathbf{v} = 3(2\mathbf{i} + 3\mathbf{j} - 2\mathbf{k}) = 6\mathbf{i} + 9\mathbf{j} - 6\mathbf{k}$

(b) $2\mathbf{v} - 3\mathbf{w} = 2(2\mathbf{i} + 3\mathbf{j} - 2\mathbf{k}) - 3(3\mathbf{i} - 4\mathbf{j} + 5\mathbf{k})$
$\qquad\qquad\quad = 4\mathbf{i} + 6\mathbf{j} - 4\mathbf{k} - 9\mathbf{i} + 12\mathbf{j} - 15\mathbf{k} = -5\mathbf{i} + 18\mathbf{j} - 19\mathbf{k}$

(c) $\|\mathbf{v}\| = \|2\mathbf{i} + 3\mathbf{j} - 2\mathbf{k}\| = \sqrt{2^2 + 3^2 + (-2)^2} = \sqrt{17}$

Now Work PROBLEMS 33 AND 39

Recall that a unit vector **u** is one for which $\|\mathbf{u}\| = 1$. In many applications, it is useful to be able to find a unit vector **u** that has the same direction as a given vector **v**.

> **THEOREM Unit Vector in the Direction of v**
>
> For any nonzero vector **v**, the vector
>
> $$\mathbf{u} = \frac{\mathbf{v}}{\|\mathbf{v}\|}$$
>
> is a unit vector that has the same direction as **v**.

The following is a consequence of this theorem.

$$\mathbf{v} = \|\mathbf{v}\|\mathbf{u}$$

EXAMPLE 5 **Finding a Unit Vector**

Find the unit vector in the same direction as $\mathbf{v} = 2\mathbf{i} - 3\mathbf{j} - 6\mathbf{k}$.

Solution Find $\|\mathbf{v}\|$ first.

$$\|\mathbf{v}\| = \|2\mathbf{i} - 3\mathbf{j} - 6\mathbf{k}\| = \sqrt{4 + 9 + 36} = \sqrt{49} = 7$$

Now multiply **v** by the scalar $\frac{1}{\|\mathbf{v}\|} = \frac{1}{7}$. The result is the unit vector

$$\mathbf{u} = \frac{\mathbf{v}}{\|\mathbf{v}\|} = \frac{2\mathbf{i} - 3\mathbf{j} - 6\mathbf{k}}{7} = \frac{2}{7}\mathbf{i} - \frac{3}{7}\mathbf{j} - \frac{6}{7}\mathbf{k}$$

Now Work PROBLEM 47

4 Find the Dot Product

The definition of *dot product* in space is an extension of the definition given for vectors in a plane.

> **DEFINITION Dot Product**
>
> If $\mathbf{v} = a_1\mathbf{i} + b_1\mathbf{j} + c_1\mathbf{k}$ and $\mathbf{w} = a_2\mathbf{i} + b_2\mathbf{j} + c_2\mathbf{k}$ are two vectors, the **dot product v · w** is defined as
>
> $$\mathbf{v} \cdot \mathbf{w} = a_1a_2 + b_1b_2 + c_1c_2 \tag{3}$$

EXAMPLE 6 **Finding Dot Products**

If $\mathbf{v} = 2\mathbf{i} - 3\mathbf{j} + 6\mathbf{k}$ and $\mathbf{w} = 5\mathbf{i} + 3\mathbf{j} - \mathbf{k}$, find:

(a) $\mathbf{v} \cdot \mathbf{w}$ (b) $\mathbf{w} \cdot \mathbf{v}$ (c) $\mathbf{v} \cdot \mathbf{v}$

(d) $\mathbf{w} \cdot \mathbf{w}$ (e) $\|\mathbf{v}\|$ (f) $\|\mathbf{w}\|$

Solution
(a) $\mathbf{v} \cdot \mathbf{w} = 2 \cdot 5 + (-3)3 + 6(-1) = -5$
(b) $\mathbf{w} \cdot \mathbf{v} = 5 \cdot 2 + 3(-3) + (-1) \cdot 6 = -5$
(c) $\mathbf{v} \cdot \mathbf{v} = 2 \cdot 2 + (-3)(-3) + 6 \cdot 6 = 49$
(d) $\mathbf{w} \cdot \mathbf{w} = 5 \cdot 5 + 3 \cdot 3 + (-1)(-1) = 35$
(e) $\|\mathbf{v}\| = \sqrt{2^2 + (-3)^2 + 6^2} = \sqrt{49} = 7$
(f) $\|\mathbf{w}\| = \sqrt{5^2 + 3^2 + (-1)^2} = \sqrt{35}$

The dot product in space has the same properties as the dot product in the plane.

THEOREM **Properties of the Dot Product**

If $\mathbf{u}, \mathbf{v}$, and $\mathbf{w}$ are vectors, then

Commutative Property

$$\mathbf{u} \cdot \mathbf{v} = \mathbf{v} \cdot \mathbf{u}$$

Distributive Property

$$\mathbf{u} \cdot (\mathbf{v} + \mathbf{w}) = \mathbf{u} \cdot \mathbf{v} + \mathbf{u} \cdot \mathbf{w}$$

$$\mathbf{v} \cdot \mathbf{v} = \|\mathbf{v}\|^2$$
$$\mathbf{0} \cdot \mathbf{v} = 0$$

---

5 Find the Angle between Two Vectors

The angle θ between two vectors in space follows the same formula as for two vectors in the plane.

THEOREM **Angle between Vectors**

If $\mathbf{u}$ and $\mathbf{v}$ are two nonzero vectors, the angle θ, $0 \le \theta \le \pi$, between $\mathbf{u}$ and $\mathbf{v}$ is determined by the formula

$$\cos \theta = \frac{\mathbf{u} \cdot \mathbf{v}}{\|\mathbf{u}\| \, \|\mathbf{v}\|} \tag{4}$$

EXAMPLE 7   **Finding the Angle between Two Vectors**

Find the angle θ between $\mathbf{u} = 2\mathbf{i} - 3\mathbf{j} + 6\mathbf{k}$ and $\mathbf{v} = 2\mathbf{i} + 5\mathbf{j} - \mathbf{k}$.

Solution   Compute the quantities $\mathbf{u} \cdot \mathbf{v}$, $\|\mathbf{u}\|$, and $\|\mathbf{v}\|$.

$$\mathbf{u} \cdot \mathbf{v} = 2 \cdot 2 + (-3) \cdot 5 + 6(-1) = -17$$
$$\|\mathbf{u}\| = \sqrt{2^2 + (-3)^2 + 6^2} = \sqrt{49} = 7$$
$$\|\mathbf{v}\| = \sqrt{2^2 + 5^2 + (-1)^2} = \sqrt{30}$$

By formula (4), if θ is the angle between $\mathbf{u}$ and $\mathbf{v}$, then

$$\cos \theta = \frac{\mathbf{u} \cdot \mathbf{v}}{\|\mathbf{u}\| \, \|\mathbf{v}\|} = \frac{-17}{7\sqrt{30}} \approx -0.443$$

So, $\theta \approx \cos^{-1}(-0.443) \approx 116.3°$.

Now Work PROBLEM 51

6 Find the Direction Angles of a Vector

A nonzero vector **v** in space can be described by specifying its magnitude and its three **direction angles** α, β, and γ. These direction angles are defined as

α = the angle between **v** and **i**, the positive x-axis, $0 \le \alpha \le \pi$

β = the angle between **v** and **j**, the positive y-axis, $0 \le \beta \le \pi$

γ = the angle between **v** and **k**, the positive z-axis, $0 \le \gamma \le \pi$

See Figure 80.

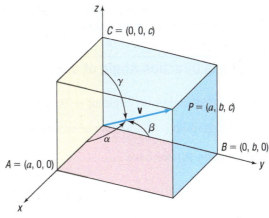

$0 \le \alpha \le \pi, 0 \le \beta \le \pi, 0 \le \gamma \le \pi$

Figure 80 Direction angles

Our first goal is to find expressions for α, β, and γ in terms of the components of a vector. Let $\mathbf{v} = a\mathbf{i} + b\mathbf{j} + c\mathbf{k}$ denote a nonzero vector. The angle α between **v** and **i**, the positive x-axis, obeys

$$\cos \alpha = \frac{\mathbf{v} \cdot \mathbf{i}}{\|\mathbf{v}\| \, \|\mathbf{i}\|} = \frac{a}{\|\mathbf{v}\|}$$

Similarly,

$$\cos \beta = \frac{b}{\|\mathbf{v}\|} \quad \text{and} \quad \cos \gamma = \frac{c}{\|\mathbf{v}\|}$$

Since $\|\mathbf{v}\| = \sqrt{a^2 + b^2 + c^2}$, the following result is obtained.

> **THEOREM** Direction Angles
>
> If $\mathbf{v} = a\mathbf{i} + b\mathbf{j} + c\mathbf{k}$ is a nonzero vector in space, the direction angles α, β, and γ obey
>
> - $\cos \alpha = \dfrac{a}{\sqrt{a^2 + b^2 + c^2}} = \dfrac{a}{\|\mathbf{v}\|}$ • $\cos \beta = \dfrac{b}{\sqrt{a^2 + b^2 + c^2}} = \dfrac{b}{\|\mathbf{v}\|}$
>
> - $\cos \gamma = \dfrac{c}{\sqrt{a^2 + b^2 + c^2}} = \dfrac{c}{\|\mathbf{v}\|}$ **(5)**

The numbers $\cos \alpha$, $\cos \beta$, and $\cos \gamma$ are called the **direction cosines** of the vector **v**.

EXAMPLE 8

Finding the Direction Angles of a Vector

Find the direction angles of $\mathbf{v} = -3\mathbf{i} + 2\mathbf{j} - 6\mathbf{k}$.

Solution $\|\mathbf{v}\| = \sqrt{(-3)^2 + 2^2 + (-6)^2} = \sqrt{49} = 7$

Using the formulas in equation (5), we have

$$\cos \alpha = \frac{-3}{7} \qquad \cos \beta = \frac{2}{7} \qquad \cos \gamma = \frac{-6}{7}$$

$$\alpha \approx 115.4° \qquad \beta \approx 73.4° \qquad \gamma \approx 149.0°$$

> **THEOREM** **Property of the Direction Cosines**
>
> If α, β, and γ are the direction angles of a nonzero vector $\mathbf{v}$ in space, then
>
> $$\cos^2 \alpha + \cos^2 \beta + \cos^2 \gamma = 1 \qquad (6)$$

The proof is a direct consequence of the equations in (5).

Based on equation (6), when two direction cosines are known, the third is determined up to its sign. Knowing two direction cosines is not sufficient to uniquely determine the direction of a vector in space.

EXAMPLE 9 **Finding a Direction Angle of a Vector**

The vector $\mathbf{v}$ makes an angle of $\alpha = \dfrac{\pi}{3}$ with the positive x-axis, an angle of $\beta = \dfrac{\pi}{3}$ with the positive y-axis, and an acute angle γ with the positive z-axis. Find γ.

Solution By equation (6), we have

$$\cos^2\left(\frac{\pi}{3}\right) + \cos^2\left(\frac{\pi}{3}\right) + \cos^2 \gamma = 1 \qquad 0 < \gamma < \frac{\pi}{2}$$

$$\left(\frac{1}{2}\right)^2 + \left(\frac{1}{2}\right)^2 + \cos^2 \gamma = 1$$

$$\cos^2 \gamma = \frac{1}{2}$$

$$\cos \gamma = \frac{\sqrt{2}}{2} \quad \text{or} \quad \cos \gamma = -\frac{\sqrt{2}}{2}$$

$$\gamma = \frac{\pi}{4} \quad \text{or} \quad \gamma = \frac{3\pi}{4}$$

Since γ must be acute, $\gamma = \dfrac{\pi}{4}$.

The direction cosines of a vector give information about only the direction of the vector; they provide no information about its magnitude. For example, *any* vector that is parallel to the xy-plane and makes an angle of $\dfrac{\pi}{4}$ radian with the positive x-axis and y-axis has direction cosines

$$\cos \alpha = \frac{\sqrt{2}}{2} \qquad \cos \beta = \frac{\sqrt{2}}{2} \qquad \cos \gamma = 0$$

However, if the direction angles *and* the magnitude of a vector are known, the vector is uniquely determined.

EXAMPLE 10 **Writing a Vector in Terms of Its Magnitude and Direction Cosines**

Show that any nonzero vector $\mathbf{v}$ in space can be written in terms of its magnitude and direction cosines as

$$\mathbf{v} = \|\mathbf{v}\|[(\cos \alpha)\mathbf{i} + (\cos \beta)\mathbf{j} + (\cos \gamma)\mathbf{k}] \qquad (7)$$

Solution Let $\mathbf{v} = a\mathbf{i} + b\mathbf{j} + c\mathbf{k}$. From the equations in (5), note that

$$a = \|\mathbf{v}\| \cos \alpha \qquad b = \|\mathbf{v}\| \cos \beta \qquad c = \|\mathbf{v}\| \cos \gamma$$

Substituting gives

$$\mathbf{v} = a\mathbf{i} + b\mathbf{j} + c\mathbf{k} = \|\mathbf{v}\|(\cos\alpha)\mathbf{i} + \|\mathbf{v}\|(\cos\beta)\mathbf{j} + \|\mathbf{v}\|(\cos\gamma)\mathbf{k}$$
$$= \|\mathbf{v}\|[(\cos\alpha)\mathbf{i} + (\cos\beta)\mathbf{j} + (\cos\gamma)\mathbf{k}]$$

Now Work PROBLEM 59

Example 10 shows that the direction cosines of a vector **v** are also the components of the unit vector in the direction of **v**.

5.6 Assess Your Understanding

'Are You Prepared?' *The answer is given at the end of these exercises. If you get the wrong answer, read the page listed in* red.

1. The distance d from $P_1 = (x_1, y_1)$ to $P_2 = (x_2, y_2)$ is $d =$ _____. (pp. 3–4)

Concepts and Vocabulary

2. If $\mathbf{v} = a\mathbf{i} + b\mathbf{j} + c\mathbf{k}$ is a vector in space, the scalars a, b, c are called the _____ of **v**.

3. The squares of the direction cosines of a vector in space add up to _____.

4. *True or False* In space, the dot product of two vectors is a positive number.

5. *True or False* A vector in space may be described by specifying its magnitude and its direction angles.

6. *Multiple Choice* In space, points of the form $(x, y, 0)$ lie in:
(a) the xy-plane (b) the xz-plane
(c) the yz-plane (d) none of these

Skill Building

In Problems 7–14, describe the set of points (x, y, z) defined by the equation(s).

7. $y = 0$ **8.** $x = 0$ **9.** $z = 2$ **10.** $y = 3$
11. $x = -4$ **12.** $z = -3$ **13.** $x = 1$ and $y = 2$ **14.** $x = 3$ and $z = 1$

In Problems 15–20, find the distance from P_1 to P_2.

15. $P_1 = (0, 0, 0)$ and $P_2 = (4, 1, 2)$ **16.** $P_1 = (0, 0, 0)$ and $P_2 = (1, -2, 3)$
17. $P_1 = (-1, 2, -3)$ and $P_2 = (0, -2, 1)$ **18.** $P_1 = (-2, 2, 3)$ and $P_2 = (4, 0, -3)$
19. $P_1 = (4, -2, -2)$ and $P_2 = (3, 2, 1)$ **20.** $P_1 = (2, -3, -3)$ and $P_2 = (4, 1, -1)$

In Problems 21–26, opposite vertices of a rectangular box whose edges are parallel to the coordinate axes are given. List the coordinates of the other six vertices of the box.

21. $(0, 0, 0)$; $(2, 1, 3)$ **22.** $(0, 0, 0)$; $(4, 2, 2)$ **23.** $(1, 2, 3)$; $(3, 4, 5)$
24. $(5, 6, 1)$; $(3, 8, 2)$ **25.** $(-1, 0, 2)$; $(4, 2, 5)$ **26.** $(-2, -3, 0)$; $(-6, 7, 1)$

*In Problems 27–32, the vector **v** has initial point P and terminal point Q. Write **v** in the form $a\mathbf{i} + b\mathbf{j} + c\mathbf{k}$; that is, find its position vector.*

27. $P = (0, 0, 0)$; $Q = (3, 4, -1)$ **28.** $P = (0, 0, 0)$; $Q = (-3, -5, 4)$
29. $P = (3, 2, -1)$; $Q = (5, 6, 0)$ **30.** $P = (-3, 2, 0)$; $Q = (6, 5, -1)$
31. $P = (-2, -1, 4)$; $Q = (6, -2, 4)$ **32.** $P = (-1, 4, -2)$; $Q = (6, 2, 2)$

In Problems 33–38, find $\|\mathbf{v}\|$.

33. $\mathbf{v} = 3\mathbf{i} - 6\mathbf{j} - 2\mathbf{k}$ **34.** $\mathbf{v} = -6\mathbf{i} + 12\mathbf{j} + 4\mathbf{k}$ **35.** $\mathbf{v} = \mathbf{i} - \mathbf{j} + \mathbf{k}$
36. $\mathbf{v} = -\mathbf{i} - \mathbf{j} + \mathbf{k}$ **37.** $\mathbf{v} = -2\mathbf{i} + 3\mathbf{j} - 3\mathbf{k}$ **38.** $\mathbf{v} = 6\mathbf{i} + 2\mathbf{j} - 2\mathbf{k}$

In Problems 39–44, find each quantity if $\mathbf{v} = 3\mathbf{i} - 5\mathbf{j} + 2\mathbf{k}$ and $\mathbf{w} = -2\mathbf{i} + 3\mathbf{j} - 2\mathbf{k}$.

39. $2\mathbf{v} + 3\mathbf{w}$ **40.** $3\mathbf{v} - 2\mathbf{w}$ **41.** $\|\mathbf{v} - \mathbf{w}\|$
42. $\|\mathbf{v} + \mathbf{w}\|$ **43.** $\|\mathbf{v}\| - \|\mathbf{w}\|$ **44.** $\|\mathbf{v}\| + \|\mathbf{w}\|$

*In Problems 45–50, find the unit vector in the same direction as **v**.*

45. $\mathbf{v} = 5\mathbf{i}$ **46.** $\mathbf{v} = -3\mathbf{j}$ **47.** $\mathbf{v} = 3\mathbf{i} - 6\mathbf{j} - 2\mathbf{k}$
48. $\mathbf{v} = -6\mathbf{i} + 12\mathbf{j} + 4\mathbf{k}$ **49.** $\mathbf{v} = \mathbf{i} + \mathbf{j} + \mathbf{k}$ **50.** $\mathbf{v} = 2\mathbf{i} - \mathbf{j} + \mathbf{k}$

*In Problems 51–58, find the dot product $\mathbf{v} \cdot \mathbf{w}$ and the angle between **v** and **w**.*

51. $\mathbf{v} = \mathbf{i} - \mathbf{j}$, $\mathbf{w} = \mathbf{i} + \mathbf{j} + \mathbf{k}$ **52.** $\mathbf{v} = \mathbf{i} + \mathbf{j}$, $\mathbf{w} = -\mathbf{i} + \mathbf{j} - \mathbf{k}$
53. $\mathbf{v} = 2\mathbf{i} + \mathbf{j} - 3\mathbf{k}$, $\mathbf{w} = \mathbf{i} + 2\mathbf{j} + 2\mathbf{k}$ **54.** $\mathbf{v} = 2\mathbf{i} + 2\mathbf{j} - \mathbf{k}$, $\mathbf{w} = \mathbf{i} + 2\mathbf{j} + 3\mathbf{k}$

55. $\mathbf{v} = 3\mathbf{i} - \mathbf{j} + 2\mathbf{k},\quad \mathbf{w} = \mathbf{i} + \mathbf{j} - \mathbf{k}$

56. $\mathbf{v} = \mathbf{i} + 3\mathbf{j} + 2\mathbf{k},\quad \mathbf{w} = \mathbf{i} - \mathbf{j} + \mathbf{k}$

57. $\mathbf{v} = 3\mathbf{i} + 4\mathbf{j} + \mathbf{k},\quad \mathbf{w} = 6\mathbf{i} + 8\mathbf{j} + 2\mathbf{k}$

58. $\mathbf{v} = 3\mathbf{i} - 4\mathbf{j} + \mathbf{k},\quad \mathbf{w} = 6\mathbf{i} - 8\mathbf{j} + 2\mathbf{k}$

In Problems 59 – 66, find the direction angles of each vector. Write each vector in the form of equation (7).

59. $\mathbf{v} = 3\mathbf{i} - 6\mathbf{j} - 2\mathbf{k}$

60. $\mathbf{v} = -6\mathbf{i} + 12\mathbf{j} + 4\mathbf{k}$

61. $\mathbf{v} = \mathbf{i} + \mathbf{j} + \mathbf{k}$

62. $\mathbf{v} = \mathbf{i} - \mathbf{j} - \mathbf{k}$

63. $\mathbf{v} = \mathbf{i} + \mathbf{j}$

64. $\mathbf{v} = \mathbf{j} + \mathbf{k}$

65. $\mathbf{v} = 3\mathbf{i} - 5\mathbf{j} + 2\mathbf{k}$

66. $\mathbf{v} = 2\mathbf{i} + 3\mathbf{j} - 4\mathbf{k}$

Applications and Extensions

67. Robotic Arm Consider the double-jointed robotic arm shown in the figure. Let the lower arm be modeled by $\mathbf{a} = \langle 2, 3, 4 \rangle$, the middle arm be modeled by $\mathbf{b} = \langle 1, -1, 3 \rangle$, and the upper arm be modeled by $\mathbf{c} = \langle 4, -1, -2 \rangle$, where units are in feet.

(a) Find a vector $\mathbf{d}$ that represents the position of the hand.

(b) Determine the distance of the hand from the origin.

68. The Sphere In space, the collection of all points that are the same distance from some fixed point is called a **sphere**. See the illustration. The constant distance is called the **radius**, and the fixed point is the **center** of the sphere. Show that an equation of a sphere with center at (x_0, y_0, z_0) and radius r is

$$(x - x_0)^2 + (y - y_0)^2 + (z - z_0)^2 = r^2$$

[**Hint**: Use the Distance Formula (1).]

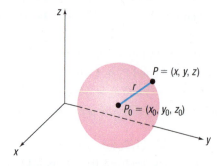

In Problems 69 and 70, find an equation of a sphere with radius r and center P_0.

69. $r = 1;\ P_0 = (3, 1, 1)$

70. $r = 2;\ P_0 = (1, 2, 2)$

In Problems 71–76, find the radius and center of each sphere.

[**Hint**: *Complete the square in each variable.*]

71. $x^2 + y^2 + z^2 + 2x - 2y = 2$

72. $x^2 + y^2 + z^2 + 2x - 2z = -1$

73. $x^2 + y^2 + z^2 - 4x + 4y + 2z = 0$

74. $x^2 + y^2 + z^2 - 4x = 0$

75. $2x^2 + 2y^2 + 2z^2 - 8x + 4z = -1$

76. $3x^2 + 3y^2 + 3z^2 + 6x - 6y = 3$

The **work** *W done by a constant force $\mathbf{F}$ in moving an object from a point A in space to a point B in space is defined as $W = \mathbf{F} \cdot \overrightarrow{AB}$. Use this definition in Problems 77–79.*

77. Work Find the work done by a force of 3 newtons acting in the direction $2\mathbf{i} + \mathbf{j} + 2\mathbf{k}$ in moving an object 2 meters from $(0, 0, 0)$ to $(0, 2, 0)$.

78. Work Find the work done by a force of 1 newton acting in the direction $2\mathbf{i} + 2\mathbf{j} + \mathbf{k}$ in moving an object 3 meters from $(0, 0, 0)$ to $(1, 2, 2)$.

79. Work Find the work done in moving an object along a vector $\mathbf{u} = 3\mathbf{i} + 2\mathbf{j} - 5\mathbf{k}$ if the applied force is $\mathbf{F} = 2\mathbf{i} - \mathbf{j} - \mathbf{k}$. Use meters for distance and newtons for force.

Retain Your Knowledge

Problems 80–88 are based on material learned earlier in the course. The purpose of these problems is to keep the material fresh in your mind so that you are better prepared for the final exam.

80. Solve: $\dfrac{3}{x - 2} \geq 5$

81. Given $f(x) = 2x - 3$, find $f(x^2 + x - 1)$.

82. Find the exact value of $\sin 80° \cos 50° - \cos 80° \sin 50°$.

83. Solve the triangle.

84. Find the distance between the points $P_1 = (-1, -2)$ and $P_2 = (9, 3)$.

85. Form a polynomial function with real coefficients having degree 4 and zeros $-i$ and $1 - 3i$.

86. The function $f(x) = \dfrac{5}{x - 8}$ is one-to-one. Find its inverse.

87. Find the average rate of change of $f(x) = 3 \tan(2x)$ from $-\dfrac{\pi}{8}$ to $\dfrac{\pi}{8}$.

88. Find the area of the region enclosed by $f(x) = \sqrt{36 - x^2}$ and $g(x) = 6 - x$.

'Are You Prepared?' Answer

1. $\sqrt{(x_2 - x_1)^2 + (y_2 - y_1)^2}$

5.7 The Cross Product

OBJECTIVES 1 Find the Cross Product of Two Vectors (p. 381)
2 Know Algebraic Properties of the Cross Product (p. 382)
3 Know Geometric Properties of the Cross Product (p. 383)
4 Find a Vector Orthogonal to Two Given Vectors (p. 384)
5 Find the Area of a Parallelogram (p. 384)

1 Find the Cross Product of Two Vectors

For vectors in space, and only for vectors in space, a second product of two vectors is defined, called the *cross product*. The cross product of two vectors in space is also a vector that has applications in both geometry and physics.

> **DEFINITION Cross Product**
>
> If $\mathbf{v} = a_1\mathbf{i} + b_1\mathbf{j} + c_1\mathbf{k}$ and $\mathbf{w} = a_2\mathbf{i} + b_2\mathbf{j} + c_2\mathbf{k}$ are two vectors in space, the **cross product $\mathbf{v} \times \mathbf{w}$** is defined as the vector
>
> $$\mathbf{v} \times \mathbf{w} = (b_1c_2 - b_2c_1)\mathbf{i} - (a_1c_2 - a_2c_1)\mathbf{j} + (a_1b_2 - a_2b_1)\mathbf{k} \qquad (1)$$

Notice that the cross product $\mathbf{v} \times \mathbf{w}$ of two vectors is a vector. Because of this, it is sometimes referred to as the **vector product**.

EXAMPLE 1   **Finding a Cross Product Using Equation (1)**

If $\mathbf{v} = 2\mathbf{i} + 3\mathbf{j} + 5\mathbf{k}$ and $\mathbf{w} = \mathbf{i} + 2\mathbf{j} + 3\mathbf{k}$, find $\mathbf{v} \times \mathbf{w}$.

Solution
$$\mathbf{v} \times \mathbf{w} = (3 \cdot 3 - 2 \cdot 5)\mathbf{i} - (2 \cdot 3 - 1 \cdot 5)\mathbf{j} + (2 \cdot 2 - 1 \cdot 3)\mathbf{k} \quad \textbf{Equation (1)}$$
$$= (9 - 10)\mathbf{i} - (6 - 5)\mathbf{j} + (4 - 3)\mathbf{k}$$
$$= -\mathbf{i} - \mathbf{j} + \mathbf{k}$$

Determinants may be used as an aid in computing cross products. **A 2 by 2 determinant**, symbolized by

$$\begin{vmatrix} a_1 & b_1 \\ a_2 & b_2 \end{vmatrix}$$

has the value $a_1b_2 - a_2b_1$; that is,

$$\begin{vmatrix} a_1 & b_1 \\ a_2 & b_2 \end{vmatrix} = a_1b_2 - a_2b_1$$

A **3 by 3 determinant** has the value

$$\begin{vmatrix} A & B & C \\ a_1 & b_1 & c_1 \\ a_2 & b_2 & c_2 \end{vmatrix} = \begin{vmatrix} b_1 & c_1 \\ b_2 & c_2 \end{vmatrix} A - \begin{vmatrix} a_1 & c_1 \\ a_2 & c_2 \end{vmatrix} B + \begin{vmatrix} a_1 & b_1 \\ a_2 & b_2 \end{vmatrix} C$$

EXAMPLE 2 | **Evaluating Determinants**

(a) $\begin{vmatrix} 2 & 3 \\ 1 & 2 \end{vmatrix} = 2 \cdot 2 - 1 \cdot 3 = 4 - 3 = 1$

(b) $\begin{vmatrix} A & B & C \\ 2 & 3 & 5 \\ 1 & 2 & 3 \end{vmatrix} = \begin{vmatrix} 3 & 5 \\ 2 & 3 \end{vmatrix} A - \begin{vmatrix} 2 & 5 \\ 1 & 3 \end{vmatrix} B + \begin{vmatrix} 2 & 3 \\ 1 & 2 \end{vmatrix} C$

$$= (9 - 10)A - (6 - 5)B + (4 - 3)C$$
$$= -A - B + C$$

Now Work PROBLEM 7

The cross product of the vectors $\mathbf{v} = a_1 \mathbf{i} + b_1 \mathbf{j} + c_1 \mathbf{k}$ and $\mathbf{w} = a_2 \mathbf{i} + b_2 \mathbf{j} + c_2 \mathbf{k}$, that is,

$$\mathbf{v} \times \mathbf{w} = (b_1 c_2 - b_2 c_1)\mathbf{i} - (a_1 c_2 - a_2 c_1)\mathbf{j} + (a_1 b_2 - a_2 b_1)\mathbf{k}$$

may be written symbolically using determinants as

$$\mathbf{v} \times \mathbf{w} = \begin{vmatrix} \mathbf{i} & \mathbf{j} & \mathbf{k} \\ a_1 & b_1 & c_1 \\ a_2 & b_2 & c_2 \end{vmatrix} = \begin{vmatrix} b_1 & c_1 \\ b_2 & c_2 \end{vmatrix}\mathbf{i} - \begin{vmatrix} a_1 & c_1 \\ a_2 & c_2 \end{vmatrix}\mathbf{j} + \begin{vmatrix} a_1 & b_1 \\ a_2 & b_2 \end{vmatrix}\mathbf{k}$$

EXAMPLE 3 | **Using Determinants to Find Cross Products**

If $\mathbf{v} = 2\mathbf{i} + 3\mathbf{j} + 5\mathbf{k}$ and $\mathbf{w} = \mathbf{i} + 2\mathbf{j} + 3\mathbf{k}$, find:

(a) $\mathbf{v} \times \mathbf{w}$ (b) $\mathbf{w} \times \mathbf{v}$ (c) $\mathbf{v} \times \mathbf{v}$ (d) $\mathbf{w} \times \mathbf{w}$

Solution

(a) $\mathbf{v} \times \mathbf{w} = \begin{vmatrix} \mathbf{i} & \mathbf{j} & \mathbf{k} \\ 2 & 3 & 5 \\ 1 & 2 & 3 \end{vmatrix} = \begin{vmatrix} 3 & 5 \\ 2 & 3 \end{vmatrix}\mathbf{i} - \begin{vmatrix} 2 & 5 \\ 1 & 3 \end{vmatrix}\mathbf{j} + \begin{vmatrix} 2 & 3 \\ 1 & 2 \end{vmatrix}\mathbf{k} = -\mathbf{i} - \mathbf{j} + \mathbf{k}$

(b) $\mathbf{w} \times \mathbf{v} = \begin{vmatrix} \mathbf{i} & \mathbf{j} & \mathbf{k} \\ 1 & 2 & 3 \\ 2 & 3 & 5 \end{vmatrix} = \begin{vmatrix} 2 & 3 \\ 3 & 5 \end{vmatrix}\mathbf{i} - \begin{vmatrix} 1 & 3 \\ 2 & 5 \end{vmatrix}\mathbf{j} + \begin{vmatrix} 1 & 2 \\ 2 & 3 \end{vmatrix}\mathbf{k} = \mathbf{i} + \mathbf{j} - \mathbf{k}$

(c) $\mathbf{v} \times \mathbf{v} = \begin{vmatrix} \mathbf{i} & \mathbf{j} & \mathbf{k} \\ 2 & 3 & 5 \\ 2 & 3 & 5 \end{vmatrix} = \begin{vmatrix} 3 & 5 \\ 3 & 5 \end{vmatrix}\mathbf{i} - \begin{vmatrix} 2 & 5 \\ 2 & 5 \end{vmatrix}\mathbf{j} + \begin{vmatrix} 2 & 3 \\ 2 & 3 \end{vmatrix}\mathbf{k} = 0\mathbf{i} - 0\mathbf{j} + 0\mathbf{k} = \mathbf{0}$

(d) $\mathbf{w} \times \mathbf{w} = \begin{vmatrix} \mathbf{i} & \mathbf{j} & \mathbf{k} \\ 1 & 2 & 3 \\ 1 & 2 & 3 \end{vmatrix}$

$= \begin{vmatrix} 2 & 3 \\ 2 & 3 \end{vmatrix}\mathbf{i} - \begin{vmatrix} 1 & 3 \\ 1 & 3 \end{vmatrix}\mathbf{j} + \begin{vmatrix} 1 & 2 \\ 1 & 2 \end{vmatrix}\mathbf{k} = 0\mathbf{i} - 0\mathbf{j} + 0\mathbf{k} = \mathbf{0}$

Now Work PROBLEM 15

2 Know Algebraic Properties of the Cross Product

Notice in Examples 3(a) and (b) that $\mathbf{v} \times \mathbf{w}$ and $\mathbf{w} \times \mathbf{v}$ are negatives of one another. From Examples 3(c) and (d), one might conjecture that the cross product of a vector with itself is the zero vector. These and other algebraic properties of the cross product are given next.

> ### THEOREM Algebraic Properties of the Cross Product
>
> If $\mathbf{u}$, $\mathbf{v}$, and $\mathbf{w}$ are vectors in space and if α is a scalar, then
>
> | | |
> |---|---|
> | • $\mathbf{u} \times \mathbf{u} = \mathbf{0}$ | (2) |
> | • $\mathbf{u} \times \mathbf{v} = -(\mathbf{v} \times \mathbf{u})$ | (3) |
> | • $\alpha(\mathbf{u} \times \mathbf{v}) = (\alpha\mathbf{u}) \times \mathbf{v} = \mathbf{u} \times (\alpha\mathbf{v})$ | (4) |
> | • $\mathbf{u} \times (\mathbf{v} + \mathbf{w}) = (\mathbf{u} \times \mathbf{v}) + (\mathbf{u} \times \mathbf{w})$ | (5) |

Proof We will prove properties (2) and (4) here and leave properties (3) and (5) as exercises (see Problems 60 and 61).

To prove property (2), let $\mathbf{u} = a_1\mathbf{i} + b_1\mathbf{j} + c_1\mathbf{k}$. Then

$$\mathbf{u} \times \mathbf{u} = \begin{vmatrix} \mathbf{i} & \mathbf{j} & \mathbf{k} \\ a_1 & b_1 & c_1 \\ a_1 & b_1 & c_1 \end{vmatrix} = \begin{vmatrix} b_1 & c_1 \\ b_1 & c_1 \end{vmatrix}\mathbf{i} - \begin{vmatrix} a_1 & c_1 \\ a_1 & c_1 \end{vmatrix}\mathbf{j} + \begin{vmatrix} a_1 & b_1 \\ a_1 & b_1 \end{vmatrix}\mathbf{k}$$

$$= 0\mathbf{i} - 0\mathbf{j} + 0\mathbf{k} = \mathbf{0}$$

To prove property (4), let $\mathbf{u} = a_1\mathbf{i} + b_1\mathbf{j} + c_1\mathbf{k}$ and $\mathbf{v} = a_2\mathbf{i} + b_2\mathbf{j} + c_2\mathbf{k}$. Then

$$\alpha(\mathbf{u} \times \mathbf{v}) \underset{\underset{\text{Apply (1).}}{\uparrow}}{=} \alpha[\,(b_1c_2 - b_2c_1)\mathbf{i} - (a_1c_2 - a_2c_1)\mathbf{j} + (a_1b_2 - a_2b_1)\mathbf{k}\,]$$

$$= \alpha(b_1c_2 - b_2c_1)\mathbf{i} - \alpha(a_1c_2 - a_2c_1)\mathbf{j} + \alpha(a_1b_2 - a_2b_1)\mathbf{k} \tag{6}$$

Since $\alpha\mathbf{u} = \alpha a_1\mathbf{i} + \alpha b_1\mathbf{j} + \alpha c_1\mathbf{k}$, we have

$$(\alpha\mathbf{u}) \times \mathbf{v} = (\alpha b_1c_2 - b_2\alpha c_1)\mathbf{i} - (\alpha a_1c_2 - a_2\alpha c_1)\mathbf{j} + (\alpha a_1b_2 - a_2\alpha b_1)\mathbf{k}$$

$$= \alpha(b_1c_2 - b_2c_1)\mathbf{i} - \alpha(a_1c_2 - a_2c_1)\mathbf{j} + \alpha(a_1b_2 - a_2b_1)\mathbf{k} \tag{7}$$

Based on equations (6) and (7), the first part of property (4) follows. The second part can be proved in like fashion. ∎

 Now Work PROBLEM 17

3 Know Geometric Properties of the Cross Product

> ### THEOREM Geometric Properties of the Cross Product
>
> Let $\mathbf{u}$ and $\mathbf{v}$ be vectors in space.
>
> | | |
> |---|---|
> | • $\mathbf{u} \times \mathbf{v}$ is orthogonal to both $\mathbf{u}$ and $\mathbf{v}$. | (8) |
> | • $\|\mathbf{u} \times \mathbf{v}\| = \|\mathbf{u}\|\,\|\mathbf{v}\| \sin\theta,$ | (9) |
> | where θ is the angle between $\mathbf{u}$ and $\mathbf{v}$. | |
> | • $\|\mathbf{u} \times \mathbf{v}\|$ is the area of the parallelogram | |
> | having $\mathbf{u} \neq \mathbf{0}$ and $\mathbf{v} \neq \mathbf{0}$ as adjacent sides. | (10) |
> | • $\mathbf{u} \times \mathbf{v} = \mathbf{0}$ if and only if $\mathbf{u}$ and $\mathbf{v}$ are parallel. | (11) |

Figure 81

Proof of Property (8) Let $\mathbf{u} = a_1\mathbf{i} + b_1\mathbf{j} + c_1\mathbf{k}$ and $\mathbf{v} = a_2\mathbf{i} + b_2\mathbf{j} + c_2\mathbf{k}$. Then

$$\mathbf{u} \times \mathbf{v} = (b_1 c_2 - b_2 c_1)\mathbf{i} - (a_1 c_2 - a_2 c_1)\mathbf{j} + (a_1 b_2 - a_2 b_1)\mathbf{k}$$

Now compute the dot product $\mathbf{u} \cdot (\mathbf{u} \times \mathbf{v})$.

$$\begin{aligned} \mathbf{u} \cdot (\mathbf{u} \times \mathbf{v}) &= (a_1\mathbf{i} + b_1\mathbf{j} + c_1\mathbf{k}) \cdot [(b_1 c_2 - b_2 c_1)\mathbf{i} - (a_1 c_2 - a_2 c_1)\mathbf{j} + (a_1 b_2 - a_2 b_1)\mathbf{k}] \\ &= a_1(b_1 c_2 - b_2 c_1) - b_1(a_1 c_2 - a_2 c_1) + c_1(a_1 b_2 - a_2 b_1) = 0 \end{aligned}$$

Since two vectors are orthogonal if their dot product is zero, it follows that $\mathbf{u}$ and $\mathbf{u} \times \mathbf{v}$ are orthogonal. Similarly, $\mathbf{v} \cdot (\mathbf{u} \times \mathbf{v}) = 0$, so $\mathbf{v}$ and $\mathbf{u} \times \mathbf{v}$ are orthogonal. ∎

4 Find a Vector Orthogonal to Two Given Vectors

Figure 82

As long as the vectors $\mathbf{u}$ and $\mathbf{v}$ are not parallel, they will form a plane in space. See Figure 81. Based on property (8), the vector $\mathbf{u} \times \mathbf{v}$ is normal to this plane. As Figure 81 illustrates, there are essentially (without regard to magnitude) two vectors normal to the plane containing $\mathbf{u}$ and $\mathbf{v}$. It can be shown that the vector $\mathbf{u} \times \mathbf{v}$ is the one determined by the thumb of the right hand when the other fingers of the right hand are cupped so that they point in a direction from $\mathbf{u}$ to $\mathbf{v}$. See Figure 82.*

EXAMPLE 4 | **Finding a Vector Orthogonal to Two Given Vectors**

Find a vector that is orthogonal to $\mathbf{u} = 3\mathbf{i} - 2\mathbf{j} + \mathbf{k}$ and $\mathbf{v} = -\mathbf{i} + 3\mathbf{j} - \mathbf{k}$.

Solution Based on property (8), such a vector is $\mathbf{u} \times \mathbf{v}$.

$$\mathbf{u} \times \mathbf{v} = \begin{vmatrix} \mathbf{i} & \mathbf{j} & \mathbf{k} \\ 3 & -2 & 1 \\ -1 & 3 & -1 \end{vmatrix} = (2 - 3)\mathbf{i} - [-3 - (-1)]\mathbf{j} + (9 - 2)\mathbf{k} = -\mathbf{i} + 2\mathbf{j} + 7\mathbf{k}$$

The vector $-\mathbf{i} + 2\mathbf{j} + 7\mathbf{k}$ is orthogonal to both $\mathbf{u}$ and $\mathbf{v}$.

✓ **Check:** Two vectors are orthogonal if their dot product is zero.

$$\mathbf{u} \cdot (-\mathbf{i} + 2\mathbf{j} + 7\mathbf{k}) = (3\mathbf{i} - 2\mathbf{j} + \mathbf{k}) \cdot (-\mathbf{i} + 2\mathbf{j} + 7\mathbf{k}) = -3 - 4 + 7 = 0$$
$$\mathbf{v} \cdot (-\mathbf{i} + 2\mathbf{j} + 7\mathbf{k}) = (-\mathbf{i} + 3\mathbf{j} - \mathbf{k}) \cdot (-\mathbf{i} + 2\mathbf{j} + 7\mathbf{k}) = 1 + 6 - 7 = 0$$

 Now Work PROBLEM 41

The proof of property (9) is left as an exercise. See Problem 62.

5 Find the Area of a Parallelogram

Proof of Property (10) Suppose that $\mathbf{u}$ and $\mathbf{v}$ are adjacent sides of a parallelogram. See Figure 83. Then the lengths of these sides are $\|\mathbf{u}\|$ and $\|\mathbf{v}\|$. If θ is the angle between $\mathbf{u}$ and $\mathbf{v}$, then the height of the parallelogram is $\|\mathbf{v}\| \sin\theta$ and its area is

$$\text{Area of parallelogram} = \text{Base} \times \text{Height} = \|\mathbf{u}\| [\|\mathbf{v}\| \sin\theta] = \|\mathbf{u} \times \mathbf{v}\|$$

$$\uparrow$$
Property (9) ∎

Figure 83

EXAMPLE 5 | **Finding the Area of a Parallelogram**

Find the area of the parallelogram whose vertices are $P_1 = (0, 0, 0)$, $P_2 = (3, -2, 1)$, $P_3 = (-1, 3, -1)$, and $P_4 = (2, 1, 0)$.

---

*This is a consequence of using a "right-handed" coordinate system.

Solution

Two adjacent sides of this parallelogram are
$$\mathbf{u} = \overrightarrow{P_1P_2} = 3\mathbf{i} - 2\mathbf{j} + \mathbf{k} \quad \text{and} \quad \mathbf{v} = \overrightarrow{P_1P_3} = -\mathbf{i} + 3\mathbf{j} - \mathbf{k}$$

WARNING Not all pairs of vertices give rise to a side. For example, $\overrightarrow{P_1P_4}$ is a diagonal of the parallelogram since $\overrightarrow{P_1P_3} + \overrightarrow{P_3P_4} = \overrightarrow{P_1P_4}$. Also, $\overrightarrow{P_1P_3}$ and $\overrightarrow{P_2P_4}$ are not adjacent sides; they are parallel sides. ■

Since $\mathbf{u} \times \mathbf{v} = -\mathbf{i} + 2\mathbf{j} + 7\mathbf{k}$ (Example 4), the area of the parallelogram is

Area of parallelogram $= \|\mathbf{u} \times \mathbf{v}\| = \sqrt{1 + 4 + 49} = \sqrt{54} = 3\sqrt{6}$ square units ■

Now Work PROBLEM 49

Proof of Property (11) The proof requires two parts. If $\mathbf{u}$ and $\mathbf{v}$ are parallel, then there is a scalar α such that $\mathbf{u} = \alpha\mathbf{v}$. Then

$$\mathbf{u} \times \mathbf{v} = (\alpha\mathbf{v}) \times \mathbf{v} = \alpha(\mathbf{v} \times \mathbf{v}) = \mathbf{0}$$

 Property (4)   Property (2)

If $\mathbf{u} \times \mathbf{v} = \mathbf{0}$, then, by property (9), we have

$$\|\mathbf{u} \times \mathbf{v}\| = \|\mathbf{u}\| \|\mathbf{v}\| \sin\theta = 0$$

Since $\mathbf{u} \neq \mathbf{0}$ and $\mathbf{v} \neq \mathbf{0}$, we must have $\sin\theta = 0$, so $\theta = 0$ or $\theta = \pi$. In either case, since θ is the angle between $\mathbf{u}$ and $\mathbf{v}$, then $\mathbf{u}$ and $\mathbf{v}$ are parallel. ■

5.7 Assess Your Understanding

Concepts and Vocabulary

1. *True or False* If $\mathbf{u}$ and $\mathbf{v}$ are parallel vectors, then $\mathbf{u} \times \mathbf{v} = \mathbf{0}$.
2. *True or False* For any vector $\mathbf{v}$, $\mathbf{v} \times \mathbf{v} = \mathbf{0}$.
3. *True or False* If $\mathbf{u}$ and $\mathbf{v}$ are vectors, then $\mathbf{u} \times \mathbf{v} + \mathbf{v} \times \mathbf{u} = \mathbf{0}$.
4. *True or False* $\mathbf{u} \times \mathbf{v}$ is a vector that is parallel to both $\mathbf{u}$ and $\mathbf{v}$.
5. *True or False* $\|\mathbf{u} \times \mathbf{v}\| = \|\mathbf{u}\| \|\mathbf{v}\| \cos\theta$, where θ is the angle between $\mathbf{u}$ and $\mathbf{v}$.
6. *True or False* The area of the parallelogram having $\mathbf{u}$ and $\mathbf{v}$ as adjacent sides is the magnitude of the cross product of $\mathbf{u}$ and $\mathbf{v}$.

Skill Building

In Problems 7–14, find the value of each determinant.

7. $\begin{vmatrix} 3 & 4 \\ 1 & 2 \end{vmatrix}$
8. $\begin{vmatrix} -2 & 5 \\ 2 & -3 \end{vmatrix}$
9. $\begin{vmatrix} 6 & 5 \\ -2 & -1 \end{vmatrix}$
10. $\begin{vmatrix} -4 & 0 \\ 5 & 3 \end{vmatrix}$

11. $\begin{vmatrix} A & B & C \\ 2 & 1 & 4 \\ 1 & 3 & 1 \end{vmatrix}$
12. $\begin{vmatrix} A & B & C \\ 0 & 2 & 4 \\ 3 & 1 & 3 \end{vmatrix}$
13. $\begin{vmatrix} A & B & C \\ -1 & 3 & 5 \\ 5 & 0 & -2 \end{vmatrix}$
14. $\begin{vmatrix} A & B & C \\ 1 & -2 & -3 \\ 0 & 2 & -2 \end{vmatrix}$

In Problems 15–22, find (a) $\mathbf{v} \times \mathbf{w}$, (b) $\mathbf{w} \times \mathbf{v}$, (c) $\mathbf{w} \times \mathbf{w}$, and (d) $\mathbf{v} \times \mathbf{v}$.

15. $\mathbf{v} = 2\mathbf{i} - 3\mathbf{j} + \mathbf{k}$
    $\mathbf{w} = 3\mathbf{i} - 2\mathbf{j} - \mathbf{k}$
16. $\mathbf{v} = -\mathbf{i} + 3\mathbf{j} + 2\mathbf{k}$
    $\mathbf{w} = 3\mathbf{i} - 2\mathbf{j} - \mathbf{k}$
17. $\mathbf{v} = \mathbf{i} + \mathbf{j}$
    $\mathbf{w} = 2\mathbf{i} + \mathbf{j} + \mathbf{k}$
18. $\mathbf{v} = \mathbf{i} - 4\mathbf{j} + 2\mathbf{k}$
    $\mathbf{w} = 3\mathbf{i} + 2\mathbf{j} + \mathbf{k}$

19. $\mathbf{v} = 2\mathbf{i} - \mathbf{j} + 2\mathbf{k}$
    $\mathbf{w} = \mathbf{j} - \mathbf{k}$
20. $\mathbf{v} = 3\mathbf{i} + \mathbf{j} + 3\mathbf{k}$
    $\mathbf{w} = \mathbf{i} - \mathbf{k}$
21. $\mathbf{v} = \mathbf{i} - \mathbf{j} - \mathbf{k}$
    $\mathbf{w} = 4\mathbf{i} - 3\mathbf{k}$
22. $\mathbf{v} = 2\mathbf{i} - 3\mathbf{j}$
    $\mathbf{w} = 3\mathbf{j} - 2\mathbf{k}$

In Problems 23–44, use the given vectors $\mathbf{u}$, $\mathbf{v}$, and $\mathbf{w}$ to find each expression.

$$\mathbf{u} = 2\mathbf{i} - 3\mathbf{j} + \mathbf{k} \quad \mathbf{v} = -3\mathbf{i} + 3\mathbf{j} + 2\mathbf{k} \quad \mathbf{w} = \mathbf{i} + \mathbf{j} + 3\mathbf{k}$$

23. $\mathbf{u} \times \mathbf{v}$
24. $\mathbf{v} \times \mathbf{w}$
25. $\mathbf{v} \times \mathbf{u}$
26. $\mathbf{w} \times \mathbf{v}$
27. $\mathbf{v} \times \mathbf{v}$
28. $\mathbf{w} \times \mathbf{w}$
29. $(3\mathbf{u}) \times \mathbf{v}$
30. $\mathbf{v} \times (4\mathbf{w})$
31. $\mathbf{u} \times (2\mathbf{v})$
32. $(-3\mathbf{v}) \times \mathbf{w}$
33. $\mathbf{u} \cdot (\mathbf{u} \times \mathbf{v})$
34. $\mathbf{v} \cdot (\mathbf{v} \times \mathbf{w})$
35. $\mathbf{u} \cdot (\mathbf{v} \times \mathbf{w})$
36. $(\mathbf{u} \times \mathbf{v}) \cdot \mathbf{w}$
37. $\mathbf{v} \cdot (\mathbf{u} \times \mathbf{w})$
38. $(\mathbf{v} \times \mathbf{u}) \cdot \mathbf{w}$
39. $\mathbf{u} \times (\mathbf{v} \times \mathbf{v})$
40. $(\mathbf{w} \times \mathbf{w}) \times \mathbf{v}$

41. Find a vector orthogonal to both **u** and **v**.

42. Find a vector orthogonal to both **u** and **w**.

43. Find a vector orthogonal to both **u** and **i** + **j**.

44. Find a vector orthogonal to both **u** and **j** + **k**.

In Problems 45–48, find the area of the parallelogram with one corner at P_1 and adjacent sides $\overrightarrow{P_1P_2}$ and $\overrightarrow{P_1P_3}$.

45. $P_1 = (0,0,0)$, $P_2 = (1,2,3)$, $P_3 = (-2,3,0)$

46. $P_1 = (0,0,0)$, $P_2 = (2,3,1)$, $P_3 = (-2,4,1)$

47. $P_1 = (1,2,0)$, $P_2 = (-2,3,4)$, $P_3 = (0,-2,3)$

48. $P_1 = (-2,0,2)$, $P_2 = (2,1,-1)$, $P_3 = (2,-1,2)$

In Problems 49–52, find the area of the parallelogram with vertices P_1, P_2, P_3, and P_4.

49. $P_1 = (1,1,2)$, $P_2 = (1,2,3)$, $P_3 = (-2,3,0)$,
$P_4 = (-2,4,1)$

50. $P_1 = (2,1,1)$, $P_2 = (2,3,1)$, $P_3 = (-2,4,1)$,
$P_4 = (-2,6,1)$

51. $P_1 = (1,2,-1)$, $P_2 = (4,2,-3)$, $P_3 = (6,-5,2)$,
$P_4 = (9,-5,0)$

52. $P_1 = (-1,1,1)$, $P_2 = (-1,2,2)$, $P_3 = (-3,4,-5)$,
$P_4 = (-3,5,-4)$

Applications and Extensions

53. Find a unit vector normal to the plane containing $\mathbf{v} = \mathbf{i} + 3\mathbf{j} - 2\mathbf{k}$ and $\mathbf{w} = -2\mathbf{i} + \mathbf{j} + 3\mathbf{k}$.

54. Find a unit vector normal to the plane containing $\mathbf{v} = 2\mathbf{i} + 3\mathbf{j} - \mathbf{k}$ and $\mathbf{w} = -2\mathbf{i} - 4\mathbf{j} - 3\mathbf{k}$.

55. Volume of a Parallelepiped A **parallelepiped** is a prism whose faces are all parallelograms. Let **A**, **B**, and **C** be the vectors that define the parallelepiped shown in the figure. The volume V of the parallelepiped is given by the formula $V = |(\mathbf{A} \times \mathbf{B}) \cdot \mathbf{C}|$.

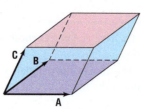

Find the volume of a parallelepiped if the defining vectors are $\mathbf{A} = 3\mathbf{i} - 2\mathbf{j} + 4\mathbf{k}$, $\mathbf{B} = 2\mathbf{i} + \mathbf{j} - 2\mathbf{k}$, and $\mathbf{C} = 3\mathbf{i} - 6\mathbf{j} - 2\mathbf{k}$.

56. Volume of a Parallelepiped Refer to Problem 55. Find the volume of a parallelepiped whose defining vectors are $\mathbf{A} = \mathbf{i} + 6\mathbf{k}$, $\mathbf{B} = 2\mathbf{i} + 3\mathbf{j} - 8\mathbf{k}$, and $\mathbf{C} = 8\mathbf{i} - 5\mathbf{j} + 6\mathbf{k}$.

57. Prove for vectors **u** and **v** that

$$\|\mathbf{u} \times \mathbf{v}\|^2 = \|\mathbf{u}\|^2\|\mathbf{v}\|^2 - (\mathbf{u} \cdot \mathbf{v})^2$$

[**Hint**: Proceed as in the proof of property (4), computing first the left side and then the right side.]

58. Show that if **u** and **v** are orthogonal, then

$$\|\mathbf{u} \times \mathbf{v}\| = \|\mathbf{u}\| \|\mathbf{v}\|$$

59. Show that if **u** and **v** are orthogonal unit vectors, then **u** × **v** is also a unit vector.

60. Prove property (3).

61. Prove property (5).

62. Prove property (9).

[**Hint**: Use the result of Problem 57 and the fact that if θ is the angle between **u** and **v**, then $\mathbf{u} \cdot \mathbf{v} = \|\mathbf{u}\| \|\mathbf{v}\| \cos \theta$.]

63. Challenge Problem Show that the vector $2\mathbf{v} \times 3\mathbf{w}$ is orthogonal to both **v** and **w**.

64. Challenge Problem If $\mathbf{v}, \mathbf{w}, \mathbf{u}$ and $\mathbf{a}, \mathbf{b}, \mathbf{c}$ are vectors for which

$$\mathbf{a} \cdot \mathbf{v} = \mathbf{b} \cdot \mathbf{w} = \mathbf{c} \cdot \mathbf{u} = 1$$

and

$$\mathbf{a} \cdot \mathbf{w} = \mathbf{a} \cdot \mathbf{u} = \mathbf{b} \cdot \mathbf{v} = \mathbf{b} \cdot \mathbf{u} = \mathbf{c} \cdot \mathbf{v} = \mathbf{c} \cdot \mathbf{w} = 0$$

show that

$$\mathbf{a} = \frac{\mathbf{w} \times \mathbf{u}}{\mathbf{v} \cdot (\mathbf{w} \times \mathbf{u})} \qquad \mathbf{b} = \frac{\mathbf{u} \times \mathbf{v}}{\mathbf{w} \cdot (\mathbf{u} \times \mathbf{v})} \qquad \mathbf{c} = \frac{\mathbf{v} \times \mathbf{w}}{\mathbf{u} \cdot (\mathbf{v} \times \mathbf{w})}$$

Discussion and Writing

65. If $\mathbf{u} \cdot \mathbf{v} = 0$ and $\mathbf{u} \times \mathbf{v} = \mathbf{0}$, what, if anything, can you conclude about **u** and **v**?

Retain Your Knowledge

Problems 66–73 are based on material learned earlier in the course. The purpose of these problems is to keep the material fresh in your mind so that you are better prepared for the final exam.

66. Find the exact value of $\cos^{-1}\left(\dfrac{1}{\sqrt{2}}\right)$.

67. Find two pairs of polar coordinates (r, θ), one with $r > 0$ and the other with $r < 0$, for the point with rectangular coordinates $(-8, -15)$. Express θ in radians.

68. Find the exact value of $\sec\left[\sin^{-1}\left(-\dfrac{\sqrt{3}}{2}\right)\right]$.

69. Find the domain of $f(x) = \dfrac{3x + 4}{x^2 - 16}$.

70. If $\cos \theta = -\dfrac{3}{8}, \dfrac{\pi}{2} < \theta < \pi$, find the exact value of $\sin\dfrac{\theta}{2}$.

71. Find the area of the triangle for which $a = 8$, $b = 9$, and $C = 60°$.

72. Rationalize the numerator: $\dfrac{\sqrt{x} - 4}{x}$

73. Find the average rate of change of $f(x) = \csc^{-1} x$ from 1 to 2.

Chapter Review

Things to Know

Polar Coordinates (pp. 316–323)

Relationship between polar coordinates (r, θ) and rectangular coordinates (x, y) (pp. 318 and 322)	$x = r \cos \theta, y = r \sin \theta$
	$r^2 = x^2 + y^2, \tan \theta = \dfrac{y}{x}$, if $x \neq 0$
	$r = y, \theta = \dfrac{\pi}{2}$, if $x = 0$

Complex Numbers and De Moivre's Theorem (pp. 340–346)

Polar form of a complex number (p. 341)	If $z = x + yi$, then $z = r(\cos \theta + i \sin \theta)$, where $r \geq 0$ and $0 \leq \theta < 2\pi$.
Exponential form of a complex number (p. 342)	$z = re^{i\theta}$, where $e^{i\theta} = \cos \theta + i \sin \theta, \theta$ in radians
De Moivre's Theorem (p. 344)	If $z = re^{i\theta}$, then $z^n = r^n e^{i(n\theta)}$, where $n \geq 1$ is an integer.
nth root of a complex number $w = re^{i\theta}, w \neq 0$ (p. 345)	$z_k = \sqrt[n]{r}\, e^{i(1/n)(\theta + 2k\pi)}, \quad k = 0, \ldots, n-1$, where $n \geq 2$ is an integer

Vectors (pp. 349–359)

	A quantity having magnitude and direction; equivalent to a directed line segment $\overrightarrow{PQ}$
Position vector (pp. 352 and 353)	A vector whose initial point is at the origin
Unit vector (pp. 352, 355 and 375)	A vector whose magnitude is 1
Direction angle of a vector $\mathbf{v}$ (p. 356)	The angle $\alpha, 0° \leq \alpha < 360°$, between $\mathbf{i}$ and $\mathbf{v}$
Dot product (pp. 364 and 375)	If $\mathbf{v} = a_1\mathbf{i} + b_1\mathbf{j}$ and $\mathbf{w} = a_2\mathbf{i} + b_2\mathbf{j}$, then $\mathbf{v} \cdot \mathbf{w} = a_1a_2 + b_1b_2$.
Angle θ between two nonzero vectors $\mathbf{u}$ and $\mathbf{v}$ (p. 365)	$\cos \theta = \dfrac{\mathbf{u} \cdot \mathbf{v}}{\|\mathbf{u}\| \|\mathbf{v}\|}, 0 \leq \theta \leq \pi$
Work (p. 368)	Work $W = $ (magnitude of force) (distance) $= \|\mathbf{F}\|\|\overrightarrow{AB}\|$ $W = \mathbf{F} \cdot \overrightarrow{AB}$
Direction angles of a vector in space (p. 377)	If $\mathbf{v} = a\mathbf{i} + b\mathbf{j} + c\mathbf{k}$, then $\mathbf{v} = \|\mathbf{v}\|[(\cos \alpha)\mathbf{i} + (\cos \beta)\mathbf{j} + (\cos \gamma)\mathbf{k}]$, where $\cos \alpha = \dfrac{a}{\|\mathbf{v}\|}, \cos \beta = \dfrac{b}{\|\mathbf{v}\|}$, and $\cos \gamma = \dfrac{c}{\|\mathbf{v}\|}$.
Cross product (p. 381)	If $\mathbf{v} = a_1\mathbf{i} + b_1\mathbf{j} + c_1\mathbf{k}$ and $\mathbf{w} = a_2\mathbf{i} + b_2\mathbf{j} + c_2\mathbf{k}$, then $\mathbf{v} \times \mathbf{w} = [b_1c_2 - b_2c_1]\mathbf{i} - [a_1c_2 - a_2c_1]\mathbf{j} + [a_1b_2 - a_2b_1]\mathbf{k}$.
Area of a parallelogram (p. 384)	$\|\mathbf{u} \times \mathbf{v}\| = \|\mathbf{u}\| \|\mathbf{v}\| \sin \theta$, where θ is the angle between the two adjacent sides $\mathbf{u}$ and $\mathbf{v}$.

Objectives

Section	You should be able to . . .	Example(s)	Review Exercises
5.1	**1** Plot points using polar coordinates (p. 316)	1–3	1–3
	2 Convert from polar coordinates to rectangular coordinates (p. 318)	4	1–3
	3 Convert from rectangular coordinates to polar coordinates (p. 320)	5–7	4–6
	4 Transform equations between polar and rectangular forms (p. 322)	8, 9	7(a)–10(a)

Section	You should be able to . . .	Example(s)	Review Exercises
5.2	**1** Identify and graph polar equations by converting to rectangular equations (p. 326)	1–6	7(b)–10(b)
	2 Test polar equations for symmetry (p. 329)	7–10	11–13
	3 Graph polar equations by plotting points (p. 330)	7–13	11–13
5.3	**1** Plot points in the complex plane (p. 340)	1	16–18
	2 Convert a complex number between rectangular form and polar form or exponential form (p. 341)	2, 3	14–18
	3 Find products and quotients of complex numbers (p. 343)	4	19–21
	4 Use De Moivre's Theorem (p. 344)	5, 6	22–25
	5 Find complex roots (p. 345)	7	26
5.4	**1** Graph vectors (p. 351)	1	27, 28
	2 Find a position vector (p. 352)	2	29, 30
	3 Add and subtract vectors algebraically (p. 354)	3	31
	4 Find a scalar multiple and the magnitude of a vector (p. 354)	4	29, 30, 32–34
	5 Find a unit vector (p. 355)	5	35
	6 Find a vector from its direction and magnitude (p. 356)	6	36, 37
	7 Model with vectors (p. 357)	8–10	59, 60
5.5	**1** Find the dot product of two vectors (p. 364)	1	46, 47
	2 Find the angle between two vectors (p. 365)	2	46, 47
	3 Determine whether two vectors are parallel (p. 366)	3	50–52
	4 Determine whether two vectors are orthogonal (p. 366)	4	50–52
	5 Decompose a vector into two orthogonal vectors (p. 366)	5, 6	53, 54, 62
	6 Compute work (p. 368)	7	61
5.6	**1** Find the distance between two points in space (p. 372)	1	38
	2 Find position vectors in space (p. 373)	2	39
	3 Perform operations on vectors (p. 374)	3–5	40–42
	4 Find the dot product (p. 375)	6	48, 49
	5 Find the angle between two vectors (p. 376)	7	48, 49
	6 Find the direction angles of a vector (p. 377)	8–10	55
5.7	**1** Find the cross product of two vectors (p. 381)	1–3	43, 44
	2 Know algebraic properties of the cross product (p. 382)	p. 383	57, 58
	3 Know geometric properties of the cross product (p. 383)	p. 383	56
	4 Find a vector orthogonal to two given vectors (p. 384)	4	45
	5 Find the area of a parallelogram (p. 384)	5	56

Review Exercises

In Problems 1–3, plot each point given in polar coordinates, and find its rectangular coordinates.

1. $\left(3, \dfrac{\pi}{6}\right)$ **2.** $\left(-2, \dfrac{4\pi}{3}\right)$ **3.** $\left(-3, -\dfrac{\pi}{2}\right)$

In Problems 4–6, the rectangular coordinates of a point are given. Find two pairs of polar coordinates (r, θ) for each point, one with $r > 0$ and the other with $r < 0$. Express θ in radians.

4. $(-3, 3)$ **5.** $(0, -2)$ **6.** $(3, 4)$

In Problems 7–10, the variables r and θ represent polar coordinates.
(a) Write each polar equation as an equation in rectangular coordinates (x, y). (b) Identify the equation and graph it.

7. $r = 2 \sin \theta$ **8.** $r = 5$ **9.** $\theta = \dfrac{\pi}{4}$ **10.** $r^2 + 4r \sin \theta - 8r \cos \theta = 5$

In Problems 11–13, graph each polar equation. Be sure to test for symmetry.

11. $r = 4\cos\theta$

12. $r = 3 - 3\sin\theta$

13. $r = 4 - \cos\theta$

In Problems 14 and 15, write each complex number in polar form and in exponential form.

14. $-1 - i$

15. $4 - 3i$

In Problems 16–18, write each complex number in rectangular form, and plot each in the complex plane.

16. $2e^{i\cdot 5\pi/6}$

17. $3\left(\cos\dfrac{2\pi}{3} + i\sin\dfrac{2\pi}{3}\right)$

18. $0.1e^{i\cdot 35\pi/18}$

In Problems 19–21, find zw and $\dfrac{z}{w}$. Write your answers in polar form and in exponential form.

19. $z = \cos\dfrac{4\pi}{9} + i\sin\dfrac{4\pi}{9}$
$w = \cos\dfrac{5\pi}{18} + i\sin\dfrac{5\pi}{18}$

20. $z = 3\left(\cos\dfrac{9\pi}{5} + i\sin\dfrac{9\pi}{5}\right)$
$w = 2\left(\cos\dfrac{\pi}{5} + i\sin\dfrac{\pi}{5}\right)$

21. $z = 5\left(\cos\dfrac{\pi}{18} + i\sin\dfrac{\pi}{18}\right)$
$w = \cos\dfrac{71\pi}{36} + i\sin\dfrac{71\pi}{36}$

In Problems 22–25, write each expression in rectangular form $x + yi$ and in exponential form $re^{i\theta}$.

22. $\left[3\left(\cos\dfrac{\pi}{9} + i\sin\dfrac{\pi}{9}\right)\right]^3$

23. $\left[\sqrt{2}\left(\cos\dfrac{5\pi}{8} + i\sin\dfrac{5\pi}{8}\right)\right]^4$

24. $\left(1 - \sqrt{3}i\right)^6$

25. $(3 + 4i)^4$

26. Find all the complex cube roots of 27.

In Problems 27 and 28, use the figure to graph each of the following:

27. $\mathbf{u} + \mathbf{v}$

28. $2\mathbf{u} + 3\mathbf{v}$

In Problems 29 and 30, the vector $\mathbf{v}$ is represented by the directed line segment $\overrightarrow{PQ}$. Write $\mathbf{v}$ in the form $a\mathbf{i} + b\mathbf{j}$ and find $\|\mathbf{v}\|$.

29. $P = (1, -2);\quad Q = (3, -6)$

30. $P = (0, -2);\quad Q = (-1, 1)$

In Problems 31–35, use the vectors $\mathbf{v} = -2\mathbf{i} + \mathbf{j}$ and $\mathbf{w} = 4\mathbf{i} - 3\mathbf{j}$ to find:

31. $\mathbf{v} + \mathbf{w}$

32. $4\mathbf{v} - 3\mathbf{w}$

33. $\|\mathbf{v}\|$

34. $\|\mathbf{v}\| + \|\mathbf{w}\|$

35. A unit vector in the same direction as $\mathbf{v}$.

36. Find the vector $\mathbf{v}$ in the xy-plane with magnitude 3 if the direction angle of $\mathbf{v}$ is $60°$.

37. Find the direction angle α of $\mathbf{v} = -\mathbf{i} + \sqrt{3}\,\mathbf{j}$.

38. Find the distance from $P_1 = (1, 3, -2)$ to $P_2 = (4, -2, 1)$.

39. A vector $\mathbf{v}$ has initial point $P = (1, 3, -2)$ and terminal point $Q = (4, -2, 1)$. Write $\mathbf{v}$ in the form $\mathbf{v} = a\mathbf{i} + b\mathbf{j} + c\mathbf{k}$.

In Problems 40–45, use the vectors $\mathbf{v} = 3\mathbf{i} + \mathbf{j} - 2\mathbf{k}$ and $\mathbf{w} = -3\mathbf{i} + 2\mathbf{j} - \mathbf{k}$ to find each expression.

40. $4\mathbf{v} - 3\mathbf{w}$

41. $\|\mathbf{v} - \mathbf{w}\|$

42. $\|\mathbf{v}\| - \|\mathbf{w}\|$

43. $\mathbf{v} \times \mathbf{w}$

44. $\mathbf{v} \cdot (\mathbf{v} \times \mathbf{w})$

45. Find a unit vector orthogonal to both $\mathbf{v}$ and $\mathbf{w}$.

In Problems 46–49, find the dot product $\mathbf{v} \cdot \mathbf{w}$ and the angle between $\mathbf{v}$ and $\mathbf{w}$.

46. $\mathbf{v} = -2\mathbf{i} + \mathbf{j},\quad \mathbf{w} = 4\mathbf{i} - 3\mathbf{j}$

47. $\mathbf{v} = \mathbf{i} - 3\mathbf{j},\quad \mathbf{w} = -\mathbf{i} + \mathbf{j}$

48. $\mathbf{v} = \mathbf{i} + \mathbf{j} + \mathbf{k},\quad \mathbf{w} = \mathbf{i} - \mathbf{j} + \mathbf{k}$

49. $\mathbf{v} = 4\mathbf{i} - \mathbf{j} + 2\mathbf{k},\quad \mathbf{w} = \mathbf{i} - 2\mathbf{j} - 3\mathbf{k}$

In Problems 50–52, determine whether $\mathbf{v}$ and $\mathbf{w}$ are parallel, orthogonal, or neither.

50. $\mathbf{v} = 2\mathbf{i} + 3\mathbf{j};\quad \mathbf{w} = -4\mathbf{i} - 6\mathbf{j}$

51. $\mathbf{v} = -2\mathbf{i} + 2\mathbf{j};\quad \mathbf{w} = -3\mathbf{i} + 2\mathbf{j}$

52. $\mathbf{v} = 3\mathbf{i} - 2\mathbf{j};\quad \mathbf{w} = 4\mathbf{i} + 6\mathbf{j}$

In Problems 53 and 54, decompose $\mathbf{v}$ into two vectors, one parallel to $\mathbf{w}$ and the other orthogonal to $\mathbf{w}$.

53. $\mathbf{v} = 2\mathbf{i} + \mathbf{j};\quad \mathbf{w} = -4\mathbf{i} + 3\mathbf{j}$

54. $\mathbf{v} = 2\mathbf{i} + 3\mathbf{j};\ \mathbf{w} = 3\mathbf{i} + \mathbf{j}$

55. Find the direction angles of the vector $\mathbf{v} = 3\mathbf{i} - 4\mathbf{j} + 2\mathbf{k}$.

56. Find the area of the parallelogram with vertices $P_1 = (1, 1, 1)$, $P_2 = (2, 3, 4)$, $P_3 = (6, 5, 2)$, and $P_4 = (7, 7, 5)$.

57. If $\mathbf{u} \times \mathbf{v} = 2\mathbf{i} - 3\mathbf{j} + \mathbf{k}$, what is $\mathbf{v} \times \mathbf{u}$?

58. Suppose that $\mathbf{u} = 3\mathbf{v}$. What is $\mathbf{u} \times \mathbf{v}$?

59. Actual Speed and Direction of a Swimmer A swimmer can maintain a constant speed of 5 miles per hour. If the swimmer heads directly across a river that has a current moving at the rate of 2 miles per hour, what is the actual speed of the swimmer? (See the figure.) If the river is 1 mile wide, how far downstream will the swimmer end up from the point directly across the river from the starting point?

Current

Swimmer's
direction

Direction
of swimmer
due to current

60. Static Equilibrium A weight of 2000 pounds is suspended from two cables, as shown in the figure. What are the tensions in the two cables?

40°   30°

2000
pounds

61. Computing Work Find the work done by a force of 5 pounds acting in the direction 60° to the horizontal in moving an object 20 feet from $(0, 0)$ to $(20, 0)$.

62. Braking Load A moving van with a gross weight of 8000 pounds is parked on a street with a 5° grade. Find the magnitude of the force required to keep the van from rolling down the hill. What is the magnitude of the force perpendicular to the hill?

Chapter Test

CHAPTER
Test Prep
VIDEOS

The Chapter Test Prep Videos include step-by-step solutions to all chapter test exercises. These videos are available in MyLab™ Math, or on this text's YouTube Channel. Refer to the Preface for a link to the YouTube channel.

In Problems 1–3, plot each point given in polar coordinates.

1. $\left(2, \dfrac{3\pi}{4}\right)$

2. $\left(3, -\dfrac{\pi}{6}\right)$

3. $\left(-4, \dfrac{\pi}{3}\right)$

4. Convert $\left(2, 2\sqrt{3}\right)$ from rectangular coordinates to polar coordinates (r, θ), where $r > 0$ and $0 \le \theta < 2\pi$.

In Problems 5–7, convert the polar equation to a rectangular equation. Graph the equation.

5. $r = 7$

6. $\tan \theta = 3$

7. $r \sin^2 \theta + 8 \sin \theta = r$

In Problems 8 and 9, test the polar equation for symmetry with respect to the pole, the polar axis, and the line $\theta = \dfrac{\pi}{2}$.

8. $r^2 \cos \theta = 5$

9. $r = 5 \sin \theta \cos^2 \theta$

In Problems 10–12, perform the given operation, where
$$z = 2\left(\cos \frac{17\pi}{36} + i \sin \frac{17\pi}{36}\right) \text{ and } w = 3\left(\cos \frac{11\pi}{90} + i \sin \frac{11\pi}{90}\right).$$

Write the answer in polar form and in exponential form.

10. $z \cdot w$

11. $\dfrac{w}{z}$

12. w^5

13. Find all the complex cube roots of $-8 + 8\sqrt{3}i$. Then plot them in the complex plane.

In Problems 14–18, $P_1 = \left(3\sqrt{2}, 7\sqrt{2}\right)$ and $P_2 = \left(8\sqrt{2}, 2\sqrt{2}\right)$.

14. Find the position vector $\mathbf{v}$ equal to $\overrightarrow{P_1 P_2}$.

15. Find $\|\mathbf{v}\|$.

16. Find the unit vector in the direction of $\mathbf{v}$.

17. Find the direction angle of $\mathbf{v}$.

18. Write the vector $\mathbf{v}$ in terms of its vertical and horizontal components.

In Problems 19–22, $\mathbf{v}_1 = 4\mathbf{i} + 6\mathbf{j}$, $\mathbf{v}_2 = -3\mathbf{i} - 6\mathbf{j}$, $\mathbf{v}_3 = -8\mathbf{i} + 4\mathbf{j}$, and $\mathbf{v}_4 = 10\mathbf{i} + 15\mathbf{j}$.

19. Find the vector $\mathbf{v}_1 + 2\mathbf{v}_2 - \mathbf{v}_3$.

20. Which two vectors are parallel?

21. Which two vectors are orthogonal?

22. Find the angle between the vectors $\mathbf{v}_1$ and $\mathbf{v}_2$.

In Problems 23–25, use the vectors $\mathbf{u} = 2\mathbf{i} - 3\mathbf{j} + \mathbf{k}$ and $\mathbf{v} = -\mathbf{i} + 3\mathbf{j} + 2\mathbf{k}$.

23. Find $\mathbf{u} \times \mathbf{v}$.

24. Find the direction angles for $\mathbf{u}$.

25. Find the area of the parallelogram that has $\mathbf{u}$ and $\mathbf{v}$ as adjacent sides.

26. A 1200-pound chandelier is to be suspended over a large ballroom; the chandelier will be hung on two cables of equal length whose ends will be attached to the ceiling, 16 feet apart. The chandelier will be free-hanging so that the ends of the cable will make equal angles with the ceiling. If the top of the chandelier is to be 16 feet from the ceiling, what is the minimum tension each cable must be able to endure?

Cumulative Review

1. Find the real solutions, if any, of the equation

$$x^2 - 9 = 0$$

2. Find an equation for the line containing the origin that makes an angle of 30° with the positive x-axis.

3. Find an equation for the circle with center at the point $(0, 1)$ and radius 3. Graph this circle.

4. What is the domain of the function $f(x) = \dfrac{1}{1 - 2x}$?

5. Test the equation $x^2 + y^3 = 2x^4$ for symmetry with respect to the x-axis, the y-axis, and the origin.

6. Graph the function $y = |\sin x|$.

7. Graph the function $y = \sin|x|$.

8. Find the exact value of $\sin^{-1}\left(-\dfrac{1}{2}\right)$.

9. Graph the equations $x = 3$ and $y = 4$ on the same set of rectangular coordinates.

10. Graph the equations $r = 2$ and $\theta = \dfrac{\pi}{3}$ on the same set of polar coordinates.

11. What are the amplitude and period of $y = -4\cos(\pi x)$?

Chapter Projects

I. **Modeling Aircraft Motion** Four aerodynamic forces act on an airplane in flight: lift, weight, thrust, and drag. While an aircraft is in flight, these four forces continuously battle each other. Weight opposes lift, and drag opposes thrust. See the diagram below. In balanced flight at constant speed, the lift and weight are equal, and the thrust and drag are equal.

1. What will happen to the aircraft if the lift is held constant while the weight is decreased (say, from burning off fuel)?

2. What will happen to the aircraft if the lift is decreased while the weight is held constant?

3. What will happen to the aircraft if the thrust is increased while the drag is held constant?

4. What will happen to the aircraft if the drag is increased while the thrust is held constant?

In 1903 the Wright brothers made the first controlled powered flight. The weight of their plane was approximately 700 pounds (lb). Newton's Second Law of Motion states that force = mass × acceleration $(F = ma)$. If the mass is measured in kilograms (kg) and acceleration in meters per second squared (m/s^2), then the force will be measured in newtons (N). [**Note:** $1\ N = 1\ kg \cdot m/s^2$.]

5. If $1\ kg = 2.205\ lb$, convert the weight of the Wright brothers' plane to kilograms.

6. If acceleration due to gravity is $a = 9.80\ m/s^2$, determine the force due to weight on the Wright brothers' plane.

7. What must be true about the lift force of the Wright brothers' plane for it to get off the ground?

8. The weight of a fully loaded Cessna 172P is 2400 lb. What lift force is required to get this plane off the ground?

9. The maximum gross weight of a Boeing 787 is 560,000 lb. What lift force is required to get this jet off the ground?

The following projects are available at the Instructor's Resource Center (IRC):

II. **Project at Motorola** *Signal Fades Due to Interference* Complex trigonometric functions are used to ensure that a cellphone has optimal reception as the user travels up and down an elevator.

III. **Compound Interest** The effect of continuously compounded interest is analyzed using polar coordinates.

IV. **Complex Equations** Analysis of complex equations illustrates the connections between complex and real equations. At times, using complex equations is more efficient for proving mathematical theorems.

6

Analytic Geometry

The Orbit of Comet Hale-Bopp

The orbits of Comet Hale-Bopp and Earth can be modeled using *ellipses*, the subject of Section 6.3. The Internet-based Project at the end of this chapter explores the possibility of Comet Hale-Bopp colliding with Earth.

 —*See the Internet-based Chapter Project I*—

Outline

⬅ A Look Back

In Chapter 1, we introduced rectangular coordinates and showed how geometry problems can be solved algebraically. We defined a circle geometrically and then used the distance formula and rectangular coordinates to obtain an equation for a circle.

A Look Ahead ➡

In this chapter, geometric definitions are given for the *conics*, and the distance formula and rectangular coordinates are used to obtain their equations.

Historically, Apollonius (200 BC) was among the first to study conics and discover some of their interesting properties. Today, conics are still studied because of their many uses. *Paraboloids of revolution* (parabolas rotated about their axes of symmetry) are used as signal collectors (the satellite dishes used with radar and dish TV, for example), as solar energy collectors, and as reflectors (telescopes, light projection, and so on). The planets circle the Sun in approximately *elliptical* orbits. Elliptical surfaces are used to reflect signals such as light and sound from one place to another. A third conic, the *hyperbola*, is used to determine the location of ships or sound sources, such as lightning strikes.

The Greeks used Euclidean geometry to study conics. However, we use the more powerful methods of analytic geometry, which uses both algebra and geometry, for our study of conics.

In Section 6.7, we introduce *parametric equations*, which allow us to represent graphs of curves that are not the graph of a function, such as a circle.

6.1 Conics

OBJECTIVE 1 Know the Names of the Conics (p. 393)

1 Know the Names of the Conics

The word *conic* derives from the word *cone*, which is a geometric figure that can be constructed in the following way: Let *a* and *g* be two distinct lines that intersect at a point *V*. Keep the line *a* fixed. Now rotate the line *g* about *a*, while maintaining the same angle between *a* and *g*. The collection of points swept out (generated) by the line *g* is called a **right circular cone**. See Figure 1. The fixed line *a* is called the **axis** of the cone; the point *V* is its **vertex**; the lines that pass through *V* and make the same angle with *a* as *g* are **generators** of the cone. Each generator is a line that lies entirely on the cone. The cone consists of two parts, called **nappes**, that intersect at the vertex.

Conics, an abbreviation for **conic sections**, are curves that result from the intersection of a right circular cone and a plane. We discuss only conics formed where the plane does not contain the vertex. These conics are **circles** when the plane is perpendicular to the axis of the cone and intersects each generator; **ellipses** when the plane is tilted slightly so that it intersects each generator, but intersects only one nappe of the cone; **parabolas** when the plane is tilted farther so that it is parallel to one (and only one) generator and intersects only one nappe of the cone; and **hyperbolas** when the plane intersects both nappes. See Figure 2.

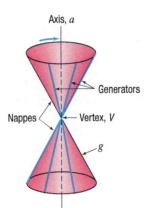

Figure 1 Right circular cone

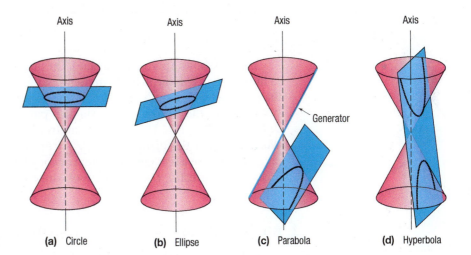

Figure 2   (a) Circle   (b) Ellipse   (c) Parabola   (d) Hyperbola

If the plane contains the vertex, the intersection of the plane and the cone is a point, a line, or a pair of intersecting lines. These are called **degenerate conics**.

Conic sections are used in modeling many applications. For example, parabolas are used in describing searchlights and telescopes (see Figures 14 and 15 on page 399). Ellipses are used to model the orbits of planets and whispering chambers (see pages 409–410). And hyperbolas are used to locate lightning strikes and model nuclear cooling towers (see Problems 76 and 77 in Section 6.4).

6.2 The Parabola

PREPARING FOR THIS SECTION *Before getting started, review the following:*

- Distance Formula (Section 1.1, pp. 3–4)
- Symmetry (Section 1.2, pp. 12–14)
- Square Root Method (Section A.4, p. A31)
- Completing the Square (Section A.3, pp. A24–A25)
- Graphing Techniques: Transformations (Section 1.6, pp. 67–77)

Now Work the 'Are You Prepared?' problems on page 400.

OBJECTIVES 1 Analyze Parabolas with Vertex at the Origin (p. 394)
2 Analyze Parabolas with Vertex at (*h, k*) (p. 397)
3 Solve Applied Problems Involving Parabolas (p. 399)

Earlier we stated that the graph of $f(x) = x^2$ is a parabola. In this section, we give a geometric definition of a parabola and use it to obtain an equation.

DEFINITION Parabola

A **parabola** is the collection of all points P in a plane that are the same distance d from a fixed point F as they are from a fixed line D. The point F is called the **focus** of the parabola, and the line D is its **directrix**. As a result, a parabola is the set of points P for which

$$d(F, P) = d(P, D) \tag{1}$$

Figure 3 shows a parabola (in blue). The line through the focus F and perpendicular to the directrix D is the **axis of symmetry** of the parabola. The point of intersection of the parabola with its axis of symmetry is the **vertex** V.

Because the vertex V lies on the parabola, it must satisfy equation (1): $d(F, V) = d(V, D)$. The vertex is midway between the focus and the directrix. We let a equal the distance $d(F, V)$ from F to V. To derive an equation for a parabola, we use a rectangular system of coordinates positioned so that the vertex V, focus F, and directrix D of the parabola are conveniently located.

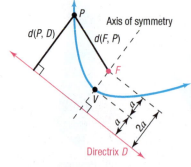

Figure 3 Parabola

1 Analyze Parabolas with Vertex at the Origin

If we locate the vertex V at the origin $(0, 0)$, we can conveniently position the focus F on either the x-axis or the y-axis. First, consider the case where the focus F is on the positive x-axis, as shown in Figure 4. Because the distance from F to V is a, the coordinates of F will be $(a, 0)$ with $a > 0$. Similarly, because the distance from V to the directrix D is also a, and because D is perpendicular to the x-axis (since the x-axis is the axis of symmetry), the equation of the directrix D is $x = -a$.

Now, if $P = (x, y)$ is any point on the parabola, then P satisfies equation (1):

$$d(F, P) = d(P, D)$$

So we have

$$\sqrt{(x-a)^2 + (y-0)^2} = |x+a| \quad \text{Use the Distance Formula.}$$
$$(x-a)^2 + y^2 = (x+a)^2 \quad \text{Square both sides.}$$
$$x^2 - 2ax + a^2 + y^2 = x^2 + 2ax + a^2 \quad \text{Multiply out.}$$
$$y^2 = 4ax \quad \text{Simplify.}$$

Figure 4

> **THEOREM** Equation of a Parabola: Vertex at $(0, 0)$, Focus at $(a, 0)$, $a > 0$
>
> The equation of a parabola with vertex at $(0, 0)$, focus at $(a, 0)$, and directrix $x = -a$, $a > 0$, is
>
> $$y^2 = 4ax \qquad (2)$$

Recall that a is the distance from the vertex to the focus of a parabola. When graphing the parabola $y^2 = 4ax$, it is helpful to determine the "opening" by finding the points that lie directly above and below the focus $(a, 0)$. Do this by substituting $x = a$ in $y^2 = 4ax$, so $y^2 = 4a \cdot a = 4a^2$, or $y = \pm 2a$. The line segment joining the two points, $(a, 2a)$ and $(a, -2a)$, is called the **latus rectum**; its length is $4a$.

EXAMPLE 1 Finding the Equation of a Parabola and Graphing It

Find an equation of the parabola with vertex at $(0, 0)$ and focus at $(3, 0)$. Graph the equation.

Solution The distance from the vertex $(0, 0)$ to the focus $(3, 0)$ is $a = 3$. Then, the equation of the parabola is

$$y^2 = 4ax \quad \text{Equation (2)}$$
$$y^2 = 12x \quad a = 3$$

To graph the parabola, find the two points that determine the latus rectum by substituting $x = 3$. Then

$$y^2 = 12x = 12 \cdot 3 = 36 \quad x = 3$$
$$y = \pm 6 \qquad\qquad \text{Solve for } y.$$

The points $(3, 6)$ and $(3, -6)$ determine the latus rectum. These points help graph the parabola because they determine the "opening." See Figure 5.

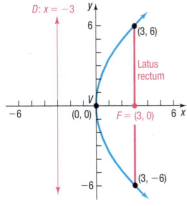

Figure 5 $y^2 = 12x$

— **Now Work** PROBLEM 21

🖥 **COMMENT** To graph the parabola $y^2 = 12x$ discussed in Example 1, graph the two functions $Y_1 = \sqrt{12x}$ and $Y_2 = -\sqrt{12x}$. Do this and compare what you see with Figure 5. ∎

By reversing the steps used to obtain equation (2), it follows that the graph of an equation of the form $y^2 = 4ax$ is a parabola; its vertex is at $(0, 0)$, its focus is at $(a, 0)$, its directrix is the line $x = -a$, and its axis of symmetry is the x-axis.

For the remainder of this section, the direction "**Analyze the equation**" means to find the vertex, focus, and directrix of the parabola and graph it.

EXAMPLE 2 Analyzing the Equation of a Parabola

Analyze the equation $y^2 = 8x$.

Solution The equation $y^2 = 8x$ is of the form $y^2 = 4ax$, where $4a = 8$, so $a = 2$. Consequently, the graph of the equation is a parabola with vertex at $(0, 0)$ and focus on the positive x-axis at $(a, 0) = (2, 0)$. The directrix is the vertical line $a = 2$ units to the left of the y-axis. That is, $x = -2$. The two points that determine the latus rectum are obtained by substituting $x = 2$ in the equation $y^2 = 8x$. Then $y^2 = 16$, so $y = \pm 4$. The points $(2, -4)$ and $(2, 4)$ determine the latus rectum. See Figure 6 for the graph.

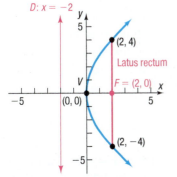

Figure 6 $y^2 = 8x$

Recall that we obtained equation (2) after placing the focus on the positive x-axis. Placing the focus on the negative x-axis, positive y-axis, or negative y-axis results in a different form of the equation for the parabola. The four forms of the equation of a parabola with vertex at $(0, 0)$ and focus on a coordinate axis a distance a from $(0, 0)$ are given in Table 1, and their graphs are given in Figure 7. Notice that each graph is symmetric with respect to its axis of symmetry.

Table 1

Equations of a Parabola: Vertex at (0, 0); Focus on an Axis; $a > 0$				
Vertex	**Focus**	**Directrix**	**Equation**	**Description**
$(0, 0)$	$(a, 0)$	$x = -a$	$y^2 = 4ax$	Axis of symmetry is the x-axis, the parabola opens right
$(0, 0)$	$(-a, 0)$	$x = a$	$y^2 = -4ax$	Axis of symmetry is the x-axis, the parabola opens left
$(0, 0)$	$(0, a)$	$y = -a$	$x^2 = 4ay$	Axis of symmetry is the y-axis, the parabola opens up (is concave up)
$(0, 0)$	$(0, -a)$	$y = a$	$x^2 = -4ay$	Axis of symmetry is the y-axis, the parabola opens down (is concave down)

  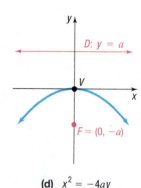

Figure 7　(a) $y^2 = 4ax$　　(b) $y^2 = -4ax$　　(c) $x^2 = 4ay$　　(d) $x^2 = -4ay$

EXAMPLE 3　**Analyzing the Equation of a Parabola**

Analyze the equation $x^2 = -12y$.

Solution　The equation $x^2 = -12y$ is of the form $x^2 = -4ay$, with $a = 3$. Consequently, the graph of the equation is a parabola with vertex at $(0, 0)$, focus at $(0, -3)$, and directrix the line $y = 3$. The parabola opens down (is concave down), and its axis of symmetry is the y-axis. To obtain the points defining the latus rectum, let $y = -3$. Then $x^2 = 36$, so $x = \pm 6$. The points $(-6, -3)$ and $(6, -3)$ determine the latus rectum. See Figure 8 for the graph.

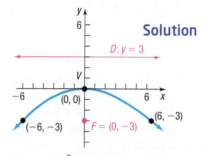

Figure 8 $x^2 = -12y$

- **Now Work** PROBLEM **41**

EXAMPLE 4　**Finding the Equation of a Parabola**

Find the equation of the parabola with focus at $(0, 4)$ and directrix the line $y = -4$. Graph the equation.

Solution　A parabola whose focus is at $(0, 4)$ and whose directrix is the horizontal line $y = -4$ has its vertex at $(0, 0)$. (Do you see why? The vertex is midway between the focus and the directrix.) Since the focus is on the positive y-axis at $(0, 4)$, the equation of the parabola is of the form $x^2 = 4ay$, with $a = 4$. That is,

$$x^2 = 4ay = 4 \cdot 4y = 16y$$
$$\uparrow$$
$$a = 4$$

Substituting $y = 4$ in the equation $x^2 = 16y$ yields $x^2 = 64$, so $x = \pm 8$. The points $(8, 4)$ and $(-8, 4)$ determine the latus rectum. Figure 9 shows the graph of $x^2 = 16y$.

Figure 9 $x^2 = 16y$

EXAMPLE 5 **Finding the Equation of a Parabola**

Find the equation of a parabola with vertex at $(0, 0)$ if its axis of symmetry is the x-axis and its graph contains the point $\left(-\frac{1}{2}, 2\right)$. Find its focus and directrix, and graph the equation.

Solution The vertex is at the origin, the axis of symmetry is the x-axis, and the graph contains a point in the second quadrant, so the parabola opens to the left. From Table 1, note that the form of the equation is

$$y^2 = -4ax$$

Because the point $\left(-\frac{1}{2}, 2\right)$ is on the parabola, the coordinates $x = -\frac{1}{2}, y = 2$ satisfy $y^2 = -4ax$. Substituting $x = -\frac{1}{2}$ and $y = 2$ into the equation leads to

$$2^2 = -4a\left(-\frac{1}{2}\right) \qquad x = -\frac{1}{2}, y = 2$$

$$a = 2 \qquad \text{Solve for } a.$$

The equation of the parabola is

$$y^2 = -4 \cdot 2x = -8x$$

The focus is $(-2, 0)$ and the directrix is the line $x = 2$. Substituting $x = -2$ in the equation $y^2 = -8x$ gives $y = \pm 4$. The points $(-2, 4)$ and $(-2, -4)$ determine the latus rectum. See Figure 10.

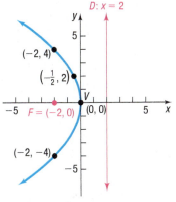

Figure 10 $y^2 = -8x$

Now Work PROBLEM 29

2 Analyze Parabolas with Vertex at (h, k)

If a parabola with vertex at the origin and axis of symmetry along a coordinate axis is shifted horizontally h units and then vertically k units, the result is a parabola with vertex at (h, k) and axis of symmetry parallel to a coordinate axis. The equations of such parabolas have the same forms as those in Table 1, but with x replaced by $x - h$ (the horizontal shift) and y replaced by $y - k$ (the vertical shift). Table 2 gives the forms of the equations of such parabolas. Figures 11(a)–(d) on page 398 illustrate the graphs for $h > 0, k > 0$.

NOTE Rather than memorizing Table 2, use transformations (shift horizontally h units, vertically k units) and the fact that a is the distance from the vertex to the focus to determine the parabola. ∎

Table 2

Equations of a Parabola: Vertex at (h, k); Axis of Symmetry Parallel to a Coordinate Axis; $a > 0$				
Vertex	**Focus**	**Directrix**	**Equation**	**Description**
(h, k)	$(h + a, k)$	$x = h - a$	$(y - k)^2 = 4a(x - h)$	Axis of symmetry is parallel to the x-axis, the parabola opens right
(h, k)	$(h - a, k)$	$x = h + a$	$(y - k)^2 = -4a(x - h)$	Axis of symmetry is parallel to the x-axis, the parabola opens left
(h, k)	$(h, k + a)$	$y = k - a$	$(x - h)^2 = 4a(y - k)$	Axis of symmetry is parallel to the y-axis, the parabola opens up (is concave up)
(h, k)	$(h, k - a)$	$y = k + a$	$(x - h)^2 = -4a(y - k)$	Axis of symmetry is parallel to the y-axis, the parabola opens down (is concave down)

(a) $(y - k)^2 = 4a(x - h)$ **(b)** $(y - k)^2 = -4a(x - h)$ **(c)** $(x - h)^2 = 4a(y - k)$ **(d)** $(x - h)^2 = -4a(y - k)$

Figure 11

EXAMPLE 6

Finding the Equation of a Parabola, Vertex Not at the Origin

Find an equation of the parabola with vertex at $(-2, 3)$ and focus at $(0, 3)$. Graph the equation.

Solution

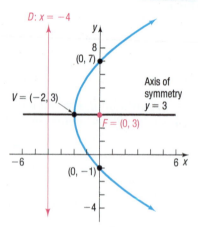

Figure 12 $(y - 3)^2 = 8(x + 2)$

The vertex $(-2, 3)$ and focus $(0, 3)$ both lie on the horizontal line $y = 3$ (the axis of symmetry). The distance a from the vertex $(-2, 3)$ to the focus $(0, 3)$ is $a = 2$. Because the focus lies to the right of the vertex, the parabola opens to the right. Consequently, the form of the equation is

$$(y - k)^2 = 4a(x - h)$$

where $(h, k) = (-2, 3)$ and $a = 2$. Therefore, the equation is

$$(y - 3)^2 = 4 \cdot 2[x - (-2)]$$
$$(y - 3)^2 = 8(x + 2)$$

Since the vertex is midway between the focus and the directrix, the line $x = -4$ is the directrix of the parabola. To find the points that define the latus rectum, substitute $x = 0$ in the equation $(y - 3)^2 = 8(x + 2)$. Then $y - 3 = \pm 4$, so $y = -1$ or $y = 7$. The points $(0, -1)$ and $(0, 7)$ determine the latus rectum. See Figure 12.

Now Work PROBLEM 31

Polynomial equations involving two variables define parabolas whenever they are quadratic in one variable and linear in the other.

EXAMPLE 7

Analyzing the Equation of a Parabola

Analyze the equation $x^2 + 4x - 4y = 0$.

Solution

Figure 13 $x^2 + 4x - 4y = 0$

To analyze the equation $x^2 + 4x - 4y = 0$, complete the square involving the variable x.

$$x^2 + 4x - 4y = 0$$
$$x^2 + 4x = 4y \qquad \text{Isolate the terms involving } x \text{ on the left side.}$$
$$x^2 + 4x + 4 = 4y + 4 \qquad \text{Complete the square on the left side.}$$
$$(x + 2)^2 = 4(y + 1) \qquad \text{Factor.}$$

The equation is of the form $(x - h)^2 = 4a(y - k)$, with $h = -2, k = -1$, and $a = 1$. The graph is a parabola with vertex at $(h, k) = (-2, -1)$ that opens up (is concave up). The focus is at $(-2, 0)$, and the directrix is the line $y = -2$. See Figure 13.

Now Work PROBLEM 49

3 Solve Applied Problems Involving Parabolas

Parabolas occur in many applications. For example, suspension bridges have cables in the shape of a parabola. Parabolas also have a reflecting property that is used in applications.

A parabola that is rotated about its axis of symmetry generates a surface called a **paraboloid of revolution**. If a light (or any other emitting source) is placed at the focus of the parabola, all the rays emanating from the light will reflect off the paraboloid of revolution in lines parallel to the axis of symmetry of the parabola. This principle is used in the design of searchlights, flashlights, certain automobile headlights, and other such devices. See Figure 14.

Conversely, suppose that rays of light (or other signals) emanate from a distant source so that they are essentially parallel. When these rays strike the surface of a parabolic mirror whose axis of symmetry is parallel to these rays, they are reflected to a single point at the focus. This principle is used in the design of some solar energy devices, satellite dishes, and the mirrors used in some types of telescopes. See Figure 15.

Figure 14 Searchlight

Figure 15 Telescope

EXAMPLE 8

Satellite Dish

A satellite dish is shaped like a paraboloid of revolution. The signals that emanate from a satellite strike the surface of the dish and are reflected to a single point, where the receiver is located. If the dish is 8 feet across at its opening and 3 feet deep at its center, at what position should the receiver be placed? That is, where is the focus?

Solution

Figure 16(a) shows the satellite dish. On a rectangular coordinate system, draw the parabola used to form the dish so that the vertex of the parabola is at the origin and its focus is on the positive y-axis. See Figure 16(b).

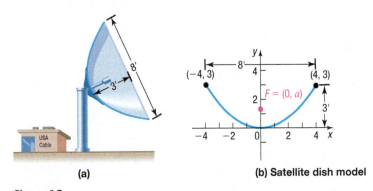

(a)

(b) Satellite dish model

Figure 16

The form of the equation of the parabola is

$$x^2 = 4ay$$

(continued)

and its focus is at $(0, a)$. Since $(4, 3)$ is a point on the graph, this gives

$$4^2 = 4a \cdot 3 \qquad x^2 = 4ay; x = 4, y = 3$$

$$a = \frac{4}{3} \qquad \text{Solve for } a.$$

The receiver should be located $\frac{4}{3}$ feet (1 foot, 4 inches) from the base of the dish, along its axis of symmetry.

 Now Work PROBLEM 69

6.2 Assess Your Understanding

'Are You Prepared?' *Answers are given at the end of these exercises. If you get a wrong answer, read the pages listed in red.*

1. The formula for the distance d from $P_1 = (x_1, y_1)$ to $P_2 = (x_2, y_2)$ is $d =$ _____. (pp. 3–4)

2. To complete the square of $x^2 - 4x$, add _____. (pp. A24–A25)

3. Use the Square Root Method to find the real solutions of $(x + 4)^2 = 9$. (p. A31)

4. The point that is symmetric with respect to the x-axis to the point $(-2, 5)$ is _____. (pp. 12–14)

5. To graph $y = (x - 3)^2 + 1$, shift the graph of $y = x^2$ to the right _____ units and then _____ 1 unit. (pp. 67–77)

Concepts and Vocabulary

6. A(n) _____ is the collection of all points in a plane that are the same distance from a fixed point as they are from a fixed line.

7. The line through the focus and perpendicular to the directrix is called the _____ of the parabola.

8. For the parabola $y^2 = 4ax$, the line segment joining the two points $(a, 2a)$ and $(a, -2a)$ is called the _____ _____.

Answer Problems 9–12 using the figure.

9. *Multiple Choice* If $a > 0$, the equation of the parabola is of the form
   (a) $(y - k)^2 = 4a(x - h)$ (b) $(y - k)^2 = -4a(x - h)$
   (c) $(x - h)^2 = 4a(y - k)$ (d) $(x - h)^2 = -4a(y - k)$

10. The coordinates of the vertex are _____.

11. *Multiple Choice* If $a = 4$, then the coordinates of the focus are _____.
    (a) $(-1, 2)$ (b) $(3, -2)$ (c) $(7, 2)$ (d) $(3, 6)$

12. *True or False* If $a = 4$, then the equation of the directrix is $x = 3$.

Skill Building

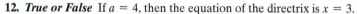

In Problems 13–20, the graph of a parabola is given. Match each graph to its equation.

(A) $y^2 = 4x$

(B) $x^2 = 4y$

(C) $y^2 = -4x$

(D) $x^2 = -4y$

(E) $(y - 1)^2 = 4(x - 1)$

(F) $(x + 1)^2 = 4(y + 1)$

(G) $(y - 1)^2 = -4(x - 1)$

(H) $(x + 1)^2 = -4(y + 1)$

13.

14.

15.

16.

17.

18.

19.

20.

In Problems 21–38, find the equation of the parabola described. Find the two points that define the latus rectum, and graph the equation.

21. Focus at $(4, 0)$ vertex at $(0, 0)$

22. Focus at $(0, 2)$; vertex at $(0, 0)$

23. Focus at $(0, -3)$; vertex at $(0, 0)$

24. Focus at $(-4, 0)$; vertex at $(0, 0)$

25. Focus at $(-2, 0)$; directrix the line $x = 2$

26. Focus at $(0, -1)$; directrix the line $y = 1$

27. Directrix the line $y = -\dfrac{1}{2}$; vertex at $(0, 0)$

28. Directrix the line $x = -\dfrac{1}{2}$; vertex at $(0, 0)$

29. Vertex at $(0, 0)$; axis of symmetry the y-axis; containing the point $(2, 3)$

30. Vertex at $(0, 0)$; axis of symmetry the x-axis; containing the point $(2, 3)$

31. Vertex at $(2, -3)$; focus at $(2, -5)$

32. Vertex at $(4, -2)$; focus at $(6, -2)$

33. Vertex at $(-1, -2)$; focus at $(0, -2)$

34. Vertex at $(3, 0)$; focus at $(3, -2)$

35. Focus at $(-3, 4)$; directrix the line $y = 2$

36. Focus at $(2, 4)$; directrix the line $x = -4$

37. Focus at $(-3, -2)$; directrix the line $x = 1$

38. Focus at $(-4, 4)$; directrix the line $y = -2$

In Problems 39–56, find the vertex, focus, and directrix of each parabola. Graph the equation.

39. $x^2 = 4y$

40. $y^2 = 8x$

41. $y^2 = -16x$

42. $x^2 = -4y$

43. $(y - 2)^2 = 8(x + 1)$

44. $(x + 4)^2 = 16(y + 2)$

45. $(x - 3)^2 = -(y + 1)$

46. $(y + 1)^2 = -4(x - 2)$

47. $(y + 3)^2 = 8(x - 2)$

48. $(x - 2)^2 = 4(y - 3)$

49. $y^2 - 4y + 4x + 4 = 0$

50. $x^2 + 6x - 4y + 1 = 0$

51. $x^2 + 8x = 4y - 8$

52. $y^2 - 2y = 8x - 1$

53. $y^2 + 2y - x = 0$

54. $x^2 - 4x = 2y$

55. $x^2 - 4x = y + 4$

56. $y^2 + 12y = -x + 1$

In Problems 57–64, write an equation for each parabola.

57.

58.

59.

60.

61.

62.

63.

64.
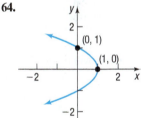

Applications and Extensions

65. Suspension Bridge The cables of a suspension bridge are in the shape of a parabola, as shown in the figure. The towers supporting the cable are 600 feet apart and 80 feet high. If the cables touch the road surface midway between the towers, what is the height of the cable from the road at a point 150 feet from the center of the bridge?

66. Suspension Bridge The cables of a suspension bridge are in the shape of a parabola. The towers supporting the cable are 400 feet apart and 100 feet high. If the cables are at a

height of 10 feet midway between the towers, what is the height of the cable at a point 50 feet from the center of the bridge?

67. Parabolic Arch Bridge A bridge is built in the shape of a parabolic arch. The arch has a span of 120 feet and a maximum height of 25 feet above the water. See the figure. Choose a suitable rectangular coordinate system and find the height of the arch at distances of 10, 30, and 50 feet from the center.

68. Parabolic Arch Bridge A bridge is to be built in the shape of a parabolic arch and is to have a span of 100 feet. The height of the arch a distance of 40 feet from the center is to be 10 feet. Find the height of the arch at its center.

69. Satellite Dish A satellite dish is shaped like a paraboloid of revolution. The signals that emanate from a satellite strike the surface of the dish and are reflected to a single point, where the receiver is located. If the dish is 10 feet across at its opening and 4 feet deep at its center, at what position should the receiver be placed?

70. Constructing a TV Dish A cable TV receiving dish is in the shape of a paraboloid of revolution. Find the location of the receiver, which is placed at the focus, if the dish is 6 feet across at its opening and 2 feet deep.

71. Constructing a Flashlight The reflector of a flashlight is in the shape of a paraboloid of revolution. Its diameter is 4 inches and its depth is 1 inch. How far from the vertex should the light bulb be placed so that the rays will be reflected parallel to the axis?

72. Constructing a Headlight A sealed-beam headlight is in the shape of a paraboloid of revolution. The bulb, which is placed at the focus, is 1 inch from the vertex. If the depth is to be 2 inches, what is the diameter of the headlight at its opening?

73. Searchlight A searchlight is shaped like a paraboloid of revolution. If the light source is located 2 feet from the base along the axis of symmetry and the opening is 5 feet across, how deep should the searchlight be?

74. Searchlight A searchlight is shaped like a paraboloid of revolution. If the light source is located 2 feet from the base along the axis of symmetry and the depth of the searchlight is 4 feet, what should the width of the opening be?

75. Solar Heat A mirror is shaped like a paraboloid of revolution and will be used to concentrate the rays of the sun at its focus, creating a heat source. See the figure. If the mirror is 20 inches across at its opening and is 6 inches deep, where will the heat source be concentrated?

76. Reflecting Telescope A reflecting telescope contains a mirror shaped like a paraboloid of revolution. If the mirror is 4 inches across at its opening and is 3 inches deep, where will the collected light be concentrated?

77. Gateway Arch The Gateway Arch in St. Louis is often mistaken to be parabolic in shape. In fact, it is a *catenary*, which has a more complicated formula than a parabola. The Arch is 630 feet high and 630 feet wide at its base.
(a) Find the equation of a parabola with the same dimensions. Let x equal the horizontal distance from the center of the arch.
(b) The table below gives the height of the Arch at various widths; find the corresponding heights for the parabola found in (a).

Width (ft)	Height (ft)
567	100
478	312.5
308	525

(c) Do the data support the notion that the Arch is in the shape of a parabola?
Source: gatewayarch.com

78. Show that an equation of the form
$$Ax^2 + Ey = 0 \qquad A \neq 0, E \neq 0$$
is the equation of a parabola with vertex at $(0, 0)$ and axis of symmetry the y-axis. Find its focus and directrix.

79. Show that an equation of the form
$$Cy^2 + Dx = 0 \qquad C \neq 0, D \neq 0$$
is the equation of a parabola with vertex at $(0, 0)$ and axis of symmetry the x-axis. Find its focus and directrix.

80. Challenge Problem Show that the graph of an equation of the form
$$Ax^2 + Dx + Ey + F = 0 \qquad A \neq 0$$
(a) Is a parabola if $E \neq 0$.
(b) Is a vertical line if $E = 0$ and $D^2 - 4AF = 0$.
(c) Is two vertical lines if $E = 0$ and $D^2 - 4AF > 0$.
(d) Contains no points if $E = 0$ and $D^2 - 4AF < 0$.

81. Challenge Problem Show that the graph of an equation of the form
$$Cy^2 + Dx + Ey + F = 0 \qquad C \neq 0$$
(a) Is a parabola if $D \neq 0$.
(b) Is a horizontal line if $D = 0$ and $E^2 - 4CF = 0$.
(c) Is two horizontal lines if $D = 0$ and $E^2 - 4CF > 0$.
(d) Contains no points if $D = 0$ and $E^2 - 4CF < 0$.

82. Challenge Problem Let A be either endpoint of the latus rectum of the parabola $y^2 - 2y - 8x + 1 = 0$, and let V be the vertex. Find the exact distance from A to V.[†]

[†]Courtesy of the Joliet Junior College Mathematics Department

---

Retain Your Knowledge

Problems 83–90 are based on material learned earlier in the course. The purpose of these problems is to keep the material fresh in your mind so that you are better prepared for the final exam.

83. For $x = 9y^2 - 36$, list the intercepts and test for symmetry.

84. Given $\tan \theta = -\dfrac{5}{8}, \dfrac{\pi}{2} < \theta < \pi$, find the exact value of each of the remaining trigonometric functions.

85. Find the exact value: $\tan\left[\cos^{-1}\left(-\dfrac{3}{7}\right)\right]$

86. Find the exact distance between the points $\left(-3,\dfrac{1}{2}\right)$ and $\left(\dfrac{2}{3},-5\right)$.

87. Find the standard form of the equation of a circle with radius $\sqrt{6}$ and center $(-12,7)$.

88. Convert $-\dfrac{7\pi}{18}$ from radians to degrees.

89. Express $\sqrt{\dfrac{1+\cos 34°}{2}}$ as a single trigonometric function.

90. Solve: $|x^2 - 5x| - 2 = 4$

'Are You Prepared?' Answers

1. $\sqrt{(x_2 - x_1)^2 + (y_2 - y_1)^2}$ **2.** 4 **3.** $\{-7, -1\}$ **4.** $(-2, -5)$ **5.** 3; up

6.3 The Ellipse

PREPARING FOR THIS SECTION *Before getting started, review the following:*

- Distance Formula (Section 1.1, pp. 3–4)
- Completing the Square (Section A.3, pp. A24– A25)
- Intercepts (Section 1.2, pp. 11–12)
- Symmetry (Section 1.2, pp. 12–14)
- Circles (Section 1.2, pp. 16–19)
- Graphing Techniques: Transformations (Section 1.6, pp. 67–77)

Now Work the 'Are You Prepared?' problems on page 410.

OBJECTIVES **1** Analyze Ellipses with Center at the Origin (p. 403)
 2 Analyze Ellipses with Center at (h, k) (p. 407)
 3 Solve Applied Problems Involving Ellipses (p. 409)

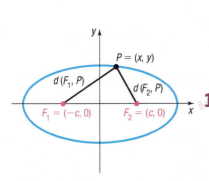

Figure 17 Ellipse

Figure 18

DEFINITION Ellipse

An **ellipse** is the collection of all points in a plane the sum of whose distances from two fixed points, called the **foci**, is a constant.

The definition contains within it a physical means for drawing an ellipse. Find a piece of string (the length of the string is the constant referred to in the definition). Then take two thumbtacks (the foci) and stick them into a piece of cardboard so that the distance between them is less than the length of the string. Now attach the ends of the string to the thumbtacks and, using the point of a pencil, pull the string taut. See Figure 17. Keeping the string taut, rotate the pencil around the two thumbtacks. The pencil traces out an ellipse, as shown in Figure 17.

In Figure 17, the foci are labeled F_1 and F_2. The line containing the foci is called the **major axis**. The midpoint of the line segment joining the foci is the **center** of the ellipse. The line through the center and perpendicular to the major axis is the **minor axis**.

The two points of intersection of the ellipse and the major axis are the **vertices**, V_1 and V_2, of the ellipse. The distance from one vertex to the other is the **length of the major axis**. The ellipse is symmetric with respect to its major axis, with respect to its minor axis, and with respect to its center.

1 Analyze Ellipses with Center at the Origin

With these ideas in mind, we are ready to find the equation of an ellipse in a rectangular coordinate system. First, place the center of the ellipse at the origin. Second, position the ellipse so that its major axis coincides with a coordinate axis, say the x-axis, as shown in Figure 18. If c is the distance from the center to a focus, one focus will be at $F_1 = (-c, 0)$ and the other at $F_2 = (c, 0)$.

As we shall see, it is convenient to let $2a$ denote the constant distance referred to in the definition. Then, if $P = (x, y)$ is any point on the ellipse,

$$d(F_1, P) + d(F_2, P) = 2a \qquad \text{The sum of the distances from } P \\ \text{to the foci equals a constant, } 2a.$$

$$\sqrt{(x + c)^2 + y^2} + \sqrt{(x - c)^2 + y^2} = 2a \qquad \text{Use the Distance Formula.}$$

$$\sqrt{(x + c)^2 + y^2} = 2a - \sqrt{(x - c)^2 + y^2} \qquad \text{Isolate one radical.}$$

$$(x + c)^2 + y^2 = 4a^2 - 4a\sqrt{(x - c)^2 + y^2} \qquad \text{Square both sides.} \\ + (x - c)^2 + y^2$$

$$x^2 + 2cx + c^2 + y^2 = 4a^2 - 4a\sqrt{(x - c)^2 + y^2} \qquad \text{Multiply out.} \\ + x^2 - 2cx + c^2 + y^2$$

$$4cx - 4a^2 = -4a\sqrt{(x - c)^2 + y^2} \qquad \text{Simplify; isolate the radical.}$$

$$cx - a^2 = -a\sqrt{(x - c)^2 + y^2} \qquad \text{Divide both sides by 4.}$$

$$(cx - a^2)^2 = a^2[(x - c)^2 + y^2] \qquad \text{Square both sides again.}$$

$$c^2x^2 - 2a^2cx + a^4 = a^2(x^2 - 2cx + c^2 + y^2) \qquad \text{Multiply out.}$$

$$(c^2 - a^2)x^2 - a^2y^2 = a^2c^2 - a^4 \qquad \text{Rearrange the terms.}$$

$$(a^2 - c^2)x^2 + a^2y^2 = a^2(a^2 - c^2) \qquad \text{Multiply both sides by } -1; \textbf{(1)} \\ \text{factor out } a^2 \text{ on the right side.}$$

To obtain points on the ellipse that are not on the major axis, we must have $a > c$. To see why, look again at Figure 18. Then

$$d(F_1, P) + d(F_2, P) > d(F_1, F_2) \qquad \text{The sum of the lengths of any two sides of a} \\ \text{triangle is greater than the length of the third side.}$$

$$2a > 2c \qquad d(F_1, P) + d(F_2, P) = 2a; d(F_1, F_2) = 2c$$

$$a > c$$

Because $a > c > 0$, this means $a^2 > c^2$, so $a^2 - c^2 > 0$. Let $b^2 = a^2 - c^2$, $b > 0$. Then $a > b$ and equation (1) can be written as

$$b^2x^2 + a^2y^2 = a^2b^2$$

$$\frac{x^2}{a^2} + \frac{y^2}{b^2} = 1 \qquad \text{Divide both sides by } a^2b^2.$$

The graph of the equation has symmetry with respect to the x-axis, the y-axis, and the origin.

Because the major axis is the x-axis, the vertices lie on the x-axis. So the vertices satisfy the equation $\dfrac{x^2}{a^2} = 1$, the solutions of which are $x = \pm a$. Consequently, the vertices of the ellipse are $V_1 = (-a, 0)$ and $V_2 = (a, 0)$. The y-intercepts of the ellipse, found by substituting $x = 0$ in the equation, have coordinates $(0, -b)$ and $(0, b)$. The four intercepts, $(a, 0)$, $(-a, 0)$, $(0, b)$, and $(0, -b)$, are used to graph the ellipse.

> **THEOREM** **Equation of an Ellipse: Center at $(0, 0)$; Major Axis along the x-Axis**
>
> An equation of the ellipse with center at $(0, 0)$, foci at $(-c, 0)$ and $(c, 0)$, and vertices at $(-a, 0)$ and $(a, 0)$ is
>
> $$\frac{x^2}{a^2} + \frac{y^2}{b^2} = 1 \qquad \text{where } a > b > 0 \text{ and } b^2 = a^2 - c^2 \qquad \textbf{(2)}$$
>
> The major axis is the x-axis. See Figure 19.

Figure 19 $\dfrac{x^2}{a^2} + \dfrac{y^2}{b^2} = 1, a > b > 0$

Notice in Figure 19 the points $(0, 0)$, $(c, 0)$, and $(0, b)$ form a right triangle. Because $b^2 = a^2 - c^2$ (or $b^2 + c^2 = a^2$), the distance from the focus at $(c, 0)$ to the point $(0, b)$ is a.

This can be seen another way. Look at the two right triangles in Figure 19. They are congruent. Do you see why? Because the sum of the distances from the foci to a point on the ellipse is $2a$, it follows that the distance from $(c, 0)$ to $(0, b)$ is a.

EXAMPLE 1

Finding an Equation of an Ellipse

Find an equation of the ellipse with center at the origin, one focus at $(3, 0)$, and a vertex at $(-4, 0)$. Graph the equation.

Solution

The ellipse has its center at the origin, and since the given focus and vertex lie on the x-axis, the major axis is the x-axis. The distance from the center, $(0, 0)$, to one of the foci, $(3, 0)$, is $c = 3$. The distance from the center, $(0, 0)$, to one of the vertices, $(-4, 0)$, is $a = 4$. From equation (2), it follows that

$$b^2 = a^2 - c^2 = 16 - 9 = 7$$

so an equation of the ellipse is

$$\frac{x^2}{16} + \frac{y^2}{7} = 1$$

Figure 20 shows the graph.

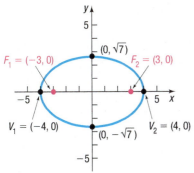

Figure 20 $\dfrac{x^2}{16} + \dfrac{y^2}{7} = 1$

In Figure 20, the intercepts of the equation are used to graph the ellipse. Following this practice makes it easier to obtain an accurate graph of an ellipse.

Now Work PROBLEM 27

COMMENT The intercepts of the ellipse also provide information about how to set the viewing rectangle for graphing an ellipse. To graph the ellipse

$$\frac{x^2}{16} + \frac{y^2}{7} = 1$$

set the viewing rectangle using a square screen that includes the intercepts, perhaps $-4.8 \leq x \leq 4.8$, $-3 \leq y \leq 3$. Then solve the equation for y:

$$\frac{x^2}{16} + \frac{y^2}{7} = 1$$

$$\frac{y^2}{7} = 1 - \frac{x^2}{16} \qquad \textbf{Subtract } \frac{x^2}{16} \textbf{ from both sides.}$$

$$y^2 = 7\left(1 - \frac{x^2}{16}\right) \qquad \textbf{Multiply both sides by 7.}$$

$$y = \pm\sqrt{7\left(1 - \frac{x^2}{16}\right)} \qquad \textbf{Use the Square Root Method.}$$

Now graph the two functions

$$Y_1 = \sqrt{7\left(1 - \frac{x^2}{16}\right)} \text{ and } Y_2 = -\sqrt{7\left(1 - \frac{x^2}{16}\right)}$$

$$Y_1 = \sqrt{7\left(1 - \frac{x^2}{16}\right)}$$

NORMAL FLOAT AUTO REAL RADIAN MP

$$Y_2 = -\sqrt{7\left(1 - \frac{x^2}{16}\right)}$$

Figure 21

Figure 21 shows the result on a TI-84 Plus C.

For the remainder of this section, the direction "**Analyze the equation**" means to find the center, major axis, foci, and vertices of the ellipse and graph it.

EXAMPLE 2   **Analyzing the Equation of an Ellipse**

Analyze the equation $\dfrac{x^2}{25} + \dfrac{y^2}{9} = 1$.

Solution   The equation is of the form of equation (2), with $a^2 = 25$ and $b^2 = 9$. The equation is that of an ellipse with center at $(0,0)$ and major axis along the x-axis. The vertices are at $(\pm a, 0) = (\pm 5, 0)$. Because $b^2 = a^2 - c^2$, this means

$$c^2 = a^2 - b^2 = 25 - 9 = 16$$

The foci are at $(\pm c, 0) = (\pm 4, 0)$. The y-intercepts are $(0, \pm b) = (0, \pm 3)$. Figure 22 shows the graph.

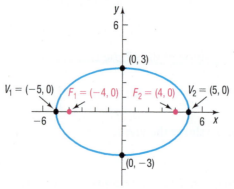

Figure 22 $\dfrac{x^2}{25} + \dfrac{y^2}{9} = 1$

Now Work PROBLEM 17

If the major axis of an ellipse with center at $(0,0)$ lies on the y-axis, the foci are at $(0, -c)$ and $(0, c)$. Using the same steps as before, the definition of an ellipse leads to the following result.

THEOREM Equation of an Ellipse: Center at (0, 0); Major Axis along the y-Axis

An equation of the ellipse with center at $(0, 0)$, foci at $(0, -c)$ and $(0, c)$, and vertices at $(0, -a)$ and $(0, a)$ is

$$\dfrac{x^2}{b^2} + \dfrac{y^2}{a^2} = 1 \qquad \text{where } a > b > 0 \text{ and } b^2 = a^2 - c^2 \tag{3}$$

The major axis is the y-axis.

Figure 23 $\dfrac{x^2}{b^2} + \dfrac{y^2}{a^2} = 1, a > b > 0$

Figure 23 illustrates the graph of such an ellipse. Again, notice the right triangle formed by the points at $(0, 0)$, $(b, 0)$, and $(0, c)$, so that $a^2 = b^2 + c^2$ (or $b^2 = a^2 - c^2$).

Look closely at equations (2) and (3). Although they may look alike, there is a difference! In equation (2), the larger number, a^2, is in the denominator of the x^2-term, so the major axis of the ellipse is along the x-axis. In equation (3), the larger number, a^2, is in the denominator of the y^2-term, so the major axis is along the y-axis.

EXAMPLE 3

Analyzing the Equation of an Ellipse

Analyze the equation $9x^2 + y^2 = 9$.

Solution

To put the equation in proper form, divide both sides by 9.

$$x^2 + \frac{y^2}{9} = 1$$

The larger denominator, 9, is in the y^2-term so, based on equation (3), this is the equation of an ellipse with center at the origin and major axis along the y-axis. Also, $a^2 = 9, b^2 = 1$, and $c^2 = a^2 - b^2 = 9 - 1 = 8$. The vertices are at $(0, \pm a) = (0, \pm 3)$, and the foci are at $(0, \pm c) = (0, \pm 2\sqrt{2})$. The x-intercepts are at $(\pm b, 0) = (\pm 1, 0)$. Figure 24 shows the graph.

 Now Work PROBLEM 21

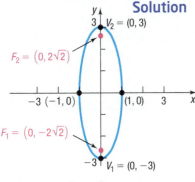

Figure 24 $9x^2 + y^2 = 9$

EXAMPLE 4

Finding an Equation of an Ellipse

Find an equation of the ellipse having one focus at $(0, 2)$ and vertices at $(0, -3)$ and $(0, 3)$. Graph the equation.

Solution

Plot the given focus and vertices, and note that the major axis is the y-axis. Because the vertices are at $(0, -3)$ and $(0, 3)$, the center of the ellipse is at their midpoint, the origin. The distance from the center, $(0, 0)$, to the given focus, $(0, 2)$, is $c = 2$. The distance from the center, $(0, 0)$, to one of the vertices, $(0, 3)$, is $a = 3$. So $b^2 = a^2 - c^2 = 9 - 4 = 5$. The form of the equation of the ellipse is given by equation (3).

$$\frac{x^2}{b^2} + \frac{y^2}{a^2} = 1$$

$$\frac{x^2}{5} + \frac{y^2}{9} = 1$$

Figure 25 shows the graph.

Figure 25 $\dfrac{x^2}{5} + \dfrac{y^2}{9} = 1$

Now Work PROBLEM 29

A circle may be considered a special kind of ellipse. To see why, let $a = b$ in equation (2) or (3). Then

$$\frac{x^2}{a^2} + \frac{y^2}{a^2} = 1$$

$$x^2 + y^2 = a^2$$

This is the equation of a circle with center at the origin and radius a. The value of c is

$$c^2 = a^2 - b^2 = 0$$
$$\uparrow$$
$$a = b$$

This indicates that the closer the two foci of an ellipse are to the center, the more the ellipse will look like a circle.

2 Analyze Ellipses with Center at (h, k)

If an ellipse with center at the origin and major axis coinciding with a coordinate axis is shifted horizontally h units and then vertically k units, the result is an ellipse with center at (h, k) and major axis parallel to a coordinate axis. The equations of such ellipses have the same forms as those given in equations (2) and (3), except that x is replaced by $x - h$ (the horizontal shift) and y is replaced by $y - k$ (the vertical shift). Table 3 (on the next page) gives the forms of the equations of such ellipses, and Figure 26 shows their graphs.

Table 3

NOTE Rather than memorizing Table 3, use transformations (shift horizontally h units, vertically k units), along with the facts that a represents the distance from the center to the vertices, c represents the distance from the center to the foci, and $b^2 = a^2 - c^2$ (or $c^2 = a^2 - b^2$). ∎

Equations of an Ellipse: Center at (h, k); Major Axis Parallel to a Coordinate Axis				
Center	Major Axis	Foci	Vertices	Equation
(h, k)	Parallel to the x-axis	$(h + c, k)$	$(h + a, k)$	$\dfrac{(x - h)^2}{a^2} + \dfrac{(y - k)^2}{b^2} = 1$
		$(h - c, k)$	$(h - a, k)$	$a > b > 0$ and $b^2 = a^2 - c^2$
(h, k)	Parallel to the y-axis	$(h, k + c)$	$(h, k + a)$	$\dfrac{(x - h)^2}{b^2} + \dfrac{(y - k)^2}{a^2} = 1$
		$(h, k - c)$	$(h, k - a)$	$a > b > 0$ and $b^2 = a^2 - c^2$

Figure 26

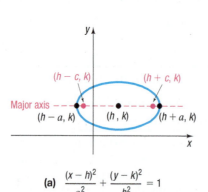

(a) $\dfrac{(x - h)^2}{a^2} + \dfrac{(y - k)^2}{b^2} = 1$

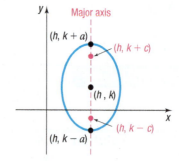

(b) $\dfrac{(x - h)^2}{b^2} + \dfrac{(y - k)^2}{a^2} = 1$

EXAMPLE 5   **Finding an Equation of an Ellipse, Center Not at the Origin**

Find an equation of the ellipse with center at $(2, -3)$, one focus at $(3, -3)$, and one vertex at $(5, -3)$. Graph the equation.

Solution   The center is at $(h, k) = (2, -3)$, so $h = 2$ and $k = -3$. Note that the center, focus, and vertex all lie on the line $y = -3$. Therefore, the major axis is parallel to the x-axis. The distance from the center $(2, -3)$ to a focus $(3, -3)$ is $c = 1$; the distance from the center $(2, -3)$ to a vertex $(5, -3)$ is $a = 3$. Then $b^2 = a^2 - c^2 = 9 - 1 = 8$. The form of the equation is

$$\frac{(x - h)^2}{a^2} + \frac{(y - k)^2}{b^2} = 1$$

$$\frac{(x - 2)^2}{9} + \frac{(y + 3)^2}{8} = 1 \quad \textcolor{blue}{h = 2, k = -3, a^2 = 9, b^2 = 8}$$

The major axis is parallel to the x-axis, so the vertices are $a = 3$ units left and right of the center $(2, -3)$. Therefore, the vertices are

$$V_1 = (2 - 3, -3) = (-1, -3) \quad \text{and} \quad V_2 = (2 + 3, -3) = (5, -3)$$

Note that the vertex $(5, -3)$ agrees with the information given in the problem. Since $c = 1$ and the major axis is parallel to the x-axis, the foci are 1 unit left and right of the center. Therefore, the foci are

$$F_1 = (2 - 1, -3) = (1, -3) \quad \text{and} \quad F_2 = (2 + 1, -3) = (3, -3)$$

Finally, use the value of $b = 2\sqrt{2}$ to find the two points above and below the center.

$$\left(2, -3 - 2\sqrt{2}\right) \quad \text{and} \quad \left(2, -3 + 2\sqrt{2}\right)$$

Use these two points and the vertices to obtain the graph. See Figure 27.

Figure 27 $\dfrac{(x - 2)^2}{9} + \dfrac{(y - 3)^2}{8} = 1$

Now Work PROBLEM 55

EXAMPLE 6   Analyzing the Equation of an Ellipse

Analyze the equation $4x^2 + y^2 - 8x + 4y + 4 = 0$.

Solution   Complete the squares in x and in y.

$$4x^2 + y^2 - 8x + 4y + 4 = 0$$

$$4x^2 - 8x + y^2 + 4y = -4 \qquad \text{Group like variables; place the constant on the right side.}$$

$$4(x^2 - 2x) + (y^2 + 4y) = -4 \qquad \text{Factor out 4 from the first two terms.}$$

$$4(x^2 - 2x + 1) + (y^2 + 4y + 4) = -4 + 4 + 4 \qquad \text{Complete each square.}$$

$$4(x - 1)^2 + (y + 2)^2 = 4 \qquad \text{Factor.}$$

$$(x - 1)^2 + \frac{(y + 2)^2}{4} = 1 \qquad \text{Divide both sides by 4.}$$

This is the equation of an ellipse with center at $(1, -2)$ and major axis parallel to the y-axis. Since $a^2 = 4$ and $b^2 = 1$, we have $c^2 = a^2 - b^2 = 4 - 1 = 3$. The vertices are at $(h, k \pm a) = (1, -2 \pm 2)$, or $(1, -4)$ and $(1, 0)$. The foci are at $(h, k \pm c) = (1, -2 \pm \sqrt{3})$, or $(1, -2 - \sqrt{3})$ and $(1, -2 + \sqrt{3})$. Figure 28 shows the graph.

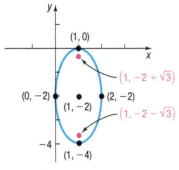

Figure 28
$4x^2 + y^2 - 8x + 4y + 4 = 0$

➤ **Now Work** PROBLEM **47**

3 Solve Applied Problems Involving Ellipses

Ellipses are found in many applications in science and engineering. For example, the orbits of the planets around the Sun are elliptical, with the Sun's position at a focus. See Figure 29.

Figure 29 Elliptical orbits

Stone and concrete bridges are often shaped as semielliptical arches. Elliptical gears are used in machinery when a variable rate of motion is required.

Ellipses also have an interesting reflection property. If a source of light (or sound) is placed at one focus, the waves transmitted by the source will reflect off the ellipse and concentrate at the other focus. This is the principle behind *whispering galleries*, which are rooms designed with elliptical ceilings. A person standing at one focus of the ellipse can whisper and be heard by a person standing at the other focus, because all the sound waves that reach the ceiling are reflected to the other person.

EXAMPLE 7   A Whispering Gallery

The whispering gallery in the Museum of Science and Industry in Chicago can be modeled by the top half of a three-dimensional ellipse, which is called an **ellipsoid**. The gallery is 47.3 feet long. The distance from the center of the room to the foci is 20.3 feet. Find an equation that describes the shape of the room. How high is the room at its center?

Source: Chicago Museum of Science and Industry Web site; www.msichicago.org

Solution Set up a rectangular coordinate system so that the center of the ellipse is at the origin and the major axis is along the x-axis. The equation of the ellipse is

$$\frac{x^2}{a^2} + \frac{y^2}{b^2} = 1$$

Since the length of the room is 47.3 feet, the distance from the center of the room to each vertex (the end of the room) will be $\dfrac{47.3}{2} = 23.65$ feet; so $a = 23.65$ feet. The distance from the center of the room to each focus is $c = 20.3$ feet. See Figure 30. Because $b^2 = a^2 - c^2$, this means that $b^2 = 23.65^2 - 20.3^2 = 147.2325$. An equation that describes the shape of the room is given by

$$\frac{x^2}{23.65^2} + \frac{y^2}{147.2325} = 1$$

The height of the room at its center is $b = \sqrt{147.2325} \approx 12.1$ feet.

Figure 30 Whispering gallery model

 Now Work PROBLEM **71**

6.3 Assess Your Understanding

'Are You Prepared?' *Answers are given at the end of these exercises. If you get a wrong answer, read the pages listed in* red.

1. The distance d from $P_1 = (2, -5)$ to $P_2 = (4, -2)$ is $d = \underline{\hspace{1cm}}$. (pp. 3–4)

2. To complete the square of $x^2 - 3x$, add $\underline{\hspace{1cm}}$. (pp. A24–25)

3. Find the intercepts of the equation $y^2 = 16 - 4x^2$. (pp. 11–12)

4. The point symmetric with respect to the y-axis to the point $(-2, 5)$ is $\underline{\hspace{1cm}}$. (pp. 12–14)

5. To graph $y = (x + 1)^2 - 4$, shift the graph of $y = x^2$ to the $\underline{\hspace{1cm}}$ unit(s) and then $\underline{\hspace{1cm}}$ unit(s). (left/right) (up/down) (pp. 67–77)

6. The standard equation of a circle with center at $(2, -3)$ and radius 1 is $\underline{\hspace{1cm}}$. (pp. 16–19)

Concepts and Vocabulary

7. A(n) $\underline{\hspace{1cm}}$ is the collection of all points in a plane the sum of whose distances from two fixed points is a constant.

8. *Multiple Choice* For an ellipse, the foci lie on a line called the $\underline{\hspace{1cm}}$.
   (a) minor axis **(b)** major axis
   (c) directrix **(d)** latus rectum

9. For the ellipse $\dfrac{x^2}{4} + \dfrac{y^2}{25} = 1$, the vertices are the points $\underline{\hspace{1cm}}$ and $\underline{\hspace{1cm}}$.

10. For the ellipse $\dfrac{x^2}{25} + \dfrac{y^2}{9} = 1$, the value of a is $\underline{\hspace{1cm}}$, the value of b is $\underline{\hspace{1cm}}$, and the major axis is the $\underline{\hspace{1cm}}$-axis.

11. If the center of an ellipse is $(2, -3)$, the major axis is parallel to the x-axis, and the distance from the center of the ellipse to a vertex is $a = 4$ units, then the coordinates of the vertices are $\underline{\hspace{1cm}}$ and $\underline{\hspace{1cm}}$.

12. *Multiple Choice* If the foci of an ellipse are $(-4, 4)$ and $(6, 4)$, then the coordinates of the center of the ellipse are $\underline{\hspace{1cm}}$.
   (a) $(1, 4)$ **(b)** $(4, 1)$
   (c) $(1, 0)$ **(d)** $(5, 4)$

Skill Building

In Problems 13–16, the graph of an ellipse is given. Match each graph to its equation.

(A) $\dfrac{x^2}{4} + y^2 = 1$ **(B)** $x^2 + \dfrac{y^2}{4} = 1$ **(C)** $\dfrac{x^2}{16} + \dfrac{y^2}{4} = 1$ **(D)** $\dfrac{x^2}{4} + \dfrac{y^2}{16} = 1$

13.

14.

15.

16.

In Problems 17–26, analyze each equation. That is, find the center, vertices, and foci of each ellipse and graph it.

17. $\dfrac{x^2}{25} + \dfrac{y^2}{4} = 1$
18. $\dfrac{x^2}{9} + \dfrac{y^2}{4} = 1$
19. $\dfrac{x^2}{9} + \dfrac{y^2}{25} = 1$
20. $x^2 + \dfrac{y^2}{16} = 1$
21. $4x^2 + y^2 = 16$

22. $x^2 + 9y^2 = 18$
23. $4y^2 + x^2 = 8$
24. $4y^2 + 9x^2 = 36$
25. $x^2 + y^2 = 16$
26. $x^2 + y^2 = 4$

In Problems 27–38, find an equation for each ellipse. Graph the equation.

27. Center at $(0, 0)$; focus at $(3, 0)$; vertex at $(5, 0)$
28. Center at $(0, 0)$; focus at $(-1, 0)$; vertex at $(3, 0)$

29. Center at $(0, 0)$; focus at $(0, -4)$; vertex at $(0, 5)$
30. Center at $(0, 0)$; focus at $(0, 1)$; vertex at $(0, -2)$

31. Foci at $(\pm 2, 0)$; length of the major axis is 6
32. Foci at $(0, \pm 2)$; length of the major axis is 8

33. Focus at $(-4, 0)$; vertices at $(\pm 5, 0)$
34. Focus at $(0, -4)$; vertices at $(0, \pm 8)$

35. Foci at $(0, \pm 3)$; x-intercepts are ± 2
36. Vertices at $(\pm 4, 0)$; y-intercepts are ± 1

37. Center at $(0, 0)$; vertex at $(0, 4)$; $b = 1$
38. Vertices at $(\pm 5, 0)$; $c = 2$

In Problems 39–42, find an equation for each ellipse.

39.

40.

41.

42.
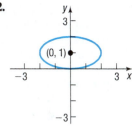

In Problems 43–54, analyze each equation; that is, find the center, foci, and vertices of each ellipse. Graph each equation.

43. $\dfrac{(x - 3)^2}{4} + \dfrac{(y + 1)^2}{9} = 1$
44. $\dfrac{(x + 4)^2}{9} + \dfrac{(y + 2)^2}{4} = 1$
45. $(x + 5)^2 + 4(y - 4)^2 = 16$

46. $9(x - 3)^2 + (y + 2)^2 = 18$
47. $x^2 + 4x + 4y^2 - 8y + 4 = 0$
48. $x^2 + 3y^2 - 12y + 9 = 0$

49. $2x^2 + 3y^2 - 8x + 6y + 5 = 0$
50. $4x^2 + 3y^2 + 8x - 6y = 5$
51. $9x^2 + 4y^2 - 18x + 16y - 11 = 0$

52. $x^2 + 9y^2 + 6x - 18y + 9 = 0$
53. $4x^2 + y^2 + 4y = 0$
54. $9x^2 + y^2 - 18x = 0$

In Problems 55–64, find an equation for each ellipse. Graph the equation.

55. Center at $(2, -2)$; vertex at $(7, -2)$; focus at $(4, -2)$
56. Center at $(-3, 1)$; vertex at $(-3, 3)$; focus at $(-3, 0)$

57. Vertices at $(4, 3)$ and $(4, 9)$; focus at $(4, 8)$
58. Foci at $(1, 2)$ and $(-3, 2)$; vertex at $(-4, 2)$

59. Foci at $(5, 1)$ and $(-1, 1)$; length of the major axis is 8
60. Vertices at $(2, 5)$ and $(2, -1)$; $c = 2$

61. Center at $(1, 2)$; focus at $(4, 2)$; contains the point $(1, 3)$
62. Center at $(1, 2)$; focus at $(1, 4)$; contains the point $(2, 2)$

63. Center at $(1, 2)$; vertex at $(4, 2)$; contains the point $(1, 5)$
64. Center at $(1, 2)$; vertex at $(1, 4)$; contains the point $\left(1 + \sqrt{3}, 3\right)$

In Problems 65–68, graph each function. Be sure to label all the intercepts. **[Hint:** *Notice that each function is half an ellipse.*]

65. $f(x) = \sqrt{16 - 4x^2}$
66. $f(x) = \sqrt{9 - 9x^2}$
67. $f(x) = -\sqrt{64 - 16x^2}$
68. $f(x) = -\sqrt{4 - 4x^2}$

Applications and Extensions

69. Semielliptical Arch Bridge An arch in the shape of the upper half of an ellipse is used to support a bridge that is to span a river 20 meters wide. The center of the arch is 6 meters above the center of the river. See the figure. Find an equation for the ellipse in which the x-axis coincides with the water level and the y-axis passes through the center of the arch.

70. Semielliptical Arch Bridge The arch of a bridge is a semiellipse with a horizontal major axis. The span is 30 feet,

and the top of the arch is 10 feet above the major axis. The roadway is horizontal and is 2 feet above the top of the arch. Find the vertical distance from the roadway to the arch at 5-foot intervals along the roadway.

71. Whispering Gallery A hall 100 feet in length is to be designed as a whispering gallery. If the foci are located 25 feet from the center, how high will the ceiling be at the center?

72. Whispering Gallery Jim, standing at one focus of a whispering gallery, is 6 feet from the nearest wall. His friend is standing at the other focus, 100 feet away. What is the length of this whispering gallery? How high is its elliptical ceiling at the center?

73. Semielliptical Arch Bridge A bridge is built in the shape of a semielliptical arch. The arch has a span of 120 feet and a maximum height of 25 feet. Choose a suitable rectangular coordinate system and find the height of the arch at distances of 10, 30, and 50 feet from the center.

74. Semielliptical Arch Bridge A bridge is to be built in the shape of a semielliptical arch and is to have a span of 100 feet. The height of the arch, at a distance of 40 feet from the center, is to be 10 feet. Find the height of the arch at its center.

75. Racetrack Design Consult the figure. A racetrack is in the shape of an ellipse 100 feet long and 50 feet wide. What is the width 10 feet from a vertex?

76. Semielliptical Arch Bridge An arch for a bridge over a highway is in the form of half an ellipse. The top of the arch is 20 feet above the ground level (the major axis). The highway has four lanes, each 12 feet wide; a center safety strip 8 feet wide; and two side strips, each 4 feet wide. What should the span of the arch be (the length of its major axis) if the height 28 feet from the center is to be 13 feet?

77. Installing a Vent Pipe A homeowner is putting in a fireplace that has a 4-inch-radius vent pipe. He needs to cut an elliptical hole in his roof to accommodate the pipe. If the pitch of his roof is $\frac{5}{4}$ (a rise of 5, run of 4), what are the dimensions of the hole?
Source: www.doe.virginia.gov

78. Volume of a Football A football is in the shape of a **prolate spheroid**, which is simply a solid obtained by rotating an ellipse $\left(\frac{x^2}{a^2} + \frac{y^2}{b^2} = 1\right)$ about its major axis. An inflated NFL football averages 11.125 inches in length and 28.25 inches in center circumference. If the volume of a prolate spheroid is $\frac{4}{3}\pi ab^2$, how much air does the football contain? (Neglect material thickness.)
Source: nfl.com

*In Problems 79–83, use the fact that the orbit of a planet about the Sun is an ellipse, with the Sun at one focus. The **aphelion** of a planet is its greatest distance from the Sun, and the **perihelion** is its shortest distance. The **mean distance** of a planet from the Sun is the length of the semimajor axis of the elliptical orbit. See the figure.*

79. Earth The mean distance of Earth from the Sun is 93 million miles. If the aphelion of Earth is 94.5 million miles, what is the perihelion? Find an equation for the orbit of Earth around the Sun.

80. Mars The mean distance of Mars from the Sun is 142 million miles. If the perihelion of Mars is 128.5 million miles, what is the aphelion? Find an equation for the orbit of Mars about the Sun.

81. Jupiter The aphelion of Jupiter is 507 million miles. If the distance from the center of its elliptical orbit to the Sun is 23.2 million miles, what is the perihelion? What is the mean distance? Find an equation for the orbit of Jupiter around the Sun.

82. Pluto The perihelion of Pluto is 4551 million miles, and the distance from the center of its elliptical orbit to the Sun is 897.5 million miles. Find the aphelion of Pluto. What is the mean distance of Pluto from the Sun? Find an equation for the orbit of Pluto about the Sun.

83. Elliptical Orbit A planet orbits a star in an elliptical orbit with the star located at one focus. The perihelion of the planet is 5 million miles. The **eccentricity** e of a conic section is $e = \frac{c}{a}$. If the eccentricity of the orbit is 0.75, find the aphelion of the planet.[†]

84. A rectangle is inscribed in an ellipse with major axis of length 14 meters and minor axis of length 4 meters. Find the maximum area of a rectangle inscribed in the ellipse. Round your answer to two decimal places.[†]

85. Lithotripsy Extracorporeal shock wave lithotripsy is a procedure that uses shockwaves to fragment kidney stones without the need for surgery. Using an elliptical reflector, a shock wave generator is placed at one focus and the kidney stone is positioned at the other focus. Shock waves pass through a water column and disintegrate the stone, allowing it to be passed out of the body naturally. If the equation of the ellipse formed by the reflector is $\frac{x^2}{324} + \frac{y^2}{100} = 1$, how far from the kidney stone does the shock wave generator need to be placed? (Units are in centimeters.)

86. Elliptical Trainer The pedals of an elliptical exercise machine travel an elliptical path as the user is exercising. If the stride length (length of the major axis) for one machine is 20 inches and the maximum vertical pedal displacement (length of the minor axis) is 9 inches, find the equation of the pedal path, assuming it is centered at the origin.

87. Challenge Problem For the ellipse, $x^2 + 5y^2 = 20$, let V be the vertex with the smaller x-coordinate and let B be the endpoint on the minor axis with the larger y-coordinate. Find the y-coordinate of the point M that is on the line $x + 5 = 0$ and is equidistant from V and B.

88 Challenge Problem Consider the circle $(x - 2)^2 + y^2 = 1$ and the ellipse with vertices at $(2, 0)$ and $(6, 0)$ and one focus at $(4 + \sqrt{3}, 0)$. Find the points of intersection of the circle and the ellipse.[†]

89. Challenge Problem Show that an equation of the form
$$Ax^2 + Cy^2 + F = 0 \qquad A \neq 0, C \neq 0, F \neq 0$$
where A and C are of the same sign and F is of opposite sign,
(a) is the equation of an ellipse with center at $(0, 0)$ if $A \neq C$.
(b) is the equation of a circle with center $(0, 0)$ if $A = C$.

90. Challenge Problem Show that the graph of an equation of the form
$$Ax^2 + Cy^2 + Dx + Ey + F = 0 \qquad A \neq 0, C \neq 0$$
where A and C are of the same sign,
(a) is an ellipse if $\frac{D^2}{4A} + \frac{E^2}{4C} - F$ is the same sign as A.
(b) is a point if $\frac{D^2}{4A} + \frac{E^2}{4C} - F = 0$.
(c) contains no points if $\frac{D^2}{4A} + \frac{E^2}{4C} - F$ is of opposite sign to A.

[†]*Courtesy of the Joliet Junior College Mathematics Department*

Discussion and Writing

91. The **eccentricity** e of an ellipse is defined as the number $\dfrac{c}{a}$, where a is the distance of a vertex from the center and c is the distance of a focus from the center. Because $a > c$, it follows that $e < 1$. Write a brief paragraph about the general shape of each of the following ellipses. Be sure to justify your conclusions.

   (a) Eccentricity close to 0   **(b)** Eccentricity = 0.5   **(c)** Eccentricity close to 1

┌ Retain Your Knowledge

Problems 92–99 are based on material learned earlier in the course. The purpose of these problems is to keep the material fresh in your mind so that you are better prepared for the final exam.

92. Find the zeros of the quadratic function $f(x) = (x - 5)^2 - 12$. What are the x-intercepts, if any, of the graph of the function?

93. Find the domain of the rational function $f(x) = \dfrac{2x - 3}{x - 5}$. Find any horizontal, vertical, or oblique asymptotes.

94. Find the work done by a force of 80 pounds acting in the direction of $50°$ to the horizontal in moving an object 12 feet from $(0, 0)$ to $(12, 0)$. Round to one decimal place.

95. Solve the right triangle shown.

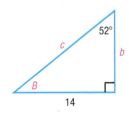

96. Solve $2\sqrt{3}\tan(5x) + 7 = 9$ for $0 \le x < \dfrac{\pi}{2}$.

97. What value does $R(x) = \dfrac{3x^2 + 14x + 8}{x^2 + x - 12}$ approach as $x \to -4$?

98. Find the difference quotient of $f(x) = 2x^2 - 7x$ as $h \to 0$.

99. Solve: $(x + 3)^2 = 20$

'Are You Prepared?' Answers

   1. $\sqrt{13}$ **2.** $\dfrac{9}{4}$ **3.** $(-2, 0), (2, 0), (0, -4), (0, 4)$ **4.** $(2, 5)$ **5.** left; 1; down; 4   **6.** $(x - 2)^2 + (y + 3)^2 = 1$

6.4 The Hyperbola

PREPARING FOR THIS SECTION *Before getting started, review the following:*

- Distance Formula (Section 1.1, pp. 3–4)
- Completing the Square (Section A.3, pp. A24–A25)
- Intercepts (Section 1.2, pp. 11–12)
- Symmetry (Section 1.2, pp. 12–14)
- Asymptotes (Section 1.6, pp. 76–77)
- Graphing Techniques: Transformations (Section 1.6, pp. 67–77)
- Square Root Method (Section A.4, p. A31)

  ✎ Now Work the 'Are You Prepared?' problems on page 423.

        OBJECTIVES **1** Analyze Hyperbolas with Center at the Origin (p. 414)
                          2 Find the Asymptotes of a Hyperbola (p. 418)
                          3 Analyze Hyperbolas with Center at (h, k) (p. 420)
                          4 Solve Applied Problems Involving Hyperbolas (p. 421)

┌ DEFINITION

A **hyperbola** is the collection of all points in a plane the difference of whose distances from two fixed points, called the **foci**, is a constant.

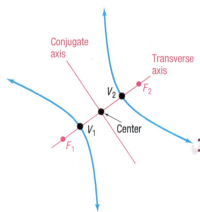

Conjugate
axis

Transverse
axis

V_2
F_2

V_1 Center

F_1

Figure 31 Hyperbola

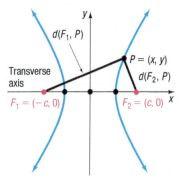

y

$d(F_1, P)$

Transverse
axis

$P = (x, y)$

$d(F_2, P)$

$F_1 = (-c, 0)$

$F_2 = (c, 0)$ x

Figure 32
$d(F_1, P) - d(F_2, P) = \pm 2a$

Figure 31 illustrates a hyperbola with foci F_1 and F_2. The line containing the foci is called the **transverse axis**. The midpoint of the line segment joining the foci is the **center** of the hyperbola. The line through the center and perpendicular to the transverse axis is the **conjugate axis**. The hyperbola consists of two separate curves, called **branches**, that are symmetric with respect to the transverse axis, conjugate axis, and center. The two points of intersection of the hyperbola and the transverse axis are the **vertices**, V_1 and V_2, of the hyperbola.

1 Analyze Hyperbolas with Center at the Origin

With these ideas in mind, we are now ready to find the equation of a hyperbola in the rectangular coordinate system. First, place the center at the origin. Next, position the hyperbola so that its transverse axis coincides with a coordinate axis. Suppose that the transverse axis coincides with the x-axis, as shown in Figure 32.

If c is the distance from the center to a focus, one focus will be at $F_1 = (-c, 0)$ and the other at $F_2 = (c, 0)$. Now we let the constant difference of the distances from any point $P = (x, y)$ on the hyperbola to the foci F_1 and F_2 be denoted by $\pm 2a$, where $a > 0$. (If P is on the right branch, the $+$ sign is used; if P is on the left branch, the $-$ sign is used.) The coordinates of P must satisfy the equation

$$d(F_1, P) - d(F_2, P) = \pm 2a \qquad \text{Difference of the distances from } P \text{ to the foci equals } \pm 2a.$$

$$\sqrt{(x + c)^2 + y^2} - \sqrt{(x - c)^2 + y^2} = \pm 2a \qquad \text{Use the Distance Formula.}$$

$$\sqrt{(x + c)^2 + y^2} = \pm 2a + \sqrt{(x - c)^2 + y^2} \qquad \text{Isolate one radical.}$$

$$(x + c)^2 + y^2 = 4a^2 \pm 4a\sqrt{(x - c)^2 + y^2} \qquad \text{Square both sides.}$$
$$+ (x - c)^2 + y^2$$

$$x^2 + 2cx + c^2 + y^2 = 4a^2 \pm 4a\sqrt{(x - c)^2 + y^2} \qquad \text{Multiply.}$$
$$+ x^2 - 2cx + c^2 + y^2$$

$$4cx - 4a^2 = \pm 4a\sqrt{(x - c)^2 + y^2} \qquad \text{Simplify; isolate the radical.}$$

$$cx - a^2 = \pm a\sqrt{(x - c)^2 + y^2} \qquad \text{Divide both sides by 4.}$$

$$(cx - a^2)^2 = a^2[(x - c)^2 + y^2] \qquad \text{Square both sides.}$$

$$c^2x^2 - 2ca^2x + a^4 = a^2(x^2 - 2cx + c^2 + y^2) \qquad \text{Multiply.}$$

$$c^2x^2 + a^4 = a^2x^2 + a^2c^2 + a^2y^2 \qquad \text{Distribute and simplify.}$$

$$(c^2 - a^2)x^2 - a^2y^2 = a^2c^2 - a^4 \qquad \text{Rearrange terms.}$$

$$(c^2 - a^2)x^2 - a^2y^2 = a^2(c^2 - a^2) \qquad \text{Factor } a^2 \text{ on the right side. } \textbf{(1)}$$

To obtain points on the hyperbola off the x-axis, we must have $a < c$. To see why, look again at Figure 32.

$$d(F_1, P) < d(F_2, P) + d(F_1, F_2) \qquad \text{Use triangle } F_1PF_2.$$

$$d(F_1, P) - d(F_2, P) < d(F_1, F_2) \qquad P \text{ is on the right branch, so}$$
$$d(F_1, P) - d(F_2, P) = 2a;$$
$$2a < 2c \qquad d(F_1, F_2) = 2c.$$

$$a < c$$

Since $a < c$, we also have $a^2 < c^2$, so $c^2 - a^2 > 0$. Let $b^2 = c^2 - a^2, b > 0$. Then equation (1) can be written as

$$b^2x^2 - a^2y^2 = a^2b^2$$

$$\frac{x^2}{a^2} - \frac{y^2}{b^2} = 1 \qquad \text{Divide both sides by } a^2b^2.$$

To find the vertices of the hyperbola, substitute $y = 0$ in the equation. The vertices satisfy the equation $\dfrac{x^2}{a^2} = 1$, the solutions of which are $x = \pm a$. Consequently, the vertices of the hyperbola are $V_1 = (-a, 0)$ and $V_2 = (a, 0)$. Notice that the distance from the center $(0, 0)$ to either vertex is a.

> **THEOREM   Equation of a Hyperbola: Center at (0, 0); Transverse Axis along the x-Axis**
>
> An equation of the hyperbola with center at $(0, 0)$, foci at $(-c, 0)$ and $(c, 0)$, and vertices at $(-a, 0)$ and $(a, 0)$ is
>
> $$\frac{x^2}{a^2} - \frac{y^2}{b^2} = 1 \qquad \text{where } b^2 = c^2 - a^2 \qquad \textbf{(2)}$$
>
> The transverse axis is the x-axis.

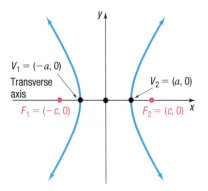

Figure 33
$$\frac{x^2}{a^2} - \frac{y^2}{b^2} = 1, \quad b^2 = c^2 - a^2$$

See Figure 33. The hyperbola defined by equation (2) is symmetric with respect to the x-axis, y-axis, and origin. To find the y-intercepts, if any, let $x = 0$ in equation (2). This results in the equation $\frac{y^2}{b^2} = -1$, which has no real solution, so the hyperbola defined by equation (2) has no y-intercepts. In fact, since $\frac{x^2}{a^2} - 1 = \frac{y^2}{b^2} \geq 0$, it follows that $\frac{x^2}{a^2} \geq 1$. There are no points on the graph for $-a < x < a$.

EXAMPLE 1

Finding and Graphing an Equation of a Hyperbola

Find an equation of the hyperbola with center at the origin, one focus at $(3, 0)$ and one vertex at $(-2, 0)$. Graph the equation.

Solution

The hyperbola has its center at the origin. Plot the center, focus, and vertex. Since they all lie on the x-axis, the transverse axis coincides with the x-axis. One focus is at $(c, 0) = (3, 0)$, so $c = 3$. One vertex is at $(-a, 0) = (-2, 0)$, so $a = 2$. From equation (2), it follows that $b^2 = c^2 - a^2 = 9 - 4 = 5$, so an equation of the hyperbola is

$$\frac{x^2}{4} - \frac{y^2}{5} = 1$$

To graph a hyperbola, it is helpful to locate and plot other points on the graph. For example, to find the points above and below the foci, let $x = \pm 3$. Then

$$\frac{x^2}{4} - \frac{y^2}{5} = 1$$

$$\frac{(\pm 3)^2}{4} - \frac{y^2}{5} = 1 \qquad \textcolor{blue}{x = \pm 3}$$

$$\frac{9}{4} - \frac{y^2}{5} = 1$$

$$\frac{y^2}{5} = \frac{5}{4}$$

$$y^2 = \frac{25}{4}$$

$$y = \pm \frac{5}{2}$$

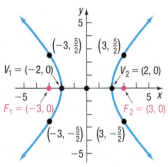

Figure 34 $\frac{x^2}{4} - \frac{y^2}{5} = 1$

The points above and below the foci are $\left(\pm 3, \frac{5}{2}\right)$ and $\left(\pm 3, -\frac{5}{2}\right)$. These points determine the "opening" of the hyperbola. See Figure 34.

📖 **COMMENT** To graph the hyperbola $\dfrac{x^2}{4} - \dfrac{y^2}{5} = 1$, graph the two functions $Y_1 = \sqrt{5}\sqrt{\dfrac{x^2}{4} - 1}$ and $Y_2 = -\sqrt{5}\sqrt{\dfrac{x^2}{4} - 1}$. Do this and compare the result with Figure 34. ∎

✏️ **Now Work** PROBLEM 19

For the next two examples, the direction "**Analyze the equation**" means to find the center, transverse axis, vertices, and foci of the hyperbola and graph it.

EXAMPLE 2

Analyzing the Equation of a Hyperbola

Analyze the equation $\dfrac{x^2}{16} - \dfrac{y^2}{4} = 1$.

Solution

The equation is of the form of equation (2), with $a^2 = 16$ and $b^2 = 4$. The graph of the equation is a hyperbola with center at $(0, 0)$ and transverse axis along the x-axis. Also, $c^2 = a^2 + b^2 = 16 + 4 = 20$. The vertices are at $(\pm a, 0) = (\pm 4, 0)$, and the foci are at $(\pm c, 0) = (\pm 2\sqrt{5}, 0)$.

To locate the points on the graph above and below the foci, let $x = \pm 2\sqrt{5}$ in the equation. Then

$$\frac{x^2}{16} - \frac{y^2}{4} = 1$$

$$\frac{\left(\pm 2\sqrt{5}\right)^2}{16} - \frac{y^2}{4} = 1 \qquad \textcolor{blue}{x = \pm 2\sqrt{5}}$$

$$\frac{20}{16} - \frac{y^2}{4} = 1$$

$$\frac{5}{4} - \frac{y^2}{4} = 1$$

$$\frac{y^2}{4} = \frac{1}{4}$$

$$y = \pm 1$$

The points above and below the foci are $(\pm 2\sqrt{5}, 1)$ and $(\pm 2\sqrt{5}, -1)$. See Figure 35.

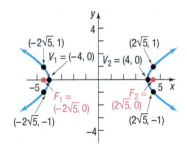

Figure 35 $\dfrac{x^2}{16} - \dfrac{y^2}{4} = 1$

THEOREM **Equation of a Hyperbola: Center at (0, 0); Transverse Axis along the y-Axis**

An equation of the hyperbola with center at $(0, 0)$, foci at $(0, -c)$ and $(0, c)$, and vertices at $(0, -a)$ and $(0, a)$ is

$$\frac{y^2}{a^2} - \frac{x^2}{b^2} = 1 \qquad \text{where } b^2 = c^2 - a^2 \qquad (3)$$

The transverse axis is the y-axis.

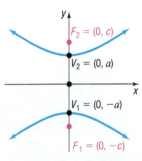

Figure 36
$\dfrac{y^2}{a^2} - \dfrac{x^2}{b^2} = 1, b^2 = c^2 - a^2$

Figure 36 shows the graph of a typical hyperbola defined by equation (3). Let's compare equations (2) and (3).

An equation of the form of equation (2), $\dfrac{x^2}{a^2} - \dfrac{y^2}{b^2} = 1$, is the equation of a hyperbola with center at the origin, foci on the x-axis at $(-c, 0)$ and $(c, 0)$, where $b^2 = c^2 - a^2$, and transverse axis along the x-axis.

An equation of the form of equation (3), $\dfrac{y^2}{a^2} - \dfrac{x^2}{b^2} = 1$, is the equation of a hyperbola with center at the origin, foci on the y-axis at $(0, -c)$ and $(0, c)$, where $b^2 = c^2 - a^2$, and transverse axis along the y-axis.

Notice the difference in the forms of equations (2) and (3). When the y^2-term is subtracted from the x^2-term, the transverse axis is along the x-axis. When the x^2-term is subtracted from the y^2-term, the transverse axis is along the y-axis.

EXAMPLE 3

Analyzing the Equation of a Hyperbola

Analyze the equation $4y^2 - x^2 = 4$.

Solution

To put the equation in proper form, divide both sides by 4:

$$y^2 - \frac{x^2}{4} = 1$$

Since the x^2-term is subtracted from the y^2-term, the equation is that of a hyperbola with center at the origin and transverse axis along the y-axis. Also, comparing the equation to equation (3), note that $a^2 = 1, b^2 = 4$, and $c^2 = a^2 + b^2 = 5$. The vertices are at $(0, \pm a) = (0, \pm 1)$, and the foci are at $(0, \pm c) = (0, \pm\sqrt{5})$.

To locate points on the graph to the left and right of the foci, let $y = \pm\sqrt{5}$ in the equation. Then

$$4y^2 - x^2 = 4$$
$$4(\pm\sqrt{5})^2 - x^2 = 4 \qquad y = \pm\sqrt{5}$$
$$20 - x^2 = 4$$
$$x^2 = 16$$
$$x = \pm 4$$

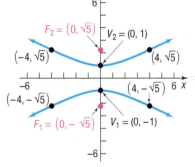

Figure 37 $4y^2 - x^2 = 4$

Four other points on the graph are $(\pm 4, \sqrt{5})$ and $(\pm 4, -\sqrt{5})$. See Figure 37.

EXAMPLE 4

Finding an Equation of a Hyperbola

Find an equation of the hyperbola that has one vertex at $(0, 2)$ and foci at $(0, -3)$ and $(0, 3)$. Graph the equation.

Solution

Since the foci are at $(0, -3)$ and $(0, 3)$, the center of the hyperbola, which is at their midpoint, is the origin. Also, the transverse axis is along the y-axis. This information tells us that $c = 3, a = 2$, and $b^2 = c^2 - a^2 = 9 - 4 = 5$. The form of the equation of the hyperbola is given by equation (3):

$$\frac{y^2}{a^2} - \frac{x^2}{b^2} = 1$$
$$\frac{y^2}{4} - \frac{x^2}{5} = 1$$

Let $y = \pm 3$ to obtain points on the graph on either side of each focus. See Figure 38.

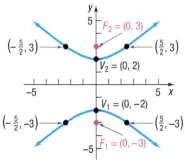

Figure 38 $\dfrac{y^2}{4} - \dfrac{x^2}{5} = 1$

Now Work PROBLEMS **15** AND **21**

Look at the equations of the hyperbolas in Examples 2 and 4. For the hyperbola in Example 2, $a^2 = 16$ and $b^2 = 4$, so $a > b$; for the hyperbola in Example 4, $a^2 = 4$ and $b^2 = 5$, so $a < b$. This indicates that for hyperbolas, there are no requirements involving the relative sizes of a and b. Contrast this situation to the case of an ellipse, in which the relative sizes of a and b dictate which axis is the major axis. Hyperbolas have another feature to distinguish them from ellipses and parabolas: hyperbolas have asymptotes.

2 Find the Asymptotes of a Hyperbola

Recall that a horizontal or oblique asymptote of a graph is a line with the property that the distance from the line to points on the graph approaches 0 as $x \to -\infty$ or as $x \to \infty$. Asymptotes provide information about the end behavior of the graph of a hyperbola.

> **THEOREM** **Asymptotes of a Hyperbola; Transverse Axis along the *x*-Axis**
>
> The hyperbola $\dfrac{x^2}{a^2} - \dfrac{y^2}{b^2} = 1$ has the two oblique asymptotes
>
> $$y = \frac{b}{a}x \quad \text{and} \quad y = -\frac{b}{a}x \tag{4}$$

Proof Begin by solving for y in the equation of the hyperbola.

$$\frac{x^2}{a^2} - \frac{y^2}{b^2} = 1$$

$$\frac{y^2}{b^2} = \frac{x^2}{a^2} - 1$$

$$y^2 = b^2\left(\frac{x^2}{a^2} - 1\right)$$

Since $x \neq 0$, the right side can be factored and rewritten as

$$y^2 = \frac{b^2 x^2}{a^2}\left(1 - \frac{a^2}{x^2}\right)$$

$$y = \pm\frac{bx}{a}\sqrt{1 - \frac{a^2}{x^2}}$$

Now, as $x \to -\infty$ or as $x \to \infty$, the term $\dfrac{a^2}{x^2}$ approaches 0, so the expression under the radical approaches 1. So as $x \to -\infty$ or as $x \to \infty$, the value of y approaches $\pm\dfrac{bx}{a}$; that is, the graph of the hyperbola approaches the lines

$$y = -\frac{b}{a}x \quad \text{and} \quad y = \frac{b}{a}x$$

These lines are oblique asymptotes of the hyperbola. ∎

The asymptotes of a hyperbola are not part of the hyperbola, but they serve as a guide for graphing the hyperbola. For example, suppose that we want to graph the equation

$$\frac{x^2}{a^2} - \frac{y^2}{b^2} = 1$$

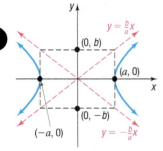

Figure 39 $\dfrac{x^2}{a^2} - \dfrac{y^2}{b^2} = 1$

Begin by plotting the vertices $(-a, 0)$ and $(a, 0)$. Then plot the points $(0, -b)$ and $(0, b)$ and use these four points to construct a rectangle, as shown in Figure 39. The diagonals of this rectangle have slopes $\dfrac{b}{a}$ and $-\dfrac{b}{a}$, and their extensions are the asymptotes of the hyperbola, $y = \dfrac{b}{a} x$ and $y = -\dfrac{b}{a} x$. If we graph the asymptotes, we can use them to establish the "opening" of the hyperbola and avoid plotting other points.

> **THEOREM Asymptotes of a Hyperbola; Transverse Axis along the y-Axis**
>
> The hyperbola $\dfrac{y^2}{a^2} - \dfrac{x^2}{b^2} = 1$ has the two oblique asymptotes
>
> $$y = \frac{a}{b} x \quad \text{and} \quad y = -\frac{a}{b} x \tag{5}$$

You are asked to prove this result in Problem 86.

For the remainder of this section, the direction "**Analyze the equation**" means to find the center, transverse axis, vertices, foci, and asymptotes of the hyperbola and graph it.

EXAMPLE 5 **Analyzing the Equation of a Hyperbola**

Analyze the equation $\dfrac{y^2}{4} - x^2 = 1$.

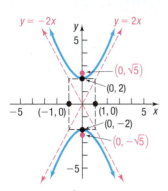

Figure 40 $\dfrac{y^2}{4} - x^2 = 1$

Solution Since the x^2-term is subtracted from the y^2-term, the equation is of the form of equation (3) and is a hyperbola with center at the origin and transverse axis along the y-axis. Comparing the equation to equation (3), note that $a^2 = 4$, $b^2 = 1$, and $c^2 = a^2 + b^2 = 5$. The vertices are at $(0, \pm a) = (0, \pm 2)$, and the foci are at $(0, \pm c) = (0, \pm \sqrt{5})$. Using equation (5) with $a = 2$ and $b = 1$, the asymptotes are the lines $y = \dfrac{a}{b} x = 2x$ and $y = -\dfrac{a}{b} x = -2x$. Form the rectangle containing the points $(0, \pm a) = (0, \pm 2)$ and $(\pm b, 0) = (\pm 1, 0)$. The extensions of the diagonals of the rectangle are the asymptotes. Now graph the asymptotes and the hyperbola. See Figure 40.

EXAMPLE 6 **Analyzing the Equation of a Hyperbola**

Analyze the equation $9x^2 - 4y^2 = 36$.

Solution Divide both sides of the equation by 36 to put the equation in proper form.

$$\frac{x^2}{4} - \frac{y^2}{9} = 1$$

The center of the hyperbola is the origin. Since the y^2-term is subtracted from the x^2-term, the transverse axis is along the x-axis, and the vertices and foci will lie on the x-axis. Using equation (2), note that $a^2 = 4$, $b^2 = 9$, and $c^2 = a^2 + b^2 = 13$. The vertices are $a = 2$ units left and right of the center at $(\pm a, 0) = (\pm 2, 0)$, the foci

(continued)

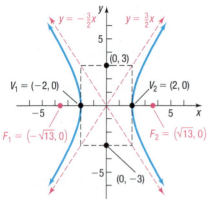

Figure 41 $9x^2 - 4y^2 = 36$

are $c = \sqrt{13}$ units left and right of the center at $(\pm c, 0) = (\pm\sqrt{13}, 0)$, and the asymptotes have the equations

$$y = \frac{b}{a}x = \frac{3}{2}x \quad \text{and} \quad y = -\frac{b}{a}x = -\frac{3}{2}x$$

To graph the hyperbola, form the rectangle containing the points $(\pm a, 0)$ and $(0, \pm b)$, that is, $(-2, 0)$, $(2, 0)$, $(0, -3)$, and $(0, 3)$. The extensions of the diagonals of the rectangle are the asymptotes. See Figure 41 for the graph.

Now Work PROBLEM 31

3 Analyze Hyperbolas with Center at (h, k)

If a hyperbola with center at the origin and transverse axis coinciding with a coordinate axis is shifted horizontally h units and then vertically k units, the result is a hyperbola with center at (h, k) and transverse axis parallel to a coordinate axis. The equations of such hyperbolas have the same forms as those given in equations (2) and (3), except that x is replaced by $x - h$ (the horizontal shift) and y is replaced by $y - k$ (the vertical shift). Table 4 gives the forms of the equations of such hyperbolas. See Figure 42 for typical graphs.

Table 4

Equations of a Hyperbola: Center at (h, k); Transverse Axis Parallel to a Coordinate Axis					
Center	**Transverse Axis**	**Foci**	**Vertices**	**Equation**	**Asymptotes**
(h, k)	Parallel to the x-axis	$(h \pm c, k)$	$(h \pm a, k)$	$\dfrac{(x-h)^2}{a^2} - \dfrac{(y-k)^2}{b^2} = 1,\ b^2 = c^2 - a^2$	$y - k = \pm\dfrac{b}{a}(x-h)$
(h, k)	Parallel to the y-axis	$(h, k \pm c)$	$(h, k \pm a)$	$\dfrac{(y-k)^2}{a^2} - \dfrac{(x-h)^2}{b^2} = 1,\ b^2 = c^2 - a^2$	$y - k = \pm\dfrac{a}{b}(x-h)$

NOTE Rather than memorize Table 4, use transformations (shift horizontally h units, vertically k units), along with the facts that a represents the distance from the center to the vertices, c represents the distance from the center to the foci, and $b^2 = c^2 - a^2$. ∎

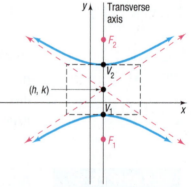

Figure 42

(a) $\dfrac{(x-h)^2}{a^2} - \dfrac{(y-k)^2}{b^2} = 1$

(b) $\dfrac{(y-k)^2}{a^2} - \dfrac{(x-h)^2}{b^2} = 1$

EXAMPLE 7

Finding an Equation of a Hyperbola, Center Not at the Origin

Find an equation for the hyperbola with center at $(1, -2)$, one focus at $(4, -2)$, and one vertex at $(3, -2)$. Graph the equation.

Solution

The center is at $(h, k) = (1, -2)$, so $h = 1$ and $k = -2$. Since the center, focus, and vertex all lie on the line $y = -2$, the transverse axis is parallel to the x-axis. The distance from the center $(1, -2)$ to the focus $(4, -2)$ is $c = 3$; the distance from the center $(1, -2)$ to the vertex $(3, -2)$ is $a = 2$. Then $b^2 = c^2 - a^2 = 9 - 4 = 5$.

The equation of the hyperbola is

$$\frac{(x-h)^2}{a^2} - \frac{(y-k)^2}{b^2} = 1$$

$$\frac{(x-1)^2}{4} - \frac{(y+2)^2}{5} = 1$$

Form the rectangle containing the vertices: $(-1, -2)$, $(3, -2)$, and the points $(h, k \pm b)$: $(1, -2 - \sqrt{5})$, $(1, -2 + \sqrt{5})$. Extend the diagonals of the rectangle to obtain the asymptotes. See Figure 43.

Figure 43 $\dfrac{(x-1)^2}{4} - \dfrac{(y+2)^2}{5} = 1$

Now Work PROBLEM 41

EXAMPLE 8	**Analyzing the Equation of a Hyperbola**

Analyze the equation $-x^2 + 4y^2 - 2x - 16y + 11 = 0$.

Solution

Complete the squares in x and in y.

$$-x^2 + 4y^2 - 2x - 16y + 11 = 0$$

$$-(x^2 + 2x) + 4(y^2 - 4y) = -11 \qquad \textbf{Group terms.}$$

$$-(x^2 + 2x + 1) + 4(y^2 - 4y + 4) = -11 - 1 + 16 \qquad \textbf{Complete each square.}$$

$$-(x+1)^2 + 4(y-2)^2 = 4 \qquad \textbf{Factor.}$$

$$(y-2)^2 - \frac{(x+1)^2}{4} = 1 \qquad \textbf{Divide both sides by 4.}$$

This is the equation of a hyperbola with center at $(-1, 2)$ and transverse axis parallel to the y-axis. Also, $a^2 = 1$ and $b^2 = 4$, so $c^2 = a^2 + b^2 = 5$. Since the transverse axis is parallel to the y-axis, the vertices and foci are located a and c units above and below the center, respectively. The vertices are at $(h, k \pm a) = (-1, 2 \pm 1)$, or $(-1, 1)$ and $(-1, 3)$. The foci are at $(h, k \pm c) = (-1, 2 \pm \sqrt{5})$. The asymptotes are $y - 2 = \frac{1}{2}(x+1)$ and $y - 2 = -\frac{1}{2}(x+1)$. Figure 44 shows the graph.

Figure 44
$-x^2 + 4y^2 - 2x - 16y + 11 = 0$

Now Work PROBLEM 55

4 Solve Applied Problems Involving Hyperbolas

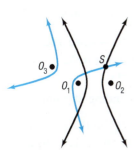

Figure 45

Look at Figure 45. Suppose that three microphones are located at points O_1, O_2, and O_3 (the foci of the two hyperbolas). In addition, suppose that a gun is fired at S and the microphone at O_1 records the gunshot 1 second after the microphone at O_2. Because sound travels at about 1100 feet per second, we conclude that the microphone at O_1 is 1100 feet farther from the gunshot than O_2. This situation is modeled by placing S on a branch of a hyperbola with foci at O_1 and O_2. (Do you see why? The difference of the distances from S to O_1 and from S to O_2 is the constant 1100.) If the third microphone at O_3 records the gunshot 2 seconds after O_1, then S lies on a branch of a second hyperbola with foci at O_1 and O_3. In this case, the constant difference will be 2200. The intersection of the two hyperbolas identifies the location of S.

EXAMPLE 9	**Lightning Strikes**

Suppose that two people standing 1 mile apart both see a flash of lightning. After a period of time, the person standing at point A hears the thunder. One second later, the person standing at point B hears the thunder. If the person at B is due west of the person at A and the lightning strike is known to occur due north of the person standing at A, where did the lightning strike occur?

Solution   See Figure 46, where the point (x, y) represents the location of the lightning strike. Sound travels at 1100 feet per second, so the person at point A is 1100 feet closer to the lightning strike than the person at point B. Since the difference of the distance from (x, y) to B and the distance from (x, y) to A is the constant 1100, the point (x, y) lies on a hyperbola whose foci are at A and B.

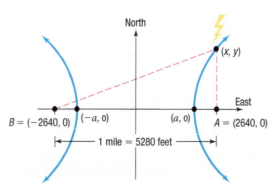

Figure 46 Lightning strike model

An equation of the hyperbola is

$$\frac{x^2}{a^2} - \frac{y^2}{b^2} = 1$$

where $2a = 1100$, or $a = 550$.

Because the distance between the two people is 1 mile (5280 feet) and each person is at a focus of the hyperbola, then

$$2c = 5280$$
$$c = \frac{5280}{2} = 2640$$

Since $b^2 = c^2 - a^2 = 2640^2 - 550^2 = 6{,}667{,}100$, the equation of the hyperbola that describes the location of the lightning strike is

$$\frac{x^2}{550^2} - \frac{y^2}{6{,}667{,}100} = 1$$

Refer to Figure 46. Since the lightning strike occurred due north of the individual at the point $A = (2640, 0)$, let $x = 2640$ and solve the resulting equation.

$$\frac{2640^2}{550^2} - \frac{y^2}{6{,}667{,}100} = 1 \qquad \text{\color{blue}{x = 2640}}$$

$$-\frac{y^2}{6{,}667{,}100} = -22.04 \qquad \text{\color{blue}{Subtract } \frac{2640^2}{550^2} \text{ from both sides.}}$$

$$y^2 = 146{,}942{,}884 \qquad \text{\color{blue}{Multiply both sides by } -6{,}667{,}100.}$$

$$y = 12{,}122 \qquad \text{\color{blue}{y > 0 since the lightning strike occurred in quadrant I.}}$$

The lightning strike occurred 12,122 feet north of the person standing at point A.

✓**Check:** The difference between the distance from $(2640, 12122)$ to the person at the point $B = (-2640, 0)$ and the distance from $(2640, 12122)$ to the person at the point $A = (2640, 0)$ should be 1100. Using the distance formula, the difference of the distances is

$$\sqrt{[2640 - (-2640)]^2 + (12{,}122 - 0)^2} - \sqrt{(2640 - 2640)^2 + (12{,}122 - 0)^2} = 1100$$

as required.

Now Work PROBLEM 75

6.4 Assess Your Understanding

'Are You Prepared?' *Answers are given at the end of these exercises. If you get a wrong answer, read the pages listed in* red.

1. The distance d from $P_1 = (3, -4)$ to $P_2 = (-2, 1)$ is
   $d =$ ____ . (pp. 3–4)

2. To complete the square of $x^2 + 5x$, add ____ . (pp. A24–A25)

3. Find the intercepts of the equation $y^2 = 9 + 4x^2$. (pp. 11–12)

4. **True or False** The equation $y^2 = 9 + x^2$ is symmetric with respect to the x-axis, the y-axis, and the origin. (pp. 12–14)

5. To graph $y = (x - 5)^3 - 4$, shift the graph of $y = x^3$ to the _____ _____ unit(s) and then _____ _____ unit(s).
   (left/right)   (up/down)
   (pp. 67–77)

6. Find the vertical asymptotes, if any, and the horizontal or oblique asymptote, if any, of $y = \dfrac{x^2 - 9}{x^2 - 4}$. (pp. 76–77)

Concepts and Vocabulary

7. A(n) _____ is the collection of points in a plane, the difference of whose distances from two fixed points is a constant.

8. For a hyperbola, the foci lie on a line called the _____ _____.

Answer Problems 9–11 using the figure to the right.

9. **Multiple Choice** The equation of the hyperbola is of the form

   (a) $\dfrac{(x - h)^2}{a^2} - \dfrac{(y - k)^2}{b^2} = 1$ (b) $\dfrac{(y - k)^2}{a^2} - \dfrac{(x - h)^2}{b^2} = 1$

   (c) $\dfrac{(x - h)^2}{a^2} + \dfrac{(y - k)^2}{b^2} = 1$ (d) $\dfrac{(x - h)^2}{b^2} + \dfrac{(y - k)^2}{a^2} = 1$

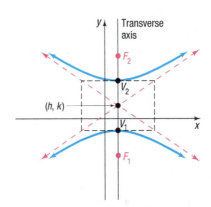

10. If the center of the hyperbola is $(2, 1)$ and $a = 3$, then the coordinates of the vertices are _____ and _____ .

11. If the center of the hyperbola is $(2, 1)$ and $c = 5$, then the coordinates of the foci are _____ and _____ .

12. **Multiple Choice** In a hyperbola, if $a = 3$ and $c = 5$, then $b =$ _____ .

    (a) 1   (b) 2   (c) 4   (d) 8

13. For the hyperbola $\dfrac{x^2}{4} - \dfrac{y^2}{9} = 1$, the value of a is ____ , the value of b is ____ , and the transverse axis is the ____ -axis.

14. For the hyperbola $\dfrac{y^2}{16} - \dfrac{x^2}{81} = 1$, the asymptotes are _____ and _____ .

Skill Building

In Problems 15–18, the graph of a hyperbola is given. Match each graph to its equation.

   (A) $\dfrac{x^2}{4} - y^2 = 1$ (B) $x^2 - \dfrac{y^2}{4} = 1$ (C) $\dfrac{y^2}{4} - x^2 = 1$ (D) $y^2 - \dfrac{x^2}{4} = 1$

15.

16.

17.

18.
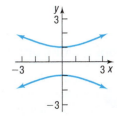

In Problems 19–28, find an equation for the hyperbola described. Graph the equation.

19. Center at $(0, 0)$; focus at $(3, 0)$; vertex at $(1, 0)$

20. Center at $(0, 0)$; focus at $(0, 5)$; vertex at $(0, 3)$

21. Center at $(0, 0)$; focus at $(0, -6)$; vertex at $(0, 4)$

22. Center at $(0, 0)$; focus at $(-3, 0)$; vertex at $(2, 0)$

23. Foci at $(-5, 0)$ and $(5, 0)$; vertex at $(3, 0)$

24. Focus at $(0, 6)$; vertices at $(0, -2)$ and $(0, 2)$

25. Vertices at $(0, -6)$ and $(0, 6)$; asymptote the line $y = 2x$

26. Vertices at $(-4, 0)$ and $(4, 0)$; asymptote the line $y = 2x$

27. Foci at $(-4, 0)$ and $(4, 0)$; asymptote the line $y = -x$

28. Foci at $(0, -2)$ and $(0, 2)$; asymptote the line $y = -x$

In Problems 29–36, find the center, transverse axis, vertices, foci, and asymptotes. Graph each equation.

29. $\dfrac{x^2}{25} - \dfrac{y^2}{9} = 1$ **30.** $\dfrac{y^2}{16} - \dfrac{x^2}{4} = 1$ **31.** $4x^2 - y^2 = 16$ **32.** $4y^2 - x^2 = 16$

33. $y^2 - 9x^2 = 9$ **34.** $x^2 - y^2 = 4$ **35.** $y^2 - x^2 = 25$ **36.** $2x^2 - y^2 = 4$

In Problems 37–40, find an equation for each hyperbola.

37.   **38.**   **39.**   **40.**

In Problems 41–48, find an equation for the hyperbola described. Graph the equation.

41. Center at $(4, -1)$; focus at $(7, -1)$; vertex at $(6, -1)$

42. Center at $(-3, 1)$; focus at $(-3, 6)$; vertex at $(-3, 4)$

43. Center at $(-3, -4)$; focus at $(-3, -8)$; vertex at $(-3, -2)$

44. Center at $(1, 4)$; focus at $(-2, 4)$; vertex at $(0, 4)$

45. Foci at $(3, 7)$ and $(7, 7)$; vertex at $(6, 7)$

46. Focus at $(-4, 0)$; vertices at $(-4, 4)$ and $(-4, 2)$

47. Vertices at $(-1, -1)$ and $(3, -1)$;

   asymptote the line $y + 1 = \dfrac{3}{2}(x - 1)$

48. Vertices at $(1, -3)$ and $(1, 1)$;

   asymptote the line $y + 1 = \dfrac{3}{2}(x - 1)$

In Problems 49–62, find the center, transverse axis, vertices, foci, and asymptotes. Graph each equation.

49. $\dfrac{(x-2)^2}{4} - \dfrac{(y+3)^2}{9} = 1$ **50.** $\dfrac{(y+3)^2}{4} - \dfrac{(x-2)^2}{9} = 1$ **51.** $(y-2)^2 - 4(x+2)^2 = 4$

52. $(x+4)^2 - 9(y-3)^2 = 9$ **53.** $(x+1)^2 - (y+2)^2 = 4$ **54.** $(y-3)^2 - (x+2)^2 = 4$

55. $x^2 - y^2 - 2x - 2y - 1 = 0$ **56.** $y^2 - x^2 - 4y + 4x - 1 = 0$ **57.** $y^2 - 4x^2 - 4y - 8x - 4 = 0$

58. $2x^2 - y^2 + 4x + 4y - 4 = 0$ **59.** $4x^2 - y^2 - 24x - 4y + 16 = 0$ **60.** $2y^2 - x^2 + 2x + 8y + 3 = 0$

61. $y^2 - 4x^2 - 16x - 2y - 19 = 0$ **62.** $x^2 - 3y^2 + 8x - 6y + 4 = 0$

In Problems 63–66, graph each function. Be sure to label any intercepts. [**Hint**: *Notice that each function is half a hyperbola.*]

63. $f(x) = \sqrt{16 + 4x^2}$ **64.** $f(x) = -\sqrt{9 + 9x^2}$ **65.** $f(x) = -\sqrt{-25 + x^2}$ **66.** $f(x) = \sqrt{-1 + x^2}$

Mixed Practice *In Problems 67–74, analyze each equation.*

67. $\dfrac{(x-3)^2}{4} - \dfrac{y^2}{25} = 1$ **68.** $\dfrac{(y+2)^2}{16} - \dfrac{(x-2)^2}{4} = 1$ **69.** $x^2 = 16(y - 3)$

70. $y^2 = -12(x + 1)$ **71.** $25x^2 + 9y^2 - 250x + 400 = 0$ **72.** $x^2 + 36y^2 - 2x + 288y + 541 = 0$

73. $x^2 - 6x - 8y - 31 = 0$ **74.** $9x^2 - y^2 - 18x - 8y - 88 = 0$

Applications and Extensions

75. Fireworks Display Suppose that two people standing 2 miles apart both see the burst from a fireworks display. After a period of time the first person, standing at point A, hears the burst. One second later the second person, standing at point B, hears the burst. If the person at point B is due west of the person at point A, and if the display is known to occur due north of the person at point A, where did the fireworks display occur?

76. Lightning Strikes Suppose that two people standing 1 mile apart both see a flash of lightning. After a period of time the first person, standing at point A, hears the thunder. Two seconds later the second person, standing at point B,

hears the thunder. If the person at point B is due west of the person at point A, and if the lightning strike is known to occur due north of the person standing at point A, where did the lightning strike occur?

77. Nuclear Power Plant Some nuclear power plants utilize "natural draft" cooling towers in the shape of a **hyperboloid**, a solid obtained by rotating a hyperbola about its conjugate axis. Suppose that such a cooling tower has a base diameter of 400 feet, and the diameter at its narrowest point, 360 feet above the ground, is 200 feet. If the diameter at the top of the tower is 300 feet, how tall is the tower?

Source: Bay Area Air Quality Management District

78. An Explosion Two recording devices are set 2400 feet apart, with the device at point A to the west of the device at point B. At a point between the devices 300 feet from point B, a small amount of explosive is detonated. The recording devices record the time until the sound reaches each. How far directly north of point B should a second explosion be done so that the measured time difference recorded by the devices is the same as that for the first detonation?

79. Rutherford's Experiment In May 1911, Ernest Rutherford published a paper in *Philosophical Magazine*. In this article, he described the motion of alpha particles as they are shot at a piece of gold foil 0.00004 cm thick. Before conducting this experiment, Rutherford expected that the alpha particles would shoot through the foil just as a bullet would shoot through snow. Instead, a small fraction of the alpha particles bounced off the foil. This led to the conclusion that the nucleus of an atom is dense, while the remainder of the atom is sparse. Only the density of the nucleus could cause the alpha particles to deviate from their path. The figure shows a diagram from Rutherford's paper that indicates that the deflected alpha particles follow the path of one branch of a hyperbola.

(a) Find an equation of the asymptotes under this scenario.
(b) If the vertex of the path of the alpha particles is 10 cm from the center of the hyperbola, find a model that describes the path of the particle.

80. Hyperbolic Mirrors Hyperbolas have interesting reflective properties that make them useful for lenses and mirrors. For example, if a ray of light strikes a convex hyperbolic mirror on a line that would (theoretically) pass through its rear focus, it is reflected through the front focus. This property, and that of the parabola, were used to develop the *Cassegrain* telescope in 1672. The focus of the parabolic mirror and the rear focus of the hyperbolic mirror are the same point. The rays are collected by the parabolic mirror, then reflected toward the (common) focus, and thus are reflected by the hyperbolic mirror through the opening to its front focus, where the eyepiece is located. If the equation of the hyperbola is $\dfrac{y^2}{9} - \dfrac{x^2}{16} = 1$ and the focal length (distance from the vertex to the focus) of the parabola is 6, find the equation of the parabola.

Source: www.enchantedlearning.com

81. Lamp Shadow The light from a lamp creates a shadow on a wall with a hyperbolic border. Find the equation of the border if the distance between the vertices is 18 inches and the foci are 4 inches from the vertices. Assume the center of the hyperbola is at the origin.

82. Sonic Boom Aircraft such as fighter jets routinely go supersonic (faster than the speed of sound). An aircraft moving faster than the speed of sound produces a cone-shaped shock wave that "booms" as it trails the vehicle. The wave intersects the ground in the shape of one half of a hyperbola and the area over which the "boom" is audible is called the "boom carpet." If an aircraft creates a shock wave that intersects the ground in the shape of the hyperbola $\dfrac{x^2}{484} - \dfrac{y^2}{100} = 1$ (units in miles), how wide is the "boom carpet" 32 miles behind the aircraft?

83. The **eccentricity** e of a hyperbola is defined as the number $\dfrac{c}{a}$, where a is the distance of a vertex from the center and c is the distance of a focus from the center. Because $c > a$, it follows that $e > 1$. Describe the general shape of a hyperbola whose eccentricity is close to 1. What is the shape if e is very large?

84. A hyperbola for which $a = b$ is called an **equilateral hyperbola**. Find the eccentricity e of an equilateral hyperbola.

[Note: The eccentricity of a hyperbola is defined in Problem 83.]

85. Two hyperbolas that have the same set of asymptotes are called **conjugate**. Show that the hyperbolas

$$\frac{x^2}{4} - y^2 = 1 \quad \text{and} \quad y^2 - \frac{x^2}{4} = 1$$

are conjugate. Graph each hyperbola on the same set of coordinate axes.

86. Prove that the hyperbola

$$\frac{y^2}{a^2} - \frac{x^2}{b^2} = 1$$

has the two oblique asymptotes

$$y = \frac{a}{b}x \quad \text{and} \quad y = -\frac{a}{b}x$$

87. Challenge Problem Show that the graph of an equation of the form

$$Ax^2 + Cy^2 + F = 0 \qquad A \neq 0, C \neq 0, F \neq 0$$

where A and C are opposite in sign, is a hyperbola with center at $(0, 0)$.

88. Challenge Problem Show that the graph of an equation of the form

$$Ax^2 + Cy^2 + Dx + Ey + F = 0 \qquad A \neq 0, C \neq 0$$

where A and C are opposite in sign,

(a) is a hyperbola if $\dfrac{D^2}{4A} + \dfrac{E^2}{4C} - F \neq 0$.

(b) is two intersecting lines if $\dfrac{D^2}{4A} + \dfrac{E^2}{4C} - F = 0$.

Retain Your Knowledge

Problems 89–96 are based on material learned earlier in the course. The purpose of these problems is to keep the material fresh in your mind so that you are better prepared for the final exam.

89. For $f(x) = -\dfrac{1}{2}\sin(3x + \pi) + 5$, find the amplitude, period, phase shift, and vertical shift.

90. Solve the triangle described: $a = 7$, $b = 10$, and $C = 100°$

91. Find the rectangular coordinates of the point with the polar coordinates $\left(12, -\dfrac{\pi}{3}\right)$.

92. Transform the polar equation $r = 6\sin\theta$ to an equation in rectangular coordinates. Then identify and graph the equation.

93. Find the area of the region enclosed by the graphs of $y = \sqrt{9 - x^2}$ and $y = x + 3$.

94. Solve $(2x + 3)^2 + x^2 = 5x(2 + x) + 1$.

95. Find the midpoint of the line segment connecting the points $(3, -8)$ and $(-2, 5)$.

96. Evaluate $\cos\left(\sin^{-1}\left(\dfrac{x}{4}\right)\right)$.

'Are You Prepared?' Answers

1. $5\sqrt{2}$ **2.** $\dfrac{25}{4}$ **3.** $(0, -3), (0, 3)$ **4.** True   **5.** right; 5; down; 4   **6.** Vertical: $x = -2, x = 2$; horizontal: $y = 1$

6.5 Rotation of Axes; General Form of a Conic

PREPARING FOR THIS SECTION *Before getting started, review the following:*

- Sum Formulas for Sine and Cosine (Section 3.5, pp. 227 and 230)
- Half-angle Formulas for Sine and Cosine (Section 3.6, p. 244)
- Double-angle Formulas for Sine and Cosine (Section 3.6, p. 241)

Now Work the 'Are You Prepared?' problems on page 432.

OBJECTIVES **1** Identify a Conic (p. 426)
2 Use a Rotation of Axes to Transform Equations (p. 427)
3 Analyze an Equation Using a Rotation of Axes (p. 430)
4 Identify Conics without Rotating the Axes (p. 432)

In this section, we show that the graph of a general second-degree polynomial equation containing two variables x and y—that is, an equation of the form

$$Ax^2 + Bxy + Cy^2 + Dx + Ey + F = 0 \qquad (1)$$

where A, B, and C are not all 0—is a conic. We are not concerned with the degenerate cases of equation (1), such as $x^2 + y^2 = 0$, whose graph is a single point $(0, 0)$; or $x^2 + 3y^2 + 3 = 0$, whose graph contains no points; or $x^2 - 4y^2 = 0$, whose graph is two lines, $x - 2y = 0$ and $x + 2y = 0$.

We begin with the case where $B = 0$. In this case, the term containing xy is not present, so equation (1) has the form

$$Ax^2 + Cy^2 + Dx + Ey + F = 0 \qquad (2)$$

where either $A \neq 0$ or $C \neq 0$.

1 Identify a Conic

We have already discussed how to identify the graph of an equation of this form. We complete the squares of the quadratic expressions in x or y, or both. Then the conic can be identified by comparing it to one of the forms studied in Sections 6.2 through 6.4.

But the conic can be identified directly from its equation without completing the squares.

> **THEOREM** **Identifying Conics without Completing the Squares**
>
> Excluding degenerate cases, the equation
>
> $$Ax^2 + Cy^2 + Dx + Ey + F = 0$$
>
> where A and C are not both equal to zero:
> - Defines a parabola if $AC = 0$.
> - Defines an ellipse (or a circle) if $AC > 0$.
> - Defines a hyperbola if $AC < 0$.

Proof

- If $AC = 0$, then either $A = 0$ or $C = 0$, but not both, so the form of equation (2) is either

$$Ax^2 + Dx + Ey + F = 0 \qquad A \neq 0$$

or

$$Cy^2 + Dx + Ey + F = 0 \qquad C \neq 0$$

Using the results of Problems 80 and 81 at the end of Section 6.2, it follows that, except for the degenerate cases, the equation is a parabola.

- If $AC > 0$, then A and C have the same sign. Using the results of Problem 90 at the end of Section 6.3, except for the degenerate cases, the equation is an ellipse.

- If $AC < 0$, then A and C have opposite signs. Using the results of Problem 88 at the end of Section 6.4, except for the degenerate cases, the equation is a hyperbola. ∎

Although we are not studying the degenerate cases of equation (2), in practice, you should be alert to the possibility of degeneracy.

EXAMPLE 1

Identifying a Conic without Completing the Squares

Identify the graph of each equation without completing the squares.

(a) $3x^2 + 6y^2 + 6x - 12y = 0$ (b) $2x^2 - 3y^2 + 6y + 4 = 0$

(c) $y^2 - 2x + 4 = 0$

Solution

(a) Note that $A = 3$ and $C = 6$. Since $AC = 18 > 0$, the equation defines an ellipse.

(b) Here $A = 2$ and $C = -3$, so $AC = -6 < 0$. The equation defines a hyperbola.

(c) Here $A = 0$ and $C = 1$, so $AC = 0$. The equation defines a parabola.

 ✏️ **Now Work** PROBLEM 11

Although we can now identify the type of conic represented by any general second-degree equation of the form of equation (2) without completing the squares, we still need to complete the squares if we desire additional information about the conic, such as its graph.

2 Use a Rotation of Axes to Transform Equations

Now let's suppose that $B \neq 0$. To discuss this case, we introduce a new procedure: *rotation of axes*.

In a **rotation of axes**, the origin remains fixed while the x-axis and y-axis are rotated through an angle θ to a new position; the new positions of the x-axis and the y-axis are denoted by x' and y', respectively, as shown in Figure 47(a).

Now look at Figure 47(b). There the point P has the coordinates (x, y) relative to the xy-plane, while the same point P has coordinates (x', y') relative to the $x'y'$-plane. We need relationships that enable us to express x and y in terms of x', y', and θ.

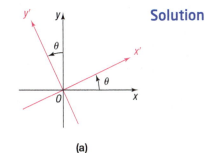

(a)

(b)

Figure 47 Rotation of axes

As Figure 47(b) shows, r denotes the distance from the origin O to the point P, and α denotes the angle between the positive x'-axis and the ray from O through P. Then, using the definitions of sine and cosine, we have

$$x' = r \cos \alpha \qquad\qquad y' = r \sin \alpha \qquad\qquad \textbf{(3)}$$

$$x = r \cos(\theta + \alpha) \qquad y = r \sin(\theta + \alpha) \qquad \textbf{(4)}$$

Now

$$x = r \cos(\theta + \alpha)$$

$$= r(\cos \theta \cos \alpha - \sin \theta \sin \alpha) \qquad \textcolor{blue}{\textbf{Use the Sum Formula for cosine.}}$$

$$= (r \cos \alpha) \cdot \cos \theta - (r \sin \alpha) \cdot \sin \theta$$

$$= x' \cos \theta - y' \sin \theta \qquad\qquad \textcolor{blue}{\textbf{By equation (3)}}$$

Similarly,

$$y = r \sin(\theta + \alpha)$$

$$= r(\sin \theta \cos \alpha + \cos \theta \sin \alpha) \qquad \textcolor{blue}{\textbf{Use the Sum Formula for sine.}}$$

$$= x' \sin \theta + y' \cos \theta \qquad\qquad \textcolor{blue}{\textbf{By equation (3)}}$$

THEOREM   Rotation Formulas

If the x- and y-axes are rotated through an angle θ, the coordinates (x, y) of a point P relative to the xy-plane and the coordinates (x', y') of the same point relative to the new x'- and y'-axes are related by the formulas

$$x = x' \cos \theta - y' \sin \theta \qquad y = x' \sin \theta + y' \cos \theta \qquad \textbf{(5)}$$

EXAMPLE 2   **Rotating Axes**

Express the equation $xy = 1$ in terms of new $x'y'$-coordinates by rotating the axes through a 45° angle. Discuss the new equation.

Solution   Let $\theta = 45°$ in formulas (5). Then

$$x = x' \cos 45° - y' \sin 45° = x' \frac{\sqrt{2}}{2} - y' \frac{\sqrt{2}}{2} = \frac{\sqrt{2}}{2}(x' - y')$$

$$y = x' \sin 45° + y' \cos 45° = x' \frac{\sqrt{2}}{2} + y' \frac{\sqrt{2}}{2} = \frac{\sqrt{2}}{2}(x' + y')$$

Substituting these expressions for x and y in $xy = 1$ gives

$$\left[\frac{\sqrt{2}}{2}(x' - y') \right] \cdot \left[\frac{\sqrt{2}}{2}(x' + y') \right] = 1$$

$$\frac{1}{2}(x'^2 - y'^2) = 1$$

$$\frac{x'^2}{2} - \frac{y'^2}{2} = 1$$

This is the equation of a hyperbola with center at $(0, 0)$ and transverse axis along the x'-axis. The vertices are at $(\pm\sqrt{2}, 0)$ on the x'-axis; the asymptotes are $y' = x'$ and $y' = -x'$ (which correspond to the original x- and y-axes).

See Figure 48 for the graph of $xy = 1$.

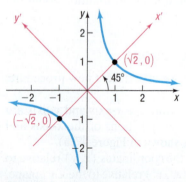

Figure 48 $xy = 1$

As Example 2 illustrates, a rotation of axes through an appropriate angle can transform a second-degree equation in x and y containing an xy-term into one in x' and y' in which no $x'y'$-term appears. In fact, a rotation of axes through an appropriate angle will transform any equation of the form of equation (1) into an equation in x' and y' without an $x'y'$-term.

To find the formula for choosing an appropriate angle θ through which to rotate the axes, begin with equation (1),

$$Ax^2 + Bxy + Cy^2 + Dx + Ey + F = 0 \quad B \neq 0$$

Now rotate the x- and y-axes through an angle θ using the rotation formulas (5).

$$A(x' \cos \theta - y' \sin \theta)^2 + B(x' \cos \theta - y' \sin \theta)(x' \sin \theta + y' \cos \theta)$$
$$+ C(x' \sin \theta + y' \cos \theta)^2 + D(x' \cos \theta - y' \sin \theta)$$
$$+ E(x' \sin \theta + y' \cos \theta) + F = 0$$

Expanding and collecting like terms gives

$$(A \cos^2 \theta + B \sin \theta \cos \theta + C \sin^2 \theta)x'^2 + [B(\cos^2 \theta - \sin^2 \theta) + 2(C - A)(\sin \theta \cos \theta)]x'y'$$
$$+ (A \sin^2 \theta - B \sin \theta \cos \theta + C \cos^2 \theta)y'^2$$
$$+ (D \cos \theta + E \sin \theta)x'$$
$$+ (-D \sin \theta + E \cos \theta)y' + F = 0 \qquad (6)$$

In equation (6), the coefficient of $x'y'$ is

$$B(\cos^2 \theta - \sin^2 \theta) + 2(C - A)(\sin \theta \cos \theta)$$

To eliminate the $x'y'$-term, select an angle θ so that this coefficient is 0.

$$B(\cos^2 \theta - \sin^2 \theta) + 2(C - A)(\sin \theta \cos \theta) = 0$$
$$B \cos(2\theta) + (C - A)\sin(2\theta) = 0 \quad \text{\textcolor{blue}{Double-angle Formulas}}$$
$$B \cos(2\theta) = (A - C)\sin(2\theta)$$
$$\cot(2\theta) = \frac{A - C}{B} \quad B \neq 0$$

THEOREM Transformation Angle

To transform the equation

$$Ax^2 + Bxy + Cy^2 + Dx + Ey + F = 0 \quad B \neq 0$$

into an equation in x' and y' without an $x'y'$-term, rotate the axes through an angle θ that satisfies the equation

$$\cot(2\theta) = \frac{A - C}{B} \qquad (7)$$

WARNING Be careful if you use a calculator to solve equation (7).

- If $\cot(2\theta) = 0$, then $2\theta = 90°$ and $\theta = 45°$.

- If $\cot(2\theta) \neq 0$, first find $\cos(2\theta)$. Then use the inverse cosine function key(s) to obtain 2θ, $0° < 2\theta < 180°$. Finally, divide by 2 to obtain the correct acute angle θ. ∎

Equation (7) has an infinite number of solutions for θ. We follow the convention of choosing the acute angle θ that satisfies (7). There are two possibilities:

If $\cot(2\theta) \geq 0$, then $0° < 2\theta \leq 90°$, so $0° < \theta \leq 45°$.
If $\cot(2\theta) < 0$, then $90° < 2\theta < 180°$, so $45° < \theta < 90°$.

Each of these results in a counterclockwise rotation of the axes through an acute angle θ.*

---

*Any rotation through an angle θ that satisfies $\cot(2\theta) = \dfrac{A - C}{B}$ will eliminate the $x'y'$-term.

However, the final form of the transformed equation may be different (but equivalent), depending on the angle chosen.

3 Analyze an Equation Using a Rotation of Axes

For the remainder of this section, the direction "**Analyze the equation**" means to transform the given equation so that it contains no xy-term and to graph the equation.

EXAMPLE 3

Analyzing an Equation Using a Rotation of Axes

Analyze the equation $x^2 + \sqrt{3}\, xy + 2y^2 - 10 = 0$.

Solution

Since an xy-term is present, rotate the axes. Using $A = 1$, $B = \sqrt{3}$, and $C = 2$ in equation (7), the acute angle θ of rotation satisfies the equation

$$\cot(2\theta) = \frac{A - C}{B} = \frac{-1}{\sqrt{3}} = -\frac{\sqrt{3}}{3} \qquad 0° < 2\theta < 180°$$

Since $\cot(2\theta) = -\dfrac{\sqrt{3}}{3}$, this means $2\theta = 120°$, so $\theta = 60°$. Using formulas (5),

$$x = x'\cos 60° - y'\sin 60° = \frac{1}{2}x' - \frac{\sqrt{3}}{2}y' = \frac{1}{2}\left(x' - \sqrt{3}\, y'\right)$$

$$y = x'\sin 60° + y'\cos 60° = \frac{\sqrt{3}}{2}x' + \frac{1}{2}y' = \frac{1}{2}\left(\sqrt{3}\, x' + y'\right)$$

Substituting these values into the original equation and simplifying gives

$$x^2 + \sqrt{3}\, xy + 2y^2 - 10 = 0$$

$$\frac{1}{4}\left(x' - \sqrt{3}\, y'\right)^2 + \sqrt{3} \cdot \left[\frac{1}{2}\left(x' - \sqrt{3}\, y'\right)\right] \cdot \left[\frac{1}{2}\left(\sqrt{3}\, x' + y'\right)\right] + 2 \cdot \left[\frac{1}{4}\left(\sqrt{3}\, x' + y'\right)^2\right] = 10$$

Multiply both sides by 4 and expand to obtain

$$x'^2 - 2\sqrt{3}\, x'y' + 3y'^2 + \sqrt{3}\left(\sqrt{3}\, x'^2 - 2x'y' - \sqrt{3}\, y'^2\right) + 2\left(3x'^2 + 2\sqrt{3}\, x'y' + y'^2\right) = 40$$

$$10x'^2 + 2y'^2 = 40$$

$$\frac{x'^2}{4} + \frac{y'^2}{20} = 1$$

This is the equation of an ellipse with center at $(0, 0)$ and major axis along the y'-axis. The vertices are at $(0, \pm 2\sqrt{5})$ on the y'-axis. See Figure 49 for the graph. ⌡

🖉 **Now Work** PROBLEMS **21 AND 31**

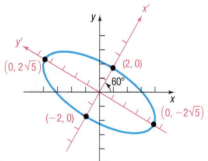

Figure 49 $\dfrac{x'^2}{4} + \dfrac{y'^2}{20} = 1$

In Example 3, the acute angle θ of rotation was easy to find because of the numbers used in the given equation. In general, the equation $\cot(2\theta) = \dfrac{A - C}{B}$ does not have such a "nice" solution. As the next example shows, we can still find the appropriate rotation formulas without a calculator approximation by using Half-angle Formulas.

EXAMPLE 4

Analyzing an Equation Using a Rotation of Axes

Analyze the equation $4x^2 - 4xy + y^2 + 5\sqrt{5}\, x + 5 = 0$.

Solution   Letting $A = 4$, $B = -4$, and $C = 1$ in equation (7), the acute angle θ to rotate the axes satisfies

$$\cot(2\theta) = \frac{A - C}{B} = \frac{3}{-4} = -\frac{3}{4}$$

To use the rotation formulas (5), we need to know the values of $\sin\theta$ and $\cos\theta$. Because the angle θ is acute, we know that $\sin\theta > 0$ and $\cos\theta > 0$. Use the Half-angle Formulas in the form

$$\sin\theta = \sqrt{\frac{1 - \cos(2\theta)}{2}} \qquad \cos\theta = \sqrt{\frac{1 + \cos(2\theta)}{2}}$$

Figure 50 $\cos(2\theta) = \dfrac{x}{r} = \dfrac{-3}{5}$

See Figure 50. Because $\cot(2\theta) = -\dfrac{3}{4}$, then $90° < 2\theta < 180°$, so $\cos(2\theta) = -\dfrac{3}{5}$. Then

$$\sin\theta = \sqrt{\frac{1 - \cos(2\theta)}{2}} = \sqrt{\frac{1 - \left(-\frac{3}{5}\right)}{2}} = \sqrt{\frac{4}{5}} = \frac{2}{\sqrt{5}} = \frac{2\sqrt{5}}{5}$$

$$\cos\theta = \sqrt{\frac{1 + \cos(2\theta)}{2}} = \sqrt{\frac{1 + \left(-\frac{3}{5}\right)}{2}} = \sqrt{\frac{1}{5}} = \frac{1}{\sqrt{5}} = \frac{\sqrt{5}}{5}$$

With these values, the rotation formulas (5) are

$$x = \frac{\sqrt{5}}{5}x' - \frac{2\sqrt{5}}{5}y' = \frac{\sqrt{5}}{5}(x' - 2y')$$

$$y = \frac{2\sqrt{5}}{5}x' + \frac{\sqrt{5}}{5}y' = \frac{\sqrt{5}}{5}(2x' + y')$$

Substituting these values in the original equation and simplifying gives

$$4x^2 - 4xy + y^2 + 5\sqrt{5}x + 5 = 0$$

$$4\left[\frac{\sqrt{5}}{5}(x' - 2y')\right]^2 - 4\left[\frac{\sqrt{5}}{5}(x' - 2y')\right]\left[\frac{\sqrt{5}}{5}(2x' + y')\right]$$

$$+ \left[\frac{\sqrt{5}}{5}(2x' + y')\right]^2 + 5\sqrt{5}\left[\frac{\sqrt{5}}{5}(x' - 2y')\right] = -5$$

Multiply both sides by 5 and expand to obtain

$$4(x'^2 - 4x'y' + 4y'^2) - 4(2x'^2 - 3x'y' - 2y'^2)$$

$$+ 4x'^2 + 4x'y' + y'^2 + 25(x' - 2y') = -25$$

$$25y'^2 - 50y' + 25x' = -25 \quad \text{\color{blue}Combine like terms.}$$

$$y'^2 - 2y' + x' = -1 \quad \text{\color{blue}Divide both sides by 25.}$$

$$y'^2 - 2y' + 1 = -x' \quad \text{\color{blue}Complete the square in } y'.$$

$$(y' - 1)^2 = -x'$$

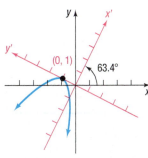

Figure 51 $(y' - 1)^2 = -x'$

This is the equation of a parabola with vertex at $(0, 1)$ in the $x'y'$-plane. The axis of symmetry is parallel to the x'-axis. Use a calculator to solve $\sin\theta = \dfrac{2\sqrt{5}}{5}$, and find that $\theta \approx 63.4°$. See Figure 51 for the graph.

━━━━ **Now Work** PROBLEM 37

4 Identify Conics without Rotating the Axes

Suppose that we are required only to identify (rather than analyze) the graph of an equation of the form

$$Ax^2 + Bxy + Cy^2 + Dx + Ey + F = 0 \qquad B \neq 0 \tag{8}$$

Applying the rotation formulas (5) to this equation gives an equation of the form

$$A'x'^2 + B'x'y' + C'y'^2 + D'x' + E'y' + F' = 0 \tag{9}$$

where $A', B', C', D', E',$ and F' can be expressed in terms of A, B, C, D, E, F and the angle θ of rotation (see Problem 55). It can be shown that the value of $B^2 - 4AC$ in equation (8) and the value of $B'^2 - 4A'C'$ in equation (9) are equal no matter what angle θ of rotation is chosen (see Problem 57). In particular, if the angle θ of rotation satisfies equation (7), then $B' = 0$ in equation (9), and $B^2 - 4AC = -4A'C'$. Since equation (9) then has the form of equation (2),

$$A'x'^2 + C'y'^2 + D'x' + E'y' + F' = 0$$

we can identify its graph without completing the squares, as we did in the beginning of this section. In fact, now we can identify the conic described by any equation of the form of equation (8) without rotating the axes.

> **THEOREM** Identifying Conics without Rotating the Axes
>
> Except for degenerate cases, the equation
>
> $$Ax^2 + Bxy + Cy^2 + Dx + Ey + F = 0 \tag{10}$$
>
> - Defines a parabola if $B^2 - 4AC = 0$.
> - Defines an ellipse (or a circle) if $B^2 - 4AC < 0$.
> - Defines a hyperbola if $B^2 - 4AC > 0$.

You are asked to prove this theorem in Problem 58. Because of the above theorem, equation (10) is called the **general equation of a conic**.

EXAMPLE 5 **Identifying a Conic without Rotating the Axes**

Identify the graph of the equation $8x^2 - 12xy + 17y^2 - 4\sqrt{5}x - 2\sqrt{5}y - 15 = 0$.

Solution Here $A = 8, B = -12,$ and $C = 17$, so $B^2 - 4AC = -400$. Since $B^2 - 4AC < 0$, the equation defines an ellipse.

> ✏️ Now Work PROBLEM 43

6.5 Assess Your Understanding

'Are You Prepared?' *Answers are given at the end of these exercises. If you get a wrong answer, read the pages listed in* red.

1. The sum formula for the sine function is $\sin(A + B) = $ _____. (p. 230)

2. The Double-angle Formula for the sine function is $\sin(2\theta) = $ _____. (p. 241)

3. If θ is acute, the Half-angle Formula for the sine function is $\sin\dfrac{\theta}{2} = $ _____. (p. 244)

4. If θ is acute, the Half-angle Formula for the cosine function is $\cos\dfrac{\theta}{2} = $ _____. (p. 244)

Concepts and Vocabulary

5. To transform the equation
$$Ax^2 + Bxy + Cy^2 + Dx + Ey + F = 0 \qquad B \neq 0$$
into one in x' and y' without an $x'y'$-term, rotate the axes through an acute angle θ that satisfies the equation _____.

6. *Multiple Choice* Except for degenerate cases, the equation
$$Ax^2 + Bxy + Cy^2 + Dx + Ey + F = 0$$
defines a(n) _____ if $B^2 - 4AC = 0$.
   (a) circle   **(b)** ellipse   **(c)** hyperbola   **(d)** parabola

7. Except for degenerate cases, the equation
$$Ax^2 + Bxy + Cy^2 + Dx + Ey + F = 0$$
defines an ellipse if _____.

8. *Multiple Choice* The equation $ax^2 + 6y^2 + 12y = 0$ defines an ellipse if _____.
   (a) $a < 0$ **(b)** $a = 0$ **(c)** $a > 0$ **(d)** a is any real number

9. *True or False* The equation $3x^2 + Bxy + 12y^2 = 10$ defines a parabola if $B = -12$.

10. *True or False* To eliminate the xy-term from the equation $x^2 - 2xy + y^2 - 2x + 3y + 5 = 0$, rotate the axes through an angle θ, where $\cot \theta = B^2 - 4AC$.

Skill Building

In Problems 11–20, identify the graph of each equation without completing the squares.

11. $x^2 + 4x + y + 3 = 0$
12. $2y^2 - 3y + 3x = 0$
13. $6x^2 + 3y^2 - 12x + 6y = 0$
14. $2x^2 + y^2 - 8x + 4y + 2 = 0$
15. $3x^2 - 2y^2 + 6x + 4 = 0$
16. $4x^2 - 3y^2 - 8x + 6y + 1 = 0$
17. $2y^2 - x^2 - y + x = 0$
18. $y^2 - 8x^2 - 2x - y = 0$
19. $x^2 + y^2 - 8x + 4y = 0$
20. $2x^2 + 2y^2 - 8x + 8y = 0$

In Problems 21–30, determine the appropriate rotation formulas to use so that the new equation contains no xy-term.

21. $x^2 + 4xy + y^2 - 3 = 0$
22. $x^2 - 4xy + y^2 - 3 = 0$
23. $5x^2 + 6xy + 5y^2 - 8 = 0$
24. $3x^2 - 10xy + 3y^2 - 32 = 0$
25. $13x^2 - 6\sqrt{3}xy + 7y^2 - 16 = 0$
26. $11x^2 + 10\sqrt{3}xy + y^2 - 4 = 0$
27. $4x^2 - 4xy + y^2 - 8\sqrt{5}x - 16\sqrt{5}y = 0$
28. $x^2 + 4xy + 4y^2 + 5\sqrt{5}y + 5 = 0$
29. $25x^2 - 36xy + 40y^2 - 12\sqrt{13}x - 8\sqrt{13}y = 0$
30. $34x^2 - 24xy + 41y^2 - 25 = 0$

In Problems 31–42, rotate the axes so that the new equation contains no xy-term. Analyze and graph the new equation. Refer to Problems 21–30 for Problems 31–40.

31. $x^2 + 4xy + y^2 - 3 = 0$
32. $x^2 - 4xy + y^2 - 3 = 0$
33. $5x^2 + 6xy + 5y^2 - 8 = 0$
34. $3x^2 - 10xy + 3y^2 - 32 = 0$
35. $13x^2 - 6\sqrt{3}xy + 7y^2 - 16 = 0$
36. $11x^2 + 10\sqrt{3}xy + y^2 - 4 = 0$
37. $4x^2 - 4xy + y^2 - 8\sqrt{5}x - 16\sqrt{5}y = 0$
38. $x^2 + 4xy + 4y^2 + 5\sqrt{5}y + 5 = 0$
39. $25x^2 - 36xy + 40y^2 - 12\sqrt{13}x - 8\sqrt{13}y = 0$
40. $34x^2 - 24xy + 41y^2 - 25 = 0$
41. $16x^2 + 24xy + 9y^2 - 130x + 90y = 0$
42. $16x^2 + 24xy + 9y^2 - 60x + 80y = 0$

In Problems 43–52, identify the graph of each equation without applying a rotation of axes.

43. $x^2 + 3xy - 2y^2 + 3x + 2y + 5 = 0$
44. $2x^2 - 3xy + 4y^2 + 2x + 3y - 5 = 0$
45. $x^2 - 7xy + 3y^2 - y - 10 = 0$
46. $2x^2 - 3xy + 2y^2 - 4x - 2 = 0$
47. $9x^2 + 12xy + 4y^2 - x - y = 0$
48. $10x^2 + 12xy + 4y^2 - x - y + 10 = 0$
49. $10x^2 - 12xy + 4y^2 - x - y - 10 = 0$
50. $4x^2 + 12xy + 9y^2 - x - y = 0$
51. $3x^2 - 2xy + y^2 + 4x + 2y - 1 = 0$
52. $3x^2 + 2xy + y^2 + 4x - 2y + 10 = 0$

Applications and Extensions

53. **Satellite Receiver** A parabolic satellite receiver is initially positioned so its axis of symmetry is parallel to the x-axis. A motor allows the receiver to rotate and track the satellite signal. If the rotated receiver has the equation
$$4x^2 - 4\sqrt{21}xy - \frac{21\sqrt{21}}{2}x + 21y^2 - 171y + 324 = 0$$
through what acute angle did the receiver rotate, to the nearest tenth of a degree?

54. **Elliptical Trainer** A runner on an elliptical trainer inclines the machine to increase the difficulty of her workout. If the initial elliptic path of the pedals had a horizontal major axis, and the inclined path has the equation
$$20x^2 - 10xy - \frac{19}{2}x + 89y^2 - 89y + \frac{48}{5} = 0$$
through what acute angle did the runner incline the machine, to the nearest tenth of a degree?

In Problems 55–58, apply the rotation formulas (5) to

$$Ax^2 + Bxy + Cy^2 + Dx + Ey + F = 0$$

to obtain the equation

$$A'x'^2 + B'x'y' + C'y'^2 + D'x' + E'y' + F' = 0$$

55. Express A', B', C', D', E', and F' in terms of A, B, C, D, E, F, and the angle θ of rotation. [**Hint:** Refer to equation (6).]

56. Show that $A + C = A' + C'$, which proves that $A + C$ is **invariant**; that is, its value does not change under a rotation of axes.

57. Refer to Problem 56. Show that $B^2 - 4AC$ is invariant.

58. Prove that, except for degenerate cases, the equation

$$Ax^2 + Bxy + Cy^2 + Dx + Ey + F = 0$$

 (a) Defines a parabola if $B^2 - 4AC = 0$.
 (b) Defines an ellipse (or a circle) if $B^2 - 4AC < 0$.
 (c) Defines a hyperbola if $B^2 - 4AC > 0$.

59. *Challenge Problem* Use the rotation formulas (5) to show that distance is invariant under a rotation of axes. That is, show that the distance from $P_1 = (x_1, y_1)$ to $P_2 = (x_2, y_2)$ in the xy-plane equals the distance from $P_1 = (x_1', y_1')$ to $P_2 = (x_2', y_2')$ in the $x'y'$-plane.

60. *Challenge Problem* Show that the graph of the equation $x^{1/2} + y^{1/2} = a^{1/2}, a > 0$, is part of the graph of a parabola.

Explaining Concepts: Discussion and Writing

61. Formulate a strategy for analyzing and graphing an equation of the form

$$Ax^2 + Cy^2 + Dx + Ey + F = 0$$

62. Explain how your strategy presented in Problem 61 changes if the equation is of the form

$$Ax^2 + Bxy + Cy^2 + Dx + Ey + F = 0$$

Retain Your Knowledge

Problems 63–69 are based on material learned earlier in the course. The purpose of these problems is to keep the material fresh in your mind so that you are better prepared for the final exam.

63. Solve the triangle: $a = 7$, $b = 9$, and $c = 11$

64. Find the area of the triangle: $a = 14$, $b = 11$, $C = 30°$

65. Transform the equation $xy = 1$ from rectangular coordinates to polar coordinates.

66. Write the complex number $2 - 5i$ in polar form.

67. Simplify

$$\frac{(4x^2 - 1)^8 \cdot 3(2x + 3)^2 \cdot 2 - (2x + 3)^3 \cdot 8(4x^2 - 1)^7 \cdot 8x}{[(4x - 1)^8]^2}$$

68. The graph of $f(x) = \sqrt{x^2 + 25} + \dfrac{1}{2}(8 - x)$ has an absolute minimum when $\dfrac{x}{\sqrt{x^2 + 25}} - \dfrac{1}{2} = 0$. What is the minimum value rounded to two decimal places?

69. If $g(x) = \sqrt{x - 7} + 2$, find $g^{-1}(3)$.

'Are You Prepared?' Answers

1. $\sin A \cos B + \cos A \sin B$ **2.** $2 \sin \theta \cos \theta$ **3.** $\sqrt{\dfrac{1 - \cos \theta}{2}}$ **4.** $\sqrt{\dfrac{1 + \cos \theta}{2}}$

6.6 Polar Equations of Conics

PREPARING FOR THIS SECTION *Before getting started, review the following:*

• Polar Coordinates (Section 5.1, pp. 316–323)

Now Work the 'Are You Prepared?' problems on page 439.

OBJECTIVES **1** Analyze and Graph Polar Equations of Conics (p. 434)
 2 Convert the Polar Equation of a Conic to a Rectangular Equation (p. 438)

1 Analyze and Graph Polar Equations of Conics

In Sections 6.2 through 6.4, we gave individual definitions for a parabola, ellipse, and hyperbola based on geometric properties and the distance formula. This section presents an alternative definition that simultaneously defines *all* these conics and is well suited to polar coordinate representation. (Refer to Section 5.1.)

> **DEFINITION Conic**
>
> Let D denote a fixed line called the **directrix**; let F denote a fixed point called the **focus**, which is not on D; and let e be a fixed positive number called the **eccentricity**. A **conic** is the set of points P in a plane for which the ratio of the distance from F to P to the distance from D to P equals e. That is, a conic is the collection of points P for which
>
> $$\frac{d(F, P)}{d(D, P)} = e \qquad \text{(1)}$$
>
> - If $e = 1$, the conic is a **parabola**.
> - If $e < 1$, the conic is an **ellipse**.
> - If $e > 1$, the conic is a **hyperbola**.

Observe that if $e = 1$, the definition of a parabola in equation (1) is exactly the same as the definition used earlier in Section 6.2.

In the case of an ellipse, the **major axis** is a line through the focus perpendicular to the directrix. In the case of a hyperbola, the **transverse axis** is a line through the focus perpendicular to the directrix. For both an ellipse and a hyperbola, the eccentricity e satisfies

$$e = \frac{c}{a} \qquad \text{(2)}$$

where c is the distance from the center to the focus, and a is the distance from the center to a vertex.

Just as we did earlier using rectangular coordinates, we derive equations for the conics in polar coordinates by choosing a convenient position for the focus F and the directrix D. The focus F is positioned at the pole, and the directrix D is either parallel or perpendicular to the polar axis.

Suppose that we start with the directrix D perpendicular to the polar axis at a distance p units to the left of the pole (the focus F). See Figure 52.

If $P = (r, \theta)$ is any point on the conic, then, by equation (1),

$$\frac{d(F, P)}{d(D, P)} = e \quad \text{or} \quad d(F, P) = e \cdot d(D, P) \qquad \text{(3)}$$

Now use the point Q obtained by dropping the perpendicular from P to the polar axis to calculate $d(D, P)$.

$$d(D, P) = p + d(O, Q) = p + r \cos \theta \qquad \text{(4)}$$

Since the focus F is at the pole (origin), it follows that

$$d(F, P) = d(O, P) = r \qquad \text{(5)}$$

Use the results in equations (4) and (5) in equation (3). Then

$$d(F, P) = e \cdot d(D, P) \qquad \text{Equation (3)}$$
$$r = e(p + r \cos \theta) \qquad d(F, P) = r; d(D, P) = p + r \cos \theta$$
$$r = ep + er \cos \theta$$
$$r - er \cos \theta = ep$$
$$r(1 - e \cos \theta) = ep$$
$$r = \frac{ep}{1 - e \cos \theta}$$

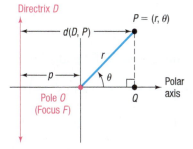

Directrix D

$P = (r, \theta)$

$d(D, P)$

p

Pole O (Focus F)

Q

Polar axis

Figure 52

> **THEOREM** **Polar Equation of a Conic; Focus at the Pole; Directrix Perpendicular to the Polar Axis a Distance p to the Left of the Pole**
>
> The polar equation of a conic with focus at the pole and directrix perpendicular to the polar axis at a distance p to the left of the pole is
>
> $$r = \frac{ep}{1 - e\cos\theta} \qquad (6)$$
>
> where e is the eccentricity of the conic.

EXAMPLE 1

Analyzing and Graphing the Polar Equation of a Conic

Analyze and graph the equation $r = \dfrac{4}{2 - \cos\theta}$.

Solution

The equation is not quite in the form of equation (6), since the first term in the denominator is 2 instead of 1. Divide the numerator and denominator by 2 to obtain

$$r = \frac{2}{1 - \frac{1}{2}\cos\theta} \qquad r = \frac{ep}{1 - e\cos\theta}$$

This equation is in the form of equation (6), with

$$e = \frac{1}{2} \quad \text{and} \quad ep = 2$$

Then

$$\frac{1}{2}p = 2, \quad \text{so} \quad p = 4$$

Since $e = \dfrac{1}{2} < 1$, the conic is an ellipse. One focus is at the pole, and the directrix is perpendicular to the polar axis, $p = 4$ units to the left of the pole. The major axis is along the polar axis. To find the vertices, let $\theta = 0$ and $\theta = \pi$. The vertices of the ellipse are $(4, 0)$ and $\left(\dfrac{4}{3}, \pi\right)$. The center of the ellipse is the midpoint of the vertices, namely, $\left(\dfrac{4}{3}, 0\right)$. [Do you see why? The vertices $(4, 0)$ and $\left(\dfrac{4}{3}, \pi\right)$ in polar coordinates are $(4, 0)$ and $\left(-\dfrac{4}{3}, 0\right)$ in rectangular coordinates. The midpoint in rectangular coordinates is $\left(\dfrac{4}{3}, 0\right)$, which is also $\left(\dfrac{4}{3}, 0\right)$ in polar coordinates.]

Then $a = $ distance from the center to a vertex $= \dfrac{8}{3}$. Using $a = \dfrac{8}{3}$ and $e = \dfrac{1}{2}$ in equation (2), $e = \dfrac{c}{a}$, yields $c = ae = \dfrac{4}{3}$. Finally, using $a = \dfrac{8}{3}$ and $c = \dfrac{4}{3}$ in $b^2 = a^2 - c^2$ yields

$$b^2 = a^2 - c^2 = \frac{64}{9} - \frac{16}{9} = \frac{48}{9}$$

$$b = \frac{4\sqrt{3}}{3}$$

Figure 53 shows the graph.

Figure 53 $r = \dfrac{4}{2 - \cos\theta}$

━━━ **Now Work** PROBLEM 11

Exploration
Graph $r_1 = \dfrac{4}{2 + \cos \theta}$ and compare the result with Figure 53. What do you conclude? Clear the

screen and graph $r_1 = \dfrac{4}{2 - \sin \theta}$ and then $r_1 = \dfrac{4}{2 + \sin \theta}$. Compare each of these graphs with

Figure 53. What do you conclude?

Equation (6) assumes that the directrix is perpendicular to the polar axis at a distance p units to the left of the pole. If the directrix is perpendicular to the polar axis at a distance p units to the right of the pole, then

$$r = \frac{ep}{1 + e \cos \theta}$$

See Problem 43.

In Problems 44 and 45, you are asked to derive the polar equations of conics with focus at the pole and directrix parallel to the polar axis. Table 5 summarizes the polar equations of conics.

Table 5

Polar Equations of Conics (Focus at the Pole, Eccentricity e)	
Equation	Description
$r = \dfrac{ep}{1 - e \cos \theta}$	Directrix is perpendicular to the polar axis at a distance p units to the left of the pole.
$r = \dfrac{ep}{1 + e \cos \theta}$	Directrix is perpendicular to the polar axis at a distance p units to the right of the pole.
$r = \dfrac{ep}{1 + e \sin \theta}$	Directrix is parallel to the polar axis at a distance p units above the pole.
$r = \dfrac{ep}{1 - e \sin \theta}$	Directrix is parallel to the polar axis at a distance p units below the pole.

Eccentricity
- If $e = 1$, the conic is a parabola; the axis of symmetry is perpendicular to the directrix.
- If $e < 1$, the conic is an ellipse; the major axis is perpendicular to the directrix.
- If $e > 1$, the conic is a hyperbola; the transverse axis is perpendicular to the directrix.

EXAMPLE 2

Analyzing and Graphing the Polar Equation of a Conic

Analyze and graph the equation $r = \dfrac{6}{3 + 3 \sin \theta}$.

Solution To put the equation in proper form, divide the numerator and denominator by 3.

$$r = \frac{2}{1 + \sin \theta}$$

See Table 5. This conic has its directrix parallel to the polar axis, a distance p units above the pole.

$$e = 1 \quad \text{and} \quad ep = 2$$
$$p = 2 \quad e = 1$$

Since $e = 1$, the conic is a parabola with focus at the pole. The directrix is parallel to the polar axis 2 units above the pole; the axis of symmetry is perpendicular to the polar axis.

The vertex of the parabola is at $\left(1, \dfrac{\pi}{2}\right)$. (Do you see why?) See Figure 54 for the graph.

Notice that two additional points, $(2, 0)$ and $(2, \pi)$, are plotted to assist in graphing.

Figure 54 $r = \dfrac{6}{3 + 3 \sin \theta}$

━━━━ **Now Work** PROBLEM 13

EXAMPLE 3

Analyzing and Graphing the Polar Equation of a Conic

Analyze and graph the equation $r = \dfrac{3}{1 + 3 \cos \theta}$.

Solution

The conic has its directrix perpendicular to the polar axis, a distance p units to the right of the pole. See Table 5.

$$e = 3 \quad \text{and} \quad ep = 3$$
$$p = 1 \quad e = 3$$

Since $e = 3 > 1$, the conic is a hyperbola with a focus at the pole. The directrix is perpendicular to the polar axis, 1 unit to the right of the pole. The transverse axis is along the polar axis. To find the vertices, let $\theta = 0$ and $\theta = \pi$. The vertices are $\left(\dfrac{3}{4}, 0\right)$ and $\left(-\dfrac{3}{2}, \pi\right)$. The center, which is at the midpoint of $\left(\dfrac{3}{4}, 0\right)$ and $\left(-\dfrac{3}{2}, \pi\right)$, is $\left(\dfrac{9}{8}, 0\right)$. Then $c = $ distance from the center to a focus $= \dfrac{9}{8}$. Using equation (2), we get

$$3 = \dfrac{\dfrac{9}{8}}{a} \quad \text{or} \quad a = \dfrac{3}{8} \quad e = \dfrac{c}{a}$$

Then,

$$b^2 = c^2 - a^2 = \dfrac{81}{64} - \dfrac{9}{64} = \dfrac{72}{64} = \dfrac{9}{8}$$

$$b = \dfrac{3}{2\sqrt{2}} = \dfrac{3\sqrt{2}}{4}$$

Figure 55 $r = \dfrac{3}{1 + 3 \cos \theta}$

Figure 55 shows the graph. Notice two additional points, $\left(3, \dfrac{\pi}{2}\right)$ and $\left(3, \dfrac{3\pi}{2}\right)$, are plotted on the left branch and symmetry is used to obtain the right branch. The asymptotes of the hyperbola were found by constructing the rectangle shown.

━━━━ **Now Work** PROBLEM **17**

2 Convert the Polar Equation of a Conic to a Rectangular Equation

EXAMPLE 4

Converting a Polar Equation to a Rectangular Equation

Convert the polar equation

$$r = \dfrac{1}{3 - 3 \cos \theta}$$

to a rectangular equation.

Solution

The strategy here is to rearrange the equation and square both sides before converting the equation in polar coordinates to an equation in rectangular coordinates.

$$r = \dfrac{1}{3 - 3 \cos \theta}$$

$$3r - 3r \cos \theta = 1$$

$$3r = 1 + 3r \cos \theta \qquad \text{Rearrange the equation.}$$

$$9r^2 = (1 + 3r \cos \theta)^2 \qquad \text{Square both sides.}$$

$$9(x^2 + y^2) = (1 + 3x)^2 \qquad x^2 + y^2 = r^2; x = r \cos \theta$$

$$9x^2 + 9y^2 = 9x^2 + 6x + 1$$

$$9y^2 = 6x + 1$$

This is the equation of a parabola in rectangular coordinates.

━━━━ **Now Work** PROBLEM **25**

6.6 Assess Your Understanding

'Are You Prepared?' *Answers are given at the end of these exercises. If you get a wrong answer, read the pages listed in red.*

1. If (x, y) are the rectangular coordinates of a point P and (r, θ) are its polar coordinates, then $x =$ _____ and $y =$ _____. (pp. 316–323)

2. Transform the equation $r = 6 \cos \theta$ from polar coordinates to rectangular coordinates. (pp. 316–323)

Concepts and Vocabulary

3. A _____ is the set of points P in a plane for which the ratio of the distance from a fixed point F, called the _____ , to P to the distance from a fixed line D, called the _____ , to P equals a constant e.

4. A conic has eccentricity e.
If $e = 1$, the conic is a(n) _____ .
If $e < 1$, the conic is a(n) _____ .
If $e > 1$, the conic is a(n) _____ .

5. *Multiple Choice* If (r, θ) are polar coordinates, the equation
$$r = \frac{2}{2 + 3 \sin \theta}$$ defines a(an) _____ .
(a) parabola **(b)** hyperbola **(c)** ellipse **(d)** circle

6. *True or False* The eccentricity e of an ellipse is $\dfrac{c}{a}$, where a is the distance of a vertex from the center and c is the distance of a focus from the center.

Skill Building

In Problems 7–12, identify the conic defined by each polar equation. Also give the position of the directrix.

7. $r = \dfrac{1}{1 + \cos \theta}$

8. $r = \dfrac{3}{1 - \sin \theta}$

9. $r = \dfrac{4}{2 - 3 \sin \theta}$

10. $r = \dfrac{2}{1 + 2 \cos \theta}$

11. $r = \dfrac{3}{4 - 2 \cos \theta}$

12. $r = \dfrac{6}{8 + 2 \sin \theta}$

In Problems 13–24, analyze each equation and graph it.

13. $r = \dfrac{1}{1 + \cos \theta}$

14. $r = \dfrac{3}{1 - \sin \theta}$

15. $r = \dfrac{8}{4 + 3 \sin \theta}$

16. $r = \dfrac{10}{5 + 4 \cos \theta}$

17. $r = \dfrac{9}{3 - 6 \cos \theta}$

18. $r = \dfrac{12}{4 + 8 \sin \theta}$

19. $r = \dfrac{8}{2 - \sin \theta}$

20. $r = \dfrac{8}{2 + 4 \cos \theta}$

21. $r(3 - 2 \sin \theta) = 6$

22. $r(2 - \cos \theta) = 2$

23. $r = \dfrac{6 \sec \theta}{2 \sec \theta - 1}$

24. $r = \dfrac{3 \csc \theta}{\csc \theta - 1}$

In Problems 25–36, convert each polar equation to a rectangular equation.

25. $r = \dfrac{1}{1 + \cos \theta}$

26. $r = \dfrac{3}{1 - \sin \theta}$

27. $r = \dfrac{8}{4 + 3 \sin \theta}$

28. $r = \dfrac{10}{5 + 4 \cos \theta}$

29. $r = \dfrac{9}{3 - 6 \cos \theta}$

30. $r = \dfrac{12}{4 + 8 \sin \theta}$

31. $r = \dfrac{8}{2 - \sin \theta}$

32. $r = \dfrac{8}{2 + 4 \cos \theta}$

33. $r(3 - 2 \sin \theta) = 6$

34. $r(2 - \cos \theta) = 2$

35. $r = \dfrac{6 \sec \theta}{2 \sec \theta - 1}$

36. $r = \dfrac{3 \csc \theta}{\csc \theta - 1}$

In Problems 37–42, find a polar equation for each conic. For each, a focus is at the pole.

37. $e = 1$; directrix is parallel to the polar axis, 1 unit above the pole.

38. $e = 1$; directrix is parallel to the polar axis, 2 units below the pole.

39. $e = \dfrac{4}{5}$; directrix is perpendicular to the polar axis, 3 units to the left of the pole.

40. $e = \dfrac{2}{3}$; directrix is parallel to the polar axis, 3 units above the pole.

41. $e = 6$; directrix is parallel to the polar axis, 2 units below the pole.

42. $e = 5$; directrix is perpendicular to the polar axis, 5 units to the right of the pole.

Applications and Extensions

43. Derive $r = \dfrac{ep}{1 + e \cos \theta}$ from Table 5.

44. Derive $r = \dfrac{ep}{1 + e \sin \theta}$ from Table 5.

45. Derive $r = \dfrac{ep}{1 - e \sin \theta}$ from Table 5.

46. Orbit of Mercury The planet Mercury travels around the Sun in an elliptical orbit given approximately by

$$r = \frac{3.442 \times 10^7}{1 - 0.206 \cos \theta}$$

where r is measured in miles and the Sun is at the pole. Find the distance from Mercury to the Sun at **aphelion** (greatest distance from the Sun) and at **perihelion** (shortest distance from the Sun). See the figure. Use the aphelion and perihelion to graph the orbit of Mercury using a graphing utility.

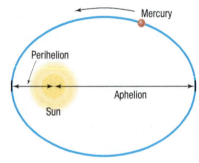

47. Halley's Comet Halley's comet travels around the Sun in an elliptical orbit given approximately by

$$r = \frac{1.155}{1 - 0.967 \cos \theta}$$

where the Sun is at the pole and r is measured in AU (astronomical units). Find the distance from Halley's comet to the Sun at aphelion and at perihelion. Use the aphelion and perihelion to graph the orbit of Halley's comet using a graphing utility.

48. Challenge Problem Water Leak A tank is punctured on its side, and water begins to stream out in a parabolic path. If the path of the water is given by

$$r = \frac{0.8}{1 + \sin \theta}$$

and the water hits the ground 4 inches away from the base of the tank, what is the height of the puncture from the base of the tank? Assume the focus is at the pole.

49. Challenge Problem Board Deflection A crate is placed at the center of a 5-meter board that is supported only at its ends. The weight of the crate causes a deflection of the board at its center. If the shape of the deflected board is a parabola given by

$$r = \frac{250}{1 - \sin \theta}$$

determine the amount of deflection at the center assuming the focus is at the pole.

50. Challenge Problem Suppose that a conic has an equation of the form $r = \dfrac{ep}{1 - e \sin \theta}$. If the polar coordinates of two points on the graph are $\left(M, \dfrac{\pi}{2} \right)$ and $\left(m, \dfrac{3\pi}{2} \right)$, show that

$$e = \frac{M - m}{M + m} \quad \text{and} \quad p = \frac{2mM}{M - m}$$

51. Challenge Problem Escape Velocity From physics, the equation for the free-flight trajectory of a satellite launched a distance r_0 from the center of the earth is given by the polar equation

$$\frac{1}{r} = \frac{1}{r_0}\left(1 - \frac{GM_e}{r_0 v_0^2} \right)\cos \theta + \frac{GM_e}{r_0^2 v_0^2}$$

where M_e is the mass of the earth, G is the gravitational constant, and v_0 is the initial velocity of the satellite. If the initial velocity is equal to the *escape velocity* (the velocity needed to overcome Earth's gravitational pull) then the resulting trajectory follows a parabolic path. What is the escape velocity?

Retain Your Knowledge

Problems 52–60 are based on material learned earlier in the course. The purpose of these problems is to keep the material fresh in your mind so that you are better prepared for the final exam.

52. Find the area of the triangle described: $a = 7, b = 8,$ and $c = 10$. Round to two decimal places.

53. Without graphing, determine the amplitude and period of $y = 4 \cos\left(\dfrac{1}{5}x\right)$.

54. Solve: $2 \cos^2 x + \cos x - 1 = 0, 0 \le x < 2\pi$

55. For $\mathbf{v} = 10\mathbf{i} - 24\mathbf{j}$, find $\|\mathbf{v}\|$.

56. If an arc length of 14 feet subtends a central angle of $105°$, what is the radius of the circle?

57. Determine where the function

$$f(x) = \begin{cases} x + 3 & \text{if } -2 \le x < -1 \\ x^2 + 1 & \text{if} \qquad x \ge -1 \end{cases}$$

is increasing, decreasing, and constant.

58. Find k so that $y = \sin(kx)$ has a period of $\dfrac{5\pi}{6}$.

59. Write the vertex form of the quadratic function whose graph has vertex $(-3, 8)$ and y-intercept 5.

60. Find the area of the region bounded by the graph of $f(x) = \dfrac{1}{2}x + 3$, the x-axis, and the vertical lines $x = 0$ and $x = 8$.

6.7 Plane Curves and Parametric Equations

PREPARING FOR THIS SECTION *Before getting started, review the following:*

• Amplitude and Period of Sinusoidal Graphs (Section 2.4, pp. 150–152)

Now Work the 'Are You Prepared?' problem on page 450.

> OBJECTIVES **1** Graph Parametric Equations (p. 441)
> **2** Find a Rectangular Equation for a Plane Curve Defined Parametrically (p. 442)
> **3** Use Time as a Parameter in Parametric Equations (p. 444)
> **4** Find Parametric Equations for Plane Curves Defined by Rectangular Equations (p. 447)

Equations of the form $y = f(x)$, where f is a function, have graphs that are intersected no more than once by any vertical line. The graphs of many of the conics and certain other, more complicated graphs do not have this characteristic. Yet each graph, like the graph of a function, is a collection of points (x, y) in the xy-plane; that is, each is a *plane curve*. This section discusses another way of representing such graphs.

> **DEFINITION** **Parametric Equations and Plane Curves**
>
> Suppose $x = x(t)$ and $y = y(t)$ are two functions of a third variable t, called the **parameter**, that are defined on the same interval I. Then the equations
>
> $$x = x(t) \qquad y = y(t)$$
>
> where t is in I, are called **parametric equations**, and the graph of the points defined by
>
> $$(x, y) = (x(t), y(t))$$
>
> is called a **plane curve**.

1 Graph Parametric Equations

Parametric equations are particularly useful in describing movement along a plane curve. Suppose that a plane curve is defined by the parametric equations

$$x = x(t) \qquad y = y(t) \qquad a \le t \le b$$

Figure 56 Plane curve

where each function is defined over the interval $a \le t \le b$. For a given value of t, the values of $x = x(t)$ and $y = y(t)$ determine a point (x, y) on the curve. In fact, as t varies over the interval from $t = a$ to $t = b$, successive values of t determine the direction of the movement along the curve. That is, the curve is traced out in a certain direction by the corresponding succession of points (x, y). See Figure 56. The arrows show the direction, or **orientation**, along the curve as t varies from a to b.

EXAMPLE 1	Graphing a Plane Curve

Graph the plane curve defined by the parametric equations

$$x(t) = 3t^2 \qquad y(t) = 2t \qquad -2 \leq t \leq 2 \qquad \textbf{(1)}$$

Solution

For each number t, $-2 \leq t \leq 2$, there corresponds a number x and a number y. For example, when $t = -2$, then $x = 3(-2)^2 = 12$ and $y = 2(-2) = -4$. When $t = 0$, then $x = 0$ and $y = 0$. Set up a table listing various choices of the parameter t and the corresponding values for x and y, as shown in Table 6. Plot these points and connect them with a smooth curve, as shown in Figure 57. The arrows in Figure 57 indicate the orientation.

COMMENT Most graphing utilities have the capability of graphing parametric equations. See Section B.9. ∎

Table 6

t	$x(t)$	$y(t)$	(x, y)
−2	12	−4	(12, −4)
−1	3	−2	(3, −2)
0	0	0	(0, 0)
1	3	2	(3, 2)
2	12	4	(12, 4)

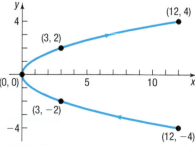

Figure 57
$x(t) = 3t^2, y(t) = 2t, -2 \leq t \leq 2$

Exploration

Graph the following parametric equations using a graphing utility with Xmin = 0, Xmax = 15, Ymin = −5, Ymax = 5, and Tstep = 0.1.

1. $x(t) = \dfrac{3t^2}{4}$, $y(t) = t$, $-4 \leq t \leq 4$

2. $x(t) = 3t^2 + 12t + 12$, $y(t) = 2t + 4$, $-4 \leq t \leq 0$

3. $x(t) = 3t^{\frac{2}{3}}$, $y(t) = 2\sqrt[3]{t}$, $-8 \leq t \leq 8$

Compare these graphs to Figure 57. Conclude that parametric equations defining a plane curve are not unique; that is, different parametric equations can represent the same graph.

---

2 Find a Rectangular Equation for a Plane Curve Defined Parametrically

The plane curve given in Example 1 should look familiar. To identify it accurately, find the corresponding rectangular equation by eliminating the parameter t from the parametric equations given in Example 1:

$$x(t) = 3t^2 \qquad y(t) = 2t \qquad -2 \leq t \leq 2$$

Solve for t in $y = 2t$, obtaining $t = \dfrac{y}{2}$, and substitute this expression in the other equation to get

$$x = 3t^2 = 3\left(\frac{y}{2}\right)^2 = \frac{3y^2}{4}$$

$$\uparrow$$
$$t = \frac{y}{2}$$

This equation, $x = \dfrac{3y^2}{4}$, is the equation of a parabola with vertex at $(0, 0)$ and axis of symmetry along the x-axis. We refer to this equation as the **rectangular equation** of the curve to distinguish it from the parametric equations.

Exploration

Graph

$$x = \frac{3y^2}{4} \left(y_1 = \sqrt{\frac{4x}{3}} \text{ and } y_2 = -\sqrt{\frac{4x}{3}} \right)$$

using a graphing utility with Xmin = 0, Xmax = 15, Ymin = −5, Ymax = 5. Compare the graph with Figure 57. Why do the graphs differ?

Note that the plane curve defined by equation (1) and shown in Figure 57 is only a part of the parabola $x = \frac{3y^2}{4}$. The graph of the rectangular equation obtained by eliminating the parameter will, in general, contain more points than the original plane curve. Care must therefore be taken when a plane curve is graphed after eliminating the parameter. Even so, eliminating the parameter t from the parametric equations to identify a plane curve accurately is sometimes a better approach than plotting points.

EXAMPLE 2 **Finding the Rectangular Equation of a Plane Curve Defined Parametrically**

Find the rectangular equation of the plane curve whose parametric equations are

$$x(t) = a \cos t \qquad y(t) = a \sin t \qquad -\infty < t < \infty$$

where $a > 0$ is a constant. Graph the plane curve, and indicate its orientation.

Solution   The presence of sines and cosines in the parametric equations suggests using a Pythagorean Identity. In fact, since

$$\cos t = \frac{x}{a} \qquad \text{and} \qquad \sin t = \frac{y}{a}$$

this means that

$$\left(\frac{x}{a}\right)^2 + \left(\frac{y}{a}\right)^2 = 1 \qquad \cos^2 t + \sin^2 t = 1$$

$$x^2 + y^2 = a^2$$

The plane curve is a circle with center at $(0, 0)$ and radius a. As the parameter t increases, say from $t = 0$ [the point $(a, 0)$] to $t = \frac{\pi}{2}$ [the point $(0, a)$] to $t = \pi$ [the point $(-a, 0)$], the corresponding points are traced in a counterclockwise direction around the circle. The orientation is as indicated in Figure 58.

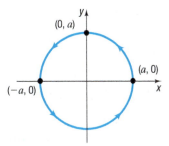

Figure 58
$x(t) = a \cos t, \quad y(t) = a \sin t$

━━━━ **Now Work** PROBLEMS 7 AND 19

Let's analyze the plane curve in Example 2 further. The domain of each parametric equation is $-\infty < t < \infty$. So, the graph in Figure 58 is repeated each time that t increases by 2π.

If we wanted the curve to consist of exactly 1 revolution in the counterclockwise direction, we could write

$$x(t) = a \cos t \qquad y(t) = a \sin t \qquad 0 \le t \le 2\pi$$

This curve starts at $t = 0$ [the point $(a, 0)$, proceeds counterclockwise around the circle, and ends at $t = 2\pi$ [also the point $(a, 0)$].

If we wanted the curve to consist of exactly three revolutions in the counterclockwise direction, we could write

$$x(t) = a \cos t \qquad y(t) = a \sin t \qquad -2\pi \le t \le 4\pi$$

or

$$x(t) = a \cos t \qquad y(t) = a \sin t \qquad 0 \le t \le 6\pi$$

or

$$x(t) = a \cos t \qquad y(t) = a \sin t \qquad 2\pi \le t \le 8\pi$$

EXAMPLE 3	**Describing Parametric Equations**

Find rectangular equations for the following plane curves defined by parametric equations. Graph each curve.

(a) $x(t) = a \cos t \quad y(t) = a \sin t \qquad 0 \le t \le \pi \quad a > 0$

(b) $x(t) = -a \sin t \quad y(t) = -a \cos t \quad 0 \le t \le \pi \quad a > 0$

Solution   (a) Eliminate the parameter t using a Pythagorean Identity.

$$\cos^2 t + \sin^2 t = 1$$

$$\left(\frac{x}{a}\right)^2 + \left(\frac{y}{a}\right)^2 = 1$$

$$x^2 + y^2 = a^2$$

The plane curve defined by these parametric equations lies on a circle with radius a and center at $(0,0)$. The curve begins at the point $(a, 0)$, when $t = 0$; passes through the point $(0, a)$, when $t = \frac{\pi}{2}$; and ends at the point $(-a, 0)$, when $t = \pi$. The parametric equations define the upper semicircle of a circle of radius a with a counterclockwise orientation. See Figure 59.

The rectangular equation is

$$y = \sqrt{a^2 - x^2} \qquad -a \le x \le a$$

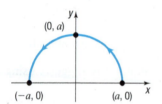

Figure 59
$x(t) = a \cos t \quad y(t) = a \sin t$
$0 \le t \le \pi \quad a > 0$

(b) Eliminate the parameter t using a Pythagorean Identity.

$$\sin^2 t + \cos^2 t = 1$$

$$\left(\frac{x}{-a}\right)^2 + \left(\frac{y}{-a}\right)^2 = 1$$

$$x^2 + y^2 = a^2$$

The plane curve defined by the parametric equations lies on a circle with radius a and center at $(0,0)$. The curve begins at the point $(0, -a)$, when $t = 0$; passes through the point $(-a, 0)$, when $t = \frac{\pi}{2}$; and ends at the point $(0, a)$, when $t = \pi$. The parametric equations define the left semicircle of a circle of radius a with a clockwise orientation. See Figure 60.

The rectangular equation is

$$x = -\sqrt{a^2 - y^2} \qquad -a \le y \le a$$

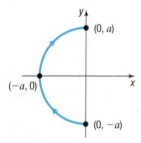

Figure 60
$x(t) = -a \sin t \quad y(t) = -a \cos t$
$0 \le t \le \pi \quad a > 0$

Example 3 illustrates the versatility of parametric equations for replacing complicated rectangular equations, while providing additional information about orientation. These characteristics make parametric equations very useful in applications, such as projectile motion.

3 Use Time as a Parameter in Parametric Equations

If we think of the parameter t as time, then the parametric equations $x = x(t)$ and $y = y(t)$ specify how the x- and y-coordinates of a moving point vary with time.

For example, we can use parametric equations to model the motion of an object, sometimes referred to as **curvilinear motion**. Using parametric equations, we can specify not only *where* the object travels—that is, its location (x, y)—but also *when* it gets there—that is, the time t.

When an object is propelled upward at an inclination θ to the horizontal with initial speed v_0, the resulting motion is called **projectile motion**.

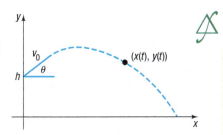

Figure 61 Projectile motion

In calculus the following result is proved.

> The parametric equations of the path of a projectile fired at an inclination θ to the horizontal, with an initial speed v_0, from a height h above the horizontal, are
>
> $$x(t) = (v_0 \cos \theta)t \qquad y(t) = -\frac{1}{2}gt^2 + (v_0 \sin \theta)t + h \qquad \textbf{(2)}$$
>
> where t is time and g is the constant acceleration due to gravity (approximately 32 ft/sec^2, or 9.8 m/sec^2).

See Figure 61.

EXAMPLE 4

Projectile Motion

Suppose that Jim hit a golf ball with an initial speed of 150 feet per second at an angle of 30° to the horizontal. See Figure 62.

(a) Find parametric equations that describe the position of the ball as a function of time.
(b) How long was the golf ball in the air?
(c) When was the ball at its maximum height? Find the maximum height of the ball.
(d) Find the distance that the ball traveled.

Figure 62

Solution

(a) We have $v_0 = 150$ ft/sec, $\theta = 30°$, $h = 0$ ft (the ball is on the ground), and $g = 32$ ft/sec^2 (since the units are in feet and seconds). Substitute these values into equations (2) to get

$$x(t) = (v_0 \cos \theta)t = (150 \cos 30°)t = 75\sqrt{3}\,t$$

$$y(t) = -\frac{1}{2}gt^2 + (v_0 \sin \theta)t + h = -\frac{1}{2}\cdot 32 \cdot t^2 + (150 \sin 30°)t + 0 = -16t^2 + 75t$$

(b) To find the length of time that the ball was in the air, solve the equation $y(t) = 0$.

$$-16t^2 + 75t = 0$$

$$t(-16t + 75) = 0$$

$$t = 0 \sec \quad \text{or} \quad t = \frac{75}{16} = 4.6875 \sec$$

The ball struck the ground after 4.6875 seconds.

Need to Review?

The vertex of a quadratic function
$y = f(x) = ax^2 + bx + c$
is the point $\left(-\dfrac{b}{2a}, f\left(-\dfrac{b}{2a}\right)\right)$.

(c) Notice that the height y of the ball is a quadratic function of t, so the maximum height of the ball can be found by determining the vertex of $y(t) = -16t^2 + 75t$. The value of t at the vertex is

$$t = \frac{-75}{-32} = 2.34375 \sec$$

The ball was at its maximum height after 2.34375 seconds. The maximum height of the ball is found by evaluating the function $y(t)$ at $t = 2.34375$ seconds.

$$\text{Maximum height} = -16 \cdot (2.34375)^2 + 75 \cdot 2.34375 \approx 87.89 \text{ feet}$$

(d) Since the ball was in the air for 4.6875 seconds, the horizontal distance that the ball traveled is

$$x = 75\sqrt{3} \cdot 4.6875 \approx 608.92 \text{ feet}$$

 Now Work PROBLEM 53

Figure 63

(a) $(t \approx 0.7)$ (b) $(t \approx 3)$ (c) $(t \approx 4)$

Now Work PROBLEM 49

A graphing utility can be used to simulate other kinds of motion as well.

EXAMPLE 5   **Simulating Motion**

Tanya, who is a long-distance runner, runs at an average speed of 8 miles per hour. Two hours after Tanya leaves your house, you leave in your Honda and follow the same route. See Figure 64. If the Honda's average speed is 40 miles per hour, how long is it before you catch up to Tanya? Use a simulation of the two motions to verify the answer.

Figure 64

Solution   Begin with two sets of parametric equations: one to describe Tanya's motion, the other to describe the motion of the Honda. We choose time $t = 0$ to be when Tanya leaves the house. If we choose $y_1 = 2$ as Tanya's path, then we can use $y_2 = 4$ as the parallel path of the Honda. The horizontal distances traversed in time t (Distance = Rate × Time) are

$$\text{Tanya: } \quad x_1(t) = 8t \qquad \text{Honda: } \quad x_2(t) = 40(t - 2)$$

You catch up to Tanya when $x_1 = x_2$.

$$8t = 40(t - 2)$$

$$8t = 40t - 80$$

$$-32t = -80$$

$$t = \frac{-80}{-32} = 2.5$$

You catch up to Tanya 2.5 hours after Tanya leaves the house.

In PARAMETRIC mode with Tstep = 0.01, simultaneously graph

Tanya: $x_1(t) = 8t$ Honda: $x_2(t) = 40(t - 2)$

$y_1(t) = 2$ $y_2(t) = 4$

for $0 \leq t \leq 3$.

Figure 65 shows the relative positions of Tanya and the Honda for $t = 0$, $t = 2$, $t = 2.25$, $t = 2.5$, and $t = 2.75$ on a TI-84 Plus C.

$t=0$

$t=2$

$t=2.25$

$t=2.5$

$t=2.75$

Figure 65

4 Find Parametric Equations for Plane Curves Defined by Rectangular Equations

If a plane curve is defined by the function $y = f(x)$, one way of finding parametric equations is to let $x = t$. Then $y = f(t)$, and

$$x(t) = t \quad y(t) = f(t) \qquad t \text{ in the domain of } f$$

are parametric equations of the plane curve.

EXAMPLE 6   **Finding Parametric Equations for a Plane Curve Defined by a Rectangular Equation**

Find two different pairs of parametric equations for the function $y = x^2 - 4$.

Solution   For the first pair of parametric equations, let $x = t$. Then the parametric equations are

$$x(t) = t \quad y(t) = t^2 - 4 \quad -\infty < t < \infty$$

A second pair of parametric equations is found by letting $x = t^3$. Then the parametric equations become

$$x(t) = t^3 \quad y(t) = t^6 - 4 \quad -\infty < t < \infty$$

Care must be taken when using the second approach in Example 6. The substitution for x must be a function that allows x to take on all the values in the domain of f. For example, letting $x(t) = t^2$ so that $y(t) = t^4 - 4$ does not result in equivalent parametric equations for $y = x^2 - 4$, since only points for which $x \geq 0$ are obtained; yet the domain of $y = x^2 - 4$ is $\{x \mid x \text{ is any real number}\}$.

Now Work PROBLEM 33

EXAMPLE 7 Finding Parametric Equations for an Object in Motion

Find parametric equations for the ellipse

$$x^2 + \frac{y^2}{9} = 1$$

where the parameter t is time (in seconds) and

(a) The motion around the ellipse is clockwise, begins at the point $(0, 3)$, and requires 1 second for a complete revolution.

(b) The motion around the ellipse is counterclockwise, begins at the point $(1, 0)$, and requires 2 seconds for a complete revolution.

Solution

(a) Figure 66 shows the graph of the ellipse. Since the motion begins at the point $(0, 3)$, we want $x = 0$ and $y = 3$ when $t = 0$. Let

$$x(t) = \sin(\omega t) \quad \text{and} \quad y(t) = 3\cos(\omega t)$$

for some constant ω. These parametric equations satisfy the equation of the ellipse. They also satisfy the requirement that when $t = 0$, then $x = 0$ and $y = 3$.

For the motion to be clockwise, the motion has to begin with the value of x increasing and the value of y decreasing as t increases. This requires that $\omega > 0$. [Do you know why? If $\omega > 0$, then $x(t) = \sin(\omega t)$ is increasing when $t \geq 0$ is near zero, and $y(t) = 3\cos(\omega t)$ is decreasing when $t \geq 0$ is near zero.] See the red part of the graph in Figure 66.

Finally, since 1 revolution takes 1 second, the period $\frac{2\pi}{\omega} = 1$, so $\omega = 2\pi$. Parametric equations that satisfy the conditions stipulated are

$$x(t) = \sin(2\pi t) \quad y(t) = 3\cos(2\pi t) \quad 0 \leq t \leq 1 \quad \textbf{(3)}$$

(b) See Figure 67. Since the motion begins at the point $(1, 0)$, we want $x = 1$ and $y = 0$ when $t = 0$. The equation is an ellipse, so begin by letting

$$x(t) = \cos(\omega t) \quad \text{and} \quad y(t) = 3\sin(\omega t)$$

for some constant ω. These parametric equations satisfy the equation of the ellipse. Furthermore, with this choice, when $t = 0$ we have $x = 1$ and $y = 0$.

For the motion to be counterclockwise, the motion has to begin with the value of x decreasing and the value of y increasing as t increases. This requires that $\omega > 0$. (Do you know why?) Finally, since 1 revolution requires 2 seconds, the period is $\frac{2\pi}{\omega} = 2$, so $\omega = \pi$. The parametric equations that satisfy the conditions stipulated are

$$x(t) = \cos(\pi t) \quad y(t) = 3\sin(\pi t) \quad 0 \leq t \leq 2 \quad \textbf{(4)}$$

Either equations (3) or equations (4) can serve as parametric equations for the ellipse $x^2 + \frac{y^2}{9} = 1$. The direction of the motion, the beginning point, and the time for 1 revolution give a particular parametric representation.

Figure 66 $x^2 + \frac{y^2}{9} = 1$ with clockwise orientation

Need to Review?
The period of a sinusoidal graph is discussed in Section 2.4, pp. 150–152.

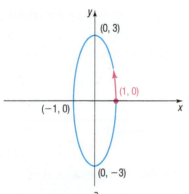

Figure 67 $x^2 + \frac{y^2}{9} = 1$ with counter-clockwise orientation

Now Work PROBLEM 39

The Cycloid

Suppose that a circle of radius *a* rolls along a horizontal line without slipping. As the circle rolls along the line, a point *P* on the circle will trace out a curve called a **cycloid** (see Figure 68). Deriving the equation of a cycloid in rectangular coordinates is difficult, but the task is relatively easy using parametric equations.

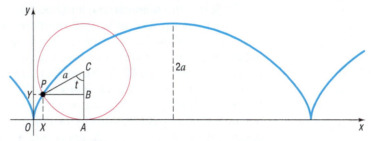

Figure 68 Cycloid

Begin with a circle of radius *a* and take the fixed line on which the circle rolls as the *x*-axis. Let the origin be one of the points at which the point *P* comes in contact with the *x*-axis. Figure 68 shows the position of this point *P* after the circle has rolled a bit. The angle *t* (in radians) measures the angle through which the circle has rolled.

Since we require no slippage, it follows that

$$\text{Arc } AP = d(O, A)$$

The length of arc *AP* is given by $s = r\theta$, where $r = a$ and $\theta = t$ radians. Then

$$at = d(O, A) \qquad s = r\,\theta, \text{where } r = a \text{ and } \theta = t$$

The *x*-coordinate of the point *P* is

$$d(O, X) = d(O, A) - d(X, A) = at - a \sin t = a(t - \sin t)$$

The *y*-coordinate of the point *P* is

$$d(O, Y) = d(A, C) - d(B, C) = a - a \cos t = a(1 - \cos t)$$

> **Exploration**
>
> Graph $x(t) = t - \sin t$, $y(t) = 1 - \cos t$, $0 \le t \le 3\pi$, using your graphing utility with Tstep $= \dfrac{\pi}{36}$ and a square screen. Compare your results with Figure 68.

> **THEOREM Parametric Equations of a Cycloid**
>
> The parametric equations of a cycloid are
>
> $$x(t) = a(t - \sin t) \qquad y(t) = a(1 - \cos t) \qquad \text{(5)}$$

Applications to Mechanics

NOTE In Greek, *brachistochrone* means "the shortest time," and *tautochrone* "equal time." ∎

If $a < 0$ in equation (5), we obtain an inverted cycloid, as shown in Figure 69(a). The inverted cycloid occurs as a result of some remarkable applications in the field of mechanics. We mention two of them: the *brachistochrone* and the *tautochrone*.

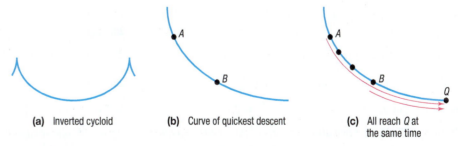

(a) Inverted cycloid

(b) Curve of quickest descent

(c) All reach *Q* at the same time

Figure 69

  The **brachistochrone** is the curve of quickest descent. If a particle is constrained to follow some path from one point *A* to a lower point *B* (not on the same vertical line) and is acted on only by gravity, the time needed to make the descent is least if the

Figure 70

Cycloid   Cycloid

Cycloid

path is an inverted cycloid. See Figure 69(b). For example, to slide packages from a loading dock onto a truck, a ramp in the shape of an inverted cycloid might be used, so the packages get to the truck in the least amount of time. This remarkable discovery, which has been attributed to many famous mathematicians (including Johann Bernoulli and Blaise Pascal), was a significant step in creating the branch of mathematics known as the *calculus of variations.*

To define the **tautochrone**, let Q be the lowest point on an inverted cycloid. If several particles placed at various positions on an inverted cycloid simultaneously begin to slide down the cycloid, they will reach the point Q at the same time, as indicated in Figure 69(c). The tautochrone property of the cycloid was used by Christiaan Huygens (1629–1695), the Dutch mathematician, physicist, and astronomer, to construct a pendulum clock with a bob that swings along a cycloid (See Figure 70). In Huygens's clock, the bob was made to swing along a cycloid by suspending the bob on a thin wire constrained by two plates shaped like cycloids. In a clock of this design, the period of the pendulum is independent of its amplitude.

6.7 Assess Your Understanding

'Are You Prepared?' *Answers are given at the end of these exercises. If you get a wrong answer, read the pages listed in* red.

1. The function $f(x) = 3 \sin(4x)$ has amplitude _____ and period _____. (pp. 150–152)

Concepts and Vocabulary

2. Suppose $x = x(t)$ and $y = y(t)$ are two functions of a third variable t that are defined on the same interval I. The graph of the collection of points defined by $(x, y) = (x(t), y(t))$ is called a (n) _____ _____. The variable t is called a (n)

_____.

3. *Multiple Choice* The parametric equations
$$x(t) = 2 \sin t \quad y(t) = 3 \cos t$$
define a (n) _____.

   (a) circle   (b) ellipse   (c) hyperbola   (d) parabola

4. *Multiple Choice* If a circle rolls along a horizontal line without slipping, a fixed point P on the circle will trace out a curve called a(n) _____.
   (a) cycloid   (b) epitrochoid
   (c) hyptrochoid   (d) pendulum

5. *True or False* Parametric equations defining a curve are unique.

6. *True or False* Plane curves defined using parametric equations have an orientation.

Skill Building

In Problems 7–26, graph the plane curve whose parametric equations are given, and show its orientation. Find the rectangular equation of each curve.

7. $x(t) = 3t + 2, \quad y(t) = t + 1; \quad 0 \le t \le 4$

8. $x(t) = t - 3, \quad y(t) = 2t + 4; \quad 0 \le t \le 2$

9. $x(t) = t + 2, \quad y(t) = \sqrt{t}; \quad t \ge 0$

10. $x(t) = \sqrt{2t}, \quad y(t) = 4t; \quad t \ge 0$

11. $x(t) = t^2 + 4, \quad y(t) = t^2 - 4; \quad -\infty < t < \infty$

12. $x(t) = \sqrt{t} + 4, \quad y(t) = \sqrt{t} - 4; \quad t \ge 0$

13. $x(t) = 3t^2, \quad y(t) = t + 1; \quad -\infty < t < \infty$

14. $x(t) = 2t - 4, \quad y(t) = 4t^2; \quad -\infty < t < \infty$

15. $x(t) = 2e^t, \quad y(t) = 1 + e^t; \quad t \ge 0$

†16. $x(t) = e^t, \quad y(t) = e^{-t}; \quad t \ge 0$

17. $x(t) = \sqrt{t}, \quad y(t) = t^{3/2}; \quad t \ge 0$

18. $x(t) = t^{3/2} + 1, \quad y(t) = \sqrt{t}; \quad t \ge 0$

19. $x(t) = 2 \cos t, \quad y(t) = 3 \sin t; \quad 0 \le t \le 2\pi$

20. $x(t) = 2 \cos t, \quad y(t) = 3 \sin t; \quad 0 \le t \le \pi$

21. $x(t) = 2 \cos t, \quad y(t) = 3 \sin t; \quad -\pi \le t \le 0$

22. $x(t) = 2 \cos t, \quad y(t) = \sin t; \quad 0 \le t \le \dfrac{\pi}{2}$

23. $x(t) = \sec t, \quad y(t) = \tan t; \quad 0 \le t \le \dfrac{\pi}{4}$

24. $x(t) = \csc t, \quad y(t) = \cot t; \quad \dfrac{\pi}{4} \le t \le \dfrac{\pi}{2}$

25. $x(t) = \sin^2 t, \quad y(t) = \cos^2 t; \quad 0 \le t \le 2\pi$

††26. $x(t) = t^2, \quad y(t) = \ln t; \quad t > 0$

In Problems 27–34, find two different pairs of parametric equations for each rectangular equation.

27. $y = 4x - 1$

28. $y = -8x + 3$

29. $y = x^2 + 1$

30. $y = -2x^2 + 1$

31. $y = x^3$

32. $y = x^4 + 1$

33. $x = y^{3/2}$

34. $x = \sqrt{y}$

†Requires Section 7.1, Exponential Functions.
††Requires Section 7.2, Logarithmic Functions.

In Problems 35–38, find parametric equations that define the plane curve shown.

35.

36.

37.

38.

In Problems 39–42, find parametric equations for an object that moves along the ellipse $\dfrac{x^2}{4} + \dfrac{y^2}{9} = 1$ with the motion described.

39. The motion begins at $(2, 0)$, is clockwise, and requires 2 seconds for a complete revolution.

40. The motion begins at $(0, 3)$, is counterclockwise, and requires 1 second for a complete revolution.

41. The motion begins at $(0, 3)$, is clockwise, and requires 1 second for a complete revolution.

42. The motion begins at $(2, 0)$, is counterclockwise, and requires 3 seconds for a complete revolution.

In Problems 43 and 44, parametric equations of four plane curves are given. Graph each of them, indicating the orientation.

43. C_1: $x(t) = t$, $y(t) = t^2$; $-4 \le t \le 4$

C_2: $x(t) = \cos t$, $y(t) = 1 - \sin^2 t$; $0 \le t \le \pi$

*C_3: $x(t) = e^t$, $y(t) = e^{2t}$; $0 \le t \le \ln 4$

C_4: $x(t) = \sqrt{t}$, $y(t) = t$; $0 \le t \le 16$

44. C_1: $x(t) = t$, $y(t) = \sqrt{1 - t^2}$; $-1 \le t \le 1$

C_2: $x(t) = \sin t$, $y(t) = \cos t$; $0 \le t \le 2\pi$

C_3: $x(t) = \cos t$, $y(t) = \sin t$; $0 \le t \le 2\pi$

C_4: $x(t) = \sqrt{1 - t^2}$, $y(t) = t$; $-1 \le t \le 1$

In Problems 45–48, use a graphing utility to graph the plane curve defined by the given parametric equations.

45. $x(t) = t \sin t$, $y(t) = t \cos t$, $t > 0$

46. $x(t) = \sin t + \cos t$, $y = \sin t - \cos t$

47. $x(t) = 4 \sin t - 2 \sin(2t)$

$y(t) = 4 \cos t - 2 \cos(2t)$

48. $x(t) = 4 \sin t + 2 \sin(2t)$

$y(t) = 4 \cos t + 2 \cos(2t)$

Applications and Extensions

49. Projectile Motion Bob throws a ball straight up with an initial speed of 50 feet per second from a height of 6 feet.
 (a) Find parametric equations that model the motion of the ball as a function of time.
 (b) How long is the ball in the air?
 (c) When is the ball at its maximum height? Determine the maximum height of the ball.
 (d) Simulate the motion of the ball by graphing the equations found in part (a).

50. Projectile Motion Alice throws a ball straight up with an initial speed of 40 feet per second from a height of 5 feet.
 (a) Find parametric equations that model the motion of the ball as a function of time.
 (b) How long is the ball in the air?
 (c) When is the ball at its maximum height? Determine the maximum height of the ball.
 (d) Simulate the motion of the ball by graphing the equations found in part (a).

51. Catching a Train Bill's train leaves at 8:06 AM and accelerates at the rate of 2 meters per second per second. Bill, who can run 5 meters per second, arrives at the train station 5 seconds after the train has left and runs for the train.
 (a) Find parametric equations that model the motions of the train and Bill as a function of time.

 [**Hint**: The position s at time t of an object having acceleration a is $s = \dfrac{1}{2}at^2$.]

 (b) Determine algebraically whether Bill will catch the train. If so, when?

 (c) Simulate the motion of the train and Bill by simultaneously graphing the equations found in part (a).

52. Catching a Bus Jodi's bus leaves at 5:30 PM and accelerates at the rate of 3 meters per second per second. Jodi, who can run 5 meters per second, arrives at the bus station 2 seconds after the bus has left and runs for the bus.
 (a) Find parametric equations that model the motions of the bus and Jodi as a function of time.

 [**Hint**: The position s at time t of an object having acceleration a is $s = \dfrac{1}{2}at^2$.]

 (b) Determine algebraically whether Jodi will catch the bus. If so, when?
 (c) Simulate the motion of the bus and Jodi by graphing simultaneously the equations found in part (a).

53. Projectile Motion Sean throws a baseball with an initial speed of 145 feet per second at an angle of 20° to the horizontal. The ball leaves Sean's hand at a height of 5 feet.
 (a) Find parametric equations that model the position of the ball as a function of time.
 (b) How long is the ball in the air?
 (c) Determine the horizontal distance that the ball travels.
 (d) When is the ball at its maximum height? Determine the maximum height of the ball.
 (e) Using a graphing utility, simultaneously graph the equations found in part (a).

*Requires Section 7.1, Exponential Functions, and Section 7.2, Logarithmic Functions.

54. Projectile Motion Billy hit a baseball with an initial speed of 125 feet per second at an angle of 40° to the horizontal. The ball was hit at a height of 3 feet above the ground.
(a) Find parametric equations that model the position of the ball as a function of time.
(b) How long was the ball in the air?
(c) Determine the horizontal distance that the ball traveled.
(d) When was the ball at its maximum height? Determine the maximum height of the ball.
(e) Using a graphing utility, simultaneously graph the equations found in part (a).

55. Projectile Motion Suppose that Adam hits a golf ball off a cliff 300 meters high with an initial speed of 40 meters per second at an angle of 45° to the horizontal.
(a) Find parametric equations that model the position of the ball as a function of time.
(b) How long is the ball in the air?
(c) Determine the horizontal distance that the ball travels.
(d) When is the ball at its maximum height? Determine the maximum height of the ball.
(e) Using a graphing utility, simultaneously graph the equations found in part (a).

56. Projectile Motion Suppose that Karla hits a golf ball off a cliff 300 meters high with an initial speed of 40 meters per second at an angle of 45° to the horizontal on the Moon (gravity on the Moon is one-sixth of that on Earth).
(a) Find parametric equations that model the position of the ball as a function of time.
(b) How long is the ball in the air?
(c) Determine the horizontal distance that the ball travels.
(d) When is the ball at its maximum height? Determine the maximum height of the ball.
(e) Using a graphing utility, simultaneously graph the equations found in part (a).

57. Uniform Motion A Toyota Camry (traveling east at 40 mph) and a Chevy Impala (traveling north at 30 mph) are heading toward the same intersection. The Camry is 5 miles from the intersection when the Impala is 4 miles from the intersection. See the figure.

(a) Find parametric equations that model the motion of the Camry and the Impala.
(b) Find a formula for the distance between the cars as a function of time.
(c) Graph the function in part (b) using a graphing utility.
(d) What is the minimum distance between the cars? When are the cars closest?
(e) Simulate the motion of the cars by simultaneously graphing the equations found in part (a).

58. Uniform Motion A Cessna (heading south at 120 mph) and a Boeing 737 (heading west at 600 mph) are flying toward the same point at the same altitude. The Cessna is 100 miles from the point where the flight patterns intersect, and the 737 is 550 miles from this intersection point. See the figure.

(a) Find parametric equations that model the motion of the Cessna and the 737.
(b) Find a formula for the distance between the planes as a function of time.
(c) Graph the function in part (b) using a graphing utility.
(d) What is the minimum distance between the planes? When are the planes closest?
(e) Simulate the motion of the planes by simultaneously graphing the equations found in part (a).

59. The Green Monster The left field wall at Fenway Park is 310 feet from home plate; the wall itself (affectionately named the Green Monster) is 37 feet high. A batted ball must clear the wall to be a home run. Suppose a ball leaves the bat 3 feet above the ground, at an angle of 45°. Use $g = 32$ ft/sec^2 as the acceleration due to gravity, and ignore any air resistance.
(a) Find parametric equations that model the position of the ball as a function of time.
(b) What is the maximum height of the ball if it leaves the bat with a speed of 90 miles per hour? Give your answer in feet.
(c) How far is the ball from home plate at its maximum height? Give your answer in feet.
(d) If the ball is hit straight down the left field line, will it clear the Green Monster? If it does, by how much does it clear the wall?

Source: The Boston Red Sox

60. Projectile Motion The position of a projectile fired with an initial velocity v_0 feet per second and at an angle θ to the horizontal at the end of t seconds is given by the parametric equations

$$x(t) = (v_0 \cos \theta)t \qquad y(t) = (v_0 \sin \theta)t - 16t^2$$

See the figure.

(a) Obtain a rectangular equation of the trajectory, and identify the curve.
(b) Show that the projectile hits the ground ($y = 0$) when $t = \dfrac{1}{16}v_0 \sin \theta$.

(c) How far has the projectile traveled (horizontally) when it strikes the ground? In other words, find the range R.

(d) Find the time t when $x = y$. Next find the horizontal distance x and the vertical distance y traveled by the projectile in this time. Then compute $\sqrt{x^2 + y^2}$. This is the distance R, the range, that the projectile travels up a plane inclined at 45° to the horizontal $(x = y)$. See the following figure. (See also Problem 105 in Section 5.6.)

61. Show that parametric equations for a line passing through the points (x_1, y_1) and (x_2, y_2) are

$$x(t) = (x_2 - x_1)t + x_1$$
$$y(t) = (y_2 - y_1)t + y_1 \quad -\infty < t < \infty$$

What is the orientation of this line?

62. Hypocycloid The hypocycloid is a plane curve defined by the parametric equations

$$x(t) = \cos^3 t \quad y(t) = \sin^3 t \quad 0 \le t \le 2\pi$$

(a) Graph the hypocycloid using a graphing utility.

(b) Find a rectangular equation of the hypocycloid.

63. *Challenge Problem* Find parametric equations for the circle $x^2 + y^2 = R^2$, using as the parameter the slope m of the line through the point $(-R, 0)$ and a general point $P = (x, y)$ on the circle.

64. *Challenge Problem* Find parametric equations for the parabola $y = x^2$, using as the parameter the slope m of the line joining the point $(1, 1)$ to a general point $P = (x, y)$ of the parabola.

Explaining Concepts: Discussion and Writing

65. In Problem 62, we graphed the hypocycloid. Now graph the rectangular equations of the hypocycloid. Did you obtain a complete graph? If not, experiment until you do.

66. Research plane curves called *hypocycloid* and *epicycloid*. Write a report on what you find. Compare and contrast them to a cycloid.

Retain Your Knowledge

Problems 67–74 are based on material learned earlier in the course. The purpose of these problems is to keep the material fresh in your mind so that you are better prepared for the final exam.

67. Graph the equation $3x - 4y = 8$ on the xy-plane.

68. Graph $y = 2\cos(2x) + \sin\left(\dfrac{x}{2}\right)$ on the xy-plane.

69. The International Space Station (ISS) orbits Earth at a height of approximately 248 miles above the surface. What is the distance, in miles, on the surface of Earth that can be observed from the ISS? Assume that Earth's radius is 3960 miles.

Source: nasa.gov

70. The displacement d (in meters) of an object at time t (in seconds) is given by $d(t) = 2\cos(4t)$.

(a) Describe the motion of the object.

(b) What is the maximum displacement of the object from its rest position?

(c) What is the time required for 1 oscillation?

(d) What is the frequency?

71. Find the oblique asymptote of $R(x) = \dfrac{4x^2 - 9x + 7}{2x + 1}$

72. Find the difference quotient of $f(x) = \dfrac{1}{x + 3}$ as $h \to 0$.

73. Find the exact value of $\cos 285°$.

74. If $f(x) = \dfrac{1}{4}x^3 + 1$ and $g(x) = \dfrac{3}{4}x^2$, find all numbers c in the interval $[0, 2]$ where $g(c)$ equals the average rate of change of f over the interval.

'Are You Prepared?' Answers

1. $3; \dfrac{\pi}{2}$

Chapter Review

Things to Know

Equations

Parabola (pp. 394–398)	See Tables 1 and 2 (pp. 396 and 397).
Ellipse (pp. 403–409)	See Table 3 (p. 408).
Hyperbola (pp. 413–421)	See Table 4 (p. 420).

General equation of a conic (p. 432) $Ax^2 + Bxy + Cy^2 + Dx + Ey + F = 0$ Parabola if $B^2 - 4AC = 0$
Ellipse (or circle) if $B^2 - 4AC < 0$
Hyperbola if $B^2 - 4AC > 0$

Polar equations of a conic with focus   See Table 5 (p. 437).
at the pole (pp. 434–438)

Parametric equations of a plane curve (p. 441) $x = x(t), y = y(t), t$ is the parameter

Definitions

Parabola (p. 394)	Set of points P in a plane for which $d(F, P) = d(P, D)$, where F is the focus and D is the directrix
Ellipse (p. 403)	Set of points P in a plane the sum of whose distances from two fixed points (the foci) is a constant
Hyperbola (p. 413)	Set of points P in a plane the difference of whose distances from two fixed points (the foci) is a constant

Conic in polar coordinates (p. 435)   The collection of points P for   Parabola if $e = 1$
which $\dfrac{d(F, P)}{d(D, P)} = e$ Ellipse if $e < 1$
Hyperbola if $e > 1$

Formulas

Rotation formulas (p. 428) $x = x' \cos \theta - y' \sin \theta$
$y = x' \sin \theta + y' \cos \theta$

Angle θ of rotation that $\cot(2\theta) = \dfrac{A - C}{B}$ $0° < \theta < 90°$
eliminates the $x'y'$-term (p. 429)

Objectives

Section		You should be able to . . .	Example(s)	Review Exercises
6.1	1	Know the names of the conics (p. 393)		1–16
6.2	1	Analyze parabolas with vertex at the origin (p. 394)	1–5	1, 11
	2	Analyze parabolas with vertex at (h, k) (p. 397)	6, 7	4, 6, 9, 14
	3	Solve applied problems involving parabolas (p. 399)	8	39
6.3	1	Analyze ellipses with center at the origin (p. 403)	1–4	3, 13
	2	Analyze ellipses with center at (h, k) (p. 407)	5, 6	8, 10, 16, 38
	3	Solve applied problems involving ellipses (p. 409)	7	40
6.4	1	Analyze hyperbolas with center at the origin (p. 414)	1–4	2, 5, 12, 37
	2	Find the asymptotes of a hyperbola (p. 418)	5, 6	2, 5, 7
	3	Analyze hyperbolas with center at (h, k) (p. 420)	7, 8	7, 15, 17, 18
	4	Solve applied problems involving hyperbolas (p. 421)	9	41
6.5	1	Identify a conic (p. 426)	1	19, 20
	2	Use a rotation of axes to transform equations (p. 427)	2	24–26
	3	Analyze an equation using a rotation of axes (p. 430)	3, 4	24–26, 44
	4	Identify conics without rotating the axes (p. 432)	5	21–23

Section	You should be able to . . .	Example(s)	Review Exercises
6.6	1 Analyze and graph polar equations of conics (p. 434)	1–3	27–29
	2 Convert the polar equation of a conic to a rectangular equation (p. 438)	4	30, 31
6.7	1 Graph parametric equations (p. 441)	1	32–34
	2 Find a rectangular equation for a plane curve defined parametrically (p. 442)	2, 3	32–34
	3 Use time as a parameter in parametric equations (p. 444)	4, 5	42, 43
	4 Find parametric equations for plane curves defined by rectangular equations (p. 447)	6, 7	35, 36

Review Exercises

In Problems 1–10, identify each equation. If it is a parabola, give its vertex, focus, and directrix; if it is an ellipse, give its center, vertices, and foci; if it is a hyperbola, give its center, vertices, foci, and asymptotes.

1. $y^2 = -16x$

2. $\dfrac{x^2}{25} - y^2 = 1$

3. $\dfrac{y^2}{25} + \dfrac{x^2}{16} = 1$

4. $x^2 + 4y = 4$

5. $4x^2 - y^2 = 8$

6. $x^2 - 4x = 2y$

7. $y^2 - 4y - 4x^2 + 8x = 4$

8. $4x^2 + 9y^2 - 16x - 18y = 11$

9. $4x^2 - 16x + 16y + 32 = 0$

10. $9x^2 + 4y^2 - 18x + 8y = 23$

In Problems 11–18, find an equation of the conic described. Graph the equation.

11. Parabola; focus at $(-2, 0)$; directrix the line $x = 2$

12. Hyperbola; center at $(0, 0)$; focus at $(0, 4)$; vertex at $(0, -2)$

13. Ellipse; foci at $(-3, 0)$ and $(3, 0)$; vertex at $(4, 0)$

14. Parabola; vertex at $(2, -3)$; focus at $(2, -4)$

15. Hyperbola; center at $(-2, -3)$; focus at $(-4, -3)$; vertex at $(-3, -3)$

16. Ellipse; foci at $(-4, 2)$ and $(-4, 8)$; vertex at $(-4, 10)$

17. Center at $(-1, 2)$; $a = 3$; $c = 4$; transverse axis parallel to the x-axis

18. Vertices at $(0, 1)$ and $(6, 1)$; asymptote the line $3y + 2x = 9$

In Problems 19–23, identify each conic without completing the squares and without applying a rotation of axes.

19. $y^2 + 4x + 3y - 8 = 0$

20. $x^2 + 2y^2 + 4x - 8y + 2 = 0$

21. $9x^2 - 12xy + 4y^2 + 8x + 12y = 0$

22. $4x^2 + 10xy + 4y^2 - 9 = 0$

23. $x^2 - 2xy + 3y^2 + 2x + 4y - 1 = 0$

In Problems 24–26, rotate the axes so that the new equation contains no xy-term. Analyze and graph the new equation.

24. $2x^2 + 5xy + 2y^2 - \dfrac{9}{2} = 0$

25. $6x^2 + 4xy + 9y^2 - 20 = 0$

26. $4x^2 - 12xy + 9y^2 + 12x + 8y = 0$

In Problems 27–29, identify the conic that each polar equation represents, and graph it.

27. $r = \dfrac{4}{1 - \cos\theta}$

28. $r = \dfrac{6}{2 - \sin\theta}$

29. $r = \dfrac{8}{4 + 8\cos\theta}$

In Problems 30 and 31, convert each polar equation to a rectangular equation.

30. $r = \dfrac{4}{1 - \cos\theta}$

31. $r = \dfrac{8}{4 + 8\cos\theta}$

In Problems 32–34, graph the plane curve whose parametric equations are given, and show its orientation. Find a rectangular equation of each curve.

32. $x(t) = 4t - 2, \quad y(t) = 1 - t; \quad -\infty < t < \infty$

33. $x(t) = 3\sin t, \quad y(t) = 4\cos t + 2; \quad 0 \le t \le 2\pi$

34. $x(t) = \sec^2 t, \quad y(t) = \tan^2 t; \quad 0 \le t \le \dfrac{\pi}{4}$

35. Find two different pairs of parametric equations for $y = -2x + 4$.

36. Find parametric equations for an object that moves along the ellipse $\dfrac{x^2}{16} + \dfrac{y^2}{9} = 1$, where the motion begins at $(4, 0)$, is counterclockwise, and requires 4 seconds for a complete revolution.

---

37. Find an equation of the hyperbola whose foci are the vertices of the ellipse $4x^2 + 9y^2 = 36$ and whose vertices are the foci of this ellipse.

38. Describe the collection of points in a plane so that the distance from each point to the point $(3, 0)$ is three-fourths of its distance from the line $x = \dfrac{16}{3}$.

39. Searchlight A searchlight is shaped like a paraboloid of revolution. If a light source is located 1 foot from the vertex along the axis of symmetry and the opening is 2 feet across, how deep should the mirror be in order to reflect the light rays parallel to the axis of symmetry?

40. Semielliptical Arch Bridge A bridge is built in the shape of a semielliptical arch. The bridge has a span of 60 feet and a maximum height of 20 feet. Find the height of the arch at distances of 5, 10, and 20 feet from the center.

41. Calibrating Instruments In a test of their recording devices, a team of seismologists positioned two devices 2000 feet apart, with the device at point A to the west of the device at point B. At a point between the devices and 200 feet from point B, a small amount of explosive was detonated and a note made of the time at which the sound reached each device. A second explosion is to be carried out at a point directly north of point B. How far north should the site of the second explosion be chosen so that the measured time difference recorded by the devices for the second detonation is the same as that recorded for the first detonation?

42. Uniform Motion Mary's train leaves at 7:15 AM and accelerates at the rate of 3 meters per second per second. Mary, who can run 6 meters per second, arrives at the train station 2 seconds after the train has left.
   (a) Find parametric equations that model the motion of the train and Mary as a function of time.
     [**Hint:** The position s at time t of an object having acceleration a is $s = \dfrac{1}{2}at^2$.]
   (b) Determine algebraically whether Mary will catch the train. If so, when?
   (c) Simulate the motions of the train and Mary by simultaneously graphing the equations found in part (a).

43. Projectile Motion Nick Foles throws a football with an initial speed of 80 feet per second at an angle of 35° to the horizontal. The ball leaves his hand at a height of 6 feet.
   (a) Find parametric equations that model the position of the ball as a function of time.
   (b) How long is the ball in the air?
   (c) When is the ball at its maximum height? Determine the maximum height of the ball.
   (d) Determine the horizontal distance that the ball travels.
   (e) Using a graphing utility, simultaneously graph the equations found in part (a).

44. Formulate a strategy for discussing and graphing an equation of the form

$$Ax^2 + Bxy + Cy^2 + Dx + Ey + F = 0$$

Chapter Test

CHAPTER
Test Prep
VIDEOS
The Chapter Test Prep Videos include step-by-step solutions to all chapter test exercises. These videos are available in MyLab™ Math, or on this text's YouTube Channel. Refer to the Preface for a link to the YouTube channel.

In Problems 1–3, identify each equation. If it is a parabola, give its vertex, focus, and directrix; if an ellipse, give its center, vertices, and foci; if a hyperbola, give its center, vertices, foci, and asymptotes.

1. $\dfrac{(x + 1)^2}{4} - \dfrac{y^2}{9} = 1$ **2.** $8y = (x - 1)^2 - 4$ **3.** $2x^2 + 3y^2 + 4x - 6y = 13$

In Problems 4–6, find an equation of the conic described; graph the equation.

4. Parabola: focus $(-1, 4.5)$, vertex $(-1, 3)$

5. Ellipse: center $(0, 0)$, vertex $(0, -4)$, focus $(0, 3)$

6. Hyperbola: center $(2, 2)$, vertex $(2, 4)$, contains the point $\left(2 + \sqrt{10}, 5\right)$

In Problems 7–9, identify each conic without completing the square or rotating axes.

7. $2x^2 + 5xy + 3y^2 + 3x - 7 = 0$ **8.** $3x^2 - xy + 2y^2 + 3y + 1 = 0$ **9.** $x^2 - 6xy + 9y^2 + 2x - 3y - 2 = 0$

10. Given the equation $41x^2 - 24xy + 34y^2 - 25 = 0$, rotate the axes so that there is no xy-term. Analyze and graph the new equation.

11. Identify the conic represented by the polar equation $r = \dfrac{3}{1 - 2\cos\theta}$. Find the rectangular equation.

12. Graph the plane curve whose parametric equations are given, and show its orientation. Find the rectangular equation for the plane curve.

$$x(t) = 3t - 2 \quad y(t) = 1 - \sqrt{t} \quad 0 \le t \le 9$$

13. A parabolic reflector (paraboloid of revolution) is used by TV crews at football games to pick up the referee's announcements, quarterback signals, and so on. A microphone is placed at the focus of the parabola. If a certain reflector is 4 feet wide and 1.5 feet deep, where should the microphone be placed?

Cumulative Review

1. For $f(x) = -3x^2 + 5x - 2$, find
$$\frac{f(x + h) - f(x)}{h} \qquad h \neq 0$$

2. Find an equation for each of the following graphs.

(a) Line:

(b) Circle:

(c) Ellipse:

(d) Parabola:

(e) Hyperbola:

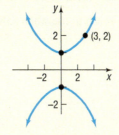

3. Find all the solutions of the equation $\sin(2\theta) = 0.5$.

4. Find a polar equation for the line containing the origin that makes an angle of $30°$ with the positive x-axis.

5. Find a polar equation for the circle with center at the point $(0, 4)$ and radius 4. Graph this circle.

6. What is the domain of the function $f(x) = \dfrac{3}{\sin x + \cos x}$?

7. Solve the equation $\cot(2\theta) = 1$, where $0° < \theta < 90°$.

8. Find the rectangular equation of the plane curve
$$x(t) = 5 \tan t \qquad y(t) = 5 \sec^2 t \qquad -\frac{\pi}{2} < t < \frac{\pi}{2}$$

Chapter Projects

Internet-based Project

I. **Comet Hale-Bopp** The orbits of planets and some comets about the Sun are ellipses, with the Sun at one focus. The **aphelion** of a planet is its greatest distance from the Sun, and the **perihelion** is its shortest distance. The **mean distance** of a planet from the Sun is the length of the semimajor axis of the elliptical orbit. See the figure.

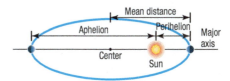

1. Research the history of Comet Hale-Bopp on the Internet. In particular, determine the aphelion and perihelion. Often these values are given in terms of astronomical units. What is an astronomical unit? What is it equivalent to in miles? In kilometers? What is the orbital period of Comet Hale-Bopp? When will it next be visible from Earth? How close does it come to Earth?

2. Find a model for the orbit of Comet Hale-Bopp around the Sun. Use the *x*-axis as the major axis.

3. Comet Hale-Bopp has an orbit that is roughly perpendicular to that of Earth. Find a model for the orbit of Earth using the *y*-axis as the major axis.

4. Use a graphing utility or some other graphing technology to graph the paths of the orbits. Based on the graphs, do the paths of the orbits intersect? Does this mean that Comet Hale-Bopp will collide with Earth?

The following projects can be found at the Instructor's Resource Center (IRC):

II. **The Orbits of Neptune and Pluto** The astronomical body known as Pluto and the planet Neptune travel around the Sun in elliptical orbits. Pluto, at times, comes closer to the Sun than Neptune, the outermost planet. This project examines and analyzes the two orbits.

III. **Project at Motorola** *Distorted Deployable Space Reflector Antennas* An engineer designs an antenna that will deploy in space to collect sunlight.

IV. **Constructing a Bridge over the East River** The size of ships using a river and fluctuations in water height due to tides or flooding must be considered when designing a bridge that will cross a major waterway.

V. **Systems of Parametric Equations** Which approach to use when solving a system of equations depends on the form of the system and on the domains of the equations.

Exponential and Logarithmic Functions

Depreciation of Cars

You are ready to buy that first new car. You know that cars lose value over time due to depreciation and that different cars have different rates of depreciation. So you will research the depreciation rates for the cars you are thinking of buying. After all, for cars that sell for about the same price, the lower the depreciation rate, the more the car will be worth each year.

 —See the Internet-based Chapter Project I—

← A Look Back

In Chapter 1, we created a library of functions, naming key functions and listing their properties, including their graphs. These functions belong to the class of **algebraic functions**—that is, functions that can be expressed in terms of sums, differences, products, quotients, powers, or roots of polynomials. Functions that are not algebraic are termed **transcendental**. That is, they transcend, or go beyond, algebraic functions.

A Look Ahead →

In this chapter, we study two transcendental functions: the exponential function and the logarithmic function. These functions occur frequently in a wide variety of applications, such as biology, chemistry, economics, and psychology.

Outline

7.1 Exponential Functions

PREPARING FOR THIS SECTION *Before getting started, review the following:*

- Exponents (Section A.1, pp. A7–A9 and Section A.7, pp. A60–A61)
- Graphing Techniques: Transformations (Section 1.6, pp. 67–76)
- Solving Equations (Section A.4, pp. A27–A34)

- Average Rate of Change (Section 1.4, pp. 50–52)
- Average Rate of Change of a Line (Section A.8, p. A65)
- Library of Functions (Section 1.5, pp. 57–61)
- Horizontal Asymptotes (Section 1.6, pp. 76–77)

Now Work the 'Are You Prepared?' problems on page 471.

> **OBJECTIVES** 1 Evaluate Exponential Functions (p. 460)
> 2 Graph Exponential Functions (p. 463)
> 3 Define the Number e (p. 467)
> 4 Solve Exponential Equations (p. 469)

1 Evaluate Exponential Functions

Section A.7 gives a definition for raising a real number a to a rational power. That discussion provides meaning to expressions of the form

$$a^r$$

where the base a is a positive real number and the exponent r is a rational number.

But what is the meaning of a^x, where the base a is a positive real number and the exponent x is an irrational number? Although a rigorous definition requires methods discussed in calculus, the basis for the definition is easy to follow: Select a rational number r that is formed by truncating (removing) all but a finite number of digits from the irrational number x. Then it is reasonable to expect that

$$a^x \approx a^r$$

For example, take the irrational number $\pi = 3.14159\ldots$. Then an approximation to a^π is

$$a^\pi \approx a^{3.14}$$

where the digits of π after the hundredths position are truncated. A better approximation is

$$a^\pi \approx a^{3.14159}$$

where the digits after the hundred-thousandths position are truncated. Continuing in this way, we can obtain approximations to a^π to any desired degree of accuracy.

Most calculators have an $\boxed{x^y}$ key or a caret key $\boxed{\wedge}$ for working with exponents. To evaluate expressions of the form a^x, enter the base a, then press the $\boxed{x^y}$ key (or the $\boxed{\wedge}$ key), enter the exponent x, and press $\boxed{=}$ (or $\boxed{\text{ENTER}}$).

EXAMPLE 1	**Using a Calculator to Evaluate Powers of 2**

Use a calculator to evaluate:

(a) $2^{1.4}$ (b) $2^{1.41}$ (c) $2^{1.414}$ (d) $2^{1.4142}$ (e) $2^{\sqrt{2}}$

Solution

(a) $2^{1.4} \approx 2.639015822$ (b) $2^{1.41} \approx 2.657371628$

(c) $2^{1.414} \approx 2.66474965$ (d) $2^{1.4142} \approx 2.665119089$

(e) $2^{\sqrt{2}} \approx 2.665144143$

 Now Work PROBLEM 17

It can be shown that the laws for rational exponents hold for real exponents.

THEOREM Laws of Exponents

If s, t, a, and b are real numbers with $a > 0$ and $b > 0$, then

$$\bullet \ a^s \cdot a^t = a^{s+t} \qquad \bullet \ (a^s)^t = a^{st} \qquad \bullet \ (ab)^s = a^s \cdot b^s$$

$$\bullet \ 1^s = 1 \qquad \bullet \ a^{-s} = \frac{1}{a^s} = \left(\frac{1}{a}\right)^s \qquad \bullet \ a^0 = 1 \tag{1}$$

Introduction to Exponential Growth

Suppose a function f has the following two properties:

- The value of f doubles with every 1-unit increase in the independent variable x.
- The value of f at $x = 0$ is 5, so $f(0) = 5$.

Table 1 shows values of the function f for $x = 0, 1, 2, 3,$ and 4.

Let's find an equation $y = f(x)$ that describes this function f. The key fact is that the value of f doubles for every 1-unit increase in x.

Table 1

x	$f(x)$
0	5
1	10
2	20
3	40
4	80

$f(0) = 5$

$f(1) = 2f(0) = 2 \cdot 5 = 5 \cdot 2^1$ **Double the value of f at 0 to get the value at 1.**

$f(2) = 2f(1) = 2(5 \cdot 2) = 5 \cdot 2^2$ **Double the value of f at 1 to get the value at 2.**

$f(3) = 2f(2) = 2(5 \cdot 2^2) = 5 \cdot 2^3$

$f(4) = 2f(3) = 2(5 \cdot 2^3) = 5 \cdot 2^4$

The pattern leads to

$$f(x) = 2f(x-1) = 2(5 \cdot 2^{x-1}) = 5 \cdot 2^x$$

DEFINITION Exponential Function

An **exponential function** is a function of the form

$$f(x) = Ca^x$$

where a is a positive real number $(a > 0)$, $a \neq 1$, and $C \neq 0$ is a real number. The domain of f is the set of all real numbers. The base a is the **growth factor**, and, because $f(0) = Ca^0 = C$, C is called the **initial value**.

In the definition of an exponential function, the base $a = 1$ is excluded because this function is simply the constant function $f(x) = C \cdot 1^x = C$. Bases that are negative are also excluded; otherwise, many values of x would have to be excluded from the domain, such as $x = \frac{1}{2}$ and $x = \frac{3}{4}$. [Recall that $(-2)^{1/2} = \sqrt{-2}, (-3)^{3/4} = \sqrt[4]{(-3)^3} = \sqrt[4]{-27}$, and so on, are not defined in the set of real numbers.]

Transformations (vertical shifts, horizontal shifts, reflections, and so on) of a function of the form $f(x) = Ca^x$ are also exponential functions. Examples of such exponential functions are

$$f(x) = 2^x \qquad F(x) = \left(\frac{1}{3}\right)^x + 5 \qquad G(x) = 2 \cdot 3^{x-3}$$

WARNING It is important to distinguish a power function, $g(x) = ax^n, n \geq 2$ an integer, from an exponential function, $f(x) = C \cdot a^x, a \neq 1, a > 0$. In a power function, the base is a variable and the exponent is a constant. In an exponential function, the base is a constant and the exponent is a variable. ∎

For each function, note that the base of the exponential expression is a constant and the exponent contains a variable.

In the function $f(x) = 5 \cdot 2^x$, notice that the ratio of consecutive outputs is constant for 1-unit increases in the input. This ratio equals the constant 2, the base of the exponential function. In other words,

$$\frac{f(1)}{f(0)} = \frac{5 \cdot 2^1}{5} = 2 \quad \frac{f(2)}{f(1)} = \frac{5 \cdot 2^2}{5 \cdot 2^1} = 2 \quad \frac{f(3)}{f(2)} = \frac{5 \cdot 2^3}{5 \cdot 2^2} = 2 \quad \text{and so on}$$

This leads to the following result.

THEOREM

For an exponential function $f(x) = Ca^x$, $a > 0$, $a \neq 1$, and $C \neq 0$, if x is any real number, then

$$\frac{f(x + 1)}{f(x)} = a \quad \text{or} \quad f(x + 1) = af(x)$$

In Words

For 1-unit changes in the input x of an exponential function $f(x) = C \cdot a^x$, the ratio of consecutive outputs is the constant a.

Proof

$$\frac{f(x + 1)}{f(x)} = \frac{Ca^{x+1}}{Ca^x} = a^{x+1-x} = a^1 = a$$

∎

EXAMPLE 2

Identifying Linear or Exponential Functions

Determine whether the given function is linear, exponential, or neither. For those that are linear, find a linear function that models the data. For those that are exponential, find an exponential function that models the data.

(a)

x	y
-1	5
0	2
1	-1
2	-4
3	-7

(b)

x	y
-1	32
0	16
1	8
2	4
3	2

(c)

x	y
-1	2
0	4
1	7
2	11
3	16

Solution

For each function, compute the average rate of change of y with respect to x and the ratio of consecutive outputs. If the average rate of change is constant, then the function is linear. If the ratio of consecutive outputs is constant, then the function is exponential.

(a) See Table 2(a). The average rate of change for every 1-unit increase in x is -3. Therefore, the function is a linear function. In a linear function the average rate of change is the slope m, so $m = -3$. The y-intercept b is the value of the function at $x = 0$, so $b = 2$. The linear function that models the data is $f(x) = mx + b = -3x + 2$.

(b) See Table 2(b). For this function, the average rate of change is not constant. So the function is not a linear function. The ratio of consecutive outputs for a 1-unit increase in the inputs is a constant, $\frac{1}{2}$. Because the ratio of consecutive outputs is constant, the function is an exponential function with growth factor $a = \dfrac{1}{2}$. The initial value C of the exponential function is $C = 16$, the value of the function at 0. Therefore, the exponential function that models the data is $g(x) = Ca^x = 16 \cdot \left(\dfrac{1}{2}\right)^x$.

(c) See Table 2(c). For this function, neither the average rate of change nor the ratio of two consecutive outputs is constant. Because the average rate of change is not constant, the function is not a linear function. Because the ratio of consecutive outputs is not a constant, the function is not an exponential function.

Table 2 (a)

x	y	Average Rate of Change	Ratio of Consecutive Outputs
−1	5	$\dfrac{\Delta y}{\Delta x} = \dfrac{2-5}{0-(-1)} = -3$	$\dfrac{2}{5}$
0	2	$\dfrac{-1-2}{1-0} = -3$	$\dfrac{-1}{2} = -\dfrac{1}{2}$
1	−1	$\dfrac{-4-(-1)}{2-1} = -3$	$\dfrac{-4}{-1} = 4$
2	−4	$\dfrac{-7-(-4)}{3-2} = -3$	$\dfrac{-7}{-4} = \dfrac{7}{4}$
3	−7		

(b)

x	y	Average Rate of Change	Ratio of Consecutive Outputs
−1	32	$\dfrac{\Delta y}{\Delta x} = \dfrac{16-32}{0-(-1)} = -16$	$\dfrac{16}{32} = \dfrac{1}{2}$
0	16	−8	$\dfrac{8}{16} = \dfrac{1}{2}$
1	8	−4	$\dfrac{4}{8} = \dfrac{1}{2}$
2	4	−2	$\dfrac{2}{4} = \dfrac{1}{2}$
3	2		

(c)

x	y	Average Rate of Change	Ratio of Consecutive Outputs
−1	2	$\dfrac{\Delta y}{\Delta x} = \dfrac{4-2}{0-(-1)} = 2$	2
0	4	3	$\dfrac{7}{4}$
1	7	4	$\dfrac{11}{7}$
2	11	5	$\dfrac{16}{11}$
3	16		

Now Work PROBLEM 29

2 Graph Exponential Functions

If we know how to graph an exponential function of the form $f(x) = a^x$, then we can use transformations (shifting, stretching, and so on) to obtain the graph of any exponential function.

First, let's graph the exponential function $f(x) = 2^x$.

| EXAMPLE 3 | **Graphing an Exponential Function** |

Graph the exponential function: $f(x) = 2^x$

Solution The domain of $f(x) = 2^x$ is the set of all real numbers. Begin by locating some points on the graph of $f(x) = 2^x$, as listed in Table 3.

Because $2^x > 0$ for all x, the graph has no x-intercepts and lies above the x-axis for all x. The y-intercept is 1.

Table 3 suggests that as $x \rightarrow -\infty$, the value of f approaches 0. Therefore, the x-axis ($y = 0$) is a horizontal asymptote of the graph of f as $x \rightarrow -\infty$. This provides the end behavior for x large and negative.

To determine the end behavior for x large and positive, look again at Table 3. As $x \rightarrow \infty$, $f(x) = 2^x$ grows very quickly, causing the graph of $f(x) = 2^x$ to rise very rapidly.

Using all this information, plot some of the points from Table 3 and connect them with a smooth, continuous curve, as shown in Figure 1. From the graph, we conclude that the range of f is $(0, \infty)$. We also conclude that f is an increasing function, and so f is one-to-one.

NOTE Recall $a^{-n} = \dfrac{1}{a^n}$. ∎

Table 3

x	$f(x) = 2^x$
-10	$2^{-10} \approx 0.00098$
-3	$2^{-3} = \dfrac{1}{8}$
-2	$2^{-2} = \dfrac{1}{4}$
-1	$2^{-1} = \dfrac{1}{2}$
0	$2^0 = 1$
1	$2^1 = 2$
2	$2^2 = 4$
3	$2^3 = 8$
10	$2^{10} = 1024$

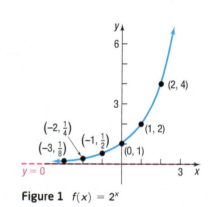

Figure 1 $f(x) = 2^x$

Graphs that look like the one in Figure 1 occur very frequently in a variety of situations. For example, the graph in Figure 2 shows the annual revenue of Amazon, Inc. from 2000 to 2017. One might conclude from this graph that Amazon's annual revenue is growing *exponentially*.

Figure 2

Source: Amazon, Inc.

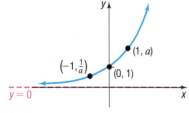

Figure 3

Later in this chapter, we discuss other situations that exhibit exponential growth. For now, we continue to explore properties of exponential functions.

The graph of $f(x) = 2^x$ in Figure 1 is typical of all exponential functions of the form $f(x) = a^x$ with $a > 1$. Such functions are increasing functions and so are one-to-one. Their graphs lie above the x-axis, contain the point $(0, 1)$, and rise rapidly as $x \to \infty$. As $x \to -\infty$, the x-axis $(y = 0)$ is a horizontal asymptote. There are no vertical asymptotes. Finally, the graphs are smooth and continuous with no corners or gaps.

Figure 3 illustrates the graphs of two other exponential functions whose bases are larger than 1. Notice that the larger the base, the steeper the graph is when $x > 0$, and when $x < 0$, the larger the base, the closer the graph is to the x-axis.

Seeing the Concept

Graph $Y_1 = 2^x$ and compare what you see to Figure 1. Clear the screen, graph $Y_1 = 3^x$ and $Y_2 = 6^x$, and compare what you see to Figure 3. Clear the screen and graph $Y_1 = 10^x$ and $Y_2 = 100^x$.

Properties of the Exponential Function $f(x) = a^x$, $a > 1$

- The domain is the set of all real numbers, or $(-\infty, \infty)$ using interval notation; the range is the set of positive real numbers, or $(0, \infty)$ using interval notation.
- There are no x-intercepts; the y-intercept is 1.
- The x-axis $(y = 0)$ is a horizontal asymptote of the graph of f as $x \to -\infty$.
- $f(x) = a^x$, $a > 1$, is an increasing function and is one-to-one.
- The graph of f contains the points $\left(-1, \dfrac{1}{a}\right)$, $(0, 1)$ and $(1, a)$.
- The graph of f is smooth and continuous, with no corners or gaps. See Figure 4.

Figure 4 $f(x) = a^x$, $a > 1$

Now consider $f(x) = a^x$ when $0 < a < 1$.

EXAMPLE 4

Graphing an Exponential Function

Graph the exponential function: $f(x) = \left(\dfrac{1}{2}\right)^x$

Solution The domain of $f(x) = \left(\dfrac{1}{2}\right)^x$ is all real numbers. As before, locate some points on the graph as shown in Table 4 on the next page . Because $\left(\dfrac{1}{2}\right)^x > 0$ for all x, the range of f is the interval $(0, \infty)$. The graph lies above the x-axis and has no x-intercepts. The y-intercept is 1. As $x \to -\infty$, $f(x) = \left(\dfrac{1}{2}\right)^x$ grows very quickly. As $x \to \infty$, the values of $f(x)$ approach 0. The x-axis $(y = 0)$ is a horizontal asymptote of the graph of f as $x \to \infty$. The function f is a decreasing function and so is one-to-one. Figure 5 on the next page shows the graph of f.

Table 4

x	$f(x) = \left(\dfrac{1}{2}\right)^x$
-10	$\left(\dfrac{1}{2}\right)^{-10} = 2^{10} = 1024$
-3	$\left(\dfrac{1}{2}\right)^{-3} = 2^3 = 8$
-2	$\left(\dfrac{1}{2}\right)^{-2} = 2^2 = 4$
-1	$\left(\dfrac{1}{2}\right)^{-1} = 2^1 = 2$
0	$\left(\dfrac{1}{2}\right)^{0} = 1$
1	$\left(\dfrac{1}{2}\right)^{1} = \dfrac{1}{2}$
2	$\left(\dfrac{1}{2}\right)^{2} = \dfrac{1}{4}$
3	$\left(\dfrac{1}{2}\right)^{3} = \dfrac{1}{8}$
10	$\left(\dfrac{1}{2}\right)^{10} \approx 0.00098$

Figure 5 $f(x) = \left(\dfrac{1}{2}\right)^x$

Need to Review?

Reflections about the y-axis are discussed in Section 1.6, p. 74.

The graph of $y = \left(\dfrac{1}{2}\right)^x$ also can be obtained from the graph of $y = 2^x$ using transformations. The graph of $y = \left(\dfrac{1}{2}\right)^x = 2^{-x}$ is a reflection about the y-axis of the graph of $y = 2^x$ (replace x by $-x$). See Figures 6(a) and (b).

Seeing the Concept

Using a graphing utility, simultaneously graph:

(a) $Y_1 = 3^x$, $Y_2 = \left(\dfrac{1}{3}\right)^x$

(b) $Y_1 = 6^x$, $Y_2 = \left(\dfrac{1}{6}\right)^x$

Conclude that the graph of $Y_2 = \left(\dfrac{1}{a}\right)^x$, for $a > 0$, is the reflection about the y-axis of the graph of $Y_1 = a^x$.

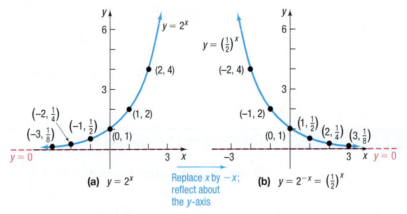

(a) $y = 2^x$

Replace x by $-x$; reflect about the y-axis

(b) $y = 2^{-x} = \left(\dfrac{1}{2}\right)^x$

Figure 6

Figure 7

The graph of $f(x) = \left(\dfrac{1}{2}\right)^x$ in Figure 5 is typical of all exponential functions of the form $f(x) = a^x$ with $0 < a < 1$. Such functions are decreasing and one-to-one. Their graphs lie above the x-axis and contain the point $(0, 1)$. The graphs rise rapidly as $x \to -\infty$. As $x \to \infty$, the x-axis ($y = 0$) is a horizontal asymptote. There are no vertical asymptotes. Finally, the graphs are smooth and continuous, with no corners or gaps.

Figure 7 illustrates the graphs of two other exponential functions whose bases are between 0 and 1. Notice that the smaller base results in a graph that is steeper when $x < 0$. When $x > 0$, the graph of the equation with the smaller base is closer to the x-axis.

> **Properties of the Exponential Function $f(x) = a^x, 0 < a < 1$**
>
>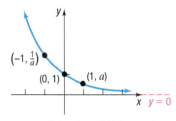
>
> - The domain is the set of all real numbers, or $(-\infty, \infty)$ using interval notation; the range is the set of positive real numbers, or $(0, \infty)$ using interval notation.
> - There are no x-intercepts; the y-intercept is 1.
> - The x-axis ($y = 0$) is a horizontal asymptote of the graph of f as $x \to \infty$.
> - $f(x) = a^x, 0 < a < 1$, is a decreasing function and is one-to-one.
> - The graph of f contains the points $\left(-1, \dfrac{1}{a}\right)$, $(0, 1)$, and $(1, a)$.
> - The graph of f is smooth and continuous, with no corners or gaps. See Figure 8.

Figure 8 $f(x) = a^x, 0 < a < 1$

EXAMPLE 5   **Graphing an Exponential Function Using Transformations**

Graph $f(x) = 2^{-x} - 3$ and determine the domain, range, horizontal asymptote, and y-intercept of f.

Solution   Begin with the graph of $y = 2^x$. Figure 9 shows the steps.

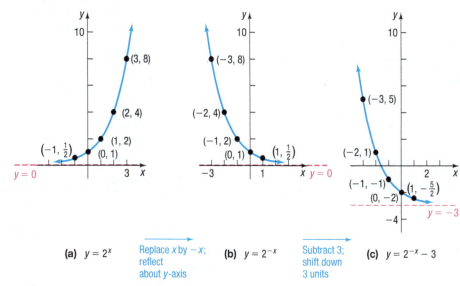

(a) $y = 2^x$ Replace x by $-x$; reflect about y-axis   **(b)** $y = 2^{-x}$ Subtract 3; shift down 3 units   **(c)** $y = 2^{-x} - 3$

Figure 9

As Figure 9(c) illustrates, the domain of $f(x) = 2^{-x} - 3$ is the interval $(-\infty, \infty)$ and the range is the interval $(-3, \infty)$. The horizontal asymptote of the graph of f is the line $y = -3$. The y-intercept is $f(0) = 2^0 - 3 = 1 - 3 = -2$.

 Now Work PROBLEM 45

---

3 Define the Number e

Many problems that occur in nature require the use of an exponential function whose base is a certain irrational number, symbolized by the letter e.

One way of arriving at this important number e is given next.

Historical Feature

The number *e* is called *Euler's number* in honor of the Swiss mathematician Leonard Euler (1707-1783). ∎

DEFINITION Number *e*

The **number *e*** is defined as the number that the expression

$$\left(1 + \frac{1}{n}\right)^n \tag{2}$$

approaches as $n \to \infty$. In calculus, this is expressed, using limit notation, as

$$e = \lim_{n \to \infty} \left(1 + \frac{1}{n}\right)^n$$

Table 5 illustrates what happens to the defining expression (2) as *n* takes on increasingly large values. The last number in the right column in the table approximates *e* correct to nine decimal places. That is, $e = 2.718281828\ldots$. Remember, the three dots indicate that the decimal places continue. Because these decimal places continue but do not repeat, *e* is an irrational number. The number *e* is often expressed as a decimal rounded to a specific number of places. For example, $e \approx 2.71828$ is rounded to five decimal places.

Table 6

x	e^x
−2	$e^{-2} \approx 0.14$
−1	$e^{-1} \approx 0.37$
0	$e^0 = 1$
1	$e^1 \approx 2.72$
2	$e^2 \approx 7.39$

Table 5

n	$\frac{1}{n}$	$1 + \frac{1}{n}$	$\left(1 + \frac{1}{n}\right)^n$
1	1	2	2
2	0.5	1.5	2.25
5	0.2	1.2	2.48832
10	0.1	1.1	2.59374246
100	0.01	1.01	2.704813829
1,000	0.001	1.001	2.716923932
10,000	0.0001	1.0001	2.718145927
100,000	0.00001	1.00001	2.718268237
1,000,000	0.000001	1.000001	2.718280469
10,000,000,000	10^{-10}	$1 + 10^{-10}$	2.718281828

The exponential function $f(x) = e^x$, whose base is the number *e*, occurs with such frequency in applications that it is usually referred to as *the* exponential function. Most calculators have the key $\boxed{e^x}$ or $\boxed{\exp(x)}$, which may be used to approximate the exponential function for a given value of *x*.*

Use your calculator to approximate e^x for $x = -2$, $x = -1$, $x = 0$, $x = 1$, and $x = 2$. See Table 6. The graph of the exponential function $f(x) = e^x$ is given in Figure 10. Since $2 < e < 3$, the graph of $y = e^x$ lies between the graphs of $y = 2^x$ and $y = 3^x$. Do you see why? (Refer to Figures 1 and 3.)

Figure 10 $y = e^x$

(2, e^2); (1, e); (0, 1); $(-1, \frac{1}{e})$; $(-2, \frac{1}{e^2})$; y = 0

Seeing the Concept

Graph $Y_1 = e^x$ and compare what you see to Figure 10. Use eVALUEate or TABLE to verify the points on the graph shown in Figure 10. Now graph $Y_2 = 2^x$ and $Y_3 = 3^x$ on the same screen as $Y_1 = e^x$. Notice that the graph of $Y_1 = e^x$ lies between these two graphs.

EXAMPLE 6 **Graphing an Exponential Function Using Transformations**

Graph $f(x) = -e^{x-3}$ and determine the domain, range, horizontal asymptote, and *y*-intercept of *f*.

*If your calculator does not have one of these keys, refer to your user's manual.

Solution   Begin with the graph of $y = e^x$. Figure 11 shows the steps.

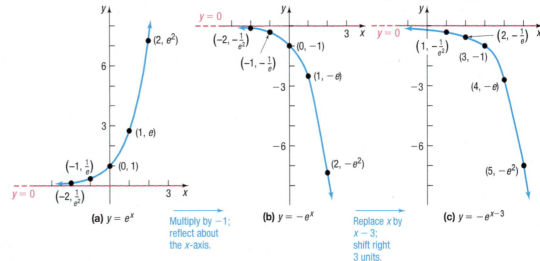

Figure 11

(a) $y = e^x$

Multiply by -1;
reflect about
the x-axis.

(b) $y = -e^x$

Replace x by
$x - 3$;
shift right
3 units.

(c) $y = -e^{x-3}$

As Figure 11(c) illustrates, the domain of $f(x) = -e^{x-3}$ is the interval $(-\infty, \infty)$, and the range is the interval $(-\infty, 0)$. The horizontal asymptote is the line $y = 0$. The y-intercept is $f(0) = -e^{0-3} = -e^{-3} \approx -0.05$.

Now Work PROBLEM 57

4 Solve Exponential Equations

Equations that involve terms of the form a^x, where $a > 0$ and $a \neq 1$, are referred to as **exponential equations**. Such equations can sometimes be solved by using the Laws of Exponents and property (3):

> If $a^u = a^v$, then $u = v$.   (3)

In Words

When two exponential expressions with the same base are equal, then their exponents are equal.

Property (3) is a consequence of the fact that exponential functions are one-to-one. To use property (3), each side of the equality must be written with the same base.

EXAMPLE 7   **Solving Exponential Equations**

Solve each exponential equation.

(a) $3^{x+1} = 81$ (b) $4^{2x-1} = 8^{x+3}$

Solution   (a) Since $81 = 3^4$, write the equation as

$$3^{x+1} = 3^4$$

Now the expressions on both sides of the equation have the same base, 3. Set the exponents equal to each other to obtain

$$x + 1 = 4$$
$$x = 3$$

The solution set is $\{3\}$.

(b) $$4^{2x-1} = 8^{x+3}$$
$$(2^2)^{(2x-1)} = (2^3)^{(x+3)} \quad 4 = 2^2; 8 = 2^3$$
$$2^{2(2x-1)} = 2^{3(x+3)} \quad (a^r)^s = a^{rs}$$
$$2(2x - 1) = 3(x + 3) \quad \text{If } a^u = a^v, \text{ then } u = v, \text{ (property (3))}.$$
$$4x - 2 = 3x + 9$$
$$x = 11$$

The solution set is $\{11\}$.

Now Work PROBLEMS 67 AND 77

EXAMPLE 8 **Solving an Exponential Equation**

Solve: $e^{-x^2} = (e^x)^2 \cdot \dfrac{1}{e^3}$

Solution Use the Laws of Exponents first to get a single expression with the base e on the right side.

$$(e^x)^2 \cdot \frac{1}{e^3} = e^{2x} \cdot e^{-3} = e^{2x-3}$$

Then,

$$e^{-x^2} = e^{2x-3}$$

$$-x^2 = 2x - 3 \qquad \text{Use property (3).}$$

$$x^2 + 2x - 3 = 0 \qquad \text{Place the quadratic equation in standard form.}$$

$$(x + 3)(x - 1) = 0 \qquad \text{Factor.}$$

$$x = -3 \quad \text{or} \quad x = 1 \qquad \text{Use the Zero-Product Property.}$$

The solution set is $\{-3, 1\}$.

 Now Work PROBLEM 83

EXAMPLE 9 **Exponential Probability**

Between 9:00 PM and 10:00 PM, cars arrive at Burger King's drive-thru at the rate of 12 cars per hour (0.2 car per minute). The following formula from probability theory can be used to determine the probability that a car will arrive within t minutes of 9:00 PM.

$$F(t) = 1 - e^{-0.2t}$$

(a) Determine the probability that a car will arrive within 5 minutes of 9 PM (that is, before 9:05 PM).

(b) Determine the probability that a car will arrive within 30 minutes of 9 PM (before 9:30 PM).

(c) Graph F.

(d) What does F approach as t increases without bound in the positive direction?

Solution (a) The probability that a car will arrive within 5 minutes is found by evaluating $F(t)$ at $t = 5$.

$$F(5) = 1 - e^{-0.2(5)} \approx 0.63212$$
$$\uparrow$$
$$\text{Use a calculator.}$$

There is a 63% probability that a car will arrive within 5 minutes.

(b) The probability that a car will arrive within 30 minutes is found by evaluating $F(t)$ at $t = 30$.

$$F(30) = 1 - e^{-0.2(30)} \approx 0.9975$$
$$\uparrow$$
$$\text{Use a calculator.}$$

There is a 99.75% probability that a car will arrive within 30 minutes.

(c) See Figure 12 for the graph of F.

(d) As time passes, the probability that a car will arrive increases. The value that F approaches can be found by letting $t \to \infty$. Since $e^{-0.2t} = \dfrac{1}{e^{0.2t}}$, it follows that $e^{-0.2t} \to 0$ as $t \to \infty$. Therefore, $F(t) = 1 - e^{-0.2t} \to 1$ as $t \to \infty$. The algebraic analysis is supported by Figure 12.

Figure 12 $F(t) = 1 - e^{-0.2t}$

 Now Work PROBLEM 115

SUMMARY

Properties of the Exponential Function

$f(x) = a^x, \quad a > 1$

- Domain: the interval $(-\infty, \infty)$; range: the interval $(0, \infty)$
- x-intercepts: none; y-intercept: 1
- Horizontal asymptote: x-axis $(y = 0)$ as $x \to -\infty$
- Increasing; one-to-one; smooth; continuous
- See Figure 4 for a typical graph.

$f(x) = a^x, \quad 0 < a < 1$

- Domain: the interval $(-\infty, \infty)$; range: the interval $(0, \infty)$
- x-intercepts: none; y-intercept: 1
- Horizontal asymptote: x-axis $(y = 0)$ as $x \to \infty$
- Decreasing; one-to-one; smooth; continuous
- See Figure 8 for a typical graph.

If $a^u = a^v$, then $u = v$.

7.1 Assess Your Understanding

'Are You Prepared?' *Answers are given at the end of these exercises. If you get a wrong answer, read the pages listed in* red.

1. $4^3 =$ _____; $8^{2/3} =$ _____; $3^{-2} =$ _____. (pp. A7–A9 and pp. A60–A61)

2. Solve: $x^2 + 3x = 4$ (pp. A30–A34)

3. *True or False* To graph $y = (x - 2)^3$, shift the graph of $y = x^3$ to the left 2 units. (pp. 67–76)

4. Find the average rate of change of $f(x) = 3x - 5$ from $x = 0$ to $x = 4$. (p. A65)

5. *True or False* The graph of the function $f(x) = \dfrac{2x}{x - 3}$ has $y = 2$ as a horizontal asymptote. (pp. 76–77)

6. Find the average rate of change of $f(x) = x^3 + 2$ from $x = -3$ to $x = 2$.

7. Where is the square function $f(x) = x^2$ increasing? Where is it decreasing? (p. 59)

Concepts and Vocabulary

8. A(n) _____ _____ is a function of the form $f(x) = Ca^x$, where $a > 0, a \neq 1$, and $C \neq 0$ are real numbers. The base a is the _____ _____ and C is the _____ _____.

9. For an exponential function $f(x) = Ca^x$, $\dfrac{f(x + 1)}{f(x)} =$ ___.

10. *True or False* The domain of the function $f(x) = a^x$, where $a > 0$ and $a \neq 1$, is the set of all real numbers.

11. *True or False* The function $f(x) = e^x$ is increasing and is one-to-one.

12. The graph of every exponential function $f(x) = a^x$, where $a > 0$ and $a \neq 1$, contains the three points: _____, _____, and _____.

13. If $3^x = 3^4$, then $x =$ ___.

14. *True or False* The graphs of $y = 3^x$ and $y = \left(\dfrac{1}{3}\right)^x$ are identical.

15. *Multiple Choice* Which exponential function is increasing?

(a) $f(x) = 0.5^x$ (b) $f(x) = \left(\dfrac{5}{2}\right)^x$

(c) $f(x) = \left(\dfrac{2}{3}\right)^x$ (d) $f(x) = 0.9^x$

16. *Multiple Choice* The range of the function $f(x) = a^x$, where $a > 0$ and $a \neq 1$, is the interval

(a) $(-\infty, \infty)$ (b) $(-\infty, 0)$ (c) $(0, \infty)$ (d) $[0, \infty)$

Skill Building

In Problems 17–28, approximate each number using a calculator. Express your answer rounded to three decimal places.

17. (a) $2^{3.14}$ **(b)** $2^{3.141}$ **(c)** $2^{3.1415}$ **(d)** 2^{π}

18. (a) $2^{2.7}$ **(b)** $2^{2.71}$ **(c)** $2^{2.718}$ **(d)** 2^{e}

19. (a) $3.1^{2.7}$ **(b)** $3.14^{2.71}$ **(c)** $3.141^{2.718}$ **(d)** π^{e}

20. (a) $2.7^{3.1}$ **(b)** $2.71^{3.14}$ **(c)** $2.718^{3.141}$ **(d)** e^{π}

21. $(1 + 0.04)^6$

22. $\left(1 + \dfrac{0.09}{12}\right)^{24}$

23. $8.4\left(\dfrac{1}{3}\right)^{2.9}$

24. $158\left(\dfrac{5}{6}\right)^{8.63}$

25. $e^{1.2}$

26. $e^{-1.3}$

27. $125e^{0.026(7)}$

28. $83.6e^{-0.157(9.5)}$

In Problems 29–36, determine whether the given function is linear, exponential, or neither. For those that are linear functions, find a linear function that models the data; for those that are exponential, find an exponential function that models the data.

29.

x	f(x)
−1	3
0	6
1	12
2	18
3	30

30.

x	g(x)
−1	2
0	5
1	8
2	11
3	14

31.

x	H(x)
−1	$\frac{1}{4}$
0	1
1	4
2	16
3	64

32.

x	F(x)
−1	$\frac{2}{3}$
0	1
1	$\frac{3}{2}$
2	$\frac{9}{4}$
3	$\frac{27}{8}$

33.

x	f(x)
−1	$\frac{3}{2}$
0	3
1	6
2	12
3	24

34.

x	g(x)
−1	6
0	1
1	0
2	3
3	10

35.

x	H(x)
−1	2
0	4
1	6
2	8
3	10

36.

x	F(x)
−1	$\frac{1}{2}$
0	$\frac{1}{4}$
1	$\frac{1}{8}$
2	$\frac{1}{16}$
3	$\frac{1}{32}$

In Problems 37–44, the graph of an exponential function is given. Match each graph to one of the following functions.

(A) $y = 3^x$ (B) $y = 3^{-x}$ (C) $y = -3^x$ (D) $y = -3^{-x}$

(E) $y = 3^x - 1$ (F) $y = 3^{x-1}$ (G) $y = 3^{1-x}$ (H) $y = 1 - 3^x$

37.

38.

39.

40.

41.

42.

43.

44.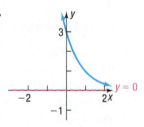

In Problems 45–56, use transformations to graph each function. Determine the domain, range, horizontal asymptote, and y-intercept of each function.

45. $f(x) = 2^x + 1$ **46.** $f(x) = 3^x - 2$ **47.** $f(x) = 3^{x-1}$ **48.** $f(x) = 2^{x+2}$

49. $f(x) = 3 \cdot \left(\frac{1}{2}\right)^x$ **50.** $f(x) = 4 \cdot \left(\frac{1}{3}\right)^x$ **51.** $f(x) = 3^{-x} - 2$ **52.** $f(x) = -3^x + 1$

53. $f(x) = 2 + 4^{x-1}$ **54.** $f(x) = 1 - 2^{x+3}$ **55.** $f(x) = 2 + 3^{x/2}$ **56.** $f(x) = 1 - 2^{-x/3}$

In Problems 57–64, begin with the graph of $y = e^x$ (Figure 10) and use transformations to graph each function. Determine the domain, range, horizontal asymptote, and y-intercept of each function.

57. $f(x) = e^{-x}$ **58.** $f(x) = -e^x$ **59.** $f(x) = e^{x+2}$ **60.** $f(x) = e^x - 1$

61. $f(x) = 5 - e^{-x}$ **62.** $f(x) = 9 - 3e^{-x}$ **63.** $f(x) = 2 - e^{-x/2}$ **64.** $f(x) = 7 - 3e^{2x}$

In Problems 65–84, solve each equation.

65. $6^x = 6^5$

66. $5^x = 5^{-6}$

67. $2^{-x} = 16$

68. $3^{-x} = 81$

69. $\left(\frac{1}{5}\right)^x = \frac{1}{25}$

70. $\left(\frac{1}{4}\right)^x = \frac{1}{64}$

71. $3^{2x-5} = 9$

72. $5^{x+3} = \frac{1}{5}$

73. $3^{x^3} = 9^x$

74. $4^{x^2} = 2^x$

75. $8^{-x+11} = 16^{2x}$

76. $9^{-x+15} = 27^x$

77. $3^{x^2-7} = 27^{2x}$

78. $5^{x^2+8} = 125^{2x}$

79. $4^x \cdot 2^{x^2} = 16^2$

80. $9^{2x} \cdot 27^{x^2} = 3^{-1}$

81. $e^{2x} = e^{5x+12}$

82. $e^{3x} = e^{2-x}$

83. $e^{x^2} = e^{3x} \cdot \frac{1}{e^2}$

84. $(e^4)^x \cdot e^{x^2} = e^{12}$

85. If $4^x = 7$, what does 4^{-2x} equal?

86. If $2^x = 3$, what does 4^{-x} equal?

87. If $3^{-x} = 2$, what does 3^{2x} equal?

88. If $5^{-x} = 3$, what does 5^{3x} equal?

89. If $9^x = 25$, what does 3^x equal?

90. If $2^{-3x} = \frac{1}{1000}$, what does 2^x equal?

In Problems 91–94, determine the exponential function whose graph is given.

91.

92.

93.

94.

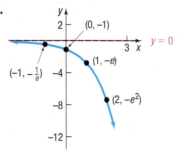

95. Find an exponential function whose graph has the horizontal asymptote $y = 2$ and contains the points $(0, 3)$ and $(1, 5)$.

96. Find an exponential function whose graph has the horizontal asymptote $y = -3$ and contains the points $(0, -2)$ and $(-2, 1)$.

97. Suppose that $f(x) = 2^x$.
  (a) What is $f(4)$? What point is on the graph of f?
  (b) If $f(x) = \frac{1}{16}$, what is x? What point is on the graph of f?

98. Suppose that $f(x) = 3^x$.
  (a) What is $f(4)$? What point is on the graph of f?
  (b) If $f(x) = \frac{1}{9}$, what is x? What point is on the graph of f?

99. Suppose that $g(x) = 4^x + 2$.
  (a) What is $g(-1)$? What point is on the graph of g?
  (b) If $g(x) = 66$, what is x? What point is on the graph of g?

100. Suppose that $g(x) = 5^x - 3$.
  (a) What is $g(-1)$? What point is on the graph of g?
  (b) If $g(x) = 122$, what is x? What point is on the graph of g?

101. Suppose that $H(x) = \left(\frac{1}{2}\right)^x - 4$.
  (a) What is $H(-6)$? What point is on the graph of H?
  (b) If $H(x) = 12$, what is x? What point is on the graph of H?
  (c) Find the zero of H.

102. Suppose that $F(x) = \left(\frac{1}{3}\right)^x - 3$.
  (a) What is $F(-5)$? What point is on the graph of F?
  (b) If $F(x) = 24$, what is x? What point is on the graph of F?
  (c) Find the zero of F.

Mixed Practice *In Problems 103–106, graph each function. Based on the graph, state the domain and the range, and find any intercepts.*

103. $f(x) = \begin{cases} e^{-x} & \text{if } x < 0 \\ e^x & \text{if } x \geq 0 \end{cases}$

104. $f(x) = \begin{cases} e^x & \text{if } x < 0 \\ e^{-x} & \text{if } x \geq 0 \end{cases}$

105. $f(x) = \begin{cases} -e^x & \text{if } x < 0 \\ -e^{-x} & \text{if } x \geq 0 \end{cases}$

106. $f(x) = \begin{cases} -e^{-x} & \text{if } x < 0 \\ -e^x & \text{if } x \geq 0 \end{cases}$

Applications and Extensions

107. Optics If a single pane of glass obliterates 3% of the light passing through it, the percent p of light that passes through n successive panes is given approximately by the function

$$p(n) = 100 \cdot 0.97^n$$

(a) What percent of light will pass through 10 panes?
(b) What percent of light will pass through 25 panes?
(c) Explain the meaning of the base 0.97 in this problem.

108. Atmospheric Pressure The atmospheric pressure p on a balloon or airplane decreases with increasing height. This pressure, measured in millimeters of mercury, is related to the height h (in kilometers) above sea level by the function

$$p(h) = 760e^{-0.145h}$$

(a) Find the atmospheric pressure at a height of 2 km (over a mile).
(b) What is it at a height of 10 kilometers (over 30,000 feet)?

109. Depreciation The price p, in dollars, of a Honda Civic EX-L sedan that is x years old is modeled by

$$p(x) = 22{,}265 \cdot 0.90^x$$

(a) How much should a 3-year-old Civic EX-L sedan cost?
(b) How much should a 9-year-old Civic EX-L sedan cost?
(c) Explain the meaning of the base 0.90 in this problem.

110. Healing of Wounds The normal healing of wounds can be modeled by an exponential function. If A_0 represents the original area of the wound and if A equals the area of the wound, then the function

$$A(n) = A_0 e^{-0.35n}$$

describes the area of a wound after n days following an injury when no infection is present to retard the healing. Suppose that a wound initially had an area of 100 square millimeters.

(a) If healing is taking place, how large will the area of the wound be after 3 days?
(b) How large will it be after 10 days?

111. Advanced-Stage Pancreatic Cancer The percentage of patients P who have survived t years after initial diagnosis of advanced-stage pancreatic cancer is modeled by the function

$$P(t) = 100 \cdot 0.3^t$$

Source: Cancer Treatment Centers of America
(a) According to the model, what percent of patients survive 1 year after initial diagnosis?
(b) What percent of patients survive 2 years after initial diagnosis?
(c) Explain the meaning of the base 0.3 in the context of this problem.

112. Endangered Species In a protected environment, the population P of a certain endangered species recovers over time t (in years) according to the model

$$P(t) = 30 \cdot 1.149^x$$

(a) What is the size of the initial population of the species?
(b) According to the model, what will be the population of the species in 5 years?

(c) According to the model, what will be the population of the species in 10 years?
(d) According to the model, what will be the population of the species in 15 years?
(e) What is happening to the population every 5 years?

113. Drug Medication The function

$$D(h) = 5e^{-0.4h}$$

can be used to find the number of milligrams D of a certain drug that is in a patient's bloodstream h hours after the drug has been administered. How many milligrams will be present after 1 hour? After 6 hours?

114. Spreading of Rumors A model for the number N of people in a college community who have heard a certain rumor is

$$N = P(1 - e^{-0.15d})$$

where P is the total population of the community and d is the number of days that have elapsed since the rumor began. In a community of 1000 students, how many students will have heard the rumor after 3 days?

115. Exponential Probability Between 12:00 PM and 1:00 PM, cars arrive at Citibank's drive-thru at the rate of 6 cars per hour (0.1 car per minute). The following formula from probability can be used to determine the probability that a car arrives within t minutes of 12:00 PM.

$$F(t) = 1 - e^{-0.1t}$$

(a) Determine the probability that a car arrives within 10 minutes of 12:00 PM (that is, before 12:10 PM).
(b) Determine the probability that a car arrives within 40 minutes of 12:00 PM (before 12:40 PM).
(c) What does F approach as t becomes unbounded in the positive direction?
(d) Graph F using a graphing utility.
(e) Using INTERSECT, determine how many minutes are needed for the probability to reach 50%.

116. Exponential Probability Between 5:00 PM and 6:00 PM, cars arrive at Jiffy Lube at the rate of 9 cars per hour (0.15 car per minute). This formula from probability can be used to determine the probability that a car arrives within t minutes of 5:00 PM:

$$F(t) = 1 - e^{-0.15t}$$

(a) Determine the probability that a car arrives within 15 minutes of 5:00 PM (that is, before 5:15 PM).
(b) Determine the probability that a car arrives within 30 minutes of 5:00 PM (before 5:30 PM).
(c) What does F approach as t becomes unbounded in the positive direction?
(d) Graph F.
(e) Using INTERSECT, determine how many minutes are needed for the probability to reach 60%.

117. Poisson Probability Between 5:00 PM and 6:00 PM, cars arrive at a McDonald's drive-thru at the rate of 20 cars per hour. The following formula from probability can be used to determine the probability that x cars arrive between 5:00 PM and 6:00 PM.

$$P(x) = \frac{20^x e^{-20}}{x!}$$

where

$$x! = x \cdot (x - 1) \cdot (x - 2) \cdots \cdot 3 \cdot 2 \cdot 1$$

(a) Determine the probability that $x = 15$ cars arrive between 5:00 PM and 6:00 PM.

(b) Determine the probability that $x = 20$ cars arrive between 5:00 PM and 6:00 PM.

118. Poisson Probability People enter a line for the *Demon Roller Coaster* at the rate of 4 per minute. The following formula from probability can be used to determine the probability that x people arrive within the next minute.

$$P(x) = \frac{4^x e^{-4}}{x!}$$

where

$$x! = x \cdot (x - 1) \cdot (x - 2) \cdots \cdot 3 \cdot 2 \cdot 1$$

(a) Determine the probability that $x = 5$ people arrive within the next minute.

(b) Determine the probability that $x = 8$ people arrive within the next minute.

119. Relative Humidity The relative humidity is the ratio (expressed as a percent) of the amount of water vapor in the air to the maximum amount that the air can hold at a specific temperature. The relative humidity, R, is found using the following formula:

$$R = 10^{\left(\frac{4221}{T+459.4} - \frac{4221}{D+459.4} + 2\right)}$$

where T is the air temperature (in °F) and D is the dew point temperature (in °F).

(a) Determine the relative humidity if the air temperature is 50° Fahrenheit and the dew point temperature is 41° Fahrenheit.

(b) Determine the relative humidity if the air temperature is 68° Fahrenheit and the dew point temperature is 59° Fahrenheit.

(c) What is the relative humidity if the air temperature and the dew point temperature are the same?

120. Learning Curve Suppose that a student has 500 vocabulary words to learn. If the student learns 15 words after 5 minutes, the function

$$L(t) = 500(1 - e^{-0.0061t})$$

approximates the number of words L that the student will have learned after t minutes.

(a) How many words will the student have learned after 30 minutes?

(b) How many words will the student have learned after 60 minutes?

121. Current in an RL Circuit The equation governing the amount of current I (in amperes) after time t (in seconds) in a single RL circuit consisting of a resistance R (in ohms), an inductance L (in henrys), and an electromotive force E (in volts) is

$$I = \frac{E}{R}\left[1 - e^{-(R/L)t}\right]$$

(a) If $E = 120$ volts, $R = 10$ ohms, and $L = 5$ henrys, how much current I_1 is flowing after 0.3 second? After 0.5 second? After 1 second?

(b) What is the maximum current?

(c) Graph this function $I = I_1(t)$, measuring I along the y-axis and t along the x-axis.

(d) If $E = 120$ volts, $R = 5$ ohms, and $L = 10$ henrys, how much current I_2 is flowing after 0.3 second? After 0.5 second? After 1 second?

(e) What is the maximum current?

(f) Graph the function $I = I_2(t)$ on the same coordinate axes as $I_1(t)$.

122. Current in an RC Circuit The equation governing the amount of current I (in amperes) after time t (in microseconds) in a single RC circuit consisting of a resistance R (in ohms), a capacitance C (in microfarads), and an electromotive force E (in volts) is

$$I = \frac{E}{R}e^{-t/(RC)}$$

(a) If $E = 120$ volts, $R = 2000$ ohms, and $C = 1.0$ microfarad, how much current I_1 is flowing initially $(t = 0)$? After 1000 microseconds? After 3000 microseconds?

(b) What is the maximum current?

(c) Graph the function $I = I_1(t)$, measuring I along the y-axis and t along the x-axis.

(d) If $E = 120$ volts, $R = 1000$ ohms, and $C = 2.0$ microfarads, how much current I_2 is flowing initially? After 1000 microseconds? After 3000 microseconds?

(e) What is the maximum current?

(f) Graph the function $I = I_2(t)$ on the same coordinate axes as $I_1(t)$.

123. If f is an exponential function of the form $f(x) = Ca^x$ with growth factor 3, and if $f(6) = 12$, what is $f(7)$?

124. Another Formula for e Use a calculator to compute the values of

$$2 + \frac{1}{2!} + \frac{1}{3!} + \cdots + \frac{1}{n!}$$

for $n = 4, 6, 8,$ and 10. Compare each result with e.

[Hint: $1! = 1, 2! = 2 \cdot 1, 3! = 3 \cdot 2 \cdot 1,$

$$n! = n(n - 1) \cdots \cdot (3)(2)(1).]$$

125. Another Formula for e Use a calculator to compute the various values of the expression. Compare the values to e.

$$2 + \cfrac{1}{1 + \cfrac{1}{2 + \cfrac{2}{3 + \cfrac{3}{4 + \cfrac{4}{\text{etc.}}}}}}$$

126. Difference Quotient If $f(x) = a^x$, show that

$$\frac{f(x+h) - f(x)}{h} = a^x \cdot \frac{a^h - 1}{h} \quad h \neq 0$$

127. If $f(x) = a^x$, show that $f(A + B) = f(A) \cdot f(B)$.

128. If $f(x) = a^x$, show that $f(-x) = \dfrac{1}{f(x)}$.

129. If $f(x) = a^x$, show that $f(\alpha x) = [f(x)]^\alpha$.

Problems 130 and 131 define two other transcendental functions.

130. The **hyperbolic sine function**, designated by $\sinh x$, is defined as

$$\sinh x = \frac{1}{2}(e^x - e^{-x})$$

   (a) Show that $f(x) = \sinh x$ is an odd function.

   (b) Graph $f(x) = \sinh x$.

131. The **hyperbolic cosine function**, designated by $\cosh x$, is defined as

$$\cosh x = \frac{1}{2}(e^x + e^{-x})$$

   (a) Show that $f(x) = \cosh x$ is an even function.

   (b) Graph $f(x) = \cosh x$.

   (c) Refer to Problem 130. Show that, for every x,

$$(\cosh x)^2 - (\sinh x)^2 = 1$$

132. Historical Problem Pierre de Fermat (1601–1665) conjectured that the function

$$f(x) = 2^{(2^x)} + 1$$

for $x = 1, 2, 3, \ldots$, would always have a value equal to a prime number. But Leonhard Euler (1707–1783) showed that this formula fails for $x = 5$. Use a calculator to determine the prime numbers produced by f for $x = 1, 2, 3, 4$. Then show that $f(5) = 641 \times 6{,}700{,}417$, which is not prime.

133. Challenge Problem Solve: $3^{2x-1} - 4 \cdot 3^x + 9 = 0$

134. Challenge Problem Solve: $2^{\frac{2}{3}x+1} - 3 \cdot 2^{\frac{1}{3}x} - 20 = 0$

Explaining Concepts: Discussion and Writing

135. The bacteria in a 4-liter container double every minute. After 60 minutes the container is full. How long did it take to fill half the container?

136. Explain in your own words what the number e is. Provide at least two applications that use this number.

137. Do you think that there is a power function that increases more rapidly than an exponential function? Explain.

138. As the base a of an exponential function $f(x) = a^x$, where $a > 1$, increases, what happens to its graph for $x > 0$? What happens to its graph for $x < 0$?

139. The graphs of $y = a^{-x}$ and $y = \left(\dfrac{1}{a}\right)^x$ are identical. Why?

Retain Your Knowledge

Problems 140–149 are based on material learned earlier in the course. The purpose of these problems is to keep the material fresh in your mind so that you are better prepared for the final exam.

140. Solve the inequality: $x^3 + 5x^2 \leq 4x + 20$.

141. Solve the inequality: $\dfrac{x+1}{x-2} \geq 1$.

142. Find the equation of the quadratic function f that has its vertex at $(3, 5)$ and contains the point $(2, 3)$.

143. Suppose $f(x) = x^2 + 2x - 3$.

   (a) Graph f by determining whether its graph is concave up or concave down and by finding its vertex, axis of symmetry, y-intercept, and x-intercepts, if any.

   (b) Find the domain and range of f.

   (c) Determine where f is increasing and where it is decreasing.

144. Solve: $13x - (5x - 6) = 2x - (8x - 27)$

145. Find an equation for the circle with center $(0, 0)$ and radius $r = 1$.

146. Solve: $x - 16\sqrt{x} + 48 = 0$

147. If \$12,000 is invested at 3.5% simple interest for 2.5 years, how much interest is earned?

148. Determine where the graph of $f(x) = x^4 - 5x^2 - 6$ is below the graph of $g(x) = 2x^2 + 12$ by solving the inequality $f(x) \leq g(x)$.

149. Find the difference quotient of $f(x) = 2x^2 - 7x$.

'Are You Prepared?' Answers

1. $64; 4; \dfrac{1}{9}$ **2.** $\{-4, 1\}$ **3.** False   **4.** 3   **5.** True   **6.** 7   **7.** $[0, \infty); (-\infty, 0]$

7.2 Logarithmic Functions

PREPARING FOR THIS SECTION *Before getting started, review the following:*

- Solving Inequalities (Section A.6, pp. A49–A52)
- Solving Equations (Section A.4, pp. A27–A34)
- One-to-One Function; Inverse Functions (Section 1.7, pp. 81–88)

Now Work the 'Are You Prepared?' problems on page 485.

OBJECTIVES 1 Change Exponential Statements to Logarithmic Statements and Logarithmic Statements to Exponential Statements (p. 477)
2 Evaluate Logarithmic Expressions (p. 478)
3 Determine the Domain of a Logarithmic Function (p. 478)
4 Graph Logarithmic Functions (p. 479)
5 Solve Logarithmic Equations (p. 483)

Recall that a one-to-one function $y = f(x)$ has an inverse function that is defined implicitly by the equation $x = f(y)$. In particular, the exponential function $y = f(x) = a^x$, where $a > 0$ and $a \neq 1$, is one-to-one, so it has an inverse function that is defined implicitly by the equation

$$x = a^y \qquad a > 0 \qquad a \neq 1$$

This inverse function is so important that it is given a name, the *logarithmic function*.

> **DEFINITION** Logarithmic Function with Base *a*
>
> The **logarithmic function with base *a***, where $a > 0$ and $a \neq 1$, is denoted by $y = \log_a x$ (read as "*y* is the logarithm with base *a* of *x*") and is defined by
>
> $$y = \log_a x \quad \text{if and only if} \quad x = a^y$$
>
> The domain of the logarithmic function $y = \log_a x$ is $x > 0$.

In Words
When you need to evaluate $\log_a x$, think to yourself "*a* raised to what power gives me x?"

As this definition illustrates, **a logarithm is a name for a certain exponent.** So $\log_a x$ equals the exponent to which *a* must be raised to obtain *x*.

EXAMPLE 1 **Relating Logarithms to Exponents**

(a) If $y = \log_3 x$, then $x = 3^y$. For example, the logarithmic statement $4 = \log_3 81$ is equivalent to the exponential statement $81 = 3^4$.

(b) If $y = \log_5 x$, then $x = 5^y$. For example, $-1 = \log_5\left(\frac{1}{5}\right)$ is equivalent to $\frac{1}{5} = 5^{-1}$.

1 Change Exponential Statements to Logarithmic Statements and Logarithmic Statements to Exponential Statements

The definition of a logarithm can be used to convert from exponential form to logarithmic form, and vice versa, as illustrated in the next two examples.

EXAMPLE 2 **Changing Exponential Statements to Logarithmic Statements**

Change each exponential statement to an equivalent statement involving a logarithm.

(a) $1.2^3 = m$ 　　　　(b) $e^b = 9$ 　　　　(c) $a^4 = 24$

Solution Use the fact that $y = \log_a x$ and $x = a^y$, where $a > 0$ and $a \neq 1$, are equivalent.

(a) If $1.2^3 = m$, then $3 = \log_{1.2} m$. 　　(b) If $e^b = 9$, then $b = \log_e 9$.
(c) If $a^4 = 24$, then $4 = \log_a 24$.

Now Work PROBLEM 11

EXAMPLE 3

Changing Logarithmic Statements to Exponential Statements

Change each logarithmic statement to an equivalent statement involving an exponent.

(a) $\log_a 4 = 5$ (b) $\log_e b = -3$ (c) $\log_3 5 = c$

Solution

(a) If $\log_a 4 = 5$, then $a^5 = 4$.

(b) If $\log_e b = -3$, then $e^{-3} = b$.

(c) If $\log_3 5 = c$, then $3^c = 5$.

━━━━━ **Now Work** PROBLEM 19

2 Evaluate Logarithmic Expressions

To find the exact value of a logarithm, write the logarithm in exponential notation using the fact that $y = \log_a x$ is equivalent to $a^y = x$, and use the fact that if $a^u = a^v$, then $u = v$.

EXAMPLE 4

Finding the Exact Value of a Logarithmic Expression

Find the exact value of:

(a) $\log_2 16$ (b) $\log_3 \dfrac{1}{27}$

Solution

(a) To find $\log_2 16$, think "2 raised to what power equals 16?" Then,

$$y = \log_2 16$$
$$2^y = 16 \quad \text{Change to exponential form.}$$
$$2^y = 2^4 \quad 16 = 2^4$$
$$y = 4 \quad \text{Equate exponents.}$$

Therefore, $\log_2 16 = 4$.

(b) To find $\log_3 \dfrac{1}{27}$, think "3 raised to what power equals $\dfrac{1}{27}$?" Then,

$$y = \log_3 \dfrac{1}{27}$$
$$3^y = \dfrac{1}{27} \quad \text{Change to exponential form.}$$
$$3^y = 3^{-3} \quad \dfrac{1}{27} = \dfrac{1}{3^3} = 3^{-3}$$
$$y = -3 \quad \text{Equate exponents.}$$

Therefore, $\log_3 \dfrac{1}{27} = -3$.

━━━━━ **Now Work** PROBLEM 27

3 Determine the Domain of a Logarithmic Function

The logarithmic function $y = \log_a x$ has been defined as the inverse of the exponential function $y = a^x$. That is, if $f(x) = a^x$, then $f^{-1}(x) = \log_a x$. Based on the discussion in Section 1.7 on inverse functions, for a function f and its inverse f^{-1},

$$\text{Domain of } f^{-1} = \text{Range of } f \quad \text{and} \quad \text{Range of } f^{-1} = \text{Domain of } f$$

Consequently, it follows that

- Domain of the logarithmic function = Range of the exponential function = $(0, \infty)$
- Range of the logarithmic function = Domain of the exponential function = $(-\infty, \infty)$

The next box summarizes some properties of the logarithmic function.

$$y = \log_a x \quad \text{if and only if} \quad x = a^y$$
$$\text{Domain: } 0 < x < \infty \quad \text{Range: } -\infty < y < \infty$$

The domain of a logarithmic function consists of the *positive* real numbers, so the argument of a logarithmic function must be greater than zero.

EXAMPLE 5   **Finding the Domain of a Logarithmic Function**

Find the domain of each logarithmic function.

(a) $F(x) = \log_2(x + 3)$ (b) $g(x) = \log_5\left(\dfrac{1 + x}{1 - x}\right)$ (c) $h(x) = \log_{1/2}|x|$

Solution   (a) The domain of F consists of all x for which $x + 3 > 0$, that is, $x > -3$. Using interval notation, the domain of F is $(-3, \infty)$.

(b) The domain of g is restricted to

$$\frac{1 + x}{1 - x} > 0$$

Solve this inequality to find that the domain of g consists of all x between -1 and 1, that is, $-1 < x < 1$, or, using interval notation, $(-1, 1)$.

(c) Since $|x| > 0$, provided that $x \neq 0$, the domain of h consists of all real numbers except zero, or, using interval notation, $(-\infty, 0) \cup (0, \infty)$.

— **Now Work** PROBLEMS **41** AND **47**

4 Graph Logarithmic Functions

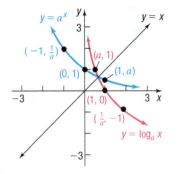

Because exponential functions and logarithmic functions are inverses of each other, the graph of the logarithmic function $y = \log_a x$ is the reflection about the line $y = x$ of the graph of the exponential function $y = a^x$, as shown in Figures 13(a) and (b).

For example, to graph $y = \log_2 x$, graph $y = 2^x$ and reflect it about the

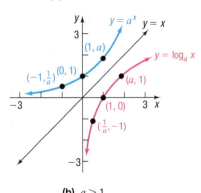

line $y = x$. See Figure 14. To graph $y = \log_{1/3} x$, graph $y = \left(\dfrac{1}{3}\right)^x$ and reflect it about the line $y = x$. See Figure 15.

(a) $0 < a < 1$

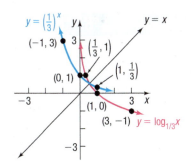

(b) $a > 1$

Figure 13   **Figure 14**   **Figure 15**

— **Now Work** PROBLEM **61**

The graphs of $y = \log_a x$ in Figures 13(a) and (b) lead to the following properties.

Properties of the Logarithmic Function $f(x) = \log_a x$; $a > 0$, $a \neq 1$

- The domain is the set of positive real numbers, or $(0, \infty)$ using interval notation; the range is the set of all real numbers, or $(-\infty, \infty)$ using interval notation.
- The x-intercept of the graph is 1. There is no y-intercept.
- The y-axis $(x = 0)$ is a vertical asymptote of the graph of f.
- A logarithmic function is decreasing if $0 < a < 1$ and is increasing if $a > 1$.
- The graph of f contains the points $(1, 0)$, $(a, 1)$, and $\left(\dfrac{1}{a}, -1\right)$.
- The graph is smooth and continuous, with no corners or gaps.

If the base of a logarithmic function is the number e, the result is the **natural logarithm function**. This function occurs so frequently in applications that it is given a special symbol, **ln** (from the Latin, *logarithmus naturalis*). That is,

$$y = \ln x \quad \text{if and only if} \quad x = e^y \qquad (1)$$

Because $y = \ln x$ and the exponential function $y = e^x$ are inverse functions, the graph of $y = \ln x$ can be obtained by reflecting the graph of $y = e^x$ about the line $y = x$. See Figure 16.

Using a calculator with an ⟨ln⟩ key, we can obtain other points on the graph of $f(x) = \ln x$. See Table 7.

Seeing the Concept
Graph $Y_1 = e^x$ and $Y_2 = \ln x$ on the same square screen. Use eVALUEate to verify the points on the graph given in Figure 16. Do you see the symmetry of the two graphs with respect to the line $y = x$?

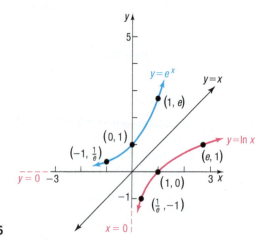

Figure 16

Table 7

x	$\ln x$
$\frac{1}{2}$	-0.69
2	0.69
3	1.10

EXAMPLE 6

Graphing a Logarithmic Function and Its Inverse

(a) Find the domain of the logarithmic function $f(x) = -\ln(x - 2)$.
(b) Graph f.
(c) From the graph, determine the range and vertical asymptote of f.
(d) Find f^{-1}, the inverse of f.
(e) Find the domain and the range of f^{-1}.
(f) Graph f^{-1}.

Solution

(a) The domain of f consists of all x for which $x - 2 > 0$, or equivalently, $x > 2$. The domain of f is $\{x \mid x > 2\}$, or $(2, \infty)$ in interval notation.

(b) To obtain the graph of $y = -\ln(x - 2)$, begin with the graph of $y = \ln x$ and use transformations. See Figure 17.

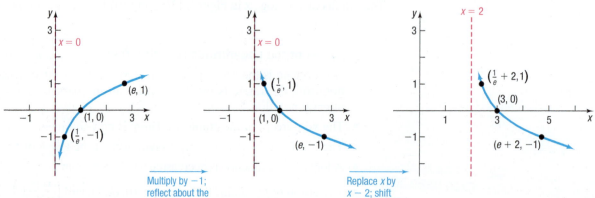

Figure 17   (a) $y = \ln x$ Multiply by -1; reflect about the x-axis.   (b) $y = -\ln x$ Replace x by $x - 2$; shift right 2 units.   (c) $y = -\ln(x - 2)$

(c) The range of $f(x) = -\ln(x - 2)$ is the set of all real numbers. The vertical asymptote is $x = 2$. [Do you see why? The original asymptote $(x = 0)$ is shifted to the right 2 units.]

(d) To find f^{-1}, begin with $y = -\ln(x - 2)$. The inverse function is defined implicitly by the equation

$$x = -\ln(y - 2)$$

Now solve for y.

$$-x = \ln(y - 2) \qquad \text{**Isolate the logarithm.**}$$
$$e^{-x} = y - 2 \qquad \text{**Change to exponential form.**}$$
$$y = e^{-x} + 2 \qquad \text{**Solve for y.**}$$

The inverse of f is $f^{-1}(x) = e^{-x} + 2$.

(e) The domain of f^{-1} equals the range of f, which is the set of all real numbers, from part (c). The range of f^{-1} is the domain of f, which is $(2, \infty)$ in interval notation.

(f) To graph f^{-1}, use the graph of f in Figure 17(c) and reflect it about the line $y = x$. See Figure 18. We could also graph $f^{-1}(x) = e^{-x} + 2$ using transformations.

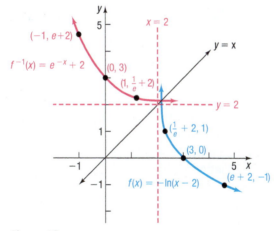

Figure 18

Now Work PROBLEM 73

If the base of a logarithmic function is the number 10, the result is the **common logarithm function**. If the base a of the logarithmic function is not indicated, it is understood to be 10. That is,

$$y = \log x \quad \text{if and only if} \quad x = 10^y$$

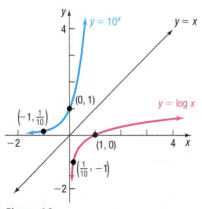

Figure 19

Because $y = \log x$ and the exponential function $y = 10^x$ are inverse functions, the graph of $y = \log x$ can be obtained by reflecting the graph of $y = 10^x$ about the line $y = x$. See Figure 19.

| EXAMPLE 7 | **Graphing a Logarithmic Function and Its Inverse** |

(a) Find the domain of the logarithmic function $f(x) = 3 \log(x - 1)$.
(b) Graph f.
(c) From the graph, determine the range and vertical asymptote of f.
(d) Find f^{-1}, the inverse of f.
(e) Find the domain and the range of f^{-1}.
(f) Graph f^{-1}.

Solution

(a) The domain of f consists of all x for which $x - 1 > 0$, or equivalently, $x > 1$. The domain of f is $\{x \mid x > 1\}$, or $(1, \infty)$ in interval notation.

(b) To obtain the graph of $y = 3 \log(x - 1)$, begin with the graph of $y = \log x$ and use transformations. See Figure 20.

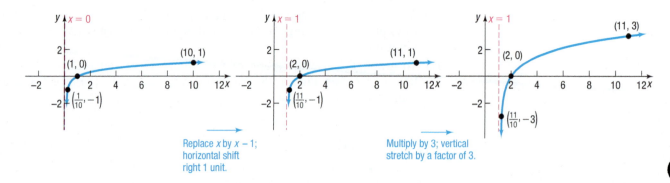

Figure 20

(a) $y = \log x$ **(b)** $y = \log(x - 1)$ **(c)** $y = 3 \log(x - 1)$

(c) The range of $f(x) = 3 \log(x - 1)$ is the set of all real numbers. The vertical asymptote is $x = 1$.

(d) Begin with $y = 3 \log(x - 1)$. The inverse function is defined implicitly by the equation

$$x = 3 \log(y - 1)$$

Solve for y.

$$\frac{x}{3} = \log(y - 1) \qquad \text{Isolate the logarithm.}$$

$$10^{x/3} = y - 1 \qquad \text{Change to exponential form.}$$

$$y = 10^{x/3} + 1 \qquad \text{Solve for } y.$$

The inverse of f is $f^{-1}(x) = 10^{x/3} + 1$.

(e) The domain of f^{-1} is the range of f, which is the set of all real numbers, from part (c). The range of f^{-1} is the domain of f, which is $(1, \infty)$ in interval notation.

(f) To graph f^{-1}, use the graph of f in Figure 20(c) and reflect it about the line $y = x$. See Figure 21.

Figure 21

We could also graph $f^{-1}(x) = 10^{x/3} + 1$ using transformations.

Now Work PROBLEM 81

5 Solve Logarithmic Equations

Equations that contain logarithms are called **logarithmic equations**. Care must be taken when solving logarithmic equations algebraically. In the expression $\log_a M$, remember that a and M are positive and $a \neq 1$. Be sure to check each apparent solution in the original equation and discard any that are extraneous.

Some logarithmic equations can be solved by changing the logarithmic equation to exponential form using the fact that $y = \log_a x$ means $a^y = x$.

EXAMPLE 8   **Solving Logarithmic Equations**

Solve:

(a) $\log_3(4x - 7) = 2$ (a) $\log_x 64 = 2$

Solution   (a) To solve, change the logarithmic equation to exponential form.

$$\log_3(4x - 7) = 2$$
$$4x - 7 = 3^2 \quad \text{Change to exponential form.}$$
$$4x - 7 = 9$$
$$4x = 16$$
$$x = 4$$

✓ **Check:** $\log_3(4x - 7) = \log_3(4 \cdot 4 - 7) = \log_3 9 = 2 \quad 3^2 = 9$
The solution set is $\{4\}$.

(b) To solve, change the logarithmic equation to exponential form.

$$\log_x 64 = 2$$
$$x^2 = 64 \qquad\qquad \text{Change to exponential form.}$$
$$x = \pm\sqrt{64} = \pm 8 \quad \text{Use the Square Root Method.}$$

Because the base of a logarithm must be positive, discard -8. Check the potential solution 8.

✓ **Check:** $\log_8 64 = 2 \quad 8^2 = 64$

The solution set is $\{8\}$.

EXAMPLE 9   **Using Logarithms to Solve an Exponential Equation**

Solve: $e^{2x} = 5$

Solution   To solve, change the exponential equation to logarithmic form.

$$e^{2x} = 5$$
$$\ln 5 = 2x \qquad \text{Change to logarithmic form.}$$
$$x = \frac{\ln 5}{2} \qquad \text{Exact solution}$$
$$\approx 0.805 \qquad \text{Approximate solution}$$

The solution set is $\left\{\dfrac{\ln 5}{2}\right\}$.

Now Work PROBLEMS **89** AND **101**

> **EXAMPLE 10**   **Alcohol and Driving**

Blood alcohol concentration (BAC) is a measure of the amount of alcohol in a person's bloodstream. A BAC of 0.04% means that a person has 4 parts alcohol per 10,000 parts blood in the body. Relative risk is defined as the likelihood of one event occurring divided by the likelihood of a second event occurring. For example, if an individual with a BAC of 0.02% is 1.4 times as likely to have a car accident as an individual who has not been drinking, the relative risk of an accident with a BAC of 0.02% is 1.4. Recent medical research suggests that the relative risk R of having an accident while driving a car can be modeled by an equation of the form

$$R = e^{kx}$$

where x is the percent concentration of alcohol in the bloodstream and k is a constant.

(a) Research indicates that the relative risk of a person having an accident with a BAC of 0.02% is 1.4. Find the constant k in the equation.

(b) Using this value of k, what is the relative risk if the BAC is 0.17%?

(c) Using this same value of k, what BAC corresponds to a relative risk of 100?

(d) If the law asserts that anyone with a relative risk of 4 or more should not have driving privileges, at what concentration of alcohol in the bloodstream should a driver be arrested and charged with DUI (driving under the influence)?

Solution   (a) For a blood alcohol concentration of 0.02% and a relative risk R of 1.4, let $x = 0.02$ and solve for k.

$$R = e^{kx}$$

$$1.4 = e^{k \cdot 0.02}$$ $R = 1.4; x = 0.02$

$$0.02k = \ln 1.4$$ **Change to a logarithmic statement.**

$$k = \frac{\ln 1.4}{0.02} \approx 16.82$$ **Solve for k.**

(b) A BAC of 0.17% means $x = 0.17$. Use $k = 16.82$ in the equation to find the relative risk R:

$$R = e^{kx} = e^{16.82 \cdot 0.17} \approx 17.5$$

For a blood alcohol concentration of 0.17%, the relative risk R of an accident is about 17.5. That is, a person with a BAC of 0.17% is 17.5 times as likely to have a car accident as a person with no alcohol in the bloodstream.

(c) A relative risk of 100 means $R = 100$. Use $k = 16.82$ in the equation $R = e^{kx}$. The blood alcohol concentration x obeys

$$100 = e^{16.82x}$$ $R = e^{kx}, R = 100, k = 16.82$

$$16.82x = \ln 100$$ **Change to a logarithmic statement.**

$$x = \frac{\ln 100}{16.82} \approx 0.27$$ **Solve for x.**

NOTE A BAC of 0.30% results in a loss of consciousness in most people. ∎

For a blood alcohol concentration of 0.27%, the relative risk R of an accident is 100.

(d) A relative risk of 4 means $R = 4$. Use $k = 16.82$ in the equation $R = e^{kx}$. The blood alcohol concentration x obeys

$$4 = e^{16.82x}$$

$$16.82x = \ln 4$$

$$x = \frac{\ln 4}{16.82} \approx 0.082$$

NOTE In most states, the blood alcohol content at which a DUI citation is given is 0.08%. ∎

A driver with a BAC of 0.082% or more should be arrested and charged with DUI.

SUMMARY

Properties of the Logarithmic Function

$f(x) = \log_a x, \quad a > 1$

$(y = \log_a x$ if and only if $x = a^y)$

- Domain: the interval $(0, \infty)$; Range: the interval $(-\infty, \infty)$
- x-intercept: 1; y-intercept: none; vertical asymptote: $x = 0$ (y-axis); increasing; one-to-one
- The graph of f is smooth and continuous. It contains the points $\left(\dfrac{1}{a}, -1\right)$, $(1, 0)$, and $(a, 1)$.
- See Figure 22(a) for a typical graph.

$f(x) = \log_a x, \quad 0 < a < 1$

$(y = \log_a x$ if and only if $x = a^y)$

- Domain: the interval $(0, \infty)$; Range: the interval $(-\infty, \infty)$
- x-intercept: 1; y-intercept: none; vertical asymptote: $x = 0$ (y-axis); decreasing; one-to-one
- The graph of f is smooth and continuous. It contains the points $(a, 1)$, $(1, 0)$, and $\left(\dfrac{1}{a}, -1\right)$.
- See Figure 22(b) for a typical graph.

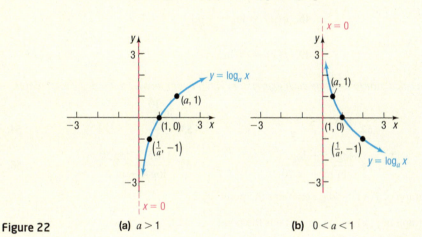

Figure 22   (a) $a > 1$ (b) $0 < a < 1$

7.2 Assess Your Understanding

'Are You Prepared?' *Answers are given at the end of these exercises. If you get a wrong answer, read the pages listed in* red.

1. Solve:
   (a) $3x - 7 \le 8 - 2x$ (pp. A49–A52)
   (b) $2x + 3 = 15$ (pp. A27–A34)

2. *True or False* A one-to-one function $y = f(x)$ has an inverse function that is defined by the equation $x = f(y)$. (pp. 81–88)

Concepts and Vocabulary

3. The domain of the logarithmic function $f(x) = \log_a x$ is _____.

4. The graph of every logarithmic function $f(x) = \log_a x$, where $a > 0$ and $a \ne 1$, contains the three points: _____, _____, and _____.

5. If the graph of a logarithmic function $f(x) = \log_a x$, where $a > 0$ and $a \ne 1$, is increasing, then its base is larger than _____.

6. *True or False* If $y = \log_a x$, then $y = a^x$.

7. *True or False* The graph of $f(x) = \log_a x$, where $a > 0$ and $a \ne 1$, has an x-intercept equal to 1 and no y-intercept.

8. *Multiple Choice* Select the answer that completes the statement: $y = \ln x$ if and only if _____.
   (a) $x = e^y$ (b) $y = e^x$ (c) $x = 10^y$ (d) $y = 10^x$

9. *Multiple Choice* The domain of $f(x) = \log_3(x + 2)$ is
   (a) $(-\infty, \infty)$ (b) $(2, \infty)$ (c) $(-2, \infty)$ (d) $(0, \infty)$

10. *Multiple Choice* $\log_3 81$ equals
    (a) 9   (b) 4   (c) 2   (d) 3

Skill Building

In Problems 11–18, change each exponential statement to an equivalent statement involving a logarithm.

11. $9 = 3^2$ **12.** $16 = 4^2$ **13.** $a^2 = 1.6$ **14.** $a^3 = 2.1$

15. $2^x = 7.2$ **16.** $3^x = 4.6$ **17.** $e^x = 8$ **18.** $e^{2.2} = M$

In Problems 19–26, change each logarithmic statement to an equivalent statement involving an exponent.

19. $\log_2 8 = 3$ **20.** $\log_3\left(\dfrac{1}{9}\right) = -2$ **21.** $\log_a 3 = 6$ **22.** $\log_b 4 = 2$

23. $\log_3 2 = x$ **24.** $\log_2 6 = x$ **25.** $\ln 4 = x$ **26.** $\ln x = 4$

In Problems 27–38, find the exact value of each logarithm without using a calculator.

27. $\log_2 1$ **28.** $\log_8 8$ **29.** $\log_7 49$ **30.** $\log_3\left(\dfrac{1}{9}\right)$

31. $\log_{1/5} 125$ **32.** $\log_{1/3} 9$ **33.** $\log \sqrt{10}$ **34.** $\log \sqrt[3]{100}$

35. $\log_{\sqrt{2}} 4$ **36.** $\log_{\sqrt{3}} 9$ **37.** $\ln \sqrt{e}$ **38.** $\ln e^3$

In Problems 39–50, find the domain of each function.

39. $f(x) = \ln(x - 3)$ **40.** $g(x) = \ln(x - 1)$ **41.** $F(x) = \log_2 x^2$

42. $H(x) = \log_5 x^3$ **43.** $f(x) = 3 - 2\log_4\left(\dfrac{x}{2} - 5\right)$ **44.** $g(x) = 8 + 5\ln(2x + 3)$

45. $f(x) = \ln\left(\dfrac{1}{x + 1}\right)$ **46.** $g(x) = \ln\left(\dfrac{1}{x - 5}\right)$ **47.** $g(x) = \log_5\left(\dfrac{x + 1}{x}\right)$

48. $h(x) = \log_3\left(\dfrac{x}{x - 1}\right)$ **49.** $f(x) = \sqrt{\ln x}$ **50.** $g(x) = \dfrac{1}{\ln x}$

In Problems 51–58, use a calculator to evaluate each expression. Round your answer to three decimal places.

51. $\ln \dfrac{5}{3}$ **52.** $\dfrac{\ln 5}{3}$ **53.** $\dfrac{\ln \dfrac{10}{3}}{0.04}$ **54.** $\dfrac{\ln \dfrac{2}{3}}{-0.1}$

55. $\dfrac{\ln 4 + \ln 2}{\log 4 + \log 2}$ **56.** $\dfrac{\log 15 + \log 20}{\ln 15 + \ln 20}$ **57.** $\dfrac{2\ln 5 + \log 50}{\log 4 - \ln 2}$ **58.** $\dfrac{3\log 80 - \ln 5}{\log 5 + \ln 20}$

59. Find a so that the graph of $f(x) = \log_a x$ contains the point $(2, 2)$.

60. Find a so that the graph of $f(x) = \log_a x$ contains the point $\left(\dfrac{1}{2}, -4\right)$.

In Problems 61–64, graph each function and its inverse on the same set of axes.

61. $f(x) = 3^x; f^{-1}(x) = \log_3 x$ **62.** $f(x) = 4^x; f^{-1}(x) = \log_4 x$

63. $f(x) = \left(\dfrac{1}{2}\right)^x; f^{-1}(x) = \log_{1/2} x$ **64.** $f(x) = \left(\dfrac{1}{3}\right)^x; f^{-1}(x) = \log_{1/3} x$

In Problems 65–72, the graph of a logarithmic function is given. Match each graph to one of the following functions:

(A) $y = \log_3 x$ (B) $y = \log_3(-x)$ (C) $y = -\log_3 x$ (D) $y = -\log_3(-x)$

(E) $y = \log_3 x - 1$ (F) $y = \log_3(x - 1)$ (G) $y = \log_3(1 - x)$ (H) $y = 1 - \log_3 x$

65.

66.

67.

68.

69.

70.

71.

72.

In Problems 73–88, use the given function f.
(a) *Find the domain of f.*   (b) *Graph f.*   (c) *From the graph, determine the range and any asymptotes of f.*
(d) *Find f^{-1}, the inverse of f.*   (e) *Find the domain and the range of f^{-1}.*   (f) *Graph f^{-1}.*

73. $f(x) = \ln(x + 4)$ **74.** $f(x) = \ln(x - 3)$ **75.** $f(x) = 2 + \ln x$ **76.** $f(x) = -\ln(-x)$

77. $f(x) = \ln(2x) - 3$ **78.** $f(x) = -2\ln(x + 1)$ **79.** $f(x) = \log(x - 4) + 2$ **80.** $f(x) = \frac{1}{2}\log x - 5$

81. $f(x) = \frac{1}{2}\log(2x)$ **82.** $f(x) = \log(-2x)$ **83.** $f(x) = 3 + \log_3(x + 2)$ **84.** $f(x) = 2 - \log_3(x + 1)$

85. $f(x) = e^{x+2} - 3$ **86.** $f(x) = 3e^x + 2$ **87.** $f(x) = 2^{x/3} + 4$ **88.** $f(x) = -3^{x+1}$

In Problems 89–112, solve each equation.

89. $\log_3 x = 2$ **90.** $\log_5 x = 3$ **91.** $\log_2(3x + 4) = 5$ **92.** $\log_3(3x - 2) = 2$

93. $\log_x 16 = 2$ **94.** $\log_x\left(\frac{1}{8}\right) = 3$ **95.** $\ln e^x = 5$ **96.** $\ln e^{-2x} = 8$

97. $\log_4 64 = x$ **98.** $\log_5 625 = x$ **99.** $\log_3 243 = 2x + 1$ **100.** $\log_6 36 = 5x + 3$

101. $e^{3x} = 10$ **102.** $e^{-2x} = \frac{1}{3}$ **103.** $e^{2x+5} = 8$ **104.** $e^{-2x+1} = 13$

105. $\log_7(x^2 + 4) = 2$ **106.** $\log_5(x^2 + x + 4) = 2$ **107.** $\log_2 8^x = -6$ **108.** $\log_3 3^x = -1$

109. $5e^{0.2x} = 7$ **110.** $8 \cdot 10^{2x-7} = 3$ **111.** $2 \cdot 10^{2-x} = 5$ **112.** $4e^{x+1} = 5$

113. Suppose that $G(x) = \log_3(2x + 1) - 2$.

   (a) What is the domain of G?
   (b) What is $G(40)$? What point is on the graph of G?
   (c) If $G(x) = 3$, what is x? What point is on the graph of G?
   (d) What is the zero of G?

114. Suppose that $F(x) = \log_2(x + 1) - 3$.

   (a) What is the domain of F?
   (b) What is $F(7)$? What point is on the graph of F?
   (c) If $F(x) = -1$, what is x? What point is on the graph of F?
   (d) What is the zero of F?

Mixed Practice In Problems 115–118, graph each function. Based on the graph, state the domain and the range, and find any intercepts.

115. $f(x) = \begin{cases} \ln(-x) & \text{if } x < 0 \\ \ln x & \text{if } x > 0 \end{cases}$ **116.** $f(x) = \begin{cases} \ln(-x) & \text{if } x \le -1 \\ -\ln(-x) & \text{if } -1 < x < 0 \end{cases}$

117. $f(x) = \begin{cases} -\ln x & \text{if } 0 < x < 1 \\ \ln x & \text{if } x \ge 1 \end{cases}$ **118.** $f(x) = \begin{cases} \ln x & \text{if } 0 < x < 1 \\ -\ln x & \text{if } x \ge 1 \end{cases}$

Applications and Extensions

119. Chemistry The pH of a chemical solution is given by the formula

$$\text{pH} = -\log_{10}[\text{H}^+]$$

where $[\text{H}^+]$ is the concentration of hydrogen ions in moles per liter. Values of pH range from 0 (acidic) to 14 (alkaline).

   (a) What is the pH of a solution for which $[\text{H}^+]$ is 0.1?
   (b) What is the pH of a solution for which $[\text{H}^+]$ is 0.01?
   (c) What is the pH of a solution for which $[\text{H}^+]$ is 0.001?
   (d) What happens to pH as the hydrogen ion concentration decreases?
   (e) Determine the hydrogen ion concentration of an orange (pH = 3.5).
   (f) Determine the hydrogen ion concentration of human blood (pH = 7.4).

120. Diversity Index Shannon's diversity index is a measure of the diversity of a population. The diversity index is given by the formula

$$H = -(p_1 \log p_1 + p_2 \log p_2 + \cdots + p_n \log p_n)$$

where p_1 is the proportion of the population that is species 1, p_2 is the proportion of the population that is species 2, and so on. In this problem, the population is people in the United States and the species is race.

   (a) According to the U.S. Census Bureau, the distribution of race in the United States in 2015 was:

Race	Proportion
White	0.617
Black or African American	0.124
American Indian and Alaska Native	0.007
Asian	0.053
Native Hawaiian and Other Pacific Islander	0.002
Hispanic	0.177
Two or More Races	0.020

Source: U.S. Census Bureau

Compute the diversity index of the United States in 2015.

   (b) The largest value of the diversity index is given by $H_{max} = \log(S)$, where S is the number of categories of race. Compute H_{max}.

   (c) The **evenness ratio** is given by $E_H = \dfrac{H}{H_{max}}$, where $0 \le E_H \le 1$. If $E_H = 1$, there is complete evenness. Compute the evenness ratio for the United States.

   (d) Obtain the distribution of race for the United States in 2010 from the Census Bureau. Compute Shannon's diversity index. Is the United States becoming more diverse? Why?

121. Atmospheric Pressure The atmospheric pressure p on an object decreases with increasing height. This pressure, measured in millimeters of mercury, is related to the height h (in kilometers) above sea level by the function

$$p(h) = 760e^{-0.145h}$$

(a) Find the height of an aircraft if the atmospheric pressure is 320 millimeters of mercury.
(b) Find the height of a mountain if the atmospheric pressure is 667 millimeters of mercury.

122. Healing of Wounds The normal healing of wounds can be modeled by an exponential function. If A_0 represents the original area of the wound, and if A equals the area of the wound, then the function

$$A(n) = A_0e^{-0.35n}$$

describes the area of a wound after n days following an injury when no infection is present to retard the healing. Suppose that a wound initially had an area of 100 square millimeters.

(a) If healing is taking place, after how many days will the wound be one-half its original size?
(b) How long before the wound is 10% of its original size?

123. Exponential Probability Between 12:00 PM and 1:00 PM, cars arrive at Citibank's drive-thru at the rate of 6 cars per hour (0.1 car per minute). The following formula from probability can be used to determine the probability that a car will arrive within t minutes of 12:00 PM.

$$F(t) = 1 - e^{-0.1t}$$

(a) Determine how many minutes are needed for the probability to reach 50%.
(b) Determine how many minutes are needed for the probability to reach 80%.
(c) Is it possible for the probability to equal 100%? Explain.

124. Exponential Probability Between 5:00 PM and 6:00 PM, cars arrive at Jiffy Lube at the rate of 9 cars per hour (0.15 car per minute). The following formula from probability can be used to determine the probability that a car will arrive within t minutes of 5:00 PM.

$$F(t) = 1 - e^{-0.15t}$$

(a) Determine how many minutes are needed for the probability to reach 50%.
(b) Determine how many minutes are needed for the probability to reach 80%.

125. Drug Medication The function

$$D(h) = 5e^{-0.4h}$$

can be used to find the number of milligrams D of a certain drug that is in a patient's bloodstream h hours after the drug was administered. When the number of milligrams reaches 2, the drug is to be administered again. What is the time between injections?

126. Spreading of Rumors A model for the number N of people in a college community who have heard a certain rumor is

$$N(d) = P(1 - e^{-0.15d})$$

where P is the total population of the community and d is the number of days that have elapsed since the rumor began. In a community of 1000 students, how many days will elapse before 450 students have heard the rumor?

127. Current in an RL Circuit The equation governing the amount of current I (in amperes) after time t (in seconds) in a simple RL circuit consisting of a resistance R (in ohms), an inductance L (in henrys), and an electromotive force E (in volts) is

$$I = \frac{E}{R}[1 - e^{-(R/L)t}]$$

If $E = 12$ volts, $R = 10$ ohms, and $L = 5$ henrys, how long does it take to obtain a current of 0.5 ampere? Of 1.0 ampere? Graph the equation.

128. Learning Curve Psychologists sometimes use the function

$$L(t) = A(1 - e^{-kt})$$

to measure the amount L learned at time t. Here A represents the amount to be learned, and the number k measures the rate of learning. Suppose that a student has an amount A of 200 vocabulary words to learn. A psychologist determines that the student has learned 20 vocabulary words after 5 minutes.

(a) Determine the rate of learning k.
(b) Approximately how many words will the student have learned after 10 minutes?
(c) After 15 minutes?
(d) How long does it take for the student to learn 180 words?

Loudness of Sound Problems 129–132 use the following discussion: The **loudness** $L(x)$, measured in decibels (dB), of a sound of intensity x, measured in watts per square meter, is defined as $L(x) = 10 \log \dfrac{x}{I_0}$, where $I_0 = 10^{-12}$ watt per square meter is the least intense sound that a human ear can detect. Determine the loudness, in decibels, of each of the following sounds.

129. Normal conversation: intensity of $x = 10^{-7}$ watt per square meter.

130. Amplified rock music: intensity of 10^{-1} watt per square meter.

131. Heavy city traffic: intensity of $x = 10^{-3}$ watt per square meter.

132. Diesel truck traveling 40 miles per hour 50 feet away: intensity 10 times that of a passenger car traveling 50 miles per hour 50 feet away, whose loudness is 70 decibels.

The Richter Scale Problems 133 and 134 use the following discussion: The **Richter scale** is one way of converting seismographic readings into numbers that provide an easy reference for measuring the magnitude M of an earthquake. All earthquakes are compared to a **zero-level earthquake** whose seismographic reading measures 0.001 millimeter at a distance of 100 kilometers from the epicenter. An earthquake whose seismographic reading measures x millimeters has **magnitude** M(x), given by

$$M(x) = \log\left(\frac{x}{x_0}\right)$$

where $x_0 = 10^{-3}$ is the reading of a zero-level earthquake the same distance from its epicenter. In Problems 133 and 134, determine the magnitude of each earthquake.

133. Magnitude of an Earthquake Mexico City in 1985: seismographic reading of 125,892 millimeters 100 kilometers from the center

134. Magnitude of an Earthquake San Francisco in 1906: seismographic reading of 50,119 millimeters 100 kilometers from the center

135. Alcohol and Driving The concentration of alcohol in a person's bloodstream is measurable. Suppose that the relative risk R of having an accident while driving a car can be modeled by an equation of the form

$$R = e^{kx}$$

where x is the percent concentration of alcohol in the bloodstream and k is a constant.
   (a) Suppose that a concentration of alcohol in the bloodstream of 0.03 percent results in a relative risk of an accident of 1.4. Find the constant k in the equation.
   (b) Using this value of k, what is the relative risk if the concentration is 0.17 percent?
   (c) Using the same value of k, what concentration of alcohol corresponds to a relative risk of 100?
   (d) If the law asserts that anyone with a relative risk of having an accident of 5 or more should not have driving privileges, at what concentration of alcohol in the bloodstream should a driver be arrested and charged with a DUI?
   (e) Compare this situation with that of Example 10. If you were a lawmaker, which situation would you support? Give your reasons.

136. The Marriage Problem There is an infamous problem from mathematics that attempts to quantify the number of potential mates one should date before choosing one's "true love." The function

$$L(x) = -x \ln x$$

represents the probability of finding the ideal mate after rejecting the first x proportion of potential mates. For example, if you reject the first 20% = 0.20 of individuals you date, the probability of finding the ideal mate is $L(0.2) \approx 0.322$. So, if you want the probability of finding the ideal mate to be greater than 0.332 and you are only willing to date up to 20 individuals, you should reject the first 0.2(20) = 4 individuals before attempting to decide on the ideal mate. Presumably, you are using those first 4 individuals to help you decide which traits you value in a mate.
   (a) Determine and interpret L(0.1).
   (b) Determine and interpret L(0.6).
   (c) What is the domain of L?
   (d) Graph L = L(x) over the domain.
   (e) Judging on the basis of the approach suggested by the model, what is the value of x that maximizes L? What is the highest probability of finding the ideal mate?

137. Challenge Problem Solve: $\log_6(\log_2 x) = 1$

138. Challenge Problem Solve: $\log_2[\log_4(\log_3 x)] = 0$

139. Challenge Problem Solve: $\log_3 9^{2x+3} = x^2 + 1$

Explaining Concepts: Discussion and Writing

140. Is there any function of the form $y = x^\alpha, 0 < \alpha < 1$, that increases more slowly than a logarithmic function whose base is greater than 1? Explain.

141. In the definition of the logarithmic function, the base a is not allowed to equal 1. Why?

142. Critical Thinking In buying a new car, one consideration might be how well the price of the car holds up over time. Different makes of cars have different depreciation rates. One way to compute a depreciation rate for a car is given here. Suppose that the current prices of a certain automobile are as shown in the table.

Age in Years					
New	1	2	3	4	5
$38,000	$36,600	$32,400	$28,750	$25,400	$21,200

Use the formula New = Old(e^{Rt}) to find R, the annual depreciation rate, for a specific time t. When might be the best time to trade in the car? Consult the NADA ("blue") book and compare two like models that you are interested in. Which has the better depreciation rate?

Retain Your Knowledge

Problems 143–152 are based on material learned earlier in the course. The purpose of these problems is to keep the material fresh in your mind so that you are better prepared for the final exam.

143. Find the real zeros of $g(x) = 4x^4 - 37x^2 + 9$. What are the x-intercepts of the graph of g?

144. Find the average rate of change of $f(x) = 9^x$ from $\frac{1}{2}$ to 1.

145. Use the Intermediate Value Theorem to show that the function $f(x) = 4x^3 - 2x^2 - 7$ has a real zero in the interval $[1, 2]$.

146. A complex polynomial function f of degree 4 with real coefficients has the zeros $-1, 2$, and $3 - i$. Find the remaining zero(s) of f. Then find a polynomial function that has the zeros.

147. Solve: $2x^2 - 7x - 1 = 0$

148. Find an equation of the line that contains the points $(0, 1)$ and $(8, -4)$. Write the equation in general form.

149. Solve: $|2x + 17| = 45$

150. Forensic Science The relationship between the height H of an adult female and the length x of her tibia, in centimeters, is estimated by the linear model $H(x) = 2.90x + 61.53$. If incomplete skeletal remains of an adult female include a tibia measuring 30.9 centimeters, estimate the height of the female. Round to the nearest tenth.

151. For $f(x) = x^3$, find $\dfrac{f(x) - f(2)}{x - 2}$.

152. Factor completely:
$$(x + 5)^4 \cdot 7(x - 3)^6 + (x - 3)^7 \cdot 4(x + 5)^3$$

'Are You Prepared?' Answers

1. (a) $x \leq 3$ **(b)** $x = 6$ **2.** True

7.3 Properties of Logarithms

OBJECTIVES **1** Work with the Properties of Logarithms (p. 490)

2 Write a Logarithmic Expression as a Sum or Difference of Logarithms (p. 492)

3 Write a Logarithmic Expression as a Single Logarithm (p. 493)

4 Evaluate Logarithms Whose Base Is Neither 10 Nor e (p. 494)

1 Work with the Properties of Logarithms

Logarithms have some very useful properties that can be derived directly from the definition and the laws of exponents.

EXAMPLE 1   **Establishing Properties of Logarithms**

(a) Show that $\log_a 1 = 0$.   (b) Show that $\log_a a = 1$.

Solution   (a) This fact was established when we graphed $y = \log_a x$ (see Figure 13 on page 479). To show the result algebraically, let $y = \log_a 1$. Then

$$y = \log_a 1$$
$$a^y = 1 \qquad \text{\textcolor{blue}{Change to exponential form.}}$$
$$a^y = a^0 \qquad \textcolor{blue}{a^0 = 1 \text{ since } a > 0, a \neq 1}$$
$$y = 0 \qquad \text{\textcolor{blue}{Equate exponents.}}$$
$$\log_a 1 = 0 \qquad \textcolor{blue}{y = \log_a 1}$$

(b) Let $y = \log_a a$. Then

$$y = \log_a a$$
$$a^y = a \qquad \text{\textcolor{blue}{Change to exponential form.}}$$
$$a^y = a^1 \qquad \textcolor{blue}{a = a^1}$$
$$y = 1 \qquad \text{\textcolor{blue}{Equate exponents.}}$$
$$\log_a a = 1 \qquad \textcolor{blue}{y = \log_a a}$$

To summarize:

$$\log_a 1 = 0 \qquad \log_a a = 1$$

THEOREM **Properties of Logarithms**

In these properties, M and a are positive real numbers, $a \neq 1$, and r is any real number.

- The number $\log_a M$ is the exponent to which a must be raised to obtain M. That is,

$$a^{\log_a M} = M \tag{1}$$

- The logarithm with base a of a raised to a power equals that power. That is,

$$\log_a a^r = r \tag{2}$$

The proof uses the fact that $y = a^x$ and $y = \log_a x$ are inverse functions.

Proof of Property (1) For inverse functions,

$$f(f^{-1}(x)) = x \quad \text{for all } x \text{ in the domain of } f^{-1}$$

Use $f(x) = a^x$ and $f^{-1}(x) = \log_a x$ to find

$$f(f^{-1}(x)) = a^{\log_a x} = x \quad \text{for } x > 0$$

Now let $x = M$ to obtain $a^{\log_a M} = M$, where $M > 0$. ∎

Proof of Property (2) For inverse functions,

$$f^{-1}(f(x)) = x \quad \text{for all } x \text{ in the domain of } f$$

Use $f(x) = a^x$ and $f^{-1}(x) = \log_a x$ to find

$$f^{-1}(f(x)) = \log_a a^x = x \quad \text{for all real numbers } x$$

Now let $x = r$ to obtain $\log_a a^r = r$, where r is any real number. ∎

EXAMPLE 2   **Using Properties of Logarithms (1) and (2)**

(a) $2^{\log_2 \pi} = \pi$ (b) $\log_{0.2} 0.2^{-\sqrt{2}} = -\sqrt{2}$ (c) $\ln e^{kt} = kt$

Now Work PROBLEM 15

Other useful properties of logarithms are given next.

THEOREM **Properties of Logarithms**

In these properties, M, N, and a are positive real numbers, $a \neq 1$, and r is any real number.

The Log of a Product Equals the Sum of the Logs

$$\log_a(MN) = \log_a M + \log_a N \tag{3}$$

The Log of a Quotient Equals the Difference of the Logs

$$\log_a\left(\frac{M}{N}\right) = \log_a M - \log_a N \tag{4}$$

The Log of a Power Equals the Product of the Power and the Log

$$\log_a M^r = r \log_a M \tag{5}$$

$$a^r = e^{r \ln a} \tag{6}$$

We prove properties (3), (5), and (6) and leave the proof of property (4) as an exercise (see Problem 109).

Proof of Property (3) Let $A = \log_a M$ and let $B = \log_a N$. These expressions are equivalent to the exponential expressions

$$a^A = M \quad \text{and} \quad a^B = N$$

Now

$$
\begin{aligned}
\log_a(MN) = \log_a(a^A a^B) = \log_a a^{A+B} &\qquad \text{\color{blue}Law of Exponents} \\
= A + B &\qquad \text{\color{blue}Property (2) of logarithms} \\
= \log_a M + \log_a N
\end{aligned}
$$

∎

Proof of Property (5) Let $A = \log_a M$. This expression is equivalent to

$$a^A = M$$

Now

$$
\begin{aligned}
\log_a M^r = \log_a(a^A)^r = \log_a a^{rA} &\qquad \text{\color{blue}Law of Exponents} \\
= rA &\qquad \text{\color{blue}Property (2) of logarithms} \\
= r\log_a M
\end{aligned}
$$

∎

Proof of Property (6) Property (1), with $a = e$, gives

$$e^{\ln M} = M$$

Now let $M = a^r$ and use property (5).

$$e^{\ln a^r} = e^{r \ln a} = a^r$$

∎

⟶ **Now Work** PROBLEM 19

2 Write a Logarithmic Expression as a Sum or Difference of Logarithms

 Logarithms can be used to transform products into sums, quotients into differences, and powers into factors. Such transformations are useful in certain calculus problems.

EXAMPLE 3 **Writing a Logarithmic Expression as a Sum of Logarithms**

Write $\log_a(x\sqrt{x^2 + 1})$, $x > 0$, as a sum of logarithms. Express all powers as factors.

Solution
$$
\begin{aligned}
\log_a(x\sqrt{x^2 + 1}) &= \log_a x + \log_a \sqrt{x^2 + 1} &\qquad \text{\color{blue}$\log_a(M \cdot N) = \log_a M + \log_a N$} \\
&= \log_a x + \log_a(x^2 + 1)^{1/2} \\
&= \log_a x + \frac{1}{2}\log_a(x^2 + 1) &\qquad \text{\color{blue}$\log_a M^r = r\log_a M$}
\end{aligned}
$$

EXAMPLE 4 **Writing a Logarithmic Expression as a Difference of Logarithms**

Write

$$\ln \frac{x^2}{(x-1)^3} \qquad x > 1$$

as a difference of logarithms. Express all powers as factors.

Solution
$$\ln \frac{x^2}{(x-1)^3} = \ln x^2 - \ln(x-1)^3 = 2\ln x - 3\ln(x-1)$$

$$\underset{\color{blue}\log_a\left(\dfrac{M}{N}\right) = \log_a M - \log_a N}{\uparrow} \qquad \underset{\color{blue}\log_a M^r = r\log_a M}{\uparrow}$$

EXAMPLE 5

Writing a Logarithmic Expression as a Sum and Difference of Logarithms

Write

$$\log_a \frac{\sqrt{x^2 + 1}}{x^3(x + 1)^4} \quad x > 0$$

as a sum and difference of logarithms. Express all powers as factors.

Solution

$$\log_a \frac{\sqrt{x^2 + 1}}{x^3(x + 1)^4} = \log_a \sqrt{x^2 + 1} - \log_a[x^3(x + 1)^4] \qquad \text{Property (4)}$$

$$= \log_a \sqrt{x^2 + 1} - [\log_a x^3 + \log_a(x + 1)^4] \qquad \text{Property (3)}$$

$$= \log_a(x^2 + 1)^{1/2} - \log_a x^3 - \log_a(x + 1)^4$$

$$= \frac{1}{2}\log_a(x^2 + 1) - 3\log_a x - 4\log_a(x + 1) \qquad \text{Property (5)}$$

WARNING In using properties (3) through (5), be careful about the values that the variable may assume. For example, the domain of the variable for $\log_a x$ is $x > 0$ and for $\log_a(x - 1)$ is $x > 1$. If these expressions are added, the domain is $x > 1$. That is, $\log_a x + \log_a(x - 1) = \log_a[x(x - 1)]$ is true only for $x > 1$. ∎

Now Work PROBLEM 51

3 Write a Logarithmic Expression as a Single Logarithm

Another use of properties (3) through (5) is to write sums and/or differences of logarithms with the same base as a single logarithm. This skill is needed to solve certain logarithmic equations discussed in the next section.

EXAMPLE 6

Writing Expressions as a Single Logarithm

Write each of the following as a single logarithm.

(a) $\log_a 7 + 4\log_a 3$ (b) $\dfrac{2}{3}\ln 8 - \ln(5^2 - 1)$

(c) $\log_a x + \log_a 9 + \log_a(x^2 + 1) - \log_a 5$

Solution

(a) $\log_a 7 + 4\log_a 3 = \log_a 7 + \log_a 3^4 \qquad r\log_a M = \log_a M^r$

$$= \log_a 7 + \log_a 81$$

$$= \log_a(7 \cdot 81) \qquad \log_a M + \log_a N = \log_a(M \cdot N)$$

$$= \log_a 567$$

(b) $\dfrac{2}{3}\ln 8 - \ln(5^2 - 1) = \ln 8^{2/3} - \ln(25 - 1) \qquad r\log_a M = \log_a M^r$

$$= \ln 4 - \ln 24 \qquad 8^{2/3} = (\sqrt[3]{8})^2 = 2^2 = 4$$

$$= \ln\left(\frac{4}{24}\right) \qquad \log_a M - \log_a N = \log_a\left(\frac{M}{N}\right)$$

$$= \ln\left(\frac{1}{6}\right)$$

$$= \ln 1 - \ln 6$$

$$= -\ln 6 \qquad \ln 1 = 0$$

(c) $\log_a x + \log_a 9 + \log_a(x^2 + 1) - \log_a 5 = \log_a(9x) + \log_a(x^2 + 1) - \log_a 5$

$$= \log_a[9x(x^2 + 1)] - \log_a 5$$

$$= \log_a\left[\frac{9x(x^2 + 1)}{5}\right]$$

WARNING A common error that some students make is to express the logarithm of a sum as the sum of logarithms.

$$\log_a(M + N) \quad \text{is not equal to} \quad \log_a M + \log_a N$$

Correct statement $\log_a(MN) = \log_a M + \log_a N$ **Property (3)**

Another common error is to express the difference of logarithms as the quotient of logarithms.

$$\log_a M - \log_a N \quad \text{is not equal to} \quad \frac{\log_a M}{\log_a N}$$

Correct statement $\log_a M - \log_a N = \log_a\left(\dfrac{M}{N}\right)$ **Property (4)**

A third common error is to express a logarithm raised to a power as the product of the power times the logarithm.

$$(\log_a M)^r \quad \text{is not equal to} \quad r\log_a M$$

Correct statement $\log_a M^r = r\log_a M$ **Property (5)** ■

Now Work PROBLEMS 57 AND 63

Two other important properties of logarithms are consequences of the fact that the logarithmic function $y = \log_a x$ is a one-to-one function.

THEOREM **Properties of Logarithms**

In these properties, M, N, and a are positive real numbers, $a \neq 1$.

- If $M = N$, then $\log_a M = \log_a N$. **(7)**
- If $\log_a M = \log_a N$, then $M = N$. **(8)**

Property (7) is used as follows: Starting with the equation $M = N$, "take the logarithm of both sides" to obtain $\log_a M = \log_a N$.

Properties (7) and (8) are useful for solving *exponential and logarithmic equations*, a topic discussed in the next section.

4 Evaluate Logarithms Whose Base Is Neither 10 Nor e

Logarithms with base 10—common logarithms—were used to facilitate arithmetic computations before the widespread use of calculators. (See the Historical Feature at the end of this section.) Natural logarithms—that is, logarithms whose base is the number e—remain very important because they arise frequently in the study of natural phenomena.

Common logarithms are usually abbreviated by writing **log**, with the base understood to be 10, just as natural logarithms are abbreviated by **ln**, with the base understood to be e.

Most calculators have both $\boxed{\log}$ and $\boxed{\ln}$ keys to calculate the common logarithm and the natural logarithm of a number, respectively. Let's look at an example to see how to approximate logarithms having a base other than 10 or e.

EXAMPLE 7 | **Approximating a Logarithm Whose Base Is Neither 10 Nor e**

Approximate $\log_2 7$. Round the answer to four decimal places.

Solution Remember, evaluating $\log_2 7$ means answering the question "2 raised to what exponent equals 7?" Let $y = \log_2 7$. Then $2^y = 7$. Because $2^2 = 4$ and $2^3 = 8$, the value of $\log_2 7$ is between 2 and 3.

$$2^y = 7$$
$$\ln 2^y = \ln 7 \qquad \textbf{Property (7)}$$
$$y \ln 2 = \ln 7 \qquad \textbf{Property (5)}$$
$$y = \frac{\ln 7}{\ln 2} \qquad \textbf{Exact value}$$
$$y \approx 2.8074 \qquad \textbf{Approximate value rounded to four decimal places}$$

Example 7 shows how to approximate a logarithm whose base is 2 by changing to logarithms involving the base e. In general, the *Change-of-Base Formula* is used.

> **THEOREM Change-of-Base Formula**
>
> If $a \neq 1, b \neq 1$, and M are positive real numbers, then
>
> $$\log_a M = \frac{\log_b M}{\log_b a} \qquad (9)$$

Proof Let $y = \log_a M$. Then

$$a^y = M$$
$$\log_b a^y = \log_b M \qquad \textbf{Property (7)}$$
$$y \log_b a = \log_b M \qquad \textbf{Property (5)}$$
$$y = \frac{\log_b M}{\log_b a} \qquad \textbf{Solve for y.}$$
$$\log_a M = \frac{\log_b M}{\log_b a} \qquad \textbf{y = log}_a\textbf{ M} \qquad ■$$

TECHNOLOGY NOTE Some calculators have features for evaluating logarithms with bases other than 10 or e. For example, the TI-84 Plus C has the logBASE function (under Math > Math > A: logBASE). Consult the user's manual for your calculator. ■

Because most calculators have keys for $\boxed{\log}$ and $\boxed{\ln}$, in practice, the Change-of-Base Formula uses either $b = 10$ or $b = e$. That is,

$$\log_a M = \frac{\log M}{\log a} \quad \text{and} \quad \log_a M = \frac{\ln M}{\ln a}$$

EXAMPLE 8 **Using the Change-of-Base Formula**

Approximate:

(a) $\log_5 89$ (b) $\log_{\sqrt{2}} \sqrt{5}$

Round answers to four decimal places.

Solution (a) $\log_5 89 = \dfrac{\log 89}{\log 5} \approx 2.7889$ (b) $\log_{\sqrt{2}} \sqrt{5} = \dfrac{\log \sqrt{5}}{\log \sqrt{2}} \approx 2.3219$

or   or

$\log_5 89 = \dfrac{\ln 89}{\ln 5} \approx 2.7889$ $\log_{\sqrt{2}} \sqrt{5} = \dfrac{\ln \sqrt{5}}{\ln \sqrt{2}} \approx 2.3219$

Now Work PROBLEMS **23** AND **71**

COMMENT We can use the Change-of-Base Formula to graph logarithmic functions when the base is different from e or 10. For example, to graph $y = \log_2 x$, graph either $y = \dfrac{\ln x}{\ln 2}$ or $y = \dfrac{\log x}{\log 2}$. ∎

— **Now Work** PROBLEM 79

SUMMARY

Properties of Logarithms

In the list that follows, $a, b, M, N,$ and r are real numbers. Also, $a > 0, a \neq 1, b > 0, b \neq 1, M > 0,$ and $N > 0$.

Definition $\qquad y = \log_a x$ if and only if $x = a^y$

Properties of logarithms
- $\log_a 1 = 0$ $\qquad$ • $\log_a a = 1$ $\qquad$ • $\log_a M^r = r \log_a M$
- $a^{\log_a M} = M$ $\qquad$ • $\log_a a^r = r$ $\qquad$ • $a^r = e^{r \ln a}$
- $\log_a (MN) = \log_a M + \log_a N$ $\qquad$ • $\log_a\left(\dfrac{M}{N}\right) = \log_a M - \log_a N$
- If $M = N$, then $\log_a M = \log_a N$. $\qquad$ • If $\log_a M = \log_a N$, then $M = N$.

Change-of-Base Formula $\qquad \log_a M = \dfrac{\log_b M}{\log_b a}$

Historical Feature

John Napier (1550–1617)

Logarithms were invented about 1590 by John Napier (1550-1617) and Joost Bürgi (1552-1632), working independently. Napier, whose work had the greater influence, was a Scottish lord, a secretive man whose neighbors were inclined to believe him to be in league with the devil. His approach to logarithms was very different from ours; it was based on the relationship between arithmetic and geometric sequences and not on the inverse function relationship of logarithms to exponential functions (described in Section 7.2).

Napier's tables, published in 1614, listed what would now be called *natural logarithms* of sines and were rather difficult to use. A London professor, Henry Briggs, became interested in the tables and visited Napier. In their conversations, they developed the idea of common logarithms, which were published in 1617. The importance of this tool for calculation was immediately recognized, and by 1650 common logarithms were being printed as far away as China. They remained an important calculation tool until the advent of inexpensive handheld calculators in about 1972, which has decreased their calculational–but not their theoretical–importance.

A side effect of the invention of logarithms was the popularization of the decimal system of notation for real numbers.

7.3 Assess Your Understanding

Concepts and Vocabulary

1. $\log_a 1 = $ _____
2. $a^{\log_a M} = $ _____
3. $\log_a a^r = $ _____
4. $\log_a (MN) = $ _____ + _____
5. $\log_a\left(\dfrac{M}{N}\right) = $ _____ − _____
6. $\log_a M^r = $ _____
7. If $\log_8 M = \dfrac{\log_5 7}{\log_5 8}$, then $M = $ _____ .
8. *True or False* $\ln(x+3) - \ln(2x) = \dfrac{\ln(x+3)}{\ln(2x)}$

9. *True or False* $\log_2(3x^4) = 4 \log_2(3x)$
10. *True or False* $\log\left(\dfrac{2}{3}\right) = \dfrac{\log 2}{\log 3}$
11. *Multiple Choice* Choose the expression equivalent to 2^x.
    (a) e^{2x} $\quad$ (b) $e^{x \ln 2}$ $\quad$ (c) $e^{\log_2 x}$ $\quad$ (d) $e^{2 \ln x}$
12. *Multiple Choice* Writing $\log_a x - \log_a y + 2 \log_a z$ as a single logarithm results in which of the following?
    (a) $\log_a (x - y + 2z)$ $\qquad$ (b) $\log_a\left(\dfrac{xz^2}{y}\right)$
    (c) $\log_a\left(\dfrac{2xz}{y}\right)$ $\qquad$ (d) $\log_a\left(\dfrac{x}{yz^2}\right)$

Skill Building

In Problems 13–28, use properties of logarithms to find the exact value of each expression. Do not use a calculator.

13. $\log_7 7^{29}$

14. $\log_2 2^{-13}$

15. $\ln e^{-4}$

16. $\ln e^{\sqrt{2}}$

17. $9^{\log_9 13}$

18. $e^{\ln 8}$

19. $\log_8 2 + \log_8 4$

20. $\log_6 9 + \log_6 4$

21. $\log_5 35 - \log_5 7$

22. $\log_8 16 - \log_8 2$

23. $\log_2 6 \cdot \log_6 8$

24. $\log_3 8 \cdot \log_8 9$

25. $4^{\log_4 6 - \log_4 5}$

26. $5^{\log_5 6 + \log_5 7}$

27. $e^{\log_{e^2} 16}$

28. $e^{\log_{e^2} 9}$

In Problems 29–36, suppose that $\ln 2 = a$ and $\ln 3 = b$. Use properties of logarithms to write each logarithm in terms of a and b.

29. $\ln 6$

30. $\ln \dfrac{2}{3}$

31. $\ln 1.5$

32. $\ln 0.5$

33. $\ln 8$

34. $\ln 27$

35. $\ln \sqrt[5]{6}$

36. $\ln \sqrt[4]{\dfrac{2}{3}}$

In Problems 37–56, write each expression as a sum and/or difference of logarithms. Express powers as factors.

37. $\log_6 36x$

38. $\log_3 \dfrac{x}{9}$

39. $\log_5 y^6$

40. $\log_7 x^5$

41. $\ln(ex)$

42. $\ln \dfrac{e}{x}$

43. $\ln \dfrac{x}{e^x}$

44. $\ln(xe^x)$

45. $\log_a(u^2 v^3) \quad u > 0, v > 0$

46. $\log_2\left(\dfrac{a}{b^2}\right) \quad a > 0, b > 0$

47. $\ln(x^2\sqrt{1-x}) \quad 0 < x < 1$

48. $\ln(x\sqrt{1+x^2}) \quad x > 0$

49. $\log_2\left(\dfrac{x^3}{x-3}\right) \quad x > 3$

50. $\log_5\left(\dfrac{\sqrt[3]{x^2+1}}{x^2-1}\right) \quad x > 1$

51. $\log\left[\dfrac{x(x+2)}{(x+3)^2}\right] \quad x > 0$

52. $\log\left[\dfrac{x^3\sqrt{x+1}}{(x-2)^2}\right] \quad x > 2$

53. $\ln\left[\dfrac{x^2-x-2}{(x+4)^2}\right]^{1/3} \quad x > 2$

54. $\ln\left[\dfrac{(x-4)^2}{x^2-1}\right]^{2/3} \quad x > 4$

55. $\ln \dfrac{5x\sqrt{1+3x}}{(x-4)^3} \quad x > 4$

56. $\ln\left[\dfrac{5x^2\sqrt[3]{1-x}}{4(x+1)^2}\right] \quad 0 < x < 1$

In Problems 57–70, write each expression as a single logarithm.

57. $3\log_5 u + 4\log_5 v$

58. $2\log_3 u - \log_3 v$

59. $\log_3 \sqrt{x} - \log_3 x^3$

60. $\log_2\left(\dfrac{1}{x}\right) + \log_2\left(\dfrac{1}{x^2}\right)$

61. $\log_4(x^2-1) - 5\log_4(x+1)$

62. $\log(x^2+3x+2) - 2\log(x+1)$

63. $\ln\left(\dfrac{x}{x-1}\right) + \ln\left(\dfrac{x+1}{x}\right) - \ln(x^2-1)$

64. $\log\left(\dfrac{x^2+2x-3}{x^2-4}\right) - \log\left(\dfrac{x^2+7x+6}{x+2}\right)$

65. $8\log_2\sqrt{3x-2} - \log_2\left(\dfrac{4}{x}\right) + \log_2 4$

66. $21\log_3\sqrt[3]{x} + \log_3(9x^2) - \log_3 9$

67. $2\log_a(5x^3) - \dfrac{1}{2}\log_a(2x+3)$

68. $\dfrac{1}{3}\log(x^3+1) + \dfrac{1}{2}\log(x^2+1)$

69. $2\log_2(x+1) - \log_2(x+3) - \log_2(x-1)$

70. $3\log_5(3x+1) - 2\log_5(2x-1) - \log_5 x$

In Problems 71–78, use the Change-of-Base Formula and a calculator to evaluate each logarithm. Round your answer to three decimal places.

71. $\log_3 21$

72. $\log_5 18$

73. $\log_{1/3} 71$

74. $\log_{1/2} 15$

75. $\log_{\sqrt{2}} 7$

76. $\log_{\sqrt{5}} 8$

77. $\log_\pi e$

78. $\log_\pi \sqrt{2}$

In Problems 79–84, graph each function using a graphing utility and the Change-of-Base Formula.

79. $y = \log_4 x$

80. $y = \log_5 x$

81. $y = \log_2(x+2)$

82. $y = \log_4(x-3)$

83. $y = \log_{x-1}(x+1)$

84. $y = \log_{x+2}(x-2)$

85. Mixed Practice If $f(x) = \ln x$, $g(x) = e^x$, and $h(x) = x^2$, find:
  (a) $(f \circ g)(x)$. What is the domain of $f \circ g$?
  (b) $(g \circ f)(x)$. What is the domain of $g \circ f$?
  (c) $(f \circ g)(5)$
  (d) $(f \circ h)(x)$. What is the domain of $f \circ h$?
  (e) $(f \circ h)(e)$

86. Mixed Practice If $f(x) = \log_2 x$, $g(x) = 2^x$, and $h(x) = 4x$, find:
  (a) $(f \circ g)(x)$. What is the domain of $f \circ g$?
  (b) $(g \circ f)(x)$. What is the domain of $g \circ f$?
  (c) $(f \circ g)(3)$
  (d) $(f \circ h)(x)$. What is the domain of $f \circ h$?
  (e) $(f \circ h)(8)$

Applications and Extensions

In Problems 87–96, express y as a function of x. The constant C is a positive number.

87. $\ln y = \ln x + \ln C$

88. $\ln y = \ln (x + C)$

89. $\ln y = \ln x + \ln (x + 1) + \ln C$

90. $\ln y = 2 \ln x - \ln (x + 1) + \ln C$

91. $\ln y = 3x + \ln C$

92. $\ln y = -2x + \ln C$

93. $\ln (y - 3) = -4x + \ln C$

94. $\ln (y + 4) = 5x + \ln C$

95. $3 \ln y = \frac{1}{2} \ln (2x + 1) - \frac{1}{3} \ln (x + 4) + \ln C$

96. $2 \ln y = -\frac{1}{2} \ln x + \frac{1}{3} \ln (x^2 + 1) + \ln C$

97. Find the value of $\log_2 3 \cdot \log_3 4 \cdot \log_4 5 \cdot \log_5 6 \cdot \log_6 7 \cdot \log_7 8$.

98. Find the value of $\log_2 4 \cdot \log_4 6 \cdot \log_6 8$.

99. Find the value of $\log_2 3 \cdot \log_3 4 \cdot \cdots \cdot \log_n (n + 1) \cdot \log_{n+1} 2$.

100. Find the value of $\log_2 2 \cdot \log_2 4 \cdot \log_2 8 \cdot \cdots \cdot \log_2 2^n$.

101. Show that $\log_a (x + \sqrt{x^2 - 1}) + \log_a (x - \sqrt{x^2 - 1}) = 0$.

102. Show that $\log_a (\sqrt{x} + \sqrt{x - 1}) + \log_a (\sqrt{x} - \sqrt{x - 1}) = 0$.

103. Show that $\ln (1 + e^{2x}) = 2x + \ln (1 + e^{-2x})$.

104. Difference Quotient If $f(x) = \log_a x$, show that $\dfrac{f(x + h) - f(x)}{h} = \log_a \left(1 + \dfrac{h}{x}\right)^{1/h}$, $h \neq 0$.

105. If $f(x) = \log_a x$, show that $-f(x) = \log_{1/a} x$.

106. If $f(x) = \log_a x$, show that $f(AB) = f(A) + f(B)$.

107. If $f(x) = \log_a x$, show that $f\left(\dfrac{1}{x}\right) = -f(x)$.

108. If $f(x) = \log_a x$, show that $f(x^\alpha) = \alpha f(x)$.

109. Show that $\log_a \left(\dfrac{M}{N}\right) = \log_a M - \log_a N$, where $a, M,$ and N are positive real numbers and $a \neq 1$.

110. Show that $\log_a \left(\dfrac{1}{N}\right) = -\log_a N$, where a and N are positive real numbers and $a \neq 1$.

111. Challenge Problem Show that $\log_a b = \dfrac{1}{\log_b a}$, where a and b are positive real numbers, $a \neq 1$, and $b \neq 1$.

112. Challenge Problem Show that $\log_{\sqrt{a}} m = \log_a m^2$, where a and m are positive real numbers and $a \neq 1$.

113. Challenge Problem Show that $\log_{a^n} b^m = \dfrac{m}{n} \log_a b$, where $a, b, m,$ and n are positive real numbers, $a \neq 1$, and $b \neq 1$.

114. Challenge Problem Find n:
$$\log_2 3 \cdot \log_3 4 \cdot \log_4 5 \cdot \ldots \cdot \log_n (n + 1) = 10$$

Explaining Concepts: Discussion and Writing

115. Graph $Y_1 = \log (x^2)$ and $Y_2 = 2 \log (x)$ using a graphing utility. Are they equivalent? What might account for any differences in the two functions?

116. Write an example that illustrates why $(\log_a x)^r \neq r \log_a x$.

117. Write an example that illustrates why
$$\log_2 (x + y) \neq \log_2 x + \log_2 y$$

118. Does $3^{\log_3 (-5)} = -5$? Why or why not?

Retain Your Knowledge

Problems 119–128 are based on material learned earlier in the course. The purpose of these problems is to keep the material fresh in your mind so that you are better prepared for the final exam.

119. Use a graphing utility to solve $x^3 - 3x^2 - 4x + 8 = 0$. Round answers to two decimal places.

120. Without solving, determine the character of the solution of the quadratic equation $4x^2 - 28x + 49 = 0$ in the complex number system.

121. Find the real zeros of
$$f(x) = 5x^5 + 44x^4 + 116x^3 + 95x^2 - 4x - 4$$

122. Graph $f(x) = \sqrt{2 - x}$ using the techniques of shifting, compressing or stretching, and reflecting. State the domain and the range of f.

123. Find the domain of $f(x) = 2\sqrt{3 - 5x} - 4$.

124. Solve: $4|x + 1| - 9 < 23$

125. Find the vertex of $f(x) = -\dfrac{1}{2}x^2 + 4x + 5$, and determine if the graph is concave up or concave down.

126. Find the center and radius of the circle
$$x^2 - 10x + y^2 + 4y = 35$$

127. Find the average rate of change $f(x) = x^3$ from -1 to 3.

128. Is the function $f(x) = \dfrac{5x^2 - 3x^4}{\sqrt[3]{x}}$ even, odd, or neither?

7.4 Logarithmic and Exponential Equations

PREPARING FOR THIS SECTION *Before getting started, review the following:*

- Using a Graphing Utility to Solve Equations (Section B.4, pp. B6–B8)
- Solving Quadratic Equations (Section A.4, pp. A30–A34)

 Now Work the 'Are You Prepared?' problems on page 503.

OBJECTIVES **1** Solve Logarithmic Equations (p. 499)
       2 Solve Exponential Equations (p. 501)
       3 Solve Logarithmic and Exponential Equations Using a Graphing Utility (p. 502)

1 Solve Logarithmic Equations

In Section 7.2, we solved logarithmic equations by changing a logarithmic expression to an exponential expression. That is, we used the definition of a logarithm:

$$y = \log_a x \quad \text{if and only if} \quad x = a^y \qquad a > 0 \quad a \neq 1$$

For example, to solve the equation $\log_2(1 - 2x) = 3$, write the logarithmic equation as an equivalent exponential equation $1 - 2x = 2^3$ and solve for x.

$$\log_2(1 - 2x) = 3$$
$$1 - 2x = 2^3 \qquad \textbf{Change to exponential form.}$$
$$-2x = 7 \qquad \textbf{Simplify.}$$
$$x = -\frac{7}{2} \qquad \textbf{Solve.}$$

You should check this solution for yourself.

For most logarithmic equations, some manipulation of the equation (usually using properties of logarithms) is required to obtain a solution. Also, to avoid extraneous solutions with logarithmic equations, determine the domain of the variable first.

Let's begin with an example of a logarithmic equation that requires using the fact that a logarithmic function is a one-to-one function:

$$\text{If } \log_a M = \log_a N, \text{ then } M = N \qquad M, N, \text{ and } a \text{ are positive and } a \neq 1$$

EXAMPLE 1

Solving a Logarithmic Equation

Solve: $2 \log_5 x = \log_5 9$

Solution

The domain of the variable in this equation is $x > 0$. Note that each logarithm has the same base, 5. Now use properties of logarithms to solve the equation.

$$2 \log_5 x = \log_5 9$$
$$\log_5 x^2 = \log_5 9 \qquad \textbf{r } \log_a M = \log_a M^r$$
$$x^2 = 9 \qquad \textbf{If } \log_a M = \log_a N, \text{ then } M = N.$$
$$x = 3 \quad \text{or} \quad x = -3$$

Since the domain of the variable is $x > 0$, -3 is extraneous and is discarded.

(continued)

✓ **Check:** $2 \log_5 3 \overset{?}{=} \log_5 9$

$\log_5 3^2 \overset{?}{=} \log_5 9 \quad r \log_a M = \log_a M^r$

$\log_5 9 = \log_5 9$

The solution set is $\{3\}$.

⟶ **Now Work** PROBLEM 17

Often one or more properties of logarithms are needed to rewrite the equation as a single logarithm. In the next example, the log of a product property is used.

EXAMPLE 2   **Solving a Logarithmic Equation**

Solve: $\log_5(x + 6) + \log_5(x + 2) = 1$

Solution   The domain of the variable requires that $x + 6 > 0$ and $x + 2 > 0$, so $x > -6$ and $x > -2$. This means any solution must satisfy $x > -2$. To obtain an exact solution, first express the left side as a single logarithm. Then change the equation to an equivalent exponential equation.

$\log_5(x + 6) + \log_5(x + 2) = 1$

$\log_5[(x + 6)(x + 2)] = 1 \qquad \log_a M + \log_a N = \log_a(MN)$

$(x + 6)(x + 2) = 5^1 = 5 \quad$ **Change to exponential form.**

$x^2 + 8x + 12 = 5 \qquad$ **Multiply out.**

$x^2 + 8x + 7 = 0 \qquad$ **Place the quadratic equation in standard form.**

$(x + 7)(x + 1) = 0 \qquad$ **Factor.**

$x = -7 \quad \text{or} \quad x = -1 \qquad$ **Use the Zero-Product Property.**

WARNING A negative solution is not automatically extraneous. You must determine whether the potential solution causes the argument of any logarithmic expression in the equation to be negative or 0. ■

Only $x = -1$ satisfies the restriction that $x > -2$, so $x = -7$ is extraneous. The solution set is $\{-1\}$, which you should check.

⟶ **Now Work** PROBLEM 25

EXAMPLE 3   **Solving a Logarithmic Equation**

Solve: $\ln x = \ln(x + 6) - \ln(x - 4)$

Solution   The domain of the variable requires that $x > 0$, $x + 6 > 0$, and $x - 4 > 0$. As a result, the domain of the variable here is $x > 4$. Begin the solution using the log of a difference property.

$\ln x = \ln(x + 6) - \ln(x - 4)$

$\ln x = \ln\left(\dfrac{x + 6}{x - 4}\right) \qquad \ln M - \ln N = \ln\left(\dfrac{M}{N}\right)$

$x = \dfrac{x + 6}{x - 4} \qquad$ **If $\ln M = \ln N$, then $M = N$.**

$x(x - 4) = x + 6 \qquad$ **Multiply both sides by $x - 4$.**

$x^2 - 4x = x + 6 \qquad$ **Distribute.**

$x^2 - 5x - 6 = 0 \qquad$ **Place the quadratic equation in standard form.**

$(x - 6)(x + 1) = 0 \qquad$ **Factor.**

$x = 6 \quad \text{or} \quad x = -1 \qquad$ **Use the Zero-Product Property.**

Because the domain of the variable is $x > 4$, discard -1 as extraneous. The solution set is $\{6\}$, which you should check.

WARNING In using properties of logarithms to solve logarithmic equations, avoid using the property $\log_a x^r = r \log_a x$, when r is even. The reason can be seen in this example:

Solve: $\log_3 x^2 = 4$

Solution: The domain of the variable x is all real numbers except 0.

(a) $\log_3 x^2 = 4$

$x^2 = 3^4 = 81$ **Change to exponential form.**

$x = -9$ or $x = 9$

(b) $\log_3 x^2 = 4$ $\log_a x^r = r \log_a x$

$2 \log_3 x = 4$ **Domain of variable is $x > 0$.**

$\log_3 x = 2$

$x = 9$

Both -9 and 9 are solutions of $\log_3 x^2 = 4$ (as you can verify). The solution in part (b) does not find the solution -9 because the domain of the variable was further restricted due to the application of the property $\log_a x^r = r \log_a x$.   ∎

 Now Work PROBLEM **35**

2 Solve Exponential Equations

In Sections 7.1 and 7.2, we solved exponential equations algebraically by expressing each side of the equation using the same base. That is, we used the one-to-one property of the exponential function:

> If $a^u = a^v$, then $u = v$ $a > 0$ $a \neq 1$

For example, to solve the exponential equation $4^{2x+1} = 16$, notice that $16 = 4^2$ and use the one-to-one property to obtain the equation $2x + 1 = 2$, from which we find $x = \dfrac{1}{2}$.

Not all exponential equations can be expressed so that each side of the equation has the same base. For such equations, other techniques often can be used to obtain exact solutions.

EXAMPLE 4

Solving Exponential Equations

Solve: (a) $2^x = 5$ (b) $8 \cdot 3^x = 5$

Solution   (a) Because 5 cannot be written as an integer power of 2 ($2^2 = 4$ and $2^3 = 8$), write the exponential equation as the equivalent logarithmic equation.

$$2^x = 5$$

$$x = \log_2 5 = \frac{\ln 5}{\ln 2}$$

↑ **Change-of-Base Formula**

Alternatively, the equation $2^x = 5$ can be solved by taking the natural logarithm (or common logarithm) of each side.

$$2^x = 5$$

$$\ln 2^x = \ln 5 \qquad \text{If } M = N, \text{ then } \ln M = \ln N.$$

$$x \ln 2 = \ln 5 \qquad \ln M^r = r \ln M$$

$$x = \frac{\ln 5}{\ln 2} \qquad \textbf{Exact solution}$$

$$\approx 2.322 \qquad \textbf{Approximate solution}$$

The solution set is $\left\{ \dfrac{\ln 5}{\ln 2} \right\}$.

(b) $8 \cdot 3^x = 5$

$$3^x = \frac{5}{8} \qquad \textbf{Solve for } 3^x.$$

(continued)

$$x = \log_3\left(\frac{5}{8}\right) = \frac{\ln\left(\frac{5}{8}\right)}{\ln 3} \qquad \text{Exact solution}$$

$$\approx -0.428 \qquad \text{Approximate solution}$$

The solution set is $\left\{\dfrac{\ln\left(\frac{5}{8}\right)}{\ln 3}\right\}$.

 Now Work PROBLEM **47**

EXAMPLE 5 **Solving an Exponential Equation**

Solve: $5^{x-2} = 3^{3x+2}$

Solution In this equation, the bases are 3 and 5, and we cannot express one as a power of the other as we did in Section 7.1 (pp. 469–470). Instead begin by taking the natural logarithm of both sides and then use the fact that $\ln M^r = r \ln M$. This results in an equation we know how to solve.

$$5^{x-2} = 3^{3x+2}$$
$$\ln 5^{x-2} = \ln 3^{3x+2} \qquad \text{If } M = N, \ln M = \ln N.$$
$$(x-2)\ln 5 = (3x+2)\ln 3 \qquad \ln M^r = r \ln M$$
$$(\ln 5)x - 2\ln 5 = (3\ln 3)x + 2\ln 3 \qquad \text{Distribute.}$$
$$(\ln 5)x - (3\ln 3)x = 2\ln 3 + 2\ln 5 \qquad \text{Place terms involving } x \text{ on the left.}$$
$$(\ln 5 - 3\ln 3)x = 2(\ln 3 + \ln 5) \qquad \text{Factor.}$$
$$x = \frac{2(\ln 3 + \ln 5)}{\ln 5 - 3\ln 3} \qquad \text{Exact solution}$$
$$\approx -3.212 \qquad \text{Approximate solution}$$

NOTE Because of the properties of logarithms, exact solutions involving logarithms often can be expressed in multiple ways. For example, the solution to $5^{x-2} = 3^{3x+2}$ from Example 5 can be expressed equivalently as $\dfrac{2\ln 15}{\ln 5 - \ln 27}$ or as $\dfrac{\ln 225}{\ln\left(\frac{5}{27}\right)}$, among others. Do you see why? ∎

The solution set is $\left\{\dfrac{2(\ln 3 + \ln 5)}{\ln 5 - 3\ln 3}\right\}$.

 Now Work PROBLEM **57**

EXAMPLE 6 **Solving an Exponential Equation That Is Quadratic in Form**

Solve: $4^x - 2^x - 12 = 0$

Solution Note that $4^x = (2^2)^x = 2^{2x} = (2^x)^2$, so the equation is quadratic in form and can be written as

$$(2^x)^2 - 2^x - 12 = 0 \qquad \text{Let } u = 2^x; \text{ then } u^2 - u - 12 = 0.$$

Now factor as usual.

$$(2^x - 4)(2^x + 3) = 0 \qquad (u - 4)(u + 3) = 0$$
$$2^x - 4 = 0 \quad \text{or} \quad 2^x + 3 = 0 \qquad u - 4 = 0 \quad \text{or} \quad u + 3 = 0$$
$$2^x = 4 \qquad\qquad 2^x = -3 \qquad u = 2^x = 4 \qquad u = 2^x = -3$$

The equation on the left has the solution $x = 2$; the equation on the right has no solution, since $2^x > 0$ for all x. The solution set is $\{2\}$.

 Now Work PROBLEM **65**

---

 3 Solve Logarithmic and Exponential Equations Using a Graphing Utility

The algebraic techniques introduced in this section to obtain exact solutions can be used to solve only certain types of logarithmic and exponential equations. Solutions for other types are generally studied in calculus, using numerical methods. For such types, we can use a graphing utility to approximate the solution.

EXAMPLE 7	Solving Equations Using a Graphing Utility

Solve: $x + e^x = 2$

Express the solution(s) rounded to two decimal places.

Solution The solution is found using a TI-84 Plus C by graphing $Y_1 = x + e^x$ and $Y_2 = 2$ as shown in Figure 23(a). Note that because Y_1 is an increasing function (do you know why?), there is only one point of intersection for Y_1 and Y_2. Using the INTERSECT command reveals that the solution is 0.44, rounded to two decimal places. Figure 23(b) shows the solution using Desmos.

Figure 23 (a) TI-84 Plus C   (b) Desmos

Now Work PROBLEM 75

7.4 Assess Your Understanding

'Are You Prepared?' *Answers are given at the end of these exercises. If you get a wrong answer, read the pages listed in* red.

1. Solve $x^2 - 7x - 30 = 0$. (pp. A30–A34)

2. Solve $(x + 3)^2 - 4(x + 3) + 3 = 0$. (pp. A30–A34)

3. Approximate the solution(s) to $x^3 = x^2 - 5$ using a graphing utility. (pp. B6–B8)

4. Approximate the solution(s) to $x^3 - 2x + 2 = 0$ using a graphing utility. (pp. B6–B8)

Skill Building

In Problems 5–44, solve each logarithmic equation. Express irrational solutions in exact form.

5. $\log_4 x = 2$

6. $\log (x + 6) = 1$

7. $\log_2(5x) = 4$

8. $\log_3(3x - 1) = 2$

9. $\log_4(x + 4) = \log_4 15$

10. $\log_5(2x + 3) = \log_5 3$

11. $\log_4|x| = 3$

12. $\log_2|x - 7| = 4$

13. $\log_5|2x - 1| = \log_5 13$

14. $\log_9|3x + 4| = \log_9|5x - 12|$

15. $\frac{1}{2}\log_7 x = 3\log_7 2$

16. $-2\log_4 x = \log_4 9$

17. $3\log_2 x = -\log_2 27$

18. $2\log_5 x = 3\log_5 4$

19. $2\log_6(x - 5) + \log_6 9 = 2$

20. $2\log_3(x + 4) - \log_3 9 = 2$

21. $\log x + \log(x + 15) = 2$

22. $\log x + \log (x - 21) = 2$

23. $\log(7x + 6) = 1 + \log(x - 1)$

24. $\log(2x) - \log(x - 3) = 1$

25. $\log_2(x + 7) + \log_2(x + 8) = 1$

26. $\log_6(x + 4) + \log_6(x + 3) = 1$

27. $\log_8(x + 6) = 1 - \log_8(x + 4)$

28. $\log_5(x + 3) = 1 - \log_5(x - 1)$

29. $\ln x + \ln(x + 2) = 4$

30. $\ln(x + 1) - \ln x = 2$

31. $\log_9(x + 8) + \log_9(x + 7) = 2$

32. $\log_2(x + 1) + \log_2(x + 7) = 3$

33. $\log_{1/3}(x^2 + x) - \log_{1/3}(x^2 - x) = -1$ 34. $\log_4(x^2 - 9) - \log_4(x + 3) = 3$

35. $\log_a(x - 1) - \log_a(x + 6) = \log_a(x - 2) - \log_a(x + 3)$

36. $\log_a x + \log_a(x - 2) = \log_a(x + 4)$

37. $2\log_5(x - 3) - \log_5 8 = \log_5 2$

38. $\log_3 x - 2\log_3 5 = \log_3(x + 1) - 2\log_3 10$

39. $2\log_6(x + 2) = 3\log_6 2 + \log_6 4$

40. $3(\log_7 x - \log_7 2) = 2\log_7 4$

41. $2\log_{13}(x + 2) = \log_{13}(4x + 7)$

42. $\log (x - 1) = \frac{1}{3}\log 2$

43. $(\log_3 x)^2 - 3\log_3 x = 10$

44. $\ln x - 3\sqrt{\ln x} + 2 = 0$

In Problems 45–72, solve each exponential equation. Express irrational solutions in exact form.

45. $2^{x-5} = 8$

46. $5^{-x} = 25$

47. $2^x = 10$

48. $3^x = 14$

49. $8^{-x} = 1.2$

50. $2^{-x} = 1.5$

51. $5(2^{3x}) = 8$

52. $0.3(4^{0.2x}) = 0.2$

53. $3^{1-2x} = 4^x$

54. $2^{x+1} = 5^{1-2x}$

55. $\left(\dfrac{3}{5}\right)^x = 7^{1-x}$

56. $\left(\dfrac{4}{3}\right)^{1-x} = 5^x$

57. $1.2^x = (0.5)^{-x}$

58. $0.3^{1+x} = 1.7^{2x-1}$

59. $\pi^{1-x} = e^x$

60. $e^{x+3} = \pi^x$

61. $2^{2x} + 2^x - 12 = 0$

62. $3^{2x} + 3^x - 2 = 0$

63. $3^{2x} + 3^{x+1} - 4 = 0$

64. $2^{2x} + 2^{x+2} - 12 = 0$

65. $16^x + 4^{x+1} - 3 = 0$

66. $9^x - 3^{x+1} + 1 = 0$

67. $25^x - 8 \cdot 5^x = -16$

68. $36^x - 6 \cdot 6^x = -9$

69. $3 \cdot 4^x + 4 \cdot 2^x + 8 = 0$

70. $2 \cdot 49^x + 11 \cdot 7^x + 5 = 0$

71. $4^x - 10 \cdot 4^{-x} = 3$

72. $3^x - 14 \cdot 3^{-x} = 5$

In Problems 73–86, use a graphing utility to solve each equation. Express your answer rounded to two decimal places.

73. $\log_5(x + 1) - \log_4(x - 2) = 1$

74. $\log_2(x - 1) - \log_6(x + 2) = 2$

75. $e^x = -x$

76. $e^{2x} = x + 2$

77. $e^x = x^2$

78. $e^x = x^3$

79. $\ln x = -x$

80. $\ln(2x) = -x + 2$

81. $\ln x = x^3 - 1$

82. $\ln x = -x^2$

83. $e^x + \ln x = 4$

84. $e^x - \ln x = 4$

85. $e^{-x} = \ln x$

86. $e^{-x} = -\ln x$

Applications and Extensions

87. $f(x) = \log_2(x + 3)$ and $g(x) = \log_2(3x + 1)$.

    (a) Solve $f(x) = 3$. What point is on the graph of f?

    (b) Solve $g(x) = 4$. What point is on the graph of g?

    (c) Solve $f(x) = g(x)$. Do the graphs of f and g intersect? If so, where?

    (d) Solve $(f + g)(x) = 7$.

    (e) Solve $(f - g)(x) = 2$.

88. $f(x) = \log_3(x + 5)$ and $g(x) = \log_3(x - 1)$.

    (a) Solve $f(x) = 2$. What point is on the graph of f?

    (b) Solve $g(x) = 3$. What point is on the graph of g?

    (c) Solve $f(x) = g(x)$. Do the graphs of f and g intersect? If so, where?

    (d) Solve $(f + g)(x) = 3$.

    (e) Solve $(f - g)(x) = 2$.

89. (a) If $f(x) = 3^{x+1}$ and $g(x) = 2^{x+2}$, graph f and g on the same Cartesian plane.

    (b) Find the point(s) of intersection of the graphs of f and g by solving $f(x) = g(x)$. Round answers to three decimal places. Label any intersection points on the graph drawn in part (a).

    (c) Based on the graph, solve $f(x) > g(x)$.

90. (a) If $f(x) = 5^{x-1}$ and $g(x) = 2^{x+1}$, graph f and g on the same Cartesian plane.

    (b) Find the point(s) of intersection of the graphs of f and g by solving $f(x) = g(x)$. Label any intersection points on the graph drawn in part (a).

    (c) Based on the graph, solve $f(x) > g(x)$.

91. (a) Graph $f(x) = 3^x$ and $g(x) = 10$ on the same Cartesian plane.

    (b) Shade the region bounded by the y-axis, $f(x) = 3^x$, and $g(x) = 10$ on the graph drawn in part (a).

    (c) Solve $f(x) = g(x)$ and label the point of intersection on the graph drawn in part (a).

92. (a) Graph $f(x) = 2^x$ and $g(x) = 12$ on the same Cartesian plane.

    (b) Shade the region bounded by the y-axis, $f(x) = 2^x$, and $g(x) = 12$ on the graph drawn in part (a).

    (c) Solve $f(x) = g(x)$ and label the point of intersection on the graph drawn in part (a).

93. (a) Graph $f(x) = 2^{x+1}$ and $g(x) = 2^{-x+2}$ on the same Cartesian plane.

    (b) Shade the region bounded by the y-axis, $f(x) = 2^{x+1}$, and $g(x) = 2^{-x+2}$ on the graph drawn in part (a).

    (c) Solve $f(x) = g(x)$ and label the point of intersection on the graph drawn in part (a).

94. (a) Graph $f(x) = 3^{-x+1}$ and $g(x) = 3^{x-2}$ on the same Cartesian plane.

    (b) Shade the region bounded by the y-axis, $f(x) = 3^{-x+1}$, and $g(x) = 3^{x-2}$ on the graph drawn in part (a).

    (c) Solve $f(x) = g(x)$ and label the point of intersection on the graph drawn in part (a).

95. (a) Graph $f(x) = 2^x - 4$.

    (b) Find the zero of f.

    (c) Based on the graph, solve $f(x) < 0$.

96. (a) Graph $g(x) = 3^x - 9$.

    (b) Find the zero of g.

    (c) Based on the graph, solve $g(x) > 0$.

97. A Population Model The resident population of the United States in 2018 was 327 million people and was growing at a rate of 0.7% per year. Assuming that this growth rate continues, the model $P(t) = 327(1.007)^{t-2018}$ represents the population P (in millions of people) in year t.

    (a) According to this model, when will the population of the United States be 415 million people?

    (b) According to this model, when will the population of the United States be 470 million people?

Source: U.S. Census Bureau

98. A Population Model The population of the world in 2018 was 7.63 billion people and was growing at a rate of 1.1% per year. Assuming that this growth rate continues, the model $P(t) = 7.63(1.011)^{t-2018}$ represents the population P (in billions of people) in year t.

    (a) According to this model, when will the population of the world be 9 billion people?

    (b) According to this model, when will the population of the world be 12.5 billion people?

Source: U.S. Census Bureau

99. Depreciation The value V of a Chevy Cruze LT that is t years old can be modeled by $V(t) = 19{,}200\,(0.82)^t$.

(a) According to the model, when will the car be worth $12,000?

(b) According to the model, when will the car be worth $9000?

(c) According to the model, when will the car be worth $3000?

Source: Kelley Blue Book

100. Depreciation The value V of a Honda Civic LX that is t years old can be modeled by $V(t) = 19{,}705\,(0.848)^t$.

(a) According to the model, when will the car be worth $14,000?

(b) According to the model, when will the car be worth $10,000?

(c) According to the model, when will the car be worth $7500?

Source: Kelley Blue Book

Challenge Problems *In Problems 101–105, solve each equation. Express irrational solutions in exact form.*

101. $\log_2(x+1) - \log_4 x = 1$

102. $(\sqrt[3]{2})^{2-x} = 2^{x^2}$

103. $\log_2 x^{\log_2 x} = 4$

104. $\ln x^2 = (\ln x)^2$

105. $\sqrt{\log x} = 2\log\sqrt{3}$

Explaining Concepts: Discussion and Writing

106. Fill in the reason for each step in the following two solutions.

Solve: $\log_3(x-1)^2 = 2$

Solution A

$\log_3(x-1)^2 = 2$

$(x-1)^2 = 3^2 = 9$ _____

$(x-1) = \pm 3$ _____

$x-1 = -3$ or $x-1 = 3$ _____

$x = -2$ or $x = 4$ _____

Solution B

$\log_3(x-1)^2 = 2$

$2\log_3(x-1) = 2$ _____

$\log_3(x-1) = 1$ _____

$x-1 = 3^1 = 3$ _____

$x = 4$ _____

Both solutions given in Solution A check. Explain what caused the solution $x = -2$ to be lost in Solution B.

Retain Your Knowledge

Problems 107–116, are based on material learned earlier in the course. The purpose of these problems is to keep the material fresh in your mind so that you are better prepared for the final exam.

107. Solve: $4x^3 + 3x^2 - 25x + 6 = 0$

108. Determine whether the function is one-to-one:

$$\{(0,-4),(2,-2),(4,0),(6,2)\}$$

109. If $f(x) = \dfrac{x}{x-2}$, find $f\!\left(\dfrac{x+5}{x-3}\right)$.

110. Find the domain of $f(x) = \sqrt{x+3} + \sqrt{x-1}$.

111. Solve: $x - \sqrt{x+7} = 5$

112. Find the real zero of the function $f(x) = 5x - 30$.

113. If $f(x) = \dfrac{x}{x-2}$ and $g(x) = \dfrac{5}{x+2}$, find $(f+g)(x)$.

114. Find the distance between the center of the circle

$$(x-2)^2 + (y+3)^2 = 25$$

and the vertex of the parabola $y = -2(x-6)^2 + 9$.

115. Find the average rate of change $f(x) = \log_2 x$ from 4 to 16.

116. Rationalize the numerator: $\dfrac{\sqrt{x+6} - \sqrt{x}}{6}$

'Are You Prepared?' Answers

1. $\{-3, 10\}$ **2.** $\{-2, 0\}$ **3.** $\{-1.43\}$ **4.** $\{-1.77\}$

7.5 Financial Models

PREPARING FOR THIS SECTION *Before getting started, review the following:*

OBJECTIVES 1 Determine the Future Value of a Lump Sum of Money (p. 506)
           2 Calculate Effective Rates of Return (p. 509)
           3 Determine the Present Value of a Lump Sum of Money (p. 510)
           4 Determine the Rate of Interest or the Time Required to Double a Lump Sum of Money (p. 511)

1 Determine the Future Value of a Lump Sum of Money

Interest is money paid for the use of money. The total amount borrowed (whether by an individual from a bank in the form of a loan or by a bank from an individual in the form of a savings account) is called the **principal**. The **rate of interest**, expressed as a percent, is the amount charged for the use of the principal for a given period of time, usually on a yearly (that is, per annum) basis.

THEOREM **Simple Interest Formula**

If a principal of P dollars is borrowed for a period of t years at a per annum interest rate r, expressed as a decimal, the interest I charged is

$$I = Prt \tag{1}$$

Interest charged according to formula (1) is called **simple interest**.

In problems involving interest, the term **payment period** is defined as follows.

Annually: Once per year   **Monthly:** 12 times per year
Semiannually: Twice per year   **Daily:** 365 times per year*
Quarterly: Four times per year

When the interest due at the end of a payment period is added to the principal so that the interest computed at the end of the next payment period is based on this new principal amount (old principal + interest), the interest is said to have been **compounded**. **Compound interest** is interest paid on the principal and on previously earned interest.

EXAMPLE 1 **Computing Compound Interest**

A credit union pays interest of 2% per annum compounded quarterly on a certain savings plan. If $1000 is deposited in the plan and the interest is left to accumulate, how much is in the account after 1 year?

Solution Use the simple interest formula, $I = Prt$. The principal P is $1000 and the rate of interest is 2% = 0.02. After the first quarter of a year, the time t is $\frac{1}{4}$ year, so the interest earned is

$$I = Prt = \$1000 \cdot 0.02 \cdot \frac{1}{4} = \$5$$

*Some banks use a 360-day "year." Why do you think they do?

The new principal is $P + I = \$1000 + \$5 = \$1005$. At the end of the second quarter, the interest on this principal is

$$I = \$1005 \cdot 0.02 \cdot \frac{1}{4} = \$5.03$$

At the end of the third quarter, the interest on the new principal of $\$1005 + \$5.03 = \$1010.03$ is

$$I = \$1010.03 \cdot 0.02 \cdot \frac{1}{4} = \$5.05$$

Finally, after the fourth quarter, the interest is

$$I = \$1015.08 \cdot 0.02 \cdot \frac{1}{4} = \$5.08$$

After 1 year the account contains $\$1015.08 + \$5.08 = \$1020.16$.

The pattern of the calculations performed in Example 1 leads to a general formula for compound interest. For this purpose, let P represent the principal to be invested at a per annum interest rate r that is compounded n times per year, so the time of each compounding period is $\frac{1}{n}$ year. (For computing purposes, r is expressed as a decimal.) The interest earned after each compounding period is given by formula (1).

$$\text{Interest} = \text{principal} \cdot \text{rate} \cdot \text{time} = P \cdot r \cdot \frac{1}{n} = P \cdot \frac{r}{n}$$

The amount A after one compounding period is

$$A = P + P \cdot \frac{r}{n} = P \cdot \left(1 + \frac{r}{n}\right)$$

After two compounding periods, the amount A, based on the new principal $P \cdot \left(1 + \frac{r}{n}\right)$, is

$$A = \underbrace{P \cdot \left(1 + \frac{r}{n}\right)}_{\substack{\textbf{New} \\ \textbf{principal}}} + \underbrace{P \cdot \left(1 + \frac{r}{n}\right) \cdot \frac{r}{n}}_{\substack{\textbf{Interest on} \\ \textbf{new principal}}} = \underbrace{P \cdot \left(1 + \frac{r}{n}\right) \cdot \left(1 + \frac{r}{n}\right)}_{\textbf{Factor out } P \cdot \left(1 + \frac{r}{n}\right)} = P \cdot \left(1 + \frac{r}{n}\right)^2$$

After three compounding periods, the amount A is

$$A = P \cdot \left(1 + \frac{r}{n}\right)^2 + P \cdot \left(1 + \frac{r}{n}\right)^2 \cdot \frac{r}{n} = P \cdot \left(1 + \frac{r}{n}\right)^2 \cdot \left(1 + \frac{r}{n}\right) = P \cdot \left(1 + \frac{r}{n}\right)^3$$

Continuing this way, after n compounding periods (1 year), the amount A is

$$A = P \cdot \left(1 + \frac{r}{n}\right)^n$$

Because t years will contain $n \cdot t$ compounding periods, the amount after t years is

$$A = P \cdot \left(1 + \frac{r}{n}\right)^{nt}$$

THEOREM Compound Interest Formula

The amount A after t years due to a principal P invested at an annual interest rate r, expressed as a decimal, compounded n times per year is

$$A = P \cdot \left(1 + \frac{r}{n}\right)^{nt} \tag{2}$$

In equation (2), the amount A is typically referred to as the **accumulated value** of the account, and P is called the **present value**.

For example, to rework Example 1, use $P = \$1000$, $r = 0.02$, $n = 4$ (quarterly compounding), and $t = 1$ year to obtain

$$A = P \cdot \left(1 + \frac{r}{n}\right)^{nt} = 1000\left(1 + \frac{0.02}{4}\right)^{4 \cdot 1} = \$1020.15$$

The result obtained here differs slightly from that obtained in Example 1 because of rounding.

Now Work PROBLEM 7

EXAMPLE 2

Comparing Investments Using Different Compounding Periods

Investing $1000 at an annual rate of 10% compounded annually, semiannually, quarterly, monthly, and daily will yield the following amounts after 1 year:

Annual compounding ($n = 1$): $A = P \cdot (1 + r)$

$$= \$1000(1 + 0.10) = \$1100.00$$

Semiannual compounding ($n = 2$): $A = P \cdot \left(1 + \dfrac{r}{2}\right)^2$

$$= \$1000(1 + 0.05)^2 = \$1102.50$$

Quarterly compounding ($n = 4$): $A = P \cdot \left(1 + \dfrac{r}{4}\right)^4$

$$= \$1000(1 + 0.025)^4 = \$1103.81$$

Monthly compounding ($n = 12$): $A = P \cdot \left(1 + \dfrac{r}{12}\right)^{12}$

$$= \$1000\left(1 + \frac{0.10}{12}\right)^{12} = \$1104.71$$

Daily compounding ($n = 365$): $A = P \cdot \left(1 + \dfrac{r}{365}\right)^{365}$

$$= \$1000\left(1 + \frac{0.10}{365}\right)^{365} = \$1105.16$$

From Example 2, note that the effect of compounding more frequently is that the amount after 1 year is higher. This leads to the following question: What would happen to the amount after 1 year if the number of times that the interest is compounded were increased without bound?

Let's find the answer. Suppose that P is the principal, r is the per annum interest rate, and n is the number of times that the interest is compounded each year. The amount A after 1 year is

$$A = P \cdot \left(1 + \frac{r}{n}\right)^n$$

Rewrite this expression as follows:

$$A = P \cdot \left(1 + \frac{r}{n}\right)^n = P \cdot \left(1 + \frac{1}{\frac{n}{r}}\right)^n = P \cdot \left[\left(1 + \frac{1}{\frac{n}{r}}\right)^{n/r}\right]^r \underset{\substack{\uparrow \\ h = \frac{n}{r}}}{=} P \cdot \left[\left(1 + \frac{1}{h}\right)^h\right]^r \quad \textbf{(3)}$$

Need to Review?
The number e is defined on page 468.

Now suppose that the number n of times that the interest is compounded per year gets larger and larger; that is, suppose that $n \to \infty$. Then $h = \dfrac{n}{r} \to \infty$, and the expression in brackets in equation (3) equals e. That is, $A \to Pe^r$.

Table 8 compares $\left(1 + \dfrac{r}{n}\right)^n$, for large values of n, to e^r for $r = 0.05$, $r = 0.10$, $r = 0.15$, and $r = 1$. As n becomes larger, the closer $\left(1 + \dfrac{r}{n}\right)^n$ gets to e^r. No matter how frequent the compounding, the amount after 1 year has the upper bound Pe^r.

Table 8

	$\left(1 + \frac{r}{n}\right)^n$			
	$n = 100$	$n = 1000$	$n = 10{,}000$	e^r
$r = 0.05$	1.0512580	1.0512698	1.051271	1.0512711
$r = 0.10$	1.1051157	1.1051654	1.1051704	1.1051709
$r = 0.15$	1.1617037	1.1618212	1.1618329	1.1618342
$r = 1$	2.7048138	2.7169239	2.7181459	2.7182818

When interest is compounded so that the amount after 1 year is Pe^r, the interest is said to be **compounded continuously**.

THEOREM **Continuous Compounding**

The amount A after t years due to a principal P invested at an annual interest rate r compounded continuously is

$$A = Pe^{rt} \tag{4}$$

EXAMPLE 3 **Using Continuous Compounding**

The amount A that results from investing a principal P of $1000 at an annual rate r of 10% compounded continuously for a time t of 1 year is

$$A = \$1000e^{0.10} = \$1000 \cdot 1.10517 = \$1105.17$$

Now Work PROBLEM 13

2 Calculate Effective Rates of Return

Suppose that you have $1000 to invest and a bank offers to pay you 3 percent annual interest compounded monthly. What simple interest rate is needed to earn an equal amount after one year? To answer this question, first determine the value after one year of the $1000 investment that earns 3 percent compounded monthly.

$$A = \$1000\left(1 + \frac{0.03}{12}\right)^{12} \quad \text{Use } A = P\left(1 + \frac{r}{n}\right)^n \text{ with } P = \$1000, r = 0.03, n = 12.$$

$$= \$1030.42$$

So the interest earned is $30.42. Using $I = Prt$ with $t = 1$, $I = \$30.42$, and $P = \$1000$, the annual simple interest rate is $0.03042 = 3.042\%$. This interest rate is known as the *effective rate of interest*.

The **effective rate of interest** is the annual simple interest rate that would yield the same amount as compounding n times per year, or continuously, after one year.

THEOREM Effective Rate of Interest

The effective rate of interest r_E of an investment earning an annual interest rate r is given by

- Compounding n times per year: $r_E = \left(1 + \dfrac{r}{n}\right)^n - 1$

- Continuous compounding: $r_E = e^r - 1$

EXAMPLE 4 **Computing the Effective Rate of Interest–Which Is the Best Deal?**

Suppose you want to buy a 5-year certificate of deposit (CD). You visit three banks to determine their CD rates. American Express offers you 2.15% annual interest compounded monthly, and First Internet Bank offers you 2.20% compounded quarterly. Discover offers 2.12% compounded daily. Determine which bank is offering the best deal.

Solution The bank that offers the best deal is the one with the highest effective interest rate.

American Express	First Internet Bank	Discover
$r_E = \left(1 + \dfrac{0.0215}{12}\right)^{12} - 1$	$r_E = \left(1 + \dfrac{0.022}{4}\right)^{4} - 1$	$r_E = \left(1 + \dfrac{0.0212}{365}\right)^{365} - 1$
$\approx 1.02171 - 1$	$\approx 1.02218 - 1$	$\approx 1.02143 - 1$
$= 0.02171$	$= 0.02218$	$= 0.02143$
$= 2.171\%$	$= 2.218\%$	$= 2.143\%$

The effective rate of interest is highest for First Internet Bank, so First Internet Bank is offering the best deal.

 Now Work PROBLEM 23

3 Determine the Present Value of a Lump Sum of Money

When people in finance speak of the "time value of money," they are usually referring to the *present value* of money. The **present value** of A dollars to be received at a future date is the principal that you would need to invest now so that it will grow to A dollars in the specified time period. The present value of money to be received at a future date is always less than the amount to be received, since the amount to be received will equal the present value (money invested now) *plus* the interest accrued over the time period.

The compound interest formula (2) is used to develop a formula for present value. If P is the present value of A dollars to be received after t years at a per annum interest rate r compounded n times per year, then, by formula (2),

$$A = P \cdot \left(1 + \frac{r}{n}\right)^{nt}$$

To solve for P, divide both sides by $\left(1 + \dfrac{r}{n}\right)^{nt}$. The result is

$$\frac{A}{\left(1 + \dfrac{r}{n}\right)^{nt}} = P \quad \text{or} \quad P = A \cdot \left(1 + \frac{r}{n}\right)^{-nt}$$

> **THEOREM Present Value Formulas**
>
> The present value P of A dollars to be received after t years, assuming a per annum interest rate r compounded n times per year, is
>
> $$P = A \cdot \left(1 + \frac{r}{n}\right)^{-nt} \qquad (5)$$
>
> If the interest is compounded continuously, then
>
> $$P = Ae^{-rt} \qquad (6)$$

To derive (6), solve $A = Pe^{rt}$ for P.

EXAMPLE 5 | **Computing the Value of a Zero-Coupon Bond**

A zero-coupon (noninterest-bearing) bond can be redeemed in 10 years for $1000. How much should you be willing to pay for it now if you want a return of

(a) 8% compounded monthly?   (b) 7% compounded continuously?

Solution   (a) To find the present value of $1000, use formula (5) with $A = \$1000$, $n = 12, r = 0.08$, and $t = 10$.

$$P = A \cdot \left(1 + \frac{r}{n}\right)^{-nt} = \$1000\left(1 + \frac{0.08}{12}\right)^{-12 \cdot 10} = \$450.52$$

For a return of 8% compounded monthly, pay $450.52 for the bond.

(b) Here use formula (6) with $A = \$1000$, $r = 0.07$, and $t = 10$.

$$P = Ae^{-rt} = \$1000e^{-0.07 \cdot 10} = \$496.59$$

For a return of 7% compounded continuously, pay $496.59 for the bond.

▬▬▶ **Now Work** PROBLEM 15

4 Determine the Rate of Interest or the Time Required to Double a Lump Sum of Money

EXAMPLE 6 | **Determining the Rate of Interest Required to Double an Investment**

What rate of interest compounded annually is needed in order to double an investment in 5 years?

Solution   If P is the principal and P is to double, then the amount A will be $2P$. Use the compound interest formula with $n = 1$ and $t = 5$ to find r.

$$A = P \cdot \left(1 + \frac{r}{n}\right)^{nt}$$

$$2P = P \cdot (1 + r)^5 \qquad \text{\color{blue}{$A = 2P, n = 1, t = 5$}}$$

$$2 = (1 + r)^5 \qquad \text{\color{blue}{Divide both sides by P.}}$$

$$1 + r = \sqrt[5]{2} \qquad \text{\color{blue}{Take the fifth root of both sides.}}$$

$$r = \sqrt[5]{2} - 1 \approx 1.148698 - 1 = 0.148698$$

The annual rate of interest needed to double the principal in 5 years is 14.87%.

▬▬▶ **Now Work** PROBLEM 31

EXAMPLE 7	**Determining the Time Required to Double or Triple an Investment**

(a) How long will it take for an investment to double in value if it earns 5% compounded continuously?

(b) How long will it take to triple at this rate?

Solution　(a) If P is the initial investment and P is to double, then the amount A will be $2P$. Use formula (4) for continuously compounded interest with $r = 0.05$.

$$A = Pe^{rt}$$
$$2P = Pe^{0.05t} \qquad \text{A = 2P, r = 0.05}$$
$$2 = e^{0.05t} \qquad \text{Cancel the P's.}$$
$$0.05t = \ln 2 \qquad \text{Rewrite as a logarithm.}$$
$$t = \frac{\ln 2}{0.05} \approx 13.86 \qquad \text{Solve for t.}$$

It will take about 14 years to double the investment.

(b) To triple the investment, let $A = 3P$ in formula (4).

$$A = Pe^{rt}$$
$$3P = Pe^{0.05t} \qquad \text{A = 3P, r = 0.05}$$
$$3 = e^{0.05t} \qquad \text{Cancel the P's.}$$
$$0.05t = \ln 3 \qquad \text{Rewrite as a logarithm.}$$
$$t = \frac{\ln 3}{0.05} \approx 21.97 \qquad \text{Solve for t.}$$

It will take about 22 years to triple the investment.

Now Work PROBLEM 35

7.5 Assess Your Understanding

Concepts and Vocabulary

1. What is the interest due when $500 is borrowed for 6 months at a simple interest rate of 6% per annum?

2. If you borrow $5000 and, after 9 months, pay off the loan in the amount of $5500, what per annum rate of interest was charged?

3. The total amount borrowed (whether by an individual from a bank in the form of a loan or by a bank from an individual in the form of a savings account) is called the _____.

4. If a principal of P dollars is borrowed for a period of t years at a per annum interest rate r, expressed as a decimal, the interest I charged is _____ = _____. Interest charged according to this formula is called _____ _____.

5. The _____ _____ ____ _____ is the annual simple interest rate that would yield the same amount as compounding n times per year, or continuously, after 1 year.

6. *Multiple Choice* The principal that must be invested now so that it will grow to a given amount in a specified time period is called the _____.
   (a) present value　　(b) future value
   (c) interest　　　　 (d) effective rate

Skill Building

In Problems 7–14, find the amount that results from each investment.

7. $100 invested at 4% compounded quarterly after a period of 2 years

8. $50 invested at 6% compounded monthly after a period of 3 years

9. $900 invested at 3% compounded semiannually after a period of $2\frac{1}{2}$ years

10. $300 invested at 12% compounded monthly after a period of $1\frac{1}{2}$ years

11. $1200 invested at 5% compounded daily after a period of 3 years

12. $700 invested at 6% compounded daily after a period of 2 years

13. $1000 invested at 11% compounded continuously after a period of 2 years

14. $400 invested at 7% compounded continuously after a period of 3 years

In Problems 15–22, find the principal needed now to get each amount; that is, find the present value.

15. To get $100 after 2 years at 6% compounded monthly

16. To get $75 after 3 years at 8% compounded quarterly

17. To get $1500 after $2\frac{1}{2}$ years at 1.5% compounded daily

18. To get $800 after $3\frac{1}{2}$ years at 7% compounded monthly

19. To get $750 after 2 years at 2.5% compounded quarterly

20. To get $300 after 4 years at 3% compounded daily

21. To get $120 after $3\frac{1}{4}$ years at 5% compounded continuously

22. To get $800 after $2\frac{1}{2}$ years at 8% compounded continuously

In Problems 23–26, find the effective rate of interest.

23. For 5% compounded quarterly

24. For 6% compounded monthly

25. For 4% compounded continuously

26. For 6% compounded continuously

In Problems 27–30, determine the rate that represents the better deal.

27. 6% compounded quarterly or $6\frac{1}{4}$% compounded annually

28. 9% compounded quarterly or $9\frac{1}{4}$% compounded annually

29. 9% compounded monthly or 8.8% compounded daily

30. 8% compounded semiannually or 7.9% compounded daily

31. What rate of interest compounded annually is required to double an investment in 3 years?

32. What rate of interest compounded annually is required to double an investment in 6 years?

33. What rate of interest compounded annually is required to triple an investment in 5 years?

34. What rate of interest compounded annually is required to triple an investment in 10 years?

35. (a) How long does it take for an investment to double in value if it is invested at 8% compounded monthly?
(b) How long does it take if the interest is compounded continuously?

36. (a) How long does it take for an investment to triple in value if it is invested at 6% compounded monthly?
(b) How long does it take if the interest is compounded continuously?

37. What rate of interest compounded quarterly will yield an effective interest rate of 7%?

38. What rate of interest compounded continuously will yield an effective interest rate of 6%?

Applications and Extensions

39. Time Required to Reach a Goal If Tanisha has $100 to invest at 4% per annum compounded monthly, how long will it be before she has $150? If the compounding is continuous, how long will it be?

40. Time Required to Reach a Goal If Angela has $100 to invest at 2.5% per annum compounded monthly, how long will it be before she has $175? If the compounding is continuous, how long will it be?

41. Time Required to Reach a Goal How many years will it take for an initial investment of $10,000 to grow to $25,000? Assume a rate of interest of 6% compounded continuously.

42. Time Required to Reach a Goal How many years will it take for an initial investment of $25,000 to grow to $80,000? Assume a rate of interest of 7% compounded continuously.

43. Price Appreciation of Homes What will a $90,000 condominium cost 5 years from now if the price appreciation for condos over that period averages 3% compounded annually?

44. Credit Card Interest A department store charges 1.25% per month on the unpaid balance for customers with charge accounts (interest is compounded monthly). A customer charges $200 and does not pay her bill for 6 months. What is the bill at that time?

45. Saving for a Car Jerome will be buying a used car for $15,000 in 3 years. How much money should he ask his parents for now so that, if he invests it at 5% compounded continuously, he will have enough to buy the car?

46. Paying off a Loan John requires $3000 in 6 months to pay off a loan that has no prepayment privileges. If he has the $3000 now, how much of it should he save in an account paying 3% compounded monthly so that in 6 months he will have exactly $3000?

47. Return on a Stock George contemplates the purchase of 100 shares of a stock selling for $15 per share. The stock pays no dividends. The history of the stock indicates that it should grow at an annual rate of 15% per year. How much should the 100 shares of stock be worth in 5 years?

48. Return on an Investment A business purchased for $650,000 in 2010 is sold in 2013 for $850,000. What is the annual rate of return for this investment?

49. Comparing Savings Plans Jim places $1000 in a bank account that pays 5.6% compounded continuously. After 1 year, will he have enough money to buy a computer system that costs $1060? If another bank will pay Jim 5.9% compounded monthly, is this a better deal?

50. Savings Plans On January 1, Kim places $1000 in a certificate of deposit that pays 6.8% compounded continuously and matures in 3 months. Then Kim places the $1000 and the interest in a passbook account that pays 5.25% compounded monthly. How much does Kim have in the passbook account on May 1?

51. Comparing IRA Investments Will invests $2000 of the money in his IRA in a bond trust that pays 9% interest compounded semiannually. His friend Henry invests $2000 in his IRA in a certificate of deposit that pays $8\frac{1}{2}$% compounded continuously. Who has more money after 20 years, Will or Henry?

52. Comparing Two Alternatives Suppose that April has access to an investment that will pay 10% interest compounded continuously. Which is better: to be given $1000 now so that she can take advantage of this investment opportunity or to be given $1325 after 3 years?

53. College Costs The average annual cost of college at 4-year private colleges was $34,740 in the 2017–2018 academic year. This was a 3.6% increase from the previous year.
   (a) If the cost of college increases by 3.6% each year, what will be the average cost of college at a 4-year private college for the 2037–2038 academic year?
   (b) College savings plans, such as a 529 plan, allow individuals to put money aside now to help pay for college later. If one such plan offers a rate of 2% compounded continuously, how much should be put in a college savings plan in 2019 to pay for 1 year of the cost of college at a 4-year private college for an incoming freshman in 2037?

   Source: The College Board

54. Analyzing Interest Rates on a Mortgage Colleen and Bill have just purchased a house for $650,000, with the seller holding a second mortgage of $100,000. They promise to pay the seller $100,000 plus all accrued interest 5 years from now. The seller offers them three interest options on the second mortgage:
   (a) Simple interest at 6% per annum
   (b) 5.5% interest compounded monthly
   (c) 5.25% interest compounded continuously

   Which option is best? That is, which results in paying the least interest on the loan?

55. Comparing Bank Accounts Two bank accounts are opened at the same time. The first has a principal of $1000 in an account earning 5% compounded monthly. The second has a principal of $2000 in an account earning 4% interest compounded annually. Determine the number of years, to the nearest tenth, at which the account balances will be equal.

56. Per Capita Federal Debt In 2018, the federal debt was about $21 trillion. In 2018, the U.S. population was about 327 million. Assuming that the federal debt is increasing about 5.5% per year and the U.S. population is increasing about 0.7% per year, determine the per capita debt (total debt divided by population) in 2030.

Problems 57–62 require the following discussion. **Inflation** *is a term used to describe the erosion of the purchasing power of money. For example, if the annual inflation rate is 3%, then $1000 worth of purchasing power now will have only $970 worth of purchasing power in 1 year because 3% of the original $1000 ($0.03 \times 1000 = 30$) has been eroded due to inflation. In general, if the rate of inflation averages r% per annum over n years, the amount A that $P will purchase after n years is*

$$A = P \cdot (1 - r)^n$$

where r is expressed as a decimal.

57. Inflation If the inflation rate averages 3%, what will be the purchasing power of $1000 in 2 years?

58. Inflation If the inflation rate averages 2%, what will be the purchasing power of $1000 in 3 years?

59. Inflation If the purchasing power of $1000 is only $950 after 2 years, what was the average inflation rate?

60. Inflation If the purchasing power of $1000 is only $930 after 2 years, what was the average inflation rate?

61. Inflation If the average inflation rate is 2%, how long is it until purchasing power is cut in half?

62. Inflation If the average inflation rate is 4%, how long is it until purchasing power is cut in half?

Problems 63–66 involve zero-coupon bonds. A **zero-coupon bond** *is a bond that is sold now at a discount and will pay its face value at the time when it matures; no interest payments are made.*

63. Zero-Coupon Bonds A zero-coupon bond can be redeemed in 20 years for $10,000. How much should you be willing to pay for it now if you want a return of:
   (a) 5% compounded monthly?
   (b) 5% compounded continuously?

64. Zero-Coupon Bonds A child's grandparents are considering buying an $80,000 face-value, zero-coupon bond at her birth so that she will have enough money for her college education 17 years later. If they want a rate of return of 6% compounded annually, what should they pay for the bond?

65. Zero-Coupon Bonds How much should a $10,000 face-value, zero-coupon bond, maturing in 10 years, be sold for now if its rate of return is to be 4.5% compounded annually?

66. Zero-Coupon Bonds If Pat pays $15,334.65 for a $25,000 face-value, zero-coupon bond that matures in 8 years, what is his annual rate of return?

67. Time to Double or Triple an Investment The formula

$$t = \frac{\ln m}{n \ln\left(1 + \dfrac{r}{n}\right)}$$

can be used to find the number of years t required to multiply an investment m times when r is the per annum interest rate compounded n times a year.
   (a) How many years will it take to double the value of an IRA that compounds annually at the rate of 6%?
   (b) How many years will it take to triple the value of a savings account that compounds quarterly at an annual rate of 5%?
   (c) Give a derivation of this formula.

68. Time to Reach an Investment Goal The formula

$$t = \frac{\ln A - \ln P}{r}$$

can be used to find the number of years t required for an investment P to grow to a value A when compounded continuously at an annual rate r.

(a) How long will it take to increase an initial investment of $1000 to $4500 at an annual rate of 5.75%?
(b) What annual rate is required to increase the value of a $2000 IRA to $30,000 in 35 years?
(c) Give a derivation of this formula.

Problems 69–72 require the following discussion. The **consumer price index (CPI)** *indicates the relative change in price over time for a fixed basket of goods and services. It is a cost-of-living index that helps measure the effect of inflation on the cost of goods and services. The CPI uses the base period 1982–1984 for comparison (the CPI for this period is 100). The CPI for March 2018 was 249.55. This means that $100 in the period 1982–1984 had the same purchasing power as $249.55 in March 2018. In general, if the rate of inflation averages r% per annum over n years, then the CPI index after n years is*

$$CPI = CPI_0\left(1 + \frac{r}{100}\right)^n$$

where CPI_0 is the CPI index at the beginning of the n-year period.
Source: *U.S. Bureau of Labor Statistics*

69. Consumer Price Index
(a) The CPI was 229.39 for 2012 and 243.80 for 2017. Assuming that annual inflation remained constant for this time period, determine the average annual inflation rate.
(b) Using the inflation rate from part (a), in what year will the CPI reach 300?

70. Consumer Price Index If the current CPI is 234.2 and the average annual inflation rate is 2.8%, what will be the CPI in 5 years?

71. Consumer Price Index If the average annual inflation rate is 3.1%, how long will it take for the CPI index to double? (A doubling of the CPI index means purchasing power is cut in half.)

72. Consumer Price Index The base period for the CPI changed in 1998. Under the previous weight and item structure, the CPI for 1995 was 456.5. If the average annual inflation rate was 5.57%, what year was used as the base period for the CPI?

Explaining Concepts: Discussion and Writing

73. Explain in your own words what the term *compound interest* means. What does *continuous compounding* mean?

74. Explain in your own words the meaning of *present value*.

75. Critical Thinking You have just contracted to buy a house and will seek financing in the amount of $100,000. You go to several banks. Bank 1 will lend you $100,000 at the rate of 4.125% amortized over 30 years with a loan origination fee of 0.45%. Bank 2 will lend you $100,000 at the rate of 3.375% amortized over 15 years with a loan origination fee of 0.95%. Bank 3 will lend you $100,000 at the rate of 4.25% amortized over 30 years with no loan origination fee. Bank 4 will lend you $100,000 at the rate of 3.625% amortized over 15 years with no loan origination fee. Which loan would you take? Why? Be sure to have sound reasons for your choice. Use the information in the table to assist you. If the amount of the monthly payment does not matter to you, which loan would you take? Again, have sound reasons for your choice. Compare your final decision with others in the class. Discuss.

	Monthly Payment	Loan Origination Fee
Bank 1	$485	$450
Bank 2	$709	$950
Bank 3	$492	$0
Bank 4	$721	$0

Retain Your Knowledge

Problems 76–85 are based on material learned earlier in the course. The purpose of these problems is to keep the material fresh in your mind so that you are better prepared for the final exam.

76. Find the remainder R when $f(x) = 6x^3 + 3x^2 + 2x - 11$ is divided by $g(x) = x - 1$. Is g a factor of f?

77. The function $f(x) = \dfrac{x}{x-2}$ is one-to-one. Find f^{-1}.

78. Find the real zeros of
$$f(x) = x^5 - x^4 - 15x^3 - 21x^2 - 16x - 20$$
Then write f in factored form.

79. Solve: $\log_2(x + 3) = 2\log_2(x - 3)$

80. Factor completely: $2x^4 + 6x^3 - 50x^2 - 150x$

81. If $f(x) = 5x^2 + 4x - 8$ and $g(x) = 3x - 1$, find $(f \circ g)(x)$.

82. Find the domain and range of $f(x) = -2x^2 - 8x + 1$.

83. For $f(x) = \dfrac{2x^2 - 5x - 4}{x - 7}$, find all vertical asymptotes, horizontal asymptotes, and oblique asymptotes, if any.

84. If $f(x) = x^2 - 4x - 3$, find an equation of the secant line containing the points $(3, f(3))$ and $(5, f(5))$.

85. Find the difference quotient for $f(x) = 3x - 5$.

7.6 Exponential Growth and Decay Models; Newton's Law; Logistic Growth and Decay Models

OBJECTIVES 1 Model Populations That Obey the Law of Uninhibited Growth (p. 516)
 2 Model Populations That Obey the Law of Uninhibited Decay (p. 518)
 3 Use Newton's Law of Cooling (p. 519)
 4 Use Logistic Models (p. 521)

1 Model Populations That Obey the Law of Uninhibited Growth

Many natural phenomena follow the law that an amount A varies with time t according to the function

$$A(t) = A_0 e^{kt} \tag{1}$$

Here A_0 is the original amount $(t = 0)$ and $k \neq 0$ is a constant.

If $k > 0$, then equation (1) states that the amount A is increasing over time; if $k < 0$, the amount A is decreasing over time. In either case, when an amount A varies over time according to equation (1), it is said to follow the **exponential law**, or the **law of uninhibited growth** $(k > 0)$ **or decay** $(k < 0)$. See Figure 24.

For example, in Section 7.5, continuously compounded interest was shown to follow the law of uninhibited growth. In this section we shall look at some additional phenomena that follow the exponential law.

Cell division is the growth process of many living organisms, such as amoebas, plants, and human skin cells. Based on an ideal situation in which no cells die and no by-products are produced, the number of cells present at a given time follows the law of uninhibited growth. Actually, however, after enough time has passed, growth at an exponential rate will cease as a consequence of factors such as lack of living space and dwindling food supply. The law of uninhibited growth accurately models only the early stages of the cell division process.

The cell division process begins with a culture containing N_0 cells. Each cell in the culture grows for a certain period of time and then divides into two identical cells. Assume that the time needed for each cell to divide in two is constant and does not change as the number of cells increases. These new cells then grow, and eventually each divides in two, and so on.

(a) $A(t) = A_0 e^{kt}, k > 0$
Exponential growth

(b) $A(t) = A_0 e^{kt}, k < 0$
Exponential decay

Figure 24

Uninhibited Growth of Cells

A model that gives the number N of cells in a culture after a time t has passed (in the early stages of growth) is

$$N(t) = N_0 e^{kt} \qquad k > 0 \tag{2}$$

where N_0 is the initial number of cells and k is a positive constant that represents the growth rate of the cells.

Using formula (2) to model the growth of cells employs a function that yields positive real numbers, even though the number of cells being counted must be an integer. This is a common practice in many applications.

EXAMPLE 1   **Bacterial Growth**

A colony of bacteria that grows according to the law of uninhibited growth is modeled by the function $N(t) = 100e^{0.045t}$, where N is measured in grams and t is measured in days.

(a) Determine the initial amount of bacteria.
(b) What is the growth rate of the bacteria?
(c) What is the population after 5 days?
(d) How long will it take for the population to reach 140 grams?
(e) What is the doubling time for the population?

Solution   (a) The initial amount of bacteria, N_0, is obtained when $t = 0$, so

$$N_0 = N(0) = 100e^{0.045 \cdot 0} = 100 \text{ grams}$$

(b) Compare $N(t) = 100e^{0.045t}$ to $N(t) = N_0 e^{kt}$. The value of k, 0.045, indicates a growth rate of 4.5%.
(c) The population after 5 days is $N(5) = 100e^{0.045 \cdot 5} \approx 125.2$ grams.
(d) To find how long it takes for the population to reach 140 grams, solve the equation $N(t) = 140$.

$$100e^{0.045t} = 140$$
$$e^{0.045t} = 1.4 \qquad \textbf{\textcolor{blue}{Divide both sides of the equation by 100.}}$$
$$0.045t = \ln 1.4 \qquad \textbf{\textcolor{blue}{Rewrite as a logarithm.}}$$
$$t = \frac{\ln 1.4}{0.045} \qquad \textbf{\textcolor{blue}{Divide both sides of the equation by 0.045.}}$$
$$\approx 7.5 \text{ days}$$

The population reaches 140 grams in about 7.5 days.

(e) The population doubles when $N(t) = 200$ grams, so the doubling time is found by solving the equation $200 = 100e^{0.045t}$ for t.

$$200 = 100e^{0.045t}$$
$$2 = e^{0.045t} \qquad \textbf{\textcolor{blue}{Divide both sides of the equation by 100.}}$$
$$\ln 2 = 0.045t \qquad \textbf{\textcolor{blue}{Rewrite as a logarithm.}}$$
$$t = \frac{\ln 2}{0.045} \qquad \textbf{\textcolor{blue}{Divide both sides of the equation by 0.045.}}$$
$$\approx 15.4 \text{ days}$$

The population doubles approximately every 15.4 days.

 Now Work PROBLEM 1

EXAMPLE 2   **Bacterial Growth**

A colony of bacteria increases according to the law of uninhibited growth.

(a) If N is the number of cells and t is the time in hours, express N as a function of t.
(b) If the number of bacteria doubles in 3 hours, find the function that gives the number of cells in the culture.
(c) How long will it take for the size of the colony to triple?
(d) How long will it take for the population to double a second time (that is, to increase four times)?

Solution   (a) Using formula (2), the number N of cells at time t is

$$N(t) = N_0 e^{kt}$$

where N_0 is the initial number of bacteria present and k is a positive number.

(continued)

(b) To find the growth rate k, note that the number of cells doubles in 3 hours, so

$$N(3) = 2N_0$$

Since $N(3) = N_0 e^{k \cdot 3}$,

$$N_0 e^{k \cdot 3} = 2N_0$$
$$e^{3k} = 2 \qquad \text{Divide both sides by } N_0.$$
$$3k = \ln 2 \qquad \text{Write the exponential equation as a logarithm.}$$
$$k = \frac{1}{3}\ln 2 \approx 0.23105$$

The function that models this growth process is therefore

$$N(t) = N_0 \, e^{0.23105t}$$

(c) The time t needed for the size of the colony to triple requires that $N = 3N_0$. Substitute $3N_0$ for N to get

$$3N_0 = N_0 \, e^{0.23105t}$$
$$3 = e^{0.23105t}$$
$$0.23105t = \ln 3$$
$$t = \frac{\ln 3}{0.23105} \approx 4.755 \text{ hours}$$

It will take about 4.755 hours, or 4 hours and 45 minutes, for the size of the colony to triple.

(d) If a population doubles in 3 hours, it will double a second time in 3 more hours, for a total time of 6 hours.

2 Model Populations That Obey the Law of Uninhibited Decay

Radioactive materials follow the law of uninhibited decay.

Uninhibited Radioactive Decay

The amount A of a radioactive material present at time t is given by

$$A(t) = A_0 e^{kt} \qquad k < 0 \tag{3}$$

where A_0 is the original amount of radioactive material and k is a negative number that represents the rate of decay.

All radioactive substances have a specific **half-life**, which is the time required for half of the radioactive substance to decay. **Carbon dating** uses the fact that all living organisms contain two kinds of carbon, carbon-12 (a stable carbon) and carbon-14 (a radioactive carbon with a half-life of 5730 years). While an organism is living, the ratio of carbon-12 to carbon-14 is constant. But when an organism dies, the original amount of carbon-12 present remains unchanged, whereas the amount of carbon-14 begins to decrease. This change in the amount of carbon-14 present relative to the amount of carbon-12 present makes it possible to calculate when the organism died.

EXAMPLE 3 **Estimating the Age of Ancient Tools**

Traces of burned wood along with ancient stone tools in an archeological dig in Chile were found to contain approximately 1.67% of the original amount of carbon-14. If the half-life of carbon-14 is 5730 years, approximately when was the tree cut and burned?

Solution   Using formula (3), the amount A of carbon-14 present at time t is

$$A(t) = A_0 e^{kt}$$

where A_0 is the original amount of carbon-14 present and k is a negative number. We first seek the number k. To find it, we use the fact that after 5730 years, half of the original amount of carbon-14 remains, so $A(5730) = \frac{1}{2}A_0$. Then

$$\frac{1}{2}A_0 = A_0 e^{k \cdot 5730}$$

$$\frac{1}{2} = e^{5730k} \qquad \text{\color{blue}\textbf{Divide both sides of the equation by } } A_0.$$

$$5730k = \ln \frac{1}{2} \qquad \text{\color{blue}\textbf{Rewrite as a logarithm.}}$$

$$k = \frac{1}{5730} \ln \frac{1}{2} \approx -0.000120968$$

Formula (3) therefore becomes

$$A(t) = A_0 \, e^{-0.000120968t}$$

If the amount A of carbon-14 now present is 1.67% of the original amount, it follows that

$$0.0167A_0 = A_0 \, e^{-0.000120968t}$$

$$0.0167 = e^{-0.000120968t} \qquad \text{\color{blue}\textbf{Divide both sides of the equation by } } A_0.$$

$$-0.000120968t = \ln 0.0167 \qquad \text{\color{blue}\textbf{Rewrite as a logarithm.}}$$

$$t = \frac{\ln 0.0167}{-0.000120968} \approx 33{,}830 \text{ years}$$

The tree was cut and burned about 33,830 years ago. Some archeologists use this conclusion to argue that humans lived in the Americas nearly 34,000 years ago, much earlier than is generally accepted.

⌐ **Now Work** PROBLEM 3

3 Use Newton's Law of Cooling

Newton's Law of Cooling* states that the temperature of a heated object decreases exponentially over time toward the temperature of the surrounding medium.

> **Newton's Law of Cooling**
>
> The temperature u of a heated object at a given time t can be modeled by the following function:
>
> $$u(t) = T + (u_0 - T)e^{kt} \qquad k < 0 \qquad (4)$$
>
> where T is the constant temperature of the surrounding medium, u_0 is the initial temperature of the heated object, and k is a negative constant.

EXAMPLE 4

Using Newton's Law of Cooling

An object is heated to 100°C (degrees Celsius) and is then allowed to cool in a room whose air temperature is 30°C.

(a) If the temperature of the object is 80°C after 5 minutes, when will its temperature be 50°C?

(b) Determine the elapsed time before the temperature of the object is 35°C.

(c) What do you notice about the temperature as time passes?

*Named after Sir Isaac Newton (1643–1727), one of the cofounders of calculus.

Solution   (a) Using formula (4) with $T = 30$ and $u_0 = 100$, the temperature $u(t)$ (in degrees Celsius) of the object at time t (in minutes) is

$$u(t) = 30 + (100 - 30)e^{kt} = 30 + 70e^{kt} \qquad \text{(5)}$$

where k is a negative constant. To find k, use the fact that $u = 80$ when $t = 5$. Then

$$
\begin{aligned}
u(t) &= 30 + 70e^{kt} & \\
80 &= 30 + 70e^{k \cdot 5} & \color{blue}{u(5) = 80} \\
50 &= 70e^{5k} & \color{blue}{\text{Simplify.}} \\
e^{5k} &= \frac{50}{70} & \color{blue}{\text{Solve for } e^{5k}.} \\
5k &= \ln\frac{5}{7} & \color{blue}{\text{Rewrite as a logarithm.}} \\
k &= \frac{1}{5}\ln\frac{5}{7} \approx -0.0673 & \color{blue}{\text{Solve for } k.}
\end{aligned}
$$

Formula (5) therefore becomes

$$u(t) = 30 + 70e^{-0.0673t} \qquad \text{(6)}$$

To find t when $u = 50°C$, solve the equation

$$
\begin{aligned}
50 &= 30 + 70e^{-0.0673t} & \\
20 &= 70e^{-0.0673t} & \color{blue}{\text{Simplify.}} \\
e^{-0.0673t} &= \frac{20}{70} & \\
-0.0673t &= \ln\frac{2}{7} & \color{blue}{\text{Rewrite as a logarithm.}} \\
t &= \frac{\ln\dfrac{2}{7}}{-0.0673} \approx 18.6 \text{ minutes} & \color{blue}{\text{Solve for } t.}
\end{aligned}
$$

The temperature of the object will be 50°C after about 18.6 minutes, or 18 minutes, 36 seconds.

(b) Use equation (6) to find t when $u = 35°C$.

$$
\begin{aligned}
35 &= 30 + 70e^{-0.0673t} & \\
5 &= 70e^{-0.0673t} & \color{blue}{\text{Simplify.}} \\
e^{-0.0673t} &= \frac{5}{70} & \\
-0.0673t &= \ln\frac{5}{70} & \color{blue}{\text{Rewrite as a logarithm.}} \\
t &= \frac{\ln\dfrac{5}{70}}{-0.0673} \approx 39.2 \text{ minutes} & \color{blue}{\text{Solve for } t.}
\end{aligned}
$$

The object will reach a temperature of 35°C after about 39.2 minutes.

(c) Look at equation (6). As t increases, the exponent $-0.0673t$ becomes unbounded in the negative direction. As a result, the value of $e^{-0.0673t}$ approaches zero, so the value of u, the temperature of the object, approaches 30°C, the air temperature of the room.

━━━▶ **Now Work** PROBLEM 13

4 Use Logistic Models

The exponential growth model $A(t) = A_0 e^{kt}$, $k > 0$, assumes uninhibited growth, meaning that the value of the function grows without limit. Recall that cell division could be modeled using this function, assuming that no cells die and no by-products are produced. However, cell division eventually is limited by factors such as living space and food supply. The **logistic model**, given next, can describe situations where the growth or decay of the dependent variable is limited.

> **Logistic Model**
>
> In a logistic model, the population P after time t is given by the function
>
> $$P(t) = \frac{c}{1 + ae^{-bt}} \qquad (7)$$
>
> where a, b, and c are constants with $a > 0$ and $c > 0$. The model is a growth model if $b > 0$; the model is a decay model if $b < 0$.

The number c is called the **carrying capacity** (for growth models) because the value $P(t)$ approaches c as t approaches infinity; that is, $\lim\limits_{t \to \infty} P(t) = c$. The number $|b|$ is the growth rate for $b > 0$ and the decay rate for $b < 0$. Figure 25(a) shows the graph of a typical logistic growth function, and Figure 25(b) shows the graph of a typical logistic decay function.

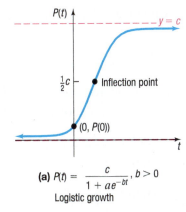

Figure 25

(a) $P(t) = \dfrac{c}{1 + ae^{-bt}}$, $b > 0$
Logistic growth

(b) $P(t) = \dfrac{c}{1 + ae^{-bt}}$, $b < 0$
Logistic decay

Based on the figures, the following properties of logistic functions emerge.

> **Properties of the Logistic Model, Equation (7)**
>
> - The domain is the set of all real numbers. The range is the interval $(0, c)$ where c is the carrying capacity.
> - There are no x-intercepts; the y-intercept is $P(0)$.
> - There are two horizontal asymptotes: $y = 0$ and $y = c$.
> - $P(t)$ is an increasing function if $b > 0$ and a decreasing function if $b < 0$.
> - There is an **inflection point** where $P(t)$ equals $\dfrac{1}{2}$ of the carrying capacity.
>
>   The inflection point is the point on the graph where the graph changes from being concave up to being concave down for growth functions, and the point where the graph changes from being concave down to being concave up for decay functions.
> - The graph is smooth and continuous, with no corners or gaps.

| EXAMPLE 5 | **Fruit Fly Population** |

Fruit flies are placed in a half-pint milk bottle with a banana (for food) and yeast plants (for food and to provide a stimulus to lay eggs). Suppose that the fruit fly population after t days is given by

$$P(t) = \frac{230}{1 + 56.5e^{-0.37t}}$$

(a) State the carrying capacity and the growth rate.
(b) Determine the initial population.
(c) What is the population after 5 days?
(d) How long does it take for the population to reach 180?
(e) Use a graphing utility to determine how long it takes for the population to reach one-half of the carrying capacity.

Solution

(a) As $t \to \infty$, $e^{-0.37t} \to 0$ and $P(t) \to \dfrac{230}{1}$. The carrying capacity of the half-pint bottle is 230 fruit flies. The growth rate is $|b| = |0.37| = 37\%$ per day.

(b) To find the initial number of fruit flies in the half-pint bottle, evaluate $P(0)$.

$$P(0) = \frac{230}{1 + 56.5e^{-0.37 \cdot 0}} = \frac{230}{1 + 56.5} = 4$$

So, initially, there were 4 fruit flies in the half-pint bottle.

(c) After 5 days the number of fruit flies in the half-pint bottle is

$$P(5) = \frac{230}{1 + 56.5e^{-0.37 \cdot 5}} \approx 23 \text{ fruit flies}$$

After 5 days, there are approximately 23 fruit flies in the bottle.

(d) To determine when the population of fruit flies will be 180, solve the equation

$$P(t) = 180$$

$$\frac{230}{1 + 56.5e^{-0.37t}} = 180$$

$$230 = 180(1 + 56.5e^{-0.37t})$$

$$0.2778 \approx 56.5e^{-0.37t} \qquad \text{Divide both sides by 180.}$$

$$0.2778 \approx 56.5e^{-0.37t} \qquad \text{Subtract 1 from both sides.}$$

$$0.0049 \approx e^{-0.37t} \qquad \text{Divide both sides by 56.5.}$$

$$\ln(0.0049) \approx -0.37t \qquad \text{Rewrite as a logarithmic expression.}$$

$$t \approx 14.4 \text{ days} \qquad \text{Divide both sides by } -0.37.$$

$Y_1 = \dfrac{230}{1 + 56.5e^{-0.37x}}$

Figure 26 TI-84 Plus C

It will take approximately 14.4 days (14 days, 10 hours) for the population to reach 180 fruit flies.

(e) One-half of the carrying capacity is 115 fruit flies. Solve $P(t) = 115$ by graphing $Y_1 = \dfrac{230}{1 + 56.5e^{-0.37t}}$ and $Y_2 = 115$, and using INTERSECT. See Figure 26.

The population will reach one-half of the carrying capacity in about 10.9 days (10 days, 22 hours).

 Look at Figure 26. Notice the point where the graph reaches 115 fruit flies (one-half of the carrying capacity): The graph changes from being concave up to being concave down. Using the language of calculus, we say the graph changes from increasing at an increasing rate to increasing at a decreasing rate. For any logistic growth function, when the population reaches one-half the carrying capacity, the population growth starts to slow down.

━━━━➤ **Now Work** PROBLEM 23

Exploration

On the same viewing rectangle, graph

$$Y_1 = \frac{500}{1 + 24e^{-0.03t}} \quad \text{and} \quad Y_2 = \frac{500}{1 + 24e^{-0.08t}}$$

What effect does the growth rate $|b|$ have on the logistic growth function?

EXAMPLE 6 | **Wood Products**

The EFISCEN wood product model classifies wood products according to their life-span. There are four classifications: short (1 year), medium short (4 years), medium long (16 years), and long (50 years). Based on data obtained from the European Forest Institute, the percentage of remaining wood products after t years for wood products with long life spans (such as those used in the building industry) is given by

$$P(t) = \frac{100.3952}{1 + 0.0316e^{0.0581t}}$$

(a) What is the decay rate?

(b) What is the percentage of remaining wood products after 10 years?

(c) How long does it take for the percentage of remaining wood products to reach 50%?

(d) Explain why the numerator given in the model is reasonable.

Solution   (a) The decay rate is $|b| = |-0.0581| = 5.81\%$ per year.

(b) Evaluate $P(10)$.

$$P(10) = \frac{100.3952}{1 + 0.0316e^{0.0581 \cdot 10}} \approx 95.0$$

So 95% of long-life-span wood products remain after 10 years.

(c) Solve the equation $P(t) = 50$.

$$\frac{100.3952}{1 + 0.0316e^{0.0581t}} = 50$$

$$100.3952 = 50(1 + 0.0316e^{0.0581t})$$

$2.0079 \approx 1 + 0.0316e^{0.0581t}$ **Divide both sides by 50.**

$1.0079 \approx 0.0316e^{0.0581t}$ **Subtract 1 from both sides.**

$31.8956 \approx e^{0.0581t}$ **Divide both sides by 0.0316.**

$\ln(31.8956) \approx 0.0581t$ **Rewrite as a logarithmic expression.**

$t \approx 59.6$ years   **Divide both sides by 0.0581.**

It will take approximately 59.6 years for the percentage of long-life-span wood products remaining to reach 50%.

(d) The numerator of 100.3952 is reasonable because the maximum percentage of wood products remaining that is possible is 100%.

7.6 Assess Your Understanding

Applications and Extensions

1. **Growth of an Insect Population** The size P of a certain insect population at time t (in days) obeys the law of uninhibited growth $P(t) = 500e^{0.02t}$.
   (a) Determine the number of insects at $t = 0$ days.
   (b) What is the growth rate of the insect population?
   (c) What is the population after 10 days?
   (d) When will the insect population reach 800?
   (e) When will the insect population double?

2. **Growth of Bacteria** The number N of bacteria present in a culture at time t (in hours) obeys the law of uninhibited growth $N(t) = 1000e^{0.01t}$.
   (a) Determine the number of bacteria at $t = 0$ hours.
   (b) What is the growth rate of the bacteria?
   (c) What is the population after 4 hours?
   (d) When will the number of bacteria reach 1700?
   (e) When will the number of bacteria double?

3. **Radioactive Decay** Strontium-90 is a radioactive material that decays according to the function $A(t) = A_0 e^{-0.0244t}$, where A_0 is the initial amount present and A is the amount present at time t (in years). Assume that a scientist has a sample of 500 grams of strontium-90.
   (a) What is the decay rate of strontium-90?
   (b) How much strontium-90 is left after 10 years?
   (c) When will 400 grams of strontium-90 be left?
   (d) What is the half-life of strontium-90?

4. **Radioactive Decay** Iodine-131 is a radioactive material that decays according to the function $A(t) = A_0 e^{-0.087t}$, where A_0 is the initial amount present and A is the amount present at time t (in days). Assume that a scientist has a sample of 100 grams of iodine-131.
   (a) What is the decay rate of iodine-131?
   (b) How much iodine-131 is left after 9 days?
   (c) When will 70 grams of iodine-131 be left?
   (d) What is the half-life of iodine-131?

5. **Growth of a Colony of Mosquitoes** The population of a colony of mosquitoes obeys the law of uninhibited growth.
   (a) If N is the population of the colony and t is the time in days, express N as a function of t.
   (b) If there are 1000 mosquitoes initially and there are 1800 after 1 day, what is the size of the colony after 3 days?
   (c) How long is it until there are 10,000 mosquitoes?

6. **Bacterial Growth** A culture of bacteria obeys the law of uninhibited growth.
   (a) If N is the number of bacteria in the culture and t is the time in hours, express N as a function of t.
   (b) If 500 bacteria are present initially and there are 800 after 1 hour, how many will be present in the culture after 5 hours?
   (c) How long is it until there are 20,000 bacteria?

7. **Population Growth** The population of a southern city is growing according to the exponential law.
   (a) If N is the population of the city and t is the time in years, express N as a function of t.
   (b) If the population doubled in size over an 18-month period and the current population is 10,000, what will the population be 2 years from now?

8. **Population Decline** The population of a midwestern city is declining according to the exponential law.
   (a) If N is the population of the city and t is the time in years, express N as a function of t.
   (b) If the population decreased from 900,000 to 800,000 from 2016 to 2018, what will the population be in 2020?

9. **Radioactive Decay** The half-life of radium is 1690 years. If 10 grams is present now, how much will be present in 50 years?

10. **Radioactive Decay** The half-life of radioactive potassium is 1.3 billion years. If 10 grams is present now, how much will be present in 100 years? In 1000 years?

11. **Estimating the Age of a Tree** A piece of charcoal is found to contain 30% of the carbon-14 that it originally had. When did the tree die from which the charcoal came? Use 5730 years as the half-life of carbon-14.

12. **Estimating Age** A fossilized leaf contains 70% of its normal amount of carbon-14. How old is the fossil?

13. **Cooling Time of a Pizza** A pizza baked at 450°F is removed from the oven at 5:00 PM and placed in a room that is a constant 70°F. After 5 minutes, the pizza is at 300°F.
    (a) At what time can you begin eating the pizza if you want its temperature to be 135°F?
    (b) Determine the time that needs to elapse before the temperature of the pizza is 160°F.
    (c) What do you notice about the temperature as time passes?

14. **Newton's Law of Cooling** A thermometer reading 72°F is placed in a refrigerator where the temperature is a constant 38°F.
    (a) If the thermometer reads 60°F after 2 minutes, what will it read after 7 minutes?
    (b) How long will it take before the thermometer reads 39°F?
    (c) Determine the time that must elapse before the thermometer reads 45°F.
    (d) What do you notice about the temperature as time passes?

15. **Newton's Law of Heating** A thermometer reading 8°C is brought into a room with a constant temperature of 35°C. If the thermometer reads 15°C after 3 minutes, what will it read after being in the room for 5 minutes? For 10 minutes?
    [**Hint:** You need to construct a formula similar to equation (4).]

16. Warming Time of a Beer Stein A beer stein has a temperature of 28°F. It is placed in a room with a constant temperature of 70°F. After 10 minutes, the temperature of the stein has risen to 35°F. What will the temperature of the stein be after 30 minutes? How long will it take the stein to reach a temperature of 45°F? (See the hint given for Problem 15.)

17. Decomposition of Chlorine in a Pool Under certain water conditions, the free chlorine (hypochlorous acid, HOCl) in a swimming pool decomposes according to the law of uninhibited decay. After shocking his pool, Ben tested the water and found the amount of free chlorine to be 2.5 parts per million (ppm). Twenty-four hours later, Ben tested the water again and found the amount of free chlorine to be 2.2 ppm. What will be the reading after 3 days (that is, 72 hours)? When the chlorine level reaches 1.0 ppm, Ben must shock the pool again. How long can Ben go before he must shock the pool again?

18. Decomposition of Dinitrogen Pentoxide At 45°C, dinitrogen pentoxide (N_2O_5) decomposes into nitrous dioxide (NO_2) and oxygen (O_2) according to the law of uninhibited decay. An initial amount of 0.25 M N_2O_5 (M is a measure of concentration known as molarity) decomposes to 0.15 M N_2O_5 in 17 minutes. What concentration of N_2O_5 will remain after 30 minutes? How long will it take until only 0.01 M N_2O_5 remains?

19. Decomposition of Sucrose Reacting with water in an acidic solution at 35°C, sucrose ($C_{12}H_{22}O_{11}$) decomposes into glucose ($C_6H_{12}O_6$) and fructose ($C_6H_{12}O_6$)* according to the law of uninhibited decay. An initial concentration of 0.40 M of sucrose decomposes to 0.36 M sucrose in 30 minutes. What concentration of sucrose will remain after 2 hours? How long will it take until only 0.10 M sucrose remains?

20. Decomposition of Salt in Water Salt (NaCl) decomposes in water into sodium (Na^+) and chloride (Cl^-) ions according to the law of uninhibited decay. If the initial amount of salt is 25 kilograms and, after 10 hours, 15 kilograms of salt is left, how much salt is left after 1 day? How long does it take until $\frac{1}{2}$ kilogram of salt is left?

21. Radioactivity from Chernobyl After the release of radioactive material into the atmosphere from a nuclear power plant at Chernobyl (Ukraine) in 1986, the hay in Austria was contaminated by iodine 131 (half-life 8 days). If it is safe to feed the hay to cows when 10% of the iodine 131 remains, how long did the farmers need to wait to use this hay?

22. Word Users According to a survey by Olsten Staffing Services, the percentage of companies reporting usage of Microsoft Word t years since 1984 is given by

$$P(t) = \frac{99.744}{1 + 3.014e^{-0.799t}}$$

(a) What is the growth rate in the percentage of Microsoft Word users?
(b) Use a graphing utility to graph $P = P(t)$.
(c) What was the percentage of Microsoft Word users in 1990?
(d) During what year did the percentage of Microsoft Word users reach 90%?
(e) Explain why the numerator given in the model is reasonable. What does it imply?

23. Home Computers The logistic model

$$P(t) = \frac{95.4993}{1 + 0.0405e^{0.1968t}}$$

represents the percentage of households that do not own a personal computer t years since 1984.
(a) Evaluate and interpret $P(0)$.
(b) Use a graphing utility to graph $P = P(t)$.
(c) What percentage of households did not own a personal computer in 1995?
(d) In what year did the percentage of households that do not own a personal computer reach 10%?
Source: U.S. Department of Commerce

24. Farmers The logistic model

$$W(t) = \frac{14,656,248}{1 + 0.059e^{0.057t}}$$

represents the number of farm workers in the United States t years after 1910.
(a) Evaluate and interpret $W(0)$.
(b) Use a graphing utility to graph $W = W(t)$.
(c) How many farm workers were there in the United States in 2010?
(d) When did the number of farm workers in the United States reach 10,000,000?
(e) According to this model, what happens to the number of farm workers in the United States as t approaches ∞? Based on this result, do you think that it is reasonable to use this model to predict the number of farm workers in the United States in 2060? Why?
Source: U.S. Department of Agriculture

25. Birthdays The logistic model

$$P(n) = \frac{113.3198}{1 + 0.115e^{0.0912n}}$$

models the probability that, in a room of n people, no two people share the same birthday.
(a) Use a graphing utility to graph $P = P(n)$.
(b) In a room of $n = 15$ people, what is the probability that no two share the same birthday?
(c) How many people must be in a room before the probability that no two people share the same birthday falls below 10%?
(d) What happens to the probability as n increases? Explain what this result means.

26. Population of an Endangered Species Environmentalists often capture an endangered species and transport the species to a controlled environment where the species can produce offspring and regenerate its population. Suppose that six American bald eagles are captured, transported to Montana, and set free. Based on experience, the environmentalists expect the population to grow according to the model

$$P(t) = \frac{500}{1 + 83.33e^{-0.162t}}$$

where t is measured in years.

*Author's Note: Surprisingly, the chemical formulas for glucose and fructose are the same: This is not a typo.   (*continued*)

(a) Determine the carrying capacity of the environment.
(b) What is the growth rate of the bald eagle?
(c) What is the population after 3 years?
(d) When will the population be 300 eagles?
(e) How long does it take for the population to reach one-half of the carrying capacity?

27. Invasive Species A habitat can be altered by invasive species that crowd out or replace native species. The logistic model

$$P(t) = \frac{431}{1 + 7.91e^{-0.017t}}$$

represents the number of invasive species present in the Great Lakes t years after 1900.

(a) Evaluate and interpret $P(0)$.
(b) What is the growth rate of invasive species?
(c) Use a graphing utility to graph $P = P(t)$.
(d) How many invasive species were present in the Great Lakes in 2000?
(e) In what year was the number of invasive species 175?

Source: NOAA

28. Social Networking The logistic model

$$P(t) = \frac{86.1}{1 + 2.12e^{-0.361t}}$$

gives the percentage of Americans who have a social media profile, where t represents the number of years after 2008.

(a) Evaluate and interpret $P(0)$.
(b) What is the growth rate?
(c) Use a graphing utility to graph $P = P(t)$.
(d) During 2017, what percentage of Americans had a social media profile?
(e) In what year did 69.3% of Americans have a social media profile?

Source: Statista, 2018

Problems 29 and 30 use the following discussion: Uninhibited growth can be modeled by exponential functions other than $A(t) = A_0e^{kt}$. For example, if an initial population P_0 requires n units of time to double, then the function $P(t) = P_0 \cdot 2^{t/n}$ models the size of the population at time t. Likewise, a population requiring n units of time to triple can be modeled by $P(t) = P_0 \cdot 3^{t/n}$.

29. Growth of an Insect Population An insect population grows exponentially.
(a) If the population triples in 20 days, and 50 insects are present initially, write an exponential function of the form $P(t) = P_0 \cdot 3^{t/n}$ that models the population.
(b) What will the population be in 47 days?
(c) When will the population reach 700?
(d) Express the model from part (a) in the form $A(t) = A_0e^{kt}$.

30. Growth of a Human Population The population of a town is growing exponentially.
(a) If its population doubled in size over an 8-year period and the current population is 25,000, write an exponential function of the form $P(t) = P_0 \cdot 2^{t/n}$ that models the population.
(b) What will the population be in 3 years?
(c) When will the population reach 80,000?
(d) Express the model from part (a) in the form $A(t) = A_0e^{kt}$.

Retain Your Knowledge

Problems 31–40 are based on material learned earlier in the course. The purpose of these problems is to keep the material fresh in your mind so that you are better prepared for the final exam.

31. Find the linear function f whose graph contains the points $(4, 1)$ and $(8, -5)$.

32. Determine whether the graphs of the linear functions $f(x) = 5x - 1$ and $g(x) = \frac{1}{5}x + 1$ are parallel, perpendicular, or neither.

33. Write the logarithmic expression $\ln\left(\frac{x^2\sqrt{y}}{z}\right)$ as the sum and/or difference of logarithms. Express powers as factors.

34. Find the domain of $f(x) = \frac{x + 3}{x^2 + 2x - 8}$.

35. If $f(x) = \frac{2x - 3}{x - 4}$ and $g(x) = \frac{3x + 1}{x - 3}$, find $(g - f)(x)$.

36. Find the x-intercept(s) and y-intercept(s) of the graph of $f(x) = 2x^2 - 5x + 1$.

37. Solve: $\frac{x + 1}{x} - \frac{x}{x + 1} = 2$

38. For the data provided, use a graphing utility to find the line of best fit. What is the correlation coefficient?

x	-4	-2	0	2	4	6
y	9	5	4	2	-1	-2

39. Use a graphing utility to graph $f(x) = x^4 - 3x^2 + 2x - 1$ over the interval $[-3, 3]$. Then, approximate any local maximum values and local minimum values, and determine where f is increasing and where f is decreasing. Round answers to two decimal places.

40. Write $\frac{10x}{3(2x + 3)^{2/3}} + 5(2x + 3)^{1/3}$ as a single quotient in which only positive exponents appear.

7.7 Building Exponential, Logarithmic, and Logistic Models from Data

OBJECTIVES **1** Build an Exponential Model from Data (p. 527)
                 2 Build a Logarithmic Model from Data (p. 529)
                 3 Build a Logistic Model from Data (p. 529)

In this section we discuss how to use a graphing utility to find equations of best fit that describe the relation between two variables when the relation is thought to be exponential $(y = ab^x)$, logarithmic $(y = a + b \ln x)$, or logistic $\left(y = \dfrac{c}{1 + ae^{-bx}} \right)$.
As before, we draw a scatter plot of the data to help to determine the appropriate model to use.

Figure 27 shows scatter plots that are typically observed for the three models.

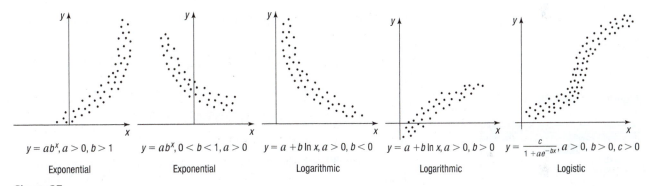

$y = ab^x, a > 0, b > 1$	$y = ab^x, 0 < b < 1, a > 0$	$y = a + b \ln x, a > 0, b < 0$	$y = a + b \ln x, a > 0, b > 0$	$y = \dfrac{c}{1 + ae^{-bx}}, a > 0, b > 0, c > 0$
Exponential	Exponential	Logarithmic	Logarithmic	Logistic

Figure 27

Most graphing utilities have REGression options that fit data to a specific type of curve. Once the data have been entered and a scatter plot obtained, the type of curve that you want to fit to the data is selected. Then that REGression option is used to obtain the curve of *best fit* of the type selected.

The correlation coefficient r will appear only if the model can be written as a linear expression. As it turns out, r will appear for the linear, power, exponential, and logarithmic models, since these models can be written as a linear expression. Remember, the closer $|r|$ is to 1, the better the fit.

1 Build an Exponential Model from Data

We saw in Section 7.5 that money earning compound interest grows exponentially, and we saw in Section 7.6 that growth and decay models also can behave exponentially. The next example shows how data can lead to an exponential model.

EXAMPLE 1

Fitting an Exponential Function to Data

Mariah deposited $20,000 into a well-diversified mutual fund 6 years ago. The data in Table 9 represent the value of the account each year for the last 7 years.

(a) Using a graphing utility, draw a scatter plot with year as the independent variable.

(b) Using a graphing utility, build an exponential model from the data.

(c) Express the function found in part (b) in the form $A = A_0 e^{kt}$.

(continued)

(d) Graph the exponential function found in part (b) or (c) on the scatter plot.

(e) Using the solution to part (b) or (c), predict the value of the account after 10 years.

(f) Interpret the value of k found in part (c).

Solution

Table 9

Year, x	Account Value, y
0	20,000
1	21,516
2	23,355
3	24,885
4	27,484
5	30,053
6	32,622

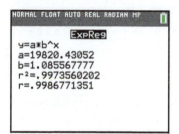

Figure 28 TI-84 Plus C

Figure 29 Exponential model using a TI-84 Plus C

(a) Enter the data into the graphing utility and draw the scatter plot as shown in Figure 28 on a TI-84 Plus C.

(b) A graphing utility fits the data in Table 9 to an exponential model of the form $y = ab^x$ using the EXPonential REGression option. Figure 29 shows that $y = ab^x = 19{,}820.43\,(1.085568)^x$ on a TI-84 Plus C. Notice that $|r| = 0.999$, which is close to 1, indicating a good fit.

(c) To express $y = ab^x$ in the form $A = A_0 e^{kt}$, where $x = t$ and $y = A$, proceed as follows:

$$ab^x = A_0 e^{kt}$$

If $x = t = 0$, then $a = A_0$. This leads to

$$b^x = e^{kt}$$
$$b^x = (e^k)^t$$
$$b = e^k \qquad \textbf{x = t}$$

Because $y = ab^x = 19{,}820.43\,(1.085568)^x$, this means that $a = 19{,}820.43$ and $b = 1.085568$.

$$a = A_0 = 19{,}820.43 \quad \text{and} \quad b = e^k = 1.085568$$

To find k, rewrite $e^k = 1.085568$ as a logarithm to obtain

$$k = \ln(1.085568) \approx 0.08210$$

As a result, $A = A_0 e^{kt} = 19{,}820.43 e^{0.08210t}$.

(d) See Figure 30 for the graph of the exponential function of best fit on a TI-84 Plus C. Figure 31 shows the exponential model using Desmos.

Figure 30

Figure 31 Exponential model using Desmos*

(e) Let $t = 10$ in the function found in part (c). The predicted value of the account after 10 years is

$$A = A_0 e^{kt} = 19{,}820.43 e^{0.08210(10)} \approx \$45{,}047$$

(f) The value of $k = 0.08210 = 8.210\%$ represents the annual growth rate of the account. It represents the rate of interest earned, assuming the account is growing continuously.

⌐

━━━ **Now Work** PROBLEM 1

---

*For this result in Desmos to agree precisely with the result of a TI-84 Plus C, the "Log Mode" option must be selected. Consult the help feature in Desmos for more information about this option.

2 Build a Logarithmic Model from Data

Some relations between variables follow a logarithmic model.

EXAMPLE 2 **Fitting a Logarithmic Function to Data**

Table 10

Atmospheric Pressure, p	Height, h
760	0
740	0.184
725	0.328
700	0.565
650	1.079
630	1.291
600	1.634
580	1.862
550	2.235

Jodi, a meteorologist, is interested in finding a function that explains the relation between the height of a weather balloon (in kilometers) and the atmospheric pressure (measured in millimeters of mercury) on the balloon. She collects the data shown in Table 10.

(a) Using a graphing utility, draw a scatter plot of the data with atmospheric pressure as the independent variable.

(b) It is known that the relation between atmospheric pressure and height follows a logarithmic model. Using a graphing utility, build a logarithmic model from the data.

(c) Graph the logarithmic function found in part (b) on the scatter plot.

(d) Use the function found in part (b) to predict the height of the weather balloon if the atmospheric pressure is 560 millimeters of mercury.

Solution

(a) Enter the data into the graphing utility, and draw the scatter plot. See Figure 32.

(b) A graphing utility fits the data in Table 10 to a logarithmic function of the form $y = a + b \ln x$ by using the LOGarithm REGression option. Figure 33 shows the result on a TI-84 Plus C. The logarithmic model from the data is

$$h(p) = 45.7863 - 6.9025 \ln p$$

Figure 32 TI-84 Plus C

where h is the height of the weather balloon and p is the atmospheric pressure. Notice that $|r|$ is close to 1, indicating a good fit.

(c) Figure 34 shows the graph of $h(p) = 45.7863 - 6.9025 \ln p$ on the scatter plot. Figure 35 shows the logarithmic model using Desmos.

Figure 33 Logarithmic model using a TI-84 Plus C

Figure 34

Figure 35 Logarithmic model using Desmos

(d) Using the function found in part (b), Jodi predicts the height of the weather balloon when the atmospheric pressure is 560 to be

$$h(560) = 45.7863 - 6.9025 \ln 560$$
$$\approx 2.108 \text{ kilometers}$$

 Now Work PROBLEM 5

3 Build a Logistic Model from Data

Logistic growth models can be used to model situations for which the value of the dependent variable is limited. Many real-world situations conform to this scenario. For example, the population of the human race is limited by the availability of natural resources, such as food and shelter. When the value of the dependent variable is limited, a logistic growth model is often appropriate.

| EXAMPLE 3 | **Fitting a Logistic Function to Data** |

The data in Table 11 represent the amount of yeast biomass in a culture after t hours.

Table 11

Time (hours)	Yeast Biomass	Time (hours)	Yeast Biomass	Time (hours)	Yeast Biomass
0	9.6	7	257.3	14	640.8
1	18.3	8	350.7	15	651.1
2	29.0	9	441.0	16	655.9
3	47.2	10	513.3	17	659.6
4	71.1	11	559.7	18	661.8
5	119.1	12	594.8		
6	174.6	13	629.4		

Source: Tor Carlson (Über Geschwindigkeit und Grösse der Hefevermehrung in Würze, Biochemische Zeitschrift, Bd. 57, pp. 313–334, 1913)

(a) Using a graphing utility, draw a scatter plot of the data with time as the independent variable.

(b) Using a graphing utility, build a logistic model from the data.

(c) Using a graphing utility, graph the function found in part (b) on the scatter plot.

(d) What is the predicted carrying capacity of the culture?

(e) Use the function found in part (b) to predict the population of the culture at $t = 19$ hours.

Solution

(a) See Figure 36 for a scatter plot of the data on a TI-84 Plus C.

(b) A graphing utility fits the data in Table 11 to a logistic growth model of the form $y = \dfrac{c}{1 + ae^{-bx}}$ by using the LOGISTIC regression option. Figure 37 shows the result on a TI-84 Plus C. The logistic model from the data is

$$y = \frac{663.0}{1 + 71.6e^{-0.5470x}}$$

where y is the amount of yeast biomass in the culture and x is the time.

(c) See Figure 38 for the graph of the logistic model on a TI-84 Plus C. Figure 39 shows the logistic model using Desmos.

Figure 36 TI-84 Plus C

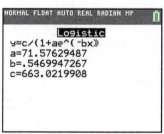

Figure 37 Logistic model using a TI-84 Plus C

Figure 38

Figure 39 Logistic model using Demos

(d) Based on the logistic growth model found in part (b), the carrying capacity of the culture is 663.

(e) Using the logistic growth model found in part (b), the predicted amount of yeast biomass at $t = 19$ hours is

$$y = \frac{663.0}{1 + 71.6e^{-0.5470 \cdot 19}} \approx 661.5$$

━━━ **Now Work** PROBLEM 7

7.7 Assess Your Understanding

Applications and Extensions

1. Biology A strain of E-coli Beu 397-recA441 is placed into a nutrient broth at 30° Celsius and allowed to grow. The following data are collected. Theory states that the number of bacteria in the petri dish will initially grow according to the law of uninhibited growth. The population is measured using an optical device in which the amount of light that passes through the petri dish is measured.

Time (hours), x	Population, y
0	0.09
2.5	0.18
3.5	0.26
4.5	0.35
6	0.50

Source: Dr. Polly Lavery, Joliet Junior College

(a) Draw a scatter plot treating time as the independent variable.
(b) Using a graphing utility, build an exponential model from the data.
(c) Express the function found in part (b) in the form $N(t) = N_0 e^{kt}$.
(d) Graph the exponential function found in part (b) or (c) on the scatter plot.
(e) Use the exponential function from part (b) or (c) to predict the population at $x = 7$ hours.
(f) Use the exponential function from part (b) or (c) to predict when the population will reach 0.75.

2. Tesla, Inc. Revenue The data in the table below represent annual revenue of Tesla, Inc. from 2010 to 2017.

Year	Revenue ($ Billion)
2010 (x = 0)	0.12
2011 (x = 1)	0.20
2012 (x = 2)	0.41
2013 (x = 3)	2.01
2014 (x = 4)	3.20
2015 (x = 5)	4.05
2016 (x = 6)	7.00
2017 (x = 7)	11.76

Source: Tesla, Inc.

(a) Using a graphing utility, draw a scatter plot of the data using 0 for 2010, 1 for 2011, and so on, as the independent variable.
(b) Using a graphing utility, build an exponential model from the data.
(c) Express the function found in part (b) in the form $A(t) = A_0 e^{kt}$.

(d) Graph the exponential function found in part (b) or (c) on the scatter plot.
(e) Use the exponential function from part (b) or (c) to predict Tesla's revenue in 2019.
(f) Interpret the meaning of k in the function found in part (c).

3. Advanced-Stage Breast Cancer The data in the table below represent the percentage of patients who have survived after diagnosis of advanced-stage breast cancer at 6-month intervals of time.

Time after Diagnosis (years)	Percentage Surviving
0.5	95.7
1	83.6
1.5	74.0
2	58.6
2.5	47.4
3	41.9
3.5	33.6

Source: Cancer Treatment Centers of America

(a) Using a graphing utility, draw a scatter plot of the data with time after diagnosis as the independent variable.
(b) Using a graphing utility, build an exponential model from the data.
(c) Express the function found in part (b) in the form $A(t) = A_0 e^{kt}$.
(d) Graph the exponential function found in part (b) or (c) on the scatter plot.
(e) Use the model to predict the percentage of patients diagnosed with advanced-stage cancer who survive for 4 years after initial diagnosis?
(f) Interpret the meaning of k in the function found in part (c).

4. Chemistry A chemist has a 100-gram sample of a radioactive material. He records the amount of radioactive material every week for 7 weeks and obtains the following data:

Week	Weight (in grams)
0	100.0
1	88.3
2	75.9
3	69.4
4	59.1
5	51.8
6	45.5

(continued)

(a) Using a graphing utility, draw a scatter plot with week as the independent variable.

(b) Using a graphing utility, build an exponential model from the data.

(c) Express the function found in part (b) in the form $A(t) = A_0 e^{kt}$.

(d) Graph the exponential function found in part (b) or (c) on the scatter plot.

(e) From the result found in part (b), determine the half-life of the radioactive material.

(f) How much radioactive material will be left after 50 weeks?

(g) When will there be 20 grams of radioactive material?

5. Milk Production The data in the table below represent the number of dairy farms (in thousands) and the amount of milk produced (in billions of pounds) in the United States for various years.

Year	Dairy Farms (thousands)	Milk Produced (billion pounds)
1980	334	128
1985	269	143
1990	193	148
1995	140	155
2000	105	167
2005	78	177
2010	63	193
2015	44	209

Source: National Agricultural Statistics Services

(a) Using a graphing utility, draw a scatter plot of the data with the number of dairy farms as the independent variable.

(b) Using a graphing utility, build a logarithmic model from the data.

(c) Graph the logarithmic function found in part (b) on the scatter plot.

(d) In 2008, there were 67 thousand dairy farms in the United States. Use the function in part (b) to predict the amount of milk produced in 2008.

(e) The actual amount of milk produced in 2008 was 190 billion pounds. How does your prediction in part (d) compare to this?

6. Social Networking The data in the table below represent the percent of U.S. citizens aged 12 and older who have a profile on at least one social network.

Year	Percent on a Social Networking Site
2008 ($x = 8$)	24
2009 ($x = 9$)	34
2010 ($x = 10$)	48
2011 ($x = 11$)	52
2012 ($x = 12$)	56
2013 ($x = 13$)	62
2014 ($x = 14$)	67
2015 ($x = 15$)	73
2016 ($x = 16$)	78
2017 ($x = 17$)	81

Source: Statista.com

(a) Using a graphing utility, draw a scatter plot of the data using 8 for 2008, 9 for 2009, and so on, as the independent variable, and percent on social networking site as the dependent variable.

(b) Using a graphing utility, build a logarithmic model from the data.

(c) Graph the logarithmic function found in part (b) on the scatter plot.

(d) Use the model to predict the percent of U.S. citizens on social networking sites in 2019.

(e) Use the model to predict the year in which 98% of U.S. citizens will be on social networking sites.

7. Population Model The following data represent the population of the United States. An ecologist is interested in building a model that describes the population of the United States.

Year	Population
1900	76,212,168
1910	92,228,496
1920	106,021,537
1930	123,202,624
1940	132,164,569
1950	151,325,798
1960	179,323,175
1970	203,302,031
1980	226,542,203
1990	248,709,873
2000	281,421,906
2010	308,745,538

Source: U.S. Census Bureau

(a) Using a graphing utility, draw a scatter plot of the data using years since 1900 as the independent variable and population as the dependent variable.

(b) Using a graphing utility, build a logistic model from the data.

(c) Using a graphing utility, graph the function found in part (b) on the scatter plot.

(d) Based on the function found in part (b), what is the carrying capacity of the United States?

(e) Use the function found in part (b) to predict the population of the United States in 2012.

(f) When will the United States population be 350,000,000?

(g) Compare actual U.S. Census figures to the predictions found in parts (e) and (f). Discuss any differences.

8. Population Model The data that follow on the next page represent world population. An ecologist is interested in building a model that describes the world population.

(a) Using a graphing utility, draw a scatter plot of the data using years since 2000 as the independent variable and population as the dependent variable.

(b) Using a graphing utility, build a logistic model from the data.

(c) Using a graphing utility, graph the function found in part (b) on the scatter plot.

(d) Based on the function found in part (b), what is the carrying capacity of the world?

(e) Use the function found in part (b) to predict the population of the world in 2025.

(f) When will world population be 10 billion?

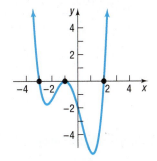

Year	Population (billions)	Year	Population (billions)
2001	6.22	2010	6.96
2002	6.30	2011	7.04
2003	6.38	2012	7.13
2004	6.46	2013	7.21
2005	6.54	2014	7.30
2006	6.62	2015	7.38
2007	6.71	2016	7.47
2008	6.79	2017	7.55
2009	6.87	2018	7.63

Source: worldometers.info

9. **Golfing** The data below represent the expected percentage of putts that will be made by professional golfers on the PGA Tour, depending on distance. For example, it is expected that 99.3% of 2-foot putts will be made.

Distance (feet)	Expected Percentage	Distance (feet)	Expected Percentage
2	99.3	14	25.0
3	94.8	15	22.0
4	85.8	16	20.0
5	74.7	17	19.0
6	64.7	18	17.0
7	55.6	19	16.0
8	48.5	20	14.0
9	43.4	21	13.0
10	38.3	22	12.0
11	34.2	23	11.0
12	30.1	24	11.0
13	27.0	25	10.0

Source: TheSandTrap.com

(a) Using a graphing utility, draw a scatter plot of the data with distance as the independent variable.

(b) Based on the scatter plot drawn in part (a), decide on a model (exponential, logarithmic, or logistic) that you think best describes the relation between distance and expected percentage. Be sure to justify your choice of model.

(c) Using a graphing utility, find the model of best fit.

(d) Graph the function found in part (c) on the scatter plot.

(e) Use the function found in part (c) to predict what percentage of 30-foot putts will be made.

Retain Your Knowledge

Problems 10–19 are based on material learned earlier in the course. The purpose of these problems is to keep the material fresh in your mind so that you are better prepared for the final exam.

10. Construct a polynomial function that might have the graph shown. (More than one answer is possible.)

11. Use the Pythagorean Theorem to find the exact length of the unlabeled side in the given right triangle.

12 Graph the equation $(x - 3)^2 + y^2 = 25$.

13. Find the midpoint of the line segment with endpoints $(-7, 5)$ and $(1, -9)$.

14. Find the function whose graph is the shape of $y = \sqrt{x}$, but shifted to the right 4 units and reflected about the x-axis.

15. Solve: $3x^2 - 4x - 5 = 0$

16. Solve: $\dfrac{x + 1}{x^2 - 25} \geq 0$

17. Use the Remainder Theorem to find the remainder when $f(x) = 3x^5 - 7x^4 - 27x^3 + 67x^2 - 36$ is divided by $x + 3$. Is $x + 3$ a factor of $f(x)$?

18. Find the average rate of change of $f(x) = 8^x$ from $\dfrac{1}{3}$ to $\dfrac{2}{3}$.

19. Use the Intermediate Value Theorem to show that $f(x) = -x^4 + 2x^3 - 5x + 1$ has a zero in the interval $[-2, -1]$. Then, approximate the zero correct to two decimal places.

Chapter Review

Things to Know

Properties of the exponential function (pp. 465, 467, 471)

$f(x) = Ca^x, \quad a > 1, C > 0$

Domain: the interval $(-\infty, \infty)$

Range: the interval $(0, \infty)$

x-intercepts: none; y-intercept: C

Horizontal asymptote: x-axis $(y = 0)$ as $x \to -\infty$

Increasing; one-to-one; smooth; continuous

See Figure 4 for a typical graph.

$f(x) = Ca^x, \quad 0 < a < 1, C > 0$

Domain: the interval $(-\infty, \infty)$

Range: the interval $(0, \infty)$

x-intercepts: none; y-intercept: C

Horizontal asymptote: x-axis $(y = 0)$ as $x \to \infty$

Decreasing; one-to-one; smooth; continuous

See Figure 8 for a typical graph.

Number e (p. 468)

Number approached by the expression $\left(1 + \dfrac{1}{n}\right)^n$ as $n \to \infty$

Property of exponents (p. 469)

If $a^u = a^v$, then $u = v$.

Natural logarithm (p. 480)

$y = \ln x$ if and only if $x = e^y$.

Properties of the logarithmic function (p. 485)

$f(x) = \log_a x, \quad a > 1$

$(y = \log_a x$ if and only if $x = a^y)$

Domain: the interval $(0, \infty)$

Range: the interval $(-\infty, \infty)$

x-intercept: 1; y-intercept: none

Vertical asymptote: $x = 0$ (y-axis)

Increasing; one-to-one; smooth; continuous

See Figure 22(a) for a typical graph.

$f(x) = \log_a x, \quad 0 < a < 1$

$(y = \log_a x$ if and only if $x = a^y)$

Domain: the interval $(0, \infty)$

Range: the interval $(-\infty, \infty)$

x-intercept: 1; y-intercept: none

Vertical asymptote: $x = 0$ (y-axis)

Decreasing; one-to-one; smooth; continuous

See Figure 22(b) for a typical graph.

Properties of logarithms (pp. 490–491, 484)

$\log_a 1 = 0 \qquad \log_a a = 1 \qquad a^{\log_a M} = M \qquad \log_a a^r = r \qquad a^r = e^{r \ln a}$

$\log_a (MN) = \log_a M + \log_a N \qquad \log_a\left(\dfrac{M}{N}\right) = \log_a M - \log_a N$

$\log_a M^r = r \log_a M$

If $M = N$, then $\log_a M = \log_a N$.

If $\log_a M = \log_a N$, then $M = N$.

Formulas

Change-of-Base Formula (p. 495)

$\log_a M = \dfrac{\log_b M}{\log_b a} \qquad a \neq 1, b \neq 1$, and M are positive real numbers

Compound Interest Formula (p. 507)

$A = P \cdot \left(1 + \dfrac{r}{n}\right)^{nt}$

Continuous compounding (p. 509)

$A = Pe^{rt}$

Effective rate of interest (p. 510)

Compounding n times per year: $r_E = \left(1 + \dfrac{r}{n}\right)^n - 1$

Continuous compounding: $r_E = e^r - 1$

Present Value Formulas (p. 511)

$P = A \cdot \left(1 + \dfrac{r}{n}\right)^{-nt} \quad$ or $\quad P = Ae^{-rt}$

Uninhibited growth or decay (pp. 516, 518)

$A(t) = A_0 e^{kt} \quad k \neq 0;$ growth, $k > 0;$ decay, $k < 0$

Newton's Law of Cooling (p. 519)

$u(t) = T + (u_0 - T)e^{kt} \quad k < 0$

Logistic model (p. 521)

$P(t) = \dfrac{c}{1 + ae^{-bt}} \quad a > 0, c > 0, b \neq 0$

Objectives

Section	You should be able to . . .	Example(s)	Review Exercises
7.1	**1** Evaluate exponential functions (p. 460)	1	1(a), (c), 34(a)
	2 Graph exponential functions (p. 463)	3–6	18–20, (a)–(c), 21(d)–(f)
	3 Define the number e (p. 467)	p. 468	
	4 Solve exponential equations (p. 469)	7, 8	22, 23, 26, 28, 34(b)
7.2	**1** Change exponential statements to logarithmic statements and logarithmic statements to exponential statements (p. 477)	2, 3	2, 3
	2 Evaluate logarithmic expressions (p. 478)	4	1(b), (d), 6, 33(b), 35(a), 36
	3 Determine the domain of a logarithmic function (p. 478)	5	4, 5, 21(a)
	4 Graph logarithmic functions (p. 479)	6, 7	18–20(d)–(f), 21(b)–(c), 33(a)
	5 Solve logarithmic equations (p. 483)	8, 9	24, 33(c), 35(b)
7.3	**1** Work with the properties of logarithms (p. 490)	1, 2	7, 8
	2 Write a logarithmic expression as a sum or difference of logarithms (p. 492)	3–5	9–12
	3 Write a logarithmic expression as a single logarithm (p. 493)	6	13–15
	4 Evaluate logarithms whose base is neither 10 nor e (p. 494)	7, 8	16, 17
7.4	**1** Solve logarithmic equations (p. 499)	1–3	24, 27, 30
	2 Solve exponential equations (p. 501)	4–6	25, 29, 31, 32
	3 Solve logarithmic and exponential equations using a graphing utility (p. 502)	7	22–32
7.5	**1** Determine the future value of a lump sum of money (p. 506)	1–3	37
	2 Calculate effective rates of return (p. 509)	4	37
	3 Determine the present value of a lump sum of money (p. 510)	5	38
	4 Determine the rate of interest or the time required to double a lump sum of money (p. 511)	6, 7	37
7.6	**1** Model populations that obey the law of uninhibited growth (p. 516)	1, 2	41
	2 Model populations that obey the law of uninhibited decay (p. 518)	3	39, 42
	3 Use Newton's Law of Cooling (p. 519)	4	40
	4 Use logistic models (p. 521)	5, 6	43
7.7	**1** Build an exponential model from data (p. 527)	1	44
	2 Build a logarithmic model from data (p. 529)	2	45
	3 Build a logistic model from data (p. 529)	3	46

Review Exercises

1. Given $f(x) = 3^x$ and $g(x) = \log_3 x$, evaluate each of the following.

    (a) $f(4)$ **(b)** $g(9)$ **(c)** $f(-2)$ **(d)** $g\left(\dfrac{1}{27}\right)$

2. Change $5^2 = z$ to an equivalent statement involving a logarithm.

3. Change $\log_5 u = 13$ to an equivalent statement involving an exponent.

In Problems 4 and 5, find the domain of each logarithmic function.

    4. $f(x) = \log(3x - 2)$ **5.** $H(x) = \log_2(x^2 - 3x + 2)$

In Problems 6–8, find the exact value of each expression. Do not use a calculator.

    6. $\log_2\left(\dfrac{1}{8}\right)$ **7.** $\ln e^{\sqrt{2}}$ **8.** $2^{\log_2 0.4}$

In Problems 9–12, write each expression as the sum and/or difference of logarithms. Express powers as factors.

    9. $\log_3\left(\dfrac{uv^2}{w}\right)$ $u > 0, v > 0, w > 0$ **10.** $\log_2\left(a^2 \sqrt{b}\right)^4$ $a > 0, \;\; b > 0$

    11. $\log\left(x^2 \sqrt{x^3 + 1}\right)$ $x > 0$ **12.** $\ln\left(\dfrac{2x + 3}{x^2 - 3x + 2}\right)^2$ $x > 2$

In Problems 13–15, write each expression as a single logarithm.

13. $3 \log_4 x^2 + \dfrac{1}{2} \log_4 \sqrt{x}$

14. $\ln\left(\dfrac{x-1}{x}\right) + \ln\left(\dfrac{x}{x+1}\right) - \ln(x^2 - 1)$

15. $\dfrac{1}{2} \ln(x^2 + 1) - 4 \ln\dfrac{1}{2} - \dfrac{1}{2}\left[\ln(x - 4) + \ln x\right]$

16. Use the Change-of-Base Formula and a calculator to evaluate $\log_4 19$. Round your answer to three decimal places.

17. Graph $y = \log_3 x$ using a graphing utility and the Change-of-Base Formula.

In Problems 18–21, for each function f:

   (a) Find the domain of f.   (b) Graph f.   (c) From the graph, determine the range and any asymptotes of f.

   (d) Find f^{-1}, the inverse function of f.   (e) Find the domain and the range of f^{-1}.   (f) Graph f^{-1}.

18. $f(x) = 2^{x-3}$ **19.** $f(x) = 1 + 3^{-x}$ **20.** $f(x) = 3e^{x-2}$ **21.** $f(x) = \dfrac{1}{2} \ln(x + 3)$

In Problems 22–32, solve each equation. Express irrational solutions in exact form.

22. $8^{6+3x} = 4$ **23.** $3^{x^2+x} = \sqrt{3}$ **24.** $\log_x 64 = -3$ **25.** $5^x = 3^{x+2}$

26. $25^{2x} = 5^{x^2-12}$ **27.** $\log_3 \sqrt{x-2} = 2$ **28.** $8 = 4^{x^2} \cdot 2^{5x}$ **29.** $2^x \cdot 5 = 10^x$

30. $\log_6(x + 3) + \log_6(x + 4) = 1$ **31.** $e^{1-x} = 5$ **32.** $9^x + 4 \cdot 3^x - 3 = 0$

---

33. Suppose that $f(x) = \log_2(x - 2) + 1$.

   (a) Graph f.

   (b) What is $f(6)$? What point is on the graph of f?

   (c) Solve $f(x) = 4$. What point is on the graph of f?

   (d) Based on the graph drawn in part (a), solve $f(x) > 0$.

   (e) Find $f^{-1}(x)$. Graph f^{-1} on the same Cartesian plane as f.

34. Amplifying Sound An amplifier's power output P (in watts) is related to its decibel voltage gain d by the formula

$$P = 25e^{0.1d}$$

   (a) Find the power output for a decibel voltage gain of 4 decibels.

   (b) For a power output of 50 watts, what is the decibel voltage gain?

35. Limiting Magnitude of a Telescope A telescope is limited in its usefulness by the brightness of the star that it is aimed at and by the diameter of its lens. One measure of a star's brightness is its *magnitude;* the dimmer the star, the larger its magnitude. A formula for the limiting magnitude L of a telescope—that is, the magnitude of the dimmest star that it can be used to view—is given by

$$L = 9 + 5.1 \log d$$

where d is the diameter (in inches) of the lens.

   (a) What is the limiting magnitude of a 3.5-inch telescope?

   (b) What diameter is required to view a star of magnitude 14?

36. Salvage Value The number of years n for a piece of machinery to depreciate to a known salvage value can be found using the formula

$$n = \frac{\log s - \log i}{\log(1 - d)}$$

where s is the salvage value of the machinery, i is its initial value, and d is the annual rate of depreciation.

   (a) How many years will it take for a piece of machinery to decline in value from \$90,000 to \$10,000 if the annual rate of depreciation is 0.20 (20%)?

   (b) How many years will it take for a piece of machinery to lose half of its value if the annual rate of depreciation is 15%?

37. Funding a College Education A child's grandparents purchase a \$10,000 bond fund that matures in 18 years to be used for her college education. The bond fund pays 4% interest compounded semiannually. How much will the bond fund be worth at maturity? What is the effective rate of interest? How long will it take the bond to double in value under these terms?

38. Funding a College Education A child's grandparents wish to purchase a bond that matures in 18 years to be used for her college education. The bond pays 4% interest compounded semiannually. How much should they pay so that the bond will be worth \$85,000 at maturity?

39. Estimating the Date When a Prehistoric Man Died The bones of a prehistoric man found in the desert of New Mexico contain approximately 5% of the original amount of carbon-14. If the half-life of carbon-14 is 5730 years, approximately how long ago did the man die?

40. Temperature of a Skillet A skillet is removed from an oven where the temperature is 450°F and placed in a room where the temperature is 70°F. After 5 minutes, the temperature of the skillet is 400°F. How long will it be until its temperature is 150°F?

41. World Population The annual growth rate of the world's population in 2018 was $k = 1.1\% = 0.011$. The population of the world in 2018 was 7,632,819,325. Letting $t = 0$ represent 2018, use the uninhibited growth model to predict the world's population in the year 2024.

Source: worldometers.info

42. Radioactive Decay The half-life of radioactive cobalt is 5.27 years. If 100 grams of radioactive cobalt is present now, how much will be present in 20 years? In 40 years?

43. Logistic Growth The logistic growth model

$$P(t) = \frac{0.8}{1 + 1.67e^{-0.16t}}$$

represents the proportion of new cars with a global positioning system (GPS). Let $t = 0$ represent 2006, $t = 1$ represent 2007, and so on.

(a) What proportion of new cars in 2006 had a GPS?
(b) Determine the maximum proportion of new cars that have a GPS.
(c) Using a graphing utility, graph $P = P(t)$.
(d) When will 75% of new cars have a GPS?

44. Rising Tuition The following data represent the average in-state tuition and fees (in 2017 dollars) at public four-year colleges and universities in the United States from the academic year 1990–91 to the academic year 2017–18.

Academic Year	Tuition and Fees (2017 dollars)
1990–91 ($x = 0$)	3580
1995–96 ($x = 5$)	4510
2000–01 ($x = 10$)	4970
2005–06 ($x = 15$)	6880
2009–10 ($x = 19$)	8040
2013–14 ($x = 23$)	9310
2017–18 ($x = 27$)	9970

Source: The College Board

(a) Using a graphing utility, draw a scatter plot with academic year as the independent variable.
(b) Using a graphing utility, build an exponential model from the data.
(c) Express the function found in part (b) in the form $A(t) = A_0 e^{kt}$.
(d) Graph the exponential function found in part (b) or (c) on the scatter plot.
(e) Predict the academic year when the average tuition will reach $16,000.

45. Wind Chill Factor The data represent the wind speed (mph) and the wind chill factor at an air temperature of 15°F.

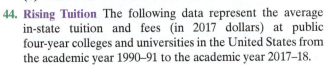

Wind Speed (mph)	Wind Chill Factor (°F)
5	7
10	3
15	0
20	−2
25	−4
30	−5
35	−7

Source: U.S. National Weather Service

(a) Using a graphing utility, draw a scatter plot with wind speed as the independent variable.
(b) Using a graphing utility, build a logarithmic model from the data.
(c) Using a graphing utility, draw the logarithmic function found in part (b) on the scatter plot.
(d) Use the function found in part (b) to predict the wind chill factor if the air temperature is 15°F and the wind speed is 23 mph.

46. Spreading of a Disease Jack and Diane live in a small town of 50 people. Unfortunately, both Jack and Diane have a cold. Those who come in contact with someone who has this cold will themselves catch the cold. The data that follow represent the number of people in the small town who have caught the cold after t days.

Days, t	Number of People with Cold, C
0	2
1	4
2	8
3	14
4	22
5	30
6	37
7	42
8	44

(a) Using a graphing utility, draw a scatter plot of the data. Comment on the type of relation that appears to exist between the number of days that have passed and the number of people with a cold.
(b) Using a graphing utility, build a logistic model from the data.
(c) Graph the function found in part (b) on the scatter plot.
(d) According to the function found in part (b), what is the maximum number of people who will catch the cold? In reality, what is the maximum number of people who could catch the cold?
(e) Sometime between the second and third day, 10 people in the town had a cold. According to the model found in part (b), when did 10 people have a cold?
(f) How long will it take for 46 people to catch the cold?

Chapter Test

CHAPTER
Test Prep
VIDEOS

The Chapter Test Prep Videos include step-by-step solutions to all chapter test exercises. These videos are available in MyLab™ Math, or on this text's YouTube Channel. Refer to the Preface for a link to the YouTube channel.

In Problems 1–3, solve each equation.

1. $3^x = 243$ **2.** $\log_b 16 = 2$ **3.** $\log_5 x = 4$

In Problems 4–7, evaluate each expression. without using a calculator.

4. $\log_6 \dfrac{1}{36}$ **5.** $\log 10{,}000$

6. $8^{\log_2 5}$ **7.** $\ln e^7$

In Problems 8 and 9, for each function f:
(a) Find the domain of f.
(b) Graph f.
(c) From the graph of f, find the range and any asymptotes.
(d) Find f^{-1}, the inverse of f.
(e) Find the domain and the range of f^{-1}.
(f) Graph f^{-1}.

8. $f(x) = 4^{x+1} - 2$ **9.** $f(x) = 1 - \log_5(x-2)$

In Problems 10–15, solve each equation. Express irrational solutions in exact form.

10. $5^{x+2} = 125$ **11.** $\log(x+9) = 2$

12. $8 - 2e^{-x} = 4$ **13.** $\log(x^2+3) = \log(x+6)$

14. $7^{x+3} = e^x$ **15.** $\log_2(x-4) + \log_2(x+4) = 3$

16. Write $\log_2\left(\dfrac{4x^3}{x^2 - 3x - 18}\right)$ as the sum and/or difference of logarithms. Express powers as factors.

17. A 50-mg sample of a radioactive substance decays to 34 mg after 30 days. How long will it take for there to be 2 mg remaining?

18. (a) If $1000 is invested at 5% compounded monthly, how much is there after 8 months?
   (b) If you want to have $1000 in 9 months, how much do you need to place in a savings account now that pays 5% compounded quarterly?
   (c) How long does it take to double your money if you can invest it at 6% compounded annually?

19. The decibel level, D, of sound is given by the equation

$$D = 10 \log\left(\frac{I}{I_0}\right)$$

where I is the intensity of the sound and $I_0 = 10^{-12}$ watt per square meter.

(a) If the shout of a single person measures 80 decibels, how loud would the sound be if two people shout at the same time? That is, how loud would the sound be if the intensity doubled?
(b) The pain threshold for sound is 125 decibels. If the Athens Olympic Stadium 2004 (Olympiako Stadio Athinas 'Spyros Louis') can seat 74,400 people, how many people in the crowd need to shout at the same time for the resulting sound level to meet or exceed the pain threshold? (Ignore any possible sound dampening.)

Cumulative Review

1. Is the following graph the graph of a function? If it is, is the function one-to-one?

2. For the function $f(x) = 2x^2 - 3x + 1$, find:
   (a) $f(3)$ (b) $f(-x)$ (c) $f(x+h)$

3. Determine which points are on the graph of $x^2 + y^2 = 1$.

   (a) $\left(\dfrac{1}{2}, \dfrac{1}{2}\right)$ (b) $\left(\dfrac{1}{2}, \dfrac{\sqrt{3}}{2}\right)$

4. Solve the equation $3(x-2) = 4(x+5)$.

5. Graph the line $2x - 4y = 16$.

6. Graph $f(x) = 3(x+1)^3 - 2$ using transformations.

7. For the function $g(x) = 3^x + 2$:
   (a) Graph g using transformations. State the domain, range, and horizontal asymptote of the graph of g.
   (b) Determine the inverse of g. State the domain, range, and vertical asymptote of the graph of g^{-1}.
   (c) On the same coordinate axes as g, graph g^{-1}.

8. Solve the equation: $4^{x-3} = 8^{2x}$

9. Solve the equation: $\log_3(x+1) + \log_3(2x-3) = \log_9 9$

10. Suppose that $f(x) = \log_3(x+2)$. Solve:
   (a) $f(x) = 0$
   (b) $f(x) > 0$
   (c) $f(x) = 3$

11. Data Analysis The following data represent the percent of all drivers by age who have been stopped by the police for any reason within the past year. The median age represents the midpoint of the upper and lower limit for the age range.

Age Range	Median Age, x	Percent Stopped, y
16–19	17.5	18.2
20–29	24.5	16.8
30–39	34.5	11.3
40–49	44.5	9.4
50–59	54.5	7.7
≥ 60	69.5	3.8

(a) Using a graphing utility, draw a scatter plot of the data treating median age, x, as the independent variable.
(b) Determine a model that best describes the relation between median age and percent stopped. You may choose from among exponential, logarithmic, and logistic models.
(c) Provide a justification for the model that you selected in part (b).

Chapter Projects

(((•))) Internet-based Project

I. **Depreciation of Cars** Kelley Blue Book is a guide that provides the current retail price of cars. You can access the Kelley Blue Book online at *www.kbb.com*.

    1. Identify three cars that you are considering purchasing, and find the Kelley Blue Book value of the cars for 0 (brand new), 1, 2, 3, 4, and 5 years of age. Online, the value of the car can be found by selecting Price New/Used. Enter the year, make, and model of the new or used car you are selecting. To be consistent, assume the cars will be driven 12,000 miles per year, so a 1-year-old car will have 12,000 miles, a 2-year-old car will have 24,000 miles, and so on. Choose the same options for each year, and select Buy from a Private Party when choosing a price type. Finally, determine the suggested retail price for cars that are in Excellent, Good, and Fair shape. You should have a total of 16 observations (1 for a brand new car, 3 for a 1-year-old car, 3 for a 2-year-old car, and so on).

    2. Draw a scatter plot of the data with age as the independent variable and value as the dependent variable using Excel, a TI-graphing calculator, or some other spreadsheet. The Chapter 3 project describes how to draw a scatter plot in Excel.

    3. Determine the exponential function of best fit. Graph the exponential function of best fit on the scatter plot. To do this in Excel, right click on any data point in the scatter plot. Now select Add Trendline. Select the Exponential radio button and select Display Equation on Chart. See

Figure 40. Move the Trendline Options window off to the side, if necessary, and you will see the exponential function of best fit displayed on the scatter plot. Do you think the function accurately describes the relation between age of the car and suggested retail price?

Figure 40

    4. The exponential function of best fit is of the form $y = Ce^{rx}$, where y is the suggested retail value of the car and x is the age of the car (in years). What does the value of C represent? What does the value of r represent? What is the depreciation rate for each car that you are considering?

    5. Write a report detailing which car you would purchase based on the depreciation rate you found for each car.

Citation: Excel © 2018 Microsoft Corporation. Used with permission from Microsoft.

The following projects are available on the Instructor's Resource Center (IRC):

II. **Hot Coffee** A fast-food restaurant wants a special container to hold coffee. The restaurant wishes the container to quickly cool the coffee from 200° to 130°F and keep the liquid between 110° and 130°F as long as possible. The restaurant has three containers to select from. Which one should be purchased?

III. **Project at Motorola** ***Thermal Fatigue of Solder Connections*** Product reliability is a major concern of a manufacturer. Here a logarithmic transformation is used to simplify the analysis of a cell phone's ability to withstand temperature change.

Review

Outline

A.1 Algebra Essentials

OBJECTIVES
1 Work with Sets (p. A1)
2 Graph Inequalities (p. A4)
3 Find Distance on the Real Number Line (p. A5)
4 Evaluate Algebraic Expressions (p. A6)
5 Determine the Domain of a Variable (p. A7)
6 Use the Laws of Exponents (p. A7)
7 Evaluate Square Roots (p. A9)
8 Use a Calculator to Evaluate Exponents (p. A10)

1 Work with Sets

A **set** is a well-defined collection of distinct objects. The objects of a set are called its **elements**. By **well-defined**, we mean that there is a rule that enables us to determine whether a given object is an element of the set. If a set has no elements, it is called the **empty set**, or **null set**, and is denoted by the symbol $\varnothing$.

For example, the set of **digits** consists of the collection of numbers 0, 1, 2, 3, 4, 5, 6, 7, 8, and 9. If we use the symbol D to denote the set of digits, then we can write

$$D = \{0, 1, 2, 3, 4, 5, 6, 7, 8, 9\}$$

In this notation, the braces $\{\ \}$ are used to enclose the objects, or **elements**, in the set. This method of denoting a set is called the **roster method**. A second way to denote a set is to use **set-builder notation**, where the set D of digits is written as

$$D = \{\quad x \quad | \quad x \text{ is a digit}\}$$

Read as "D is the set of all x such that x is a digit."

EXAMPLE 1 **Using Set-builder Notation and the Roster Method**

(a) $E = \{x | x \text{ is an even digit}\} = \{0, 2, 4, 6, 8\}$
(b) $O = \{x | x \text{ is an odd digit}\} = \{1, 3, 5, 7, 9\}$

Because the elements of a set are distinct, we never repeat elements. For example, we would never write $\{1, 2, 3, 2\}$; the correct listing is $\{1, 2, 3\}$. Because a set is a collection, the order in which the elements are listed is immaterial. $\{1, 2, 3\}, \{1, 3, 2\}, \{2, 1, 3\}$, and so on, all represent the same set.

If every element of a set A is also an element of a set B, then A is a **subset** of B, which is denoted $A \subseteq B$. If two sets A and B have the same elements, then A **equals** B, which is denoted $A = B$.

For example, $\{1, 2, 3\} \subseteq \{1, 2, 3, 4, 5\}$ and $\{1, 2, 3\} = \{2, 3, 1\}$.

DEFINITION Intersection and Union of Two Sets

If A and B are sets, the **intersection** of A with B, denoted $A \cap B$, is the set consisting of elements that belong to both A and B. The **union** of A with B, denoted $A \cup B$, is the set consisting of elements that belong to either A or B, or both.

EXAMPLE 2 **Finding the Intersection and Union of Sets**

Let $A = \{1, 3, 5, 8\}$, $B = \{3, 5, 7\}$, and $C = \{2, 4, 6, 8\}$. Find:

(a) $A \cap B$ (b) $A \cup B$ (c) $B \cap (A \cup C)$

Solution

(a) $A \cap B = \{1, 3, 5, 8\} \cap \{3, 5, 7\} = \{3, 5\}$

(b) $A \cup B = \{1, 3, 5, 8\} \cup \{3, 5, 7\} = \{1, 3, 5, 7, 8\}$

(c) $B \cap (A \cup C) = \{3, 5, 7\} \cap [\, \{1, 3, 5, 8\} \cup \{2, 4, 6, 8\} \,]$
$\qquad\qquad\quad = \{3, 5, 7\} \cap \{1, 2, 3, 4, 5, 6, 8\} = \{3, 5\}$

 Now Work PROBLEM 15

Usually, in working with sets, we designate a **universal set** U, the set consisting of all the elements that we wish to consider. Once a universal set has been designated, we can consider elements of the universal set not found in a given set.

DEFINITION Complement of a Set

If A is a set, the **complement** of A, denoted $\overline{A}$, is the set consisting of all the elements in the universal set that are not in A.[*]

EXAMPLE 3 **Finding the Complement of a Set**

If the universal set is $U = \{1, 2, 3, 4, 5, 6, 7, 8, 9\}$ and if $A = \{1, 3, 5, 7, 9\}$, then $\overline{A} = \{2, 4, 6, 8\}$.

It follows from the definition of complement that $A \cup \overline{A} = U$ and $A \cap \overline{A} = \emptyset$. Do you see why?

 Now Work PROBLEM 19

It is often helpful to draw pictures of sets. Such pictures, called **Venn diagrams**, represent sets as circles enclosed in a rectangle, which represents the universal set. Such diagrams often help us to visualize various relationships among sets. See Figure 1.

If we know that $A \subseteq B$, we might use the Venn diagram in Figure 2(a). If we know that A and B have no elements in common—that is, if $A \cap B = \emptyset$—we might use the Venn diagram in Figure 2(b). The sets A and B in Figure 2(b) are said to be **disjoint**.

Figure 1 Venn diagram

---

[*]Some texts use the notation A' or A^c for the complement of A.

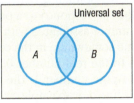

Figure 2

(a) $A \subseteq B$
subset

(b) $A \cap B = \emptyset$
disjoint sets

Figures 3(a), 3(b), and 3(c) use Venn diagrams to illustrate intersection, union, and complement, respectively.

Figure 3

(a) $A \cap B$
intersection

(b) $A \cup B$
union

(c) $\bar{A}$
complement

Real Numbers

Real numbers are represented by symbols such as

$$25, \quad 0, \quad -3, \quad \frac{1}{2}, \quad -\frac{5}{4}, \quad 0.125, \quad \sqrt{2}, \quad \pi, \quad \sqrt[3]{-2}, \quad 0.666\ldots$$

The set of **counting numbers**, or **natural numbers**, contains the numbers in the set $\{1, 2, 3, 4, \ldots\}$. (The three dots, called an **ellipsis**, indicate that the pattern continues indefinitely.) The set of **integers** contains the numbers in the set $\{\ldots, -3, -2, -1, 0, 1, 2, 3, \ldots\}$. A **rational number** is a number that can be expressed as a *quotient* $\frac{a}{b}$ of two integers, where the integer b cannot be 0. Examples of rational numbers are $\frac{3}{4}, \frac{5}{2}, \frac{0}{4}$, and $-\frac{2}{3}$. Since $\frac{a}{1} = a$ for any integer a, every integer is also a rational number. Real numbers that are not rational are called **irrational**. Examples of irrational numbers are $\sqrt{2}$ and π (the Greek letter pi), which equals the constant ratio of the circumference to the diameter of a circle. See Figure 4.

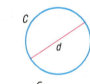

Figure 4 $\pi = \dfrac{C}{d}$

Real numbers can be represented as **decimals**. Rational real numbers have decimal representations that either **terminate** or are nonterminating with **repeating** blocks of digits. For example, $\frac{3}{4} = 0.75$, which terminates; and $\frac{2}{3} = 0.666\ldots$, in which the digit 6 repeats indefinitely. Irrational real numbers have decimal representations that neither repeat nor terminate. For example, $\sqrt{2} = 1.414213\ldots$ and $\pi = 3.14159\ldots$. In practice, the decimal representation of an irrational number is given as an approximation. We use the symbol $\approx$ (read as "approximately equal to") to write $\sqrt{2} \approx 1.4142$ and $\pi \approx 3.1416$.

Two frequently used properties of real numbers are given next.

Suppose that a, b, and c are real numbers.

Distributive Property

$$a \cdot (b + c) = ab + ac$$
$$(a + b) \cdot c = a \cdot c + b \cdot c$$

Zero-Product Property

If $ab = 0$, then $a = 0$, $b = 0$, or both equal 0.

The Distributive Property can be used to remove parentheses:

$$2(x + 3) = 2x + 2 \cdot 3 = 2x + 6$$

The Zero-Product Property will be used to solve equations (Section A.4). For example, if $2x = 0$, then $2 = 0$ or $x = 0$. Since $2 \neq 0$, it follows that $x = 0$.

The Real Number Line

Real numbers can be represented by points on a line called the **real number line**. There is a one-to-one correspondence between real numbers and points on a line. That is, every real number corresponds to a point on the line, and each point on the line has a unique real number associated with it.

Pick a point on a line somewhere in the center, and label it O. This point, called the **origin**, corresponds to the real number 0. See Figure 5. The point 1 unit to the right of O corresponds to the number 1. The distance between 0 and 1 determines the **scale** of the number line. For example, the point associated with the number 2 is twice as far from O as 1. Notice that an arrowhead on the right end of the line indicates the direction in which the numbers increase. Points to the left of the origin correspond to the real numbers -1, -2, and so on. Figure 5 also shows the points

Figure 5 Real number line

associated with the rational numbers $-\dfrac{1}{2}$ and $\dfrac{1}{2}$ and with the irrational numbers $\sqrt{2}$ and π.

DEFINITION Coordinate; Real Number Line

The real number associated with a point P is called the **coordinate** of P, and the line whose points have been assigned coordinates is called the **real number line**.

Now Work PROBLEM 23

The real number line consists of three classes of real numbers, as shown in Figure 6.

Figure 6

- The **negative real numbers** are the coordinates of points to the left of the origin O.
- The real number **zero** is the coordinate of the origin O.
- The **positive real numbers** are the coordinates of points to the right of the origin O.

2 Graph Inequalities

An important property of the real number line follows from the fact that, given two numbers a and b, either a is to the left of b, or a is at the same location as b, or a is to the right of b. See Figure 7.

If a is to the left of b, then "a is less than b," which is written $a < b$. If a is to the right of b, then "a is greater than b," which is written $a > b$. If a is at the same location as b, then $a = b$. If a is either less than or equal to b, then $a \leq b$. Similarly, $a \geq b$ means that a is either greater than or equal to b. Collectively, the symbols $<$, $>$, $\leq$, and $\geq$ are called **inequality symbols**.

Note that $a < b$ and $b > a$ mean the same thing. It does not matter whether we write $2 < 3$ or $3 > 2$.

Furthermore, if $a < b$ or if $b > a$, then the difference $b - a$ is positive. Do you see why?

(a) $a < b$

(b) $a = b$

(c) $a > b$

Figure 7

An **inequality** is a statement in which two expressions are related by an inequality symbol. The expressions are referred to as the **sides** of the inequality. Inequalities of the form $a < b$ or $b > a$ are called **strict inequalities**, whereas inequalities of the form $a \leq b$ or $b \geq a$ are called **nonstrict inequalities**.

Based on the discussion so far, we conclude that

- $a > 0$ is equivalent to a is positive
- $a < 0$ is equivalent to a is negative

We sometimes read $a > 0$ by saying that "a is positive." If $a \geq 0$, then either $a > 0$ or $a = 0$, and we may read this as "a is nonnegative."

Now Work PROBLEMS 27 AND 37

EXAMPLE 4

Graphing Inequalities

(a) On the real number line, graph all numbers x for which $x > 4$.
(b) On the real number line, graph all numbers x for which $x \leq 5$.

Solution

Figure 8 $x > 4$

Figure 9 $x \leq 5$

(a) See Figure 8. Notice that we use a left parenthesis to indicate that the number 4 is *not* part of the graph.

(b) See Figure 9. Notice that we use a right bracket to indicate that the number 5 *is* part of the graph.

Now Work PROBLEM 43

---

3 Find Distance on the Real Number Line

Figure 10

The *absolute value* of a number a is the distance from 0 to a on the number line. For example, -4 is 4 units from 0, and 3 is 3 units from 0. See Figure 10. That is, the absolute value of -4 is 4, and the absolute value of 3 is 3.

A more formal definition of absolute value is given next.

DEFINITION Absolute Value

The **absolute value** of a real number a, denoted by the symbol $|a|$, is defined by the rules

$$|a| = a \ \text{ if } a \geq 0 \qquad \text{and} \qquad |a| = -a \ \text{ if } a < 0$$

For example, because $-4 < 0$, the second rule must be used to get

$$|-4| = -(-4) = 4$$

EXAMPLE 5

Computing Absolute Value

(a) $|8| = 8$ (b) $|0| = 0$ (c) $|-15| = -(-15) = 15$

Look again at Figure 10. The distance from -4 to 3 is 7 units. This distance is the difference $3 - (-4)$, obtained by subtracting the smaller coordinate from the larger. However, since $|3 - (-4)| = |7| = 7$ and $|-4 - 3| = |-7| = 7$, we can use absolute value to calculate the distance between two points without being concerned about which is smaller.

> **DEFINITION** **Distance Between Two Points**
>
> If P and Q are two points on the real number line with coordinates a and b, respectively, the **distance between P and Q**, denoted by $d(P, Q)$, is
>
> $$d(P, Q) = |b - a|$$

Since $|b - a| = |a - b|$, it follows that $d(P, Q) = d(Q, P)$.

EXAMPLE 6 **Finding Distance on a Number Line**

Let P, Q, and R be points on the real number line with coordinates -5, 7, and -3, respectively. Find the distance

(a) between P and Q (b) between Q and R

Solution See Figure 11.

$$d(P, Q) = |7 - (-5)| = 12$$
$$d(Q, R) = |-3 - 7| = 10$$

Figure 11

(a) $d(P, Q) = |7 - (-5)| = |12| = 12$

(b) $d(Q, R) = |-3 - 7| = |-10| = 10$

Now Work PROBLEM 49

4 Evaluate Algebraic Expressions

Remember, in algebra we use letters such as x, y, a, b, and c to represent numbers. If a letter used is to represent *any* number from a given set of numbers, it is called a **variable**. A **constant** is either a fixed number, such as 5 or $\sqrt{3}$, or a letter that represents a fixed (possibly unspecified) number.

Constants and variables are combined using the operations of addition, subtraction, multiplication, and division to form *algebraic expressions*. Examples of algebraic expressions include

$$x + 3 \qquad \frac{3}{1 - t} \qquad 7x - 2y$$

To evaluate an algebraic expression, substitute a numerical value for each variable.

EXAMPLE 7 **Evaluating an Algebraic Expression**

Evaluate each expression if $x = 3$ and $y = -1$.

(a) $x + 3y$ (b) $5xy$ (c) $\dfrac{3y}{2 - 2x}$ (d) $|-4x + y|$

Solution (a) Substitute 3 for x and -1 for y in the expression $x + 3y$.

$$x + 3y = 3 + 3(-1) = 3 + (-3) = 0$$

$$\uparrow$$
$$x = 3, y = -1$$

(b) If $x = 3$ and $y = -1$, then

$$5xy = 5 \cdot 3 \cdot (-1) = -15$$

(c) If $x = 3$ and $y = -1$, then

$$\frac{3y}{2 - 2x} = \frac{3(-1)}{2 - 2 \cdot 3} = \frac{-3}{2 - 6} = \frac{-3}{-4} = \frac{3}{4}$$

(d) If $x = 3$ and $y = -1$, then

$$|-4x + y| = |-4 \cdot 3 + (-1)| = |-12 + (-1)| = |-13| = 13$$

Now Work PROBLEMS **51** AND **59**

5 Determine the Domain of a Variable

In working with expressions or formulas involving variables, the variables may be allowed to take on values from only a certain set of numbers. For example, in the formula for the area A of a circle of radius r, $A = \pi r^2$, the variable r is necessarily restricted to the positive real numbers. In the expression $\dfrac{1}{x}$, the variable x cannot take on the value 0, since division by 0 is not defined.

> **DEFINITION** Domain of a Variable
>
> The set of values that a variable may assume is called the **domain of the variable**.

EXAMPLE 8 **Finding the Domain of a Variable**

The domain of the variable x in the expression

$$\frac{5}{x - 2}$$

is $\{x \mid x \neq 2\}$ since, if $x = 2$, the denominator becomes 0, which is not defined.

EXAMPLE 9 **Circumference of a Circle**

In the formula for the circumference C of a circle of radius r,

$$C = 2\pi r$$

the domain of the variable r, representing the radius of the circle, is the set of positive real numbers, $\{r \mid r > 0\}$. The domain of the variable C, representing the circumference of the circle, is also the set of positive real numbers, $\{C \mid C > 0\}$.

In describing the domain of a variable, we may use either set notation or words, whichever is more convenient.

Now Work PROBLEM **69**

6 Use the Laws of Exponents

Integer exponents provide a shorthand notation for representing repeated multiplications of a real number. For example,

$$3^4 = 3 \cdot 3 \cdot 3 \cdot 3 = 81$$

> **DEFINITION** a^n
>
> If a is a real number and n is a positive integer, then the symbol a^n represents the product of n factors of a. That is,
>
> $$a^n = \underbrace{a \cdot a \cdot \ldots \cdot a}_{n \text{ factors}}$$ (1)

WARNING Be careful with minus signs and exponents.

$$-2^4 = -1 \cdot 2^4 = -16$$

whereas

$$(-2)^4 = (-2)(-2)(-2)(-2) = 16 \quad \blacksquare$$

In the definition it is understood that $a^1 = a$. Furthermore, $a^2 = a \cdot a$, $a^3 = a \cdot a \cdot a$, and so on. In the expression a^n, a is called the **base** and n is called the **exponent**, or **power**. We read a^n as "a raised to the power n" or as "a to the nth power." We usually read a^2 as "a squared" and a^3 as "a cubed."

In working with exponents, the operation of *raising to a power* is performed before any other operation. As examples,

$$4 \cdot 3^2 = 4 \cdot 9 = 36 \qquad 2^2 + 3^2 = 4 + 9 = 13$$

$$-2^4 = -16 \qquad 5 \cdot 3^2 + 2 \cdot 4 = 5 \cdot 9 + 2 \cdot 4 = 45 + 8 = 53$$

Parentheses are used to indicate operations to be performed first. For example,

$$(-2)^4 = (-2)(-2)(-2)(-2) = 16 \qquad (2 + 3)^2 = 5^2 = 25$$

> **DEFINITION** a^0
>
> If $a \neq 0$, then
>
> $$a^0 = 1$$

> **DEFINITION** a^{-n}
>
> If $a \neq 0$ and if n is a positive integer, then
>
> $$a^{-n} = \frac{1}{a^n}$$

Whenever you encounter a negative exponent, think "reciprocal."

EXAMPLE 10   **Evaluating Expressions Containing Negative Exponents**

(a) $2^{-3} = \dfrac{1}{2^3} = \dfrac{1}{8}$ (b) $x^{-4} = \dfrac{1}{x^4}$ (c) $\left(\dfrac{1}{5}\right)^{-2} = \dfrac{1}{\left(\dfrac{1}{5}\right)^2} = \dfrac{1}{\dfrac{1}{25}} = 25$

Now Work PROBLEMS **87** AND **107**

The following properties, called the **Laws of Exponents**, can be proved using the preceding definitions. In the list, a and b are real numbers, and m and n are integers.

> **THEOREM** **Laws of Exponents**
>
> $$a^m a^n = a^{m+n} \qquad (a^m)^n = a^{mn} \qquad (ab)^n = a^n b^n$$
>
> $$\frac{a^m}{a^n} = a^{m-n} = \frac{1}{a^{n-m}} \text{ if } a \neq 0 \qquad \left(\frac{a}{b}\right)^n = \frac{a^n}{b^n} \text{ if } b \neq 0$$

EXAMPLE 11 **Using the Laws of Exponents**

Write each expression so that all exponents are positive.

(a) $\dfrac{x^5 y^{-2}}{x^3 y}$ $x \neq 0, \quad y \neq 0$

(b) $\left(\dfrac{x^{-3}}{3y^{-1}}\right)^{-2}$ $x \neq 0, \quad y \neq 0$

Solution (a) $\dfrac{x^5 y^{-2}}{x^3 y} = \dfrac{x^5}{x^3} \cdot \dfrac{y^{-2}}{y} = x^{5-3} \cdot y^{-2-1} = x^2 y^{-3} = x^2 \cdot \dfrac{1}{y^3} = \dfrac{x^2}{y^3}$

NOTE Always write the final answer using positive exponents. ∎

(b) $\left(\dfrac{x^{-3}}{3y^{-1}}\right)^{-2} = \dfrac{(x^{-3})^{-2}}{(3y^{-1})^{-2}} = \dfrac{x^6}{3^{-2}(y^{-1})^{-2}} = \dfrac{x^6}{\dfrac{1}{9}y^2} = \dfrac{9x^6}{y^2}$

━━━━ **Now Work** PROBLEMS **89** AND **99**

7 Evaluate Square Roots

In Words
$\sqrt{36}$ means "what is the nonnegative number whose square is 36?"

A real number is squared when it is raised to the power 2. The inverse of squaring is finding a **square root**. For example, since $6^2 = 36$ and $(-6)^2 = 36$, the numbers 6 and -6 are square roots of 36.

The symbol $\sqrt{}$, called a **radical sign**, is used to denote the **principal**, or nonnegative, square root. For example, $\sqrt{36} = 6$.

> **DEFINITION** **Principal Square Root**
>
> If a is a nonnegative real number, the nonnegative number b for which $b^2 = a$ is the **principal square root** of a, and is denoted by $b = \sqrt{a}$.

The following comments are noteworthy:

• Negative numbers do not have square roots (in the real number system), because the square of any real number is *nonnegative*. For example, $\sqrt{-4}$ is not a real number, because there is no real number whose square is -4.
• The principal square root of 0 is 0, since $0^2 = 0$. That is, $\sqrt{0} = 0$.
• The principal square root of a positive number is positive.
• If $c \geq 0$, then $\left(\sqrt{c}\right)^2 = c$. For example, $\left(\sqrt{2}\right)^2 = 2$ and $\left(\sqrt{3}\right)^2 = 3$.

EXAMPLE 12 **Evaluating Square Roots**

(a) $\sqrt{64} = 8$ (b) $\sqrt{\dfrac{1}{16}} = \dfrac{1}{4}$ (c) $\left(\sqrt{1.4}\right)^2 = 1.4$

Examples 12(a) and (b) are examples of square roots of perfect squares, since $64 = 8^2$ and $\dfrac{1}{16} = \left(\dfrac{1}{4}\right)^2$.

Consider the expression $\sqrt{a^2}$. Since $a^2 \geq 0$, the principal square root of a^2 is defined whether $a > 0$ or $a < 0$. However, since the principal square root is nonnegative, we need an absolute value to ensure the nonnegative result. That is,

$$\sqrt{a^2} = |a| \quad a \text{ any real number} \tag{2}$$

EXAMPLE 13	**Using Equation (2)**

(a) $\sqrt{(2.3)^2} = |2.3| = 2.3$ (b) $\sqrt{(-2.3)^2} = |-2.3| = 2.3$

(c) $\sqrt{x^2} = |x|$

Now Work PROBLEM 95

Calculators and Graphing Utilities

Calculators are incapable of displaying decimals that contain a large number of digits. For example, some calculators are capable of displaying only eight digits. When a number requires more than eight digits, the calculator either truncates or rounds. To see how your calculator handles decimals, divide 2 by 3. How many digits do you see? Is the last digit a 6 or a 7? If it is a 6, your calculator truncates; if it is a 7, your calculator rounds.

There are different kinds of calculators. An **arithmetic** calculator can only add, subtract, multiply, and divide numbers; therefore, this type is not adequate for this course. **Scientific** calculators have all the capabilities of arithmetic calculators and also contain **function keys** labeled ln, log, sin, cos, tan, x^y, inv, and so on. **Graphing** calculators have all the capabilities of scientific calculators and contain a screen on which graphs can be displayed. We use the term **graphing utility** to refer generically to all graphing calculators and computer software packages, and use the ⬛ symbol whenever a graphing utility needs to be used. In this text the use of a graphing utility is optional.

8 Use a Calculator to Evaluate Exponents

Your calculator has either a caret key, ⌃, or an x^y key, that is used for computations involving exponents.

EXAMPLE 14	**Exponents on a Graphing Calculator**

Evaluate: $(2.3)^5$

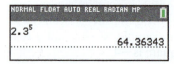

Solution Figure 12 shows the result using a TI-84 Plus C graphing calculator.

Now Work PROBLEM 125

Figure 12

A.1 Assess Your Understanding

Concepts and Vocabulary

1. A(n) _____ is a letter used in algebra to represent any number from a given set of numbers.

2. On the real number line, the real number zero is the coordinate of the _____ .

3. An inequality of the form $a > b$ is called a(n) _____ inequality.

4. In the expression 2^4, the number 2 is called the _____ and 4 is called the _____ .

5. **True or False** The distance between two distinct points on the real number line is always greater than zero.

6. **True or False** The absolute value of a real number is always greater than zero.

7. **True or False** The inverse of squaring is finding a square root.

8. **Multiple Choice** If a is a nonnegative real number, then which inequality statement best describes a?
   (a) $a < 0$ (b) $a > 0$
   (c) $a \leq 0$ (d) $a \geq 0$

9. **Multiple Choice** Let a and b be nonzero real numbers and m and n be integers. Which of the following is not a law of exponents?
   (a) $\left(\dfrac{a}{b}\right)^n = \dfrac{a^n}{b^n}$ (b) $(a^m)^n = a^{m+n}$
   (c) $\dfrac{a^m}{a^n} = a^{m-n}$ (d) $(ab)^n = a^n b^n$

10. **Multiple Choice** The set of values that a variable may assume is called the _____ of the variable.
    (a) domain   (b) range   (c) coordinate   (d) origin

Skill Building

In Problems 11–22, use U = universal set = $\{0, 1, 2, 3, 4, 5, 6, 7, 8, 9\}$, $A = \{1, 3, 4, 5, 9\}$, $B = \{2, 4, 6, 7, 8\}$, and $C = \{1, 3, 4, 6\}$ to find each set.

11. $A \cup B$

12. $A \cup C$

13. $A \cap B$

14. $A \cap C$

15. $(A \cup B) \cap C$

16. $(A \cap B) \cup C$

17. $\overline{A}$

18. $\overline{C}$

19. $\overline{A \cap B}$

20. $\overline{B \cup C}$

21. $\overline{A} \cup \overline{B}$

22. $\overline{B} \cap \overline{C}$

23. On the real number line, label the points with coordinates $0, 1, -1, \dfrac{5}{2}, -2.5, \dfrac{3}{4}$, and 0.25.

24. Repeat Problem 23 for the coordinates $0, -2, 2, -1.5, \dfrac{3}{2}, \dfrac{1}{3}$, and $\dfrac{2}{3}$.

In Problems 25–34, replace the question mark by $<$, $>$, or $=$, whichever is correct.

25. $\dfrac{1}{2}$? 0

26. 5 ? 6

27. -1 ? -2

28. -3 ? $-\dfrac{5}{2}$

29. π ? 3.14

30. $\sqrt{2}$? 1.41

31. $\dfrac{1}{2}$? 0.5

32. $\dfrac{1}{3}$? 0.33

33. $\dfrac{2}{3}$? 0.67

34. $\dfrac{1}{4}$? 0.25

In Problems 35–40, write each statement as an inequality.

35. x is positive

36. z is negative

37. x is less than 2

38. y is greater than -5

39. x is less than or equal to 1

40. x is greater than or equal to 2

In Problems 41–44, graph the numbers x on the real number line.

41. $x \geq -2$

42. $x < 4$

43. $x > -1$

44. $x \leq 7$

In Problems 45–50, use the given real number line to compute each distance.

45. $d(C, D)$

46. $d(C, A)$

47. $d(D, E)$

48. $d(C, E)$

49. $d(A, E)$

50. $d(D, B)$

In Problems 51–58, evaluate each expression if $x = -2$ and $y = 3$.

51. $x + 2y$

52. $3x + y$

53. $5xy + 2$

54. $-2x + xy$

55. $\dfrac{2x}{x - y}$

56. $\dfrac{x + y}{x - y}$

57. $\dfrac{3x + 2y}{2 + y}$

58. $\dfrac{2x - 3}{y}$

In Problems 59–68, find the value of each expression if $x = 3$ and $y = -2$.

59. $|x + y|$

60. $|x - y|$

61. $|x| + |y|$

62. $|x| - |y|$

63. $\dfrac{|x|}{x}$

64. $\dfrac{|y|}{y}$

65. $|4x - 5y|$

66. $|3x + 2y|$

67. $||4x| - |5y||$

68. $3|x| + 2|y|$

In Problems 69–76, determine which of the values (a) through (d), if any, must be excluded from the domain of the variable in each expression.

(a) $x = 3$ (b) $x = 1$ (c) $x = 0$ (d) $x = -1$

69. $\dfrac{x^2 - 1}{x}$

70. $\dfrac{x^2 + 1}{x}$

71. $\dfrac{x}{x^2 - 9}$

72. $\dfrac{x}{x^2 + 9}$

73. $\dfrac{x^2}{x^2 + 1}$

74. $\dfrac{x^3}{x^2 - 1}$

75. $\dfrac{x^2 + 5x - 10}{x^3 - x}$

76. $\dfrac{-9x^2 - x + 1}{x^3 + x}$

In Problems 77–80, determine the domain of the variable x in each expression.

77. $\dfrac{4}{x - 5}$

78. $\dfrac{-6}{x + 4}$

79. $\dfrac{x}{x + 4}$

80. $\dfrac{x - 2}{x - 6}$

In Problems 81–84, use the formula $C = \dfrac{5}{9}(F - 32)$ for converting degrees Fahrenheit into degrees Celsius to find the Celsius measure of each Fahrenheit temperature.

81. $F = 32°$ **82.** $F = 212°$ **83.** $F = 77°$ **84.** $F = -4°$

In Problems 85–96, simplify each expression.

85. $(-4)^2$ **86.** -4^2 **87.** 4^{-2} **88.** -4^{-2} **89.** $3^{-6} \cdot 3^4$ **90.** $4^{-2} \cdot 4^3$

91. $(3^{-2})^{-1}$ **92.** $(2^{-1})^{-3}$ **93.** $\sqrt{25}$ **94.** $\sqrt{36}$ **95.** $\sqrt{(-4)^2}$ **96.** $\sqrt{(-3)^2}$

In Problems 97–106, simplify each expression. Express the answer so that all exponents are positive. Whenever an exponent is 0 or negative, assume that the base is not 0.

97. $(8x^3)^2$ **98.** $(-4x^2)^{-1}$ **99.** $(x^2y^{-1})^2$ **100.** $(x^{-1}y)^3$ **101.** $\dfrac{x^2y^3}{xy^4}$

102. $\dfrac{x^{-2}y}{xy^2}$ **103.** $\dfrac{(-2)^3 x^4 (yz)^2}{3^2 xy^3 z}$ **104.** $\dfrac{4x^{-2}(yz)^{-1}}{2^3 x^4 y}$ **105.** $\left(\dfrac{3x^{-1}}{4y^{-1}}\right)^{-2}$ **106.** $\left(\dfrac{5x^{-2}}{6y^{-2}}\right)^{-3}$

In Problems 107–118, find the value of each expression if $x = 2$ and $y = -1$.

107. $2xy^{-1}$ **108.** $-3x^{-1}y$ **109.** $x^2 + y^2$ **110.** x^2y^2

111. $(xy)^2$ **112.** $(x + y)^2$ **113.** $\sqrt{x^2}$ **114.** $(\sqrt{x})^2$

115. $\sqrt{x^2 + y^2}$ **116.** $\sqrt{x^2} + \sqrt{y^2}$ **117.** x^y **118.** y^x

119. Find the value of the expression $2x^3 - 3x^2 + 5x - 4$ if $x = 2$. What is the value if $x = 1$?

120. Find the value of the expression $4x^3 + 3x^2 - x + 2$ if $x = 1$. What is the value if $x = 2$?

121. What is the value of $\dfrac{(666)^4}{(222)^4}$?   **122.** What is the value of $(0.1)^3(20)^3$?

In Problems 123–130, use a calculator to evaluate each expression. Round your answer to three decimal places.

123. $(8.2)^6$ **124.** $(3.7)^5$ **125.** $(6.1)^{-3}$ **126.** $(2.2)^{-5}$

127. $(-2.8)^6$ **128.** $-(2.8)^6$ **129.** $(-8.11)^{-4}$ **130.** $-(8.11)^{-4}$

Applications and Extensions

In Problems 131–140, express each statement as an equation involving the indicated variables.

131. Area of a Rectangle The area A of a rectangle is the product of its length l and its width w.

132. Perimeter of a Rectangle The perimeter P of a rectangle is twice the sum of its length l and its width w.

133. Circumference of a Circle The circumference C of a circle is the product of π and its diameter d.

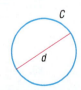

134. Area of a Triangle The area A of a triangle is one-half the product of its base b and its height h.

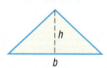

135. Area of an Equilateral Triangle The area A of an equilateral triangle is $\dfrac{\sqrt{3}}{4}$ times the square of the length x of one side.

136. Perimeter of an Equilateral Triangle The perimeter P of an equilateral triangle is 3 times the length x of one side.

137. Volume of a Sphere The volume V of a sphere is $\frac{4}{3}$ times π times the cube of the radius r.

138. Surface Area of a Sphere The surface area S of a sphere is 4 times π times the square of the radius r.

139. Volume of a Cube The volume V of a cube is the cube of the length x of a side.

140. Surface Area of a Cube The surface area S of a cube is 6 times the square of the length x of a side.

141. Manufacturing Cost The weekly production cost C of manufacturing x watches is given by the formula $C = 4000 + 2x$, where the variable C is in dollars.
   (a) What is the cost of producing 1000 watches?
   (b) What is the cost of producing 2000 watches?

142. Balancing a Checking Account At the beginning of the month, Mike had a balance of $210 in his checking account. During the next month, he deposited $80, made an ATM withdrawal for $120, made another deposit of $25, and made two electronic payments: one for $60 and the other for $32. He was also assessed a monthly service charge of $5. What was his balance at the end of the month?

In Problems 143 and 144, write an inequality using an absolute value to describe each statement.

143. x is at least 6 units from 4.

144. x is more than 5 units from 2.

145. U.S. Voltage In the United States, normal household voltage is 110 volts. It is acceptable for the actual voltage x to differ from normal by at most 5 volts. A formula that describes this is

$$|x - 110| \le 5$$

   (a) Show that a voltage of 108 volts is acceptable.
   (b) Show that a voltage of 104 volts is not acceptable.

146. Foreign Voltage In other countries, normal household voltage is 220 volts. It is acceptable for the actual voltage x to differ from normal by at most 8 volts. A formula that describes this is

$$|x - 220| \le 8$$

   (a) Show that a voltage of 214 volts is acceptable.
   (b) Show that a voltage of 209 volts is not acceptable.

147. Making Precision Ball Bearings The FireBall Company manufactures ball bearings for precision equipment. One of its products is a ball bearing with a stated radius of 3 centimeters (cm). Only ball bearings with a radius within 0.01 cm of this stated radius are acceptable. If x is the radius of a ball bearing, a formula describing this situation is

$$|x - 3| \le 0.01$$

   (a) Is a ball bearing of radius $x = 2.999$ acceptable?
   (b) Is a ball bearing of radius $x = 2.89$ acceptable?

148. Body Temperature Normal human body temperature is 98.6°F. A temperature x that differs from normal by at least 1.5°F is considered unhealthy. A formula that describes this is

$$|x - 98.6| \ge 1.5$$

   (a) Show that a temperature of 97°F is unhealthy.
   (b) Show that a temperature of 100°F is not unhealthy.

149. Does $\frac{1}{3}$ equal 0.333? If not, which is larger? By how much?

150. Does $\frac{2}{3}$ equal 0.666? If not, which is larger? By how much?

Explaining Concepts: Discussion and Writing

151. Is there a positive real number "closest" to 0?

152. Number game I'm thinking of a number! It lies between 1 and 10; its square is rational and lies between 1 and 10. The number is larger than π. Correct to two decimal places (that is, truncated to two decimal places), name the number. Now think of your own number, describe it, and challenge a fellow student to name it.

153. Write a brief paragraph that illustrates the similarities and differences between "less than" ($<$) and "less than or equal to" ($\le$).

154. Give a reason why the statement $5 < 8$ is true.

A.2 Geometry Essentials

OBJECTIVES 1 Use the Pythagorean Theorem and Its Converse (p. A14)
2 Know Geometry Formulas (p. A15)
3 Understand Congruent Triangles and Similar Triangles (p. A16)

Figure 13 A right triangle

1 Use the Pythagorean Theorem and Its Converse

The *Pythagorean Theorem* is a statement about *right triangles*. A **right triangle** is one that contains a **right angle**—that is, an angle of 90°. The side of the triangle opposite the 90° angle is called the **hypotenuse**; the remaining two sides are called **legs**. In Figure 13 we have used c to represent the length of the hypotenuse and a and b to represent the lengths of the legs. Notice the use of the symbol ⌐ to show the 90° angle. We now state the Pythagorean Theorem.

PYTHAGOREAN THEOREM

In a right triangle, the square of the length of the hypotenuse is equal to the sum of the squares of the lengths of the legs. That is, in the right triangle shown in Figure 13,

$$c^2 = a^2 + b^2 \tag{1}$$

EXAMPLE 1   Finding the Hypotenuse of a Right Triangle

In a right triangle, one leg has length 4 and the other has length 3. What is the length of the hypotenuse?

Solution   Since the triangle is a right triangle, we use the Pythagorean Theorem with $a = 4$ and $b = 3$ to find the length c of the hypotenuse. From equation (1),

$$c^2 = a^2 + b^2$$
$$c^2 = 4^2 + 3^2 = 16 + 9 = 25$$
$$c = \sqrt{25} = 5$$

Now Work PROBLEM 15

The converse of the Pythagorean Theorem is also true.

CONVERSE OF THE PYTHAGOREAN THEOREM

In a triangle, if the square of the length of one side equals the sum of the squares of the lengths of the other two sides, the triangle is a right triangle. The 90° angle is opposite the longest side.

EXAMPLE 2   Verifying That a Triangle Is a Right Triangle

Show that a triangle whose sides are of lengths 5, 12, and 13 is a right triangle. Identify the hypotenuse.

Solution   Square the lengths of the sides.

$$5^2 = 25 \qquad 12^2 = 144 \qquad 13^2 = 169$$

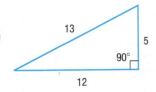

Figure 14

Notice that the sum of the first two squares (25 and 144) equals the third square (169). That is, because $5^2 + 12^2 = 13^2$, the triangle is a right triangle. The longest side, 13, is the hypotenuse. See Figure 14.

Now Work PROBLEM 23

EXAMPLE 3

Applying the Pythagorean Theorem

The tallest building in the world is Burj Khalifa in Dubai, United Arab Emirates, at 2717 feet and 163 floors. The observation deck is 1483 feet above ground level. How far can a person standing on the observation deck see (with the aid of a telescope)? Use 3960 miles for the radius of Earth, and assume Earth is a sphere.

Source: *Council on Tall Buildings and Urban Habitat*

Solution   From the center of Earth, draw two radii: one through Burj Khalifa and the other to the farthest point a person can see from the observation deck. See Figure 15. Apply the Pythagorean Theorem to the right triangle.

Since 1 mile = 5280 feet, 1483 feet = $\dfrac{1483}{5280}$ mile. Then

$$d^2 + (3960)^2 = \left(3960 + \frac{1483}{5280}\right)^2$$

$$d^2 = \left(3960 + \frac{1483}{5280}\right)^2 - (3960)^2 \approx 2224.58$$

$$d \approx 47.17$$

A person can see more than 47 miles from the observation deck.

1483 ft d

3960 mi

Figure 15

Now Work PROBLEM 55

2 Know Geometry Formulas

Certain formulas from geometry are useful in solving algebra problems.
For a rectangle of length l and width w,

Area = lw Perimeter = $2l + 2w$

For a triangle with base b and altitude h,

Area = $\dfrac{1}{2}bh$

For a circle of radius r (diameter $d = 2r$),

$$\boxed{\text{Area} = \pi r^2 \qquad \text{Circumference} = 2\pi r = \pi d}$$

For a closed rectangular box of length l, width w, and height h,

$$\boxed{\text{Volume} = lwh \qquad \text{Surface area} = 2lh + 2wh + 2lw}$$

For a sphere of radius r,

$$\boxed{\text{Volume} = \frac{4}{3}\pi r^3 \qquad \text{Surface area} = 4\pi r^2}$$

For a closed right circular cylinder of height h and radius r,

$$\boxed{\text{Volume} = \pi r^2 h \qquad \text{Surface area} = 2\pi r^2 + 2\pi r h}$$

➤ **Now Work** PROBLEM 31

EXAMPLE 4

Using Geometry Formulas

A Christmas tree ornament is in the shape of a semicircle on top of a triangle. How many square centimeters (cm^2) of copper is required to make the ornament if the height of the triangle is 6 cm and the base is 4 cm?

Solution

Figure 16

See Figure 16. The amount of copper required equals the shaded area. This area is the sum of the areas of the triangle and the semicircle. The triangle has height $h = 6$ and base $b = 4$. The semicircle has diameter $d = 4$, so its radius is $r = 2$.

$$\text{Total area} = \text{Area of triangle} + \text{Area of semicircle}$$
$$= \frac{1}{2}bh + \frac{1}{2}\pi r^2 = \frac{1}{2}\cdot 4\cdot 6 + \frac{1}{2}\pi\cdot 2^2 \qquad b = 4; h = 6; r = 2$$
$$= 12 + 2\pi \approx 18.28 \text{ cm}^2$$

About 18.28 cm^2 of copper is required.

➤ **Now Work** PROBLEM 49

---

3 Understand Congruent Triangles and Similar Triangles

Throughout the text we make reference to triangles. We begin with a discussion of *congruent* triangles. According to dictionary.com, the word **congruent** means "coinciding exactly when superimposed." For example, two angles are congruent if they have the same measure, and two line segments are congruent if they have the same length.

In Words

Two triangles are congruent if they have the same size and shape.

DEFINITION Congruent Triangles

Two triangles are **congruent** if each pair of corresponding angles have the same measure and each pair of corresponding sides are the same length.

In Figure 17, corresponding angles are equal and the corresponding sides are equal in length: $a = d$, $b = e$, and $c = f$. As a result, these triangles are congruent.

Figure 17 Congruent triangles

It is not necessary to verify that all three angles and all three sides are the same measure to determine whether two triangles are congruent.

Determining Congruent Triangles

- **Angle–Side–Angle Case** Two triangles are congruent if two of the angles are equal and the lengths of the corresponding sides between the two angles are equal.

  For example, in Figure 18(a), the two triangles are congruent because two angles and the included side are equal.
- **Side–Side–Side Case** Two triangles are congruent if the lengths of the corresponding sides of the triangles are equal.

  For example, in Figure 18(b), the two triangles are congruent because the three corresponding sides are all equal.
- **Side–Angle–Side Case** Two triangles are congruent if the lengths of two corresponding sides are equal and the angles between the two sides are the same.

  For example, in Figure 18(c), the two triangles are congruent because two sides and the included angle are equal.

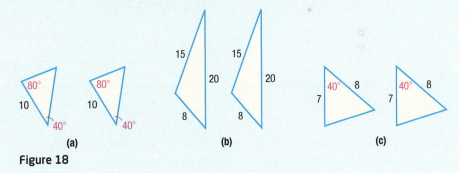

Figure 18

We contrast congruent triangles with *similar* triangles.

DEFINITION Similar Triangles

Two triangles are **similar** if the corresponding angles are equal and the lengths of the corresponding sides are proportional.

For example, the triangles in Figure 19 (on the next page) are similar because the corresponding angles are equal. In addition, the lengths of the corresponding sides are proportional because each side in the triangle on the right is twice as long as each corresponding side in the triangle on the left. That is, the ratio of the corresponding sides is a constant: $\dfrac{d}{a} = \dfrac{e}{b} = \dfrac{f}{c} = 2.$

In Words
Two triangles are similar if they have the same shape, but (possibly) different sizes.

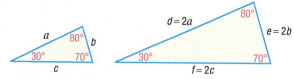

Figure 19 Similar triangles

It is not necessary to verify that all three angles are equal and all three sides are proportional to determine whether two triangles are similar.

Determining Similar Triangles

- **Angle–Angle Case** Two triangles are similar if two of the corresponding angles are equal.

  For example, in Figure 20(a), the two triangles are similar because two angles are equal.

- **Side–Side–Side Case** Two triangles are similar if the lengths of all three sides of each triangle are proportional.

  For example, in Figure 20(b), the two triangles are similar because

$$\frac{10}{30} = \frac{5}{15} = \frac{6}{18} = \frac{1}{3}$$

- **Side–Angle–Side Case** Two triangles are similar if two corresponding sides are proportional and the angles between the two sides are equal.

  For example, in Figure 20(c), the two triangles are similar because

$$\frac{4}{6} = \frac{12}{18} = \frac{2}{3}$$ and the angles between the sides are equal.

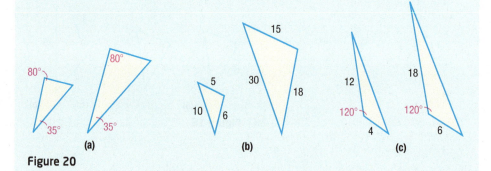

Figure 20

EXAMPLE 5

Using Similar Triangles

Given that the triangles in Figure 21 are similar, find the missing length x and the angles A, B, and C.

Figure 21

Solution Because the triangles are similar, corresponding angles are equal. So $A = 90°$, $B = 60°$, and $C = 30°$. Also, the corresponding sides are proportional. That is, $\dfrac{3}{5} = \dfrac{6}{x}$. We solve this equation for x.

$$\frac{3}{5} = \frac{6}{x}$$

$$5x \cdot \frac{3}{5} = 5x \cdot \frac{6}{x} \qquad \textbf{\textcolor{teal}{Multiply both sides by 5x}}$$

$$3x = 30 \qquad \textbf{\textcolor{teal}{Simplify.}}$$

$$x = 10 \qquad \textbf{\textcolor{teal}{Divide both sides by 3.}}$$

The missing length is 10 units.

Now Work PROBLEM 43

A.2 Assess Your Understanding

Concepts and Vocabulary

1. A(n) _____ triangle is one that contains an angle of 90 degrees. The longest side is called the _____.

2. For a triangle with base b and altitude h, a formula for the area A is _____.

3. The formula for the circumference C of a circle of radius r is _____.

4. Two triangles are _____ if corresponding angles are equal and the lengths of the corresponding sides are proportional.

5. *Multiple Choice* Which of the following is not a case for determining congruent triangles?
   (a) Angle–Side–Angle
   (b) Side–Angle–Side
   (c) Angle–Angle–Angle
   (d) Side–Side–Side

6. *Multiple Choice* Choose the formula for the volume of a sphere of radius r.

   (a) $\dfrac{4}{3}\pi r^2$ **(b)** $\dfrac{4}{3}\pi r^3$ **(c)** $4\pi r^3$ **(d)** $4\pi r^2$

7. *True or False* In a right triangle, the square of the length of the longest side equals the sum of the squares of the lengths of the other two sides.

8. *True or False* The triangle with sides of lengths 6, 8, and 10 is a right triangle.

9. *True or False* The surface area of a sphere of radius r is $\dfrac{4}{3}\pi r^2$.

10. *True or False* The triangles shown are congruent.

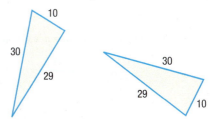

11. *True or False* The triangles shown are similar.

12. *True or False* The triangles shown are similar.

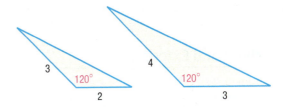

Skill Building

In Problems 13–18, the lengths of the legs of a right triangle are given. Find the hypotenuse.

13. $a = 5$, $b = 12$

14. $a = 6$, $b = 8$

15. $a = 10$, $b = 24$

16. $a = 4$, $b = 3$

17. $a = 7$, $b = 24$

18. $a = 14$, $b = 48$

In Problems 19–26, the lengths of the sides of a triangle are given. Determine which are right triangles. For those that are, identify the hypotenuse.

19. $3, 4, 5$

20. $6, 8, 10$

21. $4, 5, 6$

22. $2, 2, 3$

23. $7, 24, 25$

24. $10, 24, 26$

25. $6, 4, 3$

26. $5, 4, 7$

27. Find the area A of a rectangle with length 6 inches and width 7 inches.

28. Find the area A of a rectangle with length 9 centimeters and width 4 centimeters.

29. Find the area A of a triangle with height 14 inches and base 4 inches.

30. Find the area A of a triangle with height 9 centimeters and base 4 centimeters.

31. Find the area A and circumference C of a circle of radius 5 meters.

32. Find the area A and circumference C of a circle of radius 2 feet.

33. Find the volume V and surface area S of a closed rectangular box with length 6 feet, width 8 feet, and height 5 feet.

34. Find the volume V and surface area S of a closed rectangular box with length 9 inches, width 4 inches, and height 8 inches.

35. Find the volume V and surface area S of a sphere of radius 5 centimeters.

36. Find the volume V and surface area S of a sphere of radius 3 feet.

37. Find the volume V and surface area S of a closed right circular cylinder with radius 9 inches and height 8 inches.

38. Find the volume V and surface area S of a closed right circular cylinder with radius 8 inches and height 9 inches.

In Problems 39–42, find the area of the shaded region.

39.

40.

41.

42.
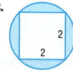

In Problems 43–46, the triangles in each pair are similar. Find the missing length x and the missing angles A, B, and C.

43.

44.

45.

46.

Applications and Extensions

47. How many feet has a wheel with a diameter of 16 inches traveled after four revolutions?

48. How many revolutions will a circular disk with a diameter of 4 feet have completed after it has rolled 20 feet?

49. In the figure shown, *ABCD* is a square, with each side of length 6 feet. The width of the border (shaded portion) between the outer square *EFGH* and *ABCD* is 2 feet. Find the area of the border.

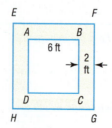

50. Refer to the figure. Square *ABCD* has an area of 100 square feet; square *BEFG* has an area of 16 square feet. What is the area of the triangle *CGF*?

51. Architecture A **Norman window** consists of a rectangle surmounted by a semicircle. Find the area of the Norman window shown in the figure. How much wood frame is needed to enclose the window?

52. Construction A circular swimming pool that is 20 feet in diameter is enclosed by a wooden deck that is 3 feet wide. What is the area of the deck? How much fence is required to enclose the deck?

53. How Tall Is the Great Pyramid? The ancient Greek philosopher Thales of Miletus is reported on one occasion to have visited Egypt and calculated the height of the Great Pyramid of Cheops by means of shadow reckoning. Thales knew that each side of the base of the pyramid was 252 paces and that his own height was 2 paces. He measured the length of the pyramid's shadow to be 114 paces and determined the length of his shadow to be 3 paces. See the figure. Using similar triangles, determine the height of the Great Pyramid in terms of the number of paces.

Source: Diggins, Julie E, *String, Straightedge, and Shadow: The Story of Geometry*, 2003, Whole Spirit Press, http://wholespiritpress.com.

54. The Bermuda Triangle Karen is doing research on the Bermuda Triangle which she defines roughly by Hamilton, Bermuda; San Juan, Puerto Rico; and Fort Lauderdale, Florida. On her atlas Karen measures the straight-line distances from Hamilton to Fort Lauderdale, Fort Lauderdale to San Juan, and San Juan to Hamilton to be approximately 57 millimeters (mm), 58 mm, and 53.5 mm, respectively. If the actual distance from Fort Lauderdale to San Juan is 1046 miles, approximate the actual distances from San Juan to Hamilton and from Hamilton to Fort Lauderdale.

In Problems 55–57, use the facts that the radius of Earth is 3960 miles and 1 mile = 5280 feet.

55. How Far Can You See? The conning tower of the U.S.S. *Silversides*, a World War II submarine now permanently stationed in Muskegon, Michigan, is approximately 20 feet above sea level. How far can you see from the conning tower?

56. How Far Can You See? A person who is 6 feet tall is standing on the beach in Fort Lauderdale, Florida, and looks out onto the Atlantic Ocean. Suddenly, a ship appears on the horizon. How far is the ship from shore?

57. How Far Can You See? The deck of a destroyer is 100 feet above sea level. How far can a person see from the deck? How far can a person see from the bridge, which is 150 feet above sea level?

58. Suppose that m and n are positive integers with $m > n$. If $a = m^2 - n^2$, $b = 2mn$, and $c = m^2 + n^2$, show that a, b, and c are the lengths of the sides of a right triangle. (This formula can be used to find the sides of a right triangle that are integers, such as 3, 4, 5; 5, 12, 13; and so on. Such triplets of integers are called **Pythagorean triples**.)

Explaining Concepts: Discussion and Writing

59. If the radius of a circle is doubled, does the area of the circle also double? Explain.

60. If the radius of a sphere is doubled, does the volume of the sphere also double? Explain.

61. You have 1000 feet of flexible pool siding and intend to construct a swimming pool. Experiment with rectangular-shaped pools with perimeters of 1000 feet. How do their areas vary? What is the shape of the rectangle with the largest area? Now compute the area enclosed by a circular pool with a perimeter (circumference) of 1000 feet. What would be your choice of shape for the pool? If rectangular, what is your preference for dimensions? Justify your choice. If your only consideration is to have a pool that encloses the most area, what shape should you use?

62. **The Gibb's Hill Lighthouse, Southampton, Bermuda,** in operation since 1846, stands 117 feet high on a hill 245 feet high, so its beam of light is 362 feet above sea level. A brochure states that the light itself can be seen on the horizon about 26 miles distant. Verify the accuracy of this information. The brochure further states that ships 40 miles away can see the light and that planes flying at 10,000 feet can see it 120 miles away. Verify the accuracy of these statements. What assumption did the brochure make about the height of the ship?

A.3 Factoring Polynomials; Completing the Square

> **OBJECTIVES** 1 Know Formulas for Special Products (p. A22)
> 2 Factor Polynomials (p. A23)
> 3 Complete the Square (p. A24)

1 Know Formulas for Special Products

Certain products, which we call **special products**, occur frequently in algebra. For example, we can find the product of two binomials using the **FOIL** (*First*, *Outer*, *Inner*, *Last*) method.

$$(ax + b)(cx + d) = \overbrace{ax \cdot cx}^{\text{First}} + \overbrace{ax \cdot d}^{\text{Outer}} + \overbrace{b \cdot cx}^{\text{Inner}} + \overbrace{b \cdot d}^{\text{Last}}$$
$$= acx^2 + adx + bcx + bd$$
$$= acx^2 + (ad + bc)x + bd$$

EXAMPLE 1 | **Using FOIL**

(a) $(x - 3)(x + 3) = x^2 + 3x - 3x - 9 = x^2 - 9$
$$\quad\quad\quad\quad\quad\quad\;\; \text{F}\quad\;\; \text{O}\quad\;\; \text{I}\quad\; \text{L}$$
(b) $(x + 2)^2 = (x + 2)(x + 2) = x^2 + 2x + 2x + 4 = x^2 + 4x + 4$

(c) $(x - 3)^2 = (x - 3)(x - 3) = x^2 - 3x - 3x + 9 = x^2 - 6x + 9$

(d) $(x + 3)(x + 1) = x^2 + x + 3x + 3 = x^2 + 4x + 3$

(e) $(2x + 1)(3x + 4) = 6x^2 + 8x + 3x + 4 = 6x^2 + 11x + 4$

Now Work PROBLEM 9

Some products have been given special names because of their form. In the list that follows, x and a are real numbers.

Difference of Two Squares

> $$(x - a)(x + a) = x^2 - a^2 \tag{1}$$

Squares of Binomials, or Perfect Squares

> $$(x + a)^2 = x^2 + 2ax + a^2 \tag{2a}$$
> $$(x - a)^2 = x^2 - 2ax + a^2 \tag{2b}$$

Cubes of Binomials, or Perfect Cubes

$$(x + a)^3 = x^3 + 3ax^2 + 3a^2x + a^3 \qquad (3a)$$
$$(x - a)^3 = x^3 - 3ax^2 + 3a^2x - a^3 \qquad (3b)$$

Difference of Two Cubes

$$(x - a)(x^2 + ax + a^2) = x^3 - a^3 \qquad (4)$$

Sum of Two Cubes

$$(x + a)(x^2 - ax + a^2) = x^3 + a^3 \qquad (5)$$

Now Work PROBLEMS 13, 17, AND 21

2 Factor Polynomials

Consider the following product:

$$(2x + 3)(x - 4) = 2x^2 - 5x - 12$$

The two polynomials on the left side are called **factors** of the polynomial on the right side. Expressing a given polynomial as a product of other polynomials—that is, finding the factors of a polynomial—is called **factoring**.

We restrict our discussion here to factoring polynomials in one variable into products of polynomials in one variable, where all coefficients are integers. We call this **factoring over the integers**.

Any polynomial can be written as the product of 1 times itself or as -1 times its additive inverse. If a polynomial cannot be written as the product of two other polynomials (excluding 1 and -1), then the polynomial is said to be **prime**. When a polynomial has been written as a product consisting only of prime factors, it is **factored completely**. Examples of prime polynomials (over the integers) are

$$2, \quad 3, \quad 5, \quad x, \quad x + 1, \quad x - 1, \quad 3x + 4, \quad x^2 + 4$$

COMMENT Over the real numbers, $3x + 4$ factors into $3\left(x + \frac{4}{3}\right)$. It is the noninteger $\frac{4}{3}$ that causes $3x + 4$ to be prime over the integers. ∎

The first factor to look for in a factoring problem is a common monomial factor present in each term of the polynomial. If one is present, use the Distributive Property to factor it out. Continue factoring out monomial factors until none are left.

EXAMPLE 2 Identifying Common Monomial Factors

Polynomial	Common Monomial Factor	Remaining Factor	Factored Form
$2x + 4$	2	$x + 2$	$2x + 4 = 2(x + 2)$
$3x - 6$	3	$x - 2$	$3x - 6 = 3(x - 2)$
$2x^2 - 4x + 8$	2	$x^2 - 2x + 4$	$2x^2 - 4x + 8 = 2(x^2 - 2x + 4)$
$8x - 12$	4	$2x - 3$	$8x - 12 = 4(2x - 3)$
$x^2 + x$	x	$x + 1$	$x^2 + x = x(x + 1)$
$x^3 - 3x^2$	x^2	$x - 3$	$x^3 - 3x^2 = x^2(x - 3)$
$6x^2 + 9x$	$3x$	$2x + 3$	$6x^2 + 9x = 3x(2x + 3)$

Notice that, once all common monomial factors have been removed from a polynomial, the remaining factor is either a prime polynomial of degree 1 or a polynomial of degree 2 or higher. (Do you see why?)

The list of special products (1) through (5) given earlier provides a list of factoring formulas when the equations are read from right to left. For example, equation (1) states that if the polynomial is the difference of two squares, $x^2 - a^2$, it can be factored into $(x - a)(x + a)$. The following example illustrates several factoring techniques.

EXAMPLE 3

Factoring Polynomials

Factor completely each polynomial.

(a) $x^4 - 16$ (b) $x^3 - 1$ (c) $9x^2 - 6x + 1$

(d) $x^2 + 4x - 12$ (e) $3x^2 + 10x - 8$ (f) $x^3 - 4x^2 + 2x - 8$

Solution

(a) $x^4 - 16 = \underset{\uparrow}{(x^2 - 4)}(x^2 + 4) = \underset{\uparrow}{(x - 2)(x + 2)}(x^2 + 4)$

              Difference of squares   **Difference of squares**

(b) $x^3 - 1 = \underset{\uparrow}{(x - 1)}(x^2 + x + 1)$

          Difference of cubes

(c) $9x^2 - 6x + 1 = \underset{\uparrow}{(3x - 1)^2}$

          Perfect square

(d) $x^2 + 4x - 12 = \underset{\uparrow}{(x + 6)(x - 2)}$

      The product of 6 and -2 is -12, and the sum of 6 and -2 is 4.

                  $12x - 2x = 10x$

(e) $3x^2 + 10x - 8 = (3x - 2)(x + 4)$

                  $3x^2$ **-8**

COMMENT The technique used in part (f) is called **factoring by grouping.** ∎

(f) $x^3 - 4x^2 + 2x - 8 = \underset{\uparrow}{(x^3 - 4x^2)} + (2x - 8)$

          Group terms

      $= \underset{\uparrow}{x^2(x - 4)} + 2(x - 4) = \underset{\uparrow}{(x^2 + 2)(x - 4)}$

          Distributive Property   **Distributive Property**

✏️ **Now Work** PROBLEMS 33, 49, AND 83

3 Complete the Square

The idea behind completing the square in one variable is to "adjust" an expression of the form $x^2 + bx$ to make it a perfect square. Perfect squares are trinomials of the form

$$x^2 + 2ax + a^2 = (x + a)^2 \quad \text{or} \quad x^2 - 2ax + a^2 = (x - a)^2$$

For example, $x^2 + 6x + 9$ is a perfect square because $x^2 + 6x + 9 = (x + 3)^2$. And $p^2 - 12p + 36$ is a perfect square because $p^2 - 12p + 36 = (p - 6)^2$.

So how do we "adjust" $x^2 + bx$ to make it a perfect square? We do it by adding a number. For example, to make $x^2 + 6x$ a perfect square, add 9. But how do we know to add 9? If we divide the coefficient of the first-degree term, 6, by 2, and then square the result, we obtain 9. This approach works in general.

> ### Completing the Square of $x^2 + bx$
> - Identify the coefficient of the first-degree term, namely b.
> - Multiply b by $\dfrac{1}{2}$ and then square the result. That is, compute $\left(\dfrac{1}{2}b\right)^2$.
> - Add $\left(\dfrac{1}{2}b\right)^2$ to $x^2 + bx$ to get $x^2 + bx + \left(\dfrac{1}{2}b\right)^2 = \left(x + \dfrac{b}{2}\right)^2$

WARNING To use $\left(\dfrac{1}{2}b\right)^2$ to complete the square, the coefficient of the x^2 term must be 1. ∎

EXAMPLE 4 **Completing the Square**

Determine the number that must be added to each expression to complete the square. Then factor the expression.

Start	Add	Result	Factored Form
$y^2 + 8y$	$\left(\dfrac{1}{2}\cdot 8\right)^2 = 16$	$y^2 + 8y + 16$	$(y + 4)^2$
$x^2 + 12x$	$\left(\dfrac{1}{2}\cdot 12\right)^2 = 36$	$x^2 + 12x + 36$	$(x + 6)^2$
$a^2 - 20a$	$\left(\dfrac{1}{2}\cdot(-20)\right)^2 = 100$	$a^2 - 20a + 100$	$(a - 10)^2$
$p^2 - 5p$	$\left(\dfrac{1}{2}\cdot(-5)\right)^2 = \dfrac{25}{4}$	$p^2 - 5p + \dfrac{25}{4}$	$\left(p - \dfrac{5}{2}\right)^2$

Notice that the factored form of a perfect square is either

$$x^2 + bx + \left(\dfrac{b}{2}\right)^2 = \left(x + \dfrac{b}{2}\right)^2 \quad \text{or} \quad x^2 - bx + \left(\dfrac{b}{2}\right)^2 = \left(x - \dfrac{b}{2}\right)^2$$

— **Now Work** PROBLEM 73

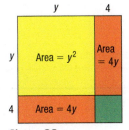

Figure 22

Are you wondering why we refer to making an expression a perfect square as "completing the square"? Look at the square in Figure 22. Its area is $(y + 4)^2$. The yellow area is y^2 and each orange area is $4y$ (for a total area of $8y$). The sum of these areas is $y^2 + 8y$. To complete the square, we need to add the area of the green region: $4 \cdot 4 = 16$. As a result, $y^2 + 8y + 16 = (y + 4)^2$.

A.3 Assess Your Understanding

Concepts and Vocabulary

1. $(x^2 - 4)(x^2 + 4) = $ _____.

2. $(x - 2)(x^2 + 2x + 4) = $ _____.

3. **True or False** The polynomial $x^2 + 4$ is prime.

4. **True or False** $3x^3 - 2x^2 - 6x + 4 = (3x - 2)(x^2 + 2)$.

5. To complete the square of the expression $x^2 + 5x$, you would

   ____ the number ____.

6. *Mutiple Choice* Choose the best description of $x^2 - 64$.
   (a) Prime
   (b) Difference of two squares
   (c) Perfect Square
   (d) Difference of two cubes

7. *Mutiple Choice* Choose the complete factorization of
   $$4x^2 - 8x - 60$$
   (a) $2(x + 3)(x - 5)$
   (b) $4(x^2 - 2x - 15)$
   (c) $(2x + 6)(2x - 10)$
   (d) $4(x + 3)(x - 5)$

8. *Mutiple Choice* To complete the square of $x^2 + bx$, use which of the following?
   (a) $(2b)^2$
   (b) $2b^2$
   (c) $\left(\dfrac{1}{2}b\right)^2$
   (d) $\dfrac{1}{2}b^2$

Skill Building

In Problems 9–24, multiply out each expression. Express your answer as a single polynomial in standard form.

9. $(x + 2)(x + 4)$
10. $(x + 3)(x + 5)$
11. $(2x + 5)(x + 2)$
12. $(3x + 1)(2x + 1)$
13. $(x - 7)(x + 7)$
14. $(x - 1)(x + 1)$
15. $(2x + 3)(2x - 3)$
16. $(3x + 2)(3x - 2)$
17. $(x + 4)^2$
18. $(x - 5)^2$
19. $(2x - 3)^2$
20. $(3x - 4)^2$
21. $(x - 2)^3$
22. $(x + 1)^3$
23. $(2x + 1)^3$
24. $(3x - 2)^3$

In Problems 25–72, factor completely each polynomial. If the polynomial cannot be factored, say it is prime.

25. $x^2 - 36$
26. $x^2 - 9$
27. $2 - 8x^2$
28. $3 - 27x^2$
29. $x^2 + 11x + 10$
30. $x^2 + 5x + 4$
31. $x^2 - 10x + 21$
32. $x^2 - 6x + 8$
33. $4x^2 - 8x + 32$
34. $3x^2 - 12x + 15$
35. $x^2 + 4x + 16$
36. $x^2 + 12x + 36$
37. $15 + 2x - x^2$
38. $14 + 6x - x^2$
39. $3x^2 - 12x - 36$
40. $x^3 + 8x^2 - 20x$
41. $y^4 + 11y^3 + 30y^2$
42. $3y^3 - 18y^2 - 48y$
43. $4x^2 + 12x + 9$
44. $9x^2 - 12x + 4$
45. $6x^2 + 8x + 2$
46. $8x^2 + 6x - 2$
47. $x^4 - 81$
48. $x^4 - 1$
49. $x^6 - 2x^3 + 1$
50. $x^6 + 2x^3 + 1$
51. $x^7 - x^5$
52. $x^8 - x^5$
53. $16x^2 + 24x + 9$
54. $9x^2 - 24x + 16$
55. $5 + 16x - 16x^2$
56. $5 + 11x - 16x^2$
57. $4y^2 - 16y + 15$
58. $9y^2 + 9y - 4$
59. $1 - 8x^2 - 9x^4$
60. $4 - 14x^2 - 8x^4$
61. $x(x + 3) - 6(x + 3)$
62. $5(3x - 7) + x(3x - 7)$
63. $(x + 2)^2 - 5(x + 2)$
64. $(x - 1)^2 - 2(x - 1)$
65. $(3x - 2)^3 - 27$
66. $(5x + 1)^3 - 1$
67. $3(x^2 + 10x + 25) - 4(x + 5)$
68. $7(x^2 - 6x + 9) + 5(x - 3)$
69. $x^3 + 2x^2 - x - 2$
70. $x^3 - 3x^2 - x + 3$
71. $x^4 - x^3 + x - 1$
72. $x^4 + x^3 + x + 1$

In Problems 73–78, determine the number that should be added to complete the square of each expression. Then factor each expression.

73. $x^2 + 10x$
74. $p^2 + 14p$
75. $y^2 - 6y$
76. $x^2 - 4x$
77. $x^2 - \dfrac{1}{2}x$
78. $x^2 + \dfrac{1}{3}x$

Applications and Extensions

In Problems 79–88, expressions that occur in calculus are given. Factor completely each expression.

79. $2(3x + 4)^2 + (2x + 3) \cdot 2(3x + 4) \cdot 3$
80. $5(2x + 1)^2 + (5x - 6) \cdot 2(2x + 1) \cdot 2$
81. $2x(2x + 5) + x^2 \cdot 2$
82. $3x^2(8x - 3) + x^3 \cdot 8$
83. $2(x + 3)(x - 2)^3 + (x + 3)^2 \cdot 3(x - 2)^2$
84. $4(x + 5)^3(x - 1)^2 + (x + 5)^4 \cdot 2(x - 1)$
85. $(4x - 3)^2 + x \cdot 2(4x - 3) \cdot 4$
86. $3x^2(3x + 4)^2 + x^3 \cdot 2(3x + 4) \cdot 3$
87. $2(3x - 5) \cdot 3(2x + 1)^3 + (3x - 5)^2 \cdot 3(2x + 1)^2 \cdot 2$
88. $3(4x + 5)^2 \cdot 4(5x + 1)^2 + (4x + 5)^3 \cdot 2(5x + 1) \cdot 5$
89. Show that $x^2 + 4$ is prime.
90. Show that $x^2 + x + 1$ is prime.

Explaining Concepts: Discussion and Writing

91. Do you prefer to memorize the rule for the square of a binomial $(x + a)^2$ or to use FOIL to obtain the product? Write a brief position paper defending your choice.

92. Make up a polynomial that factors into a perfect square.

93. Explain to a fellow student what you look for first when presented with a factoring problem. What do you do next?

A.4 Solving Equations

PREPARING FOR THIS SECTION *Before getting started, review the following:*

- Factoring Polynomials (Section A.3, pp. A23–A24)
- Zero-Product Property (Section A.1, p. A4)
- Square Roots (Section A.1, pp. A9–A10)
- Absolute Value (Section A.1, pp. A5–A6)
- Complete the Square (Section A.3, pp. A24–A25)

Now Work the 'Are You Prepared?' problems on page A34.

OBJECTIVES
1 Solve Equations by Factoring (p. A29)
2 Solve Equations Involving Absolute Value (p. A29)
3 Solve a Quadratic Equation by Factoring (p. A30)
4 Solve a Quadratic Equation by Completing the Square (p. A31)
5 Solve a Quadratic Equation Using the Quadratic Formula (p. A32)

An **equation in one variable** is a statement in which two expressions, at least one containing the variable, are equal. The expressions are called the **sides** of the equation. Since an equation is a statement, it may be true or false, depending on the value of the variable. Unless otherwise restricted, the admissible values of the variable are those in the domain of the variable. These admissible values of the variable, if any, that result in a true statement are called **solutions**, or **roots**, of the equation. To **solve an equation** means to find all the solutions of the equation.

For example, the following are all equations in one variable, x:

$$x + 5 = 9 \qquad x^2 + 5x = 2x - 2 \qquad \frac{x^2 - 4}{x + 1} = 0 \qquad \sqrt{x^2 + 9} = 5$$

The first of these statements, $x + 5 = 9$, is true when $x = 4$ and false for any other choice of x. That is, 4 is the solution of the equation $x + 5 = 9$. We also say that 4 **satisfies** the equation $x + 5 = 9$, because, when 4 is substituted for x, a true statement results.

Sometimes an equation will have more than one solution. For example, the equation

$$\frac{x^2 - 4}{x + 1} = 0$$

has -2 and 2 as solutions.

Usually, we write the solutions of an equation as a set, called the **solution set** of the equation. For example, the solution set of the equation $x^2 - 9 = 0$ is $\{-3, 3\}$.

Some equations have no real solution. For example, $x^2 + 9 = 5$ has no real solution, because there is no real number whose square, when added to 9, equals 5.

An equation that is satisfied for every value of the variable for which both sides are defined is called an **identity**. For example, the equation

$$3x + 5 = x + 3 + 2x + 2$$

is an identity, because this statement is true for any real number x.

One method for solving an equation is to replace the original equation by a succession of **equivalent equations**, equations having the same solution set, until an equation with an obvious solution is obtained.

For example, consider the following succession of equivalent equations:

$$2x + 3 = 13$$
$$2x = 10$$
$$x = 5$$

We conclude that the solution set of the original equation is $\{5\}$.

How do we obtain equivalent equations? In general, there are five ways.

Procedures That Result in Equivalent Equations

- Interchange the two sides of the equation:

  Replace $\;3 = x\;$ by $\;x = 3$

- Simplify the sides of the equation by combining like terms, eliminating parentheses, and so on:

  Replace $\quad (x + 2) + 6 = 2x + (x + 1)$

  by $\qquad\quad x + 8 = 3x + 1$

- Add or subtract the same expression on both sides of the equation:

  Replace $\qquad\quad 3x - 5 = 4$

  by $\quad (3x - 5) + 5 = 4 + 5$

- Multiply or divide both sides of the equation by the same nonzero expression:

  Replace $\qquad \dfrac{3x}{x-1} = \dfrac{6}{x-1} \qquad x \neq 1$

  by $\quad \dfrac{3x}{x-1} \cdot (x-1) = \dfrac{6}{x-1} \cdot (x-1)$

- If one side of the equation is 0 and the other side can be factored, then write it as the product of factors:

  Replace $\qquad\quad x^2 - 3x = 0$

  by $\qquad\quad x(x-3) = 0$

WARNING Squaring both sides of an equation does not necessarily lead to an equivalent equation. For example, $x = 3$ has one solution, but $x^2 = 9$ has two solutions, -3 and 3. ■

Whenever it is possible to solve an equation in your head, do so. For example,

- The solution of $2x = 8$ is 4.
- The solution of $3x - 15 = 0$ is 5.

Now Work PROBLEM 15

EXAMPLE 1   **Solving an Equation**

Solve the equation: $3x - 5 = 4$

Solution   Replace the original equation by a succession of equivalent equations.

$$3x - 5 = 4$$
$$(3x - 5) + 5 = 4 + 5 \quad \textbf{Add 5 to both sides.}$$
$$3x = 9 \qquad\qquad \textbf{Simplify.}$$
$$\frac{3x}{3} = \frac{9}{3} \qquad\qquad \textbf{Divide both sides by 3.}$$
$$x = 3 \qquad\qquad \textbf{Simplify.}$$

The last equation, $x = 3$, has the solution 3. All these equations are equivalent, so 3 is the only solution of the original equation, $3x - 5 = 4$.

✔ **Check:** Check the solution by substituting 3 for x in the original equation.

$$3x - 5 = 3 \cdot 3 - 5 = 9 - 5 = 4$$

The solution checks. The solution set is $\{3\}$.

Now Work PROBLEMS 29 AND 35

1 Solve Equations by Factoring

If an equation can be written as a product of factors equal to 0, use the Zero-Product Property to set each factor equal to 0 and solve the resulting equations.

EXAMPLE 2   ### Solving Equations by Factoring

Solve the equations: (a) $x^3 = 4x$ (b) $x^3 - x^2 - 4x + 4 = 0$

Solution   (a) Begin by collecting all terms on one side. This results in 0 on one side and an expression to be factored on the other.

$$x^3 = 4x$$

$$x^3 - 4x = 0$$

$$x(x^2 - 4) = 0 \qquad \text{Factor.}$$

$$x(x - 2)(x + 2) = 0 \qquad \text{Factor again.}$$

$$x = 0 \quad \text{or} \quad x - 2 = 0 \quad \text{or} \quad x + 2 = 0 \qquad \text{Use the Zero-Product Property.}$$

$$x = 0 \quad \text{or} \qquad x = 2 \quad \text{or} \qquad x = -2 \qquad \text{Solve for } x.$$

✓ **Check:** $x = -2$: $(-2)^3 = -8$ and $4(-2) = -8$ **−2 is a solution.**

$x = 0$: $0^3 = 0$ and $4 \cdot 0 = 0$ **0 is a solution.**

$x = 2$: $2^3 = 8$ and $4 \cdot 2 = 8$ **2 is a solution.**

The solution set is $\{-2, 0, 2\}$.

(b) Group the terms of $x^3 - x^2 - 4x + 4 = 0$ as follows:

$$(x^3 - x^2) - (4x - 4) = 0$$

Factor out x^2 from the first grouping and 4 from the second.

$$x^2(x - 1) - 4(x - 1) = 0$$

This reveals the common factor $(x - 1)$, so

$$(x^2 - 4)(x - 1) = 0$$

$$(x - 2)(x + 2)(x - 1) = 0 \qquad\qquad\qquad \text{Factor again.}$$

$$x - 2 = 0 \quad \text{or} \quad x + 2 = 0 \quad \text{or} \quad x - 1 = 0 \quad \text{Use the Zero-Product Property.}$$

$$x = 2 \qquad\qquad x = -2 \qquad\qquad x = 1 \quad \text{Solve for } x.$$

✓ **Check:**

$x = -2$: $(-2)^3 - (-2)^2 - 4(-2) + 4 = -8 - 4 + 8 + 4 = 0$ **−2 is a solution.**

$x = 1$: $1^3 - 1^2 - 4(1) + 4 = 1 - 1 - 4 + 4 = 0$ **1 is a solution.**

$x = 2$: $2^3 - 2^2 - 4(2) + 4 = 8 - 4 - 8 + 4 = 0$ **2 is a solution.**

The solution set is $\{-2, 1, 2\}$.   ⌐

── **Now Work** PROBLEMS 39 AND 45

2 Solve Equations Involving Absolute Value

On the real number line, the absolute value of a equals the distance from the origin to the point whose coordinate is a. For example, there are two points whose distance from the origin is 5 units, -5 and 5. So the equation $|x| = 5$ has the solution set $\{-5, 5\}$.

EXAMPLE 3

Solving an Equation Involving Absolute Value

Solve the equation: $|x + 4| = 13$

Solution

There are two possibilities:

$$x + 4 = 13 \quad \text{or} \quad x + 4 = -13$$
$$x = 9 \quad \text{or} \quad x = -17$$

The solution set is $\{-17, 9\}$.

Now Work PROBLEM 51

3 Solve a Quadratic Equation by Factoring

DEFINITION Quadratic Equation

A **quadratic equation** is an equation equivalent to one of the form

$$ax^2 + bx + c = 0 \tag{1}$$

where a, b, and c are real numbers and $a \neq 0$.

A quadratic equation written in the form $ax^2 + bx + c = 0$ is said to be in **standard form**.

Sometimes, a quadratic equation is called a **second-degree equation** because, when it is in standard form, the left side is a polynomial of degree 2.

When a quadratic equation is written in standard form, it may be possible to factor the expression on the left side into the product of two first-degree polynomials. Then, using the Zero-Product Property and setting each factor equal to 0, the resulting linear equations can be solved to obtain the solutions of the quadratic equation.

EXAMPLE 4

Solving a Quadratic Equation by Factoring

Solve the equation: $2x^2 = x + 3$

Solution

Put the equation $2x^2 = x + 3$ in standard form by subtracting x and 3 from both sides.

$$2x^2 = x + 3$$
$$2x^2 - x - 3 = 0 \qquad \text{Subtract x and 3 from both sides.}$$

The left side may now be factored as

$$(2x - 3)(x + 1) = 0 \qquad \text{Factor.}$$

so that

$$2x - 3 = 0 \quad \text{or} \quad x + 1 = 0 \qquad \text{Use the Zero-Product Property.}$$
$$x = \frac{3}{2} \qquad x = -1 \qquad \text{Solve.}$$

The solution set is $\left\{-1, \frac{3}{2}\right\}$.

When the left side factors into two linear equations with the same solution, the quadratic equation is said to have a **repeated solution**. This solution is also called a **root of multiplicity 2**, or a **double root**.

EXAMPLE 5 **Solving a Quadratic Equation by Factoring**

Solve the equation: $9x^2 - 6x + 1 = 0$

Solution This equation is already in standard form, and the left side can be factored.

$$9x^2 - 6x + 1 = 0$$

$$(3x - 1)(3x - 1) = 0 \quad \textcolor{blue}{\textbf{Factor.}}$$

so

$$x = \frac{1}{3} \quad \text{or} \quad x = \frac{1}{3} \quad \textcolor{blue}{\textbf{Solve for x.}}$$

This equation has only the repeated solution $\frac{1}{3}$. The solution set is $\left\{\frac{1}{3}\right\}$.

Now Work PROBLEM 69

The Square Root Method

Suppose that we wish to solve the quadratic equation

$$x^2 = p \qquad (2)$$

where $p > 0$ is a positive number. Proceeding as in the earlier examples,

$$x^2 - p = 0 \qquad \textcolor{blue}{\textbf{Put in standard form.}}$$

$$(x - \sqrt{p})(x + \sqrt{p}) = 0 \qquad \textcolor{blue}{\textbf{Factor (over the real numbers).}}$$

$$x = \sqrt{p} \quad \text{or} \quad x = -\sqrt{p} \qquad \textcolor{blue}{\textbf{Solve.}}$$

We have the following result:

If $x^2 = p$ and $p > 0$, then $x = \sqrt{p}$ or $x = -\sqrt{p}$. $\qquad (3)$

When statement (3) is used, it is called the **Square Root Method.** In statement (3), note that if $p > 0$ the equation $x^2 = p$ has two solutions, $x = \sqrt{p}$ and $x = -\sqrt{p}$. We usually abbreviate these solutions as $x = \pm\sqrt{p}$, which is read as "x equals plus or minus the square root of p."

For example, the two solutions of the equation

$$x^2 = 4$$

are

$$x = \pm\sqrt{4} \qquad \textcolor{blue}{\textbf{Use the Square Root Method.}}$$

and, since $\sqrt{4} = 2$, we have

$$x = \pm 2$$

The solution set is $\{-2, 2\}$.

Now Work PROBLEM 83

4 Solve a Quadratic Equation by Completing the Square

EXAMPLE 6 **Solving a Quadratic Equation by Completing the Square**

Solve by completing the square: $2x^2 - 8x - 5 = 0$

Solution First, rewrite the equation so that the constant is on the right side.

$$2x^2 - 8x - 5 = 0$$

$$2x^2 - 8x = 5 \qquad \textcolor{blue}{\textbf{Add 5 to both sides.}}$$

(continued)

Next, divide both sides by 2 so that the coefficient of x^2 is 1. (This enables us to complete the square at the next step.)

$$x^2 - 4x = \frac{5}{2}$$

Finally, complete the square by adding $\left[\frac{1}{2}(-4)\right]^2 = 4$ to both sides.

$$x^2 - 4x + 4 = \frac{5}{2} + 4 \qquad \text{Add 4 to both sides.}$$

$$(x - 2)^2 = \frac{13}{2} \qquad \text{Factor on the left; simplify on the right.}$$

$$x - 2 = \pm\sqrt{\frac{13}{2}} \qquad \text{Use the Square Root Method.}$$

$$x - 2 = \pm\frac{\sqrt{26}}{2} \qquad \sqrt{\frac{13}{2}} = \frac{\sqrt{13}}{\sqrt{2}} = \frac{\sqrt{13}}{\sqrt{2}} \cdot \frac{\sqrt{2}}{\sqrt{2}} = \frac{\sqrt{26}}{2}$$

$$x = 2 \pm \frac{\sqrt{26}}{2}$$

NOTE If we wanted an approximation, say rounded to two decimal places, of these solutions, we would use a calculator to get $\{-0.55, 4.55\}$. ∎

The solution set is $\left\{2 - \frac{\sqrt{26}}{2}, 2 + \frac{\sqrt{26}}{2}\right\}$.

━━ **Now Work** PROBLEM 87

5 Solve a Quadratic Equation Using the Quadratic Formula

We can use the method of completing the square to obtain a general formula for solving any quadratic equation

$$ax^2 + bx + c = 0 \qquad a \neq 0$$

As in Example 6, rearrange the terms as

$$ax^2 + bx = -c \qquad a > 0$$

NOTE There is no loss in generality to assume that $a > 0$, since if $a < 0$ we can multiply by -1 to obtain an equivalent equation with a positive leading coefficient. ∎

Since $a > 0$, divide both sides by a to get

$$x^2 + \frac{b}{a}x = -\frac{c}{a}$$

Now the coefficient of x^2 is 1. To complete the square on the left side, add the square of $\frac{1}{2}$ times the coefficient of x; that is, add

$$\left(\frac{1}{2} \cdot \frac{b}{a}\right)^2 = \frac{b^2}{4a^2}$$

to both sides. Then

$$x^2 + \frac{b}{a}x + \frac{b^2}{4a^2} = \frac{b^2}{4a^2} - \frac{c}{a}$$

$$\left(x + \frac{b}{2a}\right)^2 = \frac{b^2 - 4ac}{4a^2} \qquad \frac{b^2}{4a^2} - \frac{c}{a} = \frac{b^2}{4a^2} - \frac{4ac}{4a^2} = \frac{b^2 - 4ac}{4a^2} \qquad (4)$$

Provided that $b^2 - 4ac \geq 0$, we can now use the Square Root Method to get

$$x + \frac{b}{2a} = \pm\sqrt{\frac{b^2 - 4ac}{4a^2}}$$

$$x + \frac{b}{2a} = \frac{\pm\sqrt{b^2 - 4ac}}{2a} \qquad \text{The square root of a quotient equals the quotient of the square roots. Also, } \sqrt{4a^2} = 2a \text{ since } a > 0.$$

$$x = -\frac{b}{2a} \pm \frac{\sqrt{b^2 - 4ac}}{2a} \qquad \text{Add } -\frac{b}{2a} \text{ to both sides.}$$

$$= \frac{-b \pm \sqrt{b^2 - 4ac}}{2a} \qquad \text{Combine the quotients on the right.}$$

What if $b^2 - 4ac$ is negative? Then equation (4) states that the left expression (a real number squared) equals the right expression (a negative number). Since this is impossible for real numbers, we conclude that if $b^2 - 4ac < 0$, the quadratic equation has no *real* solution. (We discuss quadratic equations for which the quantity $b^2 - 4ac < 0$ in detail in the next section.)

THEOREM Quadratic Formula

Consider the quadratic equation

$$ax^2 + bx + c = 0 \qquad a \neq 0$$

- If $b^2 - 4ac < 0$, this equation has no real solution.
- If $b^2 - 4ac \geq 0$, the real solution(s) of this equation is (are) given by the **quadratic formula**:

$$x = \frac{-b \pm \sqrt{b^2 - 4ac}}{2a} \qquad\qquad (5)$$

The quantity $b^2 - 4ac$ is called the **discriminant** of the quadratic equation, because its value tells us whether the equation has real solutions. In fact, it also tells us how many solutions to expect.

Discriminant of a Quadratic Equation

For a quadratic equation $ax^2 + bx + c = 0, a \neq 0$:

- If $b^2 - 4ac > 0$, there are two unequal real solutions.
- If $b^2 - 4ac = 0$, there is a repeated solution, a double root.
- If $b^2 - 4ac < 0$, there is no real solution.

When asked to find the real solutions of a quadratic equation, always evaluate the discriminant first to see if there are any real solutions.

EXAMPLE 7 **Solving a Quadratic Equation Using the Quadratic Formula**

Use the quadratic formula to find the real solutions, if any, of the equation

$$3x^2 - 5x + 1 = 0$$

Solution The equation is in standard form, so compare it to $ax^2 + bx + c = 0$ to find $a, b,$ and c.

$$3x^2 - 5x + 1 = 0$$

$$ax^2 + bx + c = 0 \quad a = 3, b = -5, c = 1$$

With $a = 3, b = -5,$ and $c = 1$, evaluate the discriminant $b^2 - 4ac$.

$$b^2 - 4ac = (-5)^2 - 4(3)(1) = 25 - 12 = 13$$

(continued)

Since $b^2 - 4ac > 0$, there are two real solutions, which can be found using the quadratic formula.

$$x = \frac{-b \pm \sqrt{b^2 - 4ac}}{2a} = \frac{-(-5) \pm \sqrt{13}}{2(3)} = \frac{5 \pm \sqrt{13}}{6}$$

The solution set is $\left\{ \dfrac{5 - \sqrt{13}}{6}, \dfrac{5 + \sqrt{13}}{6} \right\}$.

EXAMPLE 8 **Solving a Quadratic Equation Using the Quadratic Formula**

Use the quadratic formula to find the real solutions, if any, of the equation

$$3x^2 + 2 = 4x$$

Solution The equation, as given, is not in standard form.

$$3x^2 + 2 = 4x$$

$$3x^2 - 4x + 2 = 0 \qquad \text{Put in standard form.}$$

$$ax^2 + bx + c = 0 \qquad \text{Compare to standard form.}$$

With $a = 3$, $b = -4$, and $c = 2$, the discriminant is

$$b^2 - 4ac = (-4)^2 - 4 \cdot 3 \cdot 2 = 16 - 24 = -8$$

Since $b^2 - 4ac < 0$, the equation has no real solution.

➤ **Now Work** PROBLEMS **93** AND **99**

SUMMARY

Steps for Solving a Quadratic Equation

To solve a quadratic equation, first put it in standard form:

$$ax^2 + bx + c = 0$$

Then:
STEP 1: Identify a, b, and c.
STEP 2: Evaluate the discriminant, $b^2 - 4ac$.
STEP 3: • If the discriminant is negative, the equation has no real solution.
  • If the discriminant is zero, the equation has one repeated real solution, a double root.
  • If the discriminant is positive, the equation has two distinct real solutions.

If you can easily spot factors, use the factoring method to solve the equation. Otherwise, use the quadratic formula or the method of completing the square.

A.4 Assess Your Understanding

'Are You Prepared?' *Answers are given at the end of these exercises. If you get a wrong answer, read the pages listed in* red.

1. Factor: $x^2 - 5x - 6$ (pp. A23–A24)

2. To complete the square of $x^2 - 4x$, add _____ to both sides. (pp. A24–A25)

3. The solution set of the equation $(x - 3)(3x + 5) = 0$ is _____. (p. A4)

4. *True or False* $\sqrt{x^2} = |x|$. (pp. A9–A10)

Concepts and Vocabulary

5. An equation that is satisfied for every choice of the variable for which both sides are defined is called a(n) _____.

6. True or False The solution of the equation $3x - 8 = 0$ is $\dfrac{3}{8}$.

7. True or False Some equations have no solution.

8. To solve the equation $x^2 + 5x = 0$ by completing the square, you would _____ the number _____ to both sides.

9. The quantity $b^2 - 4ac$ is called the _____ of a quadratic equation. If it is _____, the equation has no real solution.

10. True or False Quadratic equations always have two real solutions.

11. True or False If the discriminant of a quadratic equation is positive, then the equation has two solutions that are negatives of one another.

12. Multiple Choice An admissible value for the variable that makes the equation a true statement is called a(n) _____ of the equation.
 (a) identity   **(b)** solution   **(c)** degree   **(d)** model

13. Multiple Choice A quadratic equation is sometimes called a _____-degree equation.
 (a) first   **(b)** second   **(c)** third   **(d)** fourth

14. Multiple Choice Which of the following quadratic equations is in standard form?
 (a) $x^2 - 7x = 5$ **(b)** $9 = x^2$
 (c) $(x + 5)(x - 4) = 0$ **(d)** $0 = 5x^2 - 6x - 1$

Skill Building

In Problems 15–80, solve each equation.

15. $3x = 21$

16. $3x = -24$

17. $5x + 15 = 0$

18. $3x + 18 = 0$

19. $2x - 3 = 5$

20. $3x + 4 = -8$

21. $\dfrac{1}{3}x = \dfrac{5}{12}$

22. $\dfrac{2}{3}x = \dfrac{9}{2}$

23. $6 - x = 2x + 9$

24. $3 - 2x = 2 - x$

25. $2(3 + 2x) = 3(x - 4)$

26. $3(2 - x) = 2x - 1$

27. $8x - (2x + 1) = 3x - 10$

28. $5 - (2x - 1) = 10$

29. $\dfrac{1}{2}x - 4 = \dfrac{3}{4}x$

30. $1 - \dfrac{1}{2}x = 5$

31. $0.9t = 0.4 + 0.1t$

32. $0.9t = 1 + t$

33. $\dfrac{2}{y} + \dfrac{4}{y} = 3$

34. $\dfrac{4}{y} - 5 = \dfrac{5}{2y}$

35. $(x + 7)(x - 1) = (x + 1)^2$

36. $(x + 2)(x - 3) = (x - 3)^2$

37. $z(z^2 + 1) = 3 + z^3$

38. $w(4 - w^2) = 8 - w^3$

39. $x^2 = 9x$

40. $x^3 = x^2$

41. $t^3 - 9t^2 = 0$

42. $4z^3 - 8z^2 = 0$

43. $\dfrac{3}{2x - 3} = \dfrac{2}{x + 5}$

44. $\dfrac{-2}{x + 4} = \dfrac{-3}{x + 1}$

45. $x^3 + 2x^2 - 4x - 8 = 0$

46. $x^3 + 50 = 2x^2 + 25x$

47. $\dfrac{2}{x - 2} = \dfrac{3}{x + 5} + \dfrac{10}{(x + 5)(x - 2)}$

48. $\dfrac{1}{2x + 3} + \dfrac{1}{x - 1} = \dfrac{1}{(2x + 3)(x - 1)}$

49. $|2x| = 6$

50. $|3x| = 12$

51. $|2x + 3| = 5$

52. $|3x - 1| = 2$

53. $|1 - 4t| = 5$

54. $|1 - 2z| = 3$

55. $|-2x| = 8$

56. $|-x| = 1$

57. $|-2|x = 4$

58. $|3|x = 9$

59. $|x - 2| = -\dfrac{1}{2}$

60. $|2 - x| = -1$

61. $|x^2 - 4| = 0$

62. $|x^2 - 9| = 0$

63. $|x^2 - 2x| = 3$

64. $|x^2 + x| = 12$

65. $|x^2 + x - 1| = 1$

66. $|x^2 + 3x - 2| = 2$

67. $x^2 = 4x$

68. $x^2 = -8x$

69. $z^2 + 4z - 12 = 0$

70. $v^2 + 7v + 12 = 0$

71. $2x^2 - 5x - 3 = 0$

72. $3x^2 + 5x + 2 = 0$

73. $x(x - 7) + 12 = 0$

74. $x(x + 1) = 12$

75. $4x^2 + 9 = 12x$

76. $25x^2 + 16 = 40x$

77. $6x - 5 = \dfrac{6}{x}$

78. $x + \dfrac{12}{x} = 7$

79. $\dfrac{4(x - 2)}{x - 3} + \dfrac{3}{x} = \dfrac{-3}{x(x - 3)}$

80. $\dfrac{5}{x + 4} = 4 + \dfrac{3}{x - 2}$

In Problems 81–86, solve each equation by the Square Root Method.

81. $x^2 = 25$

82. $x^2 = 36$

83. $(x - 1)^2 = 4$

84. $(x + 2)^2 = 1$

85. $(2y + 3)^2 = 9$

86. $(3x - 2)^2 = 4$

In Problems 87–92, solve each equation by completing the square.

87. $x^2 + 4x = 21$

88. $x^2 - 6x = 13$

89. $x^2 - \dfrac{1}{2}x - \dfrac{3}{16} = 0$

90. $x^2 + \dfrac{2}{3}x - \dfrac{1}{3} = 0$

91. $3x^2 + x - \dfrac{1}{2} = 0$

92. $2x^2 - 3x - 1 = 0$

In Problems 93–104, find the real solutions, if any, of each equation. Use the quadratic formula.

93. $x^2 - 4x + 2 = 0$

94. $x^2 + 4x + 2 = 0$

95. $x^2 - 5x - 1 = 0$

96. $x^2 + 5x + 3 = 0$

97. $2x^2 - 5x + 3 = 0$

98. $2x^2 + 5x + 3 = 0$

99. $4y^2 - y + 2 = 0$

100. $4t^2 + t + 1 = 0$

101. $4x^2 = 1 - 2x$

102. $2x^2 = 1 - 2x$

103. $x^2 + \sqrt{3}\,x - 3 = 0$

104. $x^2 + \sqrt{2}\,x - 2 = 0$

In Problems 105–110, use the discriminant to determine whether each quadratic equation has two unequal real solutions, a repeated real solution, or no real solution without solving the equation.

105. $x^2 - 5x + 7 = 0$

106. $x^2 + 5x + 7 = 0$

107. $9x^2 - 30x + 25 = 0$

108. $25x^2 - 20x + 4 = 0$

109. $3x^2 + 5x - 8 = 0$

110. $2x^2 - 3x - 4 = 0$

Applications and Extensions

In Problems 111–116, solve each equation. The letters a, b, and c are constants.

111. $ax - b = c, \quad a \neq 0$

112. $1 - ax = b, \quad a \neq 0$

113. $\dfrac{x}{a} + \dfrac{x}{b} = c, \quad a \neq 0, b \neq 0, a \neq -b$

114. $\dfrac{a}{x} + \dfrac{b}{x} = c, \quad c \neq 0$

115. $\dfrac{1}{x-a} + \dfrac{1}{x+a} = \dfrac{2}{x-1}$

116. $\dfrac{b+c}{x+a} = \dfrac{b-c}{x-a}, \quad c \neq 0, \quad a \neq 0$

Problems 117–122 list some formulas that occur in applications. Solve each formula for the indicated variable.

117. Electricity $\dfrac{1}{R} = \dfrac{1}{R_1} + \dfrac{1}{R_2}$ for R

118. Finance $A = P(1 + rt)$ for r

119. Mechanics $F = \dfrac{mv^2}{R}$ for R

120. Chemistry $PV = nRT$ for T

121. Mathematics $S = \dfrac{a}{1-r}$ for r

122. Mechanics $v = -gt + v_0$ for t

123. Show that the sum of the roots of a quadratic equation is $-\dfrac{b}{a}$.

124. Show that the product of the roots of a quadratic equation is $\dfrac{c}{a}$.

125. Find k so that the equation $kx^2 + x + k = 0$ has a repeated real solution.

126. Find k so that the equation $x^2 - kx + 4 = 0$ has a repeated real solution.

127. Show that the real solutions of the equation $ax^2 + bx + c = 0$ are the negatives of the real solutions of the equation $ax^2 - bx + c = 0$. Assume that $b^2 - 4ac \geq 0$.

128. Show that the real solutions of the equation $ax^2 + bx + c = 0$ are the reciprocals of the real solutions of the equation $cx^2 + bx + a = 0$. Assume that $b^2 - 4ac \geq 0$.

Explaining Concepts: Discussion and Writing

129. Which of the following pairs of equations are equivalent? Explain.

(a) $x^2 = 9; \quad x = 3$

(b) $x = \sqrt{9}; \quad x = 3$

(c) $(x-1)(x-2) = (x-1)^2; \quad x - 2 = x - 1$

130. The equation

$$\dfrac{5}{x+3} + 3 = \dfrac{8+x}{x+3}$$

has no solution, yet when we go through the process of solving it, we obtain $x = -3$. Write a brief paragraph to explain what causes this to happen.

131. Find an equation that has no solution and give it to a fellow student to solve. Ask the fellow student to write a critique of your equation.

132. Describe three ways you might solve a quadratic equation. State your preferred method; explain why you chose it.

133. Explain the benefits of evaluating the discriminant of a quadratic equation before attempting to solve it.

134. Find three quadratic equations: one having two distinct solutions, one having no real solution, and one having exactly one real solution.

135. The word *quadratic* seems to imply four (*quad*), yet a quadratic equation is an equation that involves a polynomial of degree 2. Investigate the origin of the term *quadratic* as it is used in the expression *quadratic equation*. Write a brief essay on your findings.

'Are You Prepared?' Answers

1. $(x - 6)(x + 1)$ **2.** 4   **3.** $\left\{-\dfrac{5}{3}, 3\right\}$ **4.** True

---

A.5 Complex Numbers; Quadratic Equations in the Complex Number System*

OBJECTIVES   **1** Add, Subtract, Multiply, and Divide Complex Numbers (p. A38)
                   2 Solve Quadratic Equations in the Complex Number System (p. A42)

Complex Numbers

One property of a real number is that its square is nonnegative. For example, there is no real number x for which

$$x^2 = -1$$

To remedy this situation, we introduce a new number called the *imaginary unit*.

DEFINITION The Imaginary Unit

The **imaginary unit**, which we denote by i, is the number whose square is -1. That is,

$$i^2 = -1$$

This should not surprise you. If our universe were to consist only of integers, there would be no number x for which $2x = 1$. This was remedied by introducing numbers such as $\dfrac{1}{2}$ and $\dfrac{2}{3}$, the *rational numbers*. If our universe were to consist only of rational numbers, there would be no number whose square equals 2. That is, there would be no x for which $x^2 = 2$. To remedy this, we introduced numbers such as $\sqrt{2}$ and $\sqrt[3]{5}$, the *irrational numbers*. Recall that the *real numbers* consist of the rational numbers and the irrational numbers. Now, if our universe were to consist only of real numbers, then there would be no number x whose square is -1. To remedy this, we introduce the number i, whose square is -1.

In the progression outlined, each time we encountered a situation that was unsuitable, a new number system was introduced to remedy the situation. The number system that results from introducing the number i is called the **complex number system**.

*This section may be omitted without any loss of continuity.

> **DEFINITION** The Complex Number System
>
> **Complex numbers** are numbers of the form $a + bi$, where a and b are real numbers. The real number a is called the **real part** of the number $a + bi$; the real number b is called the **imaginary part** of $a + bi$; and i is the imaginary unit, so $i^2 = -1$.

For example, the complex number $-5 + 6i$ has the real part -5 and the imaginary part 6.

When a complex number is written in the form $a + bi$, where a and b are real numbers, it is in **standard form**. However, if the imaginary part of a complex number is negative, such as in the complex number $3 + (-2)i$, we agree to write it instead in the form $3 - 2i$.

Also, the complex number $a + 0i$ is usually written merely as a. This serves to remind us that the real numbers are a subset of the complex numbers. Similarly, the complex number $0 + bi$ is usually written as bi. Sometimes the complex number bi is called a **pure imaginary number**.

1 Add, Subtract, Multiply, and Divide Complex Numbers

Equality, addition, subtraction, and multiplication of complex numbers are defined so as to preserve the familiar rules of algebra for real numbers. Two complex numbers are equal if and only if their real parts are equal and their imaginary parts are equal.

> **Equality of Complex Numbers**
>
> $$a + bi = c + di \quad \text{if and only if} \quad a = c \text{ and } b = d \qquad (1)$$

Two complex numbers are added by forming the complex number whose real part is the sum of the real parts and whose imaginary part is the sum of the imaginary parts.

> **Sum of Complex Numbers**
>
> $$(a + bi) + (c + di) = (a + c) + (b + d)i \qquad (2)$$

To subtract two complex numbers, use this rule:

> **Difference of Complex Numbers**
>
> $$(a + bi) - (c + di) = (a - c) + (b - d)i \qquad (3)$$

EXAMPLE 1 **Adding and Subtracting Complex Numbers**

(a) $(3 + 5i) + (-2 + 3i) = [3 + (-2)] + (5 + 3)i = 1 + 8i$

(b) $(6 + 4i) - (3 + 6i) = (6 - 3) + (4 - 6)i = 3 + (-2)i = 3 - 2i$

━━━▶ **Now Work** PROBLEM 15

Products of complex numbers are calculated as illustrated in Example 2.

EXAMPLE 2

Multiplying Complex Numbers

$$(5 + 3i)(2 + 7i) = 5(2 + 7i) + 3i(2 + 7i) \qquad \text{Distributive Property}$$
$$= 10 + 35i + 6i + 21i^2 \qquad \text{Distributive Property}$$
$$= 10 + 41i + 21(-1) \qquad i^2 = -1$$
$$= -11 + 41i \qquad \text{Simplify.}$$

Based on the procedure of Example 2, the **product** of two complex numbers is defined as follows:

Product of Complex Numbers

$$(a + bi)(c + di) = (ac - bd) + (ad + bc)i \qquad \textbf{(4)}$$

Do not bother to memorize formula (4). Instead, whenever it is necessary to multiply two complex numbers, follow the usual rules for multiplying two binomials, as we did in Example 2. Just remember that $i^2 = -1$. For example,

$$(2i)(2i) = 4i^2 = 4(-1) = -4$$
$$(2 + i)(1 - i) = 2 - 2i + i - i^2 = 3 - i$$

Now Work PROBLEM 21

Algebraic properties for addition and multiplication, such as the commutative, associative, and distributive properties, hold for complex numbers. The property that every nonzero complex number has a multiplicative inverse, or reciprocal, requires a closer look.

Conjugates

DEFINITION Complex Conjugate

In Words
The conjugate of a complex number is found by changing the sign of the imaginary part.

If $z = a + bi$ is a complex number, then its **conjugate**, denoted by $\bar{z}$, is defined as

$$\bar{z} = \overline{a + bi} = a - bi$$

For example, $\overline{2 + 3i} = 2 - 3i$ and $\overline{-6 - 2i} = -6 + 2i$.

EXAMPLE 3

Multiplying a Complex Number by Its Conjugate

Find the product of the complex number $z = 3 + 4i$ and its conjugate $\bar{z}$.

Solution

Since $\bar{z} = 3 - 4i$, we have

$$z\bar{z} = (3 + 4i)(3 - 4i) = 9 - 12i + 12i - 16i^2 = 9 + 16 = 25$$

The result obtained in Example 3 has an important generalization.

THEOREM Product of Complex Conjugates

The product of a complex number and its conjugate is a nonnegative real number. That is, if $z = a + bi$, then

$$z\bar{z} = a^2 + b^2 \qquad \textbf{(5)}$$

Proof If $z = a + bi$, then

$$z\bar{z} = (a + bi)(a - bi) = a^2 - abi + abi - (bi)^2 = a^2 - b^2 i^2$$
$$= a^2 - b^2(-1) = a^2 + b^2$$

■

Now Work PROBLEM 89

To express the reciprocal of a nonzero complex number z in standard form, multiply the numerator and denominator of $\dfrac{1}{z}$ by $\bar{z}$. That is, if $z = a + bi$ is a nonzero complex number, then

$$\frac{1}{a + bi} = \frac{1}{z} = \frac{1}{z} \cdot \frac{\bar{z}}{\bar{z}} = \frac{\bar{z}}{z\bar{z}} = \frac{a - bi}{a^2 + b^2} = \frac{a}{a^2 + b^2} - \frac{b}{a^2 + b^2} i$$

$\uparrow$
Use (5).

EXAMPLE 4

Writing the Reciprocal of a Complex Number in Standard Form

Write $\dfrac{1}{3 + 4i}$ in standard form; that is, find the reciprocal of $3 + 4i$.

Solution

The idea is to multiply the numerator and denominator by the conjugate of $3 + 4i$, that is, by the complex number $3 - 4i$. The result is

$$\frac{1}{3 + 4i} = \frac{1}{3 + 4i} \cdot \frac{3 - 4i}{3 - 4i} = \frac{3 - 4i}{9 + 16} = \frac{3}{25} - \frac{4}{25} i$$

To express the quotient of two complex numbers in standard form, multiply the numerator and denominator of the quotient by the conjugate of the denominator.

EXAMPLE 5

Writing the Quotient of Two Complex Numbers in Standard Form

Write each quotient in standard form.

(a) $\dfrac{1 + 4i}{5 - 12i}$ 　　　　　　　　　　(b) $\dfrac{2 - 3i}{4 - 3i}$

Solution

(a) $\dfrac{1 + 4i}{5 - 12i} = \dfrac{1 + 4i}{5 - 12i} \cdot \dfrac{5 + 12i}{5 + 12i} = \dfrac{5 + 12i + 20i + 48i^2}{25 + 144}$

$$= \frac{-43 + 32i}{169} = -\frac{43}{169} + \frac{32}{169} i$$

(b) $\dfrac{2 - 3i}{4 - 3i} = \dfrac{2 - 3i}{4 - 3i} \cdot \dfrac{4 + 3i}{4 + 3i} = \dfrac{8 + 6i - 12i - 9i^2}{16 + 9} = \dfrac{17 - 6i}{25} = \dfrac{17}{25} - \dfrac{6}{25} i$

Now Work PROBLEM 29

EXAMPLE 6

Writing Other Expressions in Standard Form

If $z = 2 - 3i$ and $w = 5 + 2i$, write each expression in standard form.

(a) $\dfrac{z}{w}$ 　　　　　　　(b) $\overline{z + w}$ 　　　　　　　(c) $z + \bar{z}$

Solution

(a) $\dfrac{z}{w} = \dfrac{z \cdot \bar{w}}{w \cdot \bar{w}} = \dfrac{(2 - 3i)(5 - 2i)}{(5 + 2i)(5 - 2i)} = \dfrac{10 - 4i - 15i + 6i^2}{25 + 4}$

$$= \frac{4 - 19i}{29} = \frac{4}{29} - \frac{19}{29} i$$

(b) $\overline{z + w} = \overline{(2 - 3i) + (5 + 2i)} = \overline{7 - i} = 7 + i$

(c) $z + \bar{z} = (2 - 3i) + (2 + 3i) = 4$

The conjugate of a complex number has certain general properties that will be useful later.

For a real number $a = a + 0i$, the conjugate is $\bar{a} = \overline{a + 0i} = a - 0i = a$.

THEOREM

The conjugate of a real number is the real number itself.

Other properties that are direct consequences of the definition of the conjugate are given next. In each statement, z and w represent complex numbers.

THEOREMS

- The conjugate of the conjugate of a complex number is the complex number itself.

$$\bar{\bar{z}} = z \tag{6}$$

- The conjugate of the sum of two complex numbers equals the sum of their conjugates.

$$\overline{z + w} = \bar{z} + \bar{w} \tag{7}$$

- The conjugate of the product of two complex numbers equals the product of their conjugates.

$$\overline{z \cdot w} = \bar{z} \cdot \bar{w} \tag{8}$$

The proofs of equations (6), (7), and (8) are left as exercises.

Powers of i

The **powers of i** follow a pattern that is useful to know.

$i^1 = i$ $i^5 = i^4 \cdot i = 1 \cdot i = i$

$i^2 = -1$ $i^6 = i^4 \cdot i^2 = -1$

$i^3 = i^2 \cdot i = -1 \cdot i = -i$ $i^7 = i^4 \cdot i^3 = -i$

$i^4 = i^2 \cdot i^2 = (-1)(-1) = 1$ $i^8 = i^4 \cdot i^4 = 1$

And so on. The powers of i repeat with every fourth power.

EXAMPLE 7

Evaluating Powers of i

(a) $i^{27} = i^{24} \cdot i^3 = (i^4)^6 \cdot i^3 = 1^6 \cdot i^3 = -i$

(b) $i^{101} = i^{100} \cdot i^1 = (i^4)^{25} \cdot i = 1^{25} \cdot i = i$

EXAMPLE 8

Writing the Power of a Complex Number in Standard Form

Write $(2 + i)^3$ in standard form.

Solution

Use the special product formula for $(x + a)^3$.

$$(x + a)^3 = x^3 + 3ax^2 + 3a^2x + a^3$$

(continued)

NOTE Another way to find $(2 + i)^3$ is to multiply out $(2 + i)^2(2 + i)$. ∎

Using this special product formula,

$$(2 + i)^3 = 2^3 + 3 \cdot i \cdot 2^2 + 3 \cdot i^2 \cdot 2 + i^3$$

$$= 8 + 12i + 6(-1) + (-i)$$

$$= 2 + 11i$$

Now Work PROBLEMS 35 AND 43

2 Solve Quadratic Equations in the Complex Number System

Quadratic equations with a negative discriminant have no real number solution. However, if we extend our number system to allow complex numbers, quadratic equations always have solutions. Since the solution to a quadratic equation involves the square root of the discriminant, we begin with a discussion of square roots of negative numbers.

> **DEFINITION** Principal Square Root of $-N$
>
> If N is a positive real number, we define the **principal square root of** $-N$, denoted by $\sqrt{-N}$, as
>
> $$\sqrt{-N} = \sqrt{N}\,i$$
>
> where i is the imaginary unit and $i^2 = -1$.

WARNING In writing $\sqrt{-N} = \sqrt{N}\,i$, be sure to place i outside the $\sqrt{}$ symbol. ∎

EXAMPLE 9 **Evaluating the Square Root of a Negative Number**

(a) $\sqrt{-1} = \sqrt{1}\,i = i$

(b) $\sqrt{-16} = \sqrt{16}\,i = 4i$

(c) $\sqrt{-8} = \sqrt{8}\,i = 2\sqrt{2}\,i$

Now Work PROBLEM 51

EXAMPLE 10 **Solving Equations**

Solve each equation in the complex number system.

(a) $x^2 = 4$ (b) $x^2 = -9$

Solution (a) $x^2 = 4$

$$x = \pm\sqrt{4} = \pm 2$$

The equation has two solutions, -2 and 2. The solution set is $\{-2, 2\}$.

(b) $x^2 = -9$

$$x = \pm\sqrt{-9} = \pm\sqrt{9}\,i = \pm 3i$$

The equation has two solutions, $-3i$ and $3i$. The solution set is $\{-3i, 3i\}$.

Now Work PROBLEM 59

WARNING When working with square roots of negative numbers, do not set the square root of a product equal to the product of the square roots (which can be done with positive real numbers). To see why, look at this calculation: We know that $\sqrt{100} = 10$. However, it is also true that $100 = (-25)(-4)$, so

$$10 = \sqrt{100} = \sqrt{(-25)(-4)} = \sqrt{-25}\sqrt{-4} = \sqrt{25}\,i \cdot \sqrt{4}\,i = 5i \cdot 2i = 10i^2 = -10$$

↑
Here is the error.

Because we have defined the square root of a negative number, we now restate the quadratic formula without restriction.

THEOREM Quadratic Formula

In the complex number system, the solutions of the quadratic equation $ax^2 + bx + c = 0$, where $a, b,$ and c are real numbers and $a \neq 0$, are given by the formula

$$x = \frac{-b \pm \sqrt{b^2 - 4ac}}{2a} \qquad \text{(9)}$$

EXAMPLE 11

Solving a Quadratic Equation in the Complex Number System

Solve the equation $x^2 - 4x + 8 = 0$ in the complex number system.

Solution   Here $a = 1, b = -4, c = 8,$ and $b^2 - 4ac = (-4)^2 - 4 \cdot 1 \cdot 8 = -16.$ Using equation (9), we find that

$$x = \frac{-(-4) \pm \sqrt{-16}}{2 \cdot 1} = \frac{4 \pm \sqrt{16}\,i}{2} = \frac{4 \pm 4i}{2} = \frac{2(2 \pm 2i)}{2} = 2 \pm 2i$$

The equation has the solution set $\{2 - 2i, 2 + 2i\}$.

✓ **Check:** $2 + 2i$: $(2 + 2i)^2 - 4(2 + 2i) + 8 = 4 + 8i + 4i^2 - 8 - 8i + 8$

$$= 4 + 4i^2$$
$$= 4 - 4 = 0$$

$2 - 2i$: $(2 - 2i)^2 - 4(2 - 2i) + 8 = 4 - 8i + 4i^2 - 8 + 8i + 8$

$$= 4 - 4 = 0$$

Now Work PROBLEM 65

The discriminant $b^2 - 4ac$ of a quadratic equation still serves as a way to determine the character of the solutions.

Character of the Solutions of a Quadratic Equation

In the complex number system, consider a quadratic equation $ax^2 + bx + c = 0$, where $a \neq 0$ and $a, b,$ and c are real numbers.

- If $b^2 - 4ac > 0$, the equation has two unequal real solutions.
- If $b^2 - 4ac = 0$, the equation has a repeated real solution, a double root.
- If $b^2 - 4ac < 0$, the equation has two complex solutions that are not real. The solutions are conjugates of each other.

The third conclusion in the display is a consequence of the fact that if $b^2 - 4ac = -N < 0$, then by the quadratic formula, the solutions are

$$x = \frac{-b + \sqrt{b^2 - 4ac}}{2a} = \frac{-b + \sqrt{-N}}{2a} = \frac{-b + \sqrt{N}\,i}{2a} = -\frac{b}{2a} + \frac{\sqrt{N}}{2a}i$$

and

$$x = \frac{-b - \sqrt{b^2 - 4ac}}{2a} = \frac{-b - \sqrt{-N}}{2a} = \frac{-b - \sqrt{N}\,i}{2a} = -\frac{b}{2a} - \frac{\sqrt{N}}{2a}i$$

which are conjugates of each other.

EXAMPLE 12   **Determining the Character of the Solutions of a Quadratic Equation**

Without solving, determine the character of the solutions of each equation.

(a) $3x^2 + 4x + 5 = 0$ (b) $2x^2 + 4x + 1 = 0$ (c) $9x^2 - 6x + 1 = 0$

Solution   (a) Here $a = 3$, $b = 4$, and $c = 5$, so $b^2 - 4ac = 16 - 4 \cdot 3 \cdot 5 = -44$. The solutions are two complex numbers that are not real and are conjugates of each other.

(b) Here $a = 2$, $b = 4$, and $c = 1$, so $b^2 - 4ac = 16 - 8 = 8$. The solutions are two unequal real numbers.

(c) Here $a = 9$, $b = -6$, and $c = 1$, so $b^2 - 4ac = 36 - 4 \cdot 9 \cdot 1 = 0$. The solution is a repeated real number—that is, a double root.

Now Work PROBLEM 79

A.5 Assess Your Understanding

Concepts and Vocabulary

1. *True or False* The square of a complex number is sometimes negative.

2. $(2 + i)(2 - i) = $ ___ .

3. *True or False* In the complex number system, a quadratic equation has four solutions.

4. In the complex number $5 + 2i$, the number 5 is called the _____ part; the number 2 is called the _____ part; the number i is called the _____ _____.

5. *True or False* The conjugate of $2 + 5i$ is $-2 - 5i$.

6. *True or False* All real numbers are complex numbers.

7. *True or False* If $2 - 3i$ is a solution of a quadratic equation with real coefficients, then $-2 + 3i$ is also a solution.

8. *Multiple Choice* Which of the following is the principal square root of -4?
(a) $-2i$ (b) $2i$ (c) -2 (d) 2

9. *Multiple Choice* Which operation involving complex numbers requires the use of a conjugate?
(a) division   (b) multiplication
(c) subtraction   (d) addition

10. *Multiple Choice* Powers of i repeat every _____ power.
(a) second   (b) third   (c) fourth   (d) fifth

Skill Building

In Problems 11–58, perform the indicated operation, and write each expression in the standard form $a + bi$.

11. $(2 - 3i) + (6 + 8i)$ **12.** $(4 + 5i) + (-8 + 2i)$ **13.** $(-3 + 2i) - (4 - 4i)$ **14.** $(3 - 4i) - (-3 - 4i)$

15. $(2 - 5i) - (8 + 6i)$ **16.** $(-8 + 4i) - (2 - 2i)$ **17.** $3(2 - 6i)$ **18.** $-4(2 + 8i)$

19. $-3i(7 + 6i)$ **20.** $3i(-3 + 4i)$ **21.** $(3 - 4i)(2 + i)$ **22.** $(5 + 3i)(2 - i)$

23. $(-5 - i)(-5 + i)$ **24.** $(-3 + i)(3 + i)$ **25.** $\dfrac{10}{3 - 4i}$ **26.** $\dfrac{13}{5 - 12i}$

27. $\dfrac{2 + i}{i}$ **28.** $\dfrac{2 - i}{-2i}$ **29.** $\dfrac{6 - i}{1 + i}$ **30.** $\dfrac{2 + 3i}{1 - i}$

31. $\left(\dfrac{1}{2} + \dfrac{\sqrt{3}}{2}i\right)^2$ **32.** $\left(\dfrac{\sqrt{3}}{2} - \dfrac{1}{2}i\right)^2$ **33.** $(1 + i)^2$ **34.** $(1 - i)^2$

35. i^{23} **36.** i^{14} **37.** i^{-20} **38.** i^{-23} **39.** $i^6 - 5$

40. $4 + i^3$ **41.** $6i^3 - 4i^5$ **42.** $4i^3 - 2i^2 + 1$ **43.** $(1 + i)^3$ **44.** $(3i)^4 + 1$

45. $i^7(1 + i^2)$ **46.** $2i^4(1 + i^2)$ **47.** $i^8 + i^6 - i^4 - i^2$ **48.** $i^7 + i^5 + i^3 + i$ **49.** $\sqrt{-4}$

50. $\sqrt{-9}$ **51.** $\sqrt{-25}$ **52.** $\sqrt{-64}$ **53.** $\sqrt{-12}$ **54.** $\sqrt{-18}$

55. $\sqrt{-200}$ **56.** $\sqrt{-45}$ **57.** $\sqrt{(3 + 4i)(4i - 3)}$ **58.** $\sqrt{(4 + 3i)(3i - 4)}$

In Problems 59–78, solve each equation in the complex number system.

59. $x^2 + 4 = 0$ **60.** $x^2 - 4 = 0$ **61.** $x^2 - 16 = 0$ **62.** $x^2 + 25 = 0$

63. $x^2 - 6x + 13 = 0$ **64.** $x^2 + 4x + 8 = 0$ **65.** $x^2 - 6x + 10 = 0$ **66.** $x^2 - 2x + 5 = 0$

67. $25x^2 - 10x + 2 = 0$ **68.** $10x^2 + 6x + 1 = 0$ **69.** $5x^2 + 1 = 2x$ **70.** $13x^2 + 1 = 6x$

71. $x^2 + x + 1 = 0$ **72.** $x^2 - x + 1 = 0$ **73.** $x^3 - 64 = 0$ **74.** $x^3 + 27 = 0$

75. $x^4 = 16$ **76.** $x^4 = 1$ **77.** $x^4 + 13x^2 + 36 = 0$ **78.** $x^4 + 3x^2 - 4 = 0$

In Problems 79–84, without solving, determine the character of the solutions of each equation in the complex number system.

79. $3x^2 - 3x + 4 = 0$

80. $2x^2 - 4x + 1 = 0$

81. $2x^2 + 3x = 4$

82. $x^2 + 6 = 2x$

83. $9x^2 - 12x + 4 = 0$

84. $4x^2 + 12x + 9 = 0$

85. $2 + 3i$ is a solution of a quadratic equation with real coefficients. Find the other solution.

86. $4 - i$ is a solution of a quadratic equation with real coefficients. Find the other solution.

In Problems 87–90, $z = 3 - 4i$ and $w = 8 + 3i$. Write each expression in the standard form $a + bi$.

87. $z + \bar{z}$

88. $w - \bar{w}$

89. $z\bar{z}$

90. $\overline{z - w}$

Applications and Extensions

91. Electrical Circuits The impedance Z, in ohms, of a circuit element is defined as the ratio of the phasor voltage V, in volts, across the element to the phasor current I, in amperes, through the element. That is, $Z = \dfrac{V}{I}$. If the voltage across a circuit element is $18 + i$ volts and the current through the element is $3 - 4i$ amperes, find the impedance.

92. Parallel Circuits In an ac circuit with two parallel pathways, the total impedance Z, in ohms, satisfies the formula $\dfrac{1}{Z} = \dfrac{1}{Z_1} + \dfrac{1}{Z_2}$, where Z_1 is the impedance of the first pathway

and Z_2 is the impedance of the second pathway. Find the total impedance if the impedances of the two pathways are $Z_1 = 2 + i$ ohms and $Z_2 = 4 - 3i$ ohms.

93. Use $z = a + bi$ to show that $z + \bar{z} = 2a$ and $z - \bar{z} = 2bi$.

94. Use $z = a + bi$ to show that $\bar{\bar{z}} = z$.

95. Use $z = a + bi$ and $w = c + di$ to show that $\overline{z + w} = \bar{z} + \bar{w}$.

96. Use $z = a + bi$ and $w = c + di$ to show that $\overline{z \cdot w} = \bar{z} \cdot \bar{w}$.

A.6 Interval Notation; Solving Inequalities

PREPARING FOR THIS SECTION *Before getting started, review the following:*

- Algebra Essentials (Section A.1, pp. A1–A10)

Now Work the 'Are You Prepared?' problems on page A53.

OBJECTIVES
1 Use Interval Notation (p. A45)
2 Use Properties of Inequalities (p. A47)
3 Solve Inequalities (p. A49)
4 Solve Combined Inequalities (p. A50)
5 Solve Inequalities Involving Absolute Value (p. A51)

Suppose that a and b are two real numbers and $a < b$. The notation $a < x < b$ means that x is a number *between* a and b. The expression $a < x < b$ is equivalent to the two inequalities $a < x$ and $x < b$. Similarly, the expression $a \le x \le b$ is equivalent to the two inequalities $a \le x$ and $x \le b$. The remaining two possibilities, $a \le x < b$ and $a < x \le b$, are defined similarly.

Although it is acceptable to write $3 \ge x \ge 2$, it is preferable to reverse the inequality symbols and write instead $2 \le x \le 3$ so that the values go from smaller to larger, reading from left to right.

A statement such as $2 \le x \le 1$ is false because there is no number x for which $2 \le x$ and $x \le 1$. Finally, never mix inequality symbols, as in $2 \le x \ge 3$.

1 Use Interval Notation

Let a and b represent two real numbers with $a < b$.

> **DEFINITION** Intervals
>
> - An **open interval**, denoted by **(a, b)**, consists of all real numbers x for which $a < x < b$.
> - A **closed interval**, denoted by **[a, b]**, consists of all real numbers x for which $a \le x \le b$.
> - The **half-open**, or **half-closed**, **intervals** are **(a, b]**, consisting of all real numbers x for which $a < x \le b$, and **[a, b)**, consisting of all real numbers x for which $a \le x < b$.

In Words

The notation [a, b] means all real numbers between a and b, inclusive. The notation (a, b) means all real numbers between a and b, not including a and b.

In each of these definitions, a is called the **left endpoint** and b the **right endpoint** of the interval.

The symbol ∞ (read as "infinity") is not a real number, but notation used to indicate unboundedness in the positive direction. The symbol $-\infty$ (read as "negative infinity") also is not a real number, but notation used to indicate unboundedness in the negative direction. The symbols ∞ and $-\infty$ are used to define five other kinds of intervals:

> $[a, \infty)$ consists of all real numbers x for which $x \ge a$.
> (a, ∞) consists of all real numbers x for which $x > a$.
> $(-\infty, a]$ consists of all real numbers x for which $x \le a$.
> $(-\infty, a)$ consists of all real numbers x for which $x < a$.
> $(-\infty, \infty)$ consists of all real numbers.

In Words

An interval is a nonempty set of real numbers.

Note that ∞ and $-\infty$ are never included as endpoints, since neither is a real number.

Table 1 summarizes interval notation, corresponding inequality notation, and their graphs.

Table 1

Interval	Inequality	Graph
The open interval (a, b)	$a < x < b$	
The closed interval $[a, b]$	$a \le x \le b$	
The half-open interval $[a, b)$	$a \le x < b$	
The half-open interval $(a, b]$	$a < x \le b$	
The interval $[a, \infty)$	$x \ge a$	
The interval (a, ∞)	$x > a$	
The interval $(-\infty, a]$	$x \le a$	
The interval $(-\infty, a)$	$x < a$	
The interval $(-\infty, \infty)$	All real numbers	

EXAMPLE 1   **Writing Inequalities Using Interval Notation**

Write each inequality using interval notation.

(a) $1 \le x \le 3$ (b) $-4 < x < 0$ (c) $x > 5$ (d) $x \le 1$

Solution

(a) $1 \le x \le 3$ describes all real numbers x between 1 and 3, inclusive. In interval notation, we write $[1, 3]$.

(b) In interval notation, $-4 < x < 0$ is written $(-4, 0)$.

(c) In interval notation, $x > 5$ is written $(5, \infty)$.

(d) In interval notation, $x \le 1$ is written $(-\infty, 1]$.

EXAMPLE 2 **Writing Intervals Using Inequality Notation**

Write each interval as an inequality involving x.

(a) $[1, 4)$ (b) $(2, \infty)$ (c) $[2, 3]$ (d) $(-\infty, -3]$

Solution   (a) $[1, 4)$ consists of all real numbers x for which $1 \leq x < 4$.

(b) $(2, \infty)$ consists of all real numbers x for which $x > 2$.

(c) $[2, 3]$ consists of all real numbers x for which $2 \leq x \leq 3$.

(d) $(-\infty, -3]$ consists of all real numbers x for which $x \leq -3$.

 Now Work PROBLEMS 13, 25, AND 33

2 Use Properties of Inequalities

The product of two positive real numbers is positive, the product of two negative real numbers is positive, and the product of 0 and 0 is 0. For any real number a, the value of a^2 is 0 or positive; that is, a^2 is nonnegative. This is called the **nonnegative property**.

In Words
The square of a real number is never negative.

Nonnegative Property
For any real number a,

$$a^2 \geq 0 \tag{1}$$

When the same number is added to both sides of an inequality, an equivalent inequality is obtained. For example, since $3 < 5$, then $3 + 4 < 5 + 4$ or $7 < 9$. This is called the **addition property** of inequalities.

In Words
The addition property states that the sense, or direction, of an inequality remains unchanged if the same number is added to both sides.

Addition Property of Inequalities
For real numbers a, b, and c,

- If $a < b$, then $a + c < b + c$. (2a)
- If $a > b$, then $a + c > b + c$. (2b)

EXAMPLE 3 **Addition Property of Inequalities**

(a) If $x < -5$, then $x + 5 < -5 + 5$ or $x + 5 < 0$.

(b) If $x > 2$, then $x + (-2) > 2 + (-2)$ or $x - 2 > 0$.

 Now Work PROBLEM 41

Now let's see what happens when both sides of an inequality are multiplied by a nonzero number.

EXAMPLE 4 **Multiplying an Inequality by a Positive Number**

Express as an inequality the result of multiplying both sides of the inequality $3 < 7$ by 2.

Solution Begin with

$$3 < 7$$

Multiplying both sides by 2 yields the numbers 6 and 14, so we have

$$6 < 14$$

EXAMPLE 5 **Multiplying an Inequality by a Negative Number**

Express as an inequality the result of multiplying both sides of the inequality $9 > 2$ by -4.

Solution Begin with

$$9 > 2$$

Multiplying both sides by -4 yields the numbers -36 and -8, so we have

$$-36 < -8$$

In Words
Multiplying by a negative number reverses the inequality.

Note that the effect of multiplying both sides of $9 > 2$ by the negative number -4 is that the direction of the inequality symbol is reversed.

Examples 4 and 5 illustrate the following general **multiplication properties** for inequalities:

In Words
The multiplication properties state that the sense, or direction, of an inequality *remains the same* if both sides are multiplied by a *positive* real number, whereas the direction is *reversed* if both sides are multiplied by a *negative* real number.

Multiplication Properties for Inequalities
For real numbers $a, b,$ and $c,$

- If $a < b$ and if $c > 0$, then $ac < bc$.
- If $a < b$ and if $c < 0$, then $ac > bc$. **(3a)**

- If $a > b$ and if $c > 0$, then $ac > bc$.
- If $a > b$ and if $c < 0$, then $ac < bc$. **(3b)**

EXAMPLE 6 **Multiplication Property of Inequalities**

(a) If $2x < 6$, then $\frac{1}{2} \cdot 2x < \frac{1}{2} \cdot 6$ or $x < 3$.

(b) If $\frac{x}{-3} > 12$, then $-3\left(\frac{x}{-3}\right) < -3 \cdot 12$ or $x < -36$.

(c) If $-4x < -8$, then $\frac{-4x}{-4} > \frac{-8}{-4}$ or $x > 2$.

(d) If $-x > 8$, then $(-1)(-x) < (-1) \cdot 8$ or $x < -8$.

Now Work PROBLEM **47**

Reciprocal Properties for Inequalities

- If $a > 0$, then $\dfrac{1}{a} > 0$.   **(4a)**

- If $a < 0$, then $\dfrac{1}{a} < 0$.   **(4b)**

- If $b > a > 0$, then $\dfrac{1}{a} > \dfrac{1}{b} > 0$.   **(4c)**

- If $a < b < 0$, then $\dfrac{1}{b} < \dfrac{1}{a} < 0$.   **(4d)**

3 Solve Inequalities

An **inequality in one variable** is a statement involving two expressions, at least one containing the variable, separated by one of the inequality symbols: $<, \leq, >,$ or $\geq$. To **solve an inequality** means to find all values of the variable for which the statement is true. These values are called **solutions** of the inequality.

For example, the following are all inequalities involving one variable x:

$$x + 5 < 8 \qquad 2x - 3 \geq 4 \qquad x^2 - 1 \leq 3 \qquad \frac{x + 1}{x - 2} > 0$$

As with equations, one method for solving an inequality is to replace it by a series of equivalent inequalities until an inequality with an obvious solution, such as $x < 3$, is obtained. Equivalent inequalities are obtained by applying some of the same properties that are used to find equivalent equations. The addition property and the multiplication properties for inequalities form the basis for the following procedures.

Procedures That Leave the Inequality Symbol Unchanged

- Simplify both sides of the inequality by combining like terms and eliminating parentheses:

$$\text{Replace} \quad x + 2 + 6 > 2x + 5(x + 1)$$
$$\text{by} \qquad x + 8 > 7x + 5$$

- Add or subtract the same expression on both sides of the inequality:

$$\text{Replace} \qquad 3x - 5 < 4$$
$$\text{by} \quad (3x - 5) + 5 < 4 + 5$$

- Multiply or divide both sides of the inequality by the same *positive* expression:

$$\text{Replace} \quad 4x > 16 \quad \text{by} \quad \frac{4x}{4} > \frac{16}{4}$$

Procedures That Reverse the Sense or Direction of the Inequality Symbol

- Interchange the two sides of the inequality:

$$\text{Replace} \quad 3 < x \quad \text{by} \quad x > 3$$

- Multiply or divide both sides of the inequality by the same *negative* expression:

$$\text{Replace} \quad -2x > 6 \quad \text{by} \quad \frac{-2x}{-2} < \frac{6}{-2}$$

As the examples that follow illustrate, we solve inequalities using many of the same steps that are used to solve equations. In writing the solution of an inequality, either set notation or interval notation may be used, whichever is more convenient.

EXAMPLE 7   **Solving an Inequality**

Solve the inequality $4x + 7 \geq 2x - 3$, and graph the solution set.

Solution

$$4x + 7 \geq 2x - 3$$

$$4x + 7 - 7 \geq 2x - 3 - 7 \qquad \text{Subtract 7 from both sides.}$$

$$4x \geq 2x - 10 \qquad \text{Simplify.}$$

$$4x - 2x \geq 2x - 10 - 2x \qquad \text{Subtract 2x from both sides.}$$

$$2x \geq -10 \qquad \text{Simplify.}$$

$$\frac{2x}{2} \geq \frac{-10}{2} \qquad \begin{array}{l}\text{Divide both sides by 2. (The direction of the}\\ \text{inequality symbol is unchanged.)}\end{array}$$

$$x \geq -5 \qquad \text{Simplify.}$$

The solution set is $\{x \mid x \geq -5\}$ or, using interval notation, all numbers in the interval $[-5, \infty)$. See Figure 23 for the graph.

Figure 23 $x \geq -5$

Now Work PROBLEM 59

4 Solve Combined Inequalities

EXAMPLE 8   **Solving a Combined Inequality**

Solve the inequality $-5 < 3x - 2 < 1$, and graph the solution set.

Solution   Recall that the inequality

$$-5 < 3x - 2 < 1$$

is equivalent to the two inequalities

$$-5 < 3x - 2 \quad \text{and} \quad 3x - 2 < 1$$

Solve each of these inequalities separately.

$$-5 < 3x - 2 \qquad\qquad\qquad\qquad 3x - 2 < 1$$

$$-5 + 2 < 3x - 2 + 2 \quad \text{Add 2 to both sides.} \qquad 3x - 2 + 2 < 1 + 2$$

$$-3 < 3x \qquad\qquad\quad \text{Simplify.} \qquad\qquad\qquad 3x < 3$$

$$\frac{-3}{3} < \frac{3x}{3} \qquad\qquad \text{Divide both sides by 3.} \qquad \frac{3x}{3} < \frac{3}{3}$$

$$-1 < x \qquad\qquad\quad \text{Simplify.} \qquad\qquad\qquad x < 1$$

The solution set of the original pair of inequalities consists of all x for which

$$-1 < x \quad \text{and} \quad x < 1$$

This may be written more compactly as $\{x \mid -1 < x < 1\}$. In interval notation, the solution is $(-1, 1)$. See Figure 24 for the graph.

Figure 24 $-1 < x < 1$

Now Work PROBLEM 79

Observe in Example 8 that solving each of the two inequalities required exactly the same steps. A shortcut to solving the original inequality algebraically is to deal with the two inequalities at the same time, as follows:

$$
\begin{array}{rcll}
-5 < & 3x - 2 & < 1 & \\
-5 + 2 < & 3x - 2 + 2 & < 1 + 2 & \text{Add 2 to each part.} \\
-3 < & 3x & < 3 & \text{Simplify.} \\
\dfrac{-3}{3} < & \dfrac{3x}{3} & < \dfrac{3}{3} & \text{Divide each part by 3.} \\
-1 < & x & < 1 & \text{Simplify.}
\end{array}
$$

EXAMPLE 9   **Using a Reciprocal Property to Solve an Inequality**

Solve the inequality $(4x - 1)^{-1} > 0$, and graph the solution set.

Solution   Recall that $(4x - 1)^{-1} = \dfrac{1}{4x - 1}$. Reciprocal Property (4a) states that if $a > 0$, then its reciprocal is greater than zero.

$$(4x - 1)^{-1} > 0$$

$$\frac{1}{4x - 1} > 0$$

$$4x - 1 > 0 \qquad \text{Reciprocal Property (4a)}$$

$$4x > 1 \qquad \text{Add 1 to both sides.}$$

$$x > \frac{1}{4} \qquad \text{Divide both sides by 4.}$$

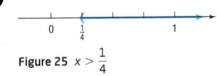

Figure 25 $x > \dfrac{1}{4}$

The solution set is $\left\{ x \mid x > \dfrac{1}{4} \right\}$, that is, all real numbers in the interval $\left(\dfrac{1}{4}, \infty \right)$. Figure 25 gives the graph.

— **Now Work** PROBLEM 89

---

5 Solve Inequalities Involving Absolute Value

EXAMPLE 10   **Solving an Inequality Involving Absolute Value**

Solve the inequality $|x| < 4$, and graph the solution set.

Solution   We are looking for all points whose coordinate x is a distance less than 4 units from the origin. See Figure 26 for the graph. Because any x between -4 and 4 satisfies the condition $|x| < 4$, the solution set consists of all numbers x for which $-4 < x < 4$, that is, all x in the interval $(-4, 4)$.

Less than 4 units from origin 0

Figure 26 $|x| < 4$

EXAMPLE 11   **Solving an Inequality Involving Absolute Value**

Solve the inequality $|x| > 3$, and graph the solution set.

Solution   We are looking for all points whose coordinate x is a distance greater than 3 units from the origin. Figure 27 gives the graph. Any number x less than -3 or greater than 3 satisfies the condition $|x| > 3$. The solution set consists of all numbers x for which $x < -3$ or $x > 3$, that is, all x in $(-\infty, -3) \cup (3, \infty)$.*

Figure 27 $|x| > 3$

*Recall that the symbol $\cup$ stands for the union of two sets. Refer to page A2 if necessary.

Examples 10 and 11 illustrate the following results:

THEOREM

If *a* is any positive number, then

- $|u| < a$ is equivalent to $-a < u < a$ **(5)**
- $|u| \leq a$ is equivalent to $-a \leq u \leq a$ **(6)**
- $|u| > a$ is equivalent to $u < -a$ or $u > a$ **(7)**
- $|u| \geq a$ is equivalent to $u \leq -a$ or $u \geq a$ **(8)**

EXAMPLE 12   **Solving an Inequality Involving Absolute Value**

Solve the inequality $|2x + 4| \leq 3$, and graph the solution set.

Solution

$$|2x + 4| \leq 3$$ This follows the form of statement (6) where $u = 2x + 4$.

$$-3 \leq \quad 2x + 4 \quad \leq 3$$ Use statement (6).

$$-3 - 4 \leq 2x + 4 - 4 \leq 3 - 4$$ Subtract 4 from each part.

$$-7 \leq \quad 2x \quad \leq -1$$ Simplify.

$$\frac{-7}{2} \leq \quad \frac{2x}{2} \quad \leq \frac{-1}{2}$$ Divide each part by 2.

$$-\frac{7}{2} \leq \quad x \quad \leq -\frac{1}{2}$$ Simplify.

Figure 28 $|2x + 4| \leq 3$

The solution set is $\left\{ x \mid -\dfrac{7}{2} \leq x \leq -\dfrac{1}{2} \right\}$, that is, all x in the interval $\left[-\dfrac{7}{2}, -\dfrac{1}{2} \right]$. See Figure 28 for a graph of the solution set.

 Now Work PROBLEM 97

EXAMPLE 13   **Solving an Inequality Involving Absolute Value**

Solve the inequality $|2x - 5| > 3$, and graph the solution set.

Solution

$$|2x - 5| > 3$$ This follows the form of statement (7) where $u = 2x - 5$.

$$2x - 5 < -3 \quad \text{or} \quad 2x - 5 > 3$$ Use statement (7).

$$2x - 5 + 5 < -3 + 5 \quad \text{or} \quad 2x - 5 + 5 > 3 + 5$$ Add 5 to each part.

$$2x < 2 \quad \text{or} \quad 2x > 8$$ Simplify.

$$\frac{2x}{2} < \frac{2}{2} \quad \text{or} \quad \frac{2x}{2} > \frac{8}{2}$$ Divide each part by 2.

$$x < 1 \quad \text{or} \quad x > 4$$ Simplify.

Figure 29 $|2x - 5| > 3$

The solution set is $\{ x \mid x < 1 \text{ or } x > 4 \}$, that is, all x in $(-\infty, 1) \cup (4, \infty)$. See Figure 29 for a graph of the solution set.

WARNING A common error to be avoided is to attempt to write the solution $x < 1$ or $x > 4$ as the combined inequality $1 > x > 4$, which is incorrect, since there are no numbers x for which $x < 1$ and $x > 4$. Another common error is to "mix" the symbols and write $1 < x > 4$, which makes no sense. ∎

Now Work PROBLEM 103

A.6 Assess Your Understanding

1. Graph the inequality: $x \geq -2$ (pp. A4–A5)

2. *True or False* $-5 > -3$ (pp. A4–A5)

Concepts and Vocabulary

3. A(n) _____ _____, denoted $[a, b]$, consists of all real numbers x for which $a \leq x \leq b$.

4. The _____ _____ state that the sense, or direction, of an inequality remains the same if both sides are multiplied by a positive number, while the direction is reversed if both sides are multiplied by a negative number.

In Problems 5–8, assume that $a < b$ and $c < 0$.

5. *True or False* $a + c < b + c$

6. *True or False* $a - c < b - c$

7. *True or False* $ac > bc$

8. *True or False* $\dfrac{a}{c} < \dfrac{b}{c}$

9. *True or False* The square of any real number is always nonnegative.

10. *True or False* A half-closed interval must have an endpoint of either $-\infty$ or ∞.

11. *Multiple Choice* Which of the following will *not* change the direction, or sense, of an inequality?
(a) Dividing both sides by a negative number
(b) Interchanging sides
(c) Taking the reciprocal of both sides
(d) Subtracting a positive number from both sides

12. *Multiple Choice* Which pair of inequalities is equivalent to $0 < x \leq 3$?
(a) $x > 0$ and $x \geq 3$
(b) $x < 0$ and $x \geq 3$
(c) $x > 0$ and $x \leq 3$
(d) $x < 0$ and $x \leq 3$

Skill Building

In Problems 13–18, express the graph shown in blue using interval notation. Also express each as an inequality involving x.

13.

14.

15.

16.

17.

18.

In Problems 19–24, an inequality is given. Write the inequality obtained by:
(a) *Adding 3 to both sides of the given inequality.*
(b) *Subtracting 5 from both sides of the given inequality.*
(c) *Multiplying both sides of the given inequality by 3.*
(d) *Multiplying both sides of the given inequality by -2.*

19. $3 < 5$ **20.** $2 > 1$ **21.** $4 > -3$ **22.** $-3 > -5$ **23.** $2x + 1 < 2$ **24.** $1 - 2x > 5$

In Problems 25–32, write each inequality using interval notation, and graph each inequality on the real number line.

25. $0 \leq x \leq 4$ **26.** $-1 < x < 5$ **27.** $4 \leq x < 6$ **28.** $-2 < x < 0$

29. $x \geq -3$ **30.** $x \leq 5$ **31.** $x < -4$ **32.** $x > 1$

In Problems 33–40, write each interval as an inequality involving x, and graph each inequality on the real number line.

33. $[2, 5]$ **34.** $(1, 2)$ **35.** $(-4, 3]$ **36.** $[0, 1)$

37. $[4, \infty)$ **38.** $(-\infty, 2]$ **39.** $(-\infty, -3)$ **40.** $(-8, \infty)$

In Problems 41–58, fill in the blank to form a correct inequality statement.

41. If $x < 5$, then $x - 5$ ____ 0.

42. If $x < -4$, then $x + 4$ ____ 0.

43. If $x > -4$, then $x + 4$ ____ 0.

44. If $x > 6$, then $x - 6$ ____ 0.

45. If $x \geq -4$, then $3x$ ____ -12.

46. If $x \leq 3$, then $2x$ ____ 6.

47. If $x > 6$, then $-2x$ ____ -12.

48. If $x > -2$, then $-4x$ ____ 8.

49. If $x \geq 5$, then $-4x$ ____ -20.

50. If $x \leq -4$, then $-3x$ ____ 12.

51. If $8x > 40$, then x ____ 5.

52. If $3x \leq 12$, then x ____ 4.

53. If $-\dfrac{1}{2}x \leq 3$, then x ____ -6.

54. If $-\dfrac{1}{4}x > 1$, then x ____ -4.

55. If $0 < 5 < x$, then $0 < \dfrac{1}{\square} < \dfrac{1}{\square}$.

56. If $x \leq -4 < 0$, then $\dfrac{1}{\square} \leq \dfrac{1}{\square} < 0$.

57. If $-5 < x < 0$, then $\dfrac{1}{\square} < \dfrac{1}{\square} < 0$.

58. If $0 < x \leq 10$, then $0 < \dfrac{1}{\square} \leq \dfrac{1}{\square}$.

In Problems 59–108, solve each inequality. Express your answer using set notation or interval notation. Graph the solution set.

59. $x + 1 < 5$

60. $x - 6 < 1$

61. $3 - 5x \leq -7$

62. $2 - 3x \leq 5$

63. $3x - 7 > 2$

64. $2x + 5 > 1$

65. $3x - 1 \geq 3 + x$

66. $2x - 2 \geq 3 + x$

67. $-2(x + 3) < 8$

68. $-3(1 - x) < 12$

69. $4 - 3(1 - x) \leq 3$

70. $8 - 4(2 - x) \leq -2x$

71. $\frac{1}{2}(x - 4) > x + 8$

72. $3x + 4 > \frac{1}{3}(x - 2)$

73. $\frac{x}{2} \geq 1 - \frac{x}{4}$

74. $\frac{x}{3} \geq 2 + \frac{x}{6}$

75. $0 < 3x - 7 \leq 5$

76. $4 \leq 2x + 2 \leq 10$

77. $-5 \leq 4 - 3x \leq 2$

78. $-3 \leq 3 - 2x \leq 9$

79. $-3 < \frac{2x - 1}{4} < 0$

80. $0 < \frac{3x + 2}{2} < 4$

81. $1 < 1 - \frac{1}{2}x < 4$

82. $0 < 1 - \frac{1}{3}x < 1$

83. $(x + 2)(x - 3) > (x - 1)(x + 1)$

84. $(x - 1)(x + 1) > (x - 3)(x + 4)$

85. $x(4x + 3) \leq (2x + 1)^2$

86. $x(9x - 5) \leq (3x - 1)^2$

87. $\frac{1}{2} \leq \frac{x + 1}{3} < \frac{3}{4}$

88. $\frac{1}{3} < \frac{x + 1}{2} \leq \frac{2}{3}$

89. $(4x + 2)^{-1} < 0$

90. $(2x - 1)^{-1} > 0$

91. $(1 - 4x)^{-1} \geq 7$

92. $2(3x + 5)^{-1} \leq -3$

93. $0 < \frac{2}{x} < \frac{3}{5}$

94. $0 < \frac{4}{x} < \frac{2}{3}$

95. $0 < (2x - 4)^{-1} < \frac{1}{2}$

96. $0 < (3x + 6)^{-1} < \frac{1}{3}$

97. $|2x| < 8$

98. $|3x| < 12$

99. $|3x| > 12$

100. $|2x| > 6$

101. $|2x - 1| \leq 1$

102. $|2x + 5| \leq 7$

103. $|1 - 2x| > 3$

104. $|2 - 3x| > 1$

105. $|-4x| + |-5| \leq 9$

106. $|-x| - |4| \leq 2$

107. $|-2x| \geq |-4|$

108. $|-x - 2| \geq 1$

Applications and Extensions

In Problems 109–118, find a and b.

109. If $-1 < x < 1$, then $a < x + 4 < b$.

110. If $-3 < x < 2$, then $a < x - 6 < b$.

111. If $2 < x < 3$, then $a < -4x < b$.

112. If $-4 < x < 0$, then $a < \frac{1}{2}x < b$.

113. If $0 < x < 4$, then $a < 2x + 3 < b$.

114. If $-3 < x < 3$, then $a < 1 - 2x < b$.

115. If $-3 < x < 0$, then $a < \frac{1}{x + 4} < b$.

116. If $2 < x < 4$, then $a < \frac{1}{x - 6} < b$.

117. If $6 < 3x < 12$, then $a < x^2 < b$.

118. If $0 < 2x < 6$, then $a < x^2 < b$.

119. What is the domain of the variable in the expression $\sqrt{3x + 6}$?

120. What is the domain of the variable in the expression $\sqrt{8 + 2x}$?

121. A young adult may be defined as someone older than 21 but less than 30 years of age. Express this statement using inequalities.

122. Middle-aged may be defined as being 40 or more and less than 60. Express this statement using inequalities.

123. Life Expectancy The Social Security Administration determined that an average 30-year-old male in 2018 could expect to live at least 52.2 more years and that an average 30-year-old female in 2018 could expect to live at least 55.8 more years.

(a) To what age could an average 30-year-old male expect to live? Express your answer as an inequality.

(b) To what age could an average 30-year-old female expect to live? Express your answer as an inequality.

(c) Who can expect to live longer, a male or a female? By how many years?

Source: Social Security Administration, 2018

124. General Chemistry For a certain ideal gas, the volume V (in cubic centimeters) equals 20 times the temperature T (in degrees Celsius). If the temperature varies from 80° to 120°C, inclusive, what is the corresponding range of the volume of the gas?

125. Real Estate A real estate agent agrees to sell an apartment complex according to the following commission schedule: $45,000 plus 25% of the selling price in excess of $900,000. Assuming that the complex will sell at some price between $900,000 and $1,100,000, inclusive, over what range does the agent's commission vary? How does the commission vary as a percent of selling price?

126. Sales Commission A used car salesperson is paid a commission of $25 plus 40% of the selling price in excess of owner's cost. The owner claims that used cars typically sell for at least owner's cost plus $200 and at most owner's cost plus $3000. For each sale made, over what range can the salesperson expect the commission to vary?

127. Federal Tax Withholding The percentage method of withholding for federal income tax (2018) states that a single person whose weekly wages, after subtracting withholding allowances, are over $815, but not over $1658, shall have $85.62 plus 22% of the excess over $815 withheld. Over what range does the amount withheld vary if the weekly wages vary from $900 to $1100, inclusive?

Source: Employer's Tax Guide. Internal Revenue Service, 2018

128. Exercising Sue wants to lose weight. For healthy weight loss, the American College of Sports Medicine (ACSM) recommends 200 to 300 minutes of exercise per week. For the first six days of the week, Sue exercised 40, 45, 0, 50, 25, and 35 minutes. How long should Sue exercise on the seventh day in order to stay within the ACSM guidelines?

129. Electricity Rates During summer months in 2018, Omaha Public Power District charged residential customers a monthly service charge of $25, plus a usage charge of 10.06¢ per kilowatt-hour (kWh). If one customer's monthly summer bills ranged from a low of $140.69 to a high of $231.23, over what range did usage vary (in kWh)?

Source: Omaha Public Power District, 2018

130. Sewer Bills The village of Oak Lawn charges homeowners $23.55 per quarter-year for sewer usage, plus $0.40 per 1000 gallons of water metered. In 2018, one homeowner's quarterly bill ranged from a high of $36.75 to a low of $30.35. Over what range did metered water usage vary?

Source: Village of Oak Lawn, Illinois, January 2018

131. Fat Content Suppose that you order a small McCafe™ chocolate shake (15 g of fat) and an Artisan Grilled Chicken Sandwich™ (7 g of fat) at McDonald's. How many oatmeal raisin cookies can you eat (5 g of fat) and still keep the total fat content of your meal to no more than 47 g?

Source: McDonald's Corporation

132. Sodium Content Suppose that you order a Garden Side Salad (95 mg of sodium) and a Strawberry Banana Smoothie (50 mg of sodium) at Burger King. How many hamburgers can you eat (380 mg of sodium) and still keep the total sodium content of your meal to no more than 1285 mg?

Source: Burger King

133. Computing Grades In your Economics 101 class, you have scores of 68, 82, 87, and 89 on the first four of five tests. To get a grade of B, the average of the first five test scores must be greater than or equal to 80 and less than 90.
   (a) Solve an inequality to find the least score you can get on the last test and still earn a B.
   (b) What score do you need if the fifth test counts double?

What do I need to get a B?

134. IQ Tests A standard intelligence test has an average score of 100. According to statistical theory, of the people who take the test, the 2.5% with the highest scores will have scores of more than 1.96σ above the average, where σ (sigma, a number called the **standard deviation**) depends on the nature of the test. If $\sigma = 12$ for this test and there is (in principle) no upper limit to the score possible on the test, write the interval of possible test scores of the people in the top 2.5%.

135. Arithmetic Mean If $a < b$, show that $a < \dfrac{a+b}{2} < b$. The number $\dfrac{a+b}{2}$ is called the **arithmetic mean** of a and b.

136. Refer to Problem 135. Show that the arithmetic mean of a and b is equidistant from a and b.

137. Geometric Mean If $0 < a < b$, show that $a < \sqrt{ab} < b$. The number $\sqrt{ab}$ is called the **geometric mean** of a and b.

138. Refer to Problems 135 and 137. Show that the geometric mean of a and b is less than the arithmetic mean of a and b.

Explaining Concepts: Discussion and Writing

139. Make up an inequality that has no solution. Make up an inequality that has exactly one solution.

140. The inequality $x^2 + 1 < -5$ has no real solution. Explain why.

141. Do you prefer to use inequality notation or interval notation to express the solution to an inequality? Give your reasons. Are there particular circumstances when you prefer one to the other? Cite examples.

'Are You Prepared?' Answers

1.

2. False

A.7 *n*th Roots; Rational Exponents

PREPARING FOR THIS SECTION *Before getting started, review the following:*

• Exponents, Square Roots (Section A.1, pp. A7–A10)

✏ **Now Work** the 'Are You Prepared?' problems on page A62.

OBJECTIVES 1 Work with *n*th Roots (p. A56)
2 Simplify Radicals (p. A57)
3 Rationalize Denominators and Numerators (p. A58)
4 Solve Radical Equations (p. A59)
5 Simplify Expressions with Rational Exponents (p. A60)

1 Work with *n*th Roots

DEFINITION Principal *n*th Root

The **principal *n*th root of a real number** $a, n \geq 2$ an integer, symbolized by $\sqrt[n]{a}$, is defined as follows:

$$\sqrt[n]{a} = b \quad \text{means} \quad a = b^n$$

where $a \geq 0$ and $b \geq 0$ if *n* is even and a, b are any real numbers if *n* is odd.

In Words
The symbol $\sqrt[n]{a}$ means "what is the number that, when raised to the power *n*, equals *a*."

Notice that if *a* is negative and *n* is even, then $\sqrt[n]{a}$ is not defined. When it is defined, the principal *n*th root of a number is unique.

The symbol $\sqrt[n]{a}$ for the principal *n*th root of *a* is called a **radical**; the integer *n* is called the **index**, and *a* is called the **radicand**. If the index of a radical is 2, we call $\sqrt[2]{a}$ the **square root** of *a* and omit the index 2 by simply writing $\sqrt{a}$. If the index is 3, we call $\sqrt[3]{a}$ the **cube root** of *a*.

EXAMPLE 1

Simplifying Principal *n*th Roots

(a) $\sqrt[3]{8} = \sqrt[3]{2^3} = 2$

(b) $\sqrt[3]{-64} = \sqrt[3]{(-4)^3} = -4$

(c) $\sqrt[4]{\dfrac{1}{16}} = \sqrt[4]{\left(\dfrac{1}{2}\right)^4} = \dfrac{1}{2}$

(d) $\sqrt[6]{(-2)^6} = |-2| = 2$

These are examples of **perfect roots**, since each simplifies to a rational number. Notice the absolute value in Example 1(d). If *n* is even, then the principal *n*th root must be nonnegative.

In general, if $n \geq 2$ is an integer and *a* is a real number, we have

$$\sqrt[n]{a^n} = a \qquad \text{if } n \geq 3 \text{ is odd} \tag{1a}$$
$$\sqrt[n]{a^n} = |a| \qquad \text{if } n \geq 2 \text{ is even} \tag{1b}$$

✏ **Now Work** PROBLEM 11

Radicals provide a way of representing many irrational real numbers. For example, it can be shown that there is no rational number whose square is 2. Using radicals, we can say that $\sqrt{2}$ is the positive number whose square is 2.

EXAMPLE 2

Using a Calculator to Approximate Roots

Use a calculator to approximate $\sqrt[5]{16}$.

Solution Figure 30 shows the result using a TI-84 Plus C graphing calculator.

Now Work PROBLEM 131

Figure 30

2 Simplify Radicals

Let $n \geq 2$ and $m \geq 2$ denote integers, and let a and b represent real numbers. Assuming that all radicals are defined, we have the following properties:

Properties of Radicals

$$\sqrt[n]{ab} = \sqrt[n]{a}\,\sqrt[n]{b} \tag{2a}$$

$$\sqrt[n]{\frac{a}{b}} = \frac{\sqrt[n]{a}}{\sqrt[n]{b}} \quad b \neq 0 \tag{2b}$$

$$\sqrt[n]{a^m} = \left(\sqrt[n]{a}\right)^m \tag{2c}$$

When used in reference to radicals, the direction to "simplify" means to remove from the radicals any perfect roots that occur as factors.

EXAMPLE 3

Simplifying Radicals

(a) $\sqrt{32} = \sqrt{16 \cdot 2} = \sqrt{16} \cdot \sqrt{2} = 4\sqrt{2}$

Factor out 16, (2a)
a perfect square.

(b) $\sqrt[3]{16} = \sqrt[3]{8 \cdot 2} = \sqrt[3]{8} \cdot \sqrt[3]{2} = \sqrt[3]{2^3} \cdot \sqrt[3]{2} = 2\sqrt[3]{2}$

Factor out 8, (2a)
a perfect cube.

(c) $\sqrt[3]{-16x^4} = \sqrt[3]{-8 \cdot 2 \cdot x^3 \cdot x} = \sqrt[3]{(-8x^3)(2x)}$

Factor perfect   Group perfect
cubes inside radical. cubes.

$= \sqrt[3]{(-2x)^3 \cdot 2x} = \sqrt[3]{(-2x)^3} \cdot \sqrt[3]{2x} = -2x\sqrt[3]{2x}$

(d) $\sqrt[4]{\frac{16x^5}{81}} = \sqrt[4]{\frac{2^4 x^4 x}{3^4}} = \sqrt[4]{\left(\frac{2x}{3}\right)^4 \cdot x} = \sqrt[4]{\left(\frac{2x}{3}\right)^4} \cdot \sqrt[4]{x} = \left|\frac{2x}{3}\right| \sqrt[4]{x}$

Now Work PROBLEMS 15 AND 27

Two or more radicals can be combined, provided that they have the same index and the same radicand. Such radicals are called **like radicals**.

EXAMPLE 4

Combining Like Radicals

(a) $-8\sqrt{12} + \sqrt{3} = -8\sqrt{4 \cdot 3} + \sqrt{3}$

$= -8 \cdot \sqrt{4}\sqrt{3} + \sqrt{3}$

$= -16\sqrt{3} + \sqrt{3} = -15\sqrt{3}$

(b) $\sqrt[3]{8x^4} + \sqrt[3]{-x} + 4\sqrt[3]{27x} = \sqrt[3]{2^3 x^3 x} + \sqrt[3]{-1 \cdot x} + 4\sqrt[3]{3^3 x}$

$= \sqrt[3]{(2x)^3} \cdot \sqrt[3]{x} + \sqrt[3]{-1} \cdot \sqrt[3]{x} + 4\sqrt[3]{3^3} \cdot \sqrt[3]{x}$

$= 2x\sqrt[3]{x} - 1 \cdot \sqrt[3]{x} + 12\sqrt[3]{x}$

$= (2x + 11)\sqrt[3]{x}$

Now Work PROBLEM 45

3 Rationalize Denominators and Numerators

When radicals occur in quotients, it is customary to rewrite the quotient so that the new denominator contains no radicals. This process is referred to as **rationalizing the denominator**.

The idea is to multiply by an appropriate expression so that the new denominator contains no radicals. For example:

If a Denominator Contains the Factor	Multiply by	To Obtain a Denominator Free of Radicals
$\sqrt{3}$	$\sqrt{3}$	$(\sqrt{3})^2 = 3$
$\sqrt{3} + 1$	$\sqrt{3} - 1$	$(\sqrt{3})^2 - 1^2 = 3 - 1 = 2$
$\sqrt{2} - 3$	$\sqrt{2} + 3$	$(\sqrt{2})^2 - 3^2 = 2 - 9 = -7$
$\sqrt{5} - \sqrt{3}$	$\sqrt{5} + \sqrt{3}$	$(\sqrt{5})^2 - (\sqrt{3})^2 = 5 - 3 = 2$
$\sqrt[3]{4}$	$\sqrt[3]{2}$	$\sqrt[3]{4} \cdot \sqrt[3]{2} = \sqrt[3]{8} = 2$

In rationalizing the denominator of a quotient, be sure to multiply both the numerator and the denominator by the expression.

EXAMPLE 5   **Rationalizing Denominators**

Rationalize the denominator of each expression:

(a) $\dfrac{1}{\sqrt{3}}$
(b) $\dfrac{5}{4\sqrt{2}}$
(c) $\dfrac{\sqrt{2}}{\sqrt{3} - 3\sqrt{2}}$

Solution   (a) The denominator contains the factor $\sqrt{3}$, so we multiply the numerator and denominator by $\sqrt{3}$ to obtain

$$\frac{1}{\sqrt{3}} = \frac{1}{\sqrt{3}} \cdot \frac{\sqrt{3}}{\sqrt{3}} = \frac{\sqrt{3}}{(\sqrt{3})^2} = \frac{\sqrt{3}}{3}$$

(b) The denominator contains the factor $\sqrt{2}$, so we multiply the numerator and denominator by $\sqrt{2}$ to obtain

$$\frac{5}{4\sqrt{2}} = \frac{5}{4\sqrt{2}} \cdot \frac{\sqrt{2}}{\sqrt{2}} = \frac{5\sqrt{2}}{4(\sqrt{2})^2} = \frac{5\sqrt{2}}{4 \cdot 2} = \frac{5\sqrt{2}}{8}$$

(c) The denominator contains the factor $\sqrt{3} - 3\sqrt{2}$, so we multiply the numerator and denominator by $\sqrt{3} + 3\sqrt{2}$ to obtain

$$\frac{\sqrt{2}}{\sqrt{3} - 3\sqrt{2}} = \frac{\sqrt{2}}{\sqrt{3} - 3\sqrt{2}} \cdot \frac{\sqrt{3} + 3\sqrt{2}}{\sqrt{3} + 3\sqrt{2}} = \frac{\sqrt{2}(\sqrt{3} + 3\sqrt{2})}{(\sqrt{3})^2 - (3\sqrt{2})^2}$$

$$= \frac{\sqrt{2}\sqrt{3} + 3(\sqrt{2})^2}{3 - 18} = \frac{\sqrt{6} + 6}{-15} = -\frac{6 + \sqrt{6}}{15}$$

━━━━━ **Now Work** PROBLEM **59**

In calculus, it is sometimes necessary to **rationalize the numerator** of a quotient. To do this, multiply by an appropriate expression so that the new numerator contains no radicals.

EXAMPLE 6   **Rationalizing Numerators**

Rationalize the numerator of the following expression.

$$\frac{\sqrt{x+h}-\sqrt{x}}{h} \quad h \neq 0$$

Solution   The numerator contains the factor $\sqrt{x+h}-\sqrt{x}$, so we multiply the numerator and denominator by $\sqrt{x+h}+\sqrt{x}$ to obtain

$$\frac{\sqrt{x+h}-\sqrt{x}}{h} = \frac{\sqrt{x+h}-\sqrt{x}}{h} \cdot \frac{\sqrt{x+h}+\sqrt{x}}{\sqrt{x+h}+\sqrt{x}} = \frac{(\sqrt{x+h})^2-(\sqrt{x})^2}{h(\sqrt{x+h}+\sqrt{x})}$$

$$= \frac{x+h-x}{h(\sqrt{x+h}+\sqrt{x})} = \frac{\cancel{h}}{\cancel{h}(\sqrt{x+h}+\sqrt{x})} = \frac{1}{\sqrt{x+h}+\sqrt{x}}$$

Now Work PROBLEM 69

4 Solve Radical Equations

When the variable in an equation occurs in a square root, cube root, and so on—that is, when it occurs in a radical—the equation is called a **radical equation**. Sometimes a suitable operation will change a radical equation to one that is linear or quadratic. A commonly used procedure is to isolate the most complicated radical on one side of the equation and then eliminate it by raising both sides to a power equal to the index of the radical. Care must be taken, however, because apparent solutions that are not, in fact, solutions of the original equation may result. These are called **extraneous solutions**. Therefore, we need to check all answers when working with radical equations, and we check them in the *original* equation.

EXAMPLE 7   **Solving a Radical Equation**

Find the real solutions of the equation: $\sqrt[3]{2x-4} - 2 = 0$

Solution   The equation contains a radical whose index is 3. Isolate it on the left side.

$$\sqrt[3]{2x-4} - 2 = 0$$
$$\sqrt[3]{2x-4} = 2 \quad \text{Add 2 to both sides.}$$

Now raise both sides to the third power (the index of the radical is 3) and solve.

$$\left(\sqrt[3]{2x-4}\right)^3 = 2^3 \quad \text{Raise both sides to the power 3.}$$
$$2x-4 = 8 \quad \text{Simplify.}$$
$$2x = 12 \quad \text{Add 4 to both sides.}$$
$$x = 6 \quad \text{Divide both sides by 2.}$$

✔ **Check:** $\sqrt[3]{2(6)-4} - 2 = \sqrt[3]{12-4} - 2 = \sqrt[3]{8} - 2 = 2 - 2 = 0$

The solution set is $\{6\}$.

Now Work PROBLEM 77

5 Simplify Expressions with Rational Exponents

Radicals are used to define rational exponents.

> **DEFINITION** $a^{1/n}$
>
> If a is a real number and $n \geq 2$ is an integer, then
>
> $$a^{1/n} = \sqrt[n]{a} \tag{3}$$
>
> provided that $\sqrt[n]{a}$ exists.

Note that if n is even and $a < 0$, then $\sqrt[n]{a}$ and $a^{1/n}$ do not exist.

EXAMPLE 8   **Writing Expressions Containing Fractional Exponents as Radicals**

(a) $4^{1/2} = \sqrt{4} = 2$ (b) $8^{1/2} = \sqrt{8} = 2\sqrt{2}$

(c) $(-27)^{1/3} = \sqrt[3]{-27} = -3$ (d) $16^{1/3} = \sqrt[3]{16} = 2\sqrt[3]{2}$

> **DEFINITION** $a^{m/n}$
>
> If a is a real number and m and n are integers containing no common factors, with $n \geq 2$, then
>
> $$a^{m/n} = \sqrt[n]{a^m} = \left(\sqrt[n]{a}\right)^m \tag{4}$$
>
> provided that $\sqrt[n]{a}$ exists.

We have two comments about equation (4):

- The exponent $\dfrac{m}{n}$ must be in lowest terms, and $n \geq 2$ must be positive.

- In simplifying the expression $a^{m/n}$, either $\sqrt[n]{a^m}$ or $\left(\sqrt[n]{a}\right)^m$ may be used, the choice depending on which is easier to simplify. Generally, taking the root first, as in $\left(\sqrt[n]{a}\right)^m$, is easier.

EXAMPLE 9   **Simplifying Expressions With Rational Exponents**

(a) $4^{3/2} = \left(\sqrt{4}\right)^3 = 2^3 = 8$ (b) $(-8)^{4/3} = \left(\sqrt[3]{-8}\right)^4 = (-2)^4 = 16$

(c) $(32)^{-2/5} = \left(\sqrt[5]{32}\right)^{-2} = 2^{-2} = \dfrac{1}{4}$ (d) $25^{6/4} = 25^{3/2} = \left(\sqrt{25}\right)^3 = 5^3 = 125$

Now Work   PROBLEM 81

It can be shown that the Laws of Exponents hold for rational exponents. The next example illustrates using the Laws of Exponents to simplify.

EXAMPLE 10   **Simplifying Expressions With Rational Exponents**

Simplify each expression. Express your answer so that only positive exponents occur. Assume that the variables are positive.

(a) $(x^{2/3}y)(x^{-2}y)^{1/2}$ (b) $\left(\dfrac{2x^{1/3}}{y^{2/3}}\right)^{-3}$ (c) $\left(\dfrac{9x^2y^{1/3}}{x^{1/3}y}\right)^{1/2}$

Solution (a) $(x^{2/3}y)\,(x^{-2}y)^{1/2} = x^{2/3}\cdot y \cdot (x^{-2})^{1/2}\cdot y^{1/2}$ $(ab)^n = a^n b^n$

$$= x^{2/3}\cdot y \cdot x^{-1}\cdot y^{1/2}\qquad (a^m)^n = a^{mn}$$

$$= x^{2/3}\cdot x^{-1}\cdot y \cdot y^{1/2}$$

$$= x^{-1/3}\cdot y^{3/2}\qquad\qquad a^m \cdot a^n = a^{m+n}$$

$$= \frac{y^{3/2}}{x^{1/3}}\qquad\qquad\qquad a^{-n} = \frac{1}{a^n}$$

(b) $\left(\dfrac{2x^{1/3}}{y^{2/3}}\right)^{-3} = \left(\dfrac{y^{2/3}}{2x^{1/3}}\right)^{3} = \dfrac{(y^{2/3})^{3}}{(2x^{1/3})^{3}} = \dfrac{y^{2}}{2^{3}(x^{1/3})^{3}} = \dfrac{y^{2}}{8x}$

(c) $\left(\dfrac{9x^{2}y^{1/3}}{x^{1/3}y}\right)^{1/2} = \left(\dfrac{9x^{2-(1/3)}}{y^{1-(1/3)}}\right)^{1/2} = \left(\dfrac{9x^{5/3}}{y^{2/3}}\right)^{1/2} = \dfrac{9^{1/2}(x^{5/3})^{1/2}}{(y^{2/3})^{1/2}} = \dfrac{3x^{5/6}}{y^{1/3}}$

—**Now Work** PROBLEM 101

The next two examples illustrate some algebra that you will need to know for certain calculus problems.

EXAMPLE 11   **Writing an Expression as a Single Quotient**

Write the following expression as a single quotient in which only positive exponents appear.

$$(x^2 + 1)^{1/2} + x\cdot\frac{1}{2}(x^2 + 1)^{-1/2}\cdot 2x$$

Solution $(x^2 + 1)^{1/2} + x\cdot\dfrac{1}{2}(x^2 + 1)^{-1/2}\cdot 2x = (x^2 + 1)^{1/2} + \dfrac{x^2}{(x^2 + 1)^{1/2}}$

$$= \frac{(x^2 + 1)^{1/2}(x^2 + 1)^{1/2} + x^2}{(x^2 + 1)^{1/2}}$$

$$= \frac{(x^2 + 1) + x^2}{(x^2 + 1)^{1/2}}$$

$$= \frac{2x^2 + 1}{(x^2 + 1)^{1/2}}$$

—**Now Work** PROBLEM 107

EXAMPLE 12   **Factoring an Expression Containing Rational Exponents**

Factor and simplify: $\dfrac{4}{3}x^{1/3}(2x + 1) + 2x^{4/3}$

Solution   Begin by writing $2x^{4/3}$ as a fraction with 3 as the denominator.

$$\frac{4}{3}x^{1/3}(2x + 1) + 2x^{4/3} = \frac{4x^{1/3}(2x + 1)}{3} + \frac{6x^{4/3}}{3} \underset{\substack{\uparrow \\ \text{Add the two fractions.}}}{=} \frac{4x^{1/3}(2x + 1) + 6x^{4/3}}{3}$$

$$\underset{\substack{\uparrow \\ 2 \text{ and } x^{1/3} \text{ are common factors.}}}{=} \frac{2x^{1/3}[\,2(2x + 1) + 3x\,]}{3} \underset{\substack{\uparrow \\ \text{Simplify.}}}{=} \frac{2x^{1/3}(7x + 2)}{3}$$

—**Now Work** PROBLEM 119

A.7 Assess Your Understanding

1. $(-3)^2 =$ ___ ; $-3^2 =$ ___ (pp. A7–A9)

2. $\sqrt{16} =$ ___ ; $\sqrt{(-4)^2} =$ ___ (pp. A9–A10)

Concepts and Vocabulary

3. In the symbol $\sqrt[n]{a}$, the integer n is called the _____.

4. We call $\sqrt[n]{a}$ the _____ _____ of a.

5. *Multiple Choice* Let $n \geq 2$ and $m \geq 2$ be integers, and let a and b be real numbers. Which of the following is not a property of radicals? Assume all radicals are defined.

(a) $\sqrt[n]{\dfrac{a}{b}} = \dfrac{\sqrt[n]{a}}{\sqrt[n]{b}}$

(b) $\sqrt[n]{a+b} = \sqrt[n]{a} + \sqrt[n]{b}$

(c) $\sqrt[n]{ab} = \sqrt[n]{a}\sqrt[n]{b}$

(d) $\sqrt[n]{a^m} = \left(\sqrt[n]{a}\right)^m$

6. *Multiple Choice* If a is a real number and $n \geq 2$ is an integer, then which of the following expressions is equivalent to $\sqrt[n]{a}$, provided that it exists?

(a) a^{-n}

(b) a^n

(c) $\dfrac{1}{a^n}$

(d) $a^{1/n}$

7. *Multiple Choice* Which of the following phrases best defines like radicals?

(a) Radical expressions that have the same index

(b) Radical expressions that have the same radicand

(c) Radical expressions that have the same index and the same radicand

(d) Radical expressions that have the same variable

8. *Multiple Choice* To rationalize the denominator of the expression $\dfrac{\sqrt{2}}{1 - \sqrt{3}}$, multiply both the numerator and the denominator by which of the following?

(a) $\sqrt{3}$　(b) $\sqrt{2}$　(c) $1 + \sqrt{3}$　(d) $1 - \sqrt{3}$

9. *True or False* $\sqrt[5]{-32} = -2$

10. *True or False* $\sqrt[4]{(-3)^4} = -3$

Skill Building

In Problems 11–54, simplify each expression. Assume that all variables are positive when they appear.

11. $\sqrt[3]{27}$

12. $\sqrt[4]{16}$

13. $\sqrt[3]{-8}$

14. $\sqrt[3]{-1}$

15. $\sqrt{8}$

16. $\sqrt{75}$

17. $\sqrt{700}$

18. $\sqrt{45x^3}$

19. $\sqrt[3]{32}$

20. $\sqrt[3]{54}$

21. $\sqrt[3]{-8x^4}$

22. $\sqrt[3]{192x^5}$

23. $\sqrt[4]{243}$

24. $\sqrt[4]{48x^5}$

25. $\sqrt[4]{x^{12}y^8}$

26. $\sqrt[5]{x^{10}y^5}$

27. $\sqrt[4]{\dfrac{x^9y^7}{xy^3}}$

28. $\sqrt[3]{\dfrac{3xy^2}{81x^4y^2}}$

29. $\sqrt{64x}$

30. $\sqrt{9x^5}$

31. $\sqrt[4]{162x^9y^{12}}$

32. $\sqrt[3]{-40x^{14}y^{10}}$

33. $\sqrt{15x^2}\sqrt{5x}$

34. $\sqrt{5x}\sqrt{20x^3}$

35. $\left(\sqrt{5}\sqrt[3]{9}\right)^2$

36. $\left(\sqrt[3]{3}\sqrt{10}\right)^4$

37. $\left(3\sqrt{6}\right)\left(2\sqrt{2}\right)$

38. $\left(5\sqrt{8}\right)\left(-3\sqrt{3}\right)$

39. $3\sqrt{2} + 4\sqrt{2}$

40. $6\sqrt{5} - 4\sqrt{5}$

41. $-\sqrt{48} + 5\sqrt{12}$

42. $2\sqrt{12} - 3\sqrt{27}$

43. $\left(\sqrt{3} + 3\right)\left(\sqrt{3} - 1\right)$

44. $\left(\sqrt{5} - 2\right)\left(\sqrt{5} + 3\right)$

45. $5\sqrt[3]{2} - 2\sqrt[3]{54}$

46. $9\sqrt[3]{24} - \sqrt[3]{81}$

47. $\left(\sqrt{x} - 1\right)^2$

48. $\left(\sqrt{x} + \sqrt{5}\right)^2$

49. $\sqrt[3]{16x^4} - \sqrt[3]{2x}$

50. $\sqrt[4]{32x} + \sqrt[4]{2x^5}$

51. $\sqrt{8x^3} - 3\sqrt{50x}$

52. $3x\sqrt{9y} + 4\sqrt{25y}$

53. $\sqrt[3]{16x^4y} - 3x\sqrt[3]{2xy} + 5\sqrt[3]{-2xy^4}$

54. $8xy - \sqrt{25x^2y^2} + \sqrt[3]{8x^3y^3}$

In Problems 55–68, rationalize the denominator of each expression. Assume that all variables are positive when they appear.

55. $\dfrac{1}{\sqrt{2}}$

56. $\dfrac{2}{\sqrt{3}}$

57. $\dfrac{-\sqrt{3}}{\sqrt{5}}$

58. $\dfrac{-\sqrt{3}}{\sqrt{8}}$

59. $\dfrac{\sqrt{3}}{5 - \sqrt{2}}$

60. $\dfrac{\sqrt{2}}{\sqrt{7} + 2}$

61. $\dfrac{2 - \sqrt{5}}{2 + 3\sqrt{5}}$

62. $\dfrac{\sqrt{3} - 1}{2\sqrt{3} + 3}$

63. $\dfrac{5}{\sqrt{2} - 1}$

64. $\dfrac{-3}{\sqrt{5} + 4}$

65. $\dfrac{5}{\sqrt[3]{2}}$

66. $\dfrac{-2}{\sqrt[3]{9}}$

67. $\dfrac{\sqrt{x+h} - \sqrt{x}}{\sqrt{x+h} + \sqrt{x}}$

68. $\dfrac{\sqrt{x+h} + \sqrt{x-h}}{\sqrt{x+h} - \sqrt{x-h}}$

In Problems 69–76, rationalize the numerator of each expression. Assume that all variables are positive when they appear.

69. $\dfrac{\sqrt{11}+1}{2}$

70. $\dfrac{5-\sqrt{43}}{3}$

71. $\dfrac{\sqrt{6}-\sqrt{15}}{\sqrt{15}}$

72. $\dfrac{\sqrt{5}+\sqrt{3}}{\sqrt{5}}$

73. $\dfrac{\sqrt{x}-\sqrt{c}}{x-c}$ $x \neq c$

74. $\dfrac{\sqrt{x}-2}{x-4}$ $x \neq 4$

75. $\dfrac{\sqrt{x-7}-1}{x-8}$ $x \neq 8$

76. $\dfrac{4-\sqrt{x-9}}{x-25}$ $x \neq 25$

In Problems 77–80, solve each equation.

77. $\sqrt[3]{2t-1}=2$

78. $\sqrt[3]{3t+1}=-2$

79. $\sqrt{15-2x}=x$

80. $\sqrt{12-x}=x$

In Problems 81–96, simplify each expression.

81. $8^{2/3}$

82. $4^{3/2}$

83. $(-64)^{1/3}$

84. $16^{3/4}$

85. $100^{3/2}$

86. $25^{3/2}$

87. $4^{-3/2}$

88. $16^{-3/2}$

89. $\left(\dfrac{9}{8}\right)^{3/2}$

90. $\left(\dfrac{27}{8}\right)^{2/3}$

91. $\left(\dfrac{8}{9}\right)^{-3/2}$

92. $\left(\dfrac{8}{27}\right)^{-2/3}$

93. $(-1000)^{-1/3}$

94. $-25^{-1/2}$

95. $\left(-\dfrac{64}{125}\right)^{-2/3}$

96. $-81^{-3/4}$

In Problems 97–104, simplify each expression. Express your answer so that only positive exponents occur. Assume that the variables are positive.

97. $x^{3/4}x^{1/3}x^{-1/2}$

98. $x^{2/3}x^{1/2}x^{-1/4}$

99. $(x^3y^6)^{1/3}$

100. $(x^4y^8)^{3/4}$

101. $\dfrac{(x^2y)^{1/3}(xy^2)^{2/3}}{x^{2/3}y^{2/3}}$

102. $\dfrac{(xy)^{1/4}(x^2y^2)^{1/2}}{(x^2y)^{3/4}}$

103. $\dfrac{(16x^2y^{-1/3})^{3/4}}{(xy^2)^{1/4}}$

104. $\dfrac{(4x^{-1}y^{1/3})^{3/2}}{(xy)^{3/2}}$

Applications and Extensions

In Problems 105–118, expressions that occur in calculus are given. Write each expression as a single quotient in which only positive exponents and/or radicals appear.

105. $\dfrac{x}{(1+x)^{1/2}}+2(1+x)^{1/2}$ $x>-1$

106. $\dfrac{1+x}{2x^{1/2}}+x^{1/2}$ $x>0$

107. $2x(x^2+1)^{1/2}+x^2\cdot\dfrac{1}{2}(x^2+1)^{-1/2}\cdot 2x$

108. $(x+1)^{1/3}+x\cdot\dfrac{1}{3}(x+1)^{-2/3}$ $x \neq -1$

109. $\sqrt{4x+3}\cdot\dfrac{1}{2\sqrt{x-5}}+\sqrt{x-5}\cdot\dfrac{1}{5\sqrt{4x+3}}$ $x>5$

110. $\dfrac{\sqrt[3]{8x+1}}{3\sqrt[3]{(x-2)^2}}+\dfrac{\sqrt[3]{x-2}}{24\sqrt[3]{(8x+1)^2}}$ $x \neq 2, x \neq -\dfrac{1}{8}$

111. $\dfrac{\sqrt{1+x}-x\cdot\dfrac{1}{2\sqrt{1+x}}}{1+x}$ $x>-1$

112. $\dfrac{\sqrt{x^2+1}-x\cdot\dfrac{2x}{2\sqrt{x^2+1}}}{x^2+1}$

113. $\dfrac{(x+4)^{1/2}-2x(x+4)^{-1/2}}{x+4}$ $x>-4$

114. $\dfrac{(9-x^2)^{1/2}+x^2(9-x^2)^{-1/2}}{9-x^2}$ $-3<x<3$

115. $\dfrac{\dfrac{x^2}{(x^2-1)^{1/2}}-(x^2-1)^{1/2}}{x^2}$ $x<-1$ or $x>1$

116. $\dfrac{(x^2+4)^{1/2}-x^2(x^2+4)^{-1/2}}{x^2+4}$

117. $\dfrac{\dfrac{1+x^2}{2\sqrt{x}}-2x\sqrt{x}}{(1+x^2)^2}$ $x>0$

118. $\dfrac{2x(1-x^2)^{1/3}+\dfrac{2}{3}x^3(1-x^2)^{-2/3}}{(1-x^2)^{2/3}}$ $x \neq -1, x \neq 1$

In Problems 119–128, expressions that occur in calculus are given. Factor each expression. Express your answer so that only positive exponents occur.

119. $(x+1)^{3/2}+x\cdot\dfrac{3}{2}(x+1)^{1/2}$ $x \geq -1$

120. $(x^2+4)^{4/3}+x\cdot\dfrac{4}{3}(x^2+4)^{1/3}\cdot 2x$

121. $6x^{1/2}(x^2+x)-8x^{3/2}-8x^{1/2}$ $x \geq 0$

122. $6x^{1/2}(2x+3)+x^{3/2}\cdot 8$ $x \geq 0$

123. $3(x^2+4)^{4/3}+x\cdot 4(x^2+4)^{1/3}\cdot 2x$

124. $2x(3x+4)^{4/3}+x^2\cdot 4(3x+4)^{1/3}$

125. $4(3x+5)^{1/3}(2x+3)^{3/2}+3(3x+5)^{4/3}(2x+3)^{1/2}$ $x \geq -\dfrac{3}{2}$

126. $6(6x+1)^{1/3}(4x-3)^{3/2}+6(6x+1)^{4/3}(4x-3)^{1/2}$ $x \geq \dfrac{3}{4}$

127. $3x^{-1/2}+\dfrac{3}{2}x^{1/2}$ $x>0$

128. $8x^{1/3}-4x^{-2/3}$ $x \neq 0$

In Problems 129–136, use a calculator to approximate each radical. Round your answer to two decimal places.

129. $\sqrt{2}$ **130.** $\sqrt{7}$ **131.** $\sqrt[3]{4}$ **132.** $\sqrt[3]{-5}$

133. $\dfrac{2 + \sqrt{3}}{3 - \sqrt{5}}$ **134.** $\dfrac{\sqrt{5} - 2}{\sqrt{2} + 4}$ **135.** $\dfrac{3\sqrt[3]{5} - \sqrt{2}}{\sqrt{3}}$ **136.** $\dfrac{2\sqrt{3} - \sqrt[3]{4}}{\sqrt{2}}$

137. Calculating the Amount of Gasoline in a Tank A Shell station stores its gasoline in underground tanks that are right circular cylinders lying on their sides. See the figure. The volume V of gasoline in the tank (in gallons) is given by the formula

$$V = 40h^2 \sqrt{\frac{96}{h} - 0.608}$$

where h is the height of the gasoline (in inches) as measured on a depth stick.
(a) If $h = 12$ inches, how many gallons of gasoline are in the tank?
(b) If $h = 1$ inch, how many gallons of gasoline are in the tank?

138. Inclined Planes The final velocity v of an object in feet per second (ft/s) after it slides down a frictionless inclined plane of height h feet is

$$v = \sqrt{64h + v_0^2}$$

where v_0 is the initial velocity (in ft/s) of the object.
(a) What is the final velocity v of an object that slides down a frictionless inclined plane of height 4 feet? Assume that the initial velocity is 0.
(b) What is the final velocity v of an object that slides down a frictionless inclined plane of height 16 feet? Assume that the initial velocity is 0.
(c) What is the final velocity v of an object that slides down a frictionless inclined plane of height 2 feet with an initial velocity of 4 ft/s?

'Are You Prepared?' Answers

1. $9; -9$ **2.** $4; 4$

A.8 Lines

OBJECTIVES
  1 Calculate and Interpret the Slope of a Line (p. A64)
  2 Graph Lines Given a Point and the Slope (p. A67)
  3 Find the Equation of a Vertical Line (p. A68)
  4 Use the Point-Slope Form of a Line; Identify Horizontal Lines (p. A68)
  5 Use the Slope-Intercept Form of a Line (p. A69)
  6 Find an Equation of a Line Given Two Points (p. A71)
  7 Graph Lines Written in General Form Using Intercepts (p. A71)
  8 Find Equations of Parallel Lines (p. A72)
  9 Find Equations of Perpendicular Lines (p. A73)

In this section we study a certain type of equation that contains two variables, called a *linear equation*, and its graph, a *line*.

1 Calculate and Interpret the Slope of a Line

Figure 31

Consider the staircase illustrated in Figure 31. Each step contains exactly the same horizontal **run** and the same vertical **rise**. The ratio of the rise to the run, called the *slope*, is a numerical measure of the steepness of the staircase. For example, if the run is increased and the rise remains the same, the staircase becomes less steep. If the run is kept the same but the rise is increased, the staircase becomes more steep. This important characteristic of a line is best defined using rectangular coordinates.

> **DEFINITION Slope**
>
> Let $P = (x_1, y_1)$ and $Q = (x_2, y_2)$ be two distinct points. If $x_1 \neq x_2$, the **slope m** of the nonvertical line L containing P and Q is defined by the formula
>
> $$m = \frac{y_2 - y_1}{x_2 - x_1} \qquad x_1 \neq x_2 \tag{1}$$
>
> If $x_1 = x_2$, then L is a **vertical line** and the slope m of L is **undefined** (since this results in division by 0).

Figure 32(a) illustrates the slope of a nonvertical line; Figure 32(b) illustrates a vertical line.

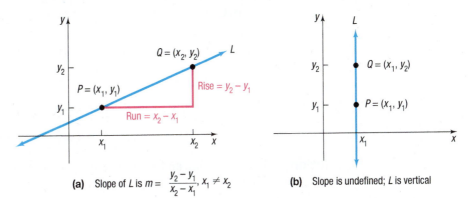

(a)   Slope of L is $m = \dfrac{y_2 - y_1}{x_2 - x_1}$, $x_1 \neq x_2$ (b)   Slope is undefined; L is vertical

Figure 32

> **In Words**
> The symbol Δ is the Greek uppercase letter delta. In mathematics, Δ is read "change in," so $\dfrac{\Delta y}{\Delta x}$ is read "change in y divided by change in x."

As Figure 32(a) illustrates, the slope m of a nonvertical line may be viewed as

$$m = \frac{y_2 - y_1}{x_2 - x_1} = \frac{\text{Rise}}{\text{Run}} \quad \text{or as} \quad m = \frac{y_2 - y_1}{x_2 - x_1} = \frac{\text{Change in } y}{\text{Change in } x} = \frac{\Delta y}{\Delta x}$$

That is, the slope m of a nonvertical line measures the amount y changes when x changes from x_1 to x_2. The expression $\dfrac{\Delta y}{\Delta x}$ is called the **average rate of change** of y with respect to x.

Two comments about computing the slope of a nonvertical line may prove helpful:

- Any two distinct points on the line can be used to compute the slope of the line. (See Figure 33 for justification.) Since any two distinct points can be used to compute the slope of a line, the average rate of change of a line is always the same number.

Figure 33
Triangles ABC and PQR are similar (equal angles), so ratios of corresponding sides are proportional. Then the slope using P and Q is

$$\frac{y_2 - y_1}{x_2 - x_1} = \frac{d(B, C)}{d(A, C)},$$

which is the slope using A and B.

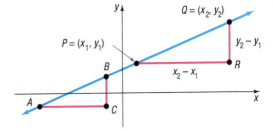

- The slope of a line may be computed from $P = (x_1, y_1)$ to $Q = (x_2, y_2)$ or from Q to P because

$$\frac{y_2 - y_1}{x_2 - x_1} = \frac{y_1 - y_2}{x_1 - x_2}$$

EXAMPLE 1   **Finding and Interpreting the Slope of a Line Given Two Points**

The slope m of the line containing the points $(1, 2)$ and $(5, -3)$ may be computed as

$$m = \frac{-3 - 2}{5 - 1} = \frac{-5}{4} = -\frac{5}{4} \quad \text{or as} \quad m = \frac{2 - (-3)}{1 - 5} = \frac{5}{-4} = -\frac{5}{4}$$

For every 4-unit change in x, y will change by -5 units. That is, if x increases by 4 units, then y will decrease by 5 units. The average rate of change of y with respect to x is $-\dfrac{5}{4}$.

— **Now Work** PROBLEMS **13** AND **19**

EXAMPLE 2

Finding the Slopes of Various Lines Containing the Same Point (2, 3)

Compute the slopes of the lines L_1, L_2, L_3, and L_4 containing the following pairs of points. Graph all four lines on the same set of coordinate axes.

$$L_1: \quad P = (2,3) \qquad Q_1 = (-1,-2)$$
$$L_2: \quad P = (2,3) \qquad Q_2 = (3,-1)$$
$$L_3: \quad P = (2,3) \qquad Q_3 = (5,3)$$
$$L_4: \quad P = (2,3) \qquad Q_4 = (2,5)$$

Solution Let m_1, m_2, m_3, and m_4 denote the slopes of the lines L_1, L_2, L_3, and L_4, respectively. Then

$$m_1 = \frac{-2-3}{-1-2} = \frac{-5}{-3} = \frac{5}{3} \quad \text{A rise of 5 divided by a run of 3}$$

$$m_2 = \frac{-1-3}{3-2} = \frac{-4}{1} = -4$$

$$m_3 = \frac{3-3}{5-2} = \frac{0}{3} = 0$$

m_4 is undefined because $x_1 = x_2 = 2$.

The graphs of these lines are in Figure 34.

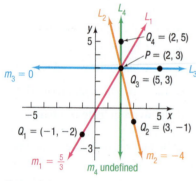

Figure 34

Figure 34 illustrates the following facts:

- When the slope of a line is positive, the line slants upward from left to right (L_1).
- When the slope of a line is negative, the line slants downward from left to right (L_2).
- When the slope is 0, the line is horizontal (L_3).
- When the slope is undefined, the line is vertical (L_4).

Seeing the Concept

On the same screen, graph the following equations:

Figure 35

$Y_1 = 0$	Slope of line is 0.
$Y_2 = \dfrac{1}{4}x$	Slope of line is $\dfrac{1}{4}$.
$Y_3 = \dfrac{1}{2}x$	Slope of line is $\dfrac{1}{2}$.
$Y_4 = x$	Slope of line is 1.
$Y_5 = 2x$	Slope of line is 2.
$Y_6 = 6x$	Slope of line is 6.

See Figure 35.

Seeing the Concept

On the same screen, graph the following equations:

Figure 36

$Y_1 = 0$ **Slope of line is 0.**

$Y_2 = -\dfrac{1}{4}x$ **Slope of line is $-\dfrac{1}{4}$.**

$Y_3 = -\dfrac{1}{2}x$ **Slope of line is $-\dfrac{1}{2}$.**

$Y_4 = -x$ **Slope of line is -1.**

$Y_5 = -2x$ **Slope of line is -2.**

$Y_6 = -6x$ **Slope of line is -6.**

See Figure 36.

Figures 35 and 36 illustrate that the closer the line is to the vertical position, the greater the magnitude of the slope.

2 Graph Lines Given a Point and the Slope

EXAMPLE 3

Graphing a Line Given a Point and a Slope

Draw a graph of the line that contains the point $(3, 2)$ and has a slope of:

(a) $\dfrac{3}{4}$ (b) $-\dfrac{4}{5}$

Solution

(a) Slope $= \dfrac{\text{Rise}}{\text{Run}}$. The slope $\dfrac{3}{4}$ means that for every horizontal change (run) of 4 units to the right, there is a vertical change (rise) of 3 units. Start at the point $(3, 2)$ and move 4 units to the right and 3 units up, arriving at the point $(7, 5)$. Drawing the line through the points $(7, 5)$ and $(3, 2)$ gives the graph. See Figure 37.

Figure 37

(b) A slope of

$$-\frac{4}{5} = \frac{-4}{5} = \frac{\text{Rise}}{\text{Run}}$$

means that for every horizontal change of 5 units to the right, there is a corresponding vertical change of -4 units (a downward movement). Start at the point $(3, 2)$ and move 5 units to the right and then 4 units down, arriving at the point $(8, -2)$. Drawing the line through these points gives the graph. See Figure 38.

Alternatively, consider that

$$-\frac{4}{5} = \frac{4}{-5} = \frac{\text{Rise}}{\text{Run}}$$

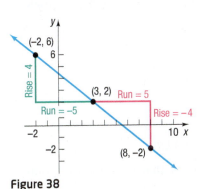

Figure 38

so for every horizontal change of -5 units (a movement to the left), there is a corresponding vertical change of 4 units (upward). This approach leads to the point $(-2, 6)$, which is also on the graph of the line in Figure 38.

Now Work PROBLEM 25

3 Find the Equation of a Vertical Line

EXAMPLE 4

Graphing a Line

Graph the equation: $x = 3$

Solution

To graph $x = 3$, we find all points (x, y) in the plane for which $x = 3$. No matter what y-coordinate is used, the corresponding x-coordinate always equals 3. Consequently, the graph of the equation $x = 3$ is a vertical line with x-intercept 3 and an undefined slope. See Figure 39 .

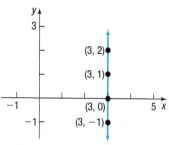

Figure 39 $x = 3$

Example 4 suggests the following result:

> **THEOREM Equation of a Vertical Line**
>
> A vertical line is given by an equation of the form
>
> $$x = a$$
>
> where a is the x-intercept.

COMMENT To graph an equation using most graphing utilities, we need to express the equation in the form $y = \{$ *expression in x* $\}$. But $x = 3$ cannot be put in this form. To overcome this, most graphing utilities have special commands for drawing vertical lines. DRAW, LINE, PLOT, and VERT are among the more common ones. Consult your manual to determine the correct methodology for your graphing utility. ∎

4 Use the Point-Slope Form of a Line; Identify Horizontal Lines

Let L be a nonvertical line with slope m that contains the point (x_1, y_1). See Figure 40. For any other point (x, y) on L, we have

$$m = \frac{y - y_1}{x - x_1} \quad \text{or} \quad y - y_1 = m(x - x_1)$$

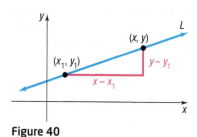

Figure 40

> **THEOREM Point-Slope Form of an Equation of a Line**
>
> An equation of a nonvertical line with slope m that contains the point (x_1, y_1) is
>
> $$y - y_1 = m(x - x_1) \tag{2}$$

EXAMPLE 5

Finding the Point-Slope Form of an Equation of a Line

Find the point-slope form of an equation of the line with slope 4, containing the point $(1, 2)$.

Solution

An equation of the line with slope 4 that contains the point $(1, 2)$ can be found by using the point-slope form with $m = 4$, $x_1 = 1$, and $y_1 = 2$.

$$y - y_1 = m(x - x_1)$$

$$y - 2 = 4(x - 1) \qquad m = 4, x_1 = 1, y_1 = 2$$

See Figure 41 for the graph.

━━━ Now Work PROBLEM 33

Figure 41 $y - 2 = 4(x - 1)$

EXAMPLE 6

Finding the Equation of a Horizontal Line

Find an equation of the horizontal line containing the point $(3, 2)$.

Solution

Because all the y-values are equal on a horizontal line, the slope of a horizontal line is 0. To get an equation, use the point-slope form with $m = 0$, $x_1 = 3$, and $y_1 = 2$.

$$y - y_1 = m(x - x_1)$$

$$y - 2 = 0 \cdot (x - 3) \qquad m = 0, x_1 = 3, y_1 = 2$$

$$y - 2 = 0$$

$$y = 2$$

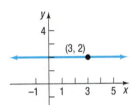

Figure 42 $y = 2$

See Figure 42 for the graph.

Example 6 suggests the following result:

> **THEOREM Equation of a Horizontal Line**
>
> A horizontal line is given by an equation of the form
>
> $$y = b$$
>
> where b is the y-intercept.

5 Use the Slope-Intercept Form of a Line

Another useful equation of a line is obtained when the slope m and y-intercept b are known. Then the point $(0, b)$ is on the line. Using the point-slope form, equation (2), we obtain

$$y - b = m(x - 0) \quad \text{or} \quad y = mx + b$$

> **THEOREM Slope-Intercept Form of an Equation of a Line**
>
> An equation of a line with slope m and y-intercept b is
>
> $$y = mx + b \qquad\qquad (3)$$

For example, if a line has slope 5 and y-intercept 2, we can write the equation in slope-intercept form as

$$y = 5x + 2$$

slope   y-intercept

━━━ Now Work PROBLEMS 53 AND 59 (express answer in slope-intercept form)

Seeing the Concept

To see the role that the slope m plays, graph the following lines on the same screen.

$Y_1 = 2$	$m = 0$
$Y_2 = x + 2$	$m = 1$
$Y_3 = -x + 2$	$m = -1$
$Y_4 = 3x + 2$	$m = 3$
$Y_5 = -3x + 2$	$m = -3$

Figure 43 displays the graphs using Desmos. What do you conclude about lines of the form $y = mx + 2$?

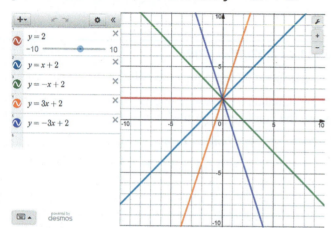

Figure 43 $y = mx + 2$

To see the role of the y-intercept b, graph the following lines on the same screen.

$Y_1 = 2x$	$b = 0$
$Y_2 = 2x + 1$	$b = 1$
$Y_3 = 2x - 1$	$b = -1$
$Y_4 = 2x + 4$	$b = 4$
$Y_5 = 2x - 4$	$b = -4$

Figure 44 displays the graphs using Desmos. What do you conclude about lines of the form $y = 2x + b$?

Figure 44 $y = 2x + b$

When the equation of a line is written in slope-intercept form, it is easy to find the slope m and y-intercept b of the line. For example, suppose that the equation of a line is

$$y = -2x + 7$$

Compare this equation to $y = mx + b$.

$$y = -2x + 7$$
$$y = mx + b$$

The slope of this line is -2 and its y-intercept is 7.

Now Work PROBLEM 79

EXAMPLE 7 | **Finding the Slope and y-Intercept of a Line**

Find the slope m and y-intercept b of the equation $2x + 4y = 8$. Graph the equation.

Solution

To find the slope and y-intercept, write the equation in slope-intercept form by solving for y.

$$2x + 4y = 8$$
$$4y = -2x + 8$$
$$y = -\frac{1}{2}x + 2 \quad y = mx + b$$

The coefficient of x, $-\frac{1}{2}$, is the slope, and the constant, 2, is the y-intercept. To graph the line with y-intercept 2 and with slope $-\frac{1}{2}$, start at the point $(0, 2)$ and move to the right 2 units and then down 1 unit to the point $(2, 1)$. Draw the line through these points. See Figure 45.

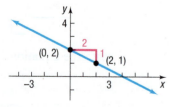

Figure 45 $y = -\frac{1}{2}x + 2$

Now Work PROBLEM 85

6 Find the Equation of a Line Given Two Points

EXAMPLE 8

Finding an Equation of a Line Given Two Points

Find an equation of the line containing the points $(2, 3)$ and $(-4, 5)$. Graph the line.

Solution

First compute the slope of the line.

$$m = \frac{5 - 3}{-4 - 2} = \frac{2}{-6} = -\frac{1}{3} \quad m = \frac{y_2 - y_1}{x_2 - x_1}$$

Use the point $(2, 3)$ and the slope $m = -\dfrac{1}{3}$ to get the point-slope form of the equation of the line.

$$y - 3 = -\frac{1}{3}(x - 2) \quad y - y_1 = m(x - x_1)$$

Continue to get the slope-intercept form.

$$y - 3 = -\frac{1}{3}x + \frac{2}{3}$$

$$y = -\frac{1}{3}x + \frac{11}{3} \quad y = mx + b$$

See Figure 46 for the graph.

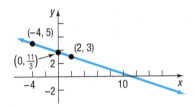

Figure 46 $y = -\dfrac{1}{3}x + \dfrac{11}{3}$

In the solution to Example 8, we could have used the other point, $(-4, 5)$, instead of the point $(2, 3)$. The equation that results, when written in slope-intercept form, is the equation that we obtained in the example. (Try it for yourself.)

Now Work PROBLEM 45

7 Graph Lines Written in General Form Using Intercepts

DEFINITION General Form

The equation of a line is in **general form*** when it is written as

$$Ax + By = C \tag{4}$$

where A, B, and C are real numbers and A and B are not both 0.

If $B = 0$ in equation (4), then $A \neq 0$ and the graph of the equation is a vertical line: $x = \dfrac{C}{A}$.

If $B \neq 0$ in equation (4), then we can solve the equation for y and write the equation in slope-intercept form as we did in Example 7.

One way to graph a line given in general form, equation (4), is to find its intercepts. Remember, the intercepts of the graph of an equation are the points where the graph crosses or touches a coordinate axis.

EXAMPLE 9

Graphing an Equation in General Form Using Its Intercepts

Graph the equation $2x + 4y = 8$ by finding its intercepts.

Solution

To obtain the x-intercept, let $y = 0$ in the equation and solve for x.

$$\begin{aligned} 2x + 4y &= 8 \\ 2x + 4 \cdot 0 &= 8 \quad \text{Let } y = 0. \\ 2x &= 8 \\ x &= 4 \quad \text{Divide both sides by 2.} \end{aligned}$$

The x-intercept is 4, and the point $(4, 0)$ is on the graph of the equation.

(continued)

*Some texts use the term **standard form**.

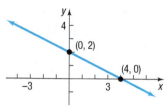

Figure 47 $2x + 4y = 8$

To obtain the y-intercept, let $x = 0$ in the equation and solve for y.

$$2x + 4y = 8$$
$$2 \cdot 0 + 4y = 8 \quad \text{Let } x = 0.$$
$$4y = 8$$
$$y = 2 \quad \text{Divide both sides by 4.}$$

The y-intercept is 2, and the point $(0, 2)$ is on the graph of the equation.

Plot the points $(4, 0)$ and $(0, 2)$ and draw the line through the points. See Figure 47.

━━━ **Now Work** PROBLEM 99

The equation of every line can be written in general form. For example, a vertical line whose equation is

$$x = a$$

can be written in the general form

$$1 \cdot x + 0 \cdot y = a \quad A = 1, B = 0, C = a$$

A horizontal line whose equation is

$$y = b$$

can be written in the general form

$$0 \cdot x + 1 \cdot y = b \quad A = 0, B = 1, C = b$$

Lines that are neither vertical nor horizontal have general equations of the form

$$Ax + By = C \quad A \neq 0 \text{ and } B \neq 0$$

Because the equation of every line can be written in general form, any equation equivalent to equation (4) is called a **linear equation**.

8 Find Equations of Parallel Lines

When two lines (in the plane) do not intersect (that is, they have no points in common), they are **parallel**. Look at Figure 48. We have drawn two parallel lines and have constructed two right triangles by drawing sides parallel to the coordinate axes. The right triangles are similar. (Do you see why? Two angles are equal.) Because the triangles are similar, the ratios of corresponding sides are equal.

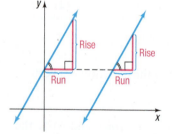

Figure 48 Parallel lines

> **THEOREM Criteria for Parallel Lines**
>
> Two nonvertical lines are parallel if and only if their slopes are equal and they have different y-intercepts.

The use of the phrase "if and only if" in the preceding theorem means that actually two statements are being made, one the converse of the other.

- If two nonvertical lines are parallel, then their slopes are equal and they have different y-intercepts.

- If two nonvertical lines have equal slopes and they have different y-intercepts, then they are parallel.

EXAMPLE 10   **Showing That Two Lines Are Parallel**

Show that the lines given by the following equations are parallel.

$$L_1: \ 2x + 3y = 6 \qquad L_2: \ 4x + 6y = 0$$

Solution   To determine whether these lines have equal slopes and different y-intercepts, write each equation in slope-intercept form.

L_1: $2x + 3y = 6$ L_2: $4x + 6y = 0$

$\qquad\qquad\quad 3y = -2x + 6$ $\qquad\qquad\qquad\qquad\qquad 6y = -4x$

$\qquad\qquad\qquad y = -\dfrac{2}{3}x + 2$ $\qquad\qquad\qquad\qquad\qquad y = -\dfrac{2}{3}x$

$\quad$ Slope $= -\dfrac{2}{3}$; y-intercept $= 2$ $\qquad\qquad$ Slope $= -\dfrac{2}{3}$; y-intercept $= 0$

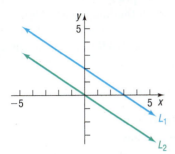

Figure 49 Parallel lines

Because these lines have the same slope, $-\dfrac{2}{3}$, but different y-intercepts, the lines are parallel. See Figure 49.

EXAMPLE 11   **Finding a Line That Is Parallel to a Given Line**

Find an equation for the line that contains the point $(2, -3)$ and is parallel to the line $2x + y = 6$.

Solution   Since the two lines are to be parallel, the slope of the line containing the point $(2, -3)$ equals the slope of the line $2x + y = 6$. Begin by writing the equation of the line $2x + y = 6$ in slope-intercept form.

$$2x + y = 6$$
$$y = -2x + 6$$

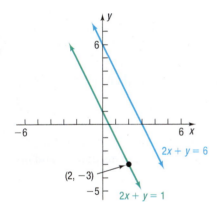

Figure 50

The slope is -2. Since the line containing the point $(2, -3)$ also has slope -2, use the point-slope form to obtain its equation.

$y - y_1 = m(x - x_1)$	**Point-slope form**
$y - (-3) = -2(x - 2)$	$m = -2, x_1 = 2, y_1 = -3$
$y + 3 = -2x + 4$	**Simplify.**
$y = -2x + 1$	**Slope-intercept form**
$2x + y = 1$	**General form**

This line is parallel to the line $2x + y = 6$ and contains the point $(2, -3)$. See Figure 50.

Now Work PROBLEM **67**

9 Find Equations of Perpendicular Lines

When two lines intersect at a right angle (90°), they are **perpendicular**. See Figure 51.

The following result gives a condition, in terms of their slopes, for two lines to be perpendicular.

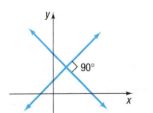

Figure 51 Perpendicular lines

> **THEOREM Criterion for Perpendicular Lines**
>
> Two nonvertical lines are perpendicular if and only if the product of their slopes is -1.

Here we prove the "only if" part of the statement:

> If two nonvertical lines are perpendicular, then the product of their slopes is -1.

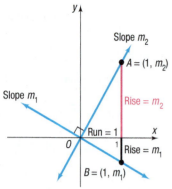

Figure 52

Proof Let m_1 and m_2 denote the slopes of the two lines. There is no loss in generality (that is, neither the angle nor the slopes are affected) if we situate the lines so that they meet at the origin. See Figure 52. The point $A = (1, m_2)$ is on the line having slope m_2, and the point $B = (1, m_1)$ is on the line having slope m_1. (Do you see why this must be true?)

Suppose that the lines are perpendicular. Then triangle OAB is a right triangle. As a result of the Pythagorean Theorem, it follows that

$$[d(O, A)]^2 + [d(O, B)]^2 = [d(A, B)]^2 \qquad (5)$$

Using the distance formula, the squares of these distances are

$$[d(O, A)]^2 = (1 - 0)^2 + (m_2 - 0)^2 = 1 + m_2^2$$
$$[d(O, B)]^2 = (1 - 0)^2 + (m_1 - 0)^2 = 1 + m_1^2$$
$$[d(A, B)]^2 = (1 - 1)^2 + (m_2 - m_1)^2 = m_2^2 - 2m_1 m_2 + m_1^2$$

Using these facts in equation (5), we get

$$(1 + m_2^2) + (1 + m_1^2) = m_2^2 - 2m_1 m_2 + m_1^2$$

which, upon simplification, can be written as

$$m_1 m_2 = -1$$

If the lines are perpendicular, the product of their slopes is -1. ∎

In Problem 136, you are asked to prove the "if" part of the theorem:

> If two nonvertical lines have slopes whose product is -1, then the lines are perpendicular.

You may find it easier to remember the condition for two nonvertical lines to be perpendicular by observing that the equality $m_1 m_2 = -1$ means that m_1 and m_2 are negative reciprocals of each other; that is,

$$m_1 = -\frac{1}{m_2} \quad \text{and} \quad m_2 = -\frac{1}{m_1}$$

EXAMPLE 12 **Finding the Slope of a Line Perpendicular to Another Line**

If a line has slope $\dfrac{3}{2}$, any line having slope $-\dfrac{2}{3}$ is perpendicular to it.

EXAMPLE 13 **Finding an Equation of a Line Perpendicular to a Given Line**

Find an equation of the line that contains the point $(1, -2)$ and is perpendicular to the line $x + 3y = 6$. Graph the two lines.

Solution First write the equation of the line $x + 3y = 6$ in slope-intercept form to find its slope.

$$x + 3y = 6$$
$$3y = -x + 6 \qquad \text{Proceed to solve for y.}$$
$$y = -\frac{1}{3}x + 2 \qquad \text{Place in the form } y = mx + b.$$

The slope of the line $x + 3y = 6$ is $-\dfrac{1}{3}$. Any line perpendicular to this line has slope 3.

Because the point $(1, -2)$ is on the line with slope 3, use the point-slope form of the equation of a line.

$$y - y_1 = m(x - x_1) \quad \text{Point-slope form}$$

$$y - (-2) = 3(x - 1) \quad m = 3, x_1 = 1, y_1 = -2$$

$$y + 2 = 3(x - 1)$$

To obtain other forms of the equation, proceed as follows:

$$y + 2 = 3x - 3 \quad \text{Simplify.}$$

$$y = 3x - 5 \quad \text{Slope-intercept form}$$

$$3x - y = 5 \quad \text{General form}$$

Figure 53

Figure 53 shows the graphs.

Now Work PROBLEM 73

WARNING Be sure to use a square screen when you use a graphing calculator to graph perpendicular lines. Otherwise, the angle between the two lines will appear distorted. A discussion of square screens is given in Section B.5. ∎

A.8 Assess Your Understanding

Concepts and Vocabulary

1. The slope of a vertical line is _____; the slope of a horizontal line is _____.

2. For the line $2x + 3y = 6$, the x-intercept is _____ and the y-intercept is _____.

3. *True or False* The equation $3x + 4y = 6$ is written in general form.

4. *True or False* The slope of the line $2y = 3x + 5$ is 3.

5. *True or False* The point $(1, 2)$ is on the line $2x + y = 4$.

6. Two nonvertical lines have slopes m_1 and m_2, respectively. The lines are parallel if _____ and the _____ are unequal; the lines are perpendicular if _____.

7. The lines $y = 2x + 3$ and $y = ax + 5$ are parallel if $a =$ _____.

8. The lines $y = 2x - 1$ and $y = ax + 2$ are perpendicular if $a =$ _____.

9. *Multiple Choice* If a line slants downward from left to right, then which of the following describes its slope?
(a) positive   (b) zero
(c) negative   (d) undefined

10. *Multiple Choice* Choose the formula for finding the slope m of a nonvertical line that contains the two distinct points (x_1, y_1) and (x_2, y_2).

(a) $m = \dfrac{y_2 - x_2}{y_1 - x_1} \quad x_1 \neq y_1$ (b) $m = \dfrac{y_2 - x_1}{x_2 - y_1} \quad y_1 \neq x_2$

(c) $m = \dfrac{x_2 - x_1}{y_2 - y_1} \quad y_1 \neq y_2$ (d) $m = \dfrac{y_2 - y_1}{x_2 - x_1} \quad x_1 \neq x_2$

11. *Multiple Choice* Choose the correct statement about the graph of the line $y = -3$.
(a) The graph is vertical with x-intercept -3.
(b) The graph is horizontal with y-intercept -3.
(c) The graph is vertical with y-intercept -3.
(d) The graph is horizontal with x-intercept -3.

12. *Multiple Choice* Choose the point-slope equation of a nonvertical line with slope m that passes through the point (x_1, y_1).

(a) $y + y_1 = mx + x_1$ (b) $y - y_1 = mx - x_1$
(c) $y + y_1 = m(x + x_1)$ (d) $y - y_1 = m(x - x_1)$

Skill Building

In Problems 13–16, (a) find the slope of the line and (b) interpret the slope.

13.

14.

15.

16.

In Problems 17–24, plot each pair of points and determine the slope of the line containing the points. Graph the line.

17. $(2, 3); (4, 0)$

18. $(4, 2); (3, 4)$

19. $(-2, 3); (2, 1)$

20. $(-1, 1); (2, 3)$

21. $(-3, -1); (2, -1)$

22. $(4, 2); (-5, 2)$

23. $(-1, 2); (-1, -2)$

24. $(2, 0); (2, 2)$

In Problems 25–32, graph the line that contains the point P and has slope m.

25. $P = (1, 2); m = 3$

26. $P = (2, 1); m = 4$

27. $P = (2, 4); m = -\dfrac{3}{4}$

28. $P = (1, 3); m = -\dfrac{2}{5}$

29. $P = (-1, 3); m = 0$

30. $P = (2, -4); m = 0$

31. $P = (0, 3)$; slope undefined

32. $P = (-2, 0)$; slope undefined

In Problems 33–38, a point on a line and its slope are given. Find the point-slope form of the equation of the line.

33. $P = (1, 2); m = 3$

34. $P = (2, 1); m = 4$

35. $P = (2, 4); m = -\dfrac{3}{4}$

36. $P = (1, 3); m = -\dfrac{2}{5}$

37. $P = (-1, 3); m = 0$

38. $P = (2, -4); m = 0$

In Problems 39–44, the slope and a point on a line are given. Use this information to locate three additional points on the line. Answers may vary. [**Hint:** *It is not necessary to find the equation of the line. See Example 3.*]

39. Slope 4; point $(1, 2)$

40. Slope 2; point $(-2, 3)$

41. Slope $-\dfrac{3}{2}$; point $(2, -4)$

42. Slope $\dfrac{4}{3}$; point $(-3, 2)$

43. Slope -2; point $(-2, -3)$

44. Slope -1; point $(4, 1)$

In Problems 45–52, find an equation of the line L.

45.

46.

47.

48.

49.

L is parallel to $y = 2x$

50.

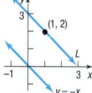

L is parallel to $y = -x$

51.

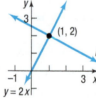

L is perpendicular to $y = 2x$

52.

L is perpendicular to $y = -x$

In Problems 53–78, find an equation for the line with the given properties. Express your answer using either the general form or the slope-intercept form of the equation of a line, whichever you prefer.

53. Slope $= 3$; containing the point $(-2, 3)$

54. Slope $= 2$; containing the point $(4, -3)$

55. Slope $= \dfrac{1}{2}$; containing the point $(3, 1)$

56. Slope $= -\dfrac{2}{3}$; containing the point $(1, -1)$

57. Containing the points $(1, 3)$ and $(-1, 2)$

58. Containing the points $(-3, 4)$ and $(2, 5)$

59. Slope $= -3$; y-intercept $= 3$

60. Slope $= -2$; y-intercept $= -2$

61. x-intercept $= -4$; y-intercept $= 4$

62. x-intercept $= 2$; y-intercept $= -1$

63. Slope undefined; containing the point $(2, 4)$

64. Slope undefined; containing the point $(3, 8)$

65. Horizontal; containing the point $(-3, 2)$

66. Vertical; containing the point $(4, -5)$

67. Parallel to the line $y = 2x$; containing the point $(-1, 2)$

68. Parallel to the line $y = -3x$; containing the point $(-1, 2)$

69. Parallel to the line $x - 2y = -5$; containing the point $(0, 0)$

70. Parallel to the line $2x - y = -2$; containing the point $(0, 0)$

71. Parallel to the line $x = 5$; containing the point $(4, 2)$

72. Parallel to the line $y = 5$; containing the point $(4, 2)$

73. Perpendicular to the line $y = \dfrac{1}{2}x + 4$; containing the point $(1, -2)$

74. Perpendicular to the line $y = 2x - 3$; containing the point $(1, -2)$

75. Perpendicular to the line $x - 2y = -5$; containing the point $(0, 4)$

76. Perpendicular to the line $2x + y = 2$; containing the point $(-3, 0)$

77. Perpendicular to the line $x = 8$; containing the point $(3, 4)$

78. Perpendicular to the line $y = 8$; containing the point $(3, 4)$

In Problems 79–98, find the slope and y-intercept of each line. Graph the line.

79. $y = 2x + 3$

80. $y = -3x + 4$

81. $\frac{1}{2}y = x - 1$

82. $\frac{1}{3}x + y = 2$

83. $y = \frac{1}{2}x + 2$

84. $y = 2x + \frac{1}{2}$

85. $x + 2y = 4$

86. $-x + 3y = 6$

87. $2x - 3y = 6$

88. $3x + 2y = 6$

89. $x + y = 1$

90. $x - y = 2$

91. $x = -4$

92. $y = -1$

93. $y = 5$

94. $x = 2$

95. $y - x = 0$

96. $x + y = 0$

97. $2y - 3x = 0$

98. $3x + 2y = 0$

In Problems 99–108, (a) find the intercepts of the graph of each equation and (b) graph the equation.

99. $2x + 3y = 6$

100. $3x - 2y = 6$

101. $-4x + 5y = 40$

102. $6x - 4y = 24$

103. $7x + 2y = 21$

104. $5x + 3y = 18$

105. $\frac{1}{2}x + \frac{1}{3}y = 1$

106. $x - \frac{2}{3}y = 4$

107. $0.2x - 0.5y = 1$

108. $-0.3x + 0.4y = 1.2$

109. Find an equation of the x-axis.

110. Find an equation of the y-axis.

In Problems 111–114, the equations of two lines are given. Determine whether the lines are parallel, perpendicular, or neither.

111. $y = 2x - 3$
$y = 2x + 4$

112. $y = \frac{1}{2}x - 3$
$y = -2x + 4$

113. $y = 4x + 5$
$y = -4x + 2$

114. $y = -2x + 3$
$y = -\frac{1}{2}x + 2$

In Problems 115–118, write an equation of each line. Express your answer using either the general form or the slope-intercept form of the equation of a line, whichever you prefer.

115.

116.

117.

118.

Applications and Extensions

119. Geometry Use slopes to show that the triangle whose vertices are $(-2, 5)$, $(1, 3)$, and $(-1, 0)$ is a right triangle.

120. Geometry Use slopes to show that the quadrilateral whose vertices are $(1, -1)$, $(4, 1)$, $(2, 2)$, and $(5, 4)$ is a parallelogram.

121. Geometry Use slopes to show that the quadrilateral whose vertices are $(-1, 0)$, $(2, 3)$, $(1, -2)$, and $(4, 1)$ is a rectangle.

122. Geometry Use slopes and the distance formula to show that the quadrilateral whose vertices are $(0, 0)$, $(1, 3)$, $(4, 2)$, and $(3, -1)$ is a square.

123. Truck Rentals A truck rental company rents a moving truck for one day by charging $39 plus $0.60 per mile. Write a linear equation that relates the cost C, in dollars, of renting the truck to the number x of miles driven. What is the cost of renting the truck if the truck is driven 110 miles? 230 miles?

124. Cost Equation The **fixed costs** of operating a business are the costs incurred regardless of the level of production. Fixed costs include rent, fixed salaries, and costs of leasing machinery. The **variable costs** of operating a business are the costs that change with the level of output. Variable costs include raw materials, hourly wages, and electricity. Suppose that a manufacturer of jeans has fixed daily costs of $1200 and variable costs of $20 for each pair of jeans manufactured. Write a linear equation that relates the daily cost C, in dollars, of manufacturing the jeans to the number x of jeans manufactured. What is the cost of manufacturing 400 pairs of jeans? 740 pairs?

125. Cost of Driving a Car The annual fixed costs of owning a small sedan are $4252, assuming the car is completely paid for. The cost to drive the car is approximately $0.14 per mile. Write a linear equation that relates the cost C and the number x of miles driven annually.

Source: AAA, 2017

126. Wages of a Car Salesperson Dan receives $525 per week for selling new and used cars. He also receives 5% of the profit on any sales. Write a linear equation that represents Dan's weekly salary S when he has sales that generate a profit of x dollars.

127. Electricity Rates in Florida Florida Power & Light Company supplies electricity to residential customers for a monthly customer charge of $8.01 plus 8.89 cents per kilowatt hour (kWh) for up to 1000 kilowatt hours.

(a) Write a linear equation that relates the monthly charge C, in dollars, to the number x of kilowatt hours used in a month, $0 \leq x \leq 1000$.

(b) Graph this equation.

(c) What is the monthly charge for using 200 kilowatt hours?

(d) What is the monthly charge for using 500 kilowatt hours?

(e) Interpret the slope of the line.

Source: Florida Power & Light Company, March 2018

128. Natural Gas Rates in Illinois Ameren Illinois supplies natural gas to residential customers for a monthly customer charge of $21.82 plus 64.9 cents per therm of heat energy.

(a) Write a linear equation that relates the monthly charge C, in dollars, to the number x of therms used in a month.

(b) What is the monthly charge for using 90 therms?

(c) What is the monthly charge for using 150 therms?

(d) Graph the equation.

(e) Interpret the slope of the line.

Source: Ameren Illinois, March 2018

129. Measuring Temperature The relationship between Celsius (°C) and Fahrenheit (°F) degrees of measuring temperature is linear. Find a linear equation relating °C and °F if 0°C corresponds to 32°F and 100°C corresponds to 212°F. Use the equation to find the Celsius measure of 70°F.

130. Measuring Temperature The Kelvin (K) scale for measuring temperature is obtained by adding 273 to the Celsius temperature.

(a) Write a linear equation relating K and °C.

(b) Write a linear equation relating K and °F (see Problem 129).

131. Access Ramp A wooden access ramp is being built to reach a platform that sits 30 inches above the floor. The ramp drops 2 inches for every 25-inch run.

(a) Write a linear equation that relates the height y of the ramp above the floor to the horizontal distance x from the platform.

(b) Find and interpret the x-intercept of the graph of your equation.

(c) Design requirements stipulate that the maximum run be 30 feet and that the maximum slope be a drop of 1 inch for each 12 inches of run. Will this ramp meet the requirements? Explain.

(d) What slopes could be used to obtain the 30-inch rise and still meet design requirements?

Source: www.adaptiveaccess.com/wood_ramps.php

132. U.S. Advertising Share A report showed that Internet ads accounted for 19% of all U.S. advertisement spending when print ads (magazines and newspapers) accounted for 26% of the spending. The report further showed that Internet ads accounted for 35% of all advertisement spending when print ads accounted for 16% of the spending.

(a) Write a linear equation that relates that percent y of print ad spending to the percent x of Internet ad spending.

(b) Find the intercepts of the graph of your equation.

(c) Do the intercepts have any meaningful interpretation?

(d) Predict the percent of print ad spending if Internet ads account for 39% of all advertisement spending in the United States.

Source: Marketing Fact Pack 2018. Ad Age Datacenter, December 18, 2017

133. Product Promotion A cereal company finds that the number of people who will buy one of its products in the first month that the product is introduced is linearly related to the amount of money it spends on advertising. If it spends $40,000 on advertising, then 100,000 boxes of cereal will be sold, and if it spends $60,000, then 200,000 boxes will be sold.

(a) Write a linear equation that relates the amount A spent on advertising to the number x of boxes the company aims to sell.

(b) How much expenditure on advertising is needed to sell 300,000 boxes of cereal?

(c) Interpret the slope.

134. The equation $2x - y = C$ defines a **family of lines**, one line for each value of C. On one set of coordinate axes, graph the members of the family when $C = -4$, $C = 0$, and $C = 2$. Can you draw a conclusion from the graph about each member of the family?

135. Find three numbers a for which the lines $x + 2y = 5$, $2x - 3y + 4 = 0$, and $ax + y = 0$ do not form a triangle.

136. Prove that if two nonvertical lines have slopes whose product is -1, then the lines are perpendicular. [**Hint:** Refer to Figure 52 and use the converse of the Pythagorean Theorem.]

Explaining Concepts: Discussion and Writing

137. Which of the following equations might have the graph shown? (More than one answer is possible.)

(a) $2x + 3y = 6$ (b) $-2x + 3y = 6$

(c) $3x - 4y = -12$ (d) $x - y = 1$

(e) $x - y = -1$ (f) $y = 3x - 5$

(g) $y = 2x + 3$ (h) $y = -3x + 3$

138. Which of the following equations might have the graph shown? (More than one answer is possible.)

(a) $2x + 3y = 6$ (b) $2x - 3y = 6$

(c) $3x + 4y = 12$ (d) $x - y = 1$

(e) $x - y = -1$ (f) $y = -2x - 1$

(g) $y = -\dfrac{1}{2}x + 10$ (h) $y = x + 4$

139. The figure shows the graph of two parallel lines. Which of the following pairs of equations might have such a graph?

(a) $x - 2y = 3$
    $x + 2y = 7$

(b) $x + y = 2$
    $x + y = -1$

(c) $x - y = -2$
    $x - y = 1$

(d) $x - y = -2$
    $2x - 2y = -4$

(e) $x + 2y = 2$
    $x + 2y = -1$

140. The figure shows the graph of two perpendicular lines. Which of the following pairs of equations might have such a graph?

(a) $y - 2x = 2$
    $y + 2x = -1$

(b) $y - 2x = 0$
    $2y + x = 0$

(c) $2y - x = 2$
    $2y + x = -2$

(d) $y - 2x = 2$
    $x + 2y = -1$

(e) $2x + y = -2$
    $2y + x = -2$

141. *m* **is for Slope** The accepted symbol used to denote the slope of a line is the letter m. Investigate the origin of this practice. Begin by consulting a French dictionary and looking up the French word *monter*. Write a brief essay on your findings.

142. **Grade of a Road** The term *grade* is used to describe the inclination of a road. How is this term related to the notion of slope of a line? Is a 4% grade very steep? Investigate the grades of some mountainous roads and determine their slopes. Write a brief essay on your findings.

143. **Carpentry** Carpenters use the term *pitch* to describe the steepness of staircases and roofs. How is pitch related to slope? Investigate typical pitches used for stairs and for roofs. Write a brief essay on your findings.

144. Can the equation of every line be written in slope-intercept form? Why?

145. Does every line have exactly one x-intercept and one y-intercept? Are there any lines that have no intercepts?

146. What can you say about two lines that have equal slopes and equal y-intercepts?

147. What can you say about two lines with the same x-intercept and the same y-intercept? Assume that the x-intercept is not 0.

148. If two distinct lines have the same slope but different x-intercepts, can they have the same y-intercept?

149. If two distinct lines have the same y-intercept but different slopes, can they have the same x-intercept?

150. Which form of the equation of a line do you prefer to use? Justify your position with an example that shows that your choice is better than another. Have reasons.

151. **What Went Wrong?** A student is asked to find the slope of the line joining $(-3, 2)$ and $(1, -4)$. He states that the slope is $\dfrac{3}{2}$. Is he correct? If not, what went wrong?

Graphing Utilities

Outline

B.1 The Viewing Rectangle

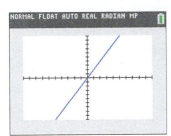

(a) $y = 2x$ on a TI-84 Plus C

(b) $y = 2x$ using Desmos

Figure 1

All graphing utilities (that is, all graphing calculators and all computer software graphing packages) graph equations by plotting points on a screen. The screen itself actually consists of small rectangles called **pixels**. The more pixels the screen has, the better the resolution. Most graphing calculators have 50 to 100 pixels per inch; most smartphones have 300 to 450 pixels per inch. When a point to be plotted lies inside a pixel, the pixel is turned on (lights up). The graph of an equation is a collection of pixels. Figure 1(a) shows how the graph of $y = 2x$ looks on a TI-84 Plus C graphing calculator, and Figure 1(b) shows the same graph using Desmos.com.

The screen of a graphing utility displays the coordinate axes of a rectangular coordinate system. However, the scale must be set on each axis. The smallest and largest values of x and y to be included in the graph must also be set. This is called **setting the viewing rectangle** or **viewing window**. Figure 2 shows a typical viewing window on a TI-84 Plus C.

To select the viewing window, values must be given to the following expressions:

Xmin:	the smallest value of x
Xmax:	the largest value of x
Xscl:	the number of units per tick mark on the x-axis
Ymin:	the smallest value of y
Ymax:	the largest value of y
Yscl:	the number of units per tick mark on the y-axis

Figure 3 illustrates these settings and their relation to the Cartesian coordinate system.

Figure 2 Viewing window on a TI-84 Plus C

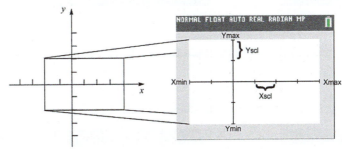

Figure 3

If the scale used on each axis is known, the minimum and maximum values of x and y shown on the screen can be determined by counting the tick marks. Look again at Figure 2. For a scale of 1 on each axis, the minimum and maximum values of x are -10 and 10, respectively; the minimum and maximum values of y are

Figure 4

also −10 and 10. If the scale is 2 on each axis, then the minimum and maximum values of *x* are −20 and 20, respectively; and the minimum and maximum values of *y* are −20 and 20, respectively.

Conversely, if the minimum and maximum values of *x* and *y* are known, the scales can be determined by counting the tick marks displayed. This text follows the practice of showing the minimum and maximum values of *x* and *y* in illustrations so that the reader will know how the viewing window was set. See Figure 4. The numbers outside of the viewing window stand for

$$X\text{min} = -3, \quad X\text{max} = 3, \quad X\text{scl} = 1$$
$$Y\text{min} = -4, \quad Y\text{max} = 4, \quad Y\text{scl} = 2$$

| EXAMPLE 1 | **Finding the Coordinates of a Point Shown on a Graphing Utility Screen** |

Find the coordinates of the point shown in Figure 5. Assume that the coordinates are integers.

Figure 5

Solution First note that the viewing window used in Figure 5 is

$$X\text{min} = -3, \quad X\text{max} = 3, \quad X\text{scl} = 1$$
$$Y\text{min} = -4, \quad Y\text{max} = 4, \quad Y\text{scl} = 2$$

The point shown is 2 tick units to the left of the origin on the horizontal axis (scale = 1) and 1 tick up on the vertical axis (scale = 2). The coordinates of the point shown are (−2, 2).

B.1 Exercises

In Problems 1–4, determine the coordinates of the points shown. Tell in which quadrant each point lies. Assume that the coordinates are integers.

1.

2.

3.

4.

In Problems 5–10, determine the viewing window used.

5.

6.

7.

8.

9.

10.

In Problems 11–16, select a setting so that each of the given points will lie within the viewing rectangle.

11. $(-10, 5), (3, -2), (4, -1)$

12. $(5, 0), (6, 8), (-2, -3)$

13. $(40, 20), (-20, -80), (10, 40)$

14. $(-80, 60), (20, -30), (-20, -40)$

15. $(0, 0), (100, 5), (5, 150)$

16. $(0, -1), (100, 50), (-10, 30)$

B.2 Using a Graphing Utility to Graph Equations

From Examples 2 and 3 in Section 1.2, recall that one way a graph can be obtained is by plotting points in a rectangular coordinate system and connecting them. Graphing utilities perform these same steps when graphing an equation. For example, the TI-84 Plus C determines 265 evenly spaced input values,* starting at Xmin and ending at Xmax; uses the equation to determine the output values; plots these points on the screen; and finally (if in the connected mode) draws a line between consecutive points.

To graph an equation in two variables x and y using a graphing utility often requires that the equation be written explicitly in the form $y = \{expression\ in\ x\}$. If the original equation is not in this form, replace it by equivalent equations until the form $y = \{expression\ in\ x\}$ is obtained.

Steps for Graphing an Equation Using a Graphing Utility

STEP 1: Solve the equation for y in terms of x.

STEP 2: Get into the graphing mode of the graphing utility. The screen will usually display $Y_1 = $ ____ , prompting you to enter the expression involving x found in Step 1. (Consult your manual for the correct way to enter the expression; for example, $y = x^2$ might be entered as $x\char`^2$ or as $x*x$ or as $x\ x^y\ 2$.)

STEP 3: Select the viewing window. Without prior knowledge about the behavior of the graph of the equation, it is common to select the **standard viewing window**** initially. The viewing window is then adjusted based on the graph that appears. In this text the standard viewing window is

$$X\text{min} = -10 \quad X\text{max} = 10 \quad X\text{scl} = 1$$
$$Y\text{min} = -10 \quad Y\text{max} = 10 \quad Y\text{scl} = 1$$

STEP 4: Graph.

STEP 5: Adjust the viewing window until a complete graph is obtained.

Figure 6 $6x^2 + 3y = 36$

NOTE Some graphing utilities allow input of implicit equations. For example, Figure 6 shows the graph of $6x^2 + 3y = 36$ using Desmos. ∎

*These input values depend on the values of Xmin and Xmax. For example, if Xmin $= -10$ and Xmax $= 10$, then the first input value will be -10 and the next input value will be $-10 + \dfrac{10 - (-10)}{264} = -9.9242$, and so on.

**Some graphing utilities have a ZOOM-STANDARD feature that automatically sets the viewing window to the standard viewing window and graphs the equation.

EXAMPLE 1   Graphing an Equation on a Graphing Utility

Graph the equation $6x^2 + 3y = 36$.

Solution   **STEP 1:** Solve for y in terms of x.

$$6x^2 + 3y = 36$$
$$3y = -6x^2 + 36 \quad \text{Subtract } 6x^2 \text{ from both sides.}$$
$$y = -2x^2 + 12 \quad \text{Divide both sides by 3 and simplify.}$$

STEP 2: From the $Y_1 =$ screen, enter the expression $-2x^2 + 12$ after the prompt.

STEP 3: Set the viewing window to the standard viewing window.

STEP 4: Graph. The screen should look like Figure 7.

STEP 5: The graph of $y = -2x^2 + 12$ is not complete. The value of Ymax must be increased so that the top portion of the graph is visible. After increasing the value of Ymax to 12, we obtain the graph in Figure 8. The graph is now complete.

Figure 7 $y = -2x^2 + 12$

Figure 8 $y = -2x^2 + 12$

Figure 9 $y = -2x^2 + 12$

Look again at Figure 8. Although a complete graph is shown, the graph might be improved by adjusting the values of Xmin and Xmax. Figure 9 shows the graph of $y = -2x^2 + 12$ using Xmin $= -4$ and Xmax $= 4$. Do you think this is a better choice for the viewing window?

EXAMPLE 2   Creating a Table and Graphing an Equation

Create a table and graph the equation $y = x^3$.

Solution   Most graphing utilities have the capability of creating a table of values for an equation. (Check your manual to see if your graphing utility has this capability.) Table 1 illustrates a table of values for $y = x^3$ on a TI-84 Plus C. See Figure 10 for the graph.

Table 1

Figure 10 $y = x^3$

B.2 Exercises

In Problems 1–16, graph each equation using the following viewing windows:

(a) Xmin $= -5$ Xmax $= 5$ Xscl $= 1$
    Ymin $= -4$ Ymax $= 4$ Yscl $= 1$

(b) Xmin $= -10$ Xmax $= 10$ Xscl $= 2$
    Ymin $= -8$ Ymax $= 8$ Yscl $= 2$

1. $y = x + 2$ **2.** $y = x - 2$ **3.** $y = -x + 2$ **4.** $y = -x - 2$

5. $y = 2x + 2$ **6.** $y = 2x - 2$ **7.** $y = -2x + 2$ **8.** $y = -2x - 2$

9. $y = x^2 + 2$ **10.** $y = x^2 - 2$ **11.** $y = -x^2 + 2$ **12.** $y = -x^2 - 2$

13. $3x + 2y = 6$ **14.** $3x - 2y = 6$ **15.** $-3x + 2y = 6$ **16.** $-3x - 2y = 6$

17–32. *For each of the equations in Problems 1–16, create a table, $-5 \leq x \leq 5$, and list points on the graph.*

B.3 Using a Graphing Utility to Locate Intercepts and Check for Symmetry

Value and Zero (or Root)

Most graphing utilities have an eVALUEate feature that, given a value of x, determines the value of y for an equation. This feature is useful for evaluating an equation at $x = 0$ to find the y-intercept. Most graphing utilities also have a ZERO (or ROOT) feature that is used to find the x-intercept(s) of an equation.

Figure 11 $y = x^3 - 8$

NOTE Some graphing utilities automatically identify key points such as intercepts and intersection points. For example, Figure 11 shows the graph of $y = x^3 - 8$ using Desmos where the intercepts are already identified. ■

EXAMPLE 1   **Finding Intercepts Using a Graphing Utility**

Use a graphing utility to find the intercepts of the equation $y = x^3 - 8$.

Solution   Figure 12(a) shows the graph of $y = x^3 - 8$ on a TI-84 Plus C graphing calculator.
  The eVALUEate feature of a TI-84 Plus C accepts as input a value of x and determines the value of y. Letting $x = 0$, we find that the y-intercept is -8. See Figure 12(b).
  The ZERO feature of a TI-84 Plus C is used to find the x-intercept(s). See Figure 12(c). The x-intercept is 2.

Figure 12   (a)   (b)   (c)

EXAMPLE 2   **Graphing the Equation $y = \dfrac{1}{x}$**

Graph the equation $y = \dfrac{1}{x}$. Based on the graph, infer information about intercepts and symmetry.

Solution   Figure 13 shows the graph. Infer from the graph that there are no intercepts; also infer that symmetry with respect to the origin is a possibility. The TABLE feature on a graphing utility provides further evidence of symmetry with respect to the origin. Using a TABLE, observe that for any ordered pair (x, y), the ordered pair $(-x, -y)$ is also a point on the graph. See Table 2.

(continued)

Figure 13 $y = \dfrac{1}{x}$

B.3 Exercises

In Problems 1–6, use ZERO (or ROOT) to approximate the smaller of the two x-intercepts of each equation. Express the answer rounded to two decimal places.

1. $y = x^2 + 4x + 2$

2. $y = x^2 + 4x - 3$

3. $y = 2x^2 + 4x + 1$

4. $y = 3x^2 + 5x + 1$

5. $y = 2x^2 - 3x - 1$

6. $y = 2x^2 - 4x - 1$

*In Problems 7–12, use ZERO (or ROOT) to approximate the **positive** x-intercepts of each equation. Express each answer rounded to two decimal places.*

7. $y = x^3 + 3.2x^2 - 16.83x - 5.31$

8. $y = x^3 + 3.2x^2 - 7.25x - 6.3$

9. $y = x^4 - 1.4x^3 - 33.71x^2 + 23.94x + 292.41$

10. $y = x^4 + 1.2x^3 - 7.46x^2 - 4.692x + 15.2881$

11. $y = x^3 + 19.5x^2 - 1021x + 1000.5$

12. $y = x^3 + 14.2x^2 - 4.8x - 12.4$

B.4 Using a Graphing Utility to Solve Equations

For many equations, there are no algebraic techniques that lead to a solution. For such equations, a graphing utility is often used to investigate possible solutions. When a graphing utility is used to solve an equation, *approximate* solutions usually are obtained. Unless otherwise stated, this text follows the practice of giving approximate solutions *rounded to two decimal places*.

The ZERO (or ROOT) feature of a graphing utility is often used to find the solutions of an equation when one side of the equation is 0. In using this feature to solve equations, make use of the fact that the x-intercepts (or zeros) of the graph of an equation are found by letting $y = 0$ and solving the equation for x. Solving an equation for x when one side of the equation is 0 is equivalent to finding where the graph of the corresponding equation crosses or touches the x-axis.

EXAMPLE 1 **Using ZERO (or ROOT) to Approximate Solutions of an Equation**

Find the solution(s) of the equation $x^2 - 6x + 7 = 0$. Round answers to two decimal places.

Solution The solutions of the equation $x^2 - 6x + 7 = 0$ are the same as the x-intercepts of the graph of $Y_1 = x^2 - 6x + 7$. Begin by graphing the equation. See Figure 14(a) for the graph using a TI-84 Plus C.

From the graph there appear to be two x-intercepts (solutions to the equation): one between 1 and 2, the other between 4 and 5.

Using the ZERO (or ROOT) feature of the graphing utility, determine that the x-intercepts, and thus the solutions to the equation, are $x = 1.59$ and $x = 4.41$, rounded to two decimal places. See Figures 14(b) and (c).

Figure 14   (a)   (b)   (c)

A second method for solving equations using a graphing utility involves the INTERSECT feature of the graphing utility. This feature is used most effectively when one side of the equation is not 0.

EXAMPLE 2

Using INTERSECT to Approximate Solutions of an Equation

Find the solution(s) of the equation $3(x - 2) = 5(x - 1)$.

Solution

Begin by graphing each side of the equation as follows: graph $Y_1 = 3(x - 2)$ and $Y_2 = 5(x - 1)$. See Figure 15(a) for the graph using a TI-84 Plus C.

At the point of intersection of the graphs, the value of the y-coordinate is the same. Conclude that the x-coordinate of the point of intersection represents the solution of the equation. Do you see why? The INTERSECT feature on a graphing utility determines the point of intersection of the graphs. Using this feature, find that the graphs intersect at $(-0.5, -7.5)$. See Figure 15(b). The solution of the equation is therefore $x = -0.5$. Figure 15(c) shows the point of intersection using Desmos.

  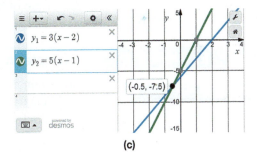

Figure 15   (a)   (b)   (c)

SUMMARY

The following steps can be used for approximating solutions of equations.

Steps for Approximating Solutions of Equations Using ZERO (or ROOT)

STEP 1: Write the equation in the form $\{expression\ in\ x\} = 0$.

STEP 2: Graph $Y_1 = \{expression\ in\ x\}$.

Be sure that the graph is complete. That is, be sure that all the intercepts are shown on the screen.

STEP 3: Use ZERO (or ROOT) to determine each x-intercept of the graph.

Steps for Approximating Solutions of Equations Using INTERSECT

STEP 1: Graph $Y_1 = \{expression\ in\ x\ on\ the\ left\ side\ of\ the\ equation\}$.

Graph $Y_2 = \{expression\ in\ x\ on\ the\ right\ side\ of\ the\ equation\}$.

STEP 2: Use INTERSECT to determine each x-coordinate of the point(s) of intersection, if any.

Be sure that the graphs are complete. That is, be sure that all the points of intersection are shown on the screen.

Figure 16 $y = \sqrt[3]{2x - 4} - 2$

EXAMPLE 3 **Solving a Radical Equation**

Find the real solutions of the equation $\sqrt[3]{2x - 4} - 2 = 0$.

Solution Figure 16 shows the graph of the equation

$$Y_1 = \sqrt[3]{2x - 4} - 2.$$

From the graph, there is one x-intercept near 6. Using ZERO (or ROOT), find that the x-intercept is 6. The only solution is $x = 6$.

B.5 Square Screens

Figure 17 $y = x$ and $y = -x$ using a standard viewing window.

Most graphing utilities have a rectangular screen. Because of this, using the same settings for both x and y results in a distorted view. For example, Figure 17 shows the graphs of $y = x$ and $y = -x$ using a TI-84 Plus C with a standard viewing window.

We expect the lines to intersect at right angles, but they do not. The selections for Xmin, Xmax, Ymin, and Ymax must be adjusted so that a **square screen** results. On the TI-84 Plus C, this is accomplished by setting the ratio of x to y at 8:5.* For example, if

$$Xmin = -16 \qquad Ymin = -10$$
$$Xmax = 16 \qquad Ymax = 10$$

then the ratio of x to y is

$$\frac{Xmax - Xmin}{Ymax - Ymin} = \frac{16 - (-16)}{10 - (-10)} = \frac{32}{20} = \frac{8}{5}$$

for a ratio of 8:5, resulting in a square screen.

Figure 18 $y = x$ and $y = -x$ with a square screen.

EXAMPLE 1 **Examples of Viewing Rectangles That Result in Square Screens**

(a) $Xmin = -8$
$Xmax = 8$
$Xscl = 1$
$Ymin = -5$
$Ymax = 5$
$Yscl = 1$

(b) $Xmin = -16$
$Xmax = 16$
$Xscl = 1$
$Ymin = -10$
$Ymax = 10$
$Yscl = 1$

(c) $Xmin = -24$
$Xmax = 24$
$Xscl = 3$
$Ymin = -15$
$Ymax = 15$
$Yscl = 3$

Figure 18 shows the graphs of $y = x$ and $y = -x$ on a square screen using the viewing rectangle given in part (b). Notice that the lines now intersect at right angles. Compare this illustration to Figure 17.

B.5 Exercises

In Problems 1–8, determine which of the given viewing rectangles result in a square screen.

1. $Xmin = -6$ $\quad Xmax = 6$ $\quad Xscl = 1$
$Ymin = -2$ $\quad Ymax = 2$ $\quad Yscl = 0.5$

2. $Xmin = -5$ $\quad Xmax = 5$ $\quad Xscl = 1$
$Ymin = -4$ $\quad Ymax = 4$ $\quad Yscl = 1$

3. $Xmin = 0$ $\quad Xmax = 16$ $\quad Xscl = 4$
$Ymin = -2$ $\quad Ymax = 8$ $\quad Yscl = 2$

4. $Xmin = -10$ $\quad Xmax = 14$ $\quad Xscl = 2$
$Ymin = -7$ $\quad Ymax = 8$ $\quad Yscl = 3$

*Some graphing utilities have a built-in function that automatically squares the screen. For example, the TI-84 Plus C has a ZSquare function that does this. Some graphing utilities require a ratio other than 8:5 to square the screen. For example, the HP 48G requires the ratio of x to y to be 2:1 for a square screen. Consult your manual.

5. If $X\text{min} = -4$, $X\text{max} = 12$, and $X\text{scl} = 1$, how should $Y\text{min}$, $Y\text{max}$, and $Y\text{scl}$ be selected so that the viewing rectangle contains the point $(4, 8)$ and the screen is square?

6. If $X\text{min} = -6$, $X\text{max} = 10$, and $X\text{scl} = 2$, how should $Y\text{min}$, $Y\text{max}$, and $Y\text{scl}$ be selected so that the viewing rectangle contains the point $(4, 8)$ and the screen is square?

B.6 Using a Graphing Utility to Graph Inequalities

EXAMPLE 1

Graphing an Inequality Using a Graphing Utility

Use a graphing utility to graph $3x + y - 6 \leq 0$.

Solution

Begin by graphing the equation $3x + y - 6 = 0$ ($Y_1 = -3x + 6$). See Figure 19.

As with graphing by hand, select test points from each region and determine whether they satisfy the inequality. To test the point $(-1, 2)$, for example, enter $3(-1) + 2 - 6 \leq 0$. See Figure 20(a). The 1 that appears indicates that the statement entered (the inequality) is true. When the point $(5, 5)$ is tested, a 0 appears, indicating that the statement entered is false. Thus, $(-1, 2)$ is a part of the graph of the inequality, and $(5, 5)$ is not. Figure 20(b) shows the graph of the inequality on a TI-84 Plus C.* Figure 20(c) shows the graph using Desmos.

Figure 19

(a)

(b)

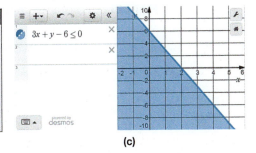
(c)

Figure 20

Steps for Graphing an Inequality Using a Graphing Utility

STEP 1: Replace the inequality symbol by an equal sign, solve the equation for y, and graph the equation.

STEP 2: In each region, select a test point P and determine whether the coordinates of P satisfy the inequality.

- If the test point satisfies the inequality, then so do all the points in the region. Indicate this by using the graphing utility to shade the region.
- If the coordinates of P do not satisfy the inequality, then neither will any of the other points in that region.

*Consult your owner's manual for shading techniques.

B.7 Using a Graphing Utility to Solve Systems of Linear Equations

Most graphing utilities have the capability to put the augmented matrix of a system of linear equations in row echelon form. The next example, Example 6 from Section 11.2, demonstrates this feature using a TI-84 Plus C graphing calculator.

EXAMPLE 1   **Solving a System of Linear Equations Using a Graphing Utility**

Solve:
$$\begin{cases} x - y + z = 8 & \textbf{(1)} \\ 2x + 3y - z = -2 & \textbf{(2)} \\ 3x - 2y - 9z = 9 & \textbf{(3)} \end{cases}$$

Solution   The augmented matrix of the system is

$$\begin{bmatrix} 1 & -1 & 1 & | & 8 \\ 2 & 3 & -1 & | & -2 \\ 3 & -2 & -9 & | & 9 \end{bmatrix}$$

Enter this matrix into a graphing utility and name it A. See Figure 21(a). Using the **ref** (row echelon form) command on matrix A, the results shown in Figure 21(b) are obtained. If the entire matrix does not fit on the screen, scroll right to see the rest of it.

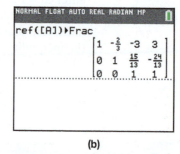

Figure 21   (a)   (b)

The system of equations represented by the matrix in row echelon form is

$$\begin{bmatrix} 1 & -\dfrac{2}{3} & -3 & | & 3 \\ 0 & 1 & \dfrac{15}{13} & | & -\dfrac{24}{13} \\ 0 & 0 & 1 & | & 1 \end{bmatrix} \quad \begin{cases} x - \dfrac{2}{3}y - 3z = 3 & \textbf{(1)} \\ y + \dfrac{15}{13}z = -\dfrac{24}{13} & \textbf{(2)} \\ z = 1 & \textbf{(3)} \end{cases}$$

Using $z = 1$, back-substitute to get

$$\begin{cases} x - \dfrac{2}{3}y - 3(1) = 3 & \textbf{(1)} \\ y + \dfrac{15}{13}(1) = -\dfrac{24}{13} & \textbf{(2)} \end{cases} \xrightarrow[\text{Simplify.}]{} \begin{cases} x - \dfrac{2}{3}y = 6 & \textbf{(1)} \\ y = -\dfrac{39}{13} = -3 & \textbf{(2)} \end{cases}$$

Solve the second equation for y to find that $y = -3$. Back-substitute $y = -3$ into $x - \dfrac{2}{3}y = 6$ to find that $x = 4$. The solution of the system is $x = 4$, $y = -3$, $z = 1$.

Figure 22

Notice that the row echelon form of the augmented matrix using the graphing utility differs from the row echelon form obtained in Example 6 of Section 11.2, yet both matrices provide the same solution! This is because the two solutions used different row operations to obtain the row echelon form. In all likelihood, the two solutions parted ways in Step 4 of the algebraic solution, where fractions were avoided by interchanging rows 2 and 3.

Most graphing utilities also have the ability to put a matrix in reduced row echelon form. Figure 22 shows the reduced row echelon form of the augmented matrix from Example 1 using the **rref** command on a TI-84 Plus C graphing calculator. Using this command, note that the solution of the system is still $x = 4$, $y = -3$, $z = 1$.

B.8 Using a Graphing Utility to Graph a Polar Equation

Most graphing utilities require the following steps in order to obtain the graph of a polar equation. Be sure to be in POLAR mode.

> **Graphing a Polar Equation Using a Graphing Utility**
>
> **STEP 1:** Set the mode to POLAR. Solve the equation for r in terms of θ.
>
> **STEP 2:** Select the viewing rectangle in polar mode. Besides setting Xmin, Xmax, Xscl, and so forth, the viewing rectangle in polar mode requires setting the minimum and maximum values for θ and an increment setting for θ (θstep). In addition, a square screen and radian measure should be used.
>
> **STEP 3:** Enter the expression involving θ that you found in Step 1. (Consult your manual for the correct way to enter the expression.)
>
> **STEP 4:** Graph.

EXAMPLE 1   **Graphing a Polar Equation Using a Graphing Utility**

Use a graphing utility to graph the polar equation $r \sin \theta = 2$.

Solution   **STEP 1:** Solve the equation for r in terms of θ.

$$r \sin \theta = 2$$

$$r = \frac{2}{\sin \theta}$$

STEP 2: From the POLAR mode, select the viewing rectangle.

$$\theta\text{min} = 0, \qquad \theta\text{max} = 2\pi, \quad \theta\text{step} = \frac{\pi}{24}$$

$$X\text{min} = -8, \quad X\text{max} = 8, \qquad X\text{scl} = 1$$

$$Y\text{min} = -5, \quad Y\text{max} = 5, \qquad Y\text{scl} = 1$$

θstep determines the number of points that the graphing utility will plot. For example, if θstep is $\frac{\pi}{24}$, the graphing utility will evaluate r at $\theta = 0\,(\theta\text{min})$, $\frac{\pi}{24}, \frac{2\pi}{24}, \frac{3\pi}{24}$, and so forth, up to $2\pi\,(\theta\text{max})$. The smaller θstep is, the more points the graphing utility will plot. Experiment with different values for θmin, θmax, and θstep to see how the graph is affected.

STEP 3: Enter the expression $\dfrac{2}{\sin \theta}$ after the prompt $r_1 = $.

STEP 4: Graph.

The graph is shown in Figure 23.

Figure 23 $r \sin \theta = 2$

B.9 Using a Graphing Utility to Graph Parametric Equations

Most graphing utilities have the capability of graphing parametric equations. The following steps are usually required to obtain the graph of parametric equations. Check your owner's manual to see how yours works.

> ### Graphing Parametric Equations Using a Graphing Utility
>
> **STEP 1:** Set the mode to PARAMETRIC. Enter $x(t)$ and $y(t)$.
>
> **STEP 2:** Select the viewing window. In addition to setting Xmin, Xmax, Xscl, and so on, the viewing window in parametric mode requires setting minimum and maximum values for the parameter t and an increment setting for t (Tstep).
>
> **STEP 3:** Graph.

EXAMPLE 1	**Graphing a Curve Defined by Parametric Equations Using a Graphing Utility**

Graph the curve defined by the parametric equations

$$x = 3t^2, \qquad y = 2t, \qquad -2 \le t \le 2$$

Solution

STEP 1: Enter the equations $x(t) = 3t^2$, $y(t) = 2t$ with the graphing utility in PARAMETRIC mode.

STEP 2: Select the viewing window. The interval is $-2 \le t \le 2$, so select the following square viewing window:

$$
\begin{aligned}
T\text{min} &= -2, & T\text{max} &= 2, & T\text{step} &= 0.1 \\
X\text{min} &= 0, & X\text{max} &= 16, & X\text{scl} &= 1 \\
Y\text{min} &= -5, & Y\text{max} &= 5, & Y\text{scl} &= 1
\end{aligned}
$$

Choose Tmin $= -2$ and Tmax $= 2$ because $-2 \le t \le 2$. Finally, the choice for Tstep will determine the number of points that the graphing utility will plot. For example, with Tstep at 0.1, the graphing utility will evaluate x and y at $t = -2, -1.9, -1.8$, and so on. The smaller the Tstep, the more points the graphing utility will plot. Experiment with different values of Tstep to see how the graph is affected.

STEP 3: Graph. Watch the direction in which the graph is drawn. This direction shows the orientation of the curve.

Using a TI-84 Plus C, the graph shown in Figure 24 is complete. ⌐

Figure 24
$x = 3t^2, y = 2t, -2 \le t \le 2$

> ### Exploration
>
> Graph the following parametric equations using a graphing utility with Xmin $= 0$, Xmax $= 16$, Ymin $= -5$, Ymax $= 5$, and Tstep $= 0.1$.
>
> **1.** $x = \dfrac{3t^2}{4}$, $y = t$, $-4 \le t \le 4$
>
> **2.** $x = 3t^2 + 12t + 12$, $y = 2t + 4$, $-4 \le t \le 0$
>
> **3.** $x = 3t^{2/3}$, $y = 2\sqrt[3]{t}$, $-8 \le t \le 8$
>
> Compare these graphs to Figure 24. Conclude that parametric equations defining a curve are not unique; that is, different parametric equations can represent the same graph.

Figure 25
$x = 3t^2, y = 2t, -2 \le t \le 2$

> ### Exploration
>
> In FUNCTION mode, graph $x = \dfrac{3y^2}{4}$ ($Y_1 = \sqrt{\dfrac{4x}{3}}$ and $Y_2 = -\sqrt{\dfrac{4x}{3}}$) with Xmin $= 0$, Xmax $= 16$, Ymin $= -5$, Ymax $= 5$. Compare this graph with Figure 24. Why do the graphs differ?

NOTE Some graphing utilities input the parametric equations as an ordered pair. For example, Figure 25 shows the graph of $x = 3t^2, y = 2t, -2 \le t \le 2$ using Desmos. ∎

Answers

CHAPTER 1 Graphs and Functions

1.1 Assess Your Understanding (page 5)

7. x-coordinate or abscissa; y-coordinate or ordinate **8.** quadrants **9.** midpoint **10.** F **11.** F **12.** T **13.** b **14.** a

15. (a) Quadrant II **(b)** x-axis   **17.** The points will be on a   **19.** $\sqrt{5}$ **33.** $d(A, B) = \sqrt{13}$
(c) Quadrant III **(d)** Quadrant I   vertical line that is 2 units **21.** $\sqrt{10}$ $\qquad d(B, C) = \sqrt{13}$
(e) y-axis **(f)** Quadrant IV   to the right of the y-axis. **23.** $2\sqrt{17}$ $\qquad d(A, C) = \sqrt{26}$
25. $\sqrt{130}$ $\qquad (\sqrt{13})^2 + (\sqrt{13})^2 = (\sqrt{26})^2$
27. $\sqrt{10}$
29. $\sqrt{6.89} \approx 2.62$ $\qquad$ Area $= \dfrac{13}{2}$ square units
31. $\sqrt{a^2 + b^2}$

35. $d(A, B) = \sqrt{130}$ **37.** $d(A, B) = 4$ **39.** $(4, 0)$ **41.** $\left(\dfrac{7}{2}, 2\right)$
$\qquad d(B, C) = \sqrt{26}$ $\qquad d(B, C) = \sqrt{41}$
$\qquad d(A, C) = 2\sqrt{26}$ $\qquad d(A, C) = 5$ **43.** $(8, -2)$ **45.** $\left(\dfrac{a}{2}, \dfrac{b}{2}\right)$
$\qquad (\sqrt{26})^2 + (2\sqrt{26})^2 = (\sqrt{130})^2$ $\qquad 4^2 + 5^2 = (\sqrt{41})^2$ **47.** $(5, 3)$ **49.** $(3, -13), (3, 11)$
$\qquad$ Area $= 26$ square units $\qquad$ Area $= 10$ square units

51. $(4 + 3\sqrt{3}, 0); (4 - 3\sqrt{3}, 0)$ **53. (a)** $(-1, 1)$ **(b)** $(0, 13)$ **55.** $(1, 2)$ **57.** $\sqrt{17}; 2\sqrt{5}; \sqrt{29}$
59. $d(P_1, P_2) = 6; d(P_2, P_3) = 4; d(P_1, P_3) = 2\sqrt{13}$; right triangle **61.** $d(P_1, P_2) = 2\sqrt{17}; d(P_2, P_3) = \sqrt{34}; d(P_1, P_3) = \sqrt{34}$; isosceles right triangle
63. $90\sqrt{2} \approx 127.28$ ft **65. (a)** $(90, 0), (90, 90), (0, 90)$ **(b)** $5\sqrt{2161} \approx 232.43$ ft **(c)** $30\sqrt{149} \approx 366.20$ ft **67.** $d = 75t$ mi

69. (a) $(2.65, 1.6)$ **(b)** Approximately 1.285 units **71.** $23,307$; a slight underestimate **73.** $\left(\dfrac{s}{2}, \dfrac{s}{2}\right)$

75. Arrange the parallelogram on the coordinate plane so that the vertices are $P_1 = (0, 0), P_2 = (a, 0), P_3 = (a + b, c)$ and $P_4 = (b, c)$.
Then the lengths of the sides are $d(P_1, P_2) = a, d(P_2, P_3) = \sqrt{b^2 + c^2}, d(P_3, P_4) = a$, and $d(P_1, P_4) = \sqrt{b^2 + c^2}$, and the lengths
of the diagonals are $d(P_1, P_3) = \sqrt{(a + b)^2 + c^2}$ and $d(P_2, P_4) = \sqrt{(a - b)^2 + c^2}$. So, the sum of the squares of the lengths of the sides is
$a^2 + (\sqrt{b^2 + c^2})^2 + a^2 + (\sqrt{b^2 + c^2})^2 = a^2 + b^2 + c^2 + a^2 + b^2 + c^2 = 2a^2 + 2b^2 + 2c^2$. The sum of the squares of the lengths of the diagonals is
$(\sqrt{(a + b)^2 + c^2})^2 + (\sqrt{(a - b)^2 + c^2})^2 = (a + b)^2 + c^2 + (a - b)^2 + c^2 = a^2 + 2ab + b^2 + c^2 + a^2 - 2ab + b^2 + c^2 = 2a^2 + 2b^2 + 2c^2$.

1.2 Assess Your Understanding (page 19)

3. intercepts **4.** y-axis **5.** $(-3, 4)$ **6.** T **7.** F **8.** radius **9.** T **10.** d **11.** F **12.** a
13. $(0, 0)$ is on the graph. **15.** $(0, 3)$ is on the graph. **17.** $(0, 2)$ and $(\sqrt{2}, \sqrt{2})$ are on the graph

19. $(-2, 0), (0, 2)$ **21.** $(-4, 0), (0, 8)$ **23.** $(-1, 0), (1, 0), (0, -1)$ **25.** $(-2, 0), (2, 0), (0, 4)$ **27.** $(3, 0), (0, 2)$

29. $(-2, 0), (2, 0), (0, 9)$ **31.** **33.** **35.** **37.**

39. **41. (a)** $(-1, 0), (1, 0)$ **43. (a)** $\left(-\dfrac{\pi}{2}, 0\right), (0, 1), \left(\dfrac{\pi}{2}, 0\right)$ **45. (a)** $(0, 0)$
$\qquad$ **(b)** Symmetric with respect to the $\qquad$ **(b)** Symmetric with respect to the y-axis $\qquad$ **(b)** Symmetric with respect to
$\qquad$ x-axis, the y-axis, and the origin $\qquad$ the x-axis
47. (a) $(-2, 0), (0, 0), (2, 0)$ **49. (a)** $(x, 0), -2 \le x \le 1$ **51. (a)** No intercepts
$\qquad$ **(b)** Symmetric with respect to the origin $\qquad$ **(b)** No symmetry $\qquad$ **(b)** Symmetric with respect to the origin

53. **55.** **57.** $(-4, 0), (0, -2), (0, 2)$; symmetric with respect to the x-axis **59.** $(0, 0)$; symmetric with
$\qquad$ respect to the origin **61.** $(0, 9), (3, 0), (-3, 0)$; symmetric with respect to the y-axis
63. $(-2, 0), (2, 0), (0, -3), (0, 3)$; symmetric with respect to the x-axis, y-axis, and origin
65. $(0, -27), (3, 0)$; no symmetry **67.** $(0, -4), (4, 0), (-1, 0)$; no symmetry **69.** $(0, 0)$;
$\qquad$ symmetric with respect to the origin **71.** $(0, 0)$; symmetric with respect to the origin

73.

75.

77. $a = -4$ or $a = 1$

79. Center $(2, 1)$; radius $= 2$; $(x - 2)^2 + (y - 1)^2 = 4$

81. Center $\left(\dfrac{5}{2}, 2\right)$; radius $= \dfrac{3}{2}$; $\left(x - \dfrac{5}{2}\right)^2 + (y - 2)^2 = \dfrac{9}{4}$

83. $x^2 + y^2 = 4$;
$x^2 + y^2 - 4 = 0$

85. $x^2 + (y - 2)^2 = 4$;
$x^2 + y^2 - 4y = 0$

87. $(x - 4)^2 + (y + 3)^2 = 25$;
$x^2 + y^2 - 8x + 6y = 0$

89. $(x + 2)^2 + (y - 1)^2 = 16$;
$x^2 + y^2 + 4x - 2y - 11 = 0$

91. $\left(x - \dfrac{1}{2}\right)^2 + y^2 = \dfrac{1}{4}$;
$x^2 + y^2 - x = 0$

93. (a) $(h, k) = (0, 0)$; $r = 2$
(b)

(c) $(\pm 2, 0)$; $(0, \pm 2)$

95. (a) $(h, k) = (3, 0)$; $r = 2$
(b)

(c) $(1, 0)$; $(5, 0)$

97. (a) $(h, k) = (1, 2)$; $r = 3$
(b)

(c) $(1 \pm \sqrt{5}, 0)$; $(0, 2 \pm 2\sqrt{2})$

99. (a) $(h, k) = (-2, 2)$; $r = 3$
(b)

(c) $(-2 \pm \sqrt{5}, 0)$; $(0, 2 \pm \sqrt{5})$

101. (a) $(h, k) = \left(\dfrac{1}{2}, -1\right)$; $r = \dfrac{1}{2}$
(b)

(c) $(0, -1)$

103. (a) $(h, k) = (3, -2)$; $r = 5$
(b)

(c) $(3 \pm \sqrt{21}, 0)$; $(0, -6)$, $(0, 2)$

105. (a) $(h, k) = (-2, 0)$; $r = 2$
(b)

(c) $(0, 0)$, $(-4, 0)$

107. $x^2 + y^2 = 13$ **109.** $(x + 1)^2 + (y - 3)^2 = 5$ **111.** $(x - 2)^2 + (y + 4)^2 = 64$ **113.** $(-1, -2)$ **115.** 4

117. (a) $(0, 0)$, $(2, 0)$, $(0, 1)$, $(0, -1)$ **(b)** x-axis symmetry **119.** 18 units^2 **121.** $x^2 + (y - 139)^2 = 15{,}625$

123. $x^2 + (y - 49.5)^2 = 3660.25$ **125.** $(x - 2)^2 + (y - 3)^2 = 9$ **127.** $(x + 1)^2 + (y - 3)^2 = 1$

129. (a) $x^2 + (mx + b)^2 = r^2$
$(1 + m^2)x^2 + 2mbx + b^2 - r^2 = 0$
The equation has one solution if and
only if the discriminant $= 0$.
$(2mb)^2 - 4(1 + m^2)(b^2 - r^2) = 0$
$-4b^2 + 4r^2 + 4m^2r^2 = 0$
$r^2(1 + m^2) = b^2$

(b) $x = \dfrac{-2mb}{2(1 + m^2)} = \dfrac{-2mb}{2b^2/r^2} = -\dfrac{r^2 m}{b}$

$y = m\left(-\dfrac{r^2 m}{b}\right) + b = -\dfrac{r^2 m^2}{b} + b = \dfrac{-r^2 m^2 + b^2}{b} = \dfrac{r^2}{b}$

(c) The slope of the tangent line $= m$.

The slope of the line joining the center to the point of tangency $= \dfrac{r^2/b}{-r^2 m/b} = -\dfrac{1}{m}$.

131. The center of the circle is $\left(\dfrac{x_1 + x_2}{2}, \dfrac{y_1 + y_2}{2}\right)$ and the radius is $\dfrac{1}{2}\sqrt{(x_1 - x_2)^2 + (y_1 - y_2)^2}$. Then the equation of the

circle is $\left(x - \dfrac{x_1 + x_2}{2}\right)^2 + \left(y - \dfrac{y_1 + y_2}{2}\right)^2 = \dfrac{1}{4}\left[(x_1 - x_2)^2 + (y_1 - y_2)^2\right]$. Expanding, gives

$$x^2 - x(x_1 + x_2) + \dfrac{(x_1 + x_2)^2}{4} + y^2 - y(y_1 + y_2) + \dfrac{(y_1 + y_2)^2}{4} = \dfrac{1}{4}\left[x_1^2 - 2x_1x_2 + x_2^2 + y_1^2 - 2y_1y_2 + y_2^2\right]$$

$$4x^2 - 4x_1x - 4x_2x + x_1^2 + 2x_1x_2 + x_2^2 + 4y^2 - 4y_1y - 4y_2y + y_1^2 + 2y_1y_2 + y_2^2 = x_1^2 - 2x_1x_2 + x_2^2 + y_1^2 - 2y_1y_2 + y_2^2$$

$$4x^2 - 4x_1x - 4x_2x + 4x_1x_2 + 4y^2 - 4y_1y - 4y_2y + 4y_1y_2 = 0$$

$$x^2 - x_1x - x_2x + x_1x_2 + y^2 - y_1y - y_2y + y_1y_2 = 0$$

$$x(x - x_1) - x_2(x - x_1) + y(y - y_1) - y_2(y - y_1) = 0$$

$$(x - x_1)(x - x_2) + (y - y_1)(y - y_2) = 0$$

133. $(0, 0)$, $(-a, 0)$, $(a, 0)$; symmetric with respect to the x-axis, the y-axis, and the origin.

135. (b), **(c)**, **(e)**, **(g)**   **137. (a)** $y = \sqrt{x^2}$ and $y = |x|$ have the same graph.   **(b)** $\sqrt{x^2} = |x|$

(c) $x \geq 0$ for $y = (\sqrt{x})^2$, while x can be any real number for $y = x$.   **(d)** $y \geq 0$ for $y = \sqrt{x^2}$

1.3 Assess Your Understanding *(page 37)*

7. independent; dependent **8.** F **9.** F **10.** F **11.** a **12.** c **13.** d **14.** a **15.** vertical **16.** 5; -3 **17.** $a = -2$ **18.** F **19.** T **20.** a

21. (a) Domain: $\{0, 22, 40, 70, 100\}$ in °C
Range: $\{1.031, 1.121, 1.229, 1.305, 1.411\}$ in kg/m^3

(b)

Temperature, °C		Density, kg/m³
0	→	1.411
22	→	1.305
40	→	1.229
70	→	1.121
100	→	1.031

(c) $\{(0, 1.411), (22, 1.305), (40, 1.229), (70, 1.121), (100, 1.031)\}$

23. Function; Domain: $\{$Elvis, Colleen, Kaleigh, Marissa$\}$; Range: $\{$January 8, March 15, September 17$\}$
25. Not a function; Domain: $\{20, 30, 40\}$; Range: $\{200, 300, 350, 425\}$ **27.** Not a function; Domain: $\{-3, 2, 4\}$; Range: $\{6, 9, 10\}$
29. Function; Domain: $\{1, 2, 3, 4\}$; Range: $\{3\}$ **31.** Not a function; Domain: $\{-2, 0, 3\}$; Range: $\{3, 4, 6, 7\}$ **33.** Function; Domain: $\{-2, -1, 0, 1\}$;
Range: $\{0, 1, 4\}$ **35.** Function **37.** Function **39.** Not a function **41.** Not a function **43.** Function **45.** Not a function
47. (a) -4 **(b)** 1 **(c)** -3 **(d)** $3x^2 - 2x - 4$ **(e)** $-3x^2 - 2x + 4$ **(f)** $3x^2 + 8x + 1$ **(g)** $12x^2 + 4x - 4$ **(h)** $3x^2 + 6xh + 3h^2 + 2x + 2h - 4$
49. (a) 0 **(b)** $\dfrac{1}{2}$ **(c)** $-\dfrac{1}{2}$ **(d)** $\dfrac{-x}{x^2 + 1}$ **(e)** $\dfrac{-x}{x^2 + 1}$ **(f)** $\dfrac{x + 1}{x^2 + 2x + 2}$ **(g)** $\dfrac{2x}{4x^2 + 1}$ **(h)** $\dfrac{x + h}{x^2 + 2xh + h^2 + 1}$
51. (a) 4 **(b)** 5 **(c)** 5 **(d)** $|x| + 4$ **(e)** $-|x| - 4$ **(f)** $|x + 1| + 4$ **(g)** $2|x| + 4$ **(h)** $|x + h| + 4$
53. (a) $-\dfrac{1}{5}$ **(b)** $-\dfrac{3}{2}$ **(c)** $\dfrac{1}{8}$ **(d)** $\dfrac{2x - 1}{3x + 5}$ **(e)** $\dfrac{-2x - 1}{3x - 5}$ **(f)** $\dfrac{2x + 3}{3x - 2}$ **(g)** $\dfrac{4x + 1}{6x - 5}$ **(h)** $\dfrac{2x + 2h + 1}{3x + 3h - 5}$ **55.** All real numbers
57. All real numbers **59.** $\{x | x \neq -4, x \neq 4\}$ **61.** $\{x | x \neq 0\}$ **63.** $\{x | x \geq 4\}$ **65.** $\{x | x > 1\}$ **67.** $\{x | x > 4\}$ **69.** $\{t | t \geq 4, t \neq 7\}$
71. 4 **73.** $2x + h$ **75.** $2x + h - 1$ **77.** $\dfrac{-(2x + h)}{x^2(x + h)^2}$ **79.** $\dfrac{6}{(x + 3)(x + h + 3)}$ **81.** $\dfrac{1}{\sqrt{x + h - 2} + \sqrt{x - 2}}$
83. (a) $f(0) = 3; f(-6) = -3$ **(b)** $f(6) = 0; f(11) = 1$ **(c)** Positive **(d)** Negative **(e)** $-3, 6,$ and 10 **(f)** $-3 < x < 6; 10 < x \leq 11$
  (g) $\{x | -6 \leq x \leq 11\}$ **(h)** $\{y | -3 \leq y \leq 4\}$ **(i)** $-3, 6, 10$ **(j)** 3 **(k)** 3 times **(l)** Once **(m)** 0, 4 **(n)** $-5, 8$
85. Not a function **87.** Function **(a)** Domain: $\{x | -\pi \leq x \leq \pi\}$; Range: $\{y | -1 \leq y \leq 1\}$ **(b)** $\left(-\dfrac{\pi}{2}, 0\right), \left(\dfrac{\pi}{2}, 0\right), (0, 1)$ **(c)** y-axis
89. Not a function **91.** Function **(a)** Domain: $\{x | 0 < x < 3\}$; Range: $\{y | y < 2\}$ **(b)** $(1, 0)$ **(c)** None
93. Function **(a)** Domain: all real numbers; Range: $\{y | y \leq 2\}$ **(b)** $(-3, 0), (3, 0), (0, 2)$ **(c)** y-axis
95. Function **(a)** Domain: all real numbers; Range: $\{y | y \geq -3\}$ **(b)** $(1, 0), (3, 0), (0, 9)$ **(c)** None
97. (a) Yes **(b)** $f(-2) = 9; (-2, 9)$ **(c)** $0, \dfrac{1}{2}; (0, -1), \left(\dfrac{1}{2}, -1\right)$ **(d)** All real numbers **(e)** $-\dfrac{1}{2}, 1$ **(f)** -1
99. (a) No **(b)** $f(4) = -3; (4, -3)$ **(c)** $14; (14, 2)$ **(d)** $\{x | x \neq 6\}$ **(e)** -2 **(f)** $-\dfrac{1}{3}$
101. (a) Yes **(b)** $f(2) = \dfrac{8}{17}; \left(2, \dfrac{8}{17}\right)$ **(c)** $-1, 1; (-1, 1), (1, 1)$ **(d)** All real numbers **(e)** 0 **(f)** 0 **103.** $\{-2, 4\}$ **105.** $A = -\dfrac{7}{2}$ **107.** $A = -4$
109. $A(x) = \dfrac{1}{2}x^2$ **111.** $G(x) = 16x$ **113. (a)** 15.1 m, 14.071 m, 12.944 m, 11.719 m **(b)** 1.01 s, 1.43 s, 1.75 s **(c)** 2.02 s
115. (a) \$222 **(b)** \$225 **(c)** \$220 **(d)** \$230
117. (a) Approximately 10.4 ft high **119. (a)** About 81.07 ft **(b)** About 129.59 ft **(c)** About 26.63 ft
  (b) Approximately 9.9 ft high **(d)** About 528.13 ft

(c)

(8, 10.4) (12, 9.9)
(15, 8.4)
10
(0, 6)
5 (20, 3.6)
(22.6, 0)
0 5 10 15 20 25 x

(d) The ball will not go through the hoop; $h(15) \approx 8.4$ ft. If $v = 30$ ft/s, $h(15) = 10$ ft.

(e)

(f) About 115.07 ft and 413.05 ft
(g) 275 ft; maximum height shown in the table is 131.8 ft
(h) 264 ft

121. (a) \$50; It costs \$50 if you use 0 gigabytes. **(b)** \$50; It costs \$50 if you use 5 gigabytes. **(c)** \$150; It costs \$150 if you use 15 gigabytes.
  (d) $\{g | 0 \leq g \leq 30\}$. There are at most 30 gigabytes in a month. **123.** $\dfrac{1}{(x + h)^{2/3} + x^{1/3}(x + h)^{1/3} + x^{2/3}}$ **125.** 12 **129.** Yes; $f(x) = 0,$
  domain $x = a$ **131. (a)** II **(b)** V **(c)** IV **(d)** III **(e)** I

133.

y
20,000 (78, 19,000)
16,000 (33, 9000)
12,000 (48, 9000)
8000
4000 (13, 4000)
(10, 2000)
0 30 60 x
Time (in minutes)
Distance (in feet)

135. (a) $(7, 7.4)$ **(b)** $(4.2, 6)$ **(c)** Increasing from 0 to 30 mi/h **(d)** 0 mi/h **(e)** 50 mi/h **(f)** $(2, 4), (4.2, 6), (7, 7.4), (7.6, 8)$
137. Yes. $f(x) = 0$ satisfies the condition that $f(x) = -f(x)$.

1.4 Assess Your Understanding *(page 52)*

6. increasing **7.** even; odd **8.** T **9.** T **10.** F **11.** c **12.** d **13.** Yes **15.** No **17.** $[-8, -2]$; $[0, 2]$; $[5, 7]$ **19.** Yes; 10 **21.** $-2, 2$; 6, 10
23. $f(-8) = -4$ **25. (a)** $(-2, 0), (0, 3), (2, 0)$ **(b)** Domain: $\{x | -4 \le x \le 4\}$ or $[-4, 4]$; Range: $\{y | 0 \le y \le 3\}$ or $[0, 3]$ **(c)** Increasing
  on $[-2, 0]$ and $[2, 4]$; Decreasing on $[-4, -2]$ and $[0, 2]$ **(d)** Even **27. (a)** $(0, 1)$ **(b)** Domain: all real numbers; Range: $\{y | y > 0\}$ or $(0, \infty)$
  (c) Increasing on $(-\infty, \infty)$ **(d)** Neither **29. (a)** $(-\pi, 0), (0, 0), (\pi, 0)$ **(b)** Domain: $\{x | -\pi \le x \le \pi\}$ or $[-\pi, \pi]$; Range: $\{y | -1 \le y \le 1\}$
  or $[-1, 1]$ **(c)** Increasing on $\left[-\dfrac{\pi}{2}, \dfrac{\pi}{2}\right]$; Decreasing on $\left[-\pi, -\dfrac{\pi}{2}\right]$ and $\left[\dfrac{\pi}{2}, \pi\right]$ **(d)** Odd **31. (a)** $\left(0, \dfrac{1}{2}\right), \left(\dfrac{1}{3}, 0\right), \left(\dfrac{5}{2}, 0\right)$
  (b) Domain: $\{x | -3 \le x \le 3\}$ or $[-3, 3]$; Range: $\{y | -1 \le y \le 2\}$ or $[-1, 2]$ **(c)** Increasing on $[2, 3]$; Decreasing on $[-1, 1]$;
  Constant on $[-3, -1]$ and $[1, 2]$ **(d)** Neither **33. (a)** $0; 3$ **(b)** $-2, 2; 0, 0$ **35. (a)** $\dfrac{\pi}{2}; 1$ **(b)** $-\dfrac{\pi}{2}; -1$ **37.** Odd **39.** Even **41.** Odd **43.** Neither
45. Even **47.** Odd **49.** Absolute maximum: $f(1) = 4$; absolute minimum: $f(5) = 1$; local maximum: $f(3) = 3$; local minimum: $f(2) = 2$
51. Absolute maximum: $f(3) = 4$; absolute minimum: $f(1) = 1$; local maximum: $f(3) = 4$; local minimum: $f(1) = 1$
53. Absolute maximum: none; absolute minimum: $f(0) = 0$; local maximum: $f(2) = 3$; local minima: $f(0) = 0$ and $f(3) = 2$
55. Absolute maximum: none; absolute minimum: none; local maximum: none; local minimum: none

57.
Increasing: $[-2, -1], [1, 2]$
Decreasing: $[-1, 1]$
Local maximum: $f(-1) = 4$
Local minimum: $f(1) = 0$

59.
Increasing: $[-2, -0.77], [0.77, 2]$
Decreasing: $[-0.77, 0.77]$
Local maximum: $f(-0.77) = 0.19$
Local minimum: $f(0.77) = -0.19$

61.
Increasing: $[-3.77, 1.77]$
Decreasing: $[-6, -3.77], [1.77, 4]$
Local maximum: $f(1.77) = -1.91$
Local minimum: $f(-3.77) = -18.89$

63.
Increasing: $[-1.87, 0], [0.97, 2]$
Decreasing: $[-3, -1.87], [0, 0.97]$
Local maximum: $f(0) = 3$
Local minima: $f(-1.87) = 0.95$,
$f(0.97) = 2.65$

65. (a) -4 **(b)** -8 **(c)** -10 **67. (a)** 15 **(b)** -3 **(c)** 9 **69. (a)** 5 **(b)** $y = 5x - 2$ **71. (a)** -1 **(b)** $y = -x$ **73. (a)** 4 **(b)** $y = 4x - 8$
75. (a) Odd **(b)** Local maximum value is 54 at $x = -3$. **77. (a)** Even **(b)** Local maximum value is 25 at $x = -2$. **(c)** 50.4 sq. units

79. (a)
(b) 10 riding lawn mowers
(c) \$239/mower

81. (a) On average, the population is increasing at a rate of 0.036 g/h from 0 to 2.5 h. **(b)** On average, from 4.5 to 6 h, the population is increasing at a rate of 0.1 g/h. **(c)** The average rate of change is increasing over time.

83. (a) 1 **(b)** 0.5 **(c)** 0.1 **(d)** 0.01 **(e)** 0.001
(f)
(g) They are getting closer to the tangent line at $(0, 0)$.
(h) They are getting closer to 0.

85. (a) 2
(b) 2; 2; 2; 2
(c) $y = 2x + 5$
(d)

87. (a) $2x + h + 2$ **(b)** 4.5; 4.1; 4.01; 4
(c) $y = 4.01x - 1.01$
(d)

89. (a) $4x + 2h - 3$ **(b)** 2; 1.2; 1.02; 1
(c) $y = 1.02x - 1.02$
(d)

91. (a) $-\dfrac{1}{(x + h)x}$
(b) $-\dfrac{2}{3}, -\dfrac{10}{11}, -\dfrac{100}{101}, -1$
(c) $y = -\dfrac{100}{101}x + \dfrac{201}{101}$
(d)

93. $\dfrac{-2 + \sqrt{19}}{3}$
97. At most one

99. Yes; the function $f(x) = 0$ is both even and odd.
101. Not necessarily. It just means $f(5) > f(2)$.

1.5 Assess Your Understanding *(page 63)*

4. $(-\infty, 0]$ **5.** piecewise-defined **6.** T **7.** F **8.** F **9.** b **10.** a **11.** C **13.** E **15.** B **17.** F

19.

21.

23.

25.

27. (a) -9 **(b)** 4 **(c)** 7 **29. (a)** 0 **(b)** 4 **(c)** 6 **(d)** 26

31. (a) All real numbers
(b) $(0, 1)$
(c)

(d) $\{y | y \neq 0\}$; $(-\infty, 0) \cup (0, \infty)$

33. (a) All real numbers
(b) $(0, 3)$
(c)

(d) $\{y | y \geq 1\}$; $[1, \infty)$

35. (a) $\{x | x \geq -2\}$; $[-2, \infty)$
(b) $(0, 3)$, $(2, 0)$
(c)

(d) $\{y | y < 4, y = 5\}$; $(-\infty, 4) \cup \{5\}$

37. (a) All real numbers
(b) $(-1, 0)$, $(0, 0)$
(c)

(d) All real numbers

39. (a) $\{x | x \geq -2, x \neq 0\}$; $[-2, 0) \cup (0, \infty)$
(b) No intercepts
(c)

(d) $\{y | y > 0\}$; $(0, \infty)$

41. (a) Domain $(0, \infty)$
(b) No intercepts
(c)

(d) Range: $(0, 7]$

43. $f(x) = \begin{cases} -x & \text{if } -1 \leq x \leq 0 \\ \frac{1}{2}x & \text{if } 0 < x \leq 2 \end{cases}$

45. $f(x) = \begin{cases} -x & \text{if } x \leq 0 \\ -x + 2 & \text{if } 0 < x \leq 2 \end{cases}$

47. (a)

(b) $[0, 6]$
(c) Absolute maximum: $f(2) = 6$;
  absolute minimum: $f(6) = -2$

49. (a) $34.99
(b) $64.99
(c) $184.99

51. (a) $47.07
(b) $119.97
(c) $C(x) = \begin{cases} 1.18172x + 23.44 & \text{if } 0 \leq x \leq 30 \\ 0.50897x + 43.6225 & \text{if } x > 30 \end{cases}$

(d)

53. $f(x) = \begin{cases} 0.10x & \text{if} & 0 < x \leq 9525 \\ 952.50 + 0.12(x - 9525) & \text{if} & 9525 < x \leq 38{,}700 \\ 4453.50 + 0.22(x - 38{,}700) & \text{if} & 38{,}700 < x \leq 82{,}500 \\ 14{,}089.50 + 0.24(x - 82{,}500) & \text{if} & 82{,}500 < x \leq 157{,}500 \\ 32{,}089.50 + 0.32(x - 157{,}500) & \text{if} & 157{,}500 < x \leq 200{,}000 \\ 45{,}689.50 + 0.35(x - 200{,}000) & \text{if} & 200{,}000 < x \leq 500{,}000 \\ 150{,}689.50 + 0.37(x - 500{,}000) & \text{if} & x > 500{,}000 \end{cases}$

55. (a)

(b) $C(x) = 10 + 0.4x$
(c) $C(x) = 70 + 0.25x$

57. (a) $C(s) = \begin{cases} 9000 & \text{if} & s \leq 659 \\ 8250 & \text{if} & 660 \leq s \leq 679 \\ 5250 & \text{if} & 680 \leq s \leq 699 \\ 3750 & \text{if} & 700 \leq s \leq 719 \\ 2250 & \text{if} & 720 \leq s \leq 739 \\ 1500 & \text{if} & s \geq 740 \end{cases}$

(b) $2250 **(c)** $8250

59. (a) $10°C$ **(b)** $4°C$ **(c)** $-3°C$ **(d)** $-4°C$
(e) The wind chill is equal to the air temperature.
(f) At wind speed greater than 20 m/s, the
  wind chill factor depends only on the air
  temperature.

61. $C(x) = \begin{cases} 1.00 & \text{if} & 0 < x \leq 1 \\ 1.21 & \text{if} & 1 < x \leq 2 \\ 1.42 & \text{if} & 2 < x \leq 3 \\ 1.63 & \text{if} & 3 < x \leq 4 \\ 1.84 & \text{if} & 4 < x \leq 5 \\ 2.05 & \text{if} & 5 < x \leq 6 \\ 2.26 & \text{if} & 6 < x \leq 7 \\ 2.47 & \text{if} & 7 < x \leq 8 \\ 2.68 & \text{if} & 8 < x \leq 9 \\ 2.89 & \text{if} & 9 < x \leq 10 \\ 3.10 & \text{if} & 10 < x \leq 11 \\ 3.31 & \text{if} & 11 < x \leq 12 \\ 3.52 & \text{if} & 12 < x \leq 13 \end{cases}$

63. $f(x) + g(x) = \begin{cases} -2x + 4 & \text{if} & x \leq 0 \\ 3x - 4 & \text{if} & 0 < x < 2 \\ x^2 + 6x - 7 & \text{if} & x \geq 2 \end{cases}$

65. Each graph is that of $y = x^2$, but shifted horizontally. If $y = (x - k)^2$, $k > 0$, the shift is right k units; if $y = (x + k)^2$, $k > 0$, the shift is left k units.

67. The graph of $y = -f(x)$ is the reflection about the x-axis of the graph of $y = f(x)$.   **69.** Yes. The graph of $y = (x - 1)^3 + 2$ is the graph
of $y = x^3$ shifted right 1 unit and up 2 units.   **71.** They all have the same general shape. All three go through the points $(-1, -1)$, $(0, 0)$, and $(1, 1)$.
As the exponent increases, the steepness of the curve increases (except near $x = 0$).

1.6 Assess Your Understanding (page 77)

1. horizontal; right   **2.** y **3.** F   **4.** T   **5.** F   **6.** d   **7.** b   **8.** T   **9.** B   **11.** H   **13.** I   **15.** L   **17.** F   **19.** G   **21.** $y = (x - 4)^3$ **23.** $y = x^3 + 4$ **25.** $y = -x^3$
27. $y = 5x^3$ **29.** $y = (2x)^3 = 8x^3$ **31.** $y = -(\sqrt{-x} + 2)$ **33.** $y = 3\sqrt{x} + 5 + 4$ **35.** c   **37.** c

39.
Domain: $(-\infty, \infty)$;
Range: $[-1, \infty)$

41.
Domain: $[0, \infty)$;
Range: $[0, \infty)$

43.
Domain: $[-2, \infty)$;
Range: $[0, \infty)$

45.
Domain: $(-\infty, \infty)$;
Range: $(-\infty, \infty)$

47.
Domain: $[0, \infty)$;
Range: $[0, \infty)$

49.
Domain: $(-\infty, \infty)$;
Range: $(-\infty, \infty)$

51.
Domain: $(-\infty, \infty)$;
Range: $[-3, \infty)$

53.
Domain: $[2, \infty)$;
Range: $[1, \infty)$

55.
Domain: $(-\infty, 0]$;
Range: $[-2, \infty)$

57.
Domain: $(-\infty, \infty)$;
Range: $(-\infty, \infty)$

59.
Domain: $(-\infty, \infty)$;
Range: $[0, \infty)$

61.
Domain: $(-\infty, 0) \cup (0, \infty)$
Range: $(-\infty, 0) \cup (0, \infty)$

63. (a) $F(x) = f(x) + 3$

(b) $G(x) = f(x + 2)$

(c) $P(x) = -f(x)$

(d) $H(x) = f(x + 1) - 2$

(e) $Q(x) = \frac{1}{2}f(x)$

(f) $g(x) = f(-x)$

(g) $h(x) = f(2x)$

65. (a) $F(x) = f(x) + 3$

(b) $G(x) = f(x + 2)$

(c) $P(x) = -f(x)$

(d) $H(x) = f(x + 1) - 2$

(e) $Q(x) = \frac{1}{2}f(x)$

(f) $g(x) = f(-x)$

(g) $h(x) = f(2x)$

67. $f(x) = (x + 1)^2 - 1$

69. $f(x) = (x - 4)^2 - 15$

71. $f(x) = 2(x - 3)^2 + 1$

73. $f(x) = -3(x + 2)^2 - 5$

75. (a) -7 and 1 **(b)** -3 and 5
(c) -5 and 3 **(d)** -3 and 5
77. (a) $[-3, 3]$ **(b)** $[4, 10]$
(c) Decreasing on $[-1, 5]$
(d) Decreasing on $[-5, 1]$

79. (a)
(b)

81. (a) $(-2, 2)$
(b) $(3, -5)$
(c) $(-1, 3)$

83. (a)
(b) 9 square units

85. (a) 72°F; 65°F
(b) The temperature decreases by 2° to 70°F during the day and 63°F overnight.

(c) The time at which the temperature adjusts between the daytime and overnight settings is moved to 1 hr sooner. It begins warming up at 5:00 AM instead of 6:00 AM, and it begins cooling down at 8:00 PM instead of 9:00 PM.

87.   **89.**

91. $[4, 8]$ **93.** The graph of $y = 4f(x)$ is a vertical stretch by a factor of 4. The graph of $y = f(4x)$ is a horizontal compression by a factor of $\frac{1}{4}$. **95.** $\frac{16}{3}$ sq. units **97.** The domain of $g(x) = \sqrt{x}$ is $[0, \infty)$. The graph of $g(x - k)$ is the graph of g shifted k units to the right, so the domain of $g(x - k)$ is $[k, \infty)$.

1.7 Assess Your Understanding *(page 89)*

5. $f(x_1) \neq f(x_2)$ **6.** one-to-one **7.** 3 **8.** $y = x$ **9.** $[4, \infty)$ **10.** T **11.** a **12.** d **13.** one-to-one **15.** not one-to-one
17. not one-to-one **19.** one-to-one **21.** one-to-one **23.** not one-to-one **25.** one-to-one

27.   **29.**   **31.**

33. $f(g(x)) = f\left(\frac{1}{3}(x - 4)\right) = 3 \cdot \frac{1}{3}(x - 4) + 4 = x - 4 + 4 = x$ **35.** $f(g(x)) = f\left(\frac{x}{4} + 2\right) = 4\left(\frac{x}{4} + 2\right) - 8 = x + 8 - 8 = x$

$g(f(x)) = g(3x + 4) = \frac{1}{3}[(3x + 4) - 4] = \frac{1}{3} \cdot 3x = x$ $g(f(x)) = g(4x - 8) = \frac{4x - 8}{4} + 2 = x - 2 + 2 = x$

37. $f(g(x)) = f(\sqrt[3]{x + 8}) = (\sqrt[3]{x + 8})^3 - 8 = x + 8 - 8 = x$ **39.** $f(g(x)) = f\left(\frac{1}{x}\right) = \frac{1}{\frac{1}{x}} = x; x \neq 0, g(f(x)) = g\left(\frac{1}{x}\right) = \frac{1}{\frac{1}{x}} = x, x \neq 0$

$g(f(x)) = g(x^3 - 8) = \sqrt[3]{(x^3 - 8) + 8} = \sqrt[3]{x^3} = x$

41. $f(g(x)) = f\left(\frac{4x - 3}{2 - x}\right) = \frac{2 \cdot \frac{4x - 3}{2 - x} + 3}{\frac{4x - 3}{2 - x} + 4} = \frac{2(4x - 3) + 3(2 - x)}{4x - 3 + 4(2 - x)} = \frac{5x}{5} = x, x \neq 2$

$g(f(x)) = g\left(\frac{2x + 3}{x + 4}\right) = \frac{4 \cdot \frac{2x + 3}{x + 4} - 3}{2 - \frac{2x + 3}{x + 4}} = \frac{4(2x + 3) - 3(x + 4)}{2(x + 4) - (2x + 3)} = \frac{5x}{5} = x, x \neq -4$

43. (a) $f^{-1}(x) = \frac{1}{3}x$ **45. (a)** $f^{-1}(x) = \frac{x}{4} - \frac{1}{2}$

$f(f^{-1}(x)) = f\left(\frac{1}{3}x\right) = 3 \cdot \frac{1}{3}x = x$ $f(f^{-1}(x)) = f\left(\frac{x}{4} - \frac{1}{2}\right) = 4\left(\frac{x}{4} - \frac{1}{2}\right) + 2 = x - 2 + 2 = x$

$f^{-1}(f(x)) = f^{-1}(3x) = \frac{1}{3} \cdot 3x = x$ $f^{-1}(f(x)) = f^{-1}(4x + 2) = \frac{4x + 2}{4} - \frac{1}{2} = x + \frac{1}{2} - \frac{1}{2} = x$

(b) Domain of f = Range of f^{-1} = All real numbers **(b)** Domain of f = Range of f^{-1} = All real numbers
Range of f = Domain of f^{-1} = All real numbers Range of f = Domain of f^{-1} = All real numbers

(c)   **(c)**

47. (a) $f^{-1}(x) = \sqrt[3]{x + 1}$ **49. (a)** $f^{-1}(x) = \sqrt{x - 4}, x \geq 4$

$f(f^{-1}(x)) = f(\sqrt[3]{x + 1}) = (\sqrt[3]{x + 1})^3 - 1 = x$ $f(f^{-1}(x)) = f(\sqrt{x - 4}) = (\sqrt{x - 4})^2 + 4 = x$

$f^{-1}(f(x)) = f^{-1}(x^3 - 1) = \sqrt[3]{(x^3 - 1) + 1} = x$ $f^{-1}(f(x)) = f^{-1}(x^2 + 4) = \sqrt{(x^2 + 4) - 4} = \sqrt{x^2} = x, x \geq 0$

(b) Domain of f = Range of f^{-1} = All real numbers **(b)** Domain of f = Range of f^{-1} = $\{x | x \geq 0\}$
Range of f = Domain of f^{-1} = All real numbers Range of f = Domain of f^{-1} = $\{x | x \geq 4\}$

(c)   **(c)**

51. (a) $f^{-1}(x) = \dfrac{4}{x}$

$$f(f^{-1}(x)) = f\left(\dfrac{4}{x}\right) = \dfrac{4}{\dfrac{4}{x}} = x$$

$$f^{-1}(f(x)) = f^{-1}\left(\dfrac{4}{x}\right) = \dfrac{4}{\dfrac{4}{x}} = x$$

(b) Domain of f = Range of $f^{-1} = \{x \mid x \neq 0\}$
Range of f = Domain of $f^{-1} = \{x \mid x \neq 0\}$

(c)

53. (a) $f^{-1}(x) = \dfrac{2x + 1}{x}$

$$f(f^{-1}(x)) = f\left(\dfrac{2x + 1}{x}\right) = \dfrac{1}{\dfrac{2x + 1}{x} - 2} = \dfrac{x}{2x + 1 - 2x} = x$$

$$f^{-1}(f(x)) = f^{-1}\left(\dfrac{1}{x - 2}\right) = \dfrac{2 \cdot \dfrac{1}{x - 2} + 1}{\dfrac{1}{x - 2}} = \dfrac{2 + x - 2}{1} = x$$

(b) Domain of f = Range of $f^{-1} = \{x \mid x \neq 2\}$
Range of f = Domain of $f^{-1} = \{x \mid x \neq 0\}$

(c)

55. (a) $f^{-1}(x) = \dfrac{2 - 3x}{x}$

$$f(f^{-1}(x)) = f\left(\dfrac{2 - 3x}{x}\right) = \dfrac{2}{3 + \dfrac{2 - 3x}{x}} = \dfrac{2x}{3x + 2 - 3x} = \dfrac{2x}{2} = x$$

$$f^{-1}(f(x)) = f^{-1}\left(\dfrac{2}{3 + x}\right) = \dfrac{2 - 3 \cdot \dfrac{2}{3 + x}}{\dfrac{2}{3 + x}} = \dfrac{2(3 + x) - 3 \cdot 2}{2} = \dfrac{2x}{2} = x$$

(b) Domain of f = Range of $f^{-1} = \{x \mid x \neq -3\}$
Range of f = Domain of $f^{-1} = \{x \mid x \neq 0\}$

57. (a) $f^{-1}(x) = \dfrac{-2x}{x - 3}$

$$f(f^{-1}(x)) = f\left(\dfrac{-2x}{x - 3}\right) = \dfrac{3 \cdot \dfrac{-2x}{x - 3}}{\dfrac{-2x}{x - 3} + 2} = \dfrac{3(-2x)}{-2x + 2(x - 3)} = \dfrac{-6x}{-6} = x$$

$$f^{-1}(f(x)) = f^{-1}\left(\dfrac{3x}{x + 2}\right) = \dfrac{-2 \cdot \dfrac{3x}{x + 2}}{\dfrac{3x}{x + 2} - 3} = \dfrac{-2 \cdot 3x}{3x - 3(x + 2)} = \dfrac{-6x}{-6} = x$$

(b) Domain of f = Range of $f^{-1} = \{x \mid x \neq -2\}$; Range of f = Domain of $f^{-1} = \{x \mid x \neq 3\}$

59. (a) $f^{-1}(x) = \dfrac{x}{3x - 2}$

$$f(f^{-1}(x)) = f\left(\dfrac{x}{3x - 2}\right) = \dfrac{2 \cdot \dfrac{x}{3x - 2}}{3 \cdot \dfrac{x}{3x - 2} - 1} = \dfrac{2x}{3x - (3x - 2)} = \dfrac{2x}{2} = x$$

$$f^{-1}(f(x)) = f^{-1}\left(\dfrac{2x}{3x - 1}\right) = \dfrac{\dfrac{2x}{3x - 1}}{3 \cdot \dfrac{2x}{3x - 1} - 2} = \dfrac{2x}{6x - 2(3x - 1)} = \dfrac{2x}{2} = x$$

(b) Domain of f = Range of $f^{-1} = \left\{x \mid x \neq \dfrac{1}{3}\right\}$; Range of f = Domain of $f^{-1} = \left\{x \mid x \neq \dfrac{2}{3}\right\}$

61. (a) $f^{-1}(x) = \dfrac{3x + 4}{2x - 3}$

$$f(f^{-1}(x)) = f\left(\dfrac{3x + 4}{2x - 3}\right) = \dfrac{3 \cdot \dfrac{3x + 4}{2x - 3} + 4}{2 \cdot \dfrac{3x + 4}{2x - 3} - 3} = \dfrac{3(3x + 4) + 4(2x - 3)}{2(3x + 4) - 3(2x - 3)} = \dfrac{17x}{17} = x$$

$$f^{-1}(f(x)) = f^{-1}\left(\dfrac{3x + 4}{2x - 3}\right) = \dfrac{3 \cdot \dfrac{3x + 4}{2x - 3} + 4}{2 \cdot \dfrac{3x + 4}{2x - 3} - 3} = \dfrac{3(3x + 4) + 4(2x - 3)}{2(3x + 4) - 3(2x - 3)} = \dfrac{17x}{17} = x$$

(b) Domain of f = Range of $f^{-1} = \left\{x \mid x \neq \dfrac{3}{2}\right\}$; Range of f = Domain of $f^{-1} = \left\{x \mid x \neq \dfrac{3}{2}\right\}$

63. (a) $f^{-1}(x) = \dfrac{-2x+3}{x-2}$

$$f(f^{-1}(x)) = f\left(\dfrac{-2x+3}{x-2}\right) = \dfrac{2\cdot\dfrac{-2x+3}{x-2}+3}{\dfrac{-2x+3}{x-2}+2} = \dfrac{2(-2x+3)+3(x-2)}{-2x+3+2(x-2)} = \dfrac{-x}{-1} = x$$

$$f^{-1}(f(x)) = f^{-1}\left(\dfrac{2x+3}{x+2}\right) = \dfrac{-2\cdot\dfrac{2x+3}{x+2}+3}{\dfrac{2x+3}{x+2}-2} = \dfrac{-2(2x+3)+3(x+2)}{2x+3-2(x+2)} = \dfrac{-x}{-1} = x$$

(b) Domain of f = Range of f^{-1} = $\{x | x \neq -2\}$; Range of f = Domain of f^{-1} = $\{x | x \neq 2\}$

65. (a) $f^{-1}(x) = \dfrac{2}{\sqrt{1-2x}}$

$$f(f^{-1}(x)) = f\left(\dfrac{2}{\sqrt{1-2x}}\right) = \dfrac{\dfrac{4}{1-2x}-4}{2\cdot\dfrac{4}{1-2x}} = \dfrac{4-4(1-2x)}{2\cdot4} = \dfrac{8x}{8} = x$$

$$f^{-1}(f(x)) = f^{-1}\left(\dfrac{x^2-4}{2x^2}\right) = \dfrac{2}{\sqrt{1-2\cdot\dfrac{x^2-4}{2x^2}}} = \dfrac{2}{\sqrt{\dfrac{4}{x^2}}} = \sqrt{x^2} = x, \text{ since } x > 0$$

(b) Domain of f = Range of f^{-1} = $\{x | x > 0\}$; Range of f = Domain of f^{-1} = $\left\{x \middle| x < \dfrac{1}{2}\right\}$

67. (a) $f^{-1}(x) = (x+4)^{\frac{3}{2}}, x \geq -4$

$$f(f^{-1}(x)) = f\left((x+4)^{\frac{3}{2}}\right) = \left((x+4)^{\frac{3}{2}}\right)^{\frac{2}{3}} - 4 = x + 4 - 4 = x$$

$$f^{-1}(f(x)) = f^{-1}\left(x^{\frac{2}{3}}-4\right) = \left(\left(x^{\frac{2}{3}}-4\right)+4\right)^{\frac{3}{2}} = \left(x^{\frac{2}{3}}\right)^{\frac{3}{2}} = x, \text{ since } x \geq 0$$

(b) Domain of f = Range of f^{-1} = $\{x | x \geq 0\}$; Domain of f^{-1} = Range of f = $\{x | x \geq -4\}$

69. (a) $f^{-1}(x) = \sqrt[5]{x^3+2}$

$$f(f^{-1}(x)) = f(\sqrt[5]{x^3+2}) = \sqrt[3]{(\sqrt[5]{x^3+2})^5 - 2} = \sqrt[3]{x^3+2-2} = \sqrt[3]{x^3} = x$$

$$f^{-1}(f(x)) = f^{-1}(\sqrt[3]{x^5-2}) = \sqrt[5]{(\sqrt[3]{x^5-2})^3 + 2} = \sqrt[5]{x^5-2+2} = \sqrt[5]{x^5} = x$$

(b) Domain of f = Range of f^{-1} = All Real Numbers; Domain of f^{-1} = Range of f = All Real Numbers

71. (a) $f^{-1}(x) = 3\sqrt{x-2}+1, x \geq 2$

$$f(f^{-1}(x)) = f(3\sqrt{x-2}+1) = \dfrac{1}{9}\left[(3\sqrt{x-2}+1)-1\right]^2 + 2 = \dfrac{1}{9}[3\sqrt{x-2}]^2 + 2 = \dfrac{1}{9}\cdot9(x-2)+2 = x-2+2 = x$$

$$f^{-1}(f(x)) = f^{-1}\left(\dfrac{1}{9}(x-1)^2+2\right) = 3\sqrt{\dfrac{1}{9}(x-1)^2+2-2}+1 = 3\sqrt{\dfrac{1}{9}(x-1)^2}+1 = 3\cdot\dfrac{1}{3}(x-1)+1 = x-1+1 = x$$

(b) Domain of f = Range of f^{-1} = $\{x | x \geq 1\}$; Domain of f^{-1} = Range of f = $\{x | x \geq 2\}$

73. (a) 0 **(b)** 2 **(c)** 0 **(d)** 1 **75.** 7 **77.** Domain of f^{-1}: $[-2, \infty)$; range of f^{-1}: $[5, \infty)$ **79.** Domain of g^{-1}: $[0, \infty)$; range of g^{-1}: $(-\infty, 0]$

81. Increasing on the interval $[f(0), f(5)]$ **83.** $f^{-1}(x) = \dfrac{1}{m}(x-b), m \neq 0$ **85.** Quadrant I

87. Possible answer: $f(x) = |x|, x \geq 0$, is one-to-one; $f^{-1}(x) = x, x \geq 0$

89. (a) $r(d) = \dfrac{d+90.39}{6.97}$

(b) $r(d(r)) = \dfrac{6.97r - 90.39 + 90.39}{6.97} = \dfrac{6.97r}{6.97} = r$

$d(r(d)) = 6.97\cdot\dfrac{d+90.39}{6.97} - 90.39 = d + 90.39 - 90.39 = d$

(c) 56 miles per hour

91. (a) 77.6 kg

(b) $h(W) = \dfrac{W-50}{2.3} + 60 = \dfrac{W+88}{2.3}$

(c) $h(W(h)) = \dfrac{50 + 2.3(h-60)+88}{2.3} = \dfrac{2.3h}{2.3} = h$

$W(h(W)) = 50 + 2.3\left(\dfrac{W+88}{2.3} - 60\right) = 50 + W + 88 - 138 = W$

(d) 73 inches

93. (a) $\{g \mid 38{,}700 < g \le 82{,}500\}$

(b) $\{T \mid 4453.50 < T \le 14{,}089.50\}$

(c) $g(T) = \dfrac{T - 4453.50}{0.22} + 38{,}700$

Domain: $\{T \mid 4453.50 < T \le 14{,}089.50\}$

Range: $\{g \mid 38{,}700 < g \le 82{,}500\}$

99. (a) Domain: $(-\infty, \infty)$; Range: $(-\infty, 3) \cup [4, \infty)$

(b) $f^{-1}(x) = \begin{cases} \dfrac{x-3}{2} & \text{if } x < 3 \\[2mm] \dfrac{x-4}{3} & \text{if } x \ge 4 \end{cases}$

(c) Domain: $(-\infty, 3) \cup [4, \infty)$; Range: $(-\infty, \infty)$

95. (a) t represents time, so $t \ge 0$.

(b) $t(H) = \sqrt{\dfrac{H - 100}{-4.9}} = \sqrt{\dfrac{100 - H}{4.9}}$

(c) 2.02 seconds

97. $f^{-1}(x) = \dfrac{-dx + b}{cx - a}; f = f^{-1}$ if $a = -d$

103. No

Review Exercises *(page 96)*

1. (a) $2\sqrt{5}$ **(b)** $(2, 1)$ **2. (a)** 5 **(b)** $\left(-\dfrac{1}{2}, 1\right)$ **3. (a)** 12 **(b)** $(4, 2)$

4.

5. $(-4, 0), (0, 2), (0, 0), (0, -2), (2, 0)$ **6.** $(0, 0)$; symmetric with respect to the x-axis

7. $(\pm 4, 0), (0, \pm 2)$; symmetric with respect to the x-axis, y-axis, and origin **8.** $(0, 1)$; symmetric with respect to the y-axis

9. $(0, 0), (\pm 1, 0)$; symmetric with respect to the origin **10.** $(0, 0), (-1, 0), (0, -2)$; no symmetry

11. $(x + 2)^2 + (y - 3)^2 = 16$ **12.** $(x + 1)^2 + (y + 2)^2 = 1$

13. Center $(0, 1)$; radius = 2

14. Center $(1, -2)$; radius = 3

15. Center $(1, -2)$; radius = $\sqrt{5}$

Intercepts: $(-\sqrt{3}, 0), (\sqrt{3}, 0),$ $(0, -1), (0, 3)$

Intercepts: $(1 - \sqrt{5}, 0), (1 + \sqrt{5}, 0),$ $(0, -2 - 2\sqrt{2}), (0, -2 + 2\sqrt{2})$

Intercepts: $(0, 0), (2, 0), (0, -4)$

16. $d(A, B) = \sqrt{(1 - 3)^2 + (1 - 4)^2} = \sqrt{13}$ and

$d(B, C) = \sqrt{(-2 - 1)^2 + (3 - 1)^2} = \sqrt{13}$ **17.** Center $(1, -2)$; Radius = $4\sqrt{2}$; $(x - 1)^2 + (y + 2)^2 = 32$

18. (a) Domain: $\{8, 16, 20, 24\}$; range: $\{\$6.30, \$12.32, \$13.99\}$ **(b)** $\{(8, \$6.30), (16, \$13.99), (20, \$12.32), (24, \$13.99)\}$

(c)

Batteries, x	Price, y
8	\$6.30
16	\$12.32
20	
24	\$13.99

(d)

19. Function; domain $\{-1, 2, 4\}$, range $\{0, 3\}$ **20.** Not a function **21. (a)** 2 **(b)** -2 **(c)** $-\dfrac{3x}{x^2 - 1}$ **(d)** $-\dfrac{3x}{x^2 - 1}$ **(e)** $\dfrac{3(x - 2)}{x^2 - 4x + 3}$ **(f)** $\dfrac{6x}{4x^2 - 1}$

22. (a) 0 **(b)** 0 **(c)** $\sqrt{x^2 - 4}$ **(d)** $-\sqrt{x^2 - 4}$ **(e)** $\sqrt{x^2 - 4x}$ **(f)** $2\sqrt{x^2 - 1}$ **23. (a)** 0 **(b)** 0 **(c)** $\dfrac{x^2 - 4}{x^2}$ **(d)** $-\dfrac{x^2 - 4}{x^2}$ **(e)** $\dfrac{x(x - 4)}{(x - 2)^2}$ **(f)** $\dfrac{x^2 - 1}{x^2}$

24. $\{x \mid x \ne -3, x \ne 3\}$ **25.** $\{x \mid x \le 2\}$ **26.** $\{x \mid x \ne 0\}$ **27.** $\{x \mid x \ne -3, x \ne 1\}$ **28.** $[-1, 2) \cup (2, \infty)$ **29.** $\{x \mid x > -8\}$, or $(-8, \infty)$

30. $-4x + 1 - 2h$ **31. (a)** Domain: $\{x \mid -4 \le x \le 3\}$; Range: $\{y \mid -3 \le y \le 3\}$

(b) $(0, 0)$ **(c)** -1 **(d)** -4 **(e)** $\{x \mid 0 < x \le 3\}$

(f)

(g)

(h)

32. (a) Domain: $\{x \mid x \le 4\}$ or $(-\infty, 4]$

Range: $\{y \mid y \le 3\}$ or $(-\infty, 3]$

(b) Increasing on $(-\infty, -2]$ and $[2, 4]$; Decreasing on $[-2, 2]$

(c) Local maximum value is 1 and occurs at $x = -2$.

Local minimum value is -1 and occurs at $x = 2$.

(d) Absolute maximum: $f(4) = 3$

Absolute minimum: none

(e) No symmetry

(f) Neither

(g) x-intercepts: $-3, 0, 3$ y-intercept: 0

33. Odd **34.** Even **35.** Neither **36.** Odd

37.

38.

39. (a) 23 **(b)** 7 **(c)** 47 **40.** −5 **41.** −17 **42.** No **43.** Yes

44.

45.

Local maximum value: 4.04 at $x = -0.91$
Local minimum value: -2.04 at $x = 0.91$
Increasing: $[-3, -0.91]$; $[0.91, 3]$
Decreasing: $[-0.91, 0.91]$

Local maximum value: 1.53 at $x = 0.41$
Local minimum values: 0.54 at
$x = -0.34$ and -3.56 at $x = 1.80$
Increasing: $[-0.34, 0.41]$; $[1.80, 3]$
Decreasing: $[-2, -0.34]$; $[0.41, 1.80]$

46.

Intercepts: $(-4, 0)$, $(4, 0)$, $(0, -4)$
Domain: all real numbers
Range: $\{y | y \geq -4\}$ or $[-4, \infty)$

47.

Intercept: $(0, 0)$
Domain: all real numbers
Range: $\{y | y \leq 0\}$ or $(-\infty, 0]$

48.

Intercept: $(1, 0)$
Domain: $\{x | x \geq 1\}$ or $[1, \infty)$
Range: $\{y | y \geq 0\}$ or $[0, \infty)$

49.

Intercepts: $(0, 1)$, $(1, 0)$
Domain: $\{x | x \leq 1\}$ or $(-\infty, 1]$
Range: $\{y | y \geq 0\}$ or $[0, \infty)$

50.

Intercept: $(0, 3)$
Domain: all real numbers
Range: $\{y | y \geq 2\}$ or $[2, \infty)$

51.

Intercepts: $(0 -24)$,
$(-2 - \sqrt[3]{4}, 0)$ or about
$(-3.6, 0)$
Domain: all real numbers
Range: all real numbers

52. (a) $\{x | x > -2\}$ or $(-2, \infty)$
(b) $(0, 0)$
(c)

(d) $\{y | y > -6\}$ or $(-6, \infty)$

53. (a) $\{x | x \geq -4\}$ or $[-4, \infty)$
(b) $(0, 1)$
(c)

(d) $\{y | -4 \leq y < 0$ or $y > 0\}$
or $[-4, 0) \cup (0, \infty)$

54. $A = 11$ **55. (a)** one-to-one **(b)** $\{(2, 1), (5, 3), (8, 5), (10, 6)\}$ **56.**

57. $f(g(x)) = f\left(\dfrac{1}{5}x + 2\right) = 5\left(\dfrac{1}{5}x + 2\right) - 10 = x + 10 - 10 = x$

$g(f(x)) = g(5x - 10) = \dfrac{1}{5}(5x - 10) + 2 = x - 2 + 2 = x$

58. $f(g(x)) = f\left(\dfrac{4}{1 - x}\right) = \dfrac{\dfrac{4}{1 - x} - 4}{\dfrac{4}{1 - x}} = \dfrac{4 - 4(1 - x)}{4} = \dfrac{4 - 4 + 4x}{4} = \dfrac{4x}{4} = x, x \neq 1$

$g(f(x)) = g\left(\dfrac{x - 4}{x}\right) = \dfrac{4}{1 - \dfrac{x - 4}{x}} = \dfrac{4x}{x - (x - 4)} = \dfrac{4x}{4} = x, x \neq 0$

59. $f^{-1}(x) = \dfrac{2x + 3}{5x - 2}$

$f(f^{-1}(x)) = \dfrac{2\left(\dfrac{2x + 3}{5x - 2}\right) + 3}{5\left(\dfrac{2x + 3}{5x - 2}\right) - 2} = x$

$f^{-1}(f(x)) = \dfrac{2\left(\dfrac{2x + 3}{5x - 2}\right) + 3}{5\left(\dfrac{2x + 3}{5x - 2}\right) - 2} = x$

Domain of f = range of f^{-1} = all real numbers except $\dfrac{2}{5}$

Range of f = domain of f^{-1} = all real numbers except $\dfrac{2}{5}$

60. $f^{-1}(x) = \dfrac{x + 1}{x}$

$f(f^{-1}(x)) = \dfrac{1}{\dfrac{x + 1}{x} - 1} = x$

$f^{-1}(f(x)) = \dfrac{\dfrac{1}{x - 1} + 1}{\dfrac{1}{x - 1}} = x$

Domain of f = range of f^{-1} = all real numbers except 1
Range of f = domain of f^{-1} = all real numbers except 0

61. $f^{-1}(x) = x^2 + 2, x \geq 0$

$f(f^{-1}(x)) = \sqrt{x^2 + 2 - 2} = |x| = x, x \geq 0$

$f^{-1}(f(x)) = (\sqrt{x - 2})^2 + 2 = x$

Domain of f = range of f^{-1} = $[2, \infty)$

Range of f = domain of f^{-1} = $[0, \infty)$

62. $f^{-1}(x) = (x - 1)^3$;

$f(f^{-1}(x)) = ((x - 1)^3)^{1/3} + 1 = x$

$f^{-1}(f(x)) = (x^{1/3} + 1 - 1)^3 = x$

Domain of f = range of f^{-1} = $(-\infty, \infty)$

Range of f = domain of f^{-1} = $(-\infty, \infty)$

Chapter Test *(page 98)*

1. $d = 2\sqrt{13}$ **2.** $(2, 1)$

3.   **4.**   **5.** Intercepts: $(-3, 0), (3, 0), (0, 9)$; symmetric with respect to the y-axis **6.** $x^2 + y^2 - 8x + 6y = 0$ **7.** Center: $(-2, 1)$; radius: 3

8. (a) Function; domain: $\{2, 4, 6, 8\}$; range: $\{5, 6, 7, 8\}$ **(b)** Not a function **(c)** Not a function **(d)** Function; domain; all real numbers; range:

$\{y|y \geq 2\}$ **9.** Domain: $\left\{x\middle|x \leq \dfrac{4}{5}\right\}$; $f(-1) = 3$ **10.** Domain: $\{x|x \neq -2\}$; $g(-1) = 1$ **11.** Domain: $\{x|x \neq -9, x \neq 4\}$; $h(-1) = \dfrac{1}{8}$

12. (a) Domain: $\{x|-5 \leq x \leq 5\}$; range: $\{y|-3 \leq y \leq 3\}$ **(b)** $(0, 2), (-2, 0)$, and $(2, 0)$ **(c)** $f(1) = 3$ **(d)** $x = -5$ and $x = 3$

(e) $\{x|-5 \leq x < -2 \text{ or } 2 < x \leq 5\}$ or $[-5, -2) \cup (2, 5]$

13. Local maxima values: $f(-0.85) \approx -0.86$; $f(2.35) \approx 15.55$; local minimum value: $f(0) = -2$; the function is increasing on the intervals $[-5, -0.85]$ and $[0, 2.35]$ and decreasing on the intervals $[-0.85, 0]$ and $[2.35, 5]$.

14. (a)   **(b)** $(0, -4), (4, 0)$ **(c)** $g(-5) = -9$ **(d)** $g(2) = -2$ **15.** 19 **16. (a)**   **(b)**

17. $f^{-1}(x) = \dfrac{2 + 5x}{3x}$, domain of $f = \left\{x\middle|x \neq \dfrac{5}{3}\right\}$, range of $f = \{y|y \neq 0\}$; domain of $f^{-1} = \{x|x \neq 0\}$; range of $f^{-1} = \left\{y\middle|y \neq \dfrac{5}{3}\right\}$

18. The point $(-5, 3)$ must be on the graph of f^{-1}.

CHAPTER 2 Trigonometric Functions

2.1 Assess Your Understanding *(page 110)*

3. standard position **4.** central angle **5.** d **6.** $r\theta$; $\dfrac{1}{2}r^2\theta$ **7.** b **8.** $\dfrac{s}{t}$; $\dfrac{\theta}{t}$ **9.** T **10.** F

11.   **13.**   **15.**   **17.**   **19.**   **21.**

23. $\dfrac{\pi}{6}$ **25.** $\dfrac{11\pi}{4}$ **27.** $-\dfrac{\pi}{3}$ **29.** 3π **31.** $-\dfrac{4\pi}{3}$ **33.** $-\dfrac{\pi}{2}$ **35.** $60°$ **37.** $-390°$ **39.** $810°$ **41.** $27°$ **43.** $-90°$ **45.** $-204°$ **47.** 0.30 **49.** -0.70 **51.** 2.18

53. $179.91°$ **55.** $401.07°$ **57.** $531.70°$ **59.** $40.17°$ **61.** $50.24°$ **63.** $9.15°$ **65.** $40°19'12''$ **67.** $18°15'18''$ **69.** $19°59'24''$ **71.** 5 m **73.** 12 ft

75. 0.9 radian **77.** $\dfrac{\pi}{3} \approx 1.047$ in. **79.** 25 m² **81.** $2\sqrt{3} \approx 3.464$ ft² **83.** 0.24 radian **85.** $\dfrac{\pi}{3} \approx 1.047$ in.² **87.** $s = 2.094$ ft; $A = 2.094$ ft²

89. $s = 14.661$ yd; $A = 87.965$ yd² **91.** $3\pi \approx 9.42$ in; $5\pi \approx 15.71$ in. **93.** $2\pi \approx 6.28$ m² **95.** $\dfrac{675\pi}{2} \approx 1060.29$ ft² **97.** $\dfrac{1075\pi}{3} \approx 1125.74$ in.²

99. $\omega = \dfrac{1}{60}$ radian/s; $v = \dfrac{1}{12}$ cm/s **101.** ≈ 23.2 mph **103.** ≈ 361.5 km/h **105.** ≈ 452.5 rpm **107.** 61.54 ft **109.** ≈ 898 mi/h **111.** ≈ 2292 mi/h

113. $\dfrac{3}{4}$ rpm **115.** ≈ 2.86 mi/h **117.** ≈ 31.47 rpm **119.** ≈ 1037 mi/h **121.** Radius ≈ 3979 mi; circumference $\approx 25,000$ mi **123.** 63π ft²

125. ≈ 2.2 rev **133.** $x = -\dfrac{7}{3}$ **134.** $\left\{-3, \dfrac{1}{5}\right\}$ **135.** $y = -|x + 3| - 4$ **136.** HA: $y = 3$; VA: $x = 7$ **137.** $c = \dfrac{5}{2}$ **138.** $\left\{\dfrac{21}{4}\right\}$

139. $\{x| x \neq -3, x \neq 3\}$ **140.** $6x^2 + 6xh + 2h^2$ **141.** $27x^3 - 54x^2 + 36x - 8$ **142.** $(-\infty, -0.99], [0.20, 0.79]$

2.2 Assess Your Understanding (page 127)

7. b   **8.** $(0, 1)$ **9.** $\left(\dfrac{\sqrt{2}}{2}, \dfrac{\sqrt{2}}{2}\right)$ **10.** F   **11.** $\dfrac{y}{r}; \dfrac{x}{r}$ **12.** a   **13.** $\sin t = \dfrac{1}{2}; \cos t = \dfrac{\sqrt{3}}{2}; \tan t = \dfrac{\sqrt{3}}{3}; \csc t = 2; \sec t = \dfrac{2\sqrt{3}}{3}; \cot t = \sqrt{3}$

15. $\sin t = \dfrac{\sqrt{21}}{5}; \cos t = -\dfrac{2}{5}; \tan t = -\dfrac{\sqrt{21}}{2}; \csc t = \dfrac{5\sqrt{21}}{21}; \sec t = -\dfrac{5}{2}; \cot t = -\dfrac{2\sqrt{21}}{21}$ **17.** $\sin t = \dfrac{\sqrt{2}}{2}; \cos t = -\dfrac{\sqrt{2}}{2};$

$\tan t = -1; \csc t = \sqrt{2}; \sec t = -\sqrt{2}; \cot t = -1$ **19.** $\sin t = -\dfrac{1}{3}; \cos t = \dfrac{2\sqrt{2}}{3}; \tan t = -\dfrac{\sqrt{2}}{4}; \csc t = -3; \sec t = \dfrac{3\sqrt{2}}{4}; \cot t = -2\sqrt{2}$

21. -1 **23.** 0 **25.** -1 **27.** 0 **29.** -1 **31.** $\dfrac{1}{2}(\sqrt{2} + 1)$ **33.** 2 **35.** $\dfrac{1}{2}$ **37.** $\sqrt{6}$ **39.** 4 **41.** 0 **43.** $2\sqrt{2} + \dfrac{4\sqrt{3}}{3}$ **45.** 1

47. $\sin \dfrac{2\pi}{3} = \dfrac{\sqrt{3}}{2}; \cos \dfrac{2\pi}{3} = -\dfrac{1}{2}; \tan \dfrac{2\pi}{3} = -\sqrt{3}; \csc \dfrac{2\pi}{3} = \dfrac{2\sqrt{3}}{3}; \sec \dfrac{2\pi}{3} = -2; \cot \dfrac{2\pi}{3} = -\dfrac{\sqrt{3}}{3}$

49. $\sin 210° = -\dfrac{1}{2}; \cos 210° = -\dfrac{\sqrt{3}}{2}; \tan 210° = \dfrac{\sqrt{3}}{3}; \csc 210° = -2; \sec 210° = -\dfrac{2\sqrt{3}}{3}; \cot 210° = \sqrt{3}$

51. $\sin \dfrac{3\pi}{4} = \dfrac{\sqrt{2}}{2}; \cos \dfrac{3\pi}{4} = -\dfrac{\sqrt{2}}{2}; \tan \dfrac{3\pi}{4} = -1; \csc \dfrac{3\pi}{4} = \sqrt{2}; \sec \dfrac{3\pi}{4} = -\sqrt{2}; \cot \dfrac{3\pi}{4} = -1$

53. $\sin \dfrac{8\pi}{3} = \dfrac{\sqrt{3}}{2}; \cos \dfrac{8\pi}{3} = -\dfrac{1}{2}; \tan \dfrac{8\pi}{3} = -\sqrt{3}; \csc \dfrac{8\pi}{3} = \dfrac{2\sqrt{3}}{3}; \sec \dfrac{8\pi}{3} = -2; \cot \dfrac{8\pi}{3} = -\dfrac{\sqrt{3}}{3}$

55. $\sin 405° = \dfrac{\sqrt{2}}{2}; \cos 405° = \dfrac{\sqrt{2}}{2}; \tan 405° = 1; \csc 405° = \sqrt{2}; \sec 405° = \sqrt{2}; \cot 405° = 1$

57. $\sin\left(-\dfrac{\pi}{6}\right) = -\dfrac{1}{2}; \cos\left(-\dfrac{\pi}{6}\right) = \dfrac{\sqrt{3}}{2}; \tan\left(-\dfrac{\pi}{6}\right) = -\dfrac{\sqrt{3}}{3}; \csc\left(-\dfrac{\pi}{6}\right) = -2; \sec\left(-\dfrac{\pi}{6}\right) = \dfrac{2\sqrt{3}}{3}; \cot\left(-\dfrac{\pi}{6}\right) = -\sqrt{3}$

59. $\sin(-135°) = -\dfrac{\sqrt{2}}{2}; \cos(-135°) = -\dfrac{\sqrt{2}}{2}; \tan(-135°) = 1; \csc(-135°) = -\sqrt{2}; \sec(-135°) = -\sqrt{2}; \cot(-135°) = 1$

61. $\sin \dfrac{5\pi}{2} = 1; \cos \dfrac{5\pi}{2} = 0; \tan \dfrac{5\pi}{2}$ is undefined; $\csc \dfrac{5\pi}{2} = 1; \sec \dfrac{5\pi}{2}$ is undefined; $\cot \dfrac{5\pi}{2} = 0$

63. $\sin\left(-\dfrac{14\pi}{3}\right) = -\dfrac{\sqrt{3}}{2}; \cos\left(-\dfrac{14\pi}{3}\right) = -\dfrac{1}{2}; \tan\left(-\dfrac{14\pi}{3}\right) = \sqrt{3}; \csc\left(-\dfrac{14\pi}{3}\right) = -\dfrac{2\sqrt{3}}{3}; \sec\left(-\dfrac{14\pi}{3}\right) = -2; \cot\left(-\dfrac{14\pi}{3}\right) = \dfrac{\sqrt{3}}{3}$

65. 0.47 **67.** 1.07 **69.** 0.32 **71.** 3.73 **73.** 0.84 **75.** 0.02 **77.** $\sin \theta = \dfrac{4}{5}; \cos \theta = -\dfrac{3}{5}; \tan \theta = -\dfrac{4}{3}; \csc \theta = \dfrac{5}{4}; \sec \theta = -\dfrac{5}{3}; \cot \theta = -\dfrac{3}{4}$

79. $\sin \theta = -\dfrac{3\sqrt{13}}{13}; \cos \theta = \dfrac{2\sqrt{13}}{13}; \tan \theta = -\dfrac{3}{2}; \csc \theta = -\dfrac{\sqrt{13}}{3}; \sec \theta = \dfrac{\sqrt{13}}{2}; \cot \theta = -\dfrac{2}{3}$

81. $\sin \theta = -\dfrac{\sqrt{2}}{2}; \cos \theta = -\dfrac{\sqrt{2}}{2}; \tan \theta = 1; \csc \theta = -\sqrt{2}; \sec \theta = -\sqrt{2}; \cot \theta = 1$

83. $\sin \theta = \dfrac{3}{5}; \cos \theta = \dfrac{4}{5}; \tan \theta = \dfrac{3}{4}; \csc \theta = \dfrac{5}{3}; \sec \theta = \dfrac{5}{4}; \cot \theta = \dfrac{4}{3}$ **85.** 0 **87.** 0 **89.** -0.1 **91.** 3 **93.** 5 **95.** $\dfrac{\sqrt{3}}{2}$ **97.** $\dfrac{1}{2}$ **99.** $\dfrac{3}{4}$

101. $\dfrac{\sqrt{3}}{2}$ **103.** $\sqrt{3}$ **105.** $-\dfrac{\sqrt{3}}{2}$ **107.** $\dfrac{1 + \sqrt{3}}{2}$ **109.** $-\dfrac{1}{2}$ **111.** $\dfrac{\sqrt{3}}{2}$ **113.** $\dfrac{\sqrt{2}}{4}$ **115. (a)** $\dfrac{\sqrt{2}}{2}; \left(\dfrac{\pi}{4}, \dfrac{\sqrt{2}}{2}\right)$ **(b)** $\left(\dfrac{\sqrt{2}}{2}, \dfrac{\pi}{4}\right)$ **(c)** $\left(\dfrac{\pi}{4}, -2\right)$

117. Answers may vary. One set of possible answers is $-\dfrac{11\pi}{3}, -\dfrac{5\pi}{3}, \dfrac{\pi}{3}, \dfrac{7\pi}{3}, \dfrac{13\pi}{3}$.

119.

θ	0.5	0.4	0.2	0.1	0.01	0.001	0.0001	0.00001
$\sin \theta$	0.4794	0.3894	0.1987	0.0998	0.0100	0.0010	0.0001	0.00001
$\dfrac{\sin \theta}{\theta}$	0.9589	0.9735	0.9933	0.9983	1.0000	1.0000	1.0000	1.0000

$\dfrac{\sin \theta}{\theta}$ approaches 1 as θ approaches 0.

121. $R \approx 310.56$ ft; $H \approx 77.64$ ft   **123.** $R \approx 19{,}541.95$ m; $H \approx 2278.14$ m   **125. (a)** 1.20 sec   **(b)** 1.12 sec   **(c)** 1.20 sec
127. (a) 1.9 hr; 0.57 hr   **(b)** 1.69 hr; 0.75 hr   **(c)** 1.63 hr; 0.86 hr   **(d)** 1.67 hr; tan 90° is undefined   **129.** 15.4 ft   **131.** 71 ft
133. (a) 466.9 ft, 587.7 ft, 612.0 ft   **(b)** $\theta \approx 0.678$ (38.8°) or $\theta \approx 1.364$ (78.2°)   **(c)** $\theta \approx 0.986$ (56.5°); this is larger than the angle needed to get the
maximum range.   **135. (a)** 16.56 ft   **(b)**   **137. (a)** values estimated to the nearest tenth:
(c) 67.5°

$\sin 1 \approx 0.8; \cos 1 \approx 0.5; \tan 1 \approx 1.6; \csc 1 \approx 1.3; \sec 1 \approx 2.0;$
$\cot 1 \approx 0.6;$ actual values to the nearest tenth: $\sin 1 \approx 0.8; \cos 1 \approx 0.5;$
$\tan 1 \approx 1.6; \csc 1 \approx 1.2; \sec 1 \approx 1.9; \cot 1 \approx 0.6$
(b) values estimated to the nearest tenth: $\sin 5.1 \approx -0.9; \cos 5.1 \approx 0.4;$
$\tan 5.1 \approx -2.3; \csc 5.1 \approx -1.1; \sec 5.1 \approx 2.5; \cot 5.1 \approx -0.4;$
actual values to the nearest tenth: $\sin 5.1 \approx -0.9; \cos 5.1 \approx 0.4;$
$\tan 5.1 \approx -2.4; \csc 5.1 \approx -1.1; \sec 5.1 \approx 2.6; \cot 5.1 \approx -0.4$

139. $y = -\dfrac{4\sqrt{3}}{7}$ **141.** $-\dfrac{12}{13}$ **146.** $4 - 3i, -5$ **147.** $R = 134$ **148.** 81π ft^2 **149.** $\dfrac{7 \pm \sqrt{157}}{6}$ **150.** $[-3, 2]$ **151.** $[-5, -4)$ **152.** $(6, -2)$ **153.** $\dfrac{x^4 + 4}{4x^2}$

2.3 Assess Your Understanding *(page 143)*

5. $2\pi; \pi$ **6.** All real numbers except odd multiples of $\dfrac{\pi}{2}$ **7.** b **8.** a **9.** 1 **10.** F **11.** $\dfrac{\sqrt{2}}{2}$ **13.** 1 **15.** 1 **17.** $\sqrt{3}$ **19.** $\dfrac{\sqrt{2}}{2}$ **21.** 0 **23.** $\sqrt{2}$ **25.** $\dfrac{\sqrt{3}}{3}$

27. II **29.** IV **31.** IV **33.** II **35.** $\tan\theta = -\dfrac{3}{4}$; $\cot\theta = -\dfrac{4}{3}$; $\sec\theta = \dfrac{5}{4}$; $\csc\theta = -\dfrac{5}{3}$ **37.** $\tan\theta = 2$; $\cot\theta = \dfrac{1}{2}$; $\sec\theta = \sqrt{5}$; $\csc\theta = \dfrac{\sqrt{5}}{2}$

39. $\tan\theta = \dfrac{\sqrt{3}}{3}$; $\cot\theta = \sqrt{3}$; $\sec\theta = \dfrac{2\sqrt{3}}{3}$; $\csc\theta = 2$ **41.** $\tan\theta = -\dfrac{\sqrt{2}}{4}$; $\cot\theta = -2\sqrt{2}$; $\sec\theta = \dfrac{3\sqrt{2}}{4}$; $\csc\theta = -3$

43. $\cos\theta = -\dfrac{5}{13}$; $\tan\theta = -\dfrac{12}{5}$; $\csc\theta = \dfrac{13}{12}$; $\sec\theta = -\dfrac{13}{5}$; $\cot\theta = -\dfrac{5}{12}$ **45.** $\sin\theta = -\dfrac{3}{5}$; $\tan\theta = \dfrac{3}{4}$; $\csc\theta = -\dfrac{5}{3}$; $\sec\theta = -\dfrac{5}{4}$; $\cot\theta = \dfrac{4}{3}$

47. $\cos\theta = -\dfrac{12}{13}$; $\tan\theta = -\dfrac{5}{12}$; $\csc\theta = \dfrac{13}{5}$; $\sec\theta = -\dfrac{13}{12}$; $\cot\theta = -\dfrac{12}{5}$ **49.** $\sin\theta = \dfrac{2\sqrt{2}}{3}$; $\tan\theta = -2\sqrt{2}$; $\csc\theta = \dfrac{3\sqrt{2}}{4}$; $\sec\theta = -3$; $\cot\theta = -\dfrac{\sqrt{2}}{4}$

51. $\cos\theta = -\dfrac{\sqrt{5}}{3}$; $\tan\theta = -\dfrac{2\sqrt{5}}{5}$; $\csc\theta = \dfrac{3}{2}$; $\sec\theta = -\dfrac{3\sqrt{5}}{5}$; $\cot\theta = -\dfrac{\sqrt{5}}{2}$ **53.** $\sin\theta = -\dfrac{\sqrt{3}}{2}$; $\cos\theta = \dfrac{1}{2}$; $\tan\theta = -\sqrt{3}$;

$\csc\theta = -\dfrac{2\sqrt{3}}{3}$; $\cot\theta = -\dfrac{\sqrt{3}}{3}$ **55.** $\sin\theta = -\dfrac{3}{5}$; $\cos\theta = -\dfrac{4}{5}$; $\csc\theta = -\dfrac{5}{3}$; $\sec\theta = -\dfrac{5}{4}$; $\cot\theta = \dfrac{4}{3}$ **57.** $\sin\theta = \dfrac{\sqrt{10}}{10}$; $\cos\theta = -\dfrac{3\sqrt{10}}{10}$;

$\csc\theta = \sqrt{10}$; $\sec\theta = -\dfrac{\sqrt{10}}{3}$; $\cot\theta = -3$ **59.** $-\dfrac{\sqrt{3}}{2}$ **61.** $-\dfrac{\sqrt{3}}{3}$ **63.** 2 **65.** -1 **67.** -1 **69.** $\dfrac{\sqrt{2}}{2}$ **71.** 0 **73.** $-\sqrt{2}$ **75.** $\dfrac{2\sqrt{3}}{3}$

77. 1 **79.** 1 **81.** 0 **83.** 1 **85.** -1 **87.** 0 **89.** 0.9 **91.** 9 **93.** 0 **95.** All real numbers **97.** Odd multiples of $\dfrac{\pi}{2}$ **99.** Odd multiples of $\dfrac{\pi}{2}$

101. $-1 \le y \le 1$ **103.** All real numbers **105.** $|y| \ge 1$ **107.** Odd; yes; origin **109.** Odd; yes; origin **111.** Even; yes; y-axis

113. (a) $-\dfrac{1}{3}$ **(b)** 1 **115. (a)** -2 **(b)** 6 **117. (a)** -4 **(b)** -12 **119.** ≈ 15.81 min **121.** 130, 90, 70

123. Let a be a real number and $P = (x, y)$ be the point on the unit circle that corresponds to t. Consider the equation $\tan t = \dfrac{y}{x} = a$. Then $y = ax$.

But $x^2 + y^2 = 1$, so $x^2 + a^2x^2 = 1$. So $x = \pm\dfrac{1}{\sqrt{1 + a^2}}$ and $y = \pm\dfrac{a}{\sqrt{1 + a^2}}$; that is, for any real number a, there is a point $P = (x, y)$ on the

unit circle for which $\tan t = a$. In other words, the range of the tangent function is the set of all real numbers.

125. Suppose that there is a number $p, 0 < p < 2\pi$, for which $\sin(\theta + p) = \sin\theta$ for all θ. If $\theta = 0$, then $\sin(0 + p) = \sin p = \sin 0 = 0$, so $p = \pi$.

If $\theta = \dfrac{\pi}{2}$, then $\sin\left(\dfrac{\pi}{2} + p\right) = \sin\left(\dfrac{\pi}{2}\right)$. But $p = \pi$. Thus, $\sin\left(\dfrac{3\pi}{2}\right) = -1 = \sin\left(\dfrac{\pi}{2}\right) = 1$. This is impossible. Therefore, the smallest positive

number p for which $\sin(\theta + p) = \sin\theta$ for all θ is 2π. **127.** $\sec\theta = \dfrac{1}{\cos\theta}$; since $\cos\theta$ has period 2π, so does $\sec\theta$.

129. If $P = (x, y)$ is the point on the unit circle corresponding to θ, then $Q = (-x, -y)$ is the point on the unit circle corresponding to $\theta + \pi$. Thus,

$\tan(\theta + \pi) = \dfrac{-y}{-x} = \dfrac{y}{x} = \tan\theta$. Suppose that there exists a number $p, 0 < p < \pi$, for which $\tan(\theta + p) = \tan\theta$ for all θ. Then, if $\theta = 0$, then

$\tan p = \tan 0 = 0$. But this means that p is a multiple of π. Since no multiple of π exists in the interval $(0, \pi)$, this is a contradiction. Therefore,

the period of $f(\theta) = \tan\theta$ is π.

131. Let $P = (x, y)$ be the point on the unit circle corresponding to θ. Then $\csc\theta = \dfrac{1}{y} = \dfrac{1}{\sin\theta}$; $\sec\theta = \dfrac{1}{x} = \dfrac{1}{\cos\theta}$; $\cot\theta = \dfrac{x}{y} = \dfrac{1}{y/x} = \dfrac{1}{\tan\theta}$.

133. $(\sin\theta\cos\phi)^2 + (\sin\theta\sin\phi)^2 + \cos^2\theta = \sin^2\theta\cos^2\phi + \sin^2\theta\sin^2\phi + \cos^2\theta = \sin^2\theta(\cos^2\phi + \sin^2\phi) + \cos^2\theta = \sin^2\theta + \cos^2\theta = 1$

135. $\dfrac{7}{5}$ **137.** $-\dfrac{7}{13}$ **143.** $f(g(x)) = x^2 - 14x + 46$ **144.**

Vertex: $(3, 5)$
axis of symmetry: $x = 3$

145. 9 **146.** $\left\{\dfrac{89}{16}\right\}$ **147.** -8 and -3

148. $c = 35$ **149.** $(0, 3), (-5, 0)$

150. $3x - 5 + \dfrac{3}{2}h$

2.4 Assess Your Understanding *(page 156)*

3. $1; \dfrac{\pi}{2}$ **4.** $3; \pi$ **5.** $3; \dfrac{\pi}{3}$ **6.** T **7.** F **8.** T **9.** d **10.** d **11. (a)** 0 **(b)** $-\dfrac{\pi}{2} \le x \le \dfrac{\pi}{2}$ **(c)** 1 **(d)** $0, \pi, 2\pi$

(e) $f(x) = 1$ for $x = -\dfrac{3\pi}{2}, \dfrac{\pi}{2}$; $f(x) = -1$ for $x = -\dfrac{\pi}{2}, \dfrac{3\pi}{2}$ **(f)** $-\dfrac{5\pi}{6}, -\dfrac{\pi}{6}, \dfrac{7\pi}{6}, \dfrac{11\pi}{6}$ **(g)** $\{x \mid x = k\pi, k \text{ an integer}\}$

13. Amplitude $= 5$; period $= 2\pi$ **15.** Amplitude $= 3$; period $= \dfrac{\pi}{2}$ **17.** Amplitude $= 6$; period $= 2$ **19.** Amplitude $= \dfrac{1}{7}$; period $= \dfrac{4\pi}{7}$

21. Amplitude $= \dfrac{10}{9}$; period $= 5$ **23.** F **25.** A **27.** H **29.** C **31.** J

33.

Domain: $(-\infty, \infty)$
Range: $[-4, 4]$

35.

Domain: $(-\infty, \infty)$
Range: $[-4, 4]$

37.

Domain: $(-\infty, \infty)$
Range: $[-1, 1]$

39.

Domain: $(-\infty, \infty)$
Range: $[-1, 1]$

41.

Domain: $(-\infty, \infty)$
Range: $[-2, 2]$

43.

Domain: $(-\infty, \infty)$
Range: $\left[-\dfrac{1}{2}, \dfrac{1}{2}\right]$

45.

Domain: $(-\infty, \infty)$
Range: $[1, 5]$

47.

Domain: $(-\infty, \infty)$
Range: $[-8, 2]$

49.

Domain: $(-\infty, \infty)$
Range: $[-2, 10]$

51.

Domain: $(-\infty, \infty)$
Range: $[2, 8]$

53.

Domain: $(-\infty, \infty)$
Range: $\left[-\dfrac{5}{3}, \dfrac{5}{3}\right]$

55.

Domain: $(-\infty, \infty)$
Range: $[-1, 2]$

57. $y = \pm 3\sin(2x)$ **59.** $y = \pm 3\sin(\pi x)$ **61.** $y = 5\cos\left(\dfrac{\pi}{4}x\right)$ **63.** $y = -3\cos\left(\dfrac{1}{2}x\right)$ **65.** $y = \dfrac{3}{4}\sin(2\pi x)$ **67.** $y = -\sin\left(\dfrac{3}{2}x\right)$

69. $y = -\cos\left(\dfrac{4\pi}{3}x\right) + 1$ **71.** $y = 3\sin\left(\dfrac{\pi}{2}x\right)$ **73.** $y = -4\cos(3x)$ **75.** $\dfrac{2}{\pi}$ **77.** $\dfrac{\sqrt{2}}{\pi}$

79. $f(g(x)) = \sin(4x)$

$g(f(x)) = 4\sin x$

81. $f(g(x)) = -2\cos x$

$g(f(x)) = \cos(-2x)$

83.

85.

87. Period $= \dfrac{1}{30}$ s
Amplitude $= 220$ amp

89. (a) $P(t) = \dfrac{[V_0 \sin(2\pi f t)]^2}{R} = \dfrac{V_0^2}{R}\sin^2(2\pi f t)$ **(b)** Since the graph of P has amplitude $\dfrac{V_0^2}{2R}$ and period $\dfrac{1}{2f}$ and is of the form $y = A\cos(\omega t) + B$,
then $A = -\dfrac{V_0^2}{2R}$ and $B = \dfrac{V_0^2}{2R}$. Since $\dfrac{1}{2f} = \dfrac{2\pi}{\omega}$, then $\omega = 4\pi f$. Therefore, $P(t) = -\dfrac{V_0^2}{2R}\cos(4\pi f t) + \dfrac{V_0^2}{2R} = \dfrac{V_0^2}{2R}[1 - \cos(4\pi f t)]$.

91. (a) 120 **(b)** 80 **(c)** 70 beats per minute **93. (a)** 75.4°F **(b)** 28.1°F **(c)** 6 months **95. (a)** 205 ft **(b)** 5 ft **(c)** 3 **(d)** 4.3 mph

97. (a) Physical potential: $\omega = \dfrac{2\pi}{23}$; emotional potential: $\omega = \dfrac{\pi}{14}$; intellectual potential: $\omega = \dfrac{2\pi}{33}$

(b) **(c)** No **(d)** Physical potential peaks at 15 days after 20th birthday. Emotional potential is 50% at 17 days, with a maximum at 10 days and a minimum at 24 days. Intellectual potential starts fairly high, drops to a minimum at 13 days, and rises to a maximum at 29 days.

99. $(0, A\cos(-BC) + A), \left(\dfrac{(2k+1)\pi}{B} + C, 0\right)$ k an integer **105.** $2x + h - 5$ **106.** $(2, 5)$ **107.** $(0, 5), \left(-\dfrac{5}{3}, 0\right), \left(-\dfrac{7}{3}, 0\right)$ **108.** $\{-8\}$

109. $y = 2x + 7$ **110.** $\left[-\dfrac{19}{12}, \infty\right)$ **111.** $(-\infty, -2] \cup \left(\dfrac{4}{3}, \infty\right)$

2.5 Assess Your Understanding *(page 167)*

3. origin; odd multiples of $\dfrac{\pi}{2}$ **4.** y-axis; odd multiples of $\dfrac{\pi}{2}$ **5.** b **6.** T **7.** 0 **9.** 1

11. $\sec x = 1$ for $x = -2\pi, 0, 2\pi$; $\sec x = -1$ for $x = -\pi, \pi$ **13.** $-\dfrac{3\pi}{2}, -\dfrac{\pi}{2}, \dfrac{\pi}{2}, \dfrac{3\pi}{2}$ **15.** $-\dfrac{3\pi}{2}, -\dfrac{\pi}{2}, \dfrac{\pi}{2}, \dfrac{3\pi}{2}$

17.

Domain: $\left\{x \,\middle|\, x \neq \dfrac{k\pi}{2}, k \text{ is an odd integer}\right\}$
Range: $(-\infty, \infty)$

19.

Domain: $\{x \mid x \neq k\pi, k \text{ is an integer}\}$
Range: $(-\infty, \infty)$

21.

Domain: $\{x \mid x \text{ does not equal an odd integer}\}$
Range: $(-\infty, \infty)$

23.

Domain: $\{x \mid x \neq 4k\pi, k \text{ is an integer}\}$
Range: $(-\infty, \infty)$

25.

Domain: $\left\{x \,\middle|\, x \neq \dfrac{k\pi}{2}, k \text{ is an odd integer}\right\}$
Range: $\{y \mid y \leq -2 \text{ or } y \geq 2\}$

27.

Domain: $\{x \mid x \neq k\pi, k \text{ is an integer}\}$
Range: $\{y \mid y \leq -3 \text{ or } y \geq 3\}$

29.

Domain: $\{x \mid x \neq k\pi, k \text{ is an odd integer}\}$
Range: $\{y \mid y \leq -4 \text{ or } y \geq 4\}$

31.

Domain: $\{x \mid x \text{ does not equal an integer}\}$
Range: $\{y \mid y \leq -2 \text{ or } y \geq 2\}$

33.

Domain: $\{x \mid x \neq 2\pi k, k \text{ is an odd integer}\}$
Range: $(-\infty, \infty)$

35.

Domain: $\left\{x \,\middle|\, x \neq \dfrac{3}{4}k, k \text{ is an odd integer}\right\}$
Range: $\{y \mid y \leq 1 \text{ or } y \geq 3\}$

37.

Domain: $\{x \mid x \neq 2\pi k, k \text{ is an odd integer}\}$
Range: $(-\infty, \infty)$

39.

Domain: $\{x \mid x \neq 3\pi k, k \text{ is an integer}\}$
Range: $\{y \mid y \leq -3 \text{ or } y \geq 1\}$

41. $\dfrac{2\sqrt{3}}{\pi}$ **43.** $\dfrac{6\sqrt{3}}{\pi}$ **45.** $f(g(x)) = \tan(4x)$ $g(f(x)) = 4\tan x$

47. $f(g(x)) = -2\cot x$ $g(f(x)) = \cot(-2x)$

49.

51. (a) $L(\theta) = \dfrac{3}{\cos\theta} + \dfrac{4}{\sin\theta}$
$= 3\sec\theta + 4\csc\theta$

(b)

(c) ≈ 0.83 (d) ≈ 9.86 ft

53.

$y = \tan x$

$y = -\cot\left(x + \dfrac{\pi}{2}\right)$

55. $(5p - 2q^2)(25p^2 + 10pq^2 + 4q^4)$ **56.** Hazel: 4 hours; Gwyneth: 6 hours

57.

58. $x + c - 3$ **59.** $(0, -2), \left(\dfrac{3}{2}, 0\right), (-2, 0)$ **60.** $\sqrt{(x+1)^2 + 5^2}$ **61.** $\left[\dfrac{2}{5}, \infty\right)$

2.6 Assess Your Understanding (page 178)

1. phase shift   **2.** False

3. Amplitude $= 4$
Period $= \pi$
Phase shift $= \dfrac{\pi}{2}$

5. Amplitude $= 2$
Period $= \dfrac{2\pi}{3}$
Phase shift $= -\dfrac{\pi}{6}$

7. Amplitude $= 3$
Period $= \pi$
Phase shift $= -\dfrac{\pi}{4}$

9. Amplitude $= 4$
Period $= 2$
Phase shift $= -\dfrac{2}{\pi}$

11. Amplitude $= 3$
Period $= 2$
Phase shift $= \dfrac{2}{\pi}$

13. Amplitude $= 3$
Period $= \pi$
Phase shift $= \dfrac{\pi}{4}$

15.

17.

19. $y = 2\sin\left[2\left(x - \dfrac{1}{2}\right)\right]$ or $y = 2\sin(2x - 1)$ **21.** $y = 3\sin\left[\dfrac{2}{3}\left(x + \dfrac{1}{3}\right)\right]$ or $y = 3\sin\left(\dfrac{2}{3}x + \dfrac{2}{9}\right)$

23. Period $= \dfrac{1}{15}$ s
Amplitude $= 120$ amp
Phase shift $= \dfrac{1}{90}$ s

25. (a)

(b) $y = 8.5\sin\left[\dfrac{2\pi}{5}\left(x - \dfrac{11}{4}\right)\right] + 24.5$ or $y = 8.5\sin\left(\dfrac{2\pi}{5}x - \dfrac{11\pi}{10}\right) + 24.5$

(c)

(d) $y = 9.46\sin(1.247x + 2.096) + 24.088$

(e)

27. (a)

(b) $y = 23.65\sin\left[\dfrac{\pi}{6}(x - 4)\right] + 51.75$ or $y = 23.65\sin\left(\dfrac{\pi}{6}x - \dfrac{2\pi}{3}\right) + 51.75$

(c)

(d) $y = 24.25\sin(0.493x - 1.927) + 51.61$

(e)

29. (a) 12:50 pm   **(b)** $y = 2.91\sin\left[\dfrac{24\pi}{149}(x + 2.688)\right] + 2.97$ or $y = 2.91\sin\left[\dfrac{24\pi}{149}x + 1.360\right] + 2.97$ **(c)** 1.48 ft

31. (a) $y = 1.615\sin\left(\dfrac{2\pi}{365}x - 1.39\right) + 12.135$
(b) 12.42 h
(c)

33. (a) $y = 6.96\sin\left(\dfrac{2\pi}{365}x - 1.39\right) + 12.41$
(b) 13.63 h
(c)

(e) The actual hours of sunlight on April 1, 2018, were 12.43 hours. This is close to the predicted amount of 12.42 hours.

(d) The actual hours of sunlight on April 1, 2018, were 13.37 hours. This is close to the predicted amount of 13.63 hours.

35. $y = 51 \sin\left[\dfrac{5\pi}{9}(t + 0.9)\right] + 55$ or $y = 51 \sin\left(\dfrac{5\pi}{9}t + \dfrac{\pi}{2}\right) + 55$ **38.** $f^{-1}(x) = \dfrac{2x - 9}{4}$ **39.** $\left\{\dfrac{7}{3}\right\}$ **40.** $64x^2 + 240xy + 225y^2$ **41.** $2\sqrt{13}$

42. $\left\{\dfrac{3}{8}, \dfrac{11}{2}\right\}$ **43.** $y = u^{3/2} - 4u^{1/2}$ or $y = (u - 4)\sqrt{u}$ **44.** $\dfrac{1}{a}\sqrt{a^2 - x^2}$ **45.** 8 ft $\times$ 19 ft   **46.** $x = -3$

Review Exercises *(page 184)*

1. $\dfrac{3\pi}{4}$ **2.** $\dfrac{\pi}{10}$ **3.** $135°$ **4.** $-450°$ **5.** $\dfrac{1}{2}$ **6.** $\dfrac{3\sqrt{2}}{2} - \dfrac{4\sqrt{3}}{3}$ **7.** $-3\sqrt{2} - 2\sqrt{3}$ **8.** 3   **9.** 0   **10.** 0   **11.** 1   **12.** 1   **13.** 1   **14.** -1 **15.** 1

16. $\cos\theta = \dfrac{3}{5}$; $\tan\theta = \dfrac{4}{3}$; $\csc\theta = \dfrac{5}{4}$; $\sec\theta = \dfrac{5}{3}$; $\cot\theta = \dfrac{3}{4}$ **17.** $\sin\theta = -\dfrac{12}{13}$; $\cos\theta = -\dfrac{5}{13}$; $\csc\theta = -\dfrac{13}{12}$; $\sec\theta = -\dfrac{13}{5}$; $\cot\theta = \dfrac{5}{12}$

18. $\sin\theta = \dfrac{3}{5}$; $\cos\theta = -\dfrac{4}{5}$; $\tan\theta = -\dfrac{3}{4}$; $\csc\theta = \dfrac{5}{3}$; $\cot\theta = -\dfrac{4}{3}$ **19.** $\cos\theta = -\dfrac{5}{13}$; $\tan\theta = -\dfrac{12}{5}$; $\csc\theta = \dfrac{13}{12}$; $\sec\theta = -\dfrac{13}{5}$; $\cot\theta = -\dfrac{5}{12}$

20. $\cos\theta = \dfrac{12}{13}$; $\tan\theta = -\dfrac{5}{12}$; $\csc\theta = -\dfrac{13}{5}$; $\sec\theta = \dfrac{13}{12}$; $\cot\theta = -\dfrac{12}{5}$ **21.** $\sin\theta = -\dfrac{\sqrt{10}}{10}$; $\cos\theta = -\dfrac{3\sqrt{10}}{10}$; $\csc\theta = -\sqrt{10}$; $\sec\theta = -\dfrac{\sqrt{10}}{3}$; $\cot\theta = 3$

22. $\sin\theta = -\dfrac{2\sqrt{2}}{3}$; $\cos\theta = \dfrac{1}{3}$; $\tan\theta = -2\sqrt{2}$; $\csc\theta = -\dfrac{3\sqrt{2}}{4}$; $\cot\theta = -\dfrac{\sqrt{2}}{4}$

23. $\sin\theta = \dfrac{\sqrt{5}}{5}$; $\cos\theta = -\dfrac{2\sqrt{5}}{5}$; $\tan\theta = -\dfrac{1}{2}$; $\csc\theta = \sqrt{5}$; $\sec\theta = -\dfrac{\sqrt{5}}{2}$

24.

Domain: $(-\infty, \infty)$
Range: $[-2, 2]$

25.

Domain: $(-\infty, \infty)$
Range: $[-3, 3]$

26.

Domain: $\left\{x \middle| x \neq \dfrac{k\pi}{2}, k \text{ is an odd integer}\right\}$
Range: $(-\infty, \infty)$

27.

Domain: $\left\{x \middle| x \neq \dfrac{\pi}{6} + k \cdot \dfrac{\pi}{3}, k \text{ is an integer}\right\}$
Range: $(-\infty, \infty)$

28.

Domain: $\left\{x \middle| x \neq -\dfrac{\pi}{4} + k\pi, k \text{ is an integer}\right\}$
Range: $(-\infty, \infty)$

29.

Domain: $\left\{x \middle| x \neq \dfrac{k\pi}{4}, k \text{ is an odd integer}\right\}$
Range: $\{y | y \leq -4 \text{ or } y \geq 4\}$

30.

Domain: $\left\{x \middle| x \neq -\dfrac{\pi}{4} + k\pi, k \text{ is an integer}\right\}$
Range: $\{y | y \leq -1 \text{ or } y \geq 1\}$

31.
Domain: $(-\infty, \infty)$
Range: $[-6, 2]$

32.

Domain: $\left\{x \middle| x \neq \dfrac{3\pi}{4} + k \cdot 3\pi, k \text{ is an integer}\right\}$
Range: $(-\infty, \infty)$

33. Amplitude = 1; Period = π **34.** Amplitude = 2; Period = $\dfrac{2}{3}$

35. Amplitude = 4
Period = $\dfrac{2\pi}{3}$
Phase shift = 0

36. Amplitude = 1
Period = 4π
Phase shift = $-\pi$

37. Amplitude = $\dfrac{1}{2}$
Period = $\dfrac{4\pi}{3}$
Phase shift = $\dfrac{2\pi}{3}$

38. Amplitude = $\dfrac{2}{3}$
Period = 2
Phase shift = $\dfrac{6}{\pi}$

39. $y = 5\cos\dfrac{x}{4}$ **40.** $y = -7\sin\left(\dfrac{\pi}{4}x\right)$ **41.** 0.38   **42.** 1.02   **43.** Sine, cosine, cosecant, and secant: negative; tangent and cotangent: positive

44. IV   **45.** $\sin\theta = \dfrac{2\sqrt{2}}{3}$; $\cos\theta = -\dfrac{1}{3}$; $\tan\theta = -2\sqrt{2}$; $\csc\theta = \dfrac{3\sqrt{2}}{4}$; $\sec\theta = -3$; $\cot\theta = -\dfrac{\sqrt{2}}{4}$

46. $\sin t = \dfrac{5\sqrt{29}}{29}$, $\cos t = -\dfrac{2\sqrt{29}}{29}$, $\tan t = -\dfrac{5}{2}$ **47.** Domain: $\left\{x \middle| x \neq \text{ odd multiple of } \dfrac{\pi}{2}\right\}$; range: $\{y | y \leq -1 \text{ or } y \geq 1\}$

48. (a) $32.34°$ **(b)** $63°10'48''$

49. $\frac{\pi}{3} \approx 1.05$ ft; $\frac{\pi}{3} \approx 1.05$ ft^2 **50.** $8\pi \approx 25.13$ in.; $\frac{16\pi}{3} \approx 16.76$ in. **51.** Approximately 114.59 revolutions/hr

52. 0.1 revolution/sec $= \frac{\pi}{5}$ radian/sec

53. (a) $\frac{1}{15}$

(b) 220

(c) $-\frac{1}{180}$

(d)

54. (a)

(b) $y = 20\sin\left[\frac{\pi}{6}(x-4)\right] + 75$ or

$y = 20\sin\left(\frac{\pi}{6}x - \frac{2\pi}{3}\right) + 75$

(c)

(d) $y = 19.81\sin(0.543x - 2.296) + 75.66$

(e)

55.

Chapter Test *(page 186)*

1. $\frac{13\pi}{9}$ **2.** $-\frac{20\pi}{9}$ **3.** $\frac{13\pi}{180}$ **4.** $-22.5°$ **5.** $810°$ **6.** $135°$ **7.** $\frac{1}{2}$ **8.** 0 **9.** $-\frac{1}{2}$ **10.** $-\frac{\sqrt{3}}{3}$ **11.** 2 **12.** $\frac{3(1-\sqrt{2})}{2}$ **13.** 0.292 **14.** 0.309

15. -1.524 **16.** 2.747 **17.**

	sin θ	cos θ	tan θ	sec θ	csc θ	cot θ
θ in QI	+	+	+	+	+	+
θ in QII	+	−	−	−	+	−
θ in QIII	−	−	+	−	−	+
θ in QIV	−	+	−	+	−	−

18. $-\frac{3}{5}$

19. $\cos\theta = -\frac{2\sqrt{6}}{7}$; $\tan\theta = -\frac{5\sqrt{6}}{12}$; $\csc\theta = \frac{7}{5}$; $\sec\theta = -\frac{7\sqrt{6}}{12}$; $\cot\theta = -\frac{2\sqrt{6}}{5}$ **20.** $\sin\theta = -\frac{\sqrt{5}}{3}$; $\tan\theta = -\frac{\sqrt{5}}{2}$; $\csc\theta = -\frac{3\sqrt{5}}{5}$;

$\sec\theta = \frac{3}{2}$; $\cot\theta = -\frac{2\sqrt{5}}{5}$ **21.** $\sin\theta = \frac{12}{13}$; $\cos\theta = -\frac{5}{13}$; $\csc\theta = \frac{13}{12}$; $\sec\theta = -\frac{13}{5}$; $\cot\theta = -\frac{5}{12}$ **22.** $\frac{7\sqrt{53}}{53}$ **23.** $-\frac{5\sqrt{146}}{146}$ **24.** $-\frac{1}{2}$

25.
$\left(-\frac{11\pi}{2}, 0\right)$ $(-4\pi, 2)$ $(2\pi, 2)$ $\left(\frac{7\pi}{2}, 0\right)$
$\left(-\frac{5\pi}{2}, 0\right)$ $(5\pi, -2)$
$(-\pi, -2)$ $\left(\frac{\pi}{2}, 0\right)$

26. $(-\pi, 3)$ $(0, 3)$

27. $y = -3\sin\left(3x + \frac{3\pi}{4}\right)$

28. 78.93 ft^2

29. 143.5 rpm

Cumulative Review *(page 187)*

1. $\left\{-1, \frac{1}{2}\right\}$ **2.** $y - 5 = -3(x+2)$ or $y = -3x - 1$ **3.** $x^2 + (y+2)^2 = 16$

4. A line; slope $\frac{2}{3}$; intercepts $(6, 0)$ and $(0, -4)$

5. A circle; center $(1, -2)$; radius 3

6.

7. (a)

(b)

(c)

(d)

8. $f^{-1}(x) = \frac{1}{3}(x+2)$ **9.** -2 **10.**

11. $3 - \frac{3\sqrt{3}}{2}$ **12.** $y = 3\cos\left(\frac{\pi}{6}x\right)$

13. (a) $f(x) = -3x - 3$;
$m = -3; (-1,0), (0,-3)$

(b) $f(x) = (x-1)^2 - 6; (0,-5)$,
$(-\sqrt{6}+1,0), (\sqrt{6}+1,0)$

CHAPTER 3 Analytic Trigonometry

3.1 Assess Your Understanding (page 199)

5. $x = \sin y$ **6.** $0 \le x \le \pi$ **7.** T   **8.** T   **9.** T   **10.** d   **11.** 0   **13.** $-\frac{\pi}{2}$ **15.** 0   **17.** $\frac{\pi}{4}$ **19.** $\frac{\pi}{3}$ **21.** $\frac{5\pi}{6}$ **23.** $\frac{\pi}{4}$ **25.** $-\frac{\pi}{6}$ **27.** 0.10   **29.** 1.37

31. 0.51   **33.** -0.38 **35.** -0.12 **37.** 1.08   **39.** $\frac{4\pi}{5}$ **41.** $-\frac{3\pi}{8}$ **43.** $-\frac{\pi}{8}$ **45.** $-\frac{\pi}{5}$ **47.** $\frac{\pi}{4}$ **49.** Not defined   **51.** $\frac{1}{4}$ **53.** 4   **55.** Not defined   **57.** π

59. $f^{-1}(x) = \sin^{-1}\frac{x-2}{5}$
Range of f = Domain of $f^{-1} = [-3,7]$
Range of $f^{-1} = \left[-\frac{\pi}{2}, \frac{\pi}{2}\right]$

61. $f^{-1}(x) = \frac{1}{3}\cos^{-1}\left(-\frac{x}{2}\right)$
Range of f = Domain of $f^{-1} = [-2,2]$
Range of $f^{-1} = \left[0, \frac{\pi}{3}\right]$

63. $f^{-1}(x) = -\tan^{-1}(x+3) - 1$
Range of f = Domain of $f^{-1} = (-\infty, \infty)$
Range of $f^{-1} = \left(-1 - \frac{\pi}{2}, \frac{\pi}{2} - 1\right)$

65. $f^{-1}(x) = \frac{1}{2}\left[\sin^{-1}\left(\frac{x}{3}\right) - 1\right]$
Range of f = Domain of $f^{-1} = [-3,3]$
Range of $f^{-1} = \left[-\frac{1}{2} - \frac{\pi}{4}, -\frac{1}{2} + \frac{\pi}{4}\right]$

67. $\left\{\frac{\sqrt{2}}{2}\right\}$ **69.** $\left\{-\frac{1}{4}\right\}$ **71.** $\{\sqrt{3}\}$ **73.** $\{-1\}$

75. (a) 13.92 h or 13 h, 55 min   **(b)** 12 h   **(c)** 13.85 h or 13 h, 51 min
77. (a) 13.3 h or 13 h, 18 min   **(b)** 12 h   **(c)** 13.26 h or 13 h, 15 min
79. (a) 12 h   **(b)** 12 h   **(c)** 12 h   **(d)** It is 12 h.   **81.** 3.35 min

83. (a) $\frac{\pi}{3}$ square units   **(b)** $\frac{5\pi}{12}$ square units   **85.** 4250 mi   **87.** $\left\{-\frac{\sqrt{7}}{4}, \frac{\sqrt{7}}{4}\right\}$ **89.** $\left[-\frac{2}{3}, 2\right]$ **90.** The graph passes the horizontal-line test.

91. $(2x+1)^{-\frac{1}{2}}(x^2+3)^{-\frac{3}{2}}(-x^2 - x + 3)$ **92.** 20 mph   **93.** $\frac{\sqrt{3}}{4}$

94. $\sin\theta = \frac{7}{25}, \tan\theta = \frac{7}{24}, \sec\theta = \frac{25}{24}, \csc\theta = \frac{25}{7}, \cot\theta = \frac{24}{7}$

95. Quadrant II   **96.** $\frac{12 - 4\sqrt{3}}{\pi}$

3.2 Assess Your Understanding (page 206)

4. $x = \sec y; \ge 1; 0; \pi$ **5.** cosine   **6.** F   **7.** T   **8.** T   **9.** $\frac{\pi}{6}$ **11.** $-\frac{\pi}{2}$ **13.** $\frac{\pi}{6}$ **15.** $\frac{2\pi}{3}$ **17.** $\frac{3\pi}{4}$ **19.** $-\frac{\pi}{4}$ **21.** 1.32   **23.** 0.46   **25.** -0.34 **27.** 2.72

29. -0.73 **31.** 2.55   **33.** $\frac{\sqrt{2}}{2}$ **35.** $-\frac{\sqrt{3}}{3}$ **37.** 2   **39.** $\sqrt{2}$ **41.** $-\frac{\sqrt{2}}{2}$ **43.** $\frac{2\sqrt{3}}{3}$ **45.** $\frac{3\pi}{4}$ **47.** $-\frac{\pi}{3}$ **49.** $\frac{\sqrt{2}}{4}$ **51.** $\frac{\sqrt{5}}{2}$ **53.** $-\frac{\sqrt{14}}{2}$

55. $-\frac{3\sqrt{10}}{10}$ **57.** $\sqrt{5}$ **59.** $-\frac{\pi}{4}$ **61.** $\frac{1}{\sqrt{1+u^2}}$ **63.** $\frac{u}{\sqrt{1-u^2}}$ **65.** $\frac{\sqrt{u^2-1}}{|u|}$ **67.** $\frac{\sqrt{u^2-1}}{|u|}$ **69.** $\frac{1}{u}$ **71.** $\frac{5}{13}$ **73.** $\frac{3\pi}{4}$ **75.** $-\frac{3}{4}$ **77.** $\frac{5}{13}$

79. $\dfrac{5\pi}{6}$ **81.** $-\sqrt{15}$ **83. (a)** $\theta = 31.89°$ **(b)** 54.64 ft in diameter   **(c)** 37.96 ft high   **85. (a)** $\theta = 22.3°$ **(b)** $v_0 = 2940.23$ ft/s   **87.** $\sqrt{2-x^2}$

90. $-5i, 5i, -2, 2$ **91.** Neither   **92.** $\dfrac{7\pi}{4}$ **93.** $\dfrac{5\pi}{2} \approx 7.85$ in.   **94. (a)** $\left(-\dfrac{5}{2}, \dfrac{31}{2}\right)$ **(b)** Concave down   **(c)** Increasing: $\left(-\infty, -\dfrac{5}{2}\right]$; decreasing: $\left[-\dfrac{5}{2}, \infty\right)$

95. $\{x \mid x \ge 3, x \ne 4, x \ne 7\}$ or $[-3, 4) \cup (4, 7) \cup (7, \infty)$ **96.** $y = 4\sin[6(x-1)]$ or $y = 4\sin(6x-6)$

97. $-\dfrac{x+c}{\sqrt{1-x^2} + \sqrt{1-c^2}}$ **98.** $-\dfrac{3\sqrt{3}+6}{5\pi}$

3.3 Assess Your Understanding *(page 214)*

7. F   **8.** T   **9.** T   **10.** F   **11.** d   **12.** a   **13.** $\left\{\dfrac{7\pi}{6}, \dfrac{11\pi}{6}\right\}$ **15.** $\left\{\dfrac{7\pi}{6}, \dfrac{11\pi}{6}\right\}$ **17.** $\left\{\dfrac{3\pi}{4}, \dfrac{7\pi}{4}\right\}$ **19.** $\left\{\dfrac{2\pi}{3}, \dfrac{4\pi}{3}\right\}$ **21.** $\left\{\dfrac{3\pi}{4}, \dfrac{5\pi}{4}\right\}$ **23.** $\left\{\dfrac{\pi}{3}, \dfrac{2\pi}{3}, \dfrac{4\pi}{3}, \dfrac{5\pi}{3}\right\}$

25. $\left\{\dfrac{\pi}{4}, \dfrac{3\pi}{4}, \dfrac{5\pi}{4}, \dfrac{7\pi}{4}\right\}$ **27.** $\left\{\dfrac{\pi}{2}, \dfrac{7\pi}{6}, \dfrac{11\pi}{6}\right\}$ **29.** $\left\{\dfrac{\pi}{3}, \dfrac{2\pi}{3}, \dfrac{4\pi}{3}, \dfrac{5\pi}{3}\right\}$ **31.** $\left\{\dfrac{4\pi}{9}, \dfrac{8\pi}{9}, \dfrac{16\pi}{9}\right\}$ **33.** $\left\{\dfrac{3\pi}{4}, \dfrac{7\pi}{4}\right\}$ **35.** $\left\{\dfrac{11\pi}{6}\right\}$

37. $\left\{\theta \mid \theta = \dfrac{\pi}{6} + 2k\pi, \theta = \dfrac{5\pi}{6} + 2k\pi\right\}$; $\dfrac{\pi}{6}, \dfrac{5\pi}{6}, \dfrac{13\pi}{6}, \dfrac{17\pi}{6}, \dfrac{25\pi}{6}, \dfrac{29\pi}{6}$ **39.** $\left\{\theta \mid \theta = \dfrac{5\pi}{6} + k\pi\right\}$; $\dfrac{5\pi}{6}, \dfrac{11\pi}{6}, \dfrac{17\pi}{6}, \dfrac{23\pi}{6}, \dfrac{29\pi}{6}, \dfrac{35\pi}{6}$

41. $\left\{\theta \mid \theta = \dfrac{\pi}{2} + 2k\pi, \theta = \dfrac{3\pi}{2} + 2k\pi\right\}$; $\dfrac{\pi}{2}, \dfrac{3\pi}{2}, \dfrac{5\pi}{2}, \dfrac{7\pi}{2}, \dfrac{9\pi}{2}, \dfrac{11\pi}{2}$ **43.** $\left\{\theta \mid \theta = \dfrac{\pi}{6} + k\pi\right\}$; $\left\{\dfrac{\pi}{6}, \dfrac{7\pi}{6}, \dfrac{13\pi}{6}, \dfrac{19\pi}{6}, \dfrac{25\pi}{6}, \dfrac{31\pi}{6}\right\}$

45. $\left\{\theta \mid \theta = \dfrac{\pi}{3} + k\pi, \theta = \dfrac{2\pi}{3} + k\pi\right\}$; $\dfrac{\pi}{3}, \dfrac{2\pi}{3}, \dfrac{4\pi}{3}, \dfrac{5\pi}{3}, \dfrac{7\pi}{3}, \dfrac{8\pi}{3}$ **47.** $\left\{\theta \mid \theta = \dfrac{8\pi}{3} + 4k\pi, \theta = \dfrac{10\pi}{3} + 4k\pi\right\}$; $\dfrac{8\pi}{3}, \dfrac{10\pi}{3}, \dfrac{20\pi}{3}, \dfrac{22\pi}{3}, \dfrac{32\pi}{3}, \dfrac{34\pi}{3}$

49. $\{0.41, 2.73\}$ **51.** $\{1.37, 4.51\}$ **53.** $\{2.69, 3.59\}$ **55.** $\{1.82, 4.46\}$ **57.** $\{2.08, 5.22\}$ **59.** $\{0.73, 2.41\}$ **61.** $\left\{\dfrac{\pi}{2}, \dfrac{2\pi}{3}, \dfrac{4\pi}{3}, \dfrac{3\pi}{2}\right\}$

63. $\left\{\dfrac{\pi}{2}, \dfrac{7\pi}{6}, \dfrac{11\pi}{6}\right\}$ **65.** $\left\{0, \dfrac{\pi}{4}, \dfrac{5\pi}{4}\right\}$ **67.** $\left\{\dfrac{\pi}{2}, \dfrac{2\pi}{3}, \dfrac{4\pi}{3}, \dfrac{3\pi}{2}\right\}$ **69.** $\{\pi\}$ **71.** $\left\{\dfrac{\pi}{4}, \dfrac{5\pi}{4}\right\}$ **73.** $\left\{0, \dfrac{\pi}{3}, \pi, \dfrac{5\pi}{3}\right\}$ **75.** $\left\{\dfrac{\pi}{6}, \dfrac{5\pi}{6}, \dfrac{3\pi}{2}\right\}$ **77.** $\left\{\dfrac{\pi}{2}\right\}$

79. $\{0\}$ **81.** $\left\{\dfrac{\pi}{3}, \dfrac{5\pi}{3}\right\}$ **83.** No real solution   **85.** $\{-1.31, 1.98, 3.84\}$ **87.** $\{0.52\}$ **89.** $\{1.26\}$ **91.** $\{-1.02, 1.02\}$

93. $\{0, 2.15\}$ **95.** $\{0.76, 1.35\}$ **97.** $\dfrac{\pi}{3}, \dfrac{2\pi}{3}, \dfrac{4\pi}{3}, \dfrac{5\pi}{3}$

99. (a) $-2\pi, -\pi, 0, \pi, 2\pi, 3\pi, 4\pi$ **(b)**

(c) $\left\{-\dfrac{11\pi}{6}, -\dfrac{7\pi}{6}, \dfrac{\pi}{6}, \dfrac{5\pi}{6}, \dfrac{13\pi}{6}, \dfrac{17\pi}{6}\right\}$

(d) $\left\{x \mid -\dfrac{11\pi}{6} < x < -\dfrac{7\pi}{6} \text{ or } \dfrac{\pi}{6} < x < \dfrac{5\pi}{6} \text{ or } \dfrac{13\pi}{6} < x < \dfrac{17\pi}{6}\right\}$

101. (a) $\left\{x \mid x = -\dfrac{\pi}{4} + k\pi, k \text{ an integer}\right\}$ **(b)** $-\dfrac{\pi}{2} < x < -\dfrac{\pi}{4}$ or $\left(-\dfrac{\pi}{2}, -\dfrac{\pi}{4}\right)$

103. (a), (d)

(b) $\left\{\dfrac{\pi}{12}, \dfrac{5\pi}{12}\right\}$ **(c)** $\left\{x \mid \dfrac{\pi}{12} < x < \dfrac{5\pi}{12}\right\}$ or $\left(\dfrac{\pi}{12}, \dfrac{5\pi}{12}\right)$

105. (a), (d)

(b) $\left\{\dfrac{2\pi}{3}, \dfrac{4\pi}{3}\right\}$ **(c)** $\left\{x \mid \dfrac{2\pi}{3} < x < \dfrac{4\pi}{3}\right\}$ or $\left(\dfrac{2\pi}{3}, \dfrac{4\pi}{3}\right)$

107. (a) 0 s, 0.43 s, 0.86 s   **(b)** 0.21 s   **(c)** $[0, 0.03] \cup [0.39, 0.43] \cup [0.86, 0.89]$

109. (a) 150 mi   **(b)** 6.06, 8.44, 15.72, 18.11 min   **(c)** Before 6.06 min, between 8.44 and 15.72 min, and after 18.11 min   **(d)** No   **111.** 2.03, 4.91

113. (a) $30°, 60°$ **(b)** 123.6 m   **115.** 28.90°   **117.** Yes; it varies from 1.25 to 1.34.   **119.** 1.47   **121.** $\theta_B = 48.8°$

(c)

123. $\left\{\theta \mid \theta = \dfrac{5\pi}{6} + k\pi, \text{ where } k \text{ is any integer}\right\}$ **127.** $\dfrac{9-\sqrt{17}}{4}, \dfrac{9+\sqrt{17}}{4}$

128. $\tan\theta = -\dfrac{1}{3}$; $\csc\theta = -\sqrt{10}$; $\sec\theta = \dfrac{\sqrt{10}}{3}$; $\cot\theta = -3$ **129.** Amplitude: 2   Period: π Phase shift: $\dfrac{\pi}{2}$

130. 3π cm ≈ 9.425 cm   **131.** $a = 6$ **132.** odd   **133.** $y = -\dfrac{5}{2}x + \dfrac{21}{2}$ **134.** $-\dfrac{2\pi}{3}$

3.4 Assess Your Understanding *(page 224)*

3. identity; conditional   4. -1 5. 0   6. T   7. F   8. T   9. c   10. b   11. $\dfrac{1}{\cos \theta}$ 13. $\dfrac{1 + \sin \theta}{\cos \theta}$ 15. $\dfrac{1}{\sin \theta \cos \theta}$ 17. 2   19. $\dfrac{3 \sin \theta + 1}{\sin \theta + 1}$

21. $\csc \theta \cdot \cos \theta = \dfrac{1}{\sin \theta} \cdot \cos \theta = \dfrac{\cos \theta}{\sin \theta} = \cot \theta$ 23. $1 + \tan^2(-\theta) = 1 + (-\tan \theta)^2 = 1 + \tan^2 \theta = \sec^2 \theta$

25. $\cos \theta (\tan \theta + \cot \theta) = \cos \theta \left(\dfrac{\sin \theta}{\cos \theta} + \dfrac{\cos \theta}{\sin \theta} \right) = \cos \theta \left(\dfrac{\sin^2 \theta + \cos^2 \theta}{\cos \theta \sin \theta} \right) = \cos \theta \left(\dfrac{1}{\cos \theta \sin \theta} \right) = \dfrac{1}{\sin \theta} = \csc \theta$

27. $\tan u \cot u - \cos^2 u = \tan u \cdot \dfrac{1}{\tan u} - \cos^2 u = 1 - \cos^2 u = \sin^2 u$ 29. $(\sec \theta - 1)(\sec \theta + 1) = \sec^2 \theta - 1 = \tan^2 \theta$

31. $(\sec \theta + \tan \theta)(\sec \theta - \tan \theta) = \sec^2 \theta - \tan^2 \theta = 1$ 33. $\cos^2 \theta (1 + \tan^2 \theta) = \cos^2 \theta \sec^2 \theta = \cos^2 \theta \cdot \dfrac{1}{\cos^2 \theta} = 1$

35. $(\sin \theta + \cos \theta)^2 + (\sin \theta - \cos \theta)^2 = \sin^2 \theta + 2 \sin \theta \cos \theta + \cos^2 \theta + \sin^2 \theta - 2 \sin \theta \cos \theta + \cos^2 \theta$
$$= \sin^2 \theta + \cos^2 \theta + \sin^2 \theta + \cos^2 \theta = 1 + 1 = 2$$

37. $\sec^4 \theta - \sec^2 \theta = \sec^2 \theta (\sec^2 \theta - 1) = (1 + \tan^2 \theta) \tan^2 \theta = \tan^4 \theta + \tan^2 \theta$

39. $\sec u - \tan u = \dfrac{1}{\cos u} - \dfrac{\sin u}{\cos u} = \dfrac{1 - \sin u}{\cos u} \cdot \dfrac{1 + \sin u}{1 + \sin u} = \dfrac{1 - \sin^2 u}{\cos u (1 + \sin u)} = \dfrac{\cos^2 u}{\cos u (1 + \sin u)} = \dfrac{\cos u}{1 + \sin u}$

41. $3 \sin^2 \theta + 4 \cos^2 \theta = 3 \sin^2 \theta + 3 \cos^2 \theta + \cos^2 \theta = 3(\sin^2 \theta + \cos^2 \theta) + \cos^2 \theta = 3 + \cos^2 \theta$

43. $1 - \dfrac{\cos^2 \theta}{1 + \sin \theta} = 1 - \dfrac{1 - \sin^2 \theta}{1 + \sin \theta} = 1 - \dfrac{(1 + \sin \theta)(1 - \sin \theta)}{1 + \sin \theta} = 1 - (1 - \sin \theta) = \sin \theta$

45. $\dfrac{1 + \tan v}{1 - \tan v} = \dfrac{1 + \dfrac{1}{\cot v}}{1 - \dfrac{1}{\cot v}} = \dfrac{\dfrac{\cot v + 1}{\cot v}}{\dfrac{\cot v - 1}{\cot v}} = \dfrac{\cot v + 1}{\cot v - 1}$ 47. $\dfrac{\sec \theta}{\csc \theta} + \dfrac{\sin \theta}{\cos \theta} = \dfrac{\dfrac{1}{\cos \theta}}{\dfrac{1}{\sin \theta}} + \tan \theta = \dfrac{\sin \theta}{\cos \theta} + \tan \theta = \tan \theta + \tan \theta = 2 \tan \theta$

49. $\dfrac{1 + \sin \theta}{1 - \sin \theta} = \dfrac{1 + \dfrac{1}{\csc \theta}}{1 - \dfrac{1}{\csc \theta}} = \dfrac{\dfrac{\csc \theta + 1}{\csc \theta}}{\dfrac{\csc \theta - 1}{\csc \theta}} = \dfrac{\csc \theta + 1}{\csc \theta - 1}$

51. $\dfrac{1 - \sin v}{\cos v} + \dfrac{\cos v}{1 - \sin v} = \dfrac{(1 - \sin v)^2 + \cos^2 v}{\cos v (1 - \sin v)} = \dfrac{1 - 2 \sin v + \sin^2 v + \cos^2 v}{\cos v (1 - \sin v)} = \dfrac{2 - 2 \sin v}{\cos v (1 - \sin v)} = \dfrac{2(1 - \sin v)}{\cos v (1 - \sin v)} = \dfrac{2}{\cos v} = 2 \sec v$

53. $\dfrac{\sin \theta}{\sin \theta - \cos \theta} = \dfrac{1}{\dfrac{\sin \theta - \cos \theta}{\sin \theta}} = \dfrac{1}{1 - \dfrac{\cos \theta}{\sin \theta}} = \dfrac{1}{1 - \cot \theta}$

55. $(\sec \theta - \tan \theta)^2 = \sec^2 \theta - 2 \sec \theta \tan \theta + \tan^2 \theta = \dfrac{1}{\cos^2 \theta} - \dfrac{2 \sin \theta}{\cos^2 \theta} + \dfrac{\sin^2 \theta}{\cos^2 \theta} = \dfrac{1 - 2 \sin \theta + \sin^2 \theta}{\cos^2 \theta} = \dfrac{(1 - \sin \theta)^2}{1 - \sin^2 \theta} = \dfrac{(1 - \sin \theta)^2}{(1 - \sin \theta)(1 + \sin \theta)}$
$$= \dfrac{1 - \sin \theta}{1 + \sin \theta}$$

57. $\dfrac{\cos \theta}{1 - \tan \theta} + \dfrac{\sin \theta}{1 - \cot \theta} = \dfrac{\cos \theta}{1 - \dfrac{\sin \theta}{\cos \theta}} + \dfrac{\sin \theta}{1 - \dfrac{\cos \theta}{\sin \theta}} = \dfrac{\cos \theta}{\dfrac{\cos \theta - \sin \theta}{\cos \theta}} + \dfrac{\sin \theta}{\dfrac{\sin \theta - \cos \theta}{\sin \theta}} = \dfrac{\cos^2 \theta}{\cos \theta - \sin \theta} + \dfrac{\sin^2 \theta}{\sin \theta - \cos \theta}$
$$= \dfrac{\cos^2 \theta - \sin^2 \theta}{\cos \theta - \sin \theta} = \dfrac{(\cos \theta - \sin \theta)(\cos \theta + \sin \theta)}{\cos \theta - \sin \theta} = \sin \theta + \cos \theta$$

59. $\tan \theta + \dfrac{\cos \theta}{1 + \sin \theta} = \dfrac{\sin \theta}{\cos \theta} + \dfrac{\cos \theta}{1 + \sin \theta} = \dfrac{\sin \theta (1 + \sin \theta) + \cos^2 \theta}{\cos \theta (1 + \sin \theta)} = \dfrac{\sin \theta + \sin^2 \theta + \cos^2 \theta}{\cos \theta (1 + \sin \theta)} = \dfrac{\sin \theta + 1}{\cos \theta (1 + \sin \theta)} = \dfrac{1}{\cos \theta} = \sec \theta$

61. $\dfrac{\tan \theta + \sec \theta - 1}{\tan \theta - \sec \theta + 1} = \dfrac{\tan \theta + (\sec \theta - 1)}{\tan \theta - (\sec \theta - 1)} \cdot \dfrac{\tan \theta + (\sec \theta - 1)}{\tan \theta + (\sec \theta - 1)} = \dfrac{\tan^2 \theta + 2 \tan \theta (\sec \theta - 1) + \sec^2 \theta - 2 \sec \theta + 1}{\tan^2 \theta - (\sec^2 \theta - 2 \sec \theta + 1)}$
$$= \dfrac{\sec^2 \theta - 1 + 2 \tan \theta (\sec \theta - 1) + \sec^2 \theta - 2 \sec \theta + 1}{\sec^2 \theta - 1 - \sec^2 \theta + 2 \sec \theta - 1} = \dfrac{2 \sec^2 \theta - 2 \sec \theta + 2 \tan \theta (\sec \theta - 1)}{-2 + 2 \sec \theta}$$
$$= \dfrac{2 \sec \theta (\sec \theta - 1) + 2 \tan \theta (\sec \theta - 1)}{2(\sec \theta - 1)} = \dfrac{2(\sec \theta - 1)(\sec \theta + \tan \theta)}{2(\sec \theta - 1)} = \tan \theta + \sec \theta$$

63. $\dfrac{\tan \theta - \cot \theta}{\tan \theta + \cot \theta} = \dfrac{\dfrac{\sin \theta}{\cos \theta} - \dfrac{\cos \theta}{\sin \theta}}{\dfrac{\sin \theta}{\cos \theta} + \dfrac{\cos \theta}{\sin \theta}} = \dfrac{\dfrac{\sin^2 \theta - \cos^2 \theta}{\cos \theta \sin \theta}}{\dfrac{\sin^2 \theta + \cos^2 \theta}{\cos \theta \sin \theta}} = \dfrac{\sin^2 \theta - \cos^2 \theta}{1} = \sin^2 \theta - \cos^2 \theta$

65. $\dfrac{\tan u - \cot u}{\tan u + \cot u} + 1 = \dfrac{\dfrac{\sin u}{\cos u} - \dfrac{\cos u}{\sin u}}{\dfrac{\sin u}{\cos u} + \dfrac{\cos u}{\sin u}} + 1 = \dfrac{\dfrac{\sin^2 u - \cos^2 u}{\cos u \sin u}}{\dfrac{\sin^2 u + \cos^2 u}{\cos u \sin u}} + 1 = \sin^2 u - \cos^2 u + 1 = \sin^2 u + (1 - \cos^2 u) = 2\sin^2 u$

67. $\dfrac{\sec\theta + \tan\theta}{\cot\theta + \cos\theta} = \dfrac{\dfrac{1}{\cos\theta} + \dfrac{\sin\theta}{\cos\theta}}{\dfrac{\cos\theta}{\sin\theta} + \cos\theta} = \dfrac{\dfrac{1+\sin\theta}{\cos\theta}}{\dfrac{\cos\theta + \cos\theta\sin\theta}{\sin\theta}} = \dfrac{1+\sin\theta}{\cos\theta}\cdot\dfrac{\sin\theta}{\cos\theta(1+\sin\theta)} = \dfrac{\sin\theta}{\cos\theta}\cdot\dfrac{1}{\cos\theta} = \tan\theta\sec\theta$

69. $\dfrac{1-\tan^2\theta}{1+\tan^2\theta} + 1 = \dfrac{1-\tan^2\theta + 1 + \tan^2\theta}{1+\tan^2\theta} = \dfrac{2}{1+\tan^2\theta} = \dfrac{2}{\sec^2\theta} = 2\cos^2\theta$

71. $\dfrac{\sec\theta - \csc\theta}{\sec\theta\csc\theta} = \dfrac{\sec\theta}{\sec\theta\csc\theta} - \dfrac{\csc\theta}{\sec\theta\csc\theta} = \dfrac{1}{\csc\theta} - \dfrac{1}{\sec\theta} = \sin\theta - \cos\theta$

73. $\sec\theta - \cos\theta = \dfrac{1}{\cos\theta} - \cos\theta = \dfrac{1-\cos^2\theta}{\cos\theta} = \dfrac{\sin^2\theta}{\cos\theta} = \sin\theta\cdot\dfrac{\sin\theta}{\cos\theta} = \sin\theta\tan\theta$

75. $\dfrac{1}{1-\sin\theta} + \dfrac{1}{1+\sin\theta} = \dfrac{1+\sin\theta + 1 - \sin\theta}{(1+\sin\theta)(1-\sin\theta)} = \dfrac{2}{1-\sin^2\theta} = \dfrac{2}{\cos^2\theta} = 2\sec^2\theta$

77. $\dfrac{\sec\theta}{1-\sin\theta} = \dfrac{\sec\theta}{1-\sin\theta}\cdot\dfrac{1+\sin\theta}{1+\sin\theta} = \dfrac{\sec\theta(1+\sin\theta)}{1-\sin^2\theta} = \dfrac{\sec\theta(1+\sin\theta)}{\cos^2\theta} = \dfrac{1+\sin\theta}{\cos^3\theta}$

79. $\dfrac{(\sec v - \tan v)^2 + 1}{\csc v(\sec v - \tan v)} = \dfrac{\sec^2 v - 2\sec v\tan v + \tan^2 v + 1}{\dfrac{1}{\sin v}\left(\dfrac{1}{\cos v} - \dfrac{\sin v}{\cos v}\right)} = \dfrac{2\sec^2 v - 2\sec v\tan v}{\dfrac{1}{\sin v}\left(\dfrac{1-\sin v}{\cos v}\right)} = \dfrac{\dfrac{2}{\cos^2 v} - \dfrac{2\sin v}{\cos^2 v}}{\dfrac{1-\sin v}{\sin v\cos v}} = \dfrac{2 - 2\sin v}{\cos^2 v}\cdot\dfrac{\sin v\cos v}{1-\sin v}$

$= \dfrac{2(1-\sin v)}{\cos v}\cdot\dfrac{\sin v}{1-\sin v} = \dfrac{2\sin v}{\cos v} = 2\tan v$

81. $\dfrac{\sin\theta + \cos\theta}{\cos\theta} - \dfrac{\sin\theta - \cos\theta}{\sin\theta} = \dfrac{\sin\theta}{\cos\theta} + 1 - 1 + \dfrac{\cos\theta}{\sin\theta} = \dfrac{\sin^2\theta + \cos^2\theta}{\cos\theta\sin\theta} = \dfrac{1}{\cos\theta\sin\theta} = \sec\theta\csc\theta$

83. $\dfrac{\sin^3\theta + \cos^3\theta}{\sin\theta + \cos\theta} = \dfrac{(\sin\theta + \cos\theta)(\sin^2\theta - \sin\theta\cos\theta + \cos^2\theta)}{\sin\theta + \cos\theta} = \sin^2\theta + \cos^2\theta - \sin\theta\cos\theta = 1 - \sin\theta\cos\theta$

85. $\dfrac{\cos^2\theta - \sin^2\theta}{1 - \tan^2\theta} = \dfrac{\cos^2\theta - \sin^2\theta}{1 - \dfrac{\sin^2\theta}{\cos^2\theta}} = \dfrac{\cos^2\theta - \sin^2\theta}{\dfrac{\cos^2\theta - \sin^2\theta}{\cos^2\theta}} = \cos^2\theta$

87. $\dfrac{(2\cos^2\theta - 1)^2}{\cos^4\theta - \sin^4\theta} = \dfrac{[2\cos^2\theta - (\sin^2\theta + \cos^2\theta)]^2}{(\cos^2\theta - \sin^2\theta)(\cos^2\theta + \sin^2\theta)} = \dfrac{(\cos^2\theta - \sin^2\theta)^2}{\cos^2\theta - \sin^2\theta} = \cos^2\theta - \sin^2\theta = (1-\sin^2\theta) - \sin^2\theta = 1 - 2\sin^2\theta$

89. $\dfrac{1+\sin\theta + \cos\theta}{1+\sin\theta - \cos\theta} = \dfrac{(1+\sin\theta) + \cos\theta}{(1+\sin\theta) - \cos\theta}\cdot\dfrac{(1+\sin\theta) + \cos\theta}{(1+\sin\theta) + \cos\theta} = \dfrac{1 + 2\sin\theta + \sin^2\theta + 2(1+\sin\theta)\cos\theta + \cos^2\theta}{1+2\sin\theta + \sin^2\theta - \cos^2\theta}$

$= \dfrac{1+2\sin\theta + \sin^2\theta + 2(1+\sin\theta)\cos\theta + (1-\sin^2\theta)}{1+2\sin\theta + \sin^2\theta - (1-\sin^2\theta)} = \dfrac{2 + 2\sin\theta + 2(1+\sin\theta)\cos\theta}{2\sin\theta + 2\sin^2\theta}$

$= \dfrac{2(1+\sin\theta) + 2(1+\sin\theta)\cos\theta}{2\sin\theta(1+\sin\theta)} = \dfrac{2(1+\sin\theta)(1+\cos\theta)}{2\sin\theta(1+\sin\theta)} = \dfrac{1+\cos\theta}{\sin\theta}$

91. $(a\sin\theta + b\cos\theta)^2 + (a\cos\theta - b\sin\theta)^2 = a^2\sin^2\theta + 2ab\sin\theta\cos\theta + b^2\cos^2\theta + a^2\cos^2\theta - 2ab\sin\theta\cos\theta + b^2\sin^2\theta$

$= a^2(\sin^2\theta + \cos^2\theta) + b^2(\cos^2\theta + \sin^2\theta) = a^2 + b^2$

93. $\dfrac{\tan\alpha + \tan\beta}{\cot\alpha + \cot\beta} = \dfrac{\tan\alpha + \tan\beta}{\dfrac{1}{\tan\alpha} + \dfrac{1}{\tan\beta}} = \dfrac{\tan\alpha + \tan\beta}{\dfrac{\tan\beta + \tan\alpha}{\tan\alpha\tan\beta}} = (\tan\alpha + \tan\beta)\cdot\dfrac{\tan\alpha\tan\beta}{\tan\alpha + \tan\beta} = \tan\alpha\tan\beta$

95. $(\sin\alpha + \cos\beta)^2 + (\cos\beta + \sin\alpha)(\cos\beta - \sin\alpha) = (\sin^2\alpha + 2\sin\alpha\cos\beta + \cos^2\beta) + (\cos^2\beta - \sin^2\alpha)$

$= 2\cos^2\beta + 2\sin\alpha\cos\beta = 2\cos\beta(\cos\beta + \sin\alpha) = 2\cos\beta(\sin\alpha + \cos\beta)$

97. $\ln|\sec\theta| = \ln|\cos\theta|^{-1} = -\ln|\cos\theta|$

99. $\ln|1+\cos\theta| + \ln|1-\cos\theta| = \ln(|1+\cos\theta||1-\cos\theta|) = \ln|1-\cos^2\theta| = \ln|\sin^2\theta| = 2\ln|\sin\theta|$

101. $g(x) = \sec x - \cos x = \dfrac{1}{\cos x} - \cos x = \dfrac{1}{\cos x} - \dfrac{\cos^2 x}{\cos x} = \dfrac{1-\cos^2 x}{\cos x} = \dfrac{\sin^2 x}{\cos x} = \sin x\cdot\dfrac{\sin x}{\cos x} = \sin x\cdot\tan x = f(x)$

103. $f(\theta) = \dfrac{1-\sin\theta}{\cos\theta} - \dfrac{\cos\theta}{1+\sin\theta} = \dfrac{1-\sin\theta}{\cos\theta}\cdot\dfrac{1+\sin\theta}{1+\sin\theta} - \dfrac{\cos\theta}{1+\sin\theta}\cdot\dfrac{\cos\theta}{\cos\theta} = \dfrac{1-\sin^2\theta}{\cos\theta(1+\sin\theta)} - \dfrac{\cos^2\theta}{\cos\theta(1+\sin\theta)}$

$= \dfrac{\cos^2\theta}{\cos\theta(1+\sin\theta)} - \dfrac{\cos^2\theta}{\cos\theta(1+\sin\theta)} = 0 = g(\theta)$

105. $\sqrt{16 + 16\tan^2\theta} = \sqrt{16}\sqrt{1 + \tan^2\theta} = 4\sqrt{\sec^2\theta} = 4\sec\theta$, since $\sec\theta > 0$ for $-\dfrac{\pi}{2} < \theta < \dfrac{\pi}{2}$

107. $1200\sec\theta\,(2\sec^2\theta - 1) = 1200\dfrac{1}{\cos\theta}\left(\dfrac{2}{\cos^2\theta} - 1\right) = 1200\dfrac{1}{\cos\theta}\left(\dfrac{2}{\cos^2\theta} - \dfrac{\cos^2\theta}{\cos^2\theta}\right) = 1200\dfrac{1}{\cos\theta}\left(\dfrac{2 - \cos^2\theta}{\cos^2\theta}\right) = \dfrac{1200\,(1 + 1 - \cos^2\theta)}{\cos^3\theta}$

$= \dfrac{1200\,(1 + \sin^2\theta)}{\cos^3\theta}$

109. Let $\theta = \sin^{-1}(-x)$. Then $-x = \sin\theta$. So, $x = -\sin\theta = \sin(-\theta)$ because the sine function is odd. This means $-\theta = \sin^{-1}x$, and $\theta = -\sin^{-1}x$. So, $\sin^{-1}(-x) = -\sin^{-1}x$.

115. Maximum, 1250   **116.** $f(3x - 4) = \dfrac{x - 1}{x - 2}$ **117.** $\sin\theta = \dfrac{5}{13}; \cos\theta = -\dfrac{12}{13}; \tan\theta = -\dfrac{5}{12}; \csc\theta = \dfrac{13}{5}; \sec\theta = -\dfrac{13}{12}; \cot\theta = -\dfrac{12}{5}$ **118.** $-\dfrac{2}{\pi}$

119. $4\sqrt{13}$ **120.** $\dfrac{48}{5}\pi\,\text{m}^2 \approx 30.159\,\text{m}^2$ **121.** 18 miles   **122.** $-\dfrac{\sqrt{89}}{8}$ **123.** $(x - 6)^2 + (y + 2)^2 = 9$

3.5 Assess Your Understanding *(page 236)*

7. (a) $-$ **(b)** $-$ **8.** F   **9.** F   **10.** T   **11.** a   **12.** d   **13.** $-\dfrac{1}{4}\left(\sqrt{2} + \sqrt{6}\right)$ **15.** $2 - \sqrt{3}$ **17.** $\dfrac{1}{4}\left(\sqrt{6} + \sqrt{2}\right)$ **19.** $\dfrac{1}{4}\left(\sqrt{2} - \sqrt{6}\right)$

21. $-\dfrac{1}{4}\left(\sqrt{6} + \sqrt{2}\right)$ **23.** $\sqrt{6} - \sqrt{2}$ **25.** $\dfrac{1}{2}$ **27.** 0   **29.** 1   **31.** -1 **33.** $\dfrac{1}{2}$ **35. (a)** $\dfrac{2\sqrt{5}}{25}$ **(b)** $\dfrac{11\sqrt{5}}{25}$ **(c)** $\dfrac{2\sqrt{5}}{5}$ **(d)** 2

37. (a) $\dfrac{4 - 3\sqrt{3}}{10}$ **(b)** $\dfrac{-3 - 4\sqrt{3}}{10}$ **(c)** $\dfrac{4 + 3\sqrt{3}}{10}$ **(d)** $\dfrac{25\sqrt{3} + 48}{39}$ **39. (a)** $-\dfrac{5 + 12\sqrt{3}}{26}$ **(b)** $\dfrac{12 - 5\sqrt{3}}{26}$ **(c)** $\dfrac{-5 + 12\sqrt{3}}{26}$ **(d)** $\dfrac{-240 + 169\sqrt{3}}{69}$

41. (a) $-\dfrac{2\sqrt{2}}{3}$ **(b)** $\dfrac{-2\sqrt{2} + \sqrt{3}}{6}$ **(c)** $\dfrac{-2\sqrt{2} + \sqrt{3}}{6}$ **(d)** $\dfrac{9 - 4\sqrt{2}}{7}$ **43.** $\dfrac{1 - 2\sqrt{6}}{6}$ **45.** $\dfrac{\sqrt{3} - 2\sqrt{2}}{6}$ **47.** $\dfrac{8\sqrt{2} - 9\sqrt{3}}{5}$

49. $\sin\left(\dfrac{\pi}{2} + \theta\right) = \sin\dfrac{\pi}{2}\cos\theta + \cos\dfrac{\pi}{2}\sin\theta = 1\cdot\cos\theta + 0\cdot\sin\theta = \cos\theta$

51. $\sin(\pi - \theta) = \sin\pi\cos\theta - \cos\pi\sin\theta = 0\cdot\cos\theta - (-1)\sin\theta = \sin\theta$

53. $\sin(\pi + \theta) = \sin\pi\cos\theta + \cos\pi\sin\theta = 0\cdot\cos\theta + (-1)\sin\theta = -\sin\theta$

55. $\tan(\pi - \theta) = \dfrac{\tan\pi - \tan\theta}{1 + \tan\pi\tan\theta} = \dfrac{0 - \tan\theta}{1 + 0\cdot\tan\theta} = -\tan\theta$

57. $\sin\left(\dfrac{3\pi}{2} + \theta\right) = \sin\dfrac{3\pi}{2}\cos\theta + \cos\dfrac{3\pi}{2}\sin\theta = -1\cdot\cos\theta + 0\cdot\sin\theta = -\cos\theta$

59. $\sin(\alpha + \beta) + \sin(\alpha - \beta) = \sin\alpha\cos\beta + \cos\alpha\sin\beta + \sin\alpha\cos\beta - \cos\alpha\sin\beta = 2\sin\alpha\cos\beta$

61. $\dfrac{\sin(\alpha + \beta)}{\sin\alpha\cos\beta} = \dfrac{\sin\alpha\cos\beta + \cos\alpha\sin\beta}{\sin\alpha\cos\beta} = \dfrac{\sin\alpha\cos\beta}{\sin\alpha\cos\beta} + \dfrac{\cos\alpha\sin\beta}{\sin\alpha\cos\beta} = 1 + \cot\alpha\tan\beta$

63. $\dfrac{\cos(\alpha + \beta)}{\cos\alpha\cos\beta} = \dfrac{\cos\alpha\cos\beta - \sin\alpha\sin\beta}{\cos\alpha\cos\beta} = \dfrac{\cos\alpha\cos\beta}{\cos\alpha\cos\beta} - \dfrac{\sin\alpha\sin\beta}{\cos\alpha\cos\beta} = 1 - \tan\alpha\tan\beta$

65. $\dfrac{\sin(\alpha + \beta)}{\sin(\alpha - \beta)} = \dfrac{\sin\alpha\cos\beta + \cos\alpha\sin\beta}{\sin\alpha\cos\beta - \cos\alpha\sin\beta} = \dfrac{\dfrac{\sin\alpha\cos\beta + \cos\alpha\sin\beta}{\cos\alpha\cos\beta}}{\dfrac{\sin\alpha\cos\beta - \cos\alpha\sin\beta}{\cos\alpha\cos\beta}} = \dfrac{\dfrac{\sin\alpha\cos\beta}{\cos\alpha\cos\beta} + \dfrac{\cos\alpha\sin\beta}{\cos\alpha\cos\beta}}{\dfrac{\sin\alpha\cos\beta}{\cos\alpha\cos\beta} - \dfrac{\cos\alpha\sin\beta}{\cos\alpha\cos\beta}} = \dfrac{\tan\alpha + \tan\beta}{\tan\alpha - \tan\beta}$

67. $\cot(\alpha + \beta) = \dfrac{\cos(\alpha + \beta)}{\sin(\alpha + \beta)} = \dfrac{\cos\alpha\cos\beta - \sin\alpha\sin\beta}{\sin\alpha\cos\beta + \cos\alpha\sin\beta} = \dfrac{\dfrac{\cos\alpha\cos\beta - \sin\alpha\sin\beta}{\sin\alpha\sin\beta}}{\dfrac{\sin\alpha\cos\beta + \cos\alpha\sin\beta}{\sin\alpha\sin\beta}} = \dfrac{\dfrac{\cos\alpha\cos\beta}{\sin\alpha\sin\beta} - \dfrac{\sin\alpha\sin\beta}{\sin\alpha\sin\beta}}{\dfrac{\sin\alpha\cos\beta}{\sin\alpha\sin\beta} + \dfrac{\cos\alpha\sin\beta}{\sin\alpha\sin\beta}} = \dfrac{\cot\alpha\cot\beta - 1}{\cot\beta + \cot\alpha}$

69. $\sec(\alpha + \beta) = \dfrac{1}{\cos(\alpha + \beta)} = \dfrac{1}{\cos\alpha\cos\beta - \sin\alpha\sin\beta} = \dfrac{\dfrac{1}{\sin\alpha\sin\beta}}{\dfrac{\cos\alpha\cos\beta - \sin\alpha\sin\beta}{\sin\alpha\sin\beta}} = \dfrac{\dfrac{1}{\sin\alpha}\cdot\dfrac{1}{\sin\beta}}{\dfrac{\cos\alpha\cos\beta}{\sin\alpha\sin\beta} - \dfrac{\sin\alpha\sin\beta}{\sin\alpha\sin\beta}} = \dfrac{\csc\alpha\csc\beta}{\cot\alpha\cot\beta - 1}$

71. $\sin(\alpha - \beta)\sin(\alpha + \beta) = (\sin\alpha\cos\beta - \cos\alpha\sin\beta)(\sin\alpha\cos\beta + \cos\alpha\sin\beta) = \sin^2\alpha\cos^2\beta - \cos^2\alpha\sin^2\beta$

$= \sin^2\alpha(1 - \sin^2\beta) - (1 - \sin^2\alpha)\sin^2\beta = \sin^2\alpha - \sin^2\beta$

73. $\sin(\theta + k\pi) = \sin\theta\cos k\pi + \cos\theta\sin k\pi = (\sin\theta)(-1)^k + (\cos\theta)(0) = (-1)^k\sin\theta$, k any integer

75. $\dfrac{\sqrt{3}}{2}$ **77.** $-\dfrac{24}{25}$ **79.** $-\dfrac{33}{65}$ **81.** $\dfrac{63}{65}$ **83.** $\dfrac{48 + 25\sqrt{3}}{39}$ **85.** $\dfrac{4}{3}$ **87.** $u\sqrt{1 - v^2} - v\sqrt{1 - u^2}: -1 \le u \le 1; -1 \le v \le 1$

89. $\dfrac{u\sqrt{1 - v^2} - v}{\sqrt{1 + u^2}}: -\infty < u < \infty; -1 \le v \le 1$ **91.** $\dfrac{uv - \sqrt{1 - u^2}\sqrt{1 - v^2}}{v\sqrt{1 - u^2} + u\sqrt{1 - v^2}}: -1 \le u \le 1; -1 \le v \le 1$ **93.** $\left\{\dfrac{\pi}{2}, \dfrac{7\pi}{6}\right\}$ **95.** $\left\{\dfrac{\pi}{4}\right\}$ **97.** $\left\{\dfrac{11\pi}{6}\right\}$

99. $\sin(\sin^{-1}v + \cos^{-1}v) = \sin(\sin^{-1}v)\cos(\cos^{-1}v) + \cos(\sin^{-1}v)\sin(\cos^{-1}v) = (v)(v) + \sqrt{1-v^2}\sqrt{1-v^2} = v^2 + 1 - v^2 = 1$

101. $\dfrac{\sin(x+h) - \sin x}{h} = \dfrac{\sin x \cos h + \cos x \sin h - \sin x}{h} = \dfrac{\cos x \sin h - \sin x(1 - \cos h)}{h} = \cos x \cdot \dfrac{\sin h}{h} - \sin x \cdot \dfrac{1 - \cos h}{h}$

103. (a) $\tan(\tan^{-1}1 + \tan^{-1}2 + \tan^{-1}3) = \tan((\tan^{-1}1 + \tan^{-1}2) + \tan^{-1}3) = \dfrac{\tan(\tan^{-1}1 + \tan^{-1}2) + \tan(\tan^{-1}3)}{1 - \tan(\tan^{-1}1 + \tan^{-1}2)\tan(\tan^{-1}3)}$

$$= \dfrac{\dfrac{\tan(\tan^{-1}1) + \tan(\tan^{-1}2)}{1 - \tan(\tan^{-1}1)\tan(\tan^{-1}2)} + 3}{1 - \dfrac{\tan(\tan^{-1}1) + \tan(\tan^{-1}2)}{1 - \tan(\tan^{-1}1)\tan(\tan^{-1}2)} \cdot 3} = \dfrac{\dfrac{1+2}{1 - 1\cdot 2} + 3}{1 - \dfrac{1+2}{1-1\cdot 2}\cdot 3} = \dfrac{\dfrac{3}{-1} + 3}{1 - \dfrac{3}{-1}\cdot 3} = \dfrac{-3+3}{1+9} = \dfrac{0}{10} = 0$$

(b) From the definition of the inverse tangent function, $0 < \tan^{-1}1 < \dfrac{\pi}{2}, 0 < \tan^{-1}2 < \dfrac{\pi}{2}$, and $0 < \tan^{-1}3 < \dfrac{\pi}{2}$,

so $0 < \tan^{-1}1 + \tan^{-1}2 + \tan^{-1}3 < \dfrac{3\pi}{2}$.

On the interval $\left(0, \dfrac{3\pi}{2}\right)$, $\tan\theta = 0$ if and only if $\theta = \pi$. Therefore, from part (a), $\tan^{-1}1 + \tan^{-1}2 + \tan^{-1}3 = \pi$.

105. (a) $(2400 - 1200\sqrt{3})$ in.^{2} **(b)** $(1350 + 675\sqrt{3})$ cm^2 **107.** $\tan\theta = \tan(\theta_2 - \theta_1) = \dfrac{\tan\theta_2 - \tan\theta_1}{1 + \tan\theta_1\tan\theta_2} = \dfrac{m_2 - m_1}{1 + m_1m_2}$

109. Let $\alpha = \tan^{-1}\dfrac{1}{v}$ and $\beta = \tan^{-1}v$. Because $v \neq 0, \alpha, \beta \neq 0$. Then $\tan\alpha = \dfrac{1}{v} = \dfrac{1}{\tan\beta} = \cot\beta$, and since

$\tan\alpha = \cot\left(\dfrac{\pi}{2} - \alpha\right), \cot\left(\dfrac{\pi}{2} - \alpha\right) = \cot\beta$. Because $v > 0, 0 < \alpha < \dfrac{\pi}{2}$, and so $\left(\dfrac{\pi}{2} - \alpha\right)$ and β both lie in $\left(0, \dfrac{\pi}{2}\right)$.

Then $\cot\left(\dfrac{\pi}{2} - \alpha\right) = \cot\beta$ implies $\dfrac{\pi}{2} - \alpha = \beta$, or $\alpha = \dfrac{\pi}{2} - \beta$.

111. $2\cot(\alpha - \beta) = \dfrac{2}{\tan(\alpha - \beta)} = 2\cdot\dfrac{1 + \tan\alpha\tan\beta}{\tan\alpha - \tan\beta} = 2\cdot\dfrac{1 + (x+1)(x-1)}{(x+1) - (x-1)} = 2\cdot\dfrac{1 + x^2 - 1}{x + 1 - x + 1} = \dfrac{2x^2}{2} = x^2$

113. $\tan\dfrac{\pi}{2}$ is not defined; $\tan\left(\dfrac{\pi}{2} - \theta\right) = \dfrac{\sin\left(\dfrac{\pi}{2} - \theta\right)}{\cos\left(\dfrac{\pi}{2} - \theta\right)} = \dfrac{\cos\theta}{\sin\theta} = \cot\theta$.

114. $\left(-\dfrac{1}{3}, -\dfrac{5}{9}\right), (-5, 1)$ **115.** $510°$ **116.** $\dfrac{9\pi}{2}$ m$^2 \approx 14.14$ m^2 **117.** $\sin\theta = -\dfrac{2\sqrt{5}}{5}; \cos\theta = \dfrac{\sqrt{5}}{5}; \csc\theta = -\dfrac{\sqrt{5}}{2}; \sec\theta = \sqrt{5}; \cot\theta = -\dfrac{1}{2}$

118. $f(x) = \dfrac{1}{4}(x+2)^2 - 3$ **119.** $144x^{18}y^{14}$ **120.** $\{2, 6\}$ **121.** $\dfrac{14x + 24}{(x+3)^{1/4}}, x > -3$

3.6 Assess Your Understanding *(page 247)*

1. $\sin^2\theta; 2\cos^2\theta; 2\sin^2\theta$ **2.** $\dfrac{\theta}{2}$ **3.** $\sin\theta$ **4.** T   **5.** F   **6.** F   **7.** b   **8.** c   **9. (a)** $\dfrac{24}{25}$ **(b)** $\dfrac{7}{25}$ **(c)** $\dfrac{\sqrt{10}}{10}$ **(d)** $\dfrac{3\sqrt{10}}{10}$

11. (a) $\dfrac{24}{25}$ **(b)** $-\dfrac{7}{25}$ **(c)** $\dfrac{2\sqrt{5}}{5}$ **(d)** $-\dfrac{\sqrt{5}}{5}$ **13. (a)** $-\dfrac{2\sqrt{2}}{3}$ **(b)** $\dfrac{1}{3}$ **(c)** $\sqrt{\dfrac{3+\sqrt{6}}{6}}$ **(d)** $\sqrt{\dfrac{3-\sqrt{6}}{6}}$

15. (a) $\dfrac{4\sqrt{2}}{9}$ **(b)** $-\dfrac{7}{9}$ **(c)** $\dfrac{\sqrt{3}}{3}$ **(d)** $\dfrac{\sqrt{6}}{3}$ **17. (a)** $-\dfrac{4}{5}$ **(b)** $\dfrac{3}{5}$ **(c)** $\sqrt{\dfrac{5 + 2\sqrt{5}}{10}}$ **(d)** $\sqrt{\dfrac{5 - 2\sqrt{5}}{10}}$

19. (a) $-\dfrac{3}{5}$ **(b)** $-\dfrac{4}{5}$ **(c)** $\dfrac{1}{2}\sqrt{\dfrac{10 - \sqrt{10}}{5}}$ **(d)** $-\dfrac{1}{2}\sqrt{\dfrac{10 + \sqrt{10}}{5}}$ **21.** $\dfrac{\sqrt{2 - \sqrt{2}}}{2}$ **23.** $1 - \sqrt{2}$ **25.** $-\dfrac{\sqrt{2 + \sqrt{3}}}{2}$

27. $\dfrac{2}{\sqrt{2 + \sqrt{2}}} = (2 - \sqrt{2})\sqrt{2 + \sqrt{2}}$ **29.** $-\dfrac{\sqrt{2 - \sqrt{2}}}{2}$ **31.** $-\dfrac{4}{5}$ **33.** $\dfrac{\sqrt{10(5 - \sqrt{5})}}{10}$ **35.** $\dfrac{4}{3}$ **37.** $-\dfrac{7}{8}$ **39.** $\dfrac{\sqrt{10}}{4}$ **41.** $-\dfrac{\sqrt{15}}{3}$

43. $\sin^4\theta = (\sin^2\theta)^2 = \left(\dfrac{1 - \cos(2\theta)}{2}\right)^2 = \dfrac{1}{4}[1 - 2\cos(2\theta) + \cos^2(2\theta)] = \dfrac{1}{4} - \dfrac{1}{2}\cos(2\theta) + \dfrac{1}{4}\cos^2(2\theta)$

$$= \dfrac{1}{4} - \dfrac{1}{2}\cos(2\theta) + \dfrac{1}{4}\cdot\dfrac{1 + \cos(4\theta)}{2} = \dfrac{1}{4} - \dfrac{1}{2}\cos(2\theta) + \dfrac{1}{8} + \dfrac{1}{8}\cos(4\theta) = \dfrac{3}{8} - \dfrac{1}{2}\cos(2\theta) + \dfrac{1}{8}\cos(4\theta)$$

45. $\sin^2\theta\cos^2\theta = \dfrac{1 - \cos(2\theta)}{2}\cdot\dfrac{1 + \cos(2\theta)}{2} = \dfrac{1}{4}[1 - \cos^2(2\theta)] = \dfrac{1}{4}\left[1 - \dfrac{1 + \cos(4\theta)}{2}\right] = \dfrac{1}{4}\left[\dfrac{1}{2} - \dfrac{1}{2}\cos(4\theta)\right] = \dfrac{1}{8} - \dfrac{1}{8}\cos(4\theta)$

47. $\cos(3\theta) = 4\cos^3\theta - 3\cos\theta$ **49.** $\sin(5\theta) = 16\sin^5\theta - 20\sin^3\theta + 5\sin\theta$ **51.** $\cos^4\theta - \sin^4\theta = (\cos^2\theta + \sin^2\theta)(\cos^2\theta - \sin^2\theta) = \cos(2\theta)$

53. $\cot(2\theta) = \dfrac{1}{\tan(2\theta)} = \dfrac{1 - \tan^2\theta}{2\tan\theta} = \dfrac{1 - \dfrac{1}{\cot^2\theta}}{2 \cdot \dfrac{1}{\cot\theta}} = \dfrac{\dfrac{\cot^2\theta - 1}{\cot^2\theta}}{\dfrac{2}{\cot\theta}} = \dfrac{\cot^2\theta - 1}{\cot^2\theta} \cdot \dfrac{\cot\theta}{2} = \dfrac{\cot^2\theta - 1}{2\cot\theta}$

55. $\sec(2\theta) = \dfrac{1}{\cos(2\theta)} = \dfrac{1}{2\cos^2\theta - 1} = \dfrac{1}{\dfrac{2}{\sec^2\theta} - 1} = \dfrac{1}{\dfrac{2 - \sec^2\theta}{\sec^2\theta}} = \dfrac{\sec^2\theta}{2 - \sec^2\theta}$ **57.** $\cos^2(2u) - \sin^2(2u) = \cos(2\cdot 2u) = \cos(4u)$

59. $\dfrac{\cos(2\theta)}{1 + \sin(2\theta)} = \dfrac{\cos^2\theta - \sin^2\theta}{1 + 2\sin\theta\cos\theta} = \dfrac{(\cos\theta - \sin\theta)(\cos\theta + \sin\theta)}{\sin^2\theta + \cos^2\theta + 2\sin\theta\cos\theta} = \dfrac{(\cos\theta - \sin\theta)(\cos\theta + \sin\theta)}{(\sin\theta + \cos\theta)(\sin\theta + \cos\theta)} = \dfrac{\cos\theta - \sin\theta}{\cos\theta + \sin\theta}$

$= \dfrac{\dfrac{\cos\theta - \sin\theta}{\sin\theta}}{\dfrac{\cos\theta + \sin\theta}{\sin\theta}} = \dfrac{\dfrac{\cos\theta}{\sin\theta} - \dfrac{\sin\theta}{\sin\theta}}{\dfrac{\cos\theta}{\sin\theta} + \dfrac{\sin\theta}{\sin\theta}} = \dfrac{\cot\theta - 1}{\cot\theta + 1}$

61. $\sec^2\dfrac{\theta}{2} = \dfrac{1}{\cos^2\left(\dfrac{\theta}{2}\right)} = \dfrac{1}{\dfrac{1 + \cos\theta}{2}} = \dfrac{2}{1 + \cos\theta}$

63. $\cot^2\dfrac{v}{2} = \dfrac{1}{\tan^2\left(\dfrac{v}{2}\right)} = \dfrac{1}{\dfrac{1 - \cos v}{1 + \cos v}} = \dfrac{1 + \cos v}{1 - \cos v} = \dfrac{1 + \dfrac{1}{\sec v}}{1 - \dfrac{1}{\sec v}} = \dfrac{\dfrac{\sec v + 1}{\sec v}}{\dfrac{\sec v - 1}{\sec v}} = \dfrac{\sec v + 1}{\sec v} \cdot \dfrac{\sec v}{\sec v - 1} = \dfrac{\sec v + 1}{\sec v - 1}$

65. $\dfrac{1 - \tan^2\left(\dfrac{\theta}{2}\right)}{1 + \tan^2\left(\dfrac{\theta}{2}\right)} = \dfrac{1 - \dfrac{1 - \cos\theta}{1 + \cos\theta}}{1 + \dfrac{1 - \cos\theta}{1 + \cos\theta}} = \dfrac{\dfrac{1 + \cos\theta - (1 - \cos\theta)}{1 + \cos\theta}}{\dfrac{1 + \cos\theta + 1 - \cos\theta}{1 + \cos\theta}} = \dfrac{2\cos\theta}{1 + \cos\theta} \cdot \dfrac{1 + \cos\theta}{2} = \cos\theta$

67. $\dfrac{\sin(3\theta)}{\sin\theta} - \dfrac{\cos(3\theta)}{\cos\theta} = \dfrac{\sin(3\theta)\cos\theta - \cos(3\theta)\sin\theta}{\sin\theta\cos\theta} = \dfrac{\sin(3\theta - \theta)}{\dfrac{1}{2}(2\sin\theta\cos\theta)} = \dfrac{2\sin(2\theta)}{\sin(2\theta)} = 2$

69. $\tan(3\theta) = \tan(\theta + 2\theta) = \dfrac{\tan\theta + \tan(2\theta)}{1 - \tan\theta\tan(2\theta)} = \dfrac{\tan\theta + \dfrac{2\tan\theta}{1 - \tan^2\theta}}{1 - \dfrac{\tan\theta(2\tan\theta)}{1 - \tan^2\theta}} = \dfrac{\tan\theta - \tan^3\theta + 2\tan\theta}{1 - \tan^2\theta - 2\tan^2\theta} = \dfrac{3\tan\theta - \tan^3\theta}{1 - 3\tan^2\theta}$

71. $\dfrac{1}{2}(\ln|1 - \cos(2\theta)| - \ln 2) = \ln\left(\dfrac{|1 - \cos(2\theta)|}{2}\right)^{1/2} = \ln|\sin^2\theta|^{1/2} = \ln|\sin\theta|$ **73.** $\left\{\dfrac{\pi}{3}, \dfrac{2\pi}{3}, \dfrac{4\pi}{3}, \dfrac{5\pi}{3}\right\}$ **75.** $\left\{0, \dfrac{2\pi}{3}, \dfrac{4\pi}{3}\right\}$

77. $\left\{0, \dfrac{\pi}{3}, \dfrac{\pi}{2}, \dfrac{2\pi}{3}, \pi, \dfrac{4\pi}{3}, \dfrac{3\pi}{2}, \dfrac{5\pi}{3}\right\}$ **79.** No real solution   **81.** $\left\{0, \dfrac{\pi}{3}, \pi, \dfrac{5\pi}{3}\right\}$ **83.** $\dfrac{\sqrt{3}}{2}$ **85.** $\dfrac{7}{25}$ **87.** $\dfrac{24}{7}$ **89.** $\dfrac{24}{25}$ **91.** $\dfrac{1}{5}$ **93.** $\dfrac{25}{7}$ **95.** $0, \dfrac{\pi}{3}, \pi, \dfrac{5\pi}{3}$

97. $\dfrac{\pi}{2}, \dfrac{3\pi}{2}$ **99.** $\dfrac{\pi}{2}, \dfrac{7\pi}{6}, \dfrac{3\pi}{2}, \dfrac{11\pi}{6}$ **101. (a)** $\left(1152\sqrt{2} - 1152\right)$ in.²   or $1152\sqrt{3 - 2\sqrt{2}}$ in.²   **(b)** $\left(162\sqrt{2} + 162\right)$ cm²   or $162\sqrt{3 + 2\sqrt{2}}$ cm²

103. (a) $W = 2D(\csc\theta - \cot\theta) = 2D\left(\dfrac{1}{\sin\theta} - \dfrac{\cos\theta}{\sin\theta}\right) = 2D\dfrac{1 - \cos\theta}{\sin\theta} = 2D\tan\dfrac{\theta}{2}$ **(b)** $\theta = 24.45°$

105. (a) $R = \dfrac{v_0^2\sqrt{2}}{16}\cos\theta\,(\sin\theta - \cos\theta)$ **(b)** $\dfrac{3\pi}{8}$ or $67.5°$ **(d)** $\theta = 67.5°$

$= \dfrac{v_0^2\sqrt{2}}{32}(2\cos\theta\sin\theta - 2\cos^2\theta)$ **(c)** $32(2 - \sqrt{2}) \approx 18.75$ ft $\left(\dfrac{3\pi}{8}\text{ radians}\right)$ makes R largest.

$= \dfrac{v_0^2\sqrt{2}}{32}[\sin(2\theta) - \cos(2\theta) - 1]$ $R = 18.75$ ft

107. $A = \dfrac{1}{2}h(\text{base}) = h\left(\dfrac{1}{2}\text{base}\right) = s\cos\dfrac{\theta}{2} \cdot s\sin\dfrac{\theta}{2} = \dfrac{1}{2}s^2\sin\theta$ **109.** $\sin(2\theta) = \dfrac{4x}{4 + x^2}$ **111.** $-\dfrac{1}{4}$

113. $\dfrac{2z}{1 + z^2} = \dfrac{2\tan\left(\dfrac{\alpha}{2}\right)}{1 + \tan^2\left(\dfrac{\alpha}{2}\right)} = \dfrac{2\tan\left(\dfrac{\alpha}{2}\right)}{\sec^2\left(\dfrac{\alpha}{2}\right)} = \dfrac{\dfrac{2\sin\left(\dfrac{\alpha}{2}\right)}{\cos\left(\dfrac{\alpha}{2}\right)}}{\dfrac{1}{\cos^2\left(\dfrac{\alpha}{2}\right)}} = 2\sin\left(\dfrac{\alpha}{2}\right)\cos\left(\dfrac{\alpha}{2}\right) = \sin\left(2\cdot\dfrac{\alpha}{2}\right) = \sin\alpha$ **115.**

117. $\sin\dfrac{\pi}{24} = \dfrac{\sqrt{2}}{4}\sqrt{4 - \sqrt{6} - \sqrt{2}}$; $\cos\dfrac{\pi}{24} = \dfrac{\sqrt{2}}{4}\sqrt{4 + \sqrt{6} + \sqrt{2}}$

119. $\sin^3\theta + \sin^3(\theta + 120°) + \sin^3(\theta + 240°) = \sin^3\theta + (\sin\theta\cos 120° + \cos\theta\sin 120°)^3 + (\sin\theta\cos 240° + \cos\theta\sin 240°)^3$

$$= \sin^3\theta + \left(-\frac{1}{2}\sin\theta + \frac{\sqrt{3}}{2}\cos\theta\right)^3 + \left(-\frac{1}{2}\sin\theta - \frac{\sqrt{3}}{2}\cos\theta\right)^3$$

$$= \sin^3\theta + \frac{1}{8}\left(3\sqrt{3}\cos^3\theta - 9\cos^2\theta\sin\theta + 3\sqrt{3}\cos\theta\sin^2\theta - \sin^3\theta\right) - \frac{1}{8}\left(\sin^3\theta + 3\sqrt{3}\sin^2\theta\cos\theta + 9\sin\theta\cos^2\theta + 3\sqrt{3}\cos^3\theta\right)$$

$$= \frac{3}{4}\sin^3\theta - \frac{9}{4}\cos^2\theta\sin\theta = \frac{3}{4}[\sin^3\theta - 3\sin\theta(1 - \sin^2\theta)] = \frac{3}{4}(4\sin^3\theta - 3\sin\theta) = -\frac{3}{4}\sin(3\theta) \text{ (from Example 2)}$$

121. $-\dfrac{1}{2}$ **123.** $y = \dfrac{1}{2}x - 4$ **124.**   **125.** $\dfrac{\sqrt{3}+1}{2}$ **126.** **127.** $f(x) = x^3 + 5x^2 - 4x - 20$ **128.** $f^{-1}(x) = \dfrac{5x+3}{2x+1}$

129. 13 **130.** $D = \dfrac{6x - 5y + 3}{5x - 4y + 4}$

3.7 Assess Your Understanding *(page 253)*

1. $\dfrac{1}{2}\left(\dfrac{\sqrt{3}}{2} - 1\right)$ **3.** $-\dfrac{1}{2}\left(\dfrac{\sqrt{3}}{2} + 1\right)$ **5.** $\dfrac{\sqrt{2}}{2}$ **7.** $\dfrac{1}{2}[\cos(2\theta) - \cos(6\theta)]$ **9.** $\dfrac{1}{2}[\sin(6\theta) + \sin(2\theta)]$ **11.** $\dfrac{1}{2}[\cos(2\theta) + \cos(8\theta)]$

13. $\dfrac{1}{2}[\cos\theta - \cos(3\theta)]$ **15.** $\dfrac{1}{2}[\sin(2\theta) + \sin\theta]$ **17.** $2\sin\theta\cos(3\theta)$ **19.** $2\cos(3\theta)\cos\theta$ **21.** $2\sin(2\theta)\cos\theta$ **23.** $2\sin\theta\sin\dfrac{\theta}{2}$

25. $\dfrac{\sin\theta + \sin(3\theta)}{2\sin(2\theta)} = \dfrac{2\sin(2\theta)\cos\theta}{2\sin(2\theta)} = \cos\theta$ **27.** $\dfrac{\sin(4\theta) + \sin(2\theta)}{\cos(4\theta) + \cos(2\theta)} = \dfrac{2\sin(3\theta)\cos\theta}{2\cos(3\theta)\cos\theta} = \dfrac{\sin(3\theta)}{\cos(3\theta)} = \tan(3\theta)$

29. $\dfrac{\cos\theta - \cos(3\theta)}{\sin\theta + \sin(3\theta)} = \dfrac{2\sin(2\theta)\sin\theta}{2\sin(2\theta)\cos\theta} = \dfrac{\sin\theta}{\cos\theta} = \tan\theta$

31. $\sin\theta[\sin\theta + \sin(3\theta)] = \sin\theta[2\sin(2\theta)\cos\theta] = \cos\theta[2\sin(2\theta)\sin\theta] = \cos\theta\left[2\cdot\dfrac{1}{2}[\cos\theta - \cos(3\theta)]\right] = \cos\theta[\cos\theta - \cos(3\theta)]$

33. $\dfrac{\sin(4\theta) + \sin(8\theta)}{\cos(4\theta) + \cos(8\theta)} = \dfrac{2\sin(6\theta)\cos(2\theta)}{2\cos(6\theta)\cos(2\theta)} = \dfrac{\sin(6\theta)}{\cos(6\theta)} = \tan(6\theta)$

35. $\dfrac{\sin(4\theta) + \sin(8\theta)}{\sin(4\theta) - \sin(8\theta)} = \dfrac{2\sin(6\theta)\cos(-2\theta)}{2\sin(-2\theta)\cos(6\theta)} = \dfrac{\sin(6\theta)}{\cos(6\theta)}\cdot\dfrac{\cos(2\theta)}{-\sin(2\theta)} = \tan(6\theta)[-\cot(2\theta)] = -\dfrac{\tan(6\theta)}{\tan(2\theta)}$

37. $\dfrac{\sin\alpha + \sin\beta}{\sin\alpha - \sin\beta} = \dfrac{2\sin\dfrac{\alpha+\beta}{2}\cos\dfrac{\alpha-\beta}{2}}{2\sin\dfrac{\alpha-\beta}{2}\cos\dfrac{\alpha+\beta}{2}} = \dfrac{\sin\dfrac{\alpha+\beta}{2}}{\cos\dfrac{\alpha+\beta}{2}}\cdot\dfrac{\cos\dfrac{\alpha-\beta}{2}}{\sin\dfrac{\alpha-\beta}{2}} = \tan\dfrac{\alpha+\beta}{2}\cot\dfrac{\alpha-\beta}{2}$

39. $\dfrac{\sin\alpha + \sin\beta}{\cos\alpha + \cos\beta} = \dfrac{2\sin\dfrac{\alpha+\beta}{2}\cos\dfrac{\alpha-\beta}{2}}{2\cos\dfrac{\alpha+\beta}{2}\cos\dfrac{\alpha-\beta}{2}} = \dfrac{\sin\dfrac{\alpha+\beta}{2}}{\cos\dfrac{\alpha+\beta}{2}} = \tan\dfrac{\alpha+\beta}{2}$

41. $1 + \cos(2\theta) + \cos(4\theta) + \cos(6\theta) = [1 + \cos(6\theta)] + [\cos(2\theta) + \cos(4\theta)] = 2\cos^2(3\theta) + 2\cos(3\theta)\cos(-\theta)$

$$= 2\cos(3\theta)[\cos(3\theta) + \cos\theta] = 2\cos(3\theta)[2\cos(2\theta)\cos\theta] = 4\cos\theta\cos(2\theta)\cos(3\theta)$$

43. $\sin^4\theta\cos^2\theta = (\sin^2\theta)^2\cos^2\theta = \left[\dfrac{1 - \cos(2\theta)}{2}\right]^2\cdot\dfrac{1 + \cos(2\theta)}{2} = \dfrac{1}{8}[1 - \cos(2\theta)]^2[1 + \cos(2\theta)]$

$$= \dfrac{1}{8}[1 - \cos(2\theta)][1 - \cos^2(2\theta)] = \dfrac{1}{8}[1 - \cos(2\theta)]\left[1 - \dfrac{1 + \cos(4\theta)}{2}\right] = \dfrac{1}{16}[1 - \cos(2\theta)][2 - (1 + \cos(4\theta))]$$

$$= \dfrac{1}{16}[1 - \cos(2\theta)][1 - \cos(4\theta)] = \dfrac{1}{16}[1 - \cos(2\theta) - \cos(4\theta) + \cos(4\theta)\cos(2\theta)]$$

$$= \dfrac{1}{16}\left\{1 - \cos(2\theta) - \cos(4\theta) + \dfrac{1}{2}[\cos(2\theta) + \cos(6\theta)]\right\} = \dfrac{1}{32}[2 - 2\cos(2\theta) - 2\cos(4\theta) + \cos(2\theta) + \cos(6\theta)]$$

$$= \dfrac{1}{32}[2 - \cos(2\theta) - 2\cos(4\theta) + \cos(6\theta)] = \dfrac{1}{16} - \dfrac{1}{32}\cos(2\theta) - \dfrac{1}{16}\cos(4\theta) + \dfrac{1}{32}\cos(6\theta)$$

45. $\sin^6\theta = (\sin^2\theta)^3 = \left[\dfrac{1 - \cos(2\theta)}{2}\right]^3 = \dfrac{1}{8}[1 - \cos(2\theta)]^3$

$$= \dfrac{1}{8}[1 - 2\cos(2\theta) + \cos^2(2\theta)][1 - \cos(2\theta)] = \dfrac{1}{8}\left[1 - 2\cos(2\theta) + \dfrac{1 + \cos(4\theta)}{2}\right][1 - \cos(2\theta)]$$

$$= \dfrac{1}{16}[2 - 4\cos(2\theta) + 1 + \cos(4\theta)][1 - \cos(2\theta)] = \dfrac{1}{16}[3 - 4\cos(2\theta) + \cos(4\theta)][1 - \cos(2\theta)]$$

$$= \dfrac{1}{16}[3 - 3\cos(2\theta) - 4\cos(2\theta) + 4\cos^2(2\theta) + \cos(4\theta) - \cos(4\theta)\cos(2\theta)]$$

$$= \dfrac{1}{16}\left\{3 - 7\cos(2\theta) + 4\cdot\dfrac{1 + \cos(4\theta)}{2} + \cos(4\theta) - \dfrac{1}{2}[\cos(2\theta) + \cos(6\theta)]\right\}$$

$$= \dfrac{1}{32}[6 - 14\cos(2\theta) + 4 + 4\cos(4\theta) + 2\cos(4\theta) - \cos(2\theta) - \cos(6\theta)]$$

$$= \dfrac{1}{32}[10 - 15\cos(2\theta) + 6\cos(4\theta) - \cos(6\theta)] = \dfrac{5}{16} - \dfrac{15}{32}\cos(2\theta) + \dfrac{3}{16}\cos(4\theta) - \dfrac{1}{32}\cos(6\theta)$$

47. $\left\{0, \dfrac{\pi}{3}, \dfrac{\pi}{2}, \dfrac{2\pi}{3}, \pi, \dfrac{4\pi}{3}, \dfrac{3\pi}{2}, \dfrac{5\pi}{3}\right\}$ **49.** $\left\{0, \dfrac{\pi}{5}, \dfrac{2\pi}{5}, \dfrac{3\pi}{5}, \dfrac{4\pi}{5}, \pi, \dfrac{6\pi}{5}, \dfrac{7\pi}{5}, \dfrac{8\pi}{5}, \dfrac{9\pi}{5}\right\}$

51. (a) $y = 2\sin(2061\pi t)\cos(357\pi t)$

(b) $y_{max} = 2$

(c)

53. $I_u = I_x \cos^2\theta + I_y \sin^2\theta - 2I_{xy}\sin\theta\cos\theta = I_x\cos^2\theta + I_y\sin^2\theta - I_{xy}\sin 2\theta$

$$= I_x\left(\dfrac{\cos 2\theta + 1}{2}\right) + I_y\left(\dfrac{1 - \cos 2\theta}{2}\right) - I_{xy}\sin 2\theta$$

$$= \dfrac{I_x}{2}\cos 2\theta + \dfrac{I_x}{2} + \dfrac{I_y}{2} - \dfrac{I_y}{2}\cos 2\theta - I_{xy}\sin 2\theta$$

$$= \dfrac{I_x + I_y}{2} + \dfrac{I_x - I_y}{2}\cos 2\theta - I_{xy}\sin 2\theta$$

$I_v = I_x\sin^2\theta + I_y\cos^2\theta + 2I_{xy}\sin\theta\cos\theta = I_x\left(\dfrac{1 - \cos 2\theta}{2}\right) + I_y\left(\dfrac{\cos 2\theta + 1}{2}\right) + I_{xy}\sin 2\theta$

$$= \dfrac{I_x}{2} - \dfrac{I_x}{2}\cos 2\theta + \dfrac{I_y}{2}\cos 2\theta + \dfrac{I_y}{2} + I_{xy}\sin 2\theta$$

$$= \dfrac{I_x + I_y}{2} - \dfrac{I_x - I_y}{2}\cos 2\theta + I_{xy}\sin 2\theta$$

55.
$$\sin(\alpha - \beta) = \sin\alpha\cos\beta - \cos\alpha\sin\beta$$
$$\sin(\alpha + \beta) = \sin\alpha\cos\beta + \cos\alpha\sin\beta$$
$$\sin(\alpha - \beta) + \sin(\alpha + \beta) = 2\sin\alpha\cos\beta$$
$$\sin\alpha\cos\beta = \dfrac{1}{2}[\sin(\alpha + \beta) + \sin(\alpha - \beta)]$$

57. $2\cos\dfrac{\alpha + \beta}{2}\cos\dfrac{\alpha - \beta}{2} = 2\cdot\dfrac{1}{2}\left[\cos\left(\dfrac{\alpha + \beta}{2} + \dfrac{\alpha - \beta}{2}\right) + \cos\left(\dfrac{\alpha + \beta}{2} - \dfrac{\alpha - \beta}{2}\right)\right] = \cos\dfrac{2\alpha}{2} + \cos\dfrac{2\beta}{2} = \cos\alpha + \cos\beta$

59. $\sin(2\alpha) + \sin(2\beta) + \sin(2\gamma) = 2\sin(\alpha + \beta)\cos(\alpha - \beta) + \sin(2\gamma) = 2\sin(\alpha + \beta)\cos(\alpha - \beta) + 2\sin\gamma\cos\gamma$

$= 2\sin(\pi - \gamma)\cos(\alpha - \beta) + 2\sin\gamma\cos\gamma = 2\sin\gamma\cos(\alpha - \beta) + 2\sin\gamma\cos\gamma = 2\sin\gamma[\cos(\alpha - \beta) + \cos\gamma]$

$= 2\sin\gamma\left(2\cos\dfrac{\alpha - \beta + \gamma}{2}\cos\dfrac{\alpha - \beta - \gamma}{2}\right) = 4\sin\gamma\cos\dfrac{\pi - 2\beta}{2}\cos\dfrac{2\alpha - \pi}{2} = 4\sin\gamma\cos\left(\dfrac{\pi}{2} - \beta\right)\cos\left(\alpha - \dfrac{\pi}{2}\right)$

$= 4\sin\gamma\sin\beta\sin\alpha = 4\sin\alpha\sin\beta\sin\gamma$

61. Amplitude: 5; Period: $\dfrac{\pi}{2}$; Phase shift: $\dfrac{\pi}{4}$ **62.** $\dfrac{2\sqrt{6}}{7}$

63. $f^{-1}(x) = \sin^{-1}\left(\dfrac{x + 5}{3}\right)$; Range of $f = $ Domain of $f^{-1} = [-8, -2]$; Range of $f^{-1} = \left[-\dfrac{\pi}{2}, \dfrac{\pi}{2}\right]$ **64.** $-\dfrac{\sqrt{3}}{3}$ **65.** $f(x) = \dfrac{1}{3}(x - 3)^2 - 5$

66. 4800 rpm   **67.** $h = \dfrac{2A}{b}$ **68.** $\left[\dfrac{1}{2}, 3\right]$ **69.** $\dfrac{2}{3}$

Review Exercises *(page 256)*

1. Domain: $\{x|-1 \le x \le 1\}$; Range: $\left\{y\left|-\dfrac{\pi}{2} \le y \le \dfrac{\pi}{2}\right.\right\}$ **2.** Domain: $\{x|-1 \le x \le 1\}$; Range: $\{y|0 \le y \le \pi\}$

3. Domain: $\{x|-\infty < x < \infty\}$; Range: $\left\{y\left|-\dfrac{\pi}{2} < y < \dfrac{\pi}{2}\right.\right\}$ **4.** Domain: $\{x||x| \ge 1\}$; Range: $\left\{y\left|0 \le y \le \pi, y \ne \dfrac{\pi}{2}\right.\right\}$

5. Domain: $\{x||x| \ge 1\}$; Range: $\left\{y\left|-\dfrac{\pi}{2} \le y \le \dfrac{\pi}{2}, y \ne 0\right.\right\}$ **6.** Domain: $\{x|-\infty < x < \infty\}$; Range: $\{y|0 < y < \pi\}$

7. $\dfrac{\pi}{2}$ **8.** $\dfrac{\pi}{2}$ **9.** $\dfrac{\pi}{4}$ **10.** $-\dfrac{\pi}{6}$ **11.** $\dfrac{5\pi}{6}$ **12.** $-\dfrac{\pi}{3}$ **13.** $\dfrac{\pi}{4}$ **14.** $\dfrac{3\pi}{4}$ **15.** $\dfrac{3\pi}{4}$ **16.** $\dfrac{3\pi}{8}$ **17.** $-\dfrac{\pi}{3}$ **18.** $\dfrac{\pi}{7}$ **19.** $-\dfrac{\pi}{9}$ **20.** 0.9   **21.** 0.6   **22.** 5   **23.** Not defined

24. $-\dfrac{\pi}{6}$ **25.** π **26.** $-\sqrt{3}$ **27.** $\dfrac{2\sqrt{3}}{3}$ **28.** $\dfrac{4}{5}$ **29.** $-\dfrac{4}{3}$ **30.** $f^{-1}(x) = \dfrac{1}{3}\sin^{-1}\left(\dfrac{x}{2}\right)$; Range of $f = $ Domain of $f^{-1} = [-2, 2]$; Range of $f^{-1} = \left[-\dfrac{\pi}{6}, \dfrac{\pi}{6}\right]$

31. $f^{-1}(x) = \cos^{-1}(3 - x)$; Range of $f = $ Domain of $f^{-1} = [2, 4]$; Range of $f^{-1} = [0, \pi]$ **32.** $\sqrt{1 - u^2}$ **33.** $\dfrac{|u|}{u\sqrt{u^2 - 1}}$

34. $\tan\theta\cot\theta - \sin^2\theta = 1 - \sin^2\theta = \cos^2\theta$ **35.** $\sin^2\theta(1 + \cot^2\theta) = \sin^2\theta\csc^2\theta = 1$

36. $5\cos^2\theta + 3\sin^2\theta = 2\cos^2\theta + 3(\cos^2\theta + \sin^2\theta) = 3 + 2\cos^2\theta$

37. $\dfrac{1 - \cos\theta}{\sin\theta} + \dfrac{\sin\theta}{1 - \cos\theta} = \dfrac{(1 - \cos\theta)^2 + \sin^2\theta}{\sin\theta(1 - \cos\theta)} = \dfrac{1 - 2\cos\theta + \cos^2\theta + \sin^2\theta}{\sin\theta(1 - \cos\theta)} = \dfrac{2(1 - \cos\theta)}{\sin\theta(1 - \cos\theta)} = 2\csc\theta$

38. $\dfrac{\cos\theta}{\cos\theta - \sin\theta} = \dfrac{\dfrac{\cos\theta}{\cos\theta}}{\dfrac{\cos\theta - \sin\theta}{\cos\theta}} = \dfrac{1}{1 - \dfrac{\sin\theta}{\cos\theta}} = \dfrac{1}{1 - \tan\theta}$ **39.** $\dfrac{\csc\theta}{1 + \csc\theta} = \dfrac{\dfrac{1}{\sin\theta}}{1 + \dfrac{1}{\sin\theta}} = \dfrac{1}{1 + \sin\theta} = \dfrac{1}{1 + \sin\theta}\cdot\dfrac{1 - \sin\theta}{1 - \sin\theta} = \dfrac{1 - \sin\theta}{1 - \sin^2\theta} = \dfrac{1 - \sin\theta}{\cos^2\theta}$

40. $\csc\theta - \sin\theta = \dfrac{1}{\sin\theta} - \sin\theta = \dfrac{1 - \sin^2\theta}{\sin\theta} = \dfrac{\cos^2\theta}{\sin\theta} = \cos\theta\cdot\dfrac{\cos\theta}{\sin\theta} = \cos\theta\cot\theta$

41. $\dfrac{1 - \sin\theta}{\sec\theta} = \cos\theta(1 - \sin\theta)\cdot\dfrac{1 + \sin\theta}{1 + \sin\theta} = \dfrac{\cos\theta(1 - \sin^2\theta)}{1 + \sin\theta} = \dfrac{\cos^3\theta}{1 + \sin\theta}$ **42.** $\cot\theta - \tan\theta = \dfrac{\cos\theta}{\sin\theta} - \dfrac{\sin\theta}{\cos\theta} = \dfrac{\cos^2\theta - \sin^2\theta}{\sin\theta\cos\theta} = \dfrac{1 - 2\sin^2\theta}{\sin\theta\cos\theta}$

43. $\dfrac{\cos(\alpha+\beta)}{\cos\alpha\sin\beta} = \dfrac{\cos\alpha\cos\beta - \sin\alpha\sin\beta}{\cos\alpha\sin\beta} = \dfrac{\cos\alpha\cos\beta}{\cos\alpha\sin\beta} - \dfrac{\sin\alpha\sin\beta}{\cos\alpha\sin\beta} = \cot\beta - \tan\alpha$

44. $\dfrac{\cos(\alpha-\beta)}{\cos\alpha\cos\beta} = \dfrac{\cos\alpha\cos\beta + \sin\alpha\sin\beta}{\cos\alpha\cos\beta} = \dfrac{\cos\alpha\cos\beta}{\cos\alpha\cos\beta} + \dfrac{\sin\alpha\sin\beta}{\cos\alpha\cos\beta} = 1 + \tan\alpha\tan\beta$

45. $(1+\cos\theta)\tan\dfrac{\theta}{2} = (1+\cos\theta)\cdot\dfrac{\sin\theta}{1+\cos\theta} = \sin\theta$

46. $2\cot\theta\cot 2\theta = 2\cdot\dfrac{\cos\theta}{\sin\theta}\cdot\dfrac{\cos 2\theta}{\sin 2\theta} = \dfrac{2\cos\theta(\cos^2\theta - \sin^2\theta)}{2\sin^2\theta\cos\theta} = \dfrac{\cos^2\theta - \sin^2\theta}{\sin^2\theta} = \cot^2\theta - 1$

47. $1 - 8\sin^2\theta\cos^2\theta = 1 - 2(2\sin\theta\cos\theta)^2 = 1 - 2\sin^2(2\theta) = \cos(4\theta)$

48. $\dfrac{\sin(3\theta)\cos\theta - \sin\theta\cos(3\theta)}{\sin(2\theta)} = \dfrac{\sin(2\theta)}{\sin(2\theta)} = 1$ **49.** $\dfrac{\sin(2\theta)+\sin(4\theta)}{\cos(2\theta)+\cos(4\theta)} = \dfrac{2\sin(3\theta)\cos(-\theta)}{2\cos(3\theta)\cos(-\theta)} = \tan(3\theta)$

50. $\dfrac{\cos(2\theta)-\cos(4\theta)}{\cos(2\theta)+\cos(4\theta)} - \tan\theta\tan(3\theta) = \dfrac{-2\sin(3\theta)\sin(-\theta)}{2\cos(3\theta)\cos(-\theta)} - \tan\theta\tan(3\theta) = \tan(3\theta)\tan\theta - \tan\theta\tan(3\theta) = 0$

51. $\dfrac{1}{4}(\sqrt{6}-\sqrt{2})$ **52.** $-2-\sqrt{3}$ **53.** $\dfrac{1}{4}(\sqrt{6}-\sqrt{2})$ **54.** $\dfrac{1}{4}(\sqrt{2}-\sqrt{6})$ **55.** $\dfrac{1}{2}$ **56.** $\dfrac{1}{2}$ **57.** $\sqrt{2}-1$ **58.** $\dfrac{\sqrt{2+\sqrt{2}}}{2}$ **59. (a)** $-\dfrac{33}{65}$ **(b)** $-\dfrac{56}{65}$

(c) $-\dfrac{63}{65}$ **(d)** $\dfrac{33}{56}$ **(e)** $\dfrac{24}{25}$ **(f)** $\dfrac{119}{169}$ **(g)** $\dfrac{5\sqrt{26}}{26}$ **(h)** $\dfrac{2\sqrt{5}}{5}$ **60. (a)** $-\dfrac{16}{65}$ **(b)** $-\dfrac{63}{65}$ **(c)** $-\dfrac{56}{65}$ **(d)** $\dfrac{16}{63}$ **(e)** $\dfrac{24}{25}$ **(f)** $\dfrac{119}{169}$ **(g)** $\dfrac{\sqrt{26}}{26}$ **(h)** $-\dfrac{\sqrt{10}}{10}$

61. (a) $-\dfrac{63}{65}$ **(b)** $\dfrac{16}{65}$ **(c)** $\dfrac{33}{65}$ **(d)** $-\dfrac{63}{16}$ **(e)** $\dfrac{24}{25}$ **(f)** $-\dfrac{119}{169}$ **(g)** $\dfrac{2\sqrt{13}}{13}$ **(h)** $-\dfrac{\sqrt{10}}{10}$ **62. (a)** $\dfrac{-\sqrt{3}-2\sqrt{2}}{6}$ **(b)** $\dfrac{1-2\sqrt{6}}{6}$ **(c)** $\dfrac{-\sqrt{3}+2\sqrt{2}}{6}$

(d) $\dfrac{8\sqrt{2}+9\sqrt{3}}{23}$ **(e)** $-\dfrac{\sqrt{3}}{2}$ **(f)** $-\dfrac{7}{9}$ **(g)** $\dfrac{\sqrt{3}}{3}$ **(h)** $\dfrac{\sqrt{3}}{2}$ **63. (a)** 1 **(b)** 0 **(c)** $-\dfrac{1}{9}$ **(d)** Not defined **(e)** $\dfrac{4\sqrt{5}}{9}$ **(f)** $-\dfrac{1}{9}$ **(g)** $\dfrac{\sqrt{30}}{6}$

(h) $-\dfrac{\sqrt{6}\sqrt{3}-\sqrt{5}}{6}$ **64.** $\dfrac{4+3\sqrt{3}}{10}$ **65.** $\dfrac{33}{65}$ **66.** $-\dfrac{48+25\sqrt{3}}{39}$ **67.** $-\dfrac{\sqrt{2}}{10}$ **68.** $-\dfrac{24}{25}$ **69.** $-\dfrac{7}{25}$ **70.** $\left\{\dfrac{\pi}{3}, \dfrac{5\pi}{3}\right\}$ **71.** $\left\{\dfrac{2\pi}{3}, \dfrac{5\pi}{3}\right\}$ **72.** $\left\{\dfrac{3\pi}{4}, \dfrac{7\pi}{4}\right\}$

73. $\left\{0, \dfrac{\pi}{2}, \pi, \dfrac{3\pi}{2}\right\}$ **74.** $\left\{\dfrac{\pi}{3}, \dfrac{2\pi}{3}, \dfrac{4\pi}{3}, \dfrac{5\pi}{3}\right\}$ **75.** $\{0.25, 2.89\}$ **76.** $\left\{0, \dfrac{2\pi}{3}, \pi, \dfrac{4\pi}{3}\right\}$ **77.** $\left\{0, \dfrac{\pi}{6}, \dfrac{5\pi}{6}\right\}$ **78.** $\left\{\dfrac{\pi}{6}, \dfrac{\pi}{2}, \dfrac{5\pi}{6}\right\}$ **79.** $\left\{\dfrac{\pi}{3}, \dfrac{5\pi}{3}\right\}$

80. $\left\{\dfrac{\pi}{4}, \dfrac{\pi}{2}, \dfrac{3\pi}{4}, \dfrac{3\pi}{2}\right\}$ **81.** $\left\{\dfrac{\pi}{2}, \pi\right\}$ **82.** 0.78 **83.** -1.11 **84.** 1.77 **85.** 1.23 **86.** 2.90 **87.** $\{1.11\}$ **88.** $\{0.87\}$ **89.** $\{2.22\}$ **90.** $\left\{-\dfrac{\sqrt{3}}{2}\right\}$ **91.** $\{0\}$

92. $\sin 15° = \sqrt{\dfrac{1-\cos 30°}{2}} = \sqrt{\dfrac{1-\dfrac{\sqrt{3}}{2}}{2}} = \sqrt{\dfrac{2-\sqrt{3}}{4}} = \dfrac{\sqrt{2-\sqrt{3}}}{2}$;

$\sin 15° = \sin(45° - 30°) = \sin 45°\cos 30° - \cos 45°\sin 30° = \dfrac{\sqrt{2}}{2}\cdot\dfrac{\sqrt{3}}{2} - \dfrac{\sqrt{2}}{2}\cdot\dfrac{1}{2} = \dfrac{\sqrt{6}}{4} - \dfrac{\sqrt{2}}{4} = \dfrac{\sqrt{6}-\sqrt{2}}{4}$;

$\left[\dfrac{\sqrt{2-\sqrt{3}}}{2}\right]^2 = \dfrac{2-\sqrt{3}}{4} = \dfrac{4(2-\sqrt{3})}{4\cdot 4} = \dfrac{8-4\sqrt{3}}{16} = \dfrac{6-2\sqrt{12}+2}{16} = \left(\dfrac{\sqrt{6}-\sqrt{2}}{4}\right)^2$

93. $\cos(2\theta) = 2\cos^2\theta - 1$

Chapter Test *(page 258)*

1. $\dfrac{\pi}{6}$ **2.** $-\dfrac{\pi}{4}$ **3.** $-\dfrac{\pi}{3}$ **4.** $\dfrac{\pi}{2}$ **5.** $\dfrac{\pi}{4}$ **6.** $-\dfrac{\pi}{6}$ **7.** $\dfrac{\pi}{5}$ **8.** $\dfrac{7}{3}$ **9.** 3 **10.** $-\dfrac{4}{3}$ **11.** 0.39 **12.** 0.78 **13.** 1.25 **14.** 0.20

15. $\dfrac{\csc\theta+\cot\theta}{\sec\theta+\tan\theta} = \dfrac{\csc\theta+\cot\theta}{\sec\theta+\tan\theta}\cdot\dfrac{\csc\theta-\cot\theta}{\csc\theta-\cot\theta} = \dfrac{\csc^2\theta-\cot^2\theta}{(\sec\theta+\tan\theta)(\csc\theta-\cot\theta)} = \dfrac{1}{(\sec\theta+\tan\theta)(\csc\theta-\cot\theta)}$

$= \dfrac{1}{(\sec\theta+\tan\theta)(\csc\theta-\cot\theta)}\cdot\dfrac{\sec\theta-\tan\theta}{\sec\theta-\tan\theta} = \dfrac{\sec\theta-\tan\theta}{(\sec^2\theta-\tan^2\theta)(\csc\theta-\cot\theta)} = \dfrac{\sec\theta-\tan\theta}{\csc\theta-\cot\theta}$

16. $\sin\theta\tan\theta + \cos\theta = \sin\theta\cdot\dfrac{\sin\theta}{\cos\theta} + \cos\theta = \dfrac{\sin^2\theta}{\cos\theta} + \dfrac{\cos^2\theta}{\cos\theta} = \dfrac{\sin^2\theta+\cos^2\theta}{\cos\theta} = \dfrac{1}{\cos\theta} = \sec\theta$

17. $\tan\theta + \cot\theta = \dfrac{\sin\theta}{\cos\theta} + \dfrac{\cos\theta}{\sin\theta} = \dfrac{\sin^2\theta}{\sin\theta\cos\theta} + \dfrac{\cos^2\theta}{\sin\theta\cos\theta} = \dfrac{\sin^2\theta+\cos^2\theta}{\sin\theta\cos\theta} = \dfrac{1}{\sin\theta\cos\theta} = \dfrac{2}{2\sin\theta\cos\theta} = \dfrac{2}{\sin(2\theta)} = 2\csc(2\theta)$

18. $\dfrac{\sin(\alpha+\beta)}{\tan\alpha+\tan\beta} = \dfrac{\sin\alpha\cos\beta + \cos\alpha\sin\beta}{\dfrac{\sin\alpha}{\cos\alpha} + \dfrac{\sin\beta}{\cos\beta}} = \dfrac{\sin\alpha\cos\beta + \cos\alpha\sin\beta}{\dfrac{\sin\alpha\cos\beta}{\cos\alpha\cos\beta} + \dfrac{\cos\alpha\sin\beta}{\cos\alpha\cos\beta}} = \dfrac{\sin\alpha\cos\beta + \cos\alpha\sin\beta}{\dfrac{\sin\alpha\cos\beta + \cos\alpha\sin\beta}{\cos\alpha\cos\beta}}$

$= \dfrac{\sin\alpha\cos\beta + \cos\alpha\sin\beta}{1}\cdot\dfrac{\cos\alpha\cos\beta}{\sin\alpha\cos\beta + \cos\alpha\sin\beta} = \cos\alpha\cos\beta$

19. $\sin(3\theta) = \sin(\theta+2\theta) = \sin\theta\cos(2\theta) + \cos\theta\sin(2\theta) = \sin\theta\cdot(\cos^2\theta - \sin^2\theta) + \cos\theta\cdot 2\sin\theta\cos\theta = \sin\theta\cos^2\theta - \sin^3\theta + 2\sin\theta\cos^2\theta$

$= 3\sin\theta\cos^2\theta - \sin^3\theta = 3\sin\theta(1-\sin^2\theta) - \sin^3\theta = 3\sin\theta - 3\sin^3\theta - \sin^3\theta = 3\sin\theta - 4\sin^3\theta$

20. $\dfrac{\tan\theta-\cot\theta}{\tan\theta+\cot\theta} = \dfrac{\dfrac{\sin\theta}{\cos\theta} - \dfrac{\cos\theta}{\sin\theta}}{\dfrac{\sin\theta}{\cos\theta} + \dfrac{\cos\theta}{\sin\theta}} = \dfrac{\dfrac{\sin^2\theta-\cos^2\theta}{\sin\theta\cos\theta}}{\dfrac{\sin^2\theta+\cos^2\theta}{\sin\theta\cos\theta}} = \dfrac{\sin^2\theta-\cos^2\theta}{\sin^2\theta+\cos^2\theta} = \dfrac{(1-\cos^2\theta)-\cos^2\theta}{1} = 1 - 2\cos^2\theta$ **21.** $\dfrac{1}{4}(\sqrt{6}+\sqrt{2})$

22. $2 + \sqrt{3}$ **23.** $\dfrac{\sqrt{5}}{5}$ **24.** $\dfrac{12\sqrt{85}}{49}$ **25.** $\dfrac{2\sqrt{13}\,(\sqrt{5} - 3)}{39}$ **26.** $\dfrac{2 + \sqrt{3}}{4}$ **27.** $\dfrac{\sqrt{6}}{2}$ **28.** $\dfrac{\sqrt{2}}{2}$ **29.** $\left\{\dfrac{\pi}{3}, \dfrac{2\pi}{3}, \dfrac{4\pi}{3}, \dfrac{5\pi}{3}\right\}$ **30.** $\{0, 1.911, \pi, 4.373\}$

31. $\left\{\dfrac{3\pi}{8}, \dfrac{7\pi}{8}, \dfrac{11\pi}{8}, \dfrac{15\pi}{8}\right\}$ **32.** $\{0.285, 3.427\}$ **33.** $\{0.253, 2.889\}$

Cumulative Review (page 259)

1. $\left\{\dfrac{-1 - \sqrt{13}}{6}, \dfrac{-1 + \sqrt{13}}{6}\right\}$ **2.** $y + 1 = -1(x - 4)$, or $x + y = 3$; $6\sqrt{2}$; $(1, 2)$ **3.** x-axis symmetry; $(0, -3), (0, 3), (3, 0)$

4.   **5.**

6. (a)   **(b)**   **(c)**

7. (a) $-\dfrac{2\sqrt{2}}{3}$ **(b)** $\dfrac{\sqrt{2}}{4}$ **(c)** $\dfrac{4\sqrt{2}}{9}$ **(d)** $\dfrac{7}{9}$ **(e)** $\sqrt{\dfrac{3 + 2\sqrt{2}}{6}}$ **(f)** $-\sqrt{\dfrac{3 - 2\sqrt{2}}{6}}$ **8.** $\dfrac{\sqrt{5}}{5}$ **9. (a)** $-\dfrac{2\sqrt{2}}{3}$ **(b)** $-\dfrac{2\sqrt{2}}{3}$ **(c)** $\dfrac{7}{9}$ **(d)** $\dfrac{4\sqrt{2}}{9}$ **(e)** $\dfrac{\sqrt{6}}{3}$

CHAPTER 4 Applications of Trigonometric Functions

4.1 Assess Your Understanding (page 269)

4. F **5.** b **6.** angle of elevation **7.** T **8.** F **9.** $\sin\theta = \dfrac{5}{13}$; $\cos\theta = \dfrac{12}{13}$; $\tan\theta = \dfrac{5}{12}$; $\cot\theta = \dfrac{12}{5}$; $\sec\theta = \dfrac{13}{12}$; $\csc\theta = \dfrac{13}{5}$

11. $\sin\theta = \dfrac{2\sqrt{13}}{13}$; $\cos\theta = \dfrac{3\sqrt{13}}{13}$; $\tan\theta = \dfrac{2}{3}$; $\cot\theta = \dfrac{3}{2}$; $\sec\theta = \dfrac{\sqrt{13}}{3}$; $\csc\theta = \dfrac{\sqrt{13}}{2}$

13. $\sin\theta = \dfrac{\sqrt{3}}{2}$; $\cos\theta = \dfrac{1}{2}$; $\tan\theta = \sqrt{3}$; $\cot\theta = \dfrac{\sqrt{3}}{3}$; $\sec\theta = 2$; $\csc\theta = \dfrac{2\sqrt{3}}{3}$

15. $\sin\theta = \dfrac{\sqrt{6}}{3}$; $\cos\theta = \dfrac{\sqrt{3}}{3}$; $\tan\theta = \sqrt{2}$; $\cot\theta = \dfrac{\sqrt{2}}{2}$; $\sec\theta = \sqrt{3}$; $\csc\theta = \dfrac{\sqrt{6}}{2}$

17. $\sin\theta = \dfrac{\sqrt{5}}{5}$; $\cos\theta = \dfrac{2\sqrt{5}}{5}$; $\tan\theta = \dfrac{1}{2}$; $\cot\theta = 2$; $\sec\theta = \dfrac{\sqrt{5}}{2}$; $\csc\theta = \sqrt{5}$

19. 0 **21.** 1 **23.** 0 **25.** 0 **27.** 1 **29.** $a \approx 13.74, c \approx 14.62, A = 70°$ **31.** $b \approx 5.03, c \approx 7.83, A = 50°$ **33.** $a \approx 0.71, c \approx 4.06, B = 80°$
35. $b \approx 10.72, c \approx 11.83, B = 65°$ **37.** $b \approx 3.08, a \approx 8.46, A = 70°$ **39.** $c \approx 5.83, A \approx 59.0°, B \approx 31.0°$ **41.** $b \approx 4.58, A \approx 23.6°, B \approx 66.4°$
43. 23.6° and 66.4° **45.** 4.59 in.; 6.55 in. **47. (a)** 5.52 in. or 11.83 in. **49.** 70.02 ft **51.** 985.91 ft **53.** 137.37 m **55.** 80.5°
57. (a) 111.96 ft/sec or 76.3 mi/h **(b)** 82.42 ft/sec or 56.2 mi/h **(c)** Under 18.8° **59. (a)** 2.4898×10^{13} miles **(b)** 0.000214° **61.** 554.52 ft
63. S76.6°E **65.** The embankment is 30.5 m high. **67.** 3.83 mi **69.** 1978.09 ft **71.** 60.27 ft **73.** The buildings are 7984 ft apart. **75.** 69.0°

77. 38.9° **79.** The white ball should hit the top cushion 3.34 ft from the upper left corner. **81.** 76.94 in. **86.** Yes **87.** $\dfrac{\sqrt{6} - \sqrt{2}}{4}$ or $\dfrac{\sqrt{2}}{4}(\sqrt{3} - 1)$

88. 0.236, 0.243, 0.248 **89.** $\left\{\dfrac{\pi}{2}, \dfrac{7\pi}{6}, \dfrac{11\pi}{6}\right\}$ **90.** $\dfrac{21}{4} = 5.25$ **91.** $3 + \sqrt{5}$ **92.** 65 **93.** $(x + 4)^2 + y^2 = 5$ **94.** $(-\infty, \infty)$

4.2 Assess Your Understanding (page 282)

5. a **6.** $\dfrac{\sin A}{a} = \dfrac{\sin B}{b} = \dfrac{\sin C}{c}$ **7.** d **8.** F **9.** F **10.** ambiguous case **11.** $a \approx 3.23, b \approx 3.55, A = 40°$ **13.** $a \approx 3.25, c \approx 4.23, B = 45°$
15. $C = 95°, c \approx 9.86, a \approx 6.36$ **17.** $A = 40°, a = 2, c \approx 3.06$ **19.** $C = 100°, b \approx 2.06, c \approx 4.81$ **21.** $A = 69°, a \approx 6.23, c \approx 4.88$
23. $B = 40°, a \approx 5.64, b \approx 3.86$ **25.** $C = 100°, a \approx 1.31, b \approx 1.31$ **27.** One triangle; $B \approx 30.7°, C \approx 99.3°, c \approx 3.86$
29. One triangle; $C \approx 23.8°, A \approx 41.2°, a \approx 6.55$ **31.** One right triangle; $B = 90°, C = 60°, c \approx 12.12$
33. Two triangles; $C_1 \approx 30.9°, A_1 \approx 129.1°, a_1 \approx 9.07$ or $C_2 \approx 149.1°, A_2 \approx 10.9°, a_2 \approx 2.20$ **35.** No triangle
37. Two triangles; $A_1 \approx 29°, B_1 \approx 139°, b_1 \approx 9.47$ or $A_2 \approx 151°, B_2 \approx 17°, b_2 \approx 4.22$
39. 1490.48 ft **41.** 335.16 ft **43.** 153.42 ft; 136.59 ft **45.** The tree is 39.39 ft high. **47.** Adam receives 100.6 more frequent flyer miles.
49. (a) Station Able is about 143.33 mi from the ship: Station Baker is about 135.58 mi from the ship. **(b)** Approximately 41 min
51. 84.7°; 183.72 ft **53.** 2.64 mi **55.** 38.5 in. **57.** 449.36 ft **59.** 187,600,000 km or 101,440,000 km **61.** The diameter is 252 ft.

63. $\dfrac{a-b}{c} = \dfrac{a}{c} - \dfrac{b}{c} = \dfrac{\sin A}{\sin C} - \dfrac{\sin B}{\sin C} = \dfrac{\sin A - \sin B}{\sin C} = \dfrac{2\sin\left(\dfrac{A-B}{2}\right)\cos\left(\dfrac{A+B}{2}\right)}{2\sin\dfrac{C}{2}\cos\dfrac{C}{2}} = \dfrac{\sin\left(\dfrac{A-B}{2}\right)\cos\left(\dfrac{\pi}{2}-\dfrac{C}{2}\right)}{\sin\dfrac{C}{2}\cos\dfrac{C}{2}} = \dfrac{\sin\left(\dfrac{A-B}{2}\right)}{\cos\dfrac{C}{2}}$

65. $\dfrac{a-b}{a+b} = \dfrac{\dfrac{a-b}{c}}{\dfrac{a+b}{c}} = \dfrac{\dfrac{\sin\left[\frac{1}{2}(A-B)\right]}{\cos\dfrac{C}{2}}}{\dfrac{\cos\left[\frac{1}{2}(A-B)\right]}{\sin\dfrac{C}{2}}} = \dfrac{\tan\left[\frac{1}{2}(A-B)\right]}{\cot\dfrac{C}{2}} = \dfrac{\tan\left[\frac{1}{2}(A-B)\right]}{\tan\left(\dfrac{\pi}{2}-\dfrac{C}{2}\right)} = \dfrac{\tan\left[\frac{1}{2}(A-B)\right]}{\tan\left[\frac{1}{2}(A+B)\right]}$

71. $\left\{-3, -\dfrac{4}{3}, 3\right\}$ **72.** $3\sqrt{5} \approx 6.71$ **73.** $-\dfrac{\sqrt{15}}{7}$ **74.**

75. $y = \dfrac{1}{3}$ **76.** $y = \dfrac{3}{2}x - 2$ **77.** Neither   **78.** $\left(\dfrac{6}{13}, \infty\right)$

4.3 Assess Your Understanding (page 289)

3. Cosines   **4.** a   **5.** b   **6.** F   **7.** F   **8.** T   **9.** $b \approx 2.95, A \approx 28.7°, C \approx 106.3°$ **11.** $c \approx 3.75, A \approx 32.1°, B \approx 52.9°$
13. $A \approx 48.5°, B \approx 38.6°, C \approx 92.9°$ **15.** $A \approx 127.2°, B \approx 32.1°, C \approx 20.7°$ **17.** $c \approx 2.57, A \approx 48.6°, B \approx 91.4°$
19. $a \approx 3.98, B \approx 29°, C \approx 76°$ **21.** $b \approx 6.46, A \approx 48.4°, C \approx 26.6°$ **23.** $c \approx 2.98, A \approx 39.1°, B \approx 70.9°$
25. $A \approx 43.6°, B \approx 90°, C \approx 46.4°$ **27.** $A = 60°, B = 60°, C = 60°$ **29.** $A \approx 29.8°, B \approx 65.8°, C \approx 84.3°$
31. $A \approx 127.1°, B \approx 43.8°, C \approx 9.2°$ **33.** $A = 85°, a = 14.56, c = 14.12$ **35.** $A = 40.8°, B = 60.6°, C = 78.6°$ **37.** $A = 80°, b = 8.74, c = 13.80$
39. $C = 90°, a = 4.93, b = 6.30$ **41.** Two triangles: $B_1 = 35.4°, C_1 = 134.6°, c_1 = 12.29; B_2 = 144.6°, C_2 = 25.4°, c_2 = 7.40$
43. $B = 24.5°, C = 95.5°, a = 10.44$
45. 165 yd   **47. (a)** 26.4°   **(b)** 30.8 h   **49. (a)** 63.7 ft   **(b)** 66.8 ft   **(c)** 92.8°   **51. (a)** 492.58 ft   **(b)** 269.26 ft   **53. (a)** 59.2 mm   **(b)** male
55. (a) $\alpha \approx 9.9°, \beta \approx 8.3°$ **(b)** 21.73 yd   **(c)** ≈ 0.36 yd or 13 in.   **57.** 342.33 ft   **59.** The footings should be 7.65 ft apart.

61. Suppose $0 < \theta < \pi$. Then, by the Law of Cosines, $d^2 = r^2 + r^2 - 2r^2\cos\theta = 4r^2\left(\dfrac{1-\cos\theta}{2}\right) \Rightarrow d = 2r\sqrt{\dfrac{1-\cos\theta}{2}} = 2r\sin\dfrac{\theta}{2}.$

Since, for any angle in $(0, \pi)$, d is strictly less than the length of the arc subtended by θ, that is, $d < r\theta$, then $2r\sin\dfrac{\theta}{2} < r\theta$, or $2\sin\dfrac{\theta}{2} < \theta$.

Since $\cos\dfrac{\theta}{2} < 1$, then, for $0 < \theta < \pi$, $\sin\theta = 2\sin\dfrac{\theta}{2}\cos\dfrac{\theta}{2} < 2\sin\dfrac{\theta}{2} < \theta$. Thus $\sin\theta < \theta$ for $0 < \theta < \pi$.

63. $\sin\dfrac{C}{2} = \sqrt{\dfrac{1-\cos C}{2}} = \sqrt{\dfrac{1 - \dfrac{a^2 + b^2 - c^2}{2ab}}{2}} = \sqrt{\dfrac{2ab - a^2 - b^2 + c^2}{4ab}} = \sqrt{\dfrac{c^2 - (a-b)^2}{4ab}} = \sqrt{\dfrac{(c+a-b)(c+b-a)}{4ab}}$
$= \sqrt{\dfrac{(2s-2b)(2s-2a)}{4ab}} = \sqrt{\dfrac{(s-a)(s-b)}{ab}}$

70.

71. $\sin\theta = \dfrac{2\sqrt{6}}{7}; \csc\theta = \dfrac{7\sqrt{6}}{12}; \sec\theta = -\dfrac{7}{5}; \cot\theta = -\dfrac{5\sqrt{6}}{12}$ **72.** $y = -3\sin(4x)$ **73.** $f^{-1}(x) = \dfrac{A - 2x}{5x}$
74. $\dfrac{2}{3}$ **75.** $\left(-\infty, \dfrac{2}{5}\right] \cup \left[\dfrac{4}{3}, \infty\right)$ **76.** $\dfrac{8\pi}{15}$ **77.** Quadrant II; 2 x-intercepts

4.4 Assess Your Understanding (page 296)

3. $\dfrac{1}{2}ab\sin C$ **4.** $\sqrt{s(s-a)(s-b)(s-c)}; \dfrac{1}{2}(a+b+c)$ **5.** 6   **6.** T   **7.** c   **8.** c   **9.** 2.83   **11.** 17.46   **13.** 13.42   **15.** 9.56   **17.** 4.60   **19.** 3.86   **21.** 2.72

23. 210   **25.** 6.93   **27.** 74.15   **29.** $K = \dfrac{1}{2}ab\sin C = \dfrac{1}{2}a\sin C \cdot \dfrac{a\sin B}{\sin A} = \dfrac{a^2\sin B\sin C}{2\sin A}$ **31.** 0.92   **33.** 2.27   **35.** 5.44   **37.** 9.03 sq ft   **39.** \$5446.38

41. 18.18 m^2 **43.** The area of home plate is about 216.5 in.^{2} **45.** $K = \dfrac{1}{2}r^2(\theta + \sin\theta)$ **47.** The ground area is 7517.4 ft^2.

49. Letting $d = 0$ gives
$$K = \sqrt{(s-a)(s-b)(s-c)(s-0)} - abc \cdot 0 \cdot \cos^2\theta = \sqrt{s(s-a)(s-b)(s-c)} \text{ where } s = \dfrac{1}{2}(a+b+c+0) = \dfrac{1}{2}(a+b+c)$$

51. (a) Area $\triangle OAC = \dfrac{1}{2}\,|OC|\,|AC| = \dfrac{1}{2}\cdot\dfrac{|OC|}{1}\cdot\dfrac{|AC|}{1} = \dfrac{1}{2}\sin\alpha\cos\alpha$

(b) Area $\triangle OCB = \dfrac{1}{2}\,|BC|\,|OC| = \dfrac{1}{2}\,|OB|^2\,\dfrac{|BC|}{|OB|}\cdot\dfrac{|OC|}{|OB|} = \dfrac{1}{2}\,|OB|^2\sin\beta\cos\beta$

(c) Area $\triangle OAB = \dfrac{1}{2}\,|BD|\,|OA| = \dfrac{1}{2}\,|OB|\,\dfrac{|BD|}{|OB|} = \dfrac{1}{2}\,|OB|\sin(\alpha+\beta)$

(d) $\dfrac{\cos\alpha}{\cos\beta} = \dfrac{\dfrac{|OC|}{1}}{\dfrac{|OC|}{|OB|}} = |OB|$

(e) Area $\triangle OAB = $ Area $\triangle OAC + $ Area $\triangle OCB$

$\dfrac{1}{2}\,|OB|\sin(\alpha+\beta) = \dfrac{1}{2}\sin\alpha\cos\alpha + \dfrac{1}{2}\,|OB|^2\sin\beta\cos\beta$

$\sin(\alpha+\beta) = \dfrac{1}{|OB|}\sin\alpha\cos\alpha + |OB|\sin\beta\cos\beta$

$\sin(\alpha+\beta) = \dfrac{\cos\beta}{\cos\alpha}\sin\alpha\cos\alpha + \dfrac{\cos\alpha}{\cos\beta}\sin\beta\cos\beta$

$\sin(\alpha+\beta) = \sin\alpha\cos\beta + \cos\alpha\sin\beta$

53. 31,145 ft^2 **55. (a)** The perimeter and area are both 36.   **(b)** The perimeter and area are both 60.

57. $K = \dfrac{1}{2}ah = \dfrac{1}{2}ab\sin C \Rightarrow h = b\sin C = \dfrac{a\sin B\sin C}{\sin A}$

59. $\angle POQ = 180° - \left(\dfrac{A}{2}+\dfrac{B}{2}\right) = 180° - \dfrac{1}{2}(180° - C) = 90° + \dfrac{C}{2}$, and $\sin\left(90° + \dfrac{C}{2}\right) = \cos\dfrac{C}{2}$. So, $r = \dfrac{c\sin\dfrac{A}{2}\sin\dfrac{B}{2}}{\sin\left(90° + \dfrac{C}{2}\right)} = \dfrac{c\sin\dfrac{A}{2}\sin\dfrac{B}{2}}{\cos\dfrac{C}{2}}$.

61. $\cot\dfrac{A}{2} + \cot\dfrac{B}{2} + \cot\dfrac{C}{2} = \dfrac{s-a}{r} + \dfrac{s-b}{r} + \dfrac{s-c}{r} = \dfrac{3s-(a+b+c)}{r} = \dfrac{3s-2s}{r} = \dfrac{s}{r}$ **66.** Maximum value; 17   **67.** $(-\infty,-3)\cup[-1,3)$

68. $\sin t = \dfrac{\sqrt{2}}{3}$, $\cos t = -\dfrac{\sqrt{7}}{3}$, $\tan t = -\dfrac{\sqrt{14}}{7}$, $\csc t = \dfrac{3\sqrt{2}}{2}$, $\sec t = -\dfrac{3\sqrt{7}}{7}$, $\cot t = -\dfrac{\sqrt{14}}{2}$

69. $\csc\theta - \sin\theta = \dfrac{1}{\sin\theta} - \sin\theta = \dfrac{1-\sin^2\theta}{\sin\theta} = \dfrac{\cos^2\theta}{\sin\theta} = \cos\theta\cdot\dfrac{\cos\theta}{\sin\theta} = \cos\theta\cot\theta$

70. $P(w) = 2w + 2\sqrt{144 - w^2}$ **71.** $\pm\dfrac{1}{2}$, $\pm\dfrac{3}{2}$, ± 1, ± 2, ± 3, ± 6 **72.** $[2.39, 2.41]$ **73.** $\{-2, 9\}$ **74.** $y = -2x$

4.5 Assess Your Understanding (page 305)

4. simple harmonic; amplitude   **5.** simple harmonic; damped   **6.** T   **7.** $d(t) = -5\cos(\pi t)$ **9.** $d(t) = -7\cos\left(\dfrac{2}{5}t\right)$

11. $d(t) = -5\sin(\pi t)$ **13.** $d(t) = -7\sin\left(\dfrac{2}{5}t\right)$ **15. (a)** Simple harmonic   **(b)** 5 m   **(c)** $\dfrac{2\pi}{3}$ s   **(d)** $\dfrac{3}{2\pi}$ oscillation/s

17. (a) Simple harmonic   **(b)** 8 m   **(c)** 1 s   **(d)** 1 oscillation/s   **19. (a)** Simple harmonic   **(b)** 9 m   **(c)** 8π s   **(d)** $\dfrac{1}{8\pi}$ oscillation/s

21. (a) Simple harmonic   **(b)** 7 m   **(c)** $\dfrac{2}{3}$ s   **(d)** $\dfrac{3}{2}$ oscillations/s

23.

25.

27.

29.

31.

33.

35. (a) $f(x) = \dfrac{1}{2}\left[\cos x - \cos(3x)\right]$

(b)

37. (a) $G(x) = \dfrac{1}{2}\left[\cos(6x) + \cos(2x)\right]$

(b)

39. (a) $H(x) = \sin(4x) + \sin(2x)$

(b)

41. (a) $d(t) = -10e^{-0.7t/50}\cos\left(\sqrt{\dfrac{4\pi^2}{25} - \dfrac{0.49}{2500}}\,t\right)$

(b)

43. (a) $d(t) = -18e^{-0.6t/60}\cos\left(\sqrt{\dfrac{\pi^2}{4} - \dfrac{0.36}{3600}}\,t\right)$

(b)

45. (a) $d(t) = -5e^{-0.8t/20} \cos\left(\sqrt{\dfrac{4\pi^2}{9} - \dfrac{0.64}{400}}\, t\right)$

(b)

47. (a) The motion is damped. The bob has mass $m = 20$ kg with a damping factor of 0.7 kg/s

(b) 20 m leftward

(c)

(d) 18.33 m leftward   **(e)** $d \to 0$

49. (a) The motion is damped. The bob has mass $m = 40$ kg with a damping factor of 0.6 kg/s

(b) 30 m leftward

(c)

(d) 28.47 m leftward   **(e)** $d \to 0$

51. (a) The motion is damped. The bob has mass $m = 15$ kg with a damping factor of 0.9 kg/s

(b) 15 m leftward

(c)

(d) 12.53 m leftward   **(e)** $d \to 0$

53. $\omega = 1040\pi$; $d(t) = 0.80 \cos(1040\pi t)$

55. $\omega = 880\pi$; $d(t) = 0.01 \sin(880\pi t)$

57. (a)

(b) At $t = 0, t = 2$; at $t = 1, t = 3$

(c) During the approximate intervals $0.35 < t < 0.67, 1.29 < t < 1.75$, and $2.19 < t \le 3$

59.

61.

65.

67. $y = \dfrac{1}{x} \sin x$

$y = \dfrac{1}{x^2} \sin x$

$y = \dfrac{1}{x^3} \sin x$

69. $f^{-1}(x) = \dfrac{4x - 3}{x - 1}$ **70. (a)** $\dfrac{3\sqrt{10}}{10}$ **(b)** $\dfrac{\sqrt{10}}{10}$ **(c)** $\dfrac{1}{3}$ **71.** $g(f(x)) = 10 - 5x$; Domain: $\left(-\infty, \dfrac{3}{5}\right]$ **72.** $-\dfrac{7\sqrt{6}}{12}$

73. $y = -\dfrac{3}{2}x + \dfrac{11}{2}$ **74.** $[-5, 8]$ **75.** $\left[-2, \dfrac{3}{5}\right]$

Review Exercises *(page 309)*

1. $\sin\theta = \dfrac{4}{5}$; $\cos\theta = \dfrac{3}{5}$; $\tan\theta = \dfrac{4}{3}$; $\cot\theta = \dfrac{3}{4}$; $\sec\theta = \dfrac{5}{3}$; $\csc\theta = \dfrac{5}{4}$ **2.** $\sin\theta = \dfrac{\sqrt{3}}{2}$; $\cos\theta = \dfrac{1}{2}$; $\tan\theta = \sqrt{3}$; $\cot\theta = \dfrac{\sqrt{3}}{3}$; $\sec\theta = 2$; $\csc\theta = \dfrac{2\sqrt{3}}{3}$

3. 0   **4.** 1   **5.** 1   **6.** $A = 70°$, $b \approx 3.42$, $a \approx 9.40$ **7.** $a \approx 4.58$, $A \approx 66.4°$, $B \approx 23.6°$ **8.** $C = 100°$, $b \approx 0.65$, $c \approx 1.29$

9. $B \approx 56.8°$, $C \approx 23.2°$, $b \approx 4.25$ **10.** No triangle   **11.** $b \approx 3.32$, $A \approx 62.8°$, $C \approx 17.2°$ **12.** $A \approx 36.2°$, $C \approx 63.8°$, $c \approx 4.55$

13. No triangle   **14.** $A \approx 83.3°$, $B \approx 44.0°$, $C \approx 52.6°$ **15.** $c \approx 2.32$, $A \approx 16.1°$, $B \approx 123.9°$ **16.** $B \approx 36.2°$, $C \approx 63.8°$, $c \approx 4.55$

17. $A \approx 39.6°$, $B \approx 18.6°$, $C \approx 121.9°$ **18.** Two triangles: $B_1 \approx 13.4°$, $C_1 \approx 156.6°$, $c_1 \approx 6.86$ or $B_2 \approx 166.6°$, $C_2 \approx 3.4°$, $c_2 \approx 1.02$

19. $b \approx 11.52$, $c \approx 10.13$, $C \approx 60°$ **20.** $a \approx 5.23$, $B \approx 46.0°$, $C \approx 64.0°$ **21.** 1.93   **22.** 18.79   **23.** 6   **24.** 3.80   **25.** 0.32   **26.** 1.92 in.^{2}

27. 48.2° and 41.8°   **28.** 23.32 ft   **29.** 2.15 mi   **30.** 132.55 ft/min   **31.** 12.7°   **32.** 29.97 ft   **33.** 6.22 mi   **34. (a)** 131.8 mi   **(b)** 23.1°   **(c)** 0.21 hr

35. 8798.67 sq ft   **36.** S4.0°E   **37. (a)** $\theta_c = 16.7°$ **(b)** 1.44 ft   **38. (a)** $\theta_c{'} = 5.7°$ **(b)** 0.50 ft   **39.** $d(t) = -3 \cos\left(\dfrac{\pi}{2}t\right)$

40. (a) Simple harmonic   **(b)** 6 ft   **(c)** π s   **(d)** $\dfrac{1}{\pi}$ oscillation/s

41. (a) Simple harmonic   **(b)** 2 ft   **(c)** 2 s   **(d)** $\dfrac{1}{2}$ oscillation/s

42. (a) $d(t) = -15e^{-0.75t/80} \cos\left(\sqrt{\dfrac{4\pi^2}{25} - \dfrac{0.5625}{6400}}\, t\right)$

(b)

43. (a) The motion is damped. The bob has mass $m = 20$ kg with a damping factor of 0.6 kg/s.

(b) 15 m leftward   **(c)**

(d) 13.92 m leftward

(e) $d \to 0$

44.

Chapter Test (*page 312*)

1. $\sin\theta = \dfrac{\sqrt5}{5}$; $\cos\theta = \dfrac{2\sqrt5}{5}$; $\tan\theta = \dfrac{1}{2}$; $\csc\theta = \sqrt5$; $\sec\theta = \dfrac{\sqrt5}{2}$; $\cot\theta = 2$ **2.** 0 **3.** $a = 15.88$, $B \approx 57.5°$, $C \approx 70.5°$
4. $b \approx 6.85$, $C = 117°$, $c \approx 16.30$ **5.** $A \approx 52.4°$, $B \approx 29.7°$, $C \approx 97.9°$ **6.** $b \approx 4.72$, $c \approx 1.67$, $B = 105°$ **7.** No triangle
8. $c \approx 7.62$, $A \approx 80.5°$, $B \approx 29.5°$ **9.** 15.04 square units **10.** 19.81 square units **11.** 61.0° **12.** 1.3° **13.** The area of the shaded region is 9.26 cm².
14. 54.15 square units **15.** Madison will have to swim about 2.23 miles. **16.** 12.63 square units **17.** The lengths of the sides are 15, 18, and 21.

18. $d(t) = 5(\sin 42°)\sin\left(\dfrac{\pi t}{3}\right)$ or $d \approx 3.346\sin\left(\dfrac{\pi t}{3}\right)$

Cumulative Review (*page 313*)

1. $\left\{\dfrac{1}{3}, 1\right\}$ **2.** $(x+5)^2 + (y-1)^2 = 9$ **3.** $\{x \mid x \le -1 \text{ or } x \ge 4\}$ **4.** **5.**

6. (a) $-\dfrac{2\sqrt5}{5}$ **(b)** $\dfrac{\sqrt5}{5}$ **(c)** $-\dfrac{4}{5}$ **(d)** $-\dfrac{3}{5}$ **(e)** $\sqrt{\dfrac{5-\sqrt5}{10}}$ **(f)** $-\sqrt{\dfrac{5+\sqrt5}{10}}$

7. (a) **(b)**

8. (a) **(b)** **(c)** **(d)** **(e)**

(f) **(g)**

9. Two triangles: $A_1 \approx 59.0°$, $B_1 \approx 81.0°$, $b_1 \approx 23.05$
or $A_2 \approx 121.0°$, $B_2 \approx 19.0°$, $b_2 \approx 7.59$
10. $\left\{0, \dfrac{2x}{3}, \dfrac{4x}{3}\right\}$ **11.** π ft ≈ 3.14 ft

CHAPTER 5 Polar Coordinates; Vectors

5.1 Assess Your Understanding (*page 323*)

7. pole; polar axis **8.** $r\cos\theta$; $r\sin\theta$ **9.** b **10.** d **11.** T **12.** F **13.** A **15.** C **17.** B **19.** A

21. **23.** **25.** **27.** **29.** **31.**

33. **35.**

(a) $\left(5, -\dfrac{4\pi}{3}\right)$ **37.**
(b) $\left(-5, \dfrac{5\pi}{3}\right)$
(c) $\left(5, \dfrac{8\pi}{3}\right)$

(a) $(2, -2\pi)$
(b) $(-2, \pi)$
(c) $(2, 2\pi)$

39.

41.

(a) $\left(1, -\dfrac{3\pi}{2}\right)$

(a) $\left(3, -\dfrac{5\pi}{4}\right)$

(b) $\left(-1, \dfrac{3\pi}{2}\right)$

(b) $\left(-3, \dfrac{7\pi}{4}\right)$

(c) $\left(1, \dfrac{5\pi}{2}\right)$

(c) $\left(3, \dfrac{11\pi}{4}\right)$

43. $(0, 3)$ **45.** $(-2, 0)$ **47.** $(-3\sqrt{3}, 3)$ **49.** $(\sqrt{2}, -\sqrt{2})$ **51.** $\left(-\dfrac{5\sqrt{3}}{2}, \dfrac{5}{2}\right)$

53. $(2, 0)$ **55.** $(-2.57, 7.05)$

57. $(-4.98, -3.85)$ **59.** $(3, 0)$ **61.** $(1, \pi)$

63. $\left(\sqrt{2}, -\dfrac{\pi}{4}\right)$ **65.** $\left(10, \dfrac{\pi}{3}\right)$ **67.** $(2.47, -1.02)$

69. $(9.30, 0.47)$ **71.** $r^2 = \dfrac{3}{2}$ or $r = \dfrac{\sqrt{6}}{2}$ **73.** $r^2 \cos^2 \theta - 4r \sin \theta = 0$ **75.** $r^2 \sin 2\theta = 1$

77. $r \cos \theta = 4$ **79.** $x^2 + y^2 - x = 0$ or $\left(x - \dfrac{1}{2}\right)^2 + y^2 = \dfrac{1}{4}$ **81.** $(x^2 + y^2)^{3/2} - x = 0$

83. $x^2 + y^2 = 4$ **85.** $y^2 = 8(x + 2)$

87. (a) $(-10, 36)$ **(b)** $\left(2\sqrt{349}, 180° + \tan^{-1}\left(-\dfrac{18}{5}\right)\right) \approx (37.36, 105.5°)$ **(c)** $(-3, -35)$ **(d)** $\left(\sqrt{1234}, 180° + \tan^{-1}\left(\dfrac{35}{3}\right)\right) \approx (35.13, 265.1°)$

89. (a) $(80, 25°)$; $(110, -5°)$ **(b)** $(72.50, 33.81)$; $(109.58, -9.59)$ **(c)** 342.5 mph   **94.** 2 or 0 positive real zeros; 1 negative real zero

95. $\left(-\dfrac{5}{4}, \dfrac{9}{2}\right)$ **96.** $(0, -11)$ **97.** $13 - 18i$ **98.** $\left\{\dfrac{\pi}{12}, \dfrac{5\pi}{12}, \dfrac{13\pi}{12}, \dfrac{17\pi}{12}\right\}$ **99.** $C = 78°$, $a \approx 9.27$, $b \approx 6.15$ **100.** $\dfrac{\sqrt{6} + \sqrt{2}}{4}$ or $\dfrac{\sqrt{2}(1 + \sqrt{3})}{4}$

101. $\sin^5 x = \sin x (\sin^2 x)^2 = \sin x (1 - \cos^2 x)^2 = \sin x (1 - 2\cos^2 x + \cos^4 x) = \sin x - 2\cos^2 x \sin x + \cos^4 x \sin x$

5.2 Assess Your Understanding *(page 337)*

7. polar equation **8.** F **9.** $-\theta$ **10.** $\pi - \theta$ **11.** $2n; n$ **12.** T **13.** c **14.** b

15. $x^2 + y^2 = 16$; circle, radius 4, center at pole

17. $y = \sqrt{3}\, x$; line through pole, making an angle of $\dfrac{\pi}{3}$ with polar axis

19. $y = 4$; horizontal line 4 units above the pole

21. $x = -2$; vertical line 2 units to the left of the pole

23. $(x - 1)^2 + y^2 = 1$; circle, radius 1, center $(1, 0)$ in rectangular coordinates

25. $x^2 + (y + 2)^2 = 4$; circle, radius 2, center $(0, -2)$ in rectangular coordinates

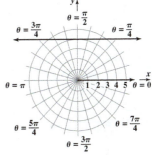

27. $(x - 2)^2 + y^2 = 4$, $x \neq 0$; circle, radius 2, center $(2, 0)$ in rectangular coordinates, hole at $(0, 0)$

29. $x^2 + (y + 1)^2 = 1$, $y \neq 0$; circle, radius 1, center $(0, -1)$ in rectangular coordinates, hole at $(0, 0)$

31. E **33.** F **35.** H **37.** D
39. Cardioid

41. Cardioid

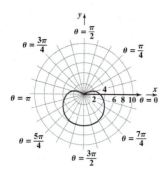

43. Limaçon without inner loop

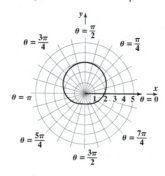

45. Limaçon without inner loop

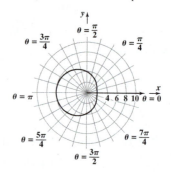

47. Limaçon with inner loop

49. Limaçon with inner loop

51. Rose

53. Rose

55. Lemniscate

57. Spiral

59. Cardioid

61. Limaçon with inner loop

63. $r = 3 + 3\cos\theta$
65. $r = 4 + \sin\theta$

67.

69.

71.

73.

75.

77.

79.

81.

83. $r \sin \theta = a$
$y = a$

85.
$r = 2a \sin \theta$
$r^2 = 2ar \sin \theta$
$x^2 + y^2 = 2ay$
$x^2 + y^2 - 2ay = 0$
$x^2 + (y - a)^2 = a^2$
Circle, radius a, center
$(0, a)$ in rectangular
coordinates

87.
$r = 2a \cos \theta$
$r^2 = 2ar \cos \theta$
$x^2 + y^2 = 2ax$
$x^2 - 2ax + y^2 = 0$
$(x - a)^2 + y^2 = a^2$
Circle, radius a, center
$(a, 0)$ in rectangular
coordinates

89. (a) 5 knots **(b)** 6 knots **(c)** 10 knots **(d)** approximately 80° to 150° **(e)** ≈9 knots; approximately 90° to 100°

91. $(x^2 + y^2)^2 = x^2 - y^2$

93. (a) $r^2 = \cos \theta; r^2 = \cos (\pi - \theta)$ **(b)** $r^2 = \sin \theta: r^2 = \sin (\pi - \theta)$
$r^2 = -\cos \theta$ $r^2 = \sin \theta$
Not equivalent; test fails.   Test works.
$(-r)^2 = \cos(-\theta) = \cos \theta$ $(-r)^2 = \sin(-\theta) = -\sin \theta$
New test works.   Not equivalent; new test fails.

97. $\{x | 3 < x \leq 8\}$, or $(3, 8]$ **98.** 420° **99.** Amplitude = 2; period = $\dfrac{2\pi}{5}$ **100.** Horizontal asymptote: $y = 0$; Vertical asymptote: $x = 4$ **101.** 3

102. $6\sqrt{30} \approx 32.86$ square units **103.** $\left\{ -\dfrac{5}{2}, \dfrac{4}{3} \right\}$ **104.** $y = -4x - 5$ **105.** $\cos^3 x = \cos x \cos^2 x = \cos x (1 - \sin^2 x) = \cos x - \sin^2 x \cos x$

Historical Problems *(page 347)*

1. (a) $1 + 4i, 1 + i$ **(b)** $-1, 2 + i$

5.3 Assess Your Understanding *(page 347)*

5. real; imaginary **6.** magnitude; modulus; argument **7.** $r_1 r_2 e^{i(\theta_1 + \theta_2)}$ **8.** F **9.** three **10.** T **11.** c **12.** a

13.

$\sqrt{2}\left(\cos \dfrac{\pi}{4} + i \sin \dfrac{\pi}{4} \right); \sqrt{2} e^{i \cdot \pi/4}$

15.

$2\left(\cos \dfrac{11\pi}{6} + i \sin \dfrac{11\pi}{6} \right); 2 e^{i \cdot 11\pi/6}$

17.

$3\left(\cos \dfrac{3\pi}{2} + i \sin \dfrac{3\pi}{2} \right); 3 e^{i \cdot 3\pi/2}$

19.

$4\sqrt{2}\left(\cos \dfrac{7\pi}{4} + i \sin \dfrac{7\pi}{4} \right); 4\sqrt{2} e^{i \cdot 7\pi/4}$

21.

$5 (\cos 5.356 + i \sin 5.356); 5 e^{i \cdot 5.356}$

23.

$\sqrt{13} (\cos 2.159 + i \sin 2.159); \sqrt{13} e^{i \cdot 2.159}$

25. $-1 + \sqrt{3}i$ **27.** $2\sqrt{2} - 2\sqrt{2}i$ **29.** $-3i$ **31.** -7 **33.** $-0.035 + 0.197i$ **35.** $1.970 + 0.347i$

37. $zw = 8\left(\cos\dfrac{\pi}{3} + i\sin\dfrac{\pi}{3}\right)$ or $8e^{i\cdot\pi/3}$; $\dfrac{z}{w} = \dfrac{1}{2}\left(\cos\dfrac{\pi}{9} + i\sin\dfrac{\pi}{9}\right)$ or $\dfrac{1}{2}e^{i\cdot\pi/9}$

39. $zw = 12\left(\cos\dfrac{2\pi}{9} + i\sin\dfrac{2\pi}{9}\right)$ or $12e^{i\cdot2\pi/9}$; $\dfrac{z}{w} = \dfrac{3}{4}\left(\cos\dfrac{11\pi}{9} + i\sin\dfrac{11\pi}{9}\right)$ or $\dfrac{3}{4}e^{i\cdot11\pi/9}$

41. $zw = 4\left(\cos\dfrac{9\pi}{40} + i\sin\dfrac{9\pi}{40}\right)$ or $4e^{i\cdot9\pi/40}$; $\dfrac{z}{w} = \cos\dfrac{\pi}{40} + i\sin\dfrac{\pi}{40}$ or $e^{i\cdot\pi/40}$

43. $zw = 4\sqrt{2}\left(\cos\dfrac{\pi}{12} + i\sin\dfrac{\pi}{12}\right)$ or $4\sqrt{2}e^{i\cdot\pi/12}$; $\dfrac{z}{w} = \sqrt{2}\left(\cos\dfrac{5\pi}{12} + i\sin\dfrac{5\pi}{12}\right)$ or $\sqrt{2}e^{i\cdot5\pi/12}$ **45.** $-32 + 32\sqrt{3}i$; $64e^{i\cdot2\pi/3}$ **47.** $32i$; $32e^{i\cdot\pi/2}$

49. $\dfrac{27}{2} + \dfrac{27\sqrt{3}}{2}i$; $\dfrac{27}{2}e^{i\cdot\pi/3}$ **51.** $-\dfrac{25\sqrt{2}}{2} + \dfrac{25\sqrt{2}}{2}i$; $25e^{i\cdot3\pi/4}$ **53.** $-4 + 4i$; $4\sqrt{2}e^{i\cdot3\pi/4}$ **55.** $-23 + 14.142i$; $27e^{i\cdot2.590}$

57. $\sqrt[6]{2}e^{i\cdot\pi/12}, \sqrt[6]{2}e^{i\cdot3\pi/4}, \sqrt[6]{2}e^{i\cdot17\pi/12}$ **59.** $\sqrt[4]{8}e^{i\cdot5\pi/12}, \sqrt[4]{8}e^{i\cdot11\pi/12}, \sqrt[4]{8}e^{i\cdot17\pi/12}, \sqrt[4]{8}e^{i\cdot23\pi/12}$

61. $2e^{i\cdot3\pi/8}, 2e^{i\cdot7\pi/8}, 2e^{i\cdot11\pi/8}, 2e^{i\cdot15\pi/8}$ **63.** $e^{i\cdot\pi/10}, e^{i\cdot\pi/2}, e^{i\cdot9\pi/10}, e^{i\cdot13\pi/10}, e^{i\cdot17\pi/10}$

65. $1, i, -1, -i$

67. Look at formula (7). $|z_k| = \sqrt[n]{r}$ for all k.

69. Look at formula (7). The z_k are spaced apart by an angle of $\dfrac{2\pi}{n}$.

71. Since the sine and cosine functions each has period 2π,
$\cos\theta = \cos(\theta + 2k\pi)$ and $\sin\theta = \sin(\theta + 2k\pi)$, k an integer. Then,
$re^{i\theta} = r(\cos\theta + i\sin\theta) = r[(\cos(\theta + 2k\pi) + i\sin(\theta + 2k\pi)] = re^{i(\theta+2k\pi)}$, k an integer.

73. Assume the theorem is true for $n \geq 1$.

For $n = 0$:
$z^0 = r^0 e^{i(0\cdot\theta)} = r^0[\cos(0\cdot\theta) + i\sin(0\cdot\theta)]$
$1 = 1\cdot[\cos 0 + i\sin 0]$
$1 = 1\cdot[1 + 0]$
$1 = 1$ True

For negative integers:
$z^{-n} = (z^n)^{-1} = [r^n e^{i(n\theta)}]^{-1} = (r^n[\cos(n\theta) + i\sin(n\theta)])^{-1}$ with $n \geq 1$
$= \dfrac{1}{r^n[\cos(n\theta) + i\sin(n\theta)]} = \dfrac{1}{r^n[\cos(n\theta) + i\sin(n\theta)]}\cdot\dfrac{\cos(n\theta) - i\sin(n\theta)}{\cos(n\theta) - i\sin(n\theta)}$
$= \dfrac{\cos(n\theta) - i\sin(n\theta)}{r^n(\cos^2(n\theta) + \sin^2(n\theta))} = \dfrac{\cos(n\theta) - i\sin(n\theta)}{r^n} = r^{-n}[\cos(n\theta) - i\sin(n\theta)]$
$= r^{-n}[\cos(-n\theta) + i\sin(-n\theta)] = r^{-n}e^{i(-n\theta)}$

Thus, De Moivre's Theorem is true for all integers.

75. ≈ 40.50 **76.** $\dfrac{4}{3}\pi$ **77.** $2y\sqrt[3]{3x^2y^2}$ **78.** Minimum: $f\left(\dfrac{6}{5}\right) = -\dfrac{16}{5}$ **79.** $A \approx 26.4°, B \approx 36.3°, C \approx 117.3°$ **80.** $f(5x^3) = 75x^6 - 20x^3$

81. $y = -\dfrac{3}{2}x + 8$ **82.** $\sqrt{16\sec^2 x - 16} = \sqrt{16(\sec^2 x - 1)} = \sqrt{16\tan^2 x} = 4\tan x$

5.4 Assess Your Understanding *(page 359)*

1. vector **2.** 0 **3.** unit **4.** position **5.** horizontal; vertical **6.** resultant **7.** T **8.** F **9.** a **10.** b

11.

13.

15.

17.

19. T **21.** F **23.** F **25.** T **27.** $\mathbf{v} = 3\mathbf{i} + 4\mathbf{j}$ **29.** $\mathbf{v} = 2\mathbf{i} + 4\mathbf{j}$ **31.** $\mathbf{v} = 8\mathbf{i} - \mathbf{j}$ **33.** $\mathbf{v} = -\mathbf{i} + \mathbf{j}$ **35.** 5 **37.** $\sqrt{2}$ **39.** $\sqrt{13}$ **41.** 1 **43.** $-\mathbf{j}$

45. $\sqrt{89}$ **47.** $\sqrt{34} - \sqrt{13}$ **49.** $\mathbf{i}$ **51.** $\dfrac{3}{5}\mathbf{i} - \dfrac{4}{5}\mathbf{j}$ **53.** $\dfrac{\sqrt{2}}{2}\mathbf{i} - \dfrac{\sqrt{2}}{2}\mathbf{j}$ **55.** 14 **57.** $\mathbf{v} = \dfrac{8\sqrt{5}}{5}\mathbf{i} + \dfrac{4\sqrt{5}}{5}\mathbf{j}$, or $\mathbf{v} = -\dfrac{8\sqrt{5}}{5}\mathbf{i} - \dfrac{4\sqrt{5}}{5}\mathbf{j}$

59. $\{-2 + \sqrt{21}, -2 - \sqrt{21}\}$ **61.** $\mathbf{v} = \dfrac{5}{2}\mathbf{i} + \dfrac{5\sqrt{3}}{2}\mathbf{j}$ **63.** $\mathbf{v} = -7\mathbf{i} + 7\sqrt{3}\mathbf{j}$ **65.** $\mathbf{v} = \dfrac{25\sqrt{3}}{2}\mathbf{i} - \dfrac{25}{2}\mathbf{j}$ **67.** 45° **69.** 150° **71.** 333.4°

73. 258.7° **75.** $\mathbf{F} = 20\sqrt{3}\mathbf{i} + 20\mathbf{j}$ **77.** $\mathbf{F} = \left(20\sqrt{3} + 30\sqrt{2}\right)\mathbf{i} + \left(20 - 30\sqrt{2}\right)\mathbf{j}$

79. (a) $\mathbf{v}_a = 550\mathbf{j}$; $\mathbf{v}_w = 50\sqrt{2}\mathbf{i} + 50\sqrt{2}\mathbf{j}$ (b) $\mathbf{v}_g = 50\sqrt{2}\mathbf{i} + (550 + 50\sqrt{2})\mathbf{j}$ (c) $\|\mathbf{v}_g\| = 624.7$ mph; N6.5°E

81. $\mathbf{v} = \left(250\sqrt{2} - 30\right)\mathbf{i} + \left(250\sqrt{2} + 30\sqrt{3}\right)\mathbf{j}$; 518.8 km/h; N38.6°E **83.** Approximately 4031 lb

85. 8.6° left of direct heading across the river; 1.52 min **87.** (a) N7.05°E (b) 12 min **89.** Tension in right cable: 1000 lb; tension in left cable: 845.2 lb

91. Tension in right part: 1088.4 lb; tension in left part: 1089.1 lb **93.** $\mu = 0.36$ **95.** 13.68 lb **97.** The truck must pull with a force of 4635.2 lb.

99. (a) $(-1, 4)$

(b)

101.

103. About 8.2° north of east.

107. $\{29\}$

108. $-3x(x+2)(x-6)$

109. $\sqrt{3}$

110. Amplitude $= \dfrac{3}{2}$; period $= \dfrac{\pi}{3}$

Phase shift $= -\dfrac{\pi}{2}$

111. 15   **112.** $(x-10)^2 + (y+2)^2 = 49$ **113.** x–intercepts: $-3, -2, 3$; y–intercept: -18 **114.** $\{5 - \sqrt{11}, 5 + \sqrt{11}\}$

115. $(x+3)(x^2+9)$ or $x^3 + 3x^2 + 9x + 27$ **116.** $f(5\sin\theta) = \sqrt{25 - (5\sin\theta)^2} = \sqrt{25 - 25\sin^2\theta} = \sqrt{25(1 - \sin^2\theta)} = \sqrt{25\cos^2\theta} = 5\cos\theta$

Historical Problem *(page 369)*

$(ai + bj) \cdot (ci + dj) = ac + bd$

Real part $[(\overline{a + bi})(c + di)] = $ real part$[(a - bi)(c + di)] = $ real part$[ac + adi - bci - bdi^2] = ac + bd$

5.5 Assess Your Understanding *(page 369)*

2. dot product   **3.** orthogonal   **4.** parallel   **5.** T   **6.** F   **7.** d   **8.** b   **9. (a)** 0   **(b)** 90°   **(c)** orthogonal   **11. (a)** 0   **(b)** 90°   **(c)** orthogonal

13. (a) $\sqrt{3} - 1$ **(b)** 75°   **(c)** neither   **15. (a)** -50 **(b)** 180°   **(c)** parallel   **17. (a)** 0   **(b)** 90°   **(c)** orthogonal   **19.** $\dfrac{2}{3}$

21. $v_1 = \dfrac{5}{2}i - \dfrac{5}{2}j, v_2 = -\dfrac{1}{2}i - \dfrac{1}{2}j$ **23.** $v_1 = -\dfrac{1}{5}i - \dfrac{2}{5}j, v_2 = \dfrac{6}{5}i - \dfrac{3}{5}j$ **25.** $v_1 = \dfrac{14}{5}i + \dfrac{7}{5}j, v_2 = \dfrac{1}{5}i - \dfrac{2}{5}j$ **27.** $12i - 9j$ or $-12i + 9j$

29. 9 ft-lb   **31. (a)** $\|I\| \approx 0.022$; the intensity of the sun's rays is approximately 0.022 W/cm². $\|A\| = 500$; the area of the solar panel is 500 cm².
(b) $W = 10$; ten watts of energy is collected.   **(c)** Vectors I and A should be parallel with the solar panels facing the sun.

33. Force required to keep the Sienna from rolling down the hill: 737.6 lb; force perpendicular to the hill: 5248.4 lb   **35.** Timmy must exert 85.5 lb.

37. 60°   **39.** Let $v = ai + bj$. Then $0 \cdot v = 0a + 0b = 0$.

41. $v = \cos\alpha\, i + \sin\alpha\, j, 0 \le \alpha \le \pi$; $w = \cos\beta\, i + \sin\beta\, j, 0 \le \beta \le \pi$. If θ is the angle between v and w, then $v \cdot w = \cos\theta$, since $\|v\| = 1$ and $\|w\| = 1$.
Now $\theta = \alpha - \beta$ or $\theta = \beta - \alpha$. Since the cosine function is even, $v \cdot w = \cos(\alpha - \beta)$. Also, $v \cdot w = \cos\alpha\cos\beta + \sin\alpha\sin\beta$.
So $\cos(\alpha - \beta) = \cos\alpha\cos\beta + \sin\alpha\sin\beta$.

43. $(\|w\|v + \|v\|w) \cdot (\|w\|v - \|v\|w) = \|w\|^2\, v \cdot v - \|w\|\|v\|v \cdot w + \|v\|\|w\|w \cdot v - \|v\|^2\, w \cdot w = \|w\|^2\, v \cdot v - \|v\|^2\, w \cdot w = \|w\|^2\|v\|^2 - \|v\|^2\|w\|^2 = 0$

45. $\dfrac{-40 + 4\sqrt{507}}{37} \approx 1.353$ **47.** $-2\sqrt{3}, 2\sqrt{3}$

49. (a) If $u = a_1 i + b_1 j$ and $v = a_2 i + b_2 j$, then, since $\|u\| = \|v\|, a_1^2 + b_1^2 = \|u\|^2 = \|v\|^2 = a_2^2 + b_2^2$,
$(u + v) \cdot (u - v) = (a_1 + a_2)(a_1 - a_2) + (b_1 + b_2)(b_1 - b_2) = (a_1^2 + b_1^2) - (a_2^2 + b_2^2) = 0$.
(b) The legs of the angle can be made to correspond to vectors $u + v$ and $u - v$.

52. 12   **53.** $\dfrac{9}{2}$ **54.** $(1 - \sin^2\theta)(1 + \tan^2\theta) = (\cos^2\theta)(\sec^2\theta) = \cos^2\theta \cdot \dfrac{1}{\cos^2\theta} = 1$

55. $V(x) = x(19 - 2x)(13 - 2x)$, or $V(x) = 4x^3 - 64x^2 + 247x$ **56.** $f(x) = \sqrt[3]{x + 4} + 9$

57. Vertical asymptotes: $x = -3, x = 5$; Horizontal asymptote: $y = 2$ **58.** $-\dfrac{\sqrt{3}}{2}$ **59.** Vertex: $(9, -44)$; concave up

57. $f(3\tan x) = \dfrac{1}{[(3\tan x)^2 + 9]^{3/2}} = \dfrac{1}{(9\tan^2 x + 9)^{3/2}} = \dfrac{1}{[9(\tan^2 x + 1)]^{3/2}} = \dfrac{1}{(9\sec^2 x)^{3/2}} = \dfrac{1}{27|\sec^3 x|}$

5.6 Assess Your Understanding *(page 379)*

2. components   **3.** 1   **4.** F   **5.** T   **6.** a   **7.** All points of the form $(x, 0, z)$ **9.** All points of the form $(x, y, 2)$ **11.** All points of the form $(-4, y, z)$

13. All points of the form $(1, 2, z)$ **15.** $\sqrt{21}$ **17.** $\sqrt{33}$ **19.** $\sqrt{26}$ **21.** $(2, 0, 0)$; $(2, 1, 0)$; $(0, 1, 0)$; $(2, 0, 3)$; $(0, 1, 3)$; $(0, 0, 3)$

23. $(1, 4, 3)$; $(3, 2, 3)$; $(3, 4, 3)$; $(3, 2, 5)$; $(1, 4, 5)$; $(1, 2, 5)$ **25.** $(-1, 2, 2)$; $(4, 0, 2)$; $(4, 2, 2)$; $(-1, 2, 5)$; $(4, 0, 5)$; $(-1, 0, 5)$ **27.** $v = 3i + 4j - k$

29. $v = 2i + 4j + k$ **31.** $v = 8i - j$ **33.** 7   **35.** $\sqrt{3}$ **37.** $\sqrt{22}$ **39.** $-j - 2k$ **41.** $\sqrt{105}$ **43.** $\sqrt{38} - \sqrt{17}$ **45.** i **47.** $\dfrac{3}{7}i - \dfrac{6}{7}j - \dfrac{2}{7}k$

49. $\dfrac{\sqrt{3}}{3}i + \dfrac{\sqrt{3}}{3}j + \dfrac{\sqrt{3}}{3}k$ **51.** $v \cdot w = 0$; $\theta = 90°$ **53.** $v \cdot w = -2, \theta \approx 100.3°$ **55.** $v \cdot w = 0$; $\theta = 90°$ **57.** $v \cdot w = 52$; $\theta = 0°$

59. $\alpha \approx 64.6°$; $\beta \approx 149.0°$; $\gamma \approx 106.6°$; $v = 7(\cos 64.6°i + \cos 149.0°j + \cos 106.6°k)$

61. $\alpha = \beta = \gamma \approx 54.7°$; $v = \sqrt{3}(\cos 54.7°i + \cos 54.7°j + \cos 54.7°k)$ **63.** $\alpha = \beta = 45°$; $\gamma = 90°$; $v = \sqrt{2}(\cos 45°i + \cos 45°j + \cos 90°k)$

65. $\alpha \approx 60.9°$; $\beta \approx 144.2°$; $\gamma \approx 71.1°$; $v = \sqrt{38}(\cos 60.9°i + \cos 144.2°j + \cos 71.1°k)$ **67. (a)** $d = a + b + c = <7, 1, 5>$ **(b)** 8.66 ft

69. $(x - 3)^2 + (y - 1)^2 + (z - 1)^2 = 1$ **71.** Radius = 2, center $(-1, 1, 0)$ **73.** Radius = 3, center $(2, -2, -1)$

75. Radius = $\dfrac{3\sqrt{2}}{2}$, center $(2, 0, -1)$ **77.** 2 newton-meters = 2 joules   **79.** 9 newton-meters = 9 joules   **80.** $\left\{x \,\middle|\, 2 < x \le \dfrac{13}{5}\right\}$ or $\left(2, \dfrac{13}{5}\right]$

81. $2x^2 + 2x - 5$ **82.** $\dfrac{1}{2}$ **83.** $c = 3\sqrt{5} \approx 6.71$; $A \approx 26.6°$; $B \approx 63.4°$ **84.** $5\sqrt{5}$ **85.** $P(x) = x^4 - 2x^3 + 11x^2 - 2x + 10$

86. $f^{-1}(x) = \dfrac{8x + 5}{x}$ **87.** $\dfrac{24}{\pi}$ **88.** $9\pi - 18$ square units

5.7 Assess Your Understanding (page 385)

1. T **2.** T **3.** T **4.** F **5.** F **6.** T **7.** 2 **9.** 4 **11.** $-11A + 2B + 5C$ **13.** $-6A + 23B - 15C$

15. (a) $5\mathbf{i} + 5\mathbf{j} + 5\mathbf{k}$ (b) $-5\mathbf{i} - 5\mathbf{j} - 5\mathbf{k}$ (c) 0 (d) 0 **17.** (a) $\mathbf{i} - \mathbf{j} - \mathbf{k}$ (b) $-\mathbf{i} + \mathbf{j} + \mathbf{k}$ (c) 0 (d) 0

19. (a) $-\mathbf{i} + 2\mathbf{j} + 2\mathbf{k}$ (b) $\mathbf{i} - 2\mathbf{j} - 2\mathbf{k}$ (c) 0 (d) 0 **21.** (a) $3\mathbf{i} - \mathbf{j} + 4\mathbf{k}$ (b) $-3\mathbf{i} + \mathbf{j} - 4\mathbf{k}$ (c) 0 (d) 0 **23.** $-9\mathbf{i} - 7\mathbf{j} - 3\mathbf{k}$

25. $9\mathbf{i} + 7\mathbf{j} + 3\mathbf{k}$ **27.** 0 **29.** $-27\mathbf{i} - 21\mathbf{j} - 9\mathbf{k}$ **31.** $-18\mathbf{i} - 14\mathbf{j} - 6\mathbf{k}$ **33.** 0 **35.** -25 **37.** 25 **39.** 0

41. Any vector of the form $c(-9\mathbf{i} - 7\mathbf{j} - 3\mathbf{k})$, where c is a nonzero scalar **43.** Any vector of the form $c(-\mathbf{i} + \mathbf{j} + 5\mathbf{k})$, where c is a nonzero scalar

45. $\sqrt{166}$ **47.** $\sqrt{555}$ **49.** $\sqrt{34}$ **51.** $\sqrt{998}$ **53.** $\dfrac{11\sqrt{19}}{57}\mathbf{i} + \dfrac{\sqrt{19}}{57}\mathbf{j} + \dfrac{7\sqrt{19}}{57}\mathbf{k}$ or $-\dfrac{11\sqrt{19}}{57}\mathbf{i} - \dfrac{\sqrt{19}}{57}\mathbf{j} - \dfrac{7\sqrt{19}}{57}\mathbf{k}$ **55.** 98 cubic units

57. $\mathbf{u} \times \mathbf{v} = \begin{vmatrix} \mathbf{i} & \mathbf{j} & \mathbf{k} \\ a_1 & b_1 & c_1 \\ a_2 & b_2 & c_2 \end{vmatrix} = (b_1c_2 - b_2c_1)\mathbf{i} - (a_1c_2 - a_2c_1)\mathbf{j} + (a_1b_2 - a_2b_1)\mathbf{k}$

$\|\mathbf{u} \times \mathbf{v}\|^2 = \sqrt{(b_1c_2 - b_2c_1)^2 + (a_1c_2 - a_2c_1)^2 + (a_1b_2 - a_2b_1)^2}\,^2$

$= b_1^2c_2^2 - 2b_1b_2c_1c_2 + b_2^2c_1^2 + a_1^2c_2^2 - 2a_1a_2c_1c_2 + a_2^2c_1^2 + a_1^2b_2^2 - 2a_1a_2b_1b_2 + a_2^2b_1^2$

$\|\mathbf{u}\|^2 = a_1^2 + b_1^2 + c_1^2, \|\mathbf{v}\|^2 = a_2^2 + b_2^2 + c_2^2$

$\|\mathbf{u}\|^2\|\mathbf{v}\|^2 = (a_1^2 + b_1^2 + c_1^2)(a_2^2 + b_2^2 + c_2^2) = a_1^2a_2^2 + a_1^2b_2^2 + a_1^2c_2^2 + b_1^2a_2^2 + b_1^2b_2^2 + b_1^2c_2^2 + a_2^2c_1^2 + b_2^2c_1^2 + c_1^2c_2^2$

$(\mathbf{u} \cdot \mathbf{v})^2 = (a_1a_2 + b_1b_2 + c_1c_2)^2 = (a_1a_2 + b_1b_2 + c_1c_2)(a_1a_2 + b_1b_2 + c_1c_2)$

$= a_1^2a_2^2 + a_1a_2b_1b_2 + a_1a_2c_1c_2 + b_1b_2c_1c_2 + b_1b_2a_1a_2 + b_1^2b_2^2 + b_1b_2c_1c_2 + a_1a_2c_1c_2 + c_1^2c_2^2$

$= a_1^2a_2^2 + b_1^2b_2^2 + c_1^2c_2^2 + 2a_1a_2b_1b_2 + 2b_1b_2c_1c_2 + 2a_1a_2c_1c_2$

$\|\mathbf{u}\|^2\|\mathbf{v}\|^2 - (\mathbf{u} \cdot \mathbf{v})^2 = a_1^2b_2^2 + a_1^2c_2^2 + b_1^2a_2^2 + a_2^2c_1^2 + b_2^2c_1^2 + b_1^2c_2^2 - 2a_1a_2b_1b_2 - 2b_1b_2c_1c_2 - 2a_1a_2c_1c_2$, which equals $\|\mathbf{u} \times \mathbf{v}\|^2$.

59. By Problem 58, since $\mathbf{u}$ and $\mathbf{v}$ are orthogonal, $\|\mathbf{u} \times \mathbf{v}\| = \|\mathbf{u}\|\,\|\mathbf{v}\|$. If, in addition, $\mathbf{u}$ and $\mathbf{v}$ are unit vectors, $\|\mathbf{u} \times \mathbf{v}\| = 1 \cdot 1 = 1$.

61. Assume that $\mathbf{u} = a\mathbf{i} + b\mathbf{j} + c\mathbf{k}$, $\mathbf{v} = d\mathbf{i} + e\mathbf{j} + f\mathbf{k}$, and $\mathbf{w} = l\mathbf{i} + m\mathbf{j} + n\mathbf{k}$. Then $\mathbf{u} \times \mathbf{v} = (bf - ec)\mathbf{i} - (af - dc)\mathbf{j} + (ae - db)\mathbf{k}$,

$\mathbf{u} \times \mathbf{w} = (bn - mc)\mathbf{i} - (an - lc)\mathbf{j} + (am - lb)\mathbf{k}$, and $\mathbf{v} + \mathbf{w} = (d + l)\mathbf{i} + (e + m)\mathbf{j} + (f + n)\mathbf{k}$.

Therefore, $(\mathbf{u} \times \mathbf{v}) + (\mathbf{u} \times \mathbf{w}) = (bf - ec + bn - mc)\mathbf{i} - (af - dc + an - lc)\mathbf{j} + (ae - db + am - lb)\mathbf{k}$ and

$\mathbf{u} \times (\mathbf{v} + \mathbf{w}) = [b(f + n) - (e + m)c]\mathbf{i} - [a(f + n) - (d + l)c]\mathbf{j} + [a(e + m) - (d + l)b]\mathbf{k}$

$= (bf - ec + bn - mc)\mathbf{i} - (af - dc + an - lc)\mathbf{j} + (ae - db + am - lb)\mathbf{k}$, which equals $(\mathbf{u} \times \mathbf{v}) + (\mathbf{u} \times \mathbf{w})$.

63. Let $\mathbf{v} = v_1\mathbf{i} + v_2\mathbf{j} + v_3\mathbf{k}$ and $\mathbf{w} = w_1\mathbf{i} + w_2\mathbf{j} + w_3\mathbf{k}$.

Then $2\mathbf{v} \times 3\mathbf{w} = 6(\mathbf{v} \times \mathbf{w}) = 6[(v_2w_3 - v_3w_2)\mathbf{i} - (v_1w_3 - v_3w_1)\mathbf{j} + (v_1w_2 - v_2w_1)\mathbf{k}]$.

$\mathbf{v} \cdot (2\mathbf{v} \times 3\mathbf{w}) = 6[(v_2w_3 - v_3w_2)\mathbf{v} \cdot \mathbf{i} - (v_1w_3 - v_3w_1)\mathbf{v} \cdot \mathbf{j} + (v_1w_2 - v_2w_1)\mathbf{v} \cdot \mathbf{k}]$

$= 6[(v_2w_3 - v_3w_2)v_1 - (v_1w_3 - v_3w_1)v_2 + (v_1w_2 - v_2w_1)v_3] = 0$

Since $\mathbf{v} \cdot (2\mathbf{v} \times 3\mathbf{w}) = 0$, $\mathbf{v}$ is orthogonal to $2\mathbf{v} \times 3\mathbf{w}$.

Similarly, $\mathbf{w} \cdot (2\mathbf{v} \times 3\mathbf{w}) = 0$, so $\mathbf{w}$ is also orthogonal to $2\mathbf{v} \times 3\mathbf{w}$.

66. $\dfrac{\pi}{4}$ **67.** $(17, 4.22)$, $(-17, 1.08)$ **68.** 2 **69.** $\{x \mid x \neq -4, x \neq 4\}$ **70.** $\dfrac{\sqrt{11}}{4}$ **71.** $18\sqrt{3}$ square units **72.** $\dfrac{x - 16}{x(\sqrt{x} + 4)}$ **73.** $-\dfrac{\pi}{3}$

Review Exercises (page 388)

1. $\left(\dfrac{3\sqrt{3}}{2}, \dfrac{3}{2}\right)$

2. $(1, \sqrt{3})$

3. $(0, 3)$

4. $\left(3\sqrt{2}, \dfrac{3\pi}{4}\right), \left(-3\sqrt{2}, -\dfrac{\pi}{4}\right)$

5. $\left(2, -\dfrac{\pi}{2}\right), \left(-2, \dfrac{\pi}{2}\right)$

6. $(5, 0.93)$, $(-5, 4.07)$

7. (a) $x^2 + (y - 1)^2 = 1$
   (b) circle, radius 1, center $(0, 1)$ in rectangular coordinates

8. (a) $x^2 + y^2 = 25$
   (b) circle, radius 5, center at pole

9. (a) $x - y = 0$
   (b) line though pole, making an angle of $\dfrac{\pi}{4}$ with polar axis

10. (a) $(x-4)^2 + (y+2)^2 = 25$
(b) circle, radius 5, center $(4, -2)$ in rectangular coordinates

11. Circle; radius 2, center $(2,0)$ in rectangular coordinates; symmetric with respect to the polar axis

12. Cardioid; symmetric with respect to the line $\theta = \dfrac{\pi}{2}$

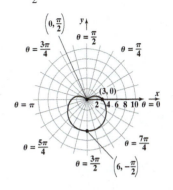

13. Limaçon without inner loop; symmetric with respect to the polar axis

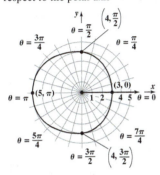

14. $\sqrt{2}\left(\cos\dfrac{5\pi}{4} + i\sin\dfrac{5\pi}{4}\right)$; $\sqrt{2}e^{i\cdot 5\pi/4}$

15. $5(\cos 5.640 + i\sin 5.640)$; $5e^{i\cdot 5.640}$

16. $-\sqrt{3} + i$

17. $-\dfrac{3}{2} + \dfrac{3\sqrt{3}}{2}i$

18. $0.10 - 0.02i$

19. $zw = \cos\dfrac{13\pi}{18} + i\sin\dfrac{13\pi}{18}$ or $e^{i\cdot 13\pi/18}$, $\dfrac{z}{w} = \cos\dfrac{\pi}{6} + i\sin\dfrac{\pi}{6}$ or $e^{i\cdot\pi/6}$

20. $zw = 6$ or $6e^{i\cdot 0}$; $\dfrac{z}{w} = \dfrac{3}{2}\left(\cos\dfrac{8\pi}{5} + i\sin\dfrac{8\pi}{5}\right)$ or $\dfrac{3}{2}e^{i\cdot 8\pi/5}$

21. $zw = 5\left(\cos\dfrac{\pi}{36} + i\sin\dfrac{\pi}{36}\right)$ or $5e^{i\cdot\pi/36}$, $\dfrac{z}{w} = 5\left(\cos\dfrac{\pi}{12} + i\sin\dfrac{\pi}{12}\right)$ or $5e^{i\cdot\pi/12}$ **22.** $\dfrac{27}{2} + \dfrac{27\sqrt{3}}{2}i$; $27e^{i\cdot\pi/3}$

23. $4i = 4e^{i\cdot\pi/2}$ **24.** 64; $64e^{i\cdot 0}$ **25.** $-527 - 336i$; $625e^{i\cdot 3.709}$ **26.** $3, 3e^{i\cdot 2\pi/3}, 3e^{i\cdot 4\pi/3}$ or $3, -\dfrac{3}{2} + \dfrac{3\sqrt{3}}{2}i, -\dfrac{3}{2} - \dfrac{3\sqrt{3}}{2}i$

27. **28.**

29. $\mathbf{v} = 2\mathbf{i} - 4\mathbf{j}$; $\|\mathbf{v}\| = 2\sqrt{5}$ **30.** $\mathbf{v} = -\mathbf{i} + 3\mathbf{j}$; $\|\mathbf{v}\| = \sqrt{10}$ **31.** $2\mathbf{i} - 2\mathbf{j}$ **32.** $-20\mathbf{i} + 13\mathbf{j}$

33. $\sqrt{5}$ **34.** $\sqrt{5} + 5 \approx 7.24$ **35.** $-\dfrac{2\sqrt{5}}{5}\mathbf{i} + \dfrac{\sqrt{5}}{5}\mathbf{j}$ **36.** $\mathbf{v} = \dfrac{3}{2}\mathbf{i} + \dfrac{3\sqrt{3}}{2}\mathbf{j}$ **37.** $120°$ **38.** $\sqrt{43} \approx 6.56$

39. $\mathbf{v} = 3\mathbf{i} - 5\mathbf{j} + 3\mathbf{k}$ **40.** $21\mathbf{i} - 2\mathbf{j} - 5\mathbf{k}$ **41.** $\sqrt{38}$ **42.** 0 **43.** $3\mathbf{i} + 9\mathbf{j} + 9\mathbf{k}$ **44.** 0

45. $\dfrac{\sqrt{19}}{19}\mathbf{i} + \dfrac{3\sqrt{19}}{19}\mathbf{j} + \dfrac{3\sqrt{19}}{19}\mathbf{k}$ or $-\dfrac{\sqrt{19}}{19}\mathbf{i} - \dfrac{3\sqrt{19}}{19}\mathbf{j} - \dfrac{3\sqrt{19}}{19}\mathbf{k}$ **46.** $\mathbf{v}\cdot\mathbf{w} = -11$; $\theta \approx 169.7°$

47. $\mathbf{v}\cdot\mathbf{w} = -4$; $\theta \approx 153.4°$ **48.** $\mathbf{v}\cdot\mathbf{w} = 1$; $\theta \approx 70.5°$ **49.** $\mathbf{v}\cdot\mathbf{w} = 0$; $\theta = 90°$ **50.** Parallel **51.** Neither

52. Orthogonal **53.** $\mathbf{v}_1 = \dfrac{4}{5}\mathbf{i} - \dfrac{3}{5}\mathbf{j}$; $\mathbf{v}_2 = \dfrac{6}{5}\mathbf{i} + \dfrac{8}{5}\mathbf{j}$ **54.** $\mathbf{v}_1 = \dfrac{9}{10}(3\mathbf{i} + \mathbf{j})$; $\mathbf{v}_2 = -\dfrac{7}{10}\mathbf{i} + \dfrac{21}{10}\mathbf{j}$ **55.** $\alpha \approx 56.1°$; $\beta \approx 138°$; $\gamma \approx 68.2°$

56. $2\sqrt{83}$ **57.** $-2\mathbf{i} + 3\mathbf{j} - \mathbf{k}$ **58.** 0 **59.** $\sqrt{29} \approx 5.39$ mph; 0.4 mi **60.** Left cable: 1843.21 lb; right cable: 1630.41 lb **61.** 50 ft-lb
62. A force of 697.2 lb is needed to keep the van from rolling down the hill. The magnitude of the force on the hill is 7969.6 lb.

Chapter Test *(page 390)*

1–3. **4.** $\left(4, \dfrac{\pi}{3}\right)$ **5.** $x^2 + y^2 = 49$ **6.** $\dfrac{y}{x} = 3$ or $y = 3x$

7. $8y = x^2$

8. $r^2 \cos \theta = 5$ is symmetric about the pole, the polar axis, and the line $\theta = \dfrac{\pi}{2}$.

9. $r = 5 \sin \theta \cos^2 \theta$ is symmetric about the line $\theta = \dfrac{\pi}{2}$. The tests for symmetry about the pole and the polar axis fail, so the graph of $r = 5 \sin \theta \cos^2 \theta$ may or may not be symmetric about the pole or polar axis.

10. $z \cdot w = 6\left(\cos\dfrac{107\pi}{180} + i \sin\dfrac{107\pi}{180}\right)$; $6e^{i \cdot 107\pi/180}$ **11.** $\dfrac{w}{z} = \dfrac{3}{2}\left(\cos\dfrac{33\pi}{20} + i \sin\dfrac{33\pi}{20}\right)$; $\dfrac{3}{2}e^{i \cdot 33\pi/20}$

12. $w^5 = 243\left(\cos\dfrac{11\pi}{18} + i \sin\dfrac{11\pi}{18}\right)$; $243e^{i \cdot 11\pi/18}$

13. $z_0 = 2\sqrt[3]{2}e^{i \cdot 2\pi/9}$, $z_1 = 2\sqrt[3]{2}e^{i \cdot 8\pi/9}$, $z_2 = 2\sqrt[3]{2}e^{i \cdot 14\pi/9}$

14. $\mathbf{v} = \langle 5\sqrt{2}, -5\sqrt{2} \rangle$ **15.** $\|\mathbf{v}\| = 10$ **16.** $\mathbf{u} = \dfrac{\mathbf{v}}{\|\mathbf{v}\|} = \left\langle \dfrac{\sqrt{2}}{2}, -\dfrac{\sqrt{2}}{2} \right\rangle$

17. 315° off the positive x-axis **18.** $\mathbf{v} = 5\sqrt{2}\mathbf{i} - 5\sqrt{2}\mathbf{j}$ **19.** $\mathbf{v}_1 + 2\mathbf{v}_2 - \mathbf{v}_3 = 6\mathbf{i} - 10\mathbf{j}$

20. Vectors $\mathbf{v}_1$ and $\mathbf{v}_4$ are parallel. **21.** Vectors $\mathbf{v}_2$ and $\mathbf{v}_3$ are orthogonal.

22. 172.87° **23.** $-9\mathbf{i} - 5\mathbf{j} + 3\mathbf{k}$ **24.** $\alpha \approx 57.7°, \beta \approx 143.3°, \gamma \approx 74.5°$ **25.** $\sqrt{115}$

26. The cable must be able to endure a tension of approximately 670.82 lb.

Cumulative Review *(page 391)*

1. $\{-3, 3\}$ **2.** $y = \dfrac{\sqrt{3}}{3}x$ **3.** $x^2 + (y - 1)^2 = 9$ **4.** $\left\{x \Big| x < \dfrac{1}{2}\right\}$ or $\left(-\infty, \dfrac{1}{2}\right)$ **5.** Symmetry with respect to the y-axis

6.   **7.**   **8.** $-\dfrac{\pi}{6}$ **9.**   **10.**   **11.** Amplitude: 4; period: 2

CHAPTER 6 Analytic Geometry

6.2 Assess Your Understanding *(page 400)*

6. parabola **7.** axis of symmetry **8.** latus rectum **9.** c **10.** $(3, 2)$ **11.** d **12.** F **13.** B **15.** E **17.** H **19.** C

21. $y^2 = 16x$

23. $x^2 = -12y$

25. $y^2 = -8x$

27. $x^2 = 2y$

29. $x^2 = \dfrac{4}{3}y$

31. $(x - 2)^2 = -8(y + 3)$

33. $(y + 2)^2 = 4(x + 1)$

35. $(x + 3)^2 = 4(y - 3)$

37. $(y + 2)^2 = -8(x + 1)$

39. Vertex: $(0, 0)$; focus: $(0, 1)$; directrix: $y = -1$

41. Vertex: $(0, 0)$; focus: $(-4, 0)$; directrix: $x = 4$

43. Vertex: $(-1, 2)$; focus: $(1, 2)$; directrix: $x = -3$

45. Vertex: $(3, -1)$; focus: $\left(3, -\dfrac{5}{4}\right)$; directrix: $y = -\dfrac{3}{4}$

47. Vertex: $(2, -3)$; focus: $(4, -3)$; directrix: $x = 0$

49. Vertex: $(0, 2)$; focus: $(-1, 2)$; directrix: $x = 1$

51. Vertex: $(-4, -2)$; focus: $(-4, -1)$; directrix: $y = -3$

53. Vertex: $(-1, -1)$; focus: $\left(-\dfrac{3}{4}, -1\right)$; directrix: $x = -\dfrac{5}{4}$

55. Vertex: $(2, -8)$; focus: $\left(2, -\dfrac{31}{4}\right)$; directrix: $y = -\dfrac{33}{4}$

57. $(y - 1)^2 = x$ **59.** $(y - 1)^2 = -(x - 2)$

61. $x^2 = 4(y - 1)$ **63.** $y^2 = \dfrac{1}{2}(x + 2)$

65. 20 ft   **67.** 24.31 ft, 18.75 ft, 7.64 ft

69. 1.5625 ft from the base of the dish, along the axis of symmetry

71. 1 in. from the vertex, along the axis of symmetry

73. 0.78125 ft

75. 4.17 in. from the base, along the axis of symmetry

77. (a) $y = -\dfrac{2}{315}x^2 + 630$

(b) 567 ft: 119.7 ft; 478 ft: 267.3 ft; 308 ft: 479.4 ft   **(c)** No

79. $Cy^2 + Dx = 0, C \neq 0, D \neq 0$
$Cy^2 = -Dx$
$y^2 = -\dfrac{D}{C}x$

This is the equation of a parabola with vertex at $(0, 0)$ and axis of symmetry the x-axis. The focus is $\left(-\dfrac{D}{4C}, 0\right)$; the directrix is the line $x = \dfrac{D}{4C}$. The parabola opens to the right if $-\dfrac{D}{C} > 0$ and to the left if $-\dfrac{D}{C} < 0$.

81. $Cy^2 + Dx + Ey + F = 0, C \neq 0$
$Cy^2 + Ey = -Dx - F$
$y^2 + \dfrac{E}{C}y = -\dfrac{D}{C}x - \dfrac{F}{C}$
$\left(y + \dfrac{E}{2C}\right)^2 = -\dfrac{D}{C}x - \dfrac{F}{C} + \dfrac{E^2}{4C^2}$
$\left(y + \dfrac{E}{2C}\right)^2 = -\dfrac{D}{C}x + \dfrac{E^2 - 4CF}{4C^2}$

(a) If $D \neq 0$, then the equation may be written as
$$\left(y + \dfrac{E}{2C}\right)^2 = -\dfrac{D}{C}\left(x - \dfrac{E^2 - 4CF}{4CD}\right).$$
This is the equation of a parabola with vertex at $\left(\dfrac{E^2 - 4CF}{4CD}, -\dfrac{E}{2C}\right)$ and axis of symmetry parallel to the x-axis.

(b)–(d) If $D = 0$, the graph of the equation contains no points if $E^2 - 4CF < 0$, is a single horizontal line if $E^2 - 4CF = 0$, and is two horizontal lines if $E^2 - 4CF > 0$.

83. $(0, 2), (0, -2), (-36, 0)$; symmetric with respect to the x-axis.   **84.** $\sin \theta = \dfrac{5\sqrt{89}}{89}$; $\cos \theta = -\dfrac{8\sqrt{89}}{89}$; $\csc \theta = \dfrac{\sqrt{89}}{5}$; $\sec \theta = -\dfrac{\sqrt{89}}{8}$; $\cot \theta = -\dfrac{8}{5}$

85. $-\dfrac{2\sqrt{10}}{3}$ **86.** $\dfrac{11\sqrt{13}}{6}$ **87.** $(x + 12)^2 + (y - 7)^2 = 6$ **88.** $-70°$ **89.** $\cos 17°$ **90.** $\{-1, 2, 3, 6\}$

6.3 Assess Your Understanding (page 410)

7. ellipse   **8.** b   **9.** $(0, -5)$; $(0, 5)$ **10.** 5; 3; x **11.** $(-2, -3)$; $(6, -3)$ **12.** a   **13.** C   **15.** B

17. Vertices: $(-5, 0), (0, 5)$
Foci: $(-\sqrt{21}, 0), (\sqrt{21}, 0)$

19. Vertices: $(0, -5), (0, 5)$
Foci: $(0, -4), (0, 4)$

21. $\dfrac{x^2}{4} + \dfrac{y^2}{16} = 1$
Vertices: $(0, -4), (0, 4)$
Foci: $(0, -2\sqrt{3}), (0, 2\sqrt{3})$

23. $\dfrac{x^2}{8} + \dfrac{y^2}{2} = 1$
Vertices: $(-2\sqrt{2}, 0), (2\sqrt{2}, 0)$
Foci: $(-\sqrt{6}, 0), (\sqrt{6}, 0)$

25. $\dfrac{x^2}{16} + \dfrac{y^2}{16} = 1$

Vertices: $(-4,0)$, $(4,0)$, $(0,-4)$, $(0,4)$; Focus: $(0,0)$

27. $\dfrac{x^2}{25} + \dfrac{y^2}{16} = 1$

29. $\dfrac{x^2}{9} + \dfrac{y^2}{25} = 1$

31. $\dfrac{x^2}{9} + \dfrac{y^2}{5} = 1$

33. $\dfrac{x^2}{25} + \dfrac{y^2}{9} = 1$

35. $\dfrac{x^2}{4} + \dfrac{y^2}{13} = 1$

37. $x^2 + \dfrac{y^2}{16} = 1$

39. $\dfrac{(x+1)^2}{4} + (y-1)^2 = 1$

41. $(x-1)^2 + \dfrac{y^2}{4} = 1$

43. Center: $(3,-1)$; vertices: $(3,-4)$, $(3,2)$; foci: $\left(3, -1 - \sqrt{5}\right)$, $\left(3, -1 + \sqrt{5}\right)$

45. $\dfrac{(x+5)^2}{16} + \dfrac{(y-4)^2}{4} = 1$

Center: $(-5,4)$; vertices: $(-9,4)$, $(-1,4)$; foci: $\left(-5 - 2\sqrt{3}, 4\right)$, $\left(-5 + 2\sqrt{3}, 4\right)$

47. $\dfrac{(x+2)^2}{4} + (y-1)^2 = 1$

Center: $(-2,1)$; vertices: $(-4,1)$, $(0,1)$; foci: $\left(-2 - \sqrt{3}, 1\right)$, $\left(-2 + \sqrt{3}, 1\right)$

49. $\dfrac{(x-2)^2}{3} + \dfrac{(y+1)^2}{2} = 1$

Center: $(2,-1)$; vertices: $\left(2 - \sqrt{3}, -1\right)$; $\left(2 + \sqrt{3}, -1\right)$; foci: $(1,-1)$, $(3,-1)$

51. $\dfrac{(x-1)^2}{4} + \dfrac{(y+2)^2}{9} = 1$

Center: $(1,-2)$; vertices: $(1,-5)$, $(1,1)$; foci: $\left(1, -2 - \sqrt{5}\right)$, $\left(1, -2 + \sqrt{5}\right)$

53. $x^2 + \dfrac{(y+2)^2}{4} = 1$

Center: $(0,-2)$; vertices: $(0,-4)$, $(0,0)$; foci: $\left(0, -2 - \sqrt{3}\right)$, $\left(0, -2 + \sqrt{3}\right)$

55. $\dfrac{(x-2)^2}{25} + \dfrac{(y+2)^2}{21} = 1$

57. $\dfrac{(x-4)^2}{5} + \dfrac{(y-6)^2}{9} = 1$

59. $\dfrac{(x-2)^2}{16} + \dfrac{(y-1)^2}{7} = 1$

61. $\dfrac{(x-1)^2}{10} + (y-2)^2 = 1$

63. $\dfrac{(x-1)^2}{9} + \dfrac{(y-2)^2}{9} = 1$

65.

67.

69. $\dfrac{x^2}{100} + \dfrac{y^2}{36} = 1$ **71.** 43.3 ft **73.** 24.65 ft, 21.65 ft, 13.82 ft **75.** 30 ft **77.** The elliptical hole will have a major axis of length $2\sqrt{41}$ in. and a minor axis of length 8 in. **79.** 91.5 million mi; $\dfrac{x^2}{(93)^2} + \dfrac{y^2}{8646.75} = 1$ **81.** Perihelion: 460.6 million mi; mean distance: 483.8 million mi; $\dfrac{x^2}{(483.8)^2} + \dfrac{y^2}{233,524.2} = 1$ **83.** 35 million mi **85.** $8\sqrt{14}$ cm $\approx$ 29.93 cm **87.** $5\sqrt{5} - 4$

89. (a) $Ax^2 + Cy^2 + F = 0$ If A and C are of the same sign and F is of opposite sign, then the equation takes the form

$Ax^2 + Cy^2 = -F$ $\dfrac{x^2}{\left(-\dfrac{F}{A}\right)} + \dfrac{y^2}{\left(-\dfrac{F}{C}\right)} = 1$, where $-\dfrac{F}{A}$ and $-\dfrac{F}{C}$ are positive. This is the equation of an ellipse with center at $(0,0)$.

(b) If $A = C$, the equation may be written as $x^2 + y^2 = -\dfrac{F}{A}$. This is the equation of a circle with center at $(0,0)$ and radius equal to $\sqrt{-\dfrac{F}{A}}$.

92. Zeros: $5 - 2\sqrt{3}, 5 + 2\sqrt{3}$; x-intercepts: $5 - 2\sqrt{3}, 5 + 2\sqrt{3}$ **93.** Domain: $\{x \mid x \neq 5\}$; Horizontal asymptote: $y = 2$; Vertical asymptote: $x = 5$

94. 617.1 ft-lb **95.** $b \approx 10.94, c \approx 17.77, B = 38°$ **96.** $\left\{\dfrac{\pi}{30}, \dfrac{7\pi}{30}, \dfrac{13\pi}{30}\right\}$ **97.** $\dfrac{10}{7}$ **98.** $4x - 7$ **99.** $\{-3 - 2\sqrt{5}, -3 + 2\sqrt{5}\}$

6.4 Assess Your Understanding *(page 423)*

7. hyperbola **8.** transverse axis **9.** b **10.** $(2,4); (2,-2)$ **11.** $(2,6); (2,-4)$ **12.** c **13.** $2; 3; x$ **14.** $y = -\dfrac{4}{9}x; y = \dfrac{4}{9}x$ **15.** B **17.** A

19. $x^2 - \dfrac{y^2}{8} = 1$

21. $\dfrac{y^2}{16} - \dfrac{x^2}{20} = 1$

23. $\dfrac{x^2}{9} - \dfrac{y^2}{16} = 1$

25. $\dfrac{y^2}{36} - \dfrac{x^2}{9} = 1$

27. $\dfrac{x^2}{8} - \dfrac{y^2}{8} = 1$

29. $\dfrac{x^2}{25} - \dfrac{y^2}{9} = 1$

Center: $(0,0)$
Transverse axis: x-axis
Vertices: $(-5,0), (5,0)$
Foci: $\left(-\sqrt{34},0\right), \left(\sqrt{34},0\right)$
Asymptotes: $y = \pm\dfrac{3}{5}x$

31. $\dfrac{x^2}{4} - \dfrac{y^2}{16} = 1$

Center: $(0,0)$
Transverse axis: x-axis
Vertices: $(-2,0), (2,0)$
Foci: $\left(-2\sqrt{5},0\right), \left(2\sqrt{5},0\right)$
Asymptotes: $y = \pm 2x$

33. $\dfrac{y^2}{9} - x^2 = 1$

Center: $(0,0)$
Transverse axis: y-axis
Vertices: $(0,-3), (0,3)$
Foci: $\left(0, -\sqrt{10}\right), \left(0, \sqrt{10}\right)$
Asymptotes: $y = \pm 3x$

35. $\dfrac{y^2}{25} - \dfrac{x^2}{25} = 1$

Center: $(0,0)$
Transverse axis: y-axis
Vertices: $(0,-5), (0,5)$
Foci: $\left(0, -5\sqrt{2}\right), \left(0, 5\sqrt{2}\right)$
Asymptotes: $y = \pm x$

37. $x^2 - y^2 = 1$

39. $\dfrac{y^2}{36} - \dfrac{x^2}{9} = 1$

41. $\dfrac{(x-4)^2}{4} - \dfrac{(y+1)^2}{5} = 1$

43. $\dfrac{(y+4)^2}{4} - \dfrac{(x+3)^2}{12} = 1$

45. $(x-5)^2 - \dfrac{(y-7)^2}{3} = 1$

47. $\dfrac{(x-1)^2}{4} - \dfrac{(y+1)^2}{9} = 1$

$y + 1 = -\frac{3}{2}(x-1)$ $y + 1 = \frac{3}{2}(x-1)$

$F_1 = (1 - \sqrt{13}, -1)$ $F_2 = (1 + \sqrt{13}, -1)$
$V_1 = (-1, -1)$ $V_2 = (3, -1)$

49. $\dfrac{(x-2)^2}{4} - \dfrac{(y+3)^2}{9} = 1$

Center: $(2, -3)$
Transverse axis: parallel to x-axis
Vertices: $(0, -3), (4, -3)$
Foci: $\left(2 - \sqrt{13}, -3\right), \left(2 + \sqrt{13}, -3\right)$
Asymptotes: $y + 3 = \pm\dfrac{3}{2}(x - 2)$

51. $\dfrac{(y-2)^2}{4} - (x+2)^2 = 1$

Center: $(-2, 2)$
Transverse axis: parallel to y-axis
Vertices: $(-2, 0), (-2, 4)$
Foci: $\left(-2, 2 - \sqrt{5}\right), \left(-2, 2 + \sqrt{5}\right)$
Asymptotes: $y - 2 = \pm 2(x + 2)$

53. $\dfrac{(x+1)^2}{4} - \dfrac{(y+2)^2}{4} = 1$

Center: $(-1, -2)$
Transverse axis: parallel to x-axis
Vertices: $(-3, -2), (1, -2)$
Foci: $\left(-1 - 2\sqrt{2}, -2\right), \left(-1 + 2\sqrt{2}, -2\right)$
Asymptotes: $y + 2 = \pm(x + 1)$

55. $(x-1)^2 - (y+1)^2 = 1$

Center: $(1, -1)$
Transverse axis: parallel to x-axis
Vertices: $(0, -1), (2, -1)$
Foci: $\left(1 - \sqrt{2}, -1\right), \left(1 + \sqrt{2}, -1\right)$
Asymptotes: $y + 1 = \pm(x - 1)$

57. $\dfrac{(y-2)^2}{4} - (x+1)^2 = 1$

Center: $(-1, 2)$
Transverse axis: parallel to y-axis
Vertices: $(-1, 0), (-1, 4)$
Foci: $\left(-1, 2 - \sqrt{5}\right), \left(-1, 2 + \sqrt{5}\right)$
Asymptotes: $y - 2 = \pm 2(x + 1)$

59. $\dfrac{(x-3)^2}{4} - \dfrac{(y+2)^2}{16} = 1$

Center: $(3, -2)$
Transverse axis: parallel to x-axis
Vertices: $(1, -2), (5, -2)$
Foci: $\left(3 - 2\sqrt{5}, -2\right), \left(3 + 2\sqrt{5}, -2\right)$
Asymptotes: $y + 2 = \pm 2(x - 3)$

61. $\dfrac{(y-1)^2}{4} - (x+2)^2 = 1$

Center: $(-2, 1)$
Transverse axis: parallel to y-axis
Vertices: $(-2, -1), (-2, 3)$
Foci: $\left(-2, 1 - \sqrt{5}\right), \left(-2, 1 + \sqrt{5}\right)$
Asymptotes: $y - 1 = \pm 2(x + 2)$

63.

65.

$y = -x$ $y = x$

67. Center: $(3, 0)$
Transverse axis: parallel to x-axis
Vertices: $(1, 0), (5, 0)$
Foci: $(3 - \sqrt{29}, 0), (3 + \sqrt{29}, 0)$
Asymptotes: $y = \pm\dfrac{5}{2}(x - 3)$

69. Vertex: $(0, 3)$; focus: $(0, 7)$;
directrix: $y = -1$

$F = (0, 7)$
$V = (0, 3)$
$D: y = -1$

71. $\dfrac{(x-5)^2}{9} + \dfrac{y^2}{25} = 1$

Center: $(5, 0)$; vertices: $(5, 5), (5, -5)$;
foci: $(5, -4), (5, 4)$

$V_1 = (5, 5)$
$F_1 = (5, 4)$
$(5, 0)$ $(8, 0)$
$(2, 0)$
$F_2 = (5, -4)$
$V_2 = (5, -5)$

73. $(x-3)^2 = 8(y+5)$
Vertex: $(3, -5)$; focus: $(3, -3)$;
directrix: $y = -7$

$F = (3, -3)$
$V = (3, -5)$
$D: y = -7$

75. The fireworks display is 50,138 ft north of the person at point A.   **77.** The tower is 592.4 ft tall.   **79. (a)** $y = \pm x$ **(b)** $\dfrac{x^2}{100} - \dfrac{y^2}{100} = 1, x \geq 0$

81. $\dfrac{y^2}{81} - \dfrac{x^2}{88} = 1$ **83.** If the eccentricity is close to 1, the "opening" of the hyperbola is very small. As e increases, the opening gets bigger.

85. $\dfrac{x^2}{4} - y^2 = 1$; asymptotes $y = \pm\dfrac{1}{2}x$

$y^2 - \dfrac{x^2}{4} = 1$; asymptotes $y = \pm\dfrac{1}{2}x$

87. $Ax^2 + Cy^2 + F = 0$

$Ax^2 + Cy^2 = -F$

If A and C are of opposite sign and $F \neq 0$, this equation may be written as $\dfrac{x^2}{\left(-\dfrac{F}{A}\right)} + \dfrac{y^2}{\left(-\dfrac{F}{C}\right)} = 1$,

where $-\dfrac{F}{A}$ and $-\dfrac{F}{C}$ are opposite in sign. This is the equation of a hyperbola with center at $(0,0)$. The transverse axis is the x-axis if $-\dfrac{F}{A} > 0$; the transverse axis is the y-axis if $-\dfrac{F}{A} < 0$.

89. Amplitude $= \dfrac{1}{2}$; Period $= \dfrac{2\pi}{3}$; Phase shift $= -\dfrac{\pi}{3}$;
Vertical shift $= 5$

90. $c \approx 13.16$, $A \approx 31.6°$, $B = 48.4°$ **91.** $\left(6, -6\sqrt{3}\right)$

92. $x^2 + (y - 3)^2 = 9$; circle, radius 3, center at $(0,3)$ in rectangular coordinates

93. $\dfrac{9\pi}{4} - \dfrac{9}{2} \approx 2.57$ sq units

94. $\{-4\}$ **95.** $\left(\dfrac{1}{2}, -\dfrac{3}{2}\right)$ **96.** $\dfrac{\sqrt{16 - x^2}}{4}$

6.5 Assess Your Understanding *(page 432)*

5. $\cot(2\theta) = \dfrac{A - C}{B}$ **6.** d   **7.** $B^2 - 4AC < 0$ **8.** c   **9.** T   **10.** F   **11.** Parabola   **13.** Ellipse   **15.** Hyperbola

17. Hyperbola   **19.** Circle   **21.** $x = \dfrac{\sqrt{2}}{2}(x' - y')$, $y = \dfrac{\sqrt{2}}{2}(x' + y')$ **23.** $x = \dfrac{\sqrt{2}}{2}(x' - y')$, $y = \dfrac{\sqrt{2}}{2}(x' + y')$

25. $x = \dfrac{1}{2}(x' - \sqrt{3}y')$, $y = \dfrac{1}{2}(\sqrt{3}x' + y')$ **27.** $x = \dfrac{\sqrt{5}}{5}(x' - 2y')$, $y = \dfrac{\sqrt{5}}{5}(2x' + y')$ **29.** $x = \dfrac{\sqrt{13}}{13}(3x' - 2y')$, $y = \dfrac{\sqrt{13}}{13}(2x' + 3y')$

31. $\theta = 45°$ (see Problem 21)

$x'^2 - \dfrac{y'^2}{3} = 1$

Hyperbola
Center at origin
Transverse axis is the x'-axis.
Vertices at $(\pm 1, 0)$

33. $\theta = 45°$ (see Problem 23)

$x'^2 + \dfrac{y'^2}{4} = 1$

Ellipse
Center at $(0,0)$
Major axis is the y'-axis.
Vertices at $(0, \pm 2)$

35. $\theta = 60°$ (see Problem 25)

$\dfrac{x'^2}{4} + y'^2 = 1$

Ellipse
Center at $(0,0)$
Major axis is the x'-axis.
Vertices at $(\pm 2, 0)$

37. $\theta \approx 63°$ (see Problem 27)
$y'^2 = 8x'$
Parabola
Vertex at $(0,0)$
Focus at $(2,0)$

39. $\theta \approx 34°$ (see Problem 29)
$\dfrac{(x' - 2)^2}{4} + y'^2 = 1$
Ellipse
Center at $(2,0)$
Major axis is the x'-axis.
Vertices at $(4,0)$ and $(0,0)$

41. $\cot(2\theta) = \dfrac{7}{24}$;

$\theta = \sin^{-1}\left(\dfrac{3}{5}\right) \approx 37°$

$(x' - 1)^2 = -6\left(y' - \dfrac{1}{6}\right)$

Parabola
Vertex at $\left(1, \dfrac{1}{6}\right)$
Focus at $\left(1, -\dfrac{4}{3}\right)$

43. Hyperbola   **45.** Hyperbola   **47.** Parabola   **49.** Ellipse   **51.** Ellipse   **53.** $23.6°$

55. Refer to equation (6): $A' = A\cos^2\theta + B\sin\theta\cos\theta + C\sin^2\theta$

$$B' = B(\cos^2\theta - \sin^2\theta) + 2(C - A)(\sin\theta\cos\theta)$$
$$C' = A\sin^2\theta - B\sin\theta\cos\theta + C\cos^2\theta$$
$$D' = D\cos\theta + E\sin\theta$$
$$E' = -D\sin\theta + E\cos\theta$$
$$F' = F$$

57. Use Problem 55 to find $B'^2 - 4A'C'$. After much cancellation, $B'^2 - 4A'C' = B^2 - 4AC$.

59. The distance between P_1 and P_2 in the $x'y'$-plane equals $\sqrt{(x_2' - x_1')^2 + (y_2' - y_1')^2}$.

Assuming that $x' = x\cos\theta - y\sin\theta$ and $y' = x\sin\theta + y\cos\theta$, then

$(x_2' - x_1')^2 = (x_2\cos\theta - y_2\sin\theta - x_1\cos\theta + y_1\sin\theta)^2$

$= \cos^2\theta(x_2 - x_1)^2 - 2\sin\theta\cos\theta(x_2 - x_1)(y_2 - y_1) + \sin^2\theta(y_2 - y_1)^2$, and

$(y_2' - y_1')^2 = (x_2\sin\theta + y_2\cos\theta - x_1\sin\theta - y_1\cos\theta)^2 = \sin^2\theta(x_2 - x_1)^2 + 2\sin\theta\cos\theta(x_2 - x_1)(y_2 - y_1) + \cos^2\theta(y_2 - y_1)^2$.

Therefore, $(x_2' - x_1')^2 + (y_2' - y_1')^2 = \cos^2\theta(x_2 - x_1)^2 + \sin^2\theta(x_2 - x_1)^2 + \sin^2\theta(y_2 - y_1)^2 + \cos^2\theta(y_2 - y_1)^2$

$= (x_2 - x_1)^2(\cos^2\theta + \sin^2\theta) + (y_2 - y_1)^2(\sin^2\theta + \cos^2\theta) = (x_2 - x_1)^2 + (y_2 - y_1)^2$.

63. $A \approx 39.4°, B \approx 54.7°, C \approx 85.9°$ **64.** 38.5   **65.** $r^2\cos\theta\sin\theta = 1$ **66.** $\sqrt{29}\left(\cos 291.8° + i\sin 291.8°\right)$

67. $-\dfrac{2(2x + 3)^2(52x^2 + 96x + 3)}{(4x^2 - 1)^9}$ **68.** 8.33   **69.** 8

6.6 Assess Your Understanding *(page 439)*

3. conic; focus; directrix   **4.** parabola; hyperbola; ellipse   **5.** b   **6.** T   **7.** Parabola; directrix is perpendicular to the polar axis, 1 unit to the right of the pole.

9. Hyperbola; directrix is parallel to the polar axis, $\dfrac{4}{3}$ units below the pole.

11. Ellipse; directrix is perpendicular to the polar axis, $\dfrac{3}{2}$ units to the left of the pole.

13. Parabola; directrix is perpendicular to the polar axis, 1 unit to the right of the pole; vertex is at $\left(\dfrac{1}{2}, 0\right)$.

15. Ellipse; directrix is parallel to the polar axis, $\dfrac{8}{3}$ units above the pole; vertices are at $\left(\dfrac{8}{7}, \dfrac{\pi}{2}\right)$ and $\left(8, \dfrac{3\pi}{2}\right)$.

17. Hyperbola; directrix is perpendicular to the polar axis, $\dfrac{3}{2}$ units to the left of the pole; vertices are at $(-3, 0)$ and $(1, \pi)$.

19. Ellipse; directrix is parallel to the polar axis, 8 units below the pole; vertices are at $\left(8, \dfrac{\pi}{2}\right)$ and $\left(\dfrac{8}{3}, \dfrac{3\pi}{2}\right)$.

21. Ellipse; directrix is parallel to the polar axis, 3 units below the pole; vertices are at $\left(6, \dfrac{\pi}{2}\right)$ and $\left(\dfrac{6}{5}, \dfrac{3\pi}{2}\right)$.

23. Ellipse; directrix is perpendicular to the polar axis, 6 units to the left of the pole; vertices are at $(6, 0)$ and $(2, \pi)$.

25. $y^2 + 2x - 1 = 0$ **27.** $16x^2 + 7y^2 + 48y - 64 = 0$ **29.** $3x^2 - y^2 + 12x + 9 = 0$ **31.** $4x^2 + 3y^2 - 16y - 64 = 0$

33. $9x^2 + 5y^2 - 24y - 36 = 0$ **35.** $3x^2 + 4y^2 - 12x - 36 = 0$ **37.** $r = \dfrac{1}{1 + \sin\theta}$ **39.** $r = \dfrac{12}{5 - 4\cos\theta}$ **41.** $r = \dfrac{12}{1 - 6\sin\theta}$

43. Use $d(D, P) = p - r\cos\theta$ in the derivation of equation (6).

45. Use $d(D, P) = p + r\sin\theta$ in the derivation of equation (6).

47. Aphelion: 35 AU; perihelion: 0.587 AU

49. 0.0125 m   **51.** $v_0 = \sqrt{\dfrac{2GM_e}{r_0}}$ **52.** 27.81   **53.** Amplitude $= 4$; Period $= 10\pi$

54. $\left\{\dfrac{\pi}{3}, \pi, \dfrac{5\pi}{3}\right\}$ **55.** 26   **56.** $r = \dfrac{24}{\pi}$ ft

57. Decreasing: $[-1, 0]$; Increasing: $[-2, -1]$ and $[0, \infty)$ **58.** $k = \dfrac{12}{5}$ **59.** $f(x) = -\dfrac{1}{3}(x + 3)^2 + 8$ **60.** 40 sq units

6.7 Assess Your Understanding (page 450)

2. plane curve; parameter **3.** b **4.** a **5.** F **6.** T

7.
$x - 3y + 1 = 0$

9.
$y = \sqrt{x - 2}$

11.
$y = x - 8$

13.
$x = 3(y - 1)^2$

15.
$2y = 2 + x$

17.
$y = x^3$

19.
$\dfrac{x^2}{4} + \dfrac{y^2}{9} = 1$

21.
$\dfrac{x^2}{4} + \dfrac{y^2}{9} = 1$

23.
$x^2 - y^2 = 1$

25.
$x + y = 1$

27. $x(t) = t$ or $x(t) = \dfrac{t + 1}{4}$
$y(t) = 4t - 1$ $y(t) = t$

29. $x(t) = t$ or $x(t) = t^3$
$y(t) = t^2 + 1$ $y(t) = t^6 + 1$

31. $x(t) = t$ or $x(t) = \sqrt[3]{t}$
$y(t) = t^3$ $y(t) = t$

33. $x(t) = t$ or $x(t) = t^3$
$y(t) = t^{2/3}, t \geq 0$ $y(t) = t^2, t \geq 0$

35. $x(t) = t + 2, y(t) = t, 0 \leq t \leq 5$ **37.** $x(t) = 3 \cos t, y(t) = 2 \sin t, 0 \leq t \leq 2\pi$
39. $x(t) = 2 \cos(\pi t), y(t) = -3 \sin(\pi t), 0 \leq t \leq 2$ **41.** $x(t) = 2 \sin(2\pi t), y(t) = 3 \cos(2\pi t), 0 \leq t \leq 1$

43.

45.

47.

49. (a) $x(t) = 3$ (or any number except 0)
$y(t) = -16t^2 + 50t + 6$
(b) 3.24 s
(c) 1.56 s; 45.06 ft
(d)

51. (a) Train: $x_1(t) = t^2, y_1(t) = 1$;
Bill: $x_2(t) = 5(t - 5), y_2(t) = 3$
(b) Bill won't catch the train.
(c)
$t = 8$

53. (a) $x(t) = (145 \cos 20°)t$
$y(t) = -16t^2 + (145 \sin 20°)t + 5$
(b) 3.20 s
(c) 435.65 ft
(d) 1.55 s; 43.43 ft
(e)

55. (a) $x(t) = (40 \cos 45°)t$
$y(t) = -4.9t^2 + (40 \sin 45°)t + 300$
(b) 11.23 s
(c) 317.52 m
(d) 2.89 s; 340.82 m
(e)

57. (a) Camry: $x(t) = 40t - 5, y(t) = 0$; Chevy Impala: $x(t) = 0, y(t) = 30t - 4$
(b) $d = \sqrt{(40t - 5)^2 + (30t - 4)^2}$
(c)
(d) 0.2 mi; 7.68 min
(e)

59. (a) $x(t) = \dfrac{\sqrt{2}}{2}v_0 t, y(t) = -16t^2 + \dfrac{\sqrt{2}}{2}v_0 t + 3$ **(b)** Maximum height is 139.1 ft. **(c)** The ball is 272.25 ft from home plate. **(d)** Yes, the ball will
clear the wall by about 99.5 ft. **61.** The orientation is from (x_1, y_1) to (x_2, y_2). **63.** $x(m) = \dfrac{R(1 - m^2)}{1 + m^2}, y(m) = \dfrac{2mR}{1 + m^2}, -\infty < m < \infty$

67.

68.
$y = \sin \dfrac{x}{2}$
$y = 2 \cos(2x)$

69. Approximately 2733 miles

70. (a) Simple harmonic **(b)** 2 m **(c)** $\dfrac{\pi}{2}$ s **(d)** $\dfrac{2}{\pi}$ oscillations/s

71. $y = 2x - \dfrac{11}{2}$ **72.** $-\dfrac{1}{(x + 3)^2}$ **73.** $\dfrac{\sqrt{6} - \sqrt{2}}{4}$ **74.** $c = \dfrac{2\sqrt{3}}{3}$

Review Exercises *(page 455)*

1. Parabola; vertex $(0, 0)$, focus $(-4, 0)$, directrix $x = 4$ **2.** Hyperbola; center $(0, 0)$, vertices $(5, 0)$ and $(-5, 0)$, foci $\left(\sqrt{26}, 0\right)$ and $\left(-\sqrt{26}, 0\right)$, asymptotes $y = \dfrac{1}{5}x$ and $y = -\dfrac{1}{5}x$ **3.** Ellipse; center $(0, 0)$, vertices $(0, 5)$ and $(0, -5)$, foci $(0, 3)$ and $(0, -3)$

4. $x^2 = -4(y - 1)$: Parabola; vertex $(0, 1)$, focus $(0, 0)$, directrix $y = 2$ **5.** $\dfrac{x^2}{2} - \dfrac{y^2}{8} = 1$: Hyperbola; center $(0, 0)$, vertices $\left(\sqrt{2}, 0\right)$ and $\left(-\sqrt{2}, 0\right)$, foci $\left(\sqrt{10}, 0\right)$ and $\left(-\sqrt{10}, 0\right)$, asymptotes $y = 2x$ and $y = -2x$ **6.** $(x - 2)^2 = 2(y + 2)$: Parabola; vertex $(2, -2)$, focus $\left(2, -\dfrac{3}{2}\right)$, directrix $y = -\dfrac{5}{2}$

7. $\dfrac{(y - 2)^2}{4} - (x - 1)^2 = 1$: Hyperbola; center $(1, 2)$, vertices $(1, 4)$ and $(1, 0)$, foci $\left(1, 2 + \sqrt{5}\right)$ and $\left(1, 2 - \sqrt{5}\right)$, asymptotes $y - 2 = \pm 2(x - 1)$

8. $\dfrac{(x - 2)^2}{9} + \dfrac{(y - 1)^2}{4} = 1$: Ellipse; center $(2, 1)$, vertices $(5, 1)$ and $(-1, 1)$, foci $\left(2 + \sqrt{5}, 1\right)$ and $\left(2 - \sqrt{5}, 1\right)$

9. $(x - 2)^2 = -4(y + 1)$: Parabola; vertex $(2, -1)$, focus $(2, -2)$, directrix $y = 0$

10. $\dfrac{(x - 1)^2}{4} + \dfrac{(y + 1)^2}{9} = 1$: Ellipse; center $(1, -1)$, vertices $(1, 2)$ and $(1, -4)$, foci $\left(1, -1 + \sqrt{5}\right)$ and $\left(1, -1 - \sqrt{5}\right)$

11. $y^2 = -8x$

12. $\dfrac{y^2}{4} - \dfrac{x^2}{12} = 1$

13. $\dfrac{x^2}{16} + \dfrac{y^2}{7} = 1$

14. $(x - 2)^2 = -4(y + 3)$

15. $(x + 2)^2 - \dfrac{(y + 3)^2}{3} = 1$

16. $\dfrac{(x + 4)^2}{16} + \dfrac{(y - 5)^2}{25} = 1$

17. $\dfrac{(x + 1)^2}{9} - \dfrac{(y - 2)^2}{7} = 1$

18. $\dfrac{(x - 3)^2}{9} - \dfrac{(y - 1)^2}{4} = 1$

19. Parabola   **20.** Ellipse
21. Parabola   **22.** Hyperbola
23. Ellipse

24. $x'^2 - \dfrac{y'^2}{9} = 1$

Hyperbola
Center at the origin
Transverse axis the x'-axis
Vertices at $(\pm 1, 0)$

25. $\dfrac{x'^2}{2} + \dfrac{y'^2}{4} = 1$

Ellipse
Center at the origin
Major axis the y'-axis
Vertices at $(0, \pm 2)$

26. $y'^2 = -\dfrac{4\sqrt{13}}{13}x'$

Parabola
Vertex at the origin
Focus on the x'-axis at $\left(-\dfrac{\sqrt{13}}{13}, 0\right)$

27. Parabola; directrix is perpendicular to the polar axis 4 units to the left of the pole; vertex is $(2, \pi)$.

28. Ellipse; directrix is parallel to the polar axis 6 units below the pole; vertices are $\left(6, \dfrac{\pi}{2}\right)$ and $\left(2, \dfrac{3\pi}{2}\right)$.

29. Hyperbola; directrix is perpendicular to the polar axis 1 unit to the right of the pole; vertices are $\left(\dfrac{2}{3}, 0\right)$ and $(-2, \pi)$.

30. $y^2 - 8x - 16 = 0$ **31.** $3x^2 - y^2 - 8x + 4 = 0$

32.

$x + 4y = 2$

33.

$$\frac{x^2}{9} + \frac{(y-2)^2}{16} = 1$$

34.

$1 + y = x$

35. $x(t) = t, y(t) = -2t + 4, -\infty < t < \infty$ **36.** $x(t) = 4\cos\left(\frac{\pi}{2}t\right), y(t) = 3\sin\left(\frac{\pi}{2}t\right), 0 \le t \le 4$ **37.** $\frac{x^2}{5} - \frac{y^2}{4} = 1$ **38.** The ellipse $\frac{x^2}{16} + \frac{y^2}{7} = 1$

$x(t) = \frac{t-4}{-2}, y(t) = t, -\infty < t < \infty$

39. $\frac{1}{4}$ ft or 3 in. **40.** 19.72 ft, 18.86 ft, 14.91 ft **41.** 450 ft

42. (a) Train: $x_1(t) = \frac{3}{2}t^2, y_1 = 1$ **(c)**
Mary: $x_2(t) = 6(t-2), y_2 = 3$
(b) Mary won't catch the train.

43. (a) $x(t) = (80\cos 35°)t$ **(e)**
$y(t) = -16t^2 + (80\sin 35°)t + 6$
(b) 2.9932 s
(c) 1.4339 s; 38.9 ft
(d) 196.15 ft

Chapter Test *(page 456)*

1. Hyperbola; center: $(-1,0)$; vertices: $(-3,0)$ and $(1,0)$; foci: $(-1-\sqrt{13},0)$ and $(-1+\sqrt{13},0)$; asymptotes: $y = -\frac{3}{2}(x+1)$ and $y = \frac{3}{2}(x+1)$

2. Parabola; vertex: $\left(1,-\frac{1}{2}\right)$; focus: $\left(1,\frac{3}{2}\right)$; directrix: $y = -\frac{5}{2}$

3. Ellipse; center: $(-1,1)$; foci: $(-1-\sqrt{3},1)$ and $(-1+\sqrt{3},1)$; vertices: $(-4,1)$ and $(2,1)$

4. $(x+1)^2 = 6(y-3)$ **5.** $\frac{x^2}{7} + \frac{y^2}{16} = 1$ **6.** $\frac{(y-2)^2}{4} - \frac{(x-2)^2}{8} = 1$

7. Hyperbola **8.** Ellipse **9.** Parabola

10. $x'^2 + 2y'^2 = 1$. This is the equation of an ellipse with center at $(0,0)$ in the $x'y'$-plane. The vertices are at $(-1,0)$ and $(1,0)$ in the $x'y'$-plane.

11. Hyperbola; $(x+2)^2 - \frac{y^2}{3} = 1$ **12.** $y = 1 - \sqrt{\frac{x+2}{3}}$

13. The microphone should be located $\frac{2}{3}$ ft from the base of the reflector, along its axis of symmetry.

Cumulative Review *(page 457)*

1. $-6x + 5 - 3h$ **2. (a)** $y = 2x - 2$ **(b)** $(x-2)^2 + y^2 = 4$ **(c)** $\frac{x^2}{9} + \frac{y^2}{4} = 1$ **(d)** $y = 2(x-1)^2$ **(e)** $y^2 - \frac{x^2}{3} = 1$

3. $\theta = \frac{\pi}{12} \pm \pi k, k$ is any integer; $\theta = \frac{5\pi}{12} \pm \pi k, k$ is any integer **4.** $\theta = \frac{\pi}{6}$ **5.** $r = 8\sin\theta$

6. $\left\{x \mid x \ne \frac{3\pi}{4} \pm \pi k, k \text{ is an integer}\right\}$ **7.** $\{22.5°\}$ **8.** $y = \frac{x^2}{5} + 5$

CHAPTER 7 Exponential and Logarithmic Functions

7.1 Assess Your Understanding *(page 471)*

8. exponential function; growth factor; initial value **9.** a **10.** T **11.** T **12.** $\left(-1, \dfrac{1}{a}\right)$; $(0,1)$; $(1,a)$ **13.** 4 **14.** F **15.** b **16.** c

17. (a) 8.815 **(b)** 8.821 **(c)** 8.824 **(d)** 8.825 **19. (a)** 21.217 **(b)** 22.217 **(c)** 22.440 **(d)** 22.459 **21.** 1.265 **23.** 0.347 **25.** 3.320 **27.** 149.952
29. Neither **31.** Exponential; $H(x) = 4^x$ **33.** Exponential; $f(x) = 3 \cdot 2^x$ **35.** Linear; $H(x) = 2x + 4$ **37.** B **39.** D **41.** A **43.** E

45.
Domain: All real numbers
Range: $\{y | y > 1\}$ or $(1, \infty)$
Horizontal asymptote: $y = 1$
y-intercept: 2

47.
Domain: All real numbers
Range: $\{y | y > 0\}$ or $(0, \infty)$
Horizontal asymptote: $y = 0$
y-intercept: $\dfrac{1}{3}$

49.
Domain: All real numbers
Range: $\{y | y > 0\}$ or $(0, \infty)$
Horizontal asymptote: $y = 0$
y-intercept: 3

51.
Domain: All real numbers
Range: $\{y | y > -2\}$ or $(-2, \infty)$
Horizontal asymptote: $y = -2$
y-intercept: -1

53.
Domain: All real numbers
Range: $\{y | y > 2\}$ or $(2, \infty)$
Horizontal asymptote: $y = 2$
y-intercept: $\dfrac{9}{4}$

55.
Domain: All real numbers
Range: $\{y | y > 2\}$ or $(2, \infty)$
Horizontal asymptote: $y = 2$
y-intercept: 3

57.
Domain: All real numbers
Range: $\{y | y > 0\}$ or $(0, \infty)$
Horizontal asymptote: $y = 0$
y-intercept: 1

59.
Domain: All real numbers
Range: $\{y | y > 0\}$ or $(0, \infty)$
Horizontal asymptote: $y = 0$
y-intercept: e^2

61.
Domain: All real numbers
Range: $\{y | y < 5\}$ or $(-\infty, 5)$
Horizontal asymptote: $y = 5$
y-intercept: 4

63.
Domain: All real numbers
Range: $\{y | y < 2\}$ or $(-\infty, 2)$
Horizontal asymptote: $y = 2$
y-intercept: 1

65. $\{5\}$ **67.** $\{-4\}$ **69.** $\{2\}$ **71.** $\left\{\dfrac{7}{2}\right\}$ **73.** $\{-\sqrt{2}, 0, \sqrt{2}\}$
75. $\{3\}$ **77.** $\{-1, 7\}$ **79.** $\{-4, 2\}$ **81.** $\{-4\}$ **83.** $\{1, 2\}$ **85.** $\dfrac{1}{49}$
87. $\dfrac{1}{4}$ **89.** 5 **91.** $f(x) = 3^x$ **93.** $f(x) = -6^x$ **95.** $f(x) = 3^x + 2$
97. (a) 16; $(4, 16)$ **(b)** -4; $\left(-4, \dfrac{1}{16}\right)$ **99. (a)** $\dfrac{9}{4}$; $\left(-1, \dfrac{9}{4}\right)$ **(b)** 3; $(3, 66)$
101. (a) 60; $(-6, 60)$ **(b)** -4; $(-4, 12)$ **(c)** -2

103.
Domain: $(-\infty, \infty)$
Range: $[1, \infty)$
Intercept: $(0, 1)$

105.
Domain: $(-\infty, \infty)$
Range: $[-1, 0)$
Intercept: $(0, -1)$

107. (a) 74% **(b)** 47% **(c)** Each pane allows only 97% of light to pass through.
109. (a) \$16,231 **(b)** \$8626 **(c)** As each year passes, the sedan is worth 90% of its value the previous year. **111. (a)** 30% **(b)** 9% **(c)** Each year only 30% of the previous survivors survive again. **113.** 3.35 mg; 0.45 mg
115. (a) 0.632 **(b)** 0.982 **(c)** 1 **(d)**   **(e)** About 7 min

117. (a) 0.0516 **(b)** 0.0888 **119. (a)** 70.95% **(b)** 72.62% **(c)** 100% **125.**
121. (a) 5.41 amp, 7.59 amp, 10.38 amp **(b)** 12 amp
(d) 3.34 amp, 5.31 amp, 9.44 amp
(e) 24 amp
(c), (f)

Final Denominator	Value of Expression	Compare Value to $e \approx 2.718281828$
$1 + 1$	2.5	$2.5 < e$
$2 + 2$	2.8	$2.8 > e$
$3 + 3$	2.7	$2.7 < e$
$4 + 4$	2.721649485	$2.721649485 > e$
$5 + 5$	2.717770035	$2.717770035 < e$
$6 + 6$	2.718348855	$2.718348855 > e$

123. 36

127. $f(A + B) = a^{A+B} = a^A \cdot a^B = f(A) \cdot f(B)$ **129.** $f(\alpha x) = a^{\alpha x} = (a^x)^\alpha = [f(x)]^\alpha$

131. (a) $f(-x) = \dfrac{1}{2}(e^{-x} + e^{-(-x)})$

$= \dfrac{1}{2}(e^{-x} + e^{x})$

$= \dfrac{1}{2}(e^{x} + e^{-x})$

$= f(x)$

(b)

(c) $(\cosh x)^2 - (\sinh x)^2$

$= \left[\dfrac{1}{2}(e^x + e^{-x})\right]^2 - \left[\dfrac{1}{2}(e^x - e^{-x})\right]^2$

$= \dfrac{1}{4}[e^{2x} + 2 + e^{-2x} - e^{2x} + 2 - e^{-2x}]$

$= \dfrac{1}{4} \cdot 4 = 1$

133. $\{1, 2\}$ **135.** 59 minutes   **139.** $a^{-x} = (a^{-1})^x = \left(\dfrac{1}{a}\right)^x$ **140.** $(-\infty, -5] \cup [-2, 2]$ **141.** $(2, \infty)$ **142.** $f(x) = -2x^2 + 12x - 13$

143. (a)

(b) Domain: $(-\infty, \infty)$; Range: $[-4, \infty)$
(c) Decreasing: $(-\infty, -1]$; Increasing: $[-1, \infty)$

144. $\left\{\dfrac{3}{2}\right\}$ **145.** $x^2 + y^2 = 1$ **146.** $\{16, 144\}$

147. \$1050   **148.** $-3 \le x \le 3$ **149.** $4x + 2h - 7$

7.2 Assess Your Understanding *(page 485)*

3. $\{x \mid x > 0\}$ or $(0, \infty)$ **4.** $\left(\dfrac{1}{a}, -1\right), (1, 0), (a, 1)$ **5.** 1   **6.** F   **7.** T   **8.** a   **9.** c   **10.** b   **11.** $2 = \log_3 9$ **13.** $2 = \log_a 1.6$ **15.** $x = \log_2 7.2$

17. $x = \ln 8$ **19.** $2^3 = 8$ **21.** $a^6 = 3$ **23.** $3^x = 2$ **25.** $e^x = 4$ **27.** 0   **29.** 2   **31.** -3 **33.** $\dfrac{1}{2}$ **35.** 4   **37.** $\dfrac{1}{2}$ **39.** $\{x \mid x > 3\}$; $(3, \infty)$

41. All real numbers except 0; $\{x \mid x \ne 0\}$; $(-\infty, 0) \cup (0, \infty)$ **43.** $\{x \mid x > 10\}$; $(10, \infty)$ **45.** $\{x \mid x > -1\}$; $(-1, \infty)$
47. $\{x \mid x < -1 \text{ or } x > 0\}$; $(-\infty, -1) \cup (0, \infty)$ **49.** $\{x \mid x \ge 1\}$; $[1, \infty)$ **51.** 0.511   **53.** 30.099   **55.** 2.303   **57.** -53.991 **59.** $\sqrt{2}$

61.

63.

65. B   **67.** D   **69.** A   **71.** E

73. (a) Domain: $(-4, \infty)$
(b)

(c) Range: $(-\infty, \infty)$
  Vertical asymptote: $x = -4$
(d) $f^{-1}(x) = e^x - 4$
(e) Domain of f^{-1}: $(-\infty, \infty)$
  Range of f^{-1}: $(-4, \infty)$
(f)

75. (a) Domain: $(0, \infty)$
(b)

(c) Range: $(-\infty, \infty)$
  Vertical asymptote: $x = 0$
(d) $f^{-1}(x) = e^{x-2}$
(e) Domain of f^{-1}: $(-\infty, \infty)$
  Range of f^{-1}: $(0, \infty)$
(f)

77. (a) Domain: $(0, \infty)$
(b)

(c) Range: $(-\infty, \infty)$
  Vertical asymptote: $x = 0$
(d) $f^{-1}(x) = \dfrac{1}{2}e^{x+3}$
(e) Domain of f^{-1}: $(-\infty, \infty)$
  Range of f^{-1}: $(0, \infty)$
(f)

79. (a) Domain: $(4, \infty)$
(b)

(c) Range: $(-\infty, \infty)$
  Vertical asymptote: $x = 4$
(d) $f^{-1}(x) = 10^{x-2} + 4$
(e) Domain of f^{-1}: $(-\infty, \infty)$
  Range of f^{-1}: $(4, \infty)$
(f)

81. (a) Domain: $(0, \infty)$
(b)

(c) Range: $(-\infty, \infty)$
  Vertical asymptote: $x = 0$
(d) $f^{-1}(x) = \dfrac{1}{2} \cdot 10^{2x}$
(e) Domain of f^{-1}: $(-\infty, \infty)$
  Range of f^{-1}: $(0, \infty)$
(f)

83. (a) Domain: $(-2, \infty)$
(b)

(c) Range: $(-\infty, \infty)$
  Vertical asymptote: $x = -2$
(d) $f^{-1}(x) = 3^{x-3} - 2$
(e) Domain of f^{-1}: $(-\infty, \infty)$
  Range of f^{-1}: $(-2, \infty)$
(f)

85. (a) Domain: $(-\infty, \infty)$
(b)

(c) Range: $(-3, \infty)$
Horizontal asymptote: $y = -3$
(d) $f^{-1}(x) = \ln(x + 3) - 2$
(e) Domain of f^{-1}: $(-3, \infty)$
Range of f^{-1}: $(-\infty, \infty)$
(f)

87. (a) Domain: $(-\infty, \infty)$
(b)

(c) Range: $(4, \infty)$
Horizontal asymptote: $y = 4$
(d) $f^{-1}(x) = 3\log_2(x - 4)$
(e) Domain of f^{-1}: $(4, \infty)$
Range of f^{-1}: $(-\infty, \infty)$
(f)

89. $\{9\}$ **91.** $\left\{\dfrac{28}{3}\right\}$ **93.** $\{4\}$ **95.** $\{5\}$

97. $\{3\}$ **99.** $\{2\}$ **101.** $\left\{\dfrac{\ln 10}{3}\right\}$

103. $\left\{\dfrac{\ln 8 - 5}{2}\right\}$ **105.** $\{-3\sqrt{5}, 3\sqrt{5}\}$

107. $\{-2\}$ **109.** $\left\{5\ln\dfrac{7}{5}\right\}$ **111.** $\left\{2 - \log\dfrac{5}{2}\right\}$

113. (a) $\left\{x \middle| x > -\dfrac{1}{2}\right\}; \left(-\dfrac{1}{2}, \infty\right)$
(b) $2; (40, 2)$ **(c)** $121; (121, 3)$ **(d)** 4

115.

Domain: $\{x | x \neq 0\}$
Range: $(-\infty, \infty)$
Intercepts: $(-1, 0), (1, 0)$

117.

Domain: $\{x | x > 0\}$
Range: $\{y | y \geq 0\}$
Intercept: $(1, 0)$

119. (a) 1   **(b)** 2   **(c)** 3
(d) It increases.   **(e)** 0.000316
(f) 3.981×10^{-8}

121. (a) 5.97 km   **(b)** 0.90 km

123. (a) 6.93 min   **(b)** 16.09 min

125. $h \approx 2.29$, so the time between
injections is about 2 h, 17 min.

127. 0.2695 s
0.8959 s

129. 50 dB   **131.** 90 dB   **133.** 8.1   **135. (a)** $k \approx 11.216$ **(b)** 6.73   **(c)** 0.41%   **(d)** 0.14%   **137.** $\{64\}$ **139.** $\{-1, 5\}$

141. Because $y = \log_1 x$ means $1^y = 1 = x$, which cannot be true for $x \neq 1$ **143.** Zeros: $-3, -\dfrac{1}{2}, \dfrac{1}{2}, 3$; x-intercepts: $-3, -\dfrac{1}{2}, \dfrac{1}{2}, 3$

144. 12   **145.** $f(1) = -5; f(2) = 17$ **146.** $3 + i; f(x) = x^4 - 7x^3 + 14x^2 + 2x - 20; a = 1$ **147.** $\left\{\dfrac{7 - \sqrt{57}}{4}, \dfrac{7 + \sqrt{57}}{4}\right\}$

148. $5x + 8y = 8$ **149.** $\{-31, 14\}$ **150.** 151.1 cm   **151.** $x^2 + 2x + 4$ **152.** $(x + 5)^3(x - 3)^6(11x + 23)$

7.3 Assess Your Understanding *(page 496)*

1. 0   **2.** M **3.** r **4.** $\log_a M + \log_a N$ **5.** $\log_a M - \log_a N$ **6.** $r\log_a M$ **7.** 7   **8.** F   **9.** F   **10.** F   **11.** b   **12.** b   **13.** 29   **15.** -4 **17.** 13   **19.** 1   **21.** 1

23. 3   **25.** $\dfrac{6}{5}$ **27.** 4   **29.** $a + b$ **31.** $b - a$ **33.** $3a$ **35.** $\dfrac{1}{5}(a + b)$ **37.** $2 + \log_6 x$ **39.** $6\log_5 y$ **41.** $1 + \ln x$ **43.** $\ln x - x$ **45.** $2\log_a u + 3\log_a v$

47. $2\ln x + \dfrac{1}{2}\ln(1 - x)$ **49.** $3\log_2 x - \log_2(x - 3)$ **51.** $\log x + \log(x + 2) - 2\log(x + 3)$ **53.** $\dfrac{1}{3}\ln(x - 2) + \dfrac{1}{3}\ln(x + 1) - \dfrac{2}{3}\ln(x + 4)$

55. $\ln 5 + \ln x + \dfrac{1}{2}\ln(1 + 3x) - 3\ln(x - 4)$ **57.** $\log_5 u^3 v^4$ **59.** $\log_3\left(\dfrac{1}{x^{5/2}}\right)$ **61.** $\log_4\left[\dfrac{x - 1}{(x + 1)^4}\right]$ **63.** $-2\ln(x - 1)$ **65.** $\log_2[x(3x - 2)^4]$

67. $\log_a\left(\dfrac{25x^6}{\sqrt{2x + 3}}\right)$ **69.** $\log_2\left[\dfrac{(x + 1)^2}{(x + 3)(x - 1)}\right]$ **71.** 2.771   **73.** -3.880 **75.** 5.615   **77.** 0.874

79. $y = \dfrac{\log x}{\log 4}$

81. $y = \dfrac{\log(x + 2)}{\log 2}$

83. $y = \dfrac{\log(x + 1)}{\log(x - 1)}$

85. (a) $(f \circ g)(x) = x$; $\{x | x \text{ is any real number}\}$ or $(-\infty, \infty)$
(b) $(g \circ f)(x) = x$; $\{x | x > 0\}$ or $(0, \infty)$ **(c)** 5
(d) $(f \circ h)(x) = \ln x^2$; $\{x | x \neq 0\}$ or $(-\infty, 0) \cup (0, \infty)$ **(e)** 2

87. $y = Cx$ **89.** $y = Cx(x + 1)$ **91.** $y = Ce^{3x}$ **93.** $y = Ce^{-4x} + 3$

95. $y = \dfrac{\sqrt[3]{C}(2x + 1)^{1/6}}{(x + 4)^{1/9}}$ **97.** 3   **99.** 1

101. $\log_a(x + \sqrt{x^2 - 1}) + \log_a(x - \sqrt{x^2 - 1}) = \log_a[(x + \sqrt{x^2 - 1})(x - \sqrt{x^2 - 1})] = \log_a[x^2 - (x^2 - 1)] = \log_a 1 = 0$

103. $\ln(1 + e^{2x}) = \ln[e^{2x}(e^{-2x} + 1)] = \ln e^{2x} + \ln(e^{-2x} + 1) = 2x + \ln(1 + e^{-2x})$

105. $y = f(x) = \log_a x; a^y = x$ implies $a^y = \left(\dfrac{1}{a}\right)^{-y} = x$, so $-y = \log_{1/a} x = -f(x)$.

107. $f(x) = \log_a x; f\left(\dfrac{1}{x}\right) = \log_a\dfrac{1}{x} = \log_a 1 - \log_a x = -f(x)$ **109.** $\log_a\dfrac{M}{N} = \log_a(M \cdot N^{-1}) = \log_a M + \log_a N^{-1} = \log_a M - \log_a N.$

111. $\log_a b = \dfrac{\log_b b}{\log_b a} = \dfrac{1}{\log_b a}$ **113.** $\log_{a^n} b^m = \dfrac{\log_a b^m}{\log_a a^n} = \dfrac{m\log_a b}{n} = \dfrac{m}{n}\log_a b$

119. $\{-1.78, 1.29, 3.49\}$ **120.** A repeated real solution (double root)   **121.** $-2, \dfrac{1}{5}, \dfrac{-5-\sqrt{21}}{2}, \dfrac{-5+\sqrt{21}}{2}$

122.

Domain: $\{x \mid x \le 2\}$ or $(-\infty, 2]$
Range: $\{y \mid y \ge 0\}$ or $[0, \infty)$

123. $\left(-\infty, \dfrac{3}{5}\right]$ **124.** $\{x \mid -9 < x < 7\}$ or $(-9, 7)$ **125.** Vertex: $(4, 13)$; concave down

126. Center: $(5, -2)$; radius: 8   **127.** 7   **128.** Odd

7.4 Assess Your Understanding *(page 503)*

5. $\{16\}$ **7.** $\left\{\dfrac{16}{5}\right\}$ **9.** $\{11\}$ **11.** $\{-64, 64\}$ **13.** $\{-6, 7\}$ **15.** $\{64\}$ **17.** $\left\{\dfrac{1}{3}\right\}$ **19.** $\{7\}$ **21.** $\{5\}$ **23.** $\left\{\dfrac{16}{3}\right\}$ **25.** $\{-6\}$ **27.** $\{-2\}$

29. $\{-1 + \sqrt{1 + e^4}\}$ **31.** $\left\{\dfrac{-15 + 5\sqrt{13}}{2}\right\}$ **33.** $\{2\}$ **35.** $\left\{\dfrac{9}{2}\right\}$ **37.** $\{7\}$ **39.** $\{-2 + 4\sqrt{2}\}$ **41.** $\{-\sqrt{3}, \sqrt{3}\}$

43. $\left\{\dfrac{1}{9}, 243\right\}$ **45.** $\{8\}$ **47.** $\{\log_2 10\} = \left\{\dfrac{1}{\log 2}\right\}$ **49.** $\{-\log_8 1.2\} = \left\{-\dfrac{\ln 1.2}{\ln 8}\right\}$ **51.** $\left\{\dfrac{1}{3}\log_2\dfrac{8}{5}\right\} = \left\{\dfrac{\ln\dfrac{8}{5}}{3\ln 2}\right\}$ **53.** $\left\{\dfrac{\ln 3}{2\ln 3 + \ln 4}\right\}$

55. $\left\{\dfrac{\ln 7}{\ln 0.6 + \ln 7}\right\}$ **57.** $\{0\}$ **59.** $\left\{\dfrac{\ln \pi}{1 + \ln \pi}\right\}$ **61.** $\left\{\dfrac{\ln 3}{\ln 2}\right\}$ **63.** $\{0\}$ **65.** $\left\{\log_4(-2 + \sqrt{7})\right\}$ **67.** $\{\log_5 4\}$ **69.** No real solution

71. $\{\log_4 5\}$ **73.** $\{2.79\}$ **75.** $\{-0.57\}$ **77.** $\{-0.70\}$ **79.** $\{0.57\}$ **81.** $\{0.39, 1.00\}$ **83.** $\{1.32\}$ **85.** $\{1.31\}$

87. (a) $\{5\}$; $(5, 3)$ **(b)** $\{5\}$; $(5, 4)$ **(c)** $\{1\}$; yes, at $(1, 2)$ **(d)** $\{5\}$ **(e)** $\left\{-\dfrac{1}{11}\right\}$

89. (a), (b)   **91. (a), (b), (c)**   **93. (a), (b), (c)**   **95. (a)**

(c) $\{x \mid x > 0.710\}$ or $(0.710, \infty)$

(b) 2   **(c)** $\{x \mid x < 2\}$ or $(-\infty, 2)$

97. (a) 2052   **(b)** 2070   **99. (a)** After 2.4 yr   **(b)** After 3.8 yr   **(c)** After 9.4 yr   **101.** $\{1\}$ **103.** $\left\{\dfrac{1}{4}, 4\right\}$ **105.** $\{3^{\log 3}\}$ **107.** $\left\{-3, \dfrac{1}{4}, 2\right\}$

108. one-to-one   **109.** $\dfrac{x + 5}{-x + 11}$ **110.** $\{x \mid x \ge 1\}$ or $[1, \infty)$ **111.** $\{9\}$ **112.** 6   **113.** $\dfrac{x^2 + 7x - 10}{(x - 2)(x + 2)}$ **114.** $4\sqrt{10}$ **115.** $\dfrac{1}{6}$ **116.** $\dfrac{1}{\sqrt{x + 6} + \sqrt{x}}$

7.5 Assess Your Understanding *(page 512)*

1. $15   **2.** $13\dfrac{1}{3}\%$ **3.** principal   **4.** I; Prt; simple interest   **5.** effective rate of interest   **6.** a   **7.** $108.29   **9.** $969.56   **11.** $1394.19   **13.** $1246.08

15. $88.72   **17.** $1444.79   **19.** $713.53   **21.** $102.00   **23.** 5.095%   **25.** 4.081%   **27.** $6\dfrac{1}{4}\%$ compounded annually   **29.** 9% compounded monthly

31. 25.992%   **33.** 24.573%   **35. (a)** About 8.69 yr   **(b)** About 8.66 yr   **37.** 6.823%   **39.** 10.15 yr; 10.14 yr   **41.** 15.27 yr or 15 yr, 3 mo   **43.** $104,335
45. $12,910.62   **47.** About $30.17 per share or $3017   **49.** Not quite. Jim will have $1057.60. The second bank gives a better deal, since Jim will have
$1060.62 after 1 yr.   **51.** Will has $11,632.73; Henry has $10,947.89.   **53. (a)** $70,473   **(b)** $49,167   **55.** 64.9 yr   **57.** $940.90   **59.** 2.53%
61. 34.31 yr   **63. (a)** $3686.45   **(b)** $3678.79   **65.** $6439.28

67. (a) 11.90 yr   **(b)** 22.11 yr   **(c)** $mP = P\left(1 + \dfrac{r}{n}\right)^{nt}$

$$m = \left(1 + \dfrac{r}{n}\right)^{nt}$$

$$\ln m = \ln\left(1 + \dfrac{r}{n}\right)^{nt} = nt\ln\left(1 + \dfrac{r}{n}\right)$$

$$t = \dfrac{\ln m}{n\ln\left(1 + \dfrac{r}{n}\right)}$$

69. (a) 1.23%   **(b)** In 2034 or after 22 yr   **71.** 22.7 yr   **76.** $R = 0$; yes

77. $f^{-1}(x) = \dfrac{2x}{x - 1}$ **78.** $-2, 5$; $f(x) = (x + 2)^2(x - 5)(x^2 + 1)$ **79.** $\{6\}$

80. $2x(x + 3)(x - 5)(x + 5)$ **81.** $(f \circ g)(x) = 45x^2 - 18x - 7$
82. Domain: $(-\infty, \infty)$; Range: $(-\infty, 9]$
83. Vertical: $x = 7$; horizontal: none; oblique: $y = 2x + 9$
84. $y = 4x - 18$ **85.** 3

7.6 Assess Your Understanding *(page 524)*

1. (a) 500 insects   **(b)** $0.02 = 2\%$ per day   **(c)** About 611 insects   **(d)** After about 23.5 days   **(e)** After about 34.7 days
3. (a) $-0.0244 = -2.44\%$ per year   **(b)** About 391.7 g   **(c)** After about 9.1 yr   **(d)** 28.4 yr   **5. (a)** $N(t) = N_0 e^{kt}$ **(b)** 5832   **(c)** 3.9 days
7. (a) $N(t) = N_0 e^{kt}$ **(b)** 25,198   **9.** 9.797 g   **11.** 9953 yr ago   **13. (a)** 5:18 PM   **(b)** About 14.3 min   **(c)** The temperature of the pizza approaches 70°F.

15. $18.63°C; 25.1°C$ **17.** 1.7 ppm; 7.17 days, or 172 hr **19.** 0.26 M; 6.58 hr, or 395 min **21.** 26.6 days

23. (a) In 1984, 91.8% of households did not own a personal computer.

25. (a)

27. (a) $P(0) \approx 48$; In 1900, about 48 invasive species were present in the Great Lakes. **(b)** 1.7%

(b)

(c)

(b) 0.78, or 78%
(c) 50 people
(d) As n increases, the probability decreases.

(c) 70.6% **(d)** During 2011

(d) About 176 **(e)** During 1999

29. (a) $P(t) = 50(3)^{t/20}$ **(b)** 661 **(c)** In 48 days **(d)** $P(t) = 50e^{0.055t}$

39.

40. $\dfrac{40x + 45}{3(2x+3)^{2/3}}$

31. $f(x) = -\dfrac{3}{2}x + 7$ **32.** Neither **33.** $2\ln x + \dfrac{1}{2}\ln y - \ln z$

34. $\{x \mid x \neq -4, x \neq 2\}$ **35.** $(g-f)(x) = \dfrac{x^2 - 2x - 13}{(x-3)(x-4)}$

36. x-intercepts: $\dfrac{5 - \sqrt{17}}{4}, \dfrac{5 + \sqrt{17}}{4}$; y-intercept: 1 **37.** $\left\{ -\dfrac{\sqrt{2}}{2}, \dfrac{\sqrt{2}}{2} \right\}$

38. $y = -1.0714x + 3.9048; r = -0.985$

Local minima: $f(-1.37) = -5.85$ and $f(1) = -1$
Local maximum: $f(0.37) = -0.65$
Increasing: $[-1.37, 0.37] \cup [1, 3]$
Decreasing: $[-3, -1.37] \cup [0.37, 1]$

7.7 Assess Your Understanding (page 531)

1. (a)   **(d)**   **3. (a)**   **(d)**

(b) $y = 0.0903(1.3384)^x$ **(e)** 0.69
(c) $N(t) = 0.0903e^{0.2915t}$ **(f)** After about 7.26 hr

(b) $y = 118.7226(0.7013)^x$ **(e)** 28.7%
(c) $A(t) = 118.7226e^{-0.3548t}$ **(f)** $k = -0.3548 = -35.48\%$ is the exponential growth rate. It represents the rate at which the percentage of patients surviving advanced-stage breast cancer is decreasing.

5. (a)   **(c)**

(b) $y = 344.5217 - 37.2566 \ln x$ **(d)** 188 billion pounds
(e) Under by 2 billion pounds

7. (a)   **(c)**

(b) $y = \dfrac{762{,}176{,}844.4}{1 + 8.7428e^{-0.0162x}}$ **(d)** $762{,}176{,}844$
(e) Approximately $315{,}203{,}288$ **(f)** 2023

9. (a)   **(d)**   **10.** $f(x) = \dfrac{1}{3}(x+3)(x+1)^2(x-2)$ **11.** $\sqrt{3}$ **12.**

(b) Exponential **(e)** 5.1%
(c) $y = 115.5779(0.9012)^x$

13. $(-3, -2)$ **14.** $f(x) = -\sqrt{x - 4}$ **15.** $\left\{ \dfrac{2 - \sqrt{19}}{3}, \dfrac{2 + \sqrt{19}}{3} \right\}$ **16.** $\{x \mid -5 < x \leq -1 \text{ or } x > 5\}$ or $(-5, -1] \cup (5, \infty)$

17. $R = f(-3) = 0$; yes **18.** 6 **19.** The function f is a polynomial so it is a continuous function, and $f(-2) = -21; f(-1) = 3$; zero: -1.32

Review Exercises (page 535)

1. (a) 81 **(b)** 2 **(c)** $\dfrac{1}{9}$ **(d)** -3 **2.** $\log_5 z = 2$ **3.** $5^{13} = u$ **4.** $\left\{ x \mid x > \dfrac{2}{3} \right\}; \left(\dfrac{2}{3}, \infty \right)$ **5.** $\{x \mid x < 1 \text{ or } x > 2\}; (-\infty, 1) \cup (2, \infty)$

6. -3 **7.** $\sqrt{2}$ **8.** 0.4 **9.** $\log_3 u + 2\log_3 v - \log_3 w$ **10.** $8\log_2 a + 2\log_2 b$ **11.** $2\log x + \dfrac{1}{2}\log(x^3 + 1)$

12. $2\ln(2x+3) - 2\ln(x-1) - 2\ln(x-2)$ **13.** $\dfrac{25}{4}\log_4 x$ **14.** $-2\ln(x+1)$ **15.** $\ln\left[\dfrac{16\sqrt{x^2 + 1}}{\sqrt{x}(x-4)} \right]$ **16.** 2.124

17.

18. (a) Domain of f: $(-\infty, \infty)$
(b)
(c) Range of f: $(0, \infty)$
Horizontal asymptote: $y = 0$
(d) $f^{-1}(x) = 3 + \log_2 x$
(e) Domain of f^{-1}: $(0, \infty)$
Range of f^{-1}: $(-\infty, \infty)$
(f)

19. (a) Domain of f: $(-\infty, \infty)$
(b)
(c) Range of f: $(1, \infty)$
Horizontal asymptote: $y = 1$
(d) $f^{-1}(x) = -\log_3(x - 1)$
(e) Domain of f^{-1}: $(1, \infty)$
Range of f^{-1}: $(-\infty, \infty)$
(f)

20. (a) Domain of f: $(-\infty, \infty)$
(b)
(c) Range of f: $(0, \infty)$
Horizontal asymptote: $y = 0$
(d) $f^{-1}(x) = 2 + \ln\left(\dfrac{x}{3}\right)$
(e) Domain of f^{-1}: $(0, \infty)$
Range of f^{-1}: $(-\infty, \infty)$
(f)

21. (a) Domain of f: $(-3, \infty)$
(b)
(c) Range of f: $(-\infty, \infty)$
Vertical asymptote: $x = -3$
(d) $f^{-1}(x) = e^{2x} - 3$
(e) Domain of f^{-1}: $(-\infty, \infty)$
Range of f^{-1}: $(-3, \infty)$
(f)

22. $\left\{-\dfrac{16}{9}\right\}$ **23.** $\left\{\dfrac{-1 - \sqrt{3}}{2}, \dfrac{-1 + \sqrt{3}}{2}\right\}$ **24.** $\left\{\dfrac{1}{4}\right\}$ **25.** $\left\{\dfrac{2\ln 3}{\ln 5 - \ln 3}\right\}$ **26.** $\{-2, 6\}$ **27.** $\{83\}$ **28.** $\left\{\dfrac{1}{2}, -3\right\}$

29. $\{1\}$ **30.** $\{-1\}$ **31.** $\{1 - \ln 5\}$ **32.** $\left\{\log_3(-2 + \sqrt{7})\right\} = \left\{\dfrac{\ln(-2 + \sqrt{7})}{\ln 3}\right\}$

33. (a), (e)

(b) $3; (6, 3)$
(c) $10; (10, 4)$
(d) $\left\{x \,\middle|\, x > \dfrac{5}{2}\right\}$ or $\left(\dfrac{5}{2}, \infty\right)$
(e) $f^{-1}(x) = 2^{x-1} + 2$

34. (a) 37.3 W **(b)** 6.9 dB **35. (a)** 11.77 **(b)** 9.56 in.
36. (a) 9.85 yr **(b)** 4.27 yr **37.** \$20,398.87; 4.04%; 17.5 yr
38. \$41,668.97 **39.** 24,765 yr ago **40.** 55.22 min, or 55 min, 13 sec
41. 8,153,581,530 **42.** 7.204 g; 0.519 g

43. (a) 0.3 **(b)** 0.8
(c)
(d) In 2026

44. (a)
(b) $y = 3610.2684(1.0406)^x$
(c) $A(t) = 3610.2684e^{0.0398t}$
(d)
(e) 2027–28

45. (a)
(b) $y = 18.921 - 7.096 \ln x$
(c)
(d) Approximately $-3°F$

46. (a)
(b) $C = \dfrac{46.9292}{1 + 21.2733e^{-0.7306t}}$
(c)
(d) About 47 people; 50 people
(e) 2.4 days; during the tenth hour of day 3
(f) 9.5 days

Chapter Test *(page 538)*

1. $x = 5$ **2.** $b = 4$ **3.** $x = 625$ **4.** -2 **5.** 4 **6.** 125 **7.** 7

8. (a) Domain of f: $\{x \mid -\infty < x < \infty\}$ or $(-\infty, \infty)$

(b)

(c) Range of f: $\{y \mid y > -2\}$ or $(-2, \infty)$
Horizontal asymptote: $y = -2$
(d) $f^{-1}(x) = \log_4(x + 2) - 1$
(e) Domain of f^{-1}: $\{x \mid x > -2\}$ or $(-2, \infty)$
Range of f^{-1}: $\{y \mid -\infty < y < \infty\}$ or $(-\infty, \infty)$

(f)

9. (a) Domain of f: $\{x \mid x > 2\}$ or $(2, \infty)$

(b)

(c) Range of f: $\{y \mid -\infty < y < \infty\}$ or $(-\infty, \infty)$
Vertical asymptote: $x = 2$
(d) $f^{-1}(x) = 5^{1-x} + 2$
(e) Domain of f^{-1}: $\{x \mid -\infty < x < \infty\}$ or $(-\infty, \infty)$
Range of f^{-1}: $\{y \mid y > 2\}$ or $(2, \infty)$

(f)

10. $\{1\}$ **11.** $\{91\}$ **12.** $\{-\ln 2\}$ **13.** $\left\{ \dfrac{1 - \sqrt{13}}{2}, \dfrac{1 + \sqrt{13}}{2} \right\}$ **14.** $\left\{ \dfrac{3 \ln 7}{1 - \ln 7} \right\}$

15. $\{2\sqrt{6}\}$ **16.** $2 + 3 \log_2 x - \log_2(x - 6) - \log_2(x + 3)$ **17.** About 250.39 days **18. (a)** \$1033.82 **(b)** \$963.42 **(c)** 11.9 yr

19. (a) About 83 dB **(b)** The pain threshold will be exceeded if 31,623 people shout at the same time.

Cumulative Review *(page 538)*

1. Yes; no **2. (a)** 10 **(b)** $2x^2 + 3x + 1$ **(c)** $2x^2 + 4xh + 2h^2 - 3x - 3h + 1$ **3.** $\left(\dfrac{1}{2}, \dfrac{\sqrt{3}}{2} \right)$ is on the graph. **4.** $\{-26\}$

5.

6.

7. (a), (c)

Domain g = range g^{-1} = $(-\infty, \infty)$
Range g = domain g^{-1} = $(2, \infty)$
(b) $g^{-1}(x) = \log_3(x - 2)$

8. $\left\{ -\dfrac{3}{2} \right\}$ **9.** $\{2\}$
10. (a) $\{-1\}$ **(b)** $\{x \mid x > -1\}$ or $(-1, \infty)$
(c) $\{25\}$
11. (a)

(b) Logarithmic; $y = 49.293 - 10.563 \ln x$
(c) Highest value of $|r|$

APPENDIX A Review

A.1 Assess Your Understanding *(page A10)*

1. variable **2.** origin **3.** strict **4.** base; exponent or power **5.** T **6.** F **7.** T **8.** d **9.** b **10.** a **11.** $\{1, 2, 3, 4, 5, 6, 7, 8, 9\}$
13. $\{4\}$ **15.** $\{1, 3, 4, 6\}$ **17.** $\{0, 2, 6, 7, 8\}$ **19.** $\{0, 1, 2, 3, 5, 6, 7, 8, 9\}$ **21.** $\{0, 1, 2, 3, 5, 6, 7, 8, 9\}$
23.

  25. $>$ **27.** $>$ **29.** $>$ **31.** $=$ **33.** $<$ **35.** $x > 0$ **37.** $x < 2$ **39.** $x \leq 1$

41.   **43.**   **45.** 1 **47.** 2 **49.** 6 **51.** 4 **53.** -28 **55.** $\dfrac{4}{5}$ **57.** 0 **59.** 1 **61.** 5 **63.** 1

65. 22 **67.** 2 **69.** $x = 0$ **71.** $x = 3$ **73.** None **75.** $x = 0, x = 1, x = -1$ **77.** $\{x \mid x \neq 5\}$ **79.** $\{x \mid x \neq -4\}$ **81.** 0°C **83.** 25°C **85.** 16

87. $\dfrac{1}{16}$ **89.** $\dfrac{1}{9}$ **91.** 9 **93.** 5 **95.** 4 **97.** $64x^6$ **99.** $\dfrac{x^4}{y^2}$ **101.** $\dfrac{x}{y}$ **103.** $-\dfrac{8x^3 z}{9y}$ **105.** $\dfrac{16x^2}{9y^2}$ **107.** -4 **109.** 5 **111.** 4 **113.** 2 **115.** $\sqrt{5}$ **117.** $\dfrac{1}{2}$ **119.** 10; 0

121. 81 **123.** 304,006.671 **125.** 0.004 **127.** 481.890 **129.** 0.000 **131.** $A = lw$ **133.** $C = \pi d$ **135.** $A = \dfrac{\sqrt{3}}{4} x^2$ **137.** $V = \dfrac{4}{3} \pi r^3$ **139.** $V = x^3$

141. (a) \$6000 **(b)** \$8000 **143.** $|x - 4| \geq 6$ **145. (a)** $2 \leq 5$ **(b)** $6 > 5$ **147. (a)** Yes **(b)** No **149.** No; $\dfrac{1}{3}$ is larger; $0.000333\ldots$ **151.** No

A.2 Assess Your Understanding *(page A19)*

1. right; hypotenuse **2.** $A = \dfrac{1}{2} bh$ **3.** $C = 2\pi r$ **4.** similar **5.** c **6.** b **7.** T **8.** T **9.** F **10.** T **11.** T **12.** F **13.** 13 **15.** 26 **17.** 25
19. Right triangle; 5 **21.** Not a right triangle **23.** Right triangle; 25 **25.** Not a right triangle **27.** 42 in.² **29.** 28 in.² **31.** $A = 25\pi$ m²; $C = 10\pi$ m

33. $V = 240$ ft^3; $S = 236$ ft^2 **35.** $V = \dfrac{500}{3}\pi$ cm^3; $S = 100\pi$ cm^2 **37.** $V = 648\pi$ in.^{3}; $S = 306\pi$ in.^{2} **39.** π square units **41.** 2π square units

43. $x = 4$ units; $A = 90°$; $B = 60°$; $C = 30°$ **45.** $x = 67.5$ units; $A = 60°$; $B = 95°$; $C = 25°$ **47.** About 16.8 ft **49.** 64 ft^2

51. $24 + 2\pi \approx 30.28$ ft^2; $16 + 2\pi \approx 22.28$ ft **53.** 160 paces **55.** About 5.477 mi **57.** From 100 ft: 12.2 mi; From 150 ft: 15.0 mi

59. No. The area quadruples.

A.3 Assess Your Understanding (*page A25*)

1. $x^4 - 16$ **2.** $x^3 - 8$ **3.** T **4.** F **5.** add; $\dfrac{25}{4}$ **6.** b **7.** d **8.** c **9.** $x^2 + 6x + 8$ **11.** $2x^2 + 9x + 10$ **13.** $x^2 - 49$ **15.** $4x^2 - 9$

17. $x^2 + 8x + 16$ **19.** $4x^2 - 12x + 9$ **21.** $x^3 - 6x^2 + 12x - 8$ **23.** $8x^3 + 12x^2 + 6x + 1$ **25.** $(x + 6)(x - 6)$ **27.** $2(1 + 2x)(1 - 2x)$

29. $(x + 1)(x + 10)$ **31.** $(x - 7)(x - 3)$ **33.** $4(x^2 - 2x + 8)$ **35.** Prime **37.** $-(x - 5)(x + 3)$ **39.** $3(x + 2)(x - 6)$

41. $y^2(y + 5)(y + 6)$ **43.** $(2x + 3)^2$ **45.** $2(3x + 1)(x + 1)$ **47.** $(x - 3)(x + 3)(x^2 + 9)$ **49.** $(x - 1)^2(x^2 + x + 1)^2$

51. $x^5(x - 1)(x + 1)$ **53.** $(4x + 3)^2$ **55.** $-(4x - 5)(4x + 1)$ **57.** $(2y - 5)(2y - 3)$ **59.** $-(3x - 1)(3x + 1)(x^2 + 1)$

61. $(x + 3)(x - 6)$ **63.** $(x + 2)(x - 3)$ **65.** $(3x - 5)(9x^2 - 3x + 7)$ **67.** $(x + 5)(3x + 11)$ **69.** $(x - 1)(x + 1)(x + 2)$

71. $(x - 1)(x + 1)(x^2 - x + 1)$ **73.** $25; (x + 5)^2$ **75.** $9; (y - 3)^2$ **77.** $\dfrac{1}{16}; \left(x - \dfrac{1}{4}\right)^2$ **79.** $2(3x + 4)(9x + 13)$

81. $2x(3x + 5)$ **83.** $5(x + 3)(x - 2)^2(x + 1)$ **85.** $3(4x - 3)(4x - 1)$ **87.** $6(3x - 5)(2x + 1)^2(5x - 4)$

89. The possibilities are $(x \pm 1)(x \pm 4) = x^2 \pm 5x + 4$ or $(x \pm 2)(x \pm 2) = x^2 \pm 4x + 4$, none of which equals $x^2 + 4$.

A.4 Assess Your Understanding (*page A34*)

5. identity **6.** F **7.** T **8.** add; $\dfrac{25}{4}$ **9.** discriminant; negative **10.** F **11.** F **12.** b **13.** b **14.** d **15.** $\{7\}$ **17.** $\{-3\}$ **19.** $\{4\}$ **21.** $\left\{\dfrac{5}{4}\right\}$ **23.** $\{-1\}$

25. $\{-18\}$ **27.** $\{-3\}$ **29.** $\{-16\}$ **31.** $\{0.5\}$ **33.** $\{2\}$ **35.** $\{2\}$ **37.** $\{3\}$ **39.** $\{0, 9\}$ **41.** $\{0, 9\}$ **43.** $\{21\}$ **45.** $\{-2, 2\}$ **47.** $\{6\}$

49. $\{-3, 3\}$ **51.** $\{-4, 1\}$ **53.** $\left\{-1, \dfrac{3}{2}\right\}$ **55.** $\{-4, 4\}$ **57.** $\{2\}$ **59.** No real solution **61.** $\{-2, 2\}$ **63.** $\{-1, 3\}$ **65.** $\{-2, -1, 0, 1\}$ **67.** $\{0, 4\}$

69. $\{-6, 2\}$ **71.** $\left\{-\dfrac{1}{2}, 3\right\}$ **73.** $\{3, 4\}$ **75.** $\left\{\dfrac{3}{2}\right\}$ **77.** $\left\{-\dfrac{2}{3}, \dfrac{3}{2}\right\}$ **79.** $\left\{-\dfrac{3}{4}, 2\right\}$ **81.** $\{-5, 5\}$ **83.** $\{-1, 3\}$ **85.** $\{-3, 0\}$ **87.** $\{-7, 3\}$

89. $\left\{-\dfrac{1}{4}, \dfrac{3}{4}\right\}$ **91.** $\left\{\dfrac{-1 - \sqrt{7}}{6}, \dfrac{-1 + \sqrt{7}}{6}\right\}$ **93.** $\{2 - \sqrt{2}, 2 + \sqrt{2}\}$ **95.** $\left\{\dfrac{5 - \sqrt{29}}{2}, \dfrac{5 + \sqrt{29}}{2}\right\}$ **97.** $\left\{1, \dfrac{3}{2}\right\}$ **99.** No real solution

101. $\left\{\dfrac{-1 - \sqrt{5}}{4}, \dfrac{-1 + \sqrt{5}}{4}\right\}$ **103.** $\left\{\dfrac{-\sqrt{3} - \sqrt{15}}{2}, \dfrac{-\sqrt{3} + \sqrt{15}}{2}\right\}$ **105.** No real solution **107.** Repeated real solution

109. Two unequal real solutions **111.** $x = \dfrac{b + c}{a}$ **113.** $x = \dfrac{abc}{a + b}$ **115.** $x = a^2$ **117.** $R = \dfrac{R_1 R_2}{R_1 + R_2}$ **119.** $R = \dfrac{mv^2}{F}$ **121.** $r = \dfrac{S - a}{S}$

123. $\dfrac{-b + \sqrt{b^2 - 4ac}}{2a} + \dfrac{-b - \sqrt{b^2 - 4ac}}{2a} = \dfrac{-2b}{2a} = \dfrac{-b}{a}$ **125.** $k = -\dfrac{1}{2}$ or $\dfrac{1}{2}$

127. The solutions of $ax^2 - bx + c = 0$ are $\dfrac{b + \sqrt{b^2 - 4ac}}{2a}$ and $\dfrac{b - \sqrt{b^2 - 4ac}}{2a}$. **129.** b

A.5 Assess Your Understanding (*page A44*)

1. T **2.** 5 **3.** F **4.** real; imaginary; imaginary unit **5.** F **6.** T **7.** F **8.** b **9.** a **10.** c **11.** $8 + 5i$ **13.** $-7 + 6i$ **15.** $-6 - 11i$ **17.** $6 - 18i$

19. $18 - 21i$ **21.** $10 - 5i$ **23.** 26 **25.** $\dfrac{6}{5} + \dfrac{8}{5}i$ **27.** $1 - 2i$ **29.** $\dfrac{5}{2} - \dfrac{7}{2}i$ **31.** $-\dfrac{1}{2} + \dfrac{\sqrt{3}}{2}i$ **33.** $2i$ **35.** $-i$ **37.** 1 **39.** -6 **41.** $-10i$ **43.** $-2 + 2i$

45. 0 **47.** 0 **49.** $2i$ **51.** $5i$ **53.** $2\sqrt{3}i$ **55.** $10\sqrt{2}i$ **57.** $5i$ **59.** $\{-2i, 2i\}$ **61.** $\{-4, 4\}$ **63.** $\{3 - 2i, 3 + 2i\}$ **65.** $\{3 - i, 3 + i\}$

67. $\left\{\dfrac{1}{5} - \dfrac{1}{5}i, \dfrac{1}{5} + \dfrac{1}{5}i\right\}$ **69.** $\left\{\dfrac{1}{5} - \dfrac{2}{5}i, \dfrac{1}{5} + \dfrac{2}{5}i\right\}$ **71.** $\left\{-\dfrac{1}{2} - \dfrac{\sqrt{3}}{2}i, -\dfrac{1}{2} + \dfrac{\sqrt{3}}{2}i\right\}$ **73.** $\{4, -2 - 2\sqrt{3}i, -2 + 2\sqrt{3}i\}$ **75.** $\{-2, 2, -2i, 2i\}$

77. $\{-3i, -2i, 2i, 3i\}$ **79.** Two complex solutions that are conjugates of each other **81.** Two unequal real solutions **83.** A repeated real solution

85. $2 - 3i$ **87.** 6 **89.** 25 **91.** $2 + 3i$ ohms **93.** $z + \bar{z} = (a + bi) + (a - bi) = 2a; z - \bar{z} = (a + bi) - (a - bi) = 2bi$

95. $\overline{z + w} = \overline{(a + bi) + (c + di)} = \overline{(a + c) + (b + d)i} = (a + c) - (b + d)i = (a - bi) + (c - di) = \bar{z} + \bar{w}$

A.6 Assess Your Understanding (*page A53*)

3. closed interval **4.** Multiplication Properties **5.** T **6.** T **7.** T **8.** F **9.** T **10.** F **11.** d **12.** c **13.** $[0, 2]; 0 \le x \le 2$ **15.** $[2, \infty); x \ge 2$

17. $[0, 3); 0 \le x < 3$ **19.** (a) $6 < 8$ (b) $-2 < 0$ (c) $9 < 15$ (d) $-6 > -10$ **21.** (a) $7 > 0$ (b) $-1 > -8$ (c) $12 > -9$ (d) $-8 < 6$

23. (a) $2x + 4 < 5$ (b) $2x - 4 < -3$ (c) $6x + 3 < 6$ (d) $-4x - 2 > -4$

25. $[0, 4]$ **27.** $[4, 6)$ **29.** $[-3, \infty)$ **31.** $(-\infty, -4)$

33. $2 \le x \le 5$ **35.** $-4 < x \le 3$ **37.** $x \ge 4$ **39.** $x < -3$

41. $<$ **43.** $>$ **45.** $\ge$ **47.** $<$ **49.** $\le$ **51.** $>$ **53.** $\ge$ **55.** $0 < \dfrac{1}{x} < \dfrac{1}{5}$ **57.** $\dfrac{1}{x} < \dfrac{1}{-5} < 0$

59. $\{x \mid x < 4\}$ or $(-\infty, 4)$ **61.** $\{x \mid x \ge 2\}$ or $[2, \infty)$ **63.** $\{x \mid x > 3\}$ or $(3, \infty)$ **65.** $\{x \mid x \ge 2\}$ or $[2, \infty)$

67. $\{x|x > -7\}$ or $(-7, \infty)$

69. $\left\{x\left|x \le \dfrac{2}{3}\right.\right\}$ or $\left(-\infty, \dfrac{2}{3}\right]$

71. $\{x|x < -20\}$ or $(-\infty, -20)$

73. $\left\{x\left|x \ge \dfrac{4}{3}\right.\right\}$ or $\left[\dfrac{4}{3}, \infty\right)$

75. $\left\{x\left|\dfrac{7}{3} < x \le 4\right.\right\}$ or $\left(\dfrac{7}{3}, 4\right]$

77. $\left\{x\left|\dfrac{2}{3} \le x \le 3\right.\right\}$ or $\left[\dfrac{2}{3}, 3\right]$

79. $\left\{x\left|-\dfrac{11}{2} < x < \dfrac{1}{2}\right.\right\}$ or $\left(-\dfrac{11}{2}, \dfrac{1}{2}\right)$

81. $\{x|-6 < x < 0\}$ or $(-6, 0)$

83. $\{x|x < -5\}$ or $(-\infty, -5)$

85. $\{x|x \ge -1\}$ or $[-1, \infty)$

87. $\left\{x\left|\dfrac{1}{2} \le x < \dfrac{5}{4}\right.\right\}$ or $\left[\dfrac{1}{2}, \dfrac{5}{4}\right)$

89. $\left\{x\left|x < -\dfrac{1}{2}\right.\right\}$ or $\left(-\infty, -\dfrac{1}{2}\right)$

91. $\left\{x\left|\dfrac{3}{14} \le x < \dfrac{1}{4}\right.\right\}$ or $\left[\dfrac{3}{14}, \dfrac{1}{4}\right)$

93. $\left\{x\left|x > \dfrac{10}{3}\right.\right\}$ or $\left(\dfrac{10}{3}, \infty\right)$

95. $\{x|x > 3\}$ or $(3, \infty)$

97. $\{x|-4 < x < 4\}$; $(-4, 4)$

99. $\{x|x < -4 \text{ or } x > 4\}$; $(-\infty, -4) \cup (4, \infty)$

101. $\{x|0 \le x \le 1\}$; $[0, 1]$

103. $\{x|x < -1 \text{ or } x > 2\}$; $(-\infty, -1) \cup (2, \infty)$

105. $\{x|-1 \le x \le 1\}$; $[-1, 1]$

107. $\{x|x \le -2 \text{ or } x \ge 2\}$; $(-\infty, -2] \cup [2, \infty)$

109. $a = 3, b = 5$ **111.** $a = -12, b = -8$ **113.** $a = 3, b = 11$ **115.** $a = \dfrac{1}{4}, b = 1$ **117.** $a = 4, b = 16$ **119.** $\{x|x \ge -2\}$ **121.** $21 < \text{Age} < 30$

123. (a) Male ≥ 82.2 years (b) Female ≥ 85.8 years (c) A female could expect to live 3.6 years longer.

125. The commission ranges from \$45,000 to \$95,000, inclusive. As a percent of selling price, the commission ranges from 5% to 8.6%, inclusive.

127. The amount withheld varies from \$104.32 to \$148.32, inclusive. **129.** The usage varied from 1150 kWh to 2050 kWh, inclusive.

131. 5 cookies **133.** (a) You need at least a 74 on the fifth test. (b) You need at least a 77 on the fifth test.

135. $\dfrac{a + b}{2} - a = \dfrac{a + b - 2a}{2} = \dfrac{b - a}{2} > 0$; therefore, $a < \dfrac{a + b}{2}$. $b - \dfrac{a + b}{2} = \dfrac{2b - a - b}{2} = \dfrac{b - a}{2} > 0$; therefore, $b > \dfrac{a + b}{2}$.

137. $\left(\sqrt{ab}\right)^2 - a^2 = ab - a^2 = a(b - a) > 0$; thus $\left(\sqrt{ab}\right)^2 > a^2$ so $\sqrt{ab} > a$.
$b^2 - \left(\sqrt{ab}\right)^2 = b^2 - ab = b(b - a) > 0$; thus $b^2 > \left(\sqrt{ab}\right)^2$ so $b > \sqrt{ab}$.

A.7 Assess Your Understanding (*page A62*)

3. index **4.** cube root **5.** b **6.** d **7.** c **8.** c **9.** T **10.** F **11.** 3 **13.** -2 **15.** $2\sqrt{2}$ **17.** $10\sqrt{7}$ **19.** $2\sqrt[3]{4}$ **21.** $-2x\sqrt[3]{x}$ **23.** $3\sqrt[4]{3}$ **25.** x^3y^2

27. x^2y **29.** $8\sqrt{x}$ **31.** $3x^2y^3\sqrt[4]{2x}$ **33.** $5x\sqrt{3x}$ **35.** $15\sqrt[3]{3}$ **37.** $12\sqrt{3}$ **39.** $7\sqrt{2}$ **41.** $6\sqrt{3}$ **43.** $2\sqrt{3}$ **45.** $-\sqrt[3]{2}$ **47.** $x - 2\sqrt{x} + 1$

49. $(2x - 1)\sqrt[3]{2x}$ **51.** $(2x - 15)\sqrt{2x}$ **53.** $-(x + 5y)\sqrt[3]{2xy}$ **55.** $\dfrac{\sqrt{2}}{2}$ **57.** $-\dfrac{\sqrt{15}}{5}$ **59.** $\dfrac{(5 + \sqrt{2})\sqrt{3}}{23}$ **61.** $\dfrac{8\sqrt{5} - 19}{41}$ **63.** $5\sqrt{2} + 5$ **65.** $\dfrac{5\sqrt[3]{4}}{2}$

67. $\dfrac{2x + h - 2\sqrt{x^2 + xh}}{h}$ **69.** $\dfrac{5}{\sqrt{11} - 1}$ **71.** $-\dfrac{3}{\sqrt{10} + 5}$ **73.** $\dfrac{1}{\sqrt{x} + \sqrt{c}}$ **75.** $\dfrac{1}{\sqrt{x - 7} + 1}$ **77.** $\left\{\dfrac{9}{2}\right\}$ **79.** $\{3\}$ **81.** 4 **83.** -4 **85.** 1000 **87.** $\dfrac{1}{8}$

89. $\dfrac{27\sqrt{2}}{32}$ **91.** $\dfrac{27\sqrt{2}}{32}$ **93.** $-\dfrac{1}{10}$ **95.** $\dfrac{25}{16}$ **97.** $x^{7/12}$ **99.** xy^2 **101.** $x^{2/3}y$ **103.** $\dfrac{8x^{5/4}}{y^{3/4}}$ **105.** $\dfrac{3x + 2}{(1 + x)^{1/2}}$ **107.** $\dfrac{x(3x^2 + 2)}{(x^2 + 1)^{1/2}}$ **109.** $\dfrac{22x + 5}{10\sqrt{x - 5}\ \sqrt{4x + 3}}$

111. $\dfrac{2 + x}{2(1 + x)^{3/2}}$ **113.** $\dfrac{4 - x}{(x + 4)^{3/2}}$ **115.** $\dfrac{1}{x^2(x^2 - 1)^{1/2}}$ **117.** $\dfrac{1 - 3x^2}{2\sqrt{x}(1 + x^2)^2}$ **119.** $\dfrac{1}{2}(5x + 2)(x + 1)^{1/2}$ **121.** $2x^{1/2}(3x - 4)(x + 1)$

123. $(x^2 + 4)^{1/3}(11x^2 + 12)$ **125.** $(3x + 5)^{1/3}(2x + 3)^{1/2}(17x + 27)$ **127.** $\dfrac{3(x + 2)}{2x^{1/2}}$ **129.** 1.41 **131.** 1.59 **133.** 4.89 **135.** 2.15

137. (a) 15,660.4 gal (b) 390.7 gal

A.8 Assess Your Understanding (*page A75*)

1. undefined; 0 **2.** 3; 2 **3.** T **4.** F **5.** T **6.** $m_1 = m_2$; y-intercepts; $m_1m_2 = -1$ **7.** 2 **8.** $-\dfrac{1}{2}$ **9.** c **10.** d **11.** b **12.** d

13. (a) Slope $= \dfrac{1}{2}$

(b) If x increases by 2 units, y will increase by 1 unit.

15. (a) Slope $= -\dfrac{1}{3}$

(b) If x increases by 3 units, y will decrease by 1 unit.

17. Slope $= -\dfrac{3}{2}$

19. Slope $= -\dfrac{1}{2}$

21. Slope $= 0$

23. Slope undefined

25.   **27.**   **29.** $P = (-1,3)$ **31.** $P = (0,3)$

33. $y - 2 = 3(x - 1)$ **35.** $y - 4 = -\dfrac{3}{4}(x - 2)$
37. $y - 3 = 0$ **39.** $(2,6); (3,10); (4,14)$
41. $(4,-7); (6,-10); (8,-13)$
43. $(-1,-5); (0,-7); (1,-9)$

45. $x - 2y = 0$ or $y = \dfrac{1}{2}x$ **47.** $x + y = 2$ or $y = -x + 2$ **49.** $2x - y = 3$ or $y = 2x - 3$ **51.** $x + 2y = 5$ or $y = -\dfrac{1}{2}x + \dfrac{5}{2}$

53. $3x - y = -9$ or $y = 3x + 9$ **55.** $x - 2y = 1$ or $y = \dfrac{1}{2}x - \dfrac{1}{2}$ **57.** $x - 2y = -5$ or $y = \dfrac{1}{2}x + \dfrac{5}{2}$ **59.** $3x + y = 3$ or $y = -3x + 3$

61. $x - y = -4$ or $y = x + 4$ **63.** $x = 2$; no slope-intercept form **65.** $y = 2$ **67.** $2x - y = -4$ or $y = 2x + 4$ **69.** $x - 2y = 0$ or $y = \dfrac{1}{2}x$

71. $x = 4$; no slope-intercept form **73.** $2x + y = 0$ or $y = -2x$ **75.** $2x + y = 4$ or $y = -2x + 4$ **77.** $y = 4$

79. Slope $= 2$; y-intercept $= 3$ **81.** Slope $= 2$; y-intercept $= -2$ **83.** Slope $= \dfrac{1}{2}$; y-intercept $= 2$ **85.** Slope $= -\dfrac{1}{2}$; y-intercept $= 2$

87. Slope $= \dfrac{2}{3}$; y-intercept $= -2$ **89.** Slope $= -1$; y-intercept $= 1$ **91.** Slope undefined; no y-intercept **93.** Slope $= 0$; y-intercept $= 5$

95. Slope $= 1$; y-intercept $= 0$ **97.** Slope $= \dfrac{3}{2}$; y-intercept $= 0$ **99. (a)** x-intercept: 3; y-intercept: 2 **101. (a)** x-intercept: -10; y-intercept: 8

   (b)   **(b)**

103. (a) x-intercept: 3; y-intercept: $\dfrac{21}{2}$ **(b)**

105. (a) x-intercept: 2; y-intercept: 3 **(b)**

107. (a) x-intercept: 5; y-intercept: -2 **(b)**

109. $y = 0$
111. Parallel **113.** Neither
115. $x - y = -2$ or $y = x + 2$
117. $x + 3y = 3$ or $y = -\dfrac{1}{3}x + 1$

119. $P_1 = (-2,5), P_2 = (1,3), m_1 = -\dfrac{2}{3}; P_2 = (1,3), P_3 = (-1,0), m_2 = \dfrac{3}{2};$ because $m_1 m_2 = -1$, the lines are perpendicular and the points $(-2,5), (1,3),$ and $(-1,0)$ are the vertices of a right triangle.

121. $P_1 = (-1,0), P_2 = (2,3), m = 1; P_3 = (1,-2), P_4 = (4,1), m = 1; P_1 = (-1,0), P_3 = (1,-2), m = -1; P_2 = (2,3), P_4 = (4,1), m = -1;$ opposite sides are parallel, and adjacent sides are perpendicular; the points are the vertices of a rectangle.

123. $C = 0.60x + 39$; \$105.00; \$177.00 **125.** $C = 0.14x + 4252$

127. (a) $C = 0.0889x + 8.01, 0 \le x \le 1000$ **(b)**
(c) \$25.79 **(d)** \$52.46
(e) Each additional kWh used adds \$0.0889 to the bill.

129. $^\circ C = \dfrac{5}{9}(^\circ F - 32)$; approximately $21.1^\circ C$

131. (a) $y = -\dfrac{2}{25}x + 30$ **(b)** x-intercept: 375; The ramp meets the floor 375 in. (31.25 ft) from the base of the platform. **(c)** The ramp does not meet design requirements. It has a run of 31.25 ft. **(d)** The only slope possible for the ramp to comply with the requirement is for it to drop 1 in. for every 12-in. run.

133. (a) $A = \dfrac{1}{5}x + 20,000$ **(b)** \$80,000 **(c)** Each additional box sold requires an additional \$0.20 in advertising.

135. $-2, -\dfrac{2}{3}, \dfrac{1}{2}$ **137.** b, c, e, g **139.** c **145.** No; no

147. They are the same line. **149.** Yes, if the y-intercept is 0.

APPENDIX B Graphing Utilities

B.1 Exercises *(page B2)*

1. $(-1, 4)$; II   **3.** $(3, 1)$; I   **5.** $X\min = -6$, $X\max = 6$, $X\text{scl} = 2$, $Y\min = -4$, $Y\max = 4$, $Y\text{scl} = 2$
7. $X\min = -6$, $X\max = 6$, $X\text{scl} = 2$, $Y\min = -1$, $Y\max = 3$, $Y\text{scl} = 1$ **9.** $X\min = 3$, $X\max = 9$, $X\text{scl} = 1$, $Y\min = 2$, $Y\max = 10$, $Y\text{scl} = 2$
11. $X\min = -11$, $X\max = 5$, $X\text{scl} = 1$, $Y\min = -3$, $Y\max = 6$, $Y\text{scl} = 1$
13. $X\min = -30$, $X\max = 50$, $X\text{scl} = 10$, $Y\min = -90$, $Y\max = 50$, $Y\text{scl} = 10$
15. $X\min = -10$, $X\max = 110$, $X\text{scl} = 10$, $Y\min = -10$, $Y\max = 160$, $Y\text{scl} = 10$

B.2 Exercises *(page B4)*

1. (a)   **(b)**   **3. (a)**   **(b)**

5. (a)   **(b)**   **7. (a)**   **(b)**

9. (a)   **(b)**   **11. (a)**   **(b)**

13. (a)   **(b)**   **15. (a)**   **(b)**

17. $Y_1 = X + 2$ **19.** $Y_1 = -X + 2$ **21.** $Y_1 = 2X + 2$ **23.** $Y_1 = -2X + 2$

25. $Y_1 = X^2 + 2$ **27.** $Y_1 = -X^2 + 2$ **29.** $Y_1 = -(3/2)X + 3$ **31.** $Y_1 = (3/2)X + 3$

B.3 Exercises *(page B6)*

1. -3.41 **3.** -1.71 **5.** -0.28 **7.** 3.00 **9.** 4.50 **11.** $1.00, 23.00$

B.5 Exercises *(page B8)*

1. No   **3.** Yes   **5.** Answers may vary. A possible answer is $Y\min = 0$, $Y\max = 10$, and $Y\text{scl} = 1$.

Photo Credits

Subject Index

118 Subject Index

Value (image) of function, 26, 29–31
Variable(s), A6
  dependent, 29
  domain of, A7
  independent, 29
    in calculus, 147
Variable costs, A77
Vector(s), 349–363
  adding, 350, 354
  algebraic, 352–353
  angle between, 365
  components of, 352, 353
    horizontal and vertical, 353
  decomposing, 366–368
  defined, 349
  difference of, 350
  direction of, 349, 355–357
  dot product of two, 364–365
  equality of, 350, 353
  finding, 356–357
  force, 356
  geometric, 349–350
  graphing, 351–352
  historical feature on, 359
  magnitudes of, 349, 352, 354–355, 356–357
  modeling with, 357–359
  multiplying by numbers geometrically, 351
  objects in static equilibrium, 358–359
  orthogonal, 366–368
  parallel, 366
  in physics, 349
  position, 352–353
  scalar multiples of, 351, 354–355, 364
  in space, 371–380
    angle between two vectors, 376
    cross product of two, 381–382
    direction angles, 377–379
    distance formula, 372–373
    dot product, 375–376
    operations on, 374–375
    orthogonal to two given vectors, 384
    position vectors, 373–374
    rectangular coordinates, 372
  subtracting, 354
  unit, 352, 355–356
  velocity, 356–357
  zero, 350
Vector projection, 367
Velocity vector, 356–357
Venn diagrams, A2
Vertex/vertices
  of cone, 393
  of ellipse, 403

of hyperbola, 414
of parabola, 394
of ray, 102
Vertical asymptote, 77
Vertical component of vector, 353
Vertical line, 327, 335, A64–A65
  equation of, A68
Vertical-line test, 34–35, 37
Vertically compressed or stretched graphs, 71–72
Vertical shifts, 67–69, 74
  horizontal shift combined with, 70–71
Viète, François, 289
Viewing angle, 130–131, 201
Viewing rectangle, 3, B1–B3
  setting, B1
Volume, formulas for, A16

Wallis, John, 347
Waves. *See also* Simple harmonic motion
  square, 308
  traveling speeds of, 218
Weight, 315, 391
Whispering galleries, 409–410
Wings, airplane, 315, 391
Work, 380
  dot product to compute, 368–369

x-axis, 2
  projection of P on the, 300
  reflections about, 73–74
  symmetry with respect to, 12, 13–14, 329
x-coordinate, 2
x-intercept, 11
xy-plane, 2, 372
xz-plane, 372

y-axis, 2
  projection of P on the, 300
  reflections about, 74
  symmetry with respect to, 13–14, 329
y-coordinate (ordinate), 2
y-intercept, 11, A70
yz-plane, 372

Zero-coupon bonds, 514
Zero-level earthquake, 489
Zero-Product Property, A4
Zero vector, 350

LIBRARY OF FUNCTIONS

Identity Function
$f(x) = x$

Square Function
$f(x) = x^2$

Cube Function
$f(x) = x^3$

Square Root Function
$f(x) = \sqrt{x}$

Reciprocal Function
$f(x) = \dfrac{1}{x}$

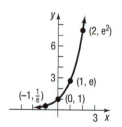

Cube Root Function
$f(x) = \sqrt[3]{x}$

Absolute Value Function
$f(x) = |x|$

Exponential Function
$f(x) = e^x$

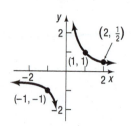

Natural Logarithm Function
$f(x) = \ln x$

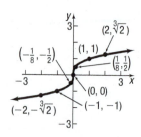

Sine Function
$f(x) = \sin x$

Cosine Function
$f(x) = \cos x$

Tangent Function
$f(x) = \tan x$

Cosecant Function
$f(x) = \csc x$

Secant Function
$f(x) = \sec x$

Cotangent Function
$f(x) = \cot x$

FORMULAS/EQUATIONS

Distance Formula

If $P_1 = (x_1, y_1)$ and $P_2 = (x_2, y_2)$, the distance from P_1 to P_2 is

$$d(P_1, P_2) = \sqrt{(x_2 - x_1)^2 + (y_2 - y_1)^2}$$

Standard Equation of a Circle

The standard equation of a circle of radius r with center (h, k) is

$$(x - h)^2 + (y - k)^2 = r^2$$

Slope Formula

The slope m of the line containing the points $P_1 = (x_1, y_1)$ and $P_2 = (x_2, y_2)$ is

$$m = \frac{y_2 - y_1}{x_2 - x_1} \qquad \text{if } x_1 \neq x_2$$

$$m \text{ is undefined} \qquad \text{if } x_1 = x_2$$

Point-Slope Equation of a Line

The equation of a line with slope m containing the point (x_1, y_1) is

$$y - y_1 = m(x - x_1)$$

Slope-Intercept Equation of a Line

The equation of a line with slope m and y-intercept b is

$$y = mx + b$$

Quadratic Formula

The solutions of the equation $ax^2 + bx + c = 0, a \neq 0$, are

$$x = \frac{-b \pm \sqrt{b^2 - 4ac}}{2a}$$

If $b^2 - 4ac > 0$, there are two unequal real solutions.
If $b^2 - 4ac = 0$, there is a repeated real solution.
If $b^2 - 4ac < 0$, there are two complex solutions that are not real.

GEOMETRY FORMULAS

Circle

$r = $ Radius, $A = $ Area, $C = $ Circumference

$$A = \pi r^2 \qquad C = 2\pi r$$

Triangle

$b = $ Base, $h = $ Altitude (Height), $A = $ area

$$A = \tfrac{1}{2} bh$$

Rectangle

$l = $ Length, $w = $ Width, $A = $ area, $P = $ perimeter

$$A = lw \qquad P = 2l + 2w$$

Rectangular Box (closed)

$l = $ Length, $w = $ Width, $h = $ Height, $V = $ Volume, $S = $ Surface area

$$V = lwh \qquad S = 2lw + 2lh + 2wh$$

Sphere

$r = $ Radius, $V = $ Volume, $S = $ Surface area

$$V = \tfrac{4}{3}\pi r^3 \qquad S = 4\pi r^2$$

Right Circular Cylinder (closed)

$r = $ Radius, $h = $ Height, $V = $ Volume, $S = $ Surface area

$$V = \pi r^2 h \qquad S = 2\pi r^2 + 2\pi rh$$

CONICS

Parabola

$$y^2 = 4ax \qquad\qquad y^2 = -4ax \qquad\qquad x^2 = 4ay \qquad\qquad x^2 = -4ay$$

Ellipse

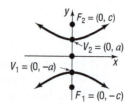

$$\frac{x^2}{a^2} + \frac{y^2}{b^2} = 1, \quad a > b, \quad c^2 = a^2 - b^2 \qquad\qquad \frac{x^2}{b^2} + \frac{y^2}{a^2} = 1, \quad a > b, \quad c^2 = a^2 - b^2$$

Hyperbola

$$\frac{x^2}{a^2} - \frac{y^2}{b^2} = 1, \quad c^2 = a^2 + b^2 \qquad\qquad \frac{y^2}{a^2} - \frac{x^2}{b^2} = 1, \quad c^2 = a^2 + b^2$$

$$\text{Asymptotes:} \quad y = \frac{b}{a}x, \quad y = -\frac{b}{a}x \qquad\qquad \text{Asymptotes:} \quad y = \frac{a}{b}x, \quad y = -\frac{a}{b}x$$

PROPERTIES OF LOGARITHMS

$$\log_a (MN) = \log_a M + \log_a N$$

$$\log_a \left(\frac{M}{N}\right) = \log_a M - \log_a N$$

$$\log_a M^r = r \log_a M$$

$$\log_a M = \frac{\log M}{\log a} = \frac{\ln M}{\ln a}$$

$$a^r = e^{r \ln a}$$

TRIGONOMETRIC FUNCTIONS

Let t be a real number and let $P = (x, y)$ be the point on the unit circle that corresponds to t.

$$\sin t = y \qquad \cos t = x \qquad \tan t = \frac{y}{x}, \quad x \neq 0$$

$$\csc t = \frac{1}{y}, \quad y \neq 0 \qquad \sec t = \frac{1}{x}, \quad x \neq 0 \qquad \cot t = \frac{x}{y}, \quad y \neq 0$$

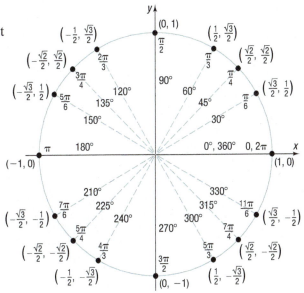

TRIGONOMETRIC IDENTITIES

Fundamental Identities

$$\tan \theta = \frac{\sin \theta}{\cos \theta} \qquad \cot \theta = \frac{\cos \theta}{\sin \theta}$$

$$\csc \theta = \frac{1}{\sin \theta} \qquad \sec \theta = \frac{1}{\cos \theta} \qquad \cot \theta = \frac{1}{\tan \theta}$$

$$\sin^2 \theta + \cos^2 \theta = 1$$
$$\tan^2 \theta + 1 = \sec^2 \theta$$
$$\cot^2 \theta + 1 = \csc^2 \theta$$

Even-Odd Identities

$$\sin(-\theta) = -\sin \theta \qquad \csc(-\theta) = -\csc \theta$$

$$\cos(-\theta) = \cos \theta \qquad \sec(-\theta) = \sec \theta$$

$$\tan(-\theta) = -\tan \theta \qquad \cot(-\theta) = -\cot \theta$$

Sum and Difference Formulas

$$\sin(\alpha + \beta) = \sin \alpha \cos \beta + \cos \alpha \sin \beta$$
$$\sin(\alpha - \beta) = \sin \alpha \cos \beta - \cos \alpha \sin \beta$$
$$\cos(\alpha + \beta) = \cos \alpha \cos \beta - \sin \alpha \sin \beta$$
$$\cos(\alpha - \beta) = \cos \alpha \cos \beta + \sin \alpha \sin \beta$$
$$\tan(\alpha + \beta) = \frac{\tan \alpha + \tan \beta}{1 - \tan \alpha \tan \beta}$$
$$\tan(\alpha - \beta) = \frac{\tan \alpha - \tan \beta}{1 + \tan \alpha \tan \beta}$$

Half-Angle Formulas

$$\sin \frac{\theta}{2} = \pm \sqrt{\frac{1 - \cos \theta}{2}}$$

$$\cos \frac{\theta}{2} = \pm \sqrt{\frac{1 + \cos \theta}{2}}$$

$$\tan \frac{\theta}{2} = \frac{1 - \cos \theta}{\sin \theta}$$

Double-Angle Formulas

$$\sin(2\theta) = 2 \sin \theta \cos \theta$$
$$\cos(2\theta) = \cos^2 \theta - \sin^2 \theta$$
$$\cos(2\theta) = 2 \cos^2 \theta - 1$$
$$\cos(2\theta) = 1 - 2 \sin^2 \theta$$
$$\tan(2\theta) = \frac{2 \tan \theta}{1 - \tan^2 \theta}$$

Product-to-Sum Formulas

$$\sin \alpha \sin \beta = \tfrac{1}{2}[\cos(\alpha - \beta) - \cos(\alpha + \beta)]$$

$$\cos \alpha \cos \beta = \tfrac{1}{2}[\cos(\alpha - \beta) + \cos(\alpha + \beta)]$$

$$\sin \alpha \cos \beta = \tfrac{1}{2}[\sin(\alpha + \beta) + \sin(\alpha - \beta)]$$

Sum-to-Product Formulas

$$\sin \alpha + \sin \beta = 2 \sin \frac{\alpha + \beta}{2} \cos \frac{\alpha - \beta}{2}$$

$$\sin \alpha - \sin \beta = 2 \sin \frac{\alpha - \beta}{2} \cos \frac{\alpha + \beta}{2}$$

$$\cos \alpha + \cos \beta = 2 \cos \frac{\alpha + \beta}{2} \cos \frac{\alpha - \beta}{2}$$

$$\cos \alpha - \cos \beta = -2 \sin \frac{\alpha + \beta}{2} \sin \frac{\alpha - \beta}{2}$$

SOLVING TRIANGLES

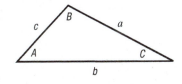

Law of Sines

$$\frac{\sin A}{a} = \frac{\sin B}{b} = \frac{\sin C}{c}$$

Law of Cosines

$$a^2 = b^2 + c^2 - 2bc \cos A$$
$$b^2 = a^2 + c^2 - 2ac \cos B$$
$$c^2 = a^2 + b^2 - 2ab \cos C$$